FUNDAMENTALS OF
MICROBIOLOGY

The Pommerville Microbiology Course Solution

More For Less? Of Course!

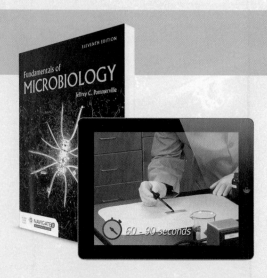

ELEVENTH EDITION

FUNDAMENTALS OF
MICROBIOLOGY

JEFFREY C. POMMERVILLE
Glendale Community College

JONES & BARTLETT
LEARNING

World Headquarters
Jones & Bartlett Learning
5 Wall Street
Burlington, MA 01803
978-443-5000
info@jblearning.com
www.jblearning.com

Jones & Bartlett Learning books and products are available through most bookstores and online booksellers. To contact Jones & Bartlett Learning directly, call 800-832-0034, fax 978-443-8000, or visit our website, www.jblearning.com.

Substantial discounts on bulk quantities of Jones & Bartlett Learning publications are available to corporations, professional associations, and other qualified organizations. For details and specific discount information, contact the special sales department at Jones & Bartlett Learning via the above contact information or send an email to specialsales@jblearning.com.

10104-1

Production Credits
VP, Executive Publisher: David D. Cella
Executive Editor: Matthew Kane
Senior Development Editor: Nancy Hoffmann
Associate Editor: Audrey Schwinn
Production Manager: Dan Stone
Senior Production Editor: Nancy Hitchcock
Marketing Manager: Lindsay White
Director of Marketing: Andrea DeFronzo
Production Services Manager: Colleen Lamy
Manufacturing and Inventory Control Supervisor:
    Amy Bacus
Composition: Cenveo® Publisher Services

Interior Design: Kristin E. Parker
Cover Design: Michael O'Donnell
Director of Rights & Media: Joanna Gallant
Rights & Media Specialist: Jamey O'Quinn
Rights & Media Specialist: Wes DeShano
Media Development Editor: Troy Liston
Cover Image (Title Page, Part Opener, Chapter Opener):
    © M.I.T. Case No. 17683C, "C Food", by Roman Stocker,
    Steven Paul Smriga, and Vicente Ignacio Fernandez
Printing and Binding: LSC Communications
Cover Printing: LSC Communications

Library of Congress Cataloging-in-Publication Data
Names: Pommerville, Jeffrey C., author.
Title: Fundamentals of microbiology / Jeffrey C. Pommerville.
Description: Eleventh edition. | Burlington, MA : Jones & Bartlett Learning,
    [2018] | Includes index.
Identifiers: LCCN 2017014168 | ISBN 9781284100952
Subjects: | MESH: Microbiological Phenomena
Classification: LCC QR41.2 | NLM QW 4 | DDC 616.9/041—dc23
LC record available at https://lccn.loc.gov/2017014168

6048

Printed in the United States of America
21 20 19 18 17   10 9 8 7 6 5 4 3 2 1

BRIEF CONTENTS

*eBook only

CONTENTS

INDEX OF BOXED FEATURES

PREFACE

■ Ebola, Zika – What's Next?

The human race has experienced and felt the effects of new infectious diseases for millennia, even well before the discovery of the infectious agents responsible for such diseases. Today, however, despite extraordinary advances to eliminate or lessen the development and spread of infectious disease, their appearance continues—and indeed, it is inevitable. Recent examples of emerging diseases include HIV/AIDS, severe acute respiratory syndrome (SARS), Ebola virus disease, and, most recently, Zika virus infection. Each of these unexpected illnesses has had a global impact on governments, economics, and society, and for Zika, even threatened the 2016 Summer Olympics.

The emergence of infectious diseases like Ebola and Zika is the result of several factors. Realize that more than 60% percent of new human infections originate in, or are transmitted by, wild animals, as is the case for all the diseases mentioned above—AIDS (apes and monkeys to humans), SARS and Ebola (bats [to other animals] to humans), and Zika (monkeys to mosquitoes to humans). Consequently, as human habitation spreads into more remote areas around the world, unknown infectious microbes in wild animals will "jump" to humans as these interacting species make contact. In addition, many of these infectious agents undergo rapid genetic changes, as exemplified by the AIDS and influenza viruses, and they can share genetic information as the influenza virus does every flu season. In some cases, this can make the infection more dangerous.

Adding to these factors is the globalized world we live in today. Airline travel makes an infectious disease outbreak in one corner of the world only a day's plane ride from almost any other destination on the globe. Accordingly, infectious diseases can "pop up" from seemingly nowhere.

The next emerging disease will have a different name and different symptoms from Ebola and Zika, and it will come from another region of the world—but it and others are coming. Therefore, what we can (and must) do is recognize and react to these "infectious events" before they can cause an outbreak or epidemic. We need to be better prepared to deal with these events by managing better the next Ebola- or Zika-like emergence, something that was not done in the most recent Ebola outbreak in West Africa and Zika outbreak in Brazil. Preventing another outbreak/epidemic of a new infectious disease can be accomplished only by aggressive vigilance, continued research for detecting new infectious agents (surveillance tools, diagnostics, drugs, and vaccines), and rapidly deploying these countermeasures.

Each new emerging disease brings unique challenges, forcing the medical community to continually adapt to these ever-shifting threats. The battle against emerging infectious diseases is a continual process in trying to get ahead and stay ahead of the next infectious agent before it can explode on the world scene.

More than likely, you are planning a career in the healthcare field. As such, it is important that you understand how new infectious diseases come about. Therefore, I am excited and honored that you are using and reading this new, eleventh edition of *Fundamentals of Microbiology*. I hope it is very useful in your studies and you come away from your course with a much better appreciation for the role that microorganisms play in the environment as well as with us. Always take time to read the sidebars (MicroFocus boxes) whether they are assigned or not. They will help in your overall microbiology experience and the realization that microorganisms do rule the world!

■ A Concept-Based Curriculum

Fundamentals of Microbiology, Eleventh Edition is written for introductory microbiology courses having an emphasis in the health sciences. It is geared toward students in health and allied health science curricula such as nursing, dental hygiene, medical assistance, sanitary science, and medical laboratory technology. It also will be an asset to students studying food science, agriculture, environmental science, and health administration. In addition, the text provides a firm foundation for advanced programs in biological sciences, as well as medicine, pharmacy, dentistry, and other health professions.

The textbook is divided into seven areas of concentration. Each area reflects the *Concept-Based Microbiology Curriculum Guidelines* as recommended by the American Society for Microbiology.

Overarching Concepts and Fundamental Statements[1]

Evolution	• Cells, organelles (e.g., mitochondria and chloroplasts), and all major metabolic pathways evolved from early prokaryotic cells. • Mutations and horizontal gene transfer, with the immense variety of microenvironments, have selected for a huge diversity of microorganisms. • Human impact on the environment influences the evolution of microorganisms. • The traditional concept of species is not readily applicable to microbes due to asexual reproduction and the frequent occurrence of horizontal gene transfer. • Evolutionary relatedness of organisms is best reflected in phylogenetic trees.
Cell Structure and Function	• The structure and function of microorganisms have been revealed by the use of microscopy. • Bacteria have unique cell structures that can be targets for antibiotics, immunity, and phage infection. • Bacteria and Archaea have specialized structures that often confer critical capabilities. • While microscopic eukaryotes carry out some of the same processes as bacteria, many of the cellular properties are fundamentally different. • The replication cycles of viruses differ among viruses and are determined by their unique structures and genomes.
Metabolic Pathways	• Bacteria and Archaea exhibit extensive, and often unique, metabolic diversity. • The interactions of microorganisms among themselves and with their environment are determined by their metabolic abilities. • The survival and growth of any microorganism in a given environment depends on its metabolic characteristics. • The growth of microorganisms can be controlled by physical, chemical, mechanical, or biological means.
Information Flow and Genetics	• Genetic variations can impact microbial functions. • Although the central dogma is universal in all cells, the processes of replication, transcription, and translation differ in Bacteria, Archaea, and Eukarya. • The regulation of gene expression is influenced by external and internal molecular cues and/or signals. • The synthesis of viral genetic material and proteins is dependent on host cells. • Cell genomes can be manipulated to alter cell function.
Microbial Systems	• Microorganisms are ubiquitous and live in diverse and dynamic ecosystems. • Most bacteria in nature live in biofilm communities. • Microorganisms and their environment interact with and modify each other. • Microorganisms, cellular and viral, can interact with both human and nonhuman hosts in beneficial, neutral, or detrimental ways.
Impact of Microorganisms	• Microbes are essential for life, as we know it, and the processes that support life. • Microorganisms provide essential models that give us fundamental knowledge about life processes. • Humans use and harness microorganisms and their products. • Because the true diversity of microbial life is largely unknown, its effects and potential benefits have not been fully explored.

[1] Merkel, S. 2012. The Development of Curricular Guidelines for Introductory Microbiology That Focus On Understanding. J Microbiol Biol Educ 13323810.1128/ jmbe.v13i1.363236537793577306 http://dx.doi.org/10.1128/jmbe.v13i1.363

■ What's New in This Edition

When you read this text, you get a global perspective on microbiology and infectious disease as found in no other similar textbook. The current edition has been updated with the latest scientific and education research and has incorporated many suggestions made by my colleagues, by emails received from microbiology instructors, and by my students. Along with these revisions, the visual aspects of the text have been improved to make the understanding of difficult concepts more approachable and the figures more engaging. What's new? Here is a summary list.

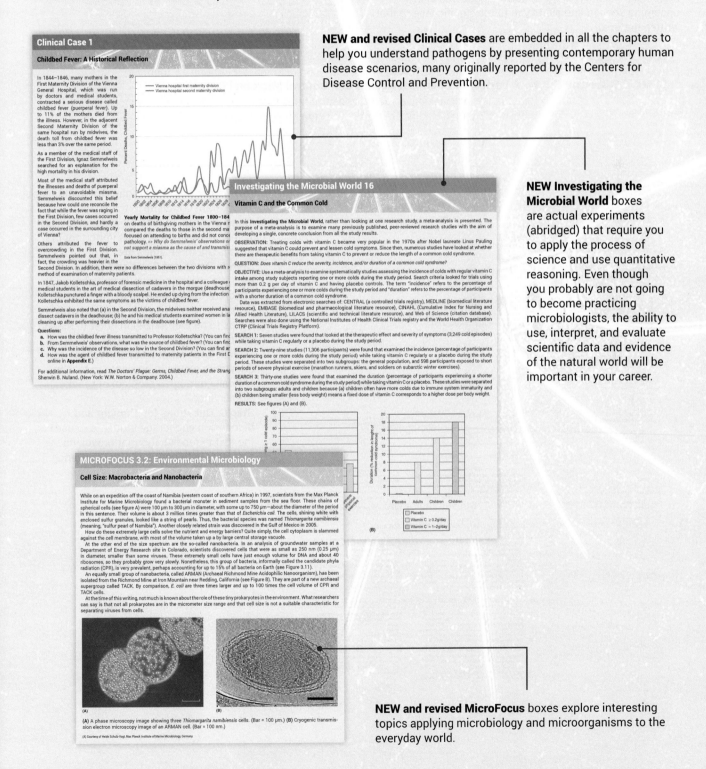

NEW and revised Clinical Cases are embedded in all the chapters to help you understand pathogens by presenting contemporary human disease scenarios, many originally reported by the Centers for Disease Control and Prevention.

NEW Investigating the Microbial World boxes are actual experiments (abridged) that require you to apply the process of science and use quantitative reasoning. Even though you probably are not going to become practicing microbiologists, the ability to use, interpret, and evaluate scientific data and evidence of the natural world will be important in your career.

NEW and revised MicroFocus boxes explore interesting topics applying microbiology and microorganisms to the everyday world.

NEW and revised MicroInquiry boxes allow you to investigate (usually interactively) some important aspect of the chapter being studied.

NEW Key Concept organization presents section statements identifying the important concepts in the upcoming section and alerts you to the significance of that written material.

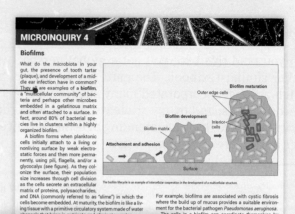

MICROINQUIRY 4

Biofilms

What do the microbiota in your gut, the presence of tooth tartar (plaque), and development of a middle ear infection have in common? They all are examples of a **biofilm**, a "multicellular community" of bacteria and perhaps other microbes embedded in a gelatinous matrix and often attached to a surface. In fact, around 80% of bacterial species live in clusters within a highly organized biofilm.

A biofilm forms when planktonic cells initially attach to a living or nonliving surface by weak electrostatic forces and then more permanently, using pili, flagella, and/or a glycocalyx (see figure). As they colonize the surface, their population size increases through cell division as the cells secrete an extracellular matrix of proteins, polysaccharides, and DNA (commonly referred to as "slime") in which the cells become embedded. At maturity, the biofilm is like a living tissue with a primitive circulatory system made of water channels that bring in nutrients and eliminate wastes.

The biofilm lifecycle is an example of intercellular cooperation in the development of a multicellular structure.

For example, biofilms are associated with cystic fibrosis where the build up of mucus provides a suitable environment for the bacterial pathogen *Pseudomonas aeruginosa*. The cells in a biofilm can coordinate themselves by [...] and "talking with" neighboring cells through [...] our electronic communications in human [...]ing through Twitter and Facebook. Thus, [...] process, called **quorum sensing (QS)**, is [...]ective decision making, wherein the cells [...]pulation numbers through the exchange of extracellular chemicals. When these molecules reach a critical threshold, the community of cells acts together and, depending on the species, gene regulation triggers a specific behavioral response.

■ **KEY CONCEPT 9.4**   **A Variety of Chemical Agents Can Control Microbial Growth**

metabolic activity, which, along with the dense and thick slime, makes the cells less susceptible to antibiotics and host immune defenses. Thus, many human wound and chronic infections are the result of biofilm development.

Step A: Review and Facts and Terms   603

■ **CHAPTER SELF-TEST**

For **Steps A–D**, you can find answers online in **Appendix D**.

■ **STEP A: REVIEW AND FACTS AND TERMS**

Multiple Choice

Read each question carefully before selecting the *one* answer that best fits the question or statement.

1. Mononucleosis is an infection of _____ cells by the _____.
   A. T; cytomegalovirus
   B. B; Epstein-Barr virus
   C. lung; cytomegalovirus
   D. red blood; Epstein-Barr virus
2. Which of the following is *not* a transmission mechanism for hepatitis B?
   A. Sexual contact
   B. Nonsterile body piercing equipment
   C. Fecal–oral route
   D. Blood-contaminated needles
3. Symptoms of headache, fever, and muscle pain lasting 3 to 5 days, followed by a 2 to 24 hour abating of symptoms are characteristics of _____.
   A. yellow fever
   B. hepatitis C
   C. dengue fever
   D. Ebola hemorrhagic fever
4. A long thread-like RNA virus is typical of the _____ viruses.
   A. hepatitis C
   B. Ebola
   C. polio
   D. West Nile
5. The reservoir for Lassa fever is _____.
   A. rats
   B. mosquitoes
   C. ticks
   D. sandflies

6. Which one of the following characteristics pertains to hepatitis A?
   A. Transmission is by the fecal–oral route.
   B. The incubation period is 2 to 4 weeks.
   C. It is an acute, inflammatory liver disease.
   D. All of the above (A–C) are correct.
7. _____ are the single most important cause of diarrhea in infants and young children admitted to American hospitals.
   A. Noroviruses
   B. Echoviruses
   C. Hepatitis A viruses
   D. Rotaviruses
8. Hydrophobia is a term applied to _____.
   A. rotavirus infections
   B. West Nile fever
   C. arboviral encephalitis
   D. rabies

True-False

Each of the following statements is true (T) or false (F). If the underlined word or phrase to make the statement true.

12. _____ The infectious agent for both yellow fever and dengue fever is a <u>DNA</u> virus transmitted by mosquitoes.
13. _____ Eighty percent of people infected by West Nile virus experience <u>flu-like symptoms</u>.

---

604   CHAPTER 17 VIRAL INFECTIONS OF THE BLOOD, LYMPHATIC, GASTROINTESTINAL

16. _____ The <u>Epstein-Barr</u> virus is the cause of infectious mononucleosis.
17. _____ The Salk and Sabin vaccines have been used for immunizations against <u>hepatitis</u>.
18. _____ <u>Hepatitis B</u> is most commonly transmitted by contact with infected semen or infected blood.

19. _____ One of the most important causes of diarrhea in infants and young children admitted to hospitals is the <u>Epstein-Barr virus</u>.
20. _____ <u>Filoviruses</u> are long, thread-like viruses that cause hemorrhagic fevers and include the marburgvirus.

■ **STEP B: CONCEPT REVIEW**

21. Compare the similarities and differences between the nature of the hepatitis B and C viruses and the illnesses they cause. **(Key Concept 17.1)**
22. Summarize the symptoms of **Ebola virus disease** and **Marburg virus disease. (Key Concept 17.2)**
23. Describe how the hepatitis A virus is spread and prevented. **(Key Concept 17.3)**

24. Explain why rotavirus infections are so deadly in children and describe how noroviruses are transmitted. **(Key Concept 17.3)**
25. Describe the outcome to someone who has been bitten by a rabid animal and recommend treatment if **rabies** symptoms have not yet appeared. **(Key Concept 17.4)**
26. Explain how the polioviruses cause disease and identify the two types of **polio** vaccines. **(Key Concept 17.4)**

■ **STEP C: APPLICATIONS AND PROBLEM SOLVING**

27. Written on some blood donor cards is the notation "CMV." What do you think the letters mean, and why are they placed there?
28. Sicilian barbers are renowned for their skill and dexterity with razors (and sometimes their singing voices). French researchers studied a group of 37 Sicilian barbers and found that 14 had antibodies against hepatitis C, despite never having been sick with the disease. By comparison, when a random group of 50 blood donors was studied, none had

the antibodies. As an epidemiologist, what might account for the high incidence of exposure to hepatitis C among these barbers?
29. As a state health inspector, you suggest all restaurant workers should be immunized with the hepatitis A vaccine. Why would restaurant owners agree or disagree with your idea?
30. An epidemiologist notes that India has a high rate of dengue fever but a very low rate of yellow fever. What might be the cause of this anomaly?

■ **STEP D: QUESTIONS FOR THOUGHT AND DISCUSSION**

31. Health authorities panicked when an outbreak of Ebola hemorrhagic disease occurred among imported macaques in a quarantine facility in Reston, Virginia, in 1989. What sparks such a dramatic response when a disease like Ebola virus breaks out?
32. A diagnostic test has been developed to detect hepatitis C in blood intended for transfusion purposes. Obviously, if the test is positive, the blood is not used. However, there is a lively controversy as to whether the blood donor should be informed of the positive result. What is your opinion? Why?
33. In the southwestern United States, abundant rain and a mild winter often bring conditions that encourage

a burgeoning rodent population. Under these circumstances, what viral disease would health officials anticipate and what precautions should they give residents?
34. Disney World and 20 swampy counties in Florida use "sentinel chickens" strategically placed on the grounds to detect any signs of viral encephalitis. Why do you suppose they use chickens? Why are Disney World and many Florida counties particularly susceptible to outbreaks of viral encephalitis? What recommendations might be offered to tourists if the disease broke out?

NEW Chapter Self-Test organization outlines the important concepts in the chapters through Bloom's Taxonomy, a classification of levels of intellectual skills important in learning. The three steps are:
- **Step A: Review of Facts and Terms** are multiple-choice questions focusing on concrete "facts" learned in the chapter. Let's face it; there is information that needs to be memorized in order to reason critically.
- **Step B: Applications and Problems** are questions requiring students to reason critically through a problem of practical significance.
- **Step C: Questions for Thought and Discussion** encourage students to use the text to resolve thought-provoking problems with contemporary relevance.

■ Chapter-By-Chapter Revisions

Each chapter of *Fundamentals of Microbiology, Eleventh Edition* has been carefully and thoroughly revised. In addition, new information pertinent to nursing and allied health has been included, while many figures and tables have been updated, revised, and/or reorganized for clarity. Here are the major changes to each chapter.

Chapter 1 Microbiology: Then and Now

- New Clinical Case study
- Modified MicroInquiry feature
- Two new and one revised and updated MicroFocus feature
- Chapter Self-Test redesigned

Chapter 2 The Chemical Building Blocks of Life

- 16 figures modified for clarity
- 2 new figures
- Chapter Self-Test redesigned

Chapter 3 Concepts and Tools for Studying Microorganisms

- New discussion on the importance of cell size
- New MicroFocus feature on very small cells
- More basic information on eukaryotic organelles
- New text material on endosymbiosis
- New Microinquiry feature on microbial identification
- Chapter Self-Test redesigned

Chapter 4 Structure and Organization of Prokaryotic Cells

- New MicroInquiry feature on biofilms
- New information on bacterial cell compartments
- Chapter Self-Test redesigned

Chapter 5 Microbial Growth and Nutrition

- New information on biofilm growth
- New and revised MicroFocus features
- New section on chemical factors influencing microbial growth
- Chapter Self-Test redesigned

Chapter 6 Microbial Metabolism

- Revised MicroInquiry feature
- New clinical case
- Revised section on cellular respiration
- Revised section on metabolic diversity
- Chapter Self-Test redesigned

Chapter 7 Microbial Genetics

- New information on bacterial genomes
- New information on organelle DNA
- Revised section on mutations
- Chapter Self-Test redesigned

Chapter 8 Gene Transfer, Genetic Engineering, and Genomics

- New Clinical Case study
- Added discussion on bioethics in biotechnology
- Several new figures
- Chapter Self-Test redesigned

Chapter 9 Control of Microorganisms: Physical Methods and Chemical Agents

- New Investigating the Microbial World
- Chapter Self-Test redesigned

Chapter 10 Control of Microorganisms: Antimicrobial Drugs and Superbugs
Formerly Chapter 24

- New Investigating the Microbial World box on antibiotic resistance
- Chapter Self-Test redesigned

Chapter 11 Airborne Bacterial Diseases

- Expanded discussion of the human respiratory microbiome
- Revised material on pertussis and tuberculosis
- Revised tables
- Chapter Self-Test redesigned

Chapter 12 Foodborne and Waterborne Bacterial Diseases

- New MicroFocus feature on probiotics
- Expanded discussion of the human gut microbiome
- New figures on oral health
- Revised tables
- Chapter Self-Test redesigned

Chapter 13 Soilborne and Arthropod-borne Bacterial Diseases

- New MicroFocus feature on insect bites
- Updated discussion of Lyme disease
- New figures on arthropod-borne diseases
- Chapter Self-Test redesigned

Chapter 14 Sexually Transmitted and Contact Transmitted Bacterial Diseases

- New MicroFocus box about rosacea
- New information on the human microbiome
- New figures and art
- Chapter Self-Test redesigned

Chapter 15 The Viruses and Virus-Like Agents

- New information of giant viruses
- New figures and art
- Chapter Self-Test redesigned

Chapter 16 Viral Infections of the Respiratory Tract and Skin

- New figures on virus families
- New Investigating the Microbial World feature
- New figure on recent mumps outbreaks
- Chapter Self-Test redesigned

Chapter 17 Viral Infections of the Blood, Lymphatic, Gastrointestinal, and Nervous Systems

- New chapter introduction (on Zika virus infection)
- New MicroInquiry feature
- Coverage of Zika virus infection
- Updated material on Ebola virus disease
- Updated material on yellow fever and dengue fever
- Chapter Self-Test redesigned

Chapter 18 Eukaryotic Microorganisms: The Fungi

- New material on the fungal mycobiome
- New figures
- Chapter Self-Test redesigned

Chapter 19 Eukaryotic Microorganisms: The Parasites

- New Clinical Case study
- Chapter Self-Test redesigned

Chapter 20 The Host-Microbe Relationship and Epidemiology

- New tables
- Several figures redesigned
- Narrative reorganized
- Chapter Self-Test redesigned

Chapter 21 Resistance and the Immune System: Innate Immunity

- New tables
- New figure for inflammation
- Chapter Self-Test redesigned

Chapter 22 Resistance and the Immune System: Adaptive Immunity

- Revised figures
- Chapter Self-Test redesigned

Chapter 23 Immunity and Serology

- Two MicroFocus features revised
- Chapter Self-Test redesigned

Chapter 24 Immunization and Serology

- Update on AIDS
- New figures
- Chapter Self-Test redesigned

Chapter 25 Applied and Industrial Microbiology

- Chapter material organized around food spoilage, food preservation, and industrial uses of microbes in food production (fermentation)
- Chapter Self-Test redesigned

Chapter 26 Environmental Microbiology

Chapter completely revised to incorporate some material from previous edition Chapter 27
- New chapter opener
- Chapter material organized around water pollution, water and sewage treatment, and microbial roles in biogeochemical recycling in the environment
- Chapter Self-Test redesigned

A GLOBAL PERSPECTIVE

Many decades ago, nursing and allied health students studying microbiology only needed to be concerned about infectious diseases as related to their community or geographic region. Today, with global travel, diseases from halfway around the world can be at our doorstep almost overnight. Therefore, students need a more global perspective of infectious disease and an understanding and familiarity with these diseases, which are presented no better than in this text.

MICROFOCUS features, such as public health articles, provide students with the information and understanding they need. Each article, such as the one about an emerging hemorrhagic fever, provides the background and significance needed for students to be informed and conversant. See page xi for the complete list of Public Health boxes.

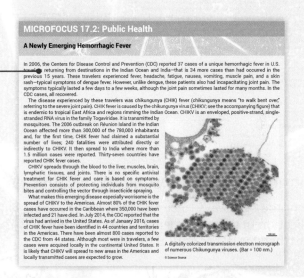

MICROFOCUS 17.2: Public Health

A Newly Emerging Hemorrhagic Fever

In 2006, the Centers for Disease Control and Prevention (CDC) reported 37 cases of a unique hemorrhagic fever in U.S. travelers returning from destinations in the Indian Ocean and India—that is 34 more cases than had occurred in the previous 15 years. These travelers experienced fever, headache, fatigue, nausea, vomiting, muscle pain, and a skin rash—typical symptoms of dengue fever. However, unlike dengue, these patients also had incapacitating joint pain. The symptoms typically lasted a few days to a few weeks, although the joint pain sometimes lasted for many months. In the CDC cases, all recovered.

The disease experienced by these travelers was chikungunya (CHIK) fever (chikungunya means "to walk bent over," referring to the severe joint pain). CHIK fever is caused by the chikungunya virus (CHIKV; see the accompanying figure) that is endemic to tropical East Africa and regions rimming the Indian Ocean. CHIKV is an enveloped, positive-strand, single-stranded RNA virus in the family Togaviridae. It is transmitted by mosquitoes. The 2006 outbreak on Réunion Island in the Indian Ocean affected more than 300,000 of the 780,000 inhabitants and, for the first time, CHIK fever had claimed a substantial number of lives; 240 fatalities were attributed directly or indirectly to CHIKV. It then spread to India where more than 1.5 million cases were reported. Thirty-seven countries have reported CHIK fever cases.

CHIKV spreads through the blood to the liver, muscles, brain, lymphatic tissues, and joints. There is no specific antiviral treatment for CHIK fever and care is based on symptoms. Prevention consists of protecting individuals from mosquito bites and controlling the vector through insecticide spraying.

What makes this emerging disease especially worrisome is the spread of CHIKV to the Americas. Almost 80% of the CHIK fever cases have occurred in the Caribbean where 350,000 have been infected and 21 have died. In July 2014, the CDC reported that the virus had arrived in the United States. As of January 2016, cases of CHIK fever have been identified in 44 countries and territories in the Americas. There have been almost 800 cases reported to the CDC from 44 states. Although most were in travelers, a few cases were acquired locally in the continental United States. It is likely that CHIKV will spread to new areas in the Americas and locally transmitted cases are expected to grow.

A digitally colorized transmission electron micrograph of numerous Chikungunya viruses. (Bar = 100 nm.)
© Science Source

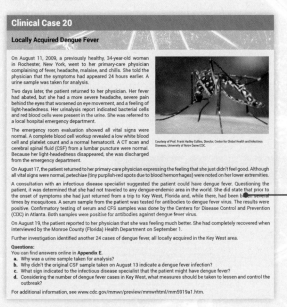

Clinical Case 20

Locally Acquired Dengue Fever

On August 11, 2009, a previously healthy, 34-year-old woman in Rochester, New York, went to her primary-care physician complaining of fever, headache, malaise, and chills. She told the physician that the symptoms had appeared 24 hours earlier. A urine sample was taken for analysis.

Two days later, the patient returned to her physician. Her fever had abated, but she had a more severe headache, severe pain behind the eyes that worsened on eye movement, and a feeling of light-headedness. Her urinalysis report indicated bacterial cells and red blood cells were present in the urine. She was referred to a local hospital emergency department.

The emergency room evaluation showed all vital signs were normal. A complete blood cell workup revealed a low white blood cell and platelet count and a normal hematocrit. A CT scan and cerebral spinal fluid (CSF) from a lumbar puncture were normal. Because her light-headedness had disappeared, she was discharged from the emergency department.

Courtesy of Prof. Frank Hadley Collins, Director, Center for Global Health and Infectious Diseases, University of Notre Dame/CDC.

On August 17, the patient returned to her primary-care physician expressing the feeling that she just didn't feel good. Although all vital signs were normal, petechiae (tiny purplish-red spots due to blood hemorrhages) were noted on her lower extremities.

A consultation with an infectious disease specialist suggested the patient could have dengue fever. Questioning the patient, it was determined that she had not traveled to any dengue-endemic area in the world. She did state that prior to the onset of symptoms she had just returned from a trip to Key West, Florida and, while there, had been bitten several times by mosquitoes. A serum sample from the patient was tested for antibodies to dengue fever virus. The results were positive. Confirmatory testing of serum and CSF samples was done by the Centers for Disease Control and Prevention (CDC) in Atlanta. Both samples were positive for antibodies against dengue fever virus.

On August 19, the patient reported to her physician that she was feeling much better. She had completely recovered when interviewed by the Monroe County (Florida) Health Department on September 1.

Further investigation identified another 24 cases of dengue fever, all locally acquired in the Key West area.

Questions:
You can find answers online in **Appendix E.**
a. Why was a urine sample taken for analysis?
b. Why didn't the original CSF sample taken on August 13 indicate a dengue fever infection?
c. What sign indicated to the infectious disease specialist that the patient might have dengue fever?
d. Considering the number of dengue fever cases in Key West, what measures should be taken to lessen and control the outbreak?

For additional information, see www.cdc.gov/mmwr/preview/mmwrhtml/mm5919a1.htm.

CLINICAL CASES also provide the global experience essential for student achievement and career success. These cases, such as the one on dengue fever, illustrate how a disease originally found in another part of the world has rapidly made it to our doorstep. See page x for the complete list of Clinical Cases.

REAL-LIFE APPLICATIONS

Some concepts and ideas in microbiology can be daunting and, at times, abstract to students studying the science. Providing students with real-life examples helps them see the significance of the concept and its application in the real world, be it their local community or worldwide.

CHAPTER CHALLENGES help students connect text material to the outside world while at the same time building their critical thinking skills. For example, foodborne illnesses are a growing concern locally, nationally, and globally. Yes, there are diseases associated with such food infections, but what about the prevention strategies? This and other chapter challenges help students see "beyond the textbook" to the real world.

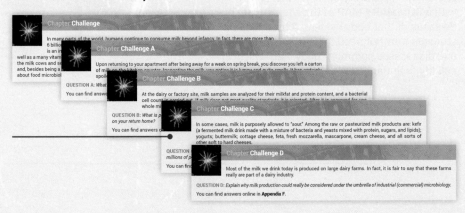

Chapter Challenge

In many parts of the world, humans continue to consume milk beyond infancy. In fact, there are more than 6 billion...

Chapter Challenge A

Upon returning to your apartment after being away for a week on spring break, you discover you left a carton of milk on the kitchen counter. Inspecting the milk, you notice it is lumpy and quite smelly. It has certainly spoiled...

QUESTION A: What...
You can answer...

Chapter Challenge B

At the dairy or factory site, milk samples are analyzed for their milkfat and protein content, and a bacterial cell count is carried out. If milk does not meet quality standards, it is rejected. After it is approved for use, whole mi...

QUESTION B: What is p...
on your return home?
You can find answers...

Chapter Challenge C

In some cases, milk is purposely allowed to "sour." Among the raw or pasteurized milk products are: kefir (a fermented milk drink made with a mixture of bacteria and yeasts mixed with protein, sugars, and lipids); yogurts; buttermilk; cottage cheese, feta, fresh mozzarella, mascarpone, cream cheese, and all sorts of other soft to hard cheeses.

QUESTION...
millions of p...
You can find...

Chapter Challenge D

Most of the milk we drink today is produced on large dairy farms. In fact, it is fair to say that these farms really are part of a dairy industry.

QUESTION D: Explain why milk production could really be considered under the umbrella of industrial (commercial) microbiology.
You can find answers online in **Appendix F.**

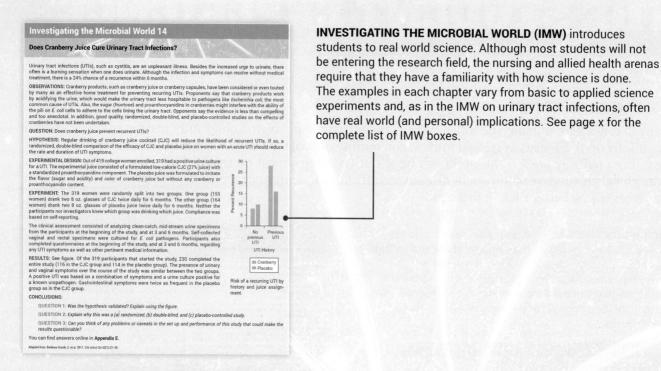

INVESTIGATING THE MICROBIAL WORLD (IMW) introduces students to real world science. Although most students will not be entering the research field, the nursing and allied health arenas require that they have a familiarity with how science is done. The examples in each chapter vary from basic to applied science experiments and, as in the IMW on urinary tract infections, often have real world (and personal) implications. See page x for the complete list of IMW boxes.

PRACTICE THROUGH THINKING AND DISCUSSING

One of the best ways to ensure mastery of a topic is through further thought and conversation. Again, the application to what a student has read will not only indicate if he or she has mastered the material, but also strengthen his or her critical thinking skills.

Many of the **MICROINQUIRY BOXES**, such as the one on smallpox, provide an opportunity for students to discuss what they have just read—and may ask for an opinion.

Sometimes the content students are trying to absorb becomes so dense they cannot "see the forest for the trees." Therefore, summary figures, diagrams, and tables can help them see the "forest." In the chapters on infectious diseases of the body systems, such as the one on the respiratory system, each ends with a **SUMMARY MAP** of the agents and diseases of that body system. Although the students may not need to know all the agents and diseases, a common "body map" will help solidify their understanding.

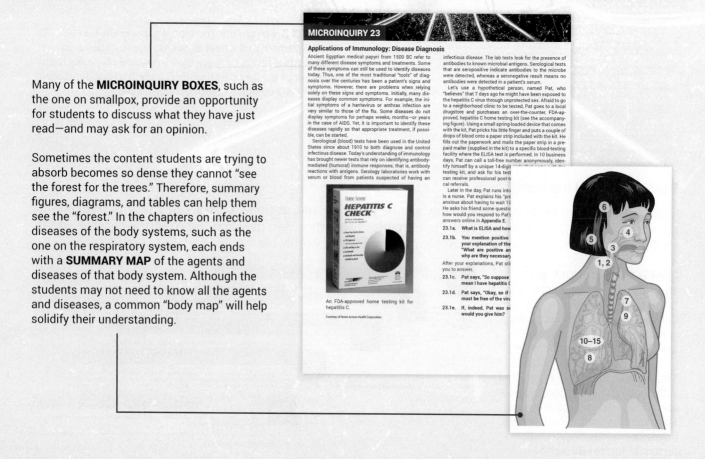

Jones & Bartlett Learning offers an assortment of supplements to assist students in mastering the concepts in this text.

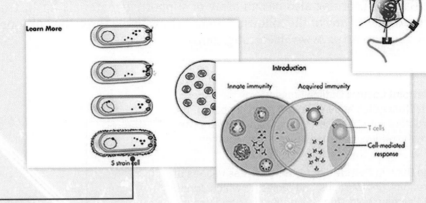

Animations: Engaging animations bring fascinating microbiology phenomena to life! Each animation guides students through microbiology processes and gauges students' progress and understanding with exercises and assessment questions introduced throughout each narrated animation.

Bonus eBook content: Two bonus chapters, "Applied and Industrial Microbiology" and "Environmental Microbiology," are available online.

Web Links: A variety of weblinks are available that present external website resources to continue your study of microbiology and keep up-to-date on what is happening in the field today.

Answer Key: Answers for the End-of-Chapter Questions, as well as the questions in the MicroInquiry, Chapter Challenge, Clinical Case, and Investigating the Microbial World feature boxes are available in the online Appendices D, E, and F (accessible with access card).

CHAPTER 25

Applied and Industrial Microbiology

CHAPTER PREVIEW

25.1 Food Spoilage Is Generally a Microbial Contamination and

25.2 Food Preservation Inhibits the of Foodborne Pathogens and Microorganisms

MicroInquiry 25: Keeping Microorganisms under Contro

25.3 Many Foods Are the Product The Microbial Ecology of Che

Investigating the Microbial W Metabolism

25.4 Microorganisms Are Used in Production of Many Industria

The idea was appealing and the price was right: a patty melt sandwich and a soft drink for lunch. The rye bread was toasted, the hamburgers were stacked and waiting to be cooked, the American cheese slices were lined up next to the grill, and the aroma from the sautéed onions was irresistible. However, on this fall day, the stage was set for one of the worst outbreaks of botulism in American history.

Between October 14 and 16, 1985, many people stopped by the restaurant and enjoyed patty melt sandwiches. Within two days, 28 individuals began experiencing the paralyzing signs of botulism. They suffered blurred vision, difficulty swallowing and chewing, and labored breathing. One by one, they called their doctors, and within a week, all were hospitalized. Twelve patients had to be placed on respirators and all but one recovered.

Investigators from the Centers for Disease Control and Prevention (CDC) soon arrived. They obtained detailed food histories from patients and from others who ate at the restaurant during the same 3-day period. First, they identified patty melt sandwiches as the probable cause (24 of 28 patients interviewed specifically recalled eating the sandwiches); then, they began a search to pinpoint the

item responsible for the illness. The in lected pointed to the onions. Investi *Clostridium botulinum* spores from the onions at the restaurant (see the ch image) and learned that after sauté were left uncovered on the warm st Furthermore, the onions were not re serving. Spores had probably germ anaerobic mounds of warm onions a tive cells deposited their deadly toxin

Incidents like this one occur to United States. The CDC estimates t 48 million Americans become sick every 6), over 128,000 are hospitali people die each year from foodbo However, each year the Foodborne D Surveillance Network (FoodNet) p the CDC reports on the changes in t

Light microscope image of gram-stained *Clostridium botulinum* cells and endospores (oval swellings).
©Michael Abbey/Science Source

2

CHAPTER 26

Environmental Microbiology

CHAPTER PREVIEW

26.1 Water Pollution Includes Biological Changes Harmful to Water Quality

26.2 Proper Treatment of Water and Sewage Ensures Safe Drinking Water

MicroInquiry 26: Doing a Standard Qualitative Water Analysis

26.3 Microbes Are Indispensable for Recycling Key Chemical Elements

Investigating the Microbial World 26: Fixated on Carbon

The year was 1887, and the farmers at a meeting in Delft, Holland all had the same problem. After planting the same crop in their fields for several years, the crop yields would drop. What could they do? At this meeting, a man named Martinus Beijerinck made a bold proposal to the assembled farmers. "Don't plant your crops in the same field as last year," he said. "Leave the field alone for the next 2 years; let it lie crop-free." Beijerinck's proposal was revolutionary because agricultural land in Holland was at a premium. Farmers could not afford to let their land lie unplanted for 2 years.

Beijerinck was a local bacteriologist. While his medical colleagues like Pasture and Koch were investigating the germ theory of disease and its implications, Beijerinck was out in the fields. He discovered that these fields were very productive when the land had just been cleared and freshly planted. These fields yielded bountiful crops when the farmer was away for a couple of years. From his investigations, Beijerinck thought he had the answer: large populations of bacterial organisms in the soil replenished needed crop nutrients.

Beijerinck was an expert on plants, but he also had something most other botanists lacked—a solid background in chemistry. He believed nitrogen was essential for plant growth, but he had no idea how nitrogen bridges the gap between atmosphere

and plant. Then, it dawned on him that bacterial organisms might be the bridge. More precisely, the little lumps and bumps on crop roots were the key. Repeatedly, he observed large communities of bacterial cells inside the little lumps and bumps ("nodules") on plant roots (see the chapter opening figure). He didn't see the nodules as often on tended crops, but they always seemed to be on wild plants growing in untended fields.

Beijerinck performed laboratory experiments to strengthen his views. He took bacterial cells from the nodules and inoculated them into seedlings of various plants. In many cases, the plants developed nodules, and when he planted the seedlings, the nitrogen content of the soil rose dramatically. Aha! The bacteria in the nodules were supplying needed nitrogen for plant growth. Thus, his advice in 1887 was straightforward: Leave the field alone for a spell; plant elsewhere; let the wild plants with nodules thrive and add nitrogen back to the soil. Then, when the field is finally planted again, the crop yield will be worth the wait.

Root nodules on soybean plant roots.
© Dan Guravich/Science Source

1

TEACHING TOOLS

Jones & Bartlett Learning also has an array of supportive materials for instructors. Additional information and review copies of any of the following items are available through your Jones & Bartlett sales representative or by going to http://www.jblearning.com.

The **PowerPoint Lecture Outline** presentation package provides lecture notes and images for each chapter of the text. Instructors with Microsoft PowerPoint software can customize the outlines, art, and order of presentation.

The **Image Bank in PowerPoint format** provides images of the illustrations, photographs, and tables (to which Jones & Bartlett Learning holds the copyright or has permission to reproduce digitally). These images are not for sale or distribution, but you can copy individual images or tables into your existing lecture presentations, test and quizzes, or other classroom materials.

An **Unlabeled Art Image Bank in PowerPoint format** provides selected images from the text with the labels removed so you can easily integrate them into your lectures, assignments, or exams.

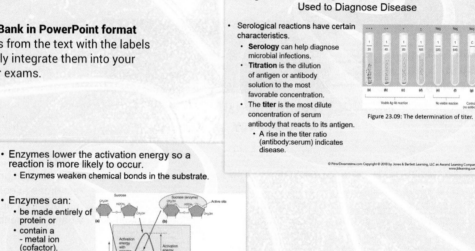

The **Instructor's Manual**, provided as a text file, includes an Instructional Overview, Instructional Objectives, Key Terms and Concepts, Chapter Teaching Points and Tips, and Essay Questions.

A robust **Test Bank**, including hundreds of assessment questions, is available.

Infectious Diseases: The *Guide to Infectious Diseases by Body Systems, Second Edition* is an excellent ancillary tool for learning about microbial diseases. Each of the fifteen body systems units presents a brief introduction to the anatomical system and the bacterial, viral, fungal, or parasitic organism infecting the system.

Encounters in Microbiology: *Encounters in Microbiology, Volume I, Second Edition*, and *Volume II* bring together "Vital Signs" articles from *Discover* magazine in which health professionals use their knowledge of microbiology in their medical cases.

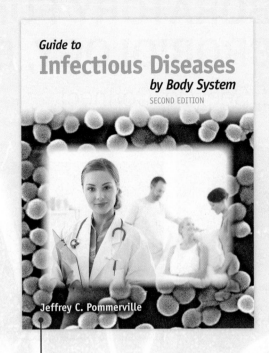

LABORATORY FUNDAMENTALS OF MICROBIOLOGY, ELEVENTH EDITION

Read. See. Do. Connect.

Packaged With Access To NEW Videos That Teach Common Lab Skills

WHAT'S NEW?

- Every new lab manual is packaged with access to over 110 minutes of NEW videos that teach common lab skills and are tied to the labs in the manual. Students clearly see how to work safely in the lab setting, how to use operating equipment, how to swab cultures, and many more valuable lab skills.

- NEW - all labs have been expanded and reorganized to fit neatly into new sections

- Features NEW introductions for each section

- Contains NEW full-color photos and micrograph examples

- Updated with NEW exercises and assessments

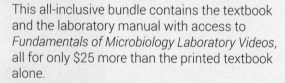

BUNDLE AND SAVE

BUNDLE *Fundamentals of Microbiology, Eleventh Edition* WITH *Laboratory Fundamentals of Microbiology, Eleventh Edition* and SAVE

This all-inclusive bundle contains the textbook and the laboratory manual with access to *Fundamentals of Microbiology Laboratory Videos*, all for only $25 more than the printed textbook alone.

Bundle ISBN: 978-1-284-14396-6

LABORATORY EXERCISES

Visit **go.jblearning.com/PommervilleLabManual** to view a sample video, sample lab exercises, and to see the complete list of 99 laboratory exercises.

FUNDAMENTALS OF MICROBIOLOGY LABORATORY VIDEOS

The Ultimate Microbiology Laboratory Experience

Welcome to the ultimate microbiology laboratory experience, with 110 minutes of instructor chosen, high-quality videos of actual students performing the most common lab skills, procedures, and techniques. From lab safety to microscopy to bacterial staining techniques, nowhere else will you find a more comprehensive and customized microbiology video program. Jeffrey Pommerville provides narration, context, and rationale for the on-screen action, and *Skills Checklists* are available online to record progress and for assignability. *Fundamentals of Microbiology Laboratory Videos* is the perfect companion to any modern microbiology laboratory course.

- 34 videos and over 110 minutes of quality production-value video!
- Visually demonstrates common skills typically covered in an introductory undergraduate microbiology lab (safety protocols, operating equipment, swabbing cultures, etc.)
- Jeffrey Pommerville provides the educational context for each lab skill, and actual students perform the techniques in a typical undergraduate lab setting
- *Skills Checklists* on PDF allows students to record progress and instructors to assign lab activities

BUNDLE AND SAVE

BUNDLE *Fundamentals of Microbiology, Eleventh Edition* WITH *Fundamentals of Microbiology Laboratory Videos* and SAVE

This bundle includes the textbook and standalone access to the lab skills videos for only $25 more than the printed textbook alone.

Bundle ISBN: 978-1-284-14435-2

STANDALONE ACCESS

Get the complete *Fundamentals of Microbiology Laboratory Videos* series for $99.95/user for one year's access.

LABORATORY VIDEOS

Section: Lab Safety
Laboratory Safety
Cleaning Up a Culture Spill

Section: Laboratory Techniques and Skills
How To Use a Gas Burner / Bacti-Cinerator to Sterilize an Inoculating Loop / Needle
Broth-To-Broth
Broth-To-Agar
Four-Way Streak Plate
Preparing the Pour Plate
Transferring a Solution with a Pipette
How To Make a Serial Dilution

Section: Microscopy
How To Use the Light Microscope: Identifying the Parts
Measuring Cell Size

Section: Bacterial Staining Techniques
How To Make a Bacterial Smear
How To Make a Simple Stain Technique
How To Make a Negative Stain Preparation
Preparing a Gram Stain
Preparing an Endospore Stain
How To Make a Capsule Stain Preparation
How To Make a Hanging Drop Preparation

Section: Control of Microorganisms
How To Prepare a Disk Diffusion Assay
How To Prepare the Kirby-Bauer Test

Section: Measuring Population Growth
How To Use a Petroff-Hausser Chamber
How To Prepare a Standard Plate Count
How To Use a Spectrophotometer

Section: Medical Microbiology
Preparing the Acid-Fast Stain
Oxidase Test
IMViC: Indole Test
IMViC: Methyl Red Test
IMViC: Voges-Proskauer Test
IMViC: Citrate Test

Section: Identification of a Bacterial Unknown
Biochemical Tests: Carbohydrate Fermentation
Starch Hydrolysis Test
Catalase Test
Hydrogen Sulfate
Urease Test

Visit **go.jblearning.com/PommervilleLabVideos** to view a sample video

ACKNOWLEDGMENTS

It takes a team of experts to put together a new edition of *Fundamentals of Microbiology*. Moreover, when the team members at Jones & Bartlett Learning are very talented professionals, my work is made easier. I wish to thank Executive Editor, Matthew Kane, who has been my "go to" person when I have had questions or I have needed guidance. As always, my Associate Editor, Audrey Schwinn, has coordinated the textbook revision with conductor-like control. Nancy Hoffmann, Senior Development Editor at Ascend Learning/Jones & Bartlett Learning, brought a new and dynamic leadership to the development of this edition. At the production end, my Production Editor, Dan Stone, managed the assembly process expertly. Jamey O'Quinn, Rights and Media Specialist, was super at locating new photos to illustrate the pages, and Troy Liston managed the revisions to the art program.

Throughout all my years of teaching at universities and colleges, I have had great fortune of working with great colleagues and outstanding students. My students keep me on my toes in the classroom, require me to always be prepared, and let me know when a topic or concept was not conveyed in as clear and understandable a way as it could (or should) be. Their suggestions and evaluations have encouraged me to continually assess my instruction, and make it the best it can be. I salute all my former students— and I hope those of you who read this text will let me know what works and what still needs improvement to make your learning effective, enjoyable, and most of all—successful.

Jeff Pommerville
Glendale Community College
Glendale, AZ

ABOUT THE AUTHOR

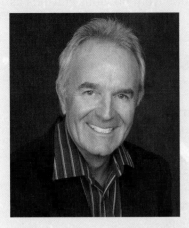

Today, I am a microbiologist, researcher, and science educator. My plans did not start with that intent. While in high school in Santa Barbara, California, I wanted to play professional baseball, study the stars, and own a '66 Corvette. None of those desires would come true—as a high school baseball player my batting average was miserable (but I was a good defensive fielder), I hated the astronomy correspondence course I took in high school, and I never bought that Corvette.

I found an interest in biology at Santa Barbara City College. After squeaking through college calculus, I transferred to the University of California at Santa Barbara (UCSB) where I received a B.S. in biology and stayed on to pursue a Ph.D. degree in the lab of Ian Ross studying cell communication and sexual pheromones in a water mold. After receiving my doctorate in cell and organismal biology, my graduation was written up in the local newspaper as a native son who was a fungal sex biologist—an image that was not lost on my three older brothers!

While in graduate school at UCSB, I rescued a secretary in distress from being licked to death by a German shepherd. Within a year, we were married (the secretary and I). When I finished my doctoral thesis, I spent several years as a postdoctoral fellow at the University of Georgia. Worried that I was involved in too many research projects, a faculty member told me something I will never forget. He said, "Jeff, it's when you can't think of a project or what to do that you need to worry." Well, I have never had to worry!

Moving to Texas A&M University, I spent 8 years in teaching and research—and telling Aggie jokes. Toward the end of this time, I realized I had a real interest in teaching and education. Leaving the sex biologist nomen behind, I headed farther west to Arizona to join the biology faculty at Glendale Community College, where I continue to teach introductory biology and microbiology.

I have been lucky to be part of several educational research projects. I was project director and lead principal investigator for a National Science Foundation grant to improve student outcomes in science through changes in curriculum and pedagogy. This culminated in my being honored with the Gustav Ohaus Award (College Division) for Innovations in Science Teaching from the National Science Teachers Association.

For 6 years I was the Perspectives Editor for the *Journal of Microbiology and Biology Education*, the education research journal of the American Society for Microbiology (ASM). I have been co-chair for the ASM Conference for Undergraduate Educators and chair of the Undergraduate Education Division of ASM. My dedication to teaching and mentoring students has been recognized by an Outstanding Instructor Award at Glendale Community College and, nationally, the Carski Foundation Distinguished Undergraduate Teaching Award for distinguished teaching of microbiology to undergraduate students and encouraging them to subsequent achievement.

I mention all this not to impress, but to show how the road of life sometimes offers opportunities in unexpected and unplanned ways. The key though is keeping your "hands on the wheel and your eyes on the prize;" then unlimited opportunities will come your way. And, hey, who knows—maybe that '66 Corvette could be in my garage yet.

◼ Dedication

This is the fifth edition of *Fundamentals of Microbiology* that I have authored. Over these 14 years, I have spent countless months revising and updating the text, often unintentionally neglecting time that should be spent with my wife. Therefore, I dedicate this eleventh edition of the textbook to my wife, Yvonne. She always has supported my passion for teaching and has encouraged me to push forward throughout the textbook revision, often providing valuable and constructive suggestions. Thanks for your support and encouragement, and enduring love through the years.

■ Reviewers for the *Eleventh Edition*

As always, it is the input, suggestions, and comments from instructors, and students alike, that evolve a textbook and make each edition an improvement on its predecessor. I thank everyone from previous editions as well as the reviewers for this edition for their time and effort with the review.

Mari Aanenson, M.S., Western Illinois University

Vasanta Lakshmi Chivukula, Ph.D., Atlanta Metropolitan State College

Heather M. Craig, Ph.D., Monterey Peninsula College

Eric DeAngelo, B.A, M.S., Lehigh Carbon Community College

Sanhita Gupta, Ph.D., Kent State University

Daniel P. Klossner, Ph.D., SUNY Delhi

Mary Ann Merz, M.Ed., MT(ASCP), West Virginia Northern Community College

Gregory Paquette, Ph.D., MLS(ASCP)SM, University of Rhode Island

Omar Perez, Ph.D., Malcolm X College

Joan Petersen, Ph.D., Queensborough Community College

Julia M. Schmitz, Ph.D., Piedmont College

Robert C. Sizemore, Ph.D., Alcorn State University

Mary Ann Smith, M.A., Marywood University

Ayodotun O. Sodipe, Ph.D., Texas Southern University

Steven J. Thurlow, M.S., Jackson College

Mary Emery Vollertsen, BGS, M.S.c. Southeast Arkansas College

Patrick M. Weir, Ph.D., Felician College

TO THE STUDENT—STUDY SMART

Your success in microbiology and any college or university course will depend on your ability to study effectively and efficiently. Therefore, this textbook was designed with you, the student, in mind. The text's organization will help you improve your learning and understanding and, ultimately, your grades. The learning design concept described in the Preface and illustrated below reflects this organization. Study it carefully, and, if you adopt the flow of study shown, you should be a big step ahead in your preparation and understanding of microbiology—and for that matter any subject you are taking.

When I was an undergraduate student, I hardly ever read the "To the Student" section (if indeed one existed) in my textbooks because the section rarely contained any information of importance. This one does, so please read on.

In college, I was a mediocre student until my junior year. Why? Mainly because I did not know how to study properly, and, important here, I did not know how to read a textbook effectively. My textbooks were filled with underlined sentences (highlighters hadn't been invented yet!) without any plan on how I would use this "emphasized" information. In fact, most textbooks assume you know how to read a textbook properly. I didn't, and you might not, either.

Reading a textbook is difficult if you are not properly prepared. So that you can take advantage of what I learned as a student and have learned from instructing thousands of students, I have worked hard to make this text user friendly with a reading style that is not threatening or complicated. Still, there is a substantial amount of information to learn and understand, so having the appropriate reading and comprehension skills is critical. Therefore, I encourage you to spend 30 minutes reading this section, as I am going to give you several tips and suggestions for acquiring those skills. Let me show you how to be an active reader. Note: the Student Study Guide also contains similar information on how to take notes from the text, how to study, how to take class (lecture) notes, how to prepare for and take exams, and perhaps most important for you, how to manage your time effectively. It all is part of this "learning design," my wish to make you a better student.

Be a Prepared Reader

Before you jump into reading a section of a chapter in this text, prepare yourself by finding the place and time and having the tools for study.

Place. Where are you right now as you read these lines? Are you in a quiet library or at home? If at home, are there any distractions, such as loud music, a blaring television, or screaming kids? Is the lighting adequate to read? Are you sitting at a desk or lounging on the living room sofa? Get where I am going? When you read for an educational purpose—that is, to learn and understand something—you need to maximize the environment for reading. Yes, it should be comfortable but not to the point that you will doze off.

Time. All of us have different times during the day when we perform some skill, be it exercising or reading, the best. The last thing you want to do is read when you are tired or simply not "in tune" for the job that needs to be done. You cannot learn and understand the information if you fall asleep or lack a positive attitude. I have kept the chapters in this text to about the same length so you can estimate the time necessary for each and plan your reading accordingly. If you have done your preliminary survey of the chapter or chapter section, you can determine about how much time you will need. If 40 minutes is needed to read—and comprehend (see below)—a section of a chapter, find the place and time that will give you 40 minutes of uninterrupted study. Brain research suggests that most people's brains cannot spend more than 45 minutes in concentrated, technical reading. Therefore, I have avoided lengthy presentations and instead have focused on smaller sections, each with its own heading. These should accommodate shorter reading periods.

Reading Tools. Lastly, as you read this, what study tools do you have at your side? Do you have a highlighter or pen for emphasizing or underlining important words or phrases? Notice, the text has wide margins, which allow you to make notes or to indicate something that needs further clarification. Do you have a pencil or pen handy to make these notes? Or, if you do not want to "deface" the text, make your notes in a notebook. Lastly, some students find having a ruler is useful to prevent your eyes from wandering on the page and to read each line without distraction.

■ Be an Explorer Before You Read

When you sit down to read a section of a chapter, do some preliminary exploring. Look at the section head and subheadings to get an idea of what is discussed. Preview any diagrams, photographs, tables, graphs, or other visuals used. They give you a better idea of what is going to occur. We have used a good deal of space in the text for these features, so use them to your advantage. They will help you learn the written information and comprehend its meaning. Do not try to understand all the visuals, but try to generate a mental "big picture" of what is to come. Familiarize yourself with any symbols or technical jargon that might be used in the visuals.

The end of each chapter contains a **Summary of Key Concepts** for that chapter. It is a good idea to read the summary before delving into the chapter. That way you will have a framework for the chapter before filling in the nitty-gritty information.

■ Be a Detective as You Read

Reading a section of a textbook is not the same as reading a novel. With a textbook, you need to uncover the important information (the terms and concepts) from the forest of words on the page. So, the first thing to do is read the complete paragraph. When you have determined the main ideas, highlight or underline them. However, I have seen students highlighting the entire paragraph in yellow, including every a, the, and and. This is an example of highlighting before knowing what is important. So, I have helped you out somewhat. Important terms and concepts are in **bold face** followed by the definition. So only highlight or underline with a pen essential ideas and key phrases—not complete sentences, if possible. By the way, the important microbiological terms and major concepts are also in the **Glossary** at the back of the text.

What if a paragraph or section has no boldfaced words? How do you find what is important here? From an English course, you may know that often the most important information is mentioned first in the paragraph. If it is followed by one or more examples, then you can backtrack and know what was important in the paragraph. In addition, I have added section "speed bumps" (called **Concept and Reasoning Checks**) to let you test your learning and understanding before getting too far ahead in the

material. These checks also are clues to what was important in the section you just read.

■ Be a Repetitious Student

Brain research has shown that each individual can only hold so much information in short-term memory. If you try to hold more, then something else needs to be removed—sort of like a full computer disk. So that you do not lose any of this important information, you need to transfer it to long-term memory—to the hard drive if you will. In reading and studying, this means retaining the term or concept; so, write it out in your notebook using your own words. Memorizing a term does not mean you have learned the term or that you understand the concept. By actively writing it out in your own words, you are forced to think and actively interact with the information. This repetition reinforces your learning.

■ Be a Patient Student

In textbooks, you cannot read at the speed that you read your e-mail or a magazine story. There are unfamiliar details to be learned and understood—and this requires being a patient, slower reader. Actually, if you are not a fast reader to begin with, as I am, it may be an advantage in your learning process. Identifying the important information from a textbook chapter requires you to slow down your reading speed. Speed-reading is of no value here.

■ Know the What, Why, and How

Have you ever read something only to say, "I have no idea what I read!" As I've already mentioned, reading a microbiology text is not the same as reading *Sports Illustrated* or *People* magazine. In these entertainment magazines, you read passively for leisure or perhaps amusement. In *Fundamentals of Microbiology, Eleventh Edition*, you must read actively for learning and understanding—that is, for comprehension. This can quickly lead to boredom unless you engage your brain as you read—that is, be an active reader. Do this by knowing the *what, why,* and *how* of your reading.

▶ *What* is the general topic or idea being discussed? This often is easy to determine because the section heading might tell you. If not, then it will appear in the first sentence or beginning part of the paragraph.

- ▶ *Why* is this information important? If I have done my job, the text section will tell you why it is important, or the examples provided will drive the importance home. These surrounding clues further explain why the main idea was important.
- ▶ *How* do I "mine" the information presented? This was discussed under being a detective.

■ A Marked-Up Reading Example

So let's put words into action. Below is a passage from the text. I have marked up the passage as if I were a student reading it for the first time. It uses many of the hints and suggestions I have provided. Remember, it is important to read the passage slowly and concentrate on the main idea (concept) and the special terms that apply.

■ KEY CONCEPT 4.4   Most Prokaryotic

The **cell envelope** is a complex structure that forms the two "wrappings"—the **cell wall** and the **cell membrane**—around the cell cytoplasm. The cell envelope helps protect the integrity of the cell. However, the cell wall is relatively porous to the diffusion of substances, so the **cell membrane** controls the transport of nutrients and metabolic products into and out of the cell.

The Bacterial Cell Wall Is a Tough and Protective External Shell

A key feature of most prokaryotic cells is the **cell wall**. By covering the entire cell surface, the cell wall acts as an exoskeleton to provide structural integrity to the cell and, along with the cytoplasmic cytoskeleton, to provide cell shape. The cell wall also anchors other molecules (i.e., pili and flagella) that extend out from the cell surface.

The cell wall prevents the cell from expanding and bursting due to the high osmotic forces pushing against the cell membrane. Most microbes live in an environment where there are more dissolved materials inside the cell than outside. The **hypertonic** condition in the cytoplasm means water diffuses inward, accounting for the increased osmotic pressure. Without a cell wall, the cell would rupture through a process called **osmotic lysis** (**FIGURE 4.9**).

■ Have a Debriefing Strategy

After reading the material, be ready to debrief. Verbally summarize what you have learned. This will start moving the short-term information into the long-term memory storage—that is, retention. Any notes you made concerning confusing material should be discussed as soon as possible with your instructor. For microbiology, allow time to draw out diagrams. Again, repetition makes for easier learning and better retention.

In many professions, such as sports or the theater, the name of the game is practice, practice, practice. The hints and suggestions I have given you form a skill that requires practice to perfect and use efficiently. Be patient, things will not happen overnight; perseverance and willingness though will pay off with practice. You might also check with your college or university academic (or learning) resource center. These folks will have more ways to help you to read a textbook better and to study well overall.

■ Concept Maps

In science as well as in other subjects you take at the college or university, there often are concepts that appear abstract or simply so complex that they are difficult to understand. A concept map is one tool to help you enhance your abilities to think and learn. Critical reasoning and the ability to make connections between complex, nonlinear information are essential to your studies and career.

Concept maps are a learning tool designed to represent complex or abstract information visually. Neurobiologists and psychologists tell us that the brain's primary function is to take incoming information and interpret it in a meaningful or practical way. They also have found that the brain has an easier time making sense of information when it is presented in a visual format. Importantly, concept maps not only present the information in "visual sentences" but also take paragraphs of material and present it in an "at-a-glance" format. Therefore, you can use concept maps to

- ▶ Communicate and organize complex ideas in a meaningful way
- ▶ Aid your learning by seeing connections within or between concepts and knowledge
- ▶ Assess your understanding or diagnose misunderstanding

▶ There are many different types of concept maps. The two most used in this textbook are the process map or flow chart and the hierarchical map. The hierarchical map starts with a general concept (the most inclusive word or phrase) at the top of the map and descends downward using more specific, less general words or terms. In several chapters in this textbook process or hierarchical maps are drawn—and you have the opportunity to construct your own hierarchical maps as well.

Concept mapping is the strategy used to produce a concept map. So, let's see how one makes a hierarchical map.

How to Construct a Concept Map

1. Print the central idea (concept or question to be mapped) in a box at the top center of a blank, unlined piece of paper. Use uppercase letters to identify the central idea.

2. Once the concept has been selected, identify the key terms (words or short phrases) that apply to or stem from the concept. Often these may be given to you as a list. If you have read a section of a text, you can extract the terms from that material, as the words are usually boldfaced or italicized.

3. Now, from this list, try to create a hierarchy for the terms you have identified; that is, list them from the most general, most inclusive to the least general, most specific. This ranking may only be approximate and subject to change as you begin mapping.

4. Construct a preliminary concept map. This can be done by writing all of the terms on Post-its®, which can be moved around easily on a large piece of paper. This is necessary as one begins to struggle with the process of building a good hierarchical organization.

5. The concept map connects terms associated with a concept in the following way:
   - The relationship between the concept and the first term(s), and between terms, is connected by an arrow pointing in the direction of the relationship (usually downward or horizontal if connecting related terms).

   - Each arrow should have a label, a very short phrase that explains the relationship with the next term. In the end, each link with a label reads like a sentence.

6. Once you have your map completed, redraw it in a more permanent form. Box in all terms that were on the sticky notes. Remember there may be more than one way to draw a good concept map, and don't be scared off if at first you have some problems mapping; mapping will become more apparent to you after you have practiced this technique a few times using the opportunities given to you in the early chapters of the textbook.

7. Now look at the map and see if it answers the following. Does it:
   - Define clearly the central idea by positioning it in the center of the page?
   - Place all the terms in a logical hierarchy and indicate clearly the relative importance of each term?
   - Allow you to figure out the relationships among the key ideas more easily?
   - Permit you to see all the information visually on one page?
   - Allow you to visualize complex relationships more easily?
   - Make recall and review more efficient?

Example

After reading the section on "Protein Synthesis," a student makes a list of the terms used and maps the concept. Using the steps outlined above, the student produces the following hierarchical map. Does it satisfy all the questions asked in (7)?

Practical Uses for Mapping

▶ Summarizing textbook readings. Use mapping to summarize a chapter section or a whole chapter in a textbook. This purpose for mapping is used many times in this text.

▶ Summarizing lectures. Although producing a concept map during the classroom period may not be the best use of the time, making a concept map or maps from the material after class will help you remember the

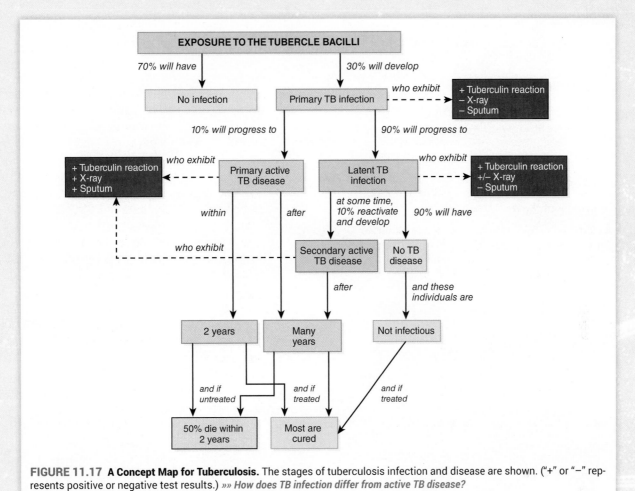

EXPOSURE TO THE TUBERCLE BACILLI

70% will have

30% will develop

No infection

Primary TB infection

who exhibit → + Tuberculin reaction
− X-ray
− Sputum

10% will progress to

90% will progress to

+ Tuberculin reaction
+ X-ray
+ Sputum

who exhibit

Primary active
TB disease

Latent TB
infection

who exhibit

+ Tuberculin reaction
+/− X-ray
− Sputum

within

after

at some time,
10% reactivate
and develop

90% will have

who exhibit

Secondary active
TB disease

No TB
disease

after

and these
individuals are

2 years

Many
years

Not infectious

and if
untreated

and if
treated

and if
treated

50% die within
2 years

Most are
cured

FIGURE 11.17 A Concept Map for Tuberculosis. The stages of tuberculosis infection and disease are shown. ("+" or "−" represents positive or negative test results.) »» *How does TB infection differ from active TB disease?*

important points and encourage high-level, critical reasoning, which is so important in university and college studies.

▶ Reviewing for an exam. Having concept maps made ahead of time can be a very useful and productive way to study for an exam, particularly if the emphasis of the course is on understanding and applying abstract, theoretical material, rather than on simply reproducing memorized information.

▶ Working on an essay. Mapping also is a powerful tool to use during the early stages of writing a course essay or term paper. Making a concept map before you write the first rough draft can help you see and ensure you have the important points and information you will want to make.

■ Send Me a Note

In closing, I would like to invite you to write me and let me know what is good about this textbook so I can build on it and what may need improvement so I can revise it. Also, I would be pleased to hear about any news of microbiology in your community, and I'd be happy to help you locate any information not covered in the text.

I wish you great success in your microbiology course. Welcome! Let's now plunge into the wonderful and often awesome world of microorganisms.

—Dr. P
Email: jeffrey.pommerville@gccaz.edu
Web site: http://www.gccaz.edu/directory/jeffrey-pommerville

The six chapters in Part I embrace several aspects of the six principal (overarching) concepts as described in the ASM Fundamental Statements for a concept-based microbiology curriculum.

PRINCIPAL CONCEPTS

Evolution	• Cells, organelles (e.g., mitochondria and chloroplasts), and all major metabolic pathways evolved from early prokaryotic cells. **Chapters 3 and 4** • The evolutionary relatedness of organisms is best reflected in phylogenetic trees. **Chapter 3**
Cell Structure and Function	• The structure and function of microorganisms have been revealed by the use of microscopy (including bright field, phase contrast, fluorescent, and electron). **Chapters 3 and 4** • Bacteria have unique cell structures that can be targets for antibiotics, immunity, and phage infection. **Chapter 4** • Bacteria and Archaea have specialized structures (e.g., flagella, endospores, and pili) that often confer critical capabilities. **Chapter 4** • While microscopic eukaryotes (for example, fungi, protozoa, and algae) carry out some of the same processes as bacteria, many of the cellular properties are fundamentally different. **Chapter 3**
Metabolic Pathways	• The interactions of microorganisms among themselves and with their environment are determined by their metabolic abilities (e.g., quorum sensing, oxygen consumption, nitrogen transformations). **Chapters 1 and 3** • The survival and growth of any microorganism in a given environment depend on its metabolic characteristics. **Chapters 2, 5, and 6**
Information Flow and Genetics	• Genetic variations can impact microbial functions (e.g., in biofilm formation, pathogenicity, and drug resistance). **Chapter 3**
Microbial Systems	• Microorganisms are ubiquitous and live in diverse and dynamic ecosystems. **Chapters 1–6** • Most bacteria in nature live in biofilm communities. **Chapters 3 and 4**
Impact of Microorganisms	• Microorganisms provide essential models that give us fundamental knowledge about life processes. **Chapter 1** • Because the true diversity of microbial life is largely unknown, its effects and potential benefits have not been fully explored. **Chapters 1 and 3**

Microbiology: Then and Now

Space. The final frontier! Really? *The* final frontier? The image on the right in the chapter opening illustration shows one of an estimated 350 billion large galaxies and more than 10^{24} stars in the visible universe. However, the invisible microbial universe on Earth consists of more than 10^{30} **microorganisms** (or **microbes** for short) distributed among an estimated 1 trillion (10^{12}) species. As the image on the left shows, microbes might be microscopic in size, but they are magnificent in their evolutionary diversity and astounding in their sheer numbers. Together, we can refer to the global community of microbes and their genes as Earth's **microbiome**.

A Day in the Life of a Microorganism

Because microorganisms exist in such diversity and vast numbers in the oceans, landmasses, and atmosphere, they must play important roles in the very survival of other organisms on the planet. Consequently, could understanding these microscopic organisms on Earth be as important as studying stars and galaxies in space? Let's uncover but a few examples of what a "day in the life of a microorganism" is like in shaping the fundamental life processes around the globe.

The Oceans

The oceans and seas cover more than 70% of planet Earth and represent the foundation that maintains our planet in a habitable condition. A critical factor in this maintenance is the marine microbiome, which, composed of some 3×10^{29} microbes, dominates ocean life. High densities of these organisms can be found anywhere from the frozen polar regions to the hot, volcanic thermal vents and the

---

Photographic image of nearby galaxies in space (right) and a light microscope image (left) of bacterial cells (larger dots), viruses (smaller dots), and a diatom (center).

(left) Courtesy of Jed Alan Fuhrman, University of Southern California. (right) Courtesy of JPL-Caltech/NASA.

cold seeps on the dark seafloor. Making up 90% of the ocean biomass, this marine microbiome is critical to regulating life on Earth.

Every day, many of the dwellers in this microbial community perform the following:

▶ Create the foundation for all marine food webs by performing photosynthesis (in sunlit areas) and chemosynthesis (in dark areas) to convert carbon into sugars and nutrients on which all fish and ocean mammals directly or indirectly depend.

▶ Provide up to 50% of the oxygen gas we breathe and many other organisms use to stay alive by performing photosynthesis (**FIGURE 1.1A**).

▶ Control atmospheric aerosols and cloud formation through the sea spray ejected into the atmosphere.

▶ Consume 50% of the dead plant and animal matter generated on Earth each year.

▶ Operate as the engines that drive and control nutrient and mineral cycling and that regulate energy flow, both of which can affect long-term climate change.

Viruses are the most abundant infectious agents on Earth. There are an estimated 10^{31} viruses in the oceans and most of these viruses remain uncharacterized. However, they probably infect marine microbes. Because many of these microbes are important in the activities listed above, viruses undoubtedly play an influential role in the global cycling of nutrients and elements, such as carbon and nutrients. In fact, infections by marine viruses are responsible for releasing and recycling some 150 gigatons of carbon per year.

Today, microbiologists around the world are discovering many marine microbes that are new to science. What are the daily occupations of these and other mysterious microbes and viruses yet to be identified and cataloged? Undoubtedly, many play useful and probably intimate roles, the consequences of which we have yet to discover. The scientific studies are just beginning.

The Land

Microbes on dry land are no less impressive in their daily activities than their marine counterparts. In fact, every time you walk on the ground, you step on billions of microbes. Moreover, like their cousins in the oceans, they can be found in every imaginable place, from the tops of the highest mountains to the deepest caves. In fact, a kilogram of moist garden soil can have up to 10^{13} microbes living in the water-filled pore spaces in the soil (**FIGURE 1.1B**).

This diverse soil microbiome is responsible for many daily activities essential to life. Here are some of the activities soil microorganisms perform:

▶ Carry out 90% of all biochemical reactions occurring in the soil.

▶ Recycle dead plant and animal material and return carbon dioxide to the atmosphere.

▶ Represent a source for many of today's antibiotics.

▶ Provide plants with important nutrients, such as nitrogen, sulfur, and phosphorus. Think of it this way—microbes are helping to feed the world.

Microbes also can degrade pesticides and other synthetic pollutants contaminating the soil. Many of these microbes have been "domesticated" for this very role through a process called **bioremediation**. As new species are identified and studied, the roles of the soil microbiome will be unraveled.

The Atmosphere

More than 315 different types of bacteria have been identified in air masses 10 kilometers above the Earth's surface. These microbes along with fungal spores account for 20% of all particles—biological and non-biological—in the atmosphere (**FIGURE 1.1C**). Although less studied than the ocean and soil microbiomes, scientists believe the atmospheric communities perform the following roles:

▶ Play an integral role with marine microbes in the formation of water vapor (clouds).

▶ Help form raindrops and snowflakes.

▶ Affect the chemical composition of the atmosphere.

▶ Influence weather cycles, alter the composition of rain and snow, and affect daily weather patterns.

Although we have much to learn and understand about microbes in the clouds, it is evident that they help stabilize the atmosphere.

The Earth's Subsurface

Consider that the oceans, soil, and air represent only those environments most familiar to us. In fact, a diverse microbial workforce is being cataloged anywhere there is an energy source and water. For example, scientists have drilled 2,400 meters into the Earth's subsurface on land or below the seafloor. In the solid rock, they have discovered another

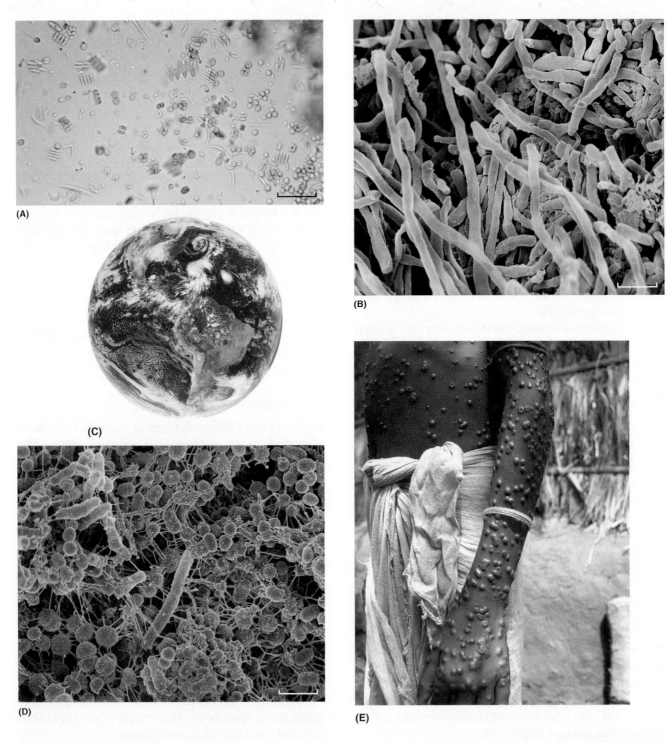

FIGURE 1.1 Daily Life in the Microbial World. Microbes play many roles. For example, **(A)** photosynthetic microbes inhabit the upper sunlit layer of almost all oceans and bodies of fresh water where they produce food molecules that sustain the aquatic food web and generate oxygen gas. (Bar = 20 μm.) **(B)** In the soil, microbes degrade dead plants and animals, form beneficial partnerships with plants, and recycle carbon, nitrogen, and sulfur. (Bar = 5 μm.) **(C)** Besides their involvement in the formation of raindrops and snowflakes, microbes in the atmosphere are important for water vapor to condense into clouds that help cool the Earth. **(D)** Large numbers of microbes can be found on and in the body where most play beneficial roles for our health. (Bar = 2 μm.) **(E)** A few microbes have played disease roles and affected world health. This 1974 photo of a Bengali boy shows the effects of smallpox, which was responsible for 300–500 million deaths during the 20th century. *»» What would happen to life on Earth if each of the examples above (A–D) was devoid of microbes?*

frontier composed of microbes that are sealed off from the rest of the world. These **intraterrestrials** (microbes living in sediment and rock) are another diverse workforce that, even buried in rock, are believed to make up more microbial biomass than that of all the microbes in water and soil. Scientists believe these intraterrestrials also are involved with the daily recycling of minerals and stabilizing the biogeochemical health of our planet.

Today, the workforce composing the global microbiome is still being cataloged and their numbers and daily activities keep growing. As Louis Pasteur, one of the fathers of microbiology, once stated, "*Life* [plant and animal] *would not long remain possible in the absence of microbes.*"

That need for microbes is evident even much closer to home. Microbial communities inhabit the bodies of all plants and animals. For animals, most every species, from termites to bees, from cows to humans, has an intimate microbiome associated with it. For example, the **human microbiome** consists of some 40 trillion (40×10^{12}) microbes, which is about equal to the number of cells (30×10^{12}) building the human body. These outwardly invisible strangers have established themselves since birth as separate and unique communities of microbes on the skin, in the respiratory tract, and especially in the gut (**FIGURE 1.1D**). In fact, the human gut microbiome is essential for a healthy life. Although some are transient, most of the species spend each day carrying out an amazing array of metabolic reactions to help us resist disease, regulate our digestion, maintain a strong immune system—and even influence our risk of obesity, asthma, and allergies. To be human and healthy, we must share our daily lives with this homegrown microbiome.

One aspect of a microbe's daily life that we have overlooked until now is its role in infectious disease. This omission was done on purpose because only a very small minority of microbes are responsible for infections and disease. Although such disease-causing agents, called **pathogens**, are rare, some, such as those causing diseases like plague, malaria, and smallpox (**FIGURE 1.1E**), throughout history have swept through cities and villages, devastated populations, killed great leaders and commoners alike, and, as a result, have transformed politics, economies, and public health on a global scale. Nevertheless, as you read through the chapters of this text, remember only a small minority of microbes are dedicated pathogens.

A major focus of this introductory chapter is to give you an introspective "first look" at microbiology—then and now. We will see how microbes were first discovered and why infectious disease preoccupied the minds and efforts of so many. Along the way, we will see how curiosity and scientific inquiry stimulated the quest to understand the microbial world just as the science of microbiology does today, as described in the box **MICROBIOLOGY PATHWAYS**. To begin our story, we reach back to the 1600s, where we encounter some very inquisitive individuals.

Chapter **Challenge**

Now that you have learned a little about the diverse global workforce of microorganisms, can you explain how the microbes were revealed and identify several challenges that some microorganisms pose for microbiology today? Let's investigate!

MICROBIOLOGY PATHWAYS

Being a Scientist

Science might not seem like the most glamorous profession. In fact, as you read many of the chapters in this text, you might wonder why many scientists have the good fortune to make key discoveries. At times, it might seem like it is the luck of the draw, but actually many scientists share a set of characteristics that puts them on the trail to success.

Robert S. Root-Bernstein, a physiology professor at Michigan State University, points out that many prominent scientists like to goof around, play games, and surround themselves

© Comstock/Thinkstock.

(continues)

MICROBIOLOGY PATHWAYS (*Continued*)

with a type of chaos aimed at revealing the unexpected. Their labs might appear to be in disorder, but they know exactly where every tube or bottle belongs. Scientists also identify intimately with the organisms or creatures they study (it is said that Louis Pasteur actually dreamed about microorganisms), and this identification brings on an intuition—a "feeling for the organism." In addition, there is the ability to recognize patterns that might bring a breakthrough. (Pasteur had studied art as a teenager and therefore had an appreciation of patterns.)

The geneticist and Nobel laureate Barbara McClintock once remarked, "I was just so interested in what I was doing I could hardly wait to get up in the morning and get at it. One of my friends, a geneticist, said I was a child, because only children can't wait to get up in the morning to get at what they want to do." Clearly, another characteristic of a scientist is having a child-like curiosity for the unknown.

Another Nobel laureate and immunologist, Peter Medawar, once said, "Scientists are people of very dissimilar temperaments doing different things in very different ways. Among scientists are collectors, classifiers, and compulsive tidiers-up; many are detectives by temperament and many are explorers; some are artists and others artisans. There are poet-scientists and philosopher-scientists and even a few mystics." In other words, scientists come from all lifestyles.

For this author, I too have found science to be an extraordinary opportunity to discover and understand something never before known. Science is fun, yet challenging—and at times arduous, tedious, and frustrating. As with most of us, we will not make the headlines for a breakthrough discovery or find a cure for a disease. However, as scientists we all hope our hard work and achievements will contribute to a better understanding of a biological (or microbiological) phenomenon and will push back the frontiers of knowledge and have a positive impact on society.

Like any profession, being a scientist is not for everyone. Besides having a bachelor's degree in biology or microbiology, you should be well read in the sciences and capable of working as part of an interdisciplinary team. Of course, you should have good quantitative and communication skills, have an inquisitive mind, and be goal oriented. If all this sounds interesting, maybe you fit the mold of a scientist. Why not consider pursuing a career in microbiology? Some possibilities are described in other **MICROBIOLOGY PATHWAYS** included in this text, but you should also visit with your instructor. Simply stop by the student union, buy two cups of coffee, and you are on your way.

■ KEY CONCEPT 1.1   The Discovery of Microbes Leads to Questioning Their Origins

As the 17th century arrived, an observational revolution was about to begin. A Dutch spectacle maker named Zacharias Janssen and others discovered that if they combined two curved lenses together, they could magnify small objects. Many individuals in Holland, England, and Italy further developed this combination of lenses that in 1625 would go by the term *microscopio* or "microscope." This new invention would be the forerunner of the modern-day instrument.

Microscopy—Discovery of the Very Small

Robert Hooke, an English natural philosopher (the term "scientist" was not coined until 1833), was one of the most inventive and ingenious minds in the history of science. As the curator of experiments for the Royal Society of London, Hooke published a book in 1665, called *Micrographia*, in which he took advantage of the magnification abilities of the early microscopes to make detailed drawings of many living objects, including fleas, lice, and peacock feathers. Perhaps his most famous description was for the structure of cork. Seeing "a great many little boxes," he called these boxes *cella* (= rooms), and from that observation today we have the word "cell" to describe the basic unit of life.

Micrographia represents one of the most important books in science history. It awakened the learned and general population of Europe to the world of the very small, revolutionized the art of scientific investigation, and showed that the microscope was an important tool for unlocking the secrets of an unseen world: the world of the cell.

At this same time, across the North Sea in Delft, Holland, Antoni van Leeuwenhoek, a successful tradesman and dry goods dealer, was using hand lenses to inspect the quality of his cloth. As such, and without any scientific training, Leeuwenhoek became skilled at grinding single pieces of glass into fine magnifying lenses. Placing such a lens between two metal plates riveted together, Leeuwenhoek's "simple microscope" could greatly out magnify Hooke's microscope (**FIGURE 1.2A, B**).

Lens

Specimen mount

Screw plate

Focusing screw

Elevating screw

(A)

(B)

(C)

FIGURE 1.2 Viewing Animalcules. (A) To view his animalcules, Leeuwenhoek placed his sample on the tip of the specimen mount that was attached to a screw plate. An elevating screw moved the specimen up or down, whereas the focusing screw pushed against the metal plate, moving the specimen toward or away from the lens. **(B)** Holding the microscope up to the bright light, Leeuwenhoek then looked through the lens to view his sample. **(C)** From such observations, Leeuwenhoek drew the animalcules he saw (bacteria in this drawing). *»» Why would Leeuwenhoek believe his living (and often moving) creatures were tiny animals?*

Aware of Hooke's *Micrographia*, Leeuwenhoek turned his microscope to the invisible world. Beginning in 1673 and lasting until his death in 1723, Leeuwenhoek communicated his microscope observations through more than 300 letters to England's Royal Society. In 1674, one of his first letters described a sample of lake water. Placing the sample before his lens, he described hundreds of tiny, moving particles he thought were minute, living animals, which he called **animalcules**. In fact, this discovery represents one of the most important observations in history because Leeuwenhoek had described and illustrated for the first time the microbial world. In 1676, he mixed pepper with a previously collected sample of snow water. After three weeks, he found more animalcules, including the first unmistakable observations of bacteria (**FIGURE 1.2C**). Among the other letters sent to the Royal Society, he described all the different types of microbes (except viruses) that we know of today. This included structural details of yeast cells, thread-like fungi, and microscopic algae and protozoa.

The process of "observation" is an important skill for all scientists and remains the cornerstone of all scientific inquiry. Hooke and Leeuwenhoek are excellent examples of individuals with sound observational skills. Unfortunately, Leeuwenhoek invited no one to work with him, nor did he show anyone how he made his lenses. Thus, naturalists at the time found it difficult to repeat and verify his observations, which also are key components of scientific inquiry. Still, Leeuwenhoek's observations of animalcules opened yet a second door to another entirely new world: the world of the microbe.

Do Animalcules Arise Spontaneously?

In the early 1600s, most naturalists were **vitalists**, individuals who thought life depended on a mysterious and pervasive "vital force" in the air. This force provided the basis for the doctrine of **spontaneous generation**, which suggested that some forms of life could arise from nonliving, often decaying matter. Others also embraced the idea, for they too witnessed toads that appeared from mud, snakes coming from the marrow of a decaying human spine, and rats arising from garbage wrapped in rags.

Resolving the reality of such bizarre beliefs would require a new form of investigation—experimentation—and a new generation of experimental naturalists arose.

Redi's Experiments

In 1668, the Italian naturalist Francesco Redi performed one of history's first biological experiments. As described in **INVESTIGATING THE MICROBIAL WORLD 1**, his experiment was designed to test the

Investigating the Microbial World 1

Can Life Arise Spontaneously?

For centuries, many people, learned and not, believed that some forms of life could arise spontaneously from nonliving, decaying matter. For example:

OBSERVATION: In the 17th century, many people believed that fly maggots (larvae) arose spontaneously from rotting meat. Francesco Redi set out to find the answer.

QUESTION: *Do fly maggots arise spontaneously from rotting meat?*

HYPOTHESIS: Redi proposed that fly maggots arise from hatched eggs laid in decaying meat by flies. If so, preventing flies from laying eggs in the rotting meat should result in no maggots being generated.

EXPERIMENTAL DESIGN: Redi obtained similar pieces of rotting meat and jars in which the meat would be placed.

EXPERIMENT: One piece of meat was placed in an open jar, whereas the other piece was placed in a similar jar that was then covered with a piece of gauze to keep out any flying insects, including flies. The meat in each jar was allowed to rot.

RESULTS: See figure.

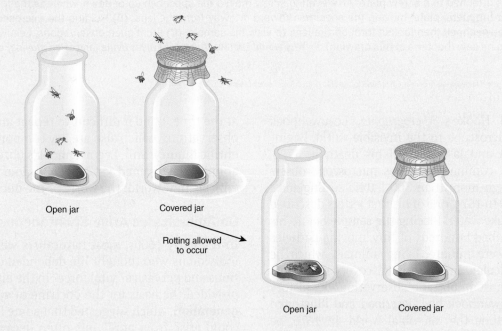

Open jar   Covered jar

Rotting allowed
to occur

Open jar   Covered jar

CONCLUSIONS:

QUESTION 1: *From Redi's experiments, was his hypothesis validated? Explain using the figure.*

QUESTION 2: *What is the control in this experiment, and why was it important to have a control?*

QUESTION 3: *Why was it important that Redi used gauze to cover the one jar and not seal it completely? Hint: Remember what people believed about the "vital force."*

You can find answers online in **Appendix E.**

Modified from Redi, F. 1688. As reprinted by Open Court Publishing Company, Chicago (1909).

belief that worm-like maggots (fly larvae) could arise spontaneously from rotting meat.

Although Redi's experiments verified that spontaneous generation did not produce larger living creatures like maggots, what about the mysterious and microscopic animalcules that appeared to straddle the boundary between the nonliving and living world? Could they arise spontaneously?

Needham's Experiments

In 1745, a British clergyman and naturalist, John Needham, proposed that the spontaneous generation of animalcules resulted from a vital force that reorganized decaying matter into life. Needham presented experiments showing that animalcules could arise spontaneously in flasks of animal broth that previously had been heated and then kept at room temperature for several days. He was convinced that the vital force provided the stimulus needed for spontaneous generation.

Spallanzani's Experiments

Experiments often can be subject to varying interpretations. As such, the Italian cleric and naturalist Lazzaro Spallanzani challenged Needham's conclusions and suggested that the animalcules came from the air and would therefore grow in the broth of the cooled flasks. To validate his claim, in 1765, he repeated Needham's experiments by boiling flasks with broth for an extended time before sealing some flasks. After 2 days, the broths in the open flasks were swarming with animalcules, whereas the sealed ones contained no animalcules. Spallanzani concluded that microbes from the air, and not spontaneous generation, accounted for the presence of animalcules in the flasks.

The controversy over spontaneous generation of animalcules continued into the mid-1800s and only deepened when Rudolf Virchow, a German pathologist, put forward, without direct evidence, the idea of **biogenesis**, which said that life arises only from life. To solve the debate concerning microbial spontaneous generation, a new experimental strategy would be needed.

Pasteur's Experiments

Louis Pasteur, a French chemist and scientist, took up the challenge of spontaneous generation, and in 1861 he designed an elegant series of experiments that were a variation of the methods of Needham and Spallanzani. The box **MICROINQUIRY 1** outlines

---

MICROINQUIRY 1

Scientific Inquiry and Spontaneous Generation

Science is a systematic way of thinking and learning about the natural world. Often we accept and integrate into our understanding new information because it appears consistent with what we believe is true. But, are we sure our ideas are always correct? In science, **scientific inquiry**, or what has been called the "scientific method," is a way to examine those ideas.

Scientific inquiry often uses **deductive reasoning**, which begins with general observations, builds a hypothesis, carries out well-designed and carefully executed experiments to test the hypothesis, and reaches a specific, logical conclusion. Pasteur used this inquiry approach in 1861 to test the idea of spontaneous generation of microorganisms. Let's see how it worked:

Observations: When studying a problem, the inquiry process usually begins with observations. For spontaneous generation, Pasteur's earlier observations suggested that organisms do not appear from nonliving matter but rather from the air in which they are located.

Question: Next comes the question, which can be asked in many ways but usually as a "what," "why," or "how" question. For example, "What accounts for the generation of microorganisms in the animal broth?"

Hypothesis: A hypothesis is a provisional but testable explanation for an observed phenomenon or observation. Pasteur's hypothesis was that "Organisms appearing in the sterile flask come from other organisms in the air." If so, air allowed to enter the sterile flask will contain organisms that will grow in such profusion that they will be seen as a cloudy liquid.

Let's analyze the experiments. In Pasteur's experimental design, only one **variable** (an adjustable condition) changed. In the accompanying figure, one flask remained intact, whereas two others were exposed to the air by

(continues)

MICROINQUIRY 1 (*Continued*)

Scientific Inquiry and Spontaneous Generation

(**A**) breaking the neck or (**B**) tipping the flask. Pasteur kept all other factors the same; that is, all of the broths were heated and cooled for the same length of time, and the flasks were identical. Thus, the experiments had a rigorous **control** (the comparative condition wherein the flask remained intact). Pasteur's finding that no organisms appeared in the intact flask is interesting but tells us very little by itself. Its significance comes by comparing it to the broken neck flask and tipped flask where organisms quickly appeared.

Conclusion: Pasteur's hypothesis was supported by the experiments.

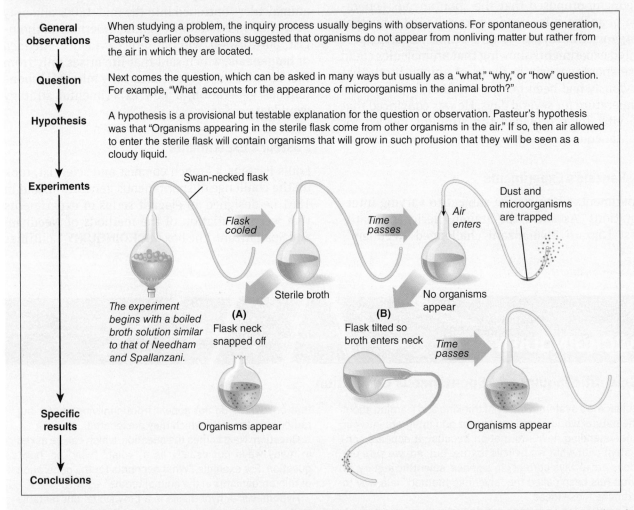

General observations — When studying a problem, the inquiry process usually begins with observations. For spontaneous generation, Pasteur's earlier observations suggested that organisms do not appear from nonliving matter but rather from the air in which they are located.

Question — Next comes the question, which can be asked in many ways but usually as a "what," "why," or "how" question. For example, "What accounts for the appearance of microorganisms in the animal broth?"

Hypothesis — A hypothesis is a provisional but testable explanation for the question or observation. Pasteur's hypothesis was that "Organisms appearing in the sterile flask come from other organisms in the air." If so, then air allowed to enter the sterile flask will contain organisms that will grow in such profusion that they will be seen as a cloudy liquid.

Experiments

Swan-necked flask

Flask cooled

Time passes

Air enters

Dust and microorganisms are trapped

Sterile broth

No organisms appear

The experiment begins with a boiled broth solution similar to that of Needham and Spallanzani.

(**A**) Flask neck snapped off

(**B**) Flask tilted so broth enters neck

Time passes

Specific results

Organisms appear

Organisms appear

Conclusions

Pasteur and the Spontaneous Generation Controversy. If broth sterilized in a swan-necked flask is left open to the air, the curvature of the neck traps dust particles and microorganisms, preventing them from reaching the broth. However, if the neck is snapped off to allow in air or the flask is tipped so broth enters the neck, organisms encounter the broth and grow.

the process of scientific inquiry and Pasteur's experiments.

Although Pasteur's experiments generated considerable debate for several years, his exacting and carefully designed experiments disproved the belief in spontaneous generation and validated the idea of biogenesis.

Today there is another form of "spontaneous generation" occurring—this time in the research laboratory, as **MICROFOCUS 1.1** relates.

MICROFOCUS 1.1: Biotechnology

Generating Life—Today

Those who believed in spontaneous generation proposed that animalcules arose from the rearrangement of molecules released from decayed organisms. Although this idea for the generation of life was incorrect, today we are getting closer to generating new life by rearranging molecules in the laboratory.

In 2002, scientists at the State University of New York, Stony Brook, reconstructed a poliovirus by assembling separate poliovirus genes and proteins. A year later, Craig Venter and his group assembled a bacteriophage—a virus that infects bacterial cells—from "off-the-shelf" biomolecules. Then, in 2010, another team, again led by Venter, synthesized from scratch the complete genetic sequence of the bacterium *Mycoplasma mycoides* and inserted the sequence into a cell of *Mycoplasma capricolum* from which its genetic information previously had been removed (see figure). This "new organism" now functioned like an *M. mycoides* cell.

However, this is not "generating new life." Rather, it is putting a new set of genes into another cell that never before had that set of genes. This new field of biology is called **synthetic biology**, and it aims to rebuild or create new "life forms" (such as bacterial cells) from scratch by recombining molecules taken from other organisms. It is like fashioning a new car by taking various parts from a Ford and Chevy and assembling them on a Toyota chassis.

Why do this? The design and construction of novel organisms, having functions very different from naturally occurring organisms, present the opportunity to expand evolution's repertoire by fabricating cells that are better at doing specific jobs. Can we, for example, design bacterial cells that are better at degrading toxic wastes, providing alternative energy sources, or making cheaper pharmaceuticals? These and many other positive benefits are envisioned as outcomes of synthetic biology and the generation of new life.

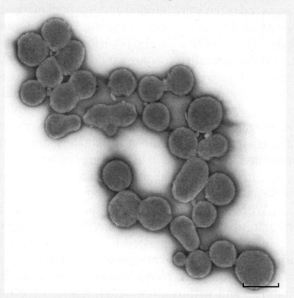

A false-color electron microscope image of synthetic *Mycoplasma* cells. (Bar = 1 μm.)

© Thomas Deerinck, NCMIR/Science Source.

Concept and Reasoning Checks 1.1

a. How were the first microbes discovered?
b. If you were alive in Leeuwenhoek's time, how would you explain the origin for the animalcules he saw with his simple microscope?
c. Evaluate the role of experimentation as an important skill to the eventual rejection of spontaneous generation as an origin for animalcules (and microbes).

▉ KEY CONCEPT 1.2   Disease Transmission Can Be Prevented

In the 13th century, people knew diseases could be transmitted between individuals, so they enforced isolations, called **quarantines**, in an attempt to prevent disease spread. Other interventions to prevent infection can be traced back to the 14th century. At that time, and throughout history, smallpox was a contagious, disfiguring, and often deadly disease that affected humans. In an effort to prevent individuals from contracting smallpox, the Chinese practiced **variolation**, which involved blowing a ground smallpox powder into the nose of individuals. By the 18th century, Europeans were inoculating dried smallpox scabs under the skin of the arm. Although some individuals did get smallpox, most contracted only a mild form of the disease and, upon recovery, were protected against subsequent smallpox exposures.

Vaccination Prevents the Spread of Smallpox

In the late 1700s, smallpox epidemics were prevalent throughout Europe. In England, smallpox epidemics were so severe that one-third of the children died before the age of three and many victims who recovered often were blinded and left pockmarked.

As an English country surgeon, Edward Jenner learned that milkmaids who occasionally contracted a mild disease called cowpox would subsequently be protected from contracting smallpox. Because cowpox was not deadly, Jenner hypothesized that intentionally giving cowpox to people should protect them against smallpox. In 1796, he took a scraping of a cowpox blister from a milkmaid's hand and scratched it into the skin of a young boy's arm. The boy soon developed a slight fever but recovered. Six weeks later, Jenner infected the boy with smallpox pus. Within days, the boy developed a reaction at the skin site but failed to show any sign of smallpox.

In 1798, Jenner repeated his experiments with others, verifying his technique of **vaccination** (*vacca* = "cow"). Prominent physicians soon confirmed his findings, and within a few years, Jenner's method of vaccination spread through Europe and abroad. However, it would not be until 1980, some 284 years after Jenner's first smallpox vaccinations, that the World Health Organization (WHO) would certify that smallpox had been eradicated globally through a massive vaccination effort that was carried out between 1966 and 1979.

In retrospect, it is remarkable that without any knowledge of microbes or disease causation, Jenner accomplished what he did. Again, hallmarks of a scientist—keen observational skills and insight—led to a therapeutic intervention against disease. His work, along with others, would lay the foundation for the field of **immunology**, the study of the body's defenses against, and responses to, foreign organisms and substances.

Disease Transmission Does Not Result from a Miasma

So, the question remains. What causes infectious disease? In the 1700s, the prevalent belief among naturalists and laypersons alike was that disease resulted from a **miasma**, which was thought to arise from a foul quality of the atmosphere or from tiny

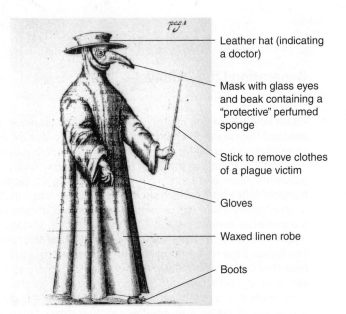

Leather hat (indicating a doctor)

Mask with glass eyes and beak containing a "protective" perfumed sponge

Stick to remove clothes of a plague victim

Gloves

Waxed linen robe

Boots

FIGURE 1.3 Dressed for Protection. This dress was thought to protect a plague doctor from the air (miasma) that caused the plague. *»» How would each item of dress offer protection?*

Courtesy of National Library of Medicine.

poisonous particles of rotting matter expelled into the air (the word "malaria" comes from *mala aria*, meaning "bad air"). To protect oneself from the black plague, for example, plague doctors in Europe often wore an elaborate costume they thought would protect them from the plague miasma (**FIGURE 1.3**).

As the 19th century unfolded, more scientists were relying on keen observations and experimentation as a way of understanding and explaining disease transmission—and challenging the idea of miasmas.

Epidemiology, as applied to infectious diseases, is another example of scientific inquiry—in this case to identify the source, cause, and mode of transmission of disease within populations. The first such epidemiological studies, carried out by Ignaz Semmelweis and John Snow, were instrumental in suggesting how diseases were transmitted and how simple measures could prevent transmission.

Semmelweis and "Cadaver Matter"

Ignaz Semmelweis was a Hungarian obstetrician who, at the age of 29, was appointed chief resident at a large maternity hospital in Vienna, Austria. Soon after his arrival, he was shocked by the numbers of women in his ward who were dying of **childbed fever** (a type of blood poisoning also called puerperal fever) following labor. The box **CLINICAL CASE 1** outlines Semmelweis' investigations into

Clinical Case 1

Childbed Fever: A Historical Reflection

In 1844–1846, many mothers in the First Maternity Division of the Vienna General Hospital, which was run by doctors and medical students, contracted a serious disease called childbed fever (puerperal fever). Up to 11% of the mothers died from the illness. However, in the adjacent Second Maternity Division of the same hospital run by midwives, the death toll from childbed fever was less than 3% over the same period.

As a member of the medical staff of the First Division, Ignaz Semmelweis searched for an explanation for the high mortality in his division.

Most of the medical staff attributed the illnesses and deaths of puerperal fever to an unavoidable miasma. Semmelweis discounted this belief because how could one reconcile the fact that while the fever was raging in the First Division, few cases occurred in the Second Division, and hardly a case occurred in the surrounding city of Vienna?

Others attributed the fever to overcrowding in the First Division. Semmelweis pointed out that, in fact, the crowding was heavier in the Second Division. In addition, there were no differences between the two divisions with regard to diet, general care, and method of examination of maternity patients.

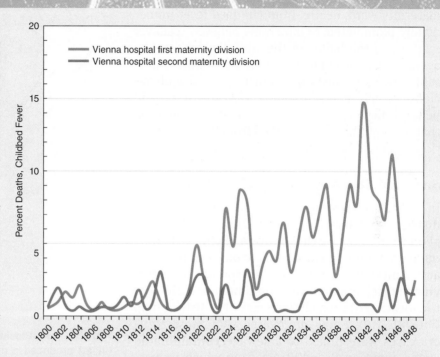

Yearly Mortality for Childbed Fever 1800–1849. Semmelweis collected data on deaths of birthgiving mothers in the Vienna maternity hospital (red line) and compared the deaths to those in the second maternity division where midwives focused on attending to births and did not concern themselves with anatomical pathology. »» *Why do Semmelweis' observations on the prevalence of childbed fever not support a miasma as the cause of and transmission for the disease?*

Data from Semmelweis (1861).

In 1847, Jakob Kolletschka, professor of forensic medicine in the hospital and a colleague of Semmelweis, was instructing medical students in the art of medical dissection of cadavers in the morgue (deadhouse). While performing an autopsy, Kolletschka punctured a finger with a bloody scalpel. He ended up dying from the infection and Semmelweis observed that Kolletschka exhibited the same symptoms as the victims of childbed fever.

Semmelweis also noted that (a) in the Second Division, the midwives neither received anatomical instruction nor did they dissect cadavers in the deadhouse; (b) he and his medical students examined women in labor in the First Division without cleaning up after performing their dissections in the deadhouse (see figure).

Questions:
a. How was the childbed fever illness transmitted to Professor Kolletschka?
b. From Semmelweis' observations, what was the source of childbed fever?
c. Why was the incidence of the disease so low in the Second Division?
d. How was the agent of childbed fever transmitted to maternity patients in the First Division?

You can find answers online in **Appendix E**.

For additional information, read *The Doctors' Plague: Germs, Childbed Fever, and the Strange Story of Ignác Semmelweis* by Sherwin B. Nuland. (New York: W.W. Norton & Company. 2004.)

the source and mode of transmission of the disease and the intervention he introduced to stop the transmission of childbed fever.

After Semmelweis directed his staff to wash their hands in chlorine water before entering the maternity ward, deaths from childbed fever immediately plummeted. Semmelweis believed "cadaver matter" on the hands of the staff doctors was the agent of disease and its transmission could be interrupted by hand washing. Unfortunately, few physicians initially heeded Semmelweis' measures and he was relieved of his position in 1849. It is believed that he might have died in 1865 from the very infectious agent he had studied.

Snow and "Organized Particles"

In 1854, a cholera epidemic hit London's Soho district. With many residents dying, English surgeon John Snow carried out one of the first thorough epidemiological studies to discover how this acute intestinal infection was spread. By interviewing sick and healthy Londoners and plotting the location of each cholera case on a district map (**FIGURE 1.4**), Snow discovered that most cholera cases were linked to a sewage-contaminated water pump on Broad Street from which numerous local residents obtained their household drinking water.

Although Snow did not know the cause of cholera, he hypothesized that the Broad Street water pump was the source of the infectious agent. Snow instituted the first known example of a public health measure to prevent disease transmission—he requested the parish Board of Guardians to remove the handle from the Broad Street pump! Cholera cases dropped and again disease spread was broken by a simple procedure.

Similar to Semmelweis' cadaver matter, Snow believed "organized particles" in the water caused cholera; this was another hypothesis that proved to be correct even though the causative agent would not be identified for another 29 years.

It is important to realize that although the miasma premise was incorrect, the fact that disease was associated with bad air and filth led to new hygiene measures, such as cleaning streets, laying new sewer lines in cities, and improving working conditions. These changes helped usher in the **Sanitary Movement** and create the infrastructure for the public health systems we have today, which is a team effort explained in MICROFOCUS 1.2.

FIGURE 1.4 Blocking Disease Transmission. John Snow (inset) produced a map plotting all the cholera cases (black rectangles) in the London Soho district and observed a cluster (circle) near the Broad Street pump (red arrow). »» *Why would removing the pump handle stop the spread of cholera?*

Courtesy of Frerichs, R. R. John Snow website: http://www.ph.ucla.edu/epi/snow.html, 2006. Inset courtesy of National Library of Medicine.

The Stage Is Set

During the early years of the 1800s, other events occurred that helped set the stage for the coming "germ revolution." In the 1830s, advances were made in microscope optics that allowed better resolution of objects. This resulted in improved and more widespread observations of tiny living organisms, many of which resembled short sticks. In fact, in 1838 the German biologist Christian Ehrenberg suggested these "rod-like" looking organisms be called **bacteria** (*bakter* = "rod").

To understand clearly the nature of infectious disease, a new concept of disease had to emerge. In doing so, it would be necessary to demonstrate that a specific infectious disease was caused by a specific "germ." This would require some very insightful work, guided by Louis Pasteur in France and Robert Koch in Germany.

MICROFOCUS 1.2: Public Health

Epidemiology—Today

On September 2, 2011, the Colorado Department of Public Health and Environment reported an abnormally high number of people admitted to local hospitals suffering from fever and muscle aches as well as diarrhea or other gastrointestinal symptoms. Doctors quickly diagnosed the illness as a bacterial infection caused by *Listeria monocytogenes*.

In the footsteps of Semmelweis and Snow, today's health experts needed to locate rapidly the source and transmission mechanism of this outbreak. This often requires an entire team of medical investigators, including those from the Centers for Disease Control and Prevention (CDC), state and local health departments, and the Coordinated Outbreak Response and Evaluation (CORE) Network, which is part of the Food and Drug Administration (FDA). See figure.

For this outbreak, the CORE team was made up of epidemiologists, microbiologists, health specialists, consumer safety officers, and policy analysts. The CORE surveillance team confirmed that *Listeria* was the infectious agent, while the response team identified locally grown *Listeria*-contaminated whole cantaloupes as the source of the outbreak. They then issued a national warning, as the number of cases had expanded beyond Colorado.

After the outbreak, the CORE post-response team discovered that the farm growing the cantaloupes had failed to follow safe food-handling practices in the facility where the whole cantaloupes were stored and packed for shipment. The farm also failed to adequately clean the processing equipment, which had been used incorrectly, leading to the contamination.

Even with this rapid response, the *Listeria* cantaloupe contamination was the deadliest foodborne disease outbreak in the United States in nearly 90 years, killing 30 people and sickening 146 in 28 states. Unlike the time of Semmelweis and Snow, today's potential for disease spread requires numerous response teams with the expertise and know-how like CORE. With an ever-present danger of emerging disease, epidemiology today is a team effort that remains a critical tool in the fight against infectious disease.

A group of FDA medical investigators form a Coordinated Outbreak Response and Evaluation (CORE) team.

Courtesy of FDA.

Concept and Reasoning Checks 1.2

a. How do the procedures of variolation and vaccination contradict the concept of miasmas?

b. How did the work of Semmelweis and Snow challenge the widely held idea of miasmas as the cause of infectious disease?

Chapter Challenge A

We are beginning to see how microbes and pathogens were discovered through the observations and studies of natural philosophers, a vaccine pioneer, and the first epidemiologists.

QUESTION A: *What three infectious diseases described in this section were thought to be caused by a miasma? What types of studies and actions suggested these diseases were not the product of a miasma?*

You can find the answers online in **Appendix F**.

■ KEY CONCEPT 1.3   The Classical Golden Age of Microbiology Reveals the Germ

Beginning around 1854, the association of microbes with the disease process blossomed. Over the next 60 years, the foundations would be laid for the maturing process that has led to the modern science of microbiology. This period is referred to as the first, or classical, Golden Age of microbiology.

Louis Pasteur Proposes That Germs Cause Infectious Disease

Trained as a chemist, Louis Pasteur (**FIGURE 1.5A**) was among the first scientists who believed that problems in science could be solved in the laboratory with the results having practical applications.

Always one to tackle big problems, Pasteur soon set out to understand the chemical process of fermentation. The prevailing theory in the 1850s held that fermentation was strictly a chemical process and the yeasts needed for fermentation were simply inert chemical "globules" catalyzing the process. Pasteur's microscope observations, however, consistently revealed large numbers of tiny yeast cells in fermented juice. When he mixed yeast in a sugar-water solution in the absence of air, the yeast multiplied and converted the sugar to alcohol. Fermentation, then, was a biological process and yeasts were living agents responsible for fermentation.

Pasteur also demonstrated that wines, beers, and vinegar each contained different and specific types of microorganisms that had specific properties. For example, in studying a local problem of wine souring, he observed that only the soured wines contained populations of bacterial cells (**FIGURE 1.5B**). Pasteur concluded that these cells must have contaminated a batch of yeast and, as another example of a biological process, produced the acids that caused the souring. Pasteur recommended a practical solution for the "wine disease" problem: heat the wine gently to kill the harmful bacterial cells but not so strongly as to affect the quality of the wine. His controlled heating technique, which would become known as **pasteurization**, soon was applied to other products. Today, pasteurization is a universal method used to kill pathogens and retard spoilage in milk and many other foods and beverages.

If wine disease was caused by tiny, living bacteria, Pasteur reasoned that human infections and disease also might be caused by other microorganisms

(A)   (B)

FIGURE 1.5 Louis Pasteur and Fermentation Bacteria. (A) Louis Pasteur as a 46-year-old professor of chemistry at the University of Paris. **(B)** A drawing of some of the bacterial cells Pasteur observed in soured wine. *»» Why would such cells exist in a wine bottle that had soured?*

in the environment—what he called **germs**. Thus, Pasteur formulated the **germ theory of disease**, which held that some microorganisms are responsible for infectious disease.

The Work of Lister and Pasteur Stimulate Disease Control and Reinforce the Germ Theory

Pasteur had reasoned that if germs were acquired from the environment, their spread could be controlled and the chain of disease transmission broken, as Semmelweis and Snow had shown years earlier.

Lister and Antisepsis

Hearing of Pasteur's germ theory, Joseph Lister, a professor of surgery at Glasgow Royal Infirmary in Scotland, wondered if germs were responsible for the large number of postoperative infections among his amputation patients. So, in 1865, knowing that carbolic acid had been effective on sewage control, Lister used a carbolic acid spray in surgery and on surgical wounds (**FIGURE 1.6**). The results were spectacular—the wounds healed without infection in almost 70% of his cases. His technique would soon not only revolutionize medicine and the practice of surgery but also lead to the practice of **antisepsis**, the use of chemical methods for disinfection of external living surfaces, such as the skin.

Therefore, germs can come from the environment—and they can be controlled.

Silkworm Disease and Cholera

Between 1857 and 1878, Pasteur sought more evidence for his germ theory idea. In 1865, Pasteur had the opportunity to study pébrine, an infectious disease of silkworms. After several setbacks and 5 years of work, he finally identified a new type of germ, unlike the bacterial cells and yeast he had observed with his microscope. These tiny germs, which he called "corpuscular parasites", were the infectious agent in silkworms and on the mulberry leaves fed to the worms. By separating the healthy silkworms from the diseased silkworms and their contaminated food, he managed to quell the spread of the silkworm disease. The identification of the pathogen was crucial to supporting the germ theory, and Pasteur would never again doubt the ability of germs to cause infectious disease.

Also in 1865, cholera engulfed Paris, killing 200 people a day. Determined to find the causative agent, Pasteur tried to capture the responsible germ by filtering the air around hospitalized patients and trapping the germs in cotton. Unfortunately, Pasteur could not grow or separate one type of germ from others because his broth cultures allowed all the organisms on the cotton to mix freely. To validate the germ theory, what was missing was the ability to isolate a

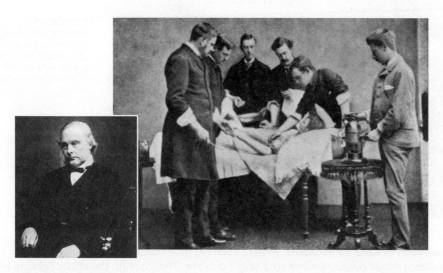

FIGURE 1.6 Lister and Antisepsis. Joseph Lister (inset) and his students used a carbolic acid spray in surgery and on surgical wounds to prevent postoperative infections. *»» Hypothesize how carbolic acid prevented surgical infections.*

specific germ from a diseased individual and demonstrate the isolated germ caused that same disease.

Robert Koch Formalizes Standards to Equate Germs with Infectious Disease

Robert Koch (**FIGURE 1.7A**) was a German country doctor who was well aware of anthrax, a deadly disease that periodically ravaged cattle and sheep and also could cause disease in humans. Was anthrax caused by a germ?

In 1875, Koch injected mice with the blood from sheep suffering anthrax. He noticed the mice soon developed the same disease signs seen in the sheep. Next, he isolated a few rod-shaped bacterial cells from the blood and, with his microscope, watched for hours as the bacterial cells multiplied, formed tangled threads, and finally reverted to spores. He then

took several spores on a sliver of wood and injected them into healthy mice. The signs of anthrax soon appeared, and when Koch autopsied the animals, he found their blood swarming with the same type of bacterial cells. Here was the first evidence that a specific germ was the causative agent of a specific disease.

Koch found that growing bacterial cells was not very convenient. Then, in 1880, he observed a slice of potato on which small masses of bacterial cells, which he termed **colonies**, were multiplying in number. Koch tried adding gelatin to his broth to prepare a similar solid culture surface. He then inoculated bacterial cells on the surface and set the dish aside to incubate. Within 24 hours, visible colonies were growing on the surface, each colony representing a **pure culture** containing only one bacterial type. By 1884, a polysaccharide called **agar** that was derived from marine

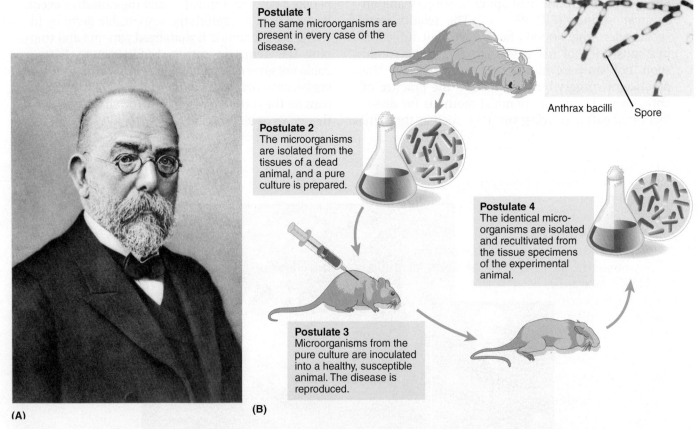

Postulate 1
The same microorganisms are present in every case of the disease.

Anthrax bacilli   Spore

Postulate 2
The microorganisms are isolated from the tissues of a dead animal, and a pure culture is prepared.

Postulate 4
The identical microorganisms are isolated and recultivated from the tissue specimens of the experimental animal.

Postulate 3
Microorganisms from the pure culture are inoculated into a healthy, susceptible animal. The disease is reproduced.

(A)   (B)

FIGURE 1.7 A Demonstration of Koch's Postulates. Robert Koch **(A)** developed what became known as Koch's postulates **(B)** that were used to relate a single microorganism to a single disease. The inset (in the upper right) is a photo of the rod-shaped anthrax bacterial cells. Many rods are swollen with spores (white ovals). »» *What is the relationship between postulate 2 and postulate 4?*

MICROFOCUS 1.3: History

Jams, Jellies, and Microorganisms

One of the major developments in microbiology was Robert Koch's use of a solid culture surface on which bacterial colonies would grow. He accomplished this by solidifying beef broth with gelatin. When inoculated onto the surface of the nutritious medium, bacterial cells grew vigorously at room temperature and produced discrete, visible colonies.

On occasion, however, Koch was dismayed to find that the gelatin turned to liquid because some bacterial species produced a chemical substance that digested the gelatin. Moreover, gelatin liquefied at the warm incubator temperatures commonly used to cultivate many bacterial species.

An associate of Koch's, Walther Hesse, mentioned the problem to his wife and laboratory assistant, Fanny Hesse. She had a possible solution. For years, she had been using a seaweed-derived powder called agar (pronounced ah'gar) to solidify her jams and jellies. Agar was valuable because it mixed easily with most liquids and once gelled, it did not liquefy, even at warm incubator temperatures.

In 1880, Walther Hesse recommend agar to Koch. Soon Koch was using it routinely to grow bacterial species, and, in 1884, he first mentioned agar in his paper on the isolation of the bacterial organism responsible for tuberculosis. It is noteworthy that Fanny Hesse is one of the first Americans (she was originally from New Jersey) to make a significant contribution to microbiology.

Another key development, the common petri dish (plate), also was invented about this time (1887) by Julius Petri, one of Koch's former assistants.

Fanny Hesse.

Courtesy of National Library of Medicine.

algae replaced gelatin as Koch's preferred solidifying agent, as MICROFOCUS 1.3 recounts.

When Koch presented his work, scientists were astonished. Here was the verification of the germ theory that had eluded Pasteur. Koch's procedures became known as **Koch's postulates** and were quickly adopted as the formalized standards for implicating a specific germ with a specific disease. FIGURE 1.7B outlines the four-step process.

By applying Koch's postulates, germ identification in the laboratory became the normal method of work. Koch's lab focused on disease causation (**etiology**), including the bacterial agents responsible

for tuberculosis and cholera. In addition, many other individuals quickly discovered other germs causing other human diseases (TABLE 1.1). Pasteur's lab, on the other hand, was more concerned with preventing disease through vaccination. This culminated in 1885, when Pasteur's rabies vaccine saved the life of a young boy who had been bitten by a rabid dog. By the turn of the century, the germ theory set a new course for studying and treating infectious disease. The detailed studies carried out by Pasteur and Koch made the discipline of **bacteriology**, the study of bacterial organisms, a well-respected field of study.

Concept and Reasoning Checks 1.3

a. How did Pasteur's studies of wine fermentation and souring suggest to him that germs might cause human infectious disease?

b. Assess Lister's antisepsis procedures and Pasteur's studies of pébrine to supporting the germ theory.

c. Explain why Koch's postulates were critical to validating the germ theory.

d. How did the development of pure cultures advance the germ theory?

TABLE 1.1 Other International Scientists Who Identified Specific Human Pathogens During the Classical Golden Age of Microbiology

Investigator (Year)	Country	Disease (Pathogenic Agent)
Otto Obermeier (1868)	Germany	Relapsing fever (bacterium)
Gerhard Hansen (1873)	Norway	Leprosy (bacterium)
Albert Neisser (1879)	Germany	Gonorrhea (bacterium)
Charles Laveran (1880)	France	Malaria (protozoan)
Karl Eberth (1880)	Germany	Typhoid fever (bacterium)
Edwin Klebs (1883)	Germany	Diphtheria (bacterium)
Arthur Nicolaier (1884)	Germany	Tetanus (bacterium)
Theodore Escherich (1885)	Germany	Infant diarrhea (bacterium)
Albert Fraenkel (1886)	Germany	Pneumonia (bacterium)
David Bruce (1887)	Australia	Undulant fever (bacterium)
Anton Weichselbaum (1887)	Austria	Cerebrospinal meningitis (bacterium)
A. A. Gärtner (1888)	Germany	Food poisoning/salmonellosis (bacterium)
William Welch and George Nuttall (1892)	United States	Gas gangrene (bacterium)
S. Kitasato and A. Yersin (1894)	Japan and France	(Independently) Bubonic plague (bacterium)
Emile van Ermengem (1896)	Belgium	Botulism (bacterium)
Kiyoshi Shiga (1898)	Japan	Bacterial dysentery (bacterium)
Walter Reed (1900)	United States	Yellow fever (virus)
Robert Forde and Joseph Everett Dutton (1902)	Great Britain	African sleeping sickness (protozoan)
Fritz Schaudinn and Erich Hoffman (1903)	Germany	Syphilis (bacterium)
Jules Bordet and Octave Gengou (1906)	France	Whooping cough/pertussis (bacterium)
George McCoy and Charles Chapin (1911)	United States	Tularemia (bacterium)

■ KEY CONCEPT 1.4   With the Discovery of Other Microbes, the Microbial World Expands

Although bacterial organisms were being discovered as the agents of some human diseases, why couldn't Koch's postulate confirm a bacterial origin for diseases such as measles, mumps, smallpox, and yellow fever? Moreover, what were the bacterial organisms being discovered in the soil doing?

Other Global Pioneers Contribute to the New Discipline of Microbiology

In the 1890s, a Russian scientist, Dimitri Ivanowsky, and Martinus Beijerinck, a Dutch investigator, independently were studying tobacco mosaic disease, which produces mottled and stunted tobacco leaves. Each independently prepared a homogenized liquid from diseased plants in the hopes of trapping the infectious agent on the filter. In fact, they discovered that when the liquid that passed through a filter was applied to healthy tobacco plants, the leaves became mottled and stunted. Ivanowsky simply assumed bacterial cells somehow had slipped through the filter, whereas Beijerinck suggested that the liquid was a "contagious, living liquid" that acted like a poison or virus (*virus* = "poison"). Then, in 1898, the causative agent for animal hoof-and-mouth disease was found to be another filterable liquid, and, in 1901, American Walter Reed, a United States Army physician, concluded that the agent responsible for yellow fever in humans also was a virus. With these discoveries, the discipline of **virology**, the study of viruses, was launched.

While scientists like Pasteur and Koch were investigating the bacterial contribution to the infectious disease process, others were identifying other types of disease-causing microbes. Fungi were found to cause plant diseases and such diseases were studied extensively by Anton de Bary in the 1860s. As already mentioned in this chapter, Pasteur identified the role of fungal yeasts (first seen by Leeuwenhoek) with fermentation. Importantly, the recognition that some fungi were linked to human skin diseases was proposed as early as 1841 when a Hungarian physician, David Gruby, discovered a fungus associated with human scalp infections. These discoveries led to the development of the field of **mycology**, the study of fungi.

The realization that infectious diseases could be caused by protozoa (again part of the animalcules first seen by Leeuwenhoek) was another major milestone in understanding infectious disease. In fact, Pasteur's "corpuscular parasites" of pébrine were protozoa. Other advances in the study of these types of microbes were dependent on studies in tropical medicine. Major advances in understanding these microbes included Charles Laveran's discovery (1880) that the protozoan parasite causing malaria could be found in human blood and David Bruce's studies (1887) that another protozoan parasite was the agent of human sleeping sickness. With these studies, the field of **protozoology** was born (**FIGURE 1.8**).

As you should now know, not all microbes are pathogens; in fact, as described in the chapter opener, relatively few cause infectious disease. Consequently, during the first Golden Age of microbiology, other scientists and microbiologists were keenly interested in microbes naturally found in the soil, and they wanted to understand the ecological importance of these nonpathogenic microbes. The Russian soil scientist Sergei Winogradsky, a student of de Bary's, and Martinus Beijerinck discovered that some bacterial organisms could convert inert atmospheric nitrogen gas (N_2) into biologically useable ammonia (NH_3) that plants need. Winogradsky and Beijerinck each developed many of the laboratory methods essential to the study of soil microbes, while uncovering the essential roles such microorganisms play in the recycling of matter on a global scale. As the founders of **microbial ecology**, Winogradsky and Beijerinck laid the foundation for what we know today about many of the so-called microbial workforce mentioned in the chapter opener.

Today, many microbiologists are still searching for, finding, and trying to understand the roles of microorganisms in the environment as well as in health and disease. In fact, less than 2% of all microorganisms on Earth have been identified and many fewer cultured, so there is still a lot to be discovered and studied in the microbial world!

The Microbial World Can Be Cataloged into Unique Groups

As time went on, more and more microbes were identified and studied. Let's briefly survey what we know about these major groups of microorganisms.

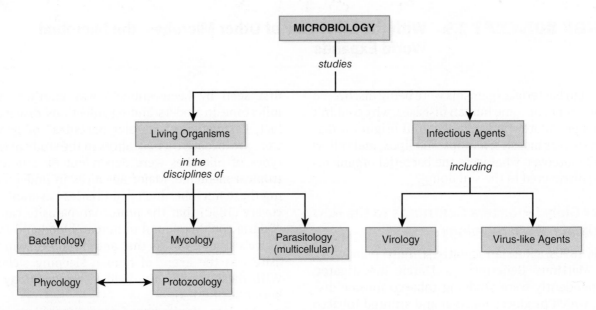

FIGURE 1.8 Microbiology Disciplines. This simple concept map shows the relationship between microbiology and its various disciplines. Phycology is the study of algae and parasitology is the study of animal parasites, which traditionally includes the parasitic protozoa and the animal parasites (worms).

Bacteria and Archaea

It is estimated that there might be more than 10 million bacterial species. Most are very small, single-celled (unicellular) organisms, although some form filaments, and the majority associate in a bacterial community called a "biofilm." Almost all bacterial cells have a rigid cell wall and many common ones are spherical, spiral, or rod-shaped (**FIGURE 1.9A**). They lack a cell nucleus and most of the typical membrane-enclosed cellular compartments typical of other microbes and multicellular organisms. Many bacterial organisms get their food from the environment, although some make their own food through photosynthesis (**FIGURE 1.9B**). Bacterial cells are found in most all environments, making up, as previously mentioned, a large percentage of the Earth's microbial workforce.

In addition to the disease-causing members, some are responsible for food spoilage, whereas others are useful in the food industry. Many bacterial members, along with several fungi, are **decomposers**, organisms that recycle nutrients from dead organisms.

Based on recent biochemical and molecular studies, many bacterial organisms have been reassigned into another unique evolutionary group, called the **Archaea**. Although they look like bacterial cells when observed with the microscope, many of the first members to be identified grew in environments that are extremely hot (such as the Yellowstone hot

springs), extremely salty (such as the Dead Sea), or of extremely low pH (such as acid mine drainage). These so-called **extremophiles** evolved many adaptations to survive in these extreme environments. Conversely, many other archaeal organisms normally grow in temperate soils and water and are an integral part of the microbiome in animal digestive tracts. No archaeal members are known to be pathogens, but like the bacterial microbes, they are an important part of Earth's microbial workforce.

Viruses

Although not correctly labeled as microorganisms, these "infectious agents" currently consist of more than 100 million known types. Probably, most every cell on Earth can be infected by some type of virus. Viruses are not cellular and cannot be grown in culture. They have a core of nucleic acid (DNA or RNA) surrounded by a protein coat. Among the features used to identify viruses are morphology (shape), genetic material (DNA, RNA), and biological properties (the organism, tissue, or cell type infected).

Viruses infect organisms for one reason only: to replicate. Viruses in the air or water, for example, cannot replicate because they need the metabolic machinery and chemical building blocks found inside living cells. Of the known viruses, only a small percentage cause disease in humans. Polio, the flu, measles, AIDS, and smallpox are examples (**FIGURE 1.9C**).

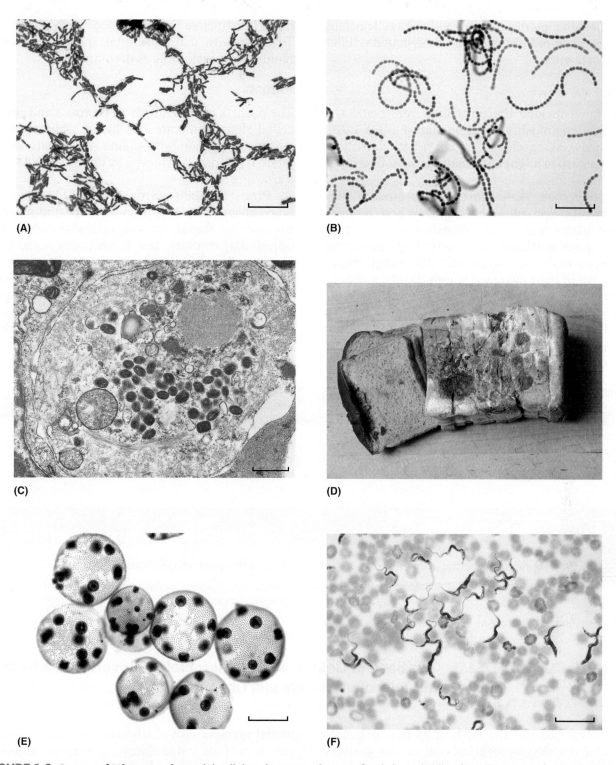

FIGURE 1.9 Groups of Microorganisms. (A) A light microscope image of rod-shaped cells of *Bacillus cereus* (stained purple), a normal inhabitant of the soil. (Bar = 10 µm.) **(B)** A light microscope image of filamentous strands of *Anabaena*, a photosynthetic bacterium. (Bar = 100 µm.) **(C)** False-color electron microscope image of smallpox viruses. (Bar = 500 nm.) **(D)** A typical blue-gray *Penicillium* mold growing on a loaf of bread. **(E)** A light microscope image of the colonial green alga *Volvox*. (Bar = 300 µm.) **(F)** A light microscope image of the ribbon-like cells of the protist *Trypanosoma*, the causative agent of African sleeping sickness. (Bar = 10 µm.) *»» Within these groups, why don't organisms like* Anabaena *and* Volvox *cause disease?*

The other groups of microbes have a cell nucleus and a variety of internal, membrane-bound cellular compartments.

Fungi

The fungi include the unicellular yeasts and the multicellular mushrooms and molds (**FIGURE 1.9D**). About 100,000 species of fungi have been described; however, there might be as many as 1.5 million species in nature.

Other than some yeasts, molds grow as filaments with rigid cell walls. Most grow best in warm, moist places and secrete digestive enzymes that break down nutrients into smaller bits that can be absorbed easily across the cell wall. Fungi, therefore, live in their own food supply. If that source of food is a human, diseases such as ringworm or vaginal yeast infections can result.

For the pharmaceutical industry, some fungi are sources for useful products, such as antibiotics. Others molds are used in the food industry to impart distinctive flavors in foods such as cheeses. Together with many bacterial species, numerous molds play a major role as decomposers.

Protists

The protists consist mostly of protozoa and single-celled algae. Some are free living, whereas others live in association with plants or animals. Movement, if present, is achieved by flagella or cilia or by a crawling motion.

Protists obtain nutrients in different ways. Some absorb nutrients from the surrounding environment or ingest smaller microorganisms. The unicellular, colonial, and filamentous algae have a rigid cell wall and can carry out photosynthesis (**FIGURE 1.9E**). The aquatic protists also provide energy and organic compounds for the lower trophic levels of the food web. Some protists (the protozoa) are capable of causing diseases in animals, including humans; these include malaria, several types of diarrhea, and sleeping sickness (**FIGURE 1.9F**).

Concept and Reasoning Checks 1.4

  a. Describe how viruses were discovered as disease-causing agents.
  b. What significant discoveries added the fungi and protozoa to the growing list of microbes?
  c. Judge the significance of the pioneering studies carried out by Winogradsky and Beijerinck.

Chapter Challenge B

By the end of the first Golden Age of microbiology (1915), all the major groups of microbes had been identified.

QUESTION B: *Construct a diagram or table that reveals how you could tell bacteria, viruses, fungi, protozoa, and algae apart from one another using structural and/or functional characteristics.*

You can find answers online in **Appendix F**.

■ KEY CONCEPT 1.5   The Second Golden Age of Microbiology Involves the Birth of Molecular Biology and Chemotherapy

The 1940s brought the birth of molecular genetics and the development of one of the greatest breakthroughs that would revolutionize medicine's ability to treat and eliminate infectious disease—antibiotics.

Molecular Biology Often Relies on Microorganisms as Model Systems

Being small, easy to grow, and having fast reproductive rates, microbes are perfect research tools or **model systems** with which to explore the molecular workings of life and to discover general principles in biology. Consequently, beginning around 1940, the second Golden Age of microbiology emerged.

Here are a few of the scientists who used microbial model systems and viruses:

▶ George Beadle and Edward Tatum (1941) used the fungus *Neurospora* to demonstrate that genes direct the synthesis of proteins.

- Salvador Luria and Max Delbrück (1943) used the bacterium *Escherichia coli* to show that mutations could occur spontaneously.
- Oswald Avery, Colin MacLeod, and Maclyn McCarty (1944) used the bacterium *Streptococcus pneumoniae* to suggest that DNA is the genetic material in all cells.
- Alfred Hershey and Martha Chase (1952) used a virus that infects bacterial cells to confirm that DNA is the genetic material.
- Francis Crick (1958) used *E. coli* and a virus to show how the DNA genetic code works to make individual proteins.

Again, microbes were at the forefront in answering fundamental questions that applied to all of biology. In addition, during this second Golden Age, other studies illuminated the basic structural organization of microbes.

Two Types of Cellular Organization Are Realized

The small size of bacterial cells hindered scientists' abilities to confirm if these cells were similar to other cellular organisms in organization or if the cells were just "bags of enzymes." In the 1940s and 1950s, a new type of microscope—the electron microscope—was developed that could magnify objects and cells thousands of times better than typical light microscopes. With the electron microscope, for the first time bacterial cells were seen as being cellular like all other microbes, plants, and animals. However, studies showed that they were organized in a structurally different way from other organisms.

It was known that animal and plant cells contained a cell nucleus that houses the genetic instructions in the form of chromosomes and that the nucleus was separated physically from other cell structures by a membrane envelope (**FIGURE 1.10A**). This type of cellular organization is called **eukaryotic** (*eu* = "true"; *karyon* = "nucleus"). Microscope observations of the protists and fungi had revealed that these organisms also have a eukaryotic organization. Thus, not only are all plants and animals **eukaryotes**, so are the microorganisms that comprise the fungi and protists.

Electron microscope studies revealed that bacterial (and archaeal) cells had few of the internal compartments typical of eukaryotic cells. They lacked a cell nucleus, indicating the DNA was not surrounded by a membrane envelope (**FIGURE 1.10B**). Therefore, members of the Bacteria and Archaea have a **prokaryotic** (*pro* = "before") type of cellular organization and represent **prokaryotes**. Importantly, there also are many genetic and molecular differences between bacterial and archaeal cells, accounting for their split into separate and unique microbial groups. Incidentally, because viruses lack a cellular organization, they are neither prokaryotes nor eukaryotes.

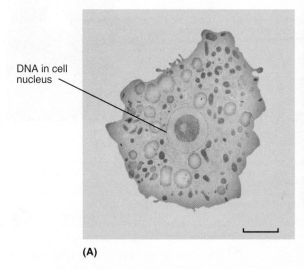
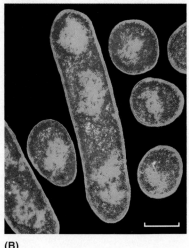

DNA in cell nucleus

(A)   (B)

FIGURE 1.10 False-Color Images of Eukaryotic and Prokaryotic Cells. (A) A false-color electron microscope image of a protist cell. All eukaryotes, including the protists and fungi, have their DNA enclosed in a cell nucleus. (Bar = 3 μm.) **(B)** A false-color electron microscope image of bacterial cells. (Bar = 0.5 μm.) *»» How is the DNA in the eukaryotic cell structurally different from the DNA in the bacterial cell?*

Antibiotics Are Used to Cure Infectious Disease

In 1910, another coworker of Koch's, Paul Ehrlich, synthesized the first chemical capable of killing pathogens without severely damaging the tissue surrounding the infection. This so-called magic bullet, an arsenic-based agent called salvarsan, could cure syphilis, a serious sexually transmitted infection. With the birth of antibacterial **chemotherapy**, the use of antimicrobial chemicals to kill microbes would become an important branch of medical microbiology and medicine.

In 1928, Alexander Fleming, a Scottish scientist, discovered a mold growing in one of his bacterial cultures (**FIGURE 1.11 A, B**). The mold, a species of *Penicillium,* killed the bacterial colonies that were near the mold. He named the antimicrobial substance penicillin and soon discovered penicillin would kill other bacterial pathogens, as well. In 1940, research headed by biochemists Howard Florey and Ernst Chain developed a way to mass produce penicillin, which would become a godsend in curing bacterial infections during World War II (**FIGURE 1.11C**). As other antibacterial chemicals were discovered, the

(A)

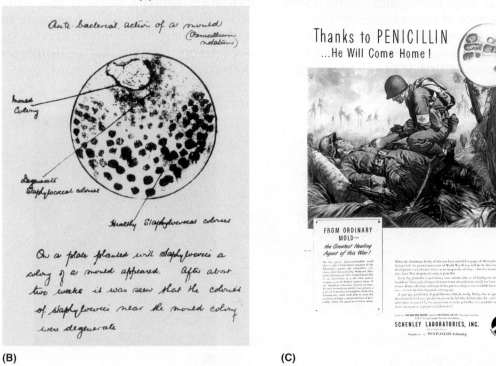

(B)   (C)

FIGURE 1.11 Fleming and Penicillin. (A) Artist's drawing of Fleming in his laboratory. **(B)** Fleming's notes and drawing on the inhibition of bacterial growth by the fungus *Penicillium.***(C)** A World War II poster touting the benefits of penicillin and illustrating the great enthusiasm in the United States for treating infectious diseases in war casualties. *»» In (B), why didn't penicillin kill the bacterial colonies farther away from the fungus?*

term **antibiotic** was coined to refer to those antimicrobial substances naturally produced by mold and bacterial species that inhibit growth or kill other microorganisms.

By the 1950s, penicillin and several other antibiotics were established treatments in medical practice. In fact, by the mid-1960s, many health experts believed all major bacterial infections would soon become rare events due to antibiotic chemotherapy. What was ignored was the mounting evidence that bacterial species were becoming resistant to or tolerant of antibiotics. Still, antibiotics represent one of the greatest breakthroughs in medicine and they have saved millions of lives since their introduction.

Concept and Reasoning Checks 1.5

a. What roles did microorganisms (and viruses) play in understanding general principles of molecular biology?
b. Distinguish between prokaryotic and eukaryotic cells.
c. Assess the importance of Fleming's discovery and Florey and Chain's production of penicillin.

KEY CONCEPT 1.6   A Third Golden Age of Microbiology Is Now

Today, microbiology again finds itself on the world stage, in part from the new age of exploration in which unique and diverse microbes are being discovered in abundance around the world, as described in the chapter opener. In addition, biotechnology has made use of the natural and genetically engineered abilities of microbial agents to carry out biological processes for industrial/commercial/medical applications. In the latest Golden Age, microbiology again is making important contributions to the life sciences and humanity.

However, the third Golden Age of microbiology also faces several challenges, many of which still concern infectious diseases that today are responsible for 16% of all deaths globally (**FIGURE 1.12**).

Microbiology Continues to Face Many Challenges

The resurgence of infectious disease has brought the subject back into the mainstream of epidemiology. Even in the United States, more than 100,000 people die each year from bacterial infections, making them the fourth leading cause of death. In fact, on a global scale, infectious diseases are spreading geographically faster than at any time in history.

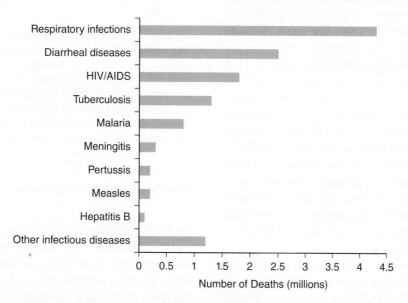

FIGURE 1.12 Global Mortality from Infectious Diseases—2015. On a global scale, some 15 million deaths annually are caused by infectious diseases. »» *Identify some of the "other infectious diseases."*

Data from *World Health Statistics 2015*: World Health Organization.

Some of these challenges are discussed briefly in the sections that follow.

Disease Transmission

Experts estimate that more than 3 billion people (9 million per day) on some 24 million flights traveled by air in 2015, making an outbreak or an epidemic in one part of the world only a few flight hours away from becoming a potentially dangerous threat in another region of the globe. It is a sobering thought to realize that since 2002, the WHO has verified more than 1,100 epidemic events worldwide. Unlike past generations, today's highly mobile, interdependent, and interconnected world provides potential opportunities for the rapid spread of infectious diseases.

Emerging and Reemerging Infectious Diseases

Infectious diseases today not only have the potential to spread faster, they also are appearing with greater frequency and intensity. Over the past few decades, new diseases have been identified at the unprecedented rate of one or more per year, such that today there are nearly 40 infectious diseases present that were unknown a generation ago. These diseases fall into one of two groups.

An **emerging infectious disease** is one that has recently surfaced in the human population for the first time. Among the more newsworthy have been AIDS, severe acute respiratory syndrome (SARS), Lyme disease, bird flu, and most recently, Ebola virus disease in West Africa and Zika virus infection in the Western Hemisphere. There is no cure for any of these, and, undoubtedly, there are more disease-causing pathogens ready to emerge.

A **reemerging infectious disease** is one that has existed in the past but is now showing resurgence in resistant forms and expansion in geographic range. Among the more prominent reemerging diseases are drug-resistant tuberculosis, cholera, dengue fever, and, for the first time in the Western Hemisphere, West Nile virus disease (**FIGURE 1.13A**). Therefore, until better surveillance of disease outbreaks is established, and more and better vaccines and antimicrobial drugs are developed, emerging and reemerging diseases will remain a challenge to public health and microbiology.

Epidemiologists and health experts estimate that about 60% of infectious diseases in humans are **zoonotic diseases**, that is, diseases that are spread from wild or domesticated animals to humans. Examples include AIDS, Ebola virus disease, Zika virus infection, and influenza. Other factors

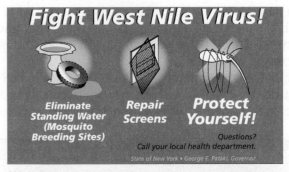

(A)

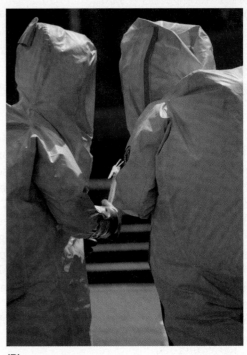

(B)

FIGURE 1.13 Emerging Disease Threats: Natural and Intentional. (A) West Nile virus (WNV) is just one of several agents responsible for emerging or reemerging diseases. Methods have been designed that individuals can use to protect themselves from mosquitoes that spread the virus. **(B)** Combating the threat of bioterrorism often requires special equipment and protection because many agents seen as possible bioweapons could be spread through the air. *»» How do these suits compare to those to protect from a miasma (see Figure 1.3)?*

(A) Reproduced with permission of the New York State Department of Health. (B) © Photodisc.

including habitat destruction, increased contact between humans and wild animals, climate change, and the trade in global agricultural products further drive disease transmission. Therefore, scientists in human, veterinary (domestic and wild), and environmental health are establishing a worldwide collaboration and communication network called **One Health**. Rather than treat each infectious disease in isolation, the strategy of One Health is to monitor and combine all information on human, animal, and environmental health in such a way that a potential outbreak of disease can be predicted before it actually happens. This challenge is daunting but doable. The success of One Health is critical to the control of infectious diseases and is indispensable for helping maintain human health.

Increased Antibiotic Resistance

Another challenge concerns our increasing inability to fight infectious disease because so many pathogens are now resistant to one or more antibiotics, and such antibiotic resistance is developing faster than new antibiotics are being discovered. Ever since it was recognized that pathogens could develop into **superbugs**, which are microbes that are resistant to multiple antibiotics, a crusade has been waged to restrain the inappropriate use of these drugs by doctors and to educate patients not to demand antibiotics in uncalled-for situations.

The challenge facing microbiologists and the medical community is to find new and effective antibiotics to which pathogens will not develop resistance quickly before the current arsenal is useless. Unfortunately, the growing threat of antibiotic resistance has been accompanied by a decline in new drug discovery and an increase in the time to develop a drug from discovery to market. Consequently, antibiotic resistance has become a major health threat and one of the most important challenges facing microbiology today.

Bioterrorism

The intentional or threatened use of biological agents to cause fear in or actually inflict death or disease upon a large population is referred to as **bioterrorism**. Most of the recognized biological agents for bioterrorism are microorganisms, viruses, or microbial toxins that are bringing diseases like anthrax, smallpox, and plague back into the human psyche (**FIGURE 1.13B**). To minimize the use of these agents to inflict mass casualties, the challenge to the scientific community and microbiologists is to improve the ways that bioterror agents are detected, discover effective measures to protect the public, and develop new and effective treatments against these pathogens for individuals or whole populations.

Climate Change and Infectious Disease

A very controversial issue is how **climate change** (including warming temperatures and altering rainfall patterns) could affect the frequency and distribution of infectious diseases around the world in the coming decades. For example, as temperatures rise in various regions of the world, mosquitoes might broaden their range and, in so doing, spread diseases like malaria and dengue fever to more temperate climates, including North America and Europe. Warming ocean waters also could provide ideal environments for the spread of waterborne diseases like cholera.

Microbiologists, epidemiologists, climate scientists, and many others are studying new strategies to limit potential disease spread before the causative agents have an opportunity to be established. In so doing, researchers need to understand better the dynamics of climate change and recognize how such changes might affect the behaviors of potentially emerging and reemerging diseases that could affect the health of humans, livestock, plants, and wildlife. Climate change coupled with the ubiquity and speed of air travel and the expanding evolution of antibiotic resistance could work synergistically to cause major epidemics, or even pandemics, if microbiologists and health authorities don't find ways to constantly survey the microbial landscape and to decrease the burden of infectious disease.

Studies in Microbial Ecology and Evolution Also Are Helping to Drive the New Golden Age

Since the time of the first Golden Age of microbiology, microbiologists have wanted to know how a microbe interacts, survives, and thrives in the environment. Today, microbiology is less concerned with a specific microbe and more concerned with the relationships among microorganisms and with their environment.

Microbial Ecology

Microbial ecology, which was started by Winogradsky and Beijerinck, involves the study of microbial dynamics in the environment and their interactions with each other and with other organisms. Traditional methods of microbial ecology required organisms from an environment be isolated and cultivated in the laboratory so that they can be characterized and identified. Unfortunately, up to 98% of microorganisms do not grow in a laboratory culture and, therefore, they cannot be studied visually. In spite of this, today many microbiologists, armed with genetic, molecular, and biotechnology tools, can study and characterize these uncultured microbes.

Today researchers are discovering that most microbes do not act as individual entities; rather, in nature they survive in complex communities called

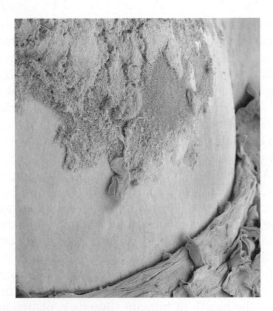

FIGURE 1.14 A Biofilm. Plaque (the false-colored pink crust) typically found on tooth surfaces is an example of a biofilm. Left untreated, it might result in tooth decay. *»» How do these examples of microbes fit into the concept of the global microbial workforce?*

© STEVE GSCHMEISSNER/Science Photo Library/Getty.

biofilms. For example, the whitish, gooey plaque on your teeth is an example of a bacterial biofilm (**FIGURE 1.14**). Microbes in biofilms act very differently than individual cells, and the biofilms can be difficult to treat when they cause chronic infectious disease.

The discovered versatility of many bacterial and archaeal species is being applied to problems that have the potential to benefit the planet. As the highly respected naturalist, E. O. Wilson has stated, *"If I could do it all over again, and relive my vision in the 21st century, I would be a microbial ecologist."*

Microbial Evolution

It was Charles Darwin—another of the scientists who combined observation with a "prepared mind"—who in 1859 first outlined the principles of evolution. Today, evolutionary theory is the unifying force in modern biology, tying together such distinct fields as genetics, ecology, medicine, and, yes, microbiology.

Like all life, microorganisms evolve. In fact, **microbial evolution** was occurring for some 2 billion years before the first truly eukaryotic cells appeared on the planet. Although challenging to study, it is possible today to "replay history" by following the accumulation of unpredictable, chance events that lead to evolutionary novelty, which is a hallmark of the microbes. In addition, microbial evolution studies are putting together new approaches to improve agricultural productivity, to monitor and assess climate change, and even to produce clean fuels and energy. The new field of **evolutionary medicine** is providing new strategies to slow the evolution of antibiotic resistance, to treat infectious diseases, and to better understand the role of microbes in human health. One has to wonder what Darwin would make of the microbial world if he were alive today.

In conclusion, microbiology (from then until now) has gone from observing the first microbes as curiosities (Leeuwenhoek) to identifying and studying individual microorganisms as pathogens (Pasteur, Koch, and others) to appreciating microbiomes as the engines that support all life on Earth (today). Yet, over this span of 300-plus years, microbiologists have only discovered perhaps 2% of all microbial species. Microbiology from then until now has come a long way but has a much longer way yet to go and many challenges to face.

Concept and Reasoning Checks 1.6

a.   Briefly, discuss four examples of the challenges facing microbiology today.

Chapter Challenge C

Many of the microbes that first occupied the minds and work of the early microbial pioneers like Pasteur and Koch still challenge microbiologists today.

QUESTION C: *Describe the natural and intentional disease threats challenging microbiology today, explaining why they are still so prominent even with all the advances in medical science and microbiology.*

You can find answers online in **Appendix F**.

SUMMARY OF KEY CONCEPTS

Concept 1.1 The Discovery of Microbes Leads to Questioning Their Origins

1. The observations with the microscope made by Hooke and especially Leeuwenhoek, who reported the existence of **animalcules** (microorganisms), sparked interest in an unknown world of microscopic life. (Figure 1.2)
2. The controversy over **spontaneous generation** initiated the need for accurate scientific experimentation, which then provided the means to refute the concept.

Concept 1.2 Disease Transmission Can Be Prevented

3. Edward Jenner established that disease (smallpox) could be prevented through **vaccination** using a similar but milder disease-causing agent.
4. Semmelweis and Snow believed that infectious disease was caused by particles transmitted from the environment (not a **miasma**) and that disease transmission could be interrupted through simple measures. (Figure 1.4)

Concept 1.3 The Classical Golden Age of Microbiology Reveals the Germ

5. Pasteur's fermentation studies suggested that microorganisms could produce chemical changes. He proposed the **germ theory of disease**, which stated that human infectious disease was due to chemical changes brought about by microorganisms infecting the body.
6. Lister's use of **antisepsis** techniques and Pasteur's studies of pébrine supported the germ theory and showed how diseases could be controlled.
7. Koch's work with anthrax and development of **pure cultures** led to a method (**Koch's postulates**) to relate a specific microorganism to a specific disease. (Figure 1.7)
8. Laboratory science arose as Pasteur and Koch hunted down the microorganisms of infectious disease. Pasteur's lab developed a vaccine for human rabies. Koch's lab isolated, cultivated, and identified the pathogens responsible for cholera and tuberculosis.

Concept 1.4 With the Discovery of Other Microbes, the Microbial World Expands

9. Ivanowsky and Beijerinck provided the first evidence for viruses as infectious agents.
10. Winogradsky and Beijerinck recognized the beneficial roles played by microorganisms found in the environment.
11. Microbes include the Bacteria and Archaea, viruses, fungi, and protists (algae and protozoa). (Figure 1.9)

Concept 1.5 A Second Golden Age of Microbiology Involves the Birth of Molecular Biology and Chemotherapy

12. Many of the advances toward understanding molecular biology and general principles in biology were based on experiments using microbial **model systems**.
13. With the advent of the electron microscope, microbiologists recognized that there were two basic types of cellular organization: **eukaryotic** and **prokaryotic**. (Figure 1.10)
14. Following from the initial work by Ehrlich, **antibiotics** were developed as "magic bullets" to cure many infectious diseases.

Concept 1.6 A Third Golden Age of Microbiology Is Now

15. In the 21st century, fighting infectious disease, identifying **emerging** and **reemerging infectious diseases**, combating increasing **antibiotic resistance**, countering the **bioterrorism** threat, and addressing the potential spread of infectious diseases due to **climate change** are challenges facing microbiology, healthcare systems, and society.
16. **Microbial ecology** is providing new clues to the roles of microorganisms in the environment. The understanding of **microbial evolution** using modern gene technologies has expanded our understanding of microorganism relationships.

CHAPTER SELF-TEST

For **Steps A–D**, you can find answers to questions and problems in **Appendix D**.

STEP A: REVIEW OF FACTS AND TERMS

Multiple Choice

Read each question carefully before selecting the *one* answer that best fits the question or statement.

1. Who was the first person to see bacterial cells with the microscope?
   - **A.** Pasteur
   - **B.** Koch
   - **C.** Leeuwenhoek
   - **D.** Hooke

2. What process was studied by Redi and Spallanzani?
   - **A.** Spontaneous generation
   - **B.** Fermentation
   - **C.** Variolation
   - **D.** Antisepsis

3. The process of _____ involved the inoculation of dried smallpox scabs under the skin.
   - **A.** vaccination
   - **B.** antisepsis
   - **C.** variolation
   - **D.** immunization

4. What is the name for the field of study established by Semmelweis and Snow in the mid-1800s?
   - **A.** Immunology
   - **B.** Bacteriology
   - **C.** Virology
   - **D.** Epidemiology

5. The process of controlled heating, called _____, was used to keep wine from spoiling.
   - **A.** curdling
   - **B.** fermentation
   - **C.** pasteurization
   - **D.** variolation

6. What surgical practice was established by Lister?
   - **A.** Antisepsis
   - **B.** Chemotherapy
   - **C.** Variolation
   - **D.** Sterilization

7. Which one of the following statements is *not* part of Koch's postulates?
   - **A.** The microorganism must be isolated from a dead animal and pure cultured.
   - **B.** The microorganism and disease can be identified from a mixed culture.
   - **C.** The pure cultured organism is inoculated into a healthy, susceptible animal.
   - **D.** The same microorganism must be present in every case of the disease.

8. Match the lab with the correct set of identified diseases.
   - **A.** Pasteur: tetanus and tuberculosis
   - **B.** Koch: anthrax and rabies
   - **C.** Koch: cholera and tuberculosis
   - **D.** Pasteur: diphtheria and typhoid fever

9. What group of microbial agents would eventually be identified from the work of Ivanowsky and Beijerinck?
   - **A.** Viruses
   - **B.** Fungi
   - **C.** Protists
   - **D.** Bacteria

10. What microbiological field was established by Winogradsky and Beijerinck?
    - **A.** Virology
    - **B.** Microbial ecology
    - **C.** Bacteriology
    - **D.** Mycology

11. What group of microorganisms has a variety of internal cell compartments and many of which act as decomposers?
    - **A.** Bacteria
    - **B.** Viruses
    - **C.** Archaea
    - **D.** Fungi

12. Which one of the following organisms was *not* a model organism related to the birth of molecular genetics?
    - **A.** *Streptococcus*
    - **B.** *Penicillium*
    - **C.** *Escherichia*
    - **D.** *Neurospora*

13. Which group of microbial agents is eukaryotic?
    - **A.** Bacteria
    - **B.** Viruses
    - **C.** Archaea
    - **D.** Algae

14. The term _____ was used to refer to antimicrobial substances naturally derived from some _____.
    - **A.** antibiotic; bacteria and fungi
    - **B.** antisepsis; other living organisms
    - **C.** antibiotic; viruses
    - **D.** fermentation; microbes

15. Which one of the following is *not* considered an emerging infectious disease?
    - **A.** Polio
    - **B.** SARS
    - **C.** Lyme disease
    - **D.** AIDS

16. What strategy is *not* part of evolutionary medicine?
    - **A.** Understanding the role of the human microbiome
    - **B.** Understanding how life began on Earth
    - **C.** Reducing the emergence of infectious diseases
    - **D.** Stimulating the development of antimicrobial drugs

True–False

Each of the following statements is **true** (**T**) or **false** (**F**). If the statement is false, substitute a word or phrase for the underlined word or phrase to make the statement true.

17. ____ <u>Leeuwenhoek</u> believed that animalcules arose spontaneously from decaying matter.
18. ____ <u>Semmelweis</u> proposed that cholera was a waterborne disease.
19. ____ Some <u>bacterial cells</u> can convert nitrogen gas (N_2) into ammonia (NH_3).
20. ____ <u>Fungi</u> are eukaryotic microorganisms, some of which are decomposers.

21. ____ <u>Koch</u> proposed the germ theory.
22. ____ Variolation involved inoculating individuals with <u>smallpox</u> scabs.
23. ____ Most microbes today exist as <u>independent, free-living cells</u>.
24. ____ Pasteur proposed that "wine disease" was a souring of wine caused by <u>yeast</u> cells.

■ STEP B: CONCEPT REVIEW

25. Describe the concept of **spontaneous generation** and distinguish between the experiments that supported and refuted the belief. (**Key Concept 1.1**)
26. Compare Jenner's work on smallpox and Pasteur's studies on rabies to the concept of preventing disease through vaccination. (**Key Concepts 1.2 and 1.3**)
27. Judge the importance of (a) the **germ theory of disease** and (b) **Koch's postulates** to the identification of microbes as agents of infectious disease. (**Key Concept 1.3**)

28. Provide evidence to support the statement: "Not all microbes cause disease; many play important roles in the environment." (**Key Concept 1.4**)
29. Distinguish between the "new generation" of scientists in the second Golden Age of microbiology that set the stage for the antibiotic revolution. (**Key Concept 1.5**)
30. Assess the importance of **microbial ecology** and **microbial evolution** to the current Golden Age of microbiology. (**Key Concept 1.6**)

■ STEP C: APPLICATIONS AND PROBLEM SOLVING

31. As a microbiologist in the 1940s, you are interested in discovering new antibiotics that will kill bacterial pathogens. You have been given a liquid sample of a chemical substance to test in order to determine if it kills bacterial cells. Drawing on the culture techniques of Robert Koch, design an experiment that would allow you to determine the killing properties of the sample substance.
32. As a microbial ecologist, you discover a new species of microbe. How could you determine if it has a prokaryotic or eukaryotic cell structure? Suppose that it has a eukaryotic structure. What information would be needed to determine if it is a member of the protista or fungi?

33. One of the foundations of scientific inquiry is proper experimental design involving the use of controls. What is the role of a control in an experiment? For each of the experiments described in the section on spontaneous generation, identify the control(s) and explain how the interpretation of the experimental results would change without such controls.
34. You isolate and pure culture a bacterial organism from ill humans that you believe causes the disease. However, you cannot find a susceptible animal for testing that contracts the disease. What would you conclude from these observations?

■ STEP D: QUESTIONS FOR THOUGHT AND DISCUSSION

35. Many people are fond of pinpointing events that alter the course of history. In your mind, which single event described in this chapter had the greatest influence on (a) the development of microbiology and (b) medicine?

36. Louis Pasteur once stated, "In the field of observation, chance favors only the prepared mind." How does this quote apply to the work done by (a) Semmelweis, (b) Snow, and (c) Fleming?

37. When you tell a friend that you are taking microbiology this semester, she asks, *"Exactly what is microbiology?"* How do you answer her?

38. As microbiologists continue to explore the microbial universe, it is becoming more apparent that microbes are "invisible emperors" that rule the world. Now that you have completed this chapter, provide examples to support the statement: Microbes rule!

39. Who would you select as the "first microbiologist"? (a) Leeuwenhoek? (b) Hooke? (c) Pasteur and Koch? Support your decision.

Concept Mapping

See pages XXXI-XXXII for tips on how to construct a concept map.

40. Construct a concept map for **Microbial Agents**, using the following terms.

Algae	Fungi	Protists
Archaea	Microorganisms	Protozoa
Bacteria	Nucleated cells	Viruses
Decomposers	Pathogens (germs)	

CHAPTER 2

The Chemical Building Blocks of Life

The origin of life is one of the great and unsolved mysteries of science. Researchers believe microbial life arose on Earth about 3.8 billion years ago, although no one can definitely say how or where life originated. In fact, there are many hypotheses. More than likely, a common primitive life form gave rise to bacterial and archaeal cells. If, in fact, this is true, did such life arise but once or was there opportunity for "life" arising more than once—and in a different chemical form?

Paul Davies is an award-winning physicist and director at Arizona State University of BEYOND: Center for Fundamental Concepts in Science. One of the "big questions" the Center is investigating is whether "alien" life might be hiding right in front of our noses. The hypothesis is that perhaps other life arose among the known life on planet Earth and still exists here today as a so-called "shadow life." To pursue this controversial idea, scientists have begun searching high and low (literally up in the air and deep in the crust of the earth) for evidence of this shadow life—specifically microorganisms that, as the result of a "second genesis," would differ chemically from all known microbial life. Because microbes and all known life employ the same chemical building blocks and operate with a nearly identical genetic code, perhaps any undiscovered life forms would look like prokaryotic cells, but an alternate and distinctive biochemistry might single them out as "alien."

Microbiologists have identified only about 2% of the microorganisms on Earth. Because we know so little about the diversity of these microbes, it is possible that there is (or are) other shadow forms dispersed or hidden among the undiscovered. Importantly, that is what **science** is all about—the field of study that builds knowledge based on testing and evidence, and that tries to describe and comprehend the nature of the universe in whole or part, wherever that might lead.

Microorganisms are found in most, if not all, habitats on Earth. Some microbes can survive in alkaline, hypersaline conditions of some lakes, like Mono Lake (see chapter opening photo), the acidic runoff from a ore mine, and high temperatures of a hot spring (**FIGURE 2.1**). In these cases, as with all environments where microbes and any shadow life

Mono Lake, California.

© Geir Olav Lyngfjell/Shutterstock.

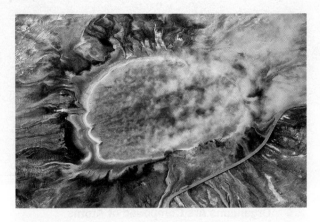

FIGURE 2.1 Survival in Earth's Extreme Environments. In the gentle flow of heated water spreading out in terraces, orange- and red-pigmented bacterial mats thrive in the warm, shallow water toward the edge. *»» What other extreme environments can you identify where microbes might survive?*

Courtesy of Jim Peaco/Yellowstone National Park.

might exist, survival depends on **cellular chemistry**, the biochemical reactions between atoms and molecules that provide for the unique metabolism found in all organisms.

Therefore, studying the basic principles of chemistry will provide you the working knowledge to understand the daily lives of the microbial workforce, whether in the environment or in your body. For example, in medicine and healthcare, understanding the biochemistry of a pathogen could guide the development of drugs, like antibiotics, and vaccines to cure or prevent infectious disease.

This chapter serves as a primer or review of the fundamental concepts of chemistry that form a foundation for the chapters ahead. We will identify the elements making up all known substances and show how these elements combine to form the major groups of organic compounds found in all "known" forms of life.

Chapter Challenge

The Earth was formed some 4.5 billion years ago. In that expanse of time, it appears that life only arose once because all life shares the same chemistry. Nonetheless, if some unrecognized, shadow form of microbial life still exists on Earth, it might contain a few unique chemical building blocks that would be used to build large substances. How would we recognize such chemicals and what are the potential markers by which "shadow life" could be detected? Let's consider some alternatives as we investigate the known chemistry of life!

■ KEY CONCEPT 2.1   Organisms Are Composed of Atoms

As far as scientists know, all matter in the physical universe—be it a rock, a tree, or a microbe—is built from substances called chemical elements. **Chemical elements** are the most basic forms of matter and they cannot be broken down into simpler substances by ordinary chemical reactions. Ninety-two naturally occurring elements have been discovered, whereas additional elements have been made in the laboratory or nuclear reactor.

Only about 25 of the 92 naturally occurring elements are essential to the survival of living organisms. Many of these are needed in relatively large amounts (TABLE 2.1). Note that just six of these elements—carbon (C), hydrogen (H), nitrogen (N), oxygen (O), phosphorus (P), and sulfur (S)—make up about 97% of the weight in both human and bacterial cells. (The acronym CHNOPS is helpful in remembering these six important elements.) Another five elements make up most of the remaining 3%.

In addition, there are a number of elements needed in much smaller amounts. These so-called "trace elements" vary from organism to organism, but often include manganese (Mn), iron (Fe), copper (Cu), and zinc (Zn).

Atoms Are Composed of Charged and Uncharged Subatomic Particles

An **atom** is the smallest unit of matter having the properties of that element; that is, if you split an atom, like carbon, into simpler parts, it no longer has the properties of carbon. Because atoms are so small, they often are illustrated using a common but overly simplified two-dimensional model.

An atom consists of a positively charged core, the **atomic nucleus** that makes up most of the atom's mass (FIGURE 2.2A). The atomic nucleus contains two kinds of tightly packed subatomic particles called **protons** and **neutrons**. Although these

TABLE 2.1 Some of the Major Elements of Bacterial and Human Cells

	Element	Symbol	Percent by Mass Bacterial Cell	Percent by Mass Human Cell	Atomic Number	Mass Number
97%	Oxygen	O	72	66	8	16
	Carbon	C	12	18	6	12
	Hydrogen	H	10	10	1	1
	Nitrogen	N	3	3	7	14
	Phosphorus	P	0.6	1	15	31
	Sulfur	S	0.3	0.3	16	32
3%	Sodium	Na	1.0	0.2	11	23
	Magnesium	Mg	0.5	0.1	12	24
	Calcium	Ca	0.5	1.5	20	40
	Potassium	K	1.0	0.4	19	39
	Chlorine	Cl	0.05	0.15	17	35

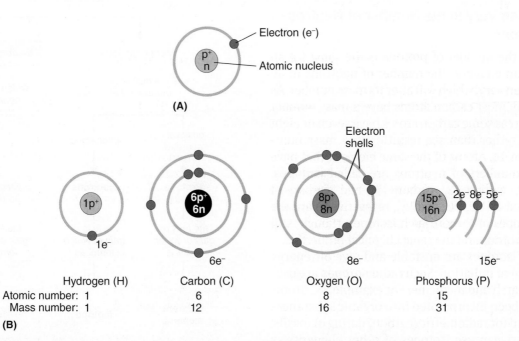

FIGURE 2.2 Atomic Structure and the Electron Configurations for Four Biologically Important Elements. Using two-dimensional models for atomic structure, **(A)** the atom is composed of protons and neutrons in the atomic nucleus, and electrons in electron shells (not drawn to scale). **(B)** The atomic structure of four biologically essential elements illustrates that the number of protons equals the number of electrons (though not necessarily equal to the number of neutrons). *»» Knowing the mass number for an element, what does that tell you about the mass of an electron?*

particles have about the same mass, each proton bears one positive electrical charge (value = +1), whereas a neutron has no charge.

The number of protons in an atom defines each element. For example, carbon atoms always have six protons. If there are seven protons, it is no longer carbon but rather the element nitrogen (see Table 2.1). The number of protons also represents the **atomic number** of the atom. As shown in Table 2.1, carbon with six protons has an atomic number of 6. The **mass number** is the total number of protons and neutrons in the nucleus. Because carbon atoms have six protons and usually six neutrons, the mass number of carbon would be 12.

Surrounding the atomic nucleus is a "cloud" of fast- moving, negatively charged **electrons** (value = −1). In uncharged atoms, the number of electrons is equal to the number of protons; that is, the atom has no net electrical charge. Because electrons move so fast, it is impossible at any moment to predict where a particular electron might be located; thus the term "electron cloud" is used. However, researchers can identify the areas within the cloud where electrons are usually found. These areas are called **electron shells**, each shell representing an area that is a different distance from the atomic nucleus and that has a specific energy level (**FIGURE 2.2B**).

Atoms Can Vary in the Number of Neutrons or Electrons

Although the number of protons is the same for all atoms of an element, the number of neutrons in an element can vary, which will alter its mass number. As mentioned, most carbon atoms have a mass number of 12, whereas some carbon atoms have seven or eight neutrons, rather than six, resulting in a mass number of 13 or 14. Atoms of the same element that have different numbers of neutrons are called **isotopes**. Therefore, carbon-12, carbon-13, and carbon-14 (symbolized as ^{12}C, ^{13}C, and ^{14}C, respectively) are the three isotopes of carbon. Such isotopes, though, still have six protons and the same chemical properties.

Some isotopes are unstable and give off energy in the form of radiation. Such **radioisotopes** are useful in research and medicine. For example, ^{14}C atoms that have been incorporated into organic substances can be used for radiometric (carbon) dating of fossils. Researchers can use isotopes of other elements as radioactive **tracers** to follow the fate of a substance (see **MicroInquiry 2** at the end of this chapter).

Should an uncharged atom gain or lose electrons, the resulting atom is called an **ion** (**FIGURE 2.3**). The

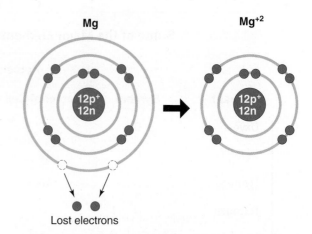

FIGURE 2.3 Formation of an Ion. Ions can be formed by the loss or gain of one or more electrons. Here, a magnesium (Mg) atom has lost two electrons to become a magnesium ion (Mg^{+2}). *»» For Mg, what does the superscript (+2) denote?*

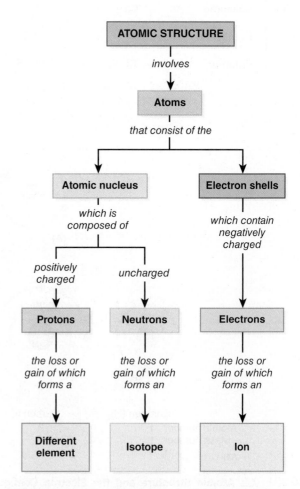

FIGURE 2.4 A Concept Map for Atomic Structure. This concept map shows the relationship between atoms, elements, isotopes, and ions. *»» Which terms in this map apply to atomic number and which apply to mass number?*

addition of one or more electrons to an atom means there is/are more negatively charged electrons than positively charged protons. Such a negatively charged ion is called an **anion**. By contrast, the loss of one or more electrons leaves the atom with fewer electrons and yields a positively charged ion, called a **cation**. As we will see, ion formation is important to some forms of chemical bonding. **FIGURE 2.4** provides a concept map that summarizes atomic structure.

Electron Distribution Determines the Chemical Properties of an Atom

Each electron shell (energy level) can hold a maximum number of electrons (see Figure 2.2B). The shell closest to the nucleus can accommodate two electrons, whereas the second and third shells each can hold eight. Any additional shells also have maximum numbers but usually no more than 18 are present in the outermost shell. Because the 25 essential elements are of lower mass number, only the first few shells are of significance to the chemical processes of life. Inner shells are filled first and,

if there are not enough electrons to fill the shell, the outermost shell will remain unfilled.

Atoms with an unfilled outer electron shell, called the **valence shell**, are unstable and the atoms tend to be chemically reactive. The valence electrons in the shell can become stable by interacting with another unstable atom. Look once again at Figure 2.2B and notice that the carbon atom, with six electrons, has two electrons in its first shell and only four in the valence shell. For this reason, carbon is extremely reactive in "finding" four more electrons and, as we will see, allows carbon to form innumerable combinations with other elements to fill the valence shell with eight electrons. Therefore, only atoms with unfilled valence shells participate in a chemical reaction.

The valence shells of a few elements normally are filled. Each of these elements, called an "inert gas," is chemically stable and unreactive. Helium (atomic number 2) and neon (atomic number 10) are examples; each has its valence shell filled with 2 and 8 electrons, respectively.

Concept and Reasoning Checks 2.1

a. Explain how the atomic number differs from the mass number.
b. Describe how an isotope differs from an ion.
c. Looking at Figure 2.2B, do these atoms have filled valence shells? Explain.

■ KEY CONCEPT 2.2   Chemical Bonds Form Between Reactive Atoms

When the valence shells of two unstable atoms come close to one another, the shells overlap, an energy exchange takes place, and each of the participating atoms assumes a more stable electron configuration. When two or more atoms are linked together, the force holding them is called a **chemical bond**. Therefore, chemical bonds are the result of atoms filling their valence shells.

The rearrangement of atoms through chemical bonding can occur in one of two ways: atoms, in the form of ions, can transfer valence electrons; or, each reactive atom can share valence electrons with one or more other reactive atoms. In both cases, the result is atoms having full valence shells.

Ionic Bonds Form Between Oppositely Charged Ions

In the formation of an **ionic bond**, one atom gives up its valence electrons to another atom. The reaction

between sodium and chlorine is an example of how these atoms become ions and then form ionic bonds when they encounter each other (**FIGURE 2.5**). With the transfer of electrons, both atoms now have their valence shell filled. Because opposite electrical charges attract, the chloride ions and sodium ions come together to form stable sodium chloride (NaCl)—table salt.

When two or more different elements interact with one another to achieve stability, they form a **compound**. Each compound has a definite formula and set of properties that distinguish it from its components. For example, sodium (Na) is an explosive metal and chlorine (Cl) is a poisonous gas, but the compound they form (NaCl) is a solid crystal of edible table salt.

Salts are typically formed through ionic bonding. Besides sodium (Na^+) and chloride (Cl^-), important salts are formed from other ions, including

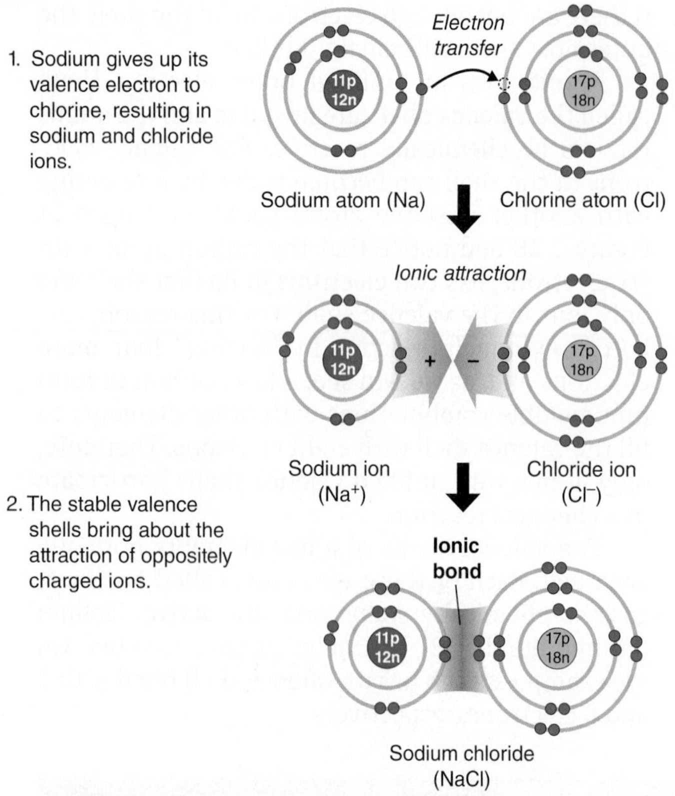

1. Sodium gives up its valence electron to chlorine, resulting in sodium and chloride ions.

Electron transfer

Sodium atom (Na)   Chlorine atom (Cl)

Ionic attraction

Sodium ion (Na$^+$)   Chloride ion (Cl$^-$)

2. The stable valence shells bring about the attraction of oppositely charged ions.

Ionic bond

Sodium chloride (NaCl)

FIGURE 2.5 Ion Formation and Ionic Bonding. The transfer of an electron from sodium to chlorine generates oppositely charged ions that are attracted to one another, forming an ionic bond through mutual attraction. *»» Which ion is a cation and which is an anion?*

calcium (Ca^{+2}), potassium (K^{+2}), magnesium (Mg^{+2}), and iron (Fe^{+2} or Fe^{+3}). In medicine, most salts represent **electrolytes**, which separate into ions when they dissolve in the fluids of the human body. Among the most important are sodium, potassium, magnesium, and phosphate (PO$_4^{-3}$). Although ionic bonds in solution are relatively weak, in large numbers they play important roles in protein structure and the reactions between antigens and antibodies in the immune response.

Covalent Bonds Share Electrons

Reactive atoms also can achieve stability by sharing valence electrons between interacting atoms, the sharing producing a **covalent bond**. Such strong bonds are very important in biology because the CHNOPS elements of life usually enter into covalent bonds with themselves and one another.

A **molecule** is two or more atoms held together by covalent bonds. Molecules can be composed of only one kind of atom, as in oxygen gas (O$_2$), or they can consist of different kinds of atoms in substances

such as water (H$_2$O), carbon dioxide (CO$_2$), and the simple sugar glucose (C$_6$H$_{12}$O$_6$). As shown by these examples, the combination and number of atoms (the subscript) in a molecule are called the **molecular formula**. (Note that the presence of one atom is represented without the subscript "1".)

Covalent bonding occurs frequently in carbon because this element has four electrons in its unfilled outer shell. The carbon atom is not strong enough to acquire four additional electrons, but it is sufficiently strong to retain the four it has. It therefore enters into a variety of covalent bonds with other reactive atoms or groups of atoms.

Nonpolar Covalent Bonds

When electrons between atoms are shared equally, a **nonpolar covalent bond** forms. An example of this type of bonding is methane (natural) gas (CH$_4$), which is a byproduct of cellulose digestion by some microbes residing in the ruminant stomach of a cow (**FIGURE 2.6A**). For methane, a reactive carbon atom shares each of its four valence electrons with the valence electron of a reactive hydrogen atom, forming four single nonpolar covalent bonds. Because of the equal sharing of electron pairs, there are no electrical charges (poles) on the molecule; methane, therefore, is a **nonpolar molecule**.

Scientists often draw chemical structures as **structural formulas**, that is, chemical diagrams showing the order and arrangement of atoms. In Figure 2.6, each line between carbon and hydrogen (C—H) represents a single nonpolar covalent bond between a pair of shared valence electrons. Other molecules, such as carbon dioxide (CO$_2$), share two pairs of valence electrons and therefore two lines are used to indicate the "double covalent bond": O=C=O. In both examples, the atoms now are stable because the outer electron shell of each atom is filled through this sharing.

Nonpolar covalent bonds form **hydrocarbons**, which are molecules consisting solely of hydrogen and carbon. Methane is the simplest hydrocarbon. When atoms are bonded covalently, they establish a geometric relationship determined largely by the electron configuration. Thus, hydrocarbons can consist of chains of carbon atoms and, in some cases, the chains can be closed to form a ring (**FIGURE 2.6B**). In all cases, the nonpolar covalent bonds are distributed equally around each carbon atom. Petroleum hydrocarbons are the primary components in oil and gasoline and can present quite a challenge to microbes (and humans) following an accidental oil spill, as described in MICROFOCUS 2.1.

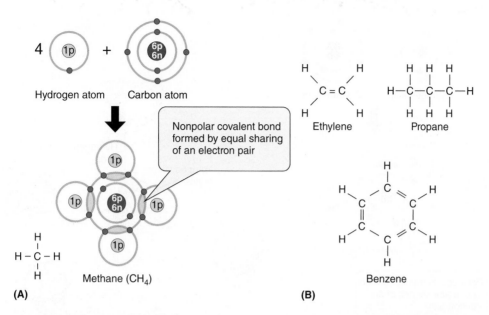

(A)

(B)

FIGURE 2.6 Chemical Bonding and Hydrocarbons. (A) A covalent bond involves the equal sharing of electron pairs between atoms, such as in this simple organic compound methane. **(B)** The molecular formulas for a few relatively simple hydrocarbons. *»» Supply the molecular formula for each of the structural formulas shown in (B).*

MICROFOCUS 2.1: Environmental Microbiology

Microbes to the Rescue!

On April 20, 2010, the Deepwater Horizon drilling rig, which was working on a well for the British Petroleum oil company, blew up in the Gulf of Mexico. For 3 months, oil spilled into the Gulf and contaminated nearby shores and wetlands, making it the largest accidental oil spill in American history. According to federal government estimates, some 5 million barrels, or 795 million liters, were spilled, and it was not until September that the well was declared sealed. Where did all those hydrocarbons go?

Petroleum or crude oil consists of a complex mixture of hydrocarbons of various sizes in liquid, gaseous, and solid forms. An oil well, such as the one in the Gulf, produces primarily crude oil, with some natural gas (primarily methane) dissolved in it. For decades, scientists have tried to use bioremediation, the breakdown (biodegradation) of contaminating compounds using microorganisms, as a natural method for cleaning up some of the environment's worst chemical hazards, including oil spills.

In the case of the ruptured Deepwater Horizon well, a natural form of bioremediation came to the rescue. The entry of oil profoundly altered the natural microbial community by significantly stimulating deep-sea, oil-hungry microbes to consume much of the smaller, dispersed hydrocarbons and methane gas (see figure). Their efficiency was amazing. With bloom sizes of 10^{23} cells and consisting of more than 50 species, within weeks, some areas of the Gulf were nearly free of oil. Overall, the microbes are estimated to have consumed 200,000 tons of oil and methane gas. Still, microbes cannot eliminate or digest all the hydrocarbons present in oil as the spill also produced tar-like hydrocarbons that were too large for microbes to digest. Still, one biogeochemist gave the microbes a "7 out of 10" for their digestive performance.

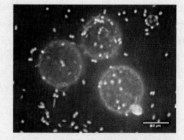

Bacteria (light-green dots) attached to and surrounding oil droplets (three large spheres).

Courtesy of Louisiana Tech University.

Polar Covalent Bonds

Not all molecules are nonpolar. In fact, water, one of the most important molecules to life, is a **polar molecule**—it has electrically charged poles (atoms)

(**FIGURE 2.7A**). Oxygen has a stronger "pull" on the shared electrons and thus has a slight negative charge. The hydrogen atoms are then left with a slight positive charge. The water molecule therefore consists of **polar covalent bonds**.

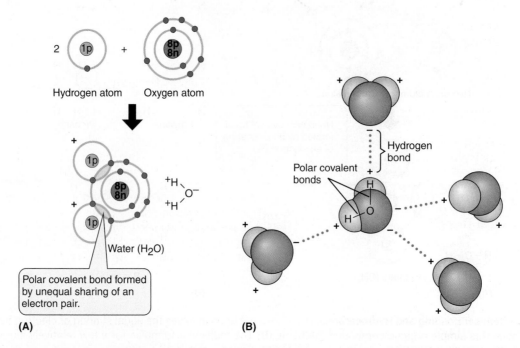

FIGURE 2.7 Chemical Bonding and Water. (A) The unequal sharing of electron pairs between oxygen and hydrogen atoms produces a polar molecule. **(B)** The charged regions of each polar water molecule allow hydrogen bonds to form between the hydrogen atom of one water molecule and the oxygen atom of another water molecule. *»» Why is there attraction between a hydrogen atom and an oxygen atom?*

Hydrogen Bonds Form Between Polar Groups or Molecules

A polar molecule, like water, has two positively charged hydrogen atoms and a negatively charged oxygen atom. Therefore, the positive hydrogen atom on one water molecule will be attracted to the negative oxygen on another water molecule. The opposing charges allow for the formation of a **hydrogen bond**. Although hydrogen bonds are much weaker than covalent bonds, hydrogen bonds provide the "glue" to hold water molecules together (**FIGURE 2.7B**). These bonds also are important to

the structure of proteins and nucleic acids, two of the major organic compounds of living cells we will soon encounter in this chapter.

TABLE 2.2 summarizes the different types of chemical bonds that have been discussed thus far.

Chemical Reactions Change Bonding Partners

A **chemical reaction** is a process in which atoms or molecules interact through making and breaking chemical bonds. Different combinations of atoms or molecules result from the reaction; that is, bonding partners change. However, the total number of

TABLE 2.2 Three Types of Chemical Bonds in Living Organisms

Type	Chemical Basis	Strength	Example
Ionic	Attraction between oppositely charged ions	Weak	Sodium chloride
Covalent	Sharing of electron pairs between atoms	Strong	Glucose
Hydrogen	Attraction of a hydrogen nucleus (a proton) to negatively charged oxygen or nitrogen atoms in the same or neighboring molecules	Weak	Water

interacting atoms remains constant. For chemical reactions, an arrow indicates in which direction the reaction will proceed. By convention, the atoms or molecules drawn to the left of the arrow are the **reactants** and those to the right are the **products** of the reaction, as demonstrated here:

$$A + B \longrightarrow A - B$$
Substrates   Product

Two types of chemical reactions are common in biological compounds: dehydration and hydrolysis. Let's take a look at them.

Dehydration Reactions

In biology, many chemical reactions are based on the assembly (synthesis) of larger compounds from smaller building blocks (**FIGURE 2.8A**). In a **dehydration reaction**, when smaller reactants are put together to build larger products, often a water molecule is formed as part of the products.

Hydrolysis Reactions

Many chemical reactions in cells also break (hydrolyze) larger reactants into smaller products (**FIGURE 2.8B**). Water is one of the reactants used to break another reactant in what is therefore referred to as a **hydrolysis reaction** (*hydro* = "water"; *lysis* =

$$C_6H_{12}O_6 + C_6H_{12}O_6 \longrightarrow C_{12}H_{22}O_{11}$$
Glucose   Glucose   Maltose
H_2O
Water

(A) Dehydration reaction

$$C_{12}H_{22}O_{11} \longrightarrow C_6H_{12}O_6 + C_6H_{12}O_6$$
Maltose   Glucose   Glucose
H_2O
Water

(B) Hydrolysis reaction

FIGURE 2.8 Dehydration and Hydrolysis Reactions. (A) Dehydration reactions result in building larger molecules with water being one of the products of the reaction. **(B)** Hydrolysis reactions break down larger molecules into smaller ones, using a water molecule as a reactant. *»» Provide proof that the numbers of atoms and types of bonds have not changed because of the chemical reaction in (A) or (B).*

"break"). The digestion of food molecules would be an example of these reactions.

In summary, the new products formed in a chemical reaction have the same number and types of atoms that were present in the reactants. In forming new products, chemical reactions only involve a change in the bonding partners. No atoms have been gained or lost from any of these reactions. The sum of all chemical reactions in cells, including dehydration and hydrolysis reactions, is referred to as **cell metabolism**.

Concept and Reasoning Checks 2.2

a.   Construct a diagram to show how the salt calcium chloride ($CaCl_2$) is formed.
b.   Why do atoms share pairs of electrons?
c.   Construct a diagram to show how hydrogen bonding occurs in a molecule of liquid ammonia (NH_3).
d.   Compare and contrast dehydration and hydrolysis reactions.

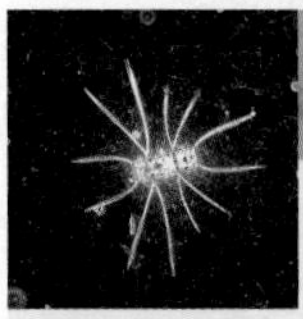

Chapter **Challenge A**

At this point, you should know that all life (as we know it) is carbon based. With four electrons to share, carbon can form many bonds with other carbon atoms and/or with other elements. Some scientists have proposed that the element silicon (Si) could replace carbon. After all, silicon is the second most abundant element in the Earth's crust (oxygen is first whereas carbon is fifteenth). In fact, microbes called diatoms (a type of alga) have a cell wall containing silica (hydrated silicon dioxide; SiO_2). A few scientists have suggested that the first life on Earth (microbial, we assume) might have been silicon based.

QUESTION A: *Knowing that silicon (Si) has an atomic number of 14 and a mass number of 28, how similar is silicon to carbon that some individuals would propose it as a potential substitute for carbon in "shadow life"? Draw the structural formula for silane, which is silicon with four hydrogen atoms covalently bonded. How similar is it to methane? As a note: In 2016, researchers were successful in creating a protein isolated from a bacterium that in the lab will bond carbon to silicon.*

You can find answers online in **Appendix F**.

■ KEY CONCEPT 2.3   All Living Organisms Depend on Water

All organisms are composed primarily of water. Human cells, as well as microbial cells, are about 70% water by mass. No organism can survive and grow without water, and many organisms, including microbes, live in watery or moist environments.

Water Has Several Unique Properties

The water molecule and its interactions with other molecules produces some unique properties.

Universal Solvent

Liquid water is the medium in which all cellular chemical reactions occur. Being a polar molecule, water molecules are attracted to other polar molecules. Take, for example, what happens when a crystal of salt is put in water. The salt is **hydrophilic**, meaning it easily dissolves in water into separate sodium and chloride ions because water molecules break the weak ionic bonds and surround each ion in a sphere of water molecules (**FIGURE 2.9**). This **aqueous solution** consists of **solutes** (salt) dissolved in water, which in biology is the **universal solvent** (**MICROFOCUS 2.2**). By contrast, substances that do not dissolve in water (lipids; see the section coming up shortly) are **hydrophobic**.

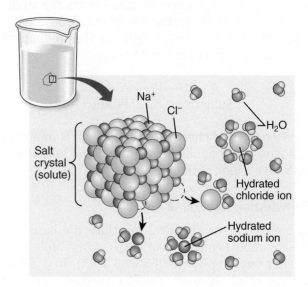

FIGURE 2.9 Solutes Dissolve in Water. Water molecules separate and surround Na⁺ and Cl⁻ ions, facilitating their dissolving into solution. *»» When dissolving, why do the H's of water surround Cl⁻ while the O's surround the Na⁺?*

Water molecules also are reactants or products in many chemical reactions, including the dehydration and hydrolysis reactions described in the previous section (see Figure 2.8).

MICROFOCUS 2.2: Tools

The Relationship Between Mass Number and Molecular Weight

Often scientists and students need to make solutions that have a specific concentration of solutes. To make these solutions, they need to know how much a particular molecule or solute weighs, which is referred to as the **molecular weight**. The calculation of the molecular weight simply consists of adding together the mass number of all the individual atoms in the substance of interest. For example, the molecular weight of a water molecule is 18 daltons,* whereas the molecular weight of a glucose molecule is 180 daltons. Other molecules can reach astonishing proportions—antibodies of the immune system can have a molecular weight of 150,000 daltons and the bacterial toxin causing botulism is over 900,000 daltons.

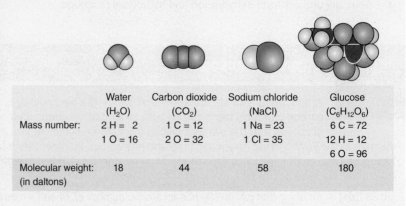

	Water (H_2O)	Carbon dioxide (CO_2)	Sodium chloride (NaCl)	Glucose ($C_6H_{12}O_6$)
Mass number:	2 H = 2 1 O = 16	1 C = 12 2 O = 32	1 Na = 23 1 Cl = 35	6 C = 72 12 H = 12 6 O = 96
Molecular weight: (in daltons)	18	44	58	180

* Dalton: The unit in biology that is used to measure the weight of atomic particles; equivalent to an atomic mass unit used in chemistry (or one-twelfth the weight of an atom of ^{12}C).

Cohesion

As you have learned, the polar nature of water molecules leads to hydrogen bonding. By forming a large number of hydrogen bonds, water molecules are held close together, which is a property called **cohesion**. However, unique to this bonding is that at freezing temperatures, the distance between water molecules expands and there now is more space than there is between water molecules at liquid temperatures. Thus, ice floats because the frozen water is less dense than liquid water. This means ice formation insulates deeper water in lakes and keeps the water from freezing and, by remaining liquid, life (including microbial life) in the water is maintained.

Temperature Modulation

Because of hydrogen bonding, it takes a large amount of heat energy to increase the temperature of water. Likewise, a large amount of heat must be lost before water temperature decreases. So, by being about 70% water, cells are bathed in a solvent that maintains a more consistent temperature even when the environmental temperatures change.

Acids and Bases Affect a Solution's pH

In an aqueous solution, water molecules can split into hydrogen ions (H^+; protons) and hydroxide ions (OH^-), only to rapidly recombine (the double arrowhead indicates a reversible reaction):

$$H_2O \leftrightarrow H^+ + OH^-$$

In addition to water, other compounds in cells can dissociate into H^+ when they dissolve in water.

For our purposes, an **acid** (such as hydrochloric acid—HCl) is a chemical substance that increases the hydrogen ion concentration ($[H^+]$) of a solution by donating H^+ to a solution.

$$HCl \leftrightarrow H^+ + Cl^-$$

By contrast, a **base** (such as sodium hydroxide—NaOH) is a substance that reduces the $[H^+]$ in solution by dissociating into hydroxide ions and cations, or by directly accepting H^+.

$$NaOH \leftrightarrow Na^+ + OH^- \qquad OH^- + H^+ \leftrightarrow H_2O$$

To indicate the concentration of H^+ in a solution, the symbol **pH** (power of hydrogen ions) and the **pH scale** are used. This numerical scale extends from 0 (extremely acidic; high $[H^+]$) to 14 (extremely basic or alkaline; low $[H^+]$) and is based on actual calculations of the number of hydrogen ions present when a substance mixes with water. A substance with a pH of 7, such as pure water, is said to be neutral; solutions that gain H^+ are said to be "acidic" and have a pH lower than 7; solutions that lose H^+ are "basic" (or alkaline) and have a pH greater than 7.

The pH scale is logarithmic; that is, every time the pH changes by one unit, the $[H^+]$ changes 10 times. For example, lemon juice (pH 2) and black coffee (pH 5) differ in $[H^+]$ by a thousandfold (10^3). **FIGURE 2.10** summarizes the pH values of several common substances. With regard to microorganisms, fungi prefer a slightly acidic environment compared to the more neutral environment preferred by most microbes—although there are some spectacular exceptions, as MICROFOCUS 2.3 points out.

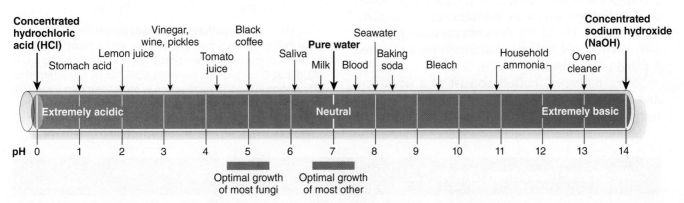

FIGURE 2.10 A Sample of pH Values for Some Common Substances. Most fungi prefer a slightly acidic pH for growth compared to most other microbes. *»» On the pH scale, notice that many of the beverages we drink (e.g., wine, tomato juice, coffee) are acidic. However, we would never normally drink equally alkaline solutions (e.g., commercial bleach, ammonia). Propose an explanation for the statement.*

MICROFOCUS 2.3: Environmental Microbiology

Just South of Chicago

All you need is a map, some pH paper, and a few collection vials. When in Chicago, use your map to find the Lake Calumet region just southeast of the city. When you arrive, pull out your pH paper, and sample some of the groundwater in the region near the Calumet River. You will be shocked to discover the pH is greater than 12—almost as alkaline as oven cleaner! In fact, this might be one of the most extreme pH environments on Earth.

How did the water get this alkaline and could anything possibly live in the groundwater?

The groundwater in the area near Lake Calumet became strongly alkaline because of the steel slag that was dumped into the area for more than 100 years. Used to fill the wetlands and lakes, water and air chemically react with the slag to produce lime [calcium hydroxide, $Ca(OH)_2$]. It is estimated that 10 trillion cubic feet of slag and the resulting lime are what have pushed the pH to such a high value.

Now use your collection vials to collect some samples of the water. Back in the lab you will be surprised to find that there are bacterial communities present in the water. Hydrogeologists who have collected such samples have discovered some bacterial species that until then had only been found in Greenland and deep gold mines of South Africa. Other identified species appear to use the hydrogen generated from the corrosion of the iron for energy.

How did these bacterial organisms get there? The hydrogeologists propose that the bacterial species have always been there and have simply adapted to the environment over the last 100 years as the slag accumulated. Otherwise, the microbes must have been imported in some way.

So, once again, provide a specific environment and they will come (or evolve)—the microbes that is.

Cell Chemistry Is Sensitive to pH Changes

As microorganisms—in fact all organisms—take up or ingest nutrients and undergo metabolism, chemical reactions occur that use up or produce H^+. Therefore, it is important for all organisms to balance the acids and bases in cells because chemical reactions and organic compounds are very sensitive to pH shifts. Proteins are especially vulnerable, as we will soon see. If the internal cellular pH is not maintained, these proteins can be destroyed. Likewise, when most microbes grow in a microbiological nutrient medium, the waste products produced can lower the pH of the medium, which could kill the organisms.

To prevent pH shifts, cells (and growth media) contain **buffers**, which are substances that maintain a specific pH. The buffer does not necessarily maintain a neutral pH, but rather whatever pH is required for that environment.

Most biological buffers consist of a weak acid and a weak base (**FIGURE 2.11**). If an excessive number of H^+ is produced (potential pH decrease), the base can absorb them. Alternatively, if there is a decrease in H^+ (potential pH increase), the weak acid can dissociate, replacing the lost H^+. Thus, the give and take of H^+ maintains the pH.

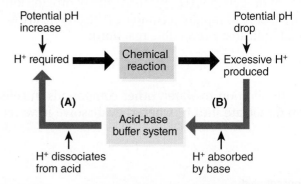

FIGURE 2.11 Hypothetical Example for pH Shifts. An acid/base buffer system can prevent pH shifts from occurring because of a chemical reaction. If the reaction is using up H^+ **(A)**, the acid component prevents a pH rise by donating H^+ to offset those used. If the reaction is producing excess H^+ **(B)**, the base can prevent a pH drop by "absorbing" them. »» *Propose what would happen if a chemical reaction continued to release excessive H^+ for a prolonged period.*

Concept and Reasoning Checks 2.3

a. Explain the difference between a solution, a solvent, and a solute.
b. What are the properties of acids and bases?
c. If the pH does drop in a cell, what does that tell you about the buffer system?

Chapter Challenge B

In trying to find "markers" to detect "shadow life," we run into the problem of microbial diversity. Microbes occupy so many diverse and often extreme habitats (many are called "extremophiles") that it is hard to know what "shadow life" really means in the context of life on Earth.

QUESTION B: *True microbial life (as we know it) can be found, for example, at extreme temperatures (including substantially above the boiling point of water), extreme pressures (in excess of 1,000 times atmospheric pressure), and extreme pHs (both exceptionally acidic and, as described in* MICROFOCUS 2.3, *tremendously alkaline environments). Are these examples of "extremophiles" representative of "shadow life"? If we do find microbes that use silicon, is that a marker for shadow life or is it just another example of the flexibility of microbes to make the best of their surroundings? If a microbe lost its carbon source, could it use silicon if it were available? What's your opinion?*

You can find answers online in **Appendix F**.

■ KEY CONCEPT 2.4   Living Organisms Are Composed of Four Types of Organic Compounds

As mentioned in the last section, a typical prokaryotic cell is composed of about 70% water. If all the water is evaporated, the predominant "dry weight" remaining consists of **organic compounds**, which are those molecules related to or having a carbon basis: the carbohydrates, lipids, proteins, and nucleic acids (**FIGURE 2.12**). Except for the lipids, each class represents a **polymer** (*poly* = "many"; *mer* = "part") built from a very large number of building blocks called **monomers** (*mono* = "one").

Functional Groups Define Molecular Behavior

Before we look at the major classes of organic compounds, we need to address one question. The monomers building carbohydrates, nucleic acids, and proteins are essentially stable, unreactive molecules because their outer shells are filled through covalent bonding. Why then should these molecules take part in chemical reactions to build polymers?

The answer is that these monomers are not entirely stable. Projecting from the carbon skeletons are groups of atoms called **functional groups**, which represent points where further chemical reactions can occur if facilitated by a specific **enzyme**. There are only a few functional groups, but their differences and placement on organic compounds make possible a large variety of chemical reactions. The important functional groups in living organisms are identified in TABLE 2.3.

Functional groups on monomers can interact to form larger molecules or polymers through

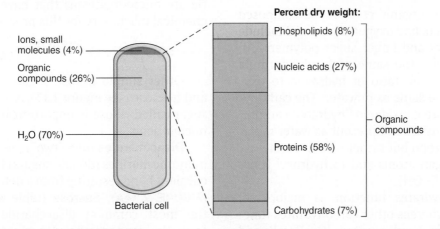

Percent dry weight:

Ions, small molecules (4%)
Organic compounds (26%)
H₂O (70%)
Bacterial cell

Phospholipids (8%)
Nucleic acids (27%)
Proteins (58%)
Carbohydrates (7%)
Organic compounds

FIGURE 2.12 Organic Compounds in Bacterial Cells. Organic compounds are abundant in cells. The approximate composition of these compounds in a bacterial cell is similar to the percentages found in other microbes. *»» Propose a reason why proteins make up almost 60% of the dry weight of a bacterial cell.*

TABLE 2.3 Common Functional Groups (and examples) on Organic Compounds

Functional Group	Structural Formula	Shorthand	Examples
Hydroxyl	—O—H	—OH	Alcohols, simple sugars, amino acids
Carboxyl	$\overset{\displaystyle O}{\underset{\displaystyle -C-OH}{\|}}$	—COOH	Fatty acids, amino acids, proteins
Carbonyl	$\overset{\displaystyle O}{\underset{\displaystyle -C-}{\|}}$	—CO—	Carbohydrates
Amino	$\overset{\displaystyle H}{\underset{\displaystyle -N-H}{\|}}$	—NH$_2$	Amino acids, proteins
Sulfhydryl	—S—H	—SH	Amino acids, proteins
Phosphate	$\overset{\displaystyle OH}{\underset{\displaystyle \overset{-O-P=O}{\underset{\displaystyle OH}{\|}}}{\|}}$	—H$_2$PO$_4$	Phospholipids, nucleic acids, ATP

dehydration reactions. In addition, functional groups can be critical for breaking larger polymers into monomers through hydrolysis reactions.

If you are unsure about the role of these functional groups, do not worry. We will see how specific functional groups interact through dehydration reactions as we now visit each of the four classes of organic compounds.

Carbohydrates Are Needed for Energy Metabolism and as Structural Materials

Carbohydrates are organic compounds composed of carbon, hydrogen, and oxygen atoms and include a variety of sugars and larger sugar polymers. All carbohydrates share the same molecular formula $(C_6H_{12}O_6)_n$, making the ratio of hydrogen to oxygen two to one, the same as in water. The carbohydrates therefore are considered "hydrated carbon." However, the atoms are not present as water molecules bound to carbon but rather carbon covalently bonded to hydrogen atoms and to hydroxyl functional groups (H–C–OH).

Some carbohydrates function as major fuel sources in cells, whereas other carbohydrates function as structural molecules present in cell walls and nucleic acids. Often the carbohydrates are termed saccharides (*sacchar* = "sugar").

Sugars

The so-called "simple sugars" are the **monosaccharides** that consist of a single sugar monomer. Glucose, a six-carbon (hexose) sugar, is the most common monosaccharide and is essential for cell metabolism (**FIGURE 2.13**). Estimates vary, but many scientists say that half the world's carbon exists in glucose. Such simple sugars are synthesized from carbon dioxide and water through the process of **photosynthesis**. Algae and cyanobacteria are microorganisms that have the cellular and chemical machinery for this process.

$$6CO_2 + 6H_2O \xrightarrow{\text{sunlight}} C_6H_{12}O_6 + 6O_2$$

Other simple hexose sugars include galactose and fructose (see Figure 2.13). A 5-carbon (pentose) sugar called ribose is important in the structure of nucleic acids.

Disaccharides (*di* = "two") are composed of two monosaccharides (double sugars) held together by a covalent bond resulting from a dehydration reaction (**FIGURE 2.14A**). Sucrose (table sugar—$C_{12}H_{22}O_{11}$), the most common disaccharide, is constructed from the monosaccharides glucose and fructose. The disaccharide is an additive to many foods and is a starting point in wine fermentations. Maltose,

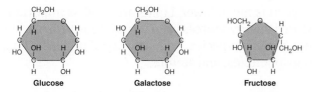

FIGURE 2.13 Structural Formulas of Common Monosaccharides. Although monosaccharides can exist in linear forms, in cells they usually occur as ring structures. »» *What is the molecular formula for all three of these monosaccharides?*

another disaccharide, is composed of two glucose monomers (**FIGURE 2.14B**). This disaccharide occurs in cereal grains, such as barley, and the sugar is fermented by yeasts when brewing beer. Lactose, a third common disaccharide, is composed of the monosaccharides glucose and galactose. Lactose is known as "milk sugar" because it is the principal disaccharide in milk. In the dairy industry, microorganisms digest the lactose in milk products, producing the lactic acid found in yogurt, sour cream, and other sour dairy products.

Polysaccharides

In the microbial world, one of the major uses of monosaccharides is as building blocks for large polymers. **Polysaccharides** (*poly* = "many") are

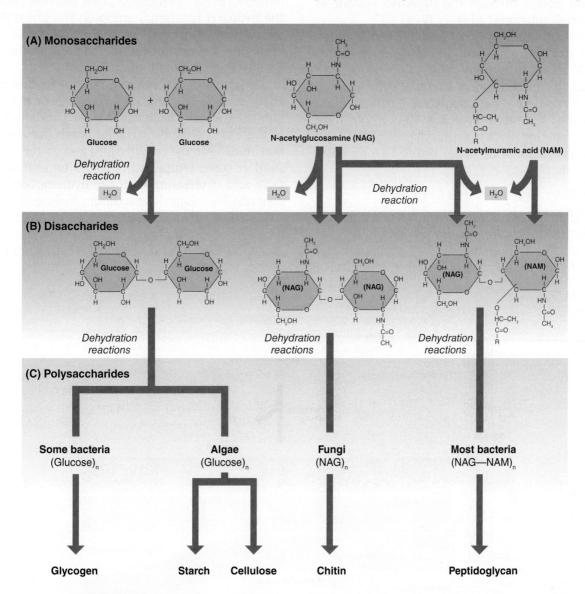

Note: The polysaccharides of glucose [(Glucose)$_n$] vary in the carbon bonding between glucose monomers and branching of the polymer chains.

FIGURE 2.14 Carbohydrate Monomers Are Built into Polymers. (A) There are many monosaccharides used by organisms that can be combined into disaccharides **(B)** or assembled into long polymers called polysaccharides **(C)**. »» *What type of chemical reaction is required to link glucose into a long polymer such as cellulose?*

"complex carbohydrates" formed by joining hundreds of thousands of similar monosaccharides, usually glucose (**FIGURE 2.14C**). Covalent bonds resulting from the dehydration reactions link the glucose units together.

Starch and **glycogen** are two common energy storage polysaccharides in algal and some bacteria cells. **Cellulose**, a structural polysaccharide, is a component of the cell walls of many single-celled algae and plants. Interestingly, no animals have the enzymes to digest cellulose. Ruminant animals (e.g., cows and sheep) depend on microbes in their gut to break down the cellulose they eat. In humans, cellulose passes through the digestive system as undigested "roughage" (dietary fiber).

Other structural polymers include **chitin**, which is built from chains of another glucose derivative called *N*-acetylglucosamine (NAG). Chitin forms the cell walls of fungi. In most bacterial cells, the cell wall is composed of **peptidoglycan**, a carbohydrate modified with protein (peptido). The carbohydrate (glycan) building block is a disaccharide of NAG and *N*-acetylmuramic acid (NAM) linked in long polymer chains.

Lipids Play Several Roles in Cells

Lipids are a broad group of nonpolar organic compounds that are **hydrophobic**; they do not mix with water. Like carbohydrates, lipids are composed of carbon, hydrogen, and oxygen, but the proportion of oxygen is much lower. Lipids are used as important stored energy sources by many microorganisms, but not bacterial species. The majority of lipids are fats, phospholipids, and sterols.

Fats

A **fat** (also called a triglyceride) consists of a three-carbon glycerol molecule and three long-chain fatty acids (**FIGURE 2.15**). Each fatty acid is a long nonpolar hydrocarbon chain containing between 16 and 18 carbon atoms, which explains why fats do not mix with water. Bonding of each fatty acid to the glycerol molecule occurs by a dehydration reaction between the hydroxyl functional group on the glycerol and the carboxyl functional group on a fatty acid. Consequently, fats are not considered polymers because a triglyceride has but four monomers.

A fatty acid is **saturated** if the hydrocarbon chain contains the maximum number of hydrogen atoms extending from the carbon backbone; that is, there are no double covalent bonds between carbon atoms. Animal fats, like lard and butter that are solid at room temperature, would be examples. A fatty acid is **unsaturated** if the hydrocarbon chain contains less than the maximum hydrogen atoms; that is, there is one or more double covalent bonds between a few carbon atoms (see Figure 2.15). Examples include plant and fish oils that are liquid at room temperature.

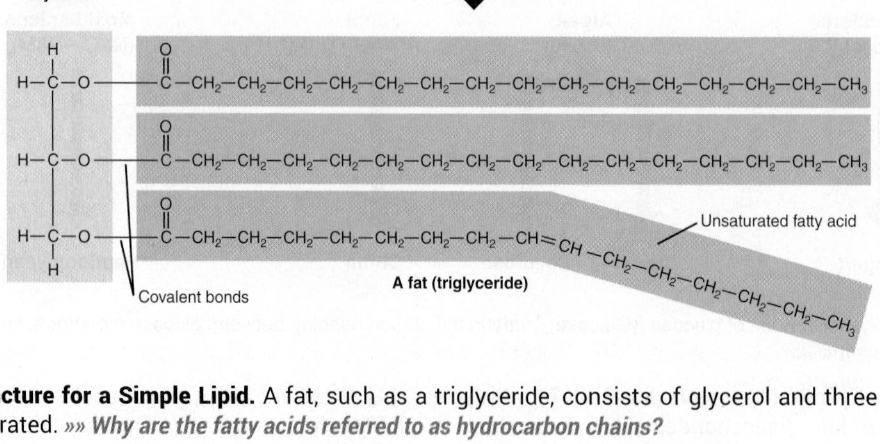

FIGURE 2.15 Structure for a Simple Lipid. A fat, such as a triglyceride, consists of glycerol and three fatty acids that are saturated or unsaturated. *»» Why are the fatty acids referred to as hydrocarbon chains?*

Phospholipids

Another type of lipid found in cell membranes is the **phospholipids**, which, unlike fats, have only two fatty acid tails attached to glycerol (**FIGURE 2.16A**). In place of the third fatty acid, there is a polar phosphate functional group, which actively interacts with other polar molecules. Some bacterial toxins are a combination of polysaccharide and lipid, as reported in **CLINICAL CASE 2**.

Because one end of the phospholipid is polar (hydrophilic) and the other end is nonpolar (hydrophobic), when mixed in water the hydrophobic ends group together, whereas the hydrophilic ends on the outside interact with water. The result is a **phospholipid bilayer** with the fatty acid tails

pointing inward. This organization forms the basic structure of the membranes found in all cells (**FIGURE 2.16B**).

Other Lipids

Besides the fats and phospholipids, other types of lipids include the waxes and sterols. Waxes are composed of long chains of fatty acids and form part of the cell wall in *Mycobacterium tuberculosis*, the bacterium causing tuberculosis. The waxy wall helps protect the cell from desiccation.

Sterols are structurally very different from the other lipids and are included with lipids solely because they too are hydrophobic molecules (**FIGURE 2.16C**). Sterols, such as ergosterol, are

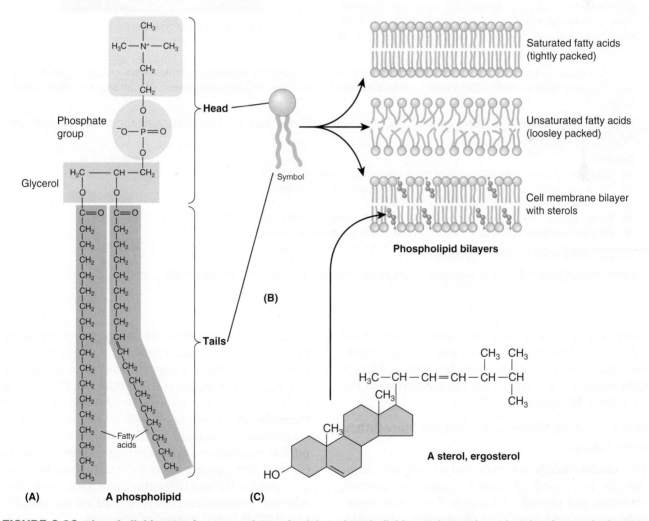

FIGURE 2.16 Phospholipids, Membranes, and Sterols. (A) A phospholipid contains a glycerol molecule attached to two fatty acids and a phosphate head group. Inset: The symbol for the structure of a phospholipid. **(B)** Phospholipids in biological membranes form a bilayer with the hydrophobic tails pointing inwards. **(C)** The sterol like ergosterol can structurally stabilize membranes. *»» Why are all these compounds considered "lipids"?*

Clinical Case 2

An Outbreak of *Salmonella* Food Poisoning

During the morning of October 17, a restaurant employee prepared a Caesar salad dressing, cracking fresh eggs into a large bowl containing olive oil. Anchovies, garlic, and warm water then were mixed into the eggs and oil.

The warm water raised the temperature of the mixture slightly before the dressing was placed in the refrigerator. Later that day, the Caesar dressing was placed at the salad bar in a cooled compartment having a temperature of about 16°C. The dressing remained at the salad bar until the restaurant closed, a period of 8 to 10 hours. During that time, many patrons helped themselves to the Caesar salad and the dressing.

Within 3 days, 15 restaurant patrons experienced gastrointestinal illness. Symptoms included diarrhea, fever, abdominal cramps, nausea, and chills. Thirteen sought medical care, and eight (all over 65 years of age) required intravenous rehydration.

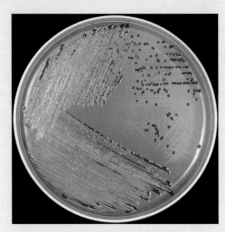

A culture plate of *Salmonella*.

Courtesy of CDC.

From the stool samples of all 13 patrons who sought medical attention, bacterial colonies were cultured (see figure) and laboratory tests identified *Salmonella enterica* serotype Enteritidis as the causative agent. All patrons recovered within 7 days.

Salmonella enterica serotype Enteritidis produces a lipopolysaccharide toxin that causes the symptoms experienced by all the affected patrons.

Questions:
a. What might have been the origin of the bacterial contamination?
b. What conditions would have encouraged the growth of *Salmonella*?
c. How could the outbreak have been prevented?
d. What types of organic compounds form the lipopolysaccharide toxin?
e. Why did so many of the older adult patrons develop a serious illness?

You can find answers online in **Appendix E**.

For additional information, see www.cdc.gov/nczved/divisions/dfbmd/diseases/salmonella_enteritidis.

---

composed of several rings of carbon atoms with side chains. Like cholesterol in animal cells, ergosterol stabilizes the membranes of protists and fungi. Other sterols are found in the cell membrane of the bacterium *Mycoplasma*.

Nucleic Acids Store and Transmit Hereditary Information

The **nucleic acids** are unbranched, organic compounds composed of carbon, hydrogen, oxygen, nitrogen, and phosphorus atoms. Two types, **deoxyribonucleic acid (DNA)** and **ribonucleic acid (RNA)**, are the basic genetic material needed to control the complex metabolism within cells of all organisms and are the genetic material in viruses.

Both DNA and RNA are composed of repeating monomers called **nucleotides**. Each nucleotide has three components: a sugar molecule, a phosphate group, and a nucleobase (**FIGURE 2.17A**). The sugar in DNA is deoxyribose, whereas in RNA it is ribose. The **nucleobases**, shown in **FIGURE 2.17B**, are nitrogen-containing molecules. In DNA, the double-ringed purine nucleobases are adenine (A) and guanine (G) and the single-ringed pyrimidine nucleobases are cytosine (C) and thymine (T). In RNA molecules, adenine, guanine, and cytosine also are present, but uracil (U) is found instead of thymine.

Nucleic acids, like polysaccharides, are polymers. In DNA and RNA, the nucleotides are covalently joined through dehydration reactions between the sugar of one nucleotide and the phosphate of the adjacent nucleotide to form a linear **polynucleotide**.

DNA Structure and Function

In 1953, James Watson, Francis Crick, Rosalind Franklin, and Maurice Wilkins published three research reports describing for the first time the structure of DNA. They described DNA as a molecule consisting of two polynucleotide strands running in opposite directions in a ladder-like arrangement (**FIGURE 2.17C**). They also reported that the nucleobases guanine and cytosine always line up opposite each other, as do thymine and adenine oppose each other in the two strands. These complementary base pairs in the double-stranded DNA are held

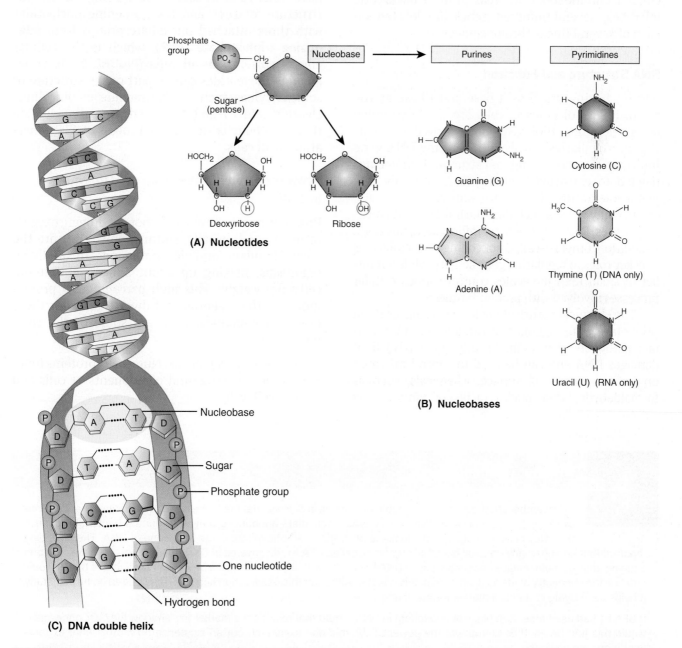

FIGURE 2.17 The Molecular Structures of Nucleotide Components and the Construction of DNA. (A) The sugars in nucleotides are ribose and deoxyribose, which are identical except for one additional oxygen atom in RNA. **(B)** For each nucleobase, note the similarities in the structures of the bases and the differences in the side groups. **(C)** The two polynucleotides of DNA forming the double helix are held together by hydrogen bonds between adenine (A) and thymine (T) and guanine (G) and cytosine (C). *»» If a segment of one strand of DNA has the bases TTAGGCACG, what would be the sequence of bases in the complementary strand?*

together by hydrogen bonds. When fully formed, the DNA twists to form a spiral arrangement called the **double helix**.

The genetic information in DNA exists in discrete units of inheritance called **genes**, which are sequences of nucleotides that encode information to regulate and synthesize proteins. In prokaryotic cells, these genes are usually found on a single, circular **chromosome**, whereas in most eukaryotic microbes, several hundred genes are located on each of several linear chromosomes.

RNA Structure and Function

Besides having uracil as a base and ribose as the sugar, RNA molecules in cells are single-stranded polynucleotides. Biologists once viewed RNAs as the intermediaries, involved in carrying DNA gene information and acting as structural molecules needed to construct proteins. This certainly is a major role for RNA but not the only roles.

In many viruses, including the influenza and measles viruses, RNA is the genetic information, not DNA. In addition, other so-called noncoding RNA molecules play key roles in regulating gene activity, while a number of small RNA molecules control various cellular processes involved with protein synthesis.

The chemical nature of nucleic acids makes them prime targets for agents that can alter or kill microbial pathogens. For example, ultraviolet (UV) light damages DNA and can be used to control microbes on an environmental surface. Chemicals, such as formaldehyde, alter nucleic acids of viruses, and microbiologists can employ some of these chemicals in the preparation of several vaccines. In addition, certain antibiotics interfere with DNA or RNA function and are used to inhibit or kill bacterial pathogens.

Other Roles for Nucleotides

It also is important to point out that nucleotides have other roles in cells besides being the subunit structure for DNA and RNA. Adenine nucleotides with three attached phosphate groups form **adenosine triphosphate** (**ATP**), which is the cellular energy currency in all cells (**FIGURE 2.18**). Other adenine nucleotides can be part of the structure of some enzymes, such as nicotinamide adenine dinucleotide (NAD^+) and flavin adenine dinucleotide (FAD). Both are critical in the process of ATP generation in all cells.

Proteins Are the "Workhorse" Polymers in Cells

Proteins are composed of carbon, hydrogen, oxygen, nitrogen, and, usually, sulfur atoms. They are the most abundant organic compounds in all living organisms, making up about 58% of a bacterial cell's dry weight. This high percentage of protein indicates the essential and diverse roles for these organic compounds. Here are some of the roles for proteins:

▶ *Structural proteins*. Numerous proteins function as structural components of cells and cell walls.

Chapter Challenge C

In November 2010, an article was published online in *Science* that made the news: *A Bacterium That Can Grow by Using Arsenic Instead of Phosphorus*. NASA scientists announced they had found a bacterial organism that they believed used arsenic in place of "traditional" phosphate in the backbone of DNA. The scientists reported that an apparently unusual bacterial organism, called GFAJ-1, discovered in California's Mono Lake (see chapter opening photo), could replace phosphorus in its DNA backbone with arsenic if no phosphate was available. If true, it would redefine the chemistry of life as discussed in this chapter. Although this observation has been discredited on further study, it helps us in trying to define a marker for our shadow life.

If GFAJ-1 had used arsenic in place of phosphate in DNA, would that represent a marker for "shadow life?" Or, once again, would this just be another example of the power of the microbe to scratch out an existence no matter what extreme conditions are present?

QUESTION C: *The atomic number for arsenic is 33 and the mass number is 75. Draw a segment of a DNA double helix (use Figure 2.17C as a template) to illustrate how arsenic could replace phosphate. What two other phosphate-containing molecules or structures discussed in this chapter might also be replaced with arsenic?*

You can find answers online in **Appendix F**.

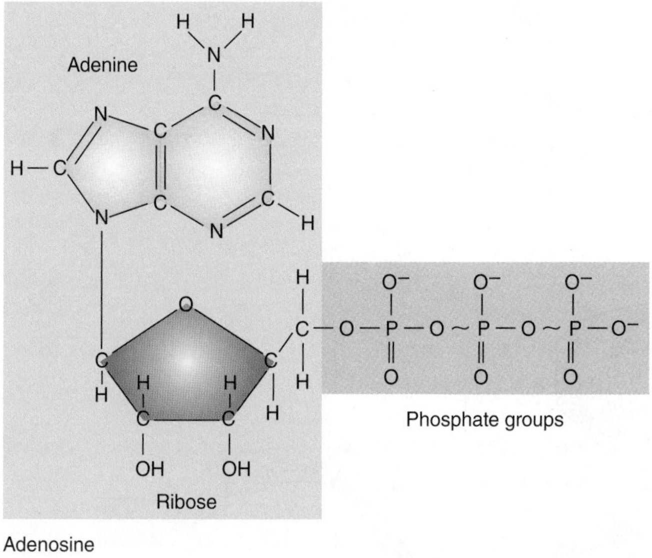

as antibodies, to defend against and eliminate invading pathogens.

▶ *Enzymes*. A large number of proteins serve as **enzymes**, protein catalysts that selectively speed up metabolic reactions.

Proteins are unbranched polymers built from nitrogen-containing monomers called **amino acids**. At the center of each amino acid is a carbon atom attached to two functional groups: an amino group ($-NH_2$) and a carboxyl group ($-COOH$) (**FIGURE 2.19**). Also attached to the carbon is a side chain, called the **R group (−R)**. Each of the 20 amino acids differs only by the atoms composing the R group. These side chains, many being functional groups, are essential in molding the final shape and function of proteins.

In forming a protein, amino acids (sometimes called peptides) are joined together by covalent (peptide) bonds where the carboxyl group of one amino acid is linked to the amino group of another amino acid through a dehydration reaction (Figure 2.19). Peptide (covalent) bonding between hundreds of amino acids produces a long polymer called a **polypeptide**. Because proteins have tremendously diverse roles, they come in many sizes and shapes. The final shape depends on the sequence of amino acids in the polypeptide.

Primary Structure

The linear sequence of amino acids in a polypeptide represents the **primary structure** (**FIGURE 2.20A**). Each protein that has a different function will have a different primary structure, as each position along the chain will have one of the 20 amino acids. The sequence of the amino acids is determined by the sequence of bases (genes) in the

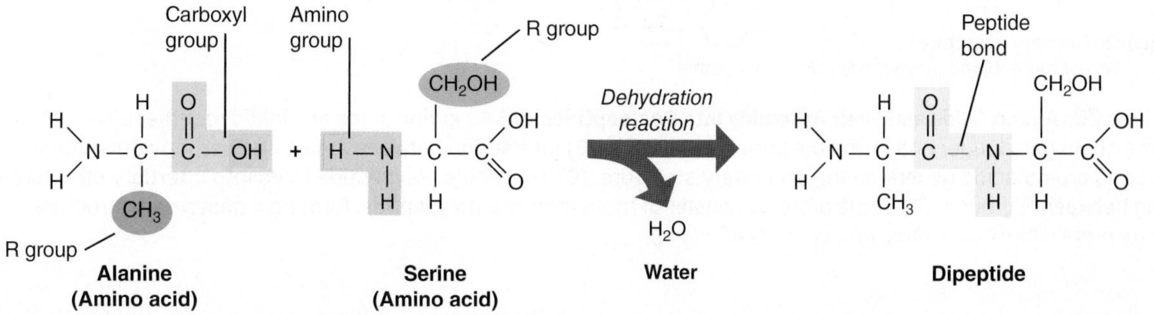

FIGURE 2.18 The Structure of ATP. The adenosine triphosphate (ATP) molecule is the main energy carrier in all cells. *»» Name the three different building blocks (monomers) that build an ATP molecule.*

▶ *Transport proteins*. Many proteins form channels or pores through membranes to facilitate the movement of materials into or out of the cell.

▶ *Regulatory proteins*. Some proteins bind to DNA and help regulate metabolic activity by switching genes on or off.

▶ *Receptor proteins*. Other proteins in cells act as sensors (receptors) that recognize chemicals in the environment.

▶ *Cell motility proteins*. Many microbes contain flagella or cilia, which are protein-containing structures involved with cell movement.

▶ *Immune system proteins*. The human immune system uses many different proteins, such

FIGURE 2.19 Amino Acids and the Dehydration Reaction. Amino acids are linked together by dehydration reactions between the carboxyl group of one amino acid and the amino group of the adjacent amino acid. *»» What type of chemical bond is the peptide bond?*

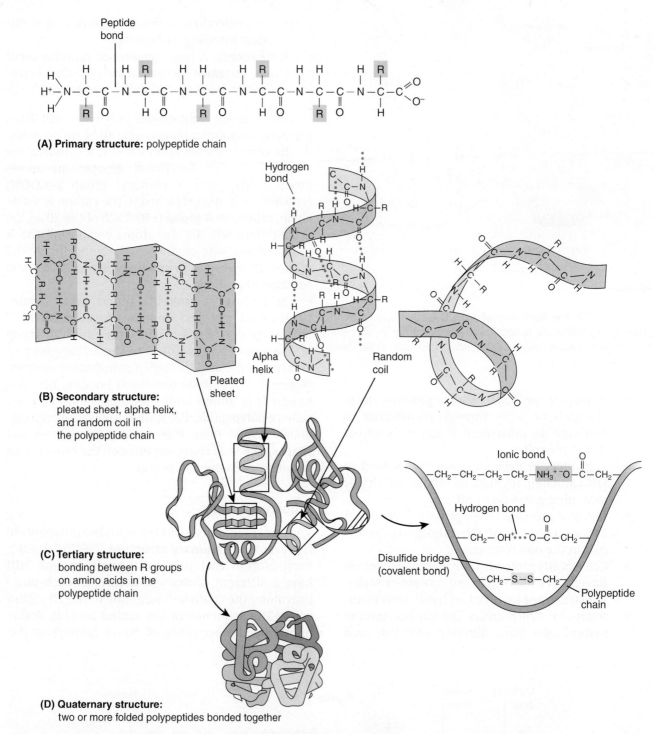

(A) Primary structure: polypeptide chain

(B) Secondary structure:
pleated sheet, alpha helix,
and random coil in
the polypeptide chain

(C) Tertiary structure:
bonding between R groups
on amino acids in the
polypeptide chain

(D) Quaternary structure:
two or more folded polypeptides bonded together

FIGURE 2.20 **Amino Acids and Their Assembly into Polypeptides. (A)** As amino acids are linked together by dehydration reactions, the chain gets longer and forms the primary structure. **(B)** Interaction between amino and carbonyl functional groups on neighboring amino acids generates the secondary structure. **(C)** The whole polypeptide folds into a tertiary structure through bonding between R groups. **(D)** Some proteins consist of more than one polypeptide, forming a quaternary structure. *»» Explain why every protein must have three or four levels of folding.*

DNA. However, the sequence of amino acids alone is not sufficient to confer function to a protein.

Secondary Structure

Most proteins have regions folded into a corkscrew shape called an **alpha helix**. These regions represent part of the protein's **secondary structure** (**FIGURE 2.20B**). Hydrogen bonds between amino groups (–NH) and carbonyl groups (–CO) on nearby amino acids maintain this structure. A secondary structure also will form when the hydrogen bonds cause portions of the polypeptide chain to zigzag in a flat plane, forming a **pleated sheet**. Other regions of the protein might not interact and remain in a "random coil."

Tertiary Structure

Many proteins also have a **tertiary structure** superimposed on the secondary structure (**FIGURE 2.20C**). Such a three-dimensional (3-D) shape of a protein is folded back on itself much like a spiral telephone cord. Ionic and hydrogen bonds between nearby R groups on amino acids help form and maintain the tertiary structure. In addition, covalent bonds, called **disulfide bridges**, between sulfhydryl (–SH) functional groups in R groups are important in stabilizing tertiary structure.

The ionic and hydrogen bonds helping hold a protein in its 3-D shape are relatively weak associations. As such, these interactions in a protein are influenced by environmental conditions. When subjected to heat, pH changes, or certain chemicals, these bonds can break, causing the polypeptide to unfold and lose its biological activity. This loss of 3-D shape is referred to as **denaturation**. For example, the white of a boiled egg is denatured egg protein (albumin) and cottage cheese is denatured milk protein. Should enzymes be denatured in living cells, the important chemical reactions they facilitate will be interrupted and death of the cell or organism might result. Viruses also can be destroyed by denaturing the proteins that build the virus' 3-D structure. So, now you should understand the importance of buffers in cells; by preventing pH shifts, they prevent protein denaturation and maintain protein function.

Quaternary Structure

Many proteins are single polypeptides. However, other proteins contain two or more polypeptides to form the complete and functional protein; this association of polypeptides is called the **quaternary structure** (**FIGURE 2.20D**). Each polypeptide chain is folded into its tertiary structure and the unique association between separate polypeptides produces the quaternary structure. The same types of chemical bonds are involved as in tertiary structure.

The four major classes of organic compounds are summarized in **FIGURE 2.21**. The box **INVESTIGATING THE MICROBIAL WORLD 2** looks at the origins of the monomers discussed in this chapter, while **MICROINQUIRY 2** shows how the radioactive attributes of two chemical elements can be used to identify whether protein or DNA is the genetic material.

In conclusion, this chapter described atoms and elements and their interactions through chemical bonding to construct compounds and molecules. The information discussed forms the underlying foundation for many of the topics we will examine in microbiology, as growth, metabolism, and genetics are based on cellular chemical reactions. In fact, many microbial pathogens that infect and damage human cells and tissues do so by means of special metabolic reactions they carry out, or by the metabolism they inhibit in the infected cells and tissues. Realize the time invested now to understand or refresh your memory about chemistry will make subsequent information easier and prepare you for a rewarding learning experience as you continue your study of the microbial world.

Concept and Reasoning Checks 2.4

a. What is common to all functional groups except the carbonyl group?

b. Discuss how stable monosaccharides are assembled into energy storage and structural polymers. Give examples.

c. Explain why lipids are not considered polymers whereas polysaccharides, polynucleotides, and polypeptides are polymers.

d. How does the structure of DNA differ from that of RNA?

e. Why does a denatured protein no longer have biological activity?

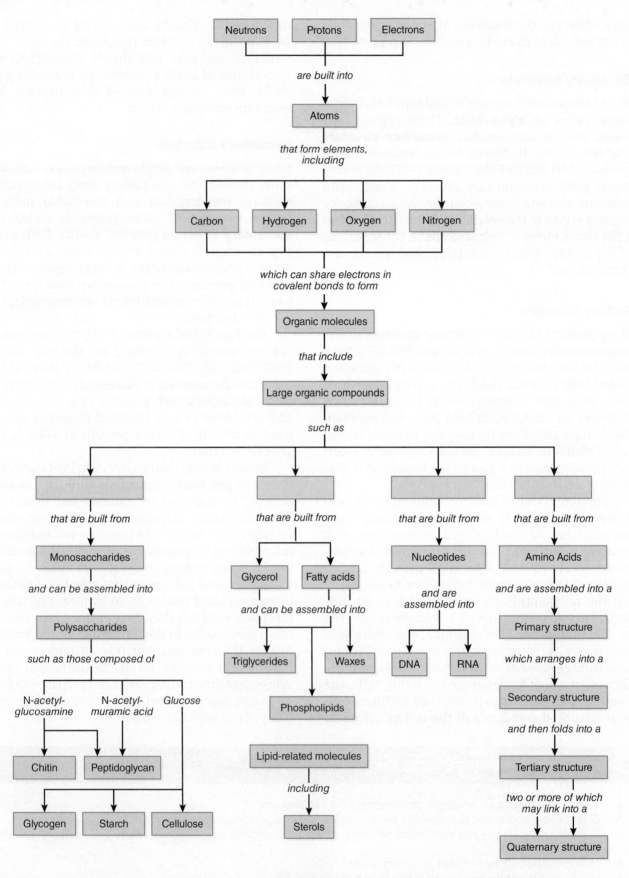

FIGURE 2.21 A Concept Map Summarizing Atoms, Elements, and Organic Compounds. »» *Finish the concept map by filling in the four empty rectangles with the correct type of organic compound.*

Investigating the Microbial World 2

Chemical Evolution

Life today is full of organic monomers that assemble into larger organic compounds that are characteristic of all life, including microorganisms. Knowing this, in the 1920s, Aleksandr Oparin, a Russian biochemist, and J. B. S. Haldane, a British geneticist, wondered whether these small monomers could have nonliving (abiotic) origins that were then used to build the first cell, a precursor to the prokaryotic cell.

OBSERVATION: Because the primitive atmosphere of Earth contained hydrogen and other substances that readily provide electrons (an atmosphere lacking oxygen gas), Oparin and Haldane suggested independently that with an appropriate supply of energy, such as lightning or ultraviolet light, a variety of organic monomers might be formed. In 1953, this suggestion for chemical evolution was put to the test by Stanley Miller, an American chemist/biologist, and Harold Urey, a physical chemist.

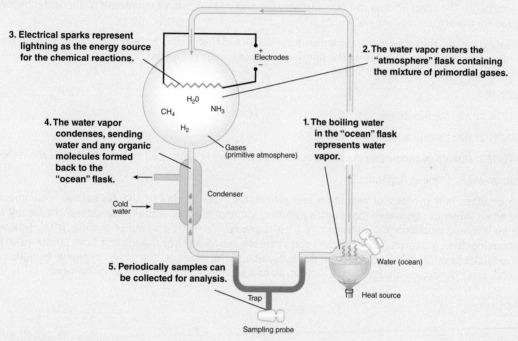

The Miller-Urey apparatus

Modified from Miller, S. L. 1953. *Science* 117(3046): 528–529; Johnson, A. P., et al. 2008. *Science* 322(5900):404.

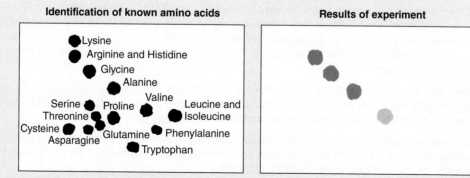

Identification of known amino acids and results of experiment.

Modified from Miller, S. L. 1953. *Science* 117(3046):528–529; Johnson, A.P., et al. 2008. *Science* 322(5900):404.

(*continues*)

Investigating the Microbial World 2 (*Continued*)

Chemical Evolution

QUESTION: *Can organic monomers be formed in an atmosphere mimicking the conditions on the primitive Earth?*

HYPOTHESIS: Organic monomers, specifically amino acids, can be formed using the prebiotic atmospheric conditions of early Earth. If so, a combination of the prebiotic atmospheric gases, along with an energy source to drive the reactions, should give rise to a variety of amino acids.

EXPERIMENTAL DESIGN: The apparatus illustrated on the previous page depicts the design of a closed environment to mimic the prebiotic conditions believed to represent Earth's primitive atmosphere [methane (CH_4), ammonia (NH_3), hydrogen (H_2), and water (H_2O)].

EXPERIMENT: The water in the closed apparatus was boiled, allowing the water vapor to mix with the three gases and the electrical discharge. The reaction was run continuously for one week. Then, the contents of the water in the apparatus were analyzed for the presence of amino acids.

RESULTS: During the run, the water in the apparatus turned pink and then a cloudy deep red. The red color was presumably due to the presence of organic molecules. The figure shows the results for amino acid detection (right figure) along with the pattern for known amino acids (left figure).

CONCLUSIONS:

QUESTION 1: *Was the hypothesis validated?*

QUESTION 2: *What amino acids were detected?*

QUESTION 3: *How do you know that the amino acids detected were not produced by living bacterial cells in the apparatus?*

You can find answers online in **Appendix E**.

Note: Science often is a process of repetition and refinement. The validity of the Miller-Urey experiment was later questioned when evidence suggested that Earth's primitive atmosphere was actually composed of different gases, ones released from volcanic eruptions [CO_2, nitrogen (N_2), hydrogen sulfide (H_2S), and sulfur dioxide (SO_2)]. However, modern experiments using these gases, along with those used in the original Miller-Urey experiment, have produced up to 22 amino acids, all five nucleobases in DNA and RNA, and the sugar ribose. So, whatever the gases, chemical evolution and abiotic synthesis of organic molecules appear to be validated as at least one source for organic monomers.

Chapter **Challenge D**

In this chapter challenge, we have been investigating how we might chemically detect whether another form of life—"shadow life"—arose in a different chemical form on Earth. If there truly is a molecule capable of making life different chemically, it might be in the amino acids used to make proteins. Most all known life uses the same 20 amino acids to synthesize proteins. However, other amino acids exist and some, like isovaline and pseudoleucine, have been identified in meteorites. Therefore, is it possible that a marker for shadow life microbes could be unusual amino acids in their proteins?

QUESTION D: *The Miller-Urey experiment described in the box* INVESTIGATING THE MICROBIAL WORLD 2 *might have produced more than the 20 traditional amino acids. If one or more of the rare amino acids were formed in the primordial Earth, and incorporated into protein, would the existing organisms be candidates for "shadow life"?*

Scientists have found some bacterial organisms like *Salmonella* can become antibiotic resistant if they synthesize and use a rare form of the amino acid lysine when making proteins. Some extremophile members of the prokaryotic world that make methane gas possess an unusual amino acid called pyrrolysine.

So, in concluding this chapter challenge, does there appear to be any evidence of shadow life on Earth? Explain.

You can find answers online in **Appendix F**.

MICROINQUIRY 2

Is Protein or DNA the Genetic Material?

In the early 1950s, there were scientists who still debated whether protein or DNA was the genetic material in cells. To settle the controversy, in 1952 Alfred Hershey and Martha Chase carried out a series of experiments to trace the fates of protein and DNA and in so doing settled the debate.

It was known that some viruses that infect bacterial cells were composed of DNA and protein and that the virus genetic material had to enter the bacterial cells to direct the production of more viruses. Because the viruses left a viral coat on the surface, what actually entered the cells—protein or DNA? Whichever did must be the genetic material.

Several biologically important elements have isotopes that are radioactive. The table to the right lists a few such elements. Hershey and Chase decided to radioactively label the viruses such that the protein and DNA could be identified by their unique radioactive profiles.

2a. Which of the radioactive elements would only label protein?

2b. Which of the radioactive elements would only label DNA?

2c. Could ^{3}H or ^{14}C have been used to label the viruses? Explain.

The Hershey and Chase experiment is outlined above. From the two experiments, they could then measure the radioactivity in the pellet (bacterial cells) and the fluid (virus coats) and determine which radioactive isotope was associated with the bacterial cells.

2d. If protein is the genetic material, which isotope should be associated with the pellet?

2e. If DNA is the genetic material, which isotope should be associated with the pellet?

2f. Did the experiments carried out by Hershey and Chase support or refute the hypothesis that DNA was the genetic material? Explain.

You can find answers online in **Appendix E**.

Some Radioactive Isotopes

Element	Common Form	Radioactive Form
Hydrogen	^{1}H	^{3}H (tritium)
Carbon	^{12}C	^{14}C
Phosphorus	^{31}P	^{32}P
Sulfur	^{32}S	^{35}S

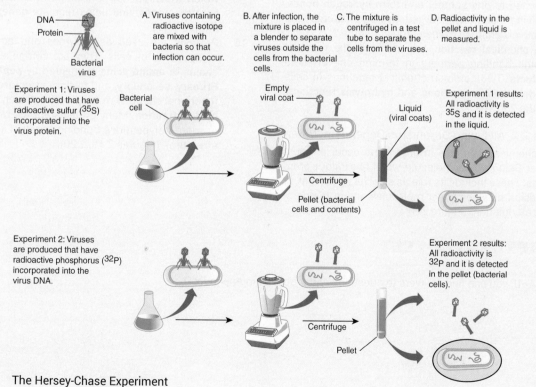

DNA — Protein —
Bacterial virus

A. Viruses containing radioactive isotope are mixed with bacteria so that infection can occur.

B. After infection, the mixture is placed in a blender to separate viruses outside the cells from the bacterial cells.

C. The mixture is centrifuged in a test tube to separate the cells from the viruses.

D. Radioactivity in the pellet and liquid is measured.

Experiment 1: Viruses are produced that have radioactive sulfur (^{35}S) incorporated into the virus protein.

Bacterial cell

Empty viral coat

Centrifuge

Pellet (bacterial cells and contents)

Liquid (viral coats)

Experiment 1 results: All radioactivity is ^{35}S and it is detected in the liquid.

Experiment 2: Viruses are produced that have radioactive phosphorus (^{32}P) incorporated into the virus DNA.

Centrifuge

Pellet

Experiment 2 results: All radioactivity is ^{32}P and it is detected in the pellet (bacterial cells).

The Hersey-Chase Experiment

■ SUMMARY OF KEY CONCEPTS

Concept 2.1 Organisms Are Composed of Atoms

1. **Atoms** consist of an **atomic nucleus** (with **neutrons** and positively charged **protons**) surrounded by a cloud of negatively charged **electrons**. (Figure 2.2)
2. **Isotopes** of an element have different numbers of neutrons. Some unstable isotopes, called **radioisotopes**, are useful in research and medicine. If an atom gains or loses electrons, it becomes an electrically charged **ion**. Many ions are important in microbial metabolism. (Figure 2.3)
3. Each electron shell holds a maximum number of electrons. If atoms have an unfilled valence shell, they chemically react with other reactive atoms.

Concept 2.2 Chemical Bonds Form Between Reactive Atoms

4. **Ionic bonds** result from the attraction of oppositely charged ions. Compounds called **salts** result. (Figure 2.5)
5. Most atoms achieve stability through a sharing of electrons, forming **covalent bonds**. The equal sharing of electrons produces **nonpolar molecules** (no net electrical charge). Atomic interactions between hydrogen and oxygen (or hydrogen and nitrogen) produce unequal sharing of electrons, which generates **polar molecules** (have electrical charges). (Figures 2.6, 2.7)
6. Separate polar molecules, like water, are electrically attracted to one another and form **hydrogen bonds**, involving positively charged hydrogen atoms and negatively charged oxygen atoms. (Figure 2.7)
7. In a **chemical reaction**, the atoms in the **reactant** change bonding partners in forming one or more **products**. Two common chemical reactions in cells are **dehydration reactions** and **hydrolysis reactions**. (Figure 2.8)

Concept 2.3 All Living Organisms Depend on Water

8. All chemical reactions in organisms occur in liquid water. Being a polar molecule, water has unique properties. These include its role as a universal **solvent**, in cohesion between water molecules and in temperature modulation. (Figure 2.9)

9. **Acids** donate hydrogen ions (H^+), whereas **bases** acquire H^+ from a solution. The **pH scale** indicates the number of H^+ in a solution and denotes the relative acidity of a solution. (Figure 2.10)
10. **Buffers** are a mixture of a weak acid and a weak base that maintain acid/base balance in cells. (Figure 2.11)

Concept 2.4 Living Organisms Are Composed of Four Types of Organic Compounds

11. The building of large **organic compounds** depends on the **functional groups** present on the **monomers**. Functional groups interact through dehydration reactions to form a covalent bond between monomers. (Table 2.3)
12. **Carbohydrates** include simple sugars (**monosaccharides**) such as glucose and double sugars (**disaccharides**) such as sucrose. Monosaccharides and disaccharides can be linked into **polysaccharides** that represent energy and structural molecules. (Figures 2.13, 2.14)
13. **Lipids** serve as energy sources and can be **saturated** or **unsaturated**. **Phospholipids** form part of the structure in cell membranes. Other lipids include the **sterols**. (Figures 2.15, 2.16)
14. The genetic instructions for living organisms are composed of two types of **nucleic acids**: deoxyribonucleic acid (**DNA**), which stores and encodes the hereditary information; and **ribonucleic acid** (**RNA**), which transmits the information to make proteins, controls genes, and helps regulate genetic activity. (Figure 2.17)
15. **Proteins** are used as enzymes and as structural components of cells. They are composed of linear chains of **amino acids** connected by **peptide bonds**. **Primary, secondary**, and **tertiary structures** form the functional shape of many proteins, which can unfold by **denaturation**. Many proteins are the result of two or more polypeptides bonding together (**quaternary structure**). (Figures 2.19, 2.20)

■ CHAPTER SELF-TEST

For **Steps A–D**, you can find answers to questions and problems in **Appendix D**.

■ STEP A: REVIEW OF FACTS AND TERMS

Multiple Choice

Read each question carefully before selecting the *one* answer that best fits the question or statement.

1. These positively charged particles are found in the atomic nucleus.
   A. Protons
   B. Electrons
   C. Protons and neutrons
   D. Neutrons

2. Atoms of the same element that have different numbers of neutrons are called _____.
   A. isotopes
   B. ions
   C. acids
   D. inert elements

3. If an element has two electrons in the first shell and seven in the second shell, the element is said to be what?
   A. Stable
   B. Unreactive
   C. Unstable
   D. Inert

4. The transfer of one or more electrons between interacting atoms results in what type of bond?
   A. Hydrogen
   B. Ionic
   C. Peptide
   D. Covalent

5. The covalent bonding of atoms forms a/an _____.
   A. molecule
   B. ion
   C. element
   D. isotope

6. The _____ bond is a weak bond that can exist between poles of adjacent molecules.
   A. hydrogen
   B. ionic
   C. polar covalent
   D. nonpolar covalent

7. In what type of chemical reaction are the products of water removed during the formation of covalent bonds?
   A. Hydrolysis
   B. Ionization
   C. Dehydration
   D. Decomposition

8. A nurse is preparing an IV drip for a patient. In the saline solution she is using, the salt represents a _____.
   A. solvent
   B. hydrophobic molecule
   C. solute
   D. nonpolar molecule

9. If the saline solution mentioned in the previous question has a pH of 5.5, it is mildly _____.
   A. buffered
   B. alkaline
   C. acidic
   D. basic

10. Which one of the following statements about buffers is false?
    A. They work inside cells.
    B. They consist of a weak acid and weak base.
    C. They prevent pH shifts.
    D. They speed up chemical reactions.

11. A functional group designated-COOH is known as a/an _____.
    A. carboxyl
    B. carbonyl
    C. amino
    D. hydroxyl

12. Which one of the following is *not* a polysaccharide?
    A. Chitin
    B. Glycogen
    C. Cellulose
    D. Triglyceride

13. How do the lipids differ from the other organic compounds?
    A. They are the largest organic compounds.
    B. They are nonpolar compounds.
    C. They are the product of hydrolysis reactions.
    D. They are not used for energy storage.

14. Both DNA and RNA are composed of _____.
    A. polynucleotides
    B. genes
    C. polysaccharides
    D. polypeptides

15. The _____ structure of a protein is the sequence of amino acids.
    A. primary
    B. secondary
    C. tertiary
    D. quaternary

■ STEP B: CONCEPT REVIEW

16. Assess the importance of the three types of atomic particles to atomic structure and reactivity. (**Key Concept 2.1**)

17. Explain the difference between **nonpolar covalent bonds, polar covalent bonds**, and **ionic bonds** in terms of electron interactions between atoms. (**Key Concept 2.2**)

18. Support the statement: "Life as we know it could not exist without water." (**Key Concept 2.3**)

19. Predict the chemical reactivity potential for a molecule having five **functional groups** versus one having only one functional group. Explain your prediction. (**Key Concept 2.4**)

20. Construct a table providing the name of each type of organic compound and its function or functions. (**Key Concept 2.4**)

■ STEP C: APPLICATIONS AND PROBLEM SOLVING

21. You want to grow an acid-loving bacterial species; that is, one that grows best in very acidic environments. Would you want to grow it in a culture that has a pH of 2.0, 6.8, or 11.5? Explain.

22. You are given two beakers of a broth growth medium. However, only one of the beakers of broth is buffered. How could you determine which beaker contains the buffered broth solution? Hint: You are provided with a bottle of concentrated HCl and pH papers that indicate a solution's pH.

23. The microbial community in a termite's gut contains the enzyme cellulase. How does this benefit the termite and the termite's microbial community?

24. Use the following list to identify the structures (in 25 i–v) drawn below.
   A. Amino acid
   B. Disaccharide
   C. Lipid
   D. Monosaccharide
   E. Nucleotide
   F. Polysaccharide
   G. Sterol

25. Identify all functional groups on each structure (i–v).

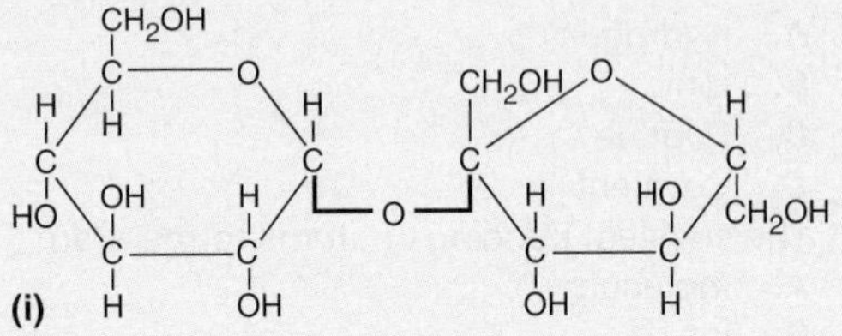

(i)

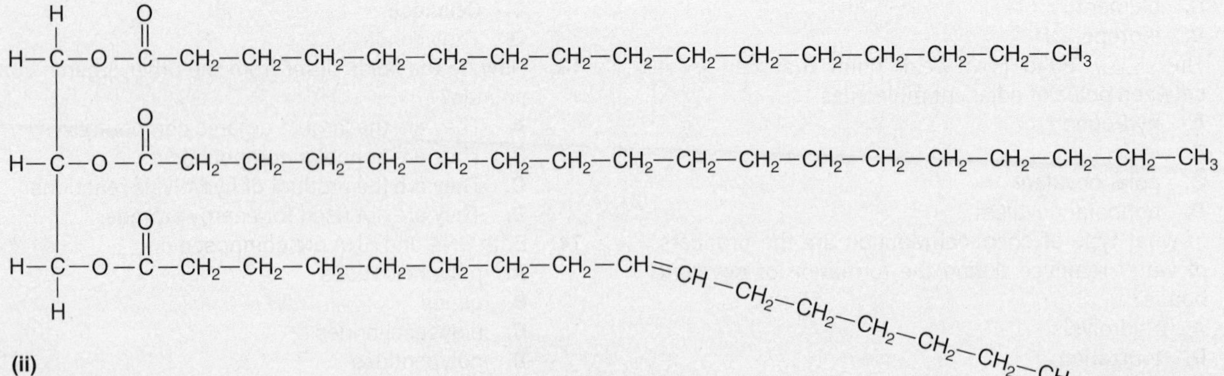

(ii)

(iii)

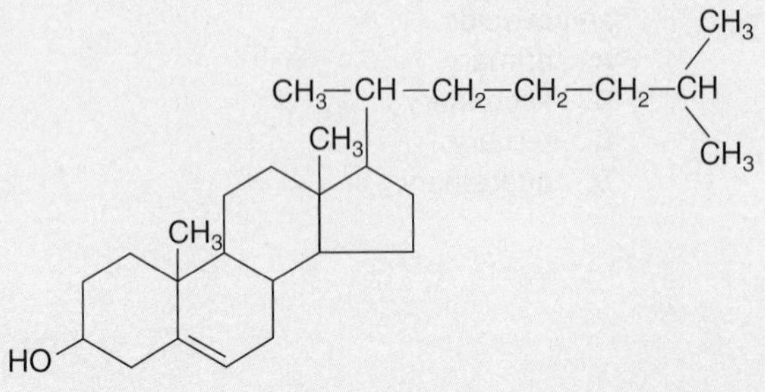

(iv)

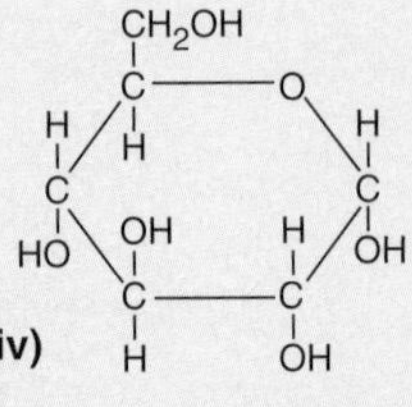

(v)

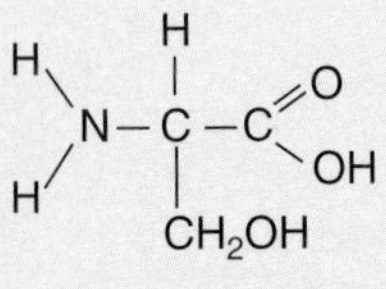

■ STEP D: QUESTIONS FOR THOUGHT AND DISCUSSION

26. Propose a reason why organic molecules tend to be so large.

27. Bacterial cells do not grow on bars of soap (that are very alkaline) even though the soap is wet and covered with bacterial organisms after one has washed. Explain this observation.

28. Suppose you had the choice of destroying one class of organic compounds in bacterial cells to prevent their spread. Which class would you choose? Why?

29. Milk production typically has the bacterium *Lactobacillus* added to the milk before it is delivered to market. In the milk, the organism produces lactic acid. (a) Why would this organism be added to the milk and (b) why was this particular bacterium chosen?

30. The toxin associated with the foodborne disease botulism is a protein. To avoid botulism, home canners are advised to heat preserved foods to boiling for at least 12 minutes. How does the heat help to avoid the disease?

31. The late Isaac Asimov, an American author and professor of biochemistry, once said, "The significant chemicals in living tissue are rickety and unstable, which is exactly what is needed for life." How does this quote apply to the atoms, molecules, and organic compounds described in this chapter?

Concept Mapping

See pages XXXI-XXXII on how to construct a concept map.

32. Construct a concept map for **Chemical Bonds**, using the following terms.

Hydrocarbon
Hydrogen bonds
Ionic bonds
Methane
NaCl
Nonpolar covalent bonds
Polar covalent bonds
Salts
Water

CHAPTER 3

Concepts and Tools for Studying Microorganisms

The oceans of the world are a teeming with an invisible forest of microorganisms and viruses. It is estimated that one liter of surface seawater contains more than 10 billion (10×10^9) microbes and 100 billion (100×10^9) viruses. Although most of these microbes are prokaryotic species, we know little about them or the marine viruses.

We do know that a substantial portion of the marine microbes represents the **phytoplankton** (*phyto* = "plant"; *plankto* = "wandering"), which are communities of suspended cyanobacteria and eukaryotic algae. Besides forming the foundation for the marine food web, the marine and freshwater phytoplankton account for 50% of the photosynthesis on Earth and, in so doing, supply about half the oxygen gas we breathe and other organisms breathe or use. Sometimes their growth is so abundant that blooms occur (see chapter opening photo).

One of the most abundant microorganisms among the marine phytoplankton is a tiny cyanobacterium called *Prochlorococcus*. Inhabiting the surface waters of tropical and subtropical oceans, a typical sample often contains up to 200,000 (2×10^5) *Prochlorococcus* cells in one drop of surface seawater (**FIGURE 3.1**). In fact, *Prochlorococcus* is the smallest and most abundant marine photosynthetic organism yet discovered.

The success of *Prochlorococcus* is due, in part, to the existence of different species subgroups called **ecotypes** that inhabit different depths within the sunlit surface layer of the oceans. For example, the high-sunlight ecotype occurs nearest the surface, whereas the low-sunlight type is found below 50 meters. This latter ecotype compensates for the decreased light by increasing the amount of cellular chlorophyll that can capture the available light.

In terms of nitrogen sources, the high-sunlight ecotype uses ammonium ions (NH_4^+). However, at increasing depth NH_4^+ is less abundant so the low-sunlight ecotype compensates by using a wider variety of nitrogen sources.

These and other attributes of *Prochlorococcus* illustrate how microbes survive and adapt to environmental change. As part of the global microbial workforce, they are important to the functioning of

Cyanobacterial bloom, Lake Erie—2011.
Courtesy of MERIS/NASA; processed by NOAA/NOS/NCCOS.

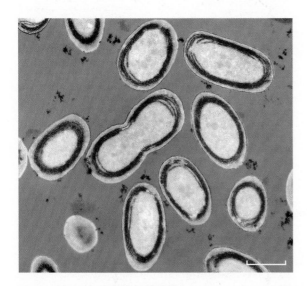

FIGURE 3.1 *Prochlorococcus*. This false-color transmission electron microscope image of a section through several *Prochlorococcus* cells reveals bands of membranes (green) that contain the chlorophyll used in photosynthesis. (Bar = 0.5 μm.) »» *Do these cells look structurally complex or simple?*

© Claire Ting/Science Source.

the biosphere and play key roles in maintaining the "health" of the planet.

This knowledge has come from the work of an interdisciplinary group of scientists who are studying how microorganisms influence life on this planet. Microbial ecologists study how the phytoplankton communities help in the natural recycling and use of chemical elements such as nitrogen. Evolutionary microbiologists look at these microorganisms to learn more about how they are related to the other marine microbes, whereas microscopists, biochemists, and geneticists study how *Prochlorococcus* cells compensate for a changing environment of sunlight and nutrients.

In this chapter, we will investigate some of the relationships between microorganisms and discover some of the many attributes they share. Along the way, we will explore the methods used to name and catalog microorganisms, and we will become familiar with one of the most basic tools used to observe the microbial world—the microscope.

Chapter **Challenge**

Look again at Figure 3.1. Bacterial cells often are described and referred to as "simple" cells, often thought of as not much more than bags of enzymes and chemicals—even if they have transformed planet Earth! Your challenge, as we proceed through this chapter, is to decide whether this perception of prokaryotic simplicity is valid and whether they are "simple" cells.

■ KEY CONCEPT 3.1 Prokaryotes Are Not Simple, Primitive Organisms

Despite their microscopic size, prokaryotes share a common set of characteristics, or emerging properties, with all living organisms. These include the following:

▶ Hereditary material in the form of DNA.
▶ Complex biochemical patterns involving growth and energy conversions.
▶ Reproduction to produce new generations of descendants.
▶ Evolutionary adaptation in response to other organisms and the environment.
▶ Complex and regulated responses to stimuli.

The last property is the focus in the first part of this chapter.

Prokaryotic Cells Exhibit Some Remarkable and Widespread Behaviors

Often, there is little cell structure seen when looking at bacterial cells with an electron microscope (**FIGURE 3.2A**). These cells appear to lack much in the way of visible cell structure. There also appears to be a minimal "pattern of organization," referring to the configuration of those structures and their relationships to one another.

On the other hand, what has been overlooked is the "cellular process," the activities that the cells carry out for the continued survival of the cell (and organism). At this level, the complexity of prokaryotic cells is just as intricate as in eukaryotic cells and they carry out many of the same cellular processes

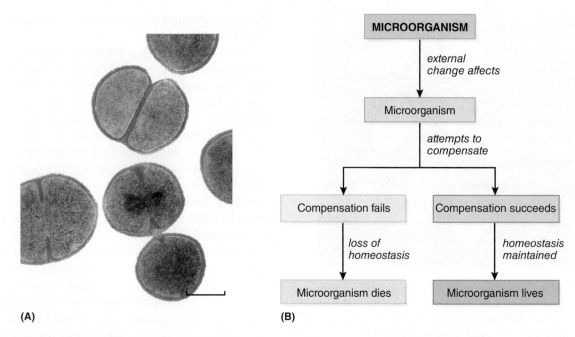

FIGURE 3.2 Homeostasis Through Compensation. (A) This false-color electron microscope image of *Staphylococcus aureus* gives the impression of simplicity in structure. (Bar = 1 μm.) **(B)** A concept map illustrating how bacterial organisms, like all microorganisms, have to compensate for environmental changes. Survival depends on such homeostatic abilities. *»» Using the concept map in (B), explain how* Prochlorococcus *"behaves" in a low-light environment.*

(A) © Biomedical Imaging Unit, Southampton General Hospital/Science Photo Library/Getty Images.

as eukaryotic cells. They just do not need the set of integrated structures typical of eukaryotic cells. Let's look at just two examples where prokaryotic cells carry out complex cellular processes.

Homeostasis

All organisms continually battle their external environment, in which factors such as temperature, sunlight, or toxic chemicals can have serious consequences. All organisms strive to maintain a stable internal state by making appropriate metabolic or structural adjustments. This ability to adjust yet maintain a relatively steady internal state is called **homeostasis** (*homeo* = "similar"; *stasis* = "state"). This is illustrated in **FIGURE 3.2B**.

The low-sunlight *Prochlorococcus* ecotype mentioned in the chapter introduction lives at depths down to 50 meters. Transmitted sunlight decreases with depth and any one nitrogen source is less accessible. The ecotype compensates for the light reduction and nitrogen limitation by (1) increasing the amount of cellular chlorophyll to maximize the capture of sunlight for photosynthesis and (2) using a more diverse set of available nitrogen sources for metabolism. These adjustments in structure and pattern organization maintain a steady internal state.

For our second example, suppose that a patient is given an antibiotic to combat a bacterial infection. In response, some of the infecting cells might compensate for the change by breaking the structure of the antibiotic. The adjustment maintains homeostasis and the reward is antibiotic resistance.

In both of these examples, if the internal environment is not maintained, the cell probably will die. However, homeostasis involves more than a simple response to the immediate environmental conditions. Compensation can be planned for ahead of time.

Microbial "Learning"

Unbelievably, bacterial cells can "think," not in the cognitive way that you and I think, but rather they can anticipate and prepare for future conditions to ensure that homeostasis is maintained. Do you remember how Pavlov conditioned his dogs to associate a meal by ringing a bell? The dogs, hearing the bell, would anticipate the coming meal by salivating even before the meal arrived. Microbiologists have shown that bacterial organisms demonstrate a similar type of associative learning. Here is one example.

When lab cultures of *Escherichia coli* cells are repeatedly shifted from 25°C and 20% oxygen to

37°C and 0% oxygen, the bacterial cells over a few weeks "learn" to anticipate the coming oxygen drop. Sensing the temperature increase, the bacterial cells altered their metabolism in preparation for the coming oxygen change. In other words, the cells had associated the change in temperature with a coming drop in oxygen. So, in the natural environment, microbes also must respond to environmental signals that precede coming events. The ability of bacterial cells to plan ahead is not something one would expect of a simple organism! In fact, it hints at the enormous flexibility and adaptability present in these microorganisms.

Prokaryotic and Eukaryotic Cells Share Similarities in Organizational Patterns

In the 1830s, Matthias Schleiden and Theodor Schwann developed part of the **cell theory**, which states, in part, that all plants and animals are composed of one or more cells, making the cell the fundamental unit of life. (Note: about 20 years later, Rudolph Virchow added to the cell theory that all cells arise from preexisting cells.) Although the concept of a microorganism was just in its infancy in the mid-1800s, the theory suggests that there are certain organizational patterns common to all cells, be they prokaryotic or eukaryotic. Researchers can identify four basic similarities in organizational patterns. Let's look at these in the sections that follow.

Genetic Organization

All organisms have a similar genetic organization, meaning the hereditary material (DNA) is communicated or expressed using an almost universal genetic code. The organizational pattern for the DNA is in the form of one or more **chromosomes**.

Structurally, eukaryotic cells have multiple, linear chromosomes enclosed by the membrane envelope of the cell nucleus. Most prokaryotic cells have a single, circular DNA molecule without an enclosing membrane (**FIGURE 3.3**).

Compartmentation

All cells have an organizational pattern separating the internal cell compartment from the surrounding environment but still allowing for the exchange

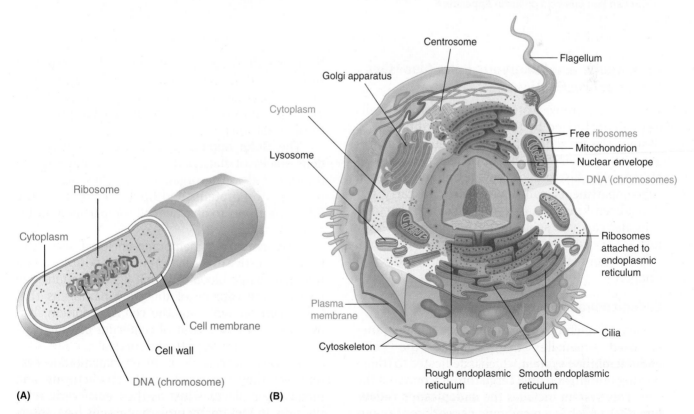

FIGURE 3.3 A Comparison of a Bacterial and Eukaryotic Cell. (A) A typical bacterial cell has relatively few visual compartments. **(B)** A protistan cell is an example of a typical eukaryotic cell. Note the variety of cellular subcompartments, many of which are discussed in the text. Universal structures are indicated in red. *»» List the ways you could microscopically distinguish a eukaryotic microbial cell from a bacterial cell.*

of solutes and wastes. The organizational pattern is the **cell membrane** (known as the **plasma membrane** in eukaryotes), where the phospholipids form the impermeable boundary to solutes and the membrane proteins are the channels through which the exchange of solutes and wastes occurs.

Metabolic Organization

The process of metabolism is a consequence of compartmentation. By being enclosed by a membrane, all cells have an internal environment in which chemical reactions occur and adenosine triphosphate (ATP), the cellular energy source, is generated. This space, called the **cytoplasm**, represents everything surrounded by the membrane and, in eukaryotic cells, exterior to the cell nucleus. If the cell structures are removed from the cytoplasm, what remains is the **cytosol**, which consists of water, salts, ions, and small organic molecules.

Protein Synthesis

All organisms must make proteins, which are the workhorses of cells and organisms. The structures capable of making (synthesizing) proteins are the **ribosomes**, tiny factories built from RNA and protein. Although the organizational pattern for protein synthesis is identical, ribosomes in prokaryotic cells are slightly smaller and composed of different-sized RNAs and a different complement of proteins than their counterparts in eukaryotic cells.

Chapter Challenge A

Up to this point, we have seen examples of the homeostatic abilities of prokaryotic cells and examined their organizational patterns.

QUESTION A: *Based on what you have read up to this point, would you consider prokaryotic cells to be simple life forms? Explain.*

You can find answers online in **Appendix F**.

Prokaryotic and Eukaryotic Cells Also Have Structural Distinctions

Historically, prokaryotic and eukaryotic cells were characterized by their structural differences. Besides the presence of a cell nucleus in eukaryotic cells, eukaryotic microbes use a variety of structurally discrete, often membrane-enclosed subcompartments called **organelles** to carry out specialized functions (see Figure 3.3). Prokaryotic cells also can have subcellular compartments—they just are not readily visible or membrane-enclosed. Details of prokaryotic structure will be described in Chapter 4.

Endomembrane System

Eukaryotic microbes have a collection of membrane-enclosed organelles that compose the cell's **endomembrane system**, which is dedicated to transporting protein and lipid cargo through or out of the cell. This system includes the **endoplasmic reticulum (ER)**, which is a membrane network continuous with the nuclear envelope (**FIGURE 3.4A**). The ER consists of flat membranes to which ribosomes are attached (the so-called rough ER). Proteins produced by these ribosomes eventually will be exported from the cell. A second form of ER forms tube-like membranes that are free of ribosomes (smooth ER). These portions of the ER are involved in lipid synthesis and transport.

The **Golgi apparatus** is a group of independent stacks of flattened membranes and vesicles (**FIGURE 3.4B**). The organelle modifies, sorts, and packages the proteins and lipids coming from the ER for transport to another cell compartment or for export from the cell.

Lysosomes are membrane-enclosed sacs containing digestive (hydrolytic) enzymes (**FIGURE 3.4C**). In human white blood cells that fight infections, the enzymes are capable of digesting pathogens that have been brought into the cell. In addition, lysosomes represent a control center for cell metabolism as well as cell growth and survival.

Prokaryotic cells lack an endomembrane system, yet they are capable of manufacturing and modifying molecules just as their eukaryotic relatives do. In fact, many prokaryotic cells have internal compartments that carry out specific chemical reactions and protect the cell from toxic byproducts from those reactions.

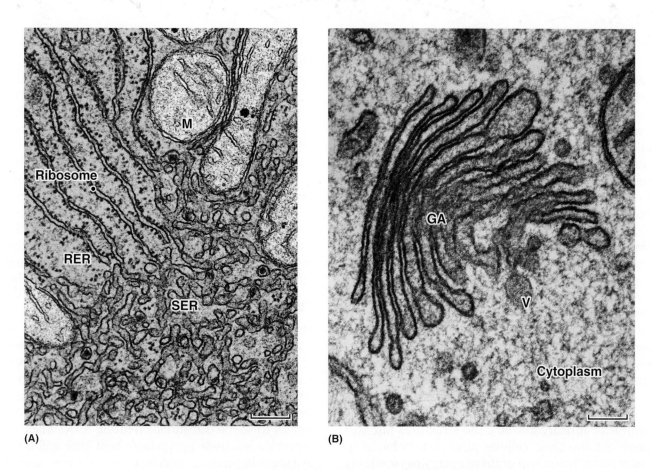

(A)

(B)

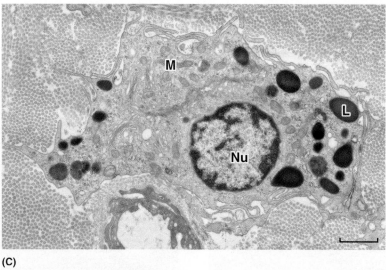

(C)

FIGURE 3.4 The Endomembrane System. These false-color electron microscope images show the various organelles of the eukaryotic endomembrane system. **(A)** In this human liver cell, the rough endoplasmic reticulum (RER) is studded with ribosomes. The smooth ER (SER) lacks ribosomes. (Bar = 1μm.) **(B)** The Golgi apparatus (GA) in this green algal cell (*Chlamydomonas*) is constructed from a stack of flat membranes that pinch off vesicles (V). (Bar = 1μm.) **(C)** The lysosomes (L) in this macrophage, a type of white blood cell contain digestive enzymes. (Bar = 1μm.) Nu = cell nucleus; M = mitochondrion. *»» Briefly explain how these three organelles are interrelated.*

Mitochondria and Chloroplasts

Cells and organisms carry out one or two types of energy transformations. Through a process called **cellular respiration**, all cells convert chemical energy into cellular energy for cellular work. In plants, animals, and eukaryotic microbes, this occurs in the cytosol and in membrane-enclosed organelles called **mitochondria** (singular: mitochondrion) (**FIGURE 3.5**). Prokaryotic cells lack mitochondria, so they use the cell membrane to complete the energy transformations that would occur in the eukaryotic mitochondrion.

A second energy transformation, **photosynthesis**, involves the conversion of light energy into chemical energy. In plants and algal protists, photosynthesis occurs in membrane-bound **chloroplasts** (see Figure 3.5). Some bacterial groups, such as the cyanobacterium *Prochlorococcus*, also carry out an almost identical set of energy transformations. Again, the prokaryotic cell membrane, or elaborations of the membrane (see Figure 3.1), represents the chemical workbench for the photosynthetic process that would occur in chloroplasts.

Importantly, in both cell respiration and photosynthesis, prokaryotes lack the structures but still carry out the same cellular processes as found in eukaryotes. The origin of these energy organelles is described later in this chapter.

Cytoskeleton

The eukaryotic **cytoskeleton** is organized into an interconnected network of cytoplasmic fibers and threads that extend throughout the cytoplasm. The cytoskeleton gives structure to the cell and assists in the transport of materials throughout the cell (**FIGURE 3.6**). Among the components of this internal cytoskeleton are microtubules, microfilaments, and intermediate filaments, with each component assembled from different protein subunits.

Prokaryotic cells also have a cytoskeleton made of proteins distantly related to those that construct the eukaryotic cytoskeleton. The prokaryotic cytoskeleton functions in determining cell shape in some cells and positioning structures in other types of bacterial cells.

Structures for Cell Motility

Many organisms, especially microorganisms, live in watery or damp environments and use the process of cell motility to "swim" from one place to another. Some animal and plant cells, as well as protists, have long, thin protein projections called **flagella** (singular: flagellum) that, covered by the plasma membrane, extend from the cell surface (**FIGURE 3.7A**). Made of microtubules, these appendages beat back and forth in a wave-like motion, which provides the mechanical force for motility.

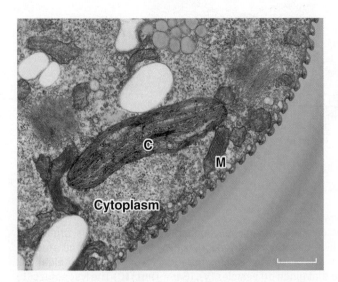

FIGURE 3.5 The Energy Organelles. In this false-color electron microscope image of the protist *Euglena*, many mitochondria (M) and one chloroplast (C) are present. (Bar = 1 μm.) »» *What are the stacked membrane organelles (olive color) to the left and upper right of the chloroplast?*

© Dennis Kunkel Microscopy/Science Source.

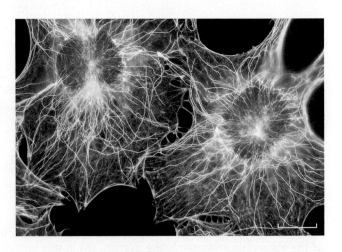

FIGURE 3.6 The Eukaryotic Cytoskeleton. This fluorescence microscope image of two human fibroblast cells shows the microtubules (yellow) and microfilaments (purple-gray) forming a network of fibers extending throughout the cells. (Bar = 1 μm.) »» *What is the large purple structure in the center of each cell?*

© Dr. Torsten Wittmann/Science Source.

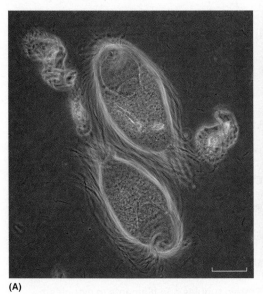
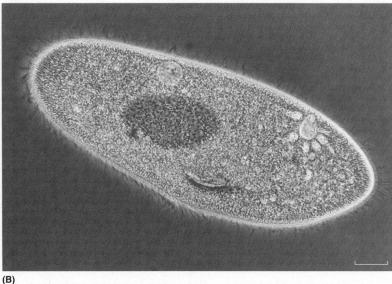

(A)   (B)

FIGURE 3.7 Structures for Cell Motility. These phase-contrast, light microscope images show **(A)** numerous flagella extending from the surface of the protist *Trichonympha* (Bar = 25 µm.) and **(B)** numerous cilia on the surface of the protist *Paramecium.* (Bar = 50 µm.) »» *What protein structures build the flagella and cilia in eukaryotic cells?*

Many prokaryotic cells also exhibit motility; however, the prokaryotic flagella are not made of microtubules and are without a membrane covering. The pattern of motility also is different, providing a rotational propeller-like force for movement.

Some animal cells and protists have other membrane-enveloped appendages called **cilia** (singular: cilium) that are more numerous than flagella and are composed of shorter microtubules (**FIGURE 3.7B**). The cilia have a wave-like synchrony that propels the cell forward. No prokaryotic cells have similar structures solely dedicated to motility.

Cell Walls

The aqueous environment in which many microorganisms live presents a situation in which the process of **diffusion** occurs, specifically the net movement of water, called **osmosis**, into the cell. Continuing unabated, the cell would eventually swell and rupture (cell lysis) because the plasma membrane does not provide the integrity to prevent swelling.

Most prokaryotic and some eukaryotic cells (fungi, algae, plants) contain a **cell wall** exterior to the cell or plasma membrane (**FIGURE 3.8**). Although the structure and organization of the wall differ between groups, all cell walls provide support for the cells, give the cells their shape, and help the cells resist cell lysis.

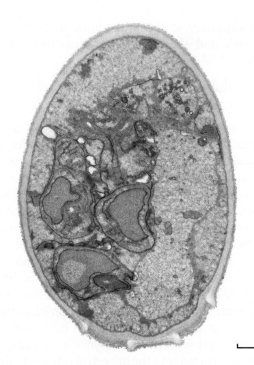

FIGURE 3.8 Yeast Cell Wall. This false-colored transmission electron micrograph image shows a cell wall (blue) of a yeast cell. The cytoplasm is colored red and a vacuole (a large storage and digestive space) is green. (Bar = 1 µm.) »» *What would happen to these yeast cells if the cell wall was removed?*

TABLE 3.1 Comparison of Prokaryotic and Eukaryotic Cell Structure

Characteristic	Cell Structure or Compartment	
	Prokaryotic	Eukaryotic
Genetic organization	Circular DNA chromosome	Linear DNA chromosomes
Cell compartmentation	Cell membrane	Plasma membrane
Metabolic organization	Cytoplasm	Cytoplasm
Protein synthesis	Ribosomes	Ribosomes
Protein/lipid transport	Cytoplasm	Endomembrane system
Energy metabolism	Cell membrane	Mitochondria and chloroplasts
Cell structure and transport	Thin protein filaments in cytoplasm	Protein tubules and filaments in cytoplasm
Cell motility	Prokaryotic flagella	Eukaryotic flagella or cilia
Water balance	Cell wall	Cell wall (algae, fungi)

A summary of the prokaryotic and eukaryotic processes and their associated structures is presented in TABLE 3.1.

The Evolution of the Eukaryotic Cell Involved Two Major Events

From what you have read so far in this chapter, all eukaryotic cells contain a variety of organelles. How did these structures evolve given that prokaryotic cells lack much of this organizational pattern? The prevailing theories suggest eukaryotic cells evolved through a series of structural changes to an ancestral archaeal cell. In this theory, two major events are proposed.

The first event involved an ancestral prokaryote developing the ability to move its plasma membrane inward, thereby forming a series of inward folds (FIGURE 3.9A). These folds might have eventually surrounded the hereditary material to form a cell nucleus as well as the internal membrane system that is present in all eukaryotic cells today. The resulting advantage to the ancestral cell was to divide metabolic processes into separate compartments in which complex chemistries would not be competing and interfering with one another. The additional membrane provided more surface or "workbench space" on which metabolic reactions could occur.

The second event in the evolution of an ancestral eukaryotic cell has more supporting evidence. Championed by the late Lynn Margulis and her colleagues, this event proposes a way ancestral eukaryotic cells acquired the energy organelles, specifically the mitochondria and chloroplasts. The theory asserts that mitochondria and chloroplasts arose through a process called endosymbiosis. A **symbiosis** is a relationship where two (or more) organisms live together and depend on each other; thus **endosymbiosis** refers to a relationship in which one organism (endosymbiont) lives inside (*endo*) the cell of another organism (the host).

The **endosymbiont theory** for the evolution of mitochondria and chloroplasts is thought to have occurred similar to what is shown in FIGURE 3.9B. About 1.5 billion years ago, an archaeal host cell engulfed an oxygen-using, nonphotosynthetic bacterial cell. Rather than the host cell destroying the bacterial cell, the host cell retained the endosymbiont in the cytoplasm and relinquished the responsibility for energy metabolism to this microbe. Over the course of evolution, the two partners became more dependent on each other for energy and survival and they became a single cell (organism) with the bacterial partner evolving into the present-day mitochondrion. With internal energy sources, the archaeal cell could develop into a larger cell with more DNA that would eventually evolve into the first eukaryotic cell.

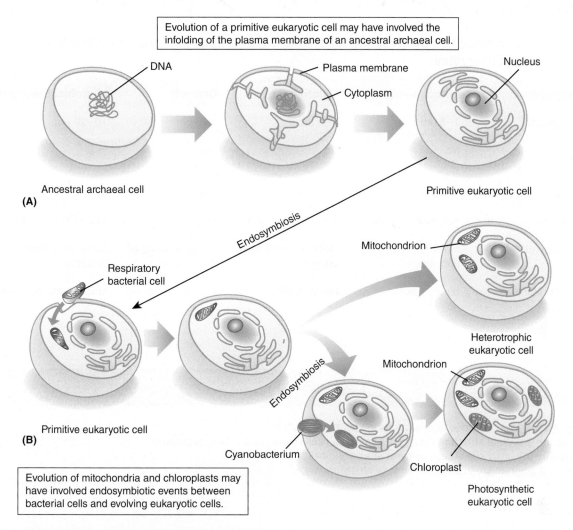

Evolution of a primitive eukaryotic cell may have involved the infolding of the plasma membrane of an ancestral archaeal cell.

DNA

Plasma membrane

Cytoplasm

Nucleus

Ancestral archaeal cell

(A)

Primitive eukaryotic cell

Endosymbiosis

Mitochondrion

Respiratory bacterial cell

Heterotrophic eukaryotic cell

Mitochondrion

Endosymbiosis

Primitive eukaryotic cell

(B)

Cyanobacterium

Chloroplast

Evolution of mitochondria and chloroplasts may have involved endosymbiotic events between bacterial cells and evolving eukaryotic cells.

Photosynthetic eukaryotic cell

FIGURE 3.9 Internal Membranes and the Endosymbiont Theory. Two major events have been proposed to explain the evolution of the first eukaryotic cell. **(A)** Some scientists have proposed that invagination of the cell membrane from an ancient archaeal cell led to the development of the cell nucleus as well as to the membranes of the endoplasmic reticulum. **(B)** Most scientists believe the mitochondria resulted from the uptake and survival (endosymbiosis) of a bacterial cell that carried out cellular respiration. A similar process, involving a bacterial cell that carried out photosynthesis, evolved into the chloroplast. *»» Which endosymbiotic event (mitochondrial or chloroplast) must have occurred first? Explain.*

Carrying the endosymbiont theory further, a slightly different scenario can be drawn for the evolution of the chloroplast found in algae and plants (see Figure 3.9B). In this case, a photosynthetic bacterial cell, perhaps a cyanobacterium, was engulfed by one of the evolving mitochondrion-containing eukaryotic cells. Tolerance of the endosymbiont permitted the symbiosis and again over the course of evolution this host cell and its photosynthetic endosymbiont merged into another "new organism" with the endosymbiont becoming the present-day chloroplasts found in the algae and plants.

Of course, unless we had a time machine and could go back 2 billion years and witness the event or events, we will never know for sure how the eukaryotic cell evolved. Still, there is considerable evidence supporting the endosymbiont theory. As TABLE 3.2 shows, the biochemical and physiological similarities between mitochondria, chloroplasts, and bacterial cells are too striking to be coincidence. All contain DNA and RNA, and the ribosomes of mitochondria and chloroplasts are somewhat similar to those in bacterial species. In addition, the organelles and bacterial cells "reproduce" themselves through a similar binary fission process. With the evolution of an endomembrane system and energy organelles, eukaryotic cells could become more complex structurally and evolve into a diverse domain of organisms.

TABLE 3.2 Similarities Between Mitochondria, Chloroplasts, Bacteria, and Microbial Eukaryotes

Characteristic	Mitochondria	Chloroplasts	Bacteria	Microbial Eukaryotes
Average size	1–5 μm	1–5 μm	1–5 μm	10–20 μm
Nuclear envelope present	No	No	No	Yes
DNA molecule shape	Circular	Circular	Circular	Linear
Ribosomes	Yes; bacterial-like	Yes; bacterial-like	Yes	Yes; eukaryotic-like
Protein synthesis	Make some of their proteins	Make some of their proteins	Many make all of their proteins	Make all of their proteins
Reproduction	Binary fission	Binary fission	Binary fission	Mitosis and cytokinesis

Concept and Reasoning Checks 3.1

a. How would you describe homeostasis in your own words?
b. List the universal structures and process that all cells possess as living organisms.
c. Identify the organelles found in most eukaryotic cells.
d. Draw a diagram illustrating the endosymbiont theory.

Chapter **Challenge B**

We have now covered the basic comparison between prokaryotic and eukaryotic cell structure and organization.

QUESTION B: *From what you have read and learned in this chapter in terms of prokaryotic "cell behavior" and now cell structure, how has your opinion changed regarding the so-called simple cells? Provide evidence to back up your opinion.*

You can find answers online in **Appendix F**.

■ KEY CONCEPT 3.2   Classifying Microorganisms Reveals Relationships Between Organisms

If you open any on-line or printed catalog, items are separated by types, styles, or uses. For example, in a fashion catalog, watches are separated from shoes and, within the shoes, men's, women's, and children's styles are separated from one another. Even the brands of shoes or their use (e.g., dress, casual, athletic) might be separated based on their degree of shared similarities.

The same reasoning holds true in biology. With such an immense diversity of organisms on planet

Earth, the human desire to catalog and order organisms into groups was originally based on shared similarities.

Classification Attempts to Catalog Organisms

In the 18th century, a Swedish scientist named Carolus Linnaeus began identifying living organisms according to similarities in form (shared characteristics) and placing these organisms in one of two "kingdoms"—Plantae and Animalia (**FIGURE 3.10**). However, as microorganisms were discovered and studied, it was clear to the German naturalist Ernst Haeckel that the microbes did not conform to the two-kingdom system. Therefore, in the mid-1800s, he created a third kingdom, the Protista, in which all the known single-celled microorganisms were placed. The bacterial organisms, which he called "moneres" (*moner* = "single") were placed near the base of the newly assembled kingdom.

With improvements in the design of light microscopes, more observations were made of the microbes in the kingdom Protista. In 1937, a French biologist named Edouard Chatton proposed that bacterial cells had unique and distinctive properties in "the prokaryotic nature of their cells" and should be separated from all other protists "which have eukaryotic cells." As a result, in 1956, Herbert Copland proposed that bacterial organisms be placed in a fourth kingdom, the Monera.

However, there was still one more problem with the kingdom Protista. A botanist named Robert H. Whittaker saw the fungi as the only eukaryotic group that must externally digest its food prior to absorption. For this and other reasons, Whittaker in 1959 added a fifth kingdom, the Fungi, to the expanding "tree of life."

This **tree of life** (**ToL**) represented the natural relatedness between organisms and reflected the evolutionary process through which organisms gradually change, as seen in the various structural and functional changes identified in different organisms. Charles Darwin, the father of evolution, called this process **natural selection**; that is, cells and organisms change through the generations to best fit (survive) in an environment. The least fit

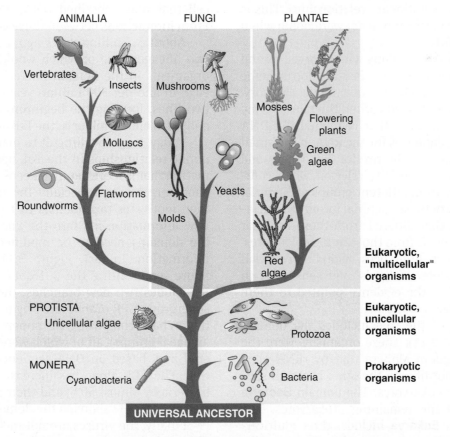

FIGURE 3.10 The Five Kingdom System for the "Tree of Life." The five-kingdom system implied an evolutionary lineage, beginning with the Monera and extending through the Protists. In the "tree of life," certain protists were believed to be ancestors of the multicellular kingdoms. *»» How many of the kingdoms consist of microorganisms?*

ones become extinct. Accordingly, the trunk of the tree suggests the Monera and Protista were the ancestral lines that gave rise to the plants, animals, and fungi. The "twigs" coming off the trunk and three main branches represent the specific groups of organisms within each kingdom.

The five-kingdom system rested safely until the late 1970s when the evolutionary biologist Carl Woese began a molecular analysis of living organisms by comparing the nucleotide sequences of genes coding for the small subunit ribosomal RNA (rRNA) found in all organisms. Woese's rRNA analysis revealed that the kingdom Monera actually contained two fundamentally unique groups that were as different from each other as they were different from the eukaryotes. Comparative analysis of rRNA genes soon became an indispensable tool in modern biology.

Kingdoms and Domains Identify Taxonomic Relationships

Taxonomy is the scientific discipline concerned with naming and classifying organisms within a framework that shows evolutionary relationships. This is what the scientists (taxonomists) were doing when developing the ToL.

The late Carl Woese, along with George Fox and coworkers, proposed a new classification scheme with a new most inclusive taxonomic category, the **domain**. As mentioned earlier, the new scheme initially came from work that compared the DNA nucleotide base sequences for the RNA in the small subunit of ribosomes. The results that Woese and Fox achieved were especially relevant when comparing sequences from different groups of bacterial organisms. Many of these bacterial forms had nucleotide sequences that differed from those in other prokaryotes as well as from those in the eukaryotes. After finding other differences, including cell wall composition, membrane lipids, and sensitivity to certain antibiotics, the evidence pointed to there being three domains to the ToL.

One branch of the tree includes the domain **Archaea** (**FIGURE 3.11**). These organisms were the ones in the kingdom Monera that by rRNA gene sequencing did not fit with the other bacterial species, or with the eukaryotes. The domain **Bacteria** encompasses all the remaining prokaryotes. The third domain, the **Eukarya**, includes three multicellular kingdoms (Plants, Fungi, and Animals). It also includes the single-celled protists that many taxonomists believe make up many kingdoms.

In the ToL, the tips of the branches represent currently living organisms, be it a species, genus, phylum (e.g., Proteobacteria) or a common biological group (e.g., slime molds). These tips represent organisms that we actually know something about. Based on DNA sequence analysis, it is possible to trace evolutionary history (the black lines in Figure 3.11) back to hypothetical common ancestors of today's organisms, which are represented by the branch points called nodes (e.g., the node at the center right represents the ancestor that gave rise to the organisms in the domain Archaea and Eukarya).

Many scientists envision the root of the tree as representing a hypothetical common ancestor called the **last common universal ancestor** (**LUCA**). LUCA presumably had all the universal features (DNA, cell membrane, cytoplasm, and ribosomes) and processes (e.g., ATP synthesis) found in living organisms today. LUCA arose through an evolutionary process involving the organization of nonliving (abiotic) building blocks and RNA (prebiotic stage) into a primitive protocell with RNA and protein. Such a cell then continued to evolve a progenote cell that now contained DNA, RNA, and protein. From interactions between these cells, LUCA arose.

About 3.7 billion years ago, LUCA gave rise to two lineages, one of which would evolve into the organisms in the domain Bacteria. The other lineage in another 500 million years would split into two lines, forming the beginnings of the Archaea and, with endosymbiosis, the Eukarya.

Today, it can be difficult to make sense of taxonomic relationships in the ToL because new information that is more detailed keeps being discovered about organisms, especially the prokaryotes. This then motivates taxonomists to figure out how the new information fits into the known ToL—or how the domains need to be modified to fit the new information. In fact, Figure 3.11 is a simplified figure representing the 2016 version for the ToL. It includes two new branches, the bacteria of the supergroup CPR (Candidate Phylum Radiation) and the archaea composing the supergroup TACK (four archaeal groups), all of which were recently discovered and, based on DNA sequences, added to the tree. The CPR alone has increased the number of bacteria by almost 50%, and their placement in the tree appears to subdivide the domain.

Finally, the viruses are not included in the ToL. This omission is intentional because the viruses do not have the cellular organization characteristic of the organisms in the three domains.

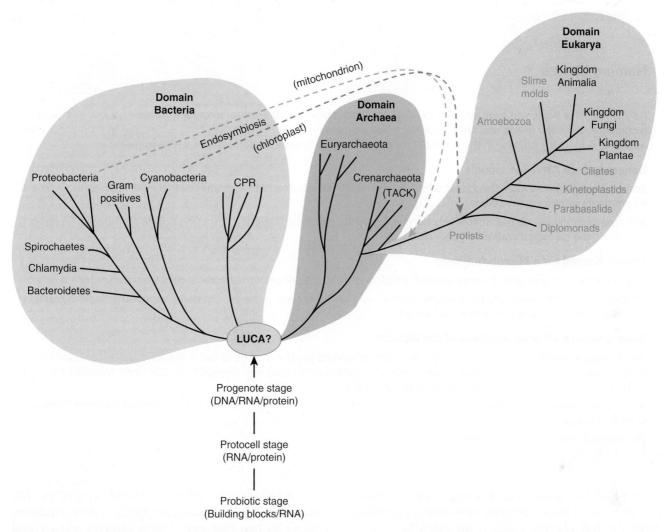

FIGURE 3.11 A Simplified View for the Current "Tree of Life." Some 3.7 billion years ago, the last universal common ancestor (LUCA) arose, which evolved through the series of stages (prebiotic, protocell, and progenote). LUCA had the universal cellular features found in all organisms today. From LUCA, presumably the Bacteria and Archaea domains (prokaryotes) arose, with the Eukarya domain (eukaryotes) branching from the Archaea lineage. Today, fundamental differences in genetic and molecular endowments are used to organize all organisms within the three domains. CPR (Candidate Phyla Radiation) and TACK (acronym for a group of four archaeal groups) are the most recent additions to the tree. All organisms in red in the domain Eukarya represent protists. **»» What portions of the "Tree of Life" represent microorganisms?**

Nomenclature Gives Scientific Names to Organisms

Another role of taxonomy is the naming of organisms and organizing them in a classification. In the late 1750s, Linnaeus published his *Systema Naturae* in which he popularized a two-word (binomial) scheme of nomenclature to identify each unique organism. The two words, usually derived from Latin or Greek stems, are the system still used today. Thus, each organism's name consists of the **genus** (plural: genera) to which the organism belongs and a **specific epithet**, a descriptor that further describes the genus. Together these two words make up the **species** name. For example,

take the common human gut bacterium *Escherichia coli*. *Escherichia* is the genus name and *coli* is the specific epithet that identifies the particular species as *Escherichia coli*. Now, you might wonder where these sometimes long, tongue-twisting names came from. MICROFOCUS 3.1 supplies the answer.

Notice in the examples that when a species name is written, only the first letter of the genus name is capitalized, while the specific epithet is all lowercase letters. In addition, both words are printed in *italics* or, if hand written, the species is underlined. After the first time a species name has been spelled out in a document or research paper, biologists usually

MICROFOCUS 3.1: Tools

Naming Names

As you read this book, you have, and will, come across many scientific names for microbes, for which a species name is a combination of the genus and specific epithet. Not only are many of these names tongue twisting to pronounce (see **Appendix C** for help), but how in the world did the organisms get those names? Here are a few examples.

Genera Named After Individuals

Escherichia coli: named after Theodore Escherich, who isolated the bacterial cells from infant feces in 1885. Being in feces, it commonly is found in the colon.

Neisseria gonorrhoeae: named after Albert Neisser, who discovered the bacterial organism in 1879. As the specific epithet points out, the disease it causes is gonorrhea.

Genera Named for a Microbe's Shape

Vibrio cholerae: vibrio means "comma-shaped," which describes the shape of the bacterial cells that cause cholera.

Staphylococcus epidermidis: *staphylo* means "cluster" and *coccus* means "spheres." So, these bacterial cells form clusters of spheres that are found on the skin surface (epidermis).

Genera Named After an Attribute of the Microbe

Saccharomyces cerevisiae: in 1837, Theodor Schwann observed yeast cells and called them *Saccharomyces* (*saccharo* = "sugar"; *myce* = "fungus") because the yeast converted grape juice (sugar) into alcohol; *cerevisiae* (from *cerevisia* = "beer") refers to the use of yeast since ancient times to make beer.

Myxococcus xanthus: *myxo* means "slime," so these are slime-producing spheres that grow as yellow (*xantho* = "yellow") colonies on agar.

Thiomargarita namibiensis: see MicroFocus 3.2.

---

abbreviate the genus name using only its initial genus letter or some accepted substitution, together with the full specific epithet, for example, *E. coli*. A cautionary note: often in magazines and newspapers, proper nomenclature is not followed, so *E. coli* often is written as Escherichia coli.

Classification Uses a Hierarchical System

Linnaeus' cataloging of plants and animals used shared and common characteristics to identify each species. Part of Linnaeus' innovation also involved organizing species and genera into higher taxa that likewise were based on shared, but broader, similarities.

Today, several similar genera are grouped into a **family**, and families with similar characteristics make up an **order**. Different orders can be placed together in a **class**, and classes are assembled together into a **phylum** (plural: phyla). All phyla would be placed together in a kingdom and/or domain, the most inclusive level of classification. TABLE 3.3 outlines the taxonomic hierarchy for three species: humans, a fungal yeast, and a bacterium.

Many microbes belong to a rank below the species level (subspecies) to indicate a special characteristic or variation within a species. For example, the cholera bacterium, *Vibrio cholerae*, exists as two biotypes: *Vibrio cholerae* classic and *Vibrio cholerae* El Tor. Other subgroup designations are subspecies, strain, ecotype, and morphotype. Serotype (*sero* = "serum") is used to distinguish between similar species by the type of antibody (serum) response generated by the immune system.

In 1923, David Hendricks Bergey devised one of the first systems of classification for bacterial species based on their structural and physiological characteristics. Today, you can find the proper taxonomic classification for the Bacteria and Archaea in *Bergey's Manual of Systematic Bacteriology*. This resource, which is based primarily on rRNA gene sequencing, consists of almost 3,500 pages separated into five volumes. It took 11 years to complete the entire manual. However, since its completion, many more prokaryotes have been identified and described.

Many Methods Are Available to Identify and Diagnose an Infection

Students taking a microbiology lab course and the staff working in the bacteriology section of a hospital's

TABLE 3.3 Taxonomic Classification of Humans, Brewer's Yeast, and a Common Bacterium

	Humans	Brewer's Yeast	*Escherichia coli*
Domain	Eukarya	Eukarya	Bacteria
Kingdom	Animalia	Fungi	
Phylum	Chordata	Ascomycota	Proteobacteria
Class	Mammalia	Saccharomycotina	Gammaproteobacteria
Order	Primates	Saccharomycetales	Enterobacteriales
Family	Hominidae	Saccharomycetaceae	Enterobacteriaceae
Genus	*Homo*	*Saccharomyces*	*Escherichia*
Species	*H. sapiens*	*S. cerevisiae*	*E. coli*

clinical microbiology lab (CML) usually are not concerned directly with bacterial classification but rather are involved with the identification of a bacterial species. Specifically, for the CML, after a doctor has confirmed that a patient has an infectious disease, it is important to identify the infectious agent that is present in a submitted sample (e.g., sputum, blood, stool, cerebrospinal fluid, or swabs from the nose or throat). To assist in the identification, the primary source of reference is *Bergey's Manual of Determinative Bacteriology*. This single-volume reference book uses morphology, other common structural and physiological characteristics, and standardized biochemical tests to identify bacterial pathogens and nonpathogens that can be cultured (**FIGURE 3.12**). Thus, after the CML tests have been run, the CML technician can refer to Bergey's manual for identification.

Usually, no single test, or type of test, can identify a particular infectious agent. Therefore, many different types of tests must be run. Often, the tests are carried out in a specific order, based on the results of previous tests. For pathogen identification purposes, one widely used technique is the **dichotomous key**. There are various forms of dichotomous keys, but one very useful construction is a flow chart in which a series of tests are listed down the page. Based on the dichotomous nature of the test (always a positive or negative result), the flow chart immediately leads to the next test and its result. Hopefully, in the end, an identification as to the bacterial agent causing the infection can be made. A simplified example is presented in **MICROINQUIRY 3**.

Genus **Escherichia**

Straight rods, 1.1–1.5 μm × 2.0–6.0 μm, occur singly or in pairs. Capsules or microcapsules occur in many strains. Gram negative. **Motile by peritrichous flagella or are nonmotile. Facultatively anaerobic.** Chemoorganotrophic, having **both a respiratory and a fermentative type of metabolism.** Optimal temperature is 37°C. D-Glucose and other carbohydrates are catabolized with the formation of acid and gas. **Oxidase negative, catalase positive, methyl red positive, Voges-Proskauer negative, and usually citrate negative. Negative for H₂S, urea hydrolysis, and lipase.** *Escherichia* species reduce nitrates. All or most strains ferment L-arabinose, maltose, D-mannitol, D-mannose, L-rhamnose, trehalose, and D-xylose. *O*-Nitrophenyl-β-D-galactopyranoside positive. Occur as normal flora in the lower part of the intestine of warm-blooded animals and, in the case of *E. blattae,* of cockroaches. *E. coli* strains that contain enterotoxins and/or other virulence factors, including invasiveness and colonization factors, cause diarrheal disease. *E. coli* is also a major cause of urinary tract infections and nosocomial infections including septicemia and meningitis. Other species, except for *E. blattae,* are rarely occurring opportunistic pathogens, usually associated with wound infections.

Type species: *Escherichia coli.*

Editorial note: *E. coli* is often subdivided serologically or by the presence of virulence factors to identify and characterize epidemiologically pathogenic strains. Com-

FIGURE 3.12 Bergey's Manual of Determinative Bacteriology. This version of *Bergey's Manual* is designed for the identification of bacteria. Here is part of the description used in the identification of the genus *Escherichia*. »» *How many different physical and biochemical characteristics can you identify in this description?*

Courtesy of Dr. Jeffrey Pommerville.

MICROINQUIRY 3

Identifying an Unknown Bacterial Species

A medical version of a taxonomic key (in the form of a dichotomous flow chart) can be used to identify very similar bacterial species based on physical and biochemical characteristics.

In this simplified scenario, an unknown bacterium has been pure cultured. In (A), Gram staining and four biochemical tests have been run. The test results are shown in the box. (B) Using the test results in (A) and the flow chart, identify the bacterial species.

In (C), the pure culture sample has been introduced into a more rapid and compact miniaturized identification system called Enterotube™ II. This system can perform 15 standard biochemical tests. (D) After 24 hours of incubation, the *positive* tests are circled and all the circled numbers in each boxed section are added to yield a 5-digit ID for the organism being tested. (E) This 5-digit number is looked up in a reference book or computer software to determine the identity of the species.

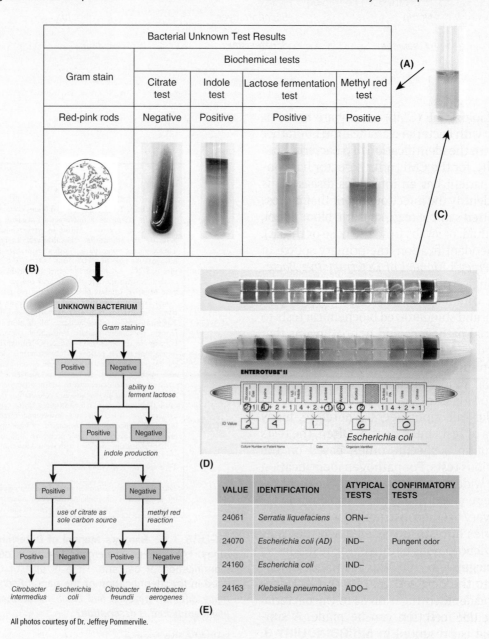

Bacterial Unknown Test Results

	Biochemical tests			
Gram stain	Citrate test	Indole test	Lactose fermentation test	Methyl red test
Red-pink rods	Negative	Positive	Positive	Positive

(A)

(B)

UNKNOWN BACTERIUM

Gram staining
→ Positive
→ Negative
ability to ferment lactose
→ Positive
→ Negative
indole production
→ Positive
→ Negative
use of citrate as sole carbon source / *methyl red reaction*
→ Positive → Negative / → Positive → Negative
Citrobacter intermedius / *Escherichia coli* / *Citrobacter freundii* / *Enterobacter aerogenes*

(C)

ENTEROTUBE® II

Glucose/Gas Lysine Ornithine H2S/Indole Adonitol Lactose Arabinose Sorbitol VP/Dulcitol/PA Urea Citrate

② 1   ④ + 2 + 1   4 + 2 + ①   ④ + ②   + 1 4 + 2 + 1

ID Value   2   4   1   6   0

Escherichia coli

(D)

VALUE	IDENTIFICATION	ATYPICAL TESTS	CONFIRMATORY TESTS
24061	*Serratia liquefaciens*	ORN–	
24070	*Escherichia coli (AD)*	IND–	Pungent odor
24160	*Escherichia coli*	IND–	
24163	*Klebsiella pneumoniae*	ADO–	

(E)

All photos courtesy of Dr. Jeffrey Pommerville.

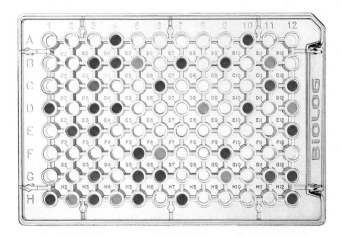

FIGURE 3.13 A Biolog MicroPlate. The Biolog system is capable of identifying hundreds of bacterial species by assessing the organism's ability to use any of 95 different substrates, each in a well of the plastic plate. The use of any substrate by the test organism is indicated by the dye in that well turning purple. The intensity of the purple coloration, which indicates the degree of substrate usage, is determined by a computer-linked automated plate reader. The first well (upper left) is a negative control with no substrate. *»» Which of the determinative methods described in this section is/are most likely to be used in this more automated system? Explain.*

Courtesy of Biolog, Inc.

With a positive identification, physicians can initiate an appropriate drug (e.g., antibiotic) or other treatment therapy.

Today, many of the tests run in the CML use automated systems (**FIGURE 3.13**). Many of the traditional tests require days to complete because the tests depend on the ability of the organism in question to grow in culture. Therefore, to make more rapid identifications, the CML is becoming more reliant on biochemical and molecular techniques, as well as nucleic acid sequencing technologies, because these methods are much faster (completed in hours) and the organism does not have to be grown in culture.

Let's briefly examine some of the more determinative bacteriological methods and a few molecular methods available to identify bacterial species.

Physical Characteristics

Because many bacterial species have physical characteristics that might help in identification, simply looking at the organism with a microscope can be sufficient for identification. The physical characteristics include cell shape (morphology) and arrangement of cells; spore-forming ability; Gram stain reactions; and motility. Unfortunately, many microbes look alike when using the microscope. If the CML can culture the organism, temperature and oxygen gas requirements can be examined. Still, many bacterial species have the same physical characteristics, so identifying biochemical features often is necessary.

Biochemical Tests

Today, a large number of biochemical tests exist that the CML can use to identify a pathogen (see **MicroInquiry 3**). These tests often include a pathogens ability to ferment carbohydrates; to use a specific substrate; and to produce specific byproducts or waste products. Nevertheless, as with the physical characteristics, often several biochemical tests are needed to differentiate between species.

Serological Tests

The immunology section within the CML often diagnoses infectious agents by detecting antigens or antibodies in a clinical sample (e.g., blood, cerebrospinal fluid) from the infected patient. Microorganisms are **antigenic**, meaning they are capable of triggering the immune system to produce **antibodies**. If the patient has been exposed to a specific microbe, antibodies might have been produced during the body's immune response to the infection. Blood tests can be run to detect these antibodies and to determine how much antibody is present.

The CML also can use antibodies to detect a pathogen. Solutions of collected antibodies, called "antisera," are commercially available for many medically important pathogens. For example, mixing a *Salmonella* antiserum with *Salmonella* cells will cause the cells to aggregate into a visible cluster. Therefore, if a foodborne illness occurs, the antiserum can be useful in identifying if *Salmonella* is the causative pathogen. More information about antibodies and serological testing is presented in Chapter 23.

Nucleic Acid Tests

Today, the fields of molecular genetics and genomics have advanced the analysis and sequencing of nucleic acids. This has given rise to a new era of **molecular taxonomy**.

If a pathogen cannot be cultured or identified by other methods, tests to identify pieces of the infectious agent's genetic material (DNA or RNA) can be completed. These nucleic acid-based tests include the polymerase chain reaction (PCR) and nucleic

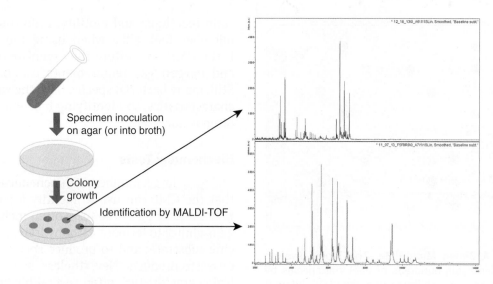

Specimen inoculation on agar (or into broth)

Colony growth

Identification by MALDI-TOF

FIGURE 3.14 MALDI-TOF Identification of Bacteria. The presence and relative intensities (heights) of biomarker peaks are unique to individual bacteria and allow rapid and accurate identification. »» *Why must the organisms be cultured before MALDI-TOF?*

Spectra courtesy of Dr. Todd Sandrin, Arizona State University-West Campus.

acid hybridization, which will be mentioned in Chapter 8.

Although PCR and other nucleic acid tests are becoming more common in CML identifications, PCR-independent methods also are being developed. One rapid and relatively low-cost method is called **matrix-assisted laser desorption/ionization time-of-flight (MALDI-TOF)** mass spectrometry. In this test, the microbes (or clinical specimen) of interest are grown on agar or in broth and prepared for MALDI-TOF. A laser beam then scatters the biological molecules from the specimen preparation (**FIGURE 3.14**). The laser-induced scattering generates a series of biomarker peaks that technicians can use to identify known prokaryotic or eukaryotic microbes down to the species level and, in some cases, to the subspecies level. As the MALDI-TOF process expands in the clinic, fast and accurate identification of pathogens will be possible.

Concept and Reasoning Checks 3.2

a. What four events changed the cataloging of microorganisms by shared characteristics?

b. With regard to the three domains, it has been said that Woese "lifted a whole submerged continent out of the ocean." What is the "submerged continent" and why is the term "lifted" used?

c. Which one of the following is a correctly written scientific name for the bacterium that causes anthrax? (i) *bacillus Anthracis*, (ii) *Bacillus Anthracis*, or (iii) *Bacillus anthracis*.

d. Why are numerous tests often needed to identify a specific bacterial species?

Chapter Challenge C

Many different types of tools and techniques can be used to classify the prokaryotes and eukaryotes into one of the three domains in the "tree of life."

QUESTION C: *Based on the two previous challenges in this chapter, what do all the similarities, differences, and "exceptions to the rule" with regard to prokaryotes and eukaryotes tell you about the nature of early life on Earth? How does LUCA factor into your judgment?*

You can find answers online in **Appendix F**.

■ KEY CONCEPT 3.3   Microscopy Is Used to Visualize the Structure of Cells

In the previous sections, microscopy was the tool that provided the images to contrast the similarities and differences in the structure of prokaryotic and eukaryotic cells. Before we examine the modern instruments used to observe the tiny microbes and viruses, we need to understand why prokaryotic cells are so small and to become familiar with the units to measure these small sizes.

Most Prokaryotic Cells Are Extremely Small

The majority of microorganisms you will examine and study in microbiology are extremely small. So, why are they so small, especially the prokaryotic cells?

Cell size in the prokaryotic world is governed by their rate of growth (reproduction). Many bacterial species reproduce extremely fast, some doubling in cell number in 15 minutes. For this rapid rate of increase in cell number, the cells must have a high metabolic rate, which necessitates the need for a constant and nearby supply of nutrients. However, as a cell grows in size, its cytoplasmic volume increases much faster than its membrane surface area. In other words, if a cell gets too large, it cannot obtain sufficient nutrients fast enough, nor can it rapidly eliminate toxic waste products. Consequently, by remaining small, the prokaryotic cell cytoplasm is always near to the cell membrane through which all nutrients enter the cell and waste products leave the cell. In eukaryotic microbes, the distance from the center of the cytoplasm to the plasma membrane can be much greater. These cells have solved the distance problem by forming subcompartments, and by having a cytoskeletal system for the rapid transport of needed nutrients and waste products.

In addition, prokaryotic cells generate all their cell energy (ATP) using their cell membrane. Because all the cytoplasm is close to the cell membrane, ATP is readily available to power cell metabolism. Conversely, if a cell grows too large, not enough ATP can move across the cell cytoplasm fast enough to supply the cell's metabolic demands. Again, the eukaryotic cells have solved this energy problem by having mitochondria scattered throughout the cell as separate sources of ATP for local distribution.

Microbes and Viruses Are Measured in Small Metric Units

Because microbial cells are so small, a convenient system of measurement is needed. The measurement system used is the metric system, where the standard unit of length is the meter, which is a little longer than a yard (see **Appendix A**). To measure microorganisms, we need to use units that are a fraction of a meter. In microbiology, the common unit for measuring length is the **micrometer (μm)**, which is a millionth (10^{-6}) of a meter.

Using these measurements, microbial agents range in size from the relatively large, almost visible protists (100 μm) down to the small bacteria (1 μm) (**FIGURE 3.15**). As a group, prokaryotic cells range from about 0.3 μm to more than 5 μm in length, whereas eukaryotic microbial cells average 5 μm to 20 μm. However, some notable exceptions to the prokaryotic size range have been discovered recently, as reported in MICROFOCUS 3.2.

Most viruses and eukaryotic cell organelles are much smaller than bacterial cells, so their size is expressed in nanometers. A **nanometer (nm)** is equivalent to a billionth (10^{-9}) of a meter; that is, 1/1,000 of a μm. Using nanometers, the size of the poliovirus, among the smallest human viruses, measures 20 nm (0.02 μm) in diameter. The human immunodeficiency virus (HIV), which is the causative agent leading to AIDS, is about 100 nm (0.10 μm) in diameter.

Light Microscopy Is Used to Observe Most Microorganisms

The basic microscope system used in the microbiology laboratory is the **light microscope**, in which visible light passes directly through the lenses and specimen (**FIGURE 3.16A**). Such an optical configuration is called **bright-field microscopy**. Visible light is projected through a condenser lens, which focuses the light into a sharp cone (**FIGURE 3.16B**). The light then passes through the opening in the stage. When hitting the glass slide containing the specimen for observation, the light is reflected or refracted as it goes through the slide and specimen. Then, the light enters the objective lens to form a magnified intermediate image inverted from that of the actual specimen. This intermediate image becomes the object magnified by the ocular lens (eyepiece) and seen by the observer.

Two important factors to consider with the microscope are magnification and resolution. Let's take a look at each.

Magnification. The increase in the apparent size of a specimen being observed is called **magnification**.

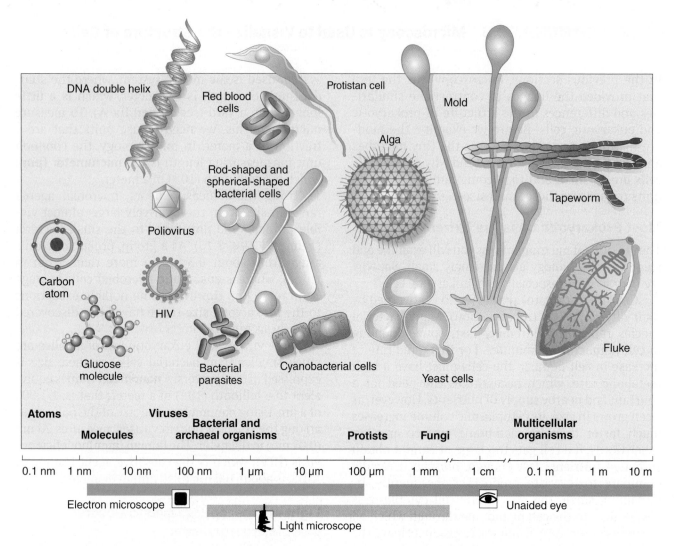

FIGURE 3.15 Size Comparisons among Various Atoms, Molecules, Viruses, and Microorganisms. (not drawn to scale) Although tapeworms and flukes usually are macroscopic, the diseases these parasites cause often are studied by microbiologists. *»» Referring to MicroFocus 3.2, where would you place the bacterial cells that were isolated from groundwater?*

A light microscope usually has at least three objective lenses that provide most of the magnification: the low-power, high-power, and oil-immersion lenses. In general, these lenses magnify an object 10, 40, and 100 times, respectively. (Magnification is represented by the multiplication sign, ×.) The ocular lens then magnifies the intermediate image produced by the objective lens by 10×. Therefore, the **total magnification** achieved is 100×, 400×, and 1,000×, respectively.

Resolution. The lens system of the microscope must have good **resolution** (also called resolving power), meaning that the observer can see a magnified object distinctly, and can distinguish closely spaced objects as separate objects. For example, a car seen

in the distance at night can appear to have a single headlight because at that distance the unaided eye lacks sufficient resolution. However, using binoculars for magnification, the two headlights can be seen clearly, because the binoculars have a higher resolving ability than does the human eye.

However, when switching from the low-power (10×) lens or high-power (40×) lens to the oil-immersion lens (100×), an observer quickly finds that the image seen through the oculars appears fuzzy. The object lacks resolution due to the refraction (bending) of light.

Both the low-power objective and high-power objective are wide enough to capture sufficient light for viewing. The lens of the oil-immersion objective, on the other hand, is so narrow that most

MICROFOCUS 3.2: Environmental Microbiology

Cell Size: Macrobacteria and Nanobacteria

While on an expedition off the coast of Namibia (western coast of southern Africa) in 1997, scientists from the Max Planck Institute for Marine Microbiology found a bacterial monster in sediment samples from the sea floor. These chains of spherical cells (see figure A) were 100 μm to 300 μm in diameter, with some up to 750 μm—about the diameter of the period in this sentence. Their volume is about 3 million times greater than that of *Escherichia coli*. The cells, shining white with enclosed sulfur granules, looked like a string of pearls. Thus, the bacterial species was named *Thiomargarita namibiensis* (meaning, "sulfur pearl of Namibia"). Another closely related strain was discovered in the Gulf of Mexico in 2005.

How do these extremely large cells solve the nutrient and energy barriers? Quite simply, the cell cytoplasm is slammed against the cell membrane, with most of the volume taken up a by large central storage vacuole.

At the other end of the size spectrum are the so-called nanobacteria. In an analysis of groundwater samples at a Department of Energy Research site in Colorado, scientists discovered cells that were as small as 250 nm (0.25 μm) in diameter, smaller than some viruses. These extremely small cells have just enough volume for DNA and about 40 ribosomes, so they probably grow very slowly. Nonetheless, this group of bacteria, informally called the candidate phyla radiation (CPR), is very prevalent, perhaps accounting for up to 15% of all bacteria on Earth (see Figure 3.11).

An equally small group of nanobacteria, called ARMAN (Archaeal Richmond Mine Acidophilic Nanoorganism), has been isolated from the Richmond Mine at Iron Mountain near Redding, California (see Figure B). They are part of a new archaeal supergroup called TACK. By comparison, *E. coli* are three times larger and up to 100 times the cell volume of CPR and TACK cells.

At the time of this writing, not much is known about the role of these tiny prokaryotes in the environment. What researchers can say is that not all prokaryotes are in the micrometer size range and that cell size is not a suitable characteristic for separating viruses from cells.

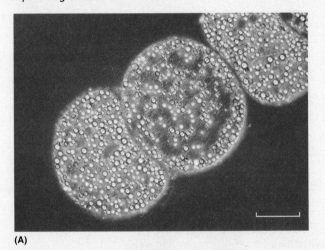

(A)

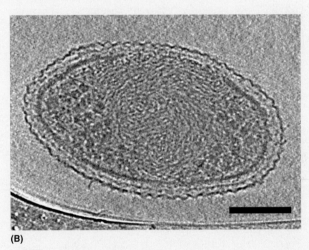

(B)

(A) A phase microscopy image showing three *Thiomargarita namibiensis* cells. (Bar = 100 μm.) **(B)** Cryogenic transmission electron microscopy image of an ARMAN cell. (Bar = 100 nm.)

light bends away from and does not pass through the objective lens (**FIGURE 3.16C**). However, by immersing the 100× lens in oil on the microscope slide, the light does not bend away from the lens as it passes from the glass slide and the specimen. The immersion oil thus provides a uniform pathway for light from the slide to the objective, and the resolution of the object increases. With the oil-immersion lens, the highest resolution possible with the light microscope is attained, which is near 0.2 μm (200 nm), as calculated in MICROFOCUS 3.3.

Staining Techniques Provide Contrast

Microbiologists commonly stain bacterial cells before viewing them with bright-field microscopy because the cell cytoplasm of bacterial cells usually lacks color, making it difficult to see the colorless cells on the bright microscope background, called

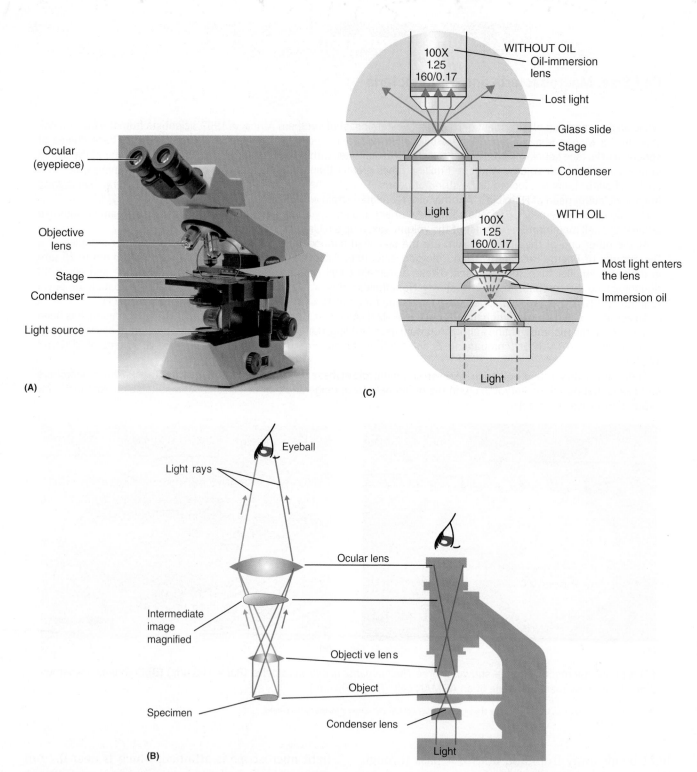

FIGURE 3.16 The Light Microscope. (A) The light microscope is used in many instructional and clinical laboratories. Note the important features of the microscope that contribute to the visualization of the object. **(B)** Image formation in the light microscope requires light to pass through the objective lens, forming a magnified intermediate image. This image serves as an object for the ocular lens, which further magnifies the image and forms the final image the eye perceives. **(C)** When using the oil immersion lens (100×), oil must be placed between and continuous with the slide and objective lens. »» *Why must oil be used with the 100× oil-immersion lens?*

MICROFOCUS 3.3: Tools

Calculating Resolving Power

The resolution or resolving power (RP) of a lens system is important in microscopy because it indicates the size of the smallest object that an observer can see clearly. The resolving power varies for each objective lens and its value is calculated by using the following formula:

$$RP = \lambda / 2 \times NA$$

In this formula, the Greek letter λ (lambda) represents the wavelength of light; for white light, it averages about 550 nm. The symbol NA stands for the numerical aperture of the lens and refers to the size of the cone of light that enters the objective lens after passing through the specimen. This number generally is printed on the side of the objective lens (see Figure 3.16C). For an oil-immersion objective with an NA of 1.25, the resolving power may be calculated as follows:

$$RP = 550 \text{ nm} / 2 \times 1.25 = 550/2.5 \text{ nm or } 0.22 \text{ } \mu m$$

Because the resolution limit for this lens system is 220 nm, any object smaller than 220 nm could not be seen as a clear, distinct object. An object larger than 220 nm would be resolved.

---

the **field** (**FIGURE 3.17A**). Several staining techniques have been developed to provide **contrast**.

Before staining the bacterial cells, a **bacterial smear** is prepared. The bacterial cells are spread on a clean glass slide and the slide air-dried. The slide then is passed briefly through a flame in a process called **heat fixation**, which bonds the cells to the slide, kills any cells still alive, and increases stain absorption.

Simple Stains

In the **simple stain technique**, the heat-fixed smear is flooded with a **basic (cationic) dye** such as methylene blue (**FIGURE 3.18A**). Because cationic dyes have a positive charge, the dye is attracted to the cytoplasm and cell wall, which primarily have negative charges. By contrasting the blue cells against the bright background, the staining procedure allows the observer to measure cell size, determine the arrangement of cells, and figure out **cell morphology**; that is, cell shape (**FIGURE 3.17B**).

Negative Stains

The **negative stain technique** works in the opposite manner (**FIGURE 3.18B**). Bacterial cells are mixed on a slide with an **acidic (anionic) dye** such as nigrosin (a black stain) or India ink (a black drawing ink). The mixture is then spread across the surface of the slide and allowed to air-dry. Because the anionic dye carries a negative charge, it is repelled from the cells and the observer sees clear or white cells on a dark or

gray background (**FIGURE 3.17C**). Because this technique avoids chemical reactions and heat fixation, the cells appear less shriveled than in a simple stain and more closely resemble their natural condition.

Differential Stains

Unlike simple staining, a **differential staining procedure** uses two contrasting colored dyes to distinguish between bacterial cells and/or cell structures when examined with the light microscope.

Gram Stain Technique. The most commonly used differential stain procedure is the **Gram stain technique**, named for Hans Christian Gram, a Danish physician who first perfected the technique in 1884.

As depicted in **FIGURE 3.18C**, a bacterial smear is (1) stained with crystal violet (a purple, basic dye), rinsed, and then (2) a special Gram's iodine solution is added that helps set (precipitate) the dye. Next, the smear is (3) rinsed with a decolorizer, such as 95% alcohol or an alcohol-acetone mixture. The alcohol will not wash away the dye-iodine complex from the so-called **gram-positive** cells because the iodine attaches the dye to those cells. The alcohol will wash away the dye-iodine complex from the so-called **gram-negative** cells because the alcohol damages the cell membrane, allowing the dye-iodine complex to escape. These cells are now colorless and difficult to contrast in the microscope. Therefore, the last step of the Gram stain technique (4) uses safranin (a red, basic dye) to counterstain the gram-negative organisms, giving them an

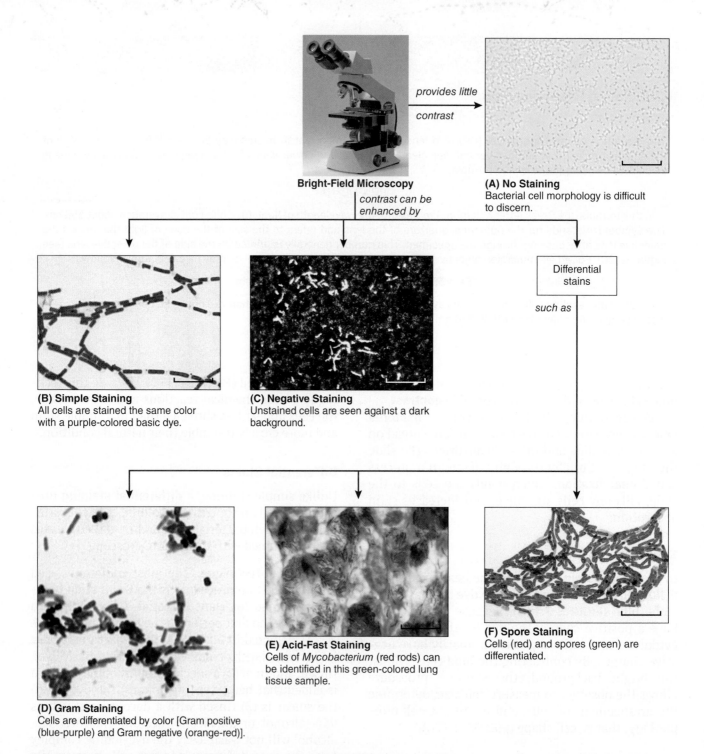

Bright-Field Microscopy

provides little contrast

(A) No Staining
Bacterial cell morphology is difficult to discern.

contrast can be enhanced by

(B) Simple Staining
All cells are stained the same color with a purple-colored basic dye.

(C) Negative Staining
Unstained cells are seen against a dark background.

Differential stains

such as

(D) Gram Staining
Cells are differentiated by color [Gram positive (blue-purple) and Gram negative (orange-red)].

(E) Acid-Fast Staining
Cells of *Mycobacterium* (red rods) can be identified in this green-colored lung tissue sample.

(F) Spore Staining
Cells (red) and spores (green) are differentiated.

FIGURE 3.17 Observing Stained Cells with Bright-Field Microscopy. Several bacterial staining techniques provide the contrast needed when observing the stained cells by bright-field microscopy. (A–F, Bar = 10 μm.) »» *What advantages does a differential stain have over a simple stain?*

(Microscope, A–D) Courtesy of Dr. Jeffrey Pommerville (E) © Dr. John D. Cunningham/Visuals Unlimited, Inc. (F) Courtesy of Larry Stauffer, Oregon State Public Health Laboratory/CDC.

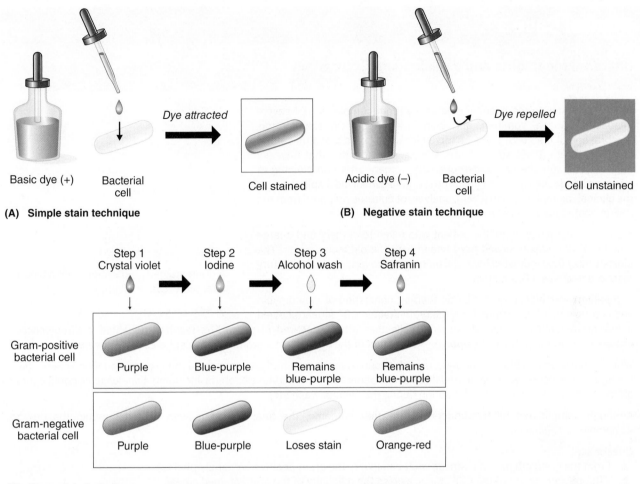

(A) Simple stain technique

Basic dye (+)   Bacterial cell   Dye attracted   Cell stained

(B) Negative stain technique

Acidic dye (−)   Bacterial cell   Dye repelled   Cell unstained

Step 1
Crystal violet

Step 2
Iodine

Step 3
Alcohol wash

Step 4
Safranin

Gram-positive bacterial cell

Purple   Blue-purple   Remains blue-purple   Remains blue-purple

Gram-negative bacterial cell

Purple   Blue-purple   Loses stain   Orange-red

(C) Gram stain technique

FIGURE 3.18 Some Important Staining Reactions in Microbiology. (A) In the simple stain technique, the cells of the smear are stained and contrasted so they can be seen against a light background. **(B)** In a negative stain, the cells remain unstained and contrasted against a dark background. **(C)** In the Gram stain, cells are stained either purple (Gram positive) or red-orange (Gram negative). *»» What three common cell characteristics can be determined by all three staining techniques?*

orange-red color. Thus, at the technique's conclusion, gram-positive cells are blue-purple, whereas gram-negative cells are orange-red. Similar to simple staining, gram staining also allows the observer to determine size and cell morphology. **FIGURE 3.17D** shows a staining outcome.

Knowing whether a bacterial cell is gram positive or gram negative is important. In the CML, the Gram stain results help the clinical technician to identify the bacterial organism using *Bergey's Manual*, if the procedure is done correctly, as described in **CLINICAL CASE 3**.

The Gram stain results also can be important for treatment decisions in a patient with a bacterial infection. Often gram-positive and gram-negative bacterial cells differ in their susceptibility to chemical substances such as antibiotics (e.g.,

gram-positive cells are more susceptible to penicillin, gram-negative cells to tetracycline), so Gram staining can be important for successful antibiotic therapy. In addition, gram-positive and gram-negative bacterial species can produce different types of disease-causing toxins.

Other Differential Staining Procedures. Another differential staining procedure is the **acid-fast technique**, which is used to identify members of the genus *Mycobacterium,* one species of which is the causative agent of tuberculosis. These bacterial cells normally are difficult to stain because the cells have very waxy walls that repel the dyes used in the Gram stain. However, the cells will stain bright red when subjected to the acid-fast procedure using the red-colored dye carbolfuchsin (**FIGURE 3.17E**).

Clinical Case 3

Bacterial Meningitis and a Misleading Gram Stain

A woman comes to the hospital emergency room complaining of severe headache, nausea, vomiting, and pain in her legs. On examination, cerebral spinal fluid (CSF) was observed leaking from a previous central nervous system (CNS) surgical site. The patient indicates that 6 weeks and 8 weeks ago she had undergone CNS surgery after complaining of migraine headaches and sinusitis. Both surgeries involved a spinal tap. In the clinical microbiology lab (CML), analysis of cultures prepared from the CSF indicated no bacterial growth.

From the emergency room, the patient was taken to surgery and a large amount of CSF was removed from underneath the old incision site. The pinkish, hazy fluid indicated bacterial meningitis, so among the laboratory tests ordered was a Gram stain.

A gram-stained preparation from a blood agar plate. (Bar = 10 μm.)

Courtesy of Dr. W. A. Clark/CDC.

The patient was placed on antibiotic therapy, consisting of vancomycin and cefotaxime. CML findings from the gram-stained CSF smear showed a few gram-positive, spherical bacterial cells that often appeared in pairs. The results suggested a *Streptococcus pneumoniae* infection. However, upon reexamination of the smear, a few gram-negative spheres were observed.

When transferred to a blood agar plate, growth occurred and a prepared smear showed many gram-negative spheres (see the figure). Further research indicated that several genera of gram-negative bacteria, including *Acinetobacter*, could appear gram-positive due to under-decolorization during the alcohol wash step.

Although complicated by the under-decolorization outcome, the final diagnosis was bacterial meningitis due to *Acinetobacter baumannii*.

Questions:
  a. From the gram-stained CSF smear, what color were the gram-positive bacterial spheres?
  b. After reexamination of the CSF smear, assess the reliability of the gram-stained smear.
  c. What reagent is used for the decolorization step in the Gram stain?

You can find answers online in **Appendix E**.

For additional information, see www.cdc.gov/HAI/organisms/acinetobacter.html.

Modified from Harrington, B. J. and Plenzler, M., 2004. Misleading Gram Stain Findings on a Smear from a Cerebrospinal Fluid Specimen. *Lab Med* 35(8):475–478.

Some gram-positive bacterial species produce endospores that can be easily differentiated from the bacterial cells by performing the **spore stain technique** (FIGURE 3.17F).

Other Light Microscopy Optics Also Enhance Contrast

The microscopist can equip a light microscope with other optical systems to improve contrast of microorganisms, with or without staining the cells. Following are three commonly employed optical systems.

Phase-contrast microscopy uses a special condenser and objective lenses. The condenser lens splits the light beam and throws the light rays slightly out of phase, which provides contrast as the rays pass through cell structures of slightly different densities. With phase-contrast microscopy, microbiologists can study the behavior of living organisms that are unstained (FIGURE 3.19A).

Dark-field microscopy also uses a special condenser lens mounted under the stage. The condenser scatters the light and causes it to hit the specimen from the side. Only light bouncing off the specimen and into the objective lens makes the specimen visible, as the surrounding area appears dark because it lacks background light.

In the CML, dark-field microscopy helps in the identification of small pathogens near the limit of resolution with the light microscope. For example,

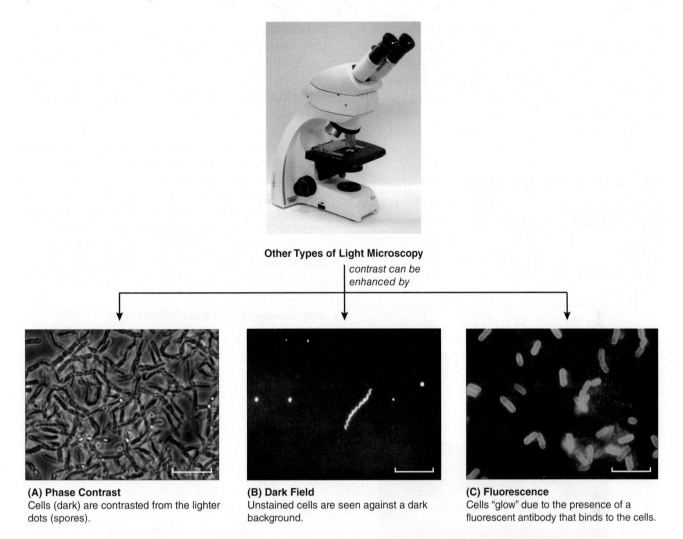

Other Types of Light Microscopy

contrast can be enhanced by

(A) Phase Contrast
Cells (dark) are contrasted from the lighter dots (spores).

(B) Dark Field
Unstained cells are seen against a dark background.

(C) Fluorescence
Cells "glow" due to the presence of a fluorescent antibody that binds to the cells.

FIGURE 3.19 Observing Cells with Other Types of Light Microscopy Optics. The three images show some common techniques for contrasting bacterial cells. These methods require special optical configurations added onto the light microscope. (A–C, Bar = 10 μm.) *»» Are the bacterial cells killed in preparing the sample for each of these three types of microscopy? Explain.*

Microscope image courtesy of Dr. Jeffrey Pommerville. (A and C) Courtesy of Larry Stauffer, Oregon State Public Health Laboratory/CDC. (B) Courtesy of Schwartz/CDC.

with dark-field microscopy, the observer can see the spiral bacterium *Treponema pallidum*, the causative agent of syphilis, which has a diameter of only about 0.15 μm (**FIGURE 3.19B**).

Fluorescence microscopy is a major asset to the CML and research laboratory. For fluorescence microscopy, fluorescent dyes, such as fluorescein are used. These dyes will coat and penetrate cells. When illuminated with an ultraviolet (UV) light source attached to the microscope, the energy in UV light excites electrons in fluorescein, causing them to give off visible light; in the case of fluorescein, a greenish yellow glow is observed in the microscope (**FIGURE 3.19C**). Other fluorescent dyes produce different colors.

Fluorescence microscopy is a major asset to clinical and research laboratories. As briefly described earlier with the serological tests, when combined with a fluorescent antibody technique, this form of microscopy has been very useful in the identification of many pathogens. Chapter 22 describes the technique.

Electron Microscopy Provides Detailed Images of Cells, Cell Parts, and Viruses

The power of electron microscopy comes from the extraordinarily short wavelength of a beam of electrons. Measured at 0.005 nm (compared to 550 nm for visible light), the short wavelength dramatically increases the resolution of the system and makes possible the visualization of viruses and detailed cellular structures, often called the **ultrastructure** of cells. The drawback of the electron microscope is that the method of specimen preparation kills the cells or organisms to be viewed.

Two types of electron microscopes are routinely used. Let's examine each one.

The Transmission Electron Microscope

The transmission electron microscope (TEM) uses magnets, rather than glass lenses, to focus a beam of electrons. The practical limit of resolution of biological samples with the TEM is about 2 nm, which is 100× better than the resolution of the light microscope (**FIGURE 3.20**).

After embedding the specimen in a suitable plastic mounting medium or freezing it, the specimen is cut into sections with a glass or diamond knife. Ultrathin sections of the prepared specimen must be used because the electron beam can penetrate matter only a very short distance. In this manner, a single

microbial cell can be sliced, like a loaf of bread, into hundreds of thin (100 nm thick) sections.

After it is cut, several of the sections are placed on a small copper grid and stained with heavy metals such as lead and osmium to provide contrast. The microscopist then inserts the grid into the vacuum tube of the microscope and focuses a 100,000-volt electron beam on one portion of one cut section at a time. As some electrons pass through while others are blocked by the specimen, an image forms on the screen below the tube or the image can be digitally recorded (**FIGURE 3.21A**). Such images recorded with the TEM are called **electron micrographs**.

The electron micrograph (maximum magnification is about 200,000×) can be enlarged to achieve a final magnification approaching 2 million times.

The Scanning Electron Microscope

The **scanning electron microscope (SEM)** also uses magnets to focus the electron beam. The resolution of the conventional SEM is about 10 nm and magnifications with the SEM are limited to about 20,000×.

(A)

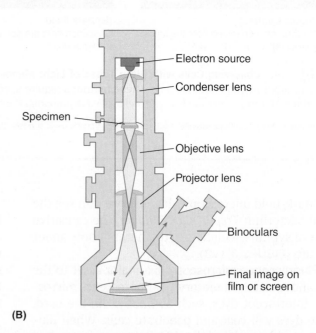

(B)

- Electron source
- Condenser lens
- Specimen
- Objective lens
- Projector lens
- Binoculars
- Final image on film or screen

FIGURE 3.20 The Transmission Electron Microscope. (A) A transmission electron microscope (TEM) is much larger than a light microscope. **(B)** A schematic of the TEM vacuum tube. A beam of electrons is emitted from the electron source and electromagnets function as lenses to focus the beam on the specimen. The image is magnified by objective and projector lenses. The final image is projected on a screen, television monitor, or electron-sensitive film. *»» How does the path of the image for the transmission electron microscope compare with that of the light microscope (Figure 3.16B)?*

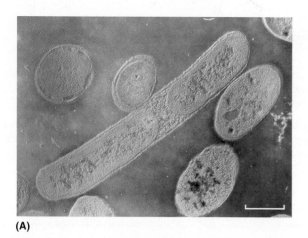

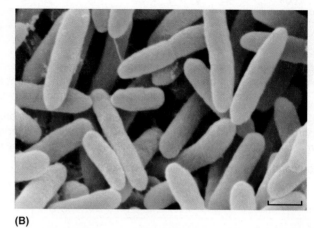

(A)   (B)

FIGURE 3.21 Transmission and Scanning Electron Microscopy Compared. The bacterium *Pseudomonas aeruginosa* (false-color images) as seen with two types of electron microscopy. **(A)** A view of sectioned cells seen with the transmission electron microscope. (Bar = 1.0 μm.) **(B)** A view of whole *P. aeruginosa* cells seen with the scanning electron microscope. (Bar = 2.0 μm.) *»» What types of information can be gathered from each of these electron micrographs?*

(A) © CNRI/Science Source. (B) © SciMAT/Science Source.

When using the SEM, a whole, dead specimen is placed in the vacuum tube and covered with a thin coat of gold. The electron beam then scans across the specimen and knocks loose showers of electrons that are captured by a detector. An image builds line by line, similar to how a television builds an image. Electrons that strike a sloping surface yield fewer electrons, thereby producing a darker contrasting spot and giving the observer a visual sense of a three-dimensional image. However, the instrument provides vivid and undistorted views of an organism's surface (**FIGURE 3.21B**).

TABLE 3.4 compares the various types of light and electron microscopy.

In conclusion, this chapter emphasized the fact that prokaryotic organisms, although they might look structurally simple, are not so simple when it comes to cellular processes and behaviors. In fact, with recent advances made to the light and electron microscopes, we now are finding more structure to prokaryotic cells than ever expected. You will discover this for yourself as you continue to learn about these organisms.

Concept and Reasoning Checks 3.3

a. If a bacterial cell is 0.75 μm in length, what is its length in nanometers?
b. Why would resolution be more important than magnification?
c. Explain the difference between a cationic dye and an anionic dye in terms of cell staining.
d. Which techniques for improving contrast do not result in killing the cells?
e. What type of electron microscope would be used to examine (1) the surface structures on a bacterial cell and (2) the organelles in an algal cell?

Chapter Challenge D

This chapter's challenge encouraged you question the widely held view that prokaryotic cells are simple and primitive. Decades of studies with the light and electron microscopes have provided extensive and rich detail into the similarities and differences between prokaryotic and eukaryotic cells.

QUESTION D: *As your final challenge, summarize how taxonomy and microscopy (two major concepts in this chapter) have contributed to both a better understanding of microbial cells (both prokaryotic and eukaryotic) and to a better realization of the organization and behavior of prokaryotic cells, as exemplified by the bacterial cell.*

You can find answers online in **Appendix F**.

TABLE 3.4 **Comparison of Various Types of Microscopy**

Type of Microscopy	Optical System	Appearance of Object	Magnification Range	Uses
Light				
Bright-field	Uses visible light to examine specimens	Stained microorganisms on clear background	100×–1,000×	Morphology and size of killed microorganisms (except viruses)
Phase-contrast	Uses "out of phase" light rays to examine specimens	Unstained microorganisms with contrasted structures	100×–1,000×	Internal structures of live, unstained eukaryotic microorganisms
Dark-field	Uses a special condenser to scatter light	Unstained microorganisms on dark background	100×–1,000×	Live, unstained microorganisms; motility of live cells
Fluorescence	Uses UV light to excite a fluorescent stain	Fluorescing microorganisms on dark background	100×–1,000×	Identification of microorganisms coated with fluorescent-tagged antibodies
Electron				
Transmission	Uses a beam of electrons to examine specimens	Alternating light and dark areas contrasting internal cell structures	100×–200,000×	Ultrathin slices of dead microorganisms, internal components, and viruses
Scanning	Uses a beam of electrons to scan specimens	Cell surfaces appear in three-dimensions	10×–20,000×	Surfaces and textures on dead microorganisms, cell components, and viruses

■ SUMMARY OF KEY CONCEPTS

Concept 3.1 Prokaryotes Are Not Simple, Primitive Organisms

1. All living organisms share the common emergent properties of life and attempt to maintain a stable internal state called **homeostasis.** (Figure 3.2)
2. Bacterial and eukaryotic cells share certain organizational patterns, including genetic organization, compartmentation, metabolic organization, and protein synthesis.
3. Although bacterial and eukaryotic cells carry out many similar processes, eukaryotic cells often contain a variety of subcellular compartments (**organelles**) to accomplish the processes. (Figure 3.3; Table 3.1)

Concept 3.2 Classifying Microorganisms Reveals Relationships Between Organisms

4. Many systems of classification have been devised to catalog organisms based on shared characteristics. (Figure 3.10)
5. Based on several molecular and biochemical differences, Woese proposed **three domains of life** that include the Bacteria, Archaea, and Eukarya, the latter consisting of the protists, fungi, plants, and animals. (Figure 3.11)
6. Part of an organism's binomial name is the **genus** name; the remaining part is the **specific epithet** that describes the genus name. Thus, a **species** name consists of the genus and specific epithet.
7. Organisms are properly classified by using a standardized hierarchical system from species (the least inclusive) to domain (the most inclusive). (Table 3.3)
8. *Bergey's Manual* is the standard reference to identify and classify bacterial species. Often a **dichotomous key** can be used to help in the identification.

Concept 3.3 Microscopy Is Used to Visualize the Structure of Cells

9. Another criterion of a microorganism is its size, a characteristic that varies among members of different groups. The **micrometer (µm)** is used to measure the dimensions of bacterial, protist, and fungal cells. The **nanometer (nm)** is commonly used to express viral sizes. (Figure 3.15)
10. The instrument most widely used to observe microorganisms is the **light microscope**. Light passes through several lens systems that magnify and resolve the object being observed. Although **magnification** is important, **resolution** is crucial. The light microscope can magnify up to 1,000× and resolve objects as small as 0.2 µm. (Figure 3.16)
11. For bacterial cells, staining generally precedes observation. The **simple, negative, Gram, acid-fast**, and other staining techniques can be used to impart contrast and determine structural or physiological properties. (Figures 3.17, 3.18)
12. Light microscopes employing **phase-contrast, dark-field**, and **fluorescence** optics have specialized uses in microbiology to contrast cells and make species identifications. (Figure 3.19)
13. To increase resolution and achieve extremely high magnification, the electron microscope employs a beam of electrons to magnify and resolve specimens. To observe internal details (**ultrastructure**), the **transmission electron microscope** is most often used; to study whole objects or surfaces, the **scanning electron microscope** is employed. (Figures 3.20, 3.21)

■ CHAPTER SELF-TEST

For **STEPS A–D**, answers to questions and problems can be found online in **Appendix D.**

■ STEP A: REVIEW OF FACTS AND TERMS

Multiple Choice

Read each question carefully before selecting the *one* answer that best fits the question or statement.

1. What is the term that describes the ability of organisms to maintain a stable internal state?
   A. Metabolism
   B. Homeostasis
   C. Biosphere
   D. Ecotype

2. Which one of the following is *not* an organizational pattern common to all organisms?
   A. Genetic organization
   B. Protein synthesis
   C. Compartmentation
   D. Endomembrane system

3. Which one of the following is *not* found in bacterial cells?
   A. Ribosomes
   B. DNA
   C. Mitochondria
   D. Cytoplasm

4. Who is considered the father of modern taxonomy?
   A. Woese
   B. Whittaker
   C. Haeckel
   D. Linnaeus

5. _____ was first used to catalog organisms into one of three domains.
   A. Photosynthesis
   B. Ribosomal RNA genes
   C. Nuclear DNA genes
   D. Cell respiration

6. Several classes of organisms would be classified into one _____.
   A. order
   B. genus
   C. phylum
   D. family

7. An important automated method used in the rapid identification of a pathogen is _____.
   A. rRNA gene sequencing
   B. polymerase chain reaction
   C. molecular taxonomy
   D. biochemical tests

8. What metric system of length is used to measure most bacterial cells?
   A. Millimeters (mm)
   B. Micrometers (µm)
   C. Nanometers (nm)
   D. Centimeters (cm)

9. Before bacterial cells are simple stained and observed with the light microscope, they must be _____.
   A. smeared on a slide
   B. heat fixed
   C. air dried
   D. All the above (**A–C**) are correct.

10. If you wanted to study the surface of a bacterial cell, you would use a _____.
    A. transmission electron microscope
    B. light microscope with phase-contrast optics
    C. scanning electron microscope
    D. light microscope with dark-field optics

Matching

Match the statement on the left to the term on the right by placing the letter of the term in the available space.

Statement

11. _____ The domain containing organisms whose cells have no cell nucleus or mitochondria in the cytoplasm.

12. _____ A member of this group of bacteria is thought to have undergone symbiosis and evolved into the chloroplast.

13. _____ Type of electron microscope for which cell sectioning is not required.

14. _____ The form of microscopy where only the light bouncing off the specimen is seen.

15. _____ The organelle, absent in bacteria, that carries out the conversion of chemical energy to cellular energy in eukaryotes.

16. _____ Domain in which fungi and protists are classified.

17. _____ Staining technique that differentiates bacterial cells into two groups.

18. _____ This form of microscopy uses a special condenser lens to split the light beam.

19. _____ The staining technique employing a single cationic dye.

20. _____ Type of microscopy using UV light to excite a dye.

Term

A. Bacteria
B. Chloroplast
C. Cyanobacteria
D. Dark-field
E. Eukarya
F. Fluorescence
G. Fungi
H. Gram
I. Mitochondrion
J. Negative
K. Phase-contrast
L. Scanning
M. Simple
N. Transmission

Label Identification

21. Identify the cell structures (a–p) indicated in the accompanying drawings **(A)** and **(B)**. What is a function for each structure?

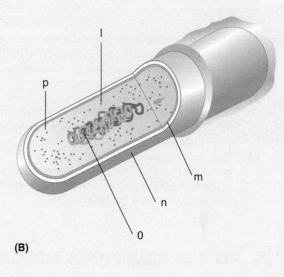

(B)

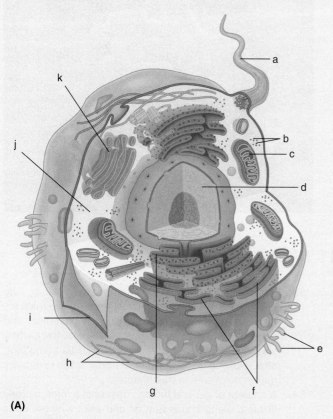

(A)

■ STEP B: CONCEPT REVIEW

22. Explain how variation in cell structure between prokaryotic and eukaryotic cells can be compatible with a similarity in cellular processes between these cells. (**Key Concept 3.1**)

23. Explain the assignment of organisms to one of the three **domains** on the "**tree of life**." (**Key Concept 3.2**)

24. Assess the importance of **magnification** and **resolution** to microscopy. (**Key Concept 3.3**)

25. Compare the uses of the **transmission** and **scanning electron microscopes**. (**Key Concept 3.3**)

◼ STEP C: APPLICATIONS AND PROBLEM SOLVING

26. A student is performing the Gram stain technique on a mixed culture of gram-positive and gram-negative bacterial cells. In reaching for the counterstain in step 4, he inadvertently takes the methylene blue bottle and proceeds with the technique. What will be the colors of gram-positive and gram-negative bacteria at the conclusion of the technique?

27. Would you obtain the best resolution with a light microscope by using red light (λ = 680 nm), green light (λ = 520 nm), or blue light (λ = 500 nm)? Explain your answer.

28. The accompanying electron micrograph shows a group of bacterial cells. The micrograph has been magnified 5,000×. Calculate in micrometers (µm) the actual length of the bacterial cells.

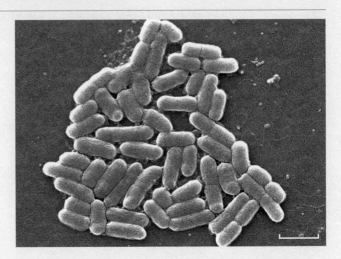

◼ STEP D: QUESTIONS FOR THOUGHT AND DISCUSSION

29. A local newspaper once contained an article about "the famous bacteria E. Coli." How many errors can you find in this phrase? Rewrite the phrase correctly.

30. Microorganisms have been described as the most chemically diverse, adaptable, and ubiquitous organisms on Earth. Although your knowledge of microorganisms still might be limited at this point, try to add to this list of "mosts."

31. Prokaryotes lack the cytoplasmic organelles commonly found in the eukaryotes. Provide a reason for this structural difference.

32. A new bacteriology section of a clinical microbiology lab is opening in your community hospital. What is one of the first books that the laboratory director will want to purchase? Why is it important to have this book?

33. Jennifer Ackerman in *Chance in the House of Fate* (2001) said, "We think we have life down; we think we understand all the conditions of its existence; and then along comes an upstart bacterium, live or fossilized, to tweak our theories or teach us something new." How does her statement apply to the taxonomy material covered in this chapter?

34. A student of general biology observes a microbiology student using immersion oil and asks why the oil is used. "To increase the magnification of the microscope" is the reply. Do you agree or disagree? Why?

35. Every state has an official animal, flower, or tree, but only Oregon has a bacterial species named in its honor: *Methanohalophilus oregonese*. The specific epithet *oregonese* should be apparent. Can you decipher the meaning of the genus name?

Concept Mapping

See pages XXXI-XXXII on how to construct a concept map.

36. Construct a concept map for **Living Organisms** using the following terms (you can use terms more than once). (**Key Concept 3.1**)

Bacterial cells	Golgi apparatus
Cell membrane	Lysosomes
Chloroplasts	Microcompartments
Cytoplasm	Mitochondria
Cytoskeleton	Nucleus
Cytosol	Rough ER
DNA region	Ribosomes
Eukaryotic cells	Smooth ER
Flagella	

37. Construct a concept map for **staining techniques** using the following terms only once. (**Key Concept 3.3**)

Acid-fast technique	Differential stain procedure
Acidic dye	Gram negative
Basic dye	Gram positive
Blue-purple cells	Gram stain technique
Cell arrangement	*Mycobacterium*
Cell shape	Negative stain technique
Cell size	Orange-red cells
Contrast	Simple stain technique

CHAPTER 4

Structure and Organization of Prokaryotic Cells

"*Double, double toil and trouble; Fire burn, and cauldron bubble*" is the refrain repeated several times by the chanting witches in Shakespeare's *Macbeth* (Act IV, Scene 1). This image of a hot, boiling cauldron actually describes the environment near which many prokaryotic species happily grow, especially archaeal members! For example, some species can be isolated from hot springs, such as the Grand Prismatic Spring (see chapter opening photo) in Yellowstone National Park in the United States.

In 1996, the late evolutionary biologist and geologist Stephen J. Gould wrote an essay for the *Washington Post Horizon* in which he said, "... bacteria are—and always have been—the dominant forms of life on Earth." He, as well as most microbiologists at the time, had no idea that embedded in these bacterial organisms was another whole domain of organisms. Thanks to the pioneering studies of Carl Woese and his colleagues, it now is quite evident there are two distinctly different groups of prokaryotes— the Bacteria and the Archaea. Many of the organisms Woese and others studied are organisms that would live a happy life in a witch's cauldron because

they can grow at high temperatures and survive in extremely acidic environments—a real cauldron by human standards! Termed **extremophiles**, these members of the domains Bacteria and Archaea have a unique genetic makeup and thrive in physically or geochemically extreme environmental conditions.

In fact, Earth's oldest fossils might have been archaeal species. Many microbiologists believe the ancestors of today's archaeal species might represent a type of organism that first inhabited planet Earth when it was a young, hot place. These unique characteristics led Woese to propose these organisms be lumped together and called the

---

The Grand Prismatic Spring in Yellowstone National Park, Wyoming.
© Martin Ruegner/Radius Images/Getty Images.

Archaebacteria (*archae* = "ancient"). Since then, the domain name has been replaced by Archaea because (1) not all members are extremophiles or related to these possible ancient ancestors and (2) they do not belong to the domain Bacteria—they are members of their own domain.

In this chapter, we will emphasize structure within this domain because of their relationships to human disease and the human microbiome. MICROFOCUS 4.1 provides a quick overview. As you will see in this chapter, a study of the structural features of bacterial cells provides a window into their activities and illustrates how the Bacteria relate to other living organisms. We will conclude the chapter with a brief survey of the domains Bacteria and Archaea.

MICROFOCUS 4.1

Bacteria in Eight Easy Lessons[1]

Mélanie Hamon, a research associate at the Institut Pasteur in Paris, says that when she introduces herself as a bacteriologist, she often is asked, "Just what does that mean?" To help explain her discipline, she gives us, in eight lessons, what she calls "some demystifying facts about bacteria."

Basic principles: Their average size is 1/25,000th of an inch. In other words, hundreds of thousands of bacteria fit into the period at the end of this sentence. In comparison, human cells are 10 to 100 times larger with a more complex inner structure. Whereas human cells have copious amounts of membrane-contained subcompartments, bacteria more closely resemble pocketless sacs. Despite their simplicity, they are self-contained living beings, unlike viruses, which depend on a host cell to carry out their life cycle.

Astonishing: Bacteria are the root of the evolutionary tree of life, the source of all living organisms. Quite successful evolutionarily speaking, they are ubiquitously distributed in soil, water, and extreme environments such as ice, acidic hot springs, or radioactive waste. In the human body, bacteria account for 2% of dry weight, populating mucosal surfaces of the oral cavity, gastrointestinal tract, urogenital tract, and surface of the skin. In fact, bacteria are so numerous on earth that scientists estimate their biomass far surpasses that of the rest of all life combined.

Crucial: It is a little known fact that most bacteria in our bodies are harmless and even essential for our survival. Inoffensive skin settlers form a protective barrier against any troublesome invader while approximately 1,000 species of gut colonizers work for our benefit, synthesizing vitamins, breaking down complex nutrients, and contributing to gut immunity. Unfortunately for babies (and parents!), we are born with a sterile gut and "colic" our way through bacterial colonization.

Tools: Besides the profitable relationship they maintain with us, bacteria have many other practical and exploitable properties, most notably, perhaps, in the production of cream, yogurt, and cheese. Less widely known are their industrial applications as antibiotic factories, insecticides, sewage processors, oil spill degraders, and so forth.

Evil: Unfortunately, not all bacteria are "good," and those that cause disease give them all an often undeserved and unpleasant reputation. If we consider the multitude of mechanisms these "bad" bacteria—pathogens—use to assail their host, it is no wonder that they get a lot of bad press. Indeed, millions of years of coevolution have shaped bacteria into organisms that "know" and "predict" their hosts' responses. Therefore, not only do bacterial toxins know their target, which is never missed, but also bacteria can predict their host's immune response and often avoid it.

Resistant: Even more worrisome than their effectiveness at targeting their host is their faculty to withstand antibiotic therapy. For close to 60 years, antibiotics have revolutionized public health in their ability to treat bacterial infections. Unfortunately, overuse and misuse of antibiotics have led to the alarming fact of resistance, which promises to be disastrous for the treatment of such diseases.

Ingenious: The appearance of antibiotic-resistant bacteria is a reflection of how adaptable they are. Thanks to their large populations, they are able to mutate their genetic makeup, or even exchange it, to find the appropriate combination that will provide them with resistance. Additionally, bacteria are able to form "biofilms," which are cellular aggregates covered in slime that allow them to tolerate antimicrobial applications that normally eradicate free-floating individual cells.

A long tradition: Although "little animalcules" were first observed in the 17th century, it was not until the 1850s that Louis Pasteur fathered modern microbiology. From this point forward, research on bacteria has developed into the flourishing field it is today. For many years to come, researchers will continue to delve into this intricate world, trying to understand how the good ones can help and how to protect ourselves from the bad ones. It is a great honor to be part of this tradition, working in the very place where it was born.

[1]Republished with permission of the author, the Institut Pasteur, and the Pasteur Foundation. The original article appeared in *Pasteur Perspectives*, Issue 20 (Spring 2007), the newsletter of the Pasteur Foundation, which you can find at www.pasteurfoundation.org. © Pasteur Foundation. Since Mélanie wrote this article, a few numbers have changed and are updated in this essay.

■ KEY CONCEPT 4.1   Prokaryotes Can Be Distinguished by Their Cell Shape and Arrangements

The domains Bacteria and Archaea are composed of prokaryotic cells (**FIGURE 4.1**). Although there are substantial differences that led to their assignment to separate domains, their cell morphology (shape) is determined by the cell wall and underlying cytoplasmic cytoskeletal proteins, which direct cell wall synthesis. When stained bacterial cells are viewed with the light microscope, many, including most of the pathogenic bacteria, appear in one of three basic shapes: the rod, the sphere, or the spiral.

Variations in Cell Shape and Cell Arrangement Exist

A prokaryotic cell with a rod shape is called a **bacillus** (plural: bacilli) and, depending on the species, can be as short as 0.5 µm or as long as 20 µm in

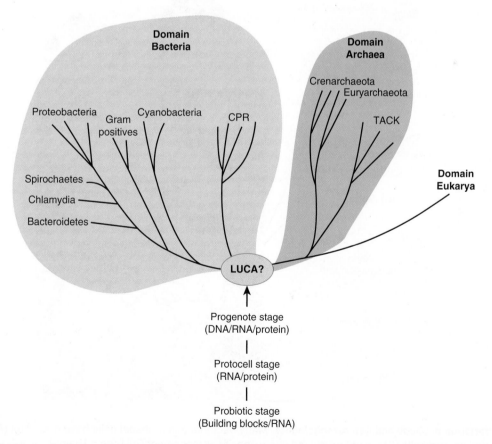

FIGURE 4.1 The Tree of Life for the Bacteria and Archaea. The tree shows several of the bacterial and archaeal lineages discussed in this chapter, including the CPR and TACK supergroups. *»» What was the original evidence that separated these prokaryotes into two domains?*

length and the rod can be slender, rectangular, or club shaped. Although some species exist as single rods, in many species the cells remain attached to one another after cell division. This results in various arrangements, including a pair of rods called **diplobacillus** or a long chain of rods called **streptobacillus** (*strepto* = "chains") (FIGURE 4.2A). You need to be aware that there are two ways to use the word "bacillus": to denote a rod-shaped bacterial cell and as a genus name (*Bacillus*).

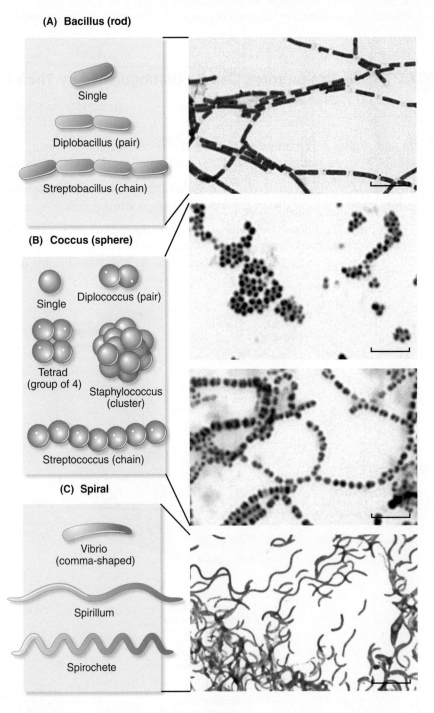

FIGURE 4.2 Variation in Shape and Cell Arrangements. Many bacterial and archaeal cells have a bacillus **(A)** or coccus **(B)** shape. Most spiral shaped-cells **(C)** are not organized into a specific arrangement (All bars = 10 μm.). *»» In Panel C, identify the vibrio and the spirillum forms in the photograph.*

All photos courtesy of Dr. Jeffrey Pommerville.

A spherically shaped bacterial cell is known as a **coccus** (plural: cocci [*kokkos* = "berry"]), which tends to be quite small, being only 0.5 μm to 1.0 μm in diameter. Although they are usually round, the coccus shape can be oval, elongated, or indented on one side. The cells of many bacterial species also remain attached after division and take on cellular arrangements characteristic of the species (**FIGURE 4.2B**). Cocci remaining in a pair represent a **diplococcus**. The causative agent of gonorrhea, *Neisseria gonorrhoeae*, and one type of bacterial meningitis (*N. meningitidis*) are diplococci. Cocci that remain in a chain represent a **streptococcus** arrangement. Certain species of streptococci are involved in strep throat (*Streptococcus pyogenes*) and tooth decay (*S. mutans*). Another arrangement of cocci is the **tetrad**, consisting of four spheres forming a square packet. A cube-like packet of eight cocci is called a **sarcina** (*sarcina* = "bundle"). *Micrococcus luteus*, a common inhabitant of the skin, is one example. Yet other species with spherical cells divide randomly and the attached cells form an irregular grape-like cluster called a **staphylococcus** (*staphylo* = "cluster") arrangement. A well-known example, *Staphylococcus aureus*, is a cause of food poisoning, toxic shock syndrome, and several skin infections. Notice again that the words "streptococcus" and "staphylococcus" can be used to describe cell arrangements or a bacterial genus (*Streptococcus* and *Staphylococcus*).

The third common morphology of many bacterial cells is the **spiral**, which do not remain attached after cell division. The spiral can take one of three forms (**FIGURE 4.2C**). The **vibrio** is a curved cell that resembles a comma. The cholera-causing organism *Vibrio cholerae* is one example. Another spiral form called **spirillum** (plural: spirilla) has a helical shape with a thick, rigid cell wall. The spiral-shaped form known as **spirochete** has a thin, flexible cell wall. The organism causing syphilis, *Treponema pallidum*, represents a spirochete. Some spirilla and spirochete forms can be up to 100 μm in length.

In addition to the bacillus, coccus, and spiral shapes, other morphologies exist. The cells of some bacterial species are spindle shaped or consist of branching filaments, whereas some archaeal species have cells with a square or star morphology. Why specific species have certain shapes is not clear. It is assumed that shape must confer some adaptive advantage. For example, a spiral cell shape appears to enhance motility, whereas cells existing as branching filaments are better adapted for buoyancy.

Concept and Reasoning Checks 4.1

a. What morphology does a bacterial cell have that is said to be coccobacillus?

Chapter Challenge A

So far, you have learned about the morphologies and arrangements of cells in the domains Bacteria and Archaea within the tree of life.

QUESTION A: *Based on what you have read up to this point, how has this diversity of life highlighted its unity?*

You can find answers online in **Appendix F**.

■ KEY CONCEPT 4.2   Bacterial and Archaeal Cells Have an Organized Structure

Bacterial and archaeal cells appear to have little visible structure when observed with a light microscope. This statement, along with their small size, gives the impression that they lack the structural organization present in eukaryotic cells.

Actually, prokaryotic cells have all the complex processes typical of eukaryotic cells. It is simply that, in most cases, the structure, and sometimes the organization, to accomplish these processes is different from the membranous organelles typical of eukaryotic species.

Cells Are Highly Ordered Externally and Internally

Studies in bacterial and archaeal cell biology have revealed that prokaryotes have a highly ordered three-dimensional organization, which is centered on three universal emergent properties of life (FIGURE 4.3).

1. **Sensing and Responding to the Surrounding Environment.** Because most prokaryotic cells are surrounded by a cell wall, some pattern of "external structures" is necessary to sense their environment and for the cells to respond to it or to other neighboring cells. External structures such as pili, flagella, and a glycocalyx provide the needed structural pattern of organization.

2. **Compartmentation of Metabolism.** Cell metabolism must be segregated from the exterior environment and yet be capable of transporting materials to and from that environment. In addition, protection from osmotic forces due to osmosis must be in place. The cell envelope (cell wall and cell membrane) fulfills those roles.

3. **Growth and Reproduction.** Cell survival and homeostasis demand a complex metabolism that occurs within the aqueous cytoplasm. These processes exist as internal structures or subcompartments localized to specific areas within the cytoplasm.

Our understanding of prokaryotic cell biology is still an developing field of study. However, there is more to cell structure than what was

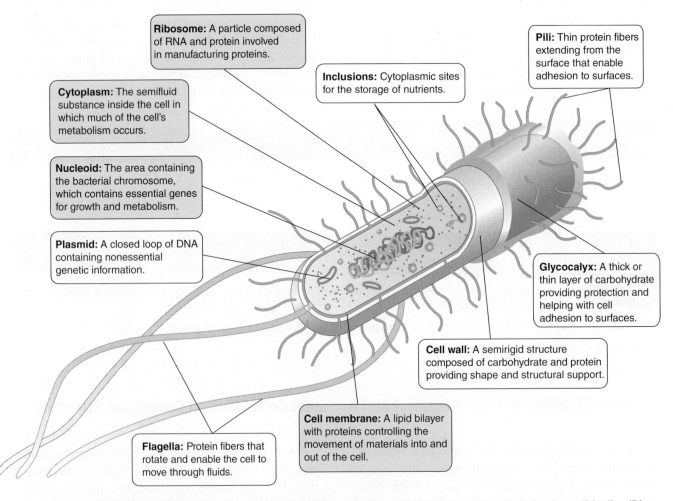

Ribosome: A particle composed of RNA and protein involved in manufacturing proteins.

Cytoplasm: The semifluid substance inside the cell in which much of the cell's metabolism occurs.

Nucleoid: The area containing the bacterial chromosome, which contains essential genes for growth and metabolism.

Plasmid: A closed loop of DNA containing nonessential genetic information.

Inclusions: Cytoplasmic sites for the storage of nutrients.

Pili: Thin protein fibers extending from the surface that enable adhesion to surfaces.

Glycocalyx: A thick or thin layer of carbohydrate providing protection and helping with cell adhesion to surfaces.

Cell wall: A semirigid structure composed of carbohydrate and protein providing shape and structural support.

Cell membrane: A lipid bilayer with proteins controlling the movement of materials into and out of the cell.

Flagella: Protein fibers that rotate and enable the cell to move through fluids.

FIGURE 4.3 Bacterial Cell Structure. The structural features, and their composition and function, existing in an "idealized" bacterial cell, such as *Escherichia coli*. Structure descriptions highlighted in blue are found in all prokaryotic cells. »» *Which structures represent (a) external structures, (b) the cell envelope, and (c) cytoplasmic structures?*

once believed—smallness does not equate with simplicity. For example, specific cellular proteins can be localized to precise regions of the cell. In *Streptococcus pyogenes*, many of the proteins that confer its pathogenic nature in leading to diseases like strep throat are secreted from a specific area of the cell surface. *Yersinia pestis*, which is the agent responsible for the black plague, contains a specialized secretion apparatus through which toxic proteins are released. This apparatus only exists on the bacterial cell surface that is in contact with the target host cells.

The cell biology studies also have important significance to clinical microbiology and the battle against infectious disease. As more is discovered about prokaryotic cells, and how they are similar and different from eukaryotic cells, the better equipped medical professionals will be to develop new antimicrobial agents capable of targeting the subcellular organization of pathogens. In an era in which there are fewer effective antibiotics to fight infections, understanding cell structure and function is extremely important.

On the following pages, we examine some of the common prokaryotic structures and their organization in an idealized bacterial cell, as no single species always contains all the structures. Our journey begins by examining the structures on or protruding from the surface of the bacterial cell. Then, we examine the cell envelope before plunging into the cell cytoplasm. Note: images of cells give the impression of a static, motionless structure. Realize such images are but a "snapshot" of a highly active living cell.

Concept and Reasoning Checks 4.2

a.   What is gained by bacterial and archaeal cells being organized into three general sets of structures—external, envelope, and cytoplasmic?

Chapter **Challenge B**

Prokaryotic cell structure is organized to efficiently carry out those functions required by the organism, that is, sensing and responding to the surrounding environment, compartmentalizing metabolism, and carrying out growth and reproduction.

QUESTION B: *Do you believe these three emergent properties of life are different from those in the eukaryotic microbial cell? Explain, keeping in mind the challenge of diversity emphasizing unity.*

You can find answers online in **Appendix F**.

■ KEY CONCEPT 4.3   Cell-Surface Structures Interact with the Environment

Prokaryotic cells need to respond to and monitor their external environment. This is made difficult by having a cell wall that "blindfolds" the cell. Therefore, many cells possess structures that extend into the environment beyond the cell surface.

Pili Are Involved with Several Cellular Processes

Numerous short, thin fibers, called **pili** (singular: pilus [*pilus* = "hair"]), protrude from the surface of most gram-negative bacteria. It should be noted that microbiologists often use the term "pili" interchangeably with "fimbriae" (singular, fimbria [*fimbria* = "fiber"]). There are several different types of pili.

Type I Pili

The pili in **FIGURE 4.4A** represent **type I pili**. They are rigid protein fibers containing specific adhesive molecules called **adhesins** that are attached at the tip. The adhesins allow the pili to attach or bind to specific cells or other objects. For example, the pili on certain pathogenic strains of *Escherichia coli* that infect the urinary tract allow the cells to attach to

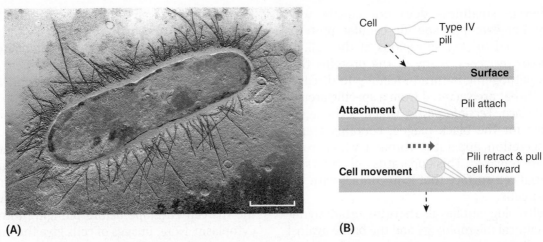

(A)   **(B)**

FIGURE 4.4   Bacterial Pili. (A) False-color transmission electron micrograph of an *Escherichia coli* cell with many type I pili. (Bar = 0.2 μm.) **(B)** Through repetitions of extension, attachment, and retraction, type IV pili pull the cell forward along a surface. *»» What is common to both type I and type IV pili?*

(A) © BSIP/Getty Images.

the epithelial cells lining the tract. In this way, the pili act as a **virulence factor**; that is, the pili facilitate colonization and biofilm formation that, in an infected individual, can lead to disease development (i.e., urinary tract infection). Could a pathogen that has lost the ability to produce pili still be dangerous? INVESTIGATING THE MICROBIAL WORLD 4 provides an answer.

Type IV Pili

Besides playing a role in attachment or adhesion to other cells and surfaces, **type IV pili** provide for a form of locomotion or cell movement called "twitching motility" or "gliding motility" (**FIGURE 4.4B**). These flexible pili extend several micrometers from the bacterial cell and attach to a surface. Then, the

Investigating the Microbial World 4

The Role of Pili

The cell surface of *Escherichia coli* (and many other bacterial species) is covered with tiny hair-like appendages called pili that function to attach the cells to human host cells—in this case, epithelial cells of the gut.

OBSERVATION: There are *E. coli* strains that normally exist in the human gut but cause no ill as long as they stay there. However, other strains, if ingested from a contaminated food source, can cause illness and disease (such as diarrhea). Therefore, what a group of researchers at Stanford University wanted to investigate is whether the pili on an enteropathogenic (intestinal disease-causing) strain of *E. coli* are necessary to initiate the disease process.

QUESTION: *Is* E. coli *attachment by pili to the intestinal epithelium required for the onset of gastrointestinal disease (diarrhea)?*

HYPOTHESIS: Pili-mediated adhesion of *E. coli* cells is necessary for the infection process leading to disease. If correct, mutants of *E. coli* that lack pili should not cause diarrhea, whereas mutants with an excessive number of pili should cause a more intense illness.

EXPERIMENTAL DESIGN: First, what was needed was a marker for infection. With the strain of *E. coli* used, this was easy—a good case of diarrhea would develop (as measured by the volume of liquid stools produced). Next, the Stanford researchers needed experimental subjects who would not complain if they developed a case of diarrhea. Therefore, a tolerant group of healthy Stanford student volunteers was enrolled in the study—and paid $300 for their "contribution" to science! The volunteers were randomized into separate groups, each group receiving a slightly different form or dose of *E. coli.* Neither the volunteers nor the Stanford researchers knew who was drinking which liquid mixture; it was a so-called double-blind study. A number of nurses and doctors were on hand to help the volunteers through their ordeal.

EXPERIMENT 1: Each of three groups drank three doses of a fruit-flavored cocktail containing either a mutant strain (muT) of *E. coli* having few pili, another mutant strain (muA) also having few pili, or the diarrhea-causing strain (wild- type; wt) with normal numbers of pili. Each group was further divided into three subgroups based on the dose they would receive (5 × 10^8, 2.5 × 10^9, or 2 × 10^{10} *E. coli* cells). Over 48 hours, the cumulative volume of liquid stool collected from each volunteer was recorded.

EXPERIMENT 2: Two additional groups drank three doses of a fruit-flavored cocktail containing a mutant strain (muF) of *E. coli* having excessive pili that causes the cells to aggregate together in culture. One group received doses of 2.5×10^9 cells, whereas the other group received doses of 2×10^{10} cells. Again, the volume of liquid stool collected from each volunteer over a 48-hour period was recorded.

RESULTS: See figure. In the figure, each vertical bar represents one volunteer and the cumulative volume of liquid stool. The short bars below the horizontal X axis represent volunteers who produced no liquid stools in the 48-hour period of the experiments.

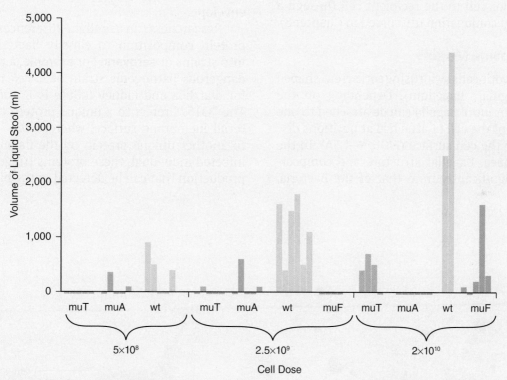

Modified from Bieber, D., et al. 1998. *Science* 280(5372):2114–2118.

CONCLUSIONS:

QUESTION 1: *Was the hypothesis validated? Explain using the figure and identify the control in experiment 1.*

QUESTION 2: *In experiment 1, explain why the volunteers drinking the cocktail with muT and muA did not suffer severe diarrhea. Were there any anomalies within these groups?*

QUESTION 3: *From experiment 2, propose a reason why the volunteers drinking the cocktail with muF did not suffer severe diarrhea. Were there any anomalies within these groups?*

You can find answers online in **Appendix E**.

Modified from Bieber, D., et al. 1998. *Science* **280**(5372):2114–2118.

pili retract (disassemble) from the base, pulling the bacterial cell forward. For example, *N. gonorrhoeae* cells (the causative agent of gonorrhea) contain type IV pili that attach the bacterial cells to the epithelial surface of the urogenital tract. Retraction of the pili pulls the cells forward, which brings the pathogens deeper into the urogenital tract. As a result, an infection is more likely.

Conjugation Pili

Some bacterial species produce flexible **conjugation pili**. These pili are longer than the other pili and only one or a few are produced on a cell. Conjugation pili establish contact with appropriate recipient bacterial cells, facilitating the transfer of DNA from the donor cell to the recipient cell through a process called conjugation (discussed in Chapter 8).

Flagella Provide Motility

Many prokaryotic cells swim using corkscrew-shaped **flagella** (singular: flagellum). Depending on the species, one or more flagella can be attached to one or both ends of the cell or attached at positions distributed over the cell surface (**FIGURE 4.5A**). In the domain Archaea, flagellar structure and composition differ significantly from that of the Bacteria.

However, this so-called archaellum also rotates, and the organelle is used for cell movement.

Bacterial flagella range in length from 10 μm to 20 μm and are many times longer than the diameter of the cell. Because they are only about 25 nm thick, they cannot be seen with the light microscope unless stained or special optics are used, such as phase-contrast or dark field.

When examined with a transmission electron microscope, each bacterial flagellum is found to be composed of three subunits: a helical filament, hook, and basal body (**FIGURE 4.5B**). The hollow filament is composed of long, rigid strands of protein, and the flagellum is not surrounded by the cell membrane. The curved, flexible hook connects the filament to a basal body (motor) anchored in the cell envelope.

Researchers can use slight differences in flagella protein composition to classify bacterial species into strains or **serovars**. For example, a particularly dangerous pathogenic strain of *E. coli* responsible for diarrhea and kidney failure is *E. coli* O157:H7. The "O157" refers to a unique protein on the bacterial membrane surface, whereas the "H7" refers to another unique protein on the flagellum. In an infected individual, these proteins trigger antibody production that can be detected in the blood.

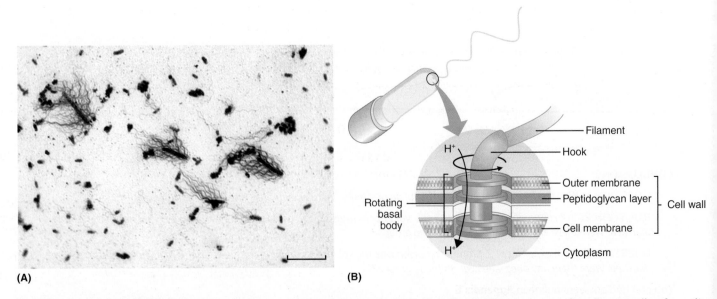

(A)   **(B)**

FIGURE 4.5 Bacterial Flagella. (A) A light micrograph of stained *Proteus vulgaris* cells showing numerous flagella extending from the cell surface. (Bar = 10 μm.) Note that the length of a flagellum is many times the width of the cell. **(B)** The flagellum on a bacterial cell such as *Escherichia coli* is attached to the cell wall and membrane by two pairs of protein rings in the basal body. *»» Why is the flagellum referred to as a rotary motor?*

(A) Courtesy of Dr. Jeffrey Pommerville.

The basal body is an assembly of more than 20 different proteins that form a central rod and set of enclosing rings embedded in the cell membrane and in the cell wall.

The basal body represents a powerful biological motor or rotary engine that generates a propeller-type rotation of the rigid filament. The energy for rotation comes from the diffusion of protons (hydrogen ions; H^+), which is facilitated by membrane proteins associated with the basal body. This energy is sufficient to generate up to 1,500 RPM by the filament and produce speeds of 50 or more cell lengths per second (the fastest human can run about 5 to 6 body lengths per second). MICROFOCUS 4.2 investigates how bacterial flagella might have evolved.

MICROFOCUS 4.2: Evolution

The Origin of the Bacterial Flagellum

Flagella are an assembly of protein parts forming a rotary engine that, like an outboard motor, propel the cell forward through its moist environment. Studies on the bacterial flagellum have shown how such a nanomachine might have evolved.

Several bacterial species, including *Yersinia pestis*, the causative agent of bubonic plague, contain structures to inject toxins into an appropriate eukaryotic host cell. These bacterial cells have a hollow tube or needle to accomplish this process, just as the bacterial flagellum and filament are hollow (see the diagram that follows). In addition, many of the flagellar proteins are similar to some of the injection proteins. Investigators discovered that *Y. pestis* cells actually contain all the genes needed for a flagellum—but the cells have lost the ability to use these genes to form a flagellum. *Y. pestis* is nonmotile and it appears that the cells use a subset of the flagellar proteins to build the injection device.

One scenario suggests that an ancient cell evolved a structure that was the progenitor of the injection and flagellar systems. In fact, many of the proteins in the basal body of flagellar and injection systems are similar to proteins involved in proton (hydrogen ion; H^+) transport. Therefore, a proton transport system might have evolved into the injection device and, through diversification events, evolved into the motility structure present on many bacterial cells today.

These investigations demonstrate that structures can evolve from other structures with a different function. It is not necessary that evolution "design" a structure from scratch but rather it can modify existing structures for other functions.

Some individuals have said that the complexity of structures, like a bacterial flagellum, is just too intricate to arise gradually through a step-by-step process. However, the investigations support the idea of modification through evolution and the bacterial flagellum almost certainly falls into that category.

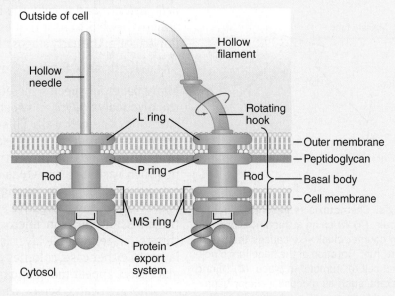

A bacterial injection device (left) compared to a bacterial flagellum (right). Both have a protein export system in the base of the basal body.

Being so small, bacterial cells struggle to move through water, which for them is equivalent to a human trying to swim in a swimming pool full of molasses. Therefore, the flagella act like propellers to provide the rotational propulsion to push the cell forward through the fluid, be it water in the ocean, broth in a culture tube, or the fluids in the human gut. When the flagellum, or a bundle of flagella, rotates counterclockwise, the cell moves forward, in what is called a "run" (**FIGURE 4.6A**). These runs can last a few seconds and the cells can move up to

10 body lengths per second. A reversal of flagellar rotation (clockwise rotation) causes the cell to "tumble" randomly for a second as the flagella become unbundled and uncoordinated. Tumbling results in a random change of direction, so when the flagella again rotate counterclockwise, another run occurs, sending the cell off in a new direction.

When searching for nutrients (e.g., simple sugars, amino acids) in an environment, a flagellated bacterial cell can manipulate its runs and tumbles so it moves toward the nutrient (attractant). This process, called **chemotaxis**, biases the cell, such that it moves up the chemical gradient by undergoing longer runs and shorter tumbles (**FIGURE 4.6B**). Similar types of motile behavior are seen in photosynthetic prokaryotes moving toward light (phototaxis) or in other cells moving toward oxygen gas (aerotaxis).

One additional type of flagellar organization is found in the spirochetes, the group of coiled bacterial species whose shape and form were described earlier in this chapter. The cells are motile by flagella that extend from one or both poles (ends) of the cell but fold back along the cell body (**FIGURE 4.7**). Such **endoflagella** lie in an area called the "periplasm" (see the discussion on the cell wall that follows). Motility results from the twisting generated on the cell by the normal rotation of the flagella; in other words, the cell swims in a corkscrew motion.

Before leaving the discussion of bacterial flagella, you should be aware of another function flagella might have, especially in a pathogenic context. In several **enteric** (intestinal) pathogens (e.g., *Salmonella enterica* and *E. coli*), the flagella can be used to adhere the cells to host tissues and thus, like pili, are important in the early stages of disease development.

The Glycocalyx Serves Several Functions

Many bacterial species secrete a sticky layer called the **glycocalyx** (*glyco* = "sweet"; *calyx* = "coat") that surrounds the cell wall. This layer, which consists of polysaccharides, or polysaccharides and small proteins, can be thick and firmly bound to the cell, in which case it is known as a **capsule**. The actual capsule can be seen by light microscopy when observing cells in a negative stain preparation or by transmission electron microscopy (**FIGURE 4.8**). A thinner, water-soluble layer is referred to as a **slime layer**. In either case, colonies containing cells with a glycocalyx appear moist and glistening.

The glycocalyx serves as an extracellular layer between the cell and the external environment. Because of its high water content, the glycocalyx can protect cells from desiccation. The glycocalyx

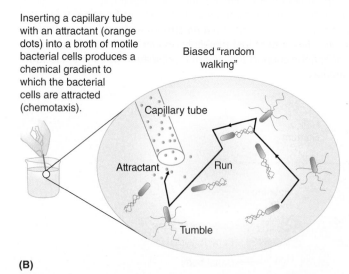

(A)

Inserting an empty capillary tube into a broth of motile bacterial cells does not cause chemotaxis.

"Random walking"

Capillary tube

No attractant

Tumble   Run

(B)

Inserting a capillary tube with an attractant (orange dots) into a broth of motile bacterial cells produces a chemical gradient to which the bacterial cells are attracted (chemotaxis).

Biased "random walking"

Capillary tube

Attractant   Run

Tumble

FIGURE 4.6 Chemotaxis. Chemotaxis represents a behavioral response to chemicals. **(A)** When no attractant is present, rotation of the flagellum counter clockwise causes the bacterial cell to swim in a short "run." Rotation of the flagellum clockwise causes the bacterial cell to "tumble" in place. **(B)** During chemotaxis to an attractant, such as glucose, cellular behavior leads to longer runs and fewer tumbles, which will result in biased movement toward the attractant. *»» Predict the behavior of a bacterial cell if it sensed a repellant, that is, a potential harmful or lethal chemical.*

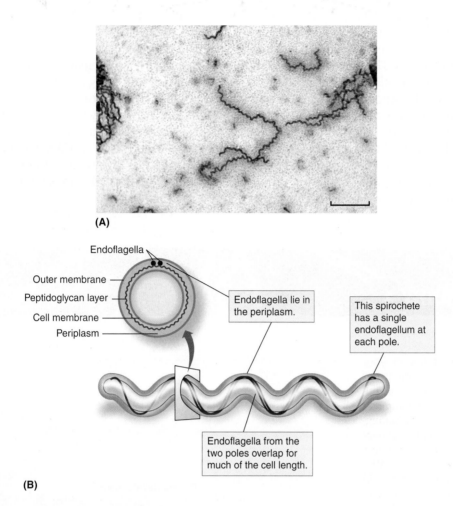

(A)

(B)

FIGURE 4.7 The Spirochete Endoflagella. (A) A light micrograph of *Treponema pallidum* shows several corkscrew-shaped spirochete cells. (Bar = 10 μm.) **(B)** Diagram showing the positioning of endoflagella in a spirochete. *»» How are endoflagella different from true bacterial flagella?*

(A) Courtesy of Dr. Jeffrey Pommerville.

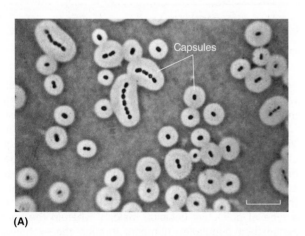

(A)

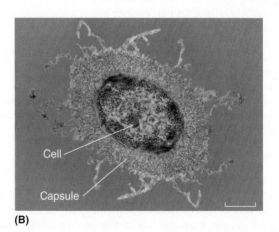

(B)

FIGURE 4.8 The Bacterial Glycocalyx. (A) The capsule surrounding *Acinetobacter* cells can be seen after negative staining. (Bar = 10 μm.) **(B)** A false-color transmission electron micrograph of *Escherichia coli.* The cell is surrounded by a thick capsule (pink). (Bar = 0.5 μm.) *»» How does the capsule provide protection for the bacterial cell?*

(A) Courtesy of Elliot Juni, Department of Microbiology and Immunology, The University of Michigan. (B) © George Musil/Visuals Unlimited.

also allows cells to stick to surfaces. The capsule of *V. cholerae*, for example, enables the cells to attach to the intestinal wall of the host. The glycocalyx of pathogens therefore represents another virulence factor.

Other encapsulated pathogens, such as *Streptococcus pneumoniae* (a principal cause of bacterial pneumonia) and *Bacillus anthracis*, evade the immune system because they cannot be easily engulfed by white blood cells during phagocytosis. Scientists believe the repulsion between pathogen and white blood cells is due to strong negative electrical charges on the capsule and the surface of the white blood cell.

S. mutans, a principal cause of tooth decay, has a slime layer containing a mass of tangled polysaccharide fibers called **dextran**. The fibers provide the bacterial cells with a way to attach to the tooth surface. This species forms part of the dental plaque, which represents a type of **biofilm** on the tooth surface. **MICROINQUIRY 4** describes the nature of a biofilm, and **CLINICAL CASE 4** details a medical consequence of a biofilm.

MICROINQUIRY 4

Biofilms

What do the microbiota in your gut, the presence of tooth tartar (plaque), and development of a middle ear infection have in common? They all are examples of a **biofilm**, a "multicellular community" of bacteria and perhaps other microbes embedded in a gelatinous matrix and often attached to a surface. In fact, around 80% of bacterial species live in clusters within a highly organized biofilm.

A biofilm forms when planktonic cells initially attach to a living or nonliving surface by weak electrostatic forces and then more permanently, using pili, flagella, and/or a glycocalyx (see figure). As they colonize the surface, their population size increases through cell division as the cells secrete an extracellular matrix of proteins, polysaccharides, and DNA (commonly referred to as "slime") in which the cells become embedded. At maturity, the biofilm is like a living tissue with a primitive circulatory system made of water channels that bring in nutrients and eliminate wastes.

Antibiotic Resistance

One of the hallmarks of a biofilm is its tolerance to chemicals, such as antibiotics and other antimicrobial compounds. Within the biofilm, there are gradients of nutrients and oxygen from the top to the bottom of the biofilm. These gradients are associated with decreased bacterial metabolic activity, which, along with the dense and thick slime, makes the cells less susceptible to antibiotics and host immune defenses. Thus, many human wound and chronic infections are the result of biofilm development.

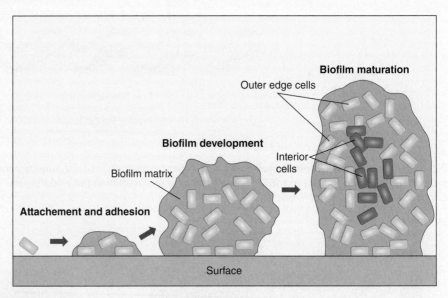

The biofilm lifecycle is an example of intercellular cooperation in the development of a multicellular structure.

For example, biofilms are associated with cystic fibrosis where the build up of mucus provides a suitable environment for the bacterial pathogen *Pseudomonas aeruginosa*.

The cells in a biofilm can coordinate themselves by "listening to" and "talking with" neighboring cells through chemical communication. In some ways, it is not that different from our electronic communications in human social networking through Twitter and Facebook. Thus, the bacterial process, called **quorum sensing** (**QS**), is a type of collective decision making, wherein the cells sense their population numbers through the exchange of extracellular chemicals. When these molecules reach a critical threshold, the community of cells acts together and, depending on the species, gene regulation triggers a specific behavioral response.

TABLE 4.1 Comparison of Bacterial Cell Wall Composition

Characteristic	Gram Positive	Gram Negative	Acid Fast
Peptidoglycan	Yes, thick layer	Yes, thin layer	Yes
Teichoic acids	Yes	No	No[1]
Outer membrane	No	Yes	Yes
Mycolic acid	No	No	Yes
Lipopolysaccharides (LPS)	No	Yes	No
Porin proteins	No	Yes	Yes
Periplasm	No	Yes	Yes

[1]Have a different type of glycopolymer.

species is the same as in bacterial species—to provide mechanical support and prevent osmotic lysis.

The Cell Membrane Represents a Selectively Permeable Barrier

A **cell (plasma) membrane** is a universal structure that separates the external environment from the internal (cytoplasmic) environment, preventing soluble materials from simply diffusing into and out of the cell.

The bacterial cell membrane, which is about 7 nm thick, is 25% phospholipid (by weight) and 75% protein. Although in illustrations the cell membrane appears very rigid (**FIGURE 4.11**), in reality it is quite fluid, having the consistency of olive oil. This means that the assortment (mosaic) of phospholipids and

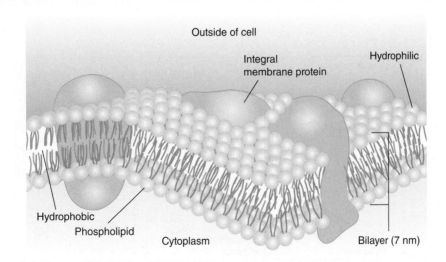

FIGURE 4.11 The Structure of the Bacterial Cell Membrane. The cell membrane of a bacterial cell consists of a phospholipid bilayer with embedded integral membrane proteins. Other proteins and ions may be associated with the integral proteins or the phospholipid heads. *»» Why is the cell membrane referred to as a fluid mosaic structure?*

proteins making up the bilayer are not trapped in place; rather, they can move laterally in the membrane. This dynamic model of membrane structure therefore is called the **fluid mosaic model**.

Phospholipids. The phospholipid molecules, typical of most biological membranes, are arranged as a bilayer and they represent the barrier function of the membrane. The phospholipids contain a charged phosphate head group attached to two hydrophobic fatty acid chains (see Chapter 2). The fatty acid "tails" are the segment that forms the permeability barrier. In contrast, the hydrophilic head groups are exposed to the aqueous external and to the cytoplasmic environments.

Several antimicrobial substances act on the cell membrane. The antibiotic polymyxin makes holes in the bilayer, whereas some detergents and alcohols dissolve the bilayer. Such actions allow the cytoplasmic contents to leak out of the cell, resulting in death through cell lysis.

Membrane Proteins. A diverse population of membrane proteins populates the phospholipid bilayer. Those membrane proteins that span the width of the bilayer are referred to as "integral membrane proteins." Other membrane proteins, called "peripheral membrane proteins," are associated with the polar heads of the bilayer or exposed parts of integral membrane proteins.

Membrane proteins carry out numerous important functions, some of which are absent from the eukaryotic plasma membrane. This includes enzymes needed for cell wall synthesis and energy metabolism. Because prokaryotic cells lack mitochondria, part of ATP generation in these cells is carried out by the cell membrane. Other membrane proteins help anchor the DNA to the membrane during replication. However, like membrane proteins in eukaryotic cells, membrane proteins in the prokaryotic cells act as receptors of chemical information, sensing changes in environment conditions and triggering appropriate behavioral responses by the affected cell.

Membrane Transport

Many integral membrane proteins are involved as transporters; that is, they control the movement of charged solutes, such as amino acids, simple sugars, and ions across the lipid bilayer. These transport proteins are highly specific and each type of transport protein only transfers a single molecular type or a very similar class of molecules. Therefore, many different transport proteins are needed to regulate the diverse molecular traffic that must flow into or out of a cell. This transport process can be passive or active.

In **facilitated diffusion**, integral membrane proteins facilitate (assist) the movement of materials along a concentration gradient, that is, from an area of higher concentration to one of lower concentration (**FIGURE 4.12**). By acting as a channel for diffusion, hydrophilic solutes can enter or leave the cell passively, that is, without the need for cellular energy (ATP).

Unlike facilitated diffusion, **active transport** allows different concentrations of solutes to be established outside or inside of the cell against the concentration gradient. These membrane proteins

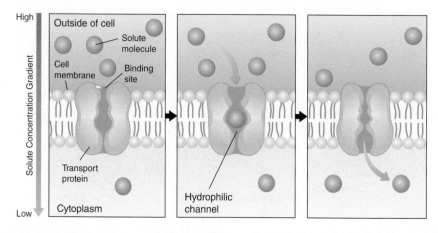

FIGURE 4.12 Facilitated Transport Through a Membrane Protein. Many transport proteins facilitate the diffusion of nutrients across the lipid bilayer. Each transport protein forms a hydrophilic channel through which a specific solute can diffuse. »» *Why wouldn't a solute simply diffuse across the lipid bilayer?*

act as "pumps" and, as such, demand an energy input from the cell. Cellular processes, such as cell energy production and flagella rotation, also depend on active transport.

Vesicle Flow

Until recently, it was believed that only eukaryotic cells produced small, membrane-bound spheres called **vesicles**, which are used to shuttle materials through the cell cytoplasm and to the plasma membrane. Recently, microbiologists discovered that most, if not all, gram-negative bacterial cells release **extracellular vesicles** (**EVs**) from the outer membrane (**FIGURE 4.13**). These EVs are 20 to 200 nm in diameter and they contain periplasmic molecules, including proteins and DNA fragments. Although their exact roles are not known yet, the EVs might transport DNA fragments between organisms and in some way be important to pathogenesis.

EVs also have been reported in gram-positive bacteria and members of the domain Archaea, which suggests that vesicle flow from a cell is a common phenomenon among organisms in all three domains of the tree of life.

Although many species within the domain Archaea have cell membranes similar in structure to those of their bacterial counterparts, other archaeal

FIGURE 4.13 Outer Membrane Vesicles. This scanning electron microscope micrograph shows vesicles "pinching off" the cell surface (arrow) of the cyanobacterium *Prochlorococcus*. (Bar = 5 μm.) »» *Would such vesicles be released from the cell surface of gram-positive bacterial cells? Explain.*

Courtesy of Dr. Steven Biller.

extremophiles have evolved a different lipid structure presumably to withstand the extreme environmental conditions (e.g., high temperatures) in which they often exist. MICROFOCUS 4.3 examines their membrane structure.

MICROFOCUS 4.3: Evolution

An Archaeal "Survival Suit"

Microbes would be unimpressed by Marvel Comics' Iron Man and his suit of armor. Take the archaeal cell membrane as one example.

A membrane surrounding the cytoplasm is a universal structure found on all prokaryotic and eukaryotic cells. As described in this chapter, this consists of phospholipids and proteins. For the bacterial and eukaryotic organisms, the cell membrane has a standard structure. However, in the domain Archaea, a unique membrane exists in many species.

One difference in the cell membrane is how the lipid tails are attached to the glycerol (figure **A**). In bacterial and eukaryotic species, the tails are bound to glycerol by "ester linkages." Archaeal species, on the other hand, have "ether linkages" to the glycerol.

The second difference concerns the actual lipid tails (figure **B**). The bacterial and eukaryotic species have fatty acid tails that form the bilayer of the membrane. Fatty acid tails are absent from the membranes of some archaeal species; instead, repeating five-carbon units are linked end-to-end to form a bilayer of lipid tails longer than the fatty acid tails found in members of the Bacteria and Eukarya. Further, in some archaeal species, the lipid bilayer has been constructed as a monolayer.

So, why the differences? Many archaeal species live in extremely hot, salty, and/or acidic locations—environments that are chemically very harsh. In these environments, ether linkages are more stable to chemical attack than are the ester-containing linkages in the Bacteria and Eukarya membranes. Further, lipid length adds rigidity to the membrane and a monolayer keeps the membrane intact in extreme conditions; the halves of a typical membrane lipid bilayer would separate under these conditions.

(continues)

MICROFOCUS 4.3: Evolution (*Continued*)

An Archaeal "Survival Suit"

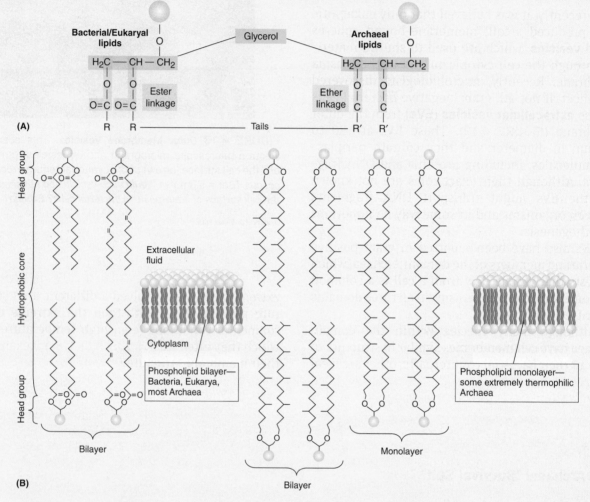

(A) Bacterial and eukaryotic cell membranes involve an ester linkage joining the glycerol to the fatty acid tails (R) while archaeal membranes have an ether linkage to the tails (R?). **(B)** Bacterial and eukaryotic membranes form a bilayer while archaeal membranes may be a bilayer or a monolayer.

As such, the archaeal cells have evolved a survival suit; they have a membrane structure that can better withstand high temperatures and resist strong acid conditions in which many archaeal species live. Therefore, if there is an environment with nutrients and energy, microbes will come and colonize, provided they have the needed "survival suit."

Concept and Reasoning Checks 4.4

a. Penicillin and lysozyme primarily affect peptidoglycan synthesis in gram-positive bacterial cells. Why are these agents less effective against gram-negative bacterial cells?

b. Justify the necessity for phospholipids and proteins in the cell membrane.

c. Draw the process by which facilitated diffusion and active transport occur, showing how the two transport processes differ.

KEY CONCEPT 4.5   The Cell Cytoplasm Is Packed with Internal Structures

The cell membrane encloses the **cytoplasm**, which is the compartment within which most growth and metabolism occurs. The cytoplasm consists of the **cytosol**, a semifluid mass of proteins, amino acids, sugars, nucleotides, salts, vitamins, and ions dissolved in water. Although prokaryotic cells lack the complex membrane compartments typical of eukaryotic cells, they still contain several bacterial structures, subcompartments, and filaments.

The Nucleoid Represents a Subcompartment Containing the Chromosome

The chromosome region in prokaryotic cells appears as a diffuse mass termed the **nucleoid** (**FIGURE 4.14**). The nucleoid is not surrounded by a membrane envelope; rather, it represents a subcompartment in the cytoplasm that is less densely packed with proteins than the cytoplasm and is devoid of ribosomes. Usually there is a single chromosome and, with few exceptions, the chromosome exists as a closed loop of DNA and protein.

The DNA contains the essential hereditary information for cell growth, metabolism, and reproduction. Because most prokaryotic cells only have one chromosome and one set of genes, the cells technically are **haploid**. Unlike eukaryotic microorganisms and other eukaryotes, the nucleoid and

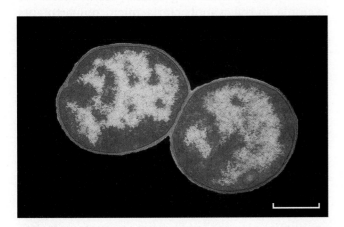

FIGURE 4.14 The Bacterial Nucleoid. In this false-color transmission electron micrograph of *Enterococcus faecalis*, nucleoids (yellow) occupy a large area in these two cells. The red area represents the cytoplasm and ribosomes. (Bar = 0.3 μm.) »» *What does the blue area surrounding each cell represent?*

© Kwangshin Kim/Science Source.

chromosome do not undergo mitosis during cell division and, having but the one set of genetic information, cannot undergo meiosis.

The complete set of genes in an organism or virus is called the **genome**. For example, the genome of *E. coli*, a bacterial species in the mid-size range, contains about 4,300 genes. In all cases, these genes determine what proteins and enzymes the cell can make, that is, what metabolic reactions, activities, and behaviors can be carried out. For *E. coli*, this equates to some 2,000 different proteins.

Plasmids Are Found in Many Prokaryotic Cells

Besides a nucleoid, many bacterial and archaeal cells contain smaller molecules of DNA called **plasmids**. About a tenth the size of the chromosome, these stable, extrachromosomal DNA molecules exist as closed loops containing 5 to 100 genes. There can be one or more plasmids in a cell and these can contain similar or different genes. Plasmids replicate independently of the chromosome and can be transferred between cells during genetic recombination. In industrial microbiology and genetic engineering, plasmids represent important vehicles to carry foreign genetic material into another cell.

Although plasmids are not essential for cellular growth, they provide a level of genetic flexibility. For example, some plasmids possess genes for disease-causing toxins and many carry genes for chemical or antibiotic resistance.

Other Subcompartments Exist in the Cell Cytoplasm

For a long time the cytoplasm and cytosol of prokaryotes were looked at as simply a formless, watery bag enclosing the genetic machinery and biochemical reactions. As more studies were carried out by using the electron microscope and via biochemical techniques, it became evident that the cytoplasm contained "more than meets the eye." Because one of the major tasks of any cell is protein synthesis, more than 30% of the molecules in the bacterial cell cytoplasm function directly or indirectly in making proteins.

Ribosomes

One of the universal structures in all cells is the **ribosome**. There are thousands of these nearly spherical

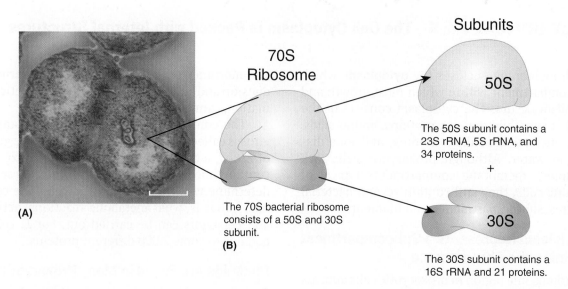

FIGURE 4.15 The Bacterial Ribosome. (A) A false-color transmission electron micrograph of *Neisseria gonorrhoeae*, showing the nucleoid (yellow) and ribosomes (blue). (Bar = 1 μm.) **(B)** The functional 70S ribosome is assembled from a small (30S) and large (50S) subunit, each of which contains dozens of proteins. *»» How many rRNA molecules and proteins construct a 70S ribosome?*

(A) © Science Source.

particles packing the cell cytosol, which gives the cytoplasm a granular appearance when viewed with the electron microscope (**FIGURE 4.15A**). In addition, there also are ribosomes loosely associated with the cell membrane. The free ribosomes make soluble proteins that are used in the cell, whereas the membrane-associated ribosomes produce proteins for the cell envelope and for secretion.

The relative size of ribosomes is measured by how fast they settle when spun in a centrifuge. Measured in Svedberg units (S), prokaryotic ribosomes represent 70S particles. Each 70S ribosome is composed of a small subunit (30S) and a large subunit (50S) (**FIGURE 4.15B**). Each subunit is associated with one or more ribosomal RNAs and dozens of proteins. For protein synthesis to occur, the two subunits come together to form the 70S functional ribosome.

Prokaryotic and eukaryotic cells differ somewhat in size and in the RNA and protein molecules they contain. Consequently, these differences enable some antibiotics, such as streptomycin and tetracycline, to bind to prokaryotic ribosomes and prevent protein synthesis; these antibiotics do not bind to eukaryotic ribosomes. In addition, the separation of prokaryotes into the domains Bacteria and Archaea was based, in part, on sequencing the ribosomal gene from the small subunit (specifically, the 16S RNA).

Micro- (and Nano-) Compartments

Some bacterial species contain **microcompartments** that appear to be unique to the domain Bacteria. They are called microcompartments because each compartment consists of a thin polyhedral, porous protein shell 100 to 200 nm in diameter (**FIGURE 4.16**). These subcompartments help optimize metabolic pathways by spatially segregating metabolism. For example, in the cyanobacteria, microcompartments called **carboxysomes** function to enhance carbon dioxide fixation. In some nonphotosynthetic species, microcompartments prevent the diffusion of volatile or toxic metabolic products from entering and damaging the cytoplasm of the cell. In general, microcompartments represent localized areas where enzymes can more directly interact with their substrates or potentially sequester harmful reaction products of metabolism.

Many bacteria also contain other protein-bound structures called **nanocompartments**. These subcompartments are only 22 to 32 nm in diameter and are unrelated to the larger microcompartments. It is believed that these structures store iron and regulate the use of iron in the cytoplasm.

Inclusion Bodies

Subcompartments called **inclusion bodies** are found in the cytosol of some prokaryotic cells. Many of

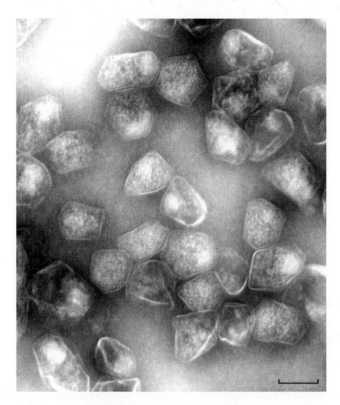

FIGURE 4.16 Microcompartments. Purified bacterial microcompartments from *Salmonella enterica* are composed of a complex protein shell that encases metabolic enzymes. (Bar = 100 nm.) »» *How do these bacterial microcompartments differ structurally from a eukaryotic organelle?*

Reprinted with permission from the American Society for Microbiology (*Microbe*, January, 2006, p. 20–24). Photo courtesy of Dr. Thomas A. Bobik, Department of Biochemistry, Biophysics and Molecular, Iowa State University.

these bodies store nutrients or the monomers for cellular structures. For example, some inclusion bodies function as nutrient reserves. They consist of aggregates or granules of polysaccharides (glycogen), globules of elemental sulfur, or lipid. Other inclusion bodies serve as important identification characters for bacterial pathogens. One example is the diphtheria bacilli that contain inclusion bodies called **metachromatic (volutin) granules**, which are deposits of polyphosphate (long chains of inorganic phosphate) along with calcium and other ions. These granules stain with dyes such as methylene blue.

Gas Vesicles

Some cyanobacteria maintain buoyancy by using **gas vesicles**, which are subcompartments built from a watertight protein shell. These vesicles are filled with a gas, which by decreasing the density of the cell, function to regulate the buoyancy and position of the cell in the water column.

Magnetosomes

The **magnetosomes**, another type of subcompartment, are an example of a truly membrane-bound organelle (**FIGURE 4.17**). Found in some aquatic, gram-negative bacterial species, each magnetosome represents an invagination of the cell membrane and contains a chain of assembled magnetic iron minerals (Fe_3O_4 or Fe_3S_4). These motile bacteria are common in aquatic sediments where they use the magnetosome as a navigation device to locate optimal areas in sediments having little or no oxygen gas.

Cytoskeletal Proteins Regulate Cell Division and Help Determine Cell Shape

Until recently, the belief was that bacterial and archaeal cells lacked a cytoskeleton, which is a common feature in eukaryotic cells. However, it is now clear that cytoskeletal proteins resembling those in the eukaryotic cytoskeleton are present in prokaryotic cells.

Some prokaryotic cytoskeletal filaments function in the regulation of cell division. During this

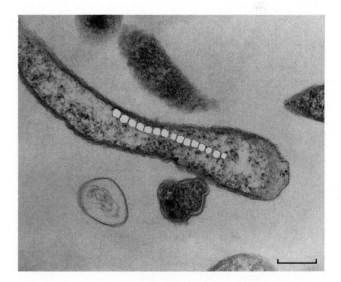

FIGURE 4.17 Bacterial Magnetosomes. Bacterial magnetosomes (yellow) are seen in this false-color transmission electron micrograph of a magnetotactic marine spirillum. (Bar = 1 μm.) »» *What is the unique feature to the composition of magnetosomes?*

© Dennis Kunkel Microscopy, Inc./Visuals Unlimited.

FIGURE 4.18 A Bacterial Cytoskeleton. Bacterial cells have proteins similar to those that form the eukaryotic cytoskeleton. The protein fibers form helical filaments that curve around the edges of these *Bacillus subtilis* cells. (Bar = 1.5 μm.) »» *Predict the shape of these cells if the cytoskeletal proteins were absent.*

Courtesy of Rut Carballido-López, INRA.

process, a cytoskeletal protein localizes around the neck of the dividing cell where it recruits other proteins needed for the deposition of a new cell wall between the dividing cells. Other filaments under the cell membrane help determine cell shape in rod-shaped cells like *E. coli* and *Bacillus subtilis* (**FIGURE 4.18**). Yet other cytoskeletal filaments are involved with chromosome segregation during cell division and magnetosome formation.

TABLE 4. 2 summarizes the structural features of prokaryotic cells.

TABLE 4.2 A Summary of the Structural Features of Prokaryotic Cells

Structure	Chemical Composition	Function	Comment
External Structures			
Pilus	Protein	Attachment to surfaces Cell movement Genetic transfer	Found primarily in gram-negative bacteria
Flagellum	Protein	Motility	Present in many rods and spirilla; few cocci; vary in number and placement
Glycocalyx	Polysaccharides and small proteins	Buffer to environment Attachment to surfaces	Capsule and slime layer Contributes to disease development Found in plaque bacteria and biofilms
Cell Envelope			
Cell wall		Cell protection Shape determination Cell lysis prevention	
Bacterial	Gram positives: thick peptidoglycan and teichoic acid Gram negatives: little peptidoglycan and an outer membrane		Site of activity of penicillin and lysozyme Gram-negative bacteria release lipid A (toxin)
Archaeal	Pseudopeptidoglycan Protein		S-layer
Cell membrane			
Bacterial	Protein and phospholipid	Cell boundary Transport into/out of cell Site of enzymatic reactions	Lipid bilayer
Archaeal			Lipid bilayer/monolayer

TABLE 4.2 A Summary of the Structural Features of Prokaryotic Cells (*Continued*)

Structure	Chemical Composition	Function	Comment
Internal Structures			
Nucleoid	DNA	Site of essential genes	Exists as single, closed loop chromosome
Plasmid	DNA	Site of nonessential genes	Might contain antibiotic resistance genes
Ribosome	RNA and protein	Protein synthesis	Inhibited by certain antibiotics
Microcompartment	Various metabolic enzymes	Carbon dioxide fixation Retention of volatile or toxic metabolites	Enzymes are enclosed in a protein shell
Nanocompartment	Iron	Regulation of iron usage	Other functions yet to be identified
Inclusion body	Glycogen, sulfur, lipid	Nutrient storage	Used as nutrients during starvation periods
Metachromatic granule	Polyphosphate	Storage of polyphosphate and calcium ions	Found in diphtheria bacilli
Gas vacuole	Protein shell	Buoyancy	Helps cells remain buoyant
Magnetosome	Magnetic iron minerals	Cell orientation	Helps locate preferred habitat
Cytoskeleton	Proteins	Cell division, chromosomal segregation, cell shape	Functionally similar to eukaryotic cytoskeletal proteins

Concept and Reasoning Checks 4.5

a. What properties distinguish the bacterial chromosome from a plasmid?
b. Provide the roles for the subcompartments found in bacterial cells.
c. Identify some of the roles for the prokaryotic cytoskeleton.

Chapter Challenge C

In the preceding three sections of the chapter, you discovered the exterior structures, studied the organization of the cell envelope, and surveyed the cytoplasmic structures in prokaryotic cells.

QUESTION C: *The chapter challenge was "Studying the diversity of life only accentuates life's unity." Summarize the validity of this statement, using cell structure as your reference.*

You can find answers online in **Appendix F**.

■ KEY CONCEPT 4.6   There Is Tremendous Diversity Among the Domains Bacteria and Archaea

The prokaryotes began their evolutionary history nearly 3.8 billion years ago. Since then, they have adapted and spread through every environment capable of supporting life. Today, there are more than 7,000 known bacterial and archaeal species and a suspected 10 million species. In this section, we highlight a few phyla and groups using the tree of life shown in Figure 4.1.

The Domain Bacteria Contains Some of the Most Studied Microbial Organisms

Today, almost 100 phyla within the domain Bacteria have been identified from culturing or nucleotide sequencing. These organisms have adapted to the diverse environments on Earth, inhabiting the air, soil, and water, and they exist in enormous numbers on the surfaces of virtually all plants and animals. They can be isolated from Arctic ice, thermal hot springs, the fringes of space, and the tissues of animals. Bacterial species, along with their archaeal relatives, have so completely colonized every part of the Earth that their mass is estimated to outweigh the mass of all plants and animals combined. As such, their impact on Earth and its life remains to be fully understood.

Let's conclude this chapter by briefly surveying some of the best-known phyla and other groups.

Phylum Proteobacteria

The **Proteobacteria** (*proteo* = "first") contains the largest and most diverse group of bacterial species. The phylum is made up of five classes, all of which are similar in having a gram-negative cell wall (**FIGURE 4.19A**). The phylum includes some of the most recognized human enteric pathogens, including genera responsible for intestinal inflammation (**gastroenteritis**) (e.g., *Escherichia*, *Shigella*, and *Salmonella*) and cholera (*Vibrio*). Other notable genera include those responsible for gonorrhea (*Neisseria*) and bubonic plague (*Yersinia*).

The Proteobacteria also include the **rickettsias**, which are called obligate, intracellular parasites because they can only reproduce after entering a host cell. Most of these tiny, gram-negative bacterial cells are pathogens and are transmitted among humans primarily by arthropods (e.g., ticks, lice, fleas, mosquitoes). Being obligate pathogens, the rickettsias are cultivated only in living tissues such as chick embryos. Different species cause a number of important human diseases, including Rocky Mountain spotted fever (*Rickettsia rickettsii*) and typhus fever (*R. prowazekii*). It is likely that the mitochondria of the Eukarya evolved through endosymbiosis from a free-living, aerobic rickettsial ancestor of the Proteobacteria.

Gram-Positive Bacteria

The gram-positive bacteria rival the Proteobacteria in diversity and the organisms are divided into two phyla.

The **Firmicutes** (*firm* = "strong"; *cuti* = "skin") are similar in having a thick peptidoglycan layer (thus the term "strong skin"). Although most are nonpathogenic, *Bacillus anthracis* and *Clostridium botulinum* are responsible for anthrax and botulism, respectively. Species within the genera *Staphylococcus* (**FIGURE 4.19B**) and *Streptococcus* are responsible for several mild to life-threatening human illnesses.

Also within the Firmicutes is the genus *Mycoplasma*, which lacks a cell wall but is otherwise taxonomically related to the gram-positive bacterial species (**FIGURE 4.19C**). Among the smallest free-living bacterial organisms, many mycoplasmas are free living in the soil. In addition, a few are pathogens.

Another phylum having gram-positive cell walls is the **Actinobacteria**. These so-called actinomycetes are soil dwellers that form a system of branched filaments containing chains of cells (**FIGURE 4.19D**). Most species are nonpathogenic and function as decomposers. In addition, the genus *Streptomyces* is the source for many important antibiotics. A medically important genus is *Mycobacterium*, one species of which is responsible for tuberculosis.

Phylum Cyanobacteria

The phylum **Cyanobacteria** is unique among bacterial groups because its members carry out photosynthesis similar to unicellular algae and green plants, using the light-trapping pigment chlorophyll. Their evolution on Earth was responsible for the "oxygen revolution" that transformed life on the young planet some 2 billion years ago. In addition, chloroplasts probably evolved through endosymbiosis from a free-living cyanobacterial ancestor.

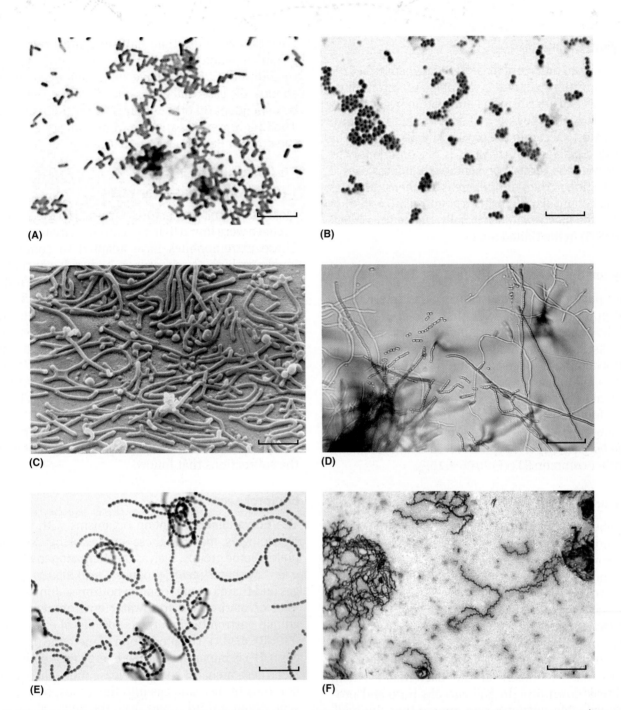

FIGURE 4.19 Members of the Domain Bacteria. (A) *Escherichia coli*. (Bar = 10 μm.) **(B)** *Staphylococcus aureus*. (Bar = 10 μm.) **(C)** *Mycoplasma* species. (Bar = 2 μm.) **(D)** *Streptomyces* species. (Bar = 20 μm.) **(E)** *Anabaena* species. (Bar = 100 μm.) **(F)** *Treponema pallidum*. (Bar = 10 μm.) All images are light micrographs except **(C)**, a false-color scanning electron micrograph. *»» What is the Gram staining result for the* E. coli *and* S. aureus *cells?*

(A, B, E, F) Courtesy of Dr. Jeffrey Pommerville. (C) © Don W. Fawcett/Science Source. (D) Courtesy of Dr. David Berd/CDC.

The members of cyanobacteria are taxonomically related to the gram-positive species and can exist as unicellular, colonial, or filamentous forms (**FIGURE 4.19E**). Their pigmentation gives them a purple, yellow, green, or red (brown) color. The periodic redness of the Red Sea, for example, is due to "blooms" of those cyanobacterial species that contain large amounts of red pigment. In the past decade, there has been an increase in the number and frequency of such blooms in Lake Erie, the Pacific coast, and the Gulf of Mexico.

Phylum Chlamydiae

Roughly half the size of the rickettsias, members of the phylum **Chlamydiae** also are obligate, intracellular parasites and are cultivated only within living animal cells. However, they are not transmitted by arthropods. Most species in the phylum are pathogens. The species *Chlamydia trachomatis* is responsible for the most common form of preventable blindness caused by infection in the world, whereas another subspecies causes chlamydial urethritis, a gonorrhea-like disease that is the most common sexually-transmitted infection (STI) in the United States.

Phylum Spirochaetes

The phylum **Spirochaetes** contains more than 340 gram-negative species that possess a unique cell body that coils into a long helix and moves in a corkscrew pattern. The ecological niches for the spirochetes are diverse, including free-living species found in mud and sediments, symbiotic species present in the digestive tracts of insects, and pathogens found in the urogenital tracts of vertebrates. Among the human diseases are Lyme disease (*Borrelia burgdorferi*) and syphilis (*Treponema pallidum*), another common STI (**FIGURE 4.19F**).

Phylum Bacteroidetes

Members of the diverse phylum **Bacteroidetes** are taxonomically related to the gram-negative bacteria. Some species have flagella for motility. Among the three large families in the phylum, the family Bacteroidaceae typically is found in the gastrointestinal tracts and mucous membranes of warm-blooded animals, including humans. In the human gastrointestinal tract, they and members of the phylum Firmicutes make up about 98% of the intestinal microbiome. Members of the Bacteroidetes help to break down protein and carbohydrate and produce valuable nutrients and energy that the body needs. They also affect immune system function and impact the overall health of the intestinal tract.

Other Phyla

There are many other phyla within the domain Bacteria. Recently, 35 new groups ("supergroup") of identified organisms, called the "Candidate Phylum Radiation" (CPR), have greatly expanded the domain Bacteria (see Figure 4.1). Although they appear to be bacterial cells, they are quite different from the other "traditional" bacterial groups. For example, more than half the genes in the CPR groups are not found in any of the other groups. In addition, CPR cells have small genomes missing supposedly essential genes, meaning that they must be dependent on other organisms for their survival. The CPR represents another whole "lifestyle" within the domain Bacteria that remains to be explored.

The Domain Archaea Contains Organisms with Diverse Characteristics

Most of the first prokaryotes assigned to the domain Archaea were found living in extreme environments. These extremophiles have adapted to conditions involving high temperatures, high salt concentrations, or extremes of pH. However, since the initial discoveries, many more species have been discovered that thrive under more modest but diverse conditions such as soil and ocean surface waters. No known species cause disease in any plants or animals. As described earlier in this chapter, although they are prokaryotes, they have many characteristics that are more related with the organisms in the domain Eukarya.

The archaeal genera can be placed into one of five phyla. The two most studied are described in the subsections that follow.

Euryarchaeota

The **Euryarchaeota** contain organisms with varying physiologies, many of these species being extremophiles. Some groups, such as the **methanogens** (*methano* = "methane"; *gen* = "produce"), are killed by oxygen gas and therefore are found in bottom sediments and mud of marine and freshwater environments (and animal gastrointestinal tracts) devoid of oxygen gas (**FIGURE 4.20A**). The production of methane (natural) gas (CH_4) is important in their energy metabolism. In fact, these archaeal species release more than 2 billion tons of methane gas into the atmosphere every year. About a third comes from the archaeal species living in the stomach (rumen) of cows.

Another group of archaeal organisms is the **extreme halophiles** (*halo* = "salt"; *phil* = "loving"). They are distinct from the methanogens in that they require oxygen gas for energy metabolism and need high concentrations of salt (up to 36% NaCl) to grow and reproduce. Therefore, they often are found in some of the saltiest places on Earth, such as the Great Salt Lake in Utah, the Dead Sea in the Middle East, and salt evaporation ponds (**FIGURE 4.20B**).

A third group in the phylum Euryarchaeota is the **extreme thermophiles** that grow optimally at temperatures above 80°C. The genus *Sulfolobus* is found in volcanic hot springs where the water is 90°C and "strain 121" reproduces in deep-sea hydrothermal vents where the water is as high as 121°C.

Crenarchaeota

The second phylum, the **Crenarchaeota**, is composed primarily of thermophilic species, typically growing in hot springs and marine hydrothermal vents (**FIGURE 4.20C**). Other species are dispersed in open oceans, often inhabiting the cold ocean waters (–3°C) of the deep-sea environments and polar seas.

Other archaeal groups perhaps more closely related to the Crenarchaeota are referred to as a supergroup called TACK (Thaumarchaeota, Aigarchaeota, Crenarchaeota, and Korarchaeota). TACK is

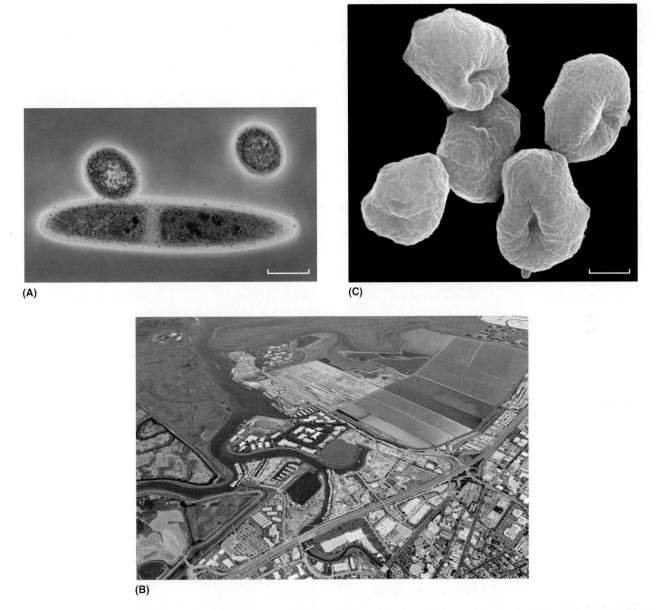

(A)

(C)

(B)

FIGURE 4.20 Members of the Domain Archaea. (A) A false-color transmission electron micrograph of the methanogen *Methanospirillum hungatei*. (Bar = 0.5 μm.) **(B)** An aerial view above Redwood City, California, of the evaporation salt ponds whose color is due to high concentrations of extreme halophiles. **(C)** A false-color scanning electron micrograph of *Sulfolobus*, a hyperthermophile that grows in waters as hot as 90°C. (Bar = 0.5 μm.) »» *What advantage is gained by these species growing in such extreme environments?*

similar to CPR in that members of both domains have small genomes and appear to lack core metabolic pathways, meaning that they too depend on other members of the microbial community for survival. Perhaps most interestingly is the discovery of the Lokiarchaea. This group is closely related to TACK, and scientists believe that the Lokiarchaea might represent the closest ancestor to the eukaryotes. Like the CPR, all these archaeal groups await more study.

TABLE 4.3 summarizes some of the characteristics among the three domains.

In conclusion, one of the take-home lessons from the discussions of cell structure and function explored in this chapter is the ability of bacterial and archaeal organisms to carry out the "complex" metabolic and biochemical processes typically associated with eukaryotic cells—usually without the need for elaborate membrane-enclosed organelles.

Concept and Reasoning Checks 4.6

a. What three unique events occurred within the Proteobacteria and Cyanobacteria that contributed to the evolution of the Eukarya and the oxygen-rich atmosphere of Earth?

b. Compared to the more moderate environments in which some archaeal species grow, why have others adapted to such extreme environments?

TABLE 4.3 A Comparison Between the Domains Bacteria, Archaea, and Eukarya

Characteristic	Bacteria	Archaea	Eukarya
Cell nucleus	No	No	Yes
Chromosome form	Single, circular	Single, circular	Multiple, linear
DNA replication origins	One	One or more	Many
Histone proteins present	No	In some species	Yes
Peptidoglycan cell wall	Yes	No	No
Membrane lipids	Unbranched, ester-linked hydrocarbons	Some branched, ether-linked hydrocarbons	Unbranched, ester-linked hydrocarbons
Ribosome sedimentation value	70S	70S	80S
Ribosome sensitivity to diphtheria toxin	No	Yes	Yes
RNA polymerase	One type	Several types	Several types
First amino acid in a protein	Formylmethionine	Methionine	Methionine
Chlorophyll-based photosynthesis	Yes (cyanobacteria)	No	Yes (algae)
Growth above 80°C	Some species	Some species	No
Growth above 100°C	No	Some species	No
Pathogens	Some species	No	Some species

SUMMARY OF KEY CONCEPTS

Concept 4.1 Prokaryotes Can Be Distinguished by Their Cell Shape and Arrangements

1. **Bacilli** have a cylindrical shape and can remain as single cells or are arranged into **diplobacilli** or chains (**streptobacilli**). **Cocci** are spherical and form a variety of arrangements, including the **diplococcus, streptococcus,** and **staphylococcus**. The spiral-shaped bacteria can be curved forms (**vibrios**) or spirals (**spirochetes** and **spirilla**). The spiral-shaped bacteria generally appear as single cells. (Figure 4.2)

Concept 4.2 Bacterial and Archaeal Cells Have an Organized Structure

2. Cell organization is centered on three specific processes: sensing and responding to environmental changes, compartmentalizing metabolism, and growing and reproducing. (Figure 4.3)

Concept 4.3 Cell-Surface Structures Interact with the Environment

3. **Pili** are short hair-like appendages facilitate attachment to a surface. **Conjugation pili** are used for genetic transfer of DNA. (Figure 4.4)
4. One or more **flagella**, found on many rods and spirals, provide for cell motility. Each flagellum consists of a **basal body** attached to the flagellar filament. In nature, flagella propel bacterial cells toward nutrient sources (**chemotaxis**). Spirochetes have **endoflagella**. (Figures 4.5, 4.7)
5. The **glycocalyx** is a sticky layer of polysaccharides that protects the cell against desiccation, attaches it to surfaces, and helps evade immune cell attack. The glycocalyx can be thick and tightly bound to the cell (**capsule**) or thinner and loosely bound (**slime layer**). (Figure 4.8)

Concept 4.4 Most Prokaryotic Cells Have a Cell Envelope

6. The **cell wall** provides structure and protects against cell lysis. Gram-positive bacteria have a thick wall of **peptidoglycan** strengthened with **teichoic acids**. Gram-negative cells have a single layer of peptidoglycan and an **outer membrane** containing **lipopolysaccharide** and **porin proteins**. (Figure 4.10)
7. Archaeal cell walls lack peptidoglycan but can have either a **pseudopeptidoglycan** or **S-layer**.

8. The **cell membrane** represents a permeability barrier and the site of transfer for nutrients and metabolites into and out of the cell. The cell membrane reflects the **fluid mosaic model** for membrane structure in that the lipids are fluid and the proteins are a mosaic that can move laterally in the bilayer. (Figures 4.11, 4.12)
9. **Extracellular vesicles** are produced by many bacterial species to transport materials between cells and to carry out other yet undiscovered functions. (Figure 4.13)

Concept 4.5 The Cell Cytoplasm Is Packed with Internal Structures

10. The DNA (**bacterial chromosome**), located in a region called the **nucleoid**, is the essential genetic information and represents most of the organism's **genome**. (Figure 4.14)
11. Bacterial and archaeal cells can contain one or more **plasmids**, which are circular pieces of nonessential DNA that replicate independently of the chromosome.
12. **Ribosomes** carry out protein synthesis, **microcompartments** and **nanocompartments** isolate species-specific processes, whereas **inclusions** store nutrients or structural building blocks. (Figures 4.15, 4.16)
13. **Gas vesicles** are used by some marine prokaryotes to regulate buoyancy in a water column.
14. **Magnetosomes** are used to locate optimal areas in a water column having little or no oxygen gas. (Figure 4.17)
15. The **cytoskeleton** helps determine cell shape, regulates cell division, and controls chromosomal segregation during cell division. (Figure 4.18)

Concept 4.6 There Is Tremendous Diversity **Among the Domains Bacteria and Archaea**

16. The **phylogenetic Tree of Life** contains many bacterial phyla, including the **Proteobacteria, gram-positive bacteria, Cyanobacteria, Chlamydiae, Spirochaetes,** and **Bacteroidetes**.
17. Many organisms in the domain Archaea live in extreme environments. The **Euryarchaeota** (**methanogens, extreme halophiles,** and **hyperthermophiles**) and the **Crenarchaeota** are two major phyla.

CHAPTER SELF-TEST

For **Steps A–D**, you can find answers to questions and problems in **Appendix D**.

■ STEP A: REVIEW OF FACTS AND TERMS

Multiple Choice

Read each question carefully before selecting the *one* answer that best fits the question or statement.

1. Spherical bacterial cells in chains would be referred to as a _____ arrangement.
   - **A.** vibrio
   - **B.** streptococcus
   - **C.** staphylococcus
   - **D.** tetrad

2. Intracellular organization in bacterial and archaeal species is centered around _____.
   - **A.** compartmentation of metabolism
   - **B.** growth and reproduction
   - **C.** sensing and responding to environment
   - **D.** All the above (A–C) are correct.

3. Which one of the following statements does *not* apply to pili?
   - **A.** Pili are made of protein.
   - **B.** Pili allow for attachment to surfaces.
   - **C.** Pili facilitate nutrient transport.
   - **D.** Pili contain adhesins.

4. Flagella are _____.
   - **A.** made of carbohydrate and lipid
   - **B.** found on all bacterial cells
   - **C.** shorter than pili
   - **D.** important for chemotaxis

5. Capsules are similar to pili because both _____.
   - **A.** contain DNA
   - **B.** are made of protein
   - **C.** contain dextran fibers
   - **D.** permit attachment to surfaces

6. Gram-negative bacterial cells would stain _____ with the Gram stain and have _____ in the wall.
   - **A.** orange-red; teichoic acid
   - **B.** orange-red; lipopolysaccharide
   - **C.** purple; peptidoglycan
   - **D.** purple; teichoic acid

7. The cell membrane of archaeal hyperthermophiles contains _____.
   - **A.** a monolayer
   - **B.** sterols
   - **C.** ester linkages
   - **D.** All the above (A–C) are correct.

8. The movement of glucose into a cell occurs by _____.
   - **A.** facilitated diffusion
   - **B.** active transport
   - **C.** simple diffusion
   - **D.** phospholipid exchange

9. When comparing bacterial and archaeal cell membranes, only archaeal cell membranes _____.
   - **A.** have three layers of phospholipids
   - **B.** have a phospholipid bilayer
   - **C.** are fluid
   - **D.** have ether linkages

10. Which one of the following statements about the nucleoid is *not* true?
    - **A.** It contains a DNA chromosome.
    - **B.** It represents a nonmembranous subcompartment.
    - **C.** It represents an area devoid of ribosomes.
    - **D.** It contains nonessential genetic information.

11. Plasmids _____.
    - **A.** replicate with the bacterial chromosome
    - **B.** contain essential growth information
    - **C.** can contain antibiotic resistance genes
    - **D.** are as large as the bacterial chromosome

12. Which one of the following is *not* a structure or subcompartment found in bacterial cells?
    - **A.** Microcompartments
    - **B.** Volutin
    - **C.** Ribosomes
    - **D.** Mitochondria

13. The bacterial cytoskeleton _____.
    - **A.** transports vesicles
    - **B.** helps determine cell shape
    - **C.** is organized identical to its eukaryotic counterpart
    - **D.** centers the nucleoid

14. Which one of the following is *not* a genus within the gram-positive bacteria?
    - **A.** *Staphylococcus*
    - **B.** *Methanogens*
    - **C.** *Mycoplasma*
    - **D.** *Bacillus* and *Clostridium*

15. The domain Archaea includes all the following groups *except* the _____.
    - **A.** mycoplasmas
    - **B.** extreme halophiles
    - **C.** Crenarchaeota
    - **D.** Euryarchaeota

Label Identification

16. Identify and label the structure on the accompanying bacterial cell from each of the following descriptions. Some separate descriptions can apply to the same structure.

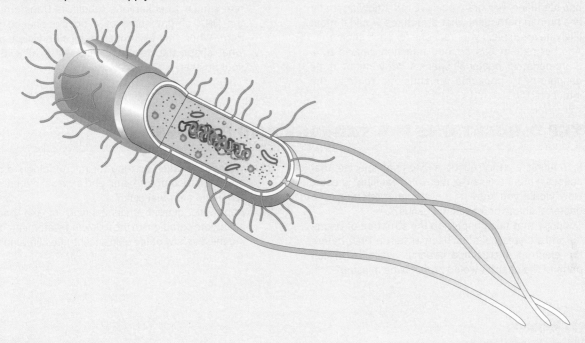

Descriptions

A. Is involved with nutrient storage.
B. An essential structure for chemotaxis, aerotaxis, or phototaxis.
C. Contains nonessential genetic information that provides genetic variability.
D. The structure that synthesizes proteins.
E. The protein structures used for attachment to surfaces.
F. Contains essential genes for metabolism and growth.

G. Prevents cell desiccation.
H. A 70S particle.
I. Contains peptidoglycan.
J. Regulates the passage of substances into and out of the cell.
K. Extrachromosomal loops of DNA.
L. Represents a capsule or slime layer.
M. The semifluid mass of proteins, amino acids, sugars, salts, and ions dissolved in water.

▮ STEP B: CONCEPT REVIEW

17. Compare the various shapes and arrangements of bacterial and archaeal cells. (**Key Concept 4.1**)
18. Summarize how the processes of sensing and responding to the environment, compartmentation of metabolism, and growth and metabolism are linked to cell structure. (**Key Concept 4.2**)
19. Assess the role of the three external cell structures to cell function and survival. (**Key Concept 4.3**)

20. Justify the need for a **cell membrane** surrounding all bacterial and archaeal cells. (**Key Concept 4.4**)
21. Describe the roles of the prokaryotic subcompartments to cell survival. (**Key Concept 4.5**)
22. Why do scientists currently believe that an organism within the domain Archaea is the ancestor of the eukaryotes? (**Key Concept 4.6**)

■ STEP C: APPLICATIONS AND PROBLEM SOLVING

23. A bacterial species has been isolated from a patient and identified as a gram-positive rod. Knowing that it is a human pathogen, what structures would it most likely have? Explain your reasons for each choice.

24. Another patient has a blood infection caused by a gram-negative bacterial species. Why might it be dangerous to prescribe an antibiotic to treat the infection?

25. In the research lab, the gene for the cytoskeletal protein similar to eukaryotic tubulin is transferred into the DNA chromosome of a coccus-shaped bacterium. When this cell undergoes cell division, predict what shape the daughter cells will exhibit. Explain your answer.

■ STEP D: QUESTIONS FOR THOUGHT AND DISCUSSION

26. In reading a story about a bacterial species that causes a human disease, the word "bacillus" is used. How would you know if the article is referring to a bacterial shape or a bacterial genus?

27. Suppose that this chapter on the structure of bacterial and archaeal cells had been written in 1940, before the electron microscope became available. Which parts of the chapter would probably be missing?

28. Why has it taken so long for microbiologists to discover microcompartments and a cytoskeleton in bacterial and archaeal cells?

29. Apply the current understanding of the bacteria/eukaryote paradigm to the following statement: "Studying the diversity of life only accentuates life's unity."

Concept Mapping

See pages XXXI–XXXII on how to construct a concept map.

30. Construct a concept map for the **Cell Envelope** using the following terms:

Active transport	Membrane proteins
Cell membrane	NAG
Cell wall	NAM
Endotoxin	Outer membrane
Facilitated transport	Peptidoglycan
Fluid-mosaic model	Phospholipids
Gram-negative wall	Polysaccharide
Gram-positive wall	Porin proteins
Lipid A	Teichoic acid
Lipopolysaccharide (LPS)	

31. Construct a concept map for the **domain Bacteria** using the following terms:

Actinobacteria	Gram-positive genera
Bacillus	*Mycoplasma*
Blooms	Proteobacteria
Chlamydiae	Rickettsias
Cyanobacteria	Spirochaetes
Escherichia	*Staphylococcus*
Firmicutes	*Streptomyces*
Gram-negative genera	*Treponema*

0 10 20 30 40 50 60 70 80 90 100 110 120 130 140 150 centimeters

CHAPTER 5

Microbial Growth and Nutrition

Books have been written about it; movies have been made; even a radio play in 1938 about it frightened thousands of Americans. What is it? Martian life. In 1877, the Italian astronomer, Giovanni Schiaparelli, observed lines on Mars, which he and others assumed were canals built by intelligent beings. It wasn't until well into the 20th century that this impression was disproved. Still, when we gaze at the Red Planet up close via a Mars rover (see chapter opening photo), we still wonder: Did life ever exist there?

We are not the only ones wondering. Astronomers, geologists, and many other scientists have asked the same question. Today, microbiologists have joined their scientific colleagues, wondering if microbial life once existed on Mars or, for that matter, elsewhere in our solar system.

Could microbes, as we know them here on Earth, survive on Mars? After all, the temperatures are far below 0°C, the atmosphere contains little oxygen gas, and ultraviolet radiation bombards the planet's surface. To test the feasibility of such ideas, researchers, using a device to simulate the Martian soil environment, placed in the soil microbes known to survive extremely cold environments here on Earth. Their results indicated that members of

the domain Archaea, specifically the methanogens, could grow in the cold, low-oxygen atmosphere, especially if they were buried just under the soil surface.

In fact, there are quite a few microbes found here on Earth that survive, and even require, living in extreme environments (TABLE 5.1)—some not so different from those of Mars (FIGURE 5.1). If life as we know it did (or does) exist on Mars, it could have been (or possibly is) composed of microbes similar to the bacterial/archaeal extremophiles found here on Earth.

In 2004, NASA sent two spacecraft to Mars to look for indirect signs of past life. Scientists monitored instruments on the Mars rovers, *Spirit* and *Opportunity*, which were designed to search for signs suggesting water once existed on the planet. Some findings suggest that there are areas where salty seas once washed over the plains of Mars, creating a life-friendly environment. Indeed, instruments on *Opportunity*

The Martian surface as recorded by NASA's Mars rover *Curiosity*.
Courtesy of JPL-CalTech/MSSS/NASA.

TABLE 5.1 Some Microbial Record Holders (and Their Taxonomic Domain)

Hottest environment (Juan de Fuca ridge)—121°C: Strain 121 (Archaea)

Coldest environment (Antarctica)—-15°C: Cryptoendoliths (Bacteria and lichens)

Highest radiation survival—5MRad, or 5,000× what kills humans: *Deinococcus radiodurans* (Bacteria)

Deepest—2.4 km underground: Many bacterial and archaeal species

Most acid environment (Iron Mountain, CA)—pH 0.0 (most life is at least 100,000× less acidic): *Ferroplasma acidarmanus* (Archaea)

Most alkaline environment (Lake Calumet, IL)—pH 12.8 (most life is at least 1,000× less basic): Proteobacteria (Bacteria)

Longest in space (NASA satellite)—6 years: *Bacillus subtilis* (Bacteria)

Highest pressure environment (Mariana Trench)—1,200× atmospheric pressure: *Moritella, Shewanella*, and others (Bacteria)

Saltiest environment (Eastern Mediterranean basin)—47% salt (15 times human blood saltiness): Several bacterial and archaeal species

found evidence for ancient shores on what is believed to have once been a sea. Scientists reported in 2008 that a more recent spacecraft, the *Phoenix Mars Lander*, detected water ice near the Martian soil surface.

Since 2012, another exploration of the Red Planet has been underway, as the Mars rover *Curiosity* is searching to see if there are any chemical signs suggesting life might have existed on the red planet.

Whether microorganisms are here on Earth in moderate or extreme environments or similar organisms on Mars, there are certain physical and chemical requirements they must possess to survive, grow, and reproduce. In this chapter, we explore the process of microbial cell reproduction, examine the physical and chemical conditions required for growth, and discover the ways that microbial growth can be measured.

FIGURE 5.1 The Martian Surface? This barren-looking landscape is not Mars but the Atacama Desert in northern Chile. It looks similar to photos taken by the Mars rovers *Spirit* and *Opportunity*. »» *Does this area look like a habitable place for life, even microbial life?*

Chapter Challenge

The United States has sent several spacecraft to Mars since the first Viking landers in 1976. Recently, an international team of scientists carried out studies suggesting terrestrial microbes could hitch a ride to Mars on such a craft and even possibly survive the journey. The team believes most spacecraft that have touched down on Mars were not thoroughly sterilized before they left Earth, so they could have carried living microbes. NASA scientists have assumed that Mars' thin atmosphere with intense ultraviolet (UV) radiation would kill any life inadvertently carried on the spacecraft. However, in laboratory tests, the scientists found that some spore-forming bacterial species could survive UV bombardment that was equivalent to that on Mars—if the spores were buried just a few millimeters in the soil. Could such an earthly extremophile survive on Mars? Could there be Martian microbes actually on the planet? Are there conditions here on Earth that might give us an idea? Let's try to hazard an educated guess as we examine the conditions that bacterial and archaeal species must possess here on Earth to survive both "normal" and "extreme" environments.

■ KEY CONCEPT 5.1   Microbial Reproduction and Growth Are Part of the Cell Cycle

Growth in the microbial world usually refers to an increase in the numbers of individuals, that is, an increase in population size. This is accomplished through **asexual reproduction**, which is the process for forming new cells while maintaining genetic constancy. In eukaryotic microbes, an elaborate interaction of microtubules and proteins with pairs of chromosomes in the cell nucleus allows for the precise events of mitosis and cytokinesis. Prokaryotic microbes divide without this elaborate division apparatus, but they still possess some similar protein filaments (cytoskeleton) to guide the cell division process.

Binary Fission Is Part of the Cell Cycle

The series of orderly events through which a cell passes during its life span is called the **cell cycle**. For a bacterial cell like *Escherichia coli*, the cycle can be divided into three stages (**FIGURE 5.2A**). Let's take a look at each of them in the following subsections.

Growth Phase

Before a bacterial cell actually divides into two cells, it goes through a phase of metabolic "growth," during which time the cell increases in mass and size. Because the cell wall is relatively inflexible, during cell enlargement enzymes break the bonds in the peptidoglycan layers so that the cell can stretch and expand. During this phase, the cell synthesizes proteins, phospholipids, and other molecules as it prepares for DNA replication.

DNA Replication Phase

When the cell reaches a critical size that makes the transport of nutrients and transfer of energy difficult, it will divide. Because all cells formed must have a complete set of genetic information, DNA replication occurs before the actual division of the cell. During this phase, DNA synthesis occurs and the bacterial chromosome is copied, resulting in two identical chromosomes in the one cell. The separation of the chromosomes after DNA replication involves the cell's cytoskeleton but, unlike eukaryotic cells, lacks the microtubular spindle to segregate replicated chromosomes.

Binary Fission Phase

As DNA replication ends, **binary fission** begins. During this phase, a partition or septum forms at midcell (**FIGURE 5.2B**). This splitting of the cell in half (**cytokinesis**) is coordinated by other cytoskeletal proteins that organize into a **fission ring apparatus**. Acting like a contractile band, the proteins ensure two nearly identical daughter cells are formed.

Following binary fission, the septum material between the daughter cells might dissolve at a slow rate, allowing pairs, chains, or clusters of cells to form that represent characteristic arrangements for many bacterial species. With other species, like *E. coli*, the cells fully separate from each other after binary fission is complete. Each of the attached or free daughter cells then enters another round of the cell cycle.

There are at least two checkpoints to make sure cell division and DNA replication occur properly (see Figure 5.2A). If cell division is incomplete, DNA replication will come to a halt within five generations. If DNA replication is not completed, or errors occur in the replication process, within three generations binary fission will stop. The inability to quickly repair the damage sends the crippled cell into an arrested (quiescent) state in which the cell remains alive but unable to resume growth or DNA replication; in other words, it no longer progresses through a cell cycle.

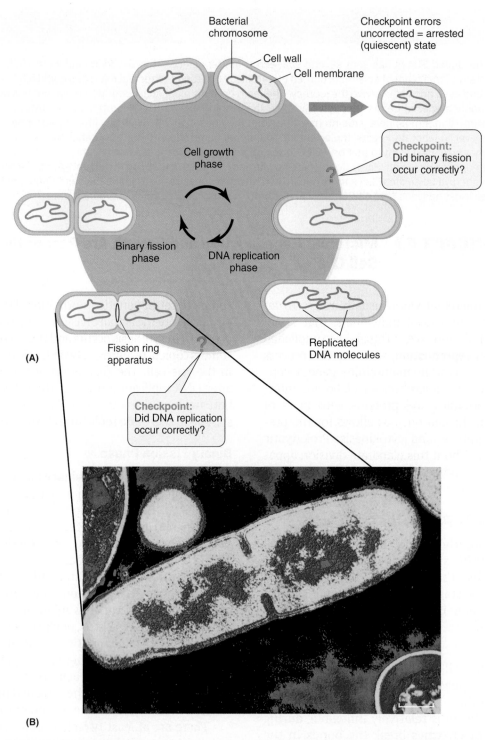

FIGURE 5.2 The Bacterial Cell Cycle. (A) The bacterial cell cycle can be separated into three phases: cell growth, DNA replication, and binary fission. **(B)** A false-color transmission electron micrograph of a *Bacillus licheniformis* cell undergoing binary fission. The inward growth of the cell envelope is evident at midcell between the segregated chromosomes (blue areas). (Bar = 0.25 μm.) »» *How would binary fission differ for a prokaryotic organism having cells arranged in chains and another forming single cells?*

(B) © Shutterstock.

Prokaryotic Cells Can Grow Exponentially

The time required for a prokaryotic cell to progress through one cell cycle and produce two daughter cells is known as the **generation time** (or doubling time). Under optimal conditions, some species have a very fast generation time (TABLE 5.2). For example, microbiologists calculate that if a single *E. coli* cell proceeds through a cell cycle every 15 minutes, within two days the mass of bacterial cells would equal the mass of the Earth! Thankfully, this will not occur because of the limitation of nutrients and the loss of essential physical and chemical factors required for growth. That means the vast majority of the 3×10^{30} bacterial cells on our planet are **oligotrophs** (*oligo* = "limited"; *troph* = "food"); that is, they spend the vast majority of their existence in a nutrient-limited state and they divide infrequently. In fact, there are some species for which the generation time is extremely slow, as MICROFOCUS 5.1 explains.

The generation time often is useful in determining the **incubation period** for an infection, that is,

the amount of time that passes from exposure to the infectious agent and detection of the first symptoms in an individual. Often, faster division times mean a shorter incubation period for a disease. Suppose that you eat an improperly refrigerated chocolate éclair that was contaminated with *Staphylococcus aureus* (FIGURE 5.3). If you ingested 100 *S. aureus* cells at 8:00 PM this evening, 200 would be present by 8:30, 400 by 9:00, and 800 by 9:30. You would have more than 25,000 by midnight. By 3:00 AM, the *S. aureus* cell population will have topped 1.6 million. Depending on the response of the immune system, under the optimal growth conditions found in the gut it is quite

TABLE 5.2 Examples of Generation Times—Common Bacterial Species Growing Under Optimal Conditions[1]

Species	Growth Medium	Generation Time (Minutes)
Escherichia coli	Chemically defined	15
Bacillus megaterium	Chemically defined	25
Staphylococcus aureus	Complex	27–30
Streptococcus lactis	Milk	26
Streptococcus lactis	Complex	48
Lactobacillus acidophilus	Milk	66–87
Mycobacterium tuberculosis	Chemically defined	792–932 (= 12–15.5 hours)
Treponema pallidum	Rabbit testes	1,980 (= 33 hours)

[1]Table modified from microblog.me.uk/138.

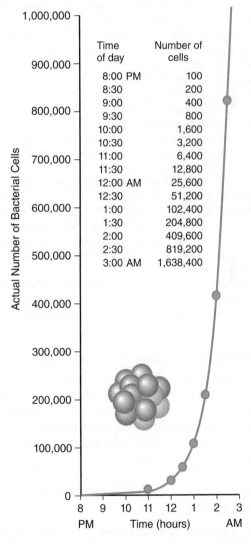

Time of day	Number of cells
8:00 PM	100
8:30	200
9:00	400
9:30	800
10:00	1,600
10:30	3,200
11:00	6,400
11:30	12,800
12:00 AM	25,600
12:30	51,200
1:00	102,400
1:30	204,800
2:00	409,600
2:30	819,200
3:00 AM	1,638,400

FIGURE 5.3 A Skyrocketing Bacterial Population. The number of *Staphylococcus aureus* cells progresses from 100 cells to almost 2 million cells in a mere 7 hours. The J-shaped growth curve gets steeper and steeper as the hours pass. Only a depletion of nutrients, buildup of waste, or some other limitation will halt the population explosion. *»» What is the generation time for* S. aureus *in this figure?*

MICROFOCUS 5.1: Evolution/Environmental Microbiology

A Microbe's Life—Zombie Style

What do *Zombieland*, *Dawn of the Dead*, *The Walking Dead*, and some extreme microbial species have in common? All are examples of a zombie state, which, for the above-mentioned big screen flicks or television series, identifies "creatures" that have died and been reanimated as unconscious, mindless monsters, often apparently arising from deep below the Earth. Well, in the microbial world, there are microbes that exist in a zombie-like state deep within the Earth.

As you already know, microbes can be found just about anywhere on and in planet Earth where an energy source and nutrients are found. This includes deep below the Earth's surface. In 2013, scientists with the Integrated Ocean Drilling Program, an international marine research program, announced they had discovered bacterial organisms in rocks 2.5 kilometers below the ocean floor. More impressive, these organisms are up to 100 million years old, and the researchers believe the bacterial cells divide only once every 10,000 years! The growth is so slow, scientists originally wondered if these cells were alive—or just undead. However, give them a nutritious microbial meal and the absorption of nutrients by the cells can be detected. Thus, the apparent snail's pace for a cell cycle and growth results from

Dr. Beth Orcutt (left front row) and researchers are trying to understand how life below the sea floor can survive and thrive.

Courtesy of Jennifer Magnusson/IODP.

the infinitesimally small amount of energy and nutrients normally available in their rocky homes. At these rates, it takes thousands of years for the single-celled intraterrestrials (living within the Earth) to produce enough energy to support binary fission. Dr. Beth Orcutt from the Bigelow Laboratory for Ocean Sciences in Maine stated, "Is it really true to call [these microbes] alive when [they're] doubling every thousands of years? It's almost like a zombie state."

Unlike the "human zombies" who attempt to disturb human society (at least in the movies), the ocean scientists believe the rock-dwelling microbial zombies could be beneficially affecting the chemistry in the rocks, the deep Earth, and the planet as a whole.

Scientifically, these bacterial communities offer an opportunity to learn how cells can survive on so few nutrients and minimal energy sources. Their very existence helps scientists to understand the line between life and death and to estimate what the bare minimum is for an organism to stay alive.

likely that sometime during the night you would know you have food poisoning, as you would be suffering from nausea, diarrhea, vomiting, and stomach cramping. With the fast generation time of *S. aureus*, usually within 6 hours, the symptoms appear.

Let's now look at the dynamics of bacterial growth in a little more detail.

A Bacterial Growth Curve Illustrates the Dynamics of Growth

A typical **bacterial growth curve** for a cell population illustrates the events occurring over time (**FIGURE 5.4**). The simplest approach for studying the dynamics of growth is to transfer a sample of bacterial cells (e.g., *E. coli*) to a tube of fresh, nutrient-rich broth. Over time, four distinct phases of growth will occur: the lag phase, logarithmic (log) phase, stationary phase, and decline phase.

The subsections that follow examine each of these phases.

The Lag Phase

During the first portion of the growth curve, called the **lag phase**, bacterial cells are adapting to their new environment and compensating for changes in nutritional conditions. In the broth, some cells might actually die from the shock of transfer or from the inability to adapt to the new environment. The actual length of the lag phase depends on the metabolic activity of the microbial population. They must grow in size, take up nutrients, and replicate their DNA—all in preparation for binary fission.

The Log Phase

The population now enters an active stage of growth called the **logarithmic (log) phase**. In the log phase,

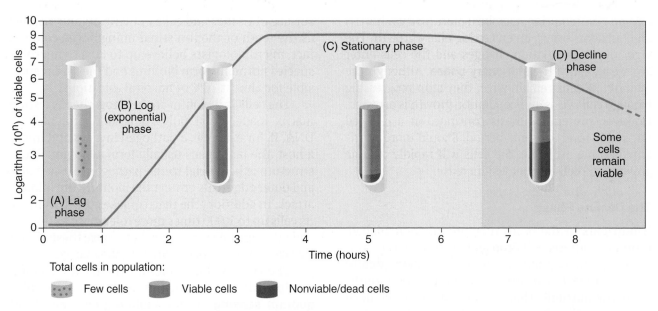

FIGURE 5.4 **The Growth Curve for a Bacterial Population.** A typical bacterial population sequentially goes through four growth phases. »» *Why would antibiotics work best to kill or inhibit cells in the log phase?*

all cells are undergoing repeated cell cycles and the generation time is dependent on the microbial species used and the environmental conditions present. For *E. coli*, the cells exhibit balanced growth because all aspects of metabolism and physiology remain constant. As time passes, the number of cells doubles and, on the bacterial growth curve, the cell number rises accordingly on a logarithmic scale.

In a broth tube, the medium becomes cloudy (turbid) due to increasing numbers of *E. coli* cells. If plated on solid growth medium, bacterial growth will be so vigorous that visible colonies appear within 24 hours and each colony will consist of billions of cells

(**FIGURE 5.5**). Vulnerability to antibiotics is also highest at this active stage of growth because many antibiotics affect vital metabolic processes like DNA replication, RNA synthesis, and protein synthesis in dividing cells.

The Stationary Phase

Because a broth tube represents a "closed system," nutrients are not available indefinitely. Therefore, after some hours, oxygen gas and nutrients like carbon and nitrogen sources become depleted as waste products accumulate. This limitation of oxygen (for

(A)

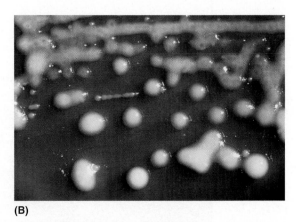

(B)

FIGURE 5.5 **Two Views of Bacterial Colonies. (A)** Bacterial colonies cultured on blood agar in a culture dish. Blood agar is a mixture of nutrient agar and blood cells. It is often used for growing bacterial colonies. **(B)** Close-up of bacterial colonies being cultured on an agar growth medium. »» *How did each colony in (a) or (b) start?*

oxygen gas–requiring microbes; more on this shortly) and nutrients brings an end to log phase growth. The vigor of the population changes and the cells enter a plateau, called the **stationary phase**. Although the majority of cells remain alive, they stop progressing through cell cycles and population growth is arrested. However, protein synthesis continues at a reduced but stable level for some period. Should more nutrients become available, the cells will rapidly resume growth, and cell number will increase.

The Decline Phase

If nutrients in the broth tube are not replenished, protein synthesis will drop and eventually the nongrowing cell population enters a **decline (death) phase**. Some *E. coli* cells can survive for months using the nutrients that are released from the dead and dying cells. Thus, the population enters a state of balanced cell death, some cells surviving for a time as other cells die.

Most Bacterial Species Exist and Grow Within a Biofilm

A **biofilm** represents a complex community of bacterial and often additional microbial species that attaches to and grows on a surface (**FIGURE 5.6**). This thin structure might form on an environmental surface like a rock, the inside surface of a drainpipe, or the surface of an indwelling or implanted medical device like a catheter. Biofilms also form on the surface of tissues and cells of living organisms, such as the teeth or the intestinal lining of the colon. In fact, microbiologists believe up to 80% of microbial species normally form biofilms and they are responsible for almost 70% of hospital infections.

The cells in a biofilm are held together by a sticky matrix made up of polysaccharides, proteins, and DNA. If, for example, a pathogen invades and infects a host, the pathogens usually form a biofilm. Such a structure is beneficial to the pathogen as the matrix and outer edge cells protect the biofilm from immune attack. In addition, the matrix makes the biofilm and its cells up to 1,000 times more tolerant to antibiotics and other antimicrobial agents because these chemicals cannot easily penetrate into the matrix.

The cells within a biofilm chemically communicate with one another through a process called **quorum sensing**. As the biofilm grows and develops, the microbial cells secrete signaling molecules that inform the other cells as to their number and cell types. When the signaling chemical reaches a threshold concentration, or "tipping point," cell behavior changes as new sets of microbial genes are turned on and others are turned off. The result can be the secretion of enzymes or toxins that trigger disease onset. Thus, by monitoring their numbers (the quorum) through chemical communication, the behavior of cells in the biofilm is modified and growth in the biofilm remains stable. The use of quorum sensing to enhance survival is clearly shown in the following two examples.

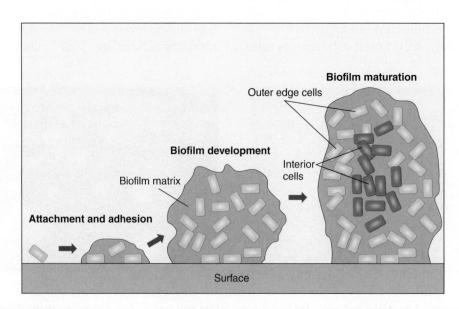

FIGURE 5.6 Biofilm Development and Maturation. Biofilm formation begins with adhesion of the cells and their attachment to a surface. The cells then divide as they lay down a sticky matrix. At maturation, some cells (yellow-orange) are nearer the edge of the biofilm while others are embedded in the interior of the biofilm (green). *»» Besides access to nutrients, what other factors may make the interior cells different from the outer edge cells?*

Metabolic Decline

As described in this chapter, the stationary phase of a bacterial growth curve is characterized by a slowing in cell population size due to the diminished availability of nutrients. As the cells in a biofilm enter stationary phase, do the cells just keep consuming the nutrients until they run out, or do they pace themselves so the nutrients last longer?

It ends up that, at least for *E. coli*, the bacterial cells can plan ahead and enter a "shortage mode." Through quorum sensing in the biofilm, the cells chemically sense when there is a reduction in nutrients. In response, the cells begin producing a chemical signal that affects gene expression. Because of the signal, the *E. coli* cells consume fewer nutrients by slowing down cell growth and protein synthesis. The cells are now programmed to survive the lean times of the stationary phase and will restart another log phase if there is a return of sufficient nutrients.

Social Conflicts

The second example concerns the microbial community's ability to resolve internal "social conflicts." Take, for example, the relative positions of cells within the biofilm (see Figure 5.6). Because of their proximity, the cells near the outer edges of the biofilm protect and shelter the cells in the interior. The cells at the outer edge also have immediate access to nutrients needed for growth. Therefore, if these peripheral cells grow rapidly, few nutrients are left to diffuse to the sheltered interior cells. Consequently, there is a social conflict here for the interior cells (protection versus starvation) and it is resolved by quorum sensing.

The interior cells normally produce ammonium ions (NH_4^+), a small metabolite that the outer edge cells need for growth and reproduction. Therefore, through quorum sensing, if the interior cells chemically sense that the cells at the outer edge are increasing in number too rapidly, the interior cells change metabolic behavior and reduce the secretion of NH_4^+. As a result, the cells at the biofilm periphery slow down their growth rate, allowing the inner cells to receive sufficient nutrients to stay alive yet remain protected by the peripheral cells. Thus, a biofilm can be viewed as a "multicellular" structure within which there is a division of labor; this is a phenomenon that we more commonly associate with multicellular plants and animals.

Some Bacterial Cells Can Exist in Metabolically Inactive States

Environmental conditions vary tremendously and often they might not be favorable to active population growth. As described earlier, such unfavorable conditions include the presence of toxic chemicals, such as antibiotics, and nutrient limitation (potential starvation). Therefore, throughout the cell cycle microbes must constantly monitor or sense their surroundings to ensure the conditions will support continued growth. For some bacterial species, no matter what the environment is like for growth (favorable or unfavorable), some cells enter a growth-arrested, dormant (quiescent) state. Two such dormancy strategies are described in the following subsections.

Persister Cell Formation

During the log phase of growth, most bacterial populations produce a subset of cells that are slow or nongrowing. Although all cells in the population are identical genetically, the so-called **persister cells** maintain a very low but stable rate of metabolism. Should the environment change for the worse (e.g., presence of an antibiotic), the persister cells survive, even though the rest of the population might be killed by the antibotic (**FIGURE 5.7**). When environmental conditions improve (e.g., antibiotic removed), the persister cells "revive" and once again begin dividing to reestablish the population.

Endospore Formation

A few medically significant gram-positive genera, especially *Bacillus* and *Clostridium*, undergo a different growth-arrested, dormancy scenario when the cells experience nutrient depletion. Species, such as *Bacillus anthracis* (the causative agent of anthrax) and *Clostridium botulinum* (the causative agent of botulism), enter the stationary phase and begin spore formation, or **sporulation**, that produces some amazing dormant structures called **endospores** (**FIGURE 5.8**). Unlike persister cells, most endospores are the result of nutrient limitation (starvation).

Spore formation requires a complex program of gene expression involving hundreds of genes. The process involves replication of the bacterial chromosome and binary fission involving an asymmetric cell division (**FIGURE 5.9**). The smaller cell, called the prespore, will become the mature endospore, while the larger mother cell will commit itself to maturation of the endospore before rupturing to free the spore. Depending on the exact asymmetry of cell division, the single endospore might develop at the end, near the end, or at the center of the mother cell. The entire process of sporulation takes about 6 to 8 hours.

After it is freed from the mother cell, the highly dehydrated free spore contains cytoplasm, highly

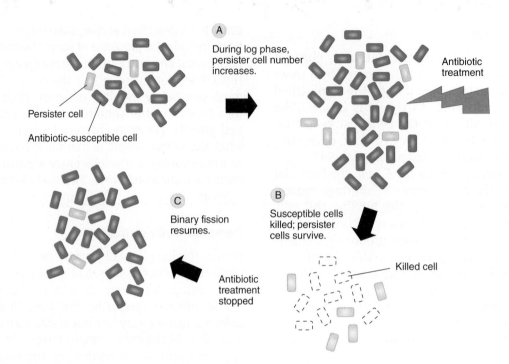

FIGURE 5.7 Persister Cells. The presence of persister cells within a bacterial cell population allows the species to survive environmental changes, in this case the presence of an antibiotic. *»» What other way could a susceptible bacterial cell population survive in the presence of an antibiotic?*

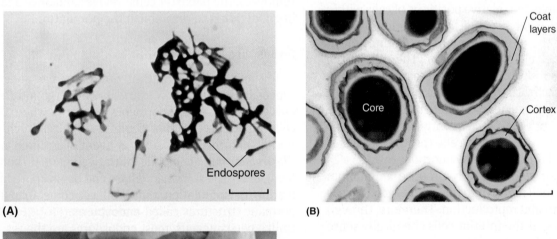

(A)

(B)

(C)

FIGURE 5.8 Three Different Views of Bacterial Endospores. (A) A light microscope image of *Clostridium* cells showing spore formation at the cell tip. Note the characteristic drumstick appearance of the spores. (Bar = 5.0 μm). **(B)** A false-color transmission electron micrograph of *Bacillus anthracis* spores showing the spore structures. (Bar = 0.5 μm). **(C)** A scanning electron microscope image of a germinating endospore of *Clostridium sporogenes*. Note that the spore coat (light yellow) has split open, allowing the vegetative cell (green) to emerge and start dividing. (Bar = 2.0 μm). *»» If an endospore is resistant to so many environmental conditions, how does a spore "know" conditions are favorable for germination?*

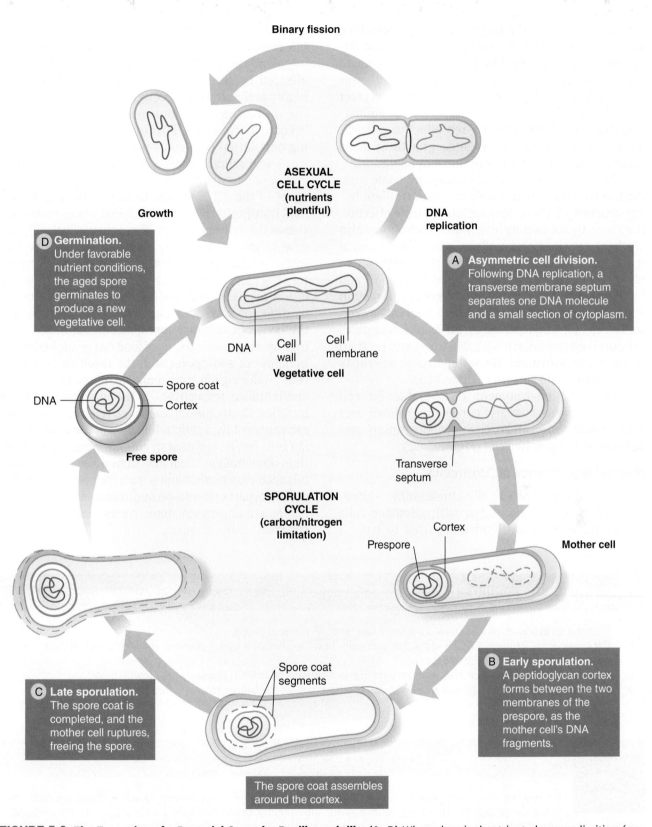

FIGURE 5.9 **The Formation of a Bacterial Spore by *Bacillus subtilis*. (A–D)** When chemical nutrients become limiting (e.g., carbon, nitrogen), endospore formers, such as *B. subtilis*, enter the sporulation cycle. *»» Hypothesize how a vegetative cell "knows" nutrient conditions are limiting.*

compacted DNA, and a large amount of **dipicolinic acid**, a unique chemical compound that helps stabilize the proteins and DNA in the spore. Thick layers of peptidoglycan form the cortex and some 70 different proteins form the layers of the spore coat to protect the cytoplasmic contents. It is important to emphasize that sporulation is not a reproductive process. Rather, the endospore represents a growth-arrested stage produced in response to nutrient depletion.

The amazing qualities of endospores come from the fact that they are probably the most resilient living structures known. Desiccation has little effect on the spore. By containing little water, endospores also are heat resistant and undergo very few chemical reactions. These properties make them difficult to eliminate from contaminated medical materials and food products. For example, endospores can remain viable in boiling water (100°C) for 2 hours. They can survive in 70% ethyl alcohol for 20 years. Endospores can survive a radiation dose 2,000 times greater than humans can withstand. No wonder some scientists believe they could survive a trip to Mars!

When the environment is favorable for cell growth, the spore coat and cortex break down over a 90-minute period and each endospore rapidly germinates as a vegetative cell (Figure 5.8C).

Medical Significance of "Dormancy"

A few serious, often life-threatening, infectious diseases are associated with persister cells and endospores. If antibiotics are used to treat a tuberculosis patient infected with *Mycobacterium tuberculosis*, any persister cells present in the infecting population will survive because the cells are not affected by the antibiotics, which target only the highly active, log phase cells (see Figure 5.7). Should the antibiotics be withdrawn later, the persister cells become active again, forming a new, actively growing population with some persister cells present.

A few serious diseases in humans are caused by endospore formers. These include *B. anthracis*, the agent of the 2001 anthrax bioterror attack that used mail transported through the United States Postal Service as the delivery system. This potentially deadly disease, originally studied by Koch and Pasteur, develops when inhaled spores germinate in the lower respiratory tract and the resulting vegetative cells secrete two deadly toxins. Tetanus and botulism are diseases caused by different species of *Clostridium*. Introduction into the body of clostridial endospores found in soil or in improperly preserved cans of food can result in the germination of endospores with the resulting vegetative cells producing life-threatening toxins causing tetanus and botulism, respectively. In hospital and healthcare facilities, *Clostridium difficile* (often referred to as *C. diff*) can cause colitis, a serious infection of the colon. The *C. diff* cells can be resistant to antibiotic treatment. More than three million *C. diff* infections occur in U.S. hospitals each year, partly coming from the germination of *C. diff* endospores present on improperly cleaned medical equipment or personal items. All these diseases will be covered in other chapters.

Concept and Reasoning Checks 5.1

a.   Propose an explanation as to how a bacterial cell "knows" when to divide.
b.   If it takes *E. coli* 7 hours to reach some 2 million cells, how long would it take *T. pallidum* to reach 2 million cells under optimal conditions? (See Table 5.2.)
c.   In a broth tube, describe the status of the bacterial cell population in each phase of the growth curve.
d.   Explain how the trigger to persister cell survival is different from the trigger to endospore formation.

Chapter Challenge A

We learned that many prokaryotic organisms on Earth survive in environments where there are minimal nutrients for growth. These oligotrophs divide infrequently. Other microbes on our planet are copiotrophs, which are organisms that grow in environments having a rich mixture of nutrients, especially carbon-based compounds.

QUESTION A: *If microbial life, as we know it, does exist on Mars, do you believe those microbes would represent oligotrophs or copiotrophs? Explain your reasoning.*

You can find answers online in **Appendix F**.

■ **KEY CONCEPT 5.2**   **Optimal Growth Is Dependent on Several Physical and Chemical Factors**

Whether in nature or in the research lab, survival and growth of microbes depend on several physical and chemical factors.

Physical Factors Significantly Affect Microbial Growth

Microbial growth can be dramatically affected by the physical characteristics in the surrounding environment. These factors include temperature, pH, and osmotic pressure. Let's take a closer look at these factors.

Temperature

Temperature is one of the most important factors affecting growth. Every microbial species has an **optimal growth temperature** at which the species grows best and has its fastest generation time. Each species also has an approximate 30°C **temperature range**, from a minimum growth temperature to a maximum growth temperature over which the cells grow, although they will have a slower generation time (**FIGURE 5.10A**). In general, most microbes can be assigned to one of three groups—psychrophiles, mesophiles, or thermophiles—based on their optimal growth temperature as well as their temperature range (**FIGURE 5.10B**).

Microbes that have their optimal growth temperature near 15°C, but can still grow at 0°C to 20°C, are called **psychrophiles** (*psychro* = "cold"; *phil* = "love"). Because about 70% of the Earth is covered by oceans having deep-water temperatures below 5°C, psychrophiles make up a large portion of the global microbial community. On the other hand, at these low temperatures, most psychrophiles could not be human pathogens because they cannot grow at the warmer 37°C typical of human body temperature.

Another group of "cold-loving" microbes is the **psychrotrophs** or **psychrotolerant** microorganisms. These species have a higher optimal growth temperature as well as a slightly broader temperature range for growth. Psychrotrophs can be found in water and soil in temperate regions of the world but are perhaps most commonly encountered as spoilage organisms as they will contaminate refrigerated foods (4°C). If these psychrotrophs produce toxins, their ingestion can cause food poisoning. **CLINICAL CASE 5** provides one example for the bacterium *Campylobacter*, the most frequently identified cause of infective diarrhea.

At the opposite extreme are the **thermophiles** (*thermo* = "heat") that grow best at temperatures around 60°C but have a growth range from 40°C to 70°C. Thermophiles are present in compost heaps and hot springs and can be contaminants in dairy products if they survive pasteurization temperatures. However, thermophiles pose little threat to human health because they do not grow well at the cooler temperature of the body.

Many archaeal species grow optimally at temperatures that exceed 80°C and can have optima near 95°C. These **hyperthermophiles** have been isolated from seawater brought up from hot-water vents along rifts on the floor of the Pacific Ocean. Because the high pressure keeps the water from boiling, some archaeal species can grow at an astonishing 121°C (see Table 5.1).

Most of the best-characterized microbial species are **mesophiles** (*meso* = "middle"), which thrive at the middle temperature range of 10°C to 45°C. This includes the pathogens that grow in warm-blooded animals, including humans, as well as those species found in aquatic and soil environments typical of the temperate and tropical regions of the world. *E. coli* and the vast majority of microbes discussed in this text are mesophiles.

pH

The cytoplasm of most microorganisms has a pH near 7.0. This means that the majority of species are **neutrophiles**, growing optimally at neutral pH and having a pH range that covers two to three pH units (1,000-fold change in H^+ concentration). However, some bacterial species, such as *Vibrio cholerae*, can tolerate acidic conditions as low as pH 2.0 and alkaline conditions as high as pH 9.5.

Acid-tolerant bacteria, called **acidophiles**, grow best at pHs below 5 and are valuable in the food and dairy industries. For example, certain species of *Lactobacillus* and *Streptococcus* produce the acid that converts milk to buttermilk and cream to sour cream. These species pose no threat to good health even when consumed in large amounts. The "active cultures" in a cup of yogurt are another example of "good" acidophilic bacterial species. **Extreme acidophiles**, preferring pHs of 1–2, are typical of

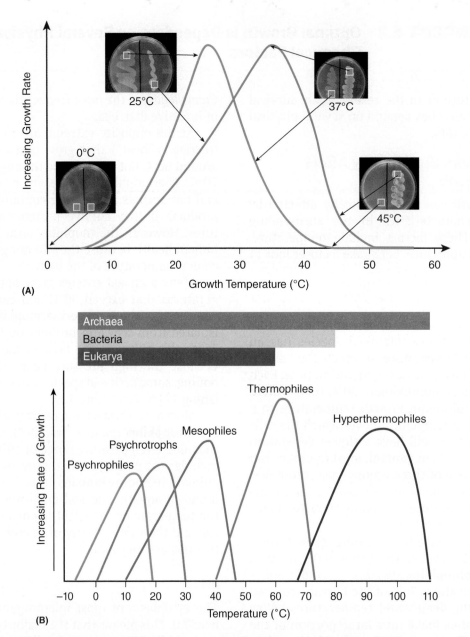

FIGURE 5.10 Growth Rates for Different Microorganisms in Response to Temperature. (A) Notice that the growth rates for these two bacterial species on nutrient agar for 24 hours are different and the rates are slower to either side of the optimal growth temperature. **(B)** Temperature optima and ranges can be used to assign microbes to one of three groups. *»» In (a), identify the temperature minimum, maximum, and optimum for growth.*

(A) Photos courtesy of Dr. Jeffrey Pommerville.

several species in the domain Archaea. At the opposite extreme are the **alkaliphiles**, which grow best at pH levels above 9.

The majority of known bacterial species, however, do not grow well under acidic conditions. Thus, the acidic environment of the stomach helps kill most bacterial pathogens, while providing a natural chemical barrier for the entry of pathogens into the intestines. In addition, you might have noted certain acidic foods such as lemons, oranges, and other citrus fruits, as well as tomatoes and many vegetables, are hardly ever contaminated by bacterial growth. However, if the produce is damaged, fungal (mold) growth can occur because many fungi grow well at a pH of 5 or lower.

Osmotic Pressure

Other microbes have adapted to the osmotic pressure exerted on cells by saline or highly saline

Clinical Case 5

An Outbreak of Food Poisoning Caused by *Campylobacter jejuni*

A cook began his day by cutting up raw chickens to be roasted for dinner. He also cut up lettuce, tomatoes, cucumbers, and other salad ingredients on the same countertop. The countertop surface where he worked was unusually small.

For lunch that day, the cook prepared sandwiches on the same countertop. Most were garnished with lettuce. Restaurant patrons enjoyed sandwiches for lunch and roasted chicken for dinner. Many patrons also had a portion of salad with their meal.

Over the next 3 days, 14 people experienced stomach cramps, nausea, and vomiting. Public health officials learned that all the affected individuals had eaten salad with lunch or dinner. *Campylobacter jejuni*, a bacterial pathogen, was isolated from their stools (see figure).

A department of health inspection concluded that the chicken was probably contaminated with *C. jejuni.* However, the cooked chicken was not the cause of the illness. Rather, inspectors identified *C. jejuni* from the raw chicken as the source.

Questions:
a. Why would the cooked chicken not be the source for the illness?
b. Why was the raw chicken identified as the source?
c. How, in fact, did the patrons become ill?

You can find answers online in **Appendix E**.

For additional information, see www.cdc.gov/mmwr/preview/mmwrhtml/00051427.htm.

Colonies of *Campylobacter jejuni* growing in agar culture.

Courtesy of Sheila Mitchell/CDC.

environments, such as the Great Salt Lake in Utah, the Dead Sea that straddles the border of Jordan and Israel, and in evaporation ponds. These **halophiles** (*halo* = "salt") are characterized by their need for hypersaline conditions for growth. Halophiles, such as *Vibrio cholerae*, grow optimally at 2% to 5% NaCl, whereas other halophiles grow optimally at 5% to 20% NaCl. **Extreme halophiles**, like *Halobacterium salinarium* and the eukaryotic green alga *Dunaliella salina*, grow optimally at 20% to 30% NaCl. In contrast, species like *E. coli* are **nonhalophiles** because they grow optimally at less than 2% NaCl and genera, such as *Staphylococcus*, are said to be **halotolerant** because they can grow in slightly saline (up to 8% NaCl) as well as nonsaline environments.

Hypersaline conditions can be deadly to microbial cells because water will be lost to the hypertonic external medium. To prevent the loss of cellular water, halophiles accumulate high cytoplasmic concentrations of solutes, such as sugars and amino acids, to balance out the osmotic difference. A major exception is for organisms like *H. salinarium*, which accumulate cytoplasmic KCl equal to the external concentration of NaCl.

The fact that microbial life can survive almost anywhere is illustrated in MICROFOCUS 5.2. FIGURE 5.11 summarizes the physical factors influencing microbial growth.

Chemical Factors for Microbial Growth Are Varied and Diverse

Besides the physical factors just described, microbes, and all organisms, have numerous chemical requirements for growth. All microbes require several core chemical elements, including carbon, oxygen, and nitrogen, as well as organic chemicals called growth factors.

MICROFOCUS 5.2: Evolution

"It's Not Toxic to Us!"

It's hard to think of oxygen as a poisonous gas considering how many organisms need it to survive today. Yet billions of years ago, oxygen was extremely toxic. One whiff by an organism and a cascade of highly destructive oxidation reactions was set into motion. Death followed quickly.

Difficult to believe? Not if you realize that ancient ancestors of the members in the domains Bacteria and Archaea relied on anaerobic chemistry for their energy needs. The atmosphere was full of methane and other gases that they could use to generate energy. But no oxygen was present. And it was that way for some 1 billion years.

Then, some 3 billion years ago, the cyanobacteria evolved. Floating on the surface of the oceans, the cyanobacteria trapped sunlight and converted it to chemical energy in the form of carbohydrates; the process was photosynthesis. But there was a downside: oxygen was a waste product of the photosynthetic process—and it was deadly because the oxygen radicals produced, such as superoxide ions (O_2^-) and peroxide anions (O_2^{-2}), could disrupt cellular metabolism in other prokaryotes by "tearing away" electrons from essential cellular molecules. However, for the next few hundred million years the amount of oxygen gas in the atmosphere was minimal and was of no great consequence to other microorganisms.

Then, about 2.4 billion years ago, there was a sudden and dramatic increase in the cyanobacteria population and in oxygen gas in the atmosphere. Unable to cope with the massive increase in oxygen radicals, enormous numbers of microbial species died and became extinct. Other species "escaped" to oxygen-free environments, such as lake and deep-sea sediments where their present-day descendants still exist. The cyanobacteria survived in the open oceans because they evolved the enzymes to safely tuck away oxygen atoms in a nontoxic form—that form was water.

Among the survivors of these first communities were gigantic, shallow-water colonies called "stromatolites." In fact, these structures—which look like rocks—still exist in a few places on Earth, such as Shark Bay off the western coast of Australia (see figure). These structures formed when ocean sediments and calcium carbonate became trapped in the microbial community (biofilm). The top few inches in the crown of a stromatolite contain the photosynthetic cyanobacteria, whereas below them are other bacterial species that can also tolerate oxygen and sunlight. Buried beneath these organisms are other bacterial species that survive the anaerobic, dark niche of the stromatolite interior where neither oxygen nor sunlight can reach. A couple of billion years would pass before one particularly well-known species of oxygen-breathing creature evolved: *Homo sapiens*.

Stromatolites, Shark Bay, Western Australia.

© Jon Nightingale/Shutterstock.

Carbon

All life on Earth is carbon-based. Therefore, the chemical element carbon is needed to build all the organic compounds (i.e., carbohydrates, lipids, nucleic acids, and proteins) found in cells. For many microbes, their source of carbon is the organic compounds (carbohydrates, lipids, and proteins) they absorb or ingest. For the photosynthetic microbes, the needed carbon comes from inorganic carbon dioxide gas (CO_2).

Oxygen

The growth of many microbes depends on the plentiful supply of oxygen gas (O_2), which makes up 21%

of the atmospheric gases. Those microbes that must have oxygen gas are called **obligate aerobes**. They use the gas in the generation of cellular energy (adenosine triphosphate [ATP]). Other microbes, termed **microaerophiles**, survive in environments where the concentration of oxygen is relatively low (2% to 10%). In the body, certain microaerophiles cause disease of the oral cavity, urinary tract, and gastrointestinal tract. For example, *Treponema pallidum*, the causative agent of syphilis, survives in the low O_2 environment of the reproductive tract. Conditions can be established in the laboratory to study these microaerophilic species (**FIGURE 5.12A**).

The **anaerobes**, by contrast, are microbes that do not or cannot use oxygen. Some are **aerotolerant**,

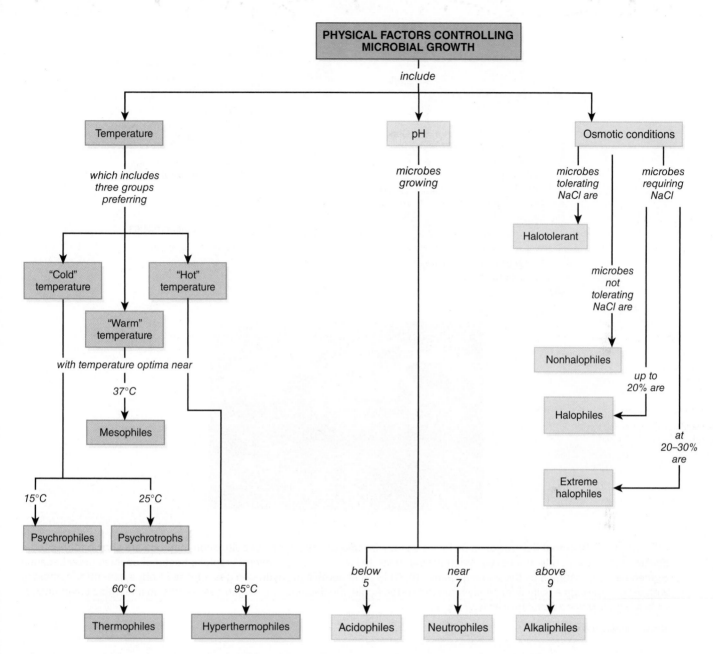

FIGURE 5.11 Types of Microbes Based on Physical Factors. This concept map shows the four major physical factors controlling microbial growth. *»» Escherichia coli is a mesophilic, facultative, nonhalophilic neutrophile. What would be the makeup of the environment where the cells would optimally grow?*

meaning they are insensitive to oxygen. Many prokaryotic species, as well as a few fungal and protistan eukaryotes, are **obligate anaerobes**, which are inhibited or killed if oxygen is present. This means they exist in environments where oxygen gas is absent and, as such, they need other ways to make ATP. Some use sulfur in their metabolic activities instead of oxygen, and therefore they produce hydrogen sulfide (H_2S) gas rather than water (H_2O) as a waste product of their metabolism. Others,

such as the ruminant microbes, produce methane (CH_4) as the byproduct of the energy conversions. In fact, life originated on Earth in an anaerobic environment consisting of methane and other gases. MICROFOCUS 5.3 details the events leading to the oxygen-rich atmosphere we have today.

The reason many anaerobes are killed by oxygen gas is that toxic forms of oxygen gas, such as superoxide radicals (O_2^-) and peroxide anions (O_2^{-2}), are produced normally during cell metabolism. Without

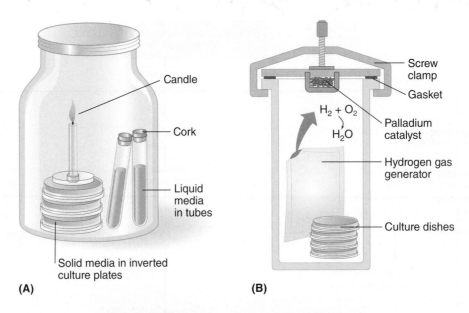

FIGURE 5.12 Bacterial Cultivation in Different Gas Environments. Two types of cultivation methods are shown for bacterial species that grow poorly in an oxygen-rich environment. **(A)** In a candle jar, microaerophilic bacterial species grow because oxygen gas is consumed by the burning candle. **(B, C)** In an anaerobic jar, hydrogen gas released from a generator combines with oxygen through a palladium catalyst to form water. Again, the depletion of oxygen gas creates an anaerobic environment. *»» In which jar would a facultative anaerobe grow?*

(C) Scott Coutts/Alamy Stock Photo.

the enzymes to detoxify these forms of oxygen, the superoxide radicals and peroxides rip off electrons from other essential cellular compounds, resulting in cell damage and possibly death.

Some anaerobic bacterial species cause disease in humans. For example, the *Clostridium* species that cause tetanus (*C. tetani*) and gas gangrene (*C. perfringens*) multiply in the dead, anaerobic tissue of a wound and produce toxins causing tissue damage. Botulism, caused by *C. botulinum*, multiplies in the oxygen-free environment of a vacuum-sealed can of food that was improperly prepared. In the food product, the pathogen produces the lethal toxins causing botulism.

Among the most widely used methods to establish anaerobic conditions in the laboratory is the GasPak™

system, in which hydrogen reacts with oxygen in the presence of a catalyst to form water, thereby creating an oxygen-free atmosphere (**FIGURE 5.12B, C**).

Finally, many microbes can grow whether oxygen gas is present or not. This group includes *E. coli*, many staphylococci and streptococci, members of the genus *Bacillus*, and the fungal yeasts. We refer to these organisms as **facultative anaerobes** because they grow best in the presence of oxygen gas but have the option of switching to anaerobic metabolism if oxygen gas is absent.

A common way to test an organism's oxygen sensitivity is to use a **thioglycollate broth**, which binds free oxygen so that only fresh oxygen entering at the top of the tube would be available (**FIGURE 5.13**).

MICROFOCUS 5.3: Environmental Microbiology

Drilling for Microbes

There are some 400 known subglacial, freshwater lakes beneath the Antarctic ice sheet. One of these, Lake Vostok, lies at the depth of more than 3 kilometers below Vostok Station, a Russian research outpost some 1,300 kilometers southeast of the South Pole (see figure). Many scientists believe that Lake Vostok, which is about the size of Lake Ontario, has been isolated from the atmosphere for 15 million years. From a microbiological perspective, the most interesting fact is that the lake is liquid water, which means there is the possibility of microbial life in the ancient lake.

For more than two decades and 57 Antarctic expeditions, Russian scientists had been carefully drilling down through the ice toward the lake to examine the water contents. Then, on Sunday, February 5, 2012, just a day before the expedition would end its season, they hit the lake! However, when the drill contacted the lake, the lake water, being under such great pressure from the ice above, shot up the borehole and froze. At this writing, the scientists have not reported if the frozen water contains any microbes.

Meanwhile, an American research group has collected ice cores from frozen water at the bottom (underside) of the glacier ice covering Lake Vostok. These cores should represent a record of the lake surface water that froze at the glacier interface and they should correspond to what is in the liquid at the surface of the lake.

In 2013, the group reported that they discovered a diverse microbial population in these cores. DNA sequence analyses from four cores identified more than 3,500 unique sequences, which presumably equates to that number of species or strains. About 95% of the sequences were associated with bacterial species (two additional sequences were related with archaeal species) and 5% linked with eukaryotic species, especially the fungi. Therefore, Lake Vostok itself might contain a complex web of organisms that has evolved and developed over the tens of millions of years of its existence.

Other subglacial lakes in Antarctica have also been examined for microbial life. In 2012, an American team reported the discovery of diverse bacterial life in Lake Vida that lies under just 27 meters of ice. This highly saline lake with a water temperature of −13°C was distinct from other saline lakes in the same area. In 2013, a different American team reported preliminary evidence for a diverse community of bacterial and archaeal cells growing in culture dishes from samples of water taken from Lake Whillans that lies 800 meters below the ice sheet.

Should we be surprised with these discoveries? Perhaps not, as we know microbes will eke out an existence anywhere an energy source and nutrients are available. That might include the ice-covered oceans beneath the surface of some moons orbiting the planets Jupiter and Saturn.

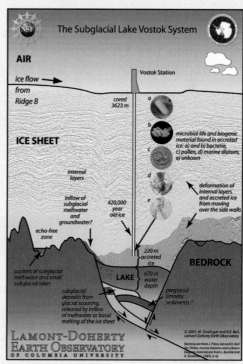

Courtesy of Michael Studinger/Lamont-Doherty Earth Observatory.

Finally, there are bacterial species said to be **capnophilic** (*capno* = "smoke"). These species require an atmosphere low in oxygen but rich in CO_2 gas. Members of the genera *Neisseria* and *Streptococcus* are capnophiles.

Other Essential Chemical Elements

Microorganisms also need a variety of other chemical elements, including nitrogen, sulfur, and phosphate, for cell growth and reproduction. Proteins contain nitrogen and sulfur atoms, and nucleic acids and the cellular energy molecule, ATP, contain nitrogen and phosphorus atoms. Most microbes get their nitrogen and sulfur by degrading proteins into amino acids and then using these amino acids to synthesize new cellular proteins. Some bacterial species also get their nitrogen from ammonium ions (NH_4^+) or nitrate ions (NO_3^-) found in organic soil materials. Phosphorus is usually obtained as phosphate ions (PO_4^{-3}).

Organic Growth Factors

Many microbial cells cannot make one or more essential organic compounds needed for growth.

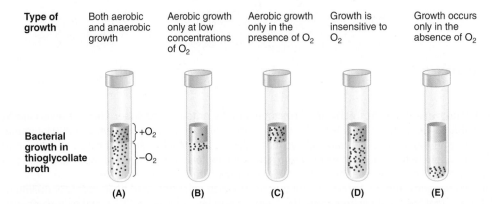

FIGURE 5.13 **The Effect of Oxygen on Microbial Growth.** Each tube contains a thioglycollate broth into which was inoculated a different bacterial species. *»» Identify the O_2 requirement in each thioglycollate tube based on the growth density [example: (A) represents facultative anaerobe species].*

Therefore, these so-called **growth factors** must be obtained from the environment where the microbes are living. For example, some microbes might be unable to synthesize some essential vitamins or amino acids. For these species, the vitamins and amino acids would be growth factors.

Concept and Reasoning Checks 5.2

a. Identify what would be extremophile-type conditions (from the human perspective) for each of the physical factors described in this section.

b. Identify some essential needs for carbon, oxygen, nitrogen, sulfur, and phosphorus in cell growth.

Chapter **Challenge B**

You have now studied many of the major physical and chemical factors that influence microbial growth and reproduction. Here on Earth, many microbes can survive these physical extremes (see Table 5.1).

QUESTION B: *Could a microbe survive the physical extremes on Mars? What kind of earthly microbe might this Martian microbe be like? Here are some useful facts about Mars:*

- Surface temperature: Estimated to be from a warm 27°C (81°F) at the equator to −143°C (−225°F) at the winter polar caps.
- Atmosphere: 95% carbon dioxide, but also present are nitrogen (2.7%), argon (1.6%), and oxygen (0.13%; 21% on Earth) gases; mean surface pressure is much lower than Earth's. UV radiation is 3 times that on Earth (note: there are earthly bacterial species that can survive 5,000 times the dose that would kill human cells).
- Soil: Existence of water ice confirmed; soil pH = about 7.7; several chemicals found that could serve as nutrients for life forms, including magnesium, potassium, and chloride.
- Salt: Dark, finger-like features could be the flow of salty water perhaps equivalent in salinity to Earth's oceans.

You can find answers online in **Appendix F**.

■ KEY CONCEPT 5.3   Culture Media Are Used to Grow Microbes and Measure Their Growth

Microorganisms can be grown (cultured) in a liquid medium (plural: media) or on a solid medium. Liquid media, called **broths**, often are contained in tubes and the broths consist of the chemical nutrients dissolved in water. After sterilization and specimen inoculation, the broth will eventually become **turbid** or cloudy as the microbial population grows in the container.

Solid media consist of the liquid media to which has been added a solidifying agent called **agar**. Agar, a cell wall polysaccharide derived from red algae, contains no nutrients but, like gelatin, it melts when heated and solidifies when cooled below 40°C. After sterilization of the nutrient agar medium, it is poured into sterile culture dishes or tubes where it will solidify on cooling. After inoculation, the specimen will grow as colonies (bacterial organisms and yeasts) or as filaments (molds) on agar or as a turbid broth in liquid form.

In the clinical microbiology lab (CML), colony characteristics on agar (e.g., color, shape, size, and elevation) can be an important step in the diagnosis of a human pathogen. Such samples for identification might have come from the environment (e.g., soils, air, or food) or directly from the affected patient. These clinical specimens could be blood, saliva, feces, or cerebrospinal fluid.

Culture Media Are of Two Basic Types

Since the time of Pasteur and Koch, microbiologists have been growing bacterial and other microbial species in artificial media, that is, in chemical nutrients designed to mimic the natural environment. Because microbial species vary in their nutritional requirements and no single medium will grow all microorganisms, today there are hundreds of different culture (nutrient) media used in the CML and research lab. However, most can be divided into one of two basic types.

A chemically undefined medium, called a **complex medium**, contains nutrients, some of which are known and others in which the exact components or their quantity is not completely known. Such media typically contain animal or plant digests (e.g., beef extract and soybean extract) or yeast extracts of an unknown nature (TABLE 5.3A). Such digests usually contain growth factors and proteins as the source of carbon and nitrogen needed for growth. Complex media are commonly used in the CML and teaching laboratory because most pathogens and microbes used for teaching purposes will grow in these nutrient-rich conditions.

The second type of growth medium is a **chemically defined medium**. In this medium, the precise chemical composition and amounts of all components are known (TABLE 5.3B). In most cases, the CML will not use a chemically defined medium because such media are more costly and the complex medium usually provides all the nutrients needed for growth of typical human pathogens. Chemically defined media are used when trying to determine an organism's specific growth requirements.

TABLE 5.3 Composition of Growth Media

Ingredient	Nutrient Supplied	Amount
A. Complex Agar Medium		
Peptone	Amino acids, peptides	5.0 g
Beef extract	Vitamins, minerals, other nutrients	3.0 g
Sodium chloride (NaCl)	Sodium and chloride ions	8.0 g
Agar		15.0 g
Water		1.0 liter
B. Chemically Defined Broth Medium		
Glucose	Simple sugar	5.0 g
Ammonium phosphate $((NH_4)_2HPO_4)$	Nitrogen and phosphate ions	1.0 g
Sodium chloride (NaCl)	Sodium and chloride ions	5.0 g
Magnesium sulfate $(MgSO_4 \cdot 7H_2O)$	Magnesium and sulphur ions	0.2 g
Potassium phosphate (K_2HPO_4)	Potassium and phosphate ions	1.0 g
Water		1.0 liter

Culture Media Can Be Modified to Identify Microbial Species

In the teaching laboratory, many bacterial species can be identified by their growth in a specific culture medium. Likewise, in the CML, the basic ingredients of nutrient media can be modified in one of three ways to provide fast and critical information for a medical diagnosis about a pathogen causing an infection or disease (TABLE 5.4).

A **selective medium** contains ingredients to inhibit the growth of certain microbes in a mixture while allowing (selecting for) the growth of others. For example, the complex medium might contain extra salt (NaCl) to select for the growth of

TABLE 5.4 A Comparison of Special Culture Media

Name	Components	Uses	Examples
Selective medium	Growth inhibitors	Selecting certain microbes (colonies) not inhibited from a mixed culture	Mannitol salt agar (MSA) for staphylococci
Differential medium	Dyes or indicator systems	Distinguishing among different microbes (colonies) growing on a medium	MacConkey agar for gram-negative bacteria
Enriched medium	Special nutrients	Promoting the growth of fastidious microbes	Blood agar for streptococci; chocolate agar for *Neisseria* species

the halotolerant pathogen *Staphylococcus aureus*. Another selective medium might contain a dye, such as crystal violet, to inhibit the growth of gram-positive bacteria and thus select for the growth of gram-negative species, which tolerate the dye. In the case for eukaryotic microbes, a selective medium might have a slightly acidic pH, which will inhibit the growth of most bacterial species and thus select for the growth of molds.

Another modification to a basic nutrient medium is the addition of one or more compounds that allow the observer to differentiate (tell the difference) between very similar species based on their colony appearance or on a visible change in growth. This **differential medium** contains specific chemicals to indicate which species possess and which species lack a particular biochemical process. For example, such indicators can make it easy to distinguish visually colonies of one organism from colonies of other similar organisms on the same differential medium. Look at **FIGURE 5.14** and see if you can determine by colony appearance which medium was used in each example. **MICROINQUIRY 5** presents another example using a visible change in a broth

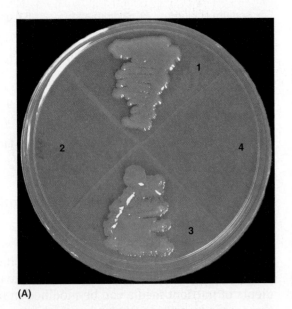
(A)

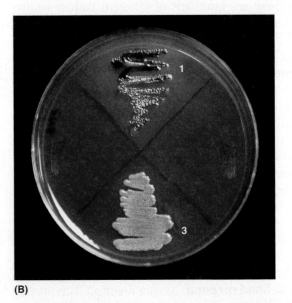
(B)

FIGURE 5.14 Special Media Formulations. (A) Four different bacterial species (1–4), two gram positive and two gram negative, were streaked onto separate sections of a MacConkey agar plate and the plates were allowed to incubate for 48 hours. MacConkey agar only supports the growth of gram-negative species. **(B)** Because the two gram-negative species cannot be visually distinguished from one another in (A), they can be streaked into an eosin methylene blue agar (EMB) plate and incubated for 48 hours. EMB allows one to distinguish between human enteric bacteria, where *Escherichia coli* (1) produces a green metallic sheen while species like *Enterobacter aerogenes* (3) produce a pink color. *»» Which special medium is selective and which is differential? Explain your reasoning.*

(A) and (B) Courtesy of Dr. Jeffrey Pommerville.

MICROINQUIRY 5

Identification of a Bacterial Species

It often is necessary to identify a bacterial species or be able to tell the difference between similar-looking species in a mixture. In microbial ecology, it might be necessary to isolate certain naturally growing species from others in a mixture. In the clinical and public health setting, microbes might be pathogens associated with disease or poor sanitation. In addition, some might be resistant to standard antibiotics normally used to treat an infection. In all these cases, identification can be accomplished by modifying the composition of a complex or synthetic growth medium. Let's go through two scenarios.

■ Suppose that you are an undergraduate student in a marine microbiology course. On a field trip, you collect some seawater samples and, now back in the lab, you want to grow only photosynthetic marine microbes.

How would you select for photosynthetic microbes? First, you know the photosynthetic organisms manufacture their own food, so their energy source will be sunlight and not the organic compounds typically found in culture media (see Table 5.3). Thus, you would need to use a chemically defined medium but leave out the glucose. In addition, you would want to add the salts typically found in ocean water to the medium. You would then inoculate a sample of the collected material into a broth tube, place the tube in the light, and incubate for one week at a temperature typical of where the organisms were collected.

5a. What would you expect to find in the broth tube after one week's incubation?

What you have used in this scenario is a selective medium, that is, one that will encourage the growth of photosynthetic microbes (light and sea salts) and suppress the growth of nonphotosynthetic microorganisms (no carbon = no energy source). Therefore, only marine photosynthetic microbes should be present.

■ As an infection disease officer in a local hospital, you routinely swab critical-care areas to determine if there are any antibiotic-resistant bacteria present. You are especially concerned about methicillin-resistant *Staphylococcus aureus* (MRSA) because it frequently can cause disease outbreaks in a hospital setting. One swab you put in a broth tube showed turbidity after 48 hours.

5b. Knowing that *Staphylococcus* species are halotolerant, how could you devise an agar medium to determine visually if any of the growth is due to *Staphylococcus aureus*?

Again, a selective medium would be used. It would be prepared by adding 7.5% salt to a complex agar medium. A sample from the broth tube would be streaked on the plate and incubated at 37°C for 48 hours.

5c. What would you expect to find on the agar plate after 48 hours?

Your selective medium contained 10 discrete colonies. You do a Gram stain and discover that all the colonies contain clusters of purple spheres; they are gram-positive. However, there are other species of *Staphylococcus* that do not cause disease. One is *S. epidermidis*, a common skin bacterium. A Gram stain therefore is of no use to differentiate *S. aureus* from *S. epidermidis*.

5d. Knowing that only *S. aureus* will produce acid in the presence of the sugar mannitol, how could you design a differential broth medium to determine if any of the colonies are *S. aureus*? (Hint: phenol red is a pH indicator that is red at neutral pH and yellow at acid pH.)

You can identify each bacterial species by taking a complex broth medium, such as nutrient broth, and adding salt and mannitol (mannitol salt broth) and phenol red. Next, you inoculate a sample of each colony into a separate tube. You inoculate the 10 tubes and incubate them for 48 hours at 37°C.

5e. The broth tubes are shown below. What do the results signify? Which tubes contain which species of *Staphylococcus*?

This method is an example of a differential medium because it allowed you to visually differentiate or distinguish between two very similar bacterial species. Knowing which colonies on the original selective medium plate are *S. aureus*, you need to determine which, if any, are resistant to the antibiotic methicillin.

5f. How could you design an agar medium to identify any MRSA colonies?

5g. If the plates are devoid of growth, what can you conclude?

Again, you have used a selective medium; the addition of methicillin will permit the growth of any MRSA bacteria and suppress the growth of staphylococci sensitive to methicillin.

You can find the answers to 5e online in **Appendix E**.

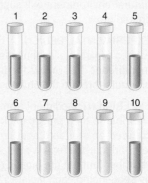

Results from differential broth tubes.

medium to identify and differentiate between similar bacterial species.

Although many microorganisms grow well in standard complex media, some organisms require an **enriched medium**, which contains extra growth factors to promote cell growth and reproduction of these often "slow growers." For example, growth of the causative agent of gonorrhea, *Neisseria gonorrhoeae*, requires powdered hemoglobin as part of the growth medium. Similarly, the growth of the bacterium causing Legionnaires' disease, *Legionella pneumophila*, requires iron and the amino acid cysteine in the culture medium.

Identification of Nongrowing Microbes

Many prokaryotic species are impossible to culture in any known culture medium. In fact, less than 2% of the species in natural water and soil samples can be cultured on agar or in broth. Consequently, it is impossible to estimate accurately microbial diversity in an environment based solely on what can be grown in a nutrient agar plate. Such nongrowing organisms are said to be in a **viable but noncultured** (**VBNC**) state, meaning that they are alive but they will not reproduce in any known culture medium. Therefore, procedures for identifying VBNC organisms include direct microscopic examination and, most commonly, amplification of diagnostic gene sequences or 16S rRNA gene sequences.

Why do these organisms remain uncultured? Microbiologists believe that part of the reason might be due to their presence in a "foreign" environment. These species have adapted to their own familiar and specific natural environment and the somewhat artificial culture medium in the lab is not their typical environment. Therefore, these species are in a growth-arrested form and do not divide; that is, they are viable but cannot be cultured. INVESTIGATING THE MICROBIAL WORLD 5 examines this anomaly in more detail.

Investigating the Microbial World 5

The Great Plate Count Anomaly

About 98% of bacterial species from the environment cannot be grown using known culture media.

OBSERVATION: Take a sediment sample from an environmental water source, mix it with saline solution (or water), and wait for the sediment to settle. Now take a drop from the liquid and place it on the surface of nutrient agar (or any type of complex medium) in a culture dish. Place another drop on a slide and add stain. On the stained slide, you will undoubtedly be able to count hundreds of cells and find dozens of different bacterial morphologies. On the plate, if you are lucky, maybe one or two colonies will appear in a few days. The majority of the cells will not grow even though they are inundated in nutrients. This is the so-called "great plate count anomaly." Standard laboratory culture techniques fail to support the growth of these viable but noncultured (VBNC) species that reside in environmental sediments.

QUESTION: *Why won't 99% of the bacterial species grow in laboratory culture media?*

HYPOTHESIS: VBNCs need some essential "nutrient" from their neighbor species in the natural environment. If so, growing the VBNCs in their natural environment will supply the needed nutrient and the uncultured should grow.

EXPERIMENTAL DESIGN: A diffusion apparatus is designed that sandwiches a microbial sediment sample in agar between two semi-permeable membranes that allow for the free diffusion of "nutrients" and waste products through the chamber (see figure).

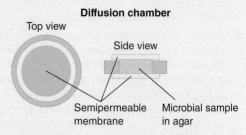

Diffusion chamber

Top view

Side view

Semipermeable membrane

Microbial sample in agar

Modified from Bollman, A., et al. 2007. *Appl Environ Microbiol* 73(20):6386–6390.

EXPERIMENT: Sediment from a freshwater pond sample is mixed with agar and placed within the diffusion chamber. The chamber is then placed back in its natural environment. Another pond sediment sample from the same environment is cultured on standard agar culture medium in a culture dish. Both the diffusion chamber culture and the lab culture are left undisturbed for 4 weeks. After the incubation period, the colonies on the culture plate and on the agar in the diffusion chamber (but not the biofilm growing on the semipermeable membrane) are identified. Phylum identification is carried out by sequencing each isolate's 16S ribosomal RNA gene.

RESULTS: See table.

Phylum Strains Obtained by Culture Dish and Diffusion-Chamber Methods

| | Number of Strains Isolated by | | | |
Phylum	Culture Dish Only	Diffusion Chamber Only	Both Methods	Total Number of Isolated Strains
Alphaproteobacteria	20	36	6	62
Betaproteobacteria	3	63	3	69
Gammaproteobacteria	2	4	1	7
Deltaproteobacteria		1		1
Bacteroidetes	5	6	1	12
Spirochaetes		4		4
Firmicutes	5	1		6
Actinobacteria	5	1		6
Total	40	116	11	167

CONCLUSIONS:

QUESTION 1: *Was the hypothesis supported? Explain using the table.*

QUESTION 2: *Are the majority of the isolates representative of gram-positive or gram-negative organisms? Explain.*

QUESTION 3: *What might have been the result if the biofilm organisms growing on the membranes were included in the analysis?*

FURTHER QUESTION: *What is it that the previously uncultured colonies in the diffusion chamber recognize? New evidence suggests that one "nutrient" is iron, which is needed for ATP generation and other biochemical processes. To grow, the cells in the uncultured colonies must "capture" the iron (presumably from their neighbors) in the bound form that is otherwise unavailable in a culture dish "environment."*

You can find answers online in **Appendix E**.

Modified from Bollman, A., et al. 2007. *Appl Environ Microbiol* 73(20):6386–6390.

Studies on VBNC prokaryotic species present a vast and as yet unexplored field (they are often referred to as "microbial dark matter"), which is important not only for detection of human pathogens but also to reveal the tremendous diversity in the microbial world.

Population Measurements Are Made Using Pure Cultures

Microorganisms rarely occur in nature as a single species. Rather, they are mixed with other species, in a so-called "mixed culture" most often as a biofilm. Therefore, to study a species that can be cultured, microbiologists and laboratory technologists must use a **pure culture**, that is, a population consisting of only one species.

Suppose that you have a mixed broth culture containing several bacterial species. How can the organisms be isolated as independent pure cultures? Two established methods are available. The first is the **pour-plate method**. With this procedure, diluted samples of the mixed culture in molten nutrient agar tubes are each poured into a sterile Petri dish where the agar solidifies. During a 24-hour to 48-hour incubation, the cells divide to form discrete colonies on and in the agar (**FIGURE 5.15**). The assumption is that each colony represents a population of cells that were derived from one original cell.

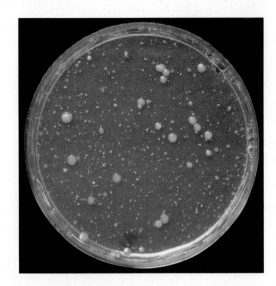

FIGURE 5.15 A Pour Plate. When mixed in molten agar, the dispersed bacterial cells grow as individual, discrete colonies. *»» By looking at this plate, how would you know the original broth culture was a mixture of bacterial species?*

Courtesy of Dr. Jeffrey Pommerville.

A second, more commonly used technique, called the **streak-plate method**, uses a single plate of sterile nutrient agar. An inoculum from a mixed culture is removed with a sterile loop, and a series of streaks is made over one area of the plate (**FIGURE 5.16A–D**). The loop is flamed, touched to the first area, and a second series of streaks is made in a second area. Similarly, streaks are made in the third and fourth areas, thereby separating the cells so that they can grow into individual colonies. After a 24-hour to 48-hour incubation, discrete colonies will be present on the plate (**FIGURE 5.16E**).

In both methods, one assumes each colony is derived from an original single cell that underwent numerous binary fissions. The researcher, medical technologist, or student can select samples of the colonies for further testing and analysis from which a pure culture can be generated.

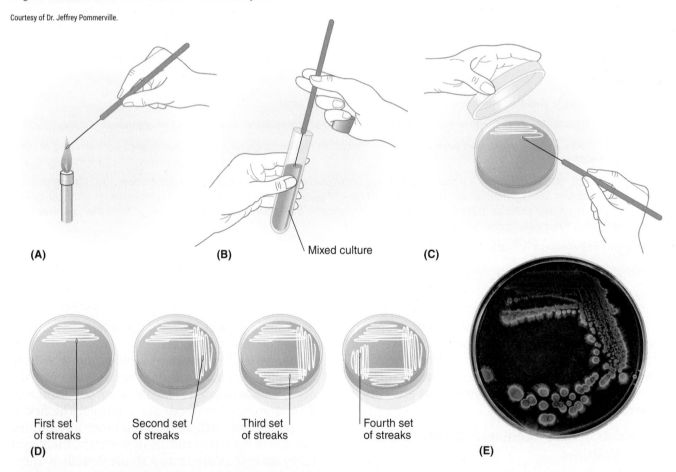

(A)   (B)   Mixed culture   (C)

First set of streaks   Second set of streaks   Third set of streaks   Fourth set of streaks

(D)   (E)

FIGURE 5.16 The Streak-Plate Method. (A) A loop is sterilized and **(B)** used to obtain a sample of cells from a mixed culture. **(C)** The cells are streaked near one edge of the agar medium. **(D)** After additional streaks are performed, the plate is incubated. **(E)** After incubation, well-isolated and defined colonies illustrate a successful isolation. *»» Justify the need to streak a mixed sample over four areas on a culture plate.*

(E) Courtesy of Dr. Jeffrey Pommerville.

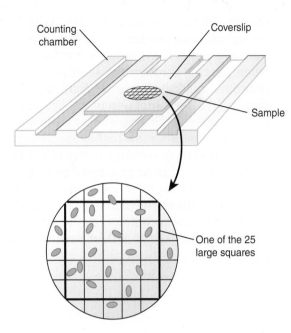

A The counting chamber is a specially marked slide containing a grid of 25 large squares of known area. The total volume of liquid held is 0.00002 ml (2×10^{-5} ml).

B The counting chamber is placed on the stage of a light microscope. The number of cells are counted in several of the large squares to determine the average number.

FIGURE 5.17 Direct Microscopic Count. This procedure can be used to estimate the total number of live and dead cells in a culture sample. *»» Suppose that the average number of cells per large square was 14. Calculate the number of cells in a 10-ml sample.*

Population Growth Can Be Measured in Several Ways

Microbial growth in a culture medium can be measured by direct and indirect methods. A few common methods are described in the subsections that follow.

Direct Methods

There are several ways to measure cell numbers directly. Scientists might want to perform a **direct microscopic count**. In this procedure, a known volume of the liquid sample is placed on a specially designed microscope slide that represents a counting chamber (**FIGURE 5.17**). However, this procedure will count both living and dead cells.

The **most probable number (MPN) test** is usually used for samples that have a low number of organisms or that will not grow on agar media. The technique has been used as a statistical estimation for bacterial cell number when measuring water quality. Undiluted microbial samples, as well as samples diluted 10 times and 100 times, are added to a set of broth tubes. After a 24-hour incubation, the presence or absence of gas or turbidity in the tubes is determined. The results are then compared to an MPN table, which gives a rough statistical estimation of the most probable cell number of living cells.

In the **standard plate count procedure**, a sample from a broth culture is placed in a sterile culture dish and melted nutrient agar is added (pour-plate method) (**FIGURE 5.18**). The assumption is that each cell will

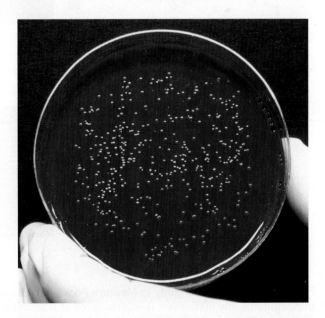

FIGURE 5.18 The Standard Plate Count. Each individual bacterial colony growing on this blood agar plate represents a colony-forming unit (CFU). *»» If a 0.1-ml sample of a 10^4 dilution contained 250 colonies, how many bacterial cells were in 10 ml of the original broth culture?*

© R.A. Longuehaye/Science Source.

undergo multiple rounds of cell division to produce separate colonies on the plate. Because two or more cells could clump together on a plate and grow as a single colony, the standard plate count is expressed as the number of **colony-forming units (CFUs)**. After incubation, the number of CFUs is used to estimate the number of living cells originally plated.

Indirect Methods

One of the simplest indirect methods for determining cell numbers in a population is to estimate the **turbidity** (cloudiness) in a broth culture medium. Turbidity is measured by using a spectrophotometer, which detects the amount of light scattered by a suspension of cells. When placed in the spectrophotometer, the cells in the broth culture will deflect or scatter a portion of the light beam passing through the sample in the spectrophotometer. The amount of light scatter (optical density [OD]) is a function of the cell number; that is, the more cells that are present, the more light is scattered, resulting in a higher absorbance reading on the spectrophotometer meter (**FIGURE 5.19**). A standard curve can be generated to serve as a measure of cell numbers. However, because more than 10 million cells are needed to get a reliable reading on the spectrophotometer, turbidity is not a useful way to study the growth of small populations of bacterial cells.

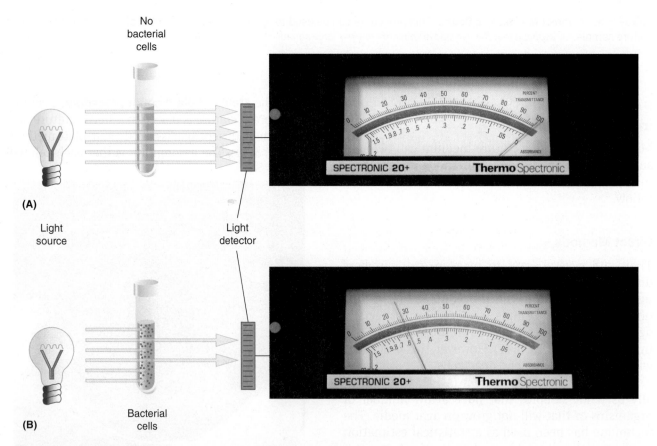

FIGURE 5.19 Using Turbidity to Measure Population Growth. (A) As light passes through a sterile broth tube in the spectrophotometer, the instrument is standardized at 0 absorbance. **(B)** As a bacterial population in another broth tube grows, the cells will scatter more of the light, which on the spectrophotometer is detected as an increase in absorbance. *»» Why do turbidity measurements represent an indirect method to measure population growth?*

Courtesy of Dr. Jeffrey Pommerville.

In conclusion, we examined the major physical and chemical factors, and the nutrient media formulations, affecting the rate at which microbial populations grow if they are not an example of a VBNC. This brings up an interesting problem when considering the diagnosis and treatment of an infectious disease. Such pathogens, including *Treponema pallidum*, have been identified from actual clinical samples using direct microscopic observation. So, one has to wonder: How many human diseases might go undiagnosed today because they are VBNC? In fact, microbiologists have wondered whether members of the Archaea can cause human disease. No species are currently known to cause human disease, yet we know from microscopic observation and DNA sequencing that archaeal organisms are present in the human body. Could some of them be VBNC pathogens?

Concept and Reasoning Checks 5.3

a. Compare and contrast complex and chemically undefined media.
b. Explain how microbiologists have figured that some 98% of the microbial world has not been "seen."
c. Why might it be more difficult to isolate a colony from a pour-plate culture than from a streak-plate culture?
d. Distinguish between direct and indirect methods to measure population growth.

Chapter Challenge C

On November 13, 1971, *Mariner 9* went into orbit around Mars. Before its arrival, we thought Mars was a dead world. Then *Mariner 9* sent back pictures of a landscape where it appeared water once flowed. However, the craft found no signs of biological material in the Martian soil. As more spacecraft have gone to Mars, the evidence for life on the Red Planet has waxed and waned—yet some are hopeful that microbial life might exist near the surface or deeper underground.

QUESTION C: *As a result, several missions to Mars are being planned by American, European, Russian, and Chinese space agencies. As an exomicrobiologist (one who looks and searches for microbial life beyond Earth), what types of experiments would you design (based on this chapter) to see if microbial life does exist on Mars? The spacecraft will not return to Earth, so the experiments need to be completed on Mars. However, communication networks are state-of-the-art.*

You can find answers online in **Appendix F**.

■ SUMMARY OF KEY CONCEPTS

Concept 5.1 Microbial Reproduction and Growth Are Part of the Cell Cycle

1. The bacterial **cell cycle** involves "metabolic" growth, DNA replication, and **binary fission** to produce genetically identical daughter cells. (Figure 5.2A)
2. A cell cycle represents the **generation time**, which may be as short as 15 minutes. (Figure 5.3)
3. The dynamics of the **bacterial growth curve** show how a microbial population grows logarithmically, reaches a certain peak and levels off, and then might decline. (Figure 5.4)
4. Growth control in a **biofilm** is highly regulated and controlled. (Figure 5.6)
5. Dormancy as related to **persister cells** and **endospores** is a response to potential or actual environmental change. (Figures 5.7, 5.8)

Concept 5.2 Optimal Growth Is Dependent on Several Physical and Chemical Factors

6. Temperature, pH, and hydrostatic/osmotic pressure are physical factors that influence microbial growth. Chemical factors include carbon, oxygen, other elements, and growth factors. (Figures 5.10, 5.13)

Concept 5. 3 Culture Media Are Used to Grow Microbes and Measure Their Growth

7. **Complex** and **chemically defined media** contain the nutrients for microbial growth.
8. Growth media can be modified to select for a desired microbial species, to differentiate between two similar species, or to enrich for species requiring special nutrients. (Figure 5.14)
9. **Pure cultures** can be produced from a mixed culture by the **pour-plate method** or the **streak-plate method**. In both cases, discrete colonies can be identified that represent only one microbial species. (Figures 5.15, 5.16)
10. Microbial growth can be measured by **direct microscopic count**, the **most probable number test**, and the **standard plate count** procedure. Indirect methods often use **turbidity** measurements. (Figures 5.17, 5.18)

■ CHAPTER SELF-TEST

For **Steps A–D**, you can find answers to questions and problems online in **Appendix D**.

■ STEP A: REVIEW OF FACTS AND TERMS

Multiple Choice

Read each question carefully before selecting the *one* answer that best fits the question or statement.

1. Which one of the following statements does *not* apply to bacterial cell division?
   A. A fission ring apparatus guides the process of cytokinesis.
   B. Septum formation occurs to separate the two cell cytoplasms.
   C. A spindle apparatus is used to separate chromosomes.
   D. Cell division is part of the cell cycle.
2. If a bacterial cell in a broth tube has a generation time of 40 minutes, how many cells will there be after 6 hours of optimal growth?
   A. 18
   B. 64
   C. 128
   D. 512

3. The generation time for a bacterial species would be determined during the _____ phase.
   A. decline
   B. lag
   C. log
   D. stationary
4. Which one of the following is *not* an example of dormancy?
   A. Log phase
   B. Endospore formation
   C. VBNCs
   D. Persister cells

5. A microbe that is a microaerophilic mesophile would grow optimally at _____ and _____.
   A. high O_2; 30°C
   B. low O_2; 20°C
   C. no O_2; 30°C
   D. low O_2; 37°C
6. If the carbon source in a growth medium is beef extract, the medium must be an example of a/an _____ medium.
   A. complex
   B. chemically defined
   C. enriched
   D. differential

7. A _____ medium would involve the addition of the antibiotic methicillin to identify methicillin-resistant bacteria.
   A. differential
   B. selective
   C. thioglycollate
   D. VBNC
8. Which one of the following is *not* part of the streak-plate method?
   A. Making four sets of streaks on a plate
   B. Diluting a mixed culture in molten agar
   C. Using a mixed culture
   D. Using a sterilized loop
9. Direct methods to measure bacterial growth would include all the following except _____.
   A. total bacterial count
   B. direct microscopic count
   C. turbidity measurements
   D. most probable number

True-False

Each of the following statements is **true (T)** or **false (F)**. If the statement is false, substitute a word or phrase for the underlined word or phrase to make the statement true.

10. _____ Endospores are produced by some gram-negative bacterial species.
11. _____ Obligate aerobes use oxygen gas as a final electron acceptor in energy production.
12. _____ The most common growth medium used in the teaching laboratory is a complex medium.
13. _____ The majority of prokaryotic organisms can be cultured in growth media.
14. _____ In attempting to culture a fastidious bacterial pathogen, a differential medium would be used.
15. _____ Acidophiles grow best at pHs greater than 7.

16. _____ Mesophiles have their optimal growth near 37°C.
17. _____ Prokaryotic cells lack a mitotic spindle to separate chromosomes.
18. _____ The fastest doubling time would be found in the lag phase of a bacterial growth curve.
19. _____ If *E. coli* cells are placed in distilled water, they will burst.
20. _____ Halophiles would dominate in marine environments.

■ STEP B: CONCEPT REVIEW

21. Describe the three phases of a **bacterial cell cycle**. (**Key Concept 5.1**)
22. Summarize the events of each phase of a **bacterial growth curve**. (**Key Concept 5.1**)
23. Explain the importance of bacterial **dormancy**. (**Key Concept 5.1**)

24. Identify the three major physical factors governing microbial growth and describe how microorganisms have adapted to these physical environments. (**Key Concept 5.2**)
25. Discuss how **selective** media and **differential media** are each constructed. (**Key Concept 5.3**)
26. Explain the procedures used in the **pour-plate** and **streak-plate** methods. (**Key Concept 5.3**)

■ STEP C: APPLICATIONS AND PROBLEM SOLVING

27. Use the log phase growth curves (1, 2, or 3) below to answer each of the following questions (a–c).

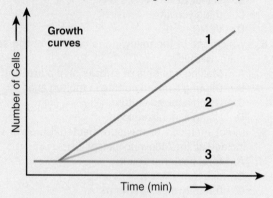

Growth curves

a. ____ Which curve (1, 2, or 3) best represents the growth curve for a mesophile incubated at 60°C?

b. ____ Which curve (1, 2, or 3) best represents a nonhalophile growing in 5% salt?

c. ____ Which curve (1, 2, or 3) best represents an acidophile growing at pH 4?

28. Consumers are advised to avoid stuffing a turkey the night before cooking, even though the turkey is refrigerated. A homemaker questions this advice and points out that the bacterial species of human disease grow mainly at warm temperatures, not in the refrigerator. What explanation might you offer to counter this argument?

29. Public health officials found that the water in a Midwestern town was contaminated with sewage bacteria. The officials suggested that homeowners boil their water for a couple of minutes before drinking it. (a) Would this treatment sterilize the water? Why? (b) Is it important that the water be sterile? Explain.

■ STEP D: QUESTIONS FOR THOUGHT AND DISCUSSION

30. To prevent decay by bacterial species and to display the mummified remains of ancient peoples, museum officials place the mummies in glass cases where oxygen has been replaced with nitrogen gas. Why do you think nitrogen is used?

31. Extremophiles are of interest to industrial corporations, who see these organisms as important sources of enzymes that function at temperatures of 100°C and pH levels of 10 (the enzymes have been dubbed "extremozymes"). What practical uses can you foresee for these enzymes?

32. During the filming of the movie *Titanic*, researchers discovered at least 20 different bacterial and archaeal species literally consuming the ship, especially a rather large piece of the midsection. What types of environmental conditions are these bacterial and archaeal species subjected to at the wreck's depth of 12,600 feet?

33. Every year news media report cases of skin and lung infections in people sitting in hot tubs. How might such infections occur in hot tubs?

Concept Mapping

See pages XXXI-XXXII on how to construct a concept map.

34. Construct a concept map for **Growth Measurements** using the following terms.

Colony-forming unit
Direct methods
Direct microscopic
   count
Indirect methods

Metabolic activity
Most probable number test
Standard plate count
Turbidity

35. Construct a concept map for microbial need for **Oxygen Gas (O_2)** using the following terms.

Aerobic
Aerotolerant
Anaerobic
Facultative anaerobes

Microaerophiles
Obligate aerobes
Obligate anaerobes

CHAPTER 6

Microbial Metabolism

Charlie Swaart had been a social drinker for years. A few beers or drinks with his pals, but no lasting alcoholic consequences. Then, in 1945, he began a nightmare that would make medical history.

One October day, while stationed in Tokyo, Japan, after World War II, Swaart suddenly became drunk for no apparent reason. He had not had any alcohol for days, but suddenly he felt like he had been partying all night. After sleeping it off, he would be fine the next day.

Unfortunately, this "behavior" returned repeatedly. For years after his return to the United States, the episodes continued—bouts of drunkenness and monumental hangovers without drinking so much as a beer! Doctors were puzzled because they couldn't detect alcohol on his breath or in his blood. Was this some type of internal metabolism gone haywire? Was it the result of a bacterial infection? It didn't seem likely. They warned him to stop drinking alcohol for fear that the alcohol would damage his liver. Swaart followed their advice to the letter; still, he experienced periods of drunkenness.

Twenty years passed before Swaart, known as the "drinkless drunk," learned of a similar case in Japan. A Japanese executive had endured years of social and professional disgrace before doctors discovered a yeast-like fungus growing in his intestine. The Japanese studies indicated that the fungal cells were fermenting carbohydrates to alcohol right there in his intestine. The fungus was identified as *Candida albicans* (see chapter opening photo).

Now, having *C. albicans* in one's intestine is rare; then again, finding fermenting *C. albicans* was historic. With the knowledge of the Japanese case, Swaart went to his doctor and asked to be tested for *C. albicans*. Sure enough, lab tests showed massive colonies of *C. albicans* in Swaart's intestines. The sugar in a cup of coffee or any carbohydrate in pasta, cake, or candy allowed the yeast to ferment sugars into ethanol, resulting in apparent drunkenness.

Swaart then learned that an antifungal drug had worked to kill the yeast cells in the Japanese executive's gut. However, to cure his own illness, Swaart had to travel back to Japan to get the effective drug.

Researchers suggested the atomic blasts of Hiroshima and Nagasaki in 1945 might have caused a normal *C. albicans* to mutate to a fermenting form, which somehow found its way into Swaart's digestive system. For Charlie Swaart, though, the nightmare was finally over.

Candida albicans (growing on agar).
© ksass/Getty Images.

The process of fermentation described here was in a eukaryotic microbe. However, many other fermentation processes also occur in prokaryotic microbes and they are but one aspect of the broad topic of microbial metabolism.

In this chapter, we examine some of the major metabolic pathways carried out by microorganisms. Much of the chapter centers on enzymes and adenosine triphosphate (ATP) because these two molecules are key to metabolic reactions. The chapter also emphasizes the role of carbohydrates in the energy conversions, so it might be worthwhile for you to review the major organic compounds (carbohydrates, lipids, and proteins).

Chapter Challenge

What do a termite's gut, a cow's rumen, and the warm, waterlogged soil in a rice paddy (field) have in common? They all produce substantial amounts of methane (CH_4) gas. In Africa, Australia, and South America, there are large areas inhabited by mound-building termites whose mounds can be up to 9 meters (nearly 30 feet) tall. The methane gas produced from the termites in these mounds contributes some 5% of the atmospheric methane. Methane produced from ruminant livestock, like dairy cattle, goats, and sheep, is currently estimated to contribute 16% of the atmospheric methane. However, rice production currently accounts for approximately 20% of global methane emissions. Therefore, methane emissions from these three "natural" sources account for more than 40% of the methane gas being released every day into the atmosphere. Where is this methane gas coming from? You can probably guess—microbes! Still, how do termite guts, cow rumens, and waterlogged soils in rice paddies produce the methane gas? Let's uncover the source!

▌ KEY CONCEPT 6.1 Enzymes and Energy Drive Cellular Metabolism

Metabolism refers to all the biochemical reactions taking place in a living organism. Because much of this metabolism requires or produces energy, cells of an organism need to manage and balance their energy resources. This balance between energy need and energy production involves two types of chemical reactions.

For some cellular reactions to occur, energy is needed. Such reactions are said to be **anabolic** because the reactions build larger organic compounds (**polymers**) from simpler building blocks (**monomers**). For example, photosynthesis is an anabolic process because the sugar glucose is built from carbon dioxide (CO_2) and water. Building reactions use energy in forming the bonds between the monomers, so anabolic reactions are said to be **endergonic** (*end* = "inner"; *ergon* = "work") **reactions** because they use more energy than they produce.

Other cellular reactions are **catabolic**, which means the reactions break down (hydrolyze) polymers into simpler molecules. A major catabolic pathway in cells is cellular respiration wherein sugars, like glucose, are broken down into CO_2 and water. Such catabolic reactions are referred to as **exergonic** (*ex* = "outside of") **reactions** because they produce more energy than they consume. TABLE 6.1

compares anabolism and catabolism, which often take place simultaneously in cells and organisms to maintain the cell's energy balance.

Enzymes Catalyze All Chemical Reactions in Cells

Microbial growth depends on metabolic processes that occur in the cell cytosol, on the cell (plasma)

TABLE 6.1 A Comparison of Two Key Aspects of Cellular Metabolism

Anabolism	Catabolism
Synthesis of larger molecules	Breakdown of large molecules
Products are large molecules	Products are small molecules
Photosynthesis	Glycolysis, citric acid cycle
Mediated by enzymes	Mediated by enzymes
Energy generally is required (endergonic)	Energy generally is released (exergonic)

membrane, in the periplasm (gram-negative bacterial cells), in eukaryotic organelles, and outside the cell. To carry out these reactions, cells need a large and diverse group of enzymes.

Enzymes usually are proteins that increase the probability of chemical reactions, while themselves remaining unchanged in the reaction. Enzymes accomplish in fractions of a second what otherwise might take hours, days, or longer to occur spontaneously. For example, even though organic molecules such as amino acids have functional groups, it is highly unlikely that they would randomly collide with one another in the precise way needed for a chemical reaction (dehydration synthesis) to occur that would bond the two amino acids together.

Enzymes have several common characteristics.

- **Enzymes are reusable**. As soon as a chemical reaction has occurred, the enzyme is released to participate in another identical reaction. In fact, the same enzyme can catalyze the same type of reaction 100 to 1 million times each second.

- **Enzymes are highly specific**. An enzyme that functions in one type of chemical reaction usually will not participate in another type of reaction. Consequently, there must be thousands of different enzymes to catalyze the thousands of different chemical reactions of metabolism occurring in a microbial cell.

- **Enzymes have an active site**. Enzyme specificity is regulated by a special pocket in the enzyme called an **active site**, which has a specific three-dimensional shape complementary to a reactant (called a **substrate**). The active site positions the substrate such that it is highly likely they will collide in such a way that a chemical reaction will occur to form one or more **products**.

- **Enzymes are required in very small amounts**. Because an enzyme can be used thousands of times to catalyze the same reaction, only tiny amounts of a particular enzyme are needed to ensure that a chemical reaction occurs.

Many enzymes can be identified by their names, which often end in "-ase." For example, "sucrase" is the enzyme that breaks down sucrose, and "ribonuclease" digests ribonucleic acid (RNA). In terms of anabolic metabolism, "polymerases" assemble polymers from monomers and "transferases" transfer a functional group from one molecule to another.

Enzymes Act Through Enzyme-Substrate Complexes

The active site of an enzyme aligns the substrate molecules in such a way that a reaction is highly favorable. In the hydrolysis reaction shown in **FIGURE 6.1**, the three-dimensional shape of the enzyme's active site recognizes and holds the substrate as an **enzyme-substrate complex**. While in the complex, chemical bonds in the substrate are stretched or weakened by the enzyme, causing the bond to break. By contrast, in a dehydration reaction the enzyme-substrate complex aligns the substrates so that they will collide in such a way to form a chemical bond. Thus, in a hydrolysis or dehydration reaction, recognition of the substrate(s) is a precisely controlled, nonrandom event.

Looking at sucrose again, the bonds holding glucose and fructose together will not break spontaneously. This is because the bond between the monosaccharides is stable and there is a substantial energy barrier preventing such a reaction (**FIGURE 6.2A**). The job of any enzyme is to bind the substrate in an enzyme-substrate complex, which lowers the energy barrier so that it is much more likely the reaction will occur. For sucrase, the bond holding glucose to fructose needs to be destabilized (i.e., stretched, weakened) while in the enzyme-substrate complex (**FIGURE 6.2B**). The energy barrier that must be overcome for a chemical reaction is called the **activation energy**.

Enzymes, then, play a key role in metabolism because they provide a reaction pathway that lowers the activation energy barrier. These catalysts assist in the destabilization of chemical bonds and the formation of new ones by separating or joining atoms in a carefully orchestrated fashion.

Some enzymes are made up entirely of protein. An example is **lysozyme**, an antimicrobial enzyme in human tears and saliva that hydrolyzes the bond between N-acetylglucosamine (NAG) and N-acetylmuramic acid (NAM) in the cell walls of gram-positive bacterial cells. Many enzymes, however, contain small, nonprotein components that participate in the catalytic reaction. If these are metal ions, such as magnesium (Mg^{+2}), iron (Fe^{+2}), or zinc (Zn^{+2}), they are called **cofactors**. If the nonprotein component is a small organic molecule, it is called a **coenzyme**, most of which are derived from vitamins. Examples of two important coenzymes are **nicotinamide adenine dinucleotide (NAD+)** and **flavin adenine dinucleotide (FAD)**. These coenzymes act as carriers of electrons by transporting the electrons

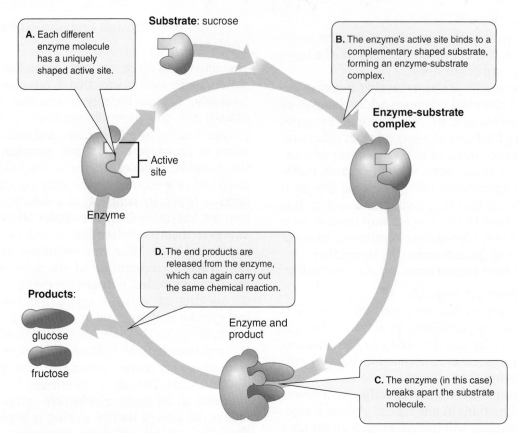

FIGURE 6.1 The Mechanism of Enzyme Action. Although this example shows an enzyme hydrolyzing a substrate (sucrose), enzymes also catalyze dehydration reactions, which combine glucose and fructose into sucrose. »» *How do enzymes recognize specific substrates?*

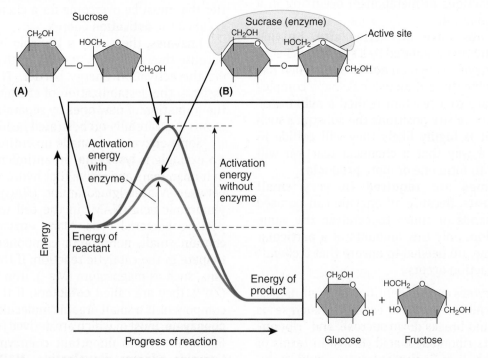

FIGURE 6.2 Enzymes and Activation Energy. Enzymes lower the activation energy barrier required for chemical reactions of metabolism. **(A)** The hydrolysis of sucrose is unlikely because of the high activation energy barrier. **(B)** When sucrase is present, the enzyme effectively lowers the activation energy barrier at the transition state **(T)**, making the hydrolysis reaction highly favorable. »» *How does the enzyme lower the activation energy of a stable substrate?*

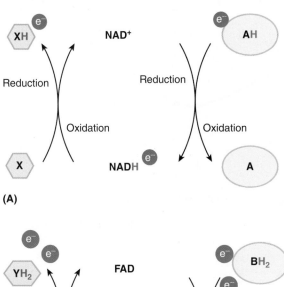

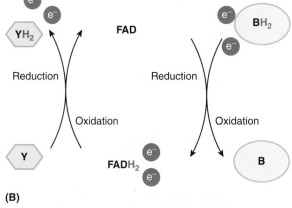

FIGURE 6.3 **Coenzymes and Redox Reactions.** Coenzymes can carry and release electrons and hydrogen atoms. **(A)** NAD^+ can pick up one electron and one hydrogen ion (proton) from a metabolic reaction and, in doing so, be reduced to NADH (reduction reaction). When the reduced form loses the electron, it is oxidized back to NAD^+ (oxidation reaction). **(B)** Likewise, FAD can be reduced to $FADH_2$ by picking up two electrons and two protons. It is oxidized back to FAD when it releases the electrons and hydrogen atoms. **»» *So, what is meant by a redox reaction?***

from a substrate and donating the electrons to another chemical reaction in so-called redox reactions, as explained in **FIGURE 6.3**. Both coenzymes play a significant role in ATP generation, and we will encounter them later in this chapter. Another coenzyme we will encounter is **nicotinamide adenine dinucleotide phosphate** (**NADP**$^+$), which plays a critical role in the anabolic reactions of photosynthesis.

Enzymes Team up in Metabolic Pathways

There are many examples, such as the sucrose example, where an enzymatic reaction is a single substrate to product reaction. However, cells more often use metabolic pathways. A **metabolic pathway** is a sequence of chemical reactions, each

reaction catalyzed by a different enzyme, in which the product (output) of one reaction serves as a substrate (input) for the next reaction (**FIGURE 6.4A**). The pathway begins with the initial substrate and finishes with the final product. The products of "in-between" stages are referred to as "intermediates."

Metabolic pathways can be anabolic, whereby larger molecules are synthesized from smaller monomers. In contrast, catabolic pathways break larger molecules into smaller ones. Such pathways can be linear, branched, or cyclic. We will see examples of these pathways in the microbial metabolism sections just ahead.

Enzyme Activity Is Affected by Several Factors

Enzyme activity can be controlled or inhibited by environmental factors and by cellular controls in metabolic pathways.

Enzyme Activity and Environmental Factors

Several physical factors affect microbial growth. One physical factor is temperature whereby the further the temperature deviates from the optimal, the slower the enzyme operates. Because most enzymes are proteins, they are sensitive to changes in temperature; indeed, high temperature can denature a protein. This loss of three-dimensional structure leads to catalytic failure, which most likely will result in the shutdown of the metabolic pathway.

Another physical factor affecting microbial growth is pH. Enzymes have an optimal pH at which they operate, so an increase or decrease in protons (hydrogen ions [H^+]) will decrease an enzyme's reaction rate, with extreme changes leading to enzyme denaturation and metabolic inhibition.

In addition, chemicals applied "environmentally" can inhibit enzyme action. Alcohols and phenol inactivate enzymes and precipitate proteins, making these chemical agents effective as antiseptics and disinfectants. Other natural or synthetic chemicals interfere with enzyme action (e.g., penicillin and sulfa drugs, respectively), making these agents effective antibiotics.

Enzyme Inhibition via Metabolic Pathway Modulation

Cells regulate enzymes so that they are present or active only when the appropriate substrate is present.

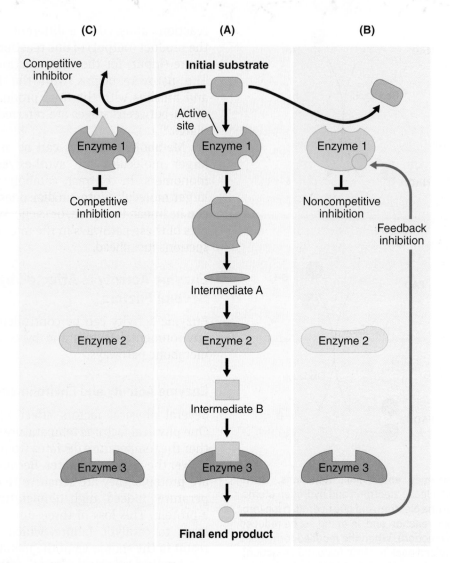

(C)   **(A)**   **(B)**

FIGURE 6.4 Metabolic Pathways and Enzyme Inhibition. (A) In a metabolic pathway, a series of enzymes transforms an initial substrate into a final product. **(B)** If excess final product accumulates, it "feeds back" on the first enzyme in the pathway and inhibits the enzyme by binding at another site on the enzyme. **(C)** In competitive inhibition, a substrate that resembles the normal substrate competes with the substrate for the enzyme's active site. Competitive inhibition would reduce the productivity of the metabolic pathway by slowing down or stopping the pathway. In both **(B)** and **(C)**, the entire pathway can become temporarily inoperative. *»» Hypothesize why most substrates cannot be converted into a final product in one enzymatic step.*

One of the most common ways of controlling enzyme activity occurs when the final product of a metabolic pathway inhibits an enzyme in that pathway (**FIGURE 6.4B**). In this example, if the first enzyme in the pathway is inhibited, no more product is available as input for the rest of the pathway. Such **feedback inhibition** stops the cell from producing more product than is necessary. In general, the final product of the metabolic pathway, or another molecule in the pathway, binds to the enzyme somewhere other than the active site. However, when this binding occurs, the shape of the active site changes and the site can no longer bind the substrate. This type of modulation is referred to as **noncompetitive inhibition**.

Another way of affecting enzyme activity is by directly blocking its active site. If a molecule closely resembles a substrate, the inhibitor (called a competitive inhibitor) blocks the active site so that the normal substrate cannot bind (**FIGURE 6.4C**). Such **competitive inhibition**, therefore, prevents the catalytic reaction because the normal substrate is kept out of the active site. Some competitive inhibitors, such as the sulfa drugs, bind reversibly and simply slow down product formation by the metabolic pathway. The metabolic rate can be so slow that the cell dies. Other competitive inhibitors bind so tightly to an enzyme's active site that the enzyme is inoperable and the metabolic pathway is shut down permanently. Competitive inhibition is uncommon in the cell, so this form of enzyme inhibition usually is the result of environmental substances.

Energy in the Form of ATP Is Required for Metabolism

Even with enzymes, metabolism would be unlikely without a way to couple the energy-requiring anabolic reactions with the energy-releasing catabolic reactions in cells. This coupling of reactions occurs using ATP (**FIGURE 6.5A**), the so-called cellular energy currency of all cells. An ATP molecule acts like a portable battery. It provides the needed energy for anabolic reactions and for energy-demanding processes like flagellar motion, active transport, and spore formation. On a more biochemical level, ATP fuels protein synthesis and macromolecular

syntheses. In fact, a major share of microbial functions depends on a continual supply of ATP. Should the supply be cut off, the cell will die very quickly because ATP cannot be stored for future use.

In prokaryotic cells, most of the ATP is generated on the cell membrane, whereas in eukaryotic cells, the reactions occur primarily in the mitochondria.

ATP molecules are relatively unstable. In Figure 6.5A, notice that the three phosphate groups all have negative charges on an oxygen atom. Similar charges repel, so the phosphate groups in ATP, being tightly packed together, are very unstable. Breaking the so-called high-energy bond holding the last phosphate group on the molecule produces a more stable **adenosine diphosphate** (**ADP**) molecule and a free phosphate group (**FIGURE 6.5B**). ATP hydrolysis is analogous to a spring compacted

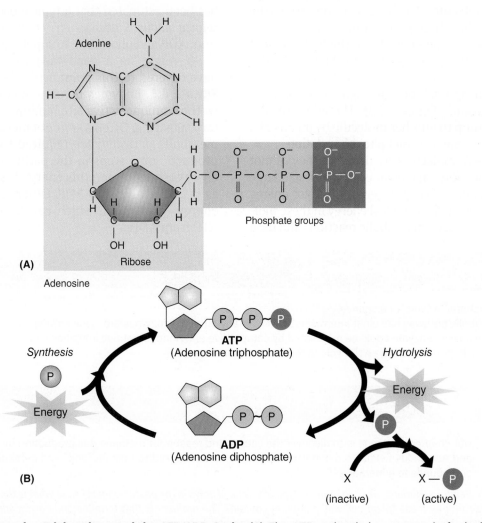

FIGURE 6.5 Adenosine Triphosphate and the ATP/ADP Cycle. (A) The ATP molecule is composed of adenine and ribose bonded to one another and to three phosphate groups. **(B)** When the ATP molecule breaks down, it releases a phosphate group and energy and becomes ADP. The freed phosphate can activate another chemical reaction through phosphorylation. For the synthesis of ATP, energy and a phosphate group must be supplied to an ADP molecule. *»» What genetic molecule closely resembles ATP?*

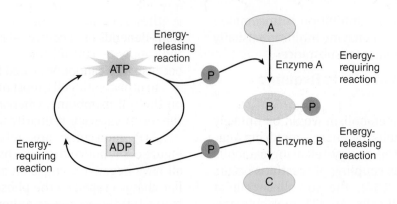

FIGURE 6.6 A Metabolic Pathway Coupled to the ATP/ADP Cycle. In this metabolic pathway, enzyme A catalyzes an energy-requiring reaction where the energy comes from ATP hydrolysis. Enzyme B converts the phosphorylated substrate to the final product C. Being an energy-releasing reaction, the free phosphate can be coupled to the reformation of ATP. *»» What are the terms for energy-requiring and energy-releasing reactions?*

in a box. Open the box (hydrolyze the phosphate group) and you have a more stable spring (a more stable ADP molecule). The release of the spring (the freeing of a phosphate group) provides the means by which work can be done. Thus, the hydrolysis of the unstable phosphate groups in ATP molecules to a more stable condition is what drives other energy-requiring reactions through the transfer of phosphate groups (Figure 6.5B). The addition of a phosphate group to another molecule by means of a transferase enzyme is called **phosphorylation**.

Because ATP molecules are unstable, they cannot be stored. Therefore, microbial cells synthesize large organic compounds like glycogen or lipids for energy storage. As needed, the chemical energy in these molecules can be released in catabolic reactions and used

to reform ATP from ADP and phosphate (Figure 6.5B). This **ATP/ADP cycle** occurs continuously in cells. It has been estimated that a typical bacterial cell must reform about 3 million ATP molecules per second from ADP and phosphate to supply its energy needs.

It might be a good idea to review what we have covered to this point, which is summarized in **FIGURE 6.6**. Enzymes regulate metabolic reactions by their unique active site binding the complementary substrate. Often a series of metabolic steps (the metabolic pathway) are required to form the final product. Some steps in the pathway might require energy (endergonic); this energy is supplied by ATP as it is hydrolyzed to ADP. Other reactions release energy (exergonic), which can be used to reform ATP from ADP.

Concept and Reasoning Checks 6.1

a. List the characteristics of enzymes.
b. If a metabolic pathway has eight intermediates, how many different enzymes are involved? Explain.
c. Describe how an enzyme could be modulated by competitive and noncompetitive inhibition.
d. Judge the importance of the ATP cycle to microbial metabolism.

Chapter **Challenge A**

Our chapter challenge is to discover the source and reason for methane gas production by termites, cows, and rice paddy fields. To get started, termites and cows need to take in "food" and catabolize the organic compounds to generate ATP.

QUESTION A: *Being vegetarians, what do termites typically eat and cows in the pasture ingest? Now, what is the basic structural polysaccharide that all these "plant cell wall products" contain? The problem is that termites and cows cannot digest these polysaccharides. Their cells of the digestive system lack the ability to synthesize the enzymes to carry out the catabolic reactions—yet termites and cows survive off these substances by breaking them down. So, if they are to get energy out of these polysaccharides, what "living agent" in their digestive tracts must be helping in the digestive process?*

You can find answers online in **Appendix F**.

■ KEY CONCEPT 6.2 Glucose Catabolism Generates Cellular Energy

Since the early part of the 20th century, the chemistry of glucose catabolism has been the subject of intense investigation by biochemists because glucose is a key source of energy for ATP production. Moreover, the process of glucose catabolism is very similar in all organisms, making this series of metabolic pathways one feature that unites all life.

Glucose Contains Stored Energy That Can Be Extracted and Transformed

In a cell, not all of the energy in an exergonic reaction is captured for metabolism. Much of the energy is lost as heat. Consequently, only about 40% of the energy in glucose is transformed to cell energy in the form of ATP.

This process of converting chemical energy in organic molecules, like glucose, into ATP is called

cellular respiration. If cells use oxygen gas (O_2) in making ATP, the process is called **aerobic respiration**. However, many microbes are anaerobic yet they still carry out cellular respiration. In these cases, the process is called **anaerobic respiration**; that is, without (O_2), another inorganic molecule will be used in the respiration process.

A form of "anaerobic metabolism," different from anaerobic respiration, is **fermentation**, which we will examine later in this chapter.

The catabolism of glucose, or another related monosaccharide or disaccharide, does not take place in one chemical reaction, nor do ATP molecules form all at once. Rather, the energy in glucose is extracted in small steps using two catabolic pathways (glycolysis and the citric acid cycle) followed by a process that forms most of the ATP molecules (oxidative phosphorylation) (**FIGURE 6.7**).

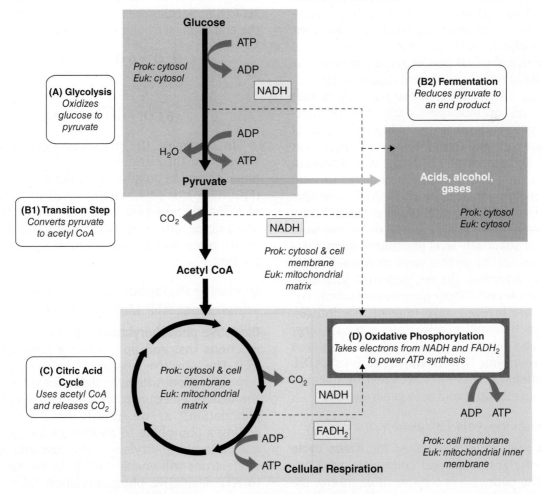

FIGURE 6.7 A Metabolic Map of Aerobic and Anaerobic Pathways for ATP Production. The production of ATP by microorganisms can be achieved through glycolysis **(A)** following a cellular respiration pathway **(B1, C, D)** or fermentation pathway **(B2)**. *»» Using this map, explain how these catabolic pathways reflect that ancient phrase: "Life is like a fire; it begins in smoke and ends in ashes."*

To begin our study of cellular respiration, we shall follow the process of aerobic respiration as it occurs in obligate aerobes, that is, in microbes that require O_2. To simplify our discussion of glucose catabolism, we shall follow the fate of one glucose molecule, as depicted in the following summary equation:

$$C_6H_{12}O_6 + 6\ O_2 + 32\ ADP + 32\ P$$
Glucose   Oxygen
$$\downarrow$$
$$6\ CO_2 + 6\ H_2O + 32\ ATP$$
Carbon   Water
dioxide

Glycolysis Is a Catabolic Pathway Used for Energy Extraction

The splitting of glucose, called **glycolysis** (*glyco* = "sweet"), occurs in the cytosol of all microorganisms and involves a catabolic pathway that converts an initial 6-carbon substrate, **glucose**, into two 3-carbon molecules (final products) called **pyruvate**. The metabolic pathway involves 10 enzyme-controlled reactions involving eight intermediates. **FIGURE 6.8** illustrates the process in detail.

Notice that the first part of glycolysis (preparatory reactions) is endergonic and requires ATP.

After an intermediate in the pathway is split into two 3-carbon molecules, each of these intermediates passes through an additional series of conversions (energy harvesting reactions) that ultimately form pyruvate. During these reactions, energy is stripped from the intermediates to form four ATP molecules. Because these ATP molecules were the result of the transfer of a phosphate from a substrate to ADP, we say these ATP molecules were the result of **substrate-level phosphorylation**. Considering two ATP molecules were consumed in the preparatory reactions, the net gain from glycolysis is two molecules of ATP per glucose consumed.

Before we proceed, take note of reaction (**6**). This enzymatic reaction releases two high-energy electrons and two protons (H^+), which are picked up by the coenzyme NAD^+, reducing each to NADH. This and similar events will have great significance shortly as an essential source to generate ATP.

The Citric Acid Cycle Extracts More Energy

The **citric acid cycle** (also called the **Krebs cycle** in honor of Hans Krebs and colleagues, who figured out the pathway) is a series of eight-enzyme controlled reactions that are referred to as a cycle because the final product formed is used as one substrate to initiate the pathway. All of the reactions take place along the cell membrane of prokaryotic cells. In cells of eukaryotic microbes, including the protists and fungi, the cycle occurs in the mitochondrial matrix.

The citric acid cycle is somewhat like a constantly turning wheel: each time the wheel comes back to the starting point, a molecule must be added to spin it for another rotation. That molecule is the pyruvate derived from glycolysis. **FIGURE 6.9** shows in detail the reactions of the citric acid cycle.

The overall purpose of the cycle is to extract more energy out of the pyruvate molecules. As shown in Figure 6.9, this includes a small amount of ATP and, importantly, the formation of 10 NADH coenzymes. Another coenzyme, FAD, is also reduced to 2 $FADH_2$ in the cycle. Finally, in several steps of the cycle (transition step and citric acid cycle), CO_2 molecules are produced.

Remember that we began with a 6-carbon glucose molecule; after one turn of the citric acid cycle, six CO_2 molecules are produced. This fulfills part of the equation (see boldface) for aerobic respiration:

$$\mathbf{C_6H_{12}O_6} + 6\ O_2 + 32\ ADP + 32\ P$$
$$\downarrow$$
$$\mathbf{6\ CO_2} + 6\ H_2O + 32\ ATP$$

In summary, the metabolic pathways of glycolysis and the citric acid cycle have extracted as much energy as possible from glucose and pyruvate (**FIGURE 6.10**). This has amounted to a small gain of ATP molecules generated from the catabolism of one glucose molecule. However, the 10 NADH and 2 $FADH_2$ molecules formed are most significant. Let's see how.

Oxidative Phosphorylation Is the Process by Which Most ATP Molecules Arise

Oxidative phosphorylation refers to a sequence of reactions that end up forming a large amount of ATP. The adjective "oxidative" is derived from the term **oxidation**, which, as defined earlier, refers to the loss of electron pairs from molecules. "Phosphorylation," as we already have seen, implies adding a phosphate to another molecule. Thus, in oxidative phosphorylation, the loss and transport of electrons will enable ADP to be phosphorylated to ATP. Oxidative phosphorylation takes place at the cell membrane in prokaryotic cells, and in the

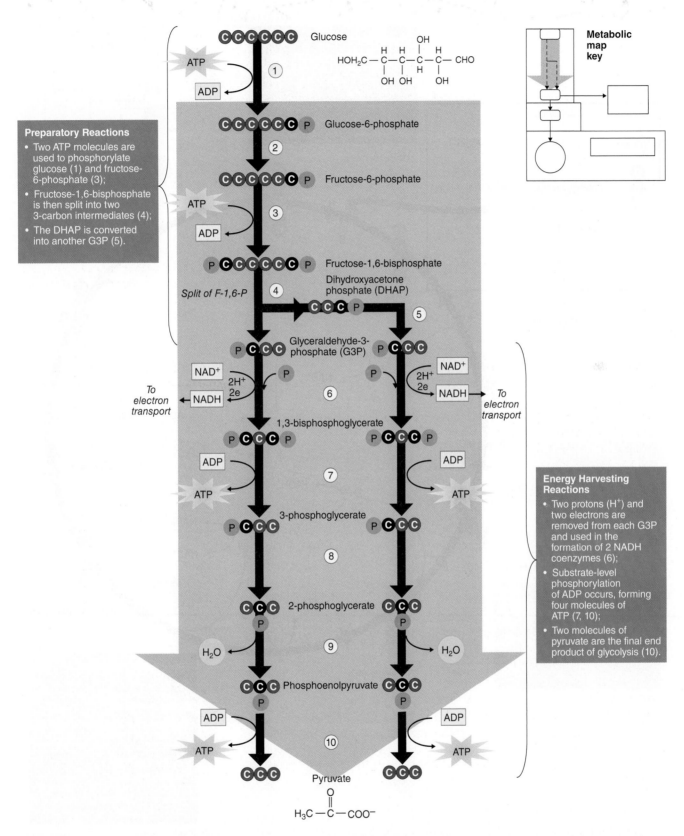

FIGURE 6.8 The Reactions of Glycolysis. Glycolysis is a metabolic pathway that converts glucose, a 6-carbon sugar, into two 3-carbon pyruvate products. In the process, two NADH coenzymes and a net gain of two ATP molecules occur. Carbon atoms are represented by circles. The dark circles represent carbon atoms bonded to phosphate groups. »» *How many substrate-level phosphorylation events occur when one glucose molecule is broken into two pyruvate molecules?*

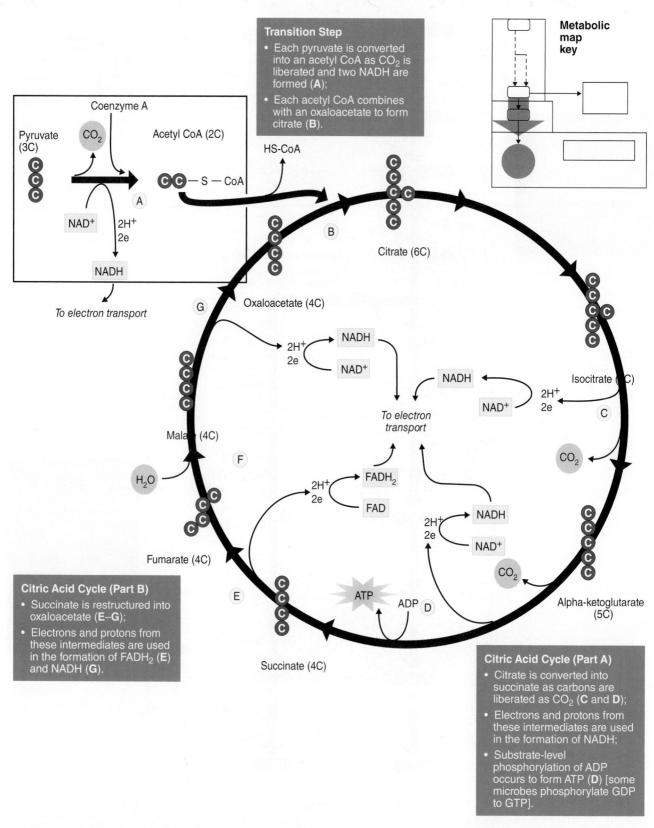

Transition Step
- Each pyruvate is converted into an acetyl CoA as CO_2 is liberated and two NADH are formed (**A**);
- Each acetyl CoA combines with an oxaloacetate to form citrate (**B**).

Metabolic map key

Pyruvate (3C)   Coenzyme A CO_2 Acetyl CoA (2C)

HS-CoA

NAD^+ $2H^+$ $2e$ NADH

To electron transport

Citrate (6C)

Oxaloacetate (4C)

$2H^+$ $2e$ NADH NAD^+

NADH NAD^+ $2H^+$ $2e$ Isocitrate (6C)   C

To electron transport CO_2

Malate (4C)   F H_2O FADH$_2$ FAD $2H^+$ $2e$

NADH NAD^+ $2H^+$ $2e$ CO_2

Fumarate (4C)   E   ATP   ADP   D   Alpha-ketoglutarate (5C)

Succinate (4C)

Citric Acid Cycle (Part B)
- Succinate is restructured into oxaloacetate (**E–G**);
- Electrons and protons from these intermediates are used in the formation of FADH$_2$ (**E**) and NADH (**G**).

Citric Acid Cycle (Part A)
- Citrate is converted into succinate as carbons are liberated as CO_2 (**C** and **D**);
- Electrons and protons from these intermediates are used in the formation of NADH;
- Substrate-level phosphorylation of ADP occurs to form ATP (**D**) [some microbes phosphorylate GDP to GTP].

FIGURE 6.9 The Reactions of the Citric Acid Cycle. Pyruvate from glycolysis combines with coenzyme A to form acetyl CoA (transition step). This molecule then joins with oxaloacetate to form citrate (citric acid cycle—Part A). Each turn of the cycle releases CO_2, produces ATP, and forms NADH and FADH$_2$ coenzymes as oxaloacetate is replaced (citric acid cycle—Part B). *»» Why are so many reactions required to extract energy out of pyruvate?*

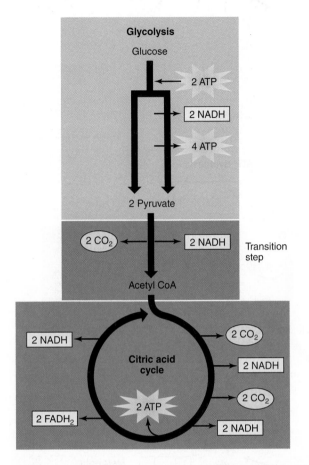

FIGURE 6.10 Summary of Glycolysis and the Citric Acid Cycle. Glycolysis and the citric acid cycle are the central metabolic pathways to extract chemical energy from glucose to generate cellular energy (ATP). *»» Every time one glucose molecule is broken down to CO_2 and water, how many ATP, NADH, and $FADH_2$ molecules are gained?*

mitochondrial inner membrane of eukaryotic cells. The process in aerobes consists of two major sets of events (**FIGURE 6.11**), which we'll look at next.

Electron Transport

Pairs of electrons carried in NADH and $FADH_2$ from glycolysis and the citric acid cycle are passed to the first in a series of **cytochromes**, which are protein compounds containing iron. Like walking along stepping stones, each electron pair is passed from one cytochrome complex (I–IV) to the next down the chain until the electron pair is finally transferred to the **final electron acceptor**, O_2, which combines with the electrons and protons (H^+) to form

water. Oxygen's role is of great significance because if O_2 was not present, there would be no way for cytochromes to "unload" their electrons and the entire cytochrome system would soon back up like a jammed conveyer belt and come to a halt.

The energy released during the passage of the electrons is used to pump protons (H^+) across the membrane into the extracellular space in prokaryotic cells or inner mitochondrial membrane in eukaryotic cells.

Chemiosmosis (ATP Synthesis)

The protons outside the membrane accumulate until a series of membrane-spanning protein channels, called **ATP synthases**, open in the membrane. Facilitated diffusion of the protons now occurs through these proteins. As the protons rush through these channels, the protons release their energy, which is used to generate ATP molecules from ADP and phosphate. This process is called **chemiosmosis**. The details of the chemiosmotic process are described in **MICROINQUIRY 6**.

If the cell membrane is damaged such that chemiosmosis cannot take place, the synthesis of ATP ceases and the organism rapidly dies. This is one reason why damage to the bacterial cell membrane, such as with antibiotics or detergent disinfectants, is so harmful.

An interesting sidelight described in **MICRO-FOCUS 6.1** concerns a novel way of using the bacterial respiratory process to generate electricity.

Based on current experimental data, 2.5 molecules of ATP can be synthesized for each pair of electrons originating from NADH; 1.5 molecules of ATP are produced for each pair of electrons from $FADH_2$ because the coenzyme interacts further down the cytochrome chain (see Figure 6.11). Therefore, as summarized in **FIGURE 6.12**, the aerobic respiration reactions can generate up to 32 ATP molecules from the oxidation of one glucose molecule to CO_2 and H_2O. It also completes the equation for aerobic respiration, accounting for oxygen being used to form water (see boldface):

$$C_6H_{12}O_6 + \textbf{6 O}_2 + \textbf{32 ADP} + \textbf{32 P}$$

$$\downarrow$$

$$6\ CO_2 + \textbf{6 H}_2\textbf{O} + \textbf{32 ATP}$$

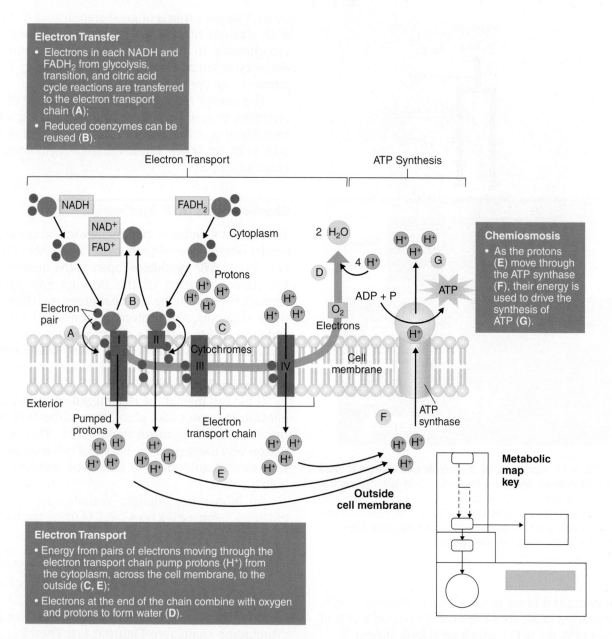

Electron Transfer
• Electrons in each NADH and FADH$_2$ from glycolysis, transition, and citric acid cycle reactions are transferred to the electron transport chain (**A**);
• Reduced coenzymes can be reused (**B**).

Electron Transport   ATP Synthesis

NADH
NAD$^+$
FAD$^+$
FADH$_2$
Cytoplasm
2 H$_2$O

Chemiosmosis
• As the protons (**E**) move through the ATP synthase (**F**), their energy is used to drive the synthesis of ATP (**G**).

Protons
Electron pair
Electrons
ADP + P
ATP
O$_2$
Cytochromes
Cell membrane
ATP synthase
Exterior
Pumped protons
Electron transport chain
Outside cell membrane

Electron Transport
• Energy from pairs of electrons moving through the electron transport chain pump protons (H$^+$) from the cytoplasm, across the cell membrane, to the outside (**C, E**);
• Electrons at the end of the chain combine with oxygen and protons to form water (**D**).

Metabolic map key

FIGURE 6.11 Oxidative Phosphorylation in Bacterial Cells. Originating in glycolysis and the citric acid cycle, coenzymes NADH and FADH$_2$ transport electron pairs to the electron transport chain in the cell membrane, which fuels the transport of protons (H$^+$) across the cell membrane. Protons then reenter the cytosol through a protein channel in the ATP synthase enzyme. ADP molecules join with phosphates as protons move through the channel, producing ATP. *»» What would happen to the oxidative phosphorylation process if this cell were deprived of oxygen?*

MICROINQUIRY 6

The Rotary Motor That Generates ATP

Every day, an adult human weighing 160 pounds uses about 80 pounds of ATP. As the ATP is broken down into ADP and phosphate, enormous amounts of energy are made available to do metabolic work. However, the body's weight neither drops by that amount nor does it even change perceptibly, because the cells are constantly regenerating ATP from the breakdown products, ADP and phosphate. This inquiry looks into how the rotary motor, an enzyme, accomplishes this task.

The Chemiosmotic Basis for ATP Synthesis

Chemiosmosis proposes that electron transport between cytochrome complexes provides the energy to "pump" protons (H^+) across the membrane; in the case of bacterial and archaeal cells, this is from the cytosol to the environment. As explained in the text, this proton gradient provides the force or potential to drive the protons back into the cell through an enzyme called ATP synthase. This flow of rapidly streaming H^+ brings together ADP and phosphate to form ATP.

How Does Proton Flow Cause ATP Synthesis?

The groundbreaking research as to how the ATP synthase works came from studies with *Escherichia coli* cells. The ATP synthase consists of many polypeptide subunits (see figure). The catalytic head faces into the cytosol and consists of several polypeptides that represent the catalytic complex for converting ADP + P to ATP. The stationary base is anchored in the cell membrane and contains the proton (H^+)-transporting channel. Linking these two polypeptide assemblies is a rotor that carries protons from the channel in the base to the cytosol.

The current model suggests the ATPase works like a rotary motor to generate ATP (see figure).

Step 1. Protons enter the channel in the stationary base.

Step 2. Binding of a proton to the rotor changes the shape of the polypeptide and causes the rotor to spin (somewhat similar to the turning of a waterwheel and very similar to that of a rotary engine).

Step 3. As each proton completes one turn (rotation) in the rotor, it leaves the rotor and passes through another channel in the stationary base that deposits the proton into the cytosol.

Step 4. The spinning of the rotor causes the stalk also to spin, which extends into the catalytic head that is stationary.

Step 5. The polypeptide subunits in the catalytic head each have an active site that allows the subunits to bind an ADP + P and catalyze the production of an ATP.

Discussion Point

Flagella also represent a rotary device and are the subject of another MicroInquiry in Chapter 4. Compare the "rotary engine" of the flagellum to that of ATP synthase. Also, what other nanomachines exist in microbial cells?

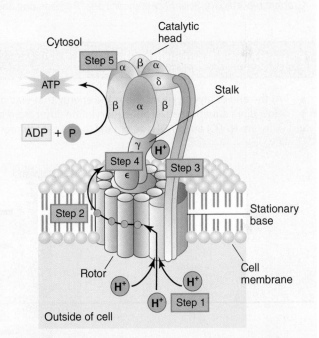

A bacterial ATP synthase enzyme consists of numerous polypeptide subunits involved in the generation of ATP.

You can find answers (and comments) online in **Appendix E**.

MICROFOCUS 6.1: Biotechnology

Bacteria Not Included

How many toys (child or adult) or electronic devices do you purchase each year for which batteries are needed to run the device? And often batteries are not included. Today, a new type of battery is being developed—one that converts sugar not into ATP but rather into electricity. The battery is one packed with bacterial cells.

Realize that cellular respiration involves the generation of minute electrical currents. During cellular respiration, electrons are transferred to cofactors like NAD^+ and then passed along a chain of cytochrome complexes during oxidative phosphorylation. Swades Chaudhuri and Derek Lovely of the University of Massachusetts at Amherst have taken this idea and applied it to developing a new type of fuel cell or battery.

The scientists mixed the bacterial species *Rhodoferax ferrireducens*, which they found in aquifer sediments in Virginia, with a variety of common sugars. When placed in a chamber with a graphite electrode, *R. ferrireducens* metabolized the sugar, stripped off the electrons, and transferred them directly to the electrode. The result: an electrical current was produced. In addition, the bacterial cells continued to grow, so a stable current could be produced with high efficiency.

Although it is still a long way from producing a reliable, long-lasting bacterial battery, the researchers believe much of the agricultural or industrial waste produced today could be the "sugar" used in making these bacterial batteries. So, as Sarah Graham reported for *ScientificAmerican.com*, "Perhaps one day electronics will be sold with the caveat 'bacteria not included.'"

Concept and Reasoning Checks 6.2

a. At the end of glycolysis, where is the energy that was originally in glucose?
b. How does substrate-level phosphorylation differ from oxidative phosphorylation?
c. Why are NADH and $FADH_2$ so critical to cell energy metabolism?

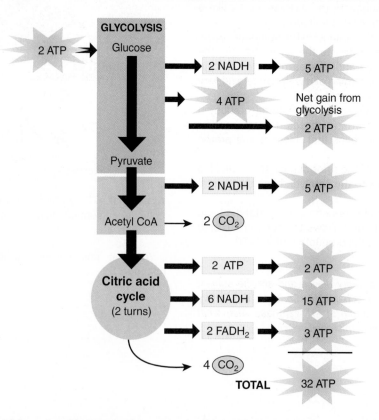

FIGURE 6.12 The ATP Yield from Aerobic Respiration. In a microbial cell, 32 molecules of ATP can result from the metabolism of a molecule of glucose. Each NADH molecule accounts for the formation of 2.5 molecules of ATP; each molecule of $FADH_2$ accounts for 1.5 ATP molecules. *»» From this diagram, what is the single most important reactant for ATP synthesis?*

Chapter Challenge B

In our Chapter Challenge, hopefully you have figured that cellulose was the structural polysaccharide of plant cell walls and that it was the common denominator in the metabolism of termites and cows; that is, cellulose is broken down by microbes that reside in the termite gut and cow rumen.

QUESTION B: *What is the building block (monomer) that builds cellulose? If this monomer is produced through a catabolic process and then used for aerobic respiration (hint: what sugar starts the cell respiration process?), what are the final products formed besides ATP? Is one of them methane gas (CH₄)? Explain. So, do the guts of termites and the rumens of cows carry out aerobic respiration? Explain. I believe we need to keep looking for the source for CH₄.*

You can find answers online in **Appendix F**.

■ **KEY CONCEPT 6.3**   **There Are Other Pathways to ATP Production**

The catabolism of glucose is a process central to the metabolism of all organisms as it provides a glimpse of how organisms obtain energy for life. In this section, we examine how cells obtain energy from other organic compounds (fats and proteins) by directing those compounds into the process of cellular respiration. We also discover how modifications to cellular respiration and glucose metabolism allow anaerobic organisms to use glucose and generate ATP without having O_2 as the final electron acceptor in electron transport.

Other Nutrients Represent Potential Energy Sources

A wide variety of organic compounds can serve as useful energy sources. All these compounds must go through a series of preparatory conversions before they are processed in glycolysis or the citric acid cycle.

Carbohydrates

Besides glucose, other monosaccharides as well as disaccharides and polysaccharides can be used to generate ATP (**FIGURE 6.13**). Sucrose, for example, is first digested by the enzyme sucrase into its two monosaccharides, glucose and fructose. The glucose molecule enters the glycolysis pathway directly, but the fructose molecule undergoes further conversions and enters glycolysis as an intermediate further into the pathway. Lactose, another disaccharide, is split into glucose and galactose by the enzyme lactase. Galactose also undergoes a series of changes before it is ready to enter glycolysis.

Stored polysaccharides, such as starch and glycogen, are metabolized by enzymes that remove one glucose unit at a time and through a series of chemical conversions are ready to enter the glycolysis pathway. Thus, carbohydrates other than glucose can be used as chemical energy sources.

Fats

The economy of metabolism is demonstrated further when we consider protein and fat catabolism. Fats are extremely valuable energy sources because their chemical bonds contain enormous amounts of chemical energy.

Fats consist of three fatty acids bonded to a glycerol molecule. To be useful for energy purposes, the fatty acids are separated from the glycerol by the enzyme lipase (Figure 6.13). After this has taken place, the glycerol portion enters as an intermediate into glycolysis. For fatty acids, there is a complex series of conversions called **beta-oxidation**, in which each long-chain fatty acid is broken by enzymes into 2-carbon units. Other enzymes then convert each unit to a molecule of acetyl CoA that enters the citric acid cycle. We previously noted that the oxidation of one glucose molecule could generate 20 molecules of ATP in the citric acid cycle. A quick calculation should illustrate the substantial energy output from a 16-carbon fatty acid (i.e., eight 2-carbon units).

Proteins

Although proteins are generally not considered energy sources, cells will use them when carbohydrates and fats are in short supply.

Proteins are broken down to amino acids. Enzymes then convert many amino acids to energy pathway components by removing the amino group and substituting a carbonyl group. This process is called **deamination** (see Figure 6.13).

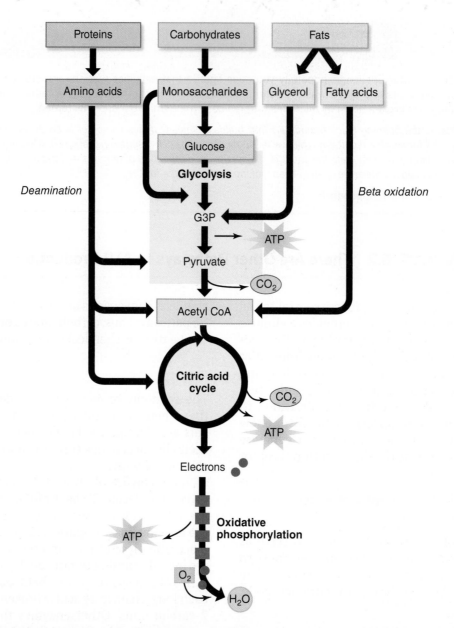

FIGURE 6.13 Carbohydrate, Protein, and Fat Metabolism. Besides glucose, other carbohydrates as well as proteins and fats can be sources of energy by providing electrons and protons for cellular respiration. The intermediates enter the pathway at various points. »» *Why would more ATP be produced from the products of one fatty acid entering the pathway at acetyl CoA than from one amino acid entering at acetyl CoA?*

For example, alanine is converted to pyruvate and aspartic acid is converted to oxaloacetate. Consequently, in most habitats, microbes can use a large and diverse set of chemical compounds as potential energy sources.

Anaerobic Respiration Produces ATP Using Other Final Electron Acceptors

Nearly all eukaryotic microbes as well as multicellular animals and plants carry out aerobic respiration,

using oxygen as the final electron acceptor in the electron transport chain. However, many bacterial and archaeal organisms exist in environments in which oxygen is scarce or absent, such as in wetland soil and water, and within human and animal digestive tracts. In these environments, the organisms have evolved a respiratory process called **anaerobic respiration** that usually relies on an inorganic final electron acceptor other than O_2 for ATP production. Considering the immense number of species that live in such anaerobic environments, anaerobic respiration is extremely important ecologically.

Many human enteric, facultative species, for example, use nitrate (NO_3^-) with which electrons combine to form nitrite (NO_2^-) or another nitrogen product during electron transport. The obligate anaerobe *Desulfovibrio* uses sulfate (SO_4^{-2}) for anaerobic respiration. The sulfate combines with the electrons from the cytochrome chain, resulting in reduced hydrogen sulfide (H_2S). This gas gives a rotten egg smell to the environment (as in a tightly compacted landfill). A final example is exhibited by the archaeal methanogens. These obligate anaerobes use CO_2 or carbonate (CO_3) as a final electron acceptor in the electron transport chain and, with hydrogen gas (H_2), form large amounts of CH_4.

As a useful side effect of anaerobic respiration, some anaerobes can "raise a stink" when competing for food resources, as MICROFOCUS 6.2 explains.

MICROFOCUS 6.2: Environmental Microbiology

Bacteria "Cause a Stink"

We are all familiar with body odor and bad breath. When one does not maintain a level of cleanliness or hygiene, bacterial species on the skin surface or in the mouth can grow out of proportion and, as they metabolize compounds like proteins, they produce noticeably unpleasant, smelly odors.

On an environmental level, there are smells that often come from decaying or rotting foods (carrion). As decomposers, these microbes also give off foul odors caused by the presence of bacterial species colonizing the dead carrion. Competing in the environment with other animal scavengers for food, are these bacterially produced odors an example of microbe warfare used to repel or deter animal species from consuming important food resources? This is what Mark Hay and collaborators at Georgia Institute of Technology wanted to know, especially with respect to marine ecosystems. Their hypothesis: Decaying food resources become repugnant and thus unappetizing to larger animal species like crabs or fish.

To test their hypothesis, the research team baited crab traps near Savannah, Georgia, with menhaden, a typical baitfish for crabs. Some traps contained bacteria-laden menhaden that had been allowed to rot for one or two days, whereas other traps contained freshly thawed carrion having relatively few bacteria. When the traps were inspected, those with fresh carrion had more than twice the number of animal scavengers per trap than did the traps with bacteria-laden carrion. Lab studies with stone crabs showed they also avoided the bacteria-laden, rotting food but readily consumed the freshly thawed (bacteria-free) menhaden.

To examine the role of bacterial organisms in the avoidance behavior by stone crabs (see figure), some menhaden was allowed to rot in water without the antibiotic chloramphenicol, whereas other samples contained the antibiotic in the water to prevent or inhibit bacterial growth. Again, the study observed that the crabs readily ate the antibiotic-incubated (bacteria-free) menhaden but avoided the menhaden without antibiotic (bacteria-laden); the bacterial organisms were in some way responsible for the aversion.

Finally, the researchers used organic extracts prepared from the bacteria-laden carrion and mixed these chemical substances with freshly thawed menhaden. Again, the crabs were repelled. Exactly what chemical compounds were responsible for the behavior were not evident from the study.

In summary, it appears that bacterial species not only act as decomposers and pathogens in the environment but also can compete very successfully with relatively large animal consumers for mutually attractive food sources.

A stone crab.

Courtesy of Catherine Billick, www.flickr.com/photos/catbcorner/.

In anaerobic respiration, the amount of ATP produced is less than in aerobic respiration. There are several reasons for this. First, only a portion of the citric acid cycle functions in anaerobic respiration, so fewer reduced coenzymes are available to the electron transport chain. In addition, not all of the cytochrome complexes function during anaerobic respiration, so the ATP yield will be less. A slower ATP generation process means these organisms will grow slower than their aerobic counterparts, which generate more ATP per glucose.

Fermentation Produces ATP Using an Organic Final Electron Acceptor

In environments that are depleted of oxygen gas (**anoxic**) and are lacking alternative final electron acceptors needed by anaerobes (e.g., NO_3^-, SO_4^{-2}, CO_3), the pyruvate from glycolysis will be catabolized through fermentation. **Fermentation**, probably the most ancient form of energy metabolism, is the enzymatic process for producing ATP by substrate-level phosphorylation, using the organic compounds of glycolysis as both electron donors and acceptors. In fermentation, the citric acid cycle and oxidative phosphorylation processes do not function.

The chemical process of fermentation makes many fewer ATP molecules than in cellular respiration. However, no matter what the end product, fermentation ensures a constant supply of NAD^+ for glycolysis and the production of two ATP molecules per glucose (**FIGURE 6.14A**). In other words, the metabolic pathways for fermentation oxidize NADH to NAD^+. For example, in the fermentation of glucose by *Streptococcus lactis*, the conversion of pyruvate to lactic acid is a way to reform the NAD^+ coenzymes that glycolysis needs to continue generating two ATP molecules for every glucose molecule consumed.

The diversity of fermentation chemistry extends to some eukaryotic microbes, as well. In yeasts like *Saccharomyces*, when pyruvate is converted to ethyl alcohol (ethanol), NAD^+ is reformed and used for the continuation of glycolysis.

The energy benefits to fermentative organisms are far less than in cellular respiration. In fermentation, each glucose passing through glycolysis yields two ATP molecules and the production of

different final products depending on the microbial species. This is in sharp contrast to the 32 molecules from cellular respiration. It is clear that cellular respiration is the better choice for energy conservation, but under anoxic conditions, there might be little alternative if life for *S. lactis*, *Saccharomyces*, or any fermentative microorganism is to continue.

Significance of Fermentation Final Products

Although the final products of fermentation are waste products for the microbes producing them, human activities and industries often make use of these end products (**FIGURE 6.14B**). Following are a few examples.

In a dairy plant, the process of **lactic acid fermentation** by *S. lactis* is carefully controlled so the acid will curdle fresh milk to make buttermilk or yogurt. The liquor industry uses the ethyl alcohol produced in **alcoholic fermentation** to make alcoholic beverages such as beer and wine.

As shown in Figure 6.14B, other industries also make use of microbial fermentations. For example, Swiss cheese develops its flavor from the propionic acid produced during fermentation and gets its holes from trapped CO_2 gas resulting from fermentation. Sauerkraut is the product of fermentation by bacteria present in cabbage. The chemical industry also has harnessed the power of fermentation in the production of acetone, butanol, and other industrial solvents. Thus, fermentation is useful not only to microorganisms as an energy-generating process but also to human consumers who enjoy or make use of the products of fermentation.

Medical Significance

The ability of microbes to carry out diverse fermentation reactions that produce different final products can be very useful in the identification of microbial species. These physiological and biochemical tests often are essential to the clinical lab microbiologist trying to identify a potential pathogen—and most likely for you in your microbiology lab class to identify an unknown bacterial species. For example, identifying a gastrointestinal pathogen might be possible by using the methyl red test and Voges-Proskauer test (**FIGURE 6.14C**).

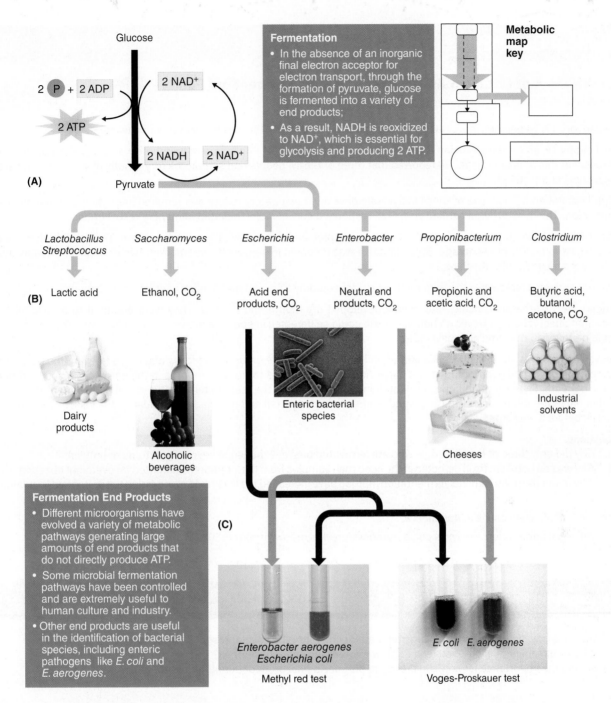

FIGURE 6.14 Microbial Fermentation. (A) Fermentation is an anaerobic process that reoxidizes NADH to NAD$^+$ by converting organic materials into final products of fermentation **(B)**. These final products of fermentation also provide a way to help identify bacterial species **(C)**. »» *How are lactic acid and alcohol fermentation identical in purpose?*

A positive methyl red test maintains a red-colored solution if a species can ferment glucose to acid final products, whereas the Voges-Proskauer test produces a solution of brownish-red color if a species forms neutral final products of glucose fermentation. Of course, additional tests would be needed to make a final identification.

In other cases, improperly fermented foods can lead to the development of potentially deadly illnesses, as described in **CLINICAL CASE 6**.

Clinical Case 6

An Outbreak of Botulism Associated with Eating Fermented Food

On January 17, 14 persons in a southwestern village in Alaska ate fermented beaver tail and paw meat.

On January 18, an Alaska physician reported to the Alaska Division of Public Health that there was a possible outbreak of botulism in this Alaskan village. He reported that three of the 14 people had symptoms (dry mouth, blurry vision, general weakness) suggesting botulism.

After hospital admission, two of these individuals developed respiratory failure and required intubation and mechanical ventilation. The third suffered cardiac arrest but responded successfully to cardiopulmonary resuscitation.

About 6 hours after the development of symptoms, all three patients received botulism antitoxin. The patients then were transferred to an ICU in Anchorage. Botulism toxin was detected in serum specimens from two of these individuals and in a stool specimen from the third.

Two patients recovered while the third required a tracheostomy tube and mechanical ventilation for 1 month.

Among the 11 other individuals, four people developed dry mouth and nausea. They were admitted to a hospital for overnight observation. However, no botulism toxin could be detected in specimens from these individuals. Traces of toxin could be found in the remaining paw meat.

Beaver is hunted in southwest Alaska and often is fermented above ground or indoors and eaten later. In this outbreak, the tail and paws had been wrapped in a paper rice sack and stored for up to 3 months in the entry to the house of one of the patients. These nontraditional fermentation methods expose the meat to temperatures that are warmer than when traditional fermentation occurs in the ground. Some of the beaver tail and paw had been added to the sack as recently as 1 week before it was eaten.

Questions:
  a. Why did only three of the 14 people who ate fermented beaver experience severe symptoms of botulism?
  b. Why was no botulism toxin detected in the specimen samples from the 11 patients admitted for overnight observation?
  c. How might the nontraditional fermentation method have increased the likelihood of ingesting the preformed botulinum toxin?

You can find answers online in **Appendix E**.

For additional information, see www.cdc.gov/mmwr/preview/mmwrhtml/mm5032a2.htm.

Concept and Reasoning Checks 6.3

  a. Describe how lipids and proteins are prepared for entry into the cellular respiration pathway.
  b. Why would obligate anaerobes tend to grow slower than obligate aerobes?
  c. Justify the importance for some microbes to produce a final product during fermentation.

Chapter Challenge C

So far, we have decided that aerobic respiration is not the source for the methane that is produced by the microbes residing in termite guts and in the rumen of cows. Yet, the structural polysaccharide we have been referring to (cellulose) still is broken down into glucose and is used as a substrate for ATP generation.

QUESTION C: *In what other catabolic pathway can glucose be used to generate ATP? If you have figured it out, what is the nature of the environment in the termite's gut and the cow's rumen? And by association, what must the waterlogged rice paddy field be like?*

You can find answers online in **Appendix F**.

■ KEY CONCEPT 6.4   Photosynthesis Converts Light Energy to Chemical Energy

Although the anabolism or synthesis of carbohydrates takes place through various mechanisms in microorganisms, the unifying feature is the requirement for energy—an anabolic process.

Photosynthesis Is a Process to Acquire Chemical Energy

Photosynthesis is the process by which light energy is converted to chemical energy that is then stored as carbohydrate or other organic compounds. In fact, photosynthesis is the most important biochemical process on the planet because it produces all of our oxygen gas and supplies most of the chemical energy many organisms need to generate ATP. In the cyanobacteria, the photosynthetic process takes place in special **thylakoid membranes**, which contain chlorophyll or chlorophyll-like pigments (**FIGURE 6.15A**).

Among eukaryotic microbes, photosynthesis occurs in the chloroplasts of such protists as diatoms, dinoflagellates, and green algae (**FIGURE 6.15B**). In all these cases, the prokaryotic and eukaryotic microbes carry out **oxygenic photosynthesis**, that is, where O_2 is a byproduct of the process. In **INVESTIGATING THE MICROBIAL WORLD 6**, we look at a historical experiment in which the use of microbes

made an important contribution toward understanding the photosynthetic process.

Using cyanobacteria as our example, the process of photosynthesis occurs in two phases (or sets of chemical reactions).

The Energy-Fixing Reactions

In the first part of photosynthesis, light energy is converted into or "fixed" as chemical energy in the form of ATP and the reduced coenzyme NADPH. Thus, the process is referred to as the **energy-fixing reactions**, or "light-dependent" reactions, because light energy is required for the process. These reactions occur on the cell membranes in cyanobacteria and the thylakoid membranes in algal chloroplasts.

Like the reactions of cellular respiration, the energy-fixing reactions of photosynthesis are dependent on electrons and protons, the details of which are shown in **FIGURE 6.16A**. The two important outcomes of these reactions are the generation of O_2 from the splitting of water and the generation of ATP and NADPH. The ability of the ancestors of modern-day cyanobacteria to generate O_2 profoundly changed the atmosphere of Earth some 2.4 billion years ago, leading to the evolution of

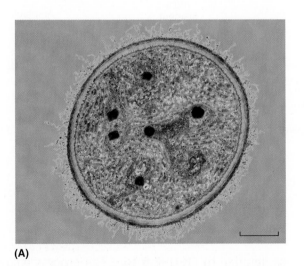
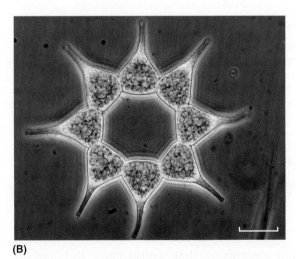

(A)   (B)

FIGURE 6.15 Photosynthetic Microbes. (A) False-color transmission electron micrograph of a cyanobacterium displaying the internal membranes (green) along which photosynthetic pigments are located. Carboxysomes are the black bodies. (Bar = 2 μm.) **(B)** A light micrograph of the colonial green alga *Pediastrum*. The green globules are chloroplasts. (Bar = 5 μm). »» *What membranes in algae are analogous to the cyanobacterial membranes?*

Investigating the Microbial World 6

Microbes Provide the Answer

It was known by the late 1700s that plants and algae produced oxygen gas during photosynthesis. In the late 1800s, Theodor W. Engelmann, a German physiologist, botanist, and microbiologist, confirmed that oxygen production was dependent on chlorophyll that is found in chloroplasts of plant and algal cells. He further observed bacterial aerotaxis, the movement of oxygen-sensing cells toward higher concentrations of oxygen gas.

OBSERVATION: In 1881, Engelmann observed that when white light was shining on the green alga *Spirogyra*, the motile bacterial cells moved toward the chloroplasts in a filament of the green alga. Knowing that the bacterial cells exhibited aerotaxis and that white light is composed of a spectrum of colors (red, orange, yellow, green, blue, indigo, and violet):

QUESTION: *Which of these colors (wavelengths) in white light stimulated photosynthesis?*

HYPOTHESIS: Photosynthesis is more efficient with certain colors within the spectrum of white light. If so, the aerotactic bacterial cells can be used to identify which colors (wavelengths) are more efficient (as measured by the production of oxygen gas).

EXPERIMENTAL DESIGN: In 1882, Engelmann had a specially designed microscope equipped with a prism that would break white light into its spectrum of colors on a microscope slide.

EXPERIMENT: A filament of the green alga *Cladophora* was placed on a slide on the special microscope, which exposed different segments of the algal filament to different colors (wavelengths) of visible light. Motile cells of the aerotactic bacterial species were then added to the slide and the algal filament was observed to see the effect that the different colors (wavelengths) of light had on the accumulation of the aerotactic bacterial cells.

RESULTS: See figure.

CONCLUSIONS:

QUESTION 1: *What could Engelmann conclude from his experiment?*

QUESTION 2: *Did the results support his hypothesis? Explain.*

QUESTION 3: *What would have been the response of the aerotactic bacteria if only green light was used to illuminate the algal filament?*

You can find answers online in **Appendix E**.

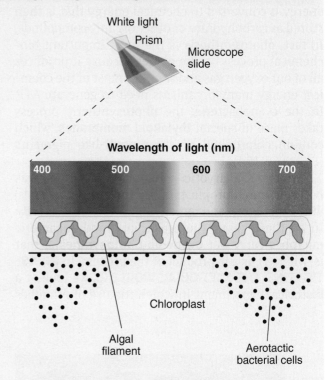

Results of the Engelmann experiment.

Modified from Engelmann, T. W. 1882. *Bot. Zeit.* 40:419–426.

aerobic organisms that would carry out aerobic respiration.

The Carbon-Fixing Reactions

In the second part of photosynthesis, another cyclic metabolic pathway is used to form carbohydrates, as detailed in **FIGURE 6.16B**. The process is known as the **carbon-fixing reactions** because the carbon CO_2 is trapped (or fixed) into carbohydrates and other organic compounds. It also is called the "Calvin-Benson cycle," named after the biologists who worked out the sequence of reactions. In

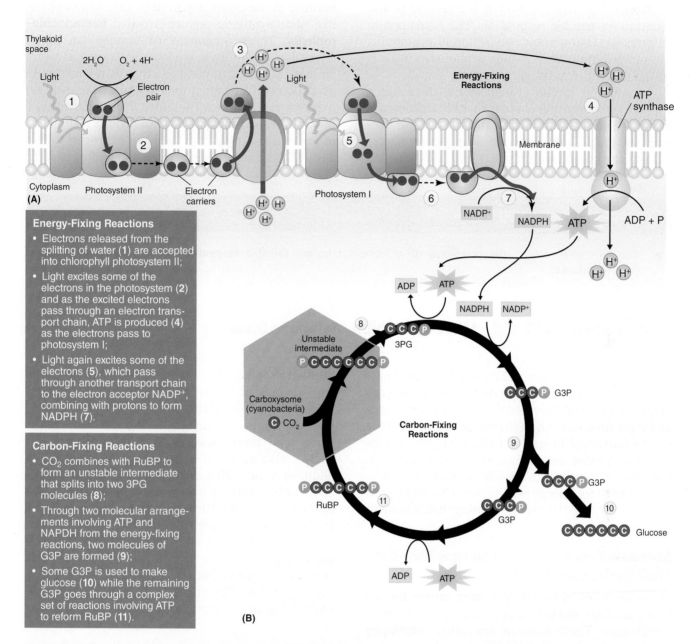

FIGURE 6.16 Photosynthesis in Cyanobacteria and Algae. (A) The energy-fixing reactions generate ATP and NADPH. **(B)** The carbon-fixing reactions unite carbon dioxide with ribulose bisphosphate (RuBP) to form two molecules of 3-phosphoglycerate (3PG) then glyceraldehyde-3-phosphate (G3P). ATP and NADPH from the energy-fixing reactions are used in the latter conversion. Condensations of two 3-carbon G3P molecules yield glucose and the remaining G3P molecules are used to form RuBP to continue the process. *»» How are the carbon-fixing reactions of photosynthesis dependent on the energy-fixing reactions?*

cyanobacteria, these reactions occur in the cytosol (in eukaryotic algae in the fluid-filled interior of chloroplasts). Importantly, in the synthesis of carbohydrate, the products of the energy-fixing reactions, ATP and NADPH, are necessary to drive the carbon-fixing reactions.

Thus, the overall formula for photosynthesis may be expressed as:

$$6\,CO_2 + 6\,H_2O + ATP$$
$$\downarrow light$$
$$C_6H_{12}O_6 + 6\,O_2 + ADP + P$$

Notice that this reaction is the reverse of the equation for aerobic respiration. The fundamental difference is that aerobic respiration is a catabolic, energy-yielding process, whereas photosynthesis is an anabolic, energy-trapping process.

In addition to the cyanobacteria, there are a few other photosynthetic groups within the domain Bacteria. These groups include the green sulfur bacteria, the purple sulfur and purple nonsulfur bacteria, and the green nonsulfur bacteria, all so named because of the colors imparted by their chlorophyll-like photosynthetic pigments called **bacteriochlorophylls**. These organisms commonly live under anaerobic conditions in environments such as sulfur springs and stagnant ponds, so they do not use water as a source of hydrogen ions and electrons, and no oxygen gas is released. Therefore, the photosynthetic process is called **anoxygenic photosynthesis**. Instead of water, these organisms use H_2 or hydrogen sulfide gas (H_2S) as a source of electrons and hydrogen ions.

Concept and Reasoning Check 6.4

a. Compare and contrast the processes of glucose catabolism (aerobic respiration) with glucose anabolism (photosynthesis).

■ KEY CONCEPT 6.5   Microbes Exhibit Metabolic Diversity

In this chapter, we examined the processes and patterns that microbes use to generate cellular energy in the form of ATP. Provided there is a carbon source and some form of energy (chemical or sunlight) that can be harvested to extract cellular energy, organisms can survive and thrive. The unique aspect is the diverse sources of organic or inorganic substances used for energy production. This we refer to as their nutritional pattern, meaning their source of carbon and energy.

Microbes Can Be Grouped on How They Get Their Carbon and Energy

Microbes differ in how they acquire carbon for use in metabolism and how they obtain energy to drive metabolism. These patterns are called **autotrophy** and **heterotrophy** (FIGURE 6.17).

Autotrophs

Organisms carrying out autotrophy synthesize their own complex organic molecules from simple inorganic carbon sources, such as CO_2, and are referred to as **autotrophs** (*auto* = "self"; *troph* = "nourish"). Those using sunlight as their energy source, such as the cyanobacteria and eukaryotic algae, are called **photoautotrophs**. They carry out photosynthesis using water and producing O_2 (oxygenic photosynthesis).

Another group of autotrophs does not use sunlight as an energy source. Instead, they use inorganic compounds of mineral origin and are called **lithoautotrophs** (*litho* = "stone"). For example, in the soil, species of *Nitrosomonas* convert ammonium ions (NH_4^+) into nitrite ions (NO_2^-) under aerobic conditions, thereby obtaining ATP. The genus *Nitrobacter* then converts NO_2^- into nitrate ions (NO_3^-), also as an ATP-generating mechanism. In addition to providing energy to both bacterial species, these reactions have great significance in the environment as a critical part of the **nitrogen cycle**. In this cycle, atmospheric nitrogen gas (N_2) is converted into nitrate or ammonia, which can then be used by green plants to form organic compounds (e.g., amino acids) that are crucial to sustain all living organisms. The nitrogen cycle is discussed in more detail in Chapter 26.

Heterotrophs

Microorganisms carrying out heterotrophy are called **heterotrophs** (*hetero* = "other"). Such heterotrophic organisms obtain their energy and carbon in different ways. The **photoheterotrophs** use sunlight as their energy source and preformed organic compounds, such as fatty acids and alcohols, as sources of carbon. Photoheterotrophs include certain green nonsulfur and purple nonsulfur bacteria.

The **chemoheterotrophs** use preformed organic compounds for both their energy and carbon sources. Glucose is the most common of these organic molecules. Those chemoheterotrophic microorganisms that feed exclusively on dead organic matter are

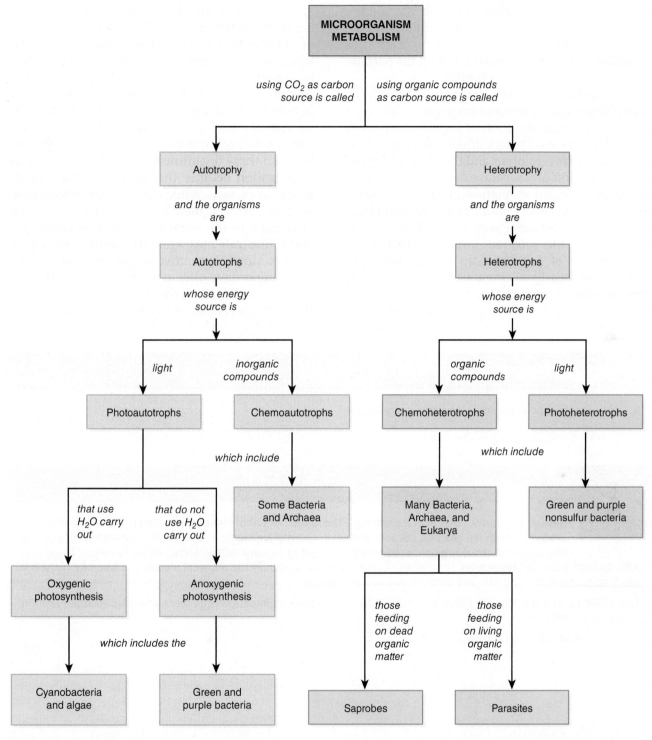

FIGURE 6.17 Microbial Metabolic Diversity. This concept map summarizes microbial metabolism based on carbon sources (CO_2 or chemical compounds) and energy sources (light or inorganic/organic compounds). *»» Why do the members of the domain Bacteria and domain Archaea show such diversity of metabolic types?*

commonly called **saprobes**. In contrast, chemo-heterotrophs that feed on living organic matter, such as human tissues, are commonly known as **parasites**. The term **pathogen** is used if the parasite causes disease in its host organism.

In conclusion, we saw that cellular respiration is an efficient energy metabolism process. Cell respiration couples the oxidation of organic compounds (such as sugars and lipids) through the central metabolic pathways (glycolysis and the citric acid cycle) to the reduction of a separate final electron acceptor (usually oxygen or another inorganic acceptor) in oxidative phosphorylation. Thus, cell respiration represents a substantial source of ATP generation compared to the substrate-level phosphorylation characteristic of fermentation. Having said this, realize that nearly all organisms (microbial and multicellular) have similar sets of chemical reactions to carry out cell respiration. The substrates and enzymes of glycolysis and the citric acid cycle are similar whether we are discussing *Escherichia coli* or a human liver cell. Even the chemical reactions by which organic compounds are assembled from their monomers are very similar in all organisms. Furthermore, oxygenic photosynthesis also is the same in cyanobacteria, algae, and green plants. Therefore, we can say that the processes of energy metabolism are almost universal. So, what is the difference between these organisms? The answer is structural organization because the diverse members of the prokaryotes lack mitochondria and chloroplasts, yet they carry out these metabolic processes almost identically to the eukaryotes. These common processes illustrate once again that prokaryotic species are not simple, primitive organisms and, perhaps more important, offer powerful evidence that the evolution of all life on the "tree of life" has occurred from a single, universal common ancestor.

Concept and Reasoning Check 6.5

a.   Why are there no autotrophic parasites or pathogens?

Chapter Challenge D

Have you figured out the chapter challenge? The reason why all of these animal and soil "environments" were producing methane gas was because the environments are highly anaerobic. The breakdown of cellulose (the polysaccharide) to glucose requires the anaerobes to use a different final electron acceptor to generate ATP through anaerobic respiration. For some of these microbes, especially the archaeal members, carbonate is the final electron acceptor, forming ATP and methane gas as a final product.

QUESTION D: *As a last step, use Figure 6.17 to trace the microorganism metabolism that these anaerobic microbes possess. Where do you end up?*

You can find answers online in **Appendix F**.

■ SUMMARY OF KEY CONCEPTS

Concept 6.1 Enzymes and Energy Drive Cellular Metabolism

1. The two major themes of microbial metabolism are **catabolism** (the breakdown of organic molecules) and **anabolism** (the synthesis of organic molecules).
2. Microorganisms and all living organisms use **enzymes**, protein molecules that speed up a chemical change, to control cellular reactions.
3. Enzymes bind to **substrates** at their **active site** (enzyme-substrate complex) where functional groups are destabilized, lowering the **activation energy**. (Figures 6.1, 6.2)
4. Many metabolic processes occur in **metabolic pathways**, where a sequence of chemical reactions is catalyzed by different enzymes. (Figure 6.4)
5. Enzymes can be inhibited by physical agents, which can denature enzymes and inhibit their action in cells. Enzymes are modulated through **feedback inhibition**. The final product often inhibits the first enzyme in the pathway.
6. Many metabolic reactions in cells require energy in the form of **adenosine triphosphate** (**ATP**). The breaking of the terminal phosphate produces enough energy to supply an **endergonic reaction** and often involves the addition of the phosphate to another molecule (**phosphorylation**). (Figure 6.5)

Concept 6.2 Glucose Catabolism Generates Cellular Energy

7. **Cellular respiration** is a series of metabolic pathways in which chemical energy is converted to cellular energy (ATP). It might require oxygen gas (**aerobic respiration**) or another inorganic final electron acceptor (**anaerobic respiration**). (Figure 6.7)
8. **Glycolysis**, the catabolism of glucose to pyruvate, extracts some energy from which two ATP and 2 NADH molecules result. (Figure 6.8)
9. The catabolism of pyruvate into carbon dioxide and water in the **citric acid cycle** extracts more energy as ATP, NADH, and $FADH_2$. Carbon dioxide gas is released. (Figure 6.9)

10. The process of **oxidative phosphorylation** involves the oxidation of NADH and FADH2, the transport of freed electrons along a cytochrome chain, the pumping of protons across the cell membrane, and the synthesis of ATP from a reversed flow of protons. These last two steps are referred to as chemiosmosis. (Figure 6.11)

Concept 6.3 There Are Other Pathways to ATP Production

11. Other carbohydrates, such as disaccharides and polysaccharides, represent energy sources that can be metabolized through cellular respiration. Besides carbohydrates, proteins and fats can be metabolized through the cellular respiratory pathways to produce ATP. (Figure 6.13)
12. The **anaerobic respiration** of glucose uses different final electron acceptors in oxidative phosphorylation. Glycolysis and parts of the citric acid cycle still function, and ATP synthesis occurs.
13. In **fermentation**, the catabolism of glucose can continue without a functional citric acid cycle or oxidative phosphorylation process. To maintain a steady supply of NAD^+ for glycolysis and ATP synthesis, pyruvate is redirected into other pathways that reoxidize NADH to NAD^+. Final products include lactic acid or ethanol. Only the two ATP molecules of glycolysis are synthesized in fermentation from each molecule of glucose. (Figure 6.14)

Concept 6.4 Photosynthesis Converts Light Energy to Chemical Energy

14. The anabolism of carbohydrates can occur by **photosynthesis**, the process whereby light energy is used to synthesize ATP, and the latter is then used to fix atmospheric CO_2 into carbohydrate molecules. (Figure 6.16)

Concept 6.5 Microbes Exhibit Metabolic Diversity

15. **Autotrophs** synthesize their own food from CO_2 and light energy (**photoautotrophs**) or CO_2 and inorganic compounds (**lithoautotrophs**). **Heterotrophs** obtain their carbon from organic compounds and energy from light (**photoheterotrophs**) or from organic compounds (**chemoheterotrophs**). (Figure 6.17)

■ CHAPTER SELF-TEST

For **Steps A–C**, you can find answers to questions and problems online in **Appendix D**.

■ STEP A: REVIEW OF FACTS AND TERMS

Multiple Choice

Read each question carefully, and then select the *one* answer that best fits the question or statement.

1. Enzymes are ____.
   A. inorganic compounds
   B. cofactors
   C. proteins
   D. vitamins

2. Enzymes combine with a ____ at the ____ site to lower the activation energy.
   A. substrate; active
   B. product; noncompetitive
   C. product; active
   D. coenzyme; active

3. Which one of the following is *not* a metabolic pathway?
   A. Citric acid cycle
   B. The carbon-fixing reactions
   C. Glycolysis
   D. Sucrose → glucose + fructose

4. If an enzyme's active site becomes deformed, ____ inhibition was likely responsible.
   A. metabolic
   B. competitive
   C. noncompetitive
   D. cellular

5. Which one of the following is *not* part of an ATP molecule?
   A. Phosphate groups
   B. Cofactor
   C. Ribose
   D. Adenine

6. The use of oxygen gas (O_2) in an exergonic pathway generating ATP is called ____.
   A. anaerobic respiration
   B. photosynthesis
   C. aerobic respiration
   D. fermentation

7. Which one of the following is *not* produced during glycolysis?
   A. ATP
   B. NADH
   C. Pyruvate
   D. Glucose

8. All the following are produced during the citric acid cycle *except* ____.
   A. CO_2
   B. O_2
   C. ATP
   D. NADH

9. The electron transport chain is directly involved with ____.
   A. ATP synthesis
   B. CO_2 production
   C. H^- pumping
   D. generating oxygen gas

10. Which one of the following macromolecules would *not* normally be used for microbial energy metabolism?
    A. DNA
    B. Proteins
    C. Carbohydrates
    D. Fats

11. Anaerobic respiration does *not* ____.
    A. use an electron transport system
    B. use oxygen gas (O_2)
    C. occur in bacterial cells
    D. generate ATP molecules

12. In fermentation, the conversion of pyruvate into a final end product is critical for the production of ____.
    A. CO_2
    B. glucose
    C. NAD^+
    D. O_2

13. Which one of the following is the correct sequence for the flow of electrons in the energy-fixing reactions of photosynthesis?
    A. Water—photosystem I—photosystem II—NADPH
    B. Photosystem I—NADPH—water—photosystem II
    C. Water—photosystem II—photosystem I—NADPH
    D. NADPH—photosystem II—photosystem I—water

14. Microorganisms that use organic compounds as energy and carbon sources are ____.
    A. chemoheterotrophs
    B. lithoautotrophs
    C. photoautotrophs
    D. photoheterotrophs

Identification

15. Identify on the metabolic map key where each reactant is used and where each product is produced in the aerobic cellular respiration summary equation.

$$C_6H_{12}O_6 + 6\ O_2 + 32\ ADP + 32\ P$$
$$\downarrow$$
$$6\ CO_2 + 6\ H_2O + 32\ ATP$$

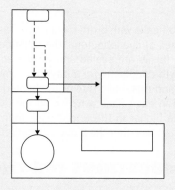

Term Selection

For each choice, circle the word or term that best completes each of the following statements.

16. The sum total of all an organism's biochemical reactions is known as (catabolism, metabolism); it includes all the (synthesis, digestion) reactions called anabolism and all the breakdown reactions known as (inactivation, catabolism).

17. Enzymes are a group of (carbohydrate, protein) molecules that generally (slow down, speed up) a chemical reaction by converting the (substrate, active site) to final products.

18. The aerobic respiration of glucose begins with the process of (oxidative phosphorylation, glycolysis) and requires that (amino acids, energy) be supplied by (ATP, NADH) molecules.

19. The process of (fermentation, the citric acid cycle) takes place in the absence of (oxygen, carbon dioxide). It begins with a molecule of (glucose, protein) and ends with molecules of (amino acids, an organic final product).

20. In oxidative phosphorylation, pairs of (protons, electrons) are passed among a series of (chromosomes, cytochromes) with the result that (oxygen, energy) is released for (NAD$^+$, ATP) synthesis.

21. In the citric acid cycle, (glucose, pyruvate) undergoes a series of changes and releases its (carbon, nitrogen) as (carbon dioxide, nitrous oxide) and its electrons to (NAD$^+$, ATP).

22. For use as energy compounds, proteins are first digested to (uric, amino) acids, which then lose their (carboxyl, amino) groups in the process of (fermentation, deamination) and become intermediates of cellular respiration.

23. Ribulose 1,5-bisphosphate bonds with (carbon monoxide, carbon dioxide) molecules during (fermentation, photosynthesis), a process that ultimately results in molecules of (pyruvate, glucose).

24. Chemoautotrophs use energy from (light, inorganic compounds) to synthesize (carbohydrates, oxygen gas) and are typified by species of (*Staphylococcus*, some Bacteria and Archaea).

25. Fats are broken down to (fatty acids, coenzymes), which are converted through (beta-oxidation, deamination) reactions to (glucose, two-carbon units) and eventually enter (cellular respiration, photosynthesis).

■ STEP B: CONCEPT REVIEW

26. State the role of an **enzyme-substrate complex** to regulating metabolism. (**Key Concept 6.1**)

27. Judge the importance of **metabolic pathways** in microbial cells. (**Key Concept 6.1**)

28. Explain the importance of **glucose** to energy metabolism. (**Key Concept 6.2**)

29. Compare and contrast **aerobic** and **anaerobic respiration**. (**Key Concept 6.3**)

30. Summarize the importance of (a) the **energy-fixing reactions** and (b) the **carbon-fixing reactions** of **photosynthesis**. (**Key Concept 6.4**)

31. Distinguish between the energy and carbon sources for the four nutritional classes of microorganisms. (**Key Concept 6.5**)

■ STEP C: APPLICATIONS AND PROBLEM SOLVING

32. You have two flasks with broth media. One flask contains a species of cyanobacteria; the other contains *E. coli*. Both flasks are sealed and incubated under optimal growth conditions for 2 days. Assuming the cell volume and metabolic rate of the bacterial cells are identical in each flask, why would the CO_2 concentration be higher in the *E. coli* flask than in the cyanobacteria flask after the 2-day incubation?

33. A stagnant pond usually has a putrid odor because hydrogen sulfide has accumulated in the water. A microbiologist recommends that tons of green sulfur bacteria be added to remove the smell. What chemical process does the microbiologist have in mind? Do you think it will work?

■ STEP D: QUESTIONS FOR THOUGHT AND DISCUSSION

34. A student goes on a college field trip and misses the microbiology exam covering microbial metabolism. Having made prior arrangements with the instructor for a make-up exam, he finds one question on the exam: "Discuss the interrelationships between anabolism and catabolism." How might you have answered this question?

35. If ATP is such an important energy source for microbes, why do you think it is not added routinely to the growth medium for these organisms?

36. One of the most important steps in the evolution of life on Earth was the appearance of certain organisms in which photosynthesis takes place. Why was this critical?

37. A population of a *Bacillus* species is growing in a soil sample. Suppose glycolysis came to a halt in these bacterial cells. Would this mean that the citric acid cycle would also stop? Why?

38. While you are taking microbiology, a friend is enrolled in a general biology course. You both are studying cell energy metabolism. Your friend looks puzzled when you tell him that the citric acid cycle and electron transport occur in the bacterial cytoplasm and cell membrane. Your friend insists these processes occur in the mitochondrion. Who is correct and why?

Concept Mapping

See pages XXXI-XXXII on how to construct a concept map.

39. Construct a concept map for **Cellular Respiration** using the following terms. Some terms can be used more than once.

Aerobic respiration	H_2O
Anaerobic respiration	H_2S
ATP	NO_2
CH_4	NO_3
CO_2	O_2
CO_3	SO_4
Glucose	

PART II The Genetics of Microorganisms

CHAPTER 7   Microbial Genetics
CHAPTER 8   Gene Transfer, Genetic Engineering, and Genomics

The two chapters in Part II embrace several aspects of the six principal (overarching) concepts as described in the ASM fundamental statements for a concept-based microbiology curriculum.

PRINCIPAL CONCEPTS

Evolution	• Cells, organelles (e.g., mitochondria and chloroplasts), and all major metabolic pathways evolved from early prokaryotic cells. **Chapters 7 and 8** • Mutations and horizontal gene transfer, with the immense variety of microenvironments, have selected for a huge diversity of microorganisms. **Chapters 7 and 8** • Human impact on the environment influences the evolution of microorganisms (e.g., emerging diseases and the selection of antibiotic resistance). **Chapters 7 and 8** • The traditional concept of species is not readily applicable to microbes due to asexual reproduction and the frequent occurrence of horizontal gene transfer. **Chapters 7 and 8**
Cell Structure and Function	• Bacteria and Archaea have specialized structures (e.g., flagella, endospores, and pili) that often confer critical capabilities. **Chapters 7 and 8** • While microscopic eukaryotes (for example, fungi, protozoa, and algae) carry out some of the same processes as bacteria, many of the cellular properties are fundamentally different. **Chapters 7 and 8**
Metabolic Pathways	• The survival and growth of any microorganism in a given environment depend on its metabolic characteristics. **Chapters 7 and 8**
Information Flow and Genetics	• Genetic variations can impact microbial functions (e.g., in biofilm formation, pathogenicity, and drug resistance). **Chapters 7 and 8** • Although the central dogma is universal in all cells, the processes of replication, transcription, and translation differ in Bacteria, Archaea, and Eukaryotes. **Chapter 7** • The regulation of gene expression is influenced by external and internal molecular cues and/or signals. **Chapter 7** • Cell genomes can be manipulated to alter cell function. **Chapters 7 and 8**
Microbial Systems	• Microorganisms and their environment interact with and modify each other. **Chapter 8**
Impact of Microorganisms	• Microorganisms provide essential models that give us fundamental knowledge about life processes. **Chapters 7 and 8** • Humans utilize and harness microorganisms and their products. **Chapter 8**

CHAPTER 7

Microbial Genetics

In our fast-paced world, we often measure time in minutes and seconds, so our minds find it difficult to imagine the colossal 4.5 billion years that the Earth has been in existence. It might help, however, if we compress Earth's history and view it from the perspective of a single year.

On January 1, the Earth and the solar system formed. During the month of January, Earth was a hot, volcanic, lifeless ball of rock bombarded by material left over from the formation of the solar system. As the earth cooled during February, water vapor condensed into oceans and seas, providing conditions more amenable for the origin of life.

Around March, prokaryotic life was already "up and running" (**FIGURE 7.1**). As life evolved in April and May, it thrived and diversified in an oxygen-free atmosphere, which was composed primarily of carbon dioxide, methane, and nitrogen. By late May, photosynthetic cyanobacteria were producing some oxygen gas. Oxygen remained scarce until mid-June (2.4 billion years ago) when geological events allowed cyanobacteria to bloom. Often encased in stromatolites similar to the present-day structures in Shark Bay, Australia (see chapter opening photo), the oxygen gas produced by the cyanobacteria

(called the "great oxidation" event) became abundant in the atmosphere.

Prokaryotes were the only organisms on Earth until August, when single-celled eukaryotes, such as the algae, emerged. These organisms flourished and they represent the ancestors of present-day species. About October, multicellular eukaryotes arose, whose descendants would evolve into diverse plants, fungi, and animals. Not until December did eukaryotic life move out of the sea onto the land. The dinosaurs were in existence from December 19 to December 25, and by December 27, the Earth showed a resemblance to modern Earth. Finally, on December 31,

Shark Bay, Western Australia.
Frans Lanting Studio/Alamy.

FIGURE 7.1 Fossil Microbes. A rock surface displaying sedimentary structures of biological origin in the 3.48 billion-year-old Dresser Formation, Pilbara region, Western Australia. Such formations support the existence of ancient microorganisms that formed carpet-like microbial mats on the former sediment surface. »» *What would it mean for the environment once oxygen-producing cyanobacteria predominated?*

Courtesy of Dr. Nora Noffke, Old Dominion University.

at 11:35 PM, humans appeared. At this scale, the average human life span is 1.25 seconds!

We take this trek through geological time to help us appreciate in evolutionary terms why

microorganisms have prospered. They have been successful primarily because they have been around the longest and have adapted well. In fact, the Bacteria and Archaea domains have been on Earth about 3.8 billion years (versus about 200,000 years for humans), as **FIGURE 7.2** shows. During this time, gene changes had been occurring regularly and nature used the microbes to test its newest genetic constructions. Evolution has eliminated the detrimental traits (together with the organisms unlucky enough to have them), whereas the beneficial traits have thrived and have been passed on to future generations—and many persist today.

Darwin's theory of natural selection and "survival of the fittest" applies not only to plants and animals but to microbes, as well. Because of their ability to reshape their genetic makeup, microbes thrive in the varied environments on Earth, whether it is the ice of the Arctic, the boiling hot volcanic vents of the ocean depths, or in the human body. No other group of organisms has this diverse capacity. Microbes have done very well in the evolutionary lottery—very well, indeed.

In this chapter, we study gene expression and consider how mutations have brought ancient microorganisms to the myriad forms we observe on Earth today. However, to understand mutations and the importance of genetic variation, we first must look at the master information molecule DNA and examine how it is replicated.

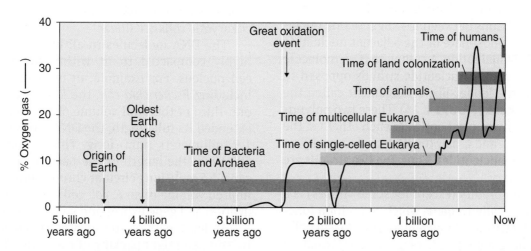

FIGURE 7.2 The Appearance of Life on Earth. This time line shows the relative amounts of time that various groups of organisms have existed on Earth. The Bacteria and Archaea have been in existence for a notably longer period than any other group, particularly humans. »» *Why did it take so long for eukaryotes (the Eukarya) to appear on Earth?*

Chapter Challenge

As we will see, there is great divergence in the number of genes that species can have. Some free-living prokaryotes have half the genes found in other free-living species like *Escherichia coli*, whereas others have twice as many genes. Even within the same genus, some species can be free living, whereas others exist as parasites. Interestingly, most obligate symbionts/parasites (referring to microbes that must live within a host cell to survive) have fewer genes than the free-living lineages. How do these microbes manage to survive—and quite well, actually—with substantially fewer genes? Let's see if we can uncover the answer.

■ KEY CONCEPT 7.1   The Hereditary Molecule in All Organisms Is DNA

Genetics is the science of heredity and the field studies the structure and expression of that genetic information, the **deoxyribonucleic acid (DNA)** molecule.

In 1953, James Watson and Francis Crick worked out DNA's double-helix structure based, in part, on the x-ray studies of Rosalind Franklin that are described in MICROFOCUS 7.1. This discovery has formed the foundation for all we know today about DNA replication, gene expression, and gene control.

Recall that DNA is a macromolecule composed of repeating monomers called **nucleotides**. Each nucleotide is composed of three components: a 5-carbon sugar molecule (deoxyribose), a phosphate group, and a nitrogen-containing nucleobase [adenine (A), guanine (G), cytosine (C), or thymine (T)]. Nucleotides are covalently joined through dehydration synthesis reactions between the sugar of one nucleotide and the phosphate of the adjacent nucleotide to form a **polynucleotide**. A complete DNA molecule consists of two polynucleotide strands opposed to each other in a ladder-like arrangement called the DNA **double helix** (FIGURE 7.3A). These two polynucleotides are hydrogen bonded to each other via the nucleobases: G and C pair together as do A and T on opposing polynucleotides. Thus, the two strands are "complementary."

A unit of heredity (except for some viruses) is a DNA segment called a **gene**, which codes for a functional product, be it ribonucleic acid (RNA) or a polypeptide. The complete set of genetic information for an organism or virus, including the organism's DNA and other information needed to maintain and reproduce the organism, is called the **genome**.

DNA Within a Cell Is Highly Compacted

Much of the genetic information in the Bacteria, Archaea, and Eukarya domains of life is contained within **chromosomes**, which are thread-like fibers composed of DNA and protein. In the prokaryotes, there is usually a single, circular chromosome, although a few species can have a single genome contained in two or more chromosomes. Such a genome with only one set of genes is called **haploid**.

The chromosome is localized in the cytosol within an area called the **nucleoid**. Remember that one of the unique features defining the nucleoid area is the absence of a surrounding **nuclear envelope**, which is typical of the cell nucleus in eukaryotic cells. In addition, eukaryotic cells have more than one chromosome and usually there are two sets of genes (**diploid**). Also, the DNA molecules in eukaryotic cells are linear.

The DNA molecules in all living cells must be highly compacted to fit within the nucleoid or cell nucleus. For example, in most bacterial cells, including *Escherichia coli*, the DNA occupies about one-third of the total volume of the cell, and when extended its full length, the DNA molecule is about 1.5 millimeters (mm) long. This is approximately 500 times the length of the bacterial cell. So, how can a 1.5-mm-long circular chromosome fit into the limited volume of an *E. coli* cell such that DNA replication and gene expression still can occur?

At least part of the answer is **supercoiling**, a twisting and tight packing of the DNA. The packing is facilitated in bacterial cells by a number of abundant **nucleoid-associated proteins (NAPs)**. The NAPs assist the coils in folding further into loops

MICROFOCUS 7.1: History

The Tortoise and the Hare

We all remember the children's fable of the tortoise and the hare. The moral of the story was those who plod along slowly and methodically (the tortoise) will win the race over those who are speedy and impetuous (the hare). The race to discover the structure of DNA is a story of collaboration and competition—a science "tortoise and the hare."

Rosalind Franklin (the tortoise) was 31 years old when she arrived at King's College in London in 1951 to work in J. T. Randell's lab. Having received a PhD in physical chemistry from Cambridge University, she moved to Paris where she learned the art of x-ray crystallography. At King's College, Franklin was part of Maurice Wilkins's group, and she was assigned the job of using x-ray crystallography to work out the structure of DNA fibers. Her training and constant pursuit of excellence allowed her to produce excellent, high-resolution x-ray images of DNA.

Meanwhile, at the Cavendish Laboratory in Cambridge, James Watson (the hare) was working with Francis Crick on the structure of DNA. Watson, who was in a rush for honor and greatness that could be gained by figuring out the structure of DNA, had a brash "bull in a china shop" attitude. This was in sharp contrast to Franklin's philosophy where you don't make conclusions until all of the experimental facts have been analyzed. Therefore, until she had all the facts, Franklin was reluctant to share her data with Wilkins—or anyone else.

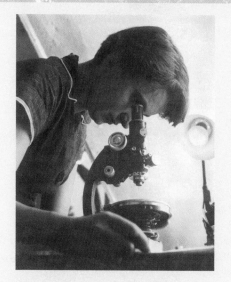

Photo of Rosalind Franklin.

© Science Source.

Feeling left out, Wilkins was more than willing to help Watson and Crick. Because Watson thought Franklin was "incompetent in interpreting x-ray photographs" and he was better able to use the data, Wilkins shared with Watson an x-ray image and report that Franklin had filed. From these materials, it was clear that DNA was a double helical molecule. It also seems clear Franklin knew this as well but, perhaps being a physical chemist, she did not grasp its importance because she was concerned with getting all the facts first and making sure everything was correct. But, looking through the report that Wilkins shared, the proverbial "light bulb" went on when Crick saw what Franklin had missed, that the two DNA strands were antiparallel. This knowledge, together with Watson's ability to work out the base pairing, led Watson and Crick to their "leap of imagination" and the structure of DNA.

In her book titled, *Rosalind Franklin: The Dark Lady of DNA* (HarperCollins, 2002), author Brenda Maddox suggests it is uncertain if Franklin could have made that leap because it was not in her character to jump beyond the data in hand. In this case, the leap of intuition won out over the methodical data collecting in research—the hare beat the tortoise this time. However, it cannot be denied that Franklin's data provided an important key from which Watson and Crick made the historical discovery.

In 1962, Watson, Crick, and Wilkins received the Nobel Prize in Physiology or Medicine for their work on the structure of DNA. Should Franklin have been included? The Nobel Prize committee does not make awards posthumously and Franklin had died 4 years earlier from ovarian cancer. So, if she had lived, did Rosalind Franklin deserve to be included in the award?

---

of about 10,000 bases (**FIGURE 7.3B**), and there are about 400 such loops in an *E. coli* chromosome. The high level of compaction is evident when the cell is broken open, releasing the DNA in a less compact form (**FIGURE 7.3C**).

In eukaryotic cells, the compaction of the DNA also is facilitated by a group of proteins called **histones** that form bead-like structures called **nucleosomes**. In the Archaea, some species have proteins similar to eukaryotic histone proteins; others lack histone-like proteins; whereas still others contain proteins similar to bacterial NAPs.

The *E. coli* chromosome and its DNA have been studied thoroughly. The genome of the common

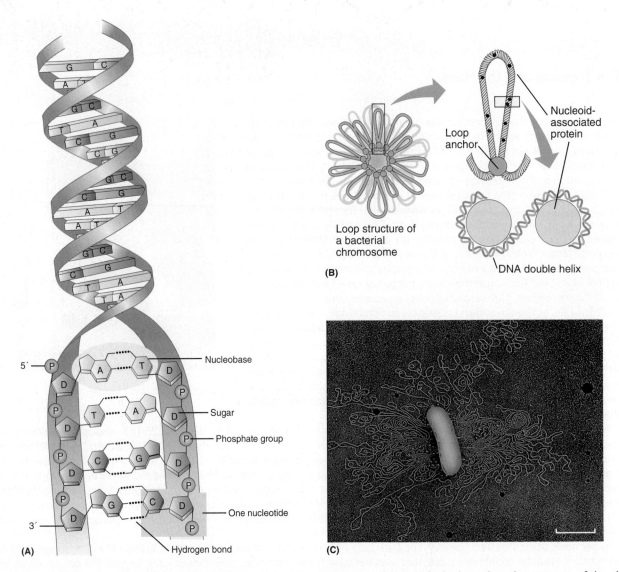

FIGURE 7.3 DNA Structure and Packing. (A) The double helical structure of DNA. **(B)** The loop domain structure of the chromosome as seen head-on. The loops in DNA help account for the compacting of a large amount of DNA in a relatively small cell. **(C)** An electron micrograph of an *E. coli* cell immediately after cell lysis. The uncoiled DNA fiber exists in loops. (Bar = 1 μm.)
»» How does plasmid structure compare to that of the chromosome in a bacterial cell?

K-12 strain has about 4,380 genes, 4,290 coding for proteins and the rest for RNAs. Such microbial genomes can be subdivided into three sets.

- ▶ **Core Genes.** There is a set of **core genes** that are essential for cell function. This portion of the genome is shared by all strains of a species and it includes those genes involved with metabolism, growth, and reproduction.
- ▶ **Variable Genes.** Another set of genes represents those found in a single strain or subset of strains within a species. Although these

variable genes are not essential for growth and reproduction, the genes can provide useful functions. For example, the O157:H7 strain of *E. coli*, which is a human pathogen, has an additional 1,346 genes not found in the K-12 strain. It is assumed that many, if not most, of these variable genes relate to increased virulence and infectious disease.
- ▶ **Unique Genes.** The third set of genes is called **unique genes** because these nucleotide sequences are unique to one strain of a species.

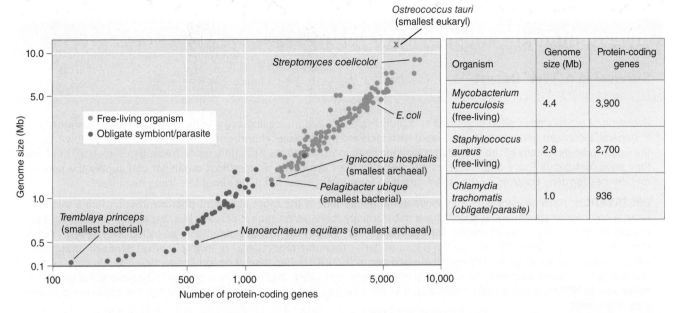

Organism	Genome size (Mb)	Protein-coding genes
Mycobacterium tuberculosis (free-living)	4.4	3,900
Staphylococcus aureus (free-living)	2.8	2,700
Chlamydia trachomatis (obligate/parasite)	1.0	936

FIGURE 7.4 Genome Size Among the Bacteria and Archaea. This graph illustrates the relationship between genome size (Mb = millions of base pairs) and number of protein-coding genes for some organisms currently sequenced. Note the extremely small size of *Tremblaya princeps*. The smallest known genome of a free-living, photosynthetic eukaryote (*Ostreococcus tauri*) is added for reference. *»» Plot on the graph the location for each of the three organisms listed in the table.*

One last part of the genome is the nucleotides that link together the core, variable, and unique genes into one continuous DNA molecule.

Many prokaryotic genomes can be larger or smaller than *E. coli*'s genome. Especially interesting are the small genomes of obligate symbionts or parasites that have a reduced number of genes (**FIGURE 7.4**). Such "genome reduction" in a free-living bacterial genus is examined in **INVESTIGATING THE MICRO-BIAL WORLD 7**. In addition, there are the much larger genomes of the eukaryotes (TABLE 7.1). Note in this table some of the archaeal/eukaryote similarities.

TABLE 7.1 Characteristics of Bacteria, Archaea, and Eukarya Chromosomes

Characteristic	Bacteria	Archaea	Eukarya
Organization	Nucleoid	Nucleoid	Nucleus
Chromosome morphology	Usually circular	Usually circular	Linear
Ploidy	Usually haploid	Usually haploid	Haploid, diploid, polyploid
Genome size (Mb)[1]	0.16–10	0.5–6	10–100,000
Average protein-coding genes	Few thousand	Few thousand	Tens of thousands
Presence of histone proteins	Histone-like	Yes	Yes
Presence of introns	No	Present in rRNA and tRNA genes	Yes
Replication	Just prior to binary fission	Just prior to binary fission	Few hours before mitosis
Replication origins	Single	Multiple	Multiple

[1]Mb = Millions of base pairs.

Investigating the Microbial World 7

Survival in a "Wet" Desert

Evolution of organisms is often described as a progression toward increasing complexity. If you look back at Figure 7.4, in general, increased genome size is equated with increased physiological and/or biochemical complexity. However, as the figure indicates, many of the obligate prokaryotic symbionts/parasites have undergone "reductive evolution" in which they have lost genes (have smaller genomes) because their symbiotic partner/host organism can supply the needed metabolites formerly supplied by the organism's own (now lost) genes. But what about free-living microbes?

OBSERVATION: *Prochlorococcus* is a cyanobacterial genus that is the most abundant photosynthetic marine microbe. From a complexity standpoint, it should have substantially more genes than it actually has, especially considering that the organism has to survive in marine surface waters that usually are low in nutrients—a so-called "wet desert." Yet in *Prochlorococcus*, evolution has resulted in a loss of genes (genome reduction). For example, when exposed to sunlight, the surface ocean water generates toxic hydrogen peroxide (HOOH). Because of genome reduction, *Prochlorococcus* has lost the gene to make catalase, the enzyme that can degrade HOOH; that is, genome reduction has made the organism vulnerable to HOOH, which should cause a loss of cell integrity and photosynthetic capacity. Yet the organism survives amazingly well!

QUESTION: *How can a HOOH-defenseless organism like Prochlorococcus thrive in a marine environment where it is continually faced with such a potentially destructive chemical?*

HYPOTHESIS: *Prochlorococcus* has evolved such that it is dependent on other marine community members ("helper" cells) to degrade HOOH in the common surrounding environment. If so, growing *Prochlorococcus* in pure culture in the presence of HOOH will kill *Prochlorococcus*, but adding a "helper" organism that can degrade HOOH will protect *Prochlorococcus*.

EXPERIMENTAL DESIGN: A strain of *Prochlorococcus* sensitive to HOOH was cultured. The "helper" organism, a strain of the heterotroph *Alteromonas* that produces catalase to degrade HOOH, was also cultured in the lab. HOOH concentrations used were similar to what would be produced naturally in a typical marine environment. Growth of *Prochlorococcus* cells was monitored by counting the number of cells/ml in the experimental culture.

EXPERIMENT 1: Does coculturing the "helper" organism with *Prochlorococcus* protect *Prochlorococcus* from HOOH damage? Two cultures containing HOOH were established: (1) *Prochlorococcus* only and (2) "helper" organism and *Prochlorococcus*. Cell growth of *Prochlorococcus* was measured over 16 days.

EXPERIMENT 2: Is removal of HOOH from the growth medium necessary and sufficient to allow *Prochlorococcus* growth? Four cultures were established. Prior to inoculation of *Prochlorococcus*: (1) the sterile growth medium was supplemented with HOOH; (2) the "helper" organism was added to the medium with HOOH; (3) the "helper" organism was filtered out after 24 h; and (4) HOOH was added back to the filtered medium of culture 3. Cell growth of *Prochlorococcus* was measured over 28 days.

EXPERIMENT 3: If growth was inhibited in experiments 1 and 2, were the *Prochlorococcus* cells actually killed or were they simply not dividing? Three cultures were established. (1) *Prochlorococcus* culture without added HOOH; (2) *Prochlorococcus* culture supplemented with HOOH; and (3) *Prochlorococcus* and the "helper" organism supplemented with HOOH. Viability of *Prochlorococcus* was monitored over 7 days using a fluorescent green dye that is taken up into dead cells only.

RESULTS: See figures.

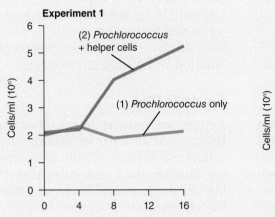

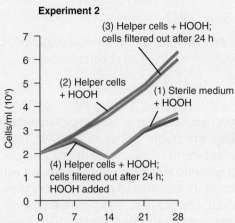

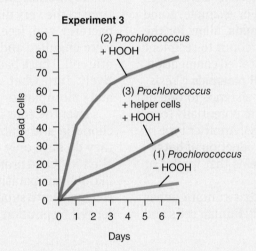

CONCLUSION: From experiment 1, it can be seen that the presence of the "helper" organism did "protect" the *Prochlorococcus* cells from cell damage by HOOH.

QUESTION 1: *From experiment 2, are the "helper" cells physically required to protect* Prochlorococcus *from HOOH damage? Explain what the "helper" cells are doing.*

QUESTION 2: *From experiment 3, does HOOH actually kill the unprotected* Prochlorococcus *cells? Explain.*

QUESTION 3: *Protection from HOOH damage is an energy-demanding process. What advantage has* Prochlorococcus *gained from genome reduction (e.g., loss of the gene to destroy HOOH)? Remember, the marine environment in which* Prochlorococcus *exists is oligotrophic, meaning available nutrients are limited and in low concentrations. It's a "wet" desert.*

You can find answers online in **Appendix E**.

Adapted from Morris, J. J. et al. 2011. *PLoS ONE* 6(2):e16805. doi:10.1371/journal.pone.0016805.

Many Microbial Cells Also Contain Extrachrosomal DNA

In many prokaryotic cells and in all eukaryotic cells, not all the DNA is restricted to chromosomes.

Plasmids

Many bacterial, archaeal, and fungal cells contain **plasmids**, which are stable, nonessential DNA molecules found outside of the nucleoid or cell nucleus. This means a plasmid could be removed from a cell without affecting its growth and division. Most plasmids are circular, contain 5 to 100 genes, and replicate as independent genetic elements in the cytosol.

Plasmids would be considered as part of the cell's variable genome and confer selective advantages by providing genetic flexibility. For example, some bacterial plasmids, called **F plasmids**, allow for the transfer of genetic material from donor to recipient through a recombination process called conjugation. Other bacterial plasmids, called **R plasmids** ("resistance" factors), carry genes for resistance to one or more antibiotics or for resistance to potentially toxic heavy metals (e.g., silver, mercury). Another type of plasmid contains genes for the production of **bacteriocins**, a group of bacterial proteins that inhibit or kill other bacterial species.

Finally, there are plasmids that contain genes coding for protein toxins affecting human cells and disease processes. The genes encoding the toxins responsible for anthrax are carried on a plasmid contained in *Bacillus anthracis*. Much more about plasmids is discussed in the chapter on gene recombination and genetic engineering.

Organelle DNA

Among the eukaryotic microbes (fungi and protists), most all have mitochondria that carry out cellular respiration. Within these energy organelles are circular DNA molecules that closely resemble the structure of bacterial DNA. In addition, the chloroplasts in algal cells also contain circular DNA molecules resembling a bacterial chromosome. In fact, biologists have put forward the **endosymbiont theory**, which states that at some time in the very distant past free-living respiratory bacteria and free-living photosynthetic bacteria were engulfed and retained by an evolving eukaryotic cell. These bacterial cells then evolved, respectively, into what we know today as mitochondria and chloroplasts.

Over the eons of time, much of the DNA in mitochondria and chloroplasts has been lost and what they do have only codes for about 5% of the organelle's RNA and protein needs. Thus, these organelles, like obligate parasites, are dependent on genes in the cell nucleus to supply the missing RNAs and proteins for cell respiration and photosynthesis, respectively.

Concept and Reasoning Checks 7.1

a. Describe the basic structure of a bacterial chromosome.
b. Justify the necessity for DNA supercoiling and looped domains.
c. Explain why plasmids are said to carry nonessential genetic information, whereas the DNA in mitochondria and chloroplasts is essential genetic information.

Chapter Challenge A

In our chapter challenge, we are examining why obligate parasites have such small genomes. Looking at Figure 7.4, there are a number of such obligate bacterial and archaeal species with extremely small genomes and very few genes. For example, *Nanoarchaeum equitans* has only some 586 protein-coding genes. Then, there is *Tremblaya princeps*, a bacterial symbiont in aphids, which has only 110 protein-coding genes. This bacterial species has the smallest genome yet discovered.

QUESTION A: *Propose why these obligate symbionts/parasites have such small genomes and thus few genes. Clue: If free-living species need substantially more genes, how do the obligate types get by with far fewer genes? (Hint: consider the genome content of modern-day mitochondria and chloroplasts.)*

You can find answers online in **Appendix F**.

■ KEY CONCEPT 7.2   DNA Replication Is Part of the Cell Cycle

Watson and Crick's 1953 paper on the structure of DNA provided a glimpse of how DNA might be copied. They concluded, "It has not escaped our notice that the specific pairing we have postulated immediately suggests a possible copying mechanism for the genetic material." In fact, the copying process for the genetic material, called **DNA replication**, occurs with such precision that the two daughter cells resulting from binary fission are genetically identical.

DNA Replication Occurs in Three Stages

One of the almost universal processes in biology is how the DNA molecule replicates itself. This anabolic process employs similar mechanisms in both prokaryotes and eukaryotes. In the following discussion, we will focus on DNA replication in bacterial organisms like *E. coli*.

DNA replication is carried out by a multiprotein complex made up of 20–30 proteins. Although it occurs in a relatively smooth, synchronized process, we can separate it into three stages: initiation, elongation, and termination. Let's look at each stage in more detail, using **FIGURE 7.5A** to guide us.

Initiation

DNA replication in *E. coli* cells starts at a fixed region on the chromosome called the **replication origin** (*ori*C). Here, initiation proteins bind to the DNA and begin to unwind the two polynucleotide strands. Subsequently, a **DNA helicase** attaches to further unwind and separate the two strands, while DNA **stabilizing proteins** keep the two strands separated. Consequently, each of the two unwinding helices is contained within a so-called **replication factory** (or **replisome**) in which each strand will be a template to synthesize a complementary strand. A V-shaped **replication fork** in each factory identifies the area active in the replication process.

Elongation

Synthesis of DNA in each factory then occurs simultaneously, using each old strand as a template for the synthesis of a new complementary strand. Many proteins are involved in DNA synthesis. A key enzyme is **DNA polymerase**. A separate polymerase binds to each template strand and then moves along each strand, catalyzing the insertion of a new complementary nucleotide to each nucleotide in the template strand (discussed in more detail a bit later). This combination of a new and an old strand is called **semiconservative replication** because in each new chromosome, one strand is the old strand of the replicated DNA and one strand is newly synthesized.

In *E. coli*, DNA synthesis takes about 40 minutes, which means that at each replication fork, DNA polymerase is adding new complementary bases at the rate of about 1,000 per second! At this pace, errors sometimes occur where an incorrect base is added. Such potential mutations could be dangerous to cell growth or reproduction, so cells have a mechanism to correct any replication errors. In fact, one subunit of the DNA polymerase can detect any mismatched nucleotides as replication is occurring. The enzyme then removes the incorrect nucleotide in the pair and adds the correct complementary nucleotide. Such "proofreading" reduces replication errors to about 1 in every 10 billion bases added. We will have more to say about mutations and DNA repair later in this chapter.

Termination

For *E. coli*, in about 40 minutes, the replication factories meet 180° from *ori*C At the **terminus** region, there are termination proteins blocking further replication, which causes the replication factories to detach. Then, the two intertwined DNA molecules (chromosomes) are separated by other enzymes. Cytoskeletal proteins reposition the two DNA molecules so that each daughter cell will inherit one complete chromosome after binary fission.

As already mentioned, the DNA in eukaryotic cells is replicated similar to that in prokaryotic cells. However, there are a few differences. Because eukaryotic chromosomes are so much larger than bacterial chromosomes, each eukaryotic DNA molecule has numerous replication origins to shorten the time to complete DNA synthesis. Eukaryotic cells also use four different DNA polymerases, three for the nuclear chromosomes and one for the organelle (mitochondrion, chloroplast) DNA.

DNA Polymerase Only Reads in the 3′ to 5′ Direction

Because DNA polymerase can "read" the template DNA only in the 3′ (pronounced "3 prime") to 5′

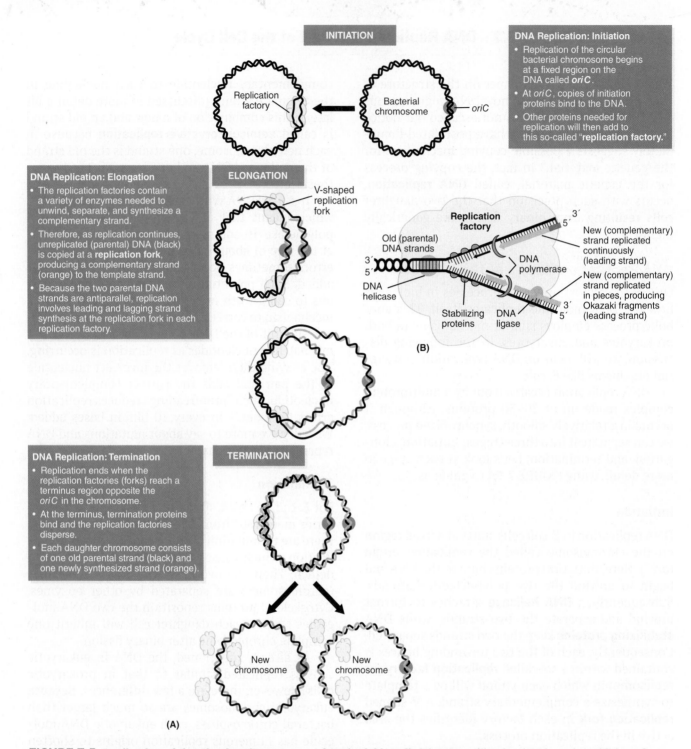

INITIATION

DNA Replication: Initiation
- Replication of the circular bacterial chromosome begins at a fixed region on the DNA called *oriC*.
- At *oriC*, copies of initiation proteins bind to the DNA.
- Other proteins needed for replication will then add to this so-called "**replication factory**."

DNA Replication: Elongation
- The replication factories contain a variety of enzymes needed to unwind, separate, and synthesize a complementary strand.
- Therefore, as replication continues, unreplicated (parental) DNA (black) is copied at a **replication fork**, producing a complementary strand (orange) to the template strand.
- Because the two parental DNA strands are antiparallel, replication involves leading and lagging strand synthesis at the replication fork in each replication factory.

ELONGATION

V-shaped replication fork

Replication factory

Old (parental) DNA strands

New (complementary) strand replicated continuously (leading strand)

DNA polymerase

New (complementary) strand replicated in pieces, producing Okazaki fragments (leading strand)

DNA helicase

Stabilizing proteins

DNA ligase

(B)

DNA Replication: Termination
- Replication ends when the replication factories (forks) reach a terminus region opposite the *oriC* in the chromosome.
- At the terminus, termination proteins bind and the replication factories disperse.
- Each daughter chromosome consists of one old parental strand (black) and one newly synthesized strand (orange).

TERMINATION

New chromosome

New chromosome

(A)

FIGURE 7.5 Replication of the Circular Chromosome of *Escherichia coli*. (A) DNA replication involves the addition of complementary bases to the parental (template) strand within replication factories. **(B)** Each replication factory contains some 15 proteins involved with DNA replication. At the replication fork, both leading and lagging strand synthesis occur. The discontinuous synthesis on the lagging strand results from the DNA polymerase moving away from the replication fork, resulting in the formation of short DNA fragments that are eventually joined by a DNA ligase. *»» Why is DNA replication considered to be semiconservative?*

("5 prime") direction (see Figure 7.3A) and the two parental (template) strands are antiparallel, at each replication fork the new complementary DNA strand is formed in opposite directions (**FIGURE 7.5B**).

One parental strand in each replication factory is the template for synthesizing a continuous complementary **leading strand**. Here the DNA polymerase reads the template in the 3' to 5' direction, bringing in triphosphate nucleotides (A, T, G, and C), forming an elongating chain of nucleotides from 5' to 3'. The other template strand in each fork of a replication factory must be read "backward" by

the DNA polymerase; that is, again in the 3' to 5' direction, which, in this case, is in a direction away from the replication fork. As a result, a discontinuous series of replication starts and stops occurs. This piecemeal strand, therefore, is called the **lagging strand**. These segments, which are about 1,000 to 2,000 nucleobases long, are called **Okazaki fragments**, after Reiji Okazaki, who discovered them. As these new polynucleotide segments are produced, the gaps between adjacent fragments are joined into a complete and elongating single strand with the help of an enzyme called **DNA ligase**.

Concept and Reasoning Checks 7.2

a. Describe the role for replication factories in DNA synthesis.
b. Why are there leading and lagging strands in each replication fork?

Chapter **Challenge B**

No matter the number of protein-coding genes found in an obligate symbiont/parasite, the genetic information contains DNA in the form of a chromosome, and the organism still needs to reproduce within the host.

QUESTION B: *When considering the obligate symbionts/parasites, would the genes required for the multiprotein complex found within the replication factories be part of the parasite's genome? Explain. Based on the text discussion of DNA replication, what would be the minimal number of genes required for DNA replication?*

You can find answers online in **Appendix F**.

■ **KEY CONCEPT 7.3**   Gene Expression Produces RNA and Protein for Cell Function

The discovery of the structure of DNA also provided a glimpse into understanding how a cell uses its genes to make proteins. The process, called **gene expression**, requires not only DNA but also RNA. TABLE 7.2 reviews the characteristics of DNA and RNA.

The process of manufacturing a protein starts when a particular gene is copied into a complementary RNA through a process called **transcription**. In the cytoplasm, the RNA carrying the gene message, along with other types of RNA, is converted into a growing protein molecule or polypeptide by linking amino acids together. This reading of the RNA bases is called **translation**, and the polypeptide produced reflects the genetic information that was originally coded in the DNA. The expression of genes through

transcription and translation is referred to as the **central dogma** (*dogma* = "truth") (**FIGURE 7.6**).

Transcription Copies Genetic Information into Complementary RNA

Transcription is the first, and perhaps most regulated, step in gene expression.

The large enzyme responsible for transcription is **RNA polymerase** (**FIGURE 7.7**). In prokaryotes, there is but one RNA polymerase with the archaeal enzyme being more similar to the eukaryotic polymerases. Like DNA polymerases, RNA polymerase "reads" the DNA template strand in the 3' to 5' direction. However, unlike DNA replication, only one of the two DNA strands within a gene is transcribed.

TABLE 7.2 A Comparison of DNA and RNA

DNA (Deoxyribonucleic Acid)	RNA (Ribonucleic Acid)
Large double-stranded molecule	Small single-stranded molecule
In Bacteria and Archaea, found in the nucleoid and plasmids; in Eukarya, found in the nucleus, mitochondria, and chloroplasts	In all organisms, found in the cytosol and in ribosomes; in Eukarya, also found in the nucleolus
Always associated with chromosome (genes); each chromosome has a fixed amount of DNA	Found mainly in combinations with proteins in ribosomes (ribosomal RNA), as messenger RNA, and as transfer RNA
Contains a 5-carbon sugar called deoxyribose	Contains a 5-carbon sugar called ribose
Contains bases adenine, guanine, cytosine, and thymine	Contains bases adenine, guanine, cytosine, and uracil
Contains phosphorus (in phosphate groups) that connects deoxyribose sugars with one another	Contains phosphorus (in phosphate groups) that connects ribose sugars with one another
Functions as the molecule of inheritance	Functions in gene expression and gene regulation

Transcription begins when RNA polymerase recognizes a sequence of bases called the **promoter**. Because this sequence is found on only one of the gene's two strands, this represents the DNA template strand recognized by the RNA polymerase. The enzyme binds to the promoter (initiation step), unwinds the helix, and separates the two strands within the gene. As the enzyme moves along the

Gene

Transcription

mRNA

Translation

Ribosome

Polypeptide (protein)

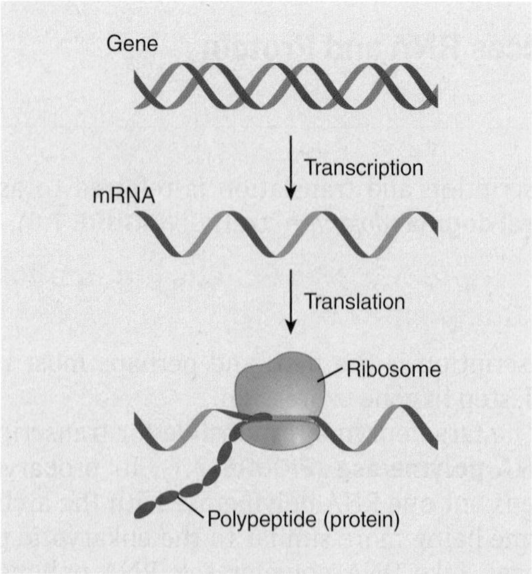

FIGURE 7.6 The Central Dogma. The flow of genetic information proceeds from DNA to RNA to protein (polypeptide). *»» Where in a bacterial cell and in a eukaryotic cell do transcription and translation occur?*

DNA template strand (elongation step), complementary pairing brings RNA triphosphate nucleotides to the template strand—guanine (G) and cytosine (C) pair with each other and thymine (T) in the DNA template pairs with adenine (A) in the RNA. However, an adenine base on the DNA template pairs with a uracil (U) base in the RNA because RNA nucleotides contain no thymine bases.

Termination of transcription occurs at a specific base sequence, called the **terminator**, on the DNA template strand. The RNA transcription product released represents a complementary base sequence to the base sequence in the DNA template strand.

Transcription produces three types of RNA that are needed for translation. In addition, a number of other types of RNA are used to regulate gene expression. Here, we focus on the three closely involved with translation.

Messenger RNA

This RNA carries the genetic information (message) that ribosomes "read" to manufacture a polypeptide. Each **messenger RNA (mRNA)** transcribed from a different gene carries a different message, that is, a different sequence of nucleotides coding for a different polypeptide. The message is encoded in a series of three-base combinations called **codons** that are found along the length of the mRNA. Each codon specifies an individual amino acid to be slotted into position during translation.

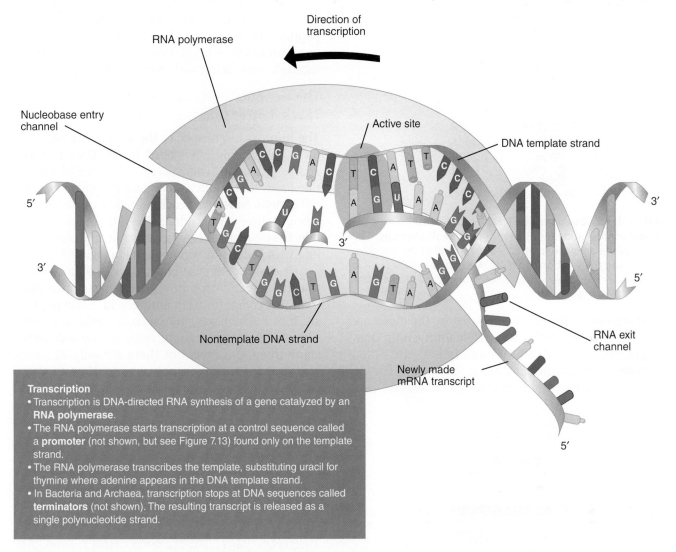

Transcription
• Transcription is DNA-directed RNA synthesis of a gene catalyzed by an **RNA polymerase**.
• The RNA polymerase starts transcription at a control sequence called a **promoter** (not shown, but see Figure 7.13) found only on the template strand.
• The RNA polymerase transcribes the template, substituting uracil for thymine where adenine appears in the DNA template strand.
• In Bacteria and Archaea, transcription stops at DNA sequences called **terminators** (not shown). The resulting transcript is released as a single polynucleotide strand.

FIGURE 7.7 The Transcription Process During Elongation. For transcription to occur, a gene must be unwound and the base pairs separated. The enzyme RNA polymerase does this as it moves along the template strand of the DNA and adds complementary RNA nucleotides. Note that the other DNA strand of the gene is not transcribed. *»» Justify the need for a promoter sequence in a gene.*

Ribosomal RNA

Three **ribosomal RNA (rRNA)** molecules are transcribed from specific regions of the DNA. Together with more than 50 proteins, these RNAs serve a structural and functional role as the framework of the ribosomes, which are the sites at which amino acids assemble into a polypeptide.

Transfer RNA

The conventional drawing for a **transfer RNA (tRNA)** is in a shape roughly like a cloverleaf (**FIGURE 7.8**). One region consists of three bases, which functions as an **anticodon**, that is, a sequence

that complementary binds to an mRNA codon. The tRNAs have a transport role in delivering amino acids to the ribosome for assembly into a polypeptide. Each tRNA has a specific amino acid attached to it. For example, the amino acid alanine binds only to the tRNA specialized to transport alanine; glycine is transported by a different tRNA.

There is one important difference between microbial RNAs. In most bacterial and some archaeal cells, all of the bases in a gene are transcribed and used to specify a particular polypeptide or RNA (**FIGURE 7.9A**). However, in other archaeal cells, and all eukaryotic cells, certain portions of the RNA coding sequence of a gene are not part of the final RNA

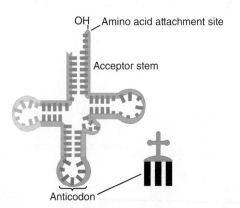

FIGURE 7.8 Structure of a Transfer RNA (tRNA). The traditional configuration for tRNA is a "cloverleaf" structure. The appropriate amino acid attaches to the end of the acceptor arm. Inset: the symbol for a tRNA. **»» *What part of the tRNA is critical for the complementary binding to a codon in an mRNA?***

and are removed from the RNA before the molecule can function (**FIGURE 7.9B**). These intervening DNA segments removed after transcription are called **introns**, whereas the remaining, amino acid-coding segments that are reattached are called **exons**.

The exons have the code for expressing a polypeptide (or protein), whereas the introns might have roles in gene regulation or in metabolic control through association with other RNAs or proteins.

The Genetic Code Consists of Three-Letter Codons

The information to specify the amino acid sequence for a polypeptide is encoded in the DNA template strand of a gene. This specific sequence of nucleotide bases that is transcribed into an mRNA is called the **genetic code**, with each sequence consisting of a string of three-letter codons. To synthesize a polypeptide, the DNA gene sequence must first be transcribed into RNA codons, as was depicted in Figure 7.7.

One of the surprising discoveries about DNA is that the genetic code in most cases contains more than one codon for each amino acid. Because there are four nucleobases, mathematics tells us 64 possible combinations can be made of the four bases, using three at a time. But there are only 20 amino acids needed to make a polypeptide. How do scientists account for the remaining 44 codons?

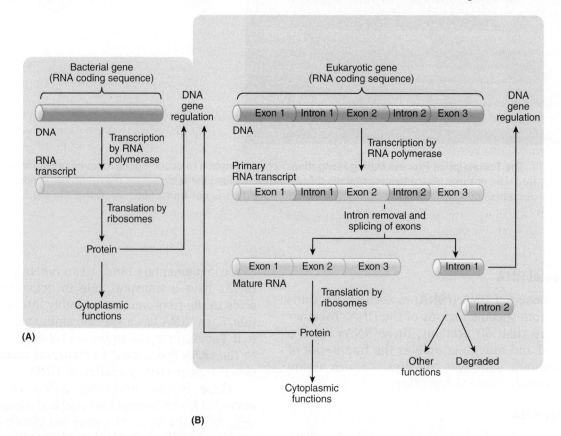

FIGURE 7.9 RNA Processing and Gene Activity. (A) In bacterial cells, gene DNA is almost entirely protein-coding information that is transcribed and translated into a protein having structural or functional roles in the cytoplasm, or gene regulation roles. **(B)** In eukaryotic cells, some of the intron RNA may be degraded, used to regulate gene function, or be associated with other RNAs or proteins in the cytoplasm. **»» *How do the roles of exon RNA and intron RNA differ?***

TABLE 7.3 The Genetic Code Decoder

The genetic code embedded in an mRNA is decoded by knowing which codon specifies which amino acid. On the far left column, find the first letter of the codon; then find the second letter from the top row; finally read up or down from the right-most column to find the third letter. The three-letter abbreviations for the amino acids are given. Note: In the Bacteria, AUG codes for formylmethionine (fMet) when starting a polypeptide.

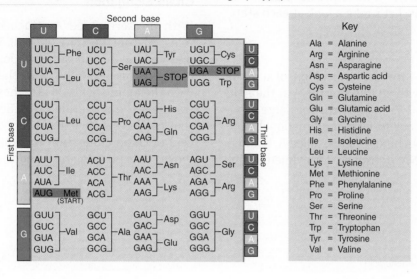

Key
Ala = Alanine
Arg = Arginine
Asn = Asparagine
Asp = Aspartic acid
Cys = Cysteine
Gln = Glutamine
Glu = Glutamic acid
Gly = Glycine
His = Histidine
Ile = Isoleucine
Leu = Leucine
Lys = Lysine
Met = Methionine
Phe = Phenylalanine
Pro = Proline
Ser = Serine
Thr = Threonine
Trp = Tryptophan
Tyr = Tyrosine
Val = Valine

In cells, 61 of the 64 codons are **sense codons** that specify an amino acid, and most of those amino acids have multiple codons (TABLE 7.3). For example, GCU, GCC, GCA, and GCG all code for the amino acid alanine (Ala). This lack of a one-to-one relationship between codon and amino acid generates the **redundancy** seen in the genetic code. In Table 7.3, notice one of the 64 codons, AUG, represents the **start codon** that sets the "reading frame" for making a polypeptide. Three additional codons, which do not code for an amino acid (UGA, UAG, and UAA), are called **stop codons** because they terminate the translation process.

Before we discuss the translation process, let's summarize the gene expression process to this point (**FIGURE 7.10**).

1. Each gene of the DNA contains information to manufacture a specific form of RNA.
2. The information can be transcribed into:
   ▶ mRNAs, which are produced from genes carrying the information as to what polypeptide will be made during translation;
   ▶ rRNAs, which form part of the structure of the ribosomes and help in the translation of the mRNA;
   ▶ tRNAs, each of which carries a specific amino acid needed for the translation process;

   ▶ Regulatory RNAs, which play important roles in controlling gene expression.
3. With the tRNAs and mRNAs present in the cytosol, they can combine within ribosomes to manufacture specific polypeptides. These polypeptides can represent a functional protein (tertiary structure) or first combine with one or more other polypeptides to form the functional protein (quaternary structure).

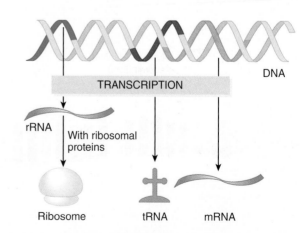

FIGURE 7.10 The Transcription of the Three Types of RNA. Genes in the DNA contain the information to produce three types of RNA needed for translation: rRNA, tRNA, and mRNA. *»» What is each type of RNA used for in a microbial cell?*

Translation Is the Process of Assembling the Polypeptide

In the process of translation, the language of the genetic code (nucleotides) is "translated" into the language of proteins (amino acids). As with DNA replication and RNA transcription, translation occurs in three steps. Let's take a look at each step.

Chain Initiation

Translation begins with the association of a small ribosomal subunit with an initiator tRNA at the AUG start codon (**FIGURE 7.11A**). Then, the large ribosomal subunit attaches to form the functional ribosome with three tRNA binding sites, called A, P, and E. At the A site, a tRNA transports formylmethionine

(fMet) in bacterial cells, whereas it transports methionine (Met) in archaeal and eukaryotic cells. Once the tRNA moves to the P site, a second tRNA can complementary bind at the A site and a ribozyme transfers the fMet to the amino acid (Ser) on the second tRNA.

Chain Elongation

With the second tRNA attached, the first tRNA is released from the E site (**FIGURE 7.11B**). Moving right one codon, the ribosome exposes the next codon (GCC in this example), and the appropriate tRNA with the amino acid Ala attaches. Again, a ribozyme transfers the dipeptide fMet–Ser to Ala. The tRNA that carried serine exits the ribosome and the process of chain elongation continues as the ribosome moves to expose the next codon at the A site.

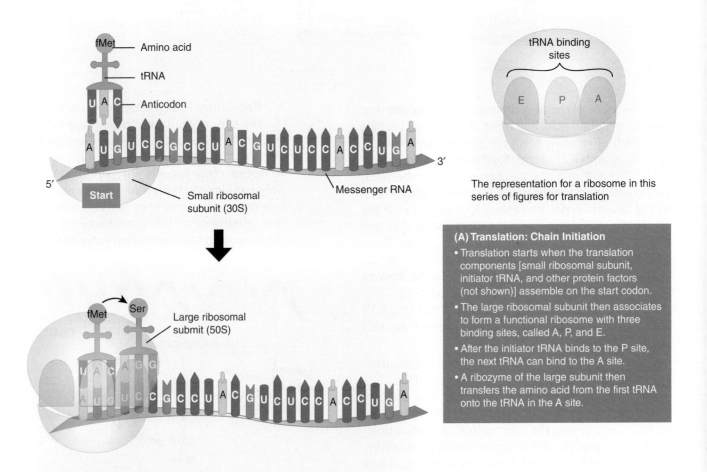

(A)

FIGURE 7.11 Protein Synthesis in a Bacterial Cell. The steps of **(A)** chain initiation, **(B)** chain elongation, and **(C)** chain termination are outlined. *»» How do the A, P, and E sites in the ribosome differ?*

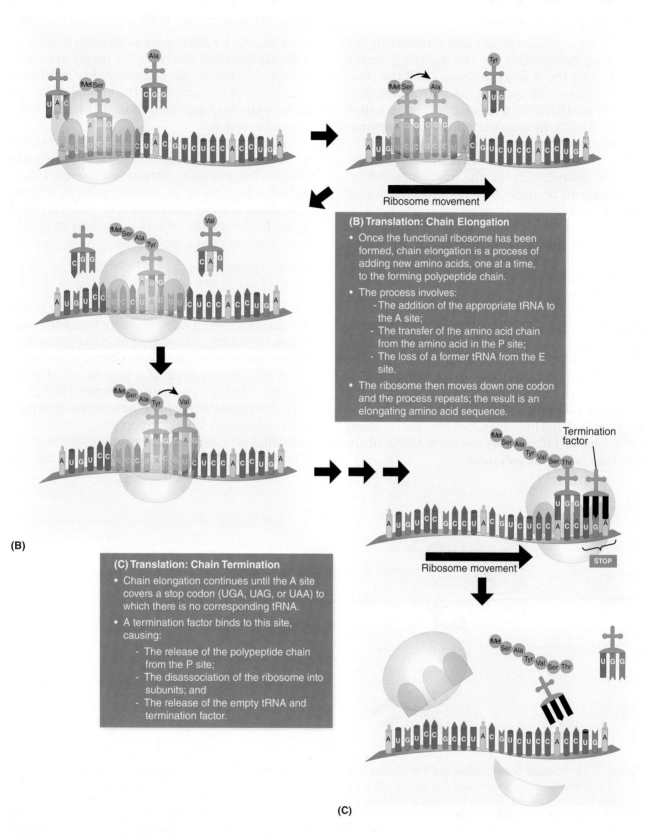

(B) Translation: Chain Elongation

- Once the functional ribosome has been formed, chain elongation is a process of adding new amino acids, one at a time, to the forming polypeptide chain.

- The process involves:
    - The addition of the appropriate tRNA to the A site;
    - The transfer of the amino acid chain from the amino acid in the P site;
    - The loss of a former tRNA from the E site.

- The ribosome then moves down one codon and the process repeats; the result is an elongating amino acid sequence.

(C) Translation: Chain Termination

- Chain elongation continues until the A site covers a stop codon (UGA, UAG, or UAA) to which there is no corresponding tRNA.

- A termination factor binds to this site, causing:
    - The release of the polypeptide chain from the P site;
    - The disassociation of the ribosome into subunits; and
    - The release of the empty tRNA and termination factor.

FIGURE 7.11   Protein Synthesis in a Bacterial Cell. (Continued)

Chain Termination/Release

The process of adding tRNAs and transferring the elongating polypeptide to the entering amino acid/tRNA at the A site continues until the ribosome reaches a stop codon (UGA in this example). There is no tRNA to recognize any stop codon (**FIGURE 7.11C**). Rather, proteins called **termination factors** bind where the tRNA would normally attach. This triggers the release of the polypeptide and disassembly of the ribosome subunits, which can be reassembled for translation of another mRNA.

During synthesis, the polypeptide already can begin to twist into its secondary and tertiary structure. For many polypeptides, groups of cytoplasmic proteins called **chaperones** ensure the folding process occurs correctly.

Cells typically make thousands of copies of each polypeptide. Producing such large amounts can be done efficiently and quickly by transcribing large numbers of identical mRNA molecules, which can each be translated simultaneously by several ribosomes (**FIGURE 7.12**). After one ribosome has moved far enough along the mRNA, another small subunit can "jump on" and initiate translation. Such a string of ribosomes all translating the same mRNA at the same time is called a **polysome**.

Antibiotics Interfere with Gene Expression

Many antibiotics affect gene expression in bacterial cells and therefore are clinically useful in treating human infections and disease. A few antibiotics interfere with transcription. For example, rifampin binds to the RNA polymerase so that transcription cannot be initiated.

A large number of antibiotics inhibit translation by binding to the bacterial 30S or 50S ribosomal subunit. For example, tetracycline prevents chain initiation by binding to the 30S subunit, whereas chloramphenicol inhibits chain elongation by binding to the 50S subunit. There is much more to learn about antibiotics in Chapter 10.

Gene Expression Can Be Controlled in Several Ways

Because transcription is the first step leading to protein manufacture in cells, one way to control what proteins and enzymes are present in a cell is to regulate the mechanisms that induce ("turn on") or repress ("turn off") transcription of a gene or set of genes. Here, we examine gene control in bacterial cells.

In 1961, two Pasteur Institute scientists, Françoise Jacob and Jacques Monod, proposed a mechanism for controlling gene expression. They suggested segments of bacterial DNA are organized into "transcriptional

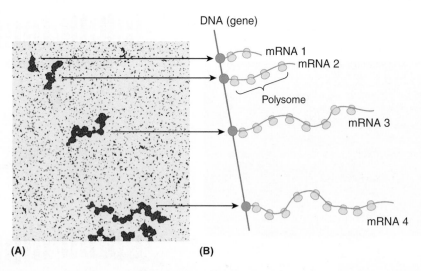

FIGURE 7.12 Coupled Transcription and Translation in *E. coli*. (A) The electron micrograph shows transcription of a gene in *E. coli* and translation of the mRNA. The dark spots are ribosomes, which coat the mRNA. An interpretation of the electron micrograph is shown in **(B)**. Each mRNA has ribosomes attached along its length. The large red dots are the RNA polymerase molecules; they are too small to be seen in the electron micrograph. The length of each mRNA is equal to the distance that each RNA polymerase has progressed from the transcription-initiation site. For clarity, the polypeptides elongating from the ribosomes are not shown. **»» *From the interpretation of the micrograph, (a) how many times has this gene been transcribed and (b) how many identical polypeptides are being translated?***

(B) Reproduced from: Miller, O. L., Hamkalo, B. A., and Thomas, C. A. 1977. *Science* **169**:392. Reprinted with permission from AAAS.

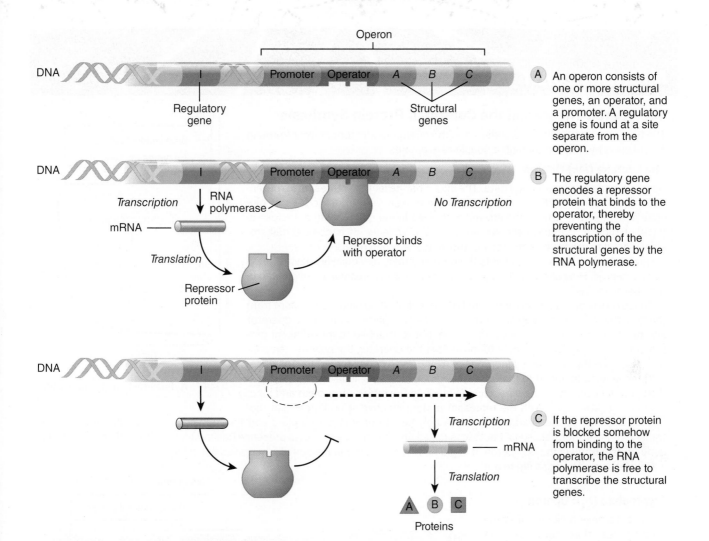

FIGURE 7.13 The Operon and Negative Control. An operon consists of a group of structural genes that are under the control of a single operator. Negative control exists if the operator prevents the RNA polymerase from transcribing the structural genes. *»» How does the operator prevent structural gene transcription in negative control?*

units" called **operons** (**FIGURE 7.13**). Their pioneering research along with more recent studies indicates that each operon consists of a cluster of **structural genes** providing genetic codes for proteins often having metabolically related functions. In this way, the cells can simultaneously regulate genes needed in the same functional or metabolic process. Adjacent to the structural genes is the **operator**, which is a sequence of bases controlling the expression (transcription) of the structural genes. Next to the operator is a **promoter**, which we mentioned represented another sequence of bases to which the RNA polymerase binds to initiate transcription. Also important, but not part of the operon, is a distant **regulatory gene** that codes for a **repressor protein**.

In the operon model, the repressor protein binds to the operator. This prevents the RNA polymerase from moving along the operon, and consequently transcription of the structural genes is prevented. This is called **negative control** because the repressor protein inhibits or "turns off" gene transcription within the operon. If the repressor in some way is prevented from binding to the operator, the RNA polymerase moves along the operon and transcribes the structural genes, which then are translated into the final polypeptides.

MICROINQUIRY 7 presents two contrasting examples of how an operon works to induce or repress gene transcription. **FIGURE 7.14** summarizes the gene expression process.

MICROINQUIRY 7

The Operon Theory and the Control of Protein Synthesis

The best way to visualize and understand the operon model for control of protein synthesis is by working through a couple of examples.

The Lactose (*lac*) Operon

Here is a piece of experimental data. The disaccharide lactose represents a potential energy source for *E. coli* cells if it can be broken into its monomers of glucose and galactose. One of the enzymes involved in the metabolism of lactose is β-galactosidase. If *E. coli* cells are grown in the absence of lactose, β-galactosidase activity cannot be detected, as shown in the graph (**FIGURE A**).

However, when lactose is added to the nutrient broth, very quickly enzyme activity is detected. How can this change from inhibition to expression be explained in the operon model?

Based on the operon theory, we would propose that when lactose is absent from the growth medium, the repressor protein for the *lac* operon binds to the operator and blocks passage of the RNA polymerase that is attached to the adjacent promoter (**FIGURE Bi**). Being unable to move past the operator, the polymerase cannot transcribe the structural genes, one of which (*lacZ*) codes for β-galactosidase.

When lactose is added to the growth medium, lactose will be transported into the bacterial cell, where the disaccharide binds to the repressor protein and inactivates it (**FIGURE Bii**). With the repressor protein inactive, it no longer can recognize and bind to the operator. The RNA polymerase now is not blocked and can translocate down the operon and transcribe the structural genes. Lactose is called an "inducer" because its presence has induced, or "turned on," structural gene transcription in the *lac* operon. It explains why β-galactosidase activity increases when lactose was present.

Now let's see if you can figure out this scenario.

Tryptophan (*trp*) Operon

E. coli cells have a cluster of structural genes that code for five enzymes in the metabolic pathway for the synthesis of the amino acid tryptophan (trp). Therefore, if *E. coli* cells are grown in a broth culture lacking trp, they continue to grow normally by synthesizing their own tryptophan, as shown in the graph (Figure A).

However, as the graph shows, when trp is added to the growth medium, new enzyme synthesis is repressed or "turned off" and cells use the trp supplied in the growth medium.

How can enzyme repression be explained by the operon model? The solution is provided online in **Appendix E**.

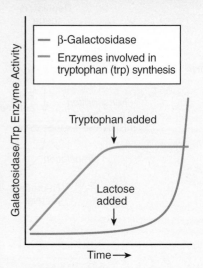

FIGURE A Enzyme activity versus time.

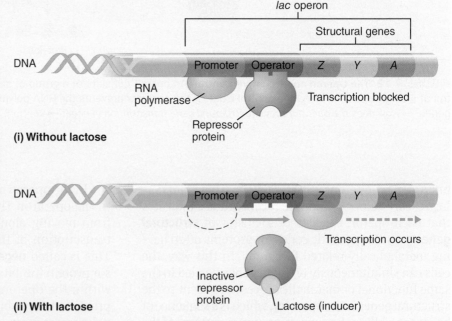

(i) **Without lactose**

(ii) **With lactose**

FIGURE B Regulation of the *lac* operon.

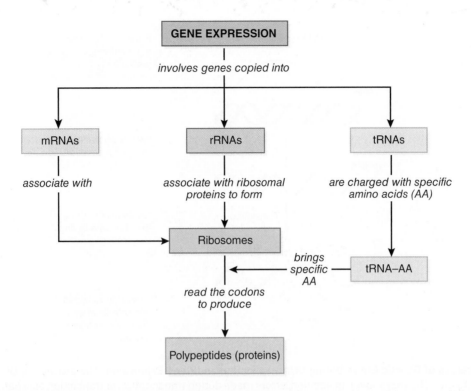

FIGURE 7.14 A Concept Map for Gene Expression. Gene expression is a combination of the transcription and translation processes. »» *Circle the part of the concept map that represents (a) transcription and (b) translation.*

Concept and Reasoning Checks 7.3

a. What is meant by the genetic code being redundant? Give two examples.
b. Explain why the ribosome can be portrayed as a "cellular translator."
c. Propose a hypothesis to explain why so many antibiotics specifically affect translation.
d. Explain what is meant by an operon being under negative control.

Chapter Challenge C

In this section of the chapter, you studied gene expression in bacterial species. This involved both transcription by an RNA polymerase and translation by ribosomes. Again, both the transcription and translation processes require many proteins to regulate and carry out the expression of a gene.

QUESTION C: *When considering the obligate symbionts/parasites, would the genes required for transcription and translation be part of the parasite's genome? Explain. Remember, a bacterial ribosome is built from four rRNAs and some 55 proteins.*

You can find answers online in **Appendix F**.

■ KEY CONCEPT 7.4   Mutations Are Heritable Changes in a Cell's DNA

The structural genes in a chromosome carry the information to build polypeptides. During DNA replication and gene expression, errors can occur that change the structure and function of the polypeptide product (**FIGURE 7.15**). Here, we will focus on those errors that might occur during the DNA replication process or result from harmful agents in the environment.

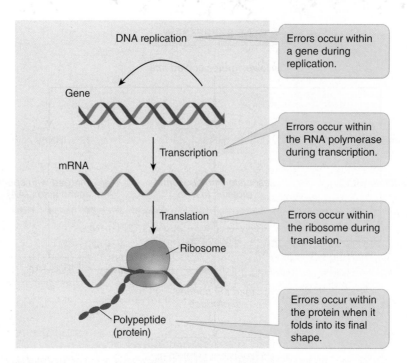

FIGURE 7.15 Sources of Genetic Errors During DNA Replication and Gene Expression. The process of DNA replication can introduce errors into the DNA (mutations). In addition, errors made during transcription or translation also can occur in the RNA transcript or protein sequence produced, respectively. Errors in folding of the final polypeptide is another possible outcome. *»» With so many potential error sites, what does that tell you about the efficiency of replication and gene expression?*

An organism's genome can be altered through a permanent and heritable change to one or more nucleotide bases. Such **mutations** usually involve a change in or disruption to a base sequence in the DNA, especially in a gene. The result is the synthesis of a miscoded mRNA and ultimately a change in one or more amino acids found in the polypeptide during translation. The majority of mutations are harmful because they alter some aspect of cellular activity in a negative way. For example, perhaps a mutation causes an enzyme to fold incorrectly, shutting down an important metabolic pathway in a cell. However, as MICROFOCUS 7.2 points out, on rare occasions, a mutation can be beneficial and give the organism a novel property, such as an increased potential of causing disease.

Mutations Can Be Spontaneous or Induced

Spontaneous mutations are heritable, random changes to the base sequence in the DNA that result from natural phenomena. These changes could result from errors made and not corrected by DNA polymerase during replication or from physical or chemical agents in the environment. It has been estimated that one such mutation can occur for every 10^6 to 10^{10} divisions in a bacterial population.

A mutant cell arising from a spontaneous mutation usually is masked by the much larger population of normal cells. However, should the environment favor the mutant, it will multiply and emerge as the predominant form. For example, for many decades doctors used penicillin to treat gonorrhea, which is caused by *Neisseria gonorrhoeae*. Then, in 1976, a penicillin-resistant strain of *N. gonorrhoeae* emerged. Many investigators suggested that the penicillin-resistant strain had existed in the *N. gonorrhoeae* population for centuries, but only with heavy penicillin use in the 1970s could the penicillin-resistant strain arise and surpass the penicillin-susceptible forms.

Most of our understanding of mutations has come from experiments involving **induced mutations**, which are produced by external physical and chemical agents called **mutagens**.

Physical Mutagens

Ultraviolet (UV) light is a **physical mutagen** whose energy causes adjacent thymine (or cytosine) bases in the DNA to covalently link together forming dimers (**FIGURE 7.16**). If these dimers occur in a protein-coding gene, the RNA polymerase cannot insert the correct bases (A–A) in mRNA molecules

MICROFOCUS 7.2: Evolution

Evolution of an Infectious Disease

Could the Black Death of the 14th century and the 25 million Europeans who succumbed to plague have been the result of a few genetic changes to a bacterial cell? Could the entire course of Western civilization have been affected by these changes?

Possibly so, according to researchers from the federal Rocky Mountain Laboratory in Montana. A research group led by Joseph Hinnebusch has identified three genes missing in the plague bacillus *Yersinia pestis* that are present in the related species *Y. pseudotuberculosis* that causes mild food poisoning. Thus, it is possible that the entire story of plague's pathogenicity revolves around a small number of gene changes to the genome of *Y. pestis*.

Bubonic, septicemic, and pneumonic plague are caused by *Y. pestis*, a rod-shaped bacterium transmitted by the rat flea. In an infected flea, the bacterial cells eventually amass in its foregut and obstruct its gastrointestinal tract. Soon the flea is starving, and it starts biting victims (humans and rodents) uncontrollably and feeding on their blood. During the bite, the flea can regurgitate some 24,000 plague bacilli into the bloodstream of the unfortunate victim.

At least three genes are important in the evolution of plague. The nonpathogenic *Y. pseudotuberculosis* bacilli have these genes, which encourage the bacilli to remain harmlessly in the midgut of the flea. Pathogenic plague bacilli, by contrast, do not have these three genes. Free of their control, the bacterial cells migrate from the midgut to the foregut, allowing the bacilli to be passed on to the victim in a fleabite.

Hinnebusch and colleagues also published evidence that another gene, carried on a plasmid, codes for an enzyme that is required for the initial survival of *Y. pestis* bacilli in the flea midgut. By acquiring this gene from another unrelated organism, *Y. pestis* made a crucial jump in its host range. It now could survive in fleas and became adapted to relying on its blood-feeding host for transmission. So, a few genetic changes to the genome might have been a key force leading to the evolution and emergence of plague. This is just another example of the flexibility that many microbes have to repackage themselves and their genomes into new and, sometimes, more dangerous agents of infectious disease.

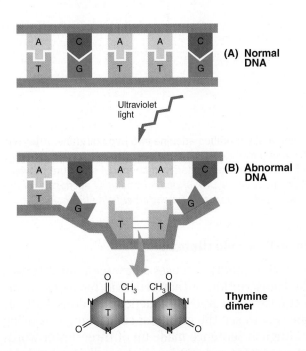

FIGURE 7.16 Ultraviolet Light and DNA. (A) When cells are irradiated with ultraviolet (UV) light either naturally or through experiment, the radiations might affect the cell's DNA. **(B)** UV light can cause adjacent thymine molecules to covalently pair (red lines) within the DNA strand to form a thymine dimer. *»» Looking at (B), how might a thymine dimer block the movement of RNA polymerase?*

where the dimers are located. In addition, ionizing radiations, such as gamma rays and X-rays, can cause physical breaks in the double-strand DNA. Loss of cellular function usually results.

Chemical Mutagens

Many chemicals are **mutagenic**; that is, they can cause mutations. Nitrous acid is an example of a chemical mutagen that converts DNA's adenine bases to hypoxanthine bases (**FIGURE 7.17A**). Adenine would normally base pair with thymine, but the presence of hypoxanthine causes a base pairing with cytosine during replication. Later, if replication occurs from the gene with the cytosine mutation, the mRNA will contain a guanine base rather than an adenine base.

Mutations also are induced by **base analogs**, which bear a close chemical similarity to a normal nucleotide. For example, acyclovir is a base analog that can substitute for guanine during virus replication (**FIGURE 7.17B**). However, incorporation of acyclovir blocks further viral replication because the analog lacks the deoxyribose sugar needed for the addition of the next nucleotide. As a result, new virus

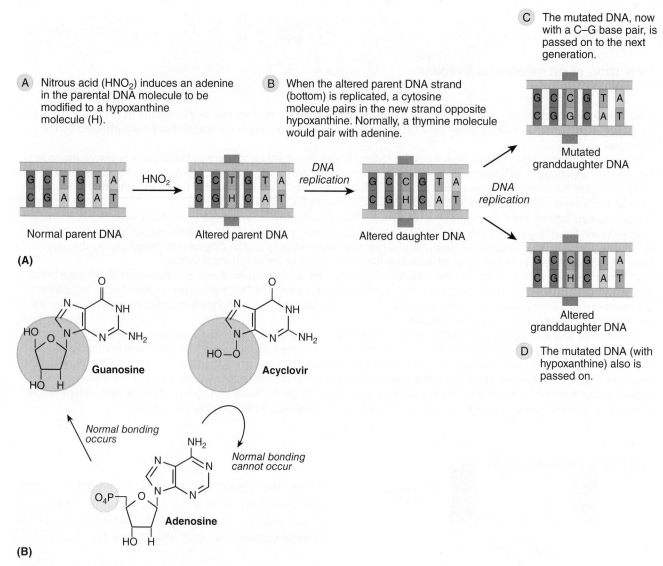

A Nitrous acid (HNO₂) induces an adenine in the parental DNA molecule to be modified to a hypoxanthine molecule (H).

B When the altered parent DNA strand (bottom) is replicated, a cytosine molecule pairs in the new strand opposite hypoxanthine. Normally, a thymine molecule would pair with adenine.

C The mutated DNA, now with a C–G base pair, is passed on to the next generation.

Normal parent DNA

Altered parent DNA

Altered daughter DNA

Mutated granddaughter DNA

Altered granddaughter DNA

D The mutated DNA (with hypoxanthine) also is passed on.

(A)

Guanosine

Acyclovir

Normal bonding occurs

Normal bonding cannot occur

Adenosine

(B)

FIGURE 7.17 The Effect of Chemical Mutagens. (A) Nitrous acid chemically modifies adenine into hypoxanthine. After replication of a hypoxanthine-containing strand, the granddaughter DNA has a mutated C–G base pair. **(B)** Base analogs induce mutations by substituting for nitrogenous bases in the synthesis of DNA. Note the similarity in chemical structure between guanosine and the base analog acyclovir. **»» If acyclovir had been incorporated into a DNA strand, why couldn't adenosine (or any other DNA nucleoside) be attached to acyclovir, while adenosine can be attached to guanoside (or any other DNA nucleoside)?**

particles cannot be produced. Acyclovir, therefore, is an effective treatment to limit infections caused by the herpes simplex 1 and 2 viruses (cold sores and genital herpes), varicella-zoster virus (shingles and chickenpox), and Epstein-Barr virus (mononucleosis).

Point Mutations Can Affect Gene Expression

Regardless of the cause of the mutation, one of the most common results is a **point mutation**, which usually affects just one point (a base pair) in a gene. Such mutations might be a change to or substitution of a different base pair or a deletion or addition of a base.

Base-Pair Substitutions

If a point mutation causes a base-pair substitution, the transcription of that gene will have one incorrect base in the mRNA sequence of codons. Perhaps one way to see the effects of such changes is using an English sentence made up of three-letter words (representing codons) in which one letter has been changed. As three-letter words, the letter substitution still reads correctly, but the sentence makes less sense.

Normal sequence: THE FAT CAT ATE THE RAT

Substitution (H for C): THE FAT HAT ATE THE RAT

As shown in **FIGURE 7.18A, B**, if the substitution, because of redundancy in the genetic code, does not alter the amino acid sequence, the change is referred to as a **silent mutation** because there is no change in protein function. If a base-pair substitution does result in a wrong amino acid in a polypeptide, the change is referred to as a **missense mutation** because the polypeptide is still assembled but might not have the correct shape. Our sentence analogy presented above is an example of a missense mutation. Finally, if the substitution causes a codon to become a stop codon, the change is called a **nonsense mutation** because translation is terminated prematurely, and any polypeptide produced probably will be nonfunctional.

Base-Pair Deletion or Insertion

Point mutations also can cause the loss or addition of a base in a gene, resulting in an inappropriate number of bases. Again, using our English sentence analogy, we can see how a deletion or insertion of one letter affects the reading frame of the three-letter word sentence.

Normal sequence: THE FAT CAT ATE THE RAT
Deletion: THE F_TC ATA TET HER AT
Insertion: THE FAT ACA TAT ETH ERA T

As you can see, the "sentence mutations" are nonsense when reading the sentence as three-letter words. The same is true in a cell. Ribosomes always read three letters (one codon) at one time, generating potentially extensive mistakes in the amino acid sequence if there are too few or too many letters (**FIGURE 7.18C**). Thus, like our English sentence, the deletion or addition of a base will cause a "reading frameshift" because the ribosome always reads the genetic code in groups of three bases. Therefore, loss or addition of a base also represents a **frameshift mutation** because it shifts the reading of the code

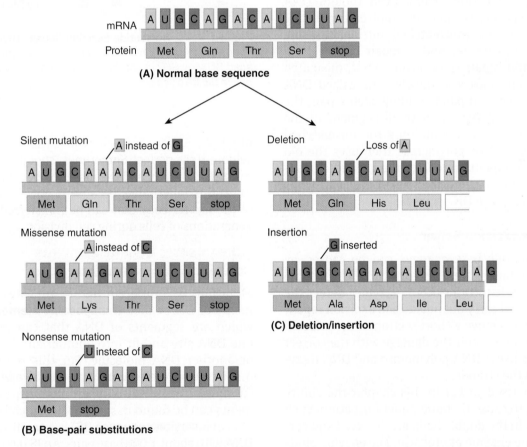

(A) Normal base sequence

(B) Base-pair substitutions

(C) Deletion/insertion

FIGURE 7.18 Categories and Results of Point Mutations. Mutations are permanent changes in DNA, but they are represented here as they are reflected in mRNA and its protein product. **(A)** The normal sequence of bases in a gene. **(B)** Base-pair substitutions can produce silent, missense, or nonsense mutations. **(C)** Deletions or insertions shift the reading frame of the ribosome. *»» Determine the normal sequence of bases in the template strand of the gene and the base change that gave rise to each of the "mutated" mRNAs.*

by one base. The result is serious sequence errors in the amino acids, which will probably produce an abnormal protein (nonsense) unable to carry out its functional role in the cell.

Repair Mechanisms Attempt to Correct Mistakes or Damage in the DNA

The fact that DNA is double stranded is not an accident. By being double stranded, one polynucleotide strand can act as the master copy or template to repair mismatches between strands or distortions within one strand. Therefore, if the damaged DNA is repaired before the cell divides, no mutation will result. Bacterial cells, such as *E. coli*, have two repair systems to mend damaged DNA.

Mismatch Repair

Realize that during the life of a microbial cell (indeed, of every cell), cellular DNA endures thousands of potentially damaging events resulting from DNA replication errors. Even though DNA polymerase is very accurate in proofreading during replication, errors (mismatched nucleotides) are sometimes missed. **Mismatch repair** can be used to detect and repair these errors. First, mismatch correction enzymes scan newly synthesized DNA for any mismatched pairs. Finding such a pair, the enzymes cut out (excise) a small segment of the polynucleotide strand containing the mismatched nucleotides. A DNA polymerase then uses the old strand as a template to replace the excised nucleotide segment with the correct set of bases, which is sealed in place by a DNA ligase.

Nucleotide Excision Repair

Some nucleotide base changes can be caused by physical mutagens that distort the DNA double helix. If such distortions occur, enzymes detect the distortions and carry out **nucleotide excision repair**. The enzymes remove a short section of the affected DNA strand and repair the damage with the correct nucleotides using DNA polymerase and DNA ligase to reattach the strands.

We discussed earlier in this chapter the ability of UV light to cause thymine dimer formation. Such distortions in the double helix are detected and corrected by the enzyme photolyase. The enzyme binds to the dimer and, when exposed to visible light (**photoreactivation**), the enzyme breaks the bond holding the thymine dimer together (**FIGURE 7.19**).

An impressive example of a very robust DNA repair system is seen in *Deinococcus radiodurans*, the subject

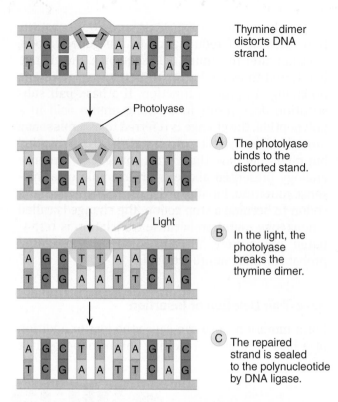

Thymine dimer distorts DNA strand.

Photolyase

A The photolyase binds to the distorted strand.

Light

B In the light, the photolyase breaks the thymine dimer.

C The repaired strand is sealed to the polynucleotide by DNA ligase.

FIGURE 7.19 Nucleotide Excision Repair. Thymine dimer distortion triggers photolyase enzymes to repair the damaged DNA. »» *How might the photolyase recognize the distortion in the DNA helix?*

of MICROFOCUS 7.3. Still, realize that DNA repair is seldom 100% perfect. Sometimes "shoddy repairs" fail to correct an error and the mutation becomes "locked in" and is inheritable, because it now exists as part of the master copy that can be transferred to future generations of cells during cell division.

Transposable Genetic Elements Can Also Cause Mutations

Mutations of a different nature are caused by segments of DNA called **transposable elements** (TEs), which are segments of DNA that can move from one DNA site and fit into another site in the same or another DNA molecule. For this reason, these mobile genetic elements have been referred to as "jumping genes." Two types of transposable elements can be found in some bacterial cells.

An **insertion sequence** (IS) is a small segment of DNA with about 1,000 base pairs. An IS has no genetic information other than for the enzyme **transposase**, which is needed to move the IS to a new location in a DNA molecule. The ends of the IS contain a number of unique "inverted repeats," that is, identical, inverted base pair sequences (**FIGURE 7.20A**).

MICROFOCUS 7.3: Evolution

Shattered Chromosomes

It has been called "Conan the Bacterium"[1] and has been listed in *The Guinness Book of World Records* as "the world's toughest bacterium." The organism is *Deinococcus radiodurans* (*deino* = "terrible"; *coccus* = "sphere"; *radio* = "ray"; *dura* = "hard"), referring to its resistance to gamma ray radiation. It is easily cultured and does not cause any known disease.

D. radiodurans can survive up to 5,000 gray (Gy; formerly called rad = 0.01 Gy) of ionizing radiation. For comparison, 10 Gy can kill a human and 100 Gy would kill *Escherichia coli*. Somehow "Conan" lives on.

So, how does *D. radiodurans* survive a dose of radiation that will cause extensive breaks in the DNA and produce "shattered chromosomes" consisting of hundreds of short DNA fragments? The key appears to be the presence of 8 to 10 copies of its genome in the cell's cytoplasm and a rapid mechanism capable of mending double-stranded breaks in its chromosomes within just a few hours. Following exposure to radiation, *D. radiodurans* undergoes massive and rapid DNA synthesis to produce a mosaic of new and old fragments with single-stranded ends that can then reconnect accurately into larger chromosomal segments. A special protein, called RecA, efficiently joins the segments into functional chromosomes. It is a remarkably efficient process occurring in an equally incredible short period.

Scientists have made use of *D. radiodurans*' talents. The bacterium has been used for bioremediation to consume and digest solvents and heavy metals, especially in highly radioactive sites. In addition, bacterial genes from *E. coli* have been introduced into *D. radiodurans* so that it can detoxify ionic mercury and toluene, chemicals often included in the radioactive waste from nuclear weapons manufacture.

A few other bacterial genera, including *Chroococcidiopsis* (phylum Cyanobacteria) and *Rubrobacter* (phylum Actinobacteria) and the archaeal species *Thermococcus gammatolerans*, also are gamma-radiation resistant. However, with its added genes for bioremediation, *D. radiodurans* rightly retains its title of "the world's toughest bacterium."

[1]Huyghe, P. 1998. Conan the bacterium. *The Sciences* 38:4, 16–19.

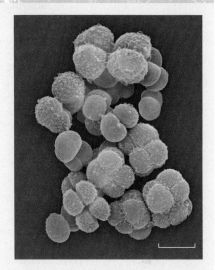

A false-color scanning electron micrograph of *Deinococcus radiodurans*. (Bar = 2 μm.)

© Dennis Kunkel Microscopy/Science Source.

A second type of TE, called a **transposon**, is larger than an IS and carries additional genes for diverse functions, such as antibiotic resistance. Like IS, the transposon has inverted repeats at each end of the element (**FIGURE 7.20B**).

The actual movement (transposition) of TEs often involves a simple "cut and paste" process. The TE is cut out of the DNA by the transposase and then, using the inverted repeats, it is inserted into a new site; nothing takes the place where the TE originated. However, other transposons are replicative; that is, the transposon remains at the original site, and a replicated copy of that transposon then jumps to a new target DNA site.

The movement of TEs can have great significance. First, when such a random transposition occurs, it can interrupt the coding sequence in an essential gene such that gene expression produces a nonfunctional protein or, more likely due to the disruption, no protein at all. Thus, TEs can be a prime force behind spontaneous mutations.

TEs can move from chromosome to chromosome, plasmid to plasmid, plasmid to chromosome, or chromosome to plasmid. Although these events are rare, they are of particular significance when transposons carrying genes for antibiotic resistance are transferred to a similar or different bacterial species (**FIGURE 7.20C**). For example, if a bacterial chromosome carries a transposon with an antibiotic resistance gene, the transposon could jump to a plasmid in that cell. If the plasmid is transferred to another bacterial cell, the transposon will move along with it, thus spreading the gene for antibiotic resistance to the recipient cell. In fact, the movement of transposons among plasmids is a major mechanism for the spread of antibiotic resistance among bacterial species, as described in **CLINICAL CASE 7.**

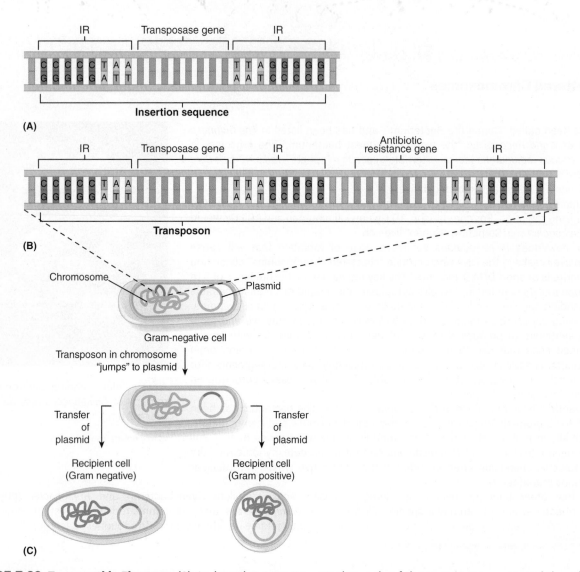

FIGURE 7.20 Transposable Elements. (A) An insertion sequence consists only of the transposase gene and the adjacent inverted repeats (IRs). **(B)** A transposon contains additional genes, such as antibiotic resistance. **(C)** Transposons can "jump" to another DNA molecule, such as a plasmid, which then can be transferred to another cell conferring new genetic capabilities (e.g., antibiotic resistance) on the recipient. *»» What is required for the transposon to "jump" to another DNA molecule?*

Chapter Challenge D

The purpose of this challenge was to try to figure out what genes obligate symbionts/parasites must contain when their survival depends on "infecting" a host cell. As you have seen, there is a great divergence in gene number and genome size among the members of the domain Bacteria and domain Archaea (see Figure 7.4). For the obligate symbionts/parasites and all free-living species, there is a set of core genes that is necessary to support any life form. This would include genes for DNA replication, transcription, and translation; genes for ribosomal subunits; and genes necessary for cell membrane biogenesis. On the other hand, genes for many aspects of metabolism might have been lost over time. In fact, it appears that the tiny genomes of symbionts/parasites are a result of outsourcing. Many genes have been transferred to the host's chromosome, or the genes have simply been lost because these organisms can rely on many of the host's own genes to supply essential cellular products. Perhaps the mitochondria and chloroplasts, once free-living bacterial cells that evolved into organelles through endosymbiosis, are the ultimate examples where symbiosis has resulted in the loss of many genes or transfer of the genes to the host.

QUESTION D: *So, to summarize, put together a list of core genes that must be part of a universal minimal genome. Scientists do not yet know what any such universal minimal genome size is (if such a genome exists). Some have estimated it could be as small as 40 genes. Mitochondria, for example, have some 40 to 80 genes.*

You can find answers online in **Appendix F**.

Clinical Case 7

Klebsiella pneumoniae in a Hospitalized Patient

A 91-year-old man with dialysis-dependent end-stage renal disease, congestive heart failure, anemia, and peptic ulcer disease was admitted to a hospital in Tel Aviv, Israel. Following amputation of the left leg below the knee due to an infected heel wound, the patient developed sepsis. The patient was treated with a variety of antibiotics, including ertapenem, a carbapenem drug.

During treatment, an acute inflammation of the gallbladder developed, so the patient had a surgical incision made in his gallbladder to help drain the infection.

During his hospital stay, the patient was screened for carbapenem-resistant Enterobacteriacae (CRE) as part of the hospital's routine infection control program aimed at limiting the spread of these often antibiotic-resistant organisms. As such, CRE is an important health challenge in healthcare settings.

From the patient, two rectal swabs were collected one week apart for analysis by the hospital's clinical microbiology lab. The first swab specimen was negative for CRE by culture. However, the second swab specimen grew colonies (see figure). Microscopy identified small, gram-stained rods, which on further analysis were shown to be facultatively anaerobic and resistant to carbapenem, as well as all cephalosporin and monobactam antibiotics. Identification was made as a carbapenem-resistant *Klebsiella pneumoniae* (CRKP) strain.

Alarm was raised, as CRKP is associated with increased mortality, particularly in patients with prolonged hospitalization. An aggressive infection control strategy was instituted in the hospital.

Two weeks later, a carbapenem-resistant *E. coli* strain was recovered from the patient's gallbladder drainage. Further infection control strategies were instituted in the hospital.

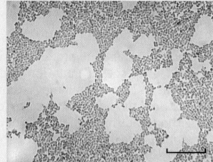

A gram-stained preparation of *Klebsiella pneumoniae.* (Bar = 10 μm.)

© Eye of Science/Science Source.

Questions:
a. From the figure, what is the Gram stain reaction for the small rods?
b. From the material covered in this chapter, propose a mechanism for the subsequent emergence of the carbapenem-resistant *E. coli* strain.
c. What infection control measures should the hospital institute?
d. Why was the presence of a carbapenem-resistant *E. coli* strain of even greater concern as a healthcare threat?

You can find answers online in **Appendix E**.

For additional information, see www.cdc.gov/eid/content/16/6/1014.htm.

Concept and Reasoning Checks 7.4

a. How do chemical mutagens interfere with DNA replication or gene expression?
b. Justify the statement: "A base-pair deletion (or insertion) potentially is more dangerous to an organism's viability than a base-pair substitution."
c. Explain why cells need at least two repair mechanisms (mismatch and excision).
d. What harm or benefit is conferred by insertion sequences and transposons?

■ KEY CONCEPT 7.5   Techniques Exist for Identifying Mutants

Any organism carrying a mutation is called a **mutant**, whereas the normal strain isolated from nature without the mutation is the **wild type**. Some mutants are easy to identify because the **phenotype**, or physical appearance, of the organism or the colony has changed from the wild type. For example, some bacterial colonies appear red because they produce a red pigment. Treat the colonies with a mutagen and, after plating on nutrient agar, mutants form colorless colonies. However, not all mutants involve an obvious phenotype change.

Plating Techniques Select for Specific Mutants or Characteristics

If a mutation does not produce a visible change, how does one "see" a mutation and then isolate the mutant cell? In microbiology, there are ways to select for (identify and isolate) a single mutant from among thousands of possible cells or colonies. Let's look at two selection techniques that make this search possible.

First, in both techniques, the chemical composition of the transfer plate is important for visual identification of the colonies being hunted. The use of a replica-plating device makes the identification possible. The device consists of a sterile velveteen cloth or filter paper mounted on a solid support. When an agar plate (master plate) with bacterial colonies is gently pressed against the surface of the velveteen, some cells from each colony stick to the velveteen. If another agar plate then is pressed against this velveteen cloth, some cells will be transferred (replicated) in the same pattern as on the master plate.

Now, suppose that you want to find a nutritional mutant that cannot grow without the amino acid histidine. This mutant (written his⁻) has lost the ability that the wild-type strain (his⁺) has to make its own histidine. Such a mutant having a nutritional requirement for growth is called an **auxotroph** (*auxo* = "increase"; *troph* = "nourishment"), whereas the wild type is a **prototroph** (*proto* = "original"). Visually, there is no difference between the two strains when they grow on a complete medium with histidine. However, you can visually identify the auxotroph by using a **negative selection** plating technique, as described in **FIGURE 7.21**.

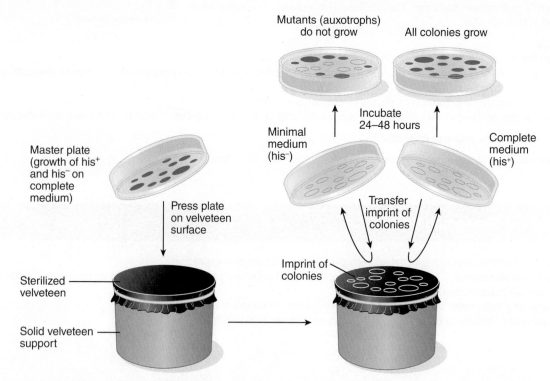

FIGURE 7.21 Negative Selection Identifies Auxotrophs. Negative selection plating techniques can be used to detect nutritional mutants (auxotrophs) that fail to grow when replica plated on minimal medium (in this example, a growth medium lacking histidine). Comparison to replica plating on complete medium visually identifies the auxotrophic mutants. *»» In this example, which colonies transferred to the complete medium represent the auxotrophs (his⁻)?*

As another example, suppose you want to see if there are any potentially dangerous carbapenem-resistant bacteria in a hospital ward. Carbapenem-resistant bacteria, including some strains of *E. coli*, have become difficult to treat because these bacteria are resistant to most all antibiotics. This includes carbapenem, which is called a "last resort" antibiotic because it is the last useful antibiotic when all others have failed in treating a patient (see Clinical Case 7). Again, phenotypically, there is no difference between those strains sensitive to carbapenem and those resistant to the antibiotic when they grow on a complete medium without carbapenem. However, a **positive selection** plating technique permits visual identification of such carbapenem-resistant mutants because they will grow on a complete medium containing carbapenem, as described in **FIGURE 7.22**.

The Ames Test Can Identify Potential Mutagens

Besides a means to identify mutants, there are ways to identify mutagens. Some years ago, scientists observed that about 90% of human **carcinogens**—agents causing tumors in humans—also induce mutations in bacterial cells. Working from this observation, Bruce Ames of the University of California developed a procedure to help identify potential

human carcinogens by determining whether the agent can mutate bacterial auxotrophs. The procedure, called the **Ames test**, uses an auxotrophic (histidine-requiring) strain (his⁻) of *Salmonella enterica* serotype Typhimurium. If inoculated onto a plate of nutrient medium lacking histidine, no colonies will appear because in this auxotrophic strain the gene inducing histidine synthesis is mutated and hence not active.

In preparation for the Ames test, the potential carcinogen is mixed with a liver enzyme preparation. The reason for doing this is because chemicals often become tumor causing and mutagenic in humans only after they have been modified by liver enzymes.

To perform the Ames test, the his⁻ strain is inoculated onto an agar plate lacking histidine (**FIGURE 7.23**). A well is cut in the middle of the agar, and the potential liver-modified carcinogen is added to the well (or a filter paper disk with the chemical is placed on the agar surface). The chemical diffuses into the agar during a 24- to 48-hour incubation. If bacterial colonies appear, one can conclude the agent mutated the bacterial his⁻ gene back to the wild type (his⁺); that is, **revertants** were generated that could again encode the enzyme needed for histidine synthesis. Because the agent is a mutagen in bacteria, it is therefore a possible

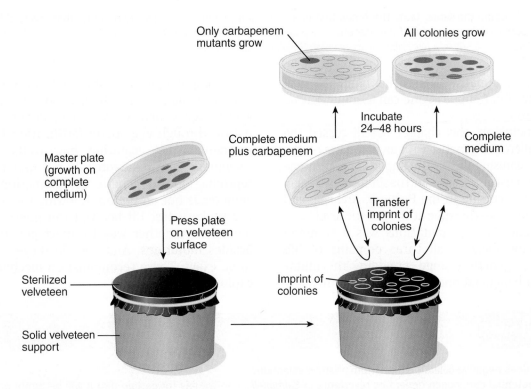

FIGURE 7.22 Positive Selection of Mutants. Positive selection plating techniques can be used to identify antibiotic-resistant mutants. »» *What do the "vacant spots" on the complete medium plus carbapenem represent?*

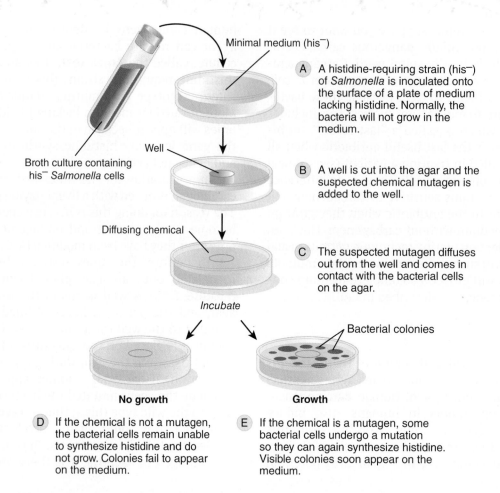

Minimal medium (his⁻)

Broth culture containing his⁻ *Salmonella* cells

Well

Diffusing chemical

Incubate

No growth

Growth

Bacterial colonies

(A) A histidine-requiring strain (his⁻) of *Salmonella* is inoculated onto the surface of a plate of medium lacking histidine. Normally, the bacteria will not grow in the medium.

(B) A well is cut into the agar and the suspected chemical mutagen is added to the well.

(C) The suspected mutagen diffuses out from the well and comes in contact with the bacterial cells on the agar.

(D) If the chemical is not a mutagen, the bacterial cells remain unable to synthesize histidine and do not grow. Colonies fail to appear on the medium.

(E) If the chemical is a mutagen, some bacterial cells undergo a mutation so they can again synthesize histidine. Visible colonies soon appear on the medium.

FIGURE 7.23 Using the Ames Test. The Ames test is a screening technique to identify mutants that reverted back to the wild type because of the presence of a mutagen. »» *Suppose, as a control, a plate with* Salmonella *(his⁻) but without mutagen is incubated. After 24–48 hours, a few colonies are seen on the plate. Explain this observation.*

carcinogen in humans. On the other hand, if bacterial colonies fail to appear, one can assume that no mutation took place.

In conclusion, this chapter has discussed the basics of microbial genetics with an emphasis on the bacterial organisms. One of the important take-home lessons emphasizes, once again, the similarity between the process occurring in the Bacteria, Archaea, and Eukarya domains—this time at the genetic level.

The generation of mutations was a common way that members of all three domains of life could affect genetic change, providing the mutation was a beneficial one. But is this the only way to accomplish genetic variation? In the eukaryotic microbes, plants, and animals, most organisms can go through sexual reproduction, so gene recombination through egg-sperm fertilization is another mechanism for acquiring new traits. However, remember the prokaryotes do not undergo sexual reproduction, so genetic recombination through gametes is not an option. Knowing the abilities of the prokaryotes, I'll bet you can guess that they must have another way to affect genetic change besides mutations. And boy, do they—three ways to be exact. These recombination mechanisms are explored in Chapter 8.

Concept and Reasoning Checks 7.5

a. How does negative selection differ from positive selection?

b. If a potential carcinogen generates revertants in *Salmonella,* can we say for certain that it will be tumor causing in humans? Explain.

■ SUMMARY OF KEY CONCEPTS

Concept 7.1 The Hereditary Molecule in All Organisms Is DNA

1. The bacterial and archaeal **chromosomes** are circular, haploid structures in the cell's **nucleoid**.
2. The DNA of a microorganism's chromosome is **supercoiled** and folded into a series of loops consisting of 10,000 bases. (Figure 7.3)
3. Many microbial cells might have one or more **plasmids**. These small closed loops of DNA carry information that can confer selective advantages (e.g., antibiotic resistance, protein toxins) to the cells.

Concept 7.2 DNA Replication Is Part of the Cell Cycle

4. DNA replicates by a **semiconservative** mechanism whereby each strand of the original DNA molecule acts as a template to synthesize a new strand. DNA replication begins with initiator proteins binding at the **replication origin**, forming two **replication "factories."** **DNA polymerase** moves along the strands inserting the correct DNA nucleotide to complementary bind with the template strand. (Figure 7.5)
5. At each **replication fork**, one of the two strands is synthesized in a continuous fashion, whereas the other strand is formed in a discontinuous fashion, forming **Okazaki fragments**, which are joined by a **DNA ligase**.

Concept 7.3 Gene Expression Produces RNA and Protein for Cell Function

6. **Transcription** occurs when the **RNA polymerase** binds to a **promoter** sequence on the DNA template strand. Various forms of RNA, including **mRNA, tRNA**, and **rRNA**, are transcribed from the DNA and are important in the translation process. (Figure 7.7)
7. The **genetic code**, a series of three nucleotides to specify a polypeptide, is redundant because often more than one **codon** can specify the same amino acid. (Table 7.3)
8. **Translation** occurs on ribosomes, which bring together mRNA and tRNAs. The ribosome "reads" the mRNA codons and inserts the correct tRNA to match the codon and **anticodon**. A **polysome** is a string of ribosomes all translating the same mRNA. (Figures 7.11, 7.12)

9. Many antibiotics interfere with protein synthesis by binding to RNA polymerase or to either the small, or large, ribosomal subunit.
10. Different control factors influence protein synthesis. The best understood is the bacterial **operon** model wherein binding of a **repressor protein** to the operon represses transcription. (Figure 7.13)

Concept 7.4 Mutations Are Heritable Changes in a Cell's DNA

11. **Mutations** are permanent and heritable changes in the cellular DNA. This can occur spontaneously in nature resulting from a replication error or the effects of natural radiation. In the laboratory, physical and chemical mutagens can induce mutations. (Figures 7.16, 7.17)
12. Base pairs in the DNA can change in one of two ways. A base-pair substitution does not change the reading frame of an mRNA but can result in a **silent, missense**, or **nonsense mutation**. A **point mutation** also can occur from the loss or gain of a base pair. Such mutations change the reading frame and often lead to loss of protein function. (Figure 7.18)
13. Replication errors or other damage done to the DNA often can be repaired. **Mismatch repair** replaces an incorrectly matched base pair with the correct pair. **Excision repair** removes a section of damaged (distorted) DNA and replaces it with the correctly paired bases. (Figure 7.19)
14. **Transposable genetic elements** exist in many microbial cells. **Insertion sequences (ISs)** only carry information to move the sequences and insert them into another location in the DNA. **Transposons** are similar to ISs except they contain additional genes, such as antibiotic resistance. (Figure 7.20)

Concept 7.5 Techniques Exist for Identifying Mutants

15. Auxotrophic mutants can be identified by **negative selection** plating techniques. **Positive selection** can be used to identify mutants having certain attributes, such as antibiotic resistance. (Figures 7.21, 7.22)
16. The **Ames test** is a method of using an auxotrophic bacterial species to identify mutagens that might be carcinogens in humans. The test is based on the ability of a potential mutagen to revert an auxotrophic mutant to its prototrophic form. (Figure 7.23)

■ CHAPTER SELF-TEST

For **Steps A–D**, you can find answers to questions and problems online in **Appendix D**.

STEP A: REVIEW OF FACTS AND TERMS

Multiple Choice

Read each question carefully, and then select the *one* answer that best fits the question or statement.

1. Which one of the following statements is *not* true of the bacterial chromosome?
   A. It is located in the nucleoid.
   B. It usually is a circular molecule.
   C. There are two copies of each chromosome.
   D. It usually is haploid.

2. DNA compaction involves _____.
   A. a twisting and packing of the DNA
   B. supercoiling
   C. loops of 10,000 bases
   D. All the above (**A–C**) are correct.

3. Plasmids are _____.
   A. another name for transposons
   B. accessory genetic information
   C. domains within a chromosome
   D. daughter chromosomes

4. The enzyme _____ adds complementary bases to the DNA template strand during replication.
   A. ligase
   B. helicase
   C. DNA polymerase
   D. RNA polymerase

5. At a chromosome replication fork, the lagging strand consists of _____ that are joined by _____.
   A. RNA sequences; DNA ligase
   B. Okazaki fragments; RNA polymerase
   C. RNA sequences; ribosomes
   D. Okazaki fragments; DNA ligase

6. In a eukaryotic microbe, those sections of a primary RNA transcript that will NOT be translated are called _____.
   A. introns
   B. anticodons
   C. "jumping genes"
   D. exons

7. Which one of the following codons would terminate translation?
   A. AUG
   B. UUU
   C. UAA
   D. UGG

8. The translation of an mRNA by multiple ribosomes is called _____ formation.
   A. Okazaki
   B. polysome
   C. plasmid
   D. transposon

9. If an antibiotic binds to a 50S subunit, what cellular process will be inhibited?
   A. DNA replication
   B. Intron excision
   C. Translation
   D. Transcription

10. Which one of the following is *not* part of an operon?
    A. Regulatory gene
    B. Operator
    C. Promoter
    D. Structural genes

11. Spontaneous mutations could arise from _____.
    A. DNA replication errors
    B. atmospheric radiation
    C. addition of insertion sequences
    D. All the above (**A–C**) are correct.

12. Which one of the following would *not* cause a change in the mRNA "reading frame"?
    A. Insertion sequence
    B. Base-pair substitution
    C. Base addition
    D. Base deletion

13. Nucleotide excision repair would correct DNA damage caused by _____.
    A. antibiotics
    B. UV light
    C. transcription
    D. a DNA replication error

14. Transposons _____.
    A. contain genes for bacterial metabolism
    B. are smaller than insertion sequences
    C. are examples of plasmids
    D. may have information for antibiotic resistance

15. Nutritional mutants are referred to as _____.
    A. prototrophs
    B. wild type
    C. revertants
    D. auxotrophs

16. The Ames test is used to _____.
    A. identify potential human carcinogens
    B. discover auxotrophic mutants
    C. find pathogenic bacterial species
    D. identify antibiotic resistant mutants

STEP B: CONCEPT REVIEW

17. Assess the role of **plasmids** in microbial cells. (**Key Concept 7.1**)
18. Identify and explain the events of the three phases of **DNA replication**. (**Key Concept 7.2**)
19. Describe the role of **RNA polymerase** in the transcription process. (**Key Concept 7.3**)
20. Label the sequences composing a bacterial **operon** and compare the functions of each sequence to the transcription process. (**Key Concept 7.3**)

21. Compare and contrast **spontaneous** and **induced mutations**, and differentiate between physical and chemical **mutagens**. (**Key Concept 7.4**)
22. Evaluate the use of the **Ames test** to identify chemicals that are potential carcinogens in humans. (**Key Concept 7.5**)

STEP C: APPLICATIONS AND PROBLEM SOLVING

Answer the following questions that pertain to (1) transcription and translation and (2) mutations. Use the genetic code (Table 7.3).

23. The following base sequence is a complete polynucleotide made in a bacterial cell.

    AUGGCGAUAGUUAAACCCGGAGGGUGA

    With this sequence, answer the following questions.

    A. Provide the sequence of nucleotide bases found in the inactive DNA strand of the gene.
    B. How many codons will be transcribed in the mRNA made from the template DNA strand?
    C. How many amino acids are coded by the mRNA made and what are the specific amino acids?
    D. Why isn't the number of codons in the template DNA the same as the number of amino acids in the polypeptide?

24. Use the base sequence to answer the following questions about mutations.

    TACACGATGGTTTTGAAGTTACGTATT

    A. Is the sequence above a single strand of DNA or RNA? Why?
    B. Using the sequence above, show the translation result if a mutation results in a C replacing the T at base 12 from the left end of the sequence. Is this an example of a silent, missense, or nonsense mutation?
    C. Using the sequence above, show the translation result if a mutation results in an A inserted between the T (base 12) and the T (base 13) from the left end of the sequence. Is this an example of a silent, missense, or nonsense mutation?

25. You are interested in identifying mutants of *E. coli*, specifically mutations occurring in the promoter and operator regions of the *lac* operon such that their respective molecules cannot bind to the region. You

have agar plates containing lactose or glucose as the energy source. How can plating the potential mutants on these media help you identify (a) promoter region mutants and (b) operator region mutants?

26. A chemical is tested with the Ames test to see if the chemical is mutagenic and therefore possibly a tumor-causing chemical in humans. On the test plate containing the chemical, no his$^+$ colonies are seen near the central well. However, many colonies are growing some distance from the well. If these colonies truly represent his$^+$ colonies, why are there no colonies closer to the central well?

27. Bioremediation is a process that uses bacteria to degrade environmental pollutants. You want to use one of these organisms to clean up a toxic waste site (biochemical refinery) that contains benzene in the soil around the refinery. Benzene (molecular formula = C_6H_6) is a major contaminant found around many of these chemical refineries. You have a culture of bacterial cells growing on a culture plate that was derived from a soil sample from the refinery area. You also have a supply of benzene. Explain how you could visually identify chemoheterotrophic bacterial colonies on agar that use benzene as their sole carbon and energy source for metabolism.

28. Suppose you now have such benzene colonies growing on agar. However, it also has been discovered that material containing radioactive phosphorus (^{32}P) is in the soil around the refinery, and this radioactive material can kill bacterial organisms. Because you want to identify colonies that might be sensitive to ^{32}P, you obtain a sample of the material containing ^{32}P. Explain how you could visually determine if any of your colonies are sensitive to ^{32}P.

■ STEP D: QUESTIONS FOR THOUGHT AND DISCUSSION

29. The author of a general biology textbook writes in reference to the development of antibiotic resistance, "The speed at which bacteria reproduce ensures that sooner or later a mutant bacterium will appear that is able to resist the poison." How might this mutant bacterial cell appear? Do you agree with the statement? Does this bode ill for the future use of antibiotics?

30. Many viruses have double-stranded DNA as their genetic information while many others have single-stranded RNA as the genetic material. Which group of viruses do you believe is more likely to efficiently repair its genetic material? Explain.

31. Some scientists suggest that mutation is the single most important event in evolution. Do you agree? Why or why not?

Concept Mapping

See pages XXXI-XXXII on how to construct a concept map.

32. Construct a concept map for **translation** using the following terms.

Amino acid	Ribosome
Chain elongation	Sense codons
Chain initiation	Small subunit
Chain termination	Start codon
Large subunit	Stop codon
mRNA	Termination factors
Polypeptide	tRNAs

CHAPTER 8

Gene Transfer, Genetic Engineering, and Genomics

One of the most devastating infectious diseases in human history was bubonic plague, commonly called the Black Death. Caused by *Yersinia pestis*, the disease took the lives of almost half of Europe's population between 1347 and 1351. In a later outbreak in London, the English writer and journalist Daniel Defoe described the dreadful incident:

> "...*Some were immediately overwhelmed with it, and came to violent fevers, vomitings, insufferable headaches, pains in the back ... Others with swellings and tumours in the neck and groin, or armpits ... while others, as I observed, were silently infected.*"

As realistic as these historical accounts are, it can be difficult from centuries-old historical records and descriptions to identify what pathogen (and disease) actually caused such human suffering and loss of life. In the outbreaks of bubonic plague, many medical experts and epidemiologists believed the historical accounts actually describe bubonic plague, whereas others questioned that identity because of the imprecise descriptions of the disease. Then, in 2015, a mass grave containing 3,500 human burials

from the 14th century was discovered in London. Did these individuals die from bubonic plague?

Today, the sequences of bases in a DNA fragment can be used like fingerprints to identify the culprit of an infectious disease (see chapter opening photo) - and the source of that DNA can come from unexpected places. Teeth from skeletons act like tiny time capsules because the tooth pulp often contains preserved DNA that can be extracted and used for **DNA sequencing** (FIGURE 8.1), the process of determining the precise order of nucleotides within the DNA. For that reason, molecular paleopathologists in Germany were sent teeth from 20 skeletons from the mass grave. From five of those skeletons, DNA was recovered from the tooth pulp. Sequencing the DNA showed that the order of DNA nucleotides closely matched that of current day *Y. pestis* bacterial DNA. The debate seems solved: the victims were infected with *Y. pestis* and the outbreaks of the 14th century were due to the Black Death.

Partial sequencing graph for a DNA fragment.
© Philippe Garo/Science Source.

FIGURE 8.1 DNA Sequencing Preparation. This DNA sequencer can rapidly sequence the bases in a DNA segment or fragment. »» *What would the scientists use as a control group to ensure the DNA they isolate truly came from bacterial DNA in the bone skeleton and teeth?*

© Gustoimages/Science Source.

Tuberculosis (TB) is another deadly infectious disease that has disfigured and killed millions throughout human history. Here again, the power of DNA sequencing can be used to answer questions about the history of infectious disease. From several archaeological sites, skeletons have been recovered that exhibit the deformities typical of TB.

After sequencing the DNA from teeth and bones of such skeletons and comparing those sequences to the DNA sequences from modern-day strains of TB, some interesting correlations have been discovered. It appears that TB has been infecting hominids for some 3 million years. In addition, many experts had thought that humans first acquired the TB bacillus, *Mycobacterium tuberculosis*, from cows thousands of years ago. However, sequencing studies now show that humans actually passed *M. tuberculosis* to cows.

These analyses are but two examples of **genomics**, which studies an organism's genome and the gene sequences that compose the genome and, as we saw here, compares genome sequences between microbes. This chapter addresses genomics and the tools of genetic engineering that make the field of biotechnology and genomics possible.

Before we can explore these topics, we need to understand the mechanisms by which bacterial cells naturally transfer genetic information from one cell to another. Understanding these mechanisms provides a unique perspective on microbial evolution, ecology, and molecular biology, while offering insights into the techniques of genetic engineering and the field of microbial genomics.

Chapter **Challenge**

The Human Genome Project (HGP) was an international scientific investigation that began in 1990. Its primary goal was to determine the sequence of nucleobases that compose human DNA and then to understand the genetic makeup of the human species. The first draft sequence of the human genome was announced in 2000 and published in 2001. In the end, it was discovered that the human genome consists of close to 20,000 genes. However, the interpretation of the genome sequence data is still being studied and evaluated. With regard to the origins of the human genome and its genes, one question being asked is: "Are there microbial genes in our genome?" Let's find out.

■ KEY CONCEPT 8.1   Bacterial Cells Can Recombine Genes in Several Ways

Traditionally, when we think about the inheritance of genetic information, we envision genes passed from parent to offspring. However, imagine being able to give genes directly to members of your own family or to one of your classmates. These mechanisms of remixing genes, called **genetic recombination**, occur in many bacterial and archaeal species, and they provide tremendous genetic and evolutionary flexibility for these organisms.

Genetic Information Can Be Transferred Vertically and Horizontally

Mutations are one of the ways by which the genetic material in a cell can be permanently altered. In bacterial cells, if a mutation occurs in the parent cell, it will be inherited by the daughter cells resulting from binary fission. Likewise, all future generations derived by binary fission also will have

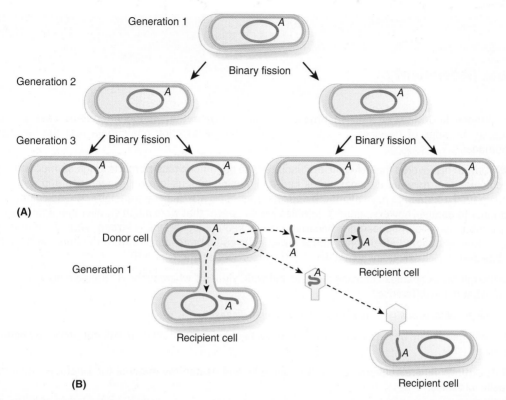

FIGURE 8.2 Gene Transfer Mechanisms. Genes can be transferred between cells in two ways. **(A)** In vertical transfer, a cell undergoes binary fission and the daughter cells of the next generation contain the identical genes found in the parent. **(B)** In horizontal transfer, genes are transferred directly to other individual cells as free DNA, by viruses, and by direct contact between cells. *»» Determine which transfer mechanism (vertical or horizontal) provides the potential for more genetic diversity.*

the mutation. This form of asexual genetic transfer, where genomes are replicated and passed to the next generation, is referred to as **vertical gene transfer** (**FIGURE 8.2A**).

Although mutation and sexual reproduction are the main drivers of genetic recombination and diversity in eukaryotic organisms, including the eukaryotic microbes, prokaryotes lack sexual reproduction as a mechanism for gene mixing. Yet, they are among the most genetically diverse species of life because they possess other ways by which they can gain new traits and from which evolutionary change can arise that are independent of sexual recombination. This occurs through the process of **horizontal gene transfer** (**HGT**), a type of genetic recombination that involves the lateral, intercellular transfer of DNA from one microbe to another (**FIGURE 8.2B**). Unlike mutation, which can be a slow and random process, HGT provides a way for organisms to quickly acquire new genes for adaptation and survival. If, for example, an antibiotic-susceptible recipient cell receives a piece of DNA for antibiotic resistance from the donor, the recipient cell is now an antibiotic-resistant cell too. In fact, the increasing resistance to antibiotics by multidrug-resistant

(MDR) pathogens is one example of how prevalent HGT can be.

HGT can occur between members of the same or related species or between very different species. Although there is mounting evidence for HGT between bacteria and eukaryotic hosts, the mechanisms for HGT are best understood within members of the domain Bacteria. There are three distinctive types of HGT: transformation, conjugation, and transduction. All three processes involve a similar four steps.

1. The donor is prepared for DNA transfer to the recipient cells.
2. The donor DNA then is transferred to the recipient cell.
3. The donor DNA is accepted by the recipient cell.
4. The donor DNA is maintained in a stable state in the recipient cell.

Let's look at each of the three types of HGT in detail.

Transformation Involves the Uptake of "Naked" DNA

Read **INVESTIGATING THE MICROBIAL WORLD 8**. It presents one of the first glimpses that would

Investigating the Microbial World 8

Spontaneous Generation?

In the 1920s, Frederick Griffith, an English physician, was studying the pathogen *Streptococcus pneumoniae*, or pneumococcus, in the hope of developing a vaccine against the devastating pneumonia that the bacterial species produced in humans.

OBSERVATION: Griffith discovered that when he first isolated the pneumococcus from the lungs of mice with pneumonia, the bacterial colonies that grew on the agar plates had a glistening, mucus-looking appearance and, with microscopy, the diplococcus pairs were seen to have a smooth capsule (the S strain). When he transferred these colonies repeatedly from one agar plate to another, however, mutant colonies would appear that were much smaller and drier looking, and the diplococci were rough looking in appearance because they lacked a capsule (the R strain). When Griffith injected mice with the S strain they contracted pneumonia, and S strain cells could be re-isolated from the infected mice (see figure, part A). Mice infected with the R strain did not develop the disease (see figure, part B).

Therefore, the S strain was capable of causing disease and was "virulent," whereas the R strain did not cause disease and was "avirulent." What is the difference?

QUESTION: *Do the pneumococcus cells need to be alive to cause disease?*

HYPOTHESIS: Cells need to be alive to cause disease. If so, heat-killing S strain cells will not produce pneumonia after injection into mice.

EXPERIMENT 1: Griffith prepared cultures of the S-strain cells and heated the material for 1 hour. He then injected the heat-treated cells into mice.

RESULTS: See figure, part C.

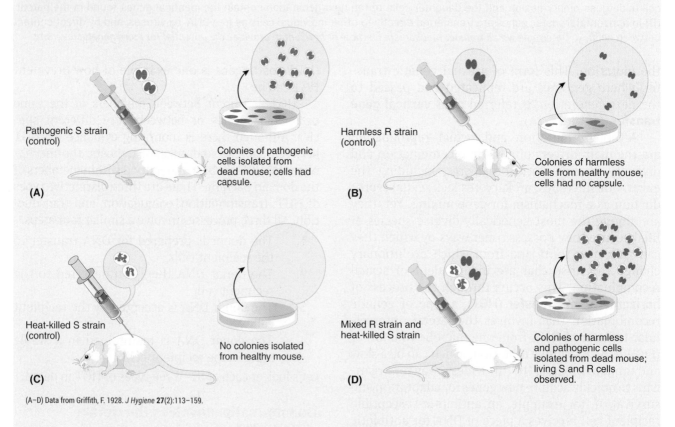

Pathogenic S strain
(control)

Colonies of pathogenic cells isolated from dead mouse; cells had capsule.

(A)

Harmless R strain
(control)

Colonies of harmless cells from healthy mouse; cells had no capsule.

(B)

Heat-killed S strain
(control)

No colonies isolated from healthy mouse.

(C)

Mixed R strain and heat-killed S strain

Colonies of harmless and pathogenic cells isolated from dead mouse; living S and R cells observed.

(D)

(A–D) Data from Griffith, F. 1928. *J Hygiene* **27**(2):113–159.

CONCLUSION:

   QUESTION 1: *What do these results suggest about the S strain cells?*

OBSERVATION: Griffith had noticed that in some of his human patients he could isolate more than one form of the pneumococcus from sputum samples.

   QUESTION 2: *What happens if a mixed culture of pneumococcus strains is injected into mice?*

HYPOTHESIS: Injecting both avirulent heat-killed S-strain cells along with living, unheated R-strain cells will produce no disease. If so, the mice should survive and not contract pneumonia.

EXPERIMENT 2: Mice were injected with unheated, live R cells along with heat-killed S cells.

RESULTS: See figure, part D.

CONCLUSIONS:

   QUESTION 3: *Did these experiments support the hypothesis? Explain.*

   QUESTION 4: *Provide an explanation as to how living S cells could be produced (spontaneous generation is not a plausible reason).*

Note: It wouldn't be until 1944 that another group of researchers would identify the transforming agent.

You can find answers online in **Appendix E**.

Adapted from Griffith, F. 1928. *J Hygiene* **27**(2):113–159.

eventually suggest that DNA is the genetic material in cells and that such information can be transferred and shared between cells or organisms. The experiments described illustrate one form of HGT called **transformation**, which involves the uptake of a free DNA fragment or plasmid from the surrounding environment and the expression of that genetic information in a recipient cell. Microbiologists regard transformation as an important genetic recombination method, even though it occurs in less than 1% of a bacterial cell population.

   The ability of some bacterial genera to undergo transformation depends on their **competence**, which refers to the ability of a recipient cell to take up an extracellular fragment of DNA that previously was released into the environment by dead bacteria. In a competent cell, transformation starts when the DNA fragment, composed of about 10 to 20 genes, binds to proteins at the cell pole of the recipient cell (**FIGURE 8.3**). Internalization of the DNA occurs and the DNA is integrated into the recipient's genome.

   Looking back at Griffith's experiments described in **INVESTIGATING THE MICROBIAL WORLD 8**, the live R-strain cells acquired the genes for capsule formation from DNA fragments released from the dead S-strain cells. With the acquired ability to form a capsule, the living recipient cells could avoid body defenses, giving the cells time to cause disease and kill the mice. Microbiologists also have demonstrated that when mildly pathogenic bacterial strains take up DNA from other mildly pathogenic strains, there is a cumulative effect, and the recipient strain often ends up being more virulent. This might explain why highly pathogenic bacterial strains appear from time to time and how transformation can contribute to the dispersal of genes for antibiotic resistance.

Bacterial Conjugation Transfers DNA via Cell-to-Cell Contact

In a second form of HGT, called **conjugation**, two bacterial cells make physical contact and the live donor cell directly transfers DNA to the live recipient cell. For example, through conjugation an antibiotic-resistant donor could transfer a gene for antibiotic resistance to an antibiotic-susceptible recipient. This transfer can occur in one of two ways.

Plasmid Transfer

In gram-negative species, like *Escherichia coli*, the process of conjugation starts with a donor cell that has an **F factor**, which is a transmissible plasmid containing about 100 genes. Most of these genes are dedicated to plasmid DNA replication and the structures needed to transfer a DNA copy to a recipient cell. The F factor–containing donor cell (designated F$^+$) produces a **conjugation pilus** that contacts a recipient cell lacking an F factor (designated F$^-$ cell) (**FIGURE 8.4**). The conjugation pilus then shortens to bring the two cells close together and a conjugation bridge formed from the F$^+$ connects the cytoplasms of the two cells (**FIGURE 8.5A**).

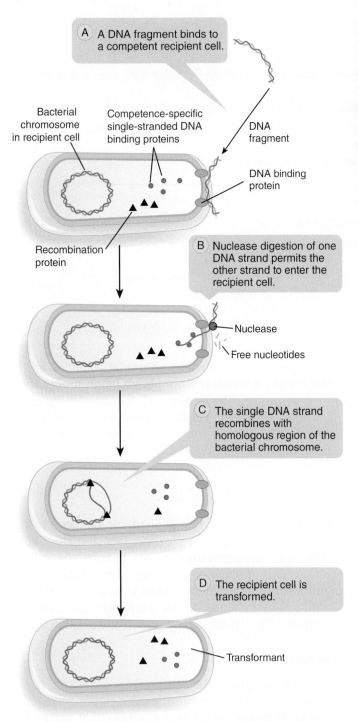

FIGURE 8.3 Transformation. Transformation (A–D) is the process in which a fragment of DNA from the environment binds to a competent recipient cell. The DNA fragment then passes into the recipient and incorporates into the recipient's chromosome. *»» Using this diagram, illustrate how Griffith's experiment with mixed heat-killed S strain and live R stain pneumococci produced live S strain encapsulated cells.*

A A DNA fragment binds to a competent recipient cell.

Bacterial chromosome in recipient cell

Competence-specific single-stranded DNA binding proteins

DNA fragment

DNA binding protein

Recombination protein

B Nuclease digestion of one DNA strand permits the other strand to enter the recipient cell.

Nuclease

Free nucleotides

C The single DNA strand recombines with homologous region of the bacterial chromosome.

D The recipient cell is transformed.

Transformant

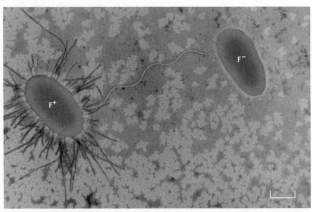

FIGURE 8.4 Bacterial Conjugation in *E. coli*. In this false-color transmission electron micrograph, the donor (F⁺ cell) (left) has produced a conjugation pilus that has contacted the recipient (F⁻ cell). Contraction of the pilus will bring donor and recipient close together for the conjugation process. (Bar = 0.5 μm.) *»» What are the other dark structures projecting from the F⁺ cell?*

© Dennis Kunkel Microscopy, Inc./Visuals Unlimited.

Following bridge formation, the F factor DNA begins to unwind, with one polynucleotide strand staying with the donor cell, while the other strand passes through the conjugation bridge to the recipient cell. This transfer takes about 5 minutes. As transfer occurs, DNA synthesis in the donor cell produces a new complementary strand to replace the transferred strand. After DNA transfer is complete, the two cells separate.

In the recipient cell, the new single-stranded DNA serves as a template for synthesis of a complementary polynucleotide strand, which then circularizes to reform an F factor. This completes the conversion of the recipient from F⁻ to F⁺ and this cell now represents a donor cell (F⁺) capable of conjugating with another F⁻ recipient. Transfer of the F factor does not involve the bacterial chromosome; therefore, the recipient does not acquire new genes other than those on the F factor.

The efficiency of DNA transfer by conjugation shows that conjugative plasmids can spread rapidly and, under laboratory conditions, can convert an entire population of F⁻ cells into plasmid-containing F⁺ cells (**FIGURE 8.6**). Indeed, conjugation appears to be the major mechanism for the transfer of several medically important traits, including antibiotic resistance, increased virulence, and toxin production.

Some gram-positive bacterial species also appear capable of conjugation. In these cases, conjugation

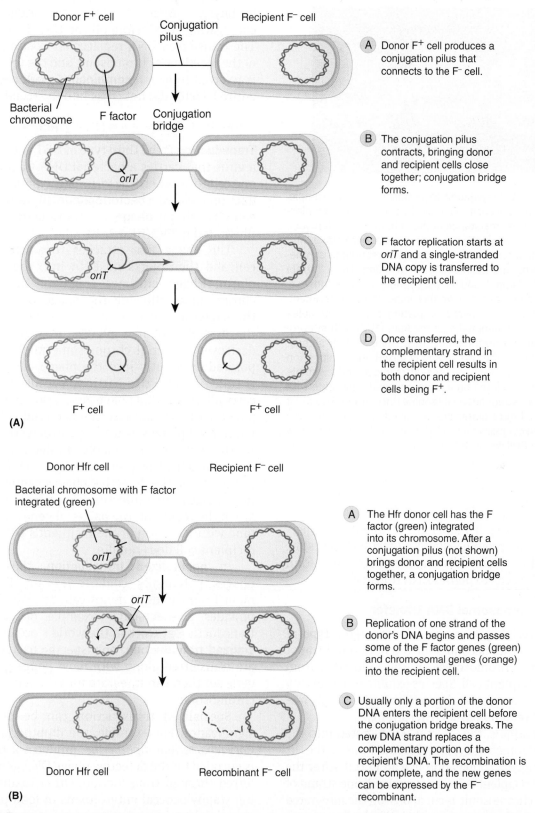

(A)

Donor F⁺ cell

Conjugation pilus

Recipient F⁻ cell

Bacterial chromosome

F factor

Conjugation bridge

oriT

oriT

F⁺ cell

F⁺ cell

A Donor F⁺ cell produces a conjugation pilus that connects to the F⁻ cell.

B The conjugation pilus contracts, bringing donor and recipient cells close together; conjugation bridge forms.

C F factor replication starts at *oriT* and a single-stranded DNA copy is transferred to the recipient cell.

D Once transferred, the complementary strand in the recipient cell results in both donor and recipient cells being F⁺.

(B)

Donor Hfr cell

Recipient F⁻ cell

Bacterial chromosome with F factor integrated (green)

oriT

oriT

Donor Hfr cell

Recombinant F⁻ cell

A The Hfr donor cell has the F factor (green) integrated into its chromosome. After a conjugation pilus (not shown) brings donor and recipient cells together, a conjugation bridge forms.

B Replication of one strand of the donor's DNA begins and passes some of the F factor genes (green) and chromosomal genes (orange) into the recipient cell.

C Usually only a portion of the donor DNA enters the recipient cell before the conjugation bridge breaks. The new DNA strand replaces a complementary portion of the recipient's DNA. The recombination is now complete, and the new genes can be expressed by the F⁻ recombinant.

FIGURE 8.5 Conjugation. (A) Conjugation between an F⁺ cell and an F⁻ cell. When the F factor is transferred from a donor (F⁺) cell to a recipient (F⁻) cell, the F⁻ cell becomes an F⁺ cell because it now possesses an F factor. **(B)** Conjugation between an Hfr and an F⁻ cell allows for the transfer of some chromosomal and plasmid DNA from donor to recipient cell. *»» Propose a hypothesis to explain why only a single-stranded DNA molecule is transferred across the conjugation pilus.*

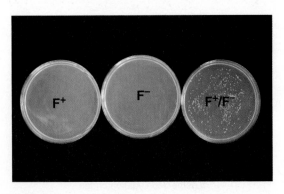

FIGURE 8.6 A Conjugation Experiment. In this example, an F⁺ strain of *Escherichia coli* contains a gene on its F plasmid that provides resistance to the antibiotic streptomycin. However, the strain lacks the gene necessary to make threonine, which is one of the amino acids essential to synthesizing proteins and maintaining metabolism. The F⁻ strain lacks an F plasmid and is susceptible to streptomycin, although it can make its own threonine. In the experiment, control plates are prepared by plating separate samples of F⁺ and F⁻ on a minimal agar medium (MM) with streptomycin (strep) but no threonine (thr). Then, another sample (experimental) of the two strains is mixed and conjugation is allowed to occur for 5 minutes. After this period, a sample of these cells is plated on another minimal agar plate with streptomycin (strep) but no threonine (thr). After 96 hours of incubation, all three plates are examined. »» *(a) Did the gene for streptomycin resistance get transferred by conjugation to the F⁻ cells? Explain. (b) Why are there no colonies on the two control plates (F⁺ and F⁻)?*

Courtesy of Dr. Jeffrey Pommerville.

does not involve pili. Rather, the recipient cell apparently secretes protein "clumping factors" that stick donor and recipient cells together. Pores then form between the cells to permit DNA transfer.

Partial Chromosomal DNA Transfer

Some bacterial species also can undergo a type of conjugation that accounts for the horizontal transfer of some chromosomal and plasmid DNA from donor to recipient cell. Species exhibiting the ability to donate chromosomal genes are called **high frequency of recombination (Hfr)** strains.

In Hfr strains, the F factor is integrated into the bacterial chromosome (**FIGURE 8.5B**). However, the Hfr cell still triggers conjugation like an F⁺ cell. After the donor and recipient cells are connected, one strand of the donor chromosome is cut at *ori*T on the integrated plasmid, and the single-stranded DNA begins to pass through the conjugation bridge into the recipient cell.

In this form of conjugation, a copy of the entire chromosome and F plasmid rarely enter the recipient because the conjugation process usually is disrupted by physical forces that break the bridge between conjugating cells before all the DNA is transferred. The transferred DNA replaces an existing segment of the recipient's chromosome and the recipient cell is referred to as a **recombinant F⁻ cell** because the entire F factor did not cross into the recipient.

Transduction Involves DNA Transfer by a Virus

Transduction is the third type of HGT and it involves a virus that carries a bacterial DNA fragment from one cell to another. The virus participating in transduction is called a **bacteriophage** (literally "bacteria eater") or simply **phage**. The transduction process is illustrated in **FIGURE 8.7**.

The phage first attaches to a live bacterial cell and then ejects its DNA into the cell cytoplasm. Phage enzymes fragment the bacterial chromosome, allowing the phage DNA to direct the bacterial cell to produce new phages. During the assembly of new phages, a fragment of the destroyed bacterial chromosome might be incorporated accidently into the phage. Such a phage is called a "defective particle" because it does not carry all the needed phage genes. However, the new phage particles can lyse the bacterial cell, releasing normal phages as well as any defective phages. All these phages are capable of infecting other live bacterial cells. However, without a complete set of phage genes, the defective phages cannot produce new phages after infection. Therefore, when they are in the recipient, the bacterial genes can recombine with a section of the recipient's DNA and the recipient cell does not die.

In **generalized transduction**, the defective phages potentially could carry any DNA fragment from the dead donor cell. MICROFOCUS 8.1 provides a spectacular example of generalized transduction at work in the world's oceans. In **specialized transduction**, only specific bacterial genes are transferred along with most phage genes. The gene set then can integrate into the recipient cell's chromosome.

Specialized transduction can be of medical significance. For example, the diphtheria bacillus, *Corynebacterium diphtheriae*, produces a toxin that is encoded on the defective phage DNA genes transferred through transduction. Other toxins, including staphylococcal enterotoxins in food poisoning, clostridial toxins causing some forms of botulism, and streptococcal toxins involved with scarlet fever, also are introduced into the otherwise harmless recipient cells by transduction. As a result, these strains become more virulent.

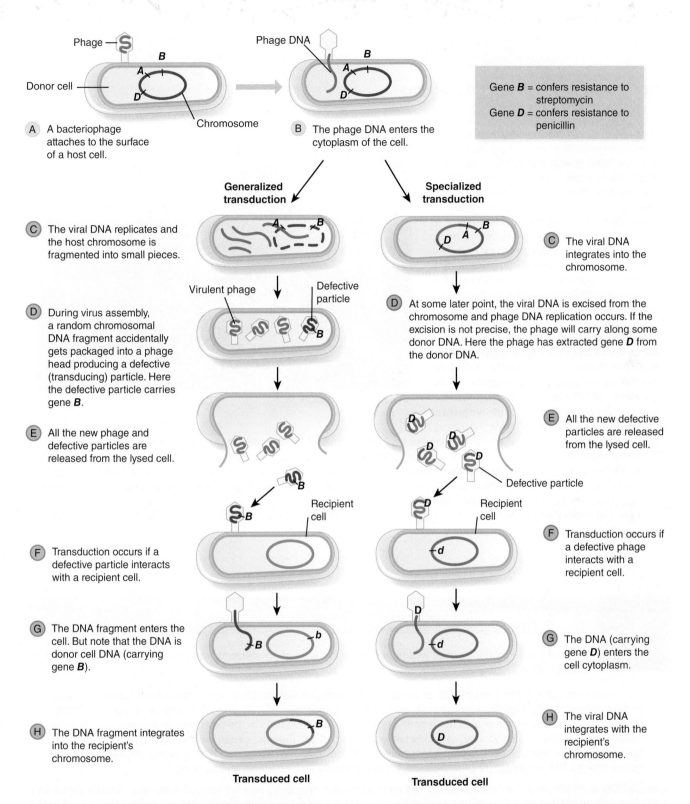

Phage

Donor cell

B
A
D

Chromosome

A A bacteriophage attaches to the surface of a host cell.

Phage DNA

B
A
D

B The phage DNA enters the cytoplasm of the cell.

Gene ***B*** = confers resistance to streptomycin
Gene ***D*** = confers resistance to penicillin

Generalized transduction

C The viral DNA replicates and the host chromosome is fragmented into small pieces.

A   *B*

Virulent phage   Defective particle

B

D During virus assembly, a random chromosomal DNA fragment accidentally gets packaged into a phage head producing a defective (transducing) particle. Here the defective particle carries gene ***B***.

E All the new phage and defective particles are released from the lysed cell.

B

B

Recipient cell

F Transduction occurs if a defective particle interacts with a recipient cell.

G The DNA fragment enters the cell. But note that the DNA is donor cell DNA (carrying gene ***B***).

B   *b*

H The DNA fragment integrates into the recipient's chromosome.

B

Transduced cell

Specialized transduction

B
D   *A*

C The viral DNA integrates into the chromosome.

D At some later point, the viral DNA is excised from the chromosome and phage DNA replication occurs. If the excision is not precise, the phage will carry along some donor DNA. Here the phage has extracted gene ***D*** from the donor DNA.

D
D
D
D

E All the new defective particles are released from the lysed cell.

D

Defective particle

D

Recipient cell

F Transduction occurs if a defective phage interacts with a recipient cell.

d

D

G The DNA (carrying gene ***D***) enters the cell cytoplasm.

d

H The viral DNA integrates with the recipient's chromosome.

D

Transduced cell

FIGURE 8.7 Generalized and Specialized Transduction. Phage can transfer bacterial genes by generalized transduction (can involve any bacterial gene) or specialized transduction (can involve genes from a region near the spot where the phage genome integrated [see B]). »» *Using genes B and D for antibiotic resistance, which form of transduction, generalized or specialized, would likely spread antibiotic resistance faster? Explain.*

MICROFOCUS 8.1: Environmental Microbiology

Gene Swapping in the World's Oceans

Many of us are familiar with the accounts of microorganisms in and around us, but we are less familiar with the massive numbers of microbes in the world's oceans and seas. For example, microbial ecologists estimate there are an estimated 10^{29} prokaryotes in the world's oceans. At the Axial Seamount, a Pacific deep-sea volcano, researchers have discovered some 3,000 archaeal species and more than 37,000 different bacterial species. In addition, there are some 10^{30} viruses called bacteriophages in the oceans that infect these oceanic microbes.

In the infection process, sometimes bacteriophages by mistake carry pieces of the bacterial chromosome (rather than viral DNA) from the infected cell to another recipient cell through generalized transduction. In the recipient cell, the new DNA fragment can be swapped for an existing part of the recipient's chromosome. It is a rare event, occurring

Courtesy of Dr. Jeffrey Pommerville.

only once in every 100 million (10^8) virus infections. That doesn't seem very significant until you consider the number of bacteriophages and susceptible bacteria existing in the oceans. Working with these numbers and the potential number of virus infections, scientists suggest that if only one in every 100 million infections brings a fragment of DNA to a recipient cell, there are about 10 million billion (that's 10,000,000,000,000,000 or 10^{16}) such gene transfers per second in the world's oceans. That is about 10^{21} transduction events per day!

We do not understand what all this gene transfer means. What we can conclude is there's an awful lot of gene swapping going on!

Chapter Challenge A

When the HGP draft sequence was published in 2001, much excitement was generated by a passage in the middle of the long manuscript.

"An interesting category is a set of 223 proteins that have significant similarity to proteins from bacteria, but no comparable similarity to proteins from yeast, worm, fly, and mustard weed, or indeed from any other (nonvertebrate) eukaryote. These sequences should not represent bacterial contamination in the draft human sequence . . . A more detailed computational analysis indicated that at least 113 of these [human] genes are widespread among bacteria, but, among eukaryotes, appear to be present only in vertebrates . . . A more [interesting] explanation is that these genes entered the vertebrate (or prevertebrate) lineage by horizontal transfer from bacteria."

QUESTION A: *If HGT did occur (and the idea is very controversial), it apparently was not through some intermediary organism. As the quote mentioned, no similar sequences could be found in yeasts, worms, flies, plants, or in any nonvertebrate eukaryote. Having studied the three types of HGT, propose which process would most likely be capable of transferring donor bacterial genes to the chromosomes in a recipient vertebrate cell.*

You can find answers online in **Appendix F**.

■ KEY CONCEPT 8.2   Genetic Engineering Involves the Deliberate Transfer of Genes Between Organisms

Prior to the 1970s, bacterial species having special or unique metabolic properties were detected by screening for organisms and their mutants possessing certain metabolic talents. Experiments in genetic recombination entered a new era in the late 1970s, when it became possible to intentionally insert a foreign gene into bacterial DNA and thereby establish a **clone**, that is, a genetically identical population of cells expressing the protein product coded by the inserted gene. **Genetic engineering** (also called **recombinant DNA technology**), uses microbial genetics, gene isolation, gene manipulation, and gene expression to manipulate genomes. The goals are illustrated in **FIGURE 8.8**.

The Tools of Genetic Engineering Make Genome Modifications Possible

The field of genetic engineering includes many techniques and depends on a group of tools to manipulate DNA. Two of the most useful are restriction enzymes and cloning vectors.

Restriction Enzymes

One of the first demonstrations of genetic engineering involved cutting open the circular DNA molecule from simian virus-40 (SV40) and then inserting (splicing) it into a bacterial chromosome. In doing so, a **recombinant DNA molecule**, a DNA molecule containing DNA segments from two or more organisms, was created.

To facilitate the process, enzymes called **restriction endonucleases** can be used. These enzymes act like molecular scissors to recognize and cut specific short stretches of nucleotides in DNA. The sequences recognized by the enzymes are called **palindromes** because the bases have the same sequence on both DNA strands when read in the

5′ to 3′ direction. Today, there is a vast array of restriction enzymes that have been isolated from prokaryotic organisms. Each restriction enzyme recognizes a specific nucleotide sequence (TABLE 8.1). Importantly, many of these enzymes leave the DNA with single-stranded extensions, called "sticky ends," that can easily attach to complementary ends protruding from another fragment of DNA. To seal these complementary DNA segments, **DNA ligase** is used.

Putting this all together, microbiologists can take a plasmid from a bacterial species such as *E. coli*

TABLE 8.1   Examples of Restriction Endonuclease Recognition Sequences

Organism	Restriction Enzyme[1]	Recognition Sequence[2]
Escherichia coli	*Eco*RI	G ↓ AATTC CTTAA ↑ G
Streptomyces albus	*Sal*I	G ↓ TCGAC CAGCT ↑ G
Haemophilus influenzae	*Hind*III	A ↓ AGCTT TTCGA ↑ A
Bacillus amyloliquefaciens	*Bam*HI	G ↓ GATC^{m}C CC^{m}TAG ↑ G
Providencia stuartii	*Pst*I	CTGCA ↓ G G ↑ ACGTC

[1] Enzyme designations are derived from the species from which they were isolated. For example, the restriction enzyme *Eco*RI stands for *Escherichia* *co*li **R**estriction enzyme **I**.
[2] Arrows indicate where the restriction enzyme cuts the two strands of the recognition sequence; C^m = methylcytosine.

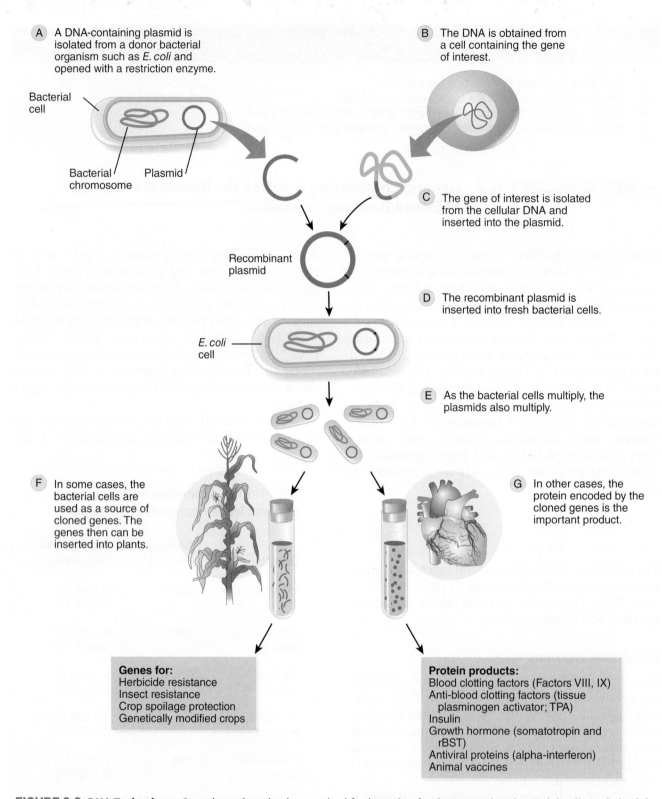

A A DNA-containing plasmid is isolated from a donor bacterial organism such as *E. coli* and opened with a restriction enzyme.

Bacterial cell

Bacterial chromosome   Plasmid

B The DNA is obtained from a cell containing the gene of interest.

C The gene of interest is isolated from the cellular DNA and inserted into the plasmid.

Recombinant plasmid

D The recombinant plasmid is inserted into fresh bacterial cells.

E. coli cell

E As the bacterial cells multiply, the plasmids also multiply.

F In some cases, the bacterial cells are used as a source of cloned genes. The genes then can be inserted into plants.

G In other cases, the protein encoded by the cloned genes is the important product.

Genes for:
Herbicide resistance
Insect resistance
Crop spoilage protection
Genetically modified crops

Protein products:
Blood clotting factors (Factors VIII, IX)
Anti-blood clotting factors (tissue plasminogen activator; TPA)
Insulin
Growth hormone (somatotropin and rBST)
Antiviral proteins (alpha-interferon)
Animal vaccines

FIGURE 8.8 DNA Technology. Genetic engineering is a method for inserting foreign genes into bacterial cells and obtaining chemically useful products. *»» How do bacterial cells that have been genetically modified to encode gene products (F) differ from those modified cells encoding protein products (G)?*

and open it with a restriction enzyme (**FIGURE 8.9**). Then, they can insert a segment of foreign DNA into the plasmid and seal the segment by using DNA ligase.

Cloning Vectors

One of the goals of genetic engineering is to insert a useful, foreign gene into another cell that will then produce the protein product of that gene. To carry this gene to the target recipient cell, a cloning vector is used. This vector can be a transposon, viral genes, or a bacterial plasmid.

The best way to understand how the vector works and to see how genetic engineering operates is to follow the steps that have been used to clone a human gene into *E. coli*.

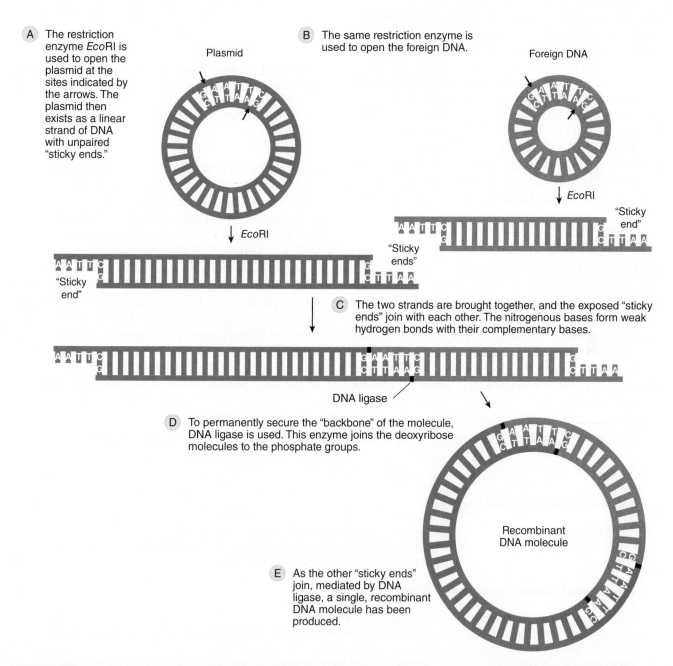

FIGURE 8.9 Construction of a Recombinant DNA Molecule. In this construction, two unrelated plasmids (loops of DNA) are united to form a single plasmid representing a recombinant DNA molecule. *»» Why is the product of genetic engineering called a recombinant DNA molecule?*

Today, more than 18 million people in the United States have **diabetes**, a group of diseases resulting from abnormally high blood glucose levels. There are more than 800,000 cases of insulin-dependent diabetes (also called juvenile or type I diabetes), requiring these diabetics to receive regular injections of **insulin** to control their blood glucose level. Before 1982, insulin was extracted and purified from the pancreas of cattle and pigs or even cadavers. However, this posed a problem because the animal insulin protein is not identical in amino acid sequence to the human insulin protein, and such an animal protein can trigger allergic reactions in the diabetic individual. In addition, the animals used to extract the insulin could contain unknown disease-causing viruses that would be isolated along with the insulin protein. The solution was to produce human insulin by using genetic engineering.

MICROINQUIRY 8 describes and illustrates one such method—by cloning the human gene for insulin into bacterial cells, which then act as tiny factories to synthesize and secrete human insulin.

Biotechnology Has Spawned Many Commercial and Practical Products

Eli Lilly marketed the first synthetic human insulin, called Humulin®, in 1982. Since then, other genetically engineered insulin products have been developed. Such commercial successes were a sign of the power of genetic engineering to modify gene expression in cells and organisms.

Biotechnology can be defined as the field that applies the techniques and tools of genetic

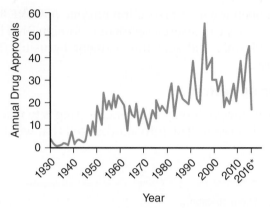

* The number of 2016 approvals is through 9/21/16.

FIGURE 8.10 FDA Approvals of New Drug Products, 1930–2016. This graph shows the number of approvals of new drug products since 1930. *»» Can you come up with a reason for the large spike in drug biopharmaceuticals in 1996?*

Data from FDA.

engineering to (1) manipulate living organisms (usually microorganisms) and (2) manufacture products important to human health and the environment. As described in **MICROBIOLOGY PATHWAYS**, today biotechnology is a very vibrant field within the biological (and interdisciplinary) sciences.

Besides insulin, many other proteins of important pharmaceutical value to humans have been produced by genetically engineered microorganisms. Some examples are listed in **TABLE 8.2**. Hundreds of biotechnology companies worldwide are working on the commercial and practical applications of genetic engineering (**FIGURE 8.10**). Many of the drug products are either proteins expressed

TABLE 8.2 Some of the Therapeutic Products of Biotechnology and Their Functions

Product	Function
Replacement Proteins and Hormones	
Factors VII, VIII, IX	Replace clotting factors missing in hemophiliacs
Growth hormone (HGH)	Replaces missing hormone in people with short stature
Insulin	Treatment of insulin-dependent diabetes
Therapeutic Proteins, Hormones, and Enzymes	
Epidermal growth factor (hEGF)	Promotes wound healing
Granulocyte colony stimulating factor (hG-CSF)	Used to stimulate white blood cell production in cancer and AIDS patients
Interferon-alpha (IFN-α)	Used with other antiviral agent to figure viral infections and some cancers
Tissue plasminogen activator (TPA)	Dissolves blood clots, prevents blood clotting after heart attacks and strokes
DNase I	Treatment of cystic fibrosis

MICROINQUIRY 8

Molecular Cloning of a Human Gene into Bacterial Cells

Genetic engineering has been used to produce pharmaceuticals of human benefit. One example concerns the need for insulin injections in people suffering from diabetes (an inability to produce the protein insulin to regulate blood glucose level). Prior to the 1980s, the only source for insulin was through a complicated and expensive extraction procedure from cattle or pig pancreases. But, what if one could isolate the human insulin gene and, through transformation, place it in bacterial cells? These cells would act as factories churning out large amounts of the pure protein that diabetics could inject.

To do molecular cloning of a gene, besides the bacterial cells, we need three ingredients: a cloning vector, the human gene of interest, and restriction enzymes. Plasmids are the "cloning vector," a genetic element used to introduce the gene of interest into the bacterial cells. Human DNA containing the insulin gene must be obtained. We will not go into detail as to how the insulin gene can be "found" from among 19,000 human genes. Suffice it to say that there are standard procedures to isolate known genes. Restriction enzymes will cut open the plasmids and cut the gene fragments that contain the insulin gene, generating complementary sticky ends.

The following description represents one procedure to genetically engineer the human insulin gene into cells of *E. coli*.

1. Plasmids often carry genes, such as antibiotic resistance. We are going to use the cloning vector shown in **FIGURE A** because it contains a gene for resistance to ampicillin (amp^R) and the *lacZ* gene that encodes the enzyme β-galactosidase (B-gal) that splits lactose into glucose and galactose. This will be important for identification of clones that have been transformed. In addition, this vector has a single restriction sequence for the restriction enzyme *Sal*I (see Table 8.1). Importantly, this cut site is within the *lacZ* gene. Also, the plasmid will replicate independently in *E. coli* cells and can be placed in the cells by transformation.

2. The vector and human DNA are cut with *Sal*I to produce complementary sticky ends on both the opened vector and the insulin gene (**FIGURE B**). The vector and the insulin gene then are mixed together in a solution with DNA ligase, which will covalently link the sticky ends. Some vectors will be recombinant plasmids, that is, plasmids containing the insulin gene. Other plasmids will close back up without incorporating the gene.

3. The plasmids are placed in *E. coli* cells by transformation. The plasmids replicate independently in the bacterial cells, but as the bacterial cells multiply, so do the plasmids. By allowing the plasmids to replicate, we have cloned the plasmids, including any that contain the insulin gene.

4. Because we do not know which plasmids contain the insulin gene, we need to screen the clones to identify what colonies contain the recombinant plasmids. This is why we selected a plasmid with amp^R and *lacZ*.

Because the *Sal*I cut site is within *lacZ*, any recombinant plasmids will have a defective *lacZ* gene and the bacterial cells carrying those plasmids will not be able to produce B-gal. Plasmids without the insulin gene have an intact *lacZ* gene and will have B-gal activity. Thus, using the positive selection technique described in the text on microbial genetics and a selective medium as described in the text on microbial growth and nutrition, we can identify which bacterial cells contain the recombinant plasmids.

Therefore, we plate all our bacterial cells onto an agar plate containing ampicillin and a substrate called X-gal that B-gal can hydrolyze.

$$X\text{-gal} \xrightarrow{\text{B-gal}} X + galactose$$

- The X product is blue in color, so any colonies having an intact *lacZ* gene will hydrolyze X-gal and appear blue on the agar plate;
- If the *lacZ* gene is inactive due to the presence of the insulin gene, then no product will be formed and those colonies on the plate will appear white.

Question a. Identify what colonies on the plate contain the insulin gene.

Question b. Explain why the bacteria without a plasmid did not grow on the plate.

You can find answers online in **Appendix E**.

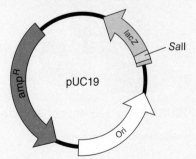

FIGURE A The Cloning Vector, pUC19.

(continues)

MICROINQUIRY 8 (*Continued*)

Molecular Cloning of a Human Gene into Bacterial Cells

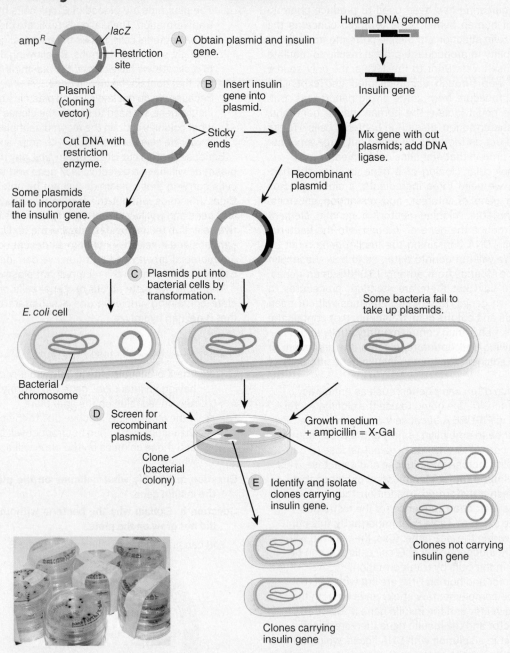

FIGURE B The Sequence of Steps to Engineer the Insulin Gene into *Escherichia coli* Cells.

Photo/inset © Martin Shields/Alamy Images.

5. These colonies with the insulin gene can now be isolated and grown in larger batches of liquid medium. These batch cultures then are inoculated into large "production vats," called bioreactors, in which the cells grow to massive numbers while secreting large quantities of insulin into the liquid.

MICROBIOLOGY PATHWAYS

Biotechnology

Industrial microbiology, or molecular biotechnology (the terms "industrial microbiology" and "biotechnology" are often one and the same), applies scientific and engineering principles to the processing of materials by microorganisms and viruses, or plant and animal cells, to create useful products or processes. Because biotechnology essentially uses the basic ingredients of life to make new products, it is both a basic technology and an applied science. Thus, by its interdisciplinary nature, biotechnology spans a number of potential career areas.

If you would like to be part of what analysts see as one of the most important fields of science, then microbiology is an excellent place to start. You should take a course in biochemistry as well as one in genetics. Courses in physiology and cell biology also are helpful. Employers will be looking for individuals with good laboratory skills, so be sure to take as many lab courses as you can.

An employer will be looking for work experience, which you can obtain by assisting a senior scientist, doing an internship, or working summers in a biotech firm. If you are at a college or university with research programs, a campus research lab is another good place to obtain work experience. It also might be a good idea to sharpen your writing skills, because you will be preparing numerous reports.

© Jon Feingersh Photogr/Blend Images/age fotostock.

You can enter the biotechnology field with an associate's or bachelor's degree. However, many of the career paths require a master's or doctoral degree. In addition, the higher degree usually correlates with a higher salary potential because there are so many levels at which individuals are hired. Careers can be at the laboratory and nonlaboratory levels.

Most laboratory positions are at the college, university, or industrial/pharmaceutical levels. Laboratory managers oversee laboratory operations and personnel. This person's job is to keep the lab running smoothly. Persons who have project responsibilities as principal investigators or research scientists often have one or more advanced degrees (a master's degree and/or PhD) in biology, microbiology, or some other allied field such as molecular biology, biochemistry, biotechnology, chemical engineering, or genetics.

Laboratory technicians represent the critical support team in a research group. They might be involved with general lab duties, equipment maintenance, or preparation of research chemicals and protocols. The technician might also have responsibility over tasks that are more specialized. Although these positions are in increasing demand, they tend to be lower-paying positions and require an associate's or bachelor's degree.

Nonlaboratory careers also occur outside of the lab. Products of biotechnology need to meet certain safety requirements and to be approved by governmental agencies such as the Food and Drug Administration. There also are opportunities in sales and marketing, which are among the highest-paying careers in the biotechnology field.

There are many opportunities in education as secondary and post-secondary teachers and educators. Each generation of scientific workforce requires a group of very knowledgeable teachers and professors to fire the imagination and train the next generation of biotechnologists.

Importantly, the novel and imaginative research that established biotechnology was founded in microbiology, and it continues to call on microbiology for its ongoing growth.

by the recombinant DNA in the bacterial cells or the recombinant DNA from the cloned plasmid (see Figure 8.8). Let's look at just a few examples to illustrate the power of genetic engineering and biotechnology.

Environmental Concerns

Many prokaryotic species of the domains Bacteria and the Archaea represent a huge, mostly untapped library of genes representing organisms with metabolically diverse processes. Consequently, a branch of biotechnology called **bioremediation** has developed in an attempt to return environmental sites to their original natural condition. The remediation process uses genetically engineered or genetically recombined microbes that have specific genes whose products will break down toxic pollutants, clean up waste materials, or degrade petroleum products in oil spills. However, we have barely scratched the surface to take advantage of the metabolic diversity offered by these microbes.

Medical and Health Concerns

Genetic engineering and biotechnology are at the heart of many pharmaceutical and therapeutic applications. In the medical field, the presence or threat of infectious disease represents a high demand for antibiotics. Although antibiotics are naturally produced by a few bacterial and fungal species, these organisms often do not produce these antimicrobial compounds in high yield. This means new antibiotic sources must be discovered and the microbes must be genetically engineered to produce larger quantities of antibiotics and/or to produce modified antibiotics to which infectious microbes have yet to show resistance.

Many proteins are being isolated and used today to fight viral infections. For example, in response to a viral infection, some human cells produce **interferon**, an antiviral protein capable of stimulating other cells to produce proteins that attempt to block viral replication in infected cells. Prior to the introduction of genetic engineering, thousands of units of human blood were needed to obtain sufficient interferon to treat a patient. With genetic engineering, much larger amounts of pure protein can be produced to aid patients suffering from hepatitis B and hepatitis C as well as some forms of cancer.

Vaccine production is safer today because of genetic engineering. For example, hepatitis B, a serious viral infection spread by contact with virus-infected blood, kills two million people globally each year. The first vaccine against hepatitis B was made by extracting the virus from infected blood of chronic hepatitis carriers and then isolating a viral surface protein. Such a procedure was complex and posed the risk of possible contamination by other infectious agents. Today, genetic engineering has developed a recombinant hepatitis B vaccine (Engerix-B®) that is made by inserting the gene for a hepatitis B viral surface protein into common baker's yeast, *Saccharomyces cerevisiae*. The protein produced by the yeast is identical to the natural viral protein and, being only a subunit of the virus, the vaccine is much safer for humans—recipients cannot contract hepatitis B from the vaccine.

A more recent recombinant vaccine, Gardasil®, also uses recombinant DNA technology by engineering a viral gene into yeast cells. The vaccine (licensed in 2006) generates protective immunity (i.e., prevents infection) against several of the papilloma viruses responsible for cervical cancer in women, penile cancer in men, and genital warts in women and men.

Lastly, developing a genetically engineered product is not always straightforward. Since 1987, scientists and genetic engineers have tried to develop protective vaccines that can be used against AIDS. However, they have had little success, not because genetic engineering can't be done; rather, it's because the virus just seems to find ways to circumvent the patient's immunity that developed from the tested vaccine. We will have much more to say about vaccines and AIDS in Chapters 23 and 24.

Medical Diagnosis

The genes of an organism contain the essential information responsible for its behavior and characteristics. Bacterial and viral pathogens, for example, contain specific genes that give the pathogen the ability to infect host organisms and cause disease. Because these nucleotide sequences are distinctive and often unique, they can be genetically manipulated and cloned as a diagnostic tool called a **DNA probe** to identify pathogens.

DNA probes are single-stranded DNA molecules capable of recognizing and binding to a distinctive nucleotide sequence of a pathogen. To make a probe, scientists first identify the segment (or gene) in the pathogen. Using this segment, they construct the single-stranded DNA probe using the **polymerase chain reaction (PCR)** outlined in MICROFOCUS 8.2 or through traditional genetic engineering

MICROFOCUS 8.2: Tools

The Polymerase Chain Reaction

The polymerase chain reaction (PCR) is a technique that takes a segment of DNA and replicates it millions of times in just a few hours.

The PCR process is a repeating three-step process (see figure). Target DNA is mixed with DNA polymerase (the enzyme that synthesizes DNA), short strands of primer DNA, and DNA nucleotides. Then, the mixture is alternately heated and cooled. During this time, the double-stranded DNA unravels, duplicates, and reforms the double strands.

The process is carried out in a highly automated PCR machine, which is the biochemist's equivalent of an office copier. Each cycle takes about 5 minutes, and each new DNA segment serves as a template for producing many additional identical copies, which in turn serve as templates for producing more identical copies. In the end, PCR produces billions of copies of a DNA sequence that are identical to the starting, target DNA sequence.

PCR is now a common and often essential tool in medical and biological research. Applications for PCR are numerous and include its use for DNA cloning, comparative DNA studies, and (as described in the text) diagnosis of infectious diseases.

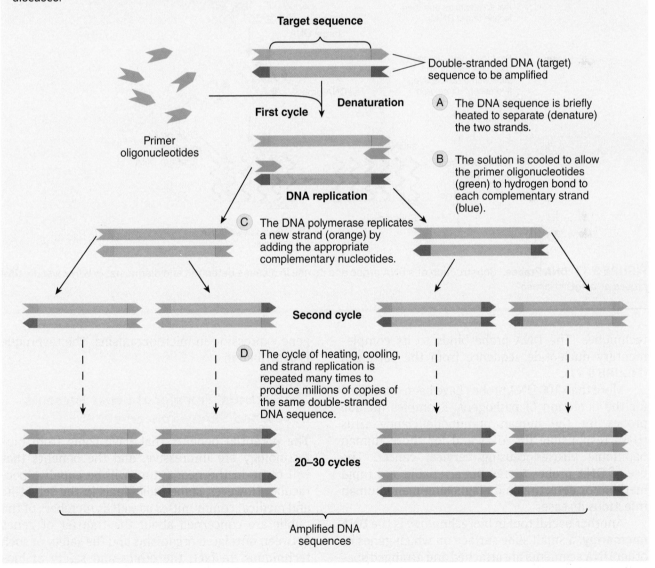

Target sequence

Double-stranded DNA (target) sequence to be amplified

First cycle

Denaturation

Ⓐ The DNA sequence is briefly heated to separate (denature) the two strands.

Ⓑ The solution is cooled to allow the primer oligonucleotides (green) to hydrogen bond to each complementary strand (blue).

Primer oligonucleotides

DNA replication

Ⓒ The DNA polymerase replicates a new strand (orange) by adding the appropriate complementary nucleotides.

Second cycle

Ⓓ The cycle of heating, cooling, and strand replication is repeated many times to produce millions of copies of the same double-stranded DNA sequence.

20–30 cycles

Amplified DNA sequences

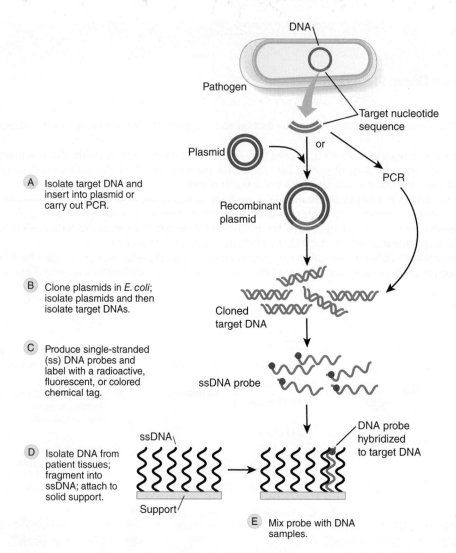

FIGURE 8.11 DNA Probes. Construction of a DNA probe and its use in disease detection and diagnosis. *»» Why must the DNA probes be single stranded?*

techniques. The DNA probe binds to its complementary nucleotide sequence from the pathogen (**FIGURE 8.11**).

More than 100 DNA probes have been developed for the detection of pathogens. Examples include probes for the human immunodeficiency virus (HIV) responsible for AIDS and for those human papilloma viruses causing cervical cancer. The use of DNA probes represents a reliable and rapid method for detecting and diagnosing many human infectious diseases.

Another useful tool in biotechnology is the **DNA microarray**, a small slide surface on which genes or other DNA segments are attached and arranged spatially in a known pattern that can be used to assess gene expression in microorganisms. The technique is described in **MICROFOCUS 8.3**.

The Potential Transfer of Genes Presents Ethical and Safety Concerns

The potential benefits that have come from biotechnology are impressive, and the benefits that will come in the near future will be equally spectacular. However, some individuals in the scientific and medical communities as well as members of the public are concerned about the transfer of genes between unrelated organisms and the safety of such techniques. In fact, the ethics and safety of biotechnology have been debated ever since scientists

MICROFOCUS 8.3: Tools

Microarrays Can Be Used to Study Gene Expression

After a microbial genome had been sequenced, scientists needed a way to discover how the genes are expressed in the organism. So, in the early 1990s, biotechnologists discovered a way to put these DNA segments on a miniaturized surface, such as a glass slide, or a plastic or silicon surface. Thousands of spots can be placed on a single microarray, often called a "DNA chip." Each spot is put in place by a mechanical robot and is unique because the spot contains many copies of a unique array of species-specific DNA sequences.

To determine which genes are being expressed, scientists use single-stranded, fluorescently labeled DNA that is complementary to a target DNA sequence. Often two different samples of probes are used, each labeled with a different-colored fluorescent dye (often red and green). When exposed to (washed over) the microarray, a probe will bind to any complementary sequences (the target), and the spot will fluoresce the color (red or green) of the bound probe. If both probes bind, they will produce a yellow fluorescence; or if one probe binds more than the other probe, the spot may be orange or light green (see figure). Therefore, a more intense fluorescence means there had been intense expression of that sequence in the organism. The results usually are scanned and analyzed by computer.

Microarrays can be used in many ways.

- **Gene expression:** Microarrays can be developed and used to study gene expression. For example, a microbiologist might want to study the effects of oxygen on gene expression in the facultative anaerobe *E. coli*. Using a microarray containing the *E. coli* genome, the microbiologist could isolate *E. coli* cells grown under aerobic conditions and another sample under anaerobic conditions. From these samples, the mRNAs (from active genes under the two conditions) would be isolated and copied (reverse transcribed) into single-stranded DNA. The DNA probes from the aerobic condition may be labeled with the red dye while the anaerobic condition may be labeled with the green dye. The two probes allow the scientist to study the same cell (genes) under two different conditions to see which genes are active with or without oxygen gas.
- **Organism detection:** Because of the constant threat posed by emerging infectious diseases and the limitations of existing approaches available to identify new pathogens, microarrays offer a rapid and accurate method for viral discovery. DNA microar-

DNA microarray showing locations of differently labeled DNA molecules.

© Alfred Pasieka/Science Source.

ray-based platforms have been developed to detect a wide range of known viruses as well as novel members of some viral families. For example, during the outbreak of severe acute respiratory syndrome (SARS) in March 2003, a viral isolate cultivated from a SARS patient was made into a DNA probe. On the microarray, it produced a spot representing the SARS virus.

A microarray consisting of human DNA can be probed with a DNA sequence from an unknown pathogen to see if that person has been infected. For example, early infection with the malarial parasite *Plasmodium falciparum* can be identified by isolating DNA from the tissues of a suspected patient, fragmenting the DNA into single-stranded DNA segments, and attaching these to the solid support. This microarray is then washed with a fluorescently labeled *P. falciparum* probe. Any fluorescent spots detected on the microarray indicate a match to *P. falciparum* DNA, confirming the patient is infected.

- **Comparative relationships:** The extent of microbial diversity in an environment can be assessed by producing a microarray, called a "Phylochip," containing oligonucleotides (short sequences of nucleotides) that are complementary to 16S rRNA gene sequence probes of different bacterial species. Any lit spot on the microarray is an indicator that that species is part of the microbial community in that environment.

began to investigate genome modifications and gene transfer.

Today, there are diverse and often conflicting beliefs concerning the recombining of genes. Thus, the public often comes away with a misunderstanding of the technology. Still, such debate is healthy and warranted.

One of the widely debated concerns about genetic engineering and biotechnology concerns the unintentional or deliberate release (or transfer) of a genetically modified organism (GMO) into the environment. Some individuals suggest that an engineered organism could become more viable than the natural organism. The result would be that the microbe spreads and damages, or destroys, the ecological balance among organisms. In addition, some opponents of biotechnology fear that a GMO could pass on genetic traits to unrelated species, such as traits for heightened virulence, leading to a harmless organism becoming a pathogen.

Debate also has occurred over genome experiments that have added new genetic content to organisms or viruses that could potentially make the agent more virulent and/or transmissible. These **"gain-of-function" (GOF)** experiments have involved infectious agents that potentially make the pathogen even more dangerous. In 2011, experiments were done with the avian flu virus to determine what changes needed to occur in the virus to make it more readily transmissible in humans. The debate over these experiments concerned whether the knowledge learned from the work justifies the risk of someone taking that information and developing a more transmissible strain of the avian flu.

To address such worries, the U.S. National Academy of Sciences has instituted standards that must be followed when doing recombinant DNA research. American research-funding organizations, such as the National Science Foundation and the National Institutes of Health, require all research proposals to include a study of any ethical and safety issues in carrying out the plan of study. To that end, in October 2013, the U.S. government stopped funding GOF studies until the biosafety and biosecurity risks are better understood.

In summary, genetic engineers have been responsible citizens in developing new microbe strains and carrying out research studies with recombinant DNA. They have thought through and taken the precautions needed to ensure that there are appropriate safeguards. Therefore, it is highly unlikely that an application or product of biotechnology could have an adverse effect, cause harm to human health, or disrupt the cycle in nature. However, the chances are not zero, given that there are groups around the world who could attempt some gene technology nightmare. Although the chances of success are extremely slim, it is imperative that there be continued oversight to reassure the public as to the safety of an engineered microorganism as well as its gene or protein product.

Concept and Reasoning Checks 8.2

a. List the steps required to form a recombinant DNA molecule.
b. Give several examples of how genetic engineering has benefited the fields of environmental biology and medicine.
c. Why are DNA probes such useful tools in the detection of disease-causing agents?
d. What are some of the health and safety concerns with regard to genetic engineering?

Chapter **Challenge B**

In this section, we have seen several examples of how genetically engineered microbes have had definite effects in trying to circumvent disease conditions arising from the human genome.

QUESTION B: *Provide a list of five ways genetically engineered microbes have helped to alleviate disease conditions caused by faulty genes in the human genome.*

You can find answers online in **Appendix F.**

■ KEY CONCEPT 8.3 Microbial Genomics Studies Genes and Genomes at the Single-Cell to Community Levels

In April 2003, exactly 50 years to the month after Watson and Crick announced that they had worked out the structure of DNA, a publicly financed, $3 billion international consortium of biologists, industrial scientists, computer experts, engineers, and ethicists completed perhaps the most ambitious project to date in the history of biology. The Human Genome Project, as it was called, had succeeded in sequencing the **human genome**—that is, the 3 billion nucleotides in a human cell were identified and their correct order was precisely determined. The completion of the project represented a scientific milestone with unimaginable potential health benefits.

DNA Sequencing Is Used to Identify Genes and Genomes in Organisms and Viruses

DNA (gene) sequencing involves figuring out the order of nucleotides (adenine, guanine, cytosine, and thymine) in an entire DNA molecule, a DNA segment, or an individual gene. To determine the sequence of nucleotides, one needs sufficient DNA. Today, this can be easily accomplished using PCR, which can amplify the needed DNA into thousands of copies. Then, the sequencing can be done. We will not discuss the details of the sequencing techniques here, except to say that the methods are now routine, allowing lab clinicians to sequence a single gene, an entire DNA molecule, or the entire genome of an organism or virus (**FIGURE 8.12**). These studies are carried out in research labs around the world and they form the field of genomics.

If the human genome were represented by a rope 2 inches in diameter, it would be 32,000 miles long. The genome of a bacterial species, like *E. coli*, at this scale would be only 1,600 miles long or about 1/20 the length of human DNA. Therefore, being substantially smaller, microbial genomes have been easier and much faster to sequence.

Many Genomes of Prokaryotes and Microbial Eukaryotes Have Been Sequenced Completely

In 1976, the first generation of sequencing technologies was used to sequence the genome of a bacteriophage. It was not until 1995 that the first complete genome of a free-living bacterium was sequenced: the 1.8 million base pairs (1.8 Mb)

composing the genome of *Haemophilus influenzae*. This was followed in the next year with the first complete genome sequence (12.5 Mb) of a eukaryotic microbe, the common baker's (or brewer's) yeast *S. cerevisiae*. Today, the field of **microbial genomics**—the discipline of sequencing and analyzing microbial genomes—has skyrocketed with the development of so-called **next-generation DNA sequencing** (**NGS**). Being able to sequence up to 3 billion sequences of DNA at one time has drastically dropped the cost and time of sequencing whole

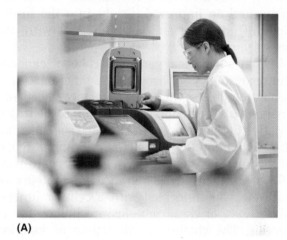

(A)

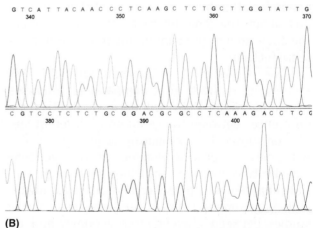

(B)

FIGURE 8.12 An Automated DNA Sequencing Machine. (A) The process of DNA sequencing is facilitated by computer analysis. **(B)** The DNA sequence represents peaks of different colors (adenine = green; guanine = black; cytosine = blue; thymine = red) seen on the computer monitor. *»» From (B), what is the sequence of nucleotides (read left to right) for the first 10 nucleotides?*

genomes. In 1995, it took 13 months and $200,000 to sequence the genome of *H. influenzae*. Today, with NGS, researchers can sequence a similar-sized prokaryotic genome in only a few hours for less than $5,000. Consequently, by the end of 2016, some 90,000 microbial sequencing projects had been completed or were being carried out (**FIGURE 8.13A**). Although many of these projects focus on clinically important bacterial pathogens, many other projects are involved with other areas of microbiology and the biological sciences (**FIGURE 8.13B**).

Comparative Genomics Brings a New Perspective to Studying Evolution

Although DNA sequences by themselves do not tell us much, sequence comparisons between organisms give us a better understanding of microbial genes and an improved insight into the interactions of microbial and viral genomes with other organisms.

After a microbial genome has been sequenced, the next step is to discover the functions for the genes composing the DNA. Sequences need to be analyzed (called **genome annotation**) to identify the location of the genes in the DNA and the function of their RNA or protein products. For example, in most of the microbial genomes sequenced to date, nearly 50% of the identified genes encoding proteins have not yet been connected with a cellular function. About 30% of these proteins are unique to each species. The challenging discipline of **functional genomics** attempts to discover what these proteins do and how the genes interact with other genes and the environment to maintain microbial growth and reproduction.

One of the most active areas involving DNA sequencing is the field of **comparative genomics**, which identifies DNA sequence similarities and differences between the genomes of different species or subspecies. Comparing sequences of similar genes can provide evidence on how genomes have evolved over time and clues to the gene relationships between microbes and other organisms, including humans. Let's examine a few comparative studies between microbe/virus genomes and the human genome.

Microbe and Human Genomes

Now that the human genome has been sequenced, are there any sequences (and genes) in the human genome that correspond to bacterial and viral gene sequences?

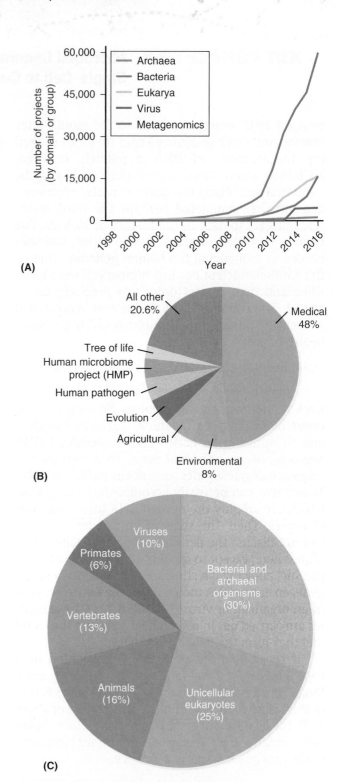

(A)

(B)

(C)

FIGURE 8.13 Genomes and Ancestry (A) The total number of genome sequencing projects from 1998–2016. Metagenomics refers to sequencing projects involving a whole community of microbes. **(B)** The relevance of bacterial projects. Almost 50% have studied medical aspects of bacteria. **(C)** Analysis of human gene sequences points to a long history of shared ancestry with microbes. *»» From (C), what percentage of genes originated in microbes and viruses?*

(A and B) Data from Genomes Online Database (https://gold.jgi.doe.gov/statistics).

Some comparisons indicate that a large percentage of our 20,000 genes originated from genes found in members of the prokaryotes and unicellular eukaryotes (**FIGURE 8.13C**). However, most of these human genes were not acquired directly from these bacterial species but rather were picked up by early animal ancestors. So important were these genes, they have been preserved and passed along through vertical gene transmission from organism to organism throughout the "tree of life" and the generations of evolution; they are life's oldest genes. For example, several researchers suggest some genes coding for brain signaling chemicals and for communication between cells did not evolve gradually in human ancestors; rather, these genes became part of the animal lineage eons ago directly from microbial organisms. This claim remains highly controversial, though, and more work is needed to sort out the relationships of the microbial genes to animal and human evolution.

Recent comparative genome studies also suggest that some 150 human genes might have infected our early primate ancestors through HGT from bacterial, unicellular eukaryotes, and viruses. We have already studied in this chapter how bacterial cells can use HGT to transfer genes. Many scientists believe this same process could have happened between microbes and early animals or in the primate ancestry leading to humans. As with all these theories, the exact time line for transfer or acquisition is unknown.

Virus and Human Genomes

Another discovery coming from comparative genomics concerns viral genomes. Several studies suggest that almost 10% of our genome (as well as parts of other animal genomes) is composed of fragments of viral DNA (see Figure 8.13C). The infectious viruses that inserted this DNA into our ancestors' genome are called the human **endogenous retroviruses** (**ERVs**). Retroviruses are RNA viruses that can reverse transcribe themselves from single-stranded RNA into double-stranded DNA, which can then insert into a human chromosome. The ERV "infections" presumably occurred in sex cells (egg and sperm) over the millions of years of primate evolution. During this period, the viral genome segments integrated into the primate lineage genome and have been transmitted vertically to future generations ever since—and these are the 100,000 remnants we see in our human DNA today (**FIGURE 8.14**).

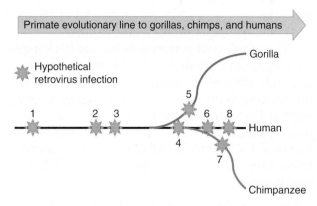

FIGURE 8.14 Simplistic Hypothesis for Retrovirus Genome Integration. This diagram presents a hypothetical example of how retroviral genes might have been integrated into the genome of the primate lineage, eventually becoming part of the ERVs. Additional integrations would have occurred earlier, prior to the branching of the primates, and also became part of the primate lineage. *»» From this figure, which integrations would be a permanent part of the human ERVs?*

Many ERVs have accumulated harmful mutations during primate evolution and no longer cause disease in the human body. However, some ERVs can "reawaken" and act like transposable elements or have evolved new functions. For example, at least six human ERV genes interact in the normal functioning of the placenta. One such ERV gene codes for a protein needed for a fertilized egg to develop into a fetus. The ERV protein allows the cells in the outer layer of the placenta to fuse together. In fact, cell fusion in general might have evolved from ERV genes that slipped into the animal genome and now function as a necessary gene for cell fusions (e.g., muscle, skin, bone, and even sperm-egg fertilization) that are part of the development of complex multicellular life. Other ERVs might play roles in cancer and neurodegenerative diseases of the spinal cord. One ERV is expressed at high levels in the brains of patients suffering from amyotrophic lateral sclerosis (ALS) and might play a role in this neurodegenerative disease of the brain and spinal cord.

Recently, even more virus "signatures" (nucleic acid sequences), very different from the ERVs, have been found in the human genome. In one scenario, these so-called **endogenous viral elements** (**EVEs**) crossed from rats and infected evolving primates. In primates, the viral genes at some point managed to infect the genomes of egg and sperm cells, allowing the viral genes to be passed through vertical transmission from generation to generation. Then, over time, the viral genes became mutated and disabled and could no longer produce new viruses.

Therefore, viral genomes in human DNA might be a legacy of long-distant infections and epidemics. Only through comparative genomics has this information been identified. Such studies strongly suggest that our genome is a hybrid of eukaryotic, prokaryotic, and viral genetic information (see Figure 8.13C). Undoubtedly, these virus integrations have affected gene expression and human evolution.

Comparative Genomics and Infectious Diseases

Humans are not the only organisms that are a hybrid of genes from various other organisms. Within the microbial world, similar hybrid genomes also are found. The genomes of nonpathogenic and pathogenic strains of *E. coli* vary by as much as 35%. Let's look at just a few specific examples where genome sequencing has provided an understanding of infectious diseases.

Genomic Islands and Infectious Diseases

Comparative genomics has revealed that all strains of a species contain a set of core genes. What separates the strains is the set of "variable genes" that diverge between strains. For example, some bacterial strains within a bacterial species contain **genomic islands**, clusters of up to 25 adjoining genes that are absent from other strains of the same species. Many of these islands came from an altogether different species, suggesting some form of HGT, such as conjugation, occurred in the past. The nonpathogenic bacterial species *Thermotoga maritima* has acquired about 25% of its genome from HGT. Sequence analysis indicates its genome is a mixture of bacterial and archaeal genes, suggesting *T. maritima* represents the modern descendants of an ancient lineage that evolved before the split of the Bacteria and Archaea domains.

One of the most interesting aspects of microbial genomics relates to infectious disease. By comparing the genomes of pathogenic and nonpathogenic bacterial species, or between pathogens with different host ranges, microbiologists are learning how pathogens evolve. Here are a couple of examples.

Gene Reduction. There are three bacterial species of *Bordetella*. *B. pertussis* causes whooping cough in humans; *B. parapertussis* causes whooping cough in infants but also infects sheep; and *B. bronchiseptica* produces respiratory infections in other animals (**FIGURE 8.15A**). Comparative genome sequence analysis of these three species reveals that *B. pertussis* and *B. parapertussis* have undergone gene loss. These two species are missing large segments of DNA (1,719 genes), which are present in *B. bronchiseptica*. In fact, of the three species, only *B. bronchiseptica* can survive independently without a host. Therefore, survival of *B. pertussis* and *B. parapertussis* depends on their ability to infect organisms that can supply them with the products of the lost genes; that is, pathogenicity has evolved from genome reduction.

Gene Expansion. A second example involves *Corynebacterium diphtheriae*, the causative agent for diphtheria (**FIGURE 8.15B**). Genome analysis suggests that an ancestor species acquired 13 genetic regions via HGT, each representing a genomic island. These islands are called **pathogenicity islands** because they encode many of the pathogenic characteristics of the bacterial species (e.g., pili formation and toxin production). In this example, pathogenicity evolved from genome expansion, specifically through the acquisition of new genes.

Pathogen Virulence

The majority of research studies in microbial genomics have focused on pathogens. These studies have provided a better understanding of how these microbes cause disease.

As this chapter has described, HGT has played a key role in the acquisition of virulence genes by some pathogens. These pathogenicity islands usually are located on mobile genetic elements (phage DNA, plasmids, and transposons). *Staphylococcus aureus* contains pathogenicity islands responsible for toxin synthesis, antibiotic resistance, and biofilm formation, all representing virulence factors that make the organism a more dangerous pathogen.

Just because an organism has some virulence gene sequences does not necessarily mean it is a pathogen. Some virulence genes have been identified by genome sequencing in avirulent strains of bacterial species in which other strains are pathogens. In these avirulent strains, the virulence genes might have separate targets. For example, bacterial toxin production in an avirulent strain might make it better able to eliminate or reduce competition for nutrients in the soil or to protect itself from predation by protists or nematodes.

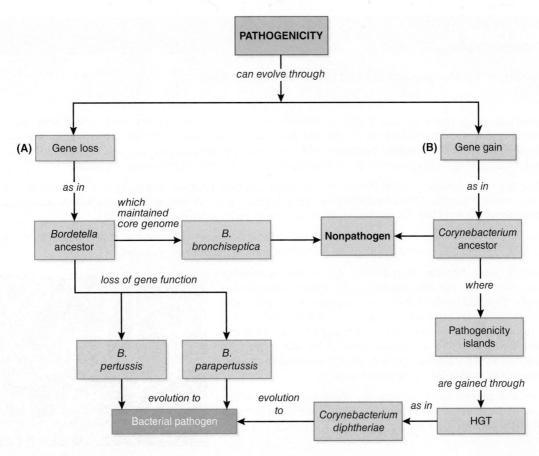

FIGURE 8.15 Comparative Genomics Suggests How Microbial Genomes Can Evolve. This concept map summarizes the evolution of three microbial genomes through loss or gain of gene function. *»» What advantage is there to losing or gaining genes or whole sets of genes (pathogenicity islands)?*

CLINICAL CASE 8 presents one last example in which the use of DNA sequencing was critical to the identification of an infectious disease.

Microbial Genomics Will Advance Our Understanding of the Microbial World

Microorganisms have existed on Earth for about 3.8 billion years, although we have known about them for little more than 300 years. Over this long period of evolution, these microbes have become established in almost every environment on Earth, make up a significant percentage of Earth's biomass, and, although they are the smallest organisms on the planet, they influence—if not control—some of the largest events. Yet, with few exceptions, we do not know a great deal about any of these microbes, and we can only culture and study in the laboratory about 2% of all microorganism species.

However, our limited knowledge is changing. With the advent of microbial genomics, microbiology

is in a third Golden Age, a time when remarkable scientific discoveries are and will continue to be made toward understanding the workings and interactions of the microbial world. Some potential consequences from the understanding of microbial genomes are outlined in the subsections that follow, especially as related to health and medicine.

A Safer Food Supply

Because microorganisms play important roles in our foods, both as contamination and spoilage agents, understanding how they get into the food product and how they produce dangerous foodborne toxins will help produce safer foods. However, a major limitation with traditional food safety surveillance is the time needed and number of tests required to identify a food pathogen.

Therefore, in 2012, the U.S. Food and Drug Administration (FDA) initiated a project to identify the microbial genomes of foodborne pathogens.

Clinical Case 8

A Diagnosis of Leptospirosis Based on DNA Sequencing

A 14-year-old boy with severe immunodeficiency (inability to mount an immune response to infection), who had undergone two previous bone marrow transplants to resolve the disorder, was admitted to the local hospital for the third time in 3 months. He presented with increased symptoms of fever, headache, nausea, and vomiting. On this third admission, his status became critical as he developed seizures that required a medically induced coma to control.

The previous summer, he had traveled to Puerto Rico, where he swam in the ocean and in a river. A fellow traveler developed a fever and blood in the urine. Within a month of his return home, the boy developed fever, headaches, and conjunctivitis. The eye symptoms returned two months later and he was treated with antimicrobial eye drops.

The following spring, he was again hospitalized twice with headaches, fever, abdominal pain, and weight loss. Lab tests, a head MRI, and an infectious diseases workup were negative and the patient was released.

Three months later, he was hospitalized again with worsening symptoms. This time a head MRI detected areas of hyperactivity in the basal ganglia. Elevated levels of angiotensin-converting enzyme were detected in the CSF that can lead to a narrowing of the blood vessels and lead to high blood pressure. On the 13th day of hospitalization, an MRI revealed meningitis, an inflammation of the meninges (the tissues that surround the brain or spinal cord). A brain autopsy showed a noncancerous inflammation in the brain tissue.

The patient was placed on IV steroids to reduce the inflammation. However, with a buildup of CSF in the skull, an extraventricular line was used to drain the fluid. The patient was intubated and placed in a drug-induced coma.

Being unable to diagnose the infection, the patient was entered into a research program involving next-generation DNA sequencing to identify the infectious pathogens. Within 100 minutes after sequencing samples from the patient's CSF, a bacterium in the family Leptospiraceae was identified. The patient was immediately started on penicillin before confirmation of the infectious agent was made. Five days later, PCR analysis identified *Leptospira santarosai* as the infectious agent (see figure).

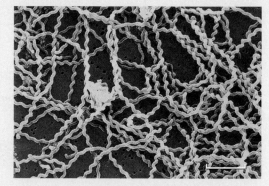

A false-color scanning electron microscope image of *Leptospira* cells. (Bar = 1 µm.)

Courtesy of the CDC/Janice Haney Carr.

The patient gradually recovered and he was finally sent home 76 days after admission.

Questions:
  a. What are the symptoms of the patient's acute infection?
  b. How did the patient most likely contract the disease?
  c. What was the significance of his fellow traveler's symptoms?
  d. Conventional testing for *Leptospira* depends on detecting antibodies in the blood against *Leptospira*. Why was this not successful in this patient?
  e. What take-home lesson can be learned from the power of next-generation DNA sequencing?

You can find answers online in **Appendix E**.

For additional information see: Wilson, M. R. et al. 2014. *New England Journal of Medicine* 370:2408–2417. http:/www .nejm.org/search?q=neuroleptospirosis.

Called the Genome-Trakr network, the FDA, in collaboration with state, federal, and international food safety labs, now has sequence information on more than 55,000 strains of bacteria. Therefore, when a foodborne outbreak occurs, the sequence information in the network will greatly speed up identification of the pathogen, identify the illness causing the outbreak, and pinpoint which ingredient in a multi-ingredient food is contaminated. For example, in 2016, a multistate foodborne outbreak of *Listeria* occurred. Using Genome-Trakr, the FDA identified a particular brand of frozen vegetables as the source of the *Listeria* contamination. Working with the Centers for Disease Control and Prevention (CDC), the FDA matched the *Listeria* genome sequence in the frozen vegetables with the *Listeria* sequence in people who had become ill.

Microbial Forensics

The advent of the anthrax bioterrorism events of 2001 and the continued threat of bioterrorism have led many researchers to look for rapid and efficient ways to detect the presence of such bioweapons. Many of the diseases caused by these potential biological agents cause no symptoms for at least several days after exposure, and when symptoms do appear, initially they are flu-like. Therefore, it can be difficult to differentiate between a natural disease outbreak and the intentional release of a potentially deadly pathogen. Genome sequencing provides a powerful tool to analyze and then respond to such incidents.

The concern with disease outbreaks and potential bioterrorism has led to a relatively new and emerging area in microbiology. Called **microbial forensics**, the discipline attempts to recognize patterns in a disease outbreak or bioterror incident; to identify the responsible pathogen; to control the pathogen's spread; and, using NGS, to discover the source of the pathogenic agent. For example:

▶ In 2009, some individuals questioned whether the swine flu pandemic was the result of an accidental release of the virus from a lab doing vaccine research. Microbial forensic investigations using NGS concluded that there was no accidental release of swine flu viruses because the genome sequences of the outbreak strain and research lab strain did not match.

▶ In 2010, a cholera outbreak occurred in Haiti. People wondered where the bacterial pathogen came from because there had not been any cholera in Haiti for more than 100 years. Some believed it was transmitted unintentionally by United Nations peacekeepers from Nepal who, earlier that year, came to Haiti as part of the U.N. relief effort after the 2010 earthquake. Proof was difficult until genomic sequencing of the cholera strains confirmed the sequence similarity between the Haitian strain and that in Nepalese peacekeepers. Nepal had been in the midst of its own cholera outbreak at the time the peacekeepers were sent to Haiti.

Microbial forensics also has been part of investigations into potentially criminal acts involving infectious agents.

▶ In 2012, a cluster of 29 hepatitis C cases occurred in a hospital. By sequencing the genomes of the viruses isolated from the outbreak and doing comparative genomics, the forensic scientists traced the origin of the outbreak to a lab technician who had hepatitis C. He was injecting himself with the clinic's narcotics and then reusing the same needles for patient medications. The accused pleaded guilty.

One last and very interesting use of microbial forensics concerns an organism's death. For example, when a human body dies and begins to decompose, its microbiome changes. Therefore, forensic pathologists have wondered if understanding the changes in microbe diversity and number could help in determining the time of death. In fact, within 2 days after death, the microbial community on a corpse changes, with some microbes increasing in number, whereas others decrease in number. After a week, the microbial community has changed again partially in response to the lack of oxygen gas. Scientists now want to develop a "microbial death clock" that could be very useful in helping law enforcement pathologists more accurately work out the time of death in homicides.

In all of these cases, tracing the infecting microbe to an institution, place, or person(s) of origin is critical. This places the fields of epidemiology, genetic engineering, biotechnology, and microbial genomics at the forefront of this emerging field.

Metagenomics Is Identifying the Previously Unseen Microbial World

Existing within, on, and around every living organism, and in most all environments on Earth, are microorganisms. Yet, some 98% of these species will not grow on any known growth medium; we refer to these organisms as **viable, but noncultured** (**VBNC**). This means that most organisms found in the soil, in oceans, and even in the human body have never been seen or named—and certainly not studied. Yet the genes of these organisms represent a wealth of genetic diversity.

There is now a microbial sequencing process for analyzing this uncultured majority (see Figure 8.13A). The process is called **metagenomics** (*meta* = "beyond"), and it refers to identifying the genome sequences within an assorted mixture of organisms in a community (the **metagenome**), that is, "beyond" what can be identified from culture. Metagenomics today is stimulating the development of advances in fields as diverse as medicine, agriculture, and energy production. Here is how the metagenomics process is carried out (**FIGURE 8.16**).

Samples of the desired community of microorganisms are collected. These samples could be from soil, water, or even the human digestive tract. The DNA is then extracted from the sample, which produces DNA fragments from all the microbes in the samples. The fragments can be amplified through PCR and sequenced as short segments. Using computer-based technologies, the segments are pieced back together, each final sequence being unique to one of the microbial species in the original sample. When completed, microbiologists have a broad spectrum of the microbial community. Using **sequence-based metagenomics**, researchers can use the reassembled genomes to compare taxonomic relationships between organisms in the sample. With **functional-based metagenomics**, researchers can use the gene products from the sequenced genes to compare metabolic relationships within the community or to search for new enzymes, vitamins, antibiotics, or other potential chemicals of therapeutic or industrial use. Thus, the role of each microbe in that community can be studied without having to grow the organism in culture.

Today, metagenomics is one of the most intensively studied fields for understanding the relationships between and within microbial communities. The field is uncovering the diversity of microbes that might help solve some of the most complex medical, environmental, agricultural, and economic challenges in today's world (**TABLE 8.3**).

In conclusion, genetic engineering and the various molecular techniques it has spun off through biotechnology have revolutionized the field of microbiology—and stimulated its third Golden Age. Examining and comparing the gene sequences of cultured and uncultured microbes (in isolation or from whole communities) have provided microbiologists with a better understanding of the processes microbes carry out and their pathogenic capabilities. With regard to taxonomic relationships, the information gathered from genome sequencing and comparative genomics has turned the "tree of life" into a somewhat "tangled bush" because of the ability of prokaryotes to undergo horizontal gene transfer in an almost promiscuous way because genes or entire genomic islands can be shared between dissimilar species. As the late American biologist and evolutionist Lynn Margulis once wrote: *"The speed of [HGT] over that of mutation is superior: it could take eukaryotic organisms a million years to adjust to a change on a worldwide scale that bacteria can accommodate in a few years."*

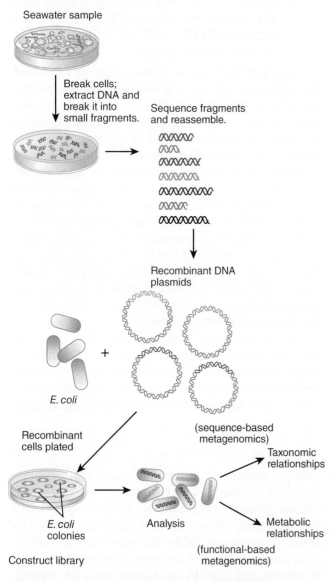

Seawater sample

Break cells; extract DNA and break it into small fragments.

Sequence fragments and reassemble.

Recombinant DNA plasmids

E. coli

+

Recombinant cells plated

E. coli colonies

Construct library

Analysis

(sequence-based metagenomics)

Taxonomic relationships

Metabolic relationships

(functional-based metagenomics)

FIGURE 8.16 The Process of Metagenomics. Metagenomics allows the simultaneous sequencing of a whole community of microorganisms without growing each species in culture. Fragments can be either sequenced or subjected to functional analysis. *»» What type of information is supplied by sequence-based metagenomics versus functional-based metagenomics?*

TABLE 8.3 Some Uses for Metagenomics

Challenge	Example
Medicine	Understanding how the microbial communities that inhabit our bodies affect human health could provide genome information that can be used to develop new strategies for diagnosing, treating, and preventing infectious diseases.
Ecology and the environment	Exploring how microbial communities in soil and the oceans affect the atmosphere and environmental conditions could help us understand, predict, and address climate change.
Energy	Harnessing the power of microbial communities might result in sustainable and ecofriendly bioenergy sources.
Bioremediation	Adding to the arsenal of microorganism-based environmental tools can help in monitoring environmental damage and cleaning up oil spills, groundwater, sewage, nuclear waste, and other biohazards.
Agriculture	Understanding the roles of beneficial microorganisms living in, on, and around domesticated plants and animals could enable detection of diseases in crops and livestock and aid in the development of improved farming practices.
Biodefense	The addition of metagenomics tools to microbial forensics will help to monitor pathogens, create more effective vaccines and therapeutics against bioterror agents, and reconstruct attacks that involve microorganisms.

Concept and Reasoning Checks 8.3

a. Why have so many bacterial species been sequenced?
b. Why do we say the human genome is a blending of vertebrate, prokaryotic, and viral genomes?
c. How is microbial genomics contributing to a better world?
d. Describe how comparative genomics helps explain some aspects of pathogenicity.
e. What does metagenomics offer that previous sequencing techniques have not been able to provide?

Chapter Challenge C

From your reading in this section, you know the human genome is a hybrid of human, prokaryotic, and viral genes. It is difficult to define what a human genome is. To many, the idea that the human genome could have evolved from viruses and other microbial genes seems somehow discomforting.

QUESTION C: *So what is your impression, reaction, or response to the following statement, which is based in the science you have learned from this text? In the human individual:*
1. There are more microbial cells in your body (human microbiome) than human cells.
2. There are 200× more microbial genes in the human microbiome than there are genes in the human genome.
3. The millions of mitochondria in all body cells have evolved from once free-living bacterial cells.
4. The human genome is littered with gene fragments from ancient viruses.

■ SUMMARY OF KEY CONCEPTS

Concept 8.1 Bacterial Cells Can Recombine Genes in Several Ways

1. Recombination implies a horizontal transfer of DNA fragments between bacterial cells and an acquisition of genes by the recipient cell. All three forms of recombination are characterized by the recombination of new genes to a recipient cell by **horizontal gene transfer**. (Figure 8.2)

2. In **transformation**, a competent recipient cell takes up "naked" DNA from the local environment. The new DNA displaces a segment of equivalent DNA in the recipient cell, and new genetic characteristics can be expressed. (Figure 8.3)

3. In one form of **conjugation**, a live donor (F$^+$) cell transfers an **F factor** (plasmid) to a recipient cell (F$^-$), which then becomes F$^+$. In another form of conjugation, **Hfr** strains contribute a portion of the donor's chromosomal genes to the recipient cell. (Figure 8.5)

4. **Transduction** involves a virus entering a bacterial cell and later replicating within it. The phage transports the DNA to a new recipient (transduced) cell. (Figure 8.7)

Concept 8.2 Genetic Engineering Involves the Deliberate Transfer of Genes Between Organisms

5. **Genetic engineering** is an outgrowth of studies in bacterial genetic recombination. The ability to construct **recombinant DNA molecules** was based on the ability of **restriction endonucleases** to form **sticky ends** on DNA fragments. (Figure 8.9)

6. **Plasmids** can be isolated from a bacterial cell, spliced with foreign genes, and inserted into fresh bacterial cells where the foreign genes are expressed as protein. The cells become biochemical factories for the synthesis of such proteins as insulin, interferon, and human growth hormone.

7. **DNA probes** can be used to detect pathogens. (Figure 8.11)

Concept 8.3 Microbial Genomics Studies Genes at the Single-Cell to Community Levels

8. Since 1995, increasing numbers of microbial genomes have been sequenced; that is, the linear sequence of bases has been identified. (Figure 8.13)

9. A comparison of bacterial genomes with the human genome has shown there might be some 200 genes in common between these organisms. Comparisons between microbial genomes indicate almost 50% of the identified genes have yet to be associated with a protein or function in the cell.

10. With the understanding of the relationships between sequenced microbial DNA molecules comes the potential for safer food production, the identification of uncultured microorganisms, a cleaner environment, and improved monitoring of pathogens through **microbial forensics**.

11. Sequencing is only the first step in understanding the behaviors and capabilities of microorganisms. **Functional genomics** attempts to determine the functions of the sequenced genes and identify how those genes interact with one another and with the environment. **Comparative genomics** compares the similarities and differences between microbial genome sequences. Such information provides an understanding of the evolutionary past and how pathogens might have arisen through the gain or loss of **pathogenicity islands**.

12. **Metagenomics** is providing new insights into the function of diverse genomes in microbial communities. (Figure 8.16)

■ CHAPTER SELF-TEST

For **Steps A–D**, you can find answers to questions and problems online in **Appendix D**.

■ STEP A: REVIEW OF FACTS AND TERMS

Multiple Choice

Read each question carefully before selecting the *one* answer that best fits the question or statement.

1. Which one of the following is *not* an example of genetic recombination?
   A. Conjugation
   B. Binary fission
   C. Transduction
   D. Transformation

2. Transformation refers to _____.
   A. using a virus to transfer DNA fragments
   B. DNA fragments transferred between live donor and recipient cells
   C. the formation of an F⁻ recombinant cell
   D. the transfer of "naked" DNA

3. An F⁻ cell is unable to initiate conjugation because it lacks _____.
   A. double-stranded DNA
   B. a prophage
   C. an F factor
   D. DNA polymerase

4. An Hfr cell _____.
   A. has a free F factor in the cytoplasm
   B. has a chromosomally integrated F factor
   C. contains a prophage for conjugation
   D. cannot conjugate with an F⁻ recombinant

5. A/An _____ is *not* associated with transduction.
   A. F plasmid
   B. phage
   C. donor
   D. recipient cell

6. Which complementary sequence would *not* be recognized by a restriction endonuclease?
   A. GAATTC
      CTTAAG
   B. AAGCTT
      TTCGAA
   C. GTCGAC
      CAGCTG
   D. AATTCC
      TTAAGG

7. A _____ seals sticky ends of recombinant DNA segments.
   A. DNA ligase
   B. restriction endonuclease
   C. protease
   D. RNA polymerase

8. _____ are single-stranded DNA molecules that can recognize and bind to a distinctive nucleotide sequence of a pathogen.
   A. Prophages
   B. Plasmids
   C. Cloning vectors
   D. DNA probes

9. The first completely sequenced genome from a free-living organism was from _____.
   A. humans
   B. *E. coli*
   C. *Haemophilus*
   D. *Bordetella*

10. Approximately what percentage of the human genome is composed of ERVs?
    A. 2%
    B. 10%
    C. 25%
    D. 50%

11. Microbial forensics attempts to _____.
    A. identify disease pathogens
    B. discover the source of a disease pathogen
    C. identify a pathogen used in a criminal act
    D. All the above (**A–C**) are correct.

12. Genomic islands are _____.
    A. gene sequences not part of the chromosomal genes
    B. adjacent gene sequences unique to one or a few strains in a species
    C. acquired by HGT
    D. Both **B** and **C** are correct.

13. A metagenome refers to _____.
    A. a large genome in an organism
    B. the collective genomes in many organisms
    C. the genome of a virus
    D. two identical genomes in different species

Fill-in

Use the following syllables to compose the term that answers each of the clues that follow. The number of letters in each term is indicated by the blank lines, and the number of syllables is shown by the number in parentheses. Each syllable is used only once.

ASE BAC CLE CON DO DROME EN GA GE HOR I I IN JU MIDS NOME NU O PAL PHAGE PLAS TAL TER TION ZON

14. Closed loops of DNA (2) __ __ __ __ __ __ __ __
15. Restriction recognition sequence (3) __ __ __ __ __ __ __ __ __ __
16. Transduction virus (5) __ __ __ __ __ __ __ __ __ __ __ __
17. Recombinant DNA enzyme (5) __ __ __ __ __ __ __ __ __ __ __ __
18. Type of HGT (4) __ __ __ __ __ __ __ __ __ __ __ __
19. Complete set of genes (2) __ __ __ __ __ __
20. Type of gene transfer (4) __ __ __ __ __ __ __ __ __ __ __

■ STEP B: CONCEPT REVIEW

21. Describe and assess the role of **transformation** as a genetic recombination mechanism. (**Key Concept 8.1**)
22. Explain how an F factor is transferred during **conjugation**. (**Key Concept 8.1**)
23. Differentiate between **genetic engineering** and **biotechnology**. (**Key Concept 8.2**)

24. Identify the role of **plasmids** and **restriction endonucleases** in the genetic engineering process. (**Key Concept 8.2**)
25. Assess the importance of **microbial genomics** in understanding the human genome. (**Key Concept 8.3**)
26. Identify how information from **metagenomics** can contribute to medicine, energy production, agriculture, and biodefense. (**Key Concept 8.3**)

■ STEP C: APPLICATIONS AND PROBLEM SOLVING

27. In 1976, an outbreak of pulmonary infections among participants at an American Legion convention in Philadelphia led to the identification of a new disease: Legionnaires' disease. The bacterial organism *Legionella pneumophila*, responsible for the disease, had never before been known to be pathogenic. From your knowledge of bacterial genetics, can you postulate how it might have acquired the ability to cause disease?

28. You are going to do a genetic engineering experiment, but the labels have fallen off the bottles containing the restriction endonucleases. One loose label says *Eco*RI and the other says *Pvu*I. How could you use the plasmid shown in MicroInquiry 8 to determine which bottle contains the *Pvu*I restriction enzyme?
29. As a research member of a genomics company, you are asked to take the lead on sequencing the genome of *L. pneumophila* (see Question 27). (a) Why is your company interested in sequencing this bacterial species, and (b) what possible applications are possible from knowing its DNA sequence?

■ STEP D: QUESTIONS FOR THOUGHT AND DISCUSSION

30. Which of the recombination processes (transformation, conjugation, or transduction) would most likely occur in the natural environment? What factors would encourage or discourage your choice from taking place?
31. Many bacterial cells can take up DNA via the transformation process. From an evolutionary perspective, what might have been the original advantage for the cells taking up naked DNA fragments from the extracellular environment?

32. The world is continually plagued by a broad series of influenza viruses that differ genetically from one another. For example, we have heard of Hong Kong flu, avian flu, and now swine flu. How might the process of transduction help explain this variability?
33. Students of microbiology often confuse the terms "reproduction" and "recombination." How do the terms differ?

34. While studying for the microbiology exam covering the material in this chapter, a friend who is a biology major asks you why genomics, and especially microbial genomics, was emphasized. How would you answer this question?

35. In 2011, researchers reported that they had discovered for the first time that bacterial cells can acquire human genes. They found that more than 10% of a *Neisseria gonorrhoeae* population, the species responsible for gonorrhea, contained a human DNA fragment. This fragment was not found in other species of *Neisseria*. How might a bacterial cell have acquired this human DNA fragment?

Concept Mapping

See pages XXXI-XXXII on how to construct a concept map.

36. Fill in the boxes in the concept map on **Horizontal Gene Transfer Types** using the Term List provided.

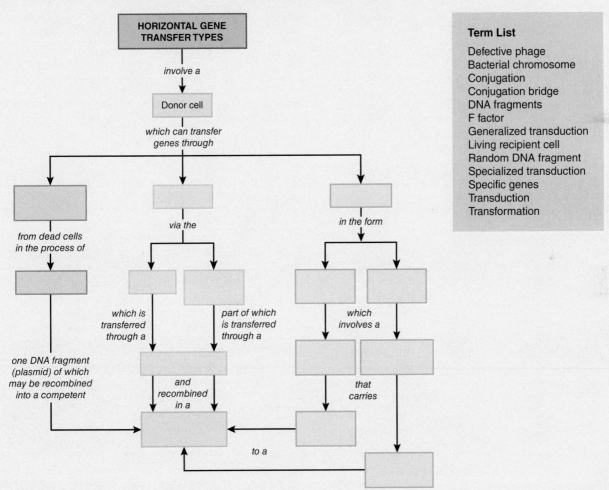

Term List

Defective phage
Bacterial chromosome
Conjugation
Conjugation bridge
DNA fragments
F factor
Generalized transduction
Living recipient cell
Random DNA fragment
Specialized transduction
Specific genes
Transduction
Transformation

PART III The Control of Microorganisms

CHAPTER 9 Control of Microorganisms:
Physical Methods and Chemical Agents

CHAPTER 10 Control of Microorganisms:
Antimicrobial Drugs and Superbugs

The two chapters in Part III embrace several aspects of the six principal (overarching) concepts as described in the ASM fundamental statements for a concept-based microbiology curriculum.

PRINCIPAL CONCEPTS

Evolution	• Human impact on the environment influences the evolution of microorganisms (e.g., emerging diseases and the selection of antibiotic resistance). **Chapter 10**
Cell Structure and Function	• Bacteria have unique cell structures that can be targets for antibiotics, immunity, and phage infection. **Chapter 10**
Metabolic Pathways	• The survival and growth of any microorganism in a given environment depends on its metabolic characteristics. **Chapters 9 and 10** • The growth of microorganisms can be controlled by physical, chemical, mechanical, or biological means. **Chapter 9**
Information Flow and Genetics	• Genetic variations can impact microbial functions (e.g., in biofilm formation, pathogenicity, and drug resistance). **Chapter 10**
Microbial Systems	• Microorganisms and their environment interact with and modify each other. **Chapter 10** • Microorganisms, cellular and viral, can interact with both human and nonhuman hosts in beneficial, neutral, or detrimental ways. **Chapter 10**
Impact of Microorganisms	• Humans utilize and harness microorganisms and their products. **Chapters 9 and 10**

CHAPTER 9

Control of Microorganisms: Physical Methods and Chemical Agents

For personal hygiene, washing our hands, taking regular showers or baths, brushing our teeth with fluoride toothpaste, and using an underarm deodorant are common practices we use to control microorganisms on our bodies. In our homes, we try to keep microbes in check by cooking and refrigerating foods, cleaning our kitchen counters and bathrooms with disinfectant chemicals, and washing our clothes with detergents.

In our attempt to be hygiene-minded consumers, sometimes we have become excessively "germophobic." The news media regularly report about this scientific study or that survey identifying places in our homes (e.g., toilets, kitchen drains) or environment (e.g., public bathrooms, drinking-water fountains) that represent infectious dangers. Consumer groups distribute pamphlets on "microbial awareness," and numerous companies have responded by manufacturing dozens of household antimicrobial products—some useful, but many unnecessary (see chapter opening photo).

Our desire to protect ourselves from microbes also stems from events beyond our doorstep. The news media again report about dangerous disease outbreaks, many of which result from a lack of sanitary controls or a lack of vigilance to maintain those controls. In our communities, we expect our drinking water to be clean. That goes for our hospitals, as well. Nowhere is this more important than in the operating rooms and surgical wards. Here, hospital personnel must maintain scrupulous levels of cleanliness and have surgical instruments that are sterile. Yet, **nosocomial** (hospital-acquired) and other **healthcare-associated infections** (**HAIs**) do occur when hygiene barriers are breached.

Microbial control also is a global endeavor. Government and health agencies in many developing nations often lack the means (financial, medical, social) to maintain proper sanitary conditions. Cholera, for example, tends to be associated with

Typical household cleaning products.

Courtesy of Dr. Jeffrey Pommerville.

poverty-stricken areas where overcrowding and inadequate sanitation practices generate contaminated water supplies (**FIGURE 9.1**).

We do need to be hygiene conscious because serious threats to health and well-being can occur if the procedures and methods to control pathogens fail or are not monitored properly. Yet, the successful control of microorganisms usually requires only simple methods and procedures. Importantly, proper hygiene and sanitation have extended our lives and have afforded us the opportunity to contribute more fully to society.

In this chapter, our study of the methods and agents commonly used to control the growth and spread of microorganisms in the environment begins by outlining some general principles and terminology. Then, physical methods of microbial control that are commonly used today will be described. The chapter also explores chemical methods for microbial control and discusses the spectrum of antiseptics and disinfectants. The use of antimicrobial chemicals, such as antibiotics, to control or eliminate pathogens within the human body will be discussed in Chapter 10.

FIGURE 9.1 Controlling Microorganisms. This shanky-town in the Phillipines has slum houses, open sewage, and littered walkways. It is not surprising that in these unsanitary conditions diseases such as cholera and typhoid are common. *»» In these examples, does controlling microorganisms mean sterilization or simply reducing the number of microbes to a safe level? Explain.*

© Flat Earth/FotoSearch.

Chapter Challenge

Most households have stocked up with various and diverse types of household cleaning agents. In addition, every day we almost unconsciously use physical methods and chemical agents to ensure that we have a microbially safe environment, both in terms of the foods we eat and the home we live in. So, just how many everyday "activities" are involved in controlling potential pathogens in our homes or daily work? Let's identify the ways as we consider the physical methods and chemical agents in this chapter.

■ KEY CONCEPT 9.1 Microbial Growth Can Be Controlled in Several Ways

The effective control of disease-causing microbes is important not only in hospitals and in healthcare facilities but also in the food industry, in restaurants, and in our own homes. In all these environments, control requires an understanding of the physical and chemical methods used to limit microbial growth and/or microbial transmission.

Sterilization and Sanitization Are Key to Good Public Health

Before we examine the types of physical and chemical methods employed in microbial control, we need to establish some basic vocabulary used by health officials—and sometimes misused by the public—when talking about microbial control in the environment.

Sterilization

In microbiology, **sterilization** involves the destruction or removal of all living microbes, spores, and viruses on an object or in an area. For example, in a surgical operation, the surgeon uses "sterile" instruments previously treated in some way to kill any microbes present on them (**FIGURE 9.2**). Agents that kill microbes are **microbicidal** (*-cide* = "kill") or more simply called "germicides." If the agent specifically kills bacterial cells, it is **bactericidal**; if it kills fungi, it is **fungicidal**. However, as soon as the sterile material or object is exposed to the air and the surrounding environment, the material or object will again become contaminated with microbes in the air or the surrounding area.

A slightly different meaning of sterilization is used in the food industry. Here, sterilization can

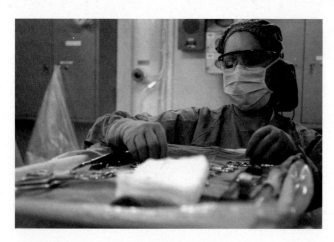

FIGURE 9.2 Sterile Surgical Instruments. Here a surgical nurse is unwrapping sterile surgical instruments. »» *How would you sterilize these instruments?*

Courtesy of Journalist 2nd Class Shane Tuc/U.S. Navy.

simply refer to the elimination of human pathogens. For example, during the canning process for vegetables and fruits, companies carry out **commercial sterilization**, which kills all the pathogens in the product. Some nonpathogenic microbial species might still be present.

Sanitization

More often, in our daily experiences we are likely to encounter materials for which microbial populations have been reduced or their growth has been inhibited.

Sanitization involves those procedures that reduce the numbers of pathogenic microbes to a level considered safe by public health standards. The process does not sterilize and, given enough time, microbes and potential pathogens will settle and grow on the sanitized object. Thus, public health depends on good sanitary practices at home and in the workplace.

Restaurants use physical and chemical procedures to sanitize cooking utensils and work surfaces. A toilet in a public restroom might have been "sanitized for your protection"; however, in a very short time, these objects can again be contaminated with microbes and pathogens from the environment or from another person. Many sanitizing agents, such as the disinfectants and antiseptics, are said to be **microbiostatic** (-*static* = "remain in place"); that

is, they reduce microbial numbers or inhibit their growth. Again, specific agents can be **bacteriostatic** or **fungistatic**.

Physical Methods and Chemical Agents Control Microbes in One or Two Ways

The physical methods and chemical agents used to control microbial growth usually exert their effect by either targeting specific cellular structures or interfering with metabolic activities. The bacterial cell envelope and microbial proteins and nucleic acids are prime targets.

The Bacterial Cell Envelope

Some control methods kill or limit the growth of bacterial cells by targeting the **cell envelope**, that is, the cell wall and cell membrane. A variety of chemical agents can damage or break the structure of the cell wall. Lacking the structural integrity of the wall, osmotic lysis will occur and the cell will burst.

The cell membrane, being composed of phospholipids and proteins, is very susceptible to physical or chemical methods that disrupt membrane structure or interfere with its function as a selectively permeable barrier. This might involve dissolving of the phospholipid bilayer or inactivating (denaturing) membrane proteins. The result is loss of permeability control, which again will lead to cell death.

Proteins and DNA

Proteins are the workhorses in all cells, both as structural components and functional factors. Recall that when heat, pH, or salt conditions vary from their normal physiological range, the three-dimensional structure of proteins is altered, and they unfold from their active state. This **denaturation** of a protein, such as an enzyme, can inactivate the protein, inhibiting cell function and perhaps leading to cell death.

Like proteins, nucleic acids are subject to structural and functional changes if environmental conditions become extreme. Several physical and chemical methods will disrupt DNA and RNA structure, interfere with cell growth and division, or prevent gene expression. Any permanent inactivation of the nucleic acids will result in death of the cell.

Concept and Reasoning Checks 9.1

a. Explain the difference between maintaining sterility and maintaining sanitary conditions.
b. Discuss two ways that physical and chemical agents affect the viability and growth of microbes.

KEY CONCEPT 9.2   There Are Various Physical Methods to Control Microbial Growth

We will now examine several physical methods, some of which have been used for centuries to kill microorganisms or inhibit their growth. In hospitals, healthcare facilities, and laboratories, these methods generally include temperature, filtration, radiation, and osmotic pressure. As we go through the methods, note the significant differences between killing microorganisms, reducing their numbers, and holding them in check.

Elevated Temperature Is a Common Physical Control Method

The killing effect of elevated temperatures on microorganisms has long been known. Heat is fast, reliable, and relatively inexpensive. All microbes exhibit a temperature range over which they grow. Above this range, proteins as well as nucleic acids in microbes can become denatured. Because heat also drives off water, and all organisms depend on liquid water for metabolism, this loss of water can lead to cell death.

The effectiveness of heat for sterilization is a function of time and temperature. For example, bacilli of *Mycobacterium tuberculosis* are destroyed in 30 minutes at 58°C but in only 2 minutes at 65°C and in a few seconds at 72°C. Thus, each microbial species has a **thermal death time**, which is the length of time needed to kill a microbial population at a specified temperature. Each species also has a **thermal death point**, which is the lowest temperature that will kill all the microbes in 10 minutes. These measurements are particularly important in the food industry, where heat is used for preservation. **MICROINQUIRY 9** examines how microbial death can be determined.

Dry Heat Has Useful Applications

The effect of dry heat on microorganisms is equivalent to that of baking. The heat denatures cell proteins and creates an arid internal environment, thereby burning microorganisms slowly. It is essential that organic matter such as oil or grease films be removed from the materials because such substances insulate against heat penetration.

Incineration

Using a direct flame can incinerate (burn to ashes) microbes very rapidly. For example, a few seconds in the heat of a lab Bunsen burner or lab bench incinerator is sufficient to sterilize the bacteriological loop before removing a sample from a culture tube (**FIGURE 9.3**).

In a healthcare setting, disposable hospital gowns, dressings, and certain plastic apparatus can be incinerated. In past centuries, the bodies of disease victims often were burned to prevent spread of the black plague. Today, it still is common practice to incinerate the carcasses of diseased cattle. The 2001 outbreak of foot-and-mouth disease in cattle in Great Britain required the mass incineration of thousands of cattle as a means to stop the spread of the disease (**FIGURE 9.4**).

Hot Air

The hot-air oven uses dry heat for sterilization. This type of energy does not penetrate materials easily; therefore, long periods of exposure to high temperatures are necessary. For example, at a temperature of 160°C, a period of 2 hours is required to sterilize most objects. The hot-air method is useful for sterilizing dry powders and water-free oily substances as well as for many types of glassware such as pipettes, flasks, and syringes.

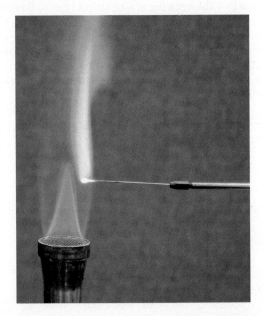

FIGURE 9.3 Use of the Direct Flame as a Sterilizing Agent. A few seconds in the flame of a laboratory Bunsen or Fisher burner is usually sufficient to effect sterilization of an inoculation loop. »» *Why is it necessary to flame an inoculation loop before using?*

Courtesy of Dr. Jeffrey Pommerville.

MICROINQUIRY 9

Microbial Death

Sterilization is a procedure that can be used to kill all life in a specific environment. Because the living and dead microbes often have the same morphology when observed with the microscope, how does one know all the microbes are dead?

Killing microbes in the environment depends on several factors. Identifying the type(s) of microbes in a product can determine whether the heating process will sterilize or eliminate only potential disease-causing species. In many foods, the microbes usually are not in water but rather in the food material. Microbes in powders or dry materials will require a different length of time to sterilize the product than microbes in organic matter.

Environmental conditions also influence the sterilization time. Microbes in acidic or alkaline materials decrease sterilization times, whereas microbes in fats and oils, which slow heat penetration, increase sterilization times. It must be remembered that sterilization times are not precise values. However, by knowing these factors, heating the product to temperatures above the maximal range for microbial growth will kill microbes rapidly and effectively.

As with cell growth, microbial death occurs in an exponential fashion. Look at the accompanying table that records the death of a microbial population by heating. Notice that the cells die at a constant rate. In this generalized example, during each minute of heating, 90% of the cells die (10% survive). Therefore, if you know the initial number of microorganisms, you can predict the **thermal death time** (TDT), which is defined as the length of time, at a specified temperature, required to kill a population of microorganisms.

The graph drawn below represents two "death curves," each representing a different bacterial species treated at the same temperature (60°C) in the same environment (curve B represents the plotted data from the table).

9a. If a food product contained 10^6 bacteria of each species, what is the TDT for each bacterial species? (Hint: $10° = 1$)

Another value of importance is the **decimal reduction time** (**DRT**) or **D value**, which is the time required at a specific temperature to kill 90% of the viable organisms. D values are usually identified by the temperature used for killing. In the graph below, the temperature was 60°C, so D is written as D60. Look at the graph again.

9b. Calculate the D60 values for the two bacterial populations depicted in curves A and B.

On another day in the canning plant, you need to sterilize a food product. However, the only information you have is a D70 = 12 minutes for the microorganism in the food product. Assuming the D value is for the volume you have to sterilize and there are 10^8 bacteria in the food product:

9c. At what temperature will you treat the food product?

9d. How long will it take to sterilize the product?

You can find answers online in **Appendix E**.

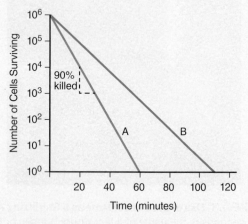

Death curves for two microbial species (A and B).

TABLE	Microbial Death Rate		
Time (minutes)	Number of Cells Surviving	% Killed
0	1,000,000	—
10	100,000	90
20	10,000	90
30	1,000	90
40	100	90
50	10	90
60	1	90

FIGURE 9.4 Incineration Is an Extreme Form of Dry Heat. Nearly 4 million hoofed animals infected with or exposed to foot-and-mouth disease in England in 2001 were incinerated and buried. *»» How is incineration similar to flame sterilization?*

© Simon Fraser/Science Source.

Moist Heat Is More Versatile Than Dry Heat

Moist heat kills the vegetative (actively metabolizing) forms of most microorganisms. There are several ways that moist heat is used to control microbes and sterilize materials.

Boiling Water

Moist heat, in the form of boiling water and steam, penetrates materials much more rapidly than dry heat because water molecules conduct heat better than air. Therefore, moist heat can be used at a lower temperature and shorter exposure time than for dry heat.

Under ordinary circumstances, with microorganisms at concentrations of less than 1 million per milliliter, most microbial species and viruses can be killed within 10 minutes. Indeed, the process might require only a few seconds. However, fungal spores, protistan cysts, and large concentrations of some viruses require a 30-minute exposure, and bacterial endospores often require many hours (**FIGURE 9.5**).

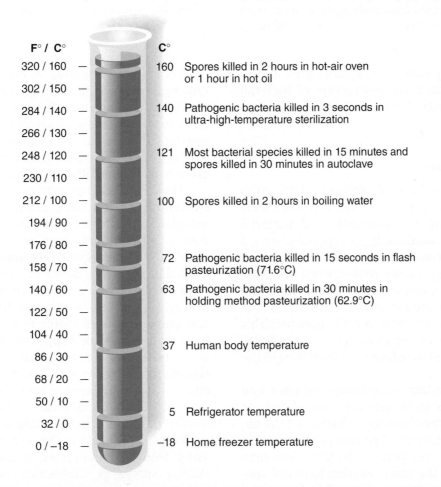

F° / C°	C°	
320 / 160	160	Spores killed in 2 hours in hot-air oven or 1 hour in hot oil
302 / 150		
284 / 140	140	Pathogenic bacteria killed in 3 seconds in ultra-high-temperature sterilization
266 / 130		
248 / 120	121	Most bacterial species killed in 15 minutes and spores killed in 30 minutes in autoclave
230 / 110		
212 / 100	100	Spores killed in 2 hours in boiling water
194 / 90		
176 / 80		
158 / 70	72	Pathogenic bacteria killed in 15 seconds in flash pasteurization (71.6°C)
140 / 60	63	Pathogenic bacteria killed in 30 minutes in holding method pasteurization (62.9°C)
122 / 50		
104 / 40		
86 / 30	37	Human body temperature
68 / 20		
50 / 10		
32 / 0	5	Refrigerator temperature
0 / −18	−18	Home freezer temperature

FIGURE 9.5 Temperature and the Physical Control of Microorganisms. Notice that materials containing bacterial endospores require longer exposure times and higher temperatures for killing. *»» Pure water boils and freezes at what temperatures on the Celsius scale?*

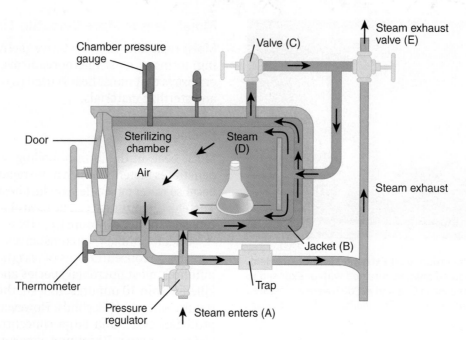

FIGURE 9.6 Operating an Autoclave. Steam enters through the port **(A)** through **(E)** and passes into the jacket **(B)**. After the air has been exhausted through the vent, a valve **(C)** opens to admit pressurized steam **(D)**, which circulates among and through the materials, thereby sterilizing them. At the conclusion of the cycle, steam is exhausted through the steam exhaust valve **(E)**. *»» Is it the steam or the pressure that kills microorganisms in an autoclave? Explain.*

Therefore, boiling water is not considered a sterilizing agent because the destruction of bacterial spores and the inactivation of viruses cannot always be assured.

Pressurized Steam

Moist heat in the form of pressurized steam is regarded as the most dependable method for sterilization, including the destruction of bacterial endospores. This method is incorporated into a machine called the **autoclave** (**FIGURE 9.6**). Working similar to a home pressure cooker, when the pressure of the steam increases, the temperature of the gas also increases proportionally. Importantly, the sterilizing agent is the heat, not the pressure. This principle is used to reduce cooking time in the home pressure cooker and to reduce sterilizing time in the autoclave.

Autoclaves contain a sterilizing chamber into which the objects to be sterilized are placed. As steam flows into the chamber, it forces out the air, increases the pressure to 15 pounds per square inch (psi), and raises the temperature to 121°C. The time for destruction of the most resistant bacterial species is about 15 minutes (see Figure 9.5). For denser objects or large volumes of liquids, more than 30 minutes of exposure might be required.

The autoclave is used in hospitals to sterilize materials that can tolerate high temperatures. Thus, blankets, bedding, utensils, instruments, intravenous solutions, and a broad variety of other heat-resistant objects can be sterilized. In the laboratory, it is used to sterilize glassware and metalware, growth media, and other solutions.

The autoclave has certain limitations. Plastic culture dishes and other plasticware will melt at autoclave temperatures. Moreover, many chemicals, such as antibiotic solutions, break down during the sterilization process, and oily substances cannot be treated because they do not mix with water.

To make certain an object or solution has been properly sterilized, autoclaves have temperature and pressure gauges visible from the outside, and most models produce a paper record of the temperature, time, and pressure. Objects often are autoclaved with autoclave tape, which turns color if the object in the autoclave has been sterilized properly. Lack of quality control can result in improperly sterilized solutions, as **CLINICAL CASE 9** illustrates.

Proper sterilization also can be monitored by using biological indicators. A nutrient vial containing spores of *Geobacillus stearothermophilus* can be included in the autoclave with the objects treated. At the conclusion of the sterilization cycle, the vial with the spores is incubated at 37°C. If the

Clinical Case 9

An Outbreak of Endophthalmitis

Three cataract patients at a local hospital in Thailand underwent cataract extraction. Within 30 hours after surgery, the three patients developed postoperative endophthalmitis, an eye inflammation affecting the ocular cavity and the adjacent structures.

The clinical microbiology lab isolated a pathogenic strain of the gram-negative bacterial species *Pseudomonas aeruginosa* from the intraocular fluid taken from two of the patients. The infected patients were treated with antibiotics. However, due to antibiotic resistance, antibiotic therapy failed, and the eyes of all three patients had to be eviscerated.

An epidemiological investigation reviewed all hospital records of all patients who had cataract surgery. No other cataract patients with postoperative endophthalmitis were identified. Further investigation identified bottles of basal salt solution (BSS) contaminated with *P. aeruginosa*. The BSS had been prepared by hospital pharmacists for use in the hospital operating rooms. To sterilize the solutions, the bottles were placed in the autoclave and left to run on its automatic cycle (see figure). The bottles were then delivered to surgery to be used to irrigate the eyes of patients undergoing cataract surgery. Some bottles were left unused.

Health investigators tested the unused bottles of BSS as well as the tubes attached to the now empty bottles. They found the identical strain of *P. aeruginosa*. Examination of the pharmacy records indicated that the bottles had been cleaned and correctly placed under an ultraviolet light before being filled with BSS and autoclaved.

However, investigators noted that the autoclave pressure had reached only 10 to 12 psi, suggesting that contaminants might have survived in the bottles and solution. Strict quality control measures were instituted.

An autoclave can be used to sterilize liquid materials.

Huntstock, Inc/Alamy.

Questions:

a. How did the cataract patients become infected with *P. aeruginosa*?

b. Because no other cases of postoperative endophthalmitis were identified, what does that say about the source of the infection in the affected patients?

c. Propose a way that the BSS bottles became contaminated.

d. What quality control measures should have been instituted?

You can find answers online in **Appendix E**.

For additional information see www.cdc.gov/mmwr/preview/mmwrhtml/00042645.htm.

---

sterilization process has been unsuccessful, the spores in the vial will germinate and grow. Their metabolism will change the color of a pH indicator in the vial's nutrient medium (**FIGURE 9.7**).

Pasteurization

The final example of moist heat involves the process of **pasteurization**, which was first introduced by Louis Pasteur in the 1860s to destroy microbes contaminating beer and wine. Today, pasteurization is used to reduce the bacterial population in liquids, such as milk, and to destroy organisms that can cause spoilage and human disease (**FIGURE 9.8**). For decades, pasteurization has been aimed at destroying *M. tuberculosis*, long considered the most heat-resistant bacterial species. More recently, attention has shifted to destruction of *Coxiella burnetii*, the agent of Q fever, because this bacterial organism can infect animal milk, and it has a higher resistance to heat. Because both organisms

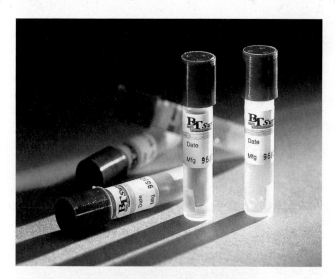

FIGURE 9.7 Monitoring the Effectiveness of Autoclave Sterilization. Vials containing spores of *Geobacillus stearothermophilus* are suspended in a growth medium containing a pH indicator. Following autoclaving, the vial is incubated at 35°C. If the autoclave run was effective, all spores will be killed and the solution will remain purple. If the spores were not killed (an ineffective sterilization process), they will germinate and grow. The acid the cells produce will turn the pH indicator yellow. »» *What do you know about this bacterial species from its scientific name?*

Courtesy of Thermo Fisher Scientific, Inc.

FIGURE 9.8 The Pasteurization of Milk. Milk is pasteurized by passing the liquid through a heat exchanger. The flow rate and temperature are monitored carefully. Following heating, the liquid is rapidly cooled. »» *Why is the liquid rapidly cooled?*

© Apply Pictures/Alamy Images.

are eliminated by pasteurization, dairy microbiologists assume that other pathogenic bacteria also are destroyed. Pasteurization also is used to eliminate any *Salmonella* and *Escherichia coli* that could contaminate fruit juices. Importantly, pasteurization does not kill all the microbes in milk. In fact, there can be up to 20,000 bacterial cells per milliliter of milk after pasteurization. The heating process also does not affect endospores.

One method for milk pasteurization, called the **holding** (or **batch**) **method**, involves heating the liquid at 63°C for 30 minutes. Today, two other methods are more commonly used. **Flash pasteurization** holds the liquid at 72°C for 15 seconds, a combination of heat and time that eliminates all pathogens. The third method, **ultra-high-temperature (UHT) sterilization** (also called ultra-pasteurization) exposes the liquid to 140°C for 3 seconds. Thus, UHT is the only method that can sterilize the liquid if done under aseptic conditions. Consequently, the products of UHT can be stored for long periods at room temperature—they can have a shelf life of six months or longer.

Refrigeration and Freezing Are Physical Control Methods That Slow Microbial Growth

In food preparation and storage, cold temperatures can be used to preserve foods by slowing microbial growth. The household refrigerator should maintain a temperature near 5°C, which will retard spoilage by lowering the metabolic rate of microbes and substantially reducing their rate of growth. Spoilage is not eliminated in refrigerated foods, however, because there are **psychrophiles**, microbes (e.g., bacteria and molds) that prefer colder temperatures that could contaminate the food material.

Freezing foods (below 0°C) also inhibits microbial metabolism and growth as the ice crystals formed can penetrate and destroy some microbial cells. Still, many bacterial pathogens can withstand subfreezing temperatures for long periods. Importantly, these organisms multiply rapidly when food thaws, which is why prompt cooking to the same degree of doneness as fresh food is recommended.

Although temperature is a valuable physical agent for controlling microorganisms, other heat-free methods also exist.

Filtration Traps Microorganisms

Filtration is a mechanical method used to remove microorganisms suspended in liquids or gases.

Several types of filters are used in microbiology laboratories and healthcare facilities. The most common is the **membrane filter**, which consists of a thin, circular pad of nitrocellulose or polycarbonate. The filters are available with pore sizes ranging from 25 µm to 0.025 µm (25 nm). When mounted in a filtering device, the unsterile fluid passes through the sterile filter, trapping organisms on the membrane because they are too large to pass through the pores (**FIGURE 9.9A, B**). The solution dripping through the filter into the receiving container is "decontaminated" or is sterile if the receiving container also is sterile. Membrane filters are used to purify such heat-sensitive liquids as beverages, some growth media, antibiotics, many pharmaceuticals, and blood solutions.

The membrane filter is particularly valuable because bacterial cells trapped on the filter will multiply and form colonies if the filter pad is placed on an agar culture plate. Microbiologists then can count the colonies growing from the bacterial cells to determine the number of cells originally present (**FIGURE 9.9C**).

Air also can be filter-purified to remove microorganisms. This might be something as simple as a surgical mask, which is worn to prevent the transmission of potential pathogens to a patient or other individual. On a larger scale is a **biological safety cabinet**, which contains a **high-efficiency particulate air (HEPA) filter** to trap particles, microorganisms, and spores (**FIGURE 9.10**). HEPA filters trap more than 99% of all particles, including microorganisms and spores with a diameter larger than 0.3 µm from escaping as aerosols into the environment. The air entering or exiting many hospital wards, surgical units, and specialized treatment facilities, such as burn units, also is HEPA filtered to exclude microorganisms and ensure air purity.

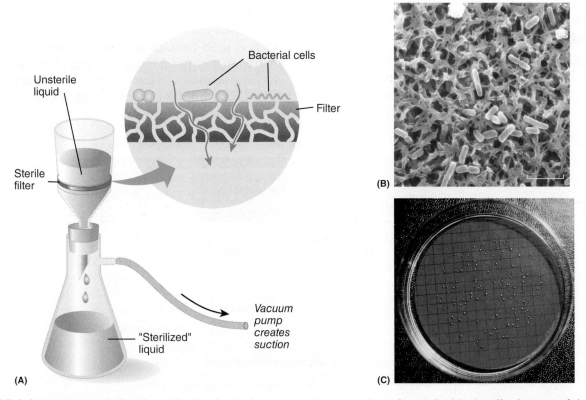

FIGURE 9.9 The Basics of Filtration. Filtration is used to remove microorganisms from a liquid. The effectiveness of the filter is proportional to the size of its pores. **(A)** A bacteria-laden liquid is poured into a filter, and a vacuum pump helps pull the liquid through and into the flask below. The bacterial cells are larger than the pores of the filter, and they become trapped. **(B)** A scanning electron micrograph of *Escherichia coli* cells trapped in the pores of a 0.45-µm membrane filter. (Bar = 5 µm.) **(C)** *E. coli* colonies growing on a membrane filter. *»» Why would most viruses not be trapped on a filter with 0.45-µm pores?*

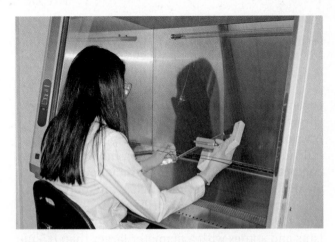

FIGURE 9.10 A Biological Safety Cabinet. The cabinet shown has a metal grid at the top that covers a HEPA filter through which air enters the cabinet. As the filtered air, free of contaminants and microbes, moves into and across the workspace, it exits out the bottom front and rear. A UV light is also positioned at the top rear to decontaminate the metal surfaces maintaining a contaminant-free workspace when the cabinet is not in use. *»» Why is air moved out of the cabinet rather than into the cabinet?*

Courtesy of Dr. Jeffrey Pommerville.

Irradiation Can Be Used to Control Microbial Growth

Radiation (or radiant energy) is the energy emitted in the form of rays or waves from an atomic source. For example, visible light is a type of radiant energy detected by the light-sensitive cells of the human eye. The wavelength of this light energy is between 400 and 750 nm. Other types of radiation have wavelengths longer and shorter than that of visible light (**FIGURE 9.11**). Two types of radiation are especially important in the control of microbes.

Ultraviolet Radiation

One type of radiation that is useful for controlling microorganisms is **ultraviolet (UV) light radiation (UVR)**. UVR is composed of UVA, UVB, and UVC types, which have a wavelength range between 100 and 400 nm. UVA is the most abundant type reaching Earth's surface and is the most damaging to living organisms. When microorganisms are subjected to UVR, cellular DNA absorbs the energy, and adjacent thymine molecules (in the same DNA strand) link together, kinking and distorting the double helix. If not repaired, the damage will ultimately cause mutations, disrupting DNA replication and transcription such that the cell or organism can no longer produce critical proteins and reproduce.

UVR has poor penetrating properties. Therefore, it is best used for disinfecting and sterilizing environmental surfaces where there is direct exposure. UVR is used to limit airborne or surface contamination in a hospital room, morgue, pharmacy, toilet facility, or food service operation. It is noteworthy that UVR from the sun can be an important factor

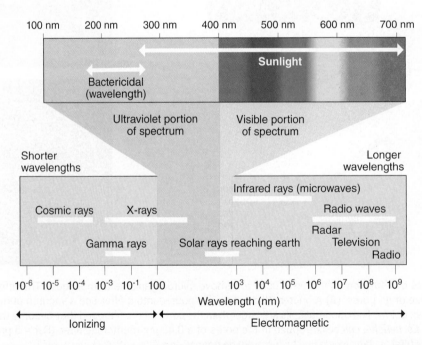

FIGURE 9.11 The Ionizing and Electromagnetic Spectrum of Energies. The complete spectrum is presented at the bottom of the chart, and the ultraviolet and visible regions are expanded at the top. *»» How does UV radiation kill bacterial cells?*

in controlling microorganisms in the air and at the soil surface.

Ionizing Radiation

There are two other forms of radiation useful for destroying microorganisms (see Figure 9.11). These are **X-rays** and **gamma rays**. Both have wavelengths shorter than the wavelength of UVR. Because X-rays and gamma rays pass through microbial molecules, they force electrons out of their shells, thereby creating ions. For this reason, the radiations are called **ionizing radiations**. The ions formed quickly combine, mainly with cellular water, and the free radicals generated affect cell metabolism and physiology and often cause mutations. Ionizing radiations currently are used to sterilize such heat-sensitive pharmaceuticals as vitamins, hormones, vaccines, and antibiotics as well as certain plastics and suture materials.

Irradiation represents the process by which an object is exposed to radiation. Ionizing radiations, in particular, have been approved for controlling microorganisms and for preserving foods, as noted in MICROFOCUS 9.1. The approval has generated some controversy, fueled by individuals concerned with the safety of factory workers and consumers. First used in 1921 to inactivate *Trichinella spiralis*, the agent of trichinellosis, irradiation now is used as a food preservation method in more than 40 countries for more than 100 food items, including potatoes, onions, cereals, flour, fresh fruit, and poultry (FIGURE 9.12A). The U.S. Food and Drug Administration (FDA) has approved cobalt-60 and cesium-137 irradiation to preserve or extend the shelf life of several foods. This includes irradiating beef, lamb, and pork products. In 2008, the FDA approved the irradiation of fresh and bagged spinach, and

MICROFOCUS 9.1: Public Health

"No, the Food Does Not Glow!"

In the United States, there are more than 76 million cases of foodborne disease accounting for more than 300,000 hospitalizations and 5,000 deaths each year. One major source of foodborne disease is agricultural produce contaminated with intestinal pathogens. Another source is from improperly cooked or handled meats or poultry harboring human intestinal pathogens, such as *E. coli* O157:H7, *Campylobacter*, *Listeria*, and *Salmonella*.

Irradiation has the potential to limit such illnesses. The Centers for Disease and Control (CDC) has estimated that if just 50% of the meat and poultry sold in the United States were to be irradiated, there would be 900,000 fewer cases of foodborne illness and 350 fewer deaths each year.

Yet, today, manufacturers continue to wrestle with the concept of food irradiation as they constantly confront a cautious public when dealing with radiation. In the United States, just 10% of the herbs and spices are irradiated and only 0.002% of fruits, vegetables, meats, and poultry are irradiated.

Food irradiation is entirely different from atomic radiation. The irradiation comes from gamma rays produced during the natural decay of cobalt-60 or cesium-137. The most common method involves electron beams (e-beams) not unlike those used in an electron microscope. None of these types of radiations produces radioactivity—the irradiated food does not glow (see figure).

Irradiated strawberries displaying the "radura" symbol, meaning they have been treated by irradiation.

© Jones & Bartlett Learning. Photographed by Kimberly Potvin.

Low doses of irradiation are used for disinfestations and extending the shelf life of packaged foods. As mentioned in the chapter narrative, a pasteurizing dose is used on meats, poultry, and other foods. Such levels do not eliminate all microbes in the food but, similar to pasteurization, help to reduce the dangers of pathogen-contaminated or cross-contaminated meats and poultry.

During the irradiation, the gamma rays or electrons penetrate the food and, just as in cooking, cause molecular changes in any contaminating microorganisms, which ultimately leads to their death.

Irradiation of foods also has its limitations. The irradiation dose will not kill bacterial endospores, inactivate viruses, or neutralize toxins. Therefore, irradiated food still must be treated in a sanitary fashion. Nutritional losses are similar to those occurring in cooking and/or freezing. Otherwise, there are virtually no known changes in the food, and there is no residue.

(A)   (B)

FIGURE 9.12 Food Irradiation. (A) The FDA has approved irradiation as a preservation method for some foods, including many fruits and vegetables, as well as poultry and red meats. **(B)** Many otherwise perishable foods eaten by NASA astronauts were prepared by irradiation. *»» Does irradiation sterilize the treated product? Explain.*

(A) © Jones & Bartlett Learning. Photographed by Kimberly Potvin. (B) Courtesy of U.S. Army Natick Soldier Center.

iceberg lettuce, to reduce potential foodborne illness. Irradiation has been used to prepare many meals for the U.S. military as well as the National Aeronautics and Space Administration (NASA) (**FIGURE 9.12B**). What is called a **pasteurizing dose** is used on meats, poultry, and other foods. Similar to heat pasteurization, the irradiation levels are not intended to eliminate all microbes in the food but rather just the pathogens. Therefore, the foods are not necessarily sterile.

Other Preservation Methods Retard Food Spoilage

Over the course of many centuries, other physical methods have been developed to retard spoilage and prolong the shelf life of foods.

Lyophilization

Another physical preservation method is **lyophilization**, more commonly called **freeze drying**. As the term suggests, this method combines both freezing and drying of the food product or substance. In this process, food is quickly frozen with liquid nitrogen (−80°C) to prevent ice crystal formation, and then a vacuum pump draws off the water. Under these conditions, the water passes directly from a solid ice phase to water vapor without passing through its liquid phase. The dry product is sealed in foil and easily reconstituted with water. Hikers and campers

find considerable value in freeze-dried food because of its light weight and durability. Lyophilization also is a useful method for storing, transporting, and preserving bacterial cultures.

Drying and Salting

Drying (**desiccation**) is useful in the preservation of various meats, fish, cereals, and other foods. Because water is necessary for life, it follows that where there is little or no water, there is virtually no life. Many nonperishable foods (such as cereals, rice, and sugar) in the kitchen pantry represent such shelf-stable products.

Preservation by salting has been used for centuries to preserve foods. The method is based on the principle of **osmotic pressure**; that is, when food is salted (usually with sodium chloride), water diffuses out of the food and out of any microorganisms associated with the food, toward the higher salt concentration and lower water concentration in the surrounding environment. This flow of water, called **osmosis**, leaves the food and microorganisms desiccated. In the case of the microbes, they might die. The same phenomenon occurs with highly sugared foods, such as syrups, jams, and jellies. However, fungal contamination (molds and yeasts) will eventually occur because the microbes can tolerate low water and high sugar concentrations.

TABLE 9.1 and **FIGURE 9.13** summarize the physical agents used for controlling microorganisms.

TABLE 9.1 A Summary of Physical Agents Used to Control Microorganisms

Physical Method	Conditions	Instrument	Examples of Uses	Comment
Incineration	A few seconds	Flame	Laboratory instruments	Object must be disposable or heat resistant
Hot air	160°C for 2 hr	Oven	Glassware Powders Oily substances	Not useful for fluid materials
Boiling water	100°C for 10 min	—	Wide variety of objects	Total immersion and precleaning necessary
	100°C for 2 hr+	—	Spores	
Pasteurization	Holding method Flash method UHT method	Pasteurizer	Dairy products Beverages	Sterilization achieved with UHT under aseptic conditions
Refrigeration/ freezing	5°C/−10°C	Refrigerator/Freezer	Numerous foods	Spoilage/food preservation
Filtration	Entrapment on membrane	Membrane filter HEPA filter	Fluids Air	Many adaptations
Ultraviolet light	265 nm energy	Generator	Surface and air sterilization	Not useful in fluids
X-rays Gamma rays	Short wavelength energy	Generator	Heat-sensitive materials	Extending food shelf life
Lyophilization (freeze drying)	Rapid freezing and drying	Freeze dryer	Numerous foods	Food preservation
Desiccation	Osmotic conditions	—	Salted and sugared foods	Food preservation

Concept and Reasoning Checks 9.2

a. How does the thermal death time differ from the thermal death point?
b. Explain how dry heat can be used to "eliminate" microorganisms.
c. Summarize the ways that moist heat controls or sterilizes materials or beverages.
d. Identify the types and uses for filtration in a healthcare setting.
e. Identify some uses for ultraviolet light radiation as a physical control method.
f. Identify some uses for ionizing radiations as a physical control method.
g. Explain how lyophilization and salting are similar as preservation methods.

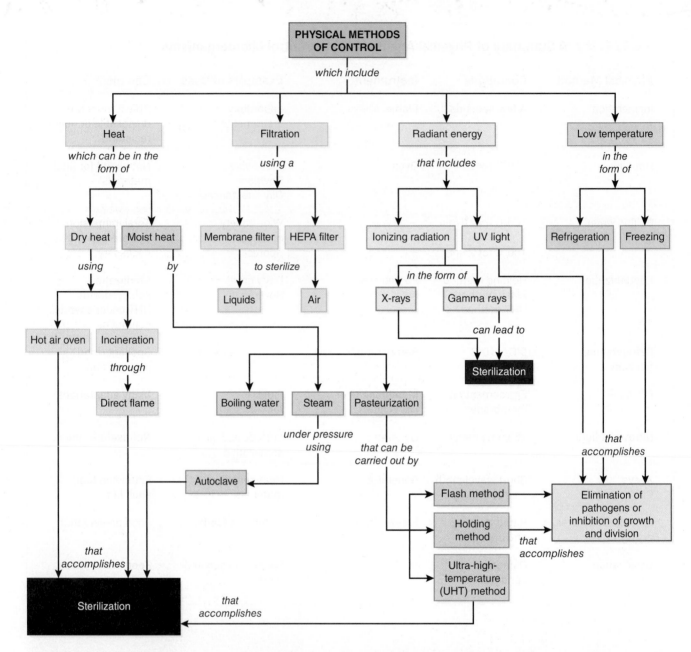

FIGURE 9.13 A Concept Map Summarizing the Physical Methods of Microbial Control. Note that some methods sterilize while others tend to inhibit growth and division. »» *What is common to most of the sterilization methods?*

Chapter Challenge A

So far, we have examined the physical methods used to control microbes. In most cases, we are primarily concerned about controlling and eliminating any potential pathogens.

QUESTION A: *Thinking about the physical methods, which of the following do you use at home or in the workplace (or in products you have purchased, such as food, etc.), and how are the methods used?*

a. Dry and moist heat
b. Filtration
c. Ultraviolet light
d. Spoilage preservation

You can find some common examples online in **Appendix F**.

■ KEY CONCEPT 9.3   Chemical Control Usually Involves Disinfection

Sanitation and disinfection methods are not unique to the modern era. The Bible refers often to cleanliness and prescribes certain dietary laws to prevent consumption of what was believed to be contaminated food. Egyptians used resins and aromatics for embalming, and ancient peoples burned sulfur for deodorizing and sanitary purposes. Arabian physicians first suggested using mercury to treat syphilis. Over the centuries, spices were used as preservatives as well as masks for foul odors, making Marco Polo's trips to Asia for new spices a necessity as well as an adventure. And fans of Western movies probably have noted that American cowboys practiced disinfection by pouring whiskey onto gunshot wounds.

Chemical Control Methods Are Dependent on the Object to Be Treated

As early as 1830, the United States Pharmacopoeia listed a solution of iodine as a valuable antiseptic, and soldiers in the Civil War used it in plentiful amounts. Joseph Lister in the 1860s established the principles of aseptic surgery by using carbolic acid (phenol) for treating wounds.

Chemical agents, as a group, rarely achieve sterilization. Some are microbicidal; they inactivate the major enzymes of most organisms and interfere with their metabolism. Other agents are microbiostatic, disrupting minor chemical reactions and slowing microbial metabolism, which results in a longer time between cell divisions.

The process of eliminating or reducing pathogens (except endospores) on inanimate objects is called **disinfection**, and the chemical agent used is called a **disinfectant**. The object is not usually sterile because there still might be nonpathogens that have survived chemical treatment. If the object treated is living, such as human skin tissue, the process is called **antisepsis**, and the chemical agent used is an **antiseptic** (**FIGURE 9.14**). The word **sepsis** (*seps* = "putrid"), often referred to as blood poisoning, refers to a condition in which microbes or their toxins are present in tissues or in the blood. For example, **septicemia** refers to bacteria growing in and spreading through the blood. Thus, the term **asepsis** means "against infection." In surgery, **aseptic technique** is critical to prevent contamination of the surgical field.

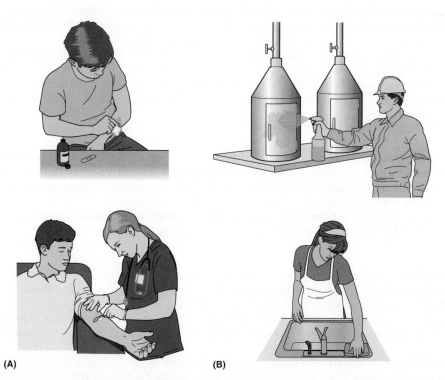

(A)   **(B)**

FIGURE 9.14 Sample Uses of Antiseptics and Disinfectants. (A) Antiseptics are used on body tissues, such as on a wound or before piercing the skin to take blood. **(B)** A disinfectant is used on inanimate objects, such as equipment used in an industrial process or tabletops and sinks. *»» Why aren't disinfectants normally used as antiseptics?*

It is important to note that many chemical agents can be used as both a disinfectant and an antiseptic. What often decides their use is the precise formulations, such that an antiseptic is not as toxic or damaging to human tissue as a disinfectant would be.

Before reading any further, you might enjoy reading MICROFOCUS 9.2, which identifies some common but surprising antiseptics.

Other expressions are associated with chemical control. As mentioned in the introduction to this chapter, sanitization of a location or object reduces the microbial population to a safe level as determined by public health standards. For example, restaurants depend on disinfectants to maintain a sanitary kitchen and work establishment. To **degerm** an object is to remove organisms mechanically from its surface. Washing one's hands with soap and water, or using alcohol to prepare the skin surface for an injection, degerms the skin surface but has little effect on microorganisms deep in the skin pores.

FIGURE 9.15 illustrates the extent of "clean-up" necessary in a hospital room prior to occupancy by the next patient. Without thorough and aggressive cleaning, often both hard and soft surfaces in hospital rooms can remain contaminated with bacterial pathogens (e.g., methicillin-resistant *Staphylococcus aureus*, *Clostridium difficile*, *E. coli*, and *Pseudomonas aeruginosa*).

MICROFOCUS 9.2: Being Skeptical

Antiseptics in Your Pantry?

Today, we live in an age when alternative and herbal medicine claims are always in the news, and these reports have generated an entire industry of health products that often make unbelievable claims. With regard to "natural products," are there some that have genuine medicinal and antiseptic properties?

Garlic

In 1858, Louis Pasteur examined the properties of garlic as an antiseptic. During World War II, when penicillin and sulfa drugs were in short supply, garlic was used as an antiseptic to disinfect open wounds and prevent gangrene. Since then, numerous scientific studies have tried to discover garlic's antiseptic powers.

Many research studies have identified a sulfur compound called allicin as one key to garlic's antiseptic properties. When a raw garlic clove is crushed or chewed, allicin gives garlic its characteristic taste and smell. Laboratory studies using garlic suggest that this compound is responsible for combating the microbes causing the common cold, flu, sore throat, sinusitis, and bronchitis. The findings indicate that the compound blocks key enzymes that bacteria and viruses need to invade and damage host cells.

Cinnamon

Cinnamon might be an antiseptic that can control pathogens, at least in fruit beverages. In studies in which cinnamon was added to commercially pasteurized apple juice purposely contaminated with typical foodborne pathogens (*Salmonella typhimurium*, *Yersinia enterocolitica*, and *Staphylococcus aureus*) and viruses, the investigators discovered the pathogens were killed more readily in the cinnamon blend than in the cinnamon-free juice.

In another study carried out by a different research group, 10% Saigon and Ceylon cinnamons deactivated 99.9% of viruses, suggesting that the spice might eliminate or prevent viruses from infecting humans.

Honey

For three decades, Professor Peter Molan, associate professor of biochemistry and director of the Waikato Honey Research Unit at the University of Waikato, New Zealand, has been studying the medicinal properties of and uses for honey. Its acidity, between 3.2 and 4.5, is low enough to inhibit many pathogens. Its low water content (15% to 21% by weight) means that it osmotically ties up free water and "drains water" from wounds, helping to deprive pathogens of an ideal environment. In addition, two proteins in honey interfere with the integrity of the bacterial envelope (cell wall and cell membrane).

Apparently, the antibacterial properties of honey depend on the kind of nectar (the plant pollen) that bees use to make honey. Manuka honey from New Zealand and honeydew from central Europe are thought to contain useful levels of antiseptic potency. In 2011, the FDA approved wound dressings containing manuka honey.

Licorice Root

Dried licorice root has been used in traditional Chinese medicine for centuries. Scientists have identified two substances in licorice root capable of killing the two most common bacterial species causing tooth decay and one species responsible for gum disease. But don't run out and start eating lots of licorice candy, because the licorice root extract originally in candies has been replaced by anise oil, which has a similar flavor but no antimicrobial properties.

Wasabi

The green, pungent, Japanese horseradish called wasabi might be more than a spicy condiment for sushi. Natural chemicals in wasabi, called isothiocyanates, inhibit the growth of *Streptococcus mutans*—one of the bacterial species that causes tooth decay. At this point, these are only test-tube laboratory studies, and the results will need to be proven in clinical trials.

The verdict? There appear to be products having genuine antimicrobial properties—and there are many more than can be described here.

Courtesy of Dr. Jeffrey Pommerville.

Chemical Agents Are Important to Laboratory and Hospital Safety

There are several factors to be considered when selecting a chemical agent to control microbes in the environment (e.g., hospital, restaurant, or at home):

▶ **Microbial susceptibility**. Pathogens generally vary greatly in their susceptibility to

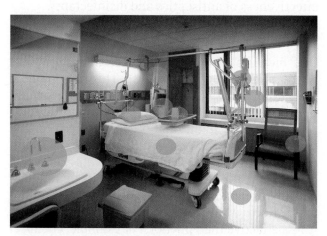

FIGURE 9.15 Potential Hotspots for Pathogen Transmission. Many potential pathogens can be difficult to eliminate even with routine cleaning of a hospital room. The red circles indicate some areas where potential pathogens might be cultured from a swab sample. *»» Should a hospital patient room be sterilized, disinfected, or sanitized before the next patient occupies the room? Explain.*

© Can Stock Photo/Babar760.

a chemical agent (**FIGURE 9.16**). Often, typical vegetative cells are killed by low-level agents, whereas killing endospores might require chemical sterilants.

▶ **Temperature and pH**. It is important to know at what temperature the treatment is to take place because a chemical reaction occurring at warmer temperatures might occur more slowly or not at all at low temperatures. Many chemical agents also have an optimal pH. For example, one agent might be most effective at neutral or slightly acidic pH, whereas another works best at an alkaline pH.

▶ **Concentration**. Chemical agents usually are more effective at higher concentrations. However, although higher concentrations might better kill or slow the growth of a pathogen, the agents might be toxic to animals or humans at the higher concentration. Therefore, it is important to follow the manufacturer's directions when preparing chemical agents.

▶ **Microbial numbers**. The greater the number of organisms in the environment (the **bioburden**), the longer disinfection will take. Microbes forming a biofilm on a surface will be more resistant to chemical attack than the individual cells.

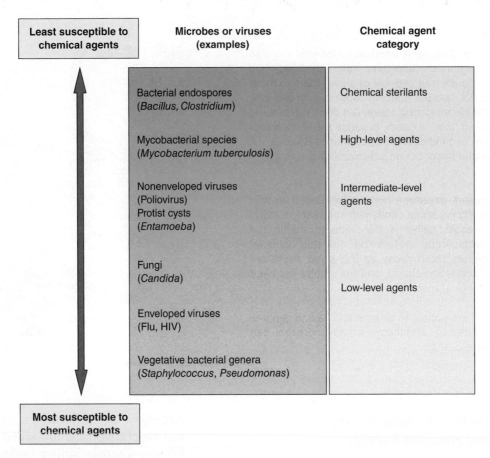

FIGURE 9.16 Microbial Susceptibility to Chemical Agents. Microbes and viruses vary in their susceptibility to chemical agents. *»» Which chemical agent category (or categories) would not destroy (kill) endospores? Polioviruses?*

▶ **Endospore formers.** Endospores are resistant to some of the chemical agents used to inhibit microbial growth, so endospore formers can evade destruction (see Figure 9.16). This can be especially important when considered in the context of the hospital environment because the elimination of bacterial spores from spore formers like *C. difficile* often requires more vigorous treatment than the removal of vegetative cells.

▶ **Other characteristics.** The chemical should be nonstaining and noncorrosive; it should be odorless, easy to obtain, and relatively inexpensive.

Antiseptics and Disinfectants Can Be Evaluated for Effectiveness

Currently, there are more than 8,000 disinfectants and antiseptics for hospital and general use (see chapter opening photo). Evaluating these chemical agents is a tedious process because of the broad diversity of conditions under which they are used.

Therefore, there are several methods to evaluate the effectiveness of antiseptics and disinfectants.

Phenol Coefficient

One measure of effectiveness of chemical agents is the **phenol coefficient (PC)**. This is a number indicating the usefulness of an antiseptic or disinfectant in comparison to phenol under identical conditions. The traditional organisms used for the **phenol coefficient test** are *S. aureus* and *Salmonella enterica* subtype Typhi (TABLE 9.2). A PC value higher than 1 indicates the chemical is more effective than phenol, and the higher the number, the more effective it is. A number less than 1 indicates it is less effective than phenol. For example, chemical A might have a PC of 78.5, whereas under identical conditions chemical B has a PC of 0.28.

The phenol coefficient test has some drawbacks because it is performed in the laboratory rather than in a real-life situation. In addition, it does not take into account many of the factors mentioned above, such as tissue toxicity and temperature variations.

TABLE 9.2 Phenol Coefficients of Some Common Antiseptics and Disinfectants[a]

Chemical Agent	Staphylococcus aureus	Salmonella enterica subtype Typhi
Phenol	1	1
Chloramine	133	100
Tincture of iodine	6.3	5.8
Lysol™	5.0	3.2
Mercury chloride	100	143
Ethyl alcohol	6.3	6.3
Formalin	0.3	0.7
Cetylpyridinium chloride	337	228

[a] Data from *Encyclopedic Dictionary of Polymers, Volume 1.* Springer-Verlag, New York. 2010.

Disk Diffusion Method

In the teaching laboratory, the **disk diffusion method** is a simple way to examine the effect of different chemical agents against a specific bacterial species. Small absorbent paper disks are each dipped in a separate solution of the chemical agents to be tested. Each damp disk is then placed on an agar plate that previously was inoculated with the bacterial species in question. Incubation of the plate for 24 to 48 hours allows the chemical to diffuse out from the disk and into the agar and possibly affect bacterial growth.

Inhibition of growth will be seen as a clear zone, called a **zone of inhibition**, around a disk (**FIGURE 9.17**). Resistant bacteria will show no zone of inhibition. Realize that zone diameters on the plate are not comparable because the chemicals might be used at different concentrations and the molecular size of the chemical agent might affect its diffusion rate through the agar (larger molecules could diffuse from the disk more slowly, producing smaller zone diameters, whereas smaller molecules could move faster and produce larger zone diameters).

Use-Dilution Test

Another method, the **use-dilution test**, uses small stainless-steel cylinders whose surfaces have been coated with the specific test microbe (usually, *P. aeruginosa, S. aureus*, or *S. choleraesuis*) before the cylinders are dried. The dried cylinders are exposed to the chemical agent at various dilutions for 10 minutes before being rinsed and placed in broth growth medium. After 48 hours, the cylinders and growth medium are examined to determine whether bacterial growth has occurred. The most effective agent is the one that shows no growth in the tubes containing cylinders having the highest dilution of chemical.

In-Use Test

A more direct way of determining the value of a chemical agent is to take swab samples from the affected area before and after the application of a chemical agent. The swabs are placed in individual tubes containing broth and incubated for 48 hours. The tubes are then examined. If there is no growth in the tube, the chemical treatment was successful.

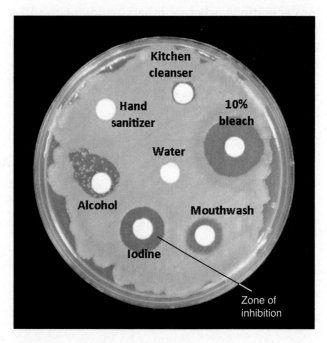

FIGURE 9.17 The Disk Diffusion Method. Paper disks containing different chemical agents can be compared by their zone diameters. *»» Hypothesize what the small bacterial colonies represent in the clear zone around the paper disk containing alcohol.*

Courtesy of Dr. Jeffrey Pommerville.

Concept and Reasoning Checks 9.3

a.   Distinguish between (1) an antiseptic and a disinfectant and (2) antisepsis and disinfection.
b.   Summarize the properties important in the selection of a disinfectant or antiseptic.
c.   Assess the need to know a chemical agent's effectiveness.

Chapter Challenge B

You might use a variety of disinfectants in your home or workplace that are designed to eliminate or control potential pathogens. Which one you use can depend on the object being treated.

QUESTION B: *Look at several disinfectants or antiseptics you have in your home or workplace. Do these products give you any information as to (a) microbial susceptibility (what is the product designed to eliminate or control) and (b) how the product should be used and stored?*

You can find answers online in **Appendix F**.

■ KEY CONCEPT 9.4   A Variety of Chemical Agents Can Control Microbial Growth

The chemical agents currently in use for controlling microorganisms range from very simple substances, such as household bleach, to very complex compounds, typified by cationic (positively charged) detergents. Many of these agents have been used for generations, whereas others represent the latest modern chemical formulations. In this section, we survey several groups of chemical agents and indicate how they are best applied in the control of microorganisms. TABLE 9.3 shows the chemical structures for many of the chemical agents described in the sections that follow.

Halogens Denature Proteins

The **halogens** are a group of highly reactive elements. Two commonly used halogens are chlorine and iodine. They represent intermediate-level chemical agents that combine with and denature certain cytoplasmic proteins, such as enzymes.

Chlorine (Cl_2) is available in a gaseous form and as both inorganic and organic liquid compounds. It is effective against a broad variety of microbes, including most bacterial and fungal species, and many viruses. At high concentrations, it can be sporicidal.

TABLE 9.3   Selected Examples of Chemical Agents and Their Structures

Chemical Agent and Structure	

Halogens: chlorine agents
• Inorganic
  ◦ Sodium hypochlorite $Na^+ [OCl]^-$
  ◦ Calcium hypochlorite $Ca^{+2} [OCl]^-_2$
• Organic
  ◦ Chloramine-T
	Phenolic compounds
	  ◦ Chlorhexidine
	
	  ◦ Triclosan

TABLE 9.3   Selected Examples of Chemical Agents and Their Structures (Continued)

Chemical Agent and Structure

Heavy metals
- Merbromin

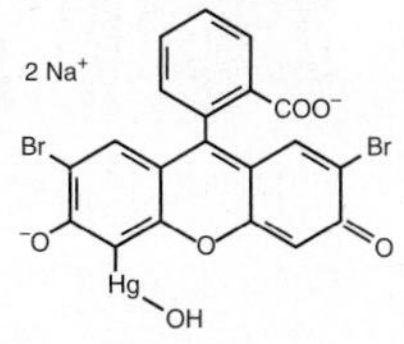

Alcohols
- Ethanol

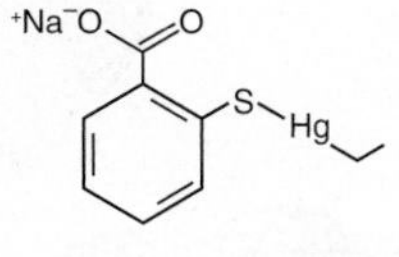

- Thimerosal

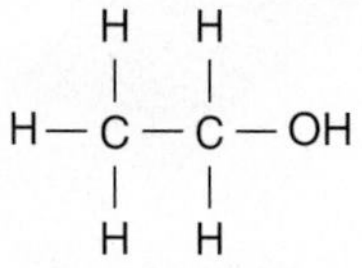

- Isopropyl

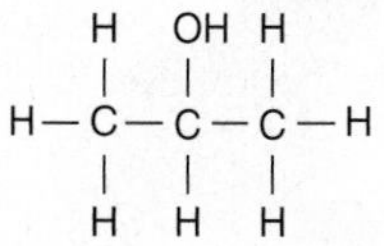

Quats
- Benzalkonium chloride

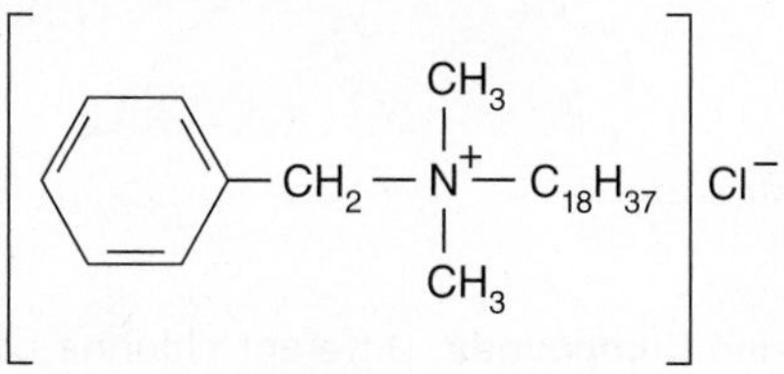

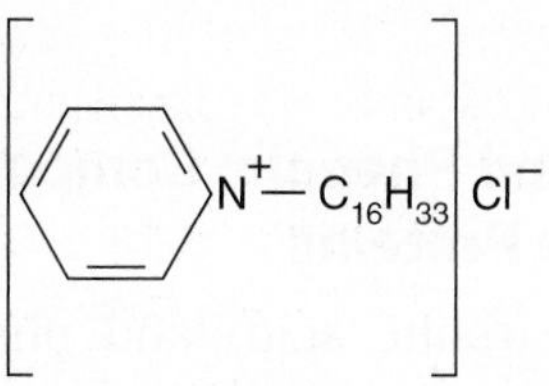

Peroxides
- Hydrogen peroxide

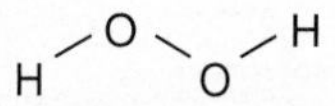

- Benzoyl peroxide

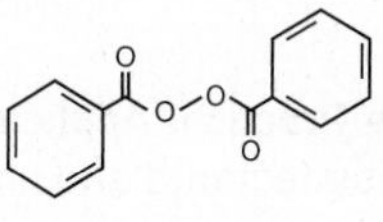

- Cetylpyridinium chloride

Aldehydes
- Formalin

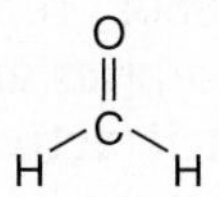

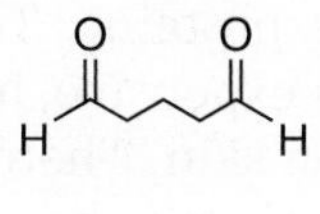

Sterilizing gses
- Ethylene dioxide

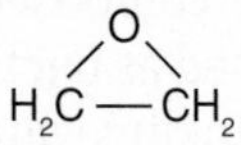

- Chlorine dioxide

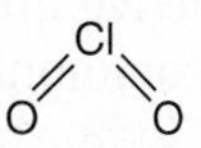

- Glutaraldehyde

Chlorine is widely used to keep microbial populations at low levels in municipal water supplies and swimming pools (**FIGURE 9.18**). Chlorine combines readily with numerous ions in water; therefore, enough chlorine must be added to ensure a residue remains for antimicrobial activity. In municipal water, the residue of chlorine is usually about 0.2 to 1.0 parts per million (ppm) of free chlorine. One ppm is equivalent to 0.0001 percent, an extremely small amount.

Inorganic forms of chlorine include sodium hypochlorite (Clorox® bleach) and calcium

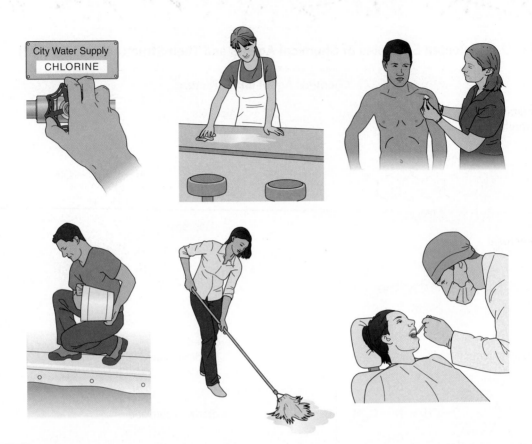

FIGURE 9.18 Some Practical Applications of Disinfection with Chlorine Compounds. Different chlorine compounds have been used as both disinfectants and antiseptics. *»» In each of the above illustrations, is the chemical agent being used as a disinfectant or an antiseptic?*

hypochlorite (see Table 9.3). The latter, also known as chlorinated lime, was used by Semmelweis in his studies on childbed fever. Hypochlorite compounds cause cellular proteins to denature, destroying their function. Hypochlorites also are useful in very dilute solutions for sanitizing dairy and restaurant equipment.

The chloramines, such as chloramine-T, are organic compounds composed of chlorine and ammonia. They are used as bactericides and used for the disinfection of drinking water.

Iodine (I_2) is slightly larger than the chlorine atom and is more reactive and more germicidal. A tincture of iodine, a commonly used antiseptic for wounds, consists of 2% iodine in alcohol. Iodine damages microbes and many endospores by reacting with enzymes and with proteins in the cell membrane and cell wall. **Iodophors** are iodine linked to a solubilizing agent, such as a detergent. These water-soluble complexes release iodine over a long period and have the added advantage of not irritating the skin. Some iodophors are used in preoperative skin preparations. An example is Betadine®, which is iodine combined with a nondetergent carrier. The combination stabilizes the iodine and releases it slowly.

Phenol and Phenolic Compounds Denature Proteins

Phenol (carbolic acid) and phenolic compounds are low- to intermediate-level chemical agents (**FIGURE 9.19**). Phenol, which has played a key role in disinfection practices since Joseph Lister first used it in the 1860s, remains a standard against which other antiseptics and disinfectants are evaluated. The agent is active against many bacterial and fungal species and some viruses. Phenol and its derivatives act by denaturing and disrupting cell membrane proteins. The drawbacks are that the chemical is expensive, has a pungent odor, and is caustic to the skin. Therefore, the usefulness of phenol as an antiseptic has diminished.

Phenol derivatives (phenolic compounds) have greater germicidal activity and lower toxicity than phenol (see Table 9.3). An important bisphenol is **chlorhexidine**. This compound is used as a surgical hand scrub and superficial skin wound cleanser. A 4% chlorhexidine solution in isopropyl alcohol is commercially available as Hibiclens®. Another bisphenol in widespread use is **triclosan**, a broad-spectrum

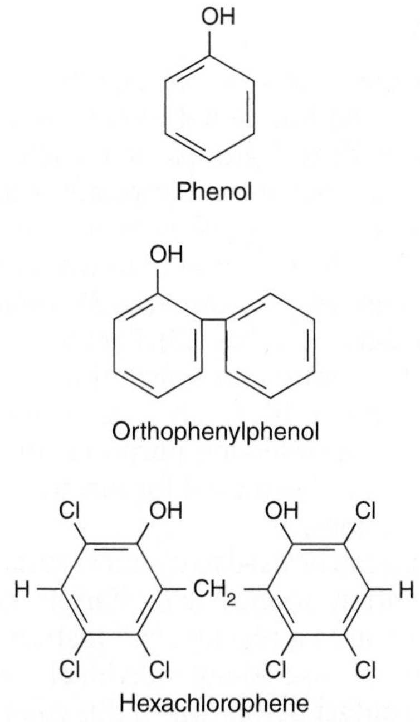

OH

Phenol

OH

Orthophenylphenol

Hexachlorophene

FIGURE 9.19 Phenol and Some Derivatives. The chemical structure of phenol and some important derivatives. »» *Why are most phenolic compounds only used as disinfectants?*

antimicrobial agent that destroys bacterial cells by disrupting cell membranes (and possibly cell walls). The chemical is included in such products as antibacterial soaps, lotions, mouthwashes, toothpastes, toys, food trays, underwear, kitchen sponges, utensils, and cutting boards. Whether triclosan actually is better than plain soap and water at preventing the spread of pathogens has been questioned. In late 2016, the FDA decided to ban triclosan and 18 other ingredients in antibacterial soaps and other home antibacterial products because of the lack of evidence supporting their effectiveness.

Heavy Metals Interfere with Microbial Metabolism

Mercury, silver, and copper are called **heavy metals** because of their large atomic mass. They are low-level chemical agents that interfere with cell metabolism by inactivating enzymes and denaturing structural proteins.

Although mercury (Hg) is very toxic to humans and other animals, the toxicity of mercury as an antimicrobial agent is reduced if other organic matter is present. In such products as merbromin (Mercurochrome) and thimerosal (Merthiolate) (see Table 9.3), mercury is combined with carrier compounds to make the metal less toxic when applied to the skin, especially after surgical incisions. Thimerosal was previously used as a preservative in vaccines. Today, only a few adult

vaccines contain the chemical agent. As a topical antiseptic, mercury compounds have been replaced by other agents, such as the iodophors.

Copper (Cu) has long been known for its antimicrobial properties. As copper sulfate ($CuSO_4$), it is incorporated into algicides, and the agent is used in swimming pools and municipal water supplies. Copper ions also are very toxic to bacterial cells and many viruses. Thus, plumbing fixtures and surfaces containing copper or copper alloys, along with effective cleaning and sanitization practices, can help control disease transmission and reduce the risk of infection by such pathogens.

In the past, state laws required hospitals to put drops of silver (Ag) (in the form of 1% silver nitrate; $AgNO_3$) into the eyes of newborns. This was done to protect the newborn from infection by *Neisseria gonorrhoeae*, which could be contracted during passage through the birth canal of a gonorrhea-infected mother. Today, silver nitrate has been replaced by antibiotic eye drops.

Alcohols Denature Proteins and Disrupt Membranes

Alcohols are intermediate-level chemical agents that are bactericidal, fungicidal, and virucidal. They have no effect on bacterial endospores or fungal spores. Alcohols denature proteins and dissolve lipids, an action leading to cell membrane destruction. Alcohols also are strong dehydrating agents.

Because alcohols react readily with organic matter, medical instruments and thermometers must be thoroughly cleaned before exposure. Alcohols are most effective when diluted in water because water prevents the rapid evaporation of alcohol. Thus, 70% alcohol is more effective than 95% alcohol in controlling microbes.

Ethyl alcohol (ethanol) is the active ingredient in many popular hand sanitizers and is used as a degerming agent to treat skin before an injection (see Table 9.3). It mechanically removes bacterial cells from the skin (degerms) and dissolves lipids. Isopropyl alcohol, or rubbing alcohol, also has high bactericidal activity.

Soaps and Detergents Act as Surfactants

Soaps are chemical compounds of fatty acids combined with potassium or sodium hydroxide. The pH of the compounds is usually about 8.0, and some microbial destruction is due to the alkaline conditions they establish on the skin. However, the major activity of soaps is as a degerming agent for the mechanical removal of microorganisms from the skin surface. Soaps, therefore, are surface-active agents called

surfactants; that is, they emulsify and solubilize particles clinging to a surface and reduce the surface tension. Soaps also remove skin oils, further reducing the surface tension and increasing the cleaning action.

Detergents are cationic, organic chemicals acting as strong surfactants. Because they are actively attracted to the phosphate groups of cellular membranes, they are bactericidal against gram-positive bacteria and are effective inhibitors of fungi and enveloped viruses. The most commonly used detergents for microbial control are derivatives of ammonium chloride. In these agents, four organic groups or hydrocarbon chains have replaced the four hydrogens in ammonia (see Table 9.3). Such compounds, often called **quaternary ammonium compounds** or, simply, **quats**, represent low-level chemical agents.

Quats have rather long, complex names, such as benzalkonium chloride in Zephiran® and cetylpyridinium chloride in Cepacol®. They are used as sanitizing agents for industrial equipment and food utensils, as antiseptics in mouthwashes and contact lens cleaners, and as disinfectants for use on hospital walls and floors.

Peroxides Damage Cellular Components

Peroxides are high-level chemical agents that contain oxygen-oxygen single bonds. Hydrogen peroxide (H_2O_2), a common household antiseptic, has been used as a rinse for wounds, scrapes, and abrasions (see Table 9.3). However, H_2O_2 applied to such areas foams and effervesces because catalase in the tissue breaks down H_2O_2 into oxygen and water. Therefore, it is not recommended as an antiseptic for open wounds. However, the furious bubbling loosens dirt, debris, and dead tissue, and the oxygen gas is effective against anaerobic bacterial species. Hydrogen peroxide decomposition also results in a reactive form of oxygen—the superoxide radical—which is highly toxic to microorganisms and viruses.

New forms of H_2O_2 are more stable than traditional forms, do not decompose spontaneously, and therefore are more stable for topical use. Such inanimate materials as soft contact lenses, utensils, heat-sensitive plastics, and food-processing equipment can be disinfected within 30 minutes. Benzoyl peroxide is an active ingredient in teeth whitening products and at low concentrations (2.5%) is used to treat acne.

Some Chemical Agents Combine with Nucleic Acids and/or Cell Proteins

The chemical agents we discussed in previous sections are used for disinfection. However, a few other chemical agents can be used as chemical sterilants.

Aldehydes

Aldehydes are highly active agents containing a –CHO (carbonyl) functional group. They cross link amino and hydroxyl groups of nucleic acids and proteins, which denature the proteins and inactivate nucleic acids in microbes, viruses, and spores.

Formaldehyde is a gas at high temperatures and a solid at room temperature. As a 37% solution, it is called **formalin** (see Table 9.3). For more than a century, formalin was used in embalming fluids for anatomical specimens (it's rarely used anymore) and by morticians for disinfecting purposes. In microbiology, formalin has been used for inactivating viruses in certain vaccines.

Formalin can be used to disinfect surgical instruments, isolation rooms, and dialysis equipment. However, it leaves a residue, and instruments must be rinsed before use. Many individuals can develop an allergic contact dermatitis to this compound.

Glutaraldehyde is a small, organic molecule that as a 2% solution can be used to disinfect objects with a 10-minute exposure (see Table 9.3). It will sterilize objects (and kill endospores) with a 10-hour exposure. Glutaraldehyde does not damage delicate objects, so it can be used to disinfect or sterilize optical equipment, such as the fiber-optic endoscopes used for arthroscopic surgery. Glutaraldehyde gives off irritating fumes, however, and instruments must be rinsed thoroughly in sterile water before use.

Sterilizing Gases

The development of plastics for use in microbiology required a suitable method for sterilizing these heat-sensitive materials. In the 1950s, research scientists discovered the antimicrobial properties of ethylene oxide, which essentially made the plastic culture dish and disposable plastic syringe possible.

Ethylene oxide (EtO) is a small molecule with excellent penetration capacity through paper and plastic (see Table 9.3). Over a 14- to 16-hour exposure, EtO denatures proteins and cross-links DNA. It is microbicidal as well as sporicidal. However, it potentially can cause cancer and is highly explosive. Its explosiveness is reduced by mixing it with Freon gas or carbon dioxide gas. EtO chambers, called "gas autoclaves," have become the chemical counterparts of heat- and pressure-based autoclaves for sterilization procedures.

EtO is used to sterilize paper, leather, wood, metal, and rubber products as well as plastics without causing any damage to the object. In medicine, it is used to sterilize catheters, artificial heart valves, heart-lung machine components, and optical equipment. NASA uses the gas for sterilization of interplanetary space capsules.

Chlorine dioxide has properties very similar to chloride gas and sodium hypochlorite, but, unlike ethylene oxide, it produces nontoxic byproducts. Chlorine dioxide can be used as a gas or liquid. In a gaseous form, with proper containment and

humidity, a 15-hour fumigation can be used to sanitize air ducts, food and meat processing plants, and hospital areas. It was the gas used to decontaminate the 2001 anthrax-contaminated mail and office buildings, as described in MICROFOCUS 9.3.

TABLE 9.4 and FIGURE 9.20 summarize the chemical agents used in controlling microorganisms.

MICROFOCUS 9.3: Tools/Bioterrorism

Decontamination of Anthrax-Contaminated Mail and Office Buildings

This chapter has examined the chemical procedures and methods used to control the numbers of microorganisms on inanimate and living objects. These control measures usually involve a level of sanitation, although some procedures can sterilize. Some examples were given for their use in the home and workplace. However, what about real instances for which large-scale and extensive decontamination have to be carried out?

In October 2001, the United States experienced a bioterrorist attack. The perpetrator(s) used anthrax spores as the bioterror agent. Four anthrax-contaminated letters were sent through the United States mail on the same day, addressed to NBC newscaster Tom Brokaw, the *New York Post*, and to two United States Senators, Senator Patrick Leahy and Senate Majority Leader Tom Daschle (see Figure A). The Centers for Disease Control and Prevention (CDC) confirmed that anthrax spores from at least the Daschle letter contaminated the Hart Senate Office Building and several post office sorting facilities in Trenton, New Jersey (from where the letters were postmarked), and Washington, D.C., areas. This resulted in the closing of the Senate building and the postal sorting facilities. With the Senate building and postal sorting facilities closed, the CDC, the U.S. Environmental Protection Agency (EPA), other governmental agencies, and commercial companies had to devise and implement a strategy to decontaminate these buildings and the mail-sorting machines (see Figure B).

As mentioned in this chapter, most sanitation procedures do not require a high technology solution. In fact, all of the methods actually used for this situation are described in this chapter.

The Hart Senate Office Building and the post office sorting facilities were contaminated with *Bacillus anthracis* endospores. Both of these buildings are large, multiroom structures with many pieces of furniture and, with the postal facility, equipment, including computers, copiers, and mail-sorting machines. To decontaminate these buildings, a gas was needed that could permeate the air ducts as well as all the office machinery and sorting machines. The gas chosen was chlorine dioxide. Essentially, the buildings were sealed as if they were going to be fumigated for termites. The gas was pumped in, and, after a time that was believed to be sufficient to kill any anthrax spores, the gas was evacuated. Similar to an in-use test, cotton swab samples were taken from the buildings. The swabs were streaked on nutrient agar plates. If any spores were still alive, they would germinate on the plates and the actively metabolizing cells would grow into visible colonies. Such results would require retreatment of the facility.

To protect the mail from similar attacks in the future, a system was devised using ultraviolet (UV) light to kill any spores that might be found in a piece of mail moving through the sorting machines.

It took months—even years—for some of the postal facilities to be declared safe and free of anthrax spores. Still, simple physical and chemical methods worked to decontaminate the buildings and equipment.

(A) Anthrax-containing letters. **(B)** Mail sorting machines.

(A) © Brian Branch-Price/AP Photo. Photo cropped from original. (B) Courtesy of the U.S. Census Bureau, Public Information Office.

TABLE 9.4 Summary Of Chemical Agents Used To Control Microorganisms

Chemical Agent	Mechanism of Activity	Applications	Limitations	Antimicrobial Spectrum
Chlorine (chlorine gas, sodium hypochlorite, chloramines)	Denatures proteins	Water treatment Skin antisepsis Equipment spraying Food processing	Inactivated by organic matter Objectionable taste, odor	Microbicidal Virucidal Sporicidal
Iodine (tincture of iodine, iodophors)	Denatures proteins	Skin antisepsis Medical/lab disinfection	Inactivated by organic matter Objectionable taste, odor	Microbicidal Virucidal Sporicidal
Phenol and derivatives	Denatures proteins Disrupts cell membranes	Surface disinfection Skin antisepsis with detergent	Toxic to tissues Disagreeable odor	Microbicidal Virucidal
Mercury (mercuric chloride, merthiolate, merbromin)	Denatures proteins	Skin antiseptics Disinfectants	Inactivated by organic matter Toxic to tissues Slow acting	Microbiostatic
Copper (copper sulfate)	Denatures proteins	Algicide in swimming pools and municipal water supplies	Inactivated by organic matter	Microbiostatic Fungistatic
Silver (silver nitrate)	Denatures proteins	Antiseptic in eyes of newborns	Skin irritation	Microbiostatic
Alcohol (70% ethyl, 75% isopropyl)	Denatures proteins Dissolves lipids Dehydrating agent	Instrument disinfectant Skin antiseptic	Precleaning necessary Skin irritation	Microbicidal Virucidal
Cationic detergents (quaternary ammonium compounds)	Disrupt lipids in cell membranes	Instrument disinfection Skin antisepsis	Neutralized by soap	Microbicidal Virucidal
Peroxides (hydrogen peroxide)	Denature proteins	Skin antisepsis Surface disinfection	Limited use	Microbicidal Virucidal Sporicidal
Formaldehyde (formalin)	Denatures proteins Inactivates nucleic acids	Embalming Vaccine production Gaseous sterilant	Poor penetration Allergenic Toxic to tissues Neutralized by organic matter	Microbicidal Virucidal Sporicidal
Glutaraldehyde	Denatures proteins Inactivates nucleic acids	Sterilization of surgical supplies	Unstable Toxic to skin Respiratory irritation	Microbicidal Virucidal Sporicidal
Ethylene oxide	Denatures proteins Inactivates nucleic acids	Sterilization of instruments, equipment, heat-sensitive objects	Explosive Toxic to skin Requires constant humidity	Microbicidal Virucidal Sporicidal
Chlorine dioxide	Denatures proteins Inactivates nucleic acids	Sanitizes equipment, rooms, buildings	Burns skin and eyes on contact	Microbicidal Virucidal Sporicidal

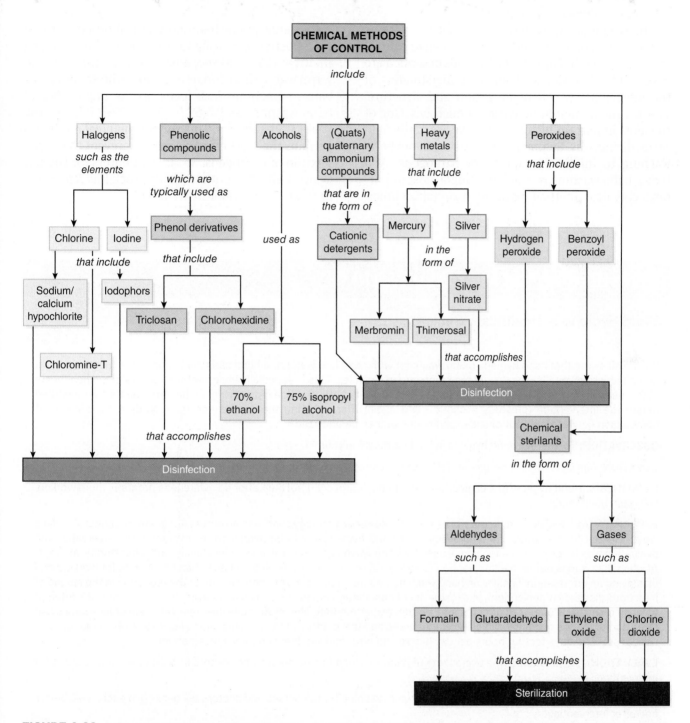

FIGURE 9.20 A Concept Map Summarizing the Chemical Methods of Microbial Control. The chemical methods predominantly disinfect, although a few can sterilize. »» *If you wanted to sanitize a kitchen counter, which chemicals might be selected?*

In conclusion, we have examined the various physical and chemical methods developed to control microbial growth. The control methods described are based on the knowledge about and understanding of the biochemistry, cell organization and structure, growth, and metabolic activities of microbes. One of the most important historical discoveries in the field of medicine was Semmelweis' insistence on hand washing to inhibit the spread of pathogens. And today, hand washing remains the single most important chemical method used in hospitals, clinics, restaurants, public restrooms, and at home. Much of the effectiveness of hand washing is simply taking the time (15 to 20 seconds) to wash with soap and warm water. Unfortunately, people who should know better sometimes lose focus on the importance of hand washing, as INVESTIGATING THE MICROBIAL WORLD 9 examines. This simple chemical method is perhaps the most effective way to prevent the transmission of pathogens. So, as the Centers for Disease Control and Prevention (CDC) proclaims *Remember to Wash—Clean Hands Save Lives!*

Investigating the Microbial World 9

Hand Hygiene in Healthcare Clinics

We are all aware that hand hygiene (hand washing) with soap and water or a hand sanitizer is one of the most important and effective ways to lessen the spread of infectious diseases. This is especially important for workers in the healthcare field. The following involves an interventional pre- and post-study, that is, a study in which the participants are monitored before the intervention (prestudy = control) and again after receiving the intervention (post-study = experiment). A comparison between the control and experiment can then be analyzed.

OBSERVATION: To improve hand hygiene in the healthcare setting.

QUESTION: *Can hand hygiene be improved through better awareness of the importance of hand washing?*

HYPOTHESIS: Hand hygiene in a clinical setting can be improved. If so, then after an awareness program, more workers will wash their hands.

EXPERIMENTAL DESIGN: Healthcare workers at an outpatient oncology clinic and an outpatient gastrointestinal (GI) clinic were selected for the study. The improvement of hand hygiene would be measured by: (a) the direct observation (no evaluation was made as to the technique used for hand washing) of healthcare workers having the "opportunity/attempt" to wash their hands before the intervention (control); (b) making available alcohol-based hand sanitizers for workers and displaying an informational poster on hand washing and the importance of hand hygiene to the reduction in the spread of infectious diseases (intervention); and (c) the direct observation of healthcare workers having the opportunity and attempt to wash their hands one week after the intervention (experimental). The "staff" consisted of an average of five nurses and three doctors in the oncology clinic and eight nurses and five doctors in the GI clinic. Individuals were observed on three nonconsecutive days for four hours per day before and after making direct contact with a patient.

EVALUATION 1: Evaluation of the overall rate of hand washing for the nurses and doctors in each clinic before and after intervention.

EVALUATION 2: Evaluation of the overall rate of hand washing for the nurses and doctors (by gender) in each clinic before and after intervention.

EVALUATION 3: Evaluation of the overall rate of hand washing for the nurses and doctors (by profession) in each clinic before and after intervention.

RESULTS: *See graphs.*

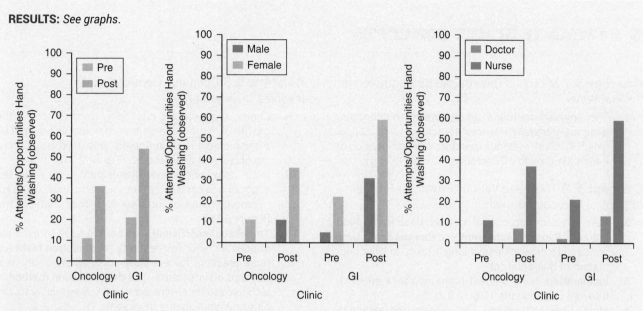

CONCLUSIONS:

> **QUESTION 1:** *Looking at graph A, was the intervention a success? Was the hypothesis supported?*
>
> **QUESTION 2:** *Looking at graph B, which gender was most likely to wash their hands?*
>
> **QUESTION 3:** *Looking at graph C, which staff members were more likely to wash their hands?*
>
> **QUESTION 4:** *From the study results, how would you assess the success of the intervention as a whole and with regard to specific groups (male versus female; doctors versus nurses)? What suggestions might you make to further improve hand hygiene?*

You can find answers online in **Appendix E**.

Modified from KuKanich, K.S. et al. 2013. *Amer J Nurs* 113(3):36–42.

Concept and Reasoning Checks 9.4

a. Compare the uses for chlorine and iodine chemical agents.
b. Explain why phenol derivatives are preferred as disinfectants and antiseptics.
c. Evaluate the use of heavy metals as antiseptics and disinfectants.
d. Why is 70% ethanol preferable to 95% ethanol as an antiseptic?
e. How do soaps differ from quats as chemical agents of control?
f. Summarize the uses for aldehydes, ethylene oxide, and chloride dioxide for sterilization.

Chapter **Challenge C**

In this last section of the chapter, you have studied several groups of chemical agents in the form of disinfectants or antiseptics.

QUESTION C: *For the products you examined in part B of the chapter challenge, as well as other personal products like a mouthwash, toothpaste, or deodorant, look at the label and especially pay attention to the "active ingredients" or "contains" section. What types of chemical agents are included in each of these products?*

You can find examples online in **Appendix F**.

■ SUMMARY OF KEY CONCEPTS

Concept 9.1 Microbial Growth Can Be Controlled in Several Ways

1. The physical methods for controlling microorganisms are generally intended to achieve sanitization, which involves methods to reduce the numbers of, or inhibit the growth of, microbes.

Concept 9.2 There Are Various Physical Methods to Control Microbial Growth

2. Heat is a common control method. Used in the food industry, **thermal death point** and **thermal death time** are used to determine how long it takes to kill a population of microbial cells.
3. **Incineration** using a direct flame achieves sterilization in a few seconds. (Figure 9.3)
4. Moist heat has several applications: the exposure to boiling water at 100°C for 2 hours can result in sterilization, but spore destruction cannot always be assured; the **autoclave** sterilizes materials in about 15 minutes at 15 psi; **pasteurization** reduces the microbial population in a liquid but the process is not intended to be a sterilization method. (Figure 9.6)
5. **Filtration** uses various materials to trap microorganisms within or on a filter. **Membrane filters** are the most common. Air can be filtered by using a **high-efficiency particulate air (HEPA) filter**. (Figures 9.9, 9.10)
6. **Ultraviolet (UV) light** is an effective way of killing microorganisms on a dry surface and in a confined air space. (Figure 9.11)
7. **X-rays** and **gamma rays** are two forms of **ionizing radiation** used to sterilize heat-sensitive objects. Irradiation also is used in the food industry to control microorganisms on perishable foods. (Figure 9.12)
8. For food preservation, drying, salting, and low temperatures can be used to control microorganisms.

Concept 9.3 Chemical Control Usually Involves Disinfection

9. Chemical agents are effectively used to control the growth of microorganisms. A chemical agent used on a living object is an **antiseptic**; one used on a nonliving object is a **disinfectant**. (Figure 9.14)
10. Both antiseptics and disinfectants are selected according to certain criteria, including an ability to kill microorganisms or interfere with their metabolism. (Figure 9.16)
11. The **phenol coefficient test** can be used to evaluate antiseptics and disinfectants. **Use dilution tests** are more practical for everyday applications of antiseptics and disinfectants. The **disk diffusion method** is also used to determine bacterial susceptibility to various chemical agents. (Figure 9.17)

Concept 9.4 A Variety of Chemical Agents Can Control Microbial Growth

12. **Halogens** (chlorine and iodine) are useful for water disinfection, wound antisepsis, and various forms of sanitation.
13. **Phenol** derivatives, such as hexachlorophene, are valuable skin antiseptics and active ingredients in presurgical scrubs. (Figure 9.19)
14. **Heavy metals** (silver and copper) are useful as antiseptics and disinfectants, respectively.
15. **Alcohol** (70% ethyl alcohol) is an effective skin antiseptic.
16. Soaps and detergents are effective degerming agents. **Quats** are more effective as a disinfectant than as an antiseptic.
17. Hydrogen peroxide acts by releasing oxygen to cause an effervescing cleansing action. It is better as a disinfectant than an antiseptic.
18. **Formalin** and **glutaraldehyde** are sterilants that alter the biochemistry of microorganisms. **Ethylene oxide** gas under controlled conditions is an effective sterilant for plasticware. **Chlorine dioxide** gas can be used to sanitize air ducts, food and meat processing plants, and hospital areas.

■ CHAPTER SELF-TEST

For **Steps A–D**, you can find answers to questions and problems in **Appendix D**.

STEP A: REVIEW OF FACTS AND TERMS

Multiple Choice

Read each question carefully before selecting the *one* answer that best fits the question or statement.

1. All the following terms apply to microbial killing *except* _____.
   A. sterilization
   B. microbicidal
   C. bactericidal
   D. fungistatic

2. The thermal death time is _____.
   A. the time to kill a microbial population at a given temperature
   B. the time to kill a microbial population in boiling water
   C. the temperature to kill all pathogens
   D. the minimal temperature needed to kill a microbial population

3. Which one of the following statements is NOT true of dry heat?
   A. The energy does not penetrate materials easily.
   B. Long periods of exposure are needed.
   C. The heat oxidizes microbial proteins.
   D. The heat corrodes sharp instruments.

4. An autoclave normally sterilizes material by heating the material to _____°C for _____ minutes at _____ psi.
   A. 100; 10; 30
   B. 121.5; 15; 15
   C. 100; 15; 0
   D. 110; 30; 5

5. Air filtration typically uses a _____ filter.
   A. HEPA
   B. membrane
   C. sand
   D. diatomaceous earth

6. For bactericidal activity, _____ has/have the ability to cause thymine dimer formation.
   A. X-rays
   B. ultraviolet (UV) light
   C. gamma rays
   D. microwaves

7. The elimination of pathogens in foods by irradiation is called _____.
   A. the D value
   B. the pasteurizing dose
   C. incineration
   D. sterilization

8. Preservation methods such as salting result in the _____ microbial cells.
   A. loss of salt from
   B. gain of water into
   C. loss of water from
   D. lysis of

9. Which one of the following statements does *not* apply to antiseptics?
   A. They are used on living objects.
   B. They usually are microbicidal.
   C. They should be useful as dilute solutions.
   D. They can sanitize objects.

10. All the following are chemical parameters considered when selecting an antiseptic or disinfectant *except* _____.
    A. dehydration
    B. temperature
    C. stability
    D. pH

11. If a chemical has a phenol coefficient (PC) of 63, it means the chemical _____.
    A. is better than one with a PC of 22
    B. will kill 63% of bacteria
    C. kills microbes at 63°C
    D. will kill all bacteria in 63 minutes

12. Which one of the following is *not* a halogen?
    A. Iodine
    B. Mercury
    C. Clorox bleach
    D. Chlorine

13. Phenolics include chemical agents _____.
    A. such as the iodophores
    B. derived from carbolic acid
    C. used as tinctures
    D. such as formaldehyde

14. Heavy metals, such as _____, work by _____.
    A. mercury; disrupting membranes
    B. copper; producing toxins
    C. iodine; denaturing proteins
    D. silver; inactivating proteins

15. Alcohols are _____.
    A. surfactants
    B. heavy metals
    C. denaturing agents
    D. detergents

16. All the following statements apply to quats *except* ____.
A. they react with cell membranes
B. they are positively charged molecules
C. they are types of soaps
D. they can be used as disinfectants

17. Hydrogen peroxide ____.
A. is an effective sterilant
B. cross-links proteins and nucleic acids
C. can emulsify and solubilize pathogens
D. is not recommended for use on open wounds

18. Ethylene oxide can be used to ____.
A. kill bacterial spores
B. clean wounds
C. sanitize work surfaces
D. treat water supplies

Fill-In: Physical Methods

Use syllables that follow to form the term that answers the clue pertaining to physical methods of control. The number of letters in the term is indicated by the dashes, and the number of syllables in the term is shown by the number in parentheses. Each syllable is used only once.

A AU CLAVE DE DER DRY HOLD ING ING LET NA O POW TION TO TRA TUR UL VI

19. Instrument for sterilization (3) ___ ___ ___ ___ ___ ___ ___ ___ ___
20. Sterilized in an oven (2) ___ ___ ___ ___ ___ ___
21. Occurs with moist heat (5) ___ ___ ___ ___ ___ ___ ___ ___ ___ ___ ___ ___ ___
22. Preserves meat, fish (2) ___ ___ ___ ___ ___ ___ ___
23. Method of pasteurization (2) ___ ___ ___ ___ ___ ___ ___
24. "Light" for air sterilization (5) ___ ___ ___ ___ ___ ___ ___ ___ ___ ___ ___

Matching: Chemical Agents

Chemical agents are a broad and diverse group, as this chapter has demonstrated. To test your knowledge over the chemical methods of control, match the chemical agent on the right to the statement on the left by placing the correct letter in the available space. A letter may be used once, more than once, or not at all.

Statement

25. ____ The halogen in bleach
26. ____ A 70% concentration is recommended
27. ____ Quaternary compounds, or quats
28. ____ Example of a heavy metal
29. ____ Triclosan is a derivative
30. ____ Used for sterilizing plastic culture dishes

Chemical agent

A. Cationic detergent
B. Chlorine
C. Ethyl alcohol
D. Ethylene oxide
E. Formaldehyde
F. Glutaraldehyde
G. Iodine
H. Phenol
I. Silver

■ STEP B: CONCEPT REVIEW

31. Distinguish between **sterilization** and **sanitization**. (**Key Concept 9.1**)
32. Summarize the **filtration** methods used to sterilize a liquid and decontaminate air. (**Key Concept 9.2**)
33. Summarize how **ultraviolet (UV) light** works to control microbial growth and explain how **X-rays** and **gamma rays** are used as physical control agents. (**Key Concept 9.2**)

34. Describe how the effectiveness of a chemical agent can be measured. (**Key Concept 9.3**)
35. Justify why **alcohol** is not a method for skin sterilization. (**Key Concept 9.4**)
36. Estimate the value of **hydrogen peroxide** as a bacteriostatic agent. (**Key Concept 9.4**)

STEP C: APPLICATIONS AND PROBLEM SOLVING

37. You need to sterilize a liquid. What methods could you devise using only the materials found in the average household?

38. You are in charge of a clinical microbiology laboratory in which instruments are routinely disinfected and equipment is sanitized. A salesperson from a disinfectant company stops in to spur your interest in a new chemical agent. What questions might you ask the salesperson about the product?

39. As a hospital infection officer, one of your duties is to ensure that the examination rooms are disinfected after patient treatment and prior to receiving new patients. Having hired new staff to carry out the examination room "cleaning," what will you tell the new employees with regard to examination room disinfection?

STEP D: QUESTIONS FOR THOUGHT AND DISCUSSION

40. Instead of saying that "food has been irradiated," manufacturers indicate that it has been "cold pasteurized." Why do you believe they must use this terminology?

41. The label on the container of a product in the dairy case proudly proclaims, "This dairy product is sterilized for your protection." However, a statement in small letters below reads, "Use within 30 days of purchase." Should this statement arouse your suspicion about the sterility of the product? Why?

42. Before taking a blood sample from the finger, the blood bank technician commonly rubs the skin with a pad soaked in alcohol. Many people think that this procedure sterilizes the skin. Explain why they are correct or incorrect in their thinking.

43. Suppose that you had just removed the thermometer from the mouth of your sick child and confirmed your suspicion of fever. Before checking the temperature of your other child, how would you treat the thermometer to disinfect it?

44. The water in your home aquarium always seems to resemble pea soup, but your friend's is clear. Not wanting to appear stupid, you avoid asking him his secret. But one day, in a moment of desperation, you break down and ask, whereupon he knowledgeably points to a few pennies among the gravel. What is the secret of the pennies?

CHAPTER 10

Control of Microorganisms: Antimicrobial Drugs and Superbugs

For centuries, physicians believed heroic measures were necessary to save patients from the ravages of infectious disease. They prescribed frightening courses of purges (bowel emptying), enormous doses of strange chemical concoctions, blood-curdling ice-water baths, deadly starvations, and bloodlettings. These treatments probably complicated an already bad situation by reducing the natural body defenses to the point of exhaustion (see chapter opening figure). In fact, George Washington might have died from a streptococcal infection of the throat, perhaps exacerbated by the bloodletting treatment that removed almost two liters of his blood within a 24-hour period.

When the germ theory of disease emerged in the 1880s, a revolution unfolded in the identification of microorganisms (germs) as the cause of infectious diseases. However, even with this new understanding of infectious disease causation, it did not change the fact that little could be done for the infected patient. Then, in the 1940s, antimicrobial agents, including antibiotics, burst on the scene, and another revolution in medicine began.

Medicine had a period of powerful, decisive growth, as doctors found they could successfully alter the course of infectious disease. Antibiotics effected a radical change in medicine and charted a new course for treating infectious disease. By 1970, many health experts believed that infectious diseases, especially ones of bacterial origin, could soon be medical history thanks to antibiotics.

15th century French illustration of patients waiting for bloodletting.
© Science Source.

Unfortunately, it soon became clear that bacterial species and other microorganisms quickly could become **superbugs**, by developing resistance to many antimicrobials. The Centers for Disease Control and Prevention (CDC) reports that each year in the United States there are more than 2 million illnesses and 23,000 deaths caused by antibiotic-resistant bacteria. In fact, the rise in **antimicrobial resistance** (**AMR**) throughout the world in large part is caused by the overuse or misuse of antibiotics by doctors and the public (**FIGURE 10.1**). Consequently, in September 2016, world leaders at the United Nations General Assembly for the first time said they were committed to "taking a broad, coordinated approach" to address the root causes of AMR affecting global health. If not, some health experts estimate that by 2050 up to 10 million people might die each year due to AMR.

In this chapter, we discuss antimicrobial drugs as the mainstays of our healthcare delivery system to treat bacterial, viral, fungal, and parasitic infections and diseases. We explore the discovery of antimicrobial drugs, learn the reasons for the rise in AMR, and examine how we, as a society, can try to reverse this trend and avoid a "post antibiotic era."

FIGURE 10.1 The Global Threat from Antibiotic Resistance. Many bacterial species around the world are becoming resistant to our antibiotic medicines. *»» Prior to reading this chapter introduction, how aware were you of the problem of antibiotic resistance?*

Courtesy of the CDC.

Chapter Challenge

Over the past few years, an increasing number of reports have warned of superbugs, microbial species that have become resistant to multiple antimicrobial agents. Now a bigger threat might loom. In 2008, a strain of *Klebsiella pneumoniae*, a bacterial species that can cause pneumonia and bloodstream infections in hospitalized patients, was isolated from the urine of a hospitalized patient who had visited India. This strain contained a transmissible genetic element, called NDM-1, which, along with other resident resistance genes, make the organism essentially resistant to all available antibiotics. Now, it is spreading globally and this "nightmare bacterium" has been reported in the United States. To make matters worse, other bacterial genera within the same taxonomic family (Enterobacteriaceae) as *Klebsiella* are being identified that have picked up the NDM-1 genetic element, turning these microbes into superbugs. So, why have superbugs become so dangerous? Let's find out as we explore antimicrobial drugs and drug resistance.

■ KEY CONCEPT 10.1   Antimicrobial Agents Are Chemical Substances Used to Treat Infectious Disease

Beginning in the late 1940s, a group of chemicals became available to assist the immune system to fight infection and disease. These **chemotherapeutic agents** were used to treat infections, diseases, and other disorders, such as cancer. In microbiology, **chemotherapy** involves those **antimicrobial agents** (**antimicrobials**) used to treat infectious disease.

The History of Chemotherapy Originated with Paul Ehrlich

In the drive to control and cure infectious disease, the efforts of microbiologists in the early 1900s were primarily directed toward enhancing the body's natural defenses. Among the leaders in the effort to control disease was an imaginative German investigator named Paul Ehrlich. Ehrlich knew that specific dyes would stain specific bacterial species. As a result, he thought there must be specific chemicals that would be toxic to specific bacterial species. This idea led Ehrlich and his coworkers to a search for chemicals that would be "magic bullets," that is, specific chemicals that in a patient would seek out and destroy pathogens in infected tissues without harming the patient.

After months of painstaking study, one of Ehrlich's collaborators, Sahachiro Hata, discovered a compound capable of destroying the syphilis spirochete *Treponema pallidum*. The drug became known as Salvarsan. Over the ensuing 20 years, German chemists continued to synthesize and manufacture dyes for fabrics and other industries, and the chemists routinely tested their new products for antimicrobial properties. In 1932, Gerhard Domagk, a German pathologist and bacteriologist, used one of these products, called Prontosil, on test animals. He found that the drug, a type of sulfa drug, had a pronounced inhibitory effect on staphylococci, streptococci, and other gram-positive bacterial species.

Many historians see this discovery as the beginning of modern antimicrobial chemotherapy.

Fleming's Work Ushered in the Era of Antibiotics

One of the first to postulate the existence and value of antibiotics was the British microbiologist Alexander Fleming (**FIGURE 10.2A**). In 1928, Fleming was performing research on staphylococci. He noted that on one of his discarded nutrient agar plates containing *Staphylococcus* a blue-green mold was contaminating the plate. Surrounding the mold, no bacterial colonies were growing (**FIGURE 10.2B**). Intrigued by the failure of staphylococci to grow near the mold, Fleming isolated the mold, identified it as a species of *Penicillium*, and found it produced a substance that kills gram-positive bacterial species. Though he failed to isolate the elusive substance, he named it "penicillin."

Then, in 1939, a group at England's Oxford University, led by pathologist Howard Florey and biochemist Ernst Chain, reisolated Fleming's penicillin, and they conducted clinical trials with highly purified samples. They discovered that penicillin was effective against a large variety of diseases, including gonorrhea, meningitis, tetanus, and diphtheria. However, England was already involved in World War II, so a group of American companies developed the techniques for the large-scale production of penicillin and made the drug available for medical use.

(A)

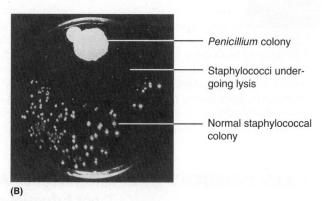

Penicillium colony

Staphylococci undergoing lysis

Normal staphylococcal colony

(B)

FIGURE 10.2 Alexander Fleming and His Culture of *Penicillium*. (A) Alexander Fleming reported the existence of penicillin in 1928 but was unable to purify it for use as an antimicrobial agent. **(B)** The actual photograph of Fleming's culture plate shows how staphylococci in the region of the *Penicillium* colony have been killed (they are "undergoing lysis") by some unknown substance produced by the mold. Fleming called the substance penicillin. *»» Why haven't the staphylococcal colonies farther from the mold also undergone lysis?*

Fleming's discovery of penicillin was a lucky discovery, something that is quite uncommon in science. However, by the early 1940s Selman Waksman's group at Rutgers University was doing a systematic screening specifically for antibacterial drugs. His group's work would revolutionize the entire pharmaceutical industry. In fact, Waksman coined the term "antibiotic."

The preceding historical account illustrates the origins for the two groups of antimicrobial agents. Those drugs like penicillin that are products of, or derived from, the metabolism of living microorganisms, represent **antibiotics**. By contrast, many antimicrobials are synthesized in the laboratory. These **synthetic drugs** include Salvarsan and Prontosil and all antiviral and many antibacterial, antifungal, and antiparasitic antimicrobials. Today, many antimicrobials have been produced by a process that chemically modifies the naturally produced drug. These are called **semisynthetic drugs**, and they often are more effective and more broadly used than the original natural antimicrobial.

The Selection of a Clinically Useful Antimicrobial Depends on Several Factors

Often antimicrobial drugs are part of the management strategy for treating patients with severe microbial infections. Therefore, dosing, knowing the appropriate drug spectrum, and using drug combinations can be key to therapeutic success. Consequently, antimicrobial agents have several important properties that need to be considered when prescribing a drug for an infection or disease.

Selective Toxicity

Ehrlich's idea of a magic bullet was based on **selective toxicity**, which states that an antimicrobial drug should harm the pathogen but do little, if any, harm to the patient. Today, two terms are used when considering the toxicity of a drug. The **toxic dose** refers to the concentration of the drug causing harm to the host. The **therapeutic dose** refers to the concentration of the drug that effectively destroys (is **microbicidal**) or inhibits (is **microbiostatic**) the pathogen. Together, these are used to formulate the **therapeutic window**, which is the concentration range or dosage of the drug in the serum that is tolerated by the host but which will eliminate the infection or disease agent (**FIGURE 10.3**). This window will differ from one patient to another, so a range (window) provides dosage flexibility. When determining dosage, it also must be kept in mind

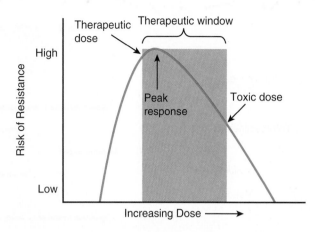

FIGURE 10.3 A Representation of the Therapeutic Window. For a patient needing antibiotic therapy, finding the correct drug dosage means considering the efficacy of the drug (therapeutic dose) and its toxicity to the patient (toxic dose). It also is important to be aware of the possibility of drug resistance development (orange curve). *»» In this example, which dose (therapeutic or toxic) would be best to treat this patient? Explain.*

what effect the administered drug will have on the development of antibiotic resistance by the bacterial pathogen causing the infection.

As we will see while progressing through this chapter, the best way to develop safe antimicrobial drugs is to identify specific and unique structural or metabolic components of microbial cells that are susceptible to the drugs. In fact, most antibiotics target the bacterial cell envelope (cell wall and cell membrane) or some aspect of bacterial metabolism. In addition, sometimes combining antimicrobials can have an effect that is greater than the effect by either drug alone. Such **synergistic effects (synergism)** will lower the therapeutic dose and make the drug combination safer for human use. We will see examples of antibiotic synergism when we examine specific antibiotics.

Antimicrobial Spectrum

Another important factor in prescribing an appropriate drug is identifying the range of pathogens for which a particular drug will work. This range of antimicrobial action is the **antimicrobial spectrum**. For example, if an antibacterial drug affects pathogens in many taxonomic groups, both gram positive and gram negative, the drug is considered a **broad-spectrum antibiotic**. One of the drawbacks of prescribing such a drug is that it will kill or inhibit

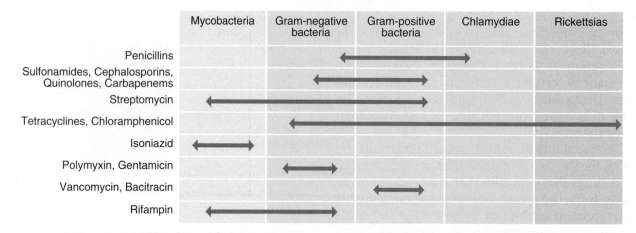

FIGURE 10.4 The Antimicrobial Spectrum of Activity. Antimicrobial drugs have a limited range of activity in the domain Bacteria. »» *Using the figure, which drugs would have a broad spectrum and which ones would have a narrow spectrum?*

not only the pathogen but also many of the bacterial species making up part of the normal human microbiome. On the other hand, if an antibacterial drug is useful against only a few specific pathogens or a specific family of microbes, the drug is a **narrow-spectrum antibiotic** (FIGURE 10.4). Penicillin G is one such example because it only affects the cell walls of gram-positive bacterial species. Note that the antimicrobial spectrum might change if a pathogen acquires AMR genes.

Antibiotics Are More Than Simply Antimicrobial Agents

Before we survey specific antibiotics, it is important to understand why some bacterial and fungal species produce antibiotics in the first place. Antibiotics represent **secondary metabolites**, which are chemical compounds not directly used by the organism for growth and reproduction. These metabolites often are produced during the stationary phase of growth, when an antibiotic would be less likely to have an adverse effect on the antibiotic producer's growth, yet when competition for limited nutrients is intense. Therefore, it makes sense to assume that in the natural environment (e.g., soil, water) there can be fierce competition between the immense numbers of microbes for the available nutrients. If a bacterial or fungal species has the ability to secrete an antibiotic, that organism might kill or inhibit the growth of those competitors sensitive to the chemical. Antibiotic production thus gives the producer a selective advantage for nutrients and space.

On the other hand, if antibiotics are so destructive, why aren't all the antibiotic-susceptible species in the soil killed by the antibiotic-producing species? Perhaps it is not always necessary to kill your neighbors. MICROFOCUS 10.1 describes some alternative roles for antibiotics.

MICROFOCUS 10.1: Evolution and Microbial Ecology

Antibiotics and Life in a Microbial Community

In the laboratory, the discovery that some bacterial or fungal species in soil can use antibiotics as a weapon to kill another microbe gave microbiologists the impression that antibiotics have but one role—kill or inhibit the growth of competitors. After all, it is a dangerous world out there and if one species can eliminate or severely limit another species' ability to obtain the limited resources available, the antibiotic-producing species should have the evolutionary advantage.

Microbiologists today are beginning to appreciate the mutually beneficial roles that low (sublethal) concentrations of antibiotics can play in the everyday life of a microbial community. Here are a few examples of what microbiologists are discovering (see figure).

Antibiotics Can Act as Cues

If exposed to low concentrations of antibiotics, these chemicals can influence the control of hundreds of genes involved in metabolism and organism behavior. This may cue the sender (species 1) to isolate itself in a biofilm, now separated from species 2.

Antibiotics Act as Synergistic Signals

Some antibiotics act as a synergistic signal, triggering cooperation between species. The result often is biofilm formation. Such signaling then benefits both the sender (species 1) and recipient (species 2).

Antibiotics Can Act as Competitive Manipulators

An antibiotic produced by one species can manipulate chemically another species to benefit the chemical producer. For example, species 1 produces a chemical that results in species 2 fleeing the environment. The manipulation of species 2 by species 1 lowers the competition without killing the competition.

So, why then do other bacterial species in the soil have antibiotic resistance genes? One hypothesis suggests that the products of these resistance genes fine tune, or modulate, the antibiotic levels to produce a more balanced cue, synergistic signal, or manipulation.

As more is learned about the diverse role of antibiotics in nature, we will certainly come to better understand and appreciate antibiotics in the daily life of a microbial community, which may not be as dangerous a place for microbes as we once thought.

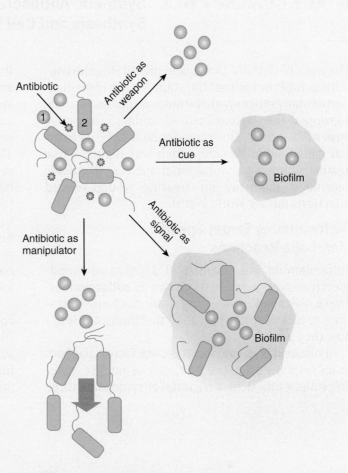

Possible roles of antibiotics.

Modified from Ratcliff, W. C. and Denison, R. F. 2011. *Science* 332(6029):547–548.

Concept and Reasoning Checks 10.1

a. Evaluate the early discoveries of and uses for chemotherapeutic agents for selective toxicity.
b. Identify how penicillin was discovered and later developed into a useful drug.
c. Assess the importance of the therapeutic window and the antimicrobial drug spectrum in prescribing treatment for an infectious disease.
d. What advantage is afforded by bacterial cells that secrete antibiotics in the soil?

Chapter Challenge A

The multidrug-resistant *K. pneumoniae* strains, combined with the genetic element NDM-1, are simply adding to the battery of resistance genes that provide resistance against antibiotics. In addition, these superbugs are gram negative (members of the Enterobacteriaceae), which, even without the presence of NDM-1, are historically difficult to kill.

QUESTION A: *(a) What is it about cell wall structure of the gram-negative species that makes them difficult to treat and (b) what does this tell you about the therapeutic window in trying to balance the therapeutic dose and toxic dose in the affected patient?*

You can find answers online in **Appendix F**.

■ KEY CONCEPT 10.2   Synthetic Antibacterial Agents Primarily Inhibit DNA Synthesis and Cell Wall Formation

The work of Gerhard Domagk in identifying the usefulness for Prontosil set the stage for the identification of many other synthetic antimicrobials. In 1935, a group at the Pasteur Institute isolated the active form of Prontosil. The group found it to be a chemical called sulfanilamide, which was highly active against gram-positive bacterial species; it quickly became a mainstay for treating wound-related infections during World War II.

Sulfonamides Target Specific Metabolic Reactions

Sulfanilamide was the first of a group of broad-spectrum synthetic agents known as **sulfonamides**. These so-called "sulfa drugs" are bacteriostatic—they interfere with bacterial metabolism. Here's how they work.

Folic acid is an important growth factor that bacterial cells need for the synthesis of nucleic acids. To produce folic acid, a bacterial enzyme joins three important substrates, one of which is **para-aminobenzoic acid (PABA)**. This molecule is similar in chemical structure to the sulfonamide called sulfamethoxazole (SMZ) (**FIGURE 10.5**). Importantly, humans do not synthesize folic acid; we acquire it from our diet. Therefore, treating patients with SMZ allows the drug to compete with PABA for the active site in the bacterial enzyme. Such **competitive inhibition** blocks folic acid synthesis, which prevents nucleic acid synthesis. Without DNA and RNA, bacterial cell metabolism will cease and cell death will occur. However, drug resistance is now quite common as mutations have arisen that allow the resistant species to absorb folic acid from outside sources.

Newer formulations of sulfonamides often combine two drugs. For example, co-trimoxazole (Bactrim®) is a combination of trimethoprim and sulfamethoxazole (TMP-SMZ) and is used to treat infections of the urinary tract, lungs (pneumonia), and ears. This combination of drugs is an example

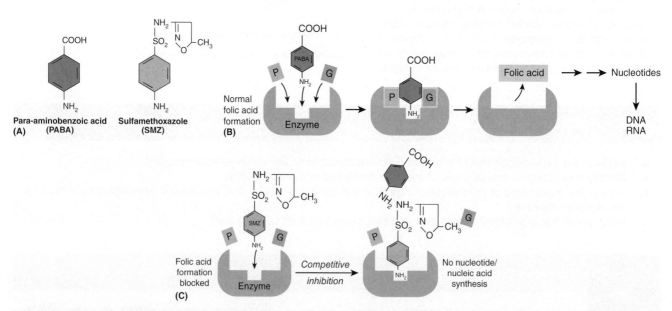

FIGURE 10.5 The Disruption of Folic Acid Synthesis by Competitive Inhibition. (A) The chemical structures of PABA and SMZ are very similar. **(B)** In the normal synthesis of folic acid, a bacterial enzyme joins three components [pteridine (P), PABA, and glutamic acid (G)] to form folic acid. However, in competitive inhibition **(C)**, SMZ competes for the active site because of its great abundance. SMZ assumes the position normally reserved for PABA, so folic acid cannot form. *»» Why would treatment with SMZ require a high therapeutic dose?*

TABLE 10.1 A Summary of Some Major Synthetic Antibacterial Drugs

Drug Class/Mode of Action	Examples	Antibacterial Spectrum/ Drug Effect	Possible Side Effects
Sulfonamides • Interfere with folic acid metabolism	Prontosil Sulfamethoxazole	Broad/bacteriostatic	Nausea or diarrhea High allergic potential
Quinolones • Inhibit DNA synthesis	Levofloxacin Ciprofloxacin	Broad/bactericidal	Nausea Central nervous system damage (rare)
Isoniazid • Interferes with mycobacterial cell wall synthesis	Isoniazid	Narrow/bactericidal	Nausea, vomiting, stomach upset

of synergism, as the two drugs together are of more benefit than either drug alone. In addition, the development of antibiotic resistance to both drugs is less likely.

Quinolones Are Synthetic Antimicrobials that Inhibit DNA Replication

Another group of synthetic, bactericidal drugs is the **quinolones**, which inhibit an enzyme needed for DNA synthesis in gram-positive and gram-negative bacterial cells. Again, the prokaryotic enzyme is different from its eukaryotic counterpart, so quinolones specifically target the bacterial pathogen.

The **fluoroquinolones**, one of the most prescribed antibiotics in the United States, are used to treat urinary tract infections, gonorrhea and chlamydia, and intestinal tract infections. Examples include levofloxacin and ciprofloxacin (Cipro). Alarmingly, the extensive use of fluoroquinolones worldwide has allowed bacteria to develop quinolone resistance.

These synthetic drugs and a synthetic antimycobacterial agent are summarized in TABLE 10.1.

Concept and Reasoning Checks 10.2

a. Explain why sulfanilamide does not interfere with DNA replication in eukaryotic cells.
b. Describe the mode of action for the fluoroquinolones.

Chapter Challenge B

Among the superbugs, AMR can be conferred by non-NDM-1 genes. One such gene many superbugs have is for the inactivation of Cipro.

QUESTION B: *A patient has a urinary tract infection (UTI) caused by* Escherichia coli. *The patient was put on TMP-SMZ, but the infection remained. So, the patient was switched to Cipro. However, the UTI was not cleared and the infection continued. What cellular process should these antibiotics have targeted if the E. coli cells were susceptible?*

You can find answers online in **Appendix F**.

■ KEY CONCEPT 10.3   Beta-Lactam Antibiotics Target Bacterial Cell Wall Synthesis

One of the most common mechanisms of antibiotic action is blocking the synthesis of the bacterial cell wall. Today, a large number of natural and semisynthetic antibiotics are available that target the assembly of the peptidoglycan component of bacterial cell walls.

Penicillin Has Remained the Most Widely Used Antibiotic

Penicillin has saved more patients with infectious diseases than any other antibiotic. Its effective therapeutic window has made it the drug of choice in eradicating many infections. All penicillins have a **beta-lactam ring** and core structure but differ in the side groups attached (**FIGURE 10.6**).

Penicillin's mode of action targets the bacterial cell wall, specifically the peptidoglycan layer and the repeating disaccharides *N*-acetylglucosamine (NAG) and *N*-acetylmuramic acid (NAM). When a cell is growing or dividing, the cell must add additional disaccharides (NAG-NAM) to the expanding peptidoglycan chains, and these must be cross-linked by short peptides between NAMs in adjacent chains. Penicillin inhibits the bacterial enzyme needed to form the peptide cross-links, which leaves the cell wall so weak that internal osmotic pressure causes the cell to swell and burst. Penicillin is therefore bactericidal in rapidly multiplying bacterial

cells. If bacterial cells are multiplying slowly, or are dormant, the drug can have either a bacteriostatic effect or no effect at all.

Natural Penicillins

Penicillins G and V are naturally produced by the fungus *Penicillium notatum*. Penicillin G has been the most popular penicillin antibiotic, and it is usually the one intended when doctors prescribe "penicillin." It is destroyed by stomach acid, so the antibiotic is given intravenously. Penicillin V is more acid resistant and can be taken orally.

The natural penicillins have a narrow drug spectrum, and they are most useful against gram-positive bacterial species. However, many bacterial species have developed resistance to the natural penicillins. These resistant species produce enzymes called **beta-lactamases** that break open the beta-lactam ring. Such a change converts penicillin G into harmless penicilloic acid (**FIGURE 10.7**). To overcome this form of AMR, semisynthetic penicillins and penicillin combinations (co-drugs) have been developed.

Semisynthetic Penicillins

Chemists have attached various chemical groups to the beta-lactam ring to create new penicillins that are more resistant to beta-lactamases and have a wider

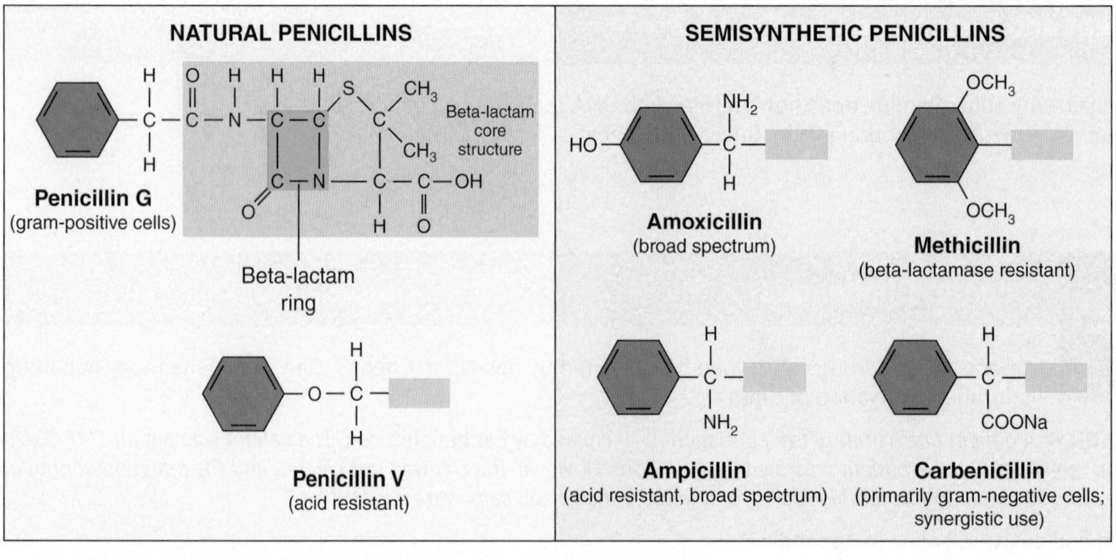

FIGURE 10.6 Some Members of the Penicillin Group of Antibiotics. The beta-lactam ring is common to all the penicillins. Different penicillins are formed by varying the side group on the molecule. »» *Why has there been such a need for so many semisynthetic penicillins?*

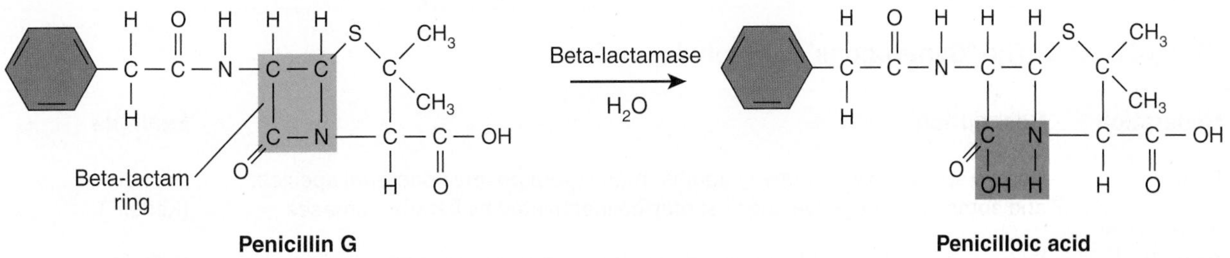

FIGURE 10.7 The Action of Beta-Lactamase on Sodium Penicillin G. The enzyme beta-lactamase converts penicillin to harmless penicilloic acid by opening the beta-lactam ring and inserting a hydroxyl group to the carbon and a hydrogen atom to the nitrogen. *»» What does the action of beta-lactamase tell you about the role of the beta-lactam ring in the structure of penicillin?*

drug spectrum. Numerous semisynthetic penicillins, such as ampicillin and amoxicillin, have been introduced (see Figure 10.6). Other semisynthetic penicillins, such as carbenicillin, are effective against an even broader range, including gram-negative bacterial species, and can be used for infections of the urinary tract. Still, bacterial species have developed resistance, as exemplified by **methicillin-resistant S. aureus (MRSA)**, which is resistant to a wide range of penicillins and other antibiotics.

Co-Drugs

Some semisynthetic penicillins are paired with a beta-lactamase inhibitor. For example, Augmentin® consists of amoxicillin and the beta-lactamase inhibitor clavulanic acid. The synergistic combination allows amoxicillin to inhibit cell wall synthesis without being destroyed by a beta-lactamase enzyme.

Other Beta-Lactam Antibiotics Also Inhibit Cell Wall Synthesis

In addition to the penicillins, there are other wall-inhibiting antibiotics with a beta-lactam ring (**FIGURE 10.8**).

Cephalosporins

A widely used group of antibiotics is the **cephalosporins**. These antibiotics, isolated from the fungus *Cephalosporium acremonium*, inhibit bacterial cell wall synthesis and resemble penicillins in chemical structure, except there are different atoms building the beta-lactam core structure. Cephalosporins are used to fight pathogens where penicillin resistance is encountered or in cases where a patient has an allergy to penicillin. Cephalosporins have a broader bactericidal spectrum against gram-negative bacterial pathogens because of the drug's ability to penetrate the porin proteins in the outer membrane.

Over the years, chemists have modified the cephalosporin structure to produce chemically related antibiotics with an improved therapeutic window and better resistance to beta-lactamases. Accordingly, the cephalosporins can be separated into "generations" by their drug spectrum (TABLE 10.2). Each succeeding generation has a greater activity against gram-negative species than the preceding generation, often with less activity against gram-positive species.

Carbapenems

Another set of beta-lactam drugs with a very broad spectrum is the **carbapenems**, which are derived from a compound produced by the bacterium *Streptomyces cattleya*. Because their structure makes them highly resistant to many beta-lactamase enzymes (see

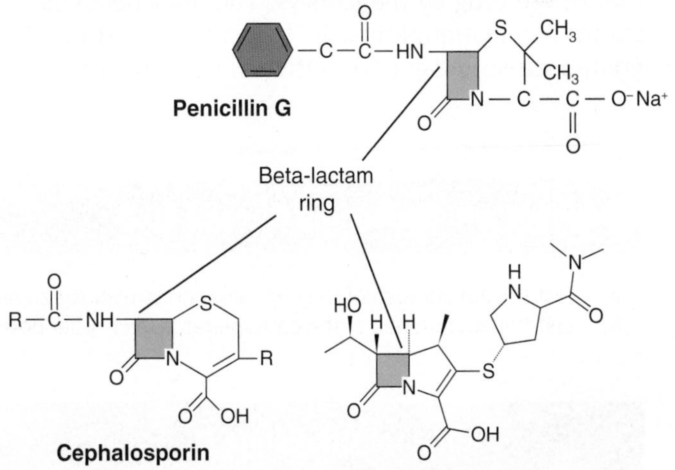

FIGURE 10.8 A Structural Comparison Between Beta-Lactam Antibiotics. Structures for three beta-lactam antibiotics (penicillin G, cephalosporin, and meropenem) contain the beta-lactam ring. *»» Some individuals with a penicillin allergy might be given meropenem. Structurally, why might meropenem be considered for treatment in a patient with a penicillin allergy?*

TABLE 10.2 The "Generations" of Cephalosporins

Generation	Description	Example
First	Narrow spectrum with activity against many gram-positive bacterial species and some gram-negative species; may be inactivated by beta-lactamases	Cephalexin (Keflex®)
Second	Expanded spectrum with increased activity against gram-negative rods; better resistance against beta-lactamases	Cefaclor
Third	Broad spectrum with more activity against gram-negative species and *Pseudomonas*; increased resistance to beta-lactamases	Ceftriaxone
Fourth	Extended spectrum with increased activity against gram-negative species resistant to third-generation drugs	Cefepime
Fifth	Useful against *Pseudomonas*, MRSA, and penicillin-resistant *Streptococcus pneumoniae*	Ceftobiprole

Figure 10.8), they have been one of the most important groups of clinically useful "last resort" antibiotics; that is, for many bacterial pathogens, such as those *E. coli* and *K. pneumoniae* strains showing resistance to other beta-lactam antibiotics. Still, there can be an inactivation problem. In patients, carbapenems, like imipenem, are normally degraded by the kidneys before they can have a therapeutic effect. Therefore, the drug usually is prescribed in synergistic combination with cilastatin, which prevents premature degradation of the drug by the kidneys. The imipenem/cilastatin combination (Primaxin®) is active against most gram-positive and gram-negative clinical isolates.

Since 2012, the medical community has been extremely worried over the spread of AMR caused by bacterial species that have acquired the genetic element **New Delhi metallo-beta-lactamase-1 (NDM-1)**. Bacterial pathogens commonly isolated from healthcare and community infections with this element, especially those in the family Enterobacteriaceae (e.g., *E. coli* as well as *Enterobacter* and *Klebsiella* species), are resistant to a broad range of antibiotics, including those in the carbapenem family. Currently, there are no new antibiotics in development for these superbugs, referred to as the **carbapenem-resistant Enterobacteriaceae (CRE)**.

Concept and Reasoning Checks 10.3

a. List the advantages of the cephalosporins over the penicillins.
b. List the advantages of the carbapenems over other beta-lactam antibiotics.

Chapter Challenge C

The original *K. pneumoniae* strain expressing the NDM-1 enzyme had the ability to destroy most known beta-lactam antibiotics, including the penicillins and cephalosporins. What was most alarming was that the strain could also hydrolyze the carbapenems, which are last resort antibiotics for treating infected patients.

QUESTION C: *Why has the alarm over this NDM-1 enzyme been heightened and of great concern to doctors who have patients with antibiotic-resistant infections caused by* E. coli *or another species in the Enterobacteriaceae?*

You can find answers online in **Appendix F**.

■ KEY CONCEPT 10.4   Other Antibiotics Target Some Aspect of Metabolism

Besides antibiotics that target folic acid synthesis and the bacterial cell wall, other antibiotics also affect the cell envelope (cell wall or cell membrane) or a specific metabolic process such as protein synthesis (translation) or nucleic acid synthesis (DNA replication, RNA transcription) (**FIGURE 10.9**).

Glycopeptide Antibiotics Inhibit Bacterial Cell Wall Synthesis

Vancomycin represents a **glycopeptide** antibiotic that is produced by *Amycolatopsis* (formerly *Streptomyces*) *orientalis*. The drug affects bacterial cell wall synthesis by blocking the addition of NAG-NAM disaccharides to the elongating peptidoglycan chain. Because the antibiotic cannot pass through porin pores in gram-negative species, vancomycin is used to treat gram-positive pathogens. It is administered by intravenous injection for severe staphylococcal diseases for which penicillin allergy or bacterial

resistance exists. In fact, vancomycin has emerged as a key treatment in therapy for MRSA, often representing the "drug of last resort." However, vancomycin-resistant *S. aureus* (VRSA) strains have recently been reported.

Polypeptide Antibiotics Affect the Cell Envelope

Both bacitracin and polymyxin B are polypeptide antibiotics produced by *Bacillus* species. These antibiotics have an unacceptable therapeutic window for internal use (renal failure, respiratory paralysis), so they generally are restricted to topical use on the skin.

Bacitracin is a cyclic polypeptide that is bactericidal and interferes with the transport of the NAG-NAM disaccharides through the cell membrane. Bacitracin is available in ointments for topical treatment of superficial skin infections caused

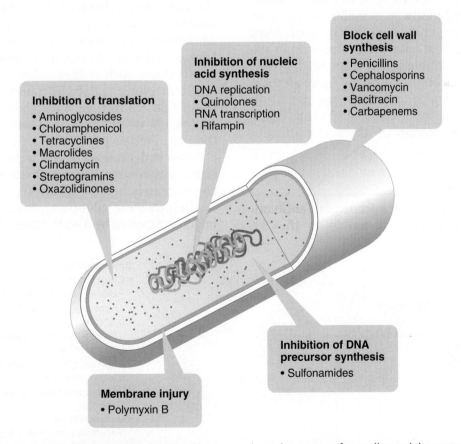

FIGURE 10.9 The Targets for Antibacterial Agents. There are six major targets for antibacterial agents: the cell wall, cell membrane, ribosomes (translation), nucleic acid synthesis (RNA transcription and DNA synthesis), and metabolic reactions. *»» Hypothesize why so many antibiotics targeted the cell wall.*

by gram-positive and gram-negative bacteria. When combined in an ointment with neomycin and polymyxin B, it is sold under the brand Neosporin®.

Polymyxins are cyclic polypeptides that insert into the cell membrane. Acting like detergents, antibiotics like polymyxin B are bactericidal, disrupting the outer membrane structure of gram-negative bacteria and leading to cell death. The two antibiotics, bacitracin and polymyxin B, are combined with gramicidin (which increases permeability of the cell membrane) in the antibiotic ointment Polysporin®.

Polymyxin E (colistin) has been valuable as a last-resort antibiotic against many superbugs. However, colistin might be losing its effectiveness in antimicrobial therapy. Studies have reported that bacteria can transfer resistance genes for colistin (*mcr-1*) through horizontal gene transfer. The gene has been identified in the food supply and clinical isolates of *mcr-1* have been found in humans in China, Europe, Canada, and the United States.

Many Antibiotics Affect Translation

Proteins often are called the "workhorses" of the cell because of their diverse functions and essential roles in cell metabolism. To manufacture proteins, ribosomes are required. Importantly, prokaryotic ribosomes differ from eukaryotic ribosomes in size and structure, so antibiotics are available that specifically target bacterial ribosomes. These antibiotics affect translation by targeting some step in the translational apparatus (**FIGURE 10.10**). Let's briefly look at the major classes of antibiotics affecting translation.

Aminoglycosides

The **aminoglycosides** are bactericidal antibiotic compounds that attach irreversibly to the 30S subunit of bacterial ribosomes, causing translational misreading of the genetic code on messenger RNA (mRNA) molecules.

In 1943, the first aminoglycoside, streptomycin, was discovered by Selman Waksman's group. At the time, the discovery was sensational because streptomycin was the first effective treatment for tuberculosis and numerous other diseases caused by gram-negative bacteria. Because streptomycin therapy can lead to nausea, vomiting, fever, and rash in patients, the drug has been largely replaced by safer alternatives, although it still is prescribed as co-drug therapy for antibiotic-susceptible cases of tuberculosis.

Newer generations of aminoglycosides, including gentamicin and tobramycin, have been isolated. However, their effectiveness is decreased because they need to be injected intramuscularly or intravenously. In addition, damage to the kidneys (renal toxicity) and to the auditory nerve (hearing loss) can occur.

Tetracyclines

The **tetracyclines** are broad-spectrum, bacteriostatic antibiotics that inhibit translation by blocking attachment of the tRNA to the 30S subunit. There are naturally occurring chlortetracyclines, isolated from species of *Streptomyces*, and semisynthetic tetracyclines, such as doxycycline (**FIGURE 10.11**).

Tetracycline antibiotics can be taken orally, a factor that led to their indiscriminate use and the subsequent development of bacterial resistance to the drug. Tetracyclines are not recommended for children under 8 years of age because the drugs can cause a yellow-gray-brown discoloration of teeth and stunted bones (see Figure 10.11).

Despite these side effects, doxycycline is used for infections caused by a wide range of gram-negative

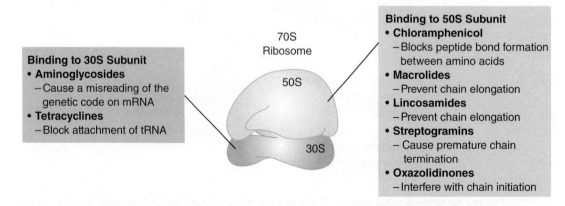

Binding to 30S Subunit
• **Aminoglycosides**
 – Cause a misreading of the genetic code on mRNA
• **Tetracyclines**
 – Block attachment of tRNA

70S Ribosome

50S

30S

Binding to 50S Subunit
• **Chloramphenicol**
 – Blocks peptide bond formation between amino acids
• **Macrolides**
 – Prevent chain elongation
• **Lincosamides**
 – Prevent chain elongation
• **Streptogramins**
 – Cause premature chain termination
• **Oxazolidinones**
 – Interfere with chain initiation

FIGURE 10.10 Antibiotics and Their Effect on Translation. Many antibiotics interfere with the translational machinery on the 30S (small) or 50S (large) ribosome subunit. **»» *Why are there so many natural antibiotics that affect translation?***

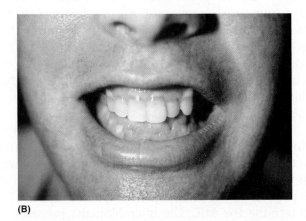

(A)

Doxycycline

(B)

FIGURE 10.11 The Tetracyclines. (A) Doxycycline consists of four hydrocarbon rings. **(B)** The staining of teeth associated with tetracycline use. *»» What chemical feature characterizes all tetracycline antibiotics?*

(B) © Kenneth E. Greer/Visuals Unlimited.

bacterial species. The antibiotic also is valuable for treating primary atypical pneumonia, syphilis, gonorrhea, and pneumococcal pneumonia.

Glycylcycline

In 2005, the U.S. Food and Drug Administration (FDA) approved a new class of antibiotics, the glycylcyclines, which are related to the tetracyclines. The new broad-spectrum drug, tigecycline, must be administered intravenously and has been effective against MRSA infections. However, by 2013, tigecycline resistance had been detected among members of the Enterobacteriaceae.

Chloramphenicol

Chloramphenicol is a bacteriostatic, broad-spectrum antibiotic that prevents peptide bond formation at specific amino acids in the forming polypeptide on the 50S ribosome subunit. It is the drug of choice in the treatment of typhoid fever and is an alternative to tetracycline for epidemic typhus and Rocky Mountain spotted fever.

Chloramphenicol usually is reserved for treating serious and life-threatening infections because the drug can produce serious side effects. It can

inhibit development of the bone marrow and result in **aplastic anemia**, which is an inability to produce red blood cells.

Macrolides

The **macrolides** are a group of bacteriostatic antibiotics that block translation by binding to the 50S subunit and inhibiting elongation of the new polypeptide chain. Erythromycin is used against gram-positive pathogens in patients with penicillin allergy, and it is placed in the eyes of newborns to prevent neonatal conjunctivitis caused by *Neisseria gonorrhoeae* and *Chlamydia trachomatis*.

Other semisynthetic macrolides with a broader spectrum include clarithromycin and azithromycin (Zithromax®), one of the world's best-selling antibiotics.

Lincosamides

The most commonly used of the **lincosamides** (derived from *Streptomyces lincolnensis*) is the semisynthetic drug clindamycin. This bacteriostatic drug inhibits polypeptide chain elongation by blocking tRNA binding to the 50S subunit. Clindamycin is an alternative antibiotic for cases in which penicillin resistance is encountered. The drug is active against aerobic, gram-positive cocci and anaerobic, gram-negative bacilli. Use of the antibiotic is limited to serious infections because the drug eliminates competing organisms from the intestines, permitting endogenous residents like *Clostridium difficile* to overgrow the area. The clostridial toxins then can induce a potentially lethal inflammation of the colon.

Streptogramins

Another group of cyclic peptides is the **streptogramins**. Synthetic versions of these antibiotics are prescribed as a combination of two cyclic peptides, called quinupristin-dalfopristin (Synercid®). Both components interfere with translation on the 50S subunit, the interference being synergistic and bactericidal. The drugs are effective against a broad range of gram-positive bacterial species, including *S. aureus* and respiratory pathogens.

Oxazolidinones

One of the new antibiotic classes is the **oxazolidinones**. One member of the group, linezolid (Zyvox®), prevents peptide bond formation at specific amino acids in the forming polypeptide on the 50S ribosome subunit. Linezolid is effective in treating gram-positive bacterial species, including MRSA. However, due to their toxicity, the oxazolidinones are drugs of last resort, being used only where other antibiotics have failed.

Some Antibiotics Inhibit Nucleic Acid Synthesis

The synthetic quinolones that inhibit bacterial DNA synthesis have already been described earlier in the chapter.

Other drugs affect transcription. **Rifampin**, a semisynthetic, bacteriostatic drug, interferes with RNA synthesis by binding to the bacterial RNA polymerase. Because mycobacterial resistance to rifampin can develop quickly, it is prescribed in combination therapy for tuberculosis and leprosy patients. It also is administered to carriers of *Neisseria* and *Haemophilus* species that cause meningitis.

FIGURE 10.12 identifies the top classes of antibiotics on the global market. TABLE 10.3 reviews those antibacterial drugs currently in use.

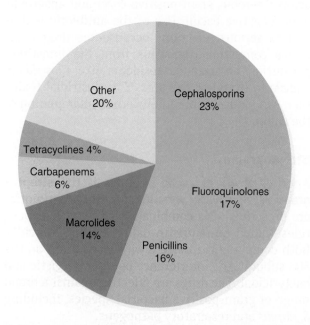

FIGURE 10.12 Global Antibiotics Market, by Class. The global antibiotics market has been dominated by four classes. *»» What percent of the antibiotic market targeted (a) bacterial cell wall synthesis, (b) translation, and (c) DNA synthesis?*

Data from BCC Research.

There Are Several Antibiotic Susceptibility Assays

The substantial variety of antibiotics developed since the 1930s necessitates that medical professionals know which ones are most effective against a pathogen. Accordingly, "antibiotic sensitivity assays" can be performed. These assays are used (1) to measure the susceptibility of a bacterial strain to one or more antibiotics and (2) to monitor the evolution of bacterial resistance. Two general methods are in common use to test the susceptibility of a bacterial pathogen to specific antibiotics: the tube dilution method and the agar strip/disk diffusion method.

The **tube (broth) dilution method** determines the lowest concentration of antibiotic that will prevent growth of the pathogen. This amount is known as the **minimum inhibitory concentration (MIC)** and is used when determining a therapeutic window. To determine the MIC, the microbiologist prepares a set of tubes (or wells) with different concentrations (dilutions) of a particular antibiotic (**FIGURE 10.13**). Each tube is inoculated with a standardized number of bacterial cells, incubated, and examined for the growth of bacterial cells. The extent of growth diminishes as the concentration of antibiotic increases, and eventually an antibiotic concentration is observed at which growth no longer occurs. This is the MIC.

The second type of antibiotic susceptibility test operates on the principle that an antibiotic will diffuse from a paper or plastic-coated strip or disk into an agar medium containing a uniformly spread test organism. In one version of the test, called the **Etest**, a strip impregnated with a marked gradient of antibiotic is placed on a nutrient agar plate that previously had been inoculated with the bacterial pathogen of interest. As the drug diffuses into the agar, the higher drug concentrations will inhibit bacterial growth (**FIGURE 10.14A**). The inhibition of growth is observed as a **zone of inhibition**, seen as a clear, oval halo on the plate. By reading the number on the strip (antibiotic concentration) where growth intersects the strip, the MIC can be determined.

In another version of the test, called the **disk diffusion method**, inhibition of bacterial growth is again observed as a failure of a susceptible bacterial species to grow on agar. A common application is the **Kirby-Bauer test**, which determines the susceptibility of a bacterial species to a group of antibiotics. To set up the test, paper disks, each having a different antibiotic, are placed on an agar plate

TABLE 10.3 A Summary of Major Antibacterial Drugs (by Mode of Action)

Antibacterial Agent	Examples	Antibacterial Spectrum/Drug Effect	Possible Side Effects
Competitive Inhibitors of DNA Precursor Synthesis (see Table 10.1)			
Inhibitors of Cell Wall Synthesis			
Penicillins	Penicillin G and V Ampicillin Amoxicillin Methicillin	Narrow (gram positive) to broad/bactericidal	Diarrhea Allergic reaction
Cephalosporins	Cephalothin Cephalexin	Broad/bactericidal	Diarrhea, nausea Allergic reaction
Carbapenems	Imipenem Meropenem	Broad/bactericidal	Diarrhea, nausea, headache, rash Allergic reaction
Glycopeptides	Vancomycin	Narrow (gram positive)/bactericidal	Hearing and kidney damage
	Bacitracin	Gram-positive bacteria, especially staphylococci	Kidney damage, if injected
Inhibitors of Cell Envelope Function			
Polypeptides	Polymyxin B	Narrow (gram positive)/bactericidal	Kidney damage
Inhibitors of Protein Synthesis (Translation)			
Aminoglycosides	Streptomycin Gentamicin Neomycin	Broad/bactericidal	Hearing defects Kidney damage
Tetracyclines Glycylcycline	Doxycycline Tigecycline	Broad/bacteriostatic Broad/bacteriostatic	Discoloration of teeth Diarrhea, nausea, vomiting Diarrhea, nausea, vomiting
Chloramphenicol	Chloramphenicol	Broad/bacteriostatic	Aplastic anemia
Macrolides	Erythromycin Clarithromycin Azithromycin	Narrow (gram positive)/bacteriostatic	Diarrhea, nausea, vomiting Liver damage
Lincosamides	Clindamycin	Narrow (gram positive)/bacteriostatic bacteria	Pseudomembranous colitis
Streptogramins	Quinupristin/dalfopristin	Narrow (gram positive)/bactericidal	Nausea, diarrhea, headache
Oxazolidinones	Linezolid	Narrow (gram positive)/bacteriostatic	Low platelet count
Inhibitors of DNA Synthesis (see Table 10.1)			
Inhibitors of RNA Synthesis (Transcription)			
Rifampin	Rifampin	Broad/bactericidal	Liver damage Reddish-orange body fluids

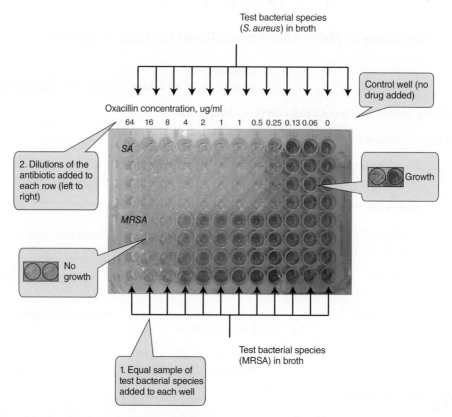

FIGURE 10.13 Determination of Minimum Inhibitory Concentration. The MICs for oxacillin against *S. aureus* and the MRSA are shown. Rather than using separate tubes, equal numbers of bacterial cells in broth were placed in each well (with four replicate rows) of a plastic plate and then dilutions of the antibiotic oxacillin added to each. The far right wells (controls) have no antibiotic added. After incubation, a colorless dye is added to each well, which changes to an orange color if the well contains living bacterial cells. From the dilutions, the MIC can be determined. *»» What are the MICs for* S. aureus *and* MRSA?

Image courtesy of Dojindo Laboratories.

Culture medium: Mueller-Hinton Broth containing various concentrations of oxacillin for 6 hours at 35°C (190 uL total volume per well). Incubation: 35°C, 2 hours (after adding 10 uL of the dye solution to each well). O.D. measurement: at 460 nm. By using the kit, the minimum inhibitory concentration (MIC) was determined in 8 hours.

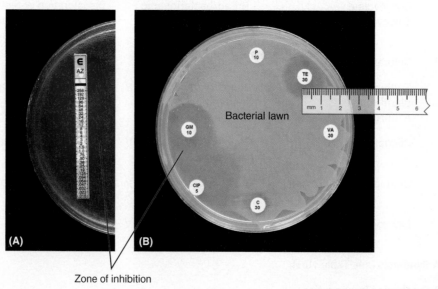

FIGURE 10.14 Antimicrobial Susceptibility Testing. (A) With the Etest®, the antibiotic-containing strip can determine categorical results (susceptible, intermediate, resistant), as well as the minimum inhibitory concentration (MIC). **(B)** The results of a disk diffusion assay (Kirby-Bauer test). *»» In (a), what is the MIC and in (b) to which antibiotics was the test bacterium susceptible?*

Courtesy of bioMerieux, Inc.

Memorial Medical Center
1 January–31 December 2007 Cumulative Antimicrobial Susceptibility Report*
Percent Susceptible

Gram-Negative Organisms	No. Strains	amikacin	ampicillin	cefazolin	cefotaxime	ceftazidime	ciprofloxacin	nitrofurantoin†	gentamicin	imipenem	piperacillin-tazobactam	trimethoprim-sulfamethoxazole	tobramycin
Acinetobacter baumannii	32	80	7	0	34	52	51	-	60	80	46	58	59
Citrobacter freundii	49	100	-	-	72	67	90	78	100	99	67	67	100
Enterobacter aerogenes	31	100	-	-	68	69	92	85	91	99	74	95	91

* %S for each organism/antimicrobial combination was generated by including the first isolate of that organism
   encountered on a given patient.
† Nitrofurantion, data from testing urine isolates only.
(-) drug not tested or drug not indicated.

FIGURE 10.15 An Antibiogram. An antibiogram is a chart or list of the antibiotic sensitivity test results for isolated pathogens and their strains. The chart lists the number of strains tested and their percent susceptibility to each antibiotic. *»» Everything else being equal, if a patient had an **Acinetobacter baumannii** infection, which antibiotic should be selected and why?*

Reproduced from: *Lab Medicine* August 2009 40(8):459–462.

containing the previously spread bacterial species. During a 24- to 48-hour incubation, each antibiotic will diffuse into the agar and might affect bacterial growth. If the bacterial cells are susceptible to a particular antibiotic, a **zone of inhibition** will surround the paper disk (**FIGURE 10.14B**). This diameter can be measured (in millimeters; mm) with a metric ruler. No halo, or a zone less than a certain diameter (using a standard table for comparison), indicates the strain was resistant to the antibiotic. A "susceptible" category indicates that the strain could be treated with the dosage used in the disk, whereas a

"resistant" category indicates the strain would not be effective at that dosage.

After completing the susceptibility tests, the clinical microbiology lab professionals can organize the results into a table called an **antibiogram** (**FIGURE 10.15**). Doctors, hospital infection control managers, and other healthcare professionals and skilled nursing facilities can consult this table (often pocket size) to help guide them in making appropriate and timely treatment decisions, which done correctly can have a significant and positive impact on patient outcomes.

Concept and Reasoning Checks 10.4

a. What problems are associated with vancomycin use?
b. Why are the polypeptide antibiotics used only topically?
c. Identify the bacterial source for the aminoglycosides, chloramphenicol, tetracyclines, macrolides, clindamycin, and the streptogramins.
d. To what enzyme does rifampin bind? As a result, what cellular process is inhibited?
e. Explain the significance of knowing a drug's minimum inhibitory concentration (MIC).

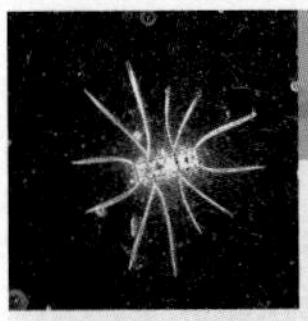

Chapter **Challenge D**

Additional antibiotic resistance in the CRE superbugs conferred by non-NDM-1 genes includes inactivation of erythromycin, rifampin, chloramphenicol, and polymyxins.

QUESTION D: *From these first four challenges (A–D), use the diagram in Figure 10.9 to identify the resistance groups (major targets) to which* K. pneumoniae *and other CRE could be resistant. Would you consider such CRE organisms as superbugs? Explain.*

You can find answers online in **Appendix F**.

■ KEY CONCEPT 10.5   Other Antimicrobial Drugs Target Viruses, Fungi, and Parasites

If you visit the CDC website that lists diseases and conditions, about 36% of the diseases have a bacterial origin, 33% have a viral origin, 6% are fungal, and 25% are due to parasites (**FIGURE 10.16**). If you look at the developing nations, the percentage of viral and especially parasitic diseases would be much higher.

Although the data are important for monitoring trends and for targeting research, prevention, and control efforts, the information does not include non-notifiable diseases, such as colds, forms of gastroenteritis, yeast infections, and so on. Still, health experts can make the following generalization: there are dozens more antibiotics in the antimicrobial arsenal to fight bacterial infections than there are drugs to fight viral, fungal, and parasitic diseases. Further, the antiviral drugs are useful against only a very limited number of viral diseases and most have been developed to fight HIV disease/AIDS. Unlike antibiotics, there are no antiviral drugs that can definitely cure a person of a viral disease (except hepatitis C). Therefore, whether it is an antiviral drug directed against HIV, herpes, or influenza, antiviral chemotherapy only lessens the effects of the infection.

The ability to use antimicrobial agents effectively against viruses and fungi presents another problem not encountered with bacterial pathogens. The major problem is finding drugs affecting the viral or fungal pathogen that are not toxic to the host tissue; that is, the therapeutic window can be very narrow for these pathogens. Because fungi are members of the Eukarya, they possess much of the cellular machinery common to animals and humans. After they make their way inside cells, viruses make use of the host cell's machinery to replicate. Thus, many drugs targeting fungi and viruses also will target host tissues and potentially be quite toxic, making it difficult to achieve selective toxicity.

Antiviral Drugs Interfere with the Viral Replication Process

A limited number of antiviral drugs are of value for treating diseases caused by herpesviruses, hepatitis B and C viruses, HIV, and influenza viruses. Antiviral drugs are designed to inhibit specific stages in the virus replication process. Let's examine a few examples.

Virus Attachment/Penetration

Some antivirals block the ability of HIV to penetrate the host cell. If a virus cannot attach to and penetrate the host cell, infection cannot occur.

Genome Replication

Several antiviral drugs target DNA replication of the herpesviruses or HIV. Some are **nucleoside (base) analogs** that are similar to the nucleosides needed for viral DNA synthesis. When the analog binds to a replicating DNA molecule, it prematurely terminates DNA synthesis. Non-nucleoside inhibitors inhibit viral DNA polymerase enzymes and therefore shut down DNA synthesis or reverse transcription in the case of HIV.

Assembly/Release

A number of antiviral drugs prevent new viruses from being formed. **Protease inhibitors** inhibit HIV assembly into new viruses. **Neuraminidase inhibitors** prevent release of flu viruses from infected host cells.

TABLE 10.4 provides some examples of antiviral drugs that target these viral activities.

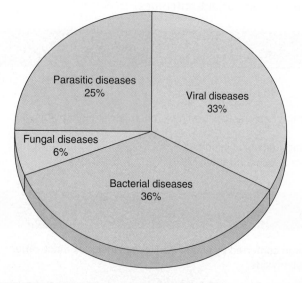

FIGURE 10.16 Types of Human Infectious Disease. The percentage of bacterial and viral disease is about equal. *»» Provide a reason why there are fewer fungal diseases than there are diseases of the other groups.*

TABLE 10.4 Examples of Antiviral Drugs (by Mode of Action)

Antiviral Drug (Trade Name)	Mode of Action	Targeted Viruses
Block Penetration		
Amantadine and rimantadine	Block viral penetration and uncoating	Influenza A viruses
Enfuvirtide	Blocks penetration/uncoating	HIV
Inhibition of Genome Replication		
Valacyclovir (Valtrex®)	Base analog (mimics guanine) inhibiting DNA polymerase	Herpesviruses (cold sores, genital herpes, shingles)
Ribavirin	Base analog (mimics guanine) inhibiting viral replication	Hepatitis B and C viruses Varicella zoster virus (shingles)
Azidothymidine (AZT; Retrovir®)	Base analog (mimics adenine) terminating DNA chain elongation	HIV
Vidarabine	Base analog (mimics adenine) terminating DNA chain elongation	Herpesviruses
Nevirapine	Non-nucleoside inhibitors of reverse transcriptase	HIV
Raltegravir	Blocks provirus integration by targeting the HIV integrase enzyme	HIV
Inhibition of Viral Assembly/Release		
Saquinavir	Inhibits HIV protease	HIV
Oseltamivir (Tamiflu®) Zanamivir (Relenza®)	Inhibit viral release	Influenza A and B viruses

Some Antifungal Drugs Cause Membrane Damage

Today, we are seeing more individuals with a reduced or compromised immune system. These **immuno-compromised** patients (especially due to AIDS) are vulnerable to many different types of secondary infections, many caused by fungal pathogens. As mentioned earlier, antimicrobial drugs used for fungal infections lack a high level of selective toxicity because fungal metabolism mimics human metabolism. Therefore, a drug affecting the fungal pathogen can also have a toxic effect on the host and generate mild to severe side effects. The main targets for antifungal drugs are the plasma membrane, cell wall, nucleic acids, and the cytoskeletal microtubules. TABLE 10.5 provides some examples of the drugs and their interactions with fungal pathogens.

The Goal of Antiprotistan Agents Is to Eradicate the Parasite

Because protists also are members of the domain Eukarya, antiprotistan drugs attempt to target unique aspects of nucleic acid synthesis, translation, or metabolic pathways found only in parasites. TABLE 10.6 identifies some of the drugs used to eradicate protistan parasites. However, as malaria is one of the most prevalent and threatening of the protistan diseases to humans, it is worth explaining what drugs have been and are being used.

Malaria is caused by any of four species of *Plasmodium*. In an infected human, the **aminoquinolines**, such as quinine, accumulate in infected red blood cells. In these cells, the drugs are toxic to the malaria parasite and interfere with the parasite's ability to break down and digest hemoglobin. The

TABLE 10.5 A Summary of Antifungal Drugs (by Mode of Action)

Antifungal Agent	Examples (brand name)	Use	Possible Side Effects
Inhibitors of Membrane Function			
Polyenes • Increase membrane permeability	Nystatin (Mycostatin®) Amphotericin B (Fungizone®)	Topical to treat *Candida* vaginal "yeast infections" To treat invasive fungal respiratory infections	Too toxic for systemic use Kidney failure
Azoles • Inhibit membrane sterol synthesis	Clotrimazole Miconazole Ketoconazole Fluconazole Itraconazole	Topical to treat *Candida* vaginal "yeast infections" and skin infections ("athlete's foot," "ringworm") Systemic fungal infections	Ketoconazole can cause liver damage
Inhibitors of Fungal Cell Wall Synthesis			
Echinocandins	Casofungin Micafungin	Used against Aspergillus and Candida infections	Narrow spectrum with few side effects
Inhibitors of Nucleic Acid Synthesis			
Flucytosine		Used to treat systemic fungal infections	Kidney and bone marrow toxicity
Inhibitors of Cell Division (Mitosis)			
Griseofulvin		Used to treat fungal infections of the skin, body, hair/beard, or nails	Severe skin reactions, liver damage

TABLE 10.6 A Summary of Antiparasitic Drugs (by Mode of Action)

Antiparasitic Agent/ Mode of Action	Examples	Use	Side Effects
Antimalarial Drugs			
Aminoquinolines	Quinine Chloroquine Mefloquine (Lariam®) Primaquine	To prevent or treat malaria caused by *Plasmodium* species	Severe neurological and behavioral side effect possible with mefloquine
Artemisinin	Artemisinin	To prevent or treat malaria caused by *Plasmodium* species	Few reports of adverse effects
Inhibitors of Nucleic Acids			
Sulfonamides • Inhibit folic acid metabolism	Sulfamethoxazole and trimethoprim	To prevent encephalitis caused by *Toxoplasma* in AIDS patients	Kidney and liver damage; allergic reactions
Nitroimidazoles • Inhibit DNA synthesis	Metronidazole	To treat *Giardia* infections of the small intestine, amebic liver abscesses, and amebic dysentery	Tumors in mice
Anti-Helminthic Drugs			
Antihelminthic agents • Alter membrane permeability • Inhibit glucose uptake • Cause parasite muscle paralysis	Praziquantel Mebendazole Ivermectin	Used for trematode infections Used for intestinal helminthic infections Used against a wide range of internal and external parasites	Diarrhea, stomach pain

MICROFOCUS 10.2: History

The Fever Tree

Rarely had a tree caused such a stir in Europe. In the 1500s, Spaniards returning from the New World told of its magical powers on malaria patients, and before long, the tree was dubbed "the fever tree." The tall evergreen grew only on the eastern slopes of the Andes Mountains (see figure). According to legend, the Countess of Chinchón, wife of the Spanish ambassador to Peru, developed malaria in 1638 and agreed to be treated with its bark. When she recovered, she spread news of the tree throughout Europe, and a century later, Linnaeus named it *Cinchona* after her.

For the next two centuries, cinchona bark remained a staple for malaria treatment. Peruvian Indians called the bark quina-quina (bark of bark), and the term *quinine* gradually evolved. In 1820, two French chemists, Pierre Pelletier and Joseph Caventou, extracted pure quinine from the bark and increased its availability still further. The ensuing rush to stockpile the chemical led to a rapid decline in the supply of cinchona trees from Peru, but Dutch farmers made new plantings in Indonesia, where the climate was similar. The island of Java eventually became the primary source of quinine for the world.

Leaves of a cinchona tree.

Photo by Forest & Kim Starr.

During World War II, Southeast Asia came under Japanese domination, and the supply of quinine to the West was drastically reduced. Scientists synthesized quinine shortly thereafter, but production costs were prohibitive. Finally, two useful substitutes (chloroquine and primaquine) were synthesized. Today, as resistance to these drugs is increasingly observed in the malarial parasites, scientists once again are looking for another "fever tree" to help control malaria.

parasite thus starves or the drug causes the accumulation of toxic products resulting from the degradation of hemoglobin in the parasite. Quinine was one of the first natural antimicrobials and was derived from the bark of the South American cinchona tree (MICROFOCUS 10.2).

Today, more effective synthetic drugs, such as chloroquine, mefloquine, and primaquine, are used. Chloroquine remains the drug of choice to treat all species of *Plasmodium*, whereas mefloquine (Lariam®) is primarily used for malaria caused by *P. falciparum*. Reports have been made that mefloquine can have rare but serious side effects, including severe depression, anxiety, paranoia, nightmares, and insomnia.

As *Plasmodium* resistance to these drugs has increased, another drug, called **artemisinin**, has become the drug of choice to treat multidrug-resistant strains of *P. falciparum*. Artemisinin was originally isolated from the shrub *Artemisia annua* (sweet wormwood); however, there are now synthetic forms more potent than the natural drug (FIGURE 10.17).

In red blood cells, artemisinin releases free radicals that destroy the malarial parasites. Unfortunately, parasite resistance to artemisinin has been detected in South East Asia. In fact, historically,

(A)

(B)   Artemisinin

FIGURE 10.17 The Source of Artemisinin. (A) *Artemisia* shrubs are grown in Africa, as indicated in this harvest by a Kenyan farmer. **(B)** Structure of artemisinin. *»» What is the mode of action of artemisinin?*

(A) Jack Barker/Alamy Images.

resistance to other antimalarial drugs has spread from South East Asia to other parts of Asia and Africa. The same scenario might come about for artemisinin resistance.

Antihelminthic Agents Target Nondividing Helminths

The helminths (flukes and tapeworms) are eukaryotic, multicellular parasites that are studied by some microbiologists. Although most antimicrobial drugs are targeted at actively dividing cells of the pathogen, the mechanism of action of most antihelminthic drugs targets the nondividing organisms.

Praziquantel is thought to change the permeability of the parasite plasma membranes of flukes and tapeworms. This leads to muscle contraction and paralysis in the parasite such that the parasites cannot feed and eventually die. **Mebendazole** inhibits uptake of glucose and other nutrients by adult and larval worms infecting the host intestine. Without the ability to carry out ATP synthesis, the parasites will die. The **avermectins** are effective against a wide variety of roundworms. The drugs affect the roundworm's nervous system, causing muscle paralysis. Table 10.6 also summarizes the characteristics of these antiparasitic agents.

Concept and Reasoning Checks 10.5

a. Why have most of the antiviral agents synthesized been targeted against viral genome replication?
b. Assess the emphasis for developing antifungal drugs targeting the fungal plasma membrane or cell wall.
c. Provide several reasons why there are fewer effective antifungal and antiparasitic drugs to treat infections.

■ KEY CONCEPT 10.6   Antimicrobial Drug Resistance Is a Growing Challenge

Between the late 1930s and the 1960s, most all major groups of antibiotics were introduced. Since then and until 2000, the intervening decades represent an "innovation gap" during which no new antibiotic classes were introduced (**FIGURE 10.18**). Importantly, during this gap, an increasing number of bacterial species became resistant to many of these antibiotics; that is, the bacterial cells could not be effectively inhibited or killed by the antibiotic, allowing the resistant cells to continue multiplying in the presence of therapeutic levels of an antibiotic. So, how do bacteria become resistant to antibiotics?

Antibiotic Resistance Can Develop and Spread in Several Ways

We have learned that some bacterial and fungal species can produce antibiotics. Therefore, it makes sense that in the natural world other bacterial species have developed the ability to become resistant to these antibiotics. This resistance to antibiotics is the result of mutations and horizontal gene transfer.

Mutations

Sometimes resistance is of its own making, such as through **mutations**, those rare, spontaneous changes in a cell's genetic material (**FIGURE 10.19A**). Because many bacterial cells have short generation times, the potential for the spread of the mutation in a bacterial cell population through binary fission and vertical gene transfer is tremendous. For example, in one study, scientists reported that a strain of *S. aureus* in an infected patient evolved resistance to vancomycin over a period of 30 months. During this period, the *S. aureus* strain accumulated 18 sequential mutations, making it resistant to vancomycin and several other antibiotics. The surprising part is that the patient was never treated with any of these antibiotics! The mutations produced "spontaneous" resistance. Today, such antibiotic-resistant organisms are increasingly responsible for human diseases of the intestinal tract, lungs, skin, and urinary tract.

Horizontal Gene Transfer

Antibiotic resistance often comes from bacterial cells picking up resistance genes (R plasmids) from other bacterial cells by means of **horizontal gene transfer** (**HGT**). Colistin resistance mentioned earlier was one example. These transfer mechanisms can allow bacterial cells to acquire antibiotic resistance genes from donor bacterial cells by undergoing **transformation**, **transduction**, or

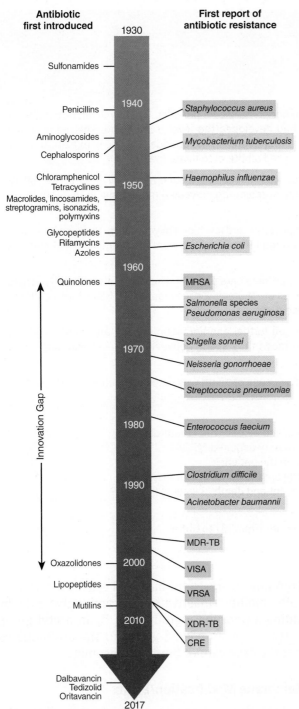

Antibiotic first introduced

Sulfonamides

Penicillins

Aminoglycosides
Cephalosporins

Chloramphenicol
Tetracyclines
Macrolides, lincosamides, streptogramins, isonazids, polymyxins
Glycopeptides
Rifamycins
Azoles

Quinolones

Oxazolidones

Lipopeptides

Mutilins

Dalbavancin
Tedizolid
Oritavancin

First report of antibiotic resistance

Staphylococcus aureus

Mycobacterium tuberculosis

Haemophilus influenzae

Escherichia coli

MRSA

Salmonella species
Pseudomonas aeruginosa

Shigella sonnei

Neisseria gonorrhoeae

Streptococcus pneumoniae

Enterococcus faecium

Clostridium difficile

Acinetobacter baumannii

MDR-TB

VISA

VRSA

XDR-TB

CRE

Innovation Gap

FIGURE 10.18 Timeline for Antibiotic Introduction and Appearance of Antibiotic Resistance. This timeline shows the time when major antibiotics were introduced and the approximate dates when some notable bacterial pathogens (pink = gram-negative species; green = gram-positive species; yellow = acid-fast species) were identified as being antibiotic resistant. MRSA (VISA, VRSA) = methicillin resistant (vancomycin intermediary resistant; vancomycin resistant) *Staphylococcus aureus*; MDR (XDR)-TB = multidrug-resistant (extensively drug-resistant) tuberculosis. *»» What are the two ways the bacterial species gained resistance?*

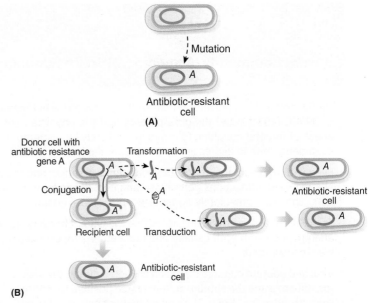

FIGURE 10.19 Mechanism of Antibiotic Resistance Acquisition. Antibiotic resistance can be the result of **(A)** a beneficial mutation and/or **(B)** transfer of an antibiotic resistance gene to a recipient cell by transformation, transduction, or conjugation. *»» Which one of the four mechanisms of gene acquisition might be the most common one for acquiring antibiotic resistance?*

conjugation (**FIGURE 10.19B**). Following acquisition of the gene(s), binary fission and vertical gene transfer will spread the resistance through a bacterial cell population.

Any bacterial cell that acquires resistance genes has the ability to resist one or more antibiotics. Because the cells can collect multiple resistance traits over time, they can become resistant to many different families of antibiotics; they can become superbugs. One tragic example is described in **CLINICAL CASE 10.**

Antibiotic Resistance Involves Different Protection Strategies

The problem of AMR, especially bacterial resistance to antibiotics, is a major global challenge facing microbiology and healthcare today. So, how do resistant bacterial species "fight off" or tolerate antibiotics? Let's briefly examine each of the resistance strategies, some of which are "offensive," whereas others are "defensive" (**FIGURE 10.20**).

Antibiotic Hydrolysis

Resistance can arise from the bacterium's ability to split apart (hydrolyze) the antibiotic.

Clinical Case 10

Severe Community-Acquired MRSA Pneumonia

On December 17, a previously healthy 8-year-old girl was taken to her primary care provider. The physician noted a fever of 39.4°C (103°F), and the girl's parents told the physician she had developed the fever 3 days ago. The girl also had spells of forceful coughing followed immediately by vomiting. She was treated in the provider's office with azithromycin, dexamethasone to decrease any swelling of the larynx, and aerosolized albuterol to dilate the bronchi.

Even with these interventions, her condition worsened, and the girl was rushed to a local emergency department, where she received intravenous ceftriaxone and nebulized albuterol. A chest radiograph revealed pneumonia in the right lower lobe. She was immediately transported to a referral hospital.

At the hospital, the physicians noted that the patient had low blood pressure and low blood oxygenation. She was intubated and provided with both cardiac and respiratory support oxygen. During intubation, she went into cardiac arrest, but she was resuscitated.

Viral and sputum cultures had been collected during the visit to the primary care provider. The viral culture tested positive for influenza and the sputum culture tested positive for methicillin-resistant *Staphylococcus aureus* (MRSA); blood cultures were negative for MRSA. The patient had not had an influenza vaccination.

Over the next 3 weeks in the hospital, the girl's situation deteriorated further and antibiotic therapy appeared hopeless. The patient suffered renal and hepatic failure, and an abscess formed in the space between the pleura and diaphragm (subpulmonary).

The patient died on January 7, 24 days after onset of symptoms. The cause of death was listed as pneumonia, respiratory distress, and MRSA sepsis.

Questions:

a. What was the point of providing azithromycin, dexamethasone, and aerosolized albuterol?
b. Why would this case study be considered an example of community-acquired MRSA?
c. Why didn't antibiotics like azithromycin and ceftriaxone work on the patient?
d. What do the viral, sputum, and blood cultures tell you about the spread of the infections?
e. What does the short duration between respiratory symptom onset and death (3 weeks) indicate?

You can find answers online in **Appendix E**.

For additional information, see www.cdc.gov/mmwr/preview/mmwrhtml/mm5614a1.htm.

The production of beta-lactamase enzymes by penicillin-resistant species is an example. By breaking the beta-lactam ring in penicillin, cephalosporin, and carbapenem antibiotics, resistant bacterial cells prevent inhibition of cell wall synthesis (see Figure 10.7). Often, the genes for beta-lactamase are carried on plasmids, so they can be moved from resistant to susceptible cells by HGT. As described earlier, the discovery of this offensive resistant mechanism led chemists to develop beta-lactamase inhibitors, such as clavulanic acid. When given with a beta-lactam antibiotic, the inhibitor prevents the beta-lactamase from hydrolyzing the antibiotic and preserves the therapeutic value of the antibiotic.

Antibiotic Modification

Another offensive mechanism involves modifying the antibiotic molecule. For example,

streptomycin-resistant bacterial cells can enzymatically modify (inactivate) the aminoglycoside by adding a phosphate group (PO_4^{-3}), an acetyl group ($C_2H_3O_2^-$), or an adenyl group to the antibiotic so that the drug cannot bind to ribosomes.

Membrane Modification/Efflux

A defensive resistance mechanism involves the ability of bacterial cells to prevent drug entry into the cytoplasm. For example, changes in membrane permeability in penicillin-resistant *Pseudomonas* prevent the antibiotic from entering the cytoplasm. Tetracycline-resistant species, such as *E. coli* and *S. aureus*, actively export (pump out) the drug. They contain cytoplasmic and membrane proteins that act as pumps to remove the antibiotic from the cell before tetracycline can affect the ribosomes in the cytoplasm. Some of the genes that code for these

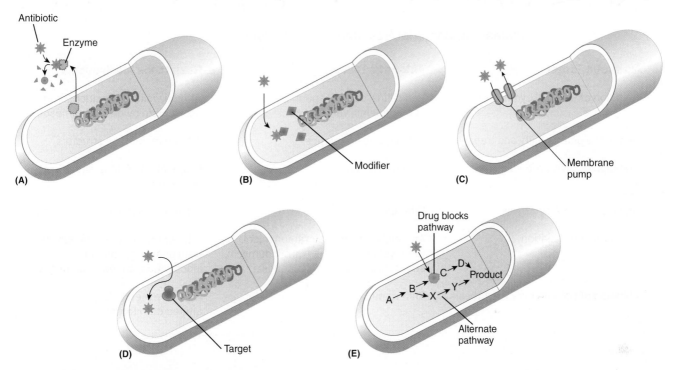

FIGURE 10.20 Antibiotic Resistance Tactics. Antibiotic resistance can develop by: **(A)** beta-lactamases breaking the structure of antibiotics like penicillin; **(B)** chemical groups inactivating an antibiotic like streptomycin; **(C)** membrane pumps pumping out an antibiotic like tetracycline; **(D)** cell targets such as ribosomes changing their structure so antibiotics like streptomycin will not bind; and **(E)** alternate pathways used to bypass a blocked metabolic pathway, such as for the sulfa drugs. *»» Which resistance mechanisms involve a chemical reaction and which involve a structural change?*

so-called efflux pumps are located on plasmids, which means they can be horizontally transferred to nonresistant recipient bacteria.

Target Modification

Another defensive mechanism involves altering the cell structure targeted by the antibiotic. Some streptomycin-resistant bacterial species have modified the structure of their ribosomes so that the antibiotic cannot bind to the organelle. However, the modified ribosomes still are functional, and they can carry out translation normally. Other targets that can be modified include RNA polymerase and enzymes involved in DNA replication.

Metabolic Pathway Alteration

Some antibiotics, such as the sulfa drugs discussed earlier in the chapter, act as competitive inhibitors. Often, bacteria have evolved alternative pathways to generate the same product as that blocked by the drug. Antibiotic resistance in this case is a matter of circumventing the blocked pathway.

TABLE 10.7 summarizes the bacterial resistance mechanisms.

Some Bacterial Species Are Tolerant of Antibiotics

Before leaving the topic of resistance mechanisms, it is important to realize that some antibiotic-susceptible species can still be tolerant of the drugs. **Antibiotic tolerance** refers to the ability of a pathogen to resist the effects of antibiotics. It is different from antibiotic resistance because tolerance is not dependent on mutations or HGT. Rather, tolerance arises from bacterial cells being in a transient, nondividing state.

Persister cells (discussed in Chapter 5) are not mutants but rather they represent a small subpopulation of regular cells that are dormant (nongrowing or extremely slow growing). They will survive the antibiotic treatment that kills the majority of the genetically identical susceptible cells in the population. Often such tolerance is seen in biofilms because these multicellular communities are impervious to antibiotic attack. The biofilm matrix retards the diffusion of the drugs into the interior of the biofilm, and the cells in the interior often have very slow metabolic rates. Consequently, diffusion limitation and the presence of slow-growing cells

TABLE 10.7 A Summary of Bacterial Resistance Mechanisms

Antibiotic Target	Antibiotic Class	Resistance Mechanisms
Inhibition of cell wall synthesis	Penicillins, cephalosporins, carbapenems, vancomycin	Altered wall composition; drug destruction by beta-lactamases
Membrane permeability and transport	Polypeptide antibiotics	Altered membrane structure
Inhibition of DNA synthesis	Fluoroquinolones	Inactivation of drug; altered drug target
Inhibition of RNA synthesis	Rifampin	Altered drug target
Inhibition of translation	Aminoglycosides, chloramphenicol, tetracyclines, macrolides, clindamycin, streptogramins, oxazolidinones	Altered membrane permeability; drug pumping; antibiotic inactivation; altered drug target
Inhibition of DNA precursor synthesis	Sulfonamides	Alternate metabolic pathway; altered drug target

make these antibiotic-susceptible cells tolerant of the drugs.

When such tolerant cells are not eliminated by antibiotic therapy and the immune system, they can persist as a source from which recurrent and chronic infections can develop. For example, persister cells and biofilm formation can be a major cause for chronic infections like tuberculosis, syphilis, and cystic fibrosis. **FIGURE 10.21** shows the consequences of persister and mutant cells during antibiotic therapy.

Antibiotic Resistance Is of Vital Concern in the Medical Community

The rise in AMR has resulted from the misuse and abuse of antimicrobials. Three factors have contributed to improper antimicrobial use.

Prescription Abuse

It is not unusual for drug companies to promote antibiotics heavily, for patients to pressure doctors for quick cures, and for physicians to sometimes misdiagnose infections or write prescriptions to avoid ordering costly tests to pinpoint the patient's illness. According to a 2014 study, 85% of antibiotic prescriptions in a skilled nursing facility were made solely from observation or without culturing. As a result, 65% of these scripts were inappropriate because the pathogens infecting the patient were resistant to the prescribed drug.

Over the past 15 years, global antibiotic consumption has increased by more than 30%. Many of these antibiotics are used in developing countries where the drugs often are available without prescription (**FIGURE 10.22**). The fact that many of these countries have substandard sanitation practices only exacerbates the demand for antibiotics.

Perscription Misuse by Individuals

People might diagnose their own illness and take "leftover" prescription antibiotics from their medicine cabinet for illnesses for which antibiotics are of no therapeutic use. Moreover, many people with a diagnosed infectious disease requiring an antibiotic tend to stop their prescription as soon as they feel better. Unfortunately, not all the pathogens might have been eliminated, and the ones remaining alive can evolve into resistant forms. The survivors proliferate well because they face reduced competition from susceptible organisms (see Figure 10.21).

Perscription Misuse by Health Organizations

Because of the considerable use of antimicrobials, hospitals are another "environment" for the emergence of AMR. In a CDC survey conducted in 183 American hospitals, 50% of the hospitalized patients received antibiotics, and about 25% of all patients received at least two different antibiotics.

In many cases, physicians overuse antibiotics, prescribing large doses (frequently broad spectrum)

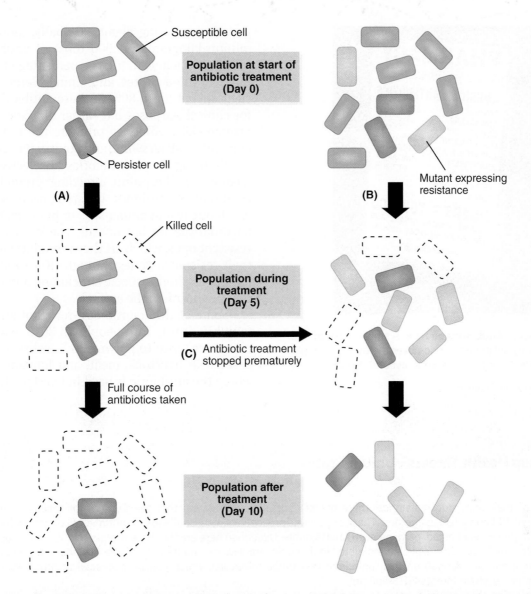

FIGURE 10.21 How Antibiotic Resistance Develops. (A) Ideally, with a complete course of antibiotics, all pathogens will be destroyed, although a few persister cells (purple) may survive. **(B)** If there are some resistant cells in the infecting population, they will survive and grow without any competition. If the person "feels better" and prematurely stops taking the antibiotic **(C)**, any mutant cells may have had the opportunity to express antibiotic resistance, survive, and again grow along with persister cells without any competition. *»» Would the persister cells be antibiotic resistant? Explain.*

to prevent infection during, and following, surgery. This misuse increases the possibility that resistant strains will replace the susceptible normal microbiome destroyed by the antibiotic (see again Figure 10.3). The result is known as a **superinfection**, meaning the pathogens are resistant to many (and perhaps most) antibiotics and now can colonize areas in the body where the normal microbiome was eliminated. In addition, antibiotic-resistant bacteria can be more adept at infecting and causing disease in hosts. Acclimating to antibiotics makes them more robust—they are super-superbugs!

Therefore, the abuse and misuse of antibiotics have allowed bacterial species to evolve and acquire more resistance to antibiotics, and many species have or are becoming multidrug resistant. MICROFOCUS 10.3 highlights the superbugs of biggest impact.

Antibiotic Abuse in Livestock

Perhaps most disturbing is the abuse and misuse of antibiotics in animal feeds. For decades, many farmers around the world have used animal feeds mixed with low, nontherapeutic amounts of antibiotics as

FIGURE 10.22 Antibiotics on Sale. A sign in front of a foreign pharmacy selling antibiotics over the counter. *»» What are the risks of buying such over-the-counter drugs?*

© Stephen Saks Photography/Alamy Images.

growth-promoting agents. Globally, more than 50 million kilograms (55,000 tons) of antibiotics are produced each year, a large percentage being used in the United States and China. Some estimates suggest that up to 80% of these antibiotics are not for clinical purposes but rather dispensed on large commercial feedlots to promote the faster growth of healthy cattle, swine, and poultry (**FIGURE 10.23**).

By using these antibiotics unselectively, many feedlots are becoming breeding grounds for the evolution of antibiotic-resistant bacterial species. Antibiotic use in animals, as in humans, kills many bacterial species but not all. The survivors are the resistant ones, which, as they reproduce, will spread their resistance genes through HGT to many similar and dissimilar species, including any pathogens that might infect the animals.

Some of these antibiotic-resistant strains have found their way into people. In 2015, bacteria carrying resistance to polymyxin E (colistin) were discovered in livestock, meat, and humans in China, where the antibiotic is used in livestock feed. Soon,

MICROFOCUS 10.3: Clinical Microbiology

Superbug Health Threats—United States

Because of their multidrug resistance, there are several microbial strains that the Centers for Disease Control and Prevention (CDC) considers dangerous. These superbugs present a clinical challenge, as antimicrobial therapy for infected patients might be very limited, perhaps even unattainable. Described here are the three threats the CDC has placed in the urgent category. All are discussed in more detail in their relevant disease chapters. In all, the CDC identifies 18 pathogens (17 bacterial and one fungal) on their antibiotic resistance threat list, which can be accessed at http://www.cdc.gov /drugresistance/pdf/ar-threats-2013-508.pdf.

C. difficile. The gram-positive, spore-forming bacterium *Clostridium difficile* (*C. diff*) has surpassed methicillin-resistant *Staphylococcus aureus* (MRSA) as a pathogen of urgent concern. Often patients receiving medical care can catch serious infections called healthcare-associated infections (HAIs). Antibiotic treatment, especially in older adults, for these HAIs can disturb the balance of the normal gut microbiota for several months. This allows time for *C. diff* to be picked up from endospore-contaminated fecal materials (e.g., contaminated surfaces or a healthcare provider's hands) and for it to multiply quickly. *C. diff* then produces disease-causing toxins, which can cause mild to severe diarrhea and even life-threatening colitis (inflammation of the bowel). Although *C. diff* is still susceptible to vancomycin, it causes an estimated 250,000 infections and 14,000 deaths each year in the United States.

Carbapenem-Resistant Enterobacteriaceae (CRE). Also considered an urgent threat by the CDC are those species of the Enterobacteriaceae (e.g., *Escherichia coli* and *Klebsiella* species) that have become resistant to nearly all antibiotics, including the carbapenems. Also included here are the bacterial species that have acquired the genetic element New Delhi metallo-beta-lactamase-1 (NDM-1), the subject of this chapter's Chapter Challenge. Healthy individuals are not usually at risk of a CRE infection. However, in healthcare facilities, there are more than 9,000 CRE cases each year among the patients. Up to half of all bloodstream infections caused by CRE result in death—approximately 600 deaths result from *E. coli* and *Klebsiella* alone. Fortunately, bloodstream infections account for a minority of all HAIs caused by Enterobacteriaceae. Infection often occurs when a doctor or a healthcare staff member touches an infected patient and then touches an uninfected patient. The bacteria also can contaminate medical instruments.

Drug-Resistant *Neisseria gonorrhoeae*. The CDC also includes this bacterial species in its urgent threat list. The species causes gonorrhea, the second most commonly reported sexually transmitted disease (more than 800,000 cases each year). It is considered a superbug because it is showing increased resistance to cephalosporins, macrolides, and tetracyclines. The CDC fears that emergence of resistance to the third-generation cephalosporin ceftriaxone would severely limit treatment options and could impair efforts to control gonorrhea spread.

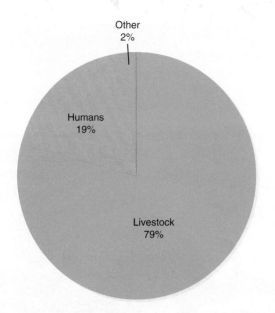

FIGURE 10.23 Antibiotic Use, United States. This chart shows the estimated annual use of antibiotics. Other = pets, aquaculture, and agricultural crops. *»» Why would antibiotics be used for pets, aquaculture (farming of aquatic organisms), and crops?*

Data from U.S. Food and Drug Administration.

colistin resistance mysteriously made its way to Denmark, where it was discovered in several samples from animals and food. Making matters worse, investigators reported that the colistin-resistant bacteria were also resistant to other antibiotics.

All this information makes it sound like antibiotic resistance originated from human abuse of antibiotics. Did antibiotic resistance exist before the modern era of antibiotic therapy, or is resistance a result of contemporary antibiotic use? INVESTIGATING THE MICROBIAL WORLD 10 looks at this important question.

New Approaches to Antimicrobial Therapy Are Needed

Combating AMR requires a two-pronged approach:

▶ **Careful Use of Antimicrobials.** First, as already described, we need to preserve the antibiotics and other antimicrobials we still have by using them prudently. The rapid evolution of resistance can be slowed through

Investigating the Microbial World 10

The Source of Antibiotic Resistance

Following the discovery and development of the first antibiotics for clinical use in the 1940s, it did not take many bacterial species long to express antibiotic resistance. But did the development and use of antibiotics in medicine produce the "driving force" for bacterial species to evolve and spread resistance genes that then transformed much of the natural microbiota? Alternatively, is resistance an ancient phenomenon that spread from the natural microbiota to clinically important pathogenic species?

OBSERVATIONS: Some surveys comparing soil samples from the pre-antibiotic era (~1940) with contemporary soil samples report that the relative abundance of antibiotic resistance genes has increased in contemporary soils. Such surveys suggest resistance genes are modern evolutionary consequences of antibiotic development and use. Other investigations analyzing microbial DNA sequences from Pleistocene permafrost sediments (dated to be 30,000 years old) report the identification of resistance genes to many different classes of antibiotics. These investigations suggest that antibiotic resistance is quite ancient, reflecting a rich and diverse reservoir of resistance genes.

QUESTION: *Is the presence of resistance genes in microbes the result of contemporary development and use of antibiotics?*

HYPOTHESIS: Antibiotic resistance is prevalent in microbial populations isolated from contact with human sources of antibiotics. If so, investigating an environment that has never been exposed to contemporary antibiotics should result in finding resistance genes among the environment's microbiome.

EXPERIMENTAL SITE AND DESIGN: A region of Lechuguilla Cave, located in Carlsbad Caverns National Park, New Mexico, was selected for study. The cave is 300 to 400 meters below the surface and is believed to have been isolated from surface water and human exposure for more than 4 million years. Of the 500 unique bacterial isolates collected from three deep, remote sample sites in the cave (see FIGURE A), 93 grew readily in tryptic soy broth. Ribosomal RNA gene sequencing classified 33% as gram-positive and 63% as gram-negative genera.

EXPERIMENT: To determine if any of the 93 strains contained antibiotic resistance, each was tested for growth in the presence and absence of up to 26 different antibiotics (20 µg/ml), representing natural, semisynthetic, and synthetic drugs.

(continues)

Investigating the Microbial World 10 (*Continued*)

The Source of Antibiotic Resistance

RESULTS:

EXPERIMENT: See Figure B. Note: Apramycin = an aminoglycoside; minocycline = a tetracycline derivative; telithromycin = macrolide; novobiocin = blocks DNA synthesis; rifampicin = rifampin; fosfomycin = inhibits DNA synthesis; daptomycin = disrupts cell membrane function.

> **QUESTION 1:** *From Figure B, do the antibiotics used in this study target the diverse bacterial cell structures and processes that are potentially susceptible to antibiotics? Explain. (Also refer to Figure 10.9.)*

> **QUESTION 2:** *Daptomycin is a new class of antibiotics that was approved for clinical use in the late 1980s. Is it surprising that almost 20% of the gram-positive isolates (see Figure B) are resistant to this drug? Does this finding favor the idea that antibiotic resistance is ancient? Explain.*

> **QUESTION 3:** *Is the hypothesis supported? Explain.*

You can find answers online in **Appendix E**.

FIGURE A Lechuguilla Cave.

Courtesy of Max Wisshak www.speleo-foto.de.

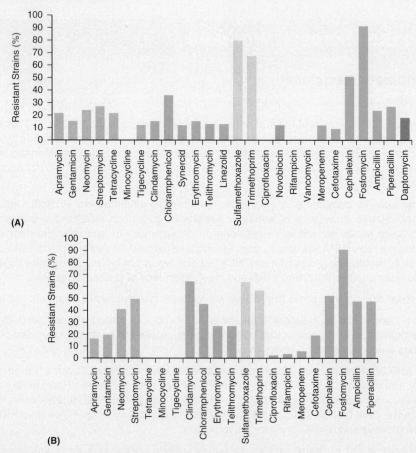

(A)

(B)

FIGURE B Resistance levels against various antibiotics. Antibiotics are grouped by their mode of action/target, where each color represents a different target. **(A)** = gram-positive isolates; **(B)** = gram-negative isolates.

Modified from Bhullar, K., et al. 2012. *PLoS ONE*, 7(4):1–11.

better education of medical personnel and people in the agricultural field so that they use antimicrobials appropriately. In 2013, the FDA put forth a voluntary plan that in 3 years would phase out the use of specific antibiotics in livestock and poultry. Compliance by drug producers and the agriculture industry will be essential for this to succeed. Everyone has a role to play. The public can help limit the spread of AMR by taking appropriate steps, as outlined in MICROFOCUS 10.4.

▶ **Search for New Antimicrobials.** The second part is a bit more involved and challenging. Due to the perception in the late 1960s that antibiotics would cure most all bacterial diseases, pharmaceutical companies and research organizations stopped or severely slowed the search for and development of new antibiotics and other antimicrobials. Unfortunately, as we have seen, many microbes and pathogens have continued to become resistant to these drugs. Yet, despite this increase in AMR, the development of new antimicrobial agents is failing in a time when there is a pressing need for these drugs. Fortunately, there has been a slight upsurge in antibiotic approvals in the last few years (**FIGURE 10.24**).

MICROFOCUS 10.4: Public Health

Preventing Antibiotic Resistance: Steps You Can Take

The Centers for Disease Control and Prevention's (CDC) **Get Smart: Know When Antibiotics Work** (http://www.cdc.gov/getsmart/) is a campaign to make the public and healthcare professional aware of ways to prevent antimicrobial resistance. The four main strategies are:

Courtesy of CDC.

A. Prevent Infection

Prevention will decrease antibiotic use—if:

   Step 1. You are vaccinated. Keeping up with all vaccinations will limit any infection by a pathogen. In addition, children should be immunized on schedule, people should get annual influenza vaccinations and keep up with other required booster shots, and those over 60 should receive a pneumococcal vaccination.

   Step 2. Avoid use of indwelling instruments, if possible. Indwelling instruments are a major source for potential infection in a healthcare setting. Request that catheters and other indwelling instruments be used only when essential and for minimal duration. Catheters should be removed aseptically when no longer essential.

   The bottom line: Maintaining high levels of immunity and minimal use of indwelling instruments will eliminate or minimize any need for antibiotics.

B. Diagnose and Treat Infection Effectively

Proper diagnosis and infection treatment will decrease antibiotic use—if:

   Step 3. The correct pathogen is identified. The clinical lab should attempt to identify the pathogen, if not known, and then prescribe the appropriate narrow-spectrum antibiotic.

   Step 4. Reliable experts are involved. For complicated or serious infections, consult an infectious disease expert. Consider a second opinion if unsatisfied with the antibiotic treatment or recovery.

   The bottom line: Appropriate diagnosis and antibiotic treatment will limit the chances of resistance development.

C. Use Antimicrobials Wisely

Proper use of antibiotics will decrease antibiotic use—if:

   Step 5. You are an informed "drug user." Minimize use of broad-spectrum antibiotics as these can lead to more rapid antibiotic resistance. Also, avoid chronic or long-term antimicrobial use.

   Step 6. You take and stop antimicrobial treatment when indicated. When cultures are negative and infection is unlikely, stop antibiotic treatment. If cultures are positive, do not stop taking the antibiotic treatment prematurely. However, once the infection has resolved with the prescribed course of antibiotic treatment, antibiotic use should stop.

   The bottom line: Appropriate use and completion of antibiotic treatment will limit potential resistance.

D. Prevent Transmission

Preventing transmission will decrease antibiotic use—if:

   Step 7. The chain of contagion is broken. Stay home when sick and cover your mouth when you cough or sneeze. Educate family members and coworkers as to proper hygiene.

   Step 8. You and others perform hand hygiene. Use alcohol-based hand rubs or wash your hands often, especially when sick. Encourage the same for all family members and coworkers.

   The bottom line: Proper hygiene will limit pathogen transmission and minimize the need for antibiotics.

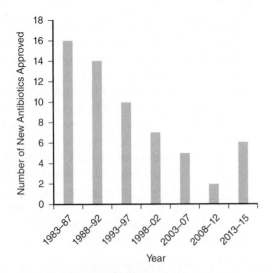

FIGURE 10.24 Antibiotic Approvals, 1983–2015. The number of antibiotic approvals has been declining over the past 30 years, although recently there has been a modest increase. *»» What are some reasons for the lack of new antibiotics coming through the development pipeline?*

Modified from: Spellberg et al, 2009. "Trends in Antimicrobial Drug Development: Implications for the Future," *Clinical Infectious Diseases* 9(38): 1279–1286.

There are several other reasons why the pharmaceutical industry has lagged behind in developing and bringing new antimicrobial drugs to market:

▶ The cost of development is staggering, requiring close to $1 billion over 10 years or more, as described in **MICROINQUIRY 10**. Pharmaceutical companies say that the relatively large research cost and time for development are lost if the experimental drug does not work as hoped when tested in clinical trials.

▶ There is a much higher financial reward to pharmaceutical companies in developing and marketing medications for the treatment of chronic illnesses (e.g., depression, hypertension, diabetes, cancer, cholesterol, arthritis) because, unlike antibiotics, which are usually given for a short 5- to 14-day period and then discontinued, medications for chronic illnesses might be taken for a lifetime.

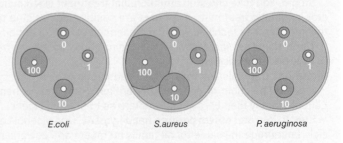

MICROINQUIRY 10

Testing Drugs—Clinical Trials

Before any antibiotic or chemotherapeutic agent can be used in therapy, it must be tested to ensure its safety and efficacy. The process runs from basic biomedical laboratory studies to approval of the product to improve healthcare. On average, it takes some 8–12 years to study the drug in the laboratory, test it in animals, and finally run human clinical trials.

In this MicroInquiry, we examine a condensed version of the steps and time estimates through which a typical antibiotic would pass. Some questions asked are followed by the answer so you can progress further in your analysis. Please try to answer these questions before moving forward through the scenario. Also, methods are used that you have studied in this and previous chapters. A real pharmaceutical company might have other, more sophisticated methods available. You can find answers online in **Appendix C**.

1. Discovery

You are head of a drug research group with a pharmaceutical company that has isolated a chemical compound (let's call it FM04) that is believed to have antibacterial properties and thus potential chemotherapeutic benefit. As head researcher, you must move this drug through the development pipeline from testing to clinical trials, evaluating, at each step, whether to proceed with further testing.

2. Preclinical Testing (4 years)

Many experiments need to be done before the chemical compound can be tested in humans. Because FM04 is "believed" to have antibacterial properties,

10.1a.   What would be the first experimental tests to be carried out?

You would experimentally test the drug on bacterial cells in culture to ascertain its relative strength and potency as an antibacterial agent. This testing might include the agar disk diffusion method described in this chapter. The figure below shows three diffusion disk plates, each plated with a different bacterial species (*Escherichia coli*, *Staphylococcus aureus*, and *Pseudomonas aeruginosa*) and disks with different concentrations of FM04 (0, 1, 10, 100 µg).

E.coli	*S.aureus*	*P. aeruginosa*

10.1b.   What can you conclude from these disk diffusion studies? Should drug testing continue? Explain.

Let's assume the bacterial studies look promising. FM04 now would be tested on human cells in culture to see if there are any toxic effects on metabolism and growth. **Table A** presents the results using the same drug concentrations from the bacterial studies. (The drug was prepared as a liquid stock solution to prepare the final drug concentration in culture.)

10.1c.   What would you conclude from the studies of human cells in culture? Should drug testing continue? Explain.

10.1d.   What was the point of including a solution with 0 mg of the drug?

When drugs are injected or ingested into the whole body, concentrations required for a desired chemotherapeutic effect may be very different from the experimental results with cells. Higher concentrations often are required and these may have toxic side effects. In animal testing, efforts should be made to use as few animals as possible and they should be subjected to humane and proper care. Let's assume the research studies of FM04 on mice used oral drug concentrations of 1, 10, 100, and 1,000 mg. **Table B** presents the simplified results.

10.2a.   What would you conclude from the mouse studies? Should drug testing continue? Explain.

Because the results look positive and higher doses can be administered without serious toxic side effects (except at 1,000 mg), additional studies would be carried out to determine how much of the drug is actually absorbed into the blood, how it is degraded and excreted in the animal, and if there are any toxic breakdown products produced. These studies could take a few years to complete. Again, for the sake of the exercise, let's say the studies with FM04 remain promising.

Although mice are quite similar to humans physiologically, they are not identical, so human clinical trials need to be done. This is the reason the U.S. Food and Drug Administration (FDA) requires all new drugs to be tested through a *three-phase clinical trial period.*

First, as head of the research team, you must provide all the preclinical and animal testing results as part of an FDA *Investigational New Drug* (IND) *Application* before human tests can begin. The application must spell out how the clinical investigations will be conducted. As head of the investigational team, you must design the protocol for these human trials.

10.2b.   How would you conduct the human tests to see if FM04 has positive therapeutic effects as an antibiotic in humans?

Briefly, there are many factors to take into consideration in designing the protocol.

(i) How will the drug be administered? Let's assume it will be oral in the form of pills.

(ii) You need to recruit groups of patients (volunteers) of similar age, weight, and health status (healthy patients as well as patients with the infectious condition to which the drug may be therapeutic). Note that the bacterial studies suggested FM04 had its most potent effect on *Staphylococcus aureus,* a gram-positive species. Perhaps patients with a staph infection would be recruited.

(iii) Eventually, volunteers (patients) need to be split into two groups, one that receives the drug and one that receives a placebo (an inactive compound that looks like the FM04 pill).

(iv) The clinical studies should be carried out as "double blind" studies. In these studies, neither the patients nor the investigators know which patients are getting the drug or placebo. This prevents patients and investigators from "wishful thinking" as to the outcome of the studies.

TABLE A Adverse Cellular Effects Based on FM04 Drug Concentration

Final Drug Concentration (µg/culture)	Adverse Cellular Effects
0	None
1	None
10	None
100	Abnormal cells and cell death

TABLE B Toxicity Results on 50 Mice Given Different Concentrations of FM04

Drug Concentration (mg)	Number of Mice	Adverse Reactions
0	10	None
1	10	None
10	10	None
100	10	None
1,000	10	Tremors and seizures in eight mice

(continues)

MICROINQUIRY 10 (*Continued*)

Testing Drugs—Clinical Trials

In actual clinical studies there are even more factors to consider, but we will assume that FM04 wins approval for clinical testing.

3. Clinical Trials (6 years)

In *Phase 1 studies*, a small number (20 to 100) of healthy volunteers are treated with the test drug at different concentrations (doses) to see if there are any adverse side effects. These studies normally take at least several months. If unfavorable side effects are minimal, Phase 2 trials begin.

Phase 2 studies use a larger population of patient volunteers (several hundred) who have the disease the drug is designed to treat. Here, it is especially important to split the patients into the test and control groups because the major purpose of these trials is to study drug effectiveness. There needs to be a control group so effectiveness can be contrasted accurately. These trials can take anywhere from several months to 2 years to complete. Provided there are no serious side effects or toxic reactions, or a lack of effectiveness, Phase 3 trials begin. There is an ethical question that can arise at this stage in the clinical trials regarding making the drug available, especially if the drug shows signs of reversing an illness that might be life threatening. Even medical experts disagree as to the answer.

10.3   Is it ethical to give ill patients placebos when effective treatment appears available?

Phase 3 studies include a few thousand patient volunteers. The trials can take several years to complete because the purpose is to evaluate safety, dosage, and effectiveness of the drug.

4. FDA Review and Approval (2 years)

If the clinical trials are positive, as head researcher you can recommend that the pharmaceutical manufacturer apply to the FDA for a New Drug Application (NDA). The application is reviewed by an internal FDA committee that examines all clinical data, proposed labeling, and manufacturing procedures. The most important questions that need to be addressed are:

(i) Were the clinical studies well controlled to provide "substantial evidence of effectiveness"?

(ii) Did the clinical trials demonstrate that the drug was safe; that is, do the benefits outweigh the risks?

Based on its findings, the committee then makes its recommendation to the FDA. If approved, an approval letter is given to the company to market the drug.

Note that very few drugs (perhaps five in 5,000 tested compounds) actually make it to human clinical trials. Then, only about one of those five are found to be safe and effective enough to reach the pharmacy.

The Infectious Disease Society of America (IDSA) and most health experts agree that it is time to focus on the search for and development of new antimicrobial drugs to replace those for which there is AMR. In fact, few drugs moving forward will treat infections due to the so-called "ESKAPE" pathogens (*Enterococcus faecium, S. aureus, K. pneumoniae, Acinetobacter baumannii, Pseudomonas aeruginosa,* and *Enterobacter* species). These pathogens, along with *C. difficile,* cause the majority of U.S. hospital infections, and they exhibit multidrug resistance, including to drugs of last resort. Consequently, the IDSA has launched a collaboration titled the "10 × 20" initiative, which aims to reverse the downward trend in antibiotic development by stimulating the development of 10 new, safe, and effective antibiotics by 2020.

In 2013, a glimmer of hope appeared. Several companies began clinically testing new classes of quinolones, tetracyclines, oxazolidinones, and cephalosporins, and they began developing novel chemical structures targeting new bacterial metabolic pathways.

Still, scientists and pharmaceutical companies need to hasten their search for new approaches to address AMR as well as identify new targets to which pathogens are susceptible. For example, the marine microbial community has barely been tapped as a source of new antimicrobials. In addition, because only a very small portion of bacterial species can be cultured, 90% to 99% of the marine metagenome has not been sampled—and there might be many beneficial antimicrobial drugs awaiting discovery.

Many microbiology labs around the world are sequencing the genomes of dozens of different microbes thought to contain genes for undiscovered antibiotics. The genus *Streptomyces* is the most prolific antibiotic producer and is the source of more than 65% of the antibiotics in current use. Undoubtedly, there are more antibiotics produced by this genus. The current explosion in microbial genome sequencing affords an opportunity to discover essential bacterial genes representing targets for antimicrobial drugs and to identify other forms of antibiotic resistance.

Although antimicrobial drugs probably are not the magic bullets once perceived by Ehrlich, new drug discovery might provide us with many new antimicrobial agents to fight pathogens and infectious disease.

Not only are the superbugs containing the NDM-1 genetic element turning up in the clinic, they are also being found in common environments in South Asia, including in tap water, water puddles, and pools where children play. As of this writing, such environmental sources have not yet caused any known outbreaks. Closer to home, CRE are being identified in many parts of the United States, including dozens of cases in Illinois. In the southeastern United States, a recent study found that community hospitals have seen a five fold increase in the number of CRE cases since 2009. The lead author of the study commented, "A CRE epidemic is fast approaching."

In conclusion, at the beginning of the 20th century, we saw that infectious diseases were among the most common causes of death in the United States and worldwide. However, by the 1950s, the "golden era" of antibiotic development and use was in full swing, with many new classes of antibiotics being introduced. These new antibiotics were viewed as being so powerful in destroying infectious disease that, by the late 1960s, some experts believed infectious bacterial diseases would soon be simply events of the past. Unfortunately, since the early 1980s, we have seen and heard about an unprecedented number of emerging and resurging infectious diseases that are now multidrug resistant. In fact, the age of antibiotics might be ending, and our declining ability to fight infections could send us back to the 1940s, when doctors were powerless to cure infectious diseases.

In 2015, the President's Council of Advisors on Science and Technology released a "National Action Plan for Combating Antibiotic-Resistant Bacteria" (https://www.whitehouse.gov/sites/default/files/docs/national_action_plan_for_combating_antibotic-resistant_bacteria.pdf). In this report, five goals were proposed to deal with the crisis of antibiotic resistance.

1. Slow the emergence of antibiotic-resistant bacteria and prevent their spread.
2. Strengthen surveillance for combating antibiotic resistance.
3. Develop and implement rapid diagnostic procedures and tests for identifying and characterizing bacterial resistance.
4. Accelerate the search for and development of new antibiotics, vaccines, and other therapeutics.
5. Improve international collaboration in dealing with the above four goals.

Perhaps the AMR crisis is best summed up by Margaret Chan, the Director-General of the World Health Organization, when she said:

"A post-antibiotic era means, in effect, an end to modern medicine as we know it. Things as common as strep throat or a child's scratched knee could once again kill . . . Some sophisticated interventions, like hip replacements, organ transplants, cancer chemotherapy, and care of preterm infants, would become far more difficult or even too dangerous to undertake . . . At a time of multiple calamities in the world, we cannot allow the loss of essential antimicrobials, essential cures for many millions of people, to become the next global crisis."

Concept and Reasoning Checks 10.6

a. How do the resistance mechanisms confer resistance to an antibiotic?
b. Summarize the misuses and abuses of antibiotics.
c. Assess the need for new and novel antibiotics and targets for those antibiotics.

Chapter Challenge E

The *K. pneumoniae* strain containing NDM-1 that was first identified in 2008 is a very dangerous superbug. By 2009, a study in India identified 22 CRE strains harboring NDM-1. Ten were *Klebsiella* species, nine were *E. coli* strains, and two were *Enterobacter* species (see MicroFocus 10.3).

QUESTION E: *The fast spread of the NDM-1 genetic element argues against mutation as the source of resistance. Explain how resistance spread so rapidly among the strains of Enterobacteriaceae.*

You can find answers online in **Appendix F**.

■ SUMMARY OF KEY CONCEPTS

Concept 10.1 Antimicrobial Agents Are Chemical Substances Used to Treat Infectious Disease

1. Ehrlich saw **antibiotics** as a chemotherapeutic approach to alleviating infectious disease. One of his students isolated the first chemotherapeutic agent, Salvarsan. Domagk used Prontosil to treat bacterial infections.
2. Penicillin was discovered by Fleming and purified and prepared for chemotherapy by Florey and Chain.
3. All antimicrobial agents represent **antibiotics** or **synthetic drugs**, demonstrate **selective toxicity**, and have a **broad** or **narrow drug spectrum**. (Figure 10.3)
4. Antibiotics are natural compounds toxic to some competitive bacterial species.

Concept 10.2 Synthetic Antibacterial Agents Primarily Inhibit DNA Synthesis and Cell Wall Formation

5. The **sulfonamides** interfere with the production of **folic acid** through **competitive inhibition**. (Figure 10.5)
6. **Quinolones** interact with DNA to inhibit replication.

Concept 10.3 Beta-Lactam Antibiotics Target Bacterial Cell Wall Synthesis

7. **Penicillin G** interferes with cell wall synthesis in gram-positive bacterial cells. Numerous semisynthetic forms of penicillin are more resistant to **beta-lactamase** activity. (Figures 10.6, 10.7)
8. Certain **cephalosporin** drugs are first-choice antibiotics for penicillin-resistant bacterial species, and a wide variety of these drugs is currently in use. The **carbapenems** inhibit cell wall synthesis and are broad-spectrum drugs. (Figure 10.8)

Concept 10.4 Other Antibiotics Target Some Aspect of Metabolism

9. **Vancomycin** inhibits cell wall synthesis.
10. **Polypeptide antibiotics** affect the permeability of the cell membrane.

11. **Aminoglycosides** inhibit translation in gram-negative bacterial cells. **Chloramphenicol** is a broad-spectrum antibiotic used against gram-positive and gram-negative bacterial species. Less severe side effects accompany **tetracycline** use, and this antibiotic is recommended against gram-negative bacterial species as well as rickettsiae and chlamydiae. The **macrolides**, **lincosamides**, and **streptogramins** also inhibit translation.
12. Besides the quinolones that inhibit DNA replication, **rifampin** interferes with transcription.

Concept 10.5 Other Antimicrobial Drugs Target Viruses, Fungi, and Parasites

13. Antiviral drugs interfere with viral entry, replication, or maturation.
14. Several antifungal drugs are valuable against fungal infections. Most affect the plasma membrane, cell wall, nucleic acids, or microtubules.
15. Antiprotistan agents include the aminoquinolines that are used to treat malaria. Today, **artemisinin** is the drug of choice for treatment.
16. Antihelminthic drugs target non-dividing organisms, affecting membrane permeability or glucose uptake or affecting the nervous system.

Concept 10.6 Antimicrobial Drug Resistance Is a Growing Challenge

17. Antibiotic assays include the **tube dilution assay**, which measures the **minimum inhibitory concentration**, and the **agar disk diffusion method**, which determines the susceptibility of a bacterial species to a series of antibiotics. (Figures 10.13, 10.14)
18. Bacterial species have developed antimicrobial resistance by altering metabolic pathways, inactivating antibiotics, reducing membrane permeability, or modifying the drug target. (Figure 10.20)
19. Arising from any of several sources such as changes in microbial biochemistry, **antibiotic resistance** threatens to put an end to the cures of infectious disease that have come to be expected from medical care today. (Figure 10.21)
20. New approaches to antibiotic resistance include the discovery of new drug targets and developing new antibiotics to fight infectious disease and antibiotic resistance.

■ CHAPTER SELF-TEST

For **Steps A–D**, you can find answers to questions and problems online in **Appendix D**.

■ STEP A: REVIEW OF FACTS AND TERMS

Multiple Choice

Read each question carefully and then select the *one* answer that best fits the question or statement.

1. Ehrlich and Hata discovered ____ that was used to treat ____.
   A. Salvarsan; syphilis
   B. penicillin; surgical wounds
   C. Salvarsan; malaria
   D. Prontosil; malaria

2. The reisolation and purification of penicillin were carried out by ____.
   A. Waksman
   B. Florey and Chain
   C. Domagk
   D. Fleming

3. The concentration of an antibiotic causing harm to the host is called the ____.
   A. toxic dose
   B. therapeutic dose
   C. minimum inhibitory concentration
   D. therapeutic window

4. In the soil, antibiotics are produced by some ____.
   A. fungal species and viruses
   B. bacterial species and viruses
   C. viruses and helminthic species
   D. fungal and bacterial species

5. Sulfamethoxazole is an example of a ____ that blocks ____ synthesis.
   A. sulfonamide; PABA
   B. penicillin; cell wall
   C. sulfonamide; folic acid
   D. macrolide; protein

6. The quinolones are ____ and have a ____ drug spectrum
   A. beta-lactams; narrow
   B. bacteriostatic; narrow
   C. natural antibiotics; broad
   D. bactericidal; broad

7. Penicillins are useful in treating ____.
   A. gram-positive infections
   B. leprosy
   C. gram-negative infections
   D. tuberculosis

8. All the following are drugs or drug classes blocking cell wall synthesis *except* ____.
   A. cephalosporins
   B. carbapenems
   C. monobactams
   D. tetracyclines

9. Vancomycin inhibits _____ synthesis.
   A. protein
   B. DNA
   C. bacterial cell wall
   D. RNA

10. Two cyclic polypeptide antibiotics are ____.
    A. vancomycin and streptomycin
    B. penicillin and cephalosporin
    C. bacitracin and polymyxins
    D. gentamicin and chloramphenicol

11. Which one of the following is *not* an inhibitor of translation?
    A. Clindamycin
    B. Macrolides
    C. Rifampin
    D. Chloramphenicol

12. Antibiotics inhibiting nucleic acid synthesis include ____.
    A. rifampin and quinolones
    B. aminoglycosides and tetracyclines
    C. lincosamides and streptogramins
    D. macrolides and aminoglycosides

13. Antiviral drugs that are base analogs inhibit ____.
    A. viral entry
    B. genome replication
    C. uncoating
    D. maturation

14. Antifungal drugs, such as ____, inhibit sterol synthesis in the ____.
    A. the azoles; plasma membrane
    B. griseofulvin; ribosomes
    C. amphotericin B; cell wall
    D. flucytosine; plasma membrane

15. All of the following antiprotistan drugs have been used to treat malaria *except* ____.
    A. metronidazole
    B. quinine
    C. mefloquine
    D. chloroquine

16. This antihelminthic agent affects the nervous system of roundworms.
    A. Mebendazole
    B. Primaquine
    C. Praziquantel
    D. Avermectins

17. The ____ is used to determine an antibiotic's minimum inhibitory concentration (MIC).
    A. Ames test
    B. tube (broth) dilution method
    C. agar disk diffusion method
    D. Kirby-Bauer test

18. The phosphorylation of an antibiotic is an example of which mechanism of resistance?
    A. Target modification
    B. Reduced permeability
    C. Antibiotic inactivation
    D. Altered metabolic pathway

19. A superinfection could arise from _____.
    A. using antibiotics when not needed
    B. using unnecessarily large doses of antibiotics
    C. stopping antibiotic treatment prematurely
    D. All of the above (A–C) are correct.

20. Antibiotic resistance can be prevented by _____.
    A. taking broad-spectrum antibiotics only until one feels better
    B. using the highest toxic dose possible to kill all the infectious pathogens
    C. taking antibiotics for any infection to prevent the pathogen's spread
    D. using narrow-spectrum antibiotics only when absolutely needed

True-False

On completing your study of these pages, test your understanding of antimicrobial drugs, deciding whether the following statements are true or false. If the statement is true, write "True" in the space. If false, substitute a word for the underlined word to make the statement true.

21. _____ Sulfonamides block folic acid formation in fungi.

22. _____ Gentamicin is an antibiotic that interferes with transcription.

23. _____ An enzyme that breaks the structure of penicillin is known as alpha-lactamase.

24. _____ Polymyxin B affects membrane permeability.

25. _____ Artemisinin is a drug that is used to treat malaria.

Identification

Identify the bacterial structure inhibited or affected by each of the antibiotics in the descriptions below.

Descriptions

26. _____ Rifampin blocks this process.
27. _____ Quinolones act here.
28. _____ Tetracyclines inhibit the function of this structure.
29. _____ Aminoglycosides inhibit this structure.
30. _____ Penicillin acts here.

■ STEP B: CONCEPT REVIEW

31. Identify the important properties of antimicrobial agents for clinical use. (**Key Concept 10.1**)
32. Explain how **sulfonamide** drugs block the **folic acid** metabolic pathway in bacterial cells. (**Key Concept 10.2**)
33. Summarize the mechanism of action of **penicillin** and its semisynthetic derivatives. (**Key Concept 10.3**)
34. Distinguish between the mechanisms of action for the polypeptide antibiotics. (**Key Concept 10.4**)

35. Describe the targets for antiviral drugs. (**Key Concept 10.5**)
36. Describe the two ways that bacterial species have developed antibiotic resistance. (**Key Concept 10.5**)
37. Describe the four mechanisms used by bacterial species to generate resistance to antibiotics. (**Key Concept 10.6**)
38. Justify the statement that **antibiotic resistance** has resulted from the misuse and abuse of antibiotics. (**Key Concept 6**)

■ STEP C: APPLICATIONS AND PROBLEM SOLVING

39. In 1877, Pasteur and his assistant Joubert observed that anthrax bacilli grew vigorously in sterile urine but failed to grow when the urine was contaminated with other bacilli. What was happening?

40. In May 1953, Edmund Hillary and Tenzing Norgay were the first to reach the summit of Mount Everest, the world's highest mountain. Since that time, more than 5,000 other mountaineers have reached the summit, and groups have gone to Nepal from all over the world on expeditions. The arrival of "civilization" has brought a drastic change to the lifestyle of Nepal's Sherpa mountain people. For example, half of all Sherpas used to die before the age of 20, but partly due to antibiotics available to fight disease, the population has grown from 9 million to more than 28 million. Medical enthusiasts are proud of this increase in the life expectancy, but population ecologists see a bleaker side. What do you suspect they foresee, and what does this tell you about the impact antibiotics have on a culture?

41. Why would a synthetically produced antibiotic be more advantageous than a naturally occurring antibiotic? Why would it be less advantageous?

■ STEP D: QUESTIONS FOR THOUGHT AND DISCUSSION

42. According to historians, 2,500 years ago the Chinese learned to treat superficial infections such as boils by applying moldy soybean curds to the skin. Can you suggest what this implies?

43. Most naturally occurring antibiotics appear to be products of the soil bacteria belonging to the genus *Streptomyces*. Can you draw any connection between the habitat of these organisms and their ability to produce antibiotics?

44. Of the thousands and thousands of types of organisms screened for antibiotics since 1940, only five bacterial genera appear capable of producing these chemicals. Does this strike you as unusual? What factors might eliminate potentially useful antibiotics?

45. Is an antibiotic that cannot be absorbed from the gastrointestinal tract necessarily useless? How about one that is rapidly expelled from the blood into the urine? Explain your answers.

46. The antibiotic resistance issue can be argued from two perspectives. Some people contend that because of side effects and microbial resistance, antibiotics will eventually be abandoned in medicine. Others see the future development of a super-antibiotic, a type of "miracle drug" that would kill or inhibit most all pathogens. What arguments can you offer for either view? Which direction in medicine do you support?

47. History shows repeatedly that creativity is a communal act, not an individual one. How does the 1945 Nobel Prize to Fleming, Florey, and Chain reflect this notion?

PART IV Bacterial Diseases of Humans

The four chapters in Part IV embrace several aspects of the six principal (overarching) concepts as described in the ASM fundamental statements for a concept-based microbiology curriculum.

PRINCIPAL CONCEPTS

Cell Structure and Function	• Bacteria have unique cell structures that can be targets for antibiotics, immunity, and phage infection. **Chapters 11–14** • Bacteria and Archaea have specialized structures (e.g., flagella, endospores, and pili) that often confer critical capabilities. **Chapters 11–14**
Metabolic Pathways	• The interactions of microorganisms among themselves and with their environment are determined by their metabolic abilities (e.g., quorum sensing, oxygen consumption, nitrogen transformations). **Chapters 11–14** • The survival and growth of any microorganism in a given environment depends on its metabolic characteristics. **Chapters 11–14** • The growth of microorganisms can be controlled by physical, chemical, mechanical, or biological means. **Chapters 11–14**
Information Flow and Genetics	• Genetic variations can impact microbial functions (e.g., in biofilm formation, pathogenicity, and drug resistance). **Chapters 11–14**
Microbial Systems	• Microorganisms are ubiquitous and live in diverse and dynamic ecosystems. **Chapters 11–14** • Most bacteria in nature live in biofilm communities. **Chapters 11–14** • Microorganisms and their environment interact with and modify each other. **Chapters 11–14** • Microorganisms, cellular and viral, can interact with both human and nonhuman hosts in beneficial, neutral, or detrimental ways. **Chapters 11–14**
Impact of Microorganisms	• Microbes are essential for life as we know it and the processes that support life (e.g., in biogeochemical cycles and plant and/or animal microflora). **Chapters 11–14** • Microorganisms provide essential models that give us fundamental knowledge about life processes. **Chapters 11–14**

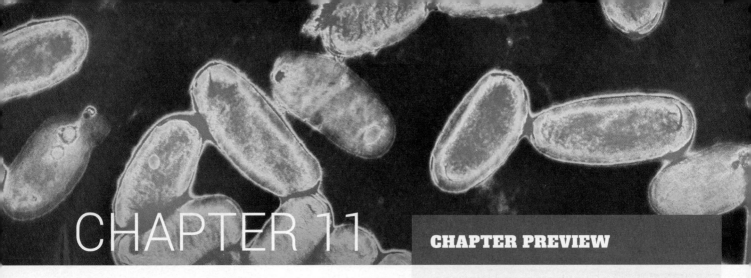

CHAPTER 11

Airborne Bacterial Diseases

In March 2010, the California Department of Public Health (CDPH) began receiving an abnormally large number of reports detailing patients with symptoms of runny nose, low-grade fever, and a mild, occasional cough that persisted for 7 to 14 days. For many, the symptoms led to a series of rapid coughs followed by a high-pitched "whoop." It appeared that an outbreak of pertussis, commonly called whooping cough, was occurring in California. Caused by *Bordetella pertussis* (**FIGURE 11.1**), this highly communicable bacterial disease was spreading throughout the state.

In the early parts of the 20th century, one of the most common childhood diseases and causes of death in the United States was pertussis. Before the introduction of a pertussis vaccine in 1940, *B. pertussis* was responsible for infection and disease in 150 out of every 100,000 people. By 1980, there was but one case in every 100,000 individuals. The vaccine had almost eliminated the pathogen.

In June of 2010, the CDPH said there had been a 418% increase in cases reported from the same period in 2009. Five infants had died, and it was obvious that a major epidemic of pertussis was in full swing in California. In fact, by the end of 2010, the CDPH had reported almost 10,000 cases of pertussis (including 10 infant deaths), making this the most cases reported in 65 years (**FIGURE 11.2**). Again, in 2014, California reported more than 11,000 cases,

False-color transmission electron microscope image of *Bordetella pertussis*.

© A. Barry Dowsett/Science Source.

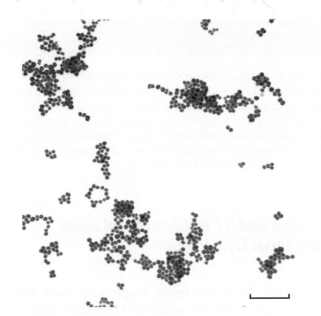

FIGURE 11.1 *Bordetella pertussis.* This Gram stain shows chains of small *B. pertussis* cells. (Bar = 20 μm.) »» *What is the Gram reaction of these stained cells?*

© Dr. George J. Wilder/Visuals Unlimited, Inc.

the highest since 1945. Several other states as well as other countries around the world, from Australia to the Netherlands, have been experiencing similar pertussis epidemics, forcing medical experts and the medical community to devise new prevention strategies. These outbreaks appear to be a combination of susceptible individuals that did not get vaccinated and a wearing off of protective immunity to pertussis.

Pertussis is but one of several bacterially caused infectious diseases affecting the respiratory system. We will divide these diseases into two general categories. The first category will include diseases of the upper respiratory tract (ears, nose, and throat), such as strep throat and diphtheria. The second category will include diseases of the lower respiratory tract (lungs), such as pertussis, tuberculosis, and pneumonia.

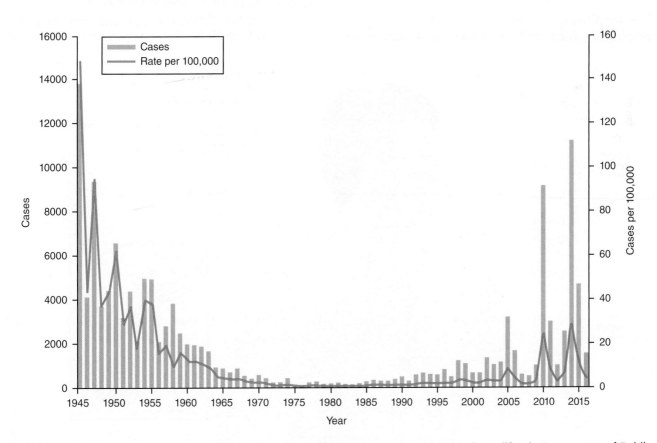

FIGURE 11.2 Reported Cases of Pertussis—California, 1945–2016. In 2010 and 2014, the California Department of Public Health reported the largest number of pertussis cases since 1945. »» *What was responsible for the reduction in pertussis cases during the 1950s to 1970s?*

Reproduced from: California Department of Public Health, Immunization Branch. Accessed 1/24/17.

Chapter Challenge

Streptococcus is a gram-positive, nonmotile, encapsulated bacterial genus in the phylum Firmicutes. Members of the genus are very diverse and widely distributed in nature. The cells are very susceptible to the environment and require close personal contact for transmission. The most common species are *S. pyogenes*, which 5% to 15% of normal healthy individuals carry as part of their normal microbiome, and *S. pneumoniae* (solely a human organism) that is carried in up to 40% of children. To confirm the presence of *S. pyogenes*, throat swabs are used, whereas sputum or blood samples plated on blood agar can be used to identify *S. pneumoniae*. So, how many illnesses are caused by these two species? Let's find out as we progress through this chapter on airborne bacterial diseases.

■ KEY CONCEPT 11.1   The Respiratory System and a Resident Microbiome Normally Impede Bacterial Colonization

The respiratory system is the most common portal of entry for many infectious agents because air typically contains microbes and viruses carried on dust and airborne droplets from sneezes and coughs.

Upper Respiratory Tract Defenses Hinder Microbe Colonization

The **respiratory system** is divided into the upper respiratory tract and the lower respiratory tract

(FIGURE 11.3). The **upper respiratory tract (URT)** is composed of the nose, sinus cavities, pharynx (throat), and larynx, whereas the **lower respiratory tract (LRT)** is composed of the trachea and a series of diverging tubes forming the bronchi and bronchioles within the lungs. The bronchioles terminate in alveoli where gas exchange occurs.

During breathing, the URT and bronchi play a critical role in defending against and filtering out

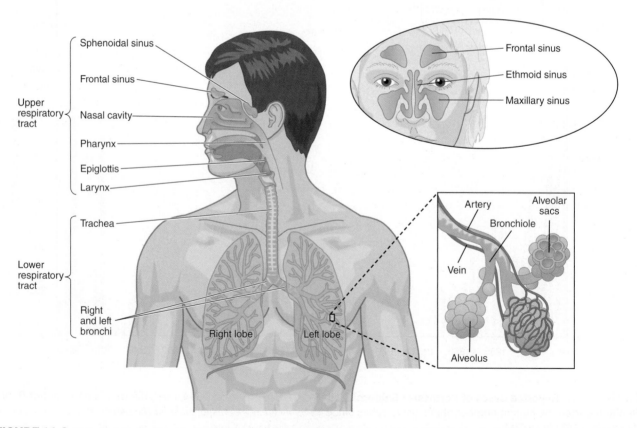

FIGURE 11.3 Respiratory System Anatomy. The major parts of the respiratory system are organized into the upper and lower respiratory tracts. Inset: the sinuses are hollow cavities within the facial bones. »» *Which part of the respiratory system would be the most susceptible to colonization and infection by airborne bacteria?*

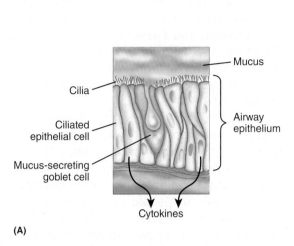

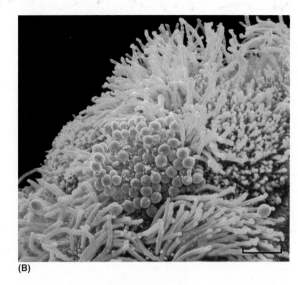

(A)

(B)

FIGURE 11.4 Defenses of the Airway Epithelium. (A) Epithelial cells provide a physical, chemical, and cellular barrier against potential pathogens. **(B)** False-color scanning electron micrograph of a *Staphylococcus* colony (yellow) on the epithelial cells of the trachea. The cilia (blue hair-like projections) attempt to keep the airways free of dust, irritants, and microbes. (Bar = 4 µm.) *»» How does the ciliated epithelium work to eliminate these bacterial cells?*

(B) © Juergen Berger/Science Source.

foreign material, such as bacterial cells and the dust particles that might carry these microbes. In particular, the airway epithelium lining the URT surfaces is involved in a defensive process called **mucociliary clearance** (**FIGURE 11.4A**). Mucus containing antimicrobial chemicals is secreted from the goblet cells in the airway epithelium. By overlying the epithelial cells, the sticky mucus traps microbes and particulate matter (**FIGURE 11.4B**). This material is moved by the ciliated epithelial cells lining the trachea, bronchi, and bronchioles toward the pharynx where the trapped material is cleared through swallowing, coughing, sneezing, or spitting. The antimicrobial substances include interferon, lysozyme, and several human defensins. In addition, immune defensive cells help protect the URT from infection.

In the LRT, epithelial cells lining the alveoli are not ciliated. However, the region is covered by alveolar fluid, which contains a number of antimicrobial components, including antibodies. In addition, immune defensive cells enter the alveoli from the pulmonary capillaries to help clear any invading pathogens.

The Healthy Respiratory Tract Harbors a Diverse Resident Microbiome

The **human microbiome** refers to the collective microbes and their genomes that reside on and in the human body. These communities of symbiotic microbes (primarily bacteria, but also including viruses, fungi, and protists) inhabit such diverse anatomic sites as the skin, mouth, reproductive organs, and intestines.

With regard to this chapter on airborne diseases, hundreds of different bacterial species representing four major phyla form the resident microbiome of the human respiratory system (TABLE 11.1; FIGURE 11.5A). Importantly, these microbes, at least in healthy individuals, cause no illness or disease. Looking more closely, unique communities of microbes are present in different microenvironments (niches) within the respiratory system.

TABLE 11.1 Some Common Resident Microbes of the Respiratory Tract

Phylum	Genera Present
Actinobacteria	*Corynebacterium, Propionibacterium*
Firmicutes	*Alloiococcus, Clostridium, Lactobacillus, Staphylococcus, Streptococcus*
Proteobacteria	*Campylobacter, Haemophilus, Neisseria, Pasteurella*
Bacteroidetes	*Bacteroides*

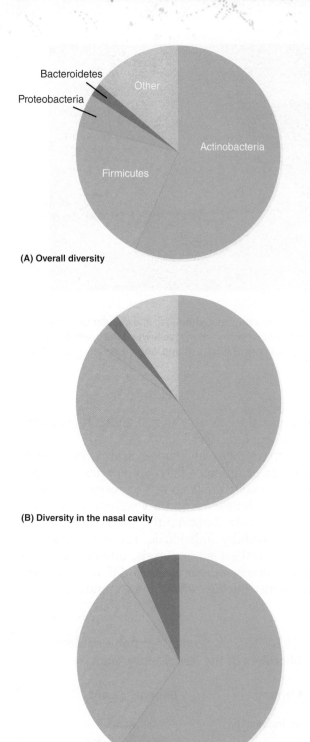

(A) Overall diversity

(B) Diversity in the nasal cavity

(C) Diversity in the oropharynx

FIGURE 11.5 The Human Respiratory Microbiome. The respiratory tract microbiome is composed of bacterial genera from four major phyla (Table 11.1). The overall compositional diversity of the respiratory tract **(A)** is compared to that for the nasal cavity **(B)** and the oropharynx **(C)**. *»» Would you expect this bacterial community to change in a person suffering from an infectious disease? Explain.*

Data from Cho, I. and Blaser, M.J. 2012. *Nature Reviews Genetics* 13:260–270.

For example, in the nasal cavity, the mucosal surfaces are mainly colonized by members of the Actinobacteria (e.g., *Corynebacterium* and *Propionibacterium*) and Firmicutes (e.g., *Staphylococcus, Streptococcus*, and *Neisseria*) (**FIGURE 11.5B**). In the oropharynx (the middle portion of the pharynx), a richer and less varied group of genera in three phyla are present (**FIGURE 11.5C**).

As in other body systems having a resident microbiome, the respiratory microbiome's role is, in part, one of **microbial antagonism**, where the resident microbial species outcompete invading pathogens for space and nutrients. In the nasal cavity, for example, 30% of people carry *S. aureus*, a potential pathogen. However, in healthy individuals the pathogen is kept under control through competition from other resident bacterial species. Should a disturbance or imbalance occur, called a **dysbiosis**, then *S. aureus* can act as an **opportunist**; that is, it can take advantage of little or no competition and cause damage to the host.

The movement of air into and mucus out of the respiratory tract makes for a dynamic and transient community of microbes (**FIGURE 11.6**). For example, inhalation of air brings 10^4 to 10^6 bacterial cells/mm^3 into the URT where most are trapped in

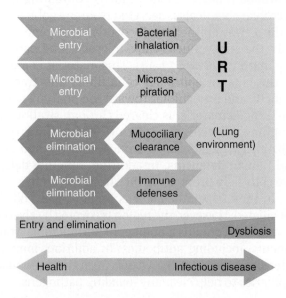

FIGURE 11.6 Dynamics of the Respiratory Microbiome. The diversity of the respiratory microbiome is dependent on microbial entry and microbial elimination. A healthy microbiome reflects the microbial composition due to entry and elimination, whereas infectious respiratory disease shows a dysbiosis in the system. *»» How might the lung microbiome become susceptible to infectious pathogens?*

Data from Dickerson, R. P. and Huffnagle, G. B. 2015. *PLOS Pathogens* (DOI:10.1371/journal .ppat.1004923).

the mucus coating the URT. Also, the respiratory environment (e.g., temperature, nutrient availability, pH, oxygen concentration, and immune defensive cell concentration) controls microbial growth. Thus, in healthy individuals, these conditions are unfavorable for pathogen colonization and infection.

The passage of fluids, secretions, and microbes, called **microaspiration**, from the oropharynx appears to be the source for the microbiome in the lungs. Microbial elimination removes microbes by mucociliary clearance and immune system activity. However, the environment in an individual suffering an infectious lung disease has too few resident bacteria, allowing for an overgrowth and dominance of those pathogens (e.g., *Bordetella* with pertussis; *Pseudomonas, Klebsiella*, and *Acinetobacter* with

pneumonia) (**FIGURE 11.7**). Such lung infections hospitalize about one million Americans every year.

Until recently, it was thought that the LRT of healthy individuals was a sterile environment and devoid of any microbes. However, numerous DNA sequence analyses indicate that the lungs do contain a resident microbial community, which appears similar to, but 1,000 times less dense than, the microbiome in the URT.

MICROBIOLOGY PATHWAYS describes working in clinical microbiology, in which some clinical microbiologists are involved with isolation and identification of bacterial pathogens as well as other infectious agents. They also might be responsible for determining the susceptibility of microorganisms to antimicrobial agents that might be needed to treat infections caused by pathogens.

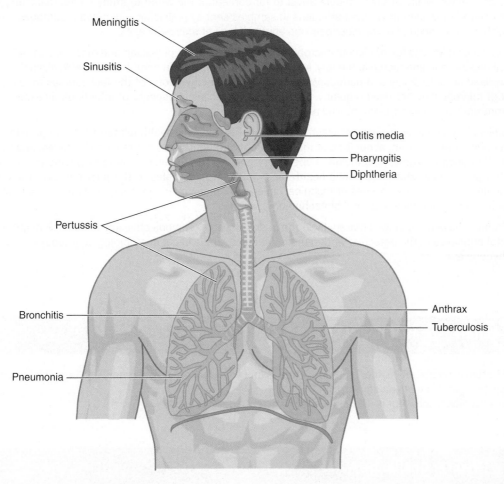

FIGURE 11.7   Major Diseases of the Respiratory System. Some pathogens target the URT or LRT, whereas others can progress from the URT to the LRT. **»» Why is the respiratory system such an accommodating environment for some pathogens?**

MICROBIOLOGY PATHWAYS

Clinical Microbiology

One of the most famous books of the 20th century was *Microbe Hunters* by Paul de Kruif. First published in the 1920s, de Kruif's book describes the joys and frustrations of Pasteur, Koch, Ehrlich, von Behring, and many of the original microbe hunters. The exploits of these scientists make for fascinating reading and help us to understand how the concepts of microbiology were formulated. I would urge you to leaf through the book (it is still in print!) at your leisure.

© Viktor Pryymachuk/Shutterstock.

Microbe hunters did not end with Pasteur, Koch, and their contemporaries, nor did the stories of the microbe hunters end with the publication of de Kruif's book. Approximately 25% of all deaths worldwide and 68% of all deaths in children younger than 5 years old are due to infectious agents. Today, clinical microbiology is concerned with the microbiology of infectious diseases, and the men and women working in hospital, public, and private laboratories are today's disease detectives. These individuals search for the pathogens of disease. Many of these people travel to far corners of the world to study microorganisms, whereas many more remain close to home, identifying the pathogens in samples sent by physicians, identifying pathogen interactions with the immune system, and working out the diagnosis and epidemiology of these diseases.

In fact, a well-developed knowledge of clinical microbiology is critical for the physician and medical staff who are faced with the concepts of disease and antimicrobial therapy. Microbiologists even work in dental clinical labs, given that many bacterial species are involved in tooth decay and periodontal disease. Microbiology is one of the few courses for which much of the fundamentals of microbiology are used regularly. This includes the clinical aspects of infectious diseases: manifestations (signs and symptoms), diagnosis, treatment, and prevention.

A career in clinical microbiology requires at least a bachelor's degree in a health-related field and preferably a master of science in infection control. In fact, about 70% of infection control practitioners are registered nurses, and 25% are medical technologists. With such a degree, jobs include supervisory positions in medical centers or private reference laboratories, infection control positions in state and public health labs, marketing and sales in the pharmaceutical and biotechnology industries, teaching at community colleges or technical colleges, or research in university and medical schools, as well as in government or industry (pharmaceutical and biotechnology) settings.

The microbe hunters have not changed materially in the past 100 years. The objectives of the search might be different, but the fundamental principles of the detective work remain the same. The clinical microbiologist is today's version of the great masters of a bygone era.

Concept and Reasoning Checks 11.1

a. The LRT often is referred to as the "respiratory tree." What would the trachea, bronchi and bronchioles, and the alveoli each represent in this tree analogy?

b. What is a dysbiosis and what is its significance to disease development?

Chapter **Challenge A**

Streptococcus can be found as part of the normal microbiome in adults.

QUESTION A: *Where in the human respiratory system are the streptococci most likely to be found?*

You can find answers online in **Appendix F**.

■ KEY CONCEPT 11.2   Several Bacterial Diseases Affect the Upper Respiratory Tract

The bacterial diseases of the URT and associated nose and ears are usually mild but can be more serious if a pathogen in the respiratory tract spreads to the blood and from there to other sensitive internal organs.

Pharyngitis Is an Inflammation of the Pharynx

A sore throat, known medically as **pharyngitis**, is an inflammation of the pharynx and sometimes the tonsils (tonsillitis). The inflammation usually is a symptom of a viral infection such as the common cold or the flu. However, about 20% of pharyngitis cases are caused by *Streptococcus pyogenes*, a gram-positive, facultative anaerobe. If the normal microbiome in the pharynx has been diminished, *S. pyogenes* dominates and can produce a potentially more dangerous form of pharyngitis called **streptococcal pharyngitis**, commonly known as **strep throat**. Each year, more than a million Americans, primarily children, suffer strep throat.

Strep throat is contagious and highly transmissible. The bacterial cells are spread by airborne respiratory droplets expelled by infected persons during coughing and sneezing or from objects contaminated with such secretions. If the cells grow and secrete toxins, inflammation of the oropharynx and tonsils can occur. Besides a sore throat, patients might develop a fever, headache, swollen lymph nodes and tonsils, and a beefy red appearance to pharyngeal tissues owing to tissue damage. Microaspiration of *S. pyogenes* cells into the LRT can result in laryngitis (inflammation of the larynx) or bronchitis (inflammation of the bronchi).

An antibiotic, such as penicillin, is often prescribed to lessen the duration and severity of the inflammation and to prevent possible complications. Good hand hygiene is the best prevention to limit pathogen transmission.

Scarlet fever is a disease arising in about 10% of children with streptococcal pharyngitis or a streptococcal skin infection. This arises because some strains of *S. pyogenes* secrete **erythrogenic** (*erythro = "red"*) **toxins** that cause a pink-red skin rash on the neck, chest, and soft-skin areas of the arms (**FIGURE 11.8A**). The rash, which usually occurs in children under 15 years of age, lasts about a week. Other symptoms include a sore throat, fever, and a "strawberry-like" inflamed tongue (**FIGURE 11.8B**).

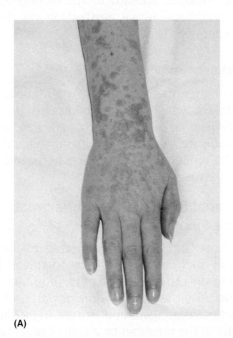

(A)   (B)

FIGURE 11.8 Scarlet Fever. Among the early symptoms of scarlet fever are **(A)** a pink-red skin rash and **(B)** a bright-red tongue with a "strawberry" appearance. Inset: Artist sketch of *S. pyrogens* cells seen with the light microscope at 1000X.
»» *What causes the skin rash seen with scarlet fever? What other symptoms are typical of scarlet fever?*

(A) Medical-on-Line/Alamy Stock Photo. (B) imagebroker/Alamy Stock Photo.

Normally, an individual experiences only one case of scarlet fever in a lifetime because recovery generates immunity. Treatment with antibiotics can shorten the duration of symptoms, although individuals with scarlet fever usually recover within 2 weeks without treatment.

A serious complication resulting from untreated streptococcal pharyngitis is **rheumatic fever**. This post-infection condition involves antibodies that were produced against *S. pyogenes*. These antibodies mistakenly bind to (cross-react with) proteins on heart muscle. The ensuing immune response causes permanent scarring and distortion of the heart valves, producing a potentially deadly condition called **rheumatic heart disease**.

Diphtheria Is an Infection of the Mucous Membranes of the Nasopharynx

Due to immunization starting in early childhood, **diphtheria** is extremely rare in the United States and other developed countries. However, low vaccine coverage in some developing nations, especially sub-Saharan Africa, makes diphtheria a health concern. In 2014, more than 7,300 cases were reported to the World Health Organization (WHO).

Causative Agent

Diphtheria is an infection of the URT that is caused by *Corynebacterium diphtheriae*, an aerobic, club-shaped (*coryne* = "club"), gram-positive rod that forms a characteristic picket-fence-like arrangement of cells called a "palisade arrangement."

Clinical Presentation

Diphtheria is acquired by inhaling airborne respiratory droplets or by direct contact with the skin from an infected person. Following a 2- to 4-day incubation, initial symptoms include a sore throat and low-grade fever. On the surface of the mucous membranes of the nasopharynx, the bacterial cells secrete a potent toxin that, once inside a cell, inhibits the translation process, resulting in cell death. Thus, a prominent feature of the disease is the accumulation of dead tissue, mucus, white blood cells, and fibrous material that forms a so-called **pseudomembrane** ("pseudo" because it does not fit the definition of a true membrane) on the tonsils or nasopharynx (**FIGURE 11.9**). Mild cases fade after a week, whereas more severe cases can persist for 2 to 6 weeks.

Complications can arise if the thickened pseudomembrane results in respiratory blockage by

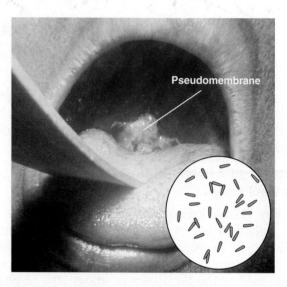

FIGURE 11.9 Diphtheria Pseudomembrane. Diphtheria is an upper-respiratory infection of mucous membranes caused by a toxigenic strain of *Corynebacterium diphtheriae*. The infection is characterized by the formation of a pseudomembrane on the tonsils or pharynx. Inset: Artist sketch of *C. diphtheria* cells seen with the light microscope at 1000X. *»» What is the composition of the pseudomembrane?*

© Science VU/CDC/Visuals Unlimited, Inc.

covering the trachea. Left untreated, diphtheria takes the lives of 5% to 10% of infected individuals. If the exotoxin spreads to the bloodstream, heart and peripheral nerve destruction can lead to an irregular heartbeat and coma.

Treatment and Prevention

Urgent treatment requires simultaneous use of antitoxins, which are antibodies with the ability to neutralize the free diphtheria toxins and antibiotics to eradicate the toxin-producing cells. Prevention can be rendered with the diphtheria-tetanus-acellular pertussis (DTaP) vaccine.

Bacterial Sinusitis Is an Inflammation of the Sinuses

The nose and ears are connected to the pharynx and, as such, allow infections to spread from one area to the other. Because of its prominent position in the URT, the nose is a major portal of entry for infectious organisms and viruses. In fact, the microbiome in the nose can be a sign of potential illness. Healthy individuals contain bacterial species primarily belonging to the Actinobacteria (68%) with many fewer Firmicutes (27%), whereas hospitalized patients exhibit a dysbiosis— Actinobacteria (27%)

and Firmicutes (71%). For example, *Staphylococcus aureus* (Firmicutes) is a normal resident in many people's noses. However, it is kept under control in the nose by other resident species (Actinobacteria and others) that produce antimicrobials that prevent colonization by *S. aureus*. Should a dysbiosis occur, *S. aureus* can colonize more effectively and lead to disease.

Bacterial sinusitis is inflammation in any of the four pairs of sinuses, the air-filled hollow cavities around the nose and nasal passages (see Figure 11.3 inset). About 10 to 15 million Americans develop a so-called "sinus infection" each year. The condition nearly always begins with an allergy or an inflammation and swelling of the nasal passages. Therefore, the condition is more correctly called **rhinosinusitis** (*rhino* = "nose").

Acute rhinosinusitis can be caused by a variety of resident microbes of the URT, especially *Streptococcus*, *Staphylococcus*, *Haemophilus*, and *Moraxella*. Trapped fluid increases the pressure in the sinuses, which causes pain, tenderness, and swelling. Yellow or green mucus might be discharged from the nose. Nasal sprays can be used for a short time, and antibiotics can be prescribed for a bacterial infection.

Acute rhinosinusitis can develop into **chronic rhinosinusitis** that lasts for more than three months. The condition might be the result of an infection or a hyperactive response by the immune system to normal, harmless microbes of the URT. The most common symptoms are nasal obstruction, nasal congestion, and postnasal drip. The treatment is the same as with acute rhinosinusitis.

In the near future, new therapies might be available to sinusitis patients. For example, in one research study, *Corynebacterium tuberculostearicum*, which is found in the noses of chronic rhinosinusitis patients, was introduced into the nasal cavities of healthy mice. These mice soon developed the symptoms of sinusitis. If the bacterium was introduced along with *Lactobacillus sakei*, which is found in the sinuses of healthy individuals, the mice did not develop sinusitis. Here again is an example of microbial antagonism, showing the protection that certain members of the human microbiome can provide. The study also suggests how treatment might alleviate the infectious condition in humans.

Middle Ear Infections Represent an Inflammation Behind the Eardrum

The ear consists of the outer, middle, and inner ear (**FIGURE 11.10**). The Eustachian tube vents the middle ear to the nasopharynx, which explains why a

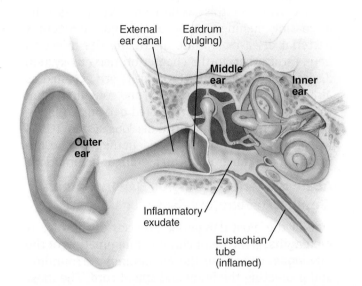

FIGURE 11.10 Ear Anatomy. The ear consists of external, middle, and inner structures. **»» *In a middle-ear infection, how do the infecting microbes reach the middle ear?***

URT infection often results in **acute otitis media** (**AOM**) (*oto* = "ear"; *itis* = "inflammation"; *media* = "middle"), an inflammation to the middle ear.

Health experts state that at least 80% of children have had at least one episode of AOM by three years of age, and 40% have had at least seven episodes by age seven. AOM often begins with a common cold or other viral infection of the URT. Inflammation of the Eustachian tube allows bacterial cells to infect the middle ear, with *S. pneumoniae*, *Haemophilus influenzae*, and *Moraxella catarrhalis* being the most common infectious agents in otherwise healthy children. Fluid buildup behind the tympanic membrane (eardrum) then provides an environment for bacterial growth and biofilm formation, which result in the middle ear becoming more inflamed. This is followed by ear pain with a red, bulging eardrum. Children with AOM can develop a fever, produce a fluid ("effusion") that drains from the ears, or have headaches. More than 80% of children with AOM recover without antibiotics. In fact, antibiotic treatment often fails to cure the infection because the bacterial pathogens form a biofilm that is unaffected by many antibiotics.

Chronic otitis media (**COM**) is a condition involving long-term infection, inflammation, and damage to the middle ear. The persistent biofilm colonization of the middle ear is a major global cause of hearing impairment and can have serious long-term effects on language, auditory and cognitive development, and educational progress.

We will end this section on URT diseases by discussing meningitis, one of the most dangerous bacterial diseases. We are discussing the infection here, rather than in the chapter on nervous system diseases because the bacterial cells are expelled as airborne droplets (e.g., sneezing and coughing) from the throat and nose of an infected person. An infection can spread to the central nervous system (CNS) where, without a protecting resident microbiome, the infection can cause a life-threatening inflammation.

Acute Bacterial Meningitis Can Be Spread Through Respiratory Secretions

One of the most dangerous diseases of the CNS is **meningitis**, an acute or chronic inflammation of the **meninges**, which are the membranes surrounding and protecting the brain and spinal cord. The most common form of meningitis, causing some 80% of all cases, is **acute bacterial meningitis (ABM)**, a rapidly developing infection affecting adults and children. Infection usually requires direct contact, such as kissing or being directly coughed or sneezed on.

ABM begins with a localized infection, which could include a sinus infection or ear infection (otitis media). The bacterial cells then can invade the epithelium of the nasopharynx and spread to the blood (**FIGURE 11.11**). When they are in the blood, the cells are capable of crossing the blood-brain barrier. The meninges then become inflamed, causing pressure on the spinal cord and brain. Testing involves doing a lumbar puncture (spinal tap), whereby a sample of the cerebrospinal fluid (CSF) is withdrawn and analyzed. The presence of bacterial cells in the CSF is indicative of bacterial meningitis.

With ABM, the individual experiences the "hallmark" signs and symptoms of meningitis, including high fever, headache, nausea and vomiting, and stiff neck. Although the infection might resolve on its own, meningitis should always be treated as a life-threatening emergency that requires urgent antibiotic therapy. Left untreated, infection through the meninges, brain, and spinal cord can be very rapid, resulting in death within hours. Even surviving ABM can result in lasting disabilities, such as deafness, blindness, and paralysis. In the United States, most cases of ABM are caused by one of three bacterial species: *Streptococcus pneumoniae, Neisseria meningitidis*, and *H. influenzae* type b.

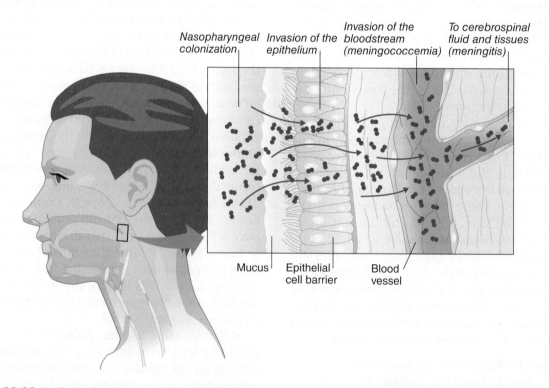

FIGURE 11.11 Pathogenic Steps Leading to Meningitis. The bacterial species capable of causing meningitis (*N. meningitidis, S. pneumoniae,* and *H. influenzae*) can colonize the nose and throat and then invade the epithelium, causing respiratory distress. They then pass into the bloodstream. Finally, they disseminate to tissues near the spinal cord, causing inflammation and meningitis. »» *What bacterial virulence factors would facilitate attachment to the nasopharynx and survival in the bloodstream?*

Pneumococcal Meningitis

By being one of the major causes of bacterial pneumonia, the gram-positive diplococci of *S. pneumoniae* (or pneumococcus) can spread and the infection develop into **pneumococcal meningitis**. This form of pneumonia is the most common ABM in infants, children, and older adults. The inflammation first occurs in the ears, nose, or lungs before entering the blood and crossing the blood-brain barrier. Pneumoccocal meningitis is responsible for about 30% of ABM cases and has a 20% to 30% mortality rate. There currently are two pneumococcal vaccines available: a pneumococcal conjugate vaccine (PCV13 or Prevnar 13®) for children under 5 years of age and adults 65 years of age and older and a pneumococcal polysaccharide vaccine (PPSV23 or Pneumovax®23) for adults 65 years of age and older.

Meningococcal Meningitis

Perhaps the most dangerous form of ABM is **meningococcal meningitis** caused by *N. meningitidis*, a small, encapsulated, aerobic, gram-negative diplococcus (**FIGURE 11.12**). After attaching by pili to the nasopharyngeal mucosa, the pathogen enters the blood, where the infection is called **meningococcemia**. Besides the classic meningitis symptoms, meningococcal meningitis produces a skin rash that begins as bright red patches and progresses to blue-black spots.

There are more than 14 serotypes of *N. meningitidis* based on the chemical nature of the capsule, with serotypes A, B, C, Y, and W-135 responsible for the majority of infections. In 2015, the Centers for Disease Control and Prevention (CDC) reported some 160 cases of meningococcal meningitis in Americans, primarily in teenagers and young adults. Globally, meningococcal meningitis has been responsible for more than 700,000 serotype A cases each year in the so-called "meningitis belt" in Africa (**FIGURE 11.13**).

Like other forms of meningitis, meningococcal meningitis is prevalent where people are in close social contact for long periods. Crowding in highly populated African countries of the meningitis belt and extended close contact in schools, daycare centers, military camps, prisons, nursing homes, and college dorms can facilitate the pathogen's spread. **CLINICAL CASE 11** describes a brief but scary meningococcal meningitis outbreak in two American universities in 2013.

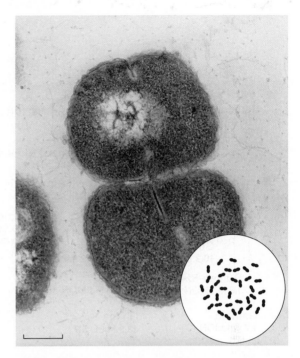

FIGURE 11.12 *Neisseria meningitidis.* A false-color transmission electron micrograph of *N. meningitidis* cells. The cytoplasm (blue) and nucleoid (pink) are shown. (Bar = 0.5 μm.) Inset: Artist sketch of *N. meningitidis* cells seen with the light microscope. *»» Living* N . meningitidis *cells have a capsule. What roles does the capsule play in the disease process?*

© CAMR/A.B. DOWSETT/Science Source.

In the United States, meningococcal meningitis is at historic lows. A meningococcal polysaccharide vaccine (Menomune®) and a conjugate vaccine (Menactra®) to serotypes A, C, Y, and W are available. In December 2010, a meningococcal vaccine (MenAfriVac®), specifically developed for African meningococcal meningitis (serotype A), was introduced, and it has had such dramatic results that medical experts are predicting the eventual elimination of serotype A throughout the meningitis belt. Two serotype B vaccines were licensed in 2014 and 2015.

Haemophilus Meningitis

The causative agent of ***Haemophilus* meningitis** is *H. influenzae* type b (Hib), an encapsulated, gram-negative coccobacillus. There are six serotypes based on the capsular material. Hib once was the most prevalent ABM causing meningitis in American children under 5 years of age. Prior to the introduction of the Hib vaccine in 1987, about 20,000 cases of *Haemophilus* meningitis and 1,000 deaths occurred

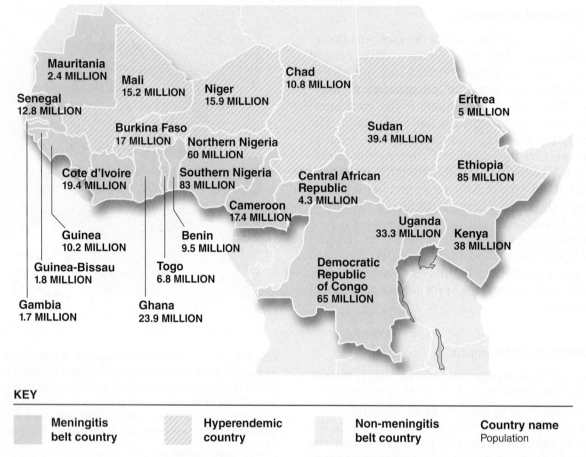

KEY

Meningitis belt country	Hyperendemic country	Non-meningitis belt country

Country name
Population

FIGURE 11.13 The African Meningitis Belt. This map of sub-Saharan Africa shows the so-called "meningitis belt" where deadly epidemics of meningitis are most prevalent. From Program for Appropriate Technology in Health (PATH)/David Simpson. »» *Why is a new meningitis vaccine, called MenAfriVac™, being initially used to immunize 1- to 29-year-olds in Mali, Burkina Faso, and Niger?*

Clinical Case 11

Invasive Meningococcal Disease—School-Based Outbreak—2013

On March 22, 2013, a female Princeton University student returning from spring break developed symptoms of meningococcal disease.

In early April, a visitor to the Princeton University campus was diagnosed with bacterial meningitis after returning home to another state. In May, two more cases of bacterial meningitis were identified in two male students. At this time, the New Jersey Department of Health declared a school-based outbreak of meningococcal disease and worked with the Centers for Disease Control and Prevention (CDC) and Princeton University Health Services to monitor the outbreak and provide antibiotics to close contacts of patients.

In late June, a Princeton male student traveling abroad developed symptoms of meningococcal disease. Meanwhile, all affected individuals were diagnosed with *Neisseria meningitidis* serogroup (type) B.

In the fall semester, additional cases of meningococcal disease developed among Princeton students. This included one female student in early October and two male students in November.

All eight patients received medical care and treatment and recovered. A serogroup B meningococcal vaccine was administered to more than 5,200 Princeton University students.

On November 18, two students at the University of California, Santa Barbara (UCSB) were diagnosed with meningococcal disease. Unfortunately, one of the male students had to have both feet amputated due to complications from the infection. This was followed by a third case on November 22 and a fourth case on December 2.

© bikeriderlondon/Shutterstock.

All four affected students at UCSB were diagnosed with *N. meningitidis* serogroup (type) B and all recovered.

Genetic analysis of bacterial samples from the two outbreaks at the two universities indicated they were different strains of serogroup B.

In January 2014, the CDC announced it would apply to make the meningococcal B vaccine available to UCSB students and staff.

Questions
   a. How does meningococcal disease spread?
   b. Why was a serogroup B meningococcal vaccine required when students had been vaccinated as adolescents with the standard MCV4 meningococcal vaccine?
   c. Why was an outbreak not announced until after the fourth identified Princeton case?
   d. Were the cases at Princeton and UCSB related? Explain.
   e. What precautions should be taken to prevent further cases of meningococcal disease at these universities?

You can find answers online in **Appendix E**.

For additional information, see http://www.cdc.gov/meningitis/index.html.

in the United States every year. The Hib vaccine protects infants as young as six weeks old and those adolescents and adults who are at increased risk for *Haemophilus* meningitis. In 2015, the CDC reported only 22 cases.

A Few Bacterial Species Cause Neonatal Meningitis

Neonatal meningitis is a rare but dangerous disease in both full-term and especially low-birth-weight neonates. About 50% of cases are caused by *Streptococcus agalactiae*. Another 20% of cases are due to the K1 strain of *Escherichia coli*, and an additional

20% are caused by *Listeria monocytogenes*. A growing number of babies who suffer from neonatal meningitis are premature and their prognosis is poor, such that even with aggressive antibiotic treatment, brain damage can result. Although transmission usually is contracted during passage through the birth canal, infection after birth can be acquired in the hospital (via contaminated inhalation therapy equipment) or in the community.

The airborne bacterial diseases of the URT are summarized in TABLE 11.2. MICROFOCUS 11.1 describes how resident, nonpathogenic microbes can prevent and trigger pathogen infections.

TABLE 11.2 **A Summary of the Major Bacterial URT Diseases**

Disease	Causative Agent	Signs and Symptoms	Transmission	Treatment	Prevention
Streptococcal pharyngitis	*Streptococcus pyogenes*	Sore throat, fever, headache, swollen lymph nodes and tonsils	Respiratory droplets	Amoxicillin Ampicillin	Practicing good hand hygiene No vaccine available
Scarlet fever	*Streptococcus pyogenes*	Pink-red rash on neck, chest, arms Strawberry-like tongue	Respiratory droplets	Penicillin Clarithromycin	Practicing good hygiene No vaccine available
Diphtheria	*Corynebacterium diphtheriae*	Pseudomembrane, sore throat, fever	Respiratory droplets	Penicillin Erythromycin	DTaP vaccination
Sinusitis	*Corynebacterium tuberculostearicum Lactobacillus sakei* Other resident species	Pain, tenderness, and swelling	Spread from infected tissues Respiratory droplets	Nasal sprays Antibiotics	Minimizing contact with individuals with colds No vaccine available
Acute otitis media (AOM)	*Streptococcus pneumoniae Haemophilus influenzae*	Ear pain Red, bulging eardrum	Infection from pharynx	Wait and see Antibiotics	Limiting time in childcare No vaccine available
Acute bacterial meningitis (ABM)	*Neisseria meningitidis Streptococcus pneumoniae Haemophilus influenzae* type b	Fever, stiff neck, severe headache, vomiting and nausea, sensitivity to light	Respiratory droplets from prolonged contact	Antibiotics	Vaccines available

Concept and Reasoning Checks 11.2

a. What makes *S. pyogenes* such a potentially dangerous pathogen in the URT?

b. In 17th century Spain, diphtheria was called *el garatillo* = "the strangler." Why was it given this name?

c. How do acute and chronic sinusitis differ?

d. In most cases with otherwise healthy children, why is the use of antibiotics not recommended for acute otitis media?

e. Describe the pathogenesis (development of disease) for acute meningitis.

f. What are the common pathogens responsible for meningococcal, pneumococcal, *Haemophilus*, and neonatal meningitis?

Chapter **Challenge B**

Both *S. pyogenes* and *S. pneumoniae* can cause infections and disease in the URT.

QUESTION B: *Identify which respiratory diseases of the URT are associated with each* Streptococcus *species.*

You can find answers online in **Appendix F**.

MICROFOCUS 11.1: Clinical Microbiology

A Yin/Yang Story

In Chinese philosophy, the concept of Yin/Yang refers to the interactions between the two forms needed to create a whole with each other. In this chapter, we have highlighted the role of the respiratory microbiome and its interactions in the infectious disease process—a microbiological yin/yang. The following are two (yin/yang) examples and their ability/inability to come together.

The yin/yang symbol.

© researcher97/Shutterstock.

Meningitis

Teenagers and young adults are very susceptible to meningococcal meningitis. In fact, many carry the pathogen *Neisseria meningitidis* in their upper respiratory tract (URT). Why do so many in this age group carry this potential pathogen?

Scientists have discovered that as children grow to teenagers and young adults, the anaerobic bacteria *Porphyromonas* and *Fusobacterium* become more prevalent as part of the microbiome in the URT. These microbes secrete a fatty acid, called propionic acid. Importantly, propionic acid acts as a growth supplement for *N. meningitidis*. Thus, the harmless anaerobes are modifying the host environment chemically, and this supports the carriage of *N. meningitidis*. Consequently, as kids grow older, their lungs more likely will be colonized by *N. meningitidis*. The yin and yang have come together and the whole might produce an infectious disease.

Sinus and Middle Ear Infections

Although *Streptococcus pneumoniae* is a major pathogen responsible for sinusitis and middle ear infections, not everyone exposed to, or harboring, the pathogen contracts one of these diseases. So, what makes one person more susceptible to *S. pneumoniae* colonization and a middle ear infection?

Well, it turns out that in children a harmless bacterial species called *Corynebacterium accolens* colonizes the noses of many children. In the nasal cavities, *C. accolens* secretes an enzyme that breaks down lipids in the nostrils. Importantly, the fatty acids produced from the hydrolysis inhibit the growth of *S. pneumoniae*. Thus, by modifying the chemical environment in the nostrils, the colonization of *S. pneumoniae* is blocked, and an ear (or sinus) infection is less likely. Again, the yin and yang have come together, but, in this example, disease development can be prevented.

These examples are just a sampling of the interactions occurring between a host microbiome and pathogens. They demonstrate how difficult it is to predict whether the convergence of the microbiome (yin) and pathogen (yang) will be beneficial or detrimental to host health.

■ KEY CONCEPT 11.3   Many Bacterial Diseases of the Lower Respiratory Tract Can Be Life Threatening

The lower respiratory tract (LRT) is not devoid of a resident microbiome. However, the number of microbes comprising this lung community is much smaller than in the URT. Therefore, depending on a person's immune status, pathogenic species can affect more easily the lung tissues and an infection can cause a potentially life-threatening disease. This includes pertussis, tuberculosis, and bacterial pneumonias.

Pertussis (Whooping Cough) Causes a Paralysis of the Cilia

Pertussis is one of the most dangerous and highly contagious bacterial diseases. Although most cases historically have occurred in children under 5 years of age, severe disease and death most often occur in infants only weeks to a few months old. Worldwide, the WHO reports that there are more than 16 million reported pertussis cases and almost 200,000 deaths annually.

Causative Agent and Epidemiology

Pertussis (*per* = "through"; *tussi* = "cough"), also known as **whooping cough**, is caused by *Bordetella pertussis*, a small, aerobic, gram-negative rod that is strictly a human pathogen. The bacilli are spread by airborne respiratory droplets, and they can infect

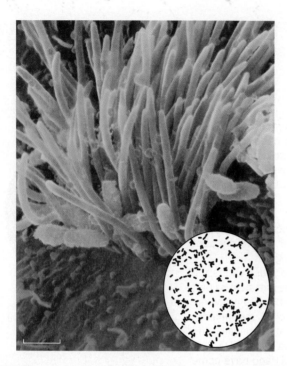

FIGURE 11.14 *Bordetella pertussis* **Associated with the Ciliated Epithelium.** False-color scanning electron micrograph of the human tracheal epithelium. The pertussis cells (yellow) cause a dramatic loss of cilia, which are essential to protect the respiratory tract from dust and particles. (Bar = 1 μm.) Inset: Artist sketch of *B. pertussis* cells seen with the light microscope seen at 1000X. *»» Explain why the loss of cilia function would be dangerous and could lead to infection.*

Photo © NIBSC/Science Source.

80% to 100% of exposed susceptible individuals. When inhaled, the bacilli adhere to and aggregate on the epithelial cells of trachea and the bronchi. The toxins produced can stimulate excessive mucus production and paralyze the cilia on the epithelial cells (**FIGURE 11.14**).

Health experts have put forth three reasons to explain the upswing in cases and outbreaks like the one in California (see the chapter opener) and several outbreaks in the United States and around the world. One factor might be that infants and children are not being vaccinated on schedule and not building up protective immunity. In fact, siblings and mothers are the most common source for the transmission of *B. pertussis* to infants. Another factor is waning immunity. It was thought that pertussis vaccination would provide life-long immunity. It now appears that immunity declines in vaccinated children between ages 8 and 12. Third, an Australian team of scientists has reported that the current outbreak strain of *B. pertussis* is a new one that is more resistant to vaccine-generated immunity. This suggests that new or modified vaccines might need to be developed.

Clinical Presentation

Following an average 9- to 10-day incubation period, typical pertussis cases occur in three stages.

1. **Catarrhal Stage.** Catarrh is an excessive discharge or buildup of mucus in a cavity or, in this case, in the respiratory airway. The catarrhal stage can last 1 to 2 weeks, during which time symptoms mimic those of a common cold, with general malaise, low-grade fever, and increasingly severe cough. It is during this stage that the individual is most contagious and carries a high bacterial load.

2. **Paroxysmal Stage.** As infected cells die over a 2- to 4-week period and disintegrate, the mucus accumulation in the airways makes breathing more labored. Spasmodic coughing occurs because the cilia are paralyzed and they cannot eliminate mucus and trapped material. Children experience multiple **paroxysms**, which consist of rapid-fire coughs all in one exhalation, followed by a forced inhalation over a partially closed glottis. This rapid inhalation produces in a high-pitched "whoop" (hence, the name whooping cough). Up to 50 paroxysms can occur daily, some being so violent that facial injury occurs (**FIGURE 11.15**). Exhaustion usually follows each paroxysm. Adolescents and adults might not experience these characteristic symptoms.

3. **Convalescent Stage.** Sporadic coughing gradually decreases over several weeks of convalescence, even after the pathogen has been eliminated.

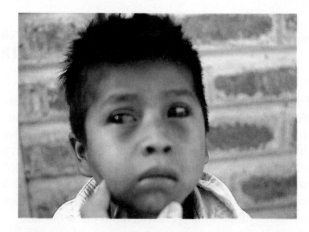

FIGURE 11.15 **A Child with Pertussis.** Pertussis can lead to broken blood vessels in the eyes and bruising on the face. *»» How could pertussis lead to the facial injuries shown in the figure?*

Courtesy of Thomas Schlenker, MD, MPH, Chief Medical Officer, Children's Hospital of Wisconsin.

Treatment and Prevention

Treatment is generally successful when antibiotics are administered during the catarrhal stage. However, antibiotic treatment only reduces the duration and severity of the illness.

Until recently, the relatively low incidence of pertussis in developed nations was the result of using the pertussis vaccine. The older vaccine (diphtheria-pertussis-tetanus, or DPT) was very effective in both preventing pertussis and stopping its transmission. However, due to vaccine reactions in some individuals, a newer acellular pertussis (aP) vaccine was developed from *B. pertussis* chemical extracts. Combined with diphtheria and tetanus toxoids, the triple vaccine has the acronym DTaP (Tripedia®). Because its ability to generate immunity slowly declines, in 2005, the US Food and Drug Administration (FDA) licensed a reduced-dose acellular pertussis vaccine. When combined with tetanus and diphtheria vaccines (Tdap), it is recommended as a booster dose for children at age 11 or 12 and as a booster every 10 years for those over 20 years of age to provide protection against pertussis and its transmission to susceptible infants.

Drug-Resistant Tuberculosis Is a Major Public Health Threat

Tuberculosis (TB) is a communicable infection that has been with us for thousands of years yet continues to evolve and resist our best drugs.

Causative Agent and Epidemiology

Tuberculosis is caused by *Mycobacterium tuberculosis*, the "tubercle" bacillus first isolated by Robert Koch in 1882. It is a small, aerobic, nonmotile rod. Its cell wall forms a waxy cell surface that greatly enhances resistance to drying, chemical disinfectants, and many antibiotics. During the first half of the 20th century, TB was called "consumption" or "white plague" because the disease wasted away the body and caused the patient to appear pale.

Over the centuries, TB has taken more lives than any other infectious disease. It remains one of the top 10 causes of death globally (**FIGURE 11.16**). For

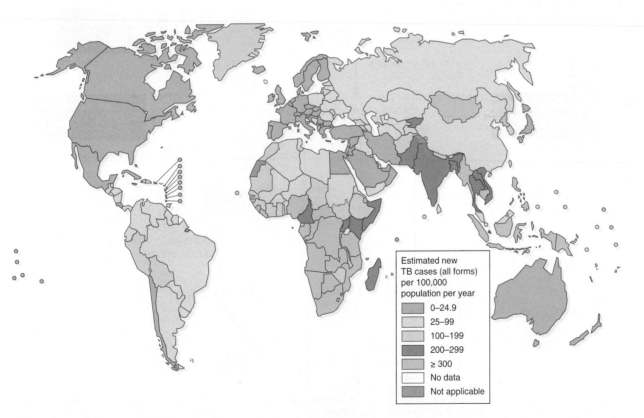

FIGURE 11.16 Estimated Global Incidence of Tuberculosis, 2015. About 85% of the new cases of TB occurred in Southeast Asia (34%), Africa (30%), and the Western Pacific (21%). **»» Where was the highest burden of new TB cases?**

2015, the WHO reported an estimated 10.4 million new TB infections and 1.8 million deaths from the disease. TB causes 25% of all deaths in AIDS individuals. For 2014, the CDC reported an all-time low of about 9,400 cases in the United States. Still, globally more people are infected with the TB bacillus than ever in history—some 2 billion people worldwide, or about 30% of the world's population.

Clinical Presentation

Tuberculosis is primarily an airborne disease and, as such, the bacilli are transmitted from person to person in small, aerosolized droplets when a person with active pulmonary disease sneezes, coughs, or even spits. Such airborne droplets can remain infectious for hours or even days. As few as 10 bacilli can initiate an infection in an individual having prolonged, frequent, or intense contact with an actively infected individual. Thus, people who live in overcrowded, urban areas are most at risk of contracting

TB. Malnutrition and a generally poor quality of life also contribute to the establishment of disease.

Unlike many other infectious diseases by which an individual becomes ill after several days or a week, the incubation period for TB is 2 to 12 weeks. Symptoms vary depending on the individual's health status and organs affected. In general, symptoms include fatigue, low-grade fever, unexplained weight loss, and loss of appetite. For lung infections, additional symptoms include persistent coughing (associated with bloody sputum) and difficulty breathing (with chest pain). The illness has two separate stages: an infection stage and a disease stage (**FIGURE 11.17**).

Primary TB Infection

If a person has a pulmonary infection (85% of infections are respiratory), the bacterial cells enter the alveoli where pathogen interactions occur (**FIGURE 11.18**). This individual has a **primary TB infection**. If tested, the person would have a positive

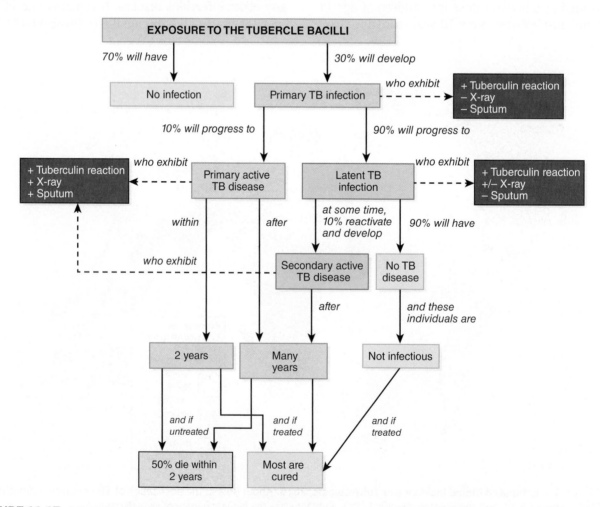

FIGURE 11.17 A Concept Map for Tuberculosis. The stages of tuberculosis infection and disease are shown. ("+" or "−" represents positive or negative test results.) *»» How does TB infection differ from active TB disease?*

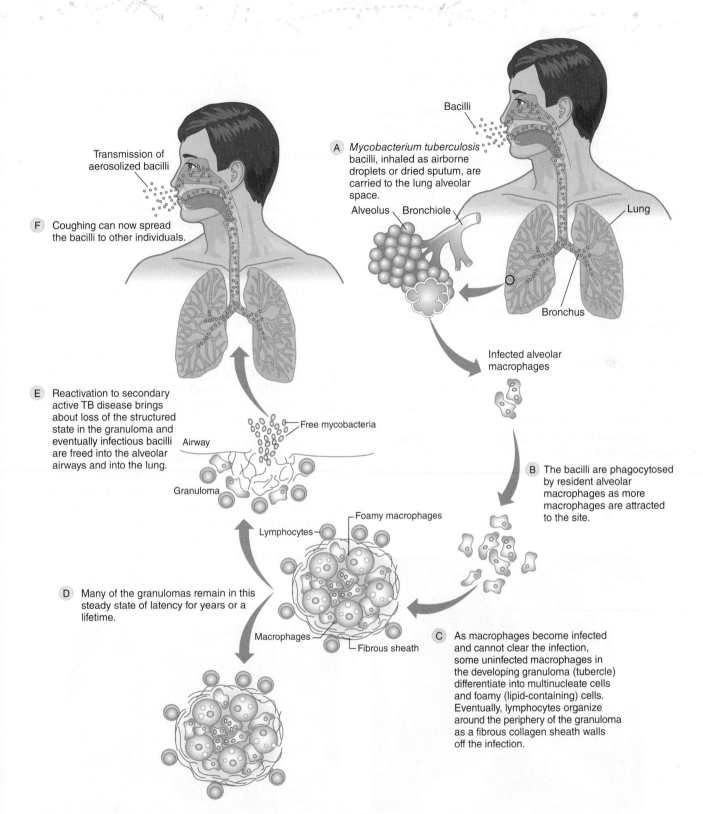

A *Mycobacterium tuberculosis* bacilli, inhaled as airborne droplets or dried sputum, are carried to the lung alveolar space.

F Coughing can now spread the bacilli to other individuals.

Transmission of aerosolized bacilli

Bacilli

Alveolus   Bronchiole   Lung

Bronchus

Infected alveolar macrophages

E Reactivation to secondary active TB disease brings about loss of the structured state in the granuloma and eventually infectious bacilli are freed into the alveolar airways and into the lung.

Free mycobacteria

Airway

Granuloma

B The bacilli are phagocytosed by resident alveolar macrophages as more macrophages are attracted to the site.

Foamy macrophages

Lymphocytes

D Many of the granulomas remain in this steady state of latency for years or a lifetime.

Macrophages

Fibrous sheath

C As macrophages become infected and cannot clear the infection, some uninfected macrophages in the developing granuloma (tubercle) differentiate into multinucleate cells and foamy (lipid-containing) cells. Eventually, lymphocytes organize around the periphery of the granuloma as a fibrous collagen sheath walls off the infection.

FIGURE 11.18   The Stages of a *Mycobacterium tuberculosis* Infection. Following inhalation of aerosolized bacilli, the bacteria invade the alveoli, where the cells are taken up by macrophages. Unable to eliminate the infection, immune cells (lymphocytes) attempt to "wall off" the bacilli. In active TB disease, the bacilli rupture the granuloma and enter the alveoli where they can be coughed up and transmitted to another susceptible host. **»» *What is the immune system attempting to do by forming tubercles?***

tuberculin reaction, but a chest X-ray often is negative and a sputum test would be negative (see the section "Disease Diagnosis"). Therefore, the person would not be contagious.

In the alveoli, macrophages respond to the infection by engulfing the bacilli through phagocytosis. Unfortunately, the bacilli often are not killed in the macrophages because of the pathogen's waxy cell wall, and they survive in the macrophage cytoplasm. As more macrophages arrive, they too phagocytize bacilli but are incapable of destroying them and an inflammatory condition ensues. After about 4 weeks, the body's immune response localizes the infection, forming a central area of large, multicellular giant cells. Recruited lymphocytes and fibroblasts surround the mass in the lung, forming a type of granuloma called a **tubercle** (hence the name tuberculosis).

Latent TB Infection

In 90% of primary TB infections, the infection becomes arrested due to immune responses and the individual usually has no clinical signs of infection. This dormant form of TB is referred to as a **latent TB infection** and is carried by 2 billion people worldwide. Of these, 90% are unlikely to develop active disease and they are not infectious.

Primary and Secondary Active TB Disease

Up to 10% of individuals who have a primary TB infection will develop **primary active TB disease** in 1 to 2 years. In addition, due to immune system dysfunction, about 10% of latent TB infections might undergo reactivation resulting in **secondary active TB disease**. Individuals will become ill within 3 months of reactivation.

In both active forms of TB disease, the immune defenses cannot keep the tubercle bacilli in check. Many of the infected macrophages die, releasing bacilli and producing a caseous (cheese-like) center in the tubercle. Live bacterial cells rupture from the tubercles, and the bacilli spread and multiply throughout the LRT. These individuals will have a positive tuberculin reaction, chest X-ray, and sputum test, and they can transmit the disease to others (**FIGURE 11.19**).

Because the bacilli in individuals with active TB disease are not killed, tubercle erosion can allow the bacterial cells to spread beyond the LRT through the blood and lymph to other organs, such as the liver, kidney, meninges, and bone. If active tubercles develop throughout the body, the disease is called **miliary (disseminated) tuberculosis** (*milium* = "seed"; in reference to the tiny lesions resembling the millet seeds in bird feed). Tubercle bacilli produce no known toxins, but growth is so unrelenting that the respiratory and other body tissues are literally consumed, a factor that gave tuberculosis its alternate name of "consumption."

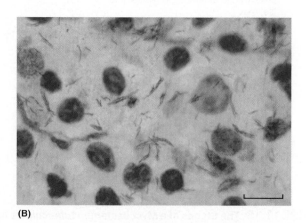

(A)

(B)

FIGURE 11.19 Pulmonary Tuberculosis. (A) This false-color chest X-ray shows the extensive fibrosis (white oval) typical in patients with advanced, active tuberculosis. **(B)** This light microscope image shows *M. tuberculosis* cells (red) stained with the acid-fast procedure. In sputum samples, the bacterial cells often exhibit growth in thick strings. (Bar = 10 μm.) »» *Why isn't the Gram stain used to identify* M. tuberculosis?

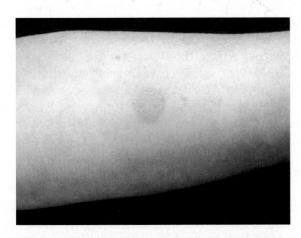

FIGURE 11.20 Tuberculin Skin Test for Tuberculosis. This is an example of a positive reaction to the Mantoux skin test. An induration of more than 15 mm is considered positive. »» *What is an induration?*

© Medical Images RM/Bob Tapper.

Disease Diagnosis

Early detection of tuberculosis is aided by the tuberculin reaction. In the **Mantoux test**, the physician injects a small amount of a purified protein derivative (PPD) of *M. tuberculosis* into the skin of the forearm. If a person has been infected with *M. tuberculosis*, the skin becomes thick, and a raised, red welt, termed an **induration**, of a defined diameter develops after 48 to 72 hours (**FIGURE 11.20**). For an individual never before exposed to *M. tuberculosis*, an induration greater than 15 mm is interpreted as a positive test. However, a positive test does not necessarily reflect the presence of active TB disease. A "false positive" test might be the result of a recent immunization, a previous tuberculin test, or past exposure to *M. tuberculosis*. It suggests a need for further tests. A chest X-ray can distinguish between active and latent TB and acid-fast staining of a sputum sample can identify the TB bacilli (see Figure 11.19). A newer TB test system is available that can identify a TB infection in less than 2 hours versus the 2 to 3 months required to verify a positive sputum test.

Treatment

Standard drug-resistant TB can be treated with such first-line antibiotics as isoniazid and rifampicin. If the antibiotics are taken daily for 2 to 6 months, cure rates can be as high as 90%.

However, there is a growing number of TB strains that are resistant to these antibiotics. Globally, in 2015, the WHO reported an estimated 600,000 cases and more than 170,000 deaths due to **multidrug-resistant tuberculosis (MDR-TB)**. This has

necessitated a switch to a group of second-line drugs, including fluoroquinolones and kanamycin. Because *M. tuberculosis* has a very slow generation time (approximately 18 hours), MDR-TB patients must undergo intensive antimicrobial drug therapy. This entails daily injections and swallowing more than 10,000 antibiotic pills over a two-year period. The side effects can be very toxic and make patients quite ill.

In all treatment therapies, patient compliance in taking the drugs is essential. This can be difficult due to the side effects and toxicity of many of the drugs. The WHO treatment strategy involves **direct observation treatment** (**DOT**), whereby a healthcare worker administers and ensures the patient has taken the drugs. Compliance makes the development of MDR-TB less likely. In 2015, only 20% of MDR-TB patients received treatment, and only half of those who started treatment were cured.

About 10% of MDR-TB cases have evolved into **extensively drug-resistant tuberculosis** (**XDR-TB**), meaning almost all antibiotics that can be used to treat TB are useless. Few treatment options remain for these individuals, and a successful treatment outcome depends upon the extent of drug resistance, the severity of the disease, and the status of the patient's immune system.

A weakened immune system is especially worrisome because TB often progresses gradually with inapparent symptoms. Thus, it can be particularly dangerous to those who have AIDS. In these coinfected patients, the immune system is severely compromised and unable to mount a response to *M. tuberculosis* as well as to other infectious agents. Unlike most other TB patients, those with HIV usually develop miliary TB in the lymph nodes, bones, liver, and numerous other organs. Ironically, AIDS patients often test negative for the tuberculin skin test, because without a robust immune system, the patients cannot produce the telltale red welt signaling exposure.

Prevention

In high-risk healthcare settings and locales with high rates of TB, appropriate precautions should be taken, which include wearing a mask to prevent TB spread. Vaccination against TB can involve intradermal injections of a weakened (attenuated) strain of *Mycobacterium bovis*, a species that causes tuberculosis in cows as well as humans. The attenuated strain is called **bacille Calmette-Guérin** (**BCG**), and the vaccine is used in parts of the world where the disease causes significant mortality and morbidity. The BCG vaccine is not recommended in the United States because it has limited effectiveness for preventing

TB in children and is ineffective in adults. The promising news is that 12 new antimicrobial compounds, consisting of subunits, molecules of DNA, and attenuated strains of mycobacteria, are currently being developed and tested as new vaccines.

Other *Mycobacterium* Species

Although most mycobacterial species are nonpathogenic, a few of the other pathogenic ones deserve mention. *M. kansasii* causes infections that are indistinguishable from those caused by *M. tuberculosis*. The group known as *M. avium* complex (MAC) causes a disseminated disease in immunocompromised individuals, such as AIDS patients and individuals on immunosuppressive chemotherapy. *M. chelonae*, frequently found in soil and water, can cause lung diseases, wound infections, arthritis, and skin abscesses. Another species, *M. haemophilum*, is a slow-growing pathogen often found in immunocompromised individuals, including those with AIDS. Cutaneous ulcerating lesions and respiratory symptoms are typical in these patients. For all the other species mentioned here, there is no evidence for spread between individuals; rather, infection comes from contacting soil or ingesting food or water contaminated with the organism.

In 2016, another species that infects 10% to 20% of patients suffering from cystic fibrosis (CF) was reported. The bacterium, *M. abscessus*, thrives in the thick mucus that builds up in the airways of CF patients, and some experts believe several different strains of the pathogen can spread from one CF treatment center to another. It has been suggested that healthy individuals with inapparent infections might be spreading the pathogens.

Infectious Bronchitis Is an Inflammation of the Bronchi

Infectious bronchitis occurs most often during the winter, and the condition can be caused by bacteria following a URT viral infection, such as the common cold. *Mycoplasma pneumoniae* and *Chlamydophila pneumoniae* often cause bacterial bronchitis in young adults, while *S. pneumoniae* and *H. influenzae* are the primary agents among middle-aged and older adults. Influenza viruses also can cause a form of viral bronchitis.

The inflammation of the bronchi generally begins with the symptoms of a common cold: runny nose, sore throat, chills, general malaise, and perhaps a slight fever. The onset of a dry cough usually signals the beginning of **acute bronchitis**, a condition that occurs when the inner walls lining the main airways of the lungs become infected and inflamed. Inflammation increases the production of mucus, which then narrows the air passages (**FIGURE 11.21**). Most cases of acute bronchitis disappear within a few days without any adverse effects, although a cough can linger for several weeks. If the condition persists for more than 3 months, it is referred to as **chronic bronchitis**. The changing of the clear or white mucus (phlegm) to a yellow or green color usually indicates a bacterial infection.

Antibiotics can be prescribed for bacterial bronchitis. Because many cases result from influenza, getting a yearly flu vaccination might reduce the risk of bacterial bronchitis. Other preventative measures include good hygiene, including hand washing, to reduce the chance of transmission.

Pneumonia Can Be Caused by Several Bacterial Species

The term **pneumonia** refers to an inflammation in the LRT. Here, the bronchioles and alveoli in the lungs become inflamed and filled with fluid. In the United States, there are 2 to 3 million pneumonia infections resulting in 45,000 deaths each year. In adults, bacteria are the most common cause of pneumonia. For discussion purposes, we can divide pneumonia into healthcare-acquired infections and community-acquired infections.

Healthcare-Acquired Pneumonia

Other than urinary tract infections, pneumonia is the most common healthcare-acquired infection. **Healthcare-acquired pneumonia (HAP)** is defined as an inflammation of one or both lungs that develops primarily in older adults and immunosuppressed patients at least 48 hours after administration to a hospital or other healthcare facility. Many of the HAP cases in the United States are associated with mechanical ventilation ("ventilator-acquired pneumonia"), which occurs when a respirator or ventilator used to provide patients with oxygen and assist patient breathing has been contaminated with bacteria or improperly cleaned. Several bacterial species are responsible for HAP, and these species are showing increased antibiotic resistance.

▶ **Staphylococcal Pneumonia.** The most common gram-positive cause of HAP results from an infection by *S. aureus*. The pathogen often is spread to others by contaminated hands and by invasive medical devices contaminated with *S. aureus*. Symptoms of **staphylococcal pneumonia** include a short

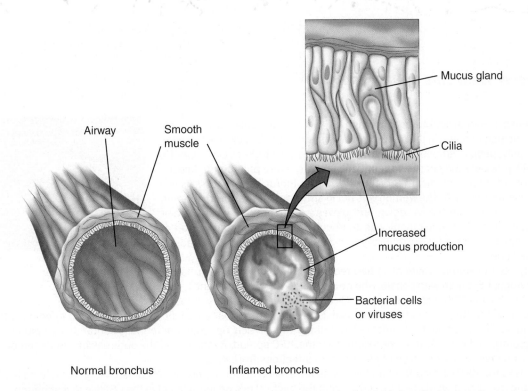

FIGURE 11.21 Bronchus Inflammation. Infectious bronchitis develops as the bronchial wall narrows due to inflammation and swelling. Increased mucus production, which contains bacterial cells or viruses, also narrows the airway. *»» How would such pathogens in the mucus be transmitted to another person?*

Data from Merck, "Acute Bronchitis: Lung and Airway Disorders," *Merck Manual Home Edition,* December 09, 2008, http://www.merck.com/mmhe/sec04/ch041/ch041a.html.

period of fever followed by rapid onset of respiratory distress, which may include rapid breathing and the appearance of bluish skin. With increased antibiotic resistance, infections can be difficult to treat and can progress to life-threatening infections.

▶ ***Klebsiella* Pneumonia.** Another form of HAP is caused by *Klebsiella pneumoniae*, a gram-negative rod with a prominent capsule. The pathogen, which is responsible for up to 15% of HAP, is acquired by airborne respiratory droplets as often it occurs naturally in the nasopharynx. ***Klebsiella* pneumonia** can be a primary infection, or a secondary infection in alcoholics or people with impaired pulmonary function. As a primary pneumonia, it is characterized by sudden onset and gelatinous reddish-brown sputum. The color results from blood seeping into the alveolar sacs of the lung as bacterial cells multiply and cause tissue damage that can result in the death of the individual.

In its secondary form, *K. pneumoniae* occurs in already ill individuals and is a hospital-acquired disease spread by such routes as clothing, intravenous solutions, foods, and the hands of healthcare workers. In addition, scientists have reported the discovery of multidrug-resistant *K. pneumoniae* strains resistant to most all available antibiotics. MICROFOCUS 11.2 relates a recent, frightening outbreak of this secondary form in one of the nation's leading research hospitals.

▶ ***Pseudomonas* Pneumonia.** HAP also can be caused by *Pseudomonas aeruginosa*, a gram-negative, aerobic rod. It is one of the most dangerous opportunistic pathogens because of its ability to cause severe or fatal healthcare-associated infections, especially in immunocompromised patients. ***Pseudomonas* pneumonia** accounts for more than 10% of all HAP cases. In addition, it is often resistant to commonly used antibiotics. The *P. aeruginosa* cells can be transmitted via aspiration from contaminated ventilator tubing or other healthcare devices, such as bronchoscopes, mechanical ventilators, and nebulizer equipment. Symptoms in immunocompromised patients include breathing problems, productive cough, fever, chills, and bluish skin. The treatment of this condition might include a combination of antibiotics.

MICROFOCUS 11.2: Public Health

"We Never Want This to Happen Again"

The 243-bed National Institutes of Health (NIH) Clinical Center, the nation's leading research hospital in the suburbs of Washington, DC, is a unique hospital, treating only people enrolled in government research studies. On June 13, 2011, a 43-year-old study participant critically ill with a rare pulmonary disease was admitted with a medical record of having multidrug-resistant *Klebsiella pneumoniae*, termed KPC. The woman was put in strict isolation, all staff entering her room donned a protective gown and gloves and rigorously washed their hands, and all medical equipment was thoroughly decontaminated. All other patients in the ICU were tested regularly to make sure KPC was not spreading. All seemed fine.

NIH Clinical Center.

Courtesy of Clinical Center/NIH.

On July 15, the woman (patient 1) had recovered, and she was sent home. Then, on August 5, a man with cancer, who never had contact with patient 1, was identified as having KPC. Later, on August 15, a woman with a primary immune deficiency fell ill to KPC. Patients 2 and 3 died. From then on, almost a patient a week was being infected with KPC. The hospital locked down patients, disinfected everything with bleach, ripped out potentially contaminated plumbing, and still patients contracted KPC. Over six worrisome months, KPC spread. By the end of the outbreak in mid-December, 19 people had been infected, and seven died of bloodstream infections from KPC.

Where was the KPC coming from and how was it traveling about the hospital? Epidemiologists were called into action to solve the mystery. It would require sequencing the bacterium's genome to solve the CSI-like investigation.

By comparing DNA sequences from all KPC patients, the isolate from the patient 1 matched the isolates taken from the other patients, making it highly likely the first patient (called the index case) was the origin of the KPC outbreak. In fact, fine differences in sequences suggested the KPC cells came from three sites on her body (lungs, groin, and throat), and so there were at least three separate transmission events. Undoubtedly, some form of intrahospital transmission had occurred, but the specific form(s) and/or vehicle(s) remain a mystery.

But where were the bacterial cells hiding out after the index patient was released from the hospital? Genetic sleuthing provided a few clues. KPC stayed alive in sink drains and even on a ventilator that had been thoroughly cleaned with bleach, providing a few potential modes of transmission for the pathogen.

Since then, the NIH Clinical Center has made changes: all ICU patients undergo extensive invasive testing, using rectal swabs to identify any "silent" pathogens; a wall now partitions off the ICU; all ICU staff, including janitors, work only in the ICU and nowhere else in the hospital; and human monitors are paid to ensure that everyone follows the rules. As Dr. Tara Palmore, deputy hospital epidemiologist at the NIH Clinical Center, said, *"We never want this to happen again."*

▶ **Acinetobacter Pneumonia.** *Acinetobacter* species are gram-negative, aerobic rods that frequently colonize the respiratory tract. Called the *Acinetobacter baumannii-calcoaceticus* (Abc) complex, the pathogen is becoming a major cause of HAP in intensive-care units. ***Acinetobacter* pneumonia** occurs in outbreaks and, like *P. aeruginosa*, is usually associated with contaminated respiratory-support equipment or fluids. The Abc complex is becoming increasingly resistant to antibiotics, presenting a significant challenge in treating these infections.

Community-Acquired Pneumonia

Globally, **community-acquired pneumonia (CAP)** is one of the most common infectious diseases and is an important cause of mortality and morbidity. For a long time, *S. pneumoniae* has been considered the major cause of CAP. However, a 2015 CDC study reported that the actual infectious agents of CAP in the United States rarely are identified. In fact, in 60% of cases, no detectable pathogen can be identified. Rather, diagnosis is made based on symptoms, a chest X-ray, and the patient's health history.

That said, let's look at what traditionally have been recognized as the most common bacterial pathogens causing CAP.

Pneumococcal Pneumonia

The most prevalent form of CAP appears to be **pneumococcal pneumonia**. It is caused by *S. pneumoniae*, a gram-positive, encapsulated chain of diplococci (**FIGURE 11.22**). The disease can occur in all age

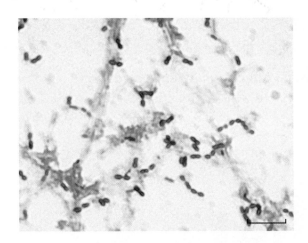

FIGURE 11.22 Streptococcus pneumoniae. Light microscope image of gram-stained *S. pneumoniae* cells (dark spheres), the causative agent for pneumococcal pneumonia. (Bar = 10 µm.) »» *How would you describe the arrangement of S.* pneumoniae *cells?*

Courtesy of Dr. Mike Miller/CDC.

groups but primarily in infants, young children, older adults, and those with underlying medical conditions.

S. pneumoniae (commonly referred to as pneumococcus) can be acquired by aerosolized droplets or through microaspiration of the bacterial cells that are a normal member of the microbiome in the URT of many individuals. However, INVESTIGATING THE MICROBIAL WORLD 11 presents another potential form of transmission.

Because mucociliary clearance is at work and the natural resistance of the body is high, pneumococcal pneumonia usually does not develop unless immune defenses are compromised. Previous viral infections as well as alcoholism, malnutrition, smoking, and treatment with immune-suppressing drugs predispose one to *S. pneumoniae* infections.

When the pneumococcus multiplies in the alveoli, the bacterial cells damage the alveolar lining,

Investigating the Microbial World 11

Can *Streptococcus pneumoniae* Survive and Remain Infectious After Desiccation?

A common inhabitant of the human nasopharynx is the gram-positive, encapsulated bacterium *Streptococcus pneumoniae*. As discussed in this chapter, its location in the upper respiratory tract makes it one of the prime candidates for meningitis, pneumonia, and middle-ear infections should it switch from a harmless existence to a pathogenic one.

OBSERVATIONS: As described in this chapter and by the Centers for Disease Control and Prevention (CDC), *S. pneumoniae* is transmitted "directly from person to person through close contact via airborne respiratory droplets" during coughing and sneezing. Yet other respiratory pathogens are also spread by nonliving objects or materials, such as environmental surfaces.

QUESTION: *Can* S. pneumoniae *also survive for a prolonged time on a dry, desiccated environmental surface?*

HYPOTHESIS: Dry, desiccated environmental surfaces are not an alternate source for the spread of *S. pneumoniae*. If so, after a "prolonged period" on a dry surface, the bacterial cells will die (no colonies will be observed when plated on blood agar) and consequently will no longer be infective.

EXPERIMENTAL DESIGN: The desiccation protocol involved growing *S. pneumoniae* on blood agar overnight, scraping the bacterial cells off the plate, and spreading a thin, even layer of cells (10^8) onto polystyrene petri dish lids. The number of viable cells per lid was determined by plating dilutions on blood agar and counting the number of colonies growing.

EXPERIMENT 1: Cell preparations were spread on lids and incubated at ambient temperature and humidity for 1 hour to 28 days. A control preparation (0 hour) was spread on a lid and immediately diluted and plated on blood agar.

EXPERIMENT 2: Because survivability could be a result of starvation on polystyrene, cell preparations for desiccation were spread on lids and incubated at ambient temperature and humidity. Cell preparations for starvation were spread on agar plates containing only phosphate-buffered saline (PBS). All lids and plates were incubated for 6, 24, or 48 hours.

EXPERIMENT 3: Because *S. pneumoniae* has a capsule, survivability could be directly related to having this type of glycocalyx. Two identical cell preparations were set up and incubated for 4 to 168 hours. One preparation used the encapsulated strain; the other used a nonencapsulated strain.

EXPERIMENT 4: If *S. pneumoniae* cells survive desiccation, they might not be infectious. Samples of 48-hour nondesiccated and desiccated cells in PBS were inoculated intranasally into mice. A control group was inoculated intranasally with only PBS. Three days postinoculation, the nasal lavage fluid was plated on blood agar plates containing the antibiotic gentamicin to which *S. pneumoniae* is naturally resistant. Colonies on agar were then counted.

(continues)

Investigating the Microbial World 11 (*Continued*)

Can *Streptococcus pneumoniae* Survive and Remain Infectious After Desiccation?

RESULTS:

EXPERIMENT 1: See figure.

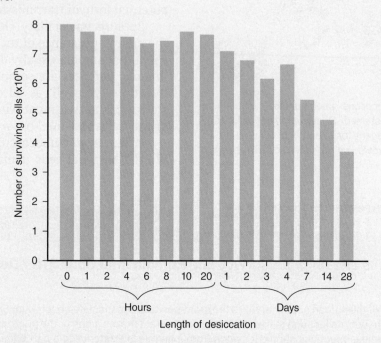

S. pneumoniae survival after desiccation. Bacteria were rehydrated and plated after 0 hours to 28 days of desiccation to determine viability. Data for the medians are shown. 0 hour = control lid (=10^8 cells).

Data from Walsh, R. L. and Camilli, A. 2011. *mBio* 2(3): e00092-11doi:10.1128/mBio.00092-11.

EXPERIMENT 2: Analysis of the desiccated and starved samples indicated that there were fewer colonies on the agar plated with starved samples than on the agar plated with desiccated samples.

EXPERIMENT 3: No significant difference was observed in bacterial viability for the nonencapsulated strain as compared to the encapsulated strain when colonies were counted on agar.

EXPERIMENT 4: No detectable *S. pneumoniae* colonies on agar could be detected from the control mice. However, 80% of mice receiving the nondesiccated sample and 75% of mice receiving the desiccated sample produced colonies on the agar plates.

CONCLUSION: The medical view has been that *S. pneumoniae* cells must be transmitted through direct contact of airborne respiratory secretions.

QUESTION 1: *Was the hypothesis concerning desiccation and infectivity supported? Explain.*

QUESTION 2: *What can you conclude from experiment 2 regarding the influence of starvation on survivability through desiccation?*

QUESTION 3: *How important is the* S. pneumoniae *capsule in providing protection against desiccation?*

QUESTION 4: *Based on the results from this work, what are your recommendations concerning the transmissibility of* S. pneumoniae *from nonliving surfaces such as handkerchiefs, utensils, and hospital surfaces?*

You can find answers online in **Appendix E**.

Modified from Walsh, R. L. and Camilli, A. 2011. *mBio* 2(3): e0009211.doi:10.1128/mBio.00092-11.

allowing fluids and blood cells to enter the sacs. Pneumonia develops as fluid fills the alveoli and reduces oxygen transfer. Patients with pneumococcal pneumonia experience high fever, sharp chest pains, difficulty breathing, and rust-colored sputum. The involvement of an entire lobe of the lung is called **lobar pneumonia**. If both left and right lungs are involved, the condition is called **double pneumonia**. Scattered patches of infection in the LRT are referred to as **bronchopneumonia**.

With appropriate antibiotic therapy, recovery is likely. Unfortunately, recovery from one

S. pneumoniae strain does not confer immunity to another strain (over 90 different strains have been identified). As mentioned earlier in the section on meningitis, there are two pneumococcal vaccines available: a pneumococcal conjugate vaccine (PCV13; Prevnar 13®) for children under 5 years of age and adults 65 years of age and older and a pneumococcal polysaccharide vaccine (PPSV23; Pneumovax®23) for adults 65 years of age and older.

Every year, some 1.6 million children worldwide die from pneumonia (MICROFOCUS 11.3). Pneumococcal pneumonia kills 800,000 children—lives

MICROFOCUS 11.3: Public Health

The Killer of Children

Global Health Magazine recently reported the following: "Chitra Kumal knows the pain of losing a child. When her daughter, Sunita, was 15 months old, she developed a respiratory infection that quickly progressed into pneumonia. With no health facilities in her Nepalese village, Kumal depended on the advice and treatment of a traditional healer or shaman. After just 3 days of fever, fast breathing, and chest indrawing, her only daughter died."

Similar stories are reported every day around the world. According to the World Health Organization (WHO), pneumonia kills 2 million children under 5 years of age each year—more than AIDS, malaria, and measles combined—accounting for nearly one in five child deaths globally (see figure). However, this number might be an underestimate because nearly half of all pneumonia cases occur in malarious parts of the world where pneumonia often is misdiagnosed as malaria.

The WHO estimates that more than 150 million episodes of pneumonia occur every year among children under 5 in developing countries, accounting for more than 95% of all new cases worldwide, and between 11 and 20 million of these episodes require hospitalization. The highest incidence of pneumonia cases among children under five occurs in South Asia and sub-Saharan Africa, accounting for more than half the total number of pneumonia episodes worldwide.

Preventing and treating childhood pneumonia obviously is critical to reducing childhood mortality. However, only about one in four caregivers knows the two key symptoms of pneumonia: fast breathing and difficult breathing (indrawing). Estimates suggest that if antibiotics were universally available and given to children with pneumonia, around 600,000 lives could be saved each year.

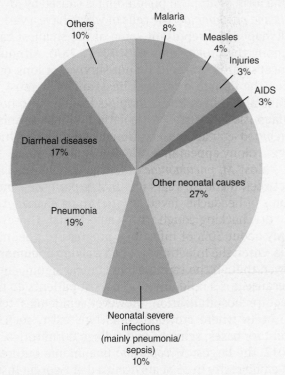

Pneumonia accounts for almost 20% of global childhood mortality.

However, this represents only about 25% of the annual cases. Clearly, other control measures are needed.

At the beginning of the 20th century, pneumonia accounted for 19% of childhood deaths in the United States, a statistic remarkably similar to the rate in developing countries today. Control in the United States was achieved largely without antibiotics and vaccines. Therefore, other control measures and strategies are needed on a global scale.

Key prevention measures include promoting balanced nutrition (including breastfeeding, vitamin A supplementation, and zinc intake), reducing environmental air pollution, and increasing immunization rates with vaccines, such as those against *Haemophilus influenzae* type b (Hib) and *Streptococcus pneumoniae* (pneumococcus). However, only about 50% of pneumonia cases in Africa and Asia are caused by these two organisms, so other vaccines need to be developed against other bacterial species (and viruses) that cause pneumonia. And, of course, hand washing, like in all areas of infectious disease, can play an important role in reducing the incidence of pneumonia.

that could be saved with an effective vaccine. Now, such vaccines are becoming available. The GAVI Alliance (formerly the Global Alliance for Vaccines and Immunization) estimates that the lives of 7 million children can be saved by 2030 through the introduction of Prevnar 13 and another vaccine called Synflorix® to the world's poorest countries, where 98% of the pneumonia deaths occur.

A more gradual form of CAP is not caused by the typical pathogens described above. These bacterial species cause **"atypical" pneumonia**. The term "atypical" is used because pneumonia caused by these species can have slightly different symptoms and respond to different antibiotics than the typical pneumonia-causing bacteria described earlier.

Mycoplasma Pneumonia

One form of atypical pneumonia is caused by *Mycoplasma pneumoniae*, the cells of which have a variety of shapes and represent some of the smallest cells among the prokaryotes (**FIGURE 11.23A**). Although encapsulated, the cells do not survive for long outside the human or animal host, and they must be transmitted in airborne droplets from host to host. Diagnosis is assisted by isolation of the organism on blood agar and observation of a distinctive "fried egg" colony appearance (**FIGURE 11.23B**).

Most *M. pneumoniae* patients, who are usually between 6 and 20 years old, first experience symptoms of headache, fever, fatigue, and a characteristic dry, hacking cough with a sore throat. However, only about 30% of infections progress to pneumonia. Often the infection is called **walking pneumonia** (even though the term has no clinical significance) because it is a mild illness and most patients do not require hospitalization. However, epidemics tend to occur where crowded conditions exist, such as military bases, schools, and college dormitories. In 2012, the largest *Mycoplasma* pneumonia outbreak at a university in 35 years occurred at Georgia Institute of Technology. Some 83 students were infected. The antibiotics erythromycin and azithromycin are commonly used as treatments.

In 2016, scientists discovered that *M. pneumoniae* in the respiratory tract can trigger a serious nervous system disease in patients with *Mycoplasma* pneumonia. When the patient's immune system produces antibodies to the infection, the antibodies not only recognize the surface of the *Mycoplasma* cells but also the myelin sheath of nerve cells. The immune attack on the nerve cells leads to an acute life-threatening

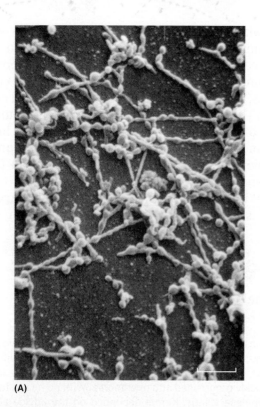

(A)

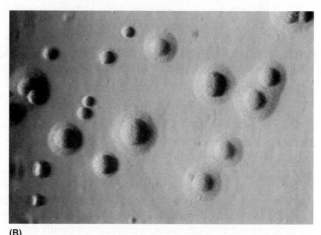

(B)

FIGURE 11.23 *Mycoplasma pneumoniae.* Two views of *M. pneumoniae,* the agent for a form of primary atypical pneumonia. **(A)** A scanning electron micrograph, demonstrating the pleomorphic shape exhibited by mycoplasmas. (Bar = 2 μm.) **(B)** Colony morphology on agar shows the typical "fried egg" appearance. *»» What structural feature is missing from the mycoplasma cells that allows for their pleomorphic shape?*

(A) © ASM/Science Source. (B) © Michael Gabridge/Visuals Unlimited, Inc.

disease called **Guillain-Barré syndrome (GBS)**. GBS can cause paralysis lasting several weeks and usually requires intensive care. Researchers now are studying this cross reactivity, as other infectious diseases also are known to trigger GBS.

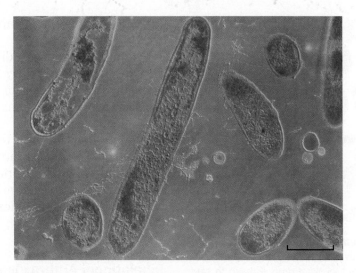

FIGURE 11.24 *Legionella pneumophila.* A false-color transmission electron micrograph of *L. pneumophila* cells. (Bar = 1 μm.)
»» What do the false-colored red and green areas in the cells represent?

© ISM/Alain POL.

Legionnaires' Disease

Another form of CAP first surfaced in July 1976, after an American Legion Convention in Philadelphia. What would become known as **Legionnaires' disease** affected 182 conventioneers and 39 other people in or near the convention hotel who suffered from headaches, fever, coughing, and shortness of breath. Thirty-four individuals died of the disease or its complications.

The causative agent of Legionnaires' disease, *Legionella pneumophila,* is an aerobic, gram-negative rod found where warm water collects, such as cooling towers, industrial air-conditioning systems, and stagnant pools (**FIGURE 11.24**). When *Legionella*-contaminated water is sprayed by air conditioning towers, fountains, or showers, the aerosolized droplets can be inhaled. In the lungs, the bacterial cells infect alveolar macrophages where the pathogen causes tissue damage and inflammation as the pathogen multiplies.

Older adults and individuals with chronic lung disease or weak immune systems are most susceptible to infection. The CDC estimates that there are 8,000 to 12,000 hospitalizations due to the disease each year. Legionnaires' disease is not transmitted person to person. In 2015, two unrelated outbreaks in New York City affected more than 133 people and killed 16. The source appeared to be *Legionella*-contaminated cooling towers.

Erythromycin is effective for treatment and prevention requires that water sources be kept chlorinated to limit possible contamination.

After *L. pneumophila* was isolated in early 1977, microbiologists realized the organism was responsible for another milder infection called **Pontiac fever**. This is an influenza-like illness that lasts 2 to 5 days but does not cause pneumonia. The term **legionellosis** encompasses both Legionnaires' disease and Pontiac fever.

CAP Also Is Caused by Intracellular Parasites

Cases of CAP can be caused by a few other bacteria. The genus *Chlamydophila* (phylum Chlamydiae) contains some of the smallest bacterial organisms known. Most are obligate, intracellular parasites, meaning that they grow only inside host cells. Most are **zoonotic diseases**, which means that the organisms are transmitted to humans from infected birds or livestock, especially dairy cows, sheep, and goats.

Psittacosis

One zoonotic form of CAP is **psittacosis**, which is caused by *Chlamydophila psittaci.* This obligate, intracellular pathogen is transmitted to humans by infected parrots, parakeets, canaries, and other members of the psittacine family of birds (*psittakos* = "parrot"). Because the disease also occurs in pigeons, chickens, turkeys, and seagulls, some microbiologists prefer to call the disease **ornithosis** (*ornith* = "bird") to reflect the more widespread occurrence in bird species. Since 1996, fewer than 50 confirmed cases have been reported to the CDC. Many more cases might occur that are not correctly diagnosed or reported.

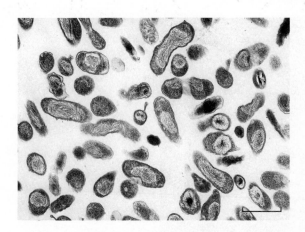

FIGURE 11.25 *Coxiella burnetii.* An electron micrograph of *C. burnetii,* the agent of Q fever. Note the oval-shaped rods of the organism. (Bar = 1 μm.) *»» What does it mean to say that Q fever is a zoonotic disease?*

Courtesy of Rocky Mountain Laboratories, NIAID, NIH.

Humans acquire *C. psittaci* by inhaling airborne dust or dried droppings from contaminated bird feces. The symptoms of psittacosis, which include fever, chills, headache, muscle aches, and a dry cough, resemble those of *Mycoplasma* pneumonia. Fever is accompanied by headaches, dry cough, and scattered patches of lung infection. Antibiotics are commonly used in therapy.

Chlamydophila Pneumonia

A mild form of CAP is caused by *Chlamydophila pneumoniae*, which is transmitted human to human by airborne respiratory droplets, principally among young adults and college students. The disease is clinically similar to bronchitis and is characterized by fever and nonproductive cough. Treatment with antibiotics hastens recovery from the infection.

Q Fever

This form of CAP is caused by *Coxiella burnetii* (phylum Proteobacteria), which is not a strict intracellular pathogen (**FIGURE 11.25**). The disease, **Q fever** (the "Q" is derived from "query," originally referring to the unknown cause of the disease), is another zoonotic disease, whereby the organisms are transmitted to humans from infected livestock, especially dairy cows, sheep, and goats. Therefore, human outbreaks can occur even some distance from where infected animals are raised, housed, or transported. In 2009, almost 4,000 cases of Q fever, including six deaths, were reported in the Netherlands where intense goat farming occurs. In 2015,

the CDC reported 128 confirmed cases of Q fever in the United States.

Being a zoonotic disease, transmission to humans occurs primarily by inhaling the organisms in dust particles or handling infected animals. In addition, humans can acquire the disease by consuming raw milk or cheese infected with *C. burnetii* or from milk that has been improperly pasteurized. Although most cases are asymptomatic, some patients experience a bronchopneumonia characterized by severe headache, high fever, dry cough, and, occasionally, lesions on the lung surface. The mortality rate is low, but antibiotic treatment should not be delayed.

Inhalational Anthrax Is an Occupational Hazard

Anthrax, another zoonotic disease, is caused by *Bacillus anthracis*, a spore-forming, aerobic, gram-positive rod. The disease is primarily associated with large, domestic herbivores (cattle, sheep, and goats).

Humans acquire **inhalational anthrax** from contaminated animal products or dust. For example, workers who process hides or wool, or shear sheep, can inhale the spores and contract a pulmonary infection as a form of pneumonia called **Woolsorters' disease**. Symptoms initially resemble a common cold (fever, chills, cough, chest pain, headache, and malaise). However, after several days, the symptoms can progress to severe breathing difficulties and shock. Inhalational anthrax is usually fatal without early antibiotic treatment. Anthrasil®, a blood plasma product containing antibodies that neutralize the toxins produced by the bacterial cells, also is available for use in combination with antibiotics for patients with inhalational anthrax.

Anthrax also is considered a biological weapon for bioterrorism and biological warfare. The seriousness of using biological agents as a means for bioterrorism was underscored in October 2001 when *B. anthracis* endospores were distributed intentionally as a powder through the U. S. mail. In all, 22 cases of anthrax (11 inhalation and 11 cutaneous) were identified. In the end, five of the individuals with inhalational anthrax died, whereas the six other individuals with inhalation anthrax and all the individuals with cutaneous anthrax recovered.

The bacterial diseases of the LRT are summarized in TABLE 11.3. **MICROINQUIRY 11** presents several case studies concerning some of the bacterial diseases discussed in this chapter.

TABLE 11.3 A Summary of the Major Bacterial LRT Diseases

Disease	Causative Agent	Signs and Symptoms	Transmission	Treatment	Prevention
Pertussis	*Bordetella pertussis*	Malaise, low-grade fever, severe cough Multiple paroxysms	Respiratory droplets	Erythromycin	Vaccinating with DTaP, Tdap
Tuberculosis	*Mycobacterium tuberculosis*	Active TB: cough, weight loss, fatigue, fever, night sweats, chills, breathing pain	Respiratory droplets	Combination therapy with antibiotics	Preventing exposure to active TB patients BCG vaccine
Infectious bronchitis	*Chlamydophila pneumoniae Haemophilus influenzae Mycoplasma pneumoniae Streptococcus pneumoniae*	Runny nose, sore throat, chills, general malaise, slight fever, and dry cough	Respiratory droplets	Antibiotics	Annual flu vaccination Good hygiene
Healthcare-acquired pneumonia	*Klebsiella pneumoniae Pseudomonas aeruginosa Staphylococcus aureus Streptococcus pneumoniae*	Chills, high fever, sweating, shortness of breath, chest pain, cough with thick, greenish or yellow sputum	Respiratory droplets	Antibiotics	Practicing good hand hygiene
Community-acquired pneumonia	*Streptococcus pneumoniae*	High fever, sharp chest pains, difficulty breathing, rust-colored sputum	Respiratory droplets	Penicillin Cefotaxime	Vaccinating Hand hygiene
"Atypical" CAP	*Legionella pneumophila Mycoplasma pneumoniae*	Headache, fever, fatigue, dry, hacking cough	Respiratory droplets Via water systems, whirlpool spas, air conditioning systems	Antibiotics	Extreme cleaning and disinfecting of water systems, pools, and spas
Psittacosis	*Chlamydophila psittaci*	Headache, fever, dry cough	Contact with infected psittacine birds	Doxycycline	Keeping susceptible birds away from the infecting agent
Chlamydial pneumonia	*Chlamydophila pneumoniae*	Headache, fever, dry cough	Respiratory droplets	Doxycycline Erythromycin	Practicing good hygiene
Q fever	*Coxiella burnetii*	Headache, fever, dry cough	Dust particles Contact with infected animals	Doxycycline	Vaccine for high-risk occupations
Inhalational anthrax	*Bacillus anthracis*	Fever, chills, cough, chest pain, headache, and malaise Severe breathing and shock can develop	Airborne endospores	Penicillin Ciprofloxacin	Avoiding contact with infected livestock and animal products

MICROINQUIRY 11

Infectious Disease Identification

Following are several descriptions of infectious diseases based on material presented in this chapter. Read the case history and then answer the questions posed. You can find answers online in **Appendix E**.

Case 1

The patient, a 33-year-old male, arrives at a local health clinic complaining that he has felt "out of sorts," has a fever, and has lost more than 10% of his body weight in the last month. He also has a cough that produced rust-colored sputum. The patient is referred to a local hospital where a chest X-ray and sputum sample are taken. Upon further questioning, the patient admits to having tested HIV positive about 1 year earlier. A tuberculin test also is ordered. Additional questioning of the patient reveals that he had been living with two roommates for 2 years. Before that, he had lived for 8 years with another roommate who had tested positive for tuberculosis about 6 months before the onset of the patient's symptoms. The sputum samples are negative for the two roommates, but both have a positive tuberculin test result. Both test negative for HIV.

11.1a.   Why was a chest X-ray ordered?

11.1b.   Why was a sputum sample taken?

11.1c.   What should a positive tuberculin skin test look like?

11.1d.   What does such a test result indicate?

11.1e.   Based on the symptoms and laboratory results, from what infectious disease does the patient suffer? What is the agent?

11.1f.   How did the patient contract the disease?

11.1g.   Why is the infectious agent more virulent in HIV-infected patients?

Following diagnosis, the patient was placed on isoniazid (INH) for 12 months.

11.1h.   Most treatment procedures call for a 6- to 8-month program. Why was the patient placed on INH for an extended period?

Case 2

Parents bring their 2-year-old daughter to the hospital emergency room. She appears to have an upper respiratory infection that her parents think started about 1 week previously. They say that their daughter had lost her appetite and appeared especially sleepy about 4 days ago. She complained of a sore throat. Examination indicates that she has a moderate fever but no chest congestion. Throat and blood cultures are taken, and their daughter is put on penicillin.

On returning to the hospital 3 days later, her throat culture shows gram-positive rods. The blood culture is negative. Her parents remark that this morning their daughter began breathing harder. Examination of her pharynx indicates the presence of a leathery membrane. On questioning the parents, it is discovered that the child has had no immunizations. The child is admitted to the hospital and treatment immediately begun.

11.2a.   What infectious agent does the child have?

11.2b.   The medical staff is concerned about the seriousness of the disease. Why does the presence of gram-positive rods in the throat cause such concern?

11.2c.   How could this disease have been prevented?

11.2d.   What is the prescribed treatment protocol?

Case 3

A 63-year-old retired steel worker who is a heavy smoker and alcoholic comes to the emergency room complaining of having a fever and shortness of breath for the past 2 days. This morning he has developed a cough with rust-colored sputum. A chest X-ray is taken, and the image shows involvement in the left lower lobe of the lungs. A sputum sample and blood sample are taken for Gram staining, and the patient is checked into the hospital where he is placed on penicillin. Two days later the patient is feeling much improved. His physician tells him that both sputum and blood cultures indicate the presence of gram-positive diplococci. The patient is released from the hospital and recovers completely after finishing antibiotic therapy.

11.3a.   What organism is responsible for the patient's infection?

11.3b.   Why is this patient at high risk for becoming infected with the bacterial organism?

11.3c.   How could the patient likely have prevented contracting the disease?

11.3d.   What bacterial factors are responsible for virulence?

11.3e.   If plated on blood agar, what type of hemolytic reaction should be seen?

In conclusion, we have seen that the respiratory system can be a major portal of entry and infection site for bacterial pathogens. In a healthy individual, the ability of such pathogens to colonize is thwarted by the immune system and the often-antagonistic responses by our respiratory microbiome. One of the most interesting examples of this dynamic is seen in people with cystic fibrosis (CF). Although this is a genetic disease, CF patients with a stable condition tend to have a diverse respiratory community of microbes. However, in CF patients with worsening disease and imminent death, a loss of bacterial diversity in the lungs is seen. CF-stable patients appear to harbor a population of dominant *Streptococcus* species that is common to the oral cavity. However, only about 50% of the CF patients contained *P. aeruginosa* as the dominant species, which is contrary to the belief that *P. aeruginosa* is the dominant species in all CF patients. These findings show just how difficult it is to understand the polymicrobial interactions among the bacteria composing a microbiome, in this case among those of the respiratory tract. With a better understanding, we might be able to design a therapy for CF patients that would use targeted antibiotics and a designed microbial community that would ensure a stable condition. And in the big picture, it promises to hold great hope for designing system microbiomes to maintain a healthy human condition.

Concept and Reasoning Checks 11.3

a. Why have the Chinese referred to pertussis as the "100-day cough"?
b. Explain how a primary or latent tuberculosis infection is different from primary or secondary tuberculosis disease.
c. What are the characteristics common to all forms of HAP?
d. Why are some pneumonia-causing species community acquired whereas other species are healthcare acquired?
e. Summarize (1) the unique characteristics and (2) the mode of transmission of the genus *Chlamydophila*.
f. What cellular factor makes *B. anthracis* a dangerous pathogen and bioterror agent?

Chapter **Challenge C**

You have now discovered that several infections and diseases can be caused by the streptococci.

QUESTION C: *Finish off your list by identifying the diseases of the LRT that are caused by* S. pneumoniae. *What age groups seem to be the most affected and why?*

You can find answers online in **Appendix F**.

■ SUMMARY OF KEY CONCEPTS

Concept 11.1 The Respiratory System and a Resident Microbiome Normally Impede Bacterial Colonization

1. The **respiratory system** is divided into the **upper respiratory tract (URT)** (nose, sinus cavities, pharynx, and larynx) and the **lower respiratory tract (LRT)** (trachea, bronchi, and lungs). The lungs contain the alveoli where gas exchange occurs. Mechanical and chemical defenses of the URT include: **mucociliary clearance** to trap microbes in a layer of **mucus** and the presence and activity of several antimicrobial substances, including lysozyme and other antimicrobial peptides, antibodies, and human defensins.

2. Among the bacteria in the resident **microbiome** of the upper respiratory tract (URT) are *Streptococcus, Neisseria, Haemophilus,* and *Staphylococcus.* The microbiome of the LRT comes from inhalation and URT microaspiration (Figures 11.3, 11.4).

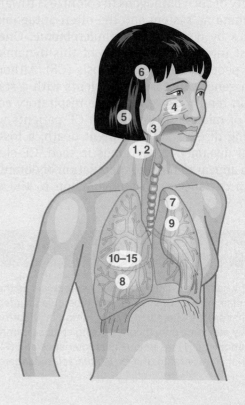

Concept 11.2 Several Bacterial Diseases Affect the Upper Respiratory Tract

▶ Streptococcal pharyngitis
   1 *Streptococcus pyogenes*
▶ Scarlet fever
   2 *Streptococcus pyogenes*
▶ Diphtheria
   3 *Corynebacterium diphtheriae*
▶ Sinusitis
   4 *Various bacterial species*
▶ Otitis media
   5 *Streptococcus pneumoniae, Haemophilus influenzae*
▶ Acute meningitis
   6 *Neisseria meningitidis, Streptococcus pneumoniae, Haemophilus influenzae type b*

Concept 11.3 Many Bacterial Diseases of the Lower Respiratory Tract Can Be Life Threatening

▶ Pertussis
   7 *Bordetella pertussis*
▶ Tuberculosis
   8 *Mycobacterium tuberculosis*
▶ Infectious bronchitis
   9 *Mycoplasma pneumoniae, Chlamydophila pneumoniae, Streptococcus pneumoniae, Haemophilus influenzae*

▶ Healthcare-acquired and community-acquired pneumonia
   10 *Streptococcus pneumoniae, Haemophilus influenzae, Staphylococcus aureus, Pseudomonas aeruginosa, Klebsiella pneumoniae*
▶ "Atypical" pneumonia
   11 *Mycoplasma pneumoniae, Legionella pneumophila*
▶ Q fever
   12 *Coxiella burnetii*
▶ Psittacosis
   13 *Chlamydophila psittaci*
▶ Chlamydial pneumonia
   14 *Chlamydophila pneumoniae*
▶ Inhalational anthrax
   15 *Bacillus anthracis*

■ CHAPTER SELF-TEST

For **Steps A–D**, you can find answers to questions and problems online in **Appendix D**.

■ STEP A: REVIEW OF FACTS AND TERMS

Multiple Choice

Read each question carefully before selecting the *one* answer that best fits the question or statement.

1. Which one of the following is *not* a part of the lower respiratory system?
   A. Lungs
   B. Pharynx
   C. Bronchi
   D. Trachea

2. A complication of streptococcal pharyngitis is _____.
   A. rheumatic fever
   B. pseudomembrane blockage
   C. strawberry tongue
   D. chest, back, and leg pain

3. A prominent feature of diphtheria is _____.
   A. meningitis
   B. a pseudomembrane
   C. pneumonia
   D. rheumatic fever

4. Which one of the following illnesses is characterized by yellow or green pus discharged from the nose?
   A. Pertussis
   B. Diphtheria
   C. Bronchitis
   D. Acute sinusitis

5. Acute meningitis _____.
   A. is an LRT infection
   B. is a disease affecting the membranes of the heart
   C. can be caused be *Corynebacterium diphtheriae*
   D. often starts as a nasopharynx infection

6. A catarrhal and paroxysmal stage is typical of _____.
   A. tuberculosis
   B. pneumonia
   C. pertussis
   D. acute otitis media

7. Which bacterial genus is typically stained using the acid-fast procedure?
   A. *Haemophilus*
   B. *Streptococcus*
   C. *Klebsiella*
   D. *Mycobacterium*

8. A person is observed to have inflamed inner walls lining the main airways of the lungs and has had a dry cough for a few days. This individual probably has _____.
   A. acute bronchitis
   B. pharyngitis
   C. pneumonia
   D. chronic bronchitis

9. Which one of the following is a gram-positive bacterial species commonly causing hospital-acquired pneumonia (HAP)?
   A. *Haemophilus influenzae*
   B. *Klebsiella pneumoniae*
   C. *Staphylococcus aureus*
   D. *Chlamydophila pneumoniae*

10. Which one of the following diseases can be acquired from the droppings of infected birds?
    A. Q fever
    B. Legionellosis
    C. Inhalational anthrax
    D. Psittacosis

Fill-in

Answer each of the following by filling in the blank with the correct word or phrase.

11. Scarlet fever is caused by a species of _____.

12. _____ is caused by a species of *Mycobacterium*.

13. An infection of the middle ear is called _____.

14. _____ is a gram-negative rod that causes whooping cough.

15. The dissemination of active tuberculosis throughout the body is called _____ tuberculosis.

16. _____ is a zoonotic disease of parrots, parakeets, and canaries.

17. A bacterial form of _____ can be caused by *Neisseria* or *Haemophilus*.

18. The cells of *Mycoplasma* species have no _____.

19. The agent of pneumococcal pneumonia is a gram-_____ organism.

20. If a person has pneumonia in both lungs, the condition is called _____.

■ STEP B: CONCEPT REVIEW

21. Explain how the URT limits colonization of the URT. (**Key Concept 11.1**)
22. Differentiate the bacterial species responsible for **acute meningitis**. (**Key Concept 11.2**)
23. Explain why **pertussis** is viewed as one of the more dangerous contagious diseases. (**Key Concept 11.3**)

24. Summarize (a) the clinical aspects of *Mycobacterium tuberculosis* as an infection and disease and (b) the problems concerning antibiotic resistance. (**Key Concept 11.3**)

■ STEP C: APPLICATIONS AND PROBLEM SOLVING

25. A patient is admitted to the hospital with high fever and a respiratory infection. Pneumococci and streptococci were eliminated as causes. Penicillin was ineffective. The most unusual sign was a continually dropping count of red blood cells. Can you make the final diagnosis?
26. One of the major world health stories of 1995 was the outbreak of diphtheria in the New Independent and Baltic States of the former Soviet Union. If you were in charge of this international public health emergency, what would be your plan to help quell the spread of *Corynebacterium diphtheriae?*

27. The CDC reports that an estimated 40,000 people in the United States die annually from pneumococcal pneumonia. Despite this high figure, only 30% of older adults who could benefit from the pneumococcal vaccine are vaccinated (compared to more than 50% who receive an influenza vaccine yearly). As an epidemiologist in charge of bringing the pneumonia vaccine to a greater percentage of older Americans, what would you do to convince older adults to be vaccinated?

■ STEP D: QUESTIONS FOR THOUGHT AND DISCUSSION

28. Between 1986 and 1996, *Haemophilus* meningitis was virtually eliminated in the United States. Indeed, at the beginning of the period, there were 18,000 cases annually, but in 1996, only 254 cases were reported. What factors probably contributed to the decline of the disease?
29. A bacterial virus is responsible for the ability of the diphtheria bacillus to produce the toxin that leads to disease. Do you believe that having the virus is advantageous to the infecting bacillus? Why or why not?

30. A children's hospital reported a dramatic increase in the number of rheumatic fever cases. Doctors were alerted to begin monitoring sore throats more carefully. Why do you suppose this protocol was recommended?

CHAPTER 12

Foodborne and Waterborne Bacterial Diseases

On October 20, 2010, an outbreak of cholera was confirmed in Haiti. This report was quite a surprise because cholera had not been reported in Haiti for decades and, because no cholera cases were reported after the January 2010 earthquake, health authorities deemed cholera as an unlikely disease to develop. Yet within a few days after the initial report of the cholera outbreak, hundreds of people became ill with stomachache, vomiting, and diarrhea so profuse that victims started dying from dehydration. Cholera treatment centers began sprouting up around the country (see chapter opening photo).

Through his pioneering epidemiological studies, John Snow, a London physician, pinpointed a water pump on Broad Street as the source of transmission for the 1854 cholera epidemic in London. Believing that the disease was being transmitted by contaminated water, Snow believed the only way to stop the spread was to remove the pump handle on the water pump so that no one could draw water from the pump. City officials took Snow's advice, and the cholera cases declined. The action supported Snow's belief that cholera was a waterborne, contagious disease. However, few at the time believed Snow's theory for a waterborne disease, and it would not be until 1883 that the cholera bacterium, *Vibrio cholerae*, was isolated.

The year 2014 marked the 160th anniversary of John Snow's landmark epidemiological studies. Unfortunately, cholera cases and deaths continue to occur in Haiti. There have been more than 750,000 cases and 9,000 deaths reported since the outbreak began in 2010. Globally, Haiti is only one of the most recent cholera epidemics to make the news. Year after year, cholera continues to cause illness and death for thousands of people in sub-Saharan Africa (**FIGURE 12.1**).

Cholera is just one of the foodborne and waterborne illnesses affecting the human digestive system. In this chapter, we will examine several of these illnesses that are caused by bacterial species transmitted through food and water.

A cholera treatment center in Haiti.

© Julie Dermansky/Science Source.

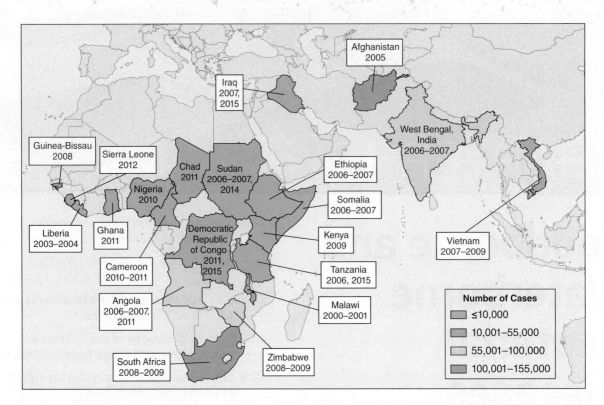

FIGURE 12.1 Recent Cholera Outbreaks—Africa and Asia, 2000–2015. There have been numerous outbreaks of cholera, especially in Africa. »» *Medical experts believe the rainy season brings the spread of the disease. Why would the rainy season spread cholera?*

Adapted from Waldor, M. K., et al. 2010. *N Engl J Med* 363(24):2279–2282; WHO, 2015.

Chapter **Challenge**

Although most experts believe that food safety in the United States is quite good, foodborne illnesses remain a major cause of illness and infectious disease. Therefore, everyone from farm to table needs to think about food safety and remain vigilant to ensure that our food and water supplies stay safe. Because foodborne and waterborne diseases still occur, let's think about how we can reduce the risk and prevent illnesses of the digestive system. As we describe many of the more common foodborne and waterborne infections, how can we prevent such infections—from the farm to the table?

■ KEY CONCEPT 12.1   The Digestive System Has Diverse Resident Microbiomes

When we eat a meal or drink fluids, our bodies take in the nutrients necessary for survival. During the ingestive and digestive processes, any pathogens that happen to be in food or liquid can upset the digestive system and lead to some form of illness. Consequently, the digestive system represents a major portal of entry for pathogens.

The Digestive System Is Composed of Two Sets of Organs

The **digestive system** consists of the **gastrointestinal (GI) tract**, which essentially is a tube running from mouth to anus, and the **accessory digestive organs** (**FIGURE 12.2A**). After food starches and some fats initially are digested by salivary enzymes in the mouth, the food passes through the pharynx, moves down the esophagus, and enters the stomach. After further digestion by hydrochloric acid (gastric juice) and stomach enzymes, the partially digested food is ready for passage into the small intestine.

Here, with an extremely large surface area, most digestion and absorption of nutrients occur. The absorbed nutrients cross the mucosa and submucosa and enter the bloodstream (**FIGURE 12.2B**). The remaining undigested materials in the small intestine move into the large intestine, or colon, where

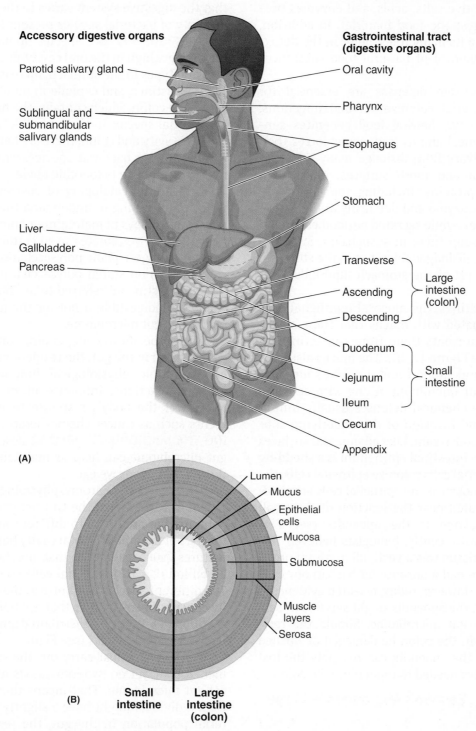

Accessory digestive organs

Parotid salivary gland

Sublingual and submandibular salivary glands

Liver

Gallbladder

Pancreas

Gastrointestinal tract (digestive organs)

Oral cavity

Pharynx

Esophagus

Stomach

Transverse

Ascending — Large intestine (colon)

Descending

Duodenum

Jejunum — Small intestine

Ileum

Cecum

Appendix

(A)

Lumen
Mucus
Epithelial cells
Mucosa
Submucosa
Muscle layers
Serosa

(B)   **Small intestine**   **Large intestine (colon)**

FIGURE 12.2 The Human Digestive System. (A) The digestive system consists of the gastrointestinal (GI) tract and accessory digestive organs. **(B)** A simplified cross section through the small and large intestine. *»» Why are organs such as the teeth, liver, and pancreas considered accessory digestive organs?*

absorption of any remaining nutrients and water occurs. The indigestible solid wastes compose the feces, and the fecal material (stool) is moved into the rectum and eliminated through the anus.

The accessory digestive organs (see Figure 12.2A) are outgrowths from and are connected to the GI tract. The digestive salts, acids, and enzymes produced help digest the food material. In addition, the liver, among its many roles, helps in the detoxification and removal of harmful food substances from the blood.

Digestive system defenses are essential for preventing bacterial colonization and infections. In the oral cavity, chewing food generates substantial mechanical and hydrodynamic forces that, along with salivary flow, dislodge many microbes from epidermal and tooth surfaces. Mucus and various other proteins, including the antimicrobial proteins lysozyme and defensins in the saliva, destroy microbes, while secreted antibodies bind to microbes and keep them in suspension. Swallowing delivers the dislodged microbes to the stomach, where the low pH of the stomach fluid kills most microbes.

In the intestines, the mucosal (epithelial cell) surfaces are coated with mucus that traps bacterial cells and prevents them from contacting the epithelium (see Figure 12.2B). The mucus also provides a structural support and nutrient source for the resident gut microbiota. Along with secreted antibodies, the chemical defenses usually inhibit colonization and invasion of the underlying submucosa and bloodstream. Like all mucosal surfaces in the body, the intestinal epithelium is a shedding surface. If bacterial cells colonize epithelial cells, the cells will be removed as the epithelial cells are shed.

Finally, located near the junction of the small and large intestine is the appendix (see Figure 12.2A). Until recently, biologists believed this small piece of tissue was a vestigial (useless) organ that might have had a purpose far back in our evolutionary past. However, today, research evidence is suggesting that the appendix might serve as a "safe house" for the gut microbiome. Should the resident microbes in the colon be damaged or flushed out (diarrhea), the appendix can resupply the lost microbial species needed to repopulate the colon.

The Digestive System's Microbiota Is Large and Diverse

Until recently, the identities of the microbial community (and their genes) inhabiting the human body, the so-called **human microbiome**, were mostly

unknown. Microbiologists knew that there was an enormous population of bacteria in the GI tract, especially in the large intestine, and that these communities provide protection through microbial antagonism. However, recent studies initiated by the **Human Gut Microbiome Initiative** (**HGMI**) suggest that the digestive system varies in the numbers and diversity of bacterial species present throughout its length (**FIGURE 12.3A**). Species number and diversity are very high in the oral cavity, drop to low numbers and types in the stomach, increase again in the small intestines, and explode in number and diversity in the colon, which contains the highest density of bacterial species in the body (**FIGURE 12.3B**). In the oral cavity and colon, there are between 500 and 1,000 different bacterial species, which are found mostly within five taxonomic phyla.

Besides the cataloging of species, microbiologists are beginning to understand the roles the gut microbiome plays in maintaining human health. We know that all vertebrates need certain microbes in the gut to digest plant polysaccharides other than starch. The non-starch polysaccharides, like cellulose and pectin, are referred to as "fibers" and they would be indigestible if not for the microbes composing the gut microbiome.

More specifically, depending on the species composition in the gut, the resident microbes influence multiple physiological functions, optimize immune reactions, influence brain development, and affect the body's response to medical treatments such as cancer chemotherapy. INVESTIGATING THE MICROBIAL WORLD 12 describes how the gut microbiome can help us maintain good health through the foods we eat.

Differences in anatomy, physiology, and organization in each part of the GI tract generate unique habitats that are home to different microbial colonizers. For example, the oral cavity has some unique features (such as teeth) that are home to some 20 billion (2×10^9) bacterial cells comprising up to 700 different species. In contrast, the colon primarily consists of soft tissues that are home to perhaps 4 trillion (4×10^{12}) cells distributed among 1,000 different bacterial species (see Figure 12.3A).

All humans must carry out the same basic gut metabolism, yet no two individuals have an identical gut microbiome. That means that even though each individual might have a slightly different bacterial population in the gut, the resident species must have overlapping or similar "duties." Very simplistically, if species A is unique to the small intestine in one healthy individual and species B is

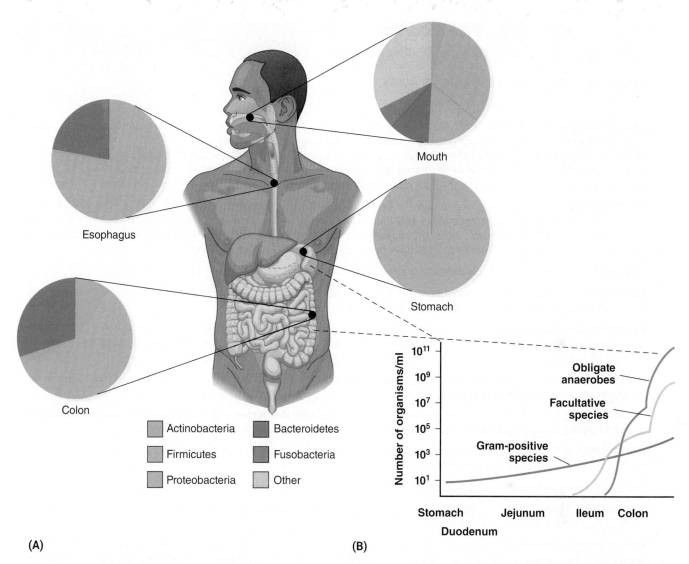

(A)   **(B)**

FIGURE 12.3 The Digestive System Microbiome. (A) At the phylum level, there can be substantial differences in the microbiome composition within an individual at different anatomical sites. Actinobacteria = gram-positive bacilli or filamentous forms; Firmicutes = gram-positive forms; Proteobacteria = anaerobic and aerobic, gram-negative forms; Bacteroidetes = anaerobic and aerobic, nonspore-forming, gram-negative bacilli; Fusobacteria = obligately anaerobic, nonspore-forming, gram-negative bacilli. **(B)** The number and diversity of bacterial species increases from the stomach to the colon. *»» What factors may influence the compositional differences at the different anatomical sites?*

(A) Modified from Cho, I. and Blaser, M. J. 2012. *Nat Rev Gen* 13(4): 260–270.

unique to that in another healthy individual, perhaps both species A and B are carrying out the same metabolic roles for each individual. Thus, different microbial members in the same part of the gut of separate individuals might still do similar tasks.

Selective pressures also influence the community of microbes, especially in the gut. Medications, like antibiotics, can eliminate or severely reduce bacterial numbers. In addition, changes in a person's diet can alter the composition and assortment of bacterial species normally in the gut. MICROFOCUS 12.1

describes one example of the effect of diet on the gut microbiome.

A loss or disruption of microbial species balance, called **dysbiosis**, can have serious physiological consequences, and the change might be associated with disease development. In fact, dysbiosis has been identified as at least partly responsible for inducing anxieties and depression, altering the risk of developing asthma or diabetes, triggering inflammatory bowel diseases, initiating colorectal cancer, and even influencing how fat or thin an individual might be.

Investigating the Microbial World 12

An Apple a Day Keeps the Doctor Away

An apple a day keeps the doctor away. Many of us remember this saying. Its origin is a mystery. Some believe it might have come from an ancient Roman proverb because the Romans thought apples could cure some illnesses. Another origin, according to the Phrase Finder (www.phrases.org.uk/), suggests that the phrase first appeared in print in *Notes and Queries*, a Welsh magazine that in February 1866 advocated, "Eat an apple on going to bed, And you'll keep the doctor from earning his bread." In any case, the idea was that apples are good for your health. Is there any truth to the maxim?

OBSERVATIONS: We continually hear about the benefits of fruits and vegetables in our diet. In addition, reports have been published suggesting that vegetables and fruits affect the bacterial populations in the gut in a positive way. The mechanisms are largely unknown, although increases in members of the gram-positive Firmicutes and decreases in numbers of the gram-negative Bacteroidetes have been correlated with positive health gains.

QUESTION: *If eating apples or apple products is so good for us, how might apples provide a health advantage?*

HYPOTHESIS: Eating apples and apple products alters the composition of the intestinal microbiome. If so, feeding rats apples or apple products should alter their gut microbiome.

EXPERIMENTAL DESIGN: Rats were used as test animals, and the changes in the bacterial microbiome in the rat cecum (an extension of the terminal portion of the small intestine and home to anaerobic bacterial species) after apple feeding (compared to a control diet) were monitored by ribosomal RNA (rRNA) gene sequencing to identify the microbial members.

EXPERIMENT 1: Rats were fed either a control diet or 10 grams of apples a day for 14 weeks and the effects on the rat cecum microbial community analyzed.

EXPERIMENT 2: Because apples contain different components, rats were given a control diet or one with various apple products (puree, pomace, pectin, or juice) and the effects on the rat cecum microbial community analyzed.

EXPERIMENT 3: Rats were given either a control diet, 10 grams of apples a day, or 7% pectin for 4 weeks and the effects on the rat cecum microbial community analyzed.

RESULTS: Apple consumption had no effect in cecal pH or relative cecal weight.

EXPERIMENT 1: Apple consumption did affect the bacterial composition in the cecum. However, it was not possible to identify specific bacterial species affected.

EXPERIMENT 2: The only apple component that showed a statistical change from the control was for butyrate, which was increased in the cecum of the rats fed 3.3% pectin. Butyrate has been shown to induce death in cancer cells and provide a fuel source for cells of the intestinal mucosa. The cecal population was altered in those rats fed either 0.33% or 3.3% pectin, although specific species could not be identified.

EXPERIMENT 3: There was an increase in butyrate production in rats fed apples or 7% pectin. Analysis of gene sequences from the cecum of rats fed pectin did identify changes in the bacterial population in the cecum (see graph).

CONCLUSIONS:

    QUESTION 1: *Was the hypothesis validated? Explain.*

    QUESTION 2: *What portion of the apple components appears to have the most influence on the bacterial population in the rat cecum?*

    QUESTION 3: *Do the results from experiment 3 (see graph) tend to support or refute the idea that apples (or apple pectin) have a health-promoting effect? Explain.*

    QUESTION 4: *How does the fact that* Clostridium *species are known butyrate producers fit into the overall picture of apples and good health?*

Rats metabolize apple components differently from humans, so we need to interpret such studies on apple or apple component consumption in rats with caution.

You can find answers online in **Appendix E**.

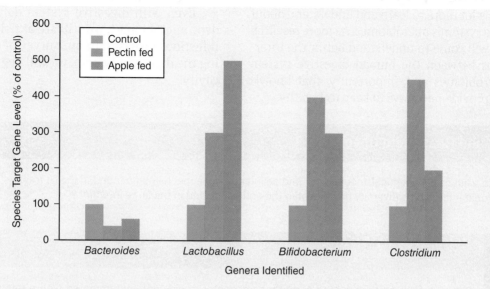

Relative amount of target genes comparing the control group, pectin-fed, and apple-fed rats.

Adapted from Licht, T. R., et al. 2010. *BMC Microbiol* 10:13.

MICROFOCUS 12.1: Being Skeptical

Probiotics to the Rescue?

In many industrialized nations, including the United States, individuals are taking in only about half the amount of recommended dietary fiber, that is, the complex carbohydrates found in many fruits, vegetables, grains, and nuts. Although the human body lacks the ability to produce the enzymes to break down most of this fiber, many of the bacteria comprising our gut microbiome produce hundreds of enzymes capable of digesting the consumed fiber. Without these gut microbes, many health experts fear intestinal physiology and immune regulation could be disrupted and disease can occur. For example, inflammatory bowel disease (IBD), which is a chronic inflammation of the intestines typical of ulcerative colitis and Crohn's disease, has been associated with a disrupted gut microbiome.

Consequently, does a diet low in dietary fiber affect the diversity (and function) of our gut microbiome and, if so, would taking probiotics help reverse the trend?

Studies carried out in mice show that when a group of mice was placed on a low dietary fiber diet, the mice lost much of their diverse gut microbiota, whereas a group on a high-fiber diet maintained the diverse microbial community. Further, if the mice on the low-fiber diet were allowed to breed within their group, their offspring also lacked a normal microbial community. Over four generations, this resulted in a profound loss of microbe diversity in the gut. Even if the offspring were fed a high-fiber diet, the "healthy" community of microbes often failed to reappear. The only way to reestablish the "healthy" community was through actually transplanting a "healthy" community of microbes from mice fed a high-fiber diet (see **microbiota-induced therapy** in this chapter). Although these studies were done in mice, some studies suggest that humans also can have a loss in diversity when on a low-fiber diet.

So, what about probiotics? When ingested, can these formulations of potentially beneficial bacteria help stabilize or have beneficial effects for the gut microbiome of people on a low-fiber diet? Some preliminary research suggests probiotics in pill form might benefit individuals with IBD, whereas other studies have found no benefit for probiotics in maintaining remission from ulcerative colitis. In the experiment described above, the mice offspring put back on a high-fiber diet (super probiotics?) often did not recover the diverse microbes.

The verdict? Traditional products like yogurt and other cultured dairy products can help maintain daily gut health. However, for probiotics, the formulations used in most of these products are extremely varied, and it is uncertain which specific bacterial species might be beneficial. Therefore, most probiotics purchased today are questionable in their ability to stabilize or maintain a good gut community in people on a low-fiber diet. For most people who do not have bowel damage, probiotics will not do any harm. Still, more research examining the roles of specific species is needed before the true value of probiotics can be evaluated.

There is a lot more to learn and understand about the digestive system's microbiomes. As more research is done, we will come to understand better the interrelationships between the human digestive system and its microbiomes and, importantly, that knowledge might provide new ways to keep us healthy.

Even with digestive system defenses and resident microbiomes, we still are at risk of contracting infectious diseases. In our survey of diseases affecting the digestive system, we will begin with the oral cavity.

Concept and Reasoning Checks 12.1

a. Prepare a list of the chemical, mechanical, and cellular defense mechanisms found in the GI tract.
b. Why would microbiome diversity be highest in the oral cavity and in the large intestine?

Chapter Challenge A

Between farm and table, food sometimes becomes accidently contaminated with a microbial pathogen. When we ingest the food, we might develop a case of food poisoning.

QUESTION A: *How is our digestive system set up to defend against entry and colonization by foodborne and waterborne pathogens?*

You can find answers online in **Appendix F**.

■ KEY CONCEPT 12.2   Bacterial Diseases of the Oral Cavity Can Affect One's Overall Health

The microbiome in the oral cavity of any individual contains some fraction of the 700 identified bacterial species (TABLE 12.1; see Figure 12.3). Some species might be in transit, moving on to the gastrointestinal tract or being aerosolized and going to the lungs. Others are permanent residents, forming different and unique communities of bacteria in the different habitats (e.g., tongue, roof of mouth, teeth, and gums) within the oral cavity. The gums, for example, even contain different bacterial populations above and below the gum line. The saliva in the mouth is yet another habitat. Researchers have discovered that it might be possible to detect pancreatic cancer based on the bacterial profile in a person's saliva. Because oral bacteria and dental infections have been implicated in diabetes, pneumonia, and stomach ulcers, an individual's oral microbiome might be one sign of his or her health status.

Like most anatomical locations on the human body, the oral cavity is populated primarily with harmless or beneficial bacterial species, many of which through microbial antagonism outcompete pathogenic invaders for nutrients and space. However, dysbiosis or poor oral hygiene can provide invaders with an opportunity to grow and proliferate, resulting in dental caries and/or periodontal disease.

Dental Caries Causes Pain and Tooth Loss in Affected Individuals

After the common cold, **dental caries** (*cario* = "rottenness"), or tooth decay, is the most prevalent

TABLE 12.1 Some Core Bacterial Genera Found in the Oral Cavity

Phylum	Core Bacterial Genera
Actinobacteria	*Actinomyces, Corynebacterium*
Firmicutes	*Lactobacillus, Streptococcus, Veillonella*
Proteobacteria	*Aggregatibacter, Haemophilus, Neisseria*
Bacteroidetes	*Capnocytophaga, Porphyromonas, Provotella*
Fusobacteria	*Fusobacterium, Leptotrichia*

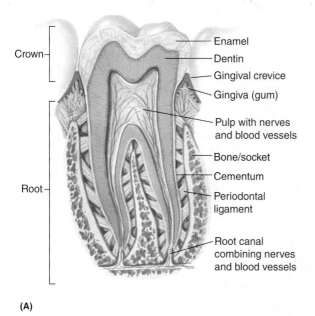

Crown
- Enamel
- Dentin
- Gingival crevice
- Gingiva (gum)
- Pulp with nerves and blood vessels
- Bone/socket

Root
- Cementum
- Periodontal ligament
- Root canal combining nerves and blood vessels

(A)

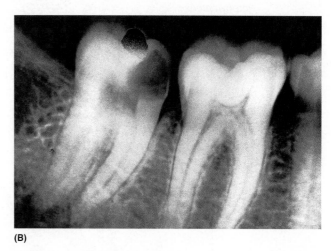

(B)

FIGURE 12.4 The Anatomy of a Tooth. (A) Different diseases can affect each part of the human molar shown in cross section. **(B)** In this false-color dental X-ray, the caries appear as a red area on the left molar. **»» *What parts of the tooth surface would be most susceptible to colorization and infection?***

(B) © BSIP/Science Source.

human infectious disease (**FIGURE 12.4**). On a clean tooth, cells of *Streptococcus* and *Corynebacterium* are among the first to colonize the surface of the enamel. As they attach, they begin to form a **biofilm**, a deposit of gelatinous material containing salivary proteins, trapped food debris, and bacterial cells and their products (**FIGURE 12.5A**). If the biofilm is not removed through flossing, brushing, and regular dental cleaning, a succession of highly organized gram-positive and gram-negative species interact and form **dental plaque** (**FIGURE 12.5B**). In the plaque, *Corynebacterium* grows outward as long filaments, providing a structural framework for the rest of the microbial community. The perimeter of the biofilm is crowned by more streptococci and other bacterial species. In fact, it has been suggested that mature dental plaque might contain up to 300 bacterial species, and some of these bacterial species might be contagious, as MICROFOCUS 12.2 explains.

By fermenting sugars, such as sucrose, in foods, plaque bacteria, including *Streptococcus mutans, S. sobrinus, Lactobacillus* species, and a few other species, produce acid end products that attack minerals in the tooth's enamel. If the plaque is not removed, a demineralization eventually results in a cavity. As the decay progresses into the pulp of the tooth, which contains nerves and blood vessels, severe pain occurs (a toothache). The immune system will

respond to the bacterial invaders by sending white blood cells to the blood vessels to fight the infection. This can result in a tooth abscess.

In summary, dental caries develops if three factors are present: a caries-susceptible tooth with a buildup of plaque; dietary carbohydrate, usually in the form of sucrose; and acidogenic (acid-producing) bacterial species (**FIGURE 12.6**).

Periodontal Disease Can Arise from Bacteria in Dental Plaque

If dental plaque is left on the teeth from a lack of proper oral hygiene, the biofilm below the gumline can result in swollen and tender gums (gingiva) that bleed when brushing. This is an outward sign of **periodontal disease** (**PD**), which in some form affects about 80% of adult Americans. One of the most common forms of early-stage PD is **gingivitis**, which develops when *Porphyromonas gingivalis* cells in the plaque secrete toxins and enzymes capable of directly or indirectly injuring the tissues surrounding and supporting the teeth (**FIGURE 12.7A**). Because there is no direct bacterial invasion of the gingival tissue, gingivitis is both treatable and preventable through professional cleaning to remove the plaque.

The presence of *P. gingivitis* in plaque provides an environment for a shift in the makeup of the oral

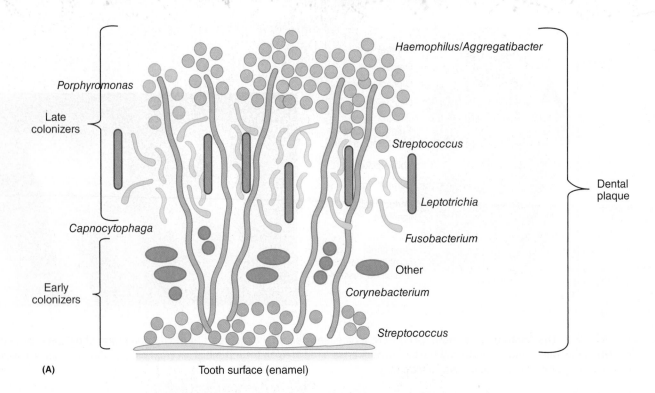

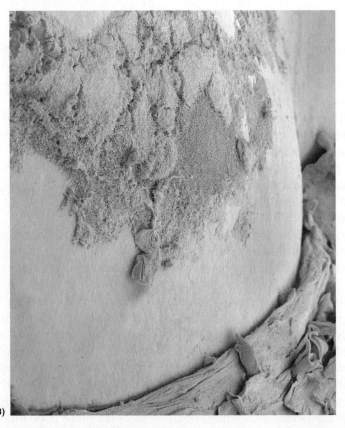

FIGURE 12.5 Dental Plaque. (A) In this model of plaque formation, an initial set of colonizing bacterial species establish a biofilm at the tooth surface. A more developed biofilm representing dental plaque occurs with the addition of late colonizers. **(B)** This false-color scanning electron micrograph of a tooth surface shows dental plaque (pink) coating the enamel of the tooth. »» *Hypothesize what would happen if the early colonizers were absent from the oral cavity and the tooth surface.*

(A) Adapted from Welch, J. L. M., et al. 2016. *PNAS* doi:10.1073/pnas.1522149113. (B) © STEVE GSCHMEISSNER/Getty Images.

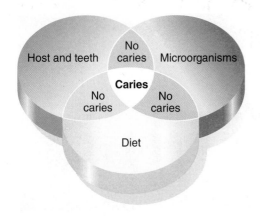

FIGURE 12.6 Dental Caries. Overlapping circles depicting the interrelationships of the three factors that lead to caries activity. **»» *What role does* S. mutans *play in the development of dental caries?***

microbiome. This dysbiosis provides an environment colonized and dominated by gram-negative anaerobes, including *Treponema denticola*, *Tannerella forsythia*, and *Fusobacterium nucleatum*. If the plaque is not removed, this community of bacteria can cause **periodontitis**, which is a serious disease of the soft tissue and bone supporting the teeth (**FIGURE 12.7B**). As a result, a tissue-destructive inflammation occurs, which can ultimately cause bone resorption and periodontal ligament loss. The result is loosening of the teeth and tooth loss.

In addition, bleeding gums provide a portal for this community of bacteria in the plaque and oral cavity to "leak" into the bloodstream, where serious systemic problems can develop. For example, PD might be a risk factor for cardiovascular disease

MICROFOCUS 12.2: Being Skeptical

I Caught a Cavity!

First, we all know that many infectious diseases are communicable—they can be passed from person to person. Just look at the number of people who get colds and the flu every year. Second, cavities and tooth decay (dental caries) rank near or at the top among the world's most common health problems. In fact, the single most common infectious childhood disease in the United States is severe tooth decay, and some see this as an epidemic problem. But are dental cavities contagious? Can you "catch" a cavity?

Dental caries is caused primarily by *Streptococcus mutans*. Research indicates that in newborns the disease-causing bacterium is usually absent. So where do the bacterial cells come from? The proposal is that children pick up *S. mutans* from an adult—usually a family member, friend, or other caretaker. Importantly, in children there appears to be a "window of infectivity" between 6 to 36 months of age, although all children maintain some level of susceptibility. That means *S. mutans* can be passed vertically from mother to child by kissing, sharing food utensils, or through any other activity involving an exchange of saliva from one mouth to another. In fact, dental studies report that many young mothers have not seen a dentist in 5 to 7 years before giving birth, meaning they carry a large "infectious dose" of *S. mutans* in their mouths that potentially can be transmitted to their children, often before the first teeth appear.

Transmission also occurs horizontally between children. For example, playmates at daycare centers can pass saliva between one another from sharing eating utensils or sweets that have been placed in the mouth (see figure). In addition, adults might be able to spread it horizontally. Dr. Margaret Mitchell, a Chicago cosmetic dentist reports, "In one instance, a patient in her 40s who had never had a cavity suddenly developed two cavities and was starting to get some gum disease," she said. Why at this age? The woman was now dating a man who hadn't seen a dentist in 18 years and had gum disease.

The verdict? Realize that dental caries involves more than simply having *S. mutans* present. As the chapter discussion describes, a person's dietary habits, fluoride use, salivary flow and composition, and oral hygiene are contributing factors. So, cavities are not being transferred. Rather, the *S. mutans* bacterial cells and perhaps other oral bacterial species that cause cavities are presumably being transmitted. So, dental decay is just another example of a bacterial infection that is no different from other infections—if the environment is right for infection, it can affect our bodies.

Sugar and sharing a lollipop. Saliva transfer?

© Carol Yepes/Moment Open/Getty.

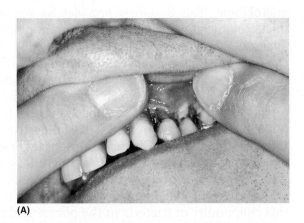

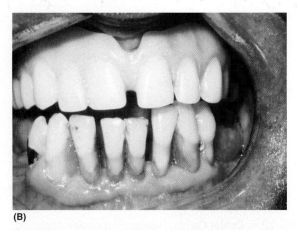

(A)

(B)

FIGURE 12.7 **Periodontal Disease. (A)** Gingivitis is an inflammation of the gums (gingiva) around the teeth, which is evident in this patient. **(B)** Periodontitis refers to conditions when gum disease affects the structures supporting the teeth. In this example, gum recession and bone loss has occurred, which can lead to loosening of the teeth and even tooth loss. *»» What are the best ways to prevent periodontal diseases?*

(A) Mediscan/Alamy Stock Photo. (B) © SPL/Science Source.

and infective endocarditis. At least in mice, the presence of plaque and oral bacteria in the blood stimulates inflammation and a narrowing and hardening of the coronary arteries. Such conditions are early indicators for possible heart attack. A specific periodontal pathogen called *Aggregatibacter actinomycetemcomitans* (see Figure 12.3) may be one risk factor in the development of infective arthritis. Periodontitis also might be a risk factor for pregnant women because women with PD are more likely to give birth to premature babies than are women with healthy gums. All this suggests that the state of one's oral health can be an indicator of one's systemic health.

TABLE 12.2 summarizes the bacterial diseases of the oral cavity.

TABLE 12.2 **A Summary of Bacterial Diseases of the Oral Cavity**

Disease	Causative Agent	Signs and Symptoms	Transmission	Treatment	Prevention
Dental Caries	*Streptococcus mutans* *Streptococcus sobrinus* *Lactobacillus*	Toothache Sensitivity and pain when drinking and eating	Due to normal microbiome	Fluoride treatment Fillings Extraction	Practicing good oral hygiene Regular dental examinations
Gingivitis	*Bacteroides, Fusobacterium, Porphyromonas, Prevotella, Peptostreptococcus*	Swollen, soft, and red gums Bleeding gums	Due to normal microbiome	Cleaning of teeth to remove plaque	Practicing good oral hygiene Regular dental examinations
Periodontitis	*Actinobacillus Porphyromonas Bacteroides Treponema*	Swollen, bright-red gums that are tender when touched Gums pulled away from teeth New spaces between teeth	Due to normal microbiome	Cleaning pockets of bacteria	Practicing good oral hygiene Regular dental examinations

Concept and Reasoning Checks 12.2

a. From the bacterial perspective, how does good oral hygiene help prevent dental caries?
b. What effects could bacterial diseases of the oral cavity have on overall human health?

Chapter **Challenge B**

Although the foods we eat are not usually contaminated with pathogens, we still can develop a disease if we do not maintain good oral hygiene after eating.

QUESTION B: *How is it possible that we can develop dental caries or periodontal disease even though the foods we eat are "clean," microbiologically speaking? Where do the bacterial species causing these diseases of the oral cavity come from?*

You can find answers online in **Appendix F**.

■ KEY CONCEPT 12.3   Bacterial Diseases of the GI Tract Are Usually Spread Through Food and Water

Infections of the GI tract have affected human societies for millennia, and they have been responsible for considerable morbidity and mortality. The Centers for Disease Control and Prevention (CDC) estimate that 48 million people in the United States suffer foodborne illnesses each year, accounting for 128,000 hospitalizations and more than 3,000 deaths—primarily infants, older adults, and immunocompromised individuals (people with weakened immune systems). Globally, the World Health Organization (WHO) reports that every year there are more than 600 million illnesses and 420,000 deaths from foodborne pathogens and another 1.7 million deaths from waterborne diseases. Most of these deaths are from diarrheal diseases and occur in children.

Diseases of the GI Tract Can Arise from Intoxications or Infections

We live in a microbial world, and there are many opportunities for food and water to become contaminated with potential pathogens. More than 250 different foodborne diseases have been described. Many of these are associated with outbreaks caused by bacterial pathogens that have contaminated a variety of food commodities (**FIGURE 12.8A, B**). These pathogens might be present in healthy animals (usually in their intestines) raised for food. However, the carcasses of cattle and poultry can become contaminated during slaughter if the processing line is exposed to small amounts of pathogen-containing intestinal contents. Likewise, water can become contaminated with animal manure or human sewage.

Many cases of food poisoning, especially prevalent with the popularity of fast food chains, are the result of transmission through the **fecal–oral route**; that is, pathogens in fecal material are passed from one infected individual and are introduced into the oral cavity of another individual. For example, in restaurants it is essential that food handlers have good hand-washing skills because they could otherwise easily spread a pathogen from the fecal material on their hands when preparing food for consumers (**FIGURE 12.8C**). Likewise, unsanitary practices (e.g., failing to cook foods or wash raw vegetables thoroughly) can lead to foodborne outbreaks.

Most foodborne and waterborne illnesses of the GI tract represent some form of **gastroenteritis**, an inflammation of the stomach and the intestines, usually with vomiting and diarrhea. Such inflammations can arise from either intoxications or infections. **Intoxications** are illnesses in which bacterial toxins in food or water are ingested. Examples are the toxins causing botulism, staphylococcal food poisoning, and clostridial food poisoning. By contrast, **infections** refer to illnesses in which live bacterial pathogens in food or water are ingested. As these pathogens grow in the body, they produce the toxins involved in the illness. Salmonellosis, shigellosis, and cholera are examples.

With many foodborne and waterborne diseases, determining the **etiology** (cause and origins) of the

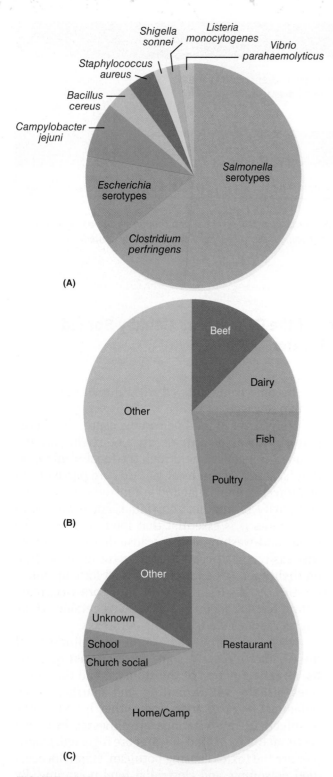

(A)

(B)

(C)

FIGURE 12.8 **Foodborne Illness Surveillance.** (A) *Salmonella* is the causative agent for more than 50% of the bacterial foodborne outbreaks. (B) Among the foodborne outbreaks, several foods are associated with illness. (C) Restaurants most often are the common source of illness due to foodborne bacterial pathogens. »» *Looking at (B), what commodities might be represented by the "other" segment?*

Data from the CDC.

illness is important to stop further spread of the disease-causing agent. In response to this problem, the CDC has established PulseNet, which is a national database designed, in part, to facilitate the identification of pathogens causing a foodborne outbreak. Although the source for an outbreak often is never determined even with this database, in those that are, *Staphylococcus aureus* and *Clostridium perfringens* account for most intoxications and *Salmonella enterica* with most infections.

Several factors are important in foodborne and waterborne outbreak investigations. Let's take a look at them in the subsections that follow.

Incubation Period

If an individual ingests and swallows a contaminated food or beverage, there is a delay, called the **incubation period**, before the symptoms appear. Depending on the bacterial species, this period can range from hours (for many intoxications) to days (for some infections).

Clinical Symptoms

The symptoms produced by an intoxication or infection depend on the specific toxin or microbe and the number of toxins (**toxic dose**) or cells (**infectious dose**) ingested. Although the intoxications and infections can have different symptoms, nausea, abdominal cramps, vomiting, and diarrhea often are common. Because these symptoms are so universal, it can be difficult to identify the microbe causing a disease unless the disease is part of a recognized outbreak, or laboratory tests are done to identify the causative agent.

Duration of Illness

Intoxications and infections can have very different lengths of time during which the symptoms persist. Some intoxications can be very abrupt and disappear in 12 to 24 hours, whereas many infections can last days to a week or longer.

Demographics

Certain individuals within a population might be more prone to infections or the effects of a toxin. Often infants, older adults, and immunocompromised patients are more at risk. In addition, populations living in unsanitary and overcrowded conditions where public health measures are lacking will be more at risk of becoming ill because there are more opportunities for food and water to become contaminated with pathogens.

Serotypes

Although there are several culture-dependent and culture-independent (nucleic acid analyses) methods to detect and identify specific species and strains of foodborne and waterborne pathogens, detection of specific bacterial proteins and carbohydrates remains an important factor for identification. Proteins and carbohydrates vary greatly from one bacterial strain to another within a bacterial species. As such, they represent antigens, because the immune system responds to these molecules by producing specific antibodies. In particular, proteins and carbohydrates in the cell wall and in the capsule, as well as proteins on the flagella of gram-negative species, can be distinguished by the antibodies produced (**FIGURE 12.9**). A strain that has antigens different from another strain of the same species is called a **serotype** (or **serovar**). We will see the importance of serotypes with several of the gram-negative foodborne and waterborne pathogens we will discuss.

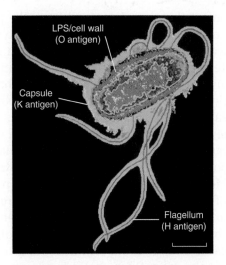

FIGURE 12.9 Bacterial Serotypes. Often bacterial structures are antigenic, and these structures can be used to "identify" different strains of a bacterial species. LPS = lipopolysaccharide. (Bar = 0.5 µm.) »» *If a strain is identified as O26:H11, what structures were used for the identification of this serotype?*

© Dr. Linda M. Stannard, University of Cape Town/Science Source.

Concept and Reasoning Checks 12.3

a. List the ways in which food and water can become contaminated with pathogens.
b. What types of etiologic information are needed to identify the cause of an intoxication or infection?

Chapter Challenge C

In this section of the chapter, the point was made that foods can become contaminated somewhere along the production line from farm to table.

QUESTION C: *Present ways that foods can be protected from pathogen contamination on the farm, during manufacturing, at a restaurant, in the grocery store, and at home. Specific examples of prevention will be examined in the rest of the chapter.*

You can find answers online in **Appendix F.**

■ KEY CONCEPT 12.4 Some Bacterial Diseases Are the Result of Foodborne Intoxications

Food intoxications often are referred to as **food poisoning**, which is a type of **noninflammatory gastroenteritis**. The illness might be linked to a single contaminated food, and the poisoning often involves a single meal and one specific food product.

Food Poisoning Can Be the Result of Enterotoxins

Individuals might consume foods contaminated with toxin-producing pathogens or consume foods already containing the toxin. As most of these toxins affect the gastrointestinal tract, the poisons are referred to as **enterotoxins** (*entero* = "intestine"). The exception is for botulism, which we'll look at separately in a moment. Most foodborne intoxications involve a brief incubation period and quick resolution. General symptoms include vomiting, diarrhea, abdominal cramps, and weakness.

One of the most common forms of food poisoning is caused by *S. aureus*, a facultatively anaerobic, gram-positive sphere (**FIGURE 12.10**). **Staphylococcal food poisoning** ranks first among reported cases of foodborne intoxications. A key reservoir of *S. aureus* in humans is the nose. Thus, an errant sneeze by a food handler can be the source of staphylococcal contamination. Studies indicate, however, that the most common mode of transmission is from boils or abscesses on the skin of an individual who then sheds staphylococci into the food product where the cocci soon begin to secrete the toxin. When investigators locate staphylococci, they can identify the organisms by growth on mannitol salt agar, Gram staining, and testing with bacterial viruses to learn the strain involved.

The incubation period is a brief 1 to 6 hours, so the individual usually can think back and pinpoint the source, which often is a protein-rich food, such as meats or fish, as well as dairy products, cream-filled pastries, and salads such as egg salad. Staphylococci grow over a broad temperature range of 8°C to 45°C, and because refrigerator temperatures are generally set at about 5°C, refrigeration is not an absolute safeguard against growth of *S. aureus* in contaminated foods.

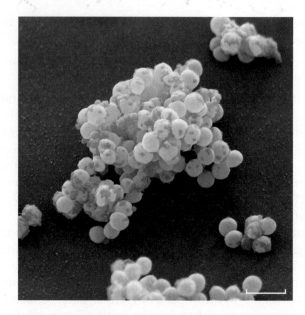

FIGURE 12.10 *Staphylococcus aureus.* A false-color scanning electron micrograph of *S. aureus* illustrating the typical grape-like cluster of cocci. (Bar = 1 µm.) »» *Why is it important to type* S. aureus *strains?*

© Eye of Science/Science Source.

Other forms of food poisoning are listed in TABLE 12.3.

Botulism Can Be a Life-Threatening Foodborne Intoxication

Of all the foodborne intoxications in humans, none is more dangerous than **botulism**. It is a rare but serious illness caused by *Clostridium botulinum*, a

TABLE 12.3 **Summary of Bacterial Foodborne Intoxications**

Intoxication	Causative Agent	Signs and Symptoms	Incubation Period	Duration	Food Examples
Staphylococcal food poisoning	*Staphylococcus aureus*	Abdominal cramps, nausea, vomiting	1–6 hours	24–48 hours	Meats, fish, egg products
Clostridial food poisoning	*Clostridium perfringens*	Abdominal cramps, watery diarrhea	8–24 hours	12–24 hours	Meat, poultry, fish
Bacillus cereus food poisoning	*Bacillus cereus*	Watery diarrhea	6–15 hours	6–24 hours	Meat, poultry, vegetables Grains
		Vomiting	½–5 hours	12–24 hours	
Botulism	*Clostridium botulinum*	Vomiting, paralysis, weakness	12–72 hours	High mortality rate if untreated	Honey, improperly preserved foods

spore-forming, obligately anaerobic, gram-positive rod. The endospores exist in the intestines of humans as well as fish, birds, and barnyard animals. They reach the soil in manure, organic fertilizers, and sewage; often, they cling to harvested products.

Most foodborne outbreaks of botulism are related to contaminated home-canned foods, such as asparagus, green beans, beets, and corn, and from other contaminated foods eaten cold. Other foods linked to botulism contamination include mushrooms, olives, salami, and sausage. In fact, the word "botulism" is derived from the Latin *botulus*, for "sausage."

Unlike most of the other diseases described in this chapter, botulism does not affect the GI tract. When *C. botulinum* spores are present in the anaerobic environment of cans or jars, they germinate to vegetative bacilli, and the bacilli produce a **neurotoxin**. The botulinum neurotoxin is one of the deadliest substances known to science. Scientists have identified seven types of *C. botulinum*, depending on the variant of toxin produced. Types A, B, and E cause most human disease, with type E being associated with most cases of foodborne transmission. The annual number of foodborne cases in the United States is low (32 cases in 2016).

The symptoms of foodborne botulism usually develop within 12 to 72 hours after ingesting the toxin-contaminated food (see Table 12.3). Patients suffer neurological manifestations, including blurred vision, slurred speech, difficulty swallowing and chewing, and labored breathing. The limbs lose their tone and become flabby, a condition called **flaccid paralysis**. These symptoms result from the toxin's effects on the nervous system. Specifically, the neurotoxin affects the nerve synapse and inhibits the release of the neurotransmitter acetylcholine. Without acetylcholine, nerve impulses cannot pass across the synapse into the muscles, and the muscles do not contract. Failure of the diaphragm and rib muscles to function leads to respiratory paralysis and death.

Because botulism is an intoxication, antibiotics are of no value as a treatment against the toxin. Instead, if treated early, large doses of specific antibodies called **antitoxins** can be administered to patients to neutralize the unbound toxins. Even with a favorable outcome, complete recovery can take up to 1 year.

Although foodborne botulism is very dangerous, other forms of botulism exist. **Wound botulism** is caused by toxins produced in the anaerobic tissue of a wound infected with *C. botulinum*. Penicillin is an effective treatment. **Infant botulism** is the most

common form of botulism in the United States, accounting for approximately 70% of the total botulism cases reported annually (115 cases reported by the CDC in 2016). Infant botulism results from the ingestion of soil or food contaminated with *C. botulinum* endospores. Thus, parents should not give their infant honey, which is the most common food triggering infant botulism. Spores that are in about 10% of honey germinate and grow in the colon where the botulinum cells release the neurotoxin. This form of botulism typically affects infants 3 to 24 months old because they have not yet established the normal balance of bowel microbes. The toxin produces lethargy and poor muscle tone and is an example of so-called **floppy baby syndrome**. Hospitalization along with mechanically assisted ventilation might be necessary, with antitoxin treatment reducing recovery time.

Today, one of the botulinum toxins has been put to practical use. In extremely tiny doses, Botox® or Dysport® (botulinum toxin type A) affects the synapses and temporarily relaxes the muscles. Therefore, the drug is useful on a number of movement disorders (so-called dystonias) caused by involuntary sustained muscle contractions. For example, the drug is used to treat strabismus, or misalignment of the eyes (crossed-eye). The toxin can be valuable in relieving stuttering, uncontrolled blinking, and musician's cramp. Most commonly, the toxin has been used to temporarily relieve facial wrinkles and frown lines (**FIGURE 12.11**). In addition, the toxin can provide temporary relief of hyperhidrosis (excessive body sweating) and chronic migraine headaches.

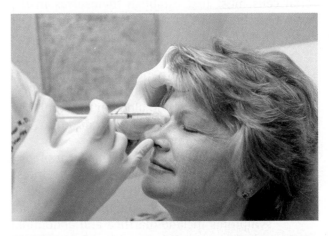

FIGURE 12.11 Cosmetic Injection of Botulinum Toxin. The botulinum toxin in controlled doses can be used to temporarily minimize wrinkles. *»» How does the toxin work to "remove wrinkles"?*

© Michael N. Paras/age fotostock.

Concept and Reasoning Checks 12.4

a. What characteristics are shared by the three foodborne intoxications of the GI tract?
b. Why is botulism considered a foodborne intoxication if it does not affect the GI tract?

Chapter Challenge D

Foodborne intoxications involve foods that inadvertently become contaminated with microbes that produce dangerous enterotoxins.

QUESTION D: *In our farm-to-table scenario, what are the most common ways that foods become contaminated with bacterial toxins or with the pathogens that produce such toxins?*

You can find answers online in **Appendix F.**

■ KEY CONCEPT 12.5   GI Infections Can Be Caused by Several Bacterial Pathogens

Foodborne infections are a substantial cause of illness in the United States. In 2015, the Foodborne Diseases Active Surveillance Network (FoodNet) reported more than 20,000 confirmed cases, 4,500 hospitalizations, and 77 deaths from foodborne infections. The FoodNet-monitored infections involve seven bacterial foodborne pathogens (*Campylobacter, Escherichia, Listeria, Salmonella, Shigella, Vibrio*, and *Yersinia*). Globally, there are more than 550 million cases and 230,000 deaths each year due to bacterial foodborne infections.

Bacterial GI infections usually have a longer incubation period than intoxications because bacterial cells must first establish themselves in the body after ingestion of the contaminated food or water. These GI infections can be one of two types:

▶ **Inflammatory Gastroenteritis.** This common form of gastroenteritis is characterized by watery diarrhea, vomiting, and usually a fever, but there is no blood in the stool. The symptoms usually are of short duration (1 to 3 days) and noninvasive.

▶ **Invasive Gastroenteritis.** This form of gastroenteritis involves damage to the intestinal mucosa caused by the invading pathogens or the secreted toxins. Signs and symptoms include diarrhea and abdominal pain but very little vomiting. Importantly, the damage to the intestines causes **dysentery**, which has a rapid onset and results in the passage of blood and mucus in the feces. Most cases resolve in 3 to 7 days.

We will examine several infections characteristic of each type of gastroenteritis.

Bacterial Gastroenteritis Often Produces an Inflammatory Condition

The most familiar illnesses causing a bacterial inflammatory gastroenteritis are cholera and traveler's diarrhea.

Cholera

No diarrheal disease can compare with the extensive diarrhea associated with **cholera** (see the chapter opener). The WHO estimates that globally there are between 1 million and 4 million cases and 21,000 to 143,000 deaths annually. The causative agent, *V. cholerae*, was first isolated by Robert Koch in 1883. Since then, seven cholera pandemics have been documented, and these pandemics have been responsible for taking the lives of millions of people. The current seventh pandemic, caused by *V. cholerae* serotype El Tor, began in 1961 in Indonesia, and the disease now involves about 35 countries, most notably in Haiti and the Democratic Republic of the Congo (see Figure 12.1).

The cells of *V. cholerae* are motile, gram-negative, curved rods (**FIGURE 12.12**). The organism naturally occurs in estuaries and marine environments, often in association with shellfish, where the cells enter the intestinal tract in contaminated food, such as raw or undercooked seafood, or in contaminated water. Because *V. cholerae* is extremely susceptible to stomach acid, most of the cells ingested are killed by the low pH and the infections usually are asymptomatic (TABLE 12.4). However, if high numbers (1 million to 100 million) are ingested, enough survive to pass into and colonize the small intestines. There, over a 1 to 3 day incubation period, the noninvasive cells stick to and colonize the entire surface of the intestinal tract, forming a thick biofilm. Then, the cells synthesize and secrete the **cholera toxin**, which disrupts the body's ability to maintain water balance. As a result, there is an unrelenting loss of fluid and electrolytes, especially from the small intestine.

Most cases of cholera are mild and they are characterized by watery diarrhea that lasts 2 to 7 days. In more severe cases, extreme dehydration might occur because a patient can lose up to 15 liters of fluid in one day. The fluid, referred to as **rice-water stools**, is colorless and watery, reflecting the conversion of the intestinal contents to a thin liquid likened to water from boiled rice. Despite continuous thirst, sufferers cannot hold fluids. Consequently, as water leaves the blood, the blood thickens, urine production ceases, and the sluggish blood flow to the brain leads to shock and coma. If untreated, an individual can die in less than 24 hours of the onset of symptoms.

For severe cases, antibiotics can be used to kill the bacterial cells. However, the key treatment is to restore the body's water and electrolyte balance. For mild to moderate cases, this entails **oral rehydration therapy**, which involves drinking solutions of electrolytes and glucose designed to restore the normal water/salt balance in the body. For the most severe cases, rehydration requires intravenous injections of the solution.

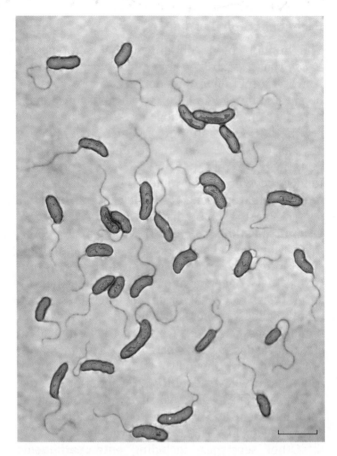

FIGURE 12.12 *Vibrio cholerae*. This light microscope image shows stained *V. cholerae* cells and their flagella. (Bar = 1 μm.) »» *What is the shape of these cells?*

© James Cavallini/Science Source.

TABLE 12.4 The Clinical Characteristics of Cholera

Characteristic	Asymptomatic Infection	Mild Infection	Severe Infection
Symptoms	None	Diarrhea similar to other forms of gastroenteritis	Prolific diarrhea and vomiting
Dehydration	None	None to mild	Moderate to severe
Stool	Normal	Watery	Rice water
Cells (per gram stool)	Up to 10^5	Up to 10^8	Up to 10^9
Treatment	None	Oral rehydration solution (ORS)	ORS, IV fluids, antibiotics

Escherichia coli Diarrhea

The human colon contains several serotypes of *E. coli* that are part of the normal gut microbiome. However, other environmental *E. coli* serotypes can be pathogenic. Their ability to cause tissue damage depends on what pathogenic traits they have acquired through mutation or gene transfer.

The cells of *E. coli* are facultatively anaerobic, gram-negative rods (**FIGURE 12.13**). Transmission follows the fecal–oral route where contaminated food or water represents the vehicle for transmission.

The most common diarrhea-causing serotype is **enterotoxigenic *E. coli*** (**ETEC**). During a 1- to 3-day incubation, the cells adhere to the epithelium of the small intestines, and the cells produce two enterotoxins that trigger significant watery diarrhea for several days to a week or longer. The illness, commonly called **traveler's diarrhea**, is the most frequent illness affecting travelers to a tropical or subtropical destination where ETEC occurs. The volume of fluid lost is usually low, but the symptoms of inflammatory gastroenteritis are common. The risk of contracting traveler's diarrhea can be reduced by careful hygiene and attention to the food and water consumed during visits to endemic areas.

Other serotypes, including **enteropathogenic *E. coli*** (**EPEC**) and **enteroaggregative *E. coli*** (**EAEC**), can cause diarrhea in infants and young children, especially where sanitation conditions are poor. As

with ETEC, careful hygiene and attention to the food and water consumed can reduce the risk of infection.

An enterohemorrhagic form of *E. coli* that causes an invasive gastroenteritis is described later in this chapter.

Clostridium difficile Infection

One of the most widespread and potentially serious illnesses involving the digestive tract and gut dysbiosis is caused by *Clostridium difficile* (or simply *C. diff*), an anaerobic, endospore-forming, gram-positive rod. ***C. diff* infections** (**CDIs**) have reached epidemic proportions around the world but especially in the United States, Canada, and Europe. The CDC estimates that there are almost 500,000 CDI cases and up to 30,000 deaths each year in the United States.

C. diff forms part of the normal microbiome in the colon of about 5% of healthy adults and 70% of healthy infants, where it remains in an inactive, noninvasive state due to microbial antagonism by the groups of harmless microbes. *C. diff* can enter the body when consuming *C. diff*-contaminated food or water. In addition, the bacterial spores are shed in feces, so any surface, instrument, or object contaminated with feces (e.g., hospital equipment, hands of healthcare workers) can serve as a source of endospores that can trigger an infection. This is particularly important because *C. diff* spores can survive for months or years on inanimate surfaces.

Healthy individuals usually do not become ill from *C. diff*, and they remain asymptomatic. Those most at risk of contracting a CDI are older adults and people under medical care who have been taking certain antibiotics for a prolonged time for another illness. Antibiotic therapy eliminates most of the harmless members of the microbiome of the colon, paving the way for proliferation and overgrowth by the antibiotic-resistant *C. diff* cells. Consequently, outbreaks tend to be associated with hospitals and outpatient facilities where *C. diff* contamination is prevalent. About 50% of CDIs develop outside the hospital or healthcare setting, while the individual is still on antibiotic therapy or has completed the therapy. Such patients having a second infection superimposed on an earlier one are said to have a **superinfection**.

Should dysbiosis occur, *C. diff* cells can colonize the epithelium of the colon and release two toxins. One toxin causes fluid loss and diarrhea, amounting to 5 to 10 clear, watery bowel movements per day. Thus, a CDI is an example of **antibiotic-induced diarrhea**. The second toxin causes mucosal injury and cell death, leading to a severe inflammation called

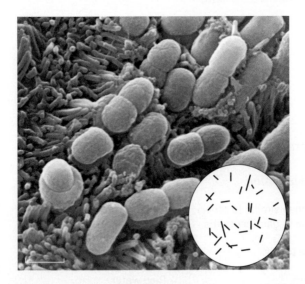

FIGURE 12.13 *Escherichia coli.* False-color scanning electron micrograph of *E. coli* cells inhabiting the surface of the intestines. (Bar = 2 μm.) Inset: Artist sketch of *E. coli* cells seen with the light microscope at 1000X »» *Why doesn't the normal intestinal* E. coli *cause gastroenteritis?*

© Stephanie Schuller/Science Source.

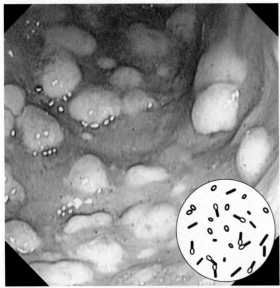

(A)

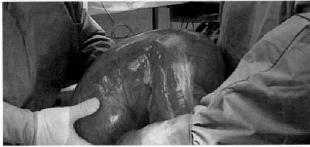

(B)

FIGURE 12.14 Pseudomembranous Colitis and Toxic Megacolon. (A) A view through an endoscope showing an inflamed colon caused by a *Clostridium difficile* infection. The white plaques represent mucus and dead cells that have built up on the colon walls. Inset: Artist sketch of *C. difficile* cells seen with the light microscope at 1000×. **(B)** Severe, untreated pseudomembranous colitis can lead to toxic megacolon. In this case, the extremely swollen sigmoid colon had to be surgically removed. *»» What factors are produced by* C. diff *that trigger pseudomembranous colitis?*

(A) © David M. Martin, MD/Science Source. (B) Courtesy of Dr. Jeffrey Pommerville.

pseudomembranous colitis (FIGURE 12.14A). In this case, adherent, white- to yellow-colored abscesses form at the injured mucosal sites. The inflammation can arise within 1 to 2 days after diarrhea symptoms begin or might not occur until several weeks after discontinuing antibiotic therapy. In about 3% of patients, the colon becomes grossly dilated and the patient is unable to expel gas and stool. This so-called **toxic megacolon** could cause the colon to rupture (**FIGURE 12.14B**). Surgical intervention usually is required.

Because most symptomatic patients exhibit only a mild diarrhea, stopping antibiotic therapy, if clinically possible, together with fluid replacement, usually results in rapid improvement. For moderate and severe cases, anticlostridial antibiotics, such as metronidazole or vancomycin, might be required.

Today, **microbiome-induced therapy** is being used to reintroduce microbes into the colon of CDI patients to stimulate the repopulation of lost resident microbes. The therapy, also called **fecal transplantation**, involves using fecal bacteria from a close relative or healthy donor. The bacterial community from the fecal sample is introduced through a colonoscope, enema, or nasogastric tube (**FIGURE 12.15**). Amazingly, the therapy has been extremely successful (95% cure rate) in reestablishing the normal gut microbiome and eliminating CDIs. Such transplants also are showing some potential for treating other bowel and gastrointestinal disorders such as irritable bowel syndrome and Crohn's disease. Thankfully, the development of a fecal pill might mean a nasogastric tube or colonoscope would no longer be needed for treatment.

The emergence of a newly discovered, hypervirulent strain of *C. diff* has been the cause of community-acquired CDIs among a wider age range of patients with no recent history of antibiotic therapy. This new strain causes a more severe illness and the hypervirulence appears to be due to amplified production of the two toxins and increased antibiotic resistance.

Vibriosis

Illnesses caused by *Vibrio* species other than *V. cholerae* are referred to as **vibriosis**. Most infections are acquired through the ingestion of contaminated raw seafood, producing symptoms of watery diarrhea, abdominal cramping, nausea, vomiting, and fever within 4 to 48 hours of ingestion. The CDC says that there are some 80,000 cases yearly, with 52,000 cases coming from eating contaminated food.

The leading cause of seafood-associated gastroenteritis in the United States is *V. parahaemolyticus,* a curved, gram-negative, halophilic rod that is naturally found in warm marine coastal seawaters. The CDC estimates that 45,000 cases of vibriosis each year come from ingestion of food contaminated with *V. parahaemolyticus*. In 2012, an outbreak was associated with consumption and inadequate cooking of shellfish harvested from Oyster Bay Harbor, New York. *V. parahaemolyticus* is now endemic along the Atlantic Coast, and the pathogen was the cause of another outbreak in 2013, constituting 104 cases reported from Connecticut to Virginia.

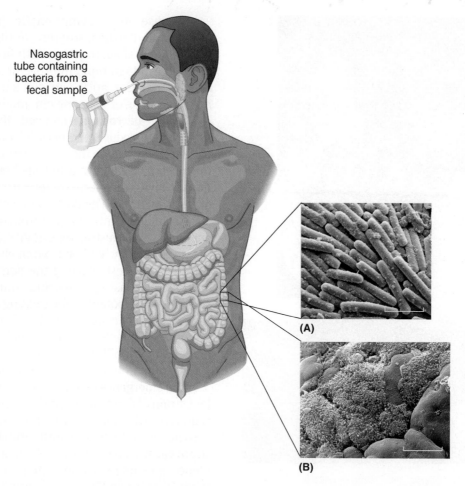

Nasogastric tube containing bacteria from a fecal sample

(A)

(B)

FIGURE 12.15 **Fecal Transplant.** For a person suffering from a severe *Clostridium difficile* infection **(A)**, a fecal transplant could be carried out. In this example, the bacterial community from a donor stool sample is introduced through a nasogastric tube to the stomach. There, the bacterial cells pass into and recolonize the small intestine, reestablishing the normal microbiota **(B)**. (A) and (B) (Bar = 2 μm.) *»» How could the transplanted bacterial cells eliminate the infection?*

Another species, *V. vulnificus*, also is found as a normal inhabitant in warm, coastal seawater. This species is the most virulent of the vibrios. People who consume contaminated raw oysters or clams are at risk, especially immunocompromised individuals and those who suffer from liver disease or low stomach acid. (Indeed, taking an antacid after a meal of any contaminated food can neutralize stomach acid and facilitate the passage of the pathogen to the bloodstream.) Exposure of an open wound to contaminated seawater results in an infection that can become systemic and produce necrotic skin ("flesh-eating") lesions and deeper tissues infections. The CDC reports as many as 100 cases, 85 hospitalizations, and 35 deaths each year, the majority of infections originating along the Gulf Coast. Early antibiotic therapy improves survival.

The bacterial GI-tract diseases causing noninvasive inflammatory gastroenteritis are summarized in TABLE 12.5.

Concept and Reasoning Checks 12.5 (PART A)

a. Explain how the mechanism of action of cholera toxin is similar to that of other enterotoxins mentioned in this chapter.
b. How is the cause of diarrhea different between ETEC and EPEC serotypes?
c. Why would *C. diff* be considered an opportunistic pathogen?
d. How might seafood become contaminated with either *V. parahaemolyticus* or *V. vulnificus*?

Chapter Challenge E

Bacterial gastroenteritis, inflammatory and invasive, can be experienced through a number of bacterial infections transmitted through food or water.

QUESTION E: *For inflammatory gastroenteritis, what types of precautions should be taken to prevent the establishment of an infection?*

You can find answers online in **Appendix F**.

Several Bacterial Species Can Cause an Invasive Gastroenteritis

Besides those bacterial species that cause inflammatory gastroenteritis, additional species or bacterial strains (serotypes) possess virulence factors that allow them to invade the submucosa. A few pathogens can become systemic and affect other body systems.

Typhoid Fever and Salmonellosis

For *Salmonella*, the first of the pathogens we will examine, there are some 2,500 unique serotypes. Different serotypes of *Salmonella* are responsible for two disease conditions: typhoid fever and salmonellosis. Let's look at each more closely.

▶ **Typhoid Fever.** Among the classical diseases that have ravaged human populations

---

TABLE 12.5 A Summary of the Bacterial Foodborne and Waterborne Infections Causing Noninvasive Inflammatory Gastroenteritis

Disease	Causative Agent	Signs and Symptoms	Toxin Involved	Transmission	Treatment	Prevention
Cholera	*Vibrio cholerae*	Severe, watery diarrhea, nausea, vomiting, muscle cramps, and dehydration	Enterotoxin	Waterborne	Oral rehydration therapy Antibiotics	Practicing good hand hygiene Avoiding untreated water
ETEC, EPEC	*Escherichia coli* serotypes	Diarrhea, vomiting, cramps, nausea, and low-grade fever	Enterotoxins	Foodborne or waterborne	Fluid replacement	Avoiding suspect foods and untreated water
Clostridium difficile infection	*Clostridium difficile*	Watery diarrhea, pseudomembranous colitis	Enterotoxin Cytotoxin	Indirect from contaminated hands or materials	Stopping antibiotic therapy Anticlostridial antibiotic therapy	Practicing good hand hygiene Keeping bathrooms and kitchens disinfected
Vibriosis	*Vibrio parahaemolyticus*	Acute abdominal pain, vomiting, watery diarrhea	Enterotoxin	Foodborne in contaminated seafood (oysters and raw shellfish)	None Antibiotic therapy for severe or prolonged illnesses	Cooking seafood thoroughly, especially oysters
	Vibrio vulnificus	Fever, nausea, severe abdominal cramps	Cytotoxin	See previous	Immediate antibiotic therapy	Avoiding raw oysters and clams

for generations is **typhoid fever**. The disease, also called enteric fever, is caused by *S. enterica* serotype Typhi (referred to as *S.* Typhi), which infects only humans. The cells are motile, nonspore-forming, gram-negative rods. They display high resistance to environmental conditions outside the body, which enhances their ability to remain alive for long periods in water, sewage, and certain foods exposed to contaminated water. The pathogen is transmitted by the five Fs (**FIGURE 12.16**): flies, food (as in raw or undercooked meats and shellfish), fingers, fecal-contaminated water, and **fomites** (non-living objects). About 85% of infections are acquired during international travel to endemic areas. Thus, in many parts of the world, typhoid fever remains a problem, with more than 27 million cases and 217,000 deaths reported each year.

The cells of *S.* Typhi are acid resistant, and with the buffering effect of food and beverages, the cells survive passage through the stomach. During the 5- to 21-day incubation period, the pathogen invades the small intestine, causing deep ulcers, bloody stools, and abdominal pain. After *S.* Typhi cells enter the submucosa, immune cells called macrophages attempt to engulf and destroy the pathogens. Unfortunately, the *S.* Typhi cells have evolved to exist and even grow while inside the macrophages. As a result, the macrophages inadvertently carry the *S.* Typhi cells through the blood and the blood invasion leads to a systemic illness.

The infected patient experiences mounting fever, lethargy, and delirium, and in about 30% of cases, the abdomen becomes covered with a faint rash (**rose spots**), indicating blood hemorrhage in the skin. Symptoms can last for 3 to 4 weeks. The gallbladder can be infected, from which the pathogen is returned to the gut, and *S.* Typhi cells then can be passed in the feces. In fact, the pathogen can persist in the gallbladder for years with the chronic carrier experiencing no symptoms. The most famous case occurred in the United States in the early 20th century, as described in MICROFOCUS 11.3.

If not treated, typhoid fever takes the lives of about 15% of infected individuals. However, treatment is generally successful with antibiotic therapy. Unfortunately, strains of *Salmonella* have become **superbugs** as they have acquired multidrug resistance. Globally, they are replacing other typhoid strains and it might not be long before antibiotic options to fight *S.* Typhi will no longer exist.

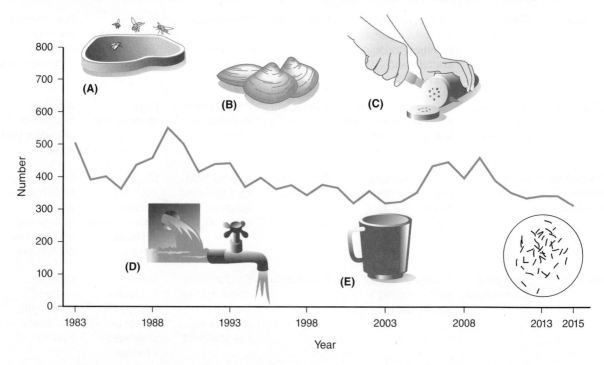

FIGURE 12.16 The Incidence of Typhoid Fever. Reported cases per 100,000 of typhoid fever in the United States by year, 1983–2015. Inset: Artist sketch of *S. typhi* cells seen with the light microscope at 1000X. **»» What do the five images (A–E) portray? Name them.**

Data from CDC, Summary of Notifiable Diseases, 2013; Notifiable Diseases and Mortality Tables, January 1, 2016.

MICROFOCUS 12.3: History

Typhoid Mary

By 1906, typhoid fever was claiming about 25,000 lives annually in the United States. During the summer of that year, a puzzling outbreak occurred in the town of Oyster Bay on Long Island, New York. One girl died of typhoid fever and five others contracted the disease, but local officials ruled out contaminated food or water as sources. Eager to find the cause, they hired George Soper, a well-known sanitary engineer from the New York City Health Department.

Soper's suspicions centered on Mary Mallon, a seemingly healthy family cook. But she had disappeared 3 weeks after the disease surfaced. Soper's investigations led him back over the 10 years' time during which Mary Mallon cooked for several households. Twenty-eight cases of typhoid fever occurred in those households, and each time, the cook left soon after the outbreak.

Soper tracked Mary Mallon through a series of leads from domestic agencies and finally came face-to-face with her in March 1907. She had assumed a false name and was now working for a family in which typhoid had broken out. Soper told her he believed she was a carrier of typhoid and pleaded with her to be tested for typhoid bacilli. When she refused to cooperate, the police forcibly brought her to a hospital for testing. Tests showed her stool was loaded with typhoid bacilli, but fearing her life was in danger, she adamantly refused the gallbladder operation (where the bacilli reside) that would eliminate them. As news of her imprisonment spread, Mary became a celebrity. Soon public sentiment led to a health department policy deploring the isolation of carriers. She was released in 1910.

But Mary's saga had not ended. In 1915, she turned up again at New York City's Sloane Hospital working as a cook under a new name. Eight people had recently died of typhoid fever, most of them doctors and nurses. Mary was arrested but again refused the gallbladder operation and vowed never to change her profession. Doctors placed her in isolation in a hospital room while trying to decide what to do. The weeks wore on.

Eventually, Mary became less incorrigible and gradually accepted her fate, even helping with routine hospital work. However, she was forced to eat in solitude and was allowed few visitors. Mary Mallon died in 1938 at the age of 70 from the effects of a stroke. She was buried without fanfare in a local cemetery.

The CDC recommends that travelers going to endemic areas with increased *S.* Typhi incidence (Africa, Asia, and Latin America) be vaccinated with one of the two available vaccines. However, both lose their effectiveness after a few years, so booster shots might be needed for future travel.

▶ **Salmonellosis.** Several nontyphoidal serotypes of *S. enterica* are associated with a gastrointestinal disease called **salmonellosis**. These strains can infect a range of animals, and they cause 94 million reported infections globally, while taking the lives of 155,000 individuals each year. The CDC estimates that there are more than 1.2 million cases of salmonellosis in the United States each year, most of which are transmitted by food consumed in the United States.

Although the most familiar serotypes involved with illness are *S. enterica* serotype Enteritidis and *S. enterica* serotype Typhimurium (**FIGURE 12.17A**), there is an increasing incidence of other serotypes, which tend to be specific to different animals and food sources. For example, in 2013, more than 280 people were sickened from an *S.* Heidelberg outbreak linked to raw chicken products. In 2015, two outbreaks involving serotypes *S.* Sandiego and *S.* Poona occurred in 16 states, both outbreaks originating in small turtles bought as pets for children. Not surprisingly, 50% of the illnesses in these latter two outbreaks were in children 5 years of age or younger.

Because salmonellosis requires a relatively large infectious dose (>100,000 cells), a broad variety of foods, most commonly beef, poultry, and eggs, must be heavily contaminated (**FIGURE 12.17B**). After an incubation period of 6 to 48 hours, the pathogen enters the mucosa and submucosa, where the inflammation causes fever, nausea, vomiting, diarrhea, and abdominal cramps. Dehydration can occur in some patients, necessitating fluid replacement. Intestinal ulceration is usually less severe than in typhoid fever, and blood invasion is uncommon as the cells are destroyed by immune cells. Symptoms typically last 3 to 7 days, and antibiotics usually are not recommended.

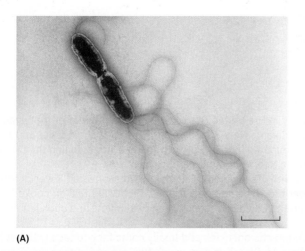

(A)

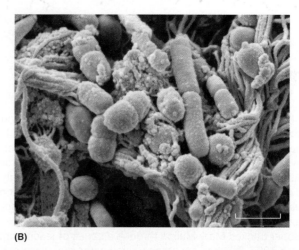

(B)

FIGURE 12.17 *Salmonella enterica.* Two views of *S. enterica.* **(A)** A light micrograph photo of *S. enterica* serotype Typhi cells. Note the long length of the flagella relative to the cell. (Bar = 3 μm.) **(B)** A false-color scanning electron micrograph of *S. enterica* serotype Typhimurium cells (blue) observed on the collagen fibers of muscle tissue from an infected chicken. (Bar = 3 μm.) *»» How might* Salmonella *cells attach to the collagen fibers?*

(A) © Kwangshin Kim/Science Source. (B) © Scimat/Science Source.

Before leaving the topic, mention should be made of another strain of *Salmonella* that is becoming a drug-resistant superbug. This is a strain of *S.* Typhimurium (ST313), and it is found in children and HIV patients in parts of sub-Saharan Africa. Instead of causing gastroenteritis typical of the non-typhoidal strains, ST313 is invasive and enters the bloodstream. The fact that it is resistant to many antibiotics makes it a challenge to treat, resulting in the death of up to 25% of the infected individuals.

Shigellosis

Another intestinal disease common among young children is **shigellosis**. It is caused by members of the genus *Shigella*, which are gram-negative, nonmotile rods closely related to *E. coli*. In the United States, shigellosis is caused primarily by *S. sonnei*, which was responsible for more than 21,000 cases in 2015. However, most infections go unreported, so the CDC estimates that there are actually about 500,000 diarrhea cases annually. About 130,000 of these cases are foodborne. In the developing world, another species, *S. dysenteriae*, causes deadly epidemic dysentery and renal failure. There are 65 million reported illnesses and 1 million deaths annually.

Humans are the primary host of *Shigella*, which is transmitted by the fecal–oral route and person to person. The pathogen also is transmitted by flies and contaminated water. Foods, such as eggs, vegetables, shellfish, and dairy products can become contaminated through unsanitary handling.

An infectious dose requires as few as 100 cells, which survive the gastric juices of the stomach and enter the small intestine where they cause diarrhea. After they enter the colon, the cells invade the mucosa, causing abscesses and bloody diarrhea (**bacillary dysentery**). Symptoms also include fever and abdominal pain. The pathogen rarely penetrates to the bloodstream as the pathogen usually is eventually destroyed by immune cells.

Most cases of shigellosis subside within a week and usually produce few complications. Patients who recover generally are carriers for a month or more and continue to shed the bacilli in their feces. Patients who lose excessive fluids must be given salt tablets, oral solutions, or intravenous injections of salt solutions for rehydration. Careful hygiene and protection of the food and water supply are most important although antibiotics are sometimes effective in reducing the duration of illness and the number of bacilli shed. However, many strains of *Shigella* are becoming more resistant to antibiotics, so prescribing antibiotics for shigellosis should be reserved for severe cases.

Hemorrhagic Colitis

Besides the ETEC and EPEC strains of *E. coli* that were described earlier in this chapter, there also are **enterohemorrhagic *E. coli* (EHEC) strains**. These strains cause disease by producing Shiga-like toxin, and the strains are called **Shiga toxin-producing *E. coli*, or STEC**. The most commonly identified and the most dangerous STEC in North America is

E. coli O157. Epidemiologists at the CDC have reported that in 2014 there were more than 6,100 cases of STEC, of which most were due to O157. Accounting for underdiagnosis, the CDC estimates there are over 96,000 cases annually. **CLINICAL CASE 12** presents one example.

The reservoir for STEC is cattle. The prevailing wisdom is that serotypes like O157 naturally exist in the intestines of healthy cattle and cause no disease in these animals. The presence of O157 in the animal's fecal material would deposit the organism in the soil, which then could be washed by rain or irrigation into neighboring agricultural fields. In addition, slaughtering might allow fecal material containing O157 to contaminate the meat-processing line. This would bring the pathogen to beef products, especially ground meats. Undercooked ground beef then could bring the pathogen to the table. Waterborne transmission also can occur by swimming in O157-contaminated water parks and swimming pools or drinking contaminated water that was inadequately chlorinated.

Clinical Case 12

Outbreak of an *E. coli* O157 Infection

On a fall day, a Connecticut family decided to take a drive in the country. Along the way, they stopped at a general store for a bite of lunch. The father and two daughters had apple cider; the mother had a soda.

Three days later, the father and children began to experience serious abdominal pains and vomiting. Moreover, there was blood in their stools. The mother had no symptoms.

One of the daughters became worse and had to be admitted to the hospital. The presence of bloody diarrhea was noted by the doctor.

The next day, the Connecticut Department of Public Health (DPH) was notified of eight more similar illnesses with disease onset during the same period as the Connecticut family. A case definition was defined, and a stool sample from the daughter was sent to the clinical lab for identification.

Meanwhile, the kidneys of the family's daughter were weakening, and the doctor advised kidney dialysis to assist kidney function.

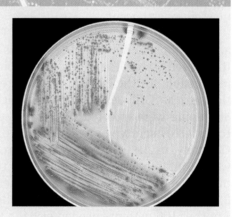

E. coli O157 growing on a culture plate.
Courtesy of the CDC.

The laboratory results identified and confirmed *E. coli* O157 as the cause of infection (see figure).

Health officials were notified, and they began a telephone survey to find out if anyone else was similarly infected. More than two dozen cases were found. All were asked about food consumption during the 7 days preceding the illness.

Based on the interviews, increased risk of illness was associated with drinking fresh apple cider from a single Connecticut juice producer.

When inspectors visited the cider plant, they were told the apples were taken from a pasture where cattle, sheep, and wild deer grazed. The apples were picked directly from the trees. What most interested investigators was hearing many apples also were picked up from the ground.

Appropriate control measures were instituted immediately to prevent further cases.

Questions

a. What would be the case definition defined by the DPH?
b. Hearing that the cause of infection was *E. coli* O157, what types of food might the DPH investigators be most interested in from the phone survey?
c. What is the infection complication exhibited by family A's daughter?
d. Why were the DPH investigators most interested in the "drop" apples collected from the soil surface of the pasture?
e. What control measures were instituted to prevent further outbreak cases?

You can find answers online in **Appendix E**.

For additional information, see: www.cdc.gov/mmwr/preview/mmwrhtml/00045558.htm.

The STEC strains are particularly pathogenic because they are acid tolerant, and less than 100 bacilli are needed to establish an infection. During an incubation period of 1 to 8 days, O157 cells attach to the lining of the colon and inject proteins directly into the adjacent host cells. An inflammatory response occurs, producing a **hemorrhagic colitis**, which is characterized by abdominal cramps, mild fever, and bloody diarrhea. In uncomplicated cases, the symptoms resolve within 5 to 7 days.

About 5% to 10% of patients, especially children, develop a potentially life-threatening complication called **hemolytic uremic syndrome** (**HUS**). The secreted Shiga-like toxins block host cell protein synthesis and kill the toxin-containing cells. This cellular damage triggers small platelet–fibrin clots to form that clog the kidneys and the filtration apparatus. Without treatment, life-threatening kidney failure can occur.

In addition to *E. coli* O157, other STEC serotypes account for another 168,000 illnesses each year in the United States. These other forms, like O26, were responsible for the 2015 *E. coli* outbreak at Chipotle Mexican Grill restaurants, in which more than 50 people were infected. One major outbreak due to the O104 serotype occurred in Europe (primarily Germany) in 2011 when fenugreek sprouts (an herb) were contaminated. By the end of the outbreak, more than 3,800 people were sickened, with some 850 exhibiting HUS that ultimately killed 54 people. Another 100 patients needed either a kidney transplant or lifelong dialysis.

The only treatment for HUS is supportive care (fluid replacement, red blood cell transfusions, and kidney dialysis). Prompt treatment usually leads to a full recovery for most patients. Antibiotics usually are not used, because they can actually increase the risk of developing HUS. Research continues on a vaccine for all *E. coli* diarrheas.

Campylobacteriosis

Since the early 1970s, **campylobacteriosis** has emerged from an obscure disease in animals to being, along with *Salmonella*, one of the most commonly reported bacterial causes of invasive gastroenteritis in the world. In the United States, the illness affects more than 2.5 million individuals and causes about 100 deaths each year.

The pathogen, *Campylobacter jejuni*, is a microaerophilic, corkscrew-shaped (*campylo* = "bending"), gram-negative rod (**FIGURE 12.18**). The pathogen has many serotypes and *C. jejuni* can be part of the

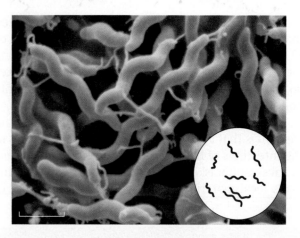

FIGURE 12.18 *Campylobacter jejuni.* A false-color scanning electron micrograph of *C. jejuni* cells. (Bar = 0.5 μm.) Inset: Artist sketch of *C. jejuni* cells seen with the light microscope at 1000X. »» ***What shape are these cells?***

Mediscan/Alamy Stock Photo.

normal microbiome in the intestinal tracts of many warm-blooded animals, including dairy cattle, chickens, and turkeys. In fact, chickens raised commercially often are colonized with *C. jejuni* by the fourth week of life. Unpasteurized dairy products, including raw milk, also can be a source of infection.

The cells of *C. jejuni* are primarily transmitted via the fecal–oral route through contact or exposure to contaminated foods or water. Chickens themselves can become ill from the pathogen, so human infections could come directly from the diseased animals. The infectious dose is about 500 cells.

During an incubation period of 2 to 7 days, the bacterial cells colonize the small or large intestine. There, invasion of the mucosa leads to inflammation, invasive disease, and occasional mild ulceration. The signs and symptoms of campylobacteriosis are typical of gastroenteritis and range from mild diarrhea to severe gastroenteritis with bloody diarrhea due to production of a cytotoxin. Most individuals recover in less than a week without treatment, but a few patients might require fluid and electrolyte replacement.

Some people might develop a rare immunoreactive condition several weeks after the diarrheal illness. This condition, called **Guillain-Barré syndrome** (**GBS**), results from an autoimmune reaction wherein antibodies attack the body's own nerves. The resulting peripheral nerve damage can cause paralysis lasting several weeks and usually requires hospitalization. Approximately 1 in every 1,000 reported cases of campylobacteriosis leads to GBS, and up to 40% of GBS cases in the United States might be caused by campylobacteriosis.

Listeriosis

An important public health concern in the United States today is **listeriosis**, a serious GI infection and foodborne illness. Listeriosis is caused by *Listeria monocytogenes*, a small, facultatively anaerobic, gram-positive rod. The CDC reported more than 630 cases in 2016. The disease primarily affects newborns, pregnant women, older adults, and those with weakened immune systems.

The organism is normally found in the gastrointestinal tracts of animals as well as in soil and water. Thus, cells of *L. monocytogenes* can contaminate foods of animal origin. However, healthy adults can consume such contaminated foods without becoming ill. Because the pathogen is a psychrotroph and can survive at low temperatures, refrigeration of contaminated food products (e.g., delicatessen cold cuts, soft cheeses, raw milk, and seafood) will not suppress growth completely. In late 2014, *Listeria*-contaminated caramel apples sickened 36 people, took the lives of seven adults, and caused the loss of one fetus. In 2015, *Listeria*-contaminated ice cream was recalled but not before three deaths occurred. In 2011, *Listeria*-contaminated whole cantaloupes sickened some 140 people in 28 states and caused the death of 30 individuals, making it the worst deadly foodborne outbreak in the United States since 1924.

The incubation period for listeriosis can be anywhere from 2 to 6 weeks. As mentioned, most healthy individuals experience no symptoms and recover completely. Almost everyone who is diagnosed with listeriosis has an invasive infection because the bacterial cells can spread beyond the gastrointestinal tract. In pregnant women, the pathogen can enter the blood and cross the placental barrier to infect the fetus. This can lead to miscarriage, stillbirth, or a life-threatening infection (meningitis or pneumonia) in newborns. In older adults and immunocompromised individuals, the pathogen can invade the intestinal lining, enter the blood, and cross the blood–brain barrier to cause an inflammation of the central nervous system. The result can be **listerial meningitis**, which has no clear symptoms or signs to distinguish it from other forms of community-acquired meningitis. Thus, listerial meningitis presents as high fever, headache, and stiff neck, and symptoms can include confusion, loss of balance, or convulsions. Listerial meningitis can be fatal in some 20% of patients or lead to a permanent disability, so early diagnosis is critical.

Treatment must be started as soon as possible in the course of the disease. Appropriate antibiotic treatment reduces the risk of serious complications or death. The general guidelines for the prevention of listeriosis include cooking raw food from animal sources thoroughly and washing hands, knives, and cutting boards used to prepare raw foods.

Yersiniosis

Another emerging foodborne illness is **yersiniosis**, caused by *Yersinia enterocolitica*. This motile, gram-negative rod is found in animals, especially pigs, which can become asymptomatic carriers of the disease (**FIGURE 12.19**). Only a few strains of *Y. enterocolitica* cause illness in humans. Infections (primarily in children) occur by consuming heavily contaminated (10^9 cells) foods that were exposed to infected domestic animals, raw or undercooked pork products, or were ingested from contaminated water or raw milk.

The *Y. enterocolitica* cells cause an invasive gastroenteritis through tissue destruction in the distal portion of the small intestine (the ileum), followed by multiplication in the associated lymphoid cells (Peyer patches). After a 1- to 2-day incubation, affected individuals experience fever, bloody diarrhea, and abdominal pain. The symptoms last 1 to 3 weeks and, unless the illness becomes systemic, antibiotic therapy is unnecessary.

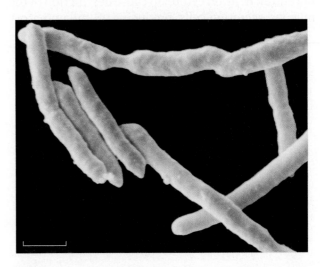

FIGURE 12.19 *Yersinia enterocolitica.* A false-color scanning electron micrograph of invasive *Y. enterocolitica.* (Bar = 0.5 μm.) »» *What types of foods or beverages might be contaminated by* Y. enterocolitica?

Mediscan/Alamy Stock Photo.

Peptic Ulcer Disease Can Be Spread Person to Person

Approximately 25 million Americans suffer from **peptic ulcers** during their lifetime, and there are approximately 500,000 new cases every year. For decades, scientists believed peptic ulcers (gastric ulcers in the stomach and duodenal ulcers in the duodenum) resulted from the production of excess stomach acid due to factors such as nervous or physiological stress, smoking, alcohol consumption, and spicy foods. However, the work of two Australian gastroenterologists, Barry Marshall and J. Robin Warren, made it clear that most gastric ulcers were of bacterial origin. Today, we know that the bacterial species *Helicobacter pylori* is primarily involved in the formation of these ulcers. This microaerophilic, motile, gram-negative curved rod infects half the world's population, yet only 2% are afflicted with peptic ulcers. Thus, the majority of infected individuals display no adverse effects.

However, it is uncertain how *H. pylori* is transmitted. Most likely, the organism is spread from an "infected" mother to her child. The way that *H. pylori* manages to survive in the intense acidity of the stomach (pH = 1.5) is interesting. In the stomach, *H. pylori* cells secrete the enzyme urease that digests urea in the area and produces ammonia as the end product (**FIGURE 12.20**). The ammonia neutralizes the stomach acid near the infection (pH = 6). That allows time for the *H. pylori* cells to penetrate the stomach mucosa that covers the stomach lining. The ammonia and an *H. pylori* cytotoxin cause destruction of the mucous-secreting cells, producing an inflammatory response (gastritis). For some people, the inflammation exposes the underlying connective tissue to the stomach acid. Over several years, an ulcer up to 12 cm in diameter appears. The pain is severe and is not relieved by food or an antacid.

One percent of infected individuals develop stomach cancer, which represents the fourth most common cancer and the second leading cause of cancer deaths globally (about 1 million deaths each year). *H. pylori* is the only bacterial agent known to be involved directly in causing cancer. In fact, the presence of *H. pylori* confers an approximately six fold increased risk of developing stomach cancer. Exactly how the pathogen contributes to cancer development remains to be identified. Some scientists suggest that infections of *H. pylori* might cause DNA damage in the stomach lining that cannot always be repaired correctly. This would then lead to mutations or cell dysfunction that damages the stomach epithelium—and possibly leading to cancer.

Thanks to Marshall and Warren, doctors have revolutionized the treatment of ulcers by prescribing antibiotics such as amoxicillin, tetracycline, or clarithromycin (Biaxin®), along with omeprazole (Prilosec®) for acid suppression. Antibiotic therapy for ulcers produces cure rates as high as 94%. Relapses of ulcers are uncommon, and the cases of stomach cancer have been declining as *H. pylori* is eliminated from people's stomachs. However, the loss of *H. pylori* might not be all good news, as the cost-benefit analysis in **MICROINQUIRY 12** reports.

TABLE 12.6 summarizes the bacterial diseases of the GI tract that cause invasive gastroenteritis.

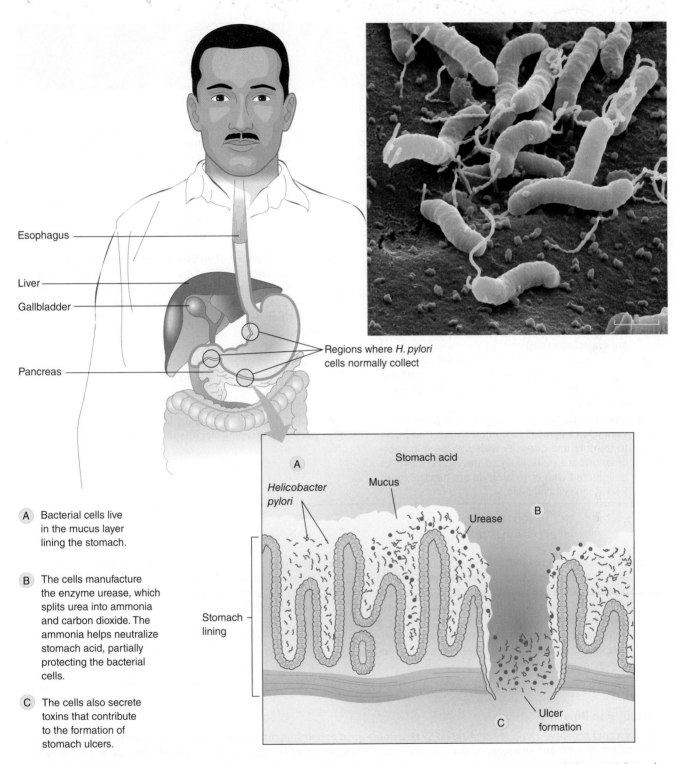

Esophagus

Liver

Gallbladder

Pancreas

Regions where *H. pylori* cells normally collect

A Bacterial cells live in the mucus layer lining the stomach.

B The cells manufacture the enzyme urease, which splits urea into ammonia and carbon dioxide. The ammonia helps neutralize stomach acid, partially protecting the bacterial cells.

C The cells also secrete toxins that contribute to the formation of stomach ulcers.

A

Stomach acid

Helicobacter pylori

Mucus

Urease

B

Stomach lining

Ulcer formation

C

FIGURE 12.20 The Progression of Gastric Ulcers. The majority of peptic ulcers are caused by *Helicobacter pylori*. Inset: False-color scanning electron microscope image of *H. pylori* cells. (Bar = 1 µm.) »» *How might peptic ulcer disease lead to stomach cancer?*

Inset © Juergen Berger/Science Source.

MICROINQUIRY 12

Helicobacter pylori: A Cost-Benefit Analysis

In the business community, a cost-benefit analysis allows a company to weigh its expected *costs* against expected *benefits* when determining the best (or most profitable) course of action for the company. We can do a similar analysis for infection with the bacterium *Helicobacter pylori*, which has been part of the human local stomach microbiome for more than 200,000 years. Specifically, what are the costs and benefits to humans of having an *H. pylori* infection? Let's do the analysis.

COST		BENEFIT
(of having *H. pylori*)		(of not having *H. pylori*)

- The bacterium *H. pylori* infects 3 billion people around the world and 2% (60 million) contract **gastric (peptic) ulcer disease**.
- About 2% of the ulcer patients (1 million) will develop **stomach cancer** (in the US: 22,000 patients are diagnosed annually of whom 10,000 are expected to die).

⟷

- With the advent of antibiotic therapy to cure peptic ulcer disease, the incidence of *H. pylori* has dropped sharply. In developed nations, infection has dropped from 80% to just a few percent. That means, many fewer people are developing peptic ulcer disease and stomach cancer.

- The presence of *H. pylori* might be linked to **adult type 2 diabetes**, which is the most common form of diabetes where the body does not respond properly to insulin.

⟷

- Elimination of *H. pylori* might lessen the chances of developing adult type 2 diabetes.

- Several studies have suggested that people with **Parkinson's disease** (a brain disorder that leads to tremors and difficulty with walking and coordination) are more likely to have ulcers and to be infected with *H. pylori* than are healthy (non-Parkinson's) individuals.

⟷

- Eradication of *H. pylori* may lessen the chances of a person developing Parkinson's disease.

COST		BENEFIT
(of not having *H. pylori*)		(of having *H. pylori*)

- Research suggests that people without *H. pylori* are at greater risk of **acid (esophageal) reflux disease** and **esophageal cancer**. These diseases have been rising dramatically as peptic ulcers have declined.

⟷

- Having *H. pylori* might lessen the chances of developing acid reflux disease and esophageal cancer. It would appear that *H. pylori* is in some way protecting the esophagus.

- Immunologists and allergy experts report that people without *H. pylori* may be more prone to **allergy-induced asthma**.

⟷

- Infection with *H. pylori* provides protection from allergy-induced asthma. It appears that somehow *H. pylori* "trains" the immune system to lower its response to triggers causing asthma.

Discussion Point

So, as a cost-benefit analysis, is colonization by H. pylori *a good or bad situation? For example, if you had a peptic ulcer, would you submit to antibiotic treatment to eliminate the infection knowing that* H. pylori *eradication could have other unhealthy or healthy consequences?*

TABLE 12.6 A Summary of the Bacterial Foodborne and Waterborne Causing Invasive, Inflammatory Gastroenteritis

Disease	Causative Agent	Signs and Symptoms	Toxin Involved	Transmission	Treatment	Prevention
Typhoid fever	*Salmonella typhi*	Bloody stools, abdominal pain, fever, lethargy, delirium	Endotoxin	Foodborne from person shedding *S. typhi* Foodborne and waterborne from contaminated sewage	Antibiotics	Avoiding risky foods and drinks Getting vaccinated
Salmonellosis	*Salmonella* serotypes	Fever, diarrhea, vomiting, and abdominal cramps	Not established	Foodborne in a broad variety of foods	Fluid replacement Antibiotic therapy	Practicing good hand hygiene and food preparation
Shigellosis	*Shigella sonnei*	Diarrhea, dysentery	Shiga enterotoxin	Foodborne and waterborne	Antibiotics Fluid and salt replacement	Practicing good hand hygiene
EHEC, STEC	*Escherichia coli* O157	Severe, bloody diarrhea	Enterotoxin Shiga-like enterotoxin	Foodborne through ingestion of undercooked meat or contaminated fruits and vegetables	For HUS: red blood cell and platelet transfusions Kidney dialysis	Practicing good hand hygiene Cooking foods thoroughly Washing fruits and vegetables; avoiding unpasteurized milk
Campylobacteriosis	*Campylobacter jejuni*	Diarrhea, fever	Enterotoxin Cytotoxin	Foodborne from contaminated foods or water	None Antibiotic therapy for severe or prolonged illnesses	Practicing good hand hygiene and food preparation
Listeriosis	*Listeria monocytogenes*	Headache, stiff neck, confusion, loss of balance, convulsions	Not established	Food contaminated with fecal matter Contaminated animal foods	Ampicillin	Practicing good hand washing Washing and preserving food properly

(continues)

TABLE 12.6 **A Summary of the Bacterial Foodborne and Waterborne Causing Invasive, Inflammatory Gastroenteritis (*Continued*)**

Disease	Causative Agent	Signs and Symptoms	Toxin Involved	Transmission	Treatment	Prevention
Yersiniosis	*Yersinia enterocolitica*	Fever, diarrhea, and abdominal pain	Enterotoxin	Foodborne from contaminated foods (pork and pork products)	None Antibiotic therapy for severe or prolonged illnesses	Avoiding raw or undercooked pork or pork products Practicing good hand hygiene
Peptic ulcer disease	*Helicobacter pylori*	Aching or burning pain in abdomen, nausea, vomiting, bloating, bloody vomit or stools	Cytotoxin	Can be transmitted person to person through direct or indirect saliva contact	Antibiotics and acid suppression medications	Practicing good hand hygiene Not sharing utensils or glasses

Concept and Reasoning Checks 12.5 (PART B)

a. Why is typhoid fever also called enteric fever?
b. What is the best way to protect oneself from salmonellosis?
c. What control measure (hygiene practice) would most likely decrease transmission rates of shigellosis?
d. Explain how *C. jejuni* could be responsible for traveler's diarrhea.
e. How does *L. monocytogenes* make its way to the brain and meninges?
f. Summarize how an *H. pylori* infection causes stomach ulceration.

Chapter Challenge F

Several bacterial species can cause some form of invasive gastroenteritis that can infect the submucosa of the intestine. One of the most common is caused by serotypes of *Salmonella enterica*. Contaminated products such as eggs, peanut butter, raw turkey burgers, ice cream, and chicken salad are among the food products that have caused outbreaks in the recent past in the United States.

QUESTION F: *Propose ways to prevent: (1) egg contamination on the farm; (2) peanut butter contamination at the manufacturing plant; (3) raw turkey contamination at the manufacturing plant; (4) ice cream contamination in delivery trucks previously used for hauling other raw products; and (5) chicken salad contamination at a local restaurant.*

You can find answers online in **Appendix F**.

In conclusion, we learned in this chapter that a diverse community of microbes in the mouth and gastrointestinal tract is extremely important to our good health. We also discovered that the intestinal microbiome could change in response to antibiotic use, diet changes, and pathogen infections. But what about at the end of a human life? What happens to the microbiome when a person dies? Studies of decomposing bodies show that the mix of bacterial species on and in a dead body also changes over time in a very predictable way. These successional changes in microbes might help pinpoint more accurately the time of death, which can be very useful in forensic investigations of murders and other criminal investigations. For example, within 20 hours of death, bacteria invade the liver, and by 58 hours, microbes have spread to the heart and other human organs. The biological clock of the dead individual might have stopped, but the microbial clock marches forward.

■ SUMMARY OF KEY CONCEPTS

Concept 12.1 The Digestive System Has Diverse Resident Microbiomes

1. The **digestive system** includes the **gastrointestinal (GI) tract** and the **accessory digestive organs** (teeth, tongue, salivary glands, liver, gallbladder, and pancreas). (Figure 12.2)
2. Digestive system defenses include: salivary mucus, lysozyme, and defensins; antibodies; stomach acid; mucus on intestinal surfaces; peristalsis; exfoliation of epithelial cells; and proteolytic enzymes.
3. The human intestinal **microbiome** contains an enormous population of normal microbiota that provides protection through microbial antagonism. (Figure 12.3)

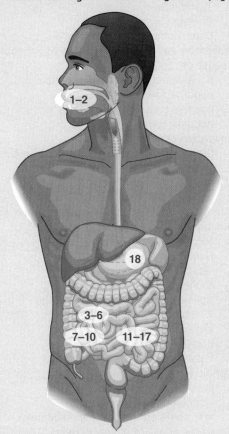

Concept 12.2 Bacterial Diseases of the Oral Cavity Can Affect One's Overall Health

▶ **Dental caries**
 1 *Streptococcus mutans, S. sobrinus, Lactobacillus species*
▶ **Gingivitis** and **periodontitis**
 2 *Bacteroides and several other genera*

Concept 12.3 Bacterial Diseases of the GI Tract Are Usually Spread Through Food and Water

▶ **Intoxications** represent a form of **noninflammatory gastroenteritis** caused by bacterial toxins, whereas **inflammatory** and **invasive gastroenteritis** are infections and diseases arising from bacterial growth in the GI tract.
▶ Contaminated food and water arising from unsanitary procedures often are the source of the intoxication or infection. (Figure 12.8)

Concept 12.4 Some Bacterial Diseases Are the Result of Foodborne Intoxications

▶ **Staphylococcal food poisoning**
 3 *Staphylococcus aureus*
▶ **Clostridial food poisoning**
 4 *Clostridium perfringens*
▶ ***Bacillus cereus* food poisoning**
 5 *Bacillus cereus*
▶ **Botulism**
 6 *Clostridium botulinum*

Concept 12.5 GI Infections Can Be Caused by Several Bacterial Pathogens

▶ **Cholera**
 7 *Vibrio cholerae*
▶ **ETEC, EPEC**
 8 *Escherichia coli*
▶ **Pseudomembranous colitis**
 9 *Clostridium difficile*
▶ **Other foodborne infections**
 10 *Vibrio parahaemolyticus, V. vulnificus*

Invasive

▶ **Typhoid fever**
 11 *Salmonella Typhi*
▶ **Salmonellosis**
 12 *Salmonella serotypes*
▶ **Shigellosis**
 13 *Shigella sonnei*
▶ **EHEC**
 14 *Escherichia coli O157*
▶ **Campylobacteriosis**
 15 *Campylobacter jejuni*
▶ **Listeriosis**
 16 *Listeria monocytogenes*
▶ **Yersiniosis**
 17 *Yersinia enterocolitica*
▶ **Gastric ulcer disease**
 18 *Helicobacter pylori*

■ CHAPTER SELF-TEST

For **Steps A–D**, you can find answers online in **Appendix D**.

■ STEP A: REVIEW OF FACTS AND TERMS

Multiple Choice

Read each question carefully, and then select the *one* answer that best fits the question or statement.

1. Which one of the following is *not* a digestive organ of the GI tract?
   A. Large intestine
   B. Oral cavity
   C. Liver
   D. Small intestine

2. What part of the GI tract contains the largest population of microorganisms (microbiota)?
   A. Colon
   B. Jejunum
   C. Duodenum
   D. Stomach

3. Which one of the following does *not* apply to dental plaque?
   A. It is an example of a biofilm.
   B. It is most noticeable on the molars.
   C. It can lead to gingivitis and periodontal disease.
   D. It is dominated by aerobic bacterial species.

4. What type of periodontal disease occurs when plaque bacteria build up between teeth and gums?
   A. Canker sore
   B. Dental caries
   C. Gingivitis
   D. Periodontitis

5. Gastroenteritis can result in _____.
   A. an intestinal inflammation
   B. an infection
   C. an intoxication
   D. All the above (**A–C**) are correct.

6. Foodborne microbes can be found in _____.
   A. cattle carcasses
   B. fresh fruits and vegetables
   C. healthy animals
   D. All the above (**A–C**) are correct.

7. Which one of the following bacterial species is *not* a cause of food poisoning (noninflammatory gastroenteritis)?
   A. *Escherichia coli*
   B. *Bacillus cereus*
   C. *Staphylococcus aureus*
   D. *Clostridium perfringens*

8. One of the most excessive diarrheas of the GI tract is associated with which of the following poisonings or diseases?
   A. Staphylococcal food poisoning
   B. Typhoid fever
   C. Cholera
   D. Campylobacteriosis

9. *Escherichia coli* is a common gram- ____ that can be a cause of ____.
   A. positive rod; hemorrhagic colitis
   B. negative rod; traveler's diarrhea
   C. positive coccus; typhoid fever
   D. negative rod; cholera

10. *Clostridium difficile* is _____.
    A. the cause of pseudomembranous colitis
    B. traveler's diarrhea
    C. meningoencephalitis
    D. undulant fever

11. This gram-positive rod is a psychrotroph that can multiply within macrophages.
    A. *Bacillus cereus*
    B. *Listeria monocytogenes*
    C. *Clostridium perfringens*
    D. *Escherichia coli*

12. This species is the most virulent of the vibrios.
    A. *V. vulnificus*
    B. *V. cholerae*
    C. *V. enterocolitica*
    D. *V. parahaemolyticus*

13. Typhoid fever is characterized by _____.
    A. the production of an exotoxin
    B. hemolytic uremic syndrome
    C. rose spots on the chest and abdomen
    D. All the above (**A–C**) are correct.

14. The symptoms of salmonellosis usually last about _____.
    A. 24 hours
    B. 48 hours
    C. 5 days
    D. 14 days

15. What is the name of the syndrome of fever, abdominal cramps, and bloody mucoid stools caused by *Shigella* species?
    A. Bacterial dysentery
    B. Typhoid fever
    C. Pseudomembranous colitis
    D. HUS
16. Enterohemorrhagic *E. coli* (EHEC) can cause _____.
    A. undulant fever
    B. hemolytic uremic syndrome
    C. Guillain-Barré syndrome
    D. stomach ulcers
17. Which of the following bacteria genera is most commonly reported as the cause of invasive bacterial gastroenteritis?
    A. *Campylobacter*
    B. *Staphylococcus*
    C. *Shigella*
    D. *Clostridium*

18. Yersiniosis is caused by _____.
    A. *Y. pestis*
    B. *Y. pseudotuberculosis*
    C. *Y. enterocolitica*
    D. All of the above (**A–C**) cause the illness.
19. Gastric ulcer disease is caused by _____.
    A. *Helicobacter pylori*
    B. *Yersinia enterocolitica*
    C. *Escherichia coli*
    D. *Salmonella* Typhi

Term Selection

For each choice, circle the word or term that best completes each of the following statements.

20. To treat patients who have botulism, large doses of (antitoxin, antibiotic) must be administered.
21. (Neurotoxins, Cytotoxins), such as those found with botulism, can cause flaccid paralysis.
22. Only a small percentage of those who recover from typhoid fever remain (carriers, free) of the bacterial cells.

23. Dental caries is caused by (*Streptococcus mutans*, *Staphylococcus aureus*), a gram-positive sphere.
24. The diarrheal disease (traveler's diarrhea, cholera) is caused by a motile, gram-negative, curved rod.
25. Botulism is an example of an (infection, intoxication).

■ STEP B: CONCEPT REVIEW

26. Summarize the digestive system defenses against pathogen colonization and infection. (**Key Concept 12.1**)
27. Describe the role played by oral bacterial species in causing **periodontal disease**. (**Key Concept 12.2**)
28. Differentiate a bacterial **intoxication** and a bacterial **infection**. (**Key Concept 12.3**)
29. Distinguish between the four bacterial species causing **food intoxications**. (**Key Concept 12.4**)

30. Summarize the clinical significance of *Salmonella* Typhi infections. (**Key Concept 12.5**)
31. Describe the consequences of **EHEC** infection. (**Key Concept 12.5**)
32. Diagram the steps involved in **gastric ulcer disease**. (**Key Concept 12.5**)

■ STEP C: APPLICATIONS AND PROBLEM SOLVING

33. You are doing the supermarket shopping for the upcoming class barbecue. What are some precautions you can take to ensure that the event is remembered for all the right reasons?
34. You read in the newspaper that botulism was diagnosed in 11 patrons of a local restaurant. The disease was subsequently traced to mushrooms bottled and preserved in the restaurant. What special cultivation practice enhances the possibility that mushrooms will be infected with the spores *Clostridium botulinum*?
35. Your roommate cuts up chickens on a wooden carving board in preparation for a summer barbecue. After running the board under water for a few seconds, he uses it to cut up tomatoes, lettuce, peppers, and other salad ingredients. What is wrong with this process?

36. The state department of health received reports of illness in 18 workers at a local pork processing plant. All the affected employees worked on the plant's "kill floor." All had gram-negative rods in their blood. Their symptoms included fever, chills, fatigue, sweats, and weight loss. Which disease was pinpointed in the workers?

37. A classmate plans to travel to a tropical country for spring break. To prevent traveler's diarrhea, she was told to take 2 ounces or 2 tablets of Pepto-Bismol four times a day for 3 weeks before travel begins. Short of turning pink, what better measures can you suggest she use to prevent traveler's diarrhea?

■ STEP D: QUESTIONS FOR THOUGHT AND DISCUSSION

38. A research report states that *Helicobacter pylori* accumulates in the gut of houseflies after the flies feed on food containing the pathogen. What are the implications of this research on human health?

39. Some years ago, the CDC noticed a puzzling trend: reported cases of salmonellosis seemed to soar in the summer months and then drop radically in September. Can you venture a guess as to why this is so?

40. A frozen-food manufacturer recalls thousands of packages of jumbo stuffed shells and cheese lasagna after a local outbreak of salmonellosis. Which parts of the pasta products would attract the attention of inspectors as possible sources of salmonellosis? Why?

CHAPTER 13

Soilborne and Arthropod-Borne Bacterial Diseases

Bubonic plague, commonly known as "the Black Death," was probably the greatest catastrophe ever to strike Europe and Asia. It swept back and forth across the continent for almost a decade, each year increasing in ferocity. By 1348, two-thirds of the European population was stricken and half of the sick had died. Houses were empty, towns were abandoned, and a dreadful solitude hung over the land. The sick died too quickly for the living to bury them, so victims often were buried in "plague pits" (FIGURE 13.1). At one point, the Rhône River was blessed as a graveyard for plague victims. Contemporary historians wrote that posterity would not believe such things could happen, because those who saw them were themselves appalled. The horror was almost impossible to imagine; a witness, Agnolo di Tura, said, *"Father abandoned child, wife husband, one brother another . . . And I, Agnolo di Tura . . . buried my five children with my own hands. . . . So many died that all believed that it was the end of the world."*

Before the century concluded, the Black Death visited Europe at least five more times in periodic reigns of infectious terror. During one epidemic in

Paris, an estimated 800 people died each day; in Siena, Italy, the population dropped from 42,000 to 15,000; and in Florence, almost 75% of the citizenry perished. Flight was the chief recourse for people who could afford it, but ironically, the escaping travelers spread the disease. Those who remained in the cities locked themselves in their homes until those infected either succumbed or recovered.

The immediate effect of the plague was a general paralysis in Europe. Trade ceased and wars stopped. Bewildered peasants who survived the disease encountered unexpected prosperity because landowners had to pay higher wages to obtain help. Land values declined and class relationships were upset, as the established social hierarchy of landowners (nobles) and peasants gradually crumbled. However, medical practices became increasingly sophisticated, with new standards of sanitation and a 40-day period of detention (quarantine) imposed

False-color transmission electron microscope image of *Yersinia pestis* cells, the causative agent of the Black Death.
© A. Barry Dowsett/Science Source.

FIGURE 13.1 **The Plague Pit.** This painting shows the unloading of dead bodies during the plague of 1665. These pits were no more than mass graves used to bury the plague victims that had been gathered up from the streets on "dead carts." »» *Why would these plague victims be buried in mass graves rather than by traditional funerals—and at night?*

© Photos.com.

on vessels docking at ports. The graveyard of plague left fertile ground for the renewal of Europe during the Renaissance. To many historians, the Black Death remains a major turning point in Western civilization.

Today, we know that the Black Death (bubonic plague) is caused by *Yersinia pestis* (see chapter opening image), a bacterial species that infects rodents and passes to fleas when they take a blood meal from the rodent. The zoonotic disease is no longer a major killer in most parts of the world, but infection and death still occur. Two recent outbreaks in Madagascar took the lives of more than 100 people, including 20 within just one week. Although most cases are curable with early antibiotic therapy,

new antibiotic-resistant strains have emerged. Consequently, it still is a disease to take seriously.

Besides plague, other diseases of bacterial origin, like typhus and relapsing fever, are transmitted by arthropods (animals having jointed appendages and segmented bodies such as ticks, lice, fleas, and mosquitoes). Neither is a major problem in our society, but a substantial number of cases of other arthropod-borne diseases, such as tularemia, Rocky Mountain spotted fever, and especially Lyme disease, are reported each year.

In this chapter, we also examine a number of soilborne diseases; these are organisms that live in soil and enter the body through a cut, wound or abrasion, or by inhalation.

Chapter Challenge

The soil is home to a diverse group of organisms. In fact, microbial ecologists estimate that up to 25% of all Earth's species inhabit the soil. Among the microorganisms, the vast majority of species is not a threat to human health; rather, these species are beneficial to the soil and the recycling of elements. However, soils do contain some human pathogens, which can be divided into two groups: those pathogens that are true soil-dwellers and those that are soil transmitted. As we progress through this chapter, let's see if we can identify to which group each of the soilborne, and even arthropod-borne, bacterial organisms belong.

■ KEY CONCEPT 13.1   Several Soilborne Bacterial Diseases Develop from Endospores

Soilborne bacterial diseases are those whose agents are transferred from the soil to the unsuspecting individual. To remain alive in the soil, the bacterial cells must resist environmental extremes and often form **endospores**, as the first three diseases illustrate.

Anthrax Is an Enzootic Disease

Anthrax was the first infectious disease shown by Koch to be caused by a germ. Anthrax is primarily an **enzootic** disease, meaning it is endemic among populations of animals such as large, domestic herbivores (cattle, sheep, and goats). Animals ingest and inhale the spores from the soil during grazing, and soon they are overwhelmed with escalating numbers of growing bacterial cells as their organs fill with bloody black fluid (*anthrac* = "coal"; the disease name is thus a reference to the blackening of the blood). About 80% of untreated animals die. The World Health Organization (WHO) estimates there are some 2,000 to 20,000 human cases of anthrax globally each year.

Anthrax is caused by *Bacillus anthracis*, a spore-forming, aerobic, gram-positive rod (**FIGURE 13.2**). Endospores can survive for years in soil rich with moist inorganic matter and calcium. When transmitted to animal tissues, the spores germinate rapidly to produce vegetative cells. The thick capsule of the cells impedes phagocytosis by immune cells, and the bacilli produce three toxins that work together to cause disease. Capsule and toxins are coded by genes carried on two plasmids.

Humans acquire anthrax from infected animal products, contaminated dust, or directly from the soil. Route of entry can be by any of three ways:

▶ **Cutaneous Anthrax.** Skin abrasions contaminated with soil or spore-contaminated animal products (e.g., violin bows [typically made of horsehair], shaving bristles, goatskin drumheads, and leather jackets) can lead to **cutaneous anthrax**. Cutaneous anthrax accounts for more than 95% of all anthrax infections. Skin infection begins as an elevated bump, called a **papule**, which resembles an insect bite. Within 5 to 7 days, the papule ulcerates, forming a black, necrotic (dying) scar called an **eschar** that soon crusts over. Lymph glands in the adjacent area might be invaded and swell. About 20% of untreated cases result in death.

A recently reported version of cutaneous anthrax is **injection anthrax**. This is when

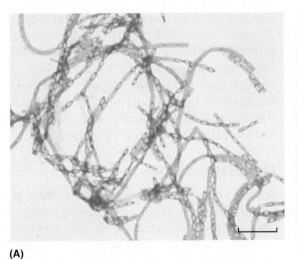

(A)

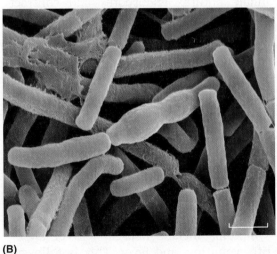

(B)

FIGURE 13.2 *Bacillus anthracis.* *B. anthracis* is the causative agent of anthrax. **(A)** Spores (white ovals) in vegetative cells can be seen in this photomicrograph. (Bar = 10 μm.) **(B)** A false-color scanning electron micrograph of vegetative cells. It is from such cells that the exotoxins are produced. Pink cells represent cells undergoing endospore formation. (Bar = 2 μm.) »» *What advantage is provided to the organism by producing endospores?*

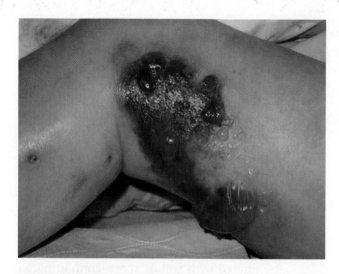

FIGURE 13.3 **An Anthrax Lesion.** This cutaneous lesion on the thigh is a result of infection with anthrax bacilli from contaminated heroin. Multiple injection sites are visible. Lesions like this one develop when anthrax spores contact the skin, germinate to vegetative cells, and multiply. *»» Why was a skin ulceration like this given the name "anthrax"?*

Courtesy of NIH.

drug users injecting heroin develop massive edema (swelling) due to heroin contaminated with anthrax spores (**FIGURE 13.3**). Although rare, a few cases have been reported in northern Europe, where the mortality rate can be up to 30%.

▶ **Inhalation Anthrax.** Workers who tan hides, shear sheep, or process wool can inhale anthrax endospores and contract **inhalation (pulmonary) anthrax** as a form of pneumonia called **woolsorter's disease**. Symptoms initially resemble a common cold (fever, chills, cough, chest pain, headache, and malaise). After several days, the symptoms can progress to severe breathing problems and shock. Inhalation anthrax is usually fatal without early treatment.

▶ **Intestinal Anthrax.** Consumption of endospore-contaminated and undercooked meat can lead to **gastrointestinal anthrax**. It is characterized by an acute inflammation of the intestinal tract. Initial signs and symptoms include nausea, loss of appetite, vomiting, and fever. This is followed by abdominal pain, vomiting of blood, and severe diarrhea. Intestinal anthrax results in death in 25% to 60% of untreated cases.

B. anthracis infections can be treated with specific antibiotics, although the survival rate from the disease can still be as low as 50%. At present, there is no vaccine for civilian use, although there is a cell-free vaccine for veterinarians and others who work with livestock. The combination of antibiotics and vaccine brings the survival rate to more than 70%.

B. anthracis is considered a potential bioterrorism and biological warfare agent, as described in MICROFOCUS 13.1. The seriousness of using biological agents as a means for bioterrorism was underscored in October 2001 when *B. anthracis* spores were distributed intentionally through the United States mail. In all, 22 cases of anthrax (11 inhalation and 11 cutaneous) were identified, making the case-fatality rate among patients with inhalation anthrax 45% (5/11). The six other individuals with inhalation anthrax and all the individuals with cutaneous anthrax recovered. Had it not been for antibiotic therapy, many more might have been stricken.

Tetanus Causes Hyperactive Muscle Contractions

Tetanus is one of the most dangerous human diseases. The organism is typically found in barnyard and garden soils containing animal manure.

Causative Agent and Epidemiology

Clostridium tetani, the bacterial species causing the disease, is an anaerobic, gram-positive rod that forms endospores. The United States has had a steady decline in the incidence of tetanus, with 25 cases confirmed in 2014. Older Americans are primarily affected because either they have not been immunized or they have not kept up their booster immunizations. Globally, neonatal tetanus accounts for the majority of cases and deaths, often the result of the umbilical stump becoming infected from nonsterile instruments or dressings.

Clinical Presentation

Spores enter the body through a deep puncture wound resulting from a fracture, gunshot, animal bite, a piece of glass, a thorn, or a rusty nail contaminated with soil. Like anthrax, even illicit drugs can contain spores. In dead, oxygen-free tissue of the wound, spores germinate into vegetative bacilli that produce several toxins. The most important of these is the **tetanus toxin (tetanospasmin)**. At the nerve-muscle synapse of the spinal cord or peripheral nerves, the neurotoxin prevents the release of

MICROFOCUS 13.1: History

The Legacy of Gruinard Island

In 1941, the specter of airborne biological warfare hung over Europe. Fearing that the Germans might launch an attack against civilian populations, British authorities performed a series of experiments to test their own biological weapons—including anthrax.

To test the possibility of using anthrax as a biowarfare agent, British investigators placed 80 tethered sheep on Gruinard Island, a 2-kilometer-long patch of land off the coast of Scotland. Bombs containing anthrax spores were exploded near the tethered sheep. Within days, all of the sheep were dead.

Warfare with biological weapons never came to reality in World War II, but the contamination of Gruinard Island remained. A series of tests in 1971 discovered anthrax spores still viable at and below the upper crust of the soil. Fearing they could be spread by earthworms, British officials posted signs warning people not to trespass on the island (see figure).

Then, a surprising protest occurred in 1981. Activists demanded that the British government decontaminate the island. They backed their demands with packages of soil taken from the island. One package actually contained spore-laden soil.

Partly because of the protests, the British government instituted a decontamination of the island in 1986. Technicians used a powerful brushwood killer, combined with burning and treatment with formalin in seawater. Finally, they managed to rid the soil of anthrax spores. By April 1987, sheep were once again grazing on the island. In 1990, the British government announced that the island was safe and that it was sold back to the heirs of the original owner for £500. A test in 2007 found no evidence of infection in island sheep.

Gruinard Island.

© Scotsman Publ Ltd.

neurotransmitters needed to inhibit muscle contraction. Without any inhibiting influence, volleys of spontaneous impulses arise in the motor neurons, causing uncontrolled, continuous muscle contraction (spasms).

Symptoms of tetanus intoxication develop rapidly, often within hours of exposure. A person first experiences generalized muscle stiffness, especially in the facial and swallowing muscles. Spasms of the jaw muscles are usually the first affected, causing the teeth to clench and bringing on a condition called **trismus**, or **lockjaw**. Severe cases are characterized by a "fixed smile" (risus sardonicus), and muscle spasms cause an arching of the back (**opisthotonus**; FIGURE 13.4). Spasmodic inhalation and seizures in the diaphragm and rib muscles lead to reduced ventilation, and patients often experience violent deaths through asphyxiation.

Treatment and Prevention

Patients are treated with sedatives and muscle relaxants and are placed in quiet, dark rooms because noise and bright light can trigger muscle spasms. Physicians prescribe antibiotics to destroy the organisms and, most important, inject tetanus antitoxin into a vein to neutralize the toxin.

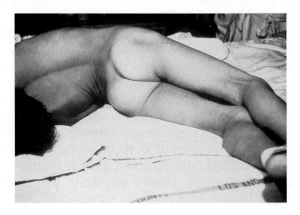

FIGURE 13.4 Opisthotonos. This patient exhibits a condition in which there is a curvature of the body posture (called opisthotonos) caused by a *Clostridium tetani* exotoxin. »» *What part of the CNS appears to be targeted in this condition?*

Courtesy of CDC.

Childhood immunizations involve injections of tetanus toxoid in the **diphtheria-tetanus-acellular pertussis (DTaP) vaccine**. Booster injections of tetanus toxoid in the **Td vaccine** (a "tetanus shot") are recommended every 10 years to keep the level of immunity to tetanus high.

Gas Gangrene Causes Massive Tissue Damage

Gangrene (*gangren* = "a sore") develops when the blood flow ceases to a part of the body, usually because of blockage by dead tissue. The body part, generally an extremity, becomes dry and shrunken, and the skin color changes to purplish or black. The gangrene can spread as enzymes from broken cells destroy other cells, which requires removal of dead, damaged, or infected tissue (**debridement**) or the body part amputated.

Gas gangrene, or **myonecrosis** (*myo* = "muscle"; *necros* = "death") is caused primarily by *Clostridium perfringens*, an anaerobic, spore-forming, gram-positive rod typically found in soil. After endospores in contaminated soil are introduced through a severe, open wound, the spores germinate, and the vegetative cells multiply rapidly in the anaerobic environment. As they grow, they ferment muscle carbohydrates and decompose the muscle proteins (thus the term "myonecrosis"). Large amounts of gas can result from this metabolism, causing a crackling sound as the gas accumulates under the skin. The gas also presses against blood vessels, thereby blocking the flow and forcing cells away from their blood supply. In the infection process, the organisms secrete at least 12 toxins. The most important is α-toxin, which damages membranes and disrupts tissues, facilitating the passage of bacterial cells into the blood (sepsis).

The symptoms of gas gangrene include a foul odor and intense pain and swelling at the wound site. Initially the body site turns dull red, then green, and finally blue black (**FIGURE 13.5**).

Treatment consists of antitoxin and antibiotic therapy as well as debridement. Amputation might be required. Hyperbaric oxygen therapy can be useful in forcing oxygen into deeper tissues, which will kill the *C. tetani* cells. However, without treatment, the disease spreads rapidly, and death frequently occurs within days of gangrene initiation.

Leptospirosis Is an Emerging Zoonotic Disease

Leptospirosis (*lepto* = "thin"; *spir* = "spiral") is the most widespread zoonotic disease in the world and

FIGURE 13.5 Gas Gangrene of the Hand. A severely infected hand showing gangrene (blackened tissue necrosis). This gangrene developed from an infection from an accident while the patient was scaling fish. In this advanced stage, amputation of the hand may be necessary. Inset: Artist sketch of *C. perfringens* cells seen with the light microscope at 1000X. *»» What properties of C. perfringens would cause the tissue necrosis?*

Courtesy of Dr. Jack Poland/CDC.

represents a reemerging disease of significant public health concern.

Causative Agent and Epidemiology

The agent of leptospirosis is *Leptospira interrogans*, a thin, aerobic, gram-negative, motile spirochete with a hook at one end resembling a question mark; hence the name interrogans (*roga* = "ask") (**FIGURE 13.6**). The undulating movements, caused by contractions of endoflagella, allow the cell to burrow into tissue.

There are more than 500,000 cases of leptospirosis reported each year, and the mortality rate can be as high as 10%. Infections are highest in warm, subtropical regions such as Southeast Asia and South America. This is of particular concern to adventure travelers who travel to these exotic locales, as described in MICROFOCUS 13.2. There are usually between 40 and 100 cases of leptospirosis reported each year in the United States (50% of cases occur in Hawaii).

Household dogs and cats as well as barnyard and wild animals and marine mammals can become infected with *L. interrogans*. In infected animals, the spirochetes can colonize the kidney tubules without causing disease. The bacterial cells are excreted in the urine to the soil or water where they can remain viable in a biofilm for several weeks. Humans acquire the pathogen by direct contact with these animals

MICROFOCUS 13.2: Public Health/Environmental Microbiology

A Real Eco-Challenge!

It started as a headache on the plane back from Borneo to the United States. Within 3 days, Steve went to a hospital emergency room in Los Angeles complaining of fever and chills, muscle aches, vomiting, and nausea.

The Eco-Challenge in Sabah, Borneo, was the site for the annual adventure race. Some 304 participants composing 76 teams from 26 countries competed in the 10-day endurance event, which was designed to push participants and their teams beyond their athletic limits. During the Eco-Challenge, teams would kayak on the open ocean, mountain bike into the rainforest, spelunk in hot caves, and swim in local rivers (see figure). Of the 76 teams starting the challenge, 44 teams finished.

A river in Borneo.

© Andrea Seemann/Shutterstock.

About the same time that Steve arrived at the emergency room, the Centers for Disease Control and Prevention (CDC) in Atlanta received calls from the Idaho Department of Health, the Los Angeles County Department of Health Services, and the GeoSentinel Network (a network of international travel clinics) reporting cases of a febrile illness similar to Steve's symptoms.

The CDC quickly carried out a phone questionnaire to 158 participants in the Eco-Challenge. Many reported symptoms similar to Steve's, including chills, fever, headache, diarrhea, and conjunctivitis. Twenty-five respondents had been hospitalized. Within a few days, antibiotic therapy had Steve recovering. In fact, all 135 affected participants completely recovered.

The similar symptoms suggested leptospirosis, and laboratory tests either confirmed the presence of *Leptospira* antibodies or identified the organism in culture from serum samples collected from ill participants.

To identify the source and the exposure risk, information was gathered from participants about various portions of the racecourse. Analysis identified swimming in and kayaking on the Sagama River as the probable source.

Several participants who did not become ill had taken antibiotics as a preventive for malaria and leptospirosis, as race organizers had advised. Unfortunately, Steve had not heeded those words and suffered a real "eco-challenge." Asked if he would participate in the next Eco-Challenge, Steve said "Heck yes! It just adds to pushing the limits."

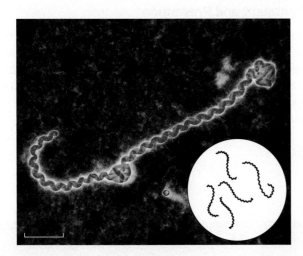

FIGURE 13.6 *Leptospira interrogans.* A false-color transmission electron micrograph of an *L. interrogans* cell, the agent of leptospirosis. Note the tightly coiling spirals. (Bar = 2 μm.) Inset: Artist sketch of *L. interrogans* cells seen with the light microscope at 1000X. *»» Besides the coil shape, what are the other notable features of this bacterial species?*

© Eye of Science/Science Source.

or indirectly from soil, food, or water contaminated with the urine from the infected animals.

Clinical Presentation

If a person swims or wades in contaminated water, *L. interrogans* can enter the human body through the mucus membranes of the eyes, nose, and mouth or through skin abrasions. In this first phase, the bacterial cells multiply rapidly, and 90% of infected individuals experience a vague febrile illness with muscle aches and eye inflammation. Episodes of fever and chills occur for 4 to 9 days but then disappear.

Depending on the serotype of infecting *L. interrogans* and the level of exposure, a more severe, second phase of leptospirosis might develop. In this phase, about 10% of infected individuals experience a short recovery period as the spirochetes directly invade and infect various organs and the immune system reacts to the infection. Bacteria would now

TABLE 13.1 A Summary of Soilborne Bacterial Diseases

Disease	Causative Agent	Signs and Symptoms	Transmission	Treatment	Prevention
Anthrax	*Bacillus anthracis*	Fever, chills, cough, chest pain, headache, and malaise Severe breathing and shock can develop	Direct contact with endospores	Penicillin Ciprofloxacin	Avoiding contact with infected livestock and animal products
Tetanus	*Clostridium tetani*	Muscle stiffness in jaw and neck, trismus	Direct contact with spores through puncture wound	Tetanus antitoxin Penicillin	Immunizing with toxoid vaccine
Gas gangrene	*Clostridium perfringens*	Foul odor and intense pain and swelling at the wound site	Direct contact with spores through puncture wound	Debridement Surgery Cephalosporin Hyperbaric oxygen	Cleaning wounds Debridement
Leptospirosis	*Leptospira interrogans*	Acute headache Muscle aches Vomiting and nausea Fever and chills	Exposure to contaminated soil or water	Antibiotics	Avoiding contaminated soil and water

be present in the urine. A fever returns, and there will be a persistent yellowing of the skin and whites of the eyes (jaundice). Aseptic meningitis can develop and last a few days or up to 2 weeks. Most important, inflammation can occur in the liver and lungs, leading to liver and kidney failure. The illness can last from a few days to more than 3 weeks. Without treatment, recovery can take several months, and complicated cases can be life threatening; in fact, the second phase carries a mortality rate of 5% to 10%.

Treatment and Prevention

Treatment of severe leptospirosis should be started as soon as possible. Despite the numerous tissues infected, antibiotic therapy provides effective treatment. Prevention involves prophylactic doxycycline for high-risk individuals and public health awareness of the occupational and recreational risks.

TABLE 13.1 summarizes the soilborne bacterial diseases.

Concept and Reasoning Checks 13.1

a. What cellular factors make *B. anthracis* a dangerous pathogen?
b. What are the definitive symptoms of tetanus?
c. What characteristic of *C. perfringens* is being "attacked" by placing a gas gangrene patient in a hyperbaric chamber?
d. What bacterial and host factors make *L. interrogans* a potent pathogen?

Chapter Challenge A

True soil-dwelling organisms are called euedaphic (*eu* = "true"; *edaph* = "soil"). In this first section of the chapter, we have discussed the bacterial species that are the causative agents for anthrax, tetanus, gas gangrene, and leptospirosis.

QUESTION A: *Which of these species represents a euedaphic pathogen, and which species is/are soil-transmitted pathogens?*

You can find answers online in **Appendix F**.

■ KEY CONCEPT 13.2   Some Bacterial Diseases Can Be Transmitted by Arthropods

A living organism such as an arthropod that transmits disease agents is called a **vector** (*vect* = "carried"). Arthropods transmit diseases to humans usually by taking a blood meal from an infected animal and themselves becoming infected. Then, they pass the infectious agents to another individual during the next blood meal. Arthropod-borne diseases occur primarily in the bloodstream, and they often are characterized by a high fever and a body rash.

Plague Can Be a Fatal Disease

Few diseases have had as rich and terrifying a history as the bubonic plague or can match the array of social and economic changes resulting from this disease, as recounted in the chapter introduction. The Justinian Plague (541–542) and the Black Death (1348–1351) that swept through Europe, North Africa, and parts of Asia killed somewhere between 125–300 million people (**FIGURE 13.7**).

Causative Agent and Epidemiology

Plague is caused by *Yersinia pestis* (*pestis* = "plague") (**FIGURE 13.8A**). This facultative, gram-negative rod stains heavily at the poles of the cell, giving it a safety-pin appearance and a characteristic called **bipolar staining** when direct smears from infected specimens are observed.

Plague remains a threat in endemic areas of the world (Africa, America, and Asia) where small mammals are infected. Between 2010 and 2015 the WHO reported more than 3,200 cases of human plague, including 584 deaths. The Centers for Disease Control and Prevention (CDC) reported 32 cases and no deaths in the United States during this same time period.

Urban plague refers to disease among rodents that live in close association with humans in urban areas. Usually, human cases of urban plague are transmitted from indoor/outdoor pets harboring infected fleas passed from rodents. The last urban case in the United States was in Los Angeles in 1924. **Sylvatic plague** occurs in rural wildlife and accounts for most American and global cases today.

In all cases of plague, if infected rodents are bitten by fleas, the plague bacilli form a biofilm in the fleas' digestive system (**FIGURE 13.8B**). When the fleas attempt to take a blood meal, the biofilm blocks the fleas' digestive system and the arthropods soon starve. This causes the fleas to become even more voracious in finding a blood meal. Then,

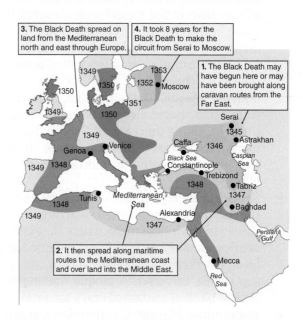

FIGURE 13.7 The Spread of Bubonic Plague (Black Death) in 14th Century Europe. Over a period of 8 years, the plague bacillus spread clockwise from the Black Sea through Europe to Moscow. »» *If such an outbreak occurred today, would it spread faster or slower than in the 14th century? Explain.*

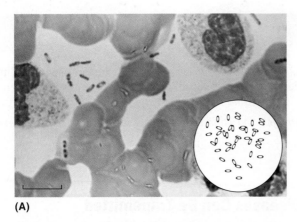

(A) **(B)**

FIGURE 13.8 *Yersinia pestis* **and the Flea Vector. (A)** This light micrograph shows the bipolar staining of the bacterial cells. (Bar = 1 μm.) Inset: Artist sketch of *Y. pestis* cells seen with the light microscope at 1000X. **(B)** *Xenopsylla cheopis* (oriental rat flea) with clotted *Y. pestis* mass (red foregut). *»» How does clotting in the foregut lead to human infections?*

(A and B) Courtesy of CDC.

as septicemic rats die, the fleas might jump to another animal, such as humans, in a futile attempt to feed. In this process, bacterial cells break free of the biofilm, and the cells are regurgitated into the human bloodstream during an attempted flea bite.

Clinical Presentation

There are three major forms of plague:

▶ **Bubonic Plague.** In an infection, *Y. pestis* cells multiply in the submucosa and eventually localize in the lymph nodes, especially those of the armpits, neck, and groin. Hemorrhaging in the lymph nodes causes painful and substantial swellings called **buboes** (*bubon* = "the groin") and thus the name **bubonic plague.** Dark, purplish splotches from hemorrhages also can be seen through the skin, giving this form of the disease the name **Black Death.** Blood pressure drops, and, without treatment, mortality reaches 60%.

▶ **Septicemic Plague.** From the lymph nodes, the bacilli in an infected individual can spread to the bloodstream where they cause **septicemic plague.** Symptoms include high fever, diarrhea, and abdominal pain and the septicemia can lead to **plague meningitis.** Nearly 100% of untreated cases are fatal.

▶ **Pneumonic Plague.** During historical epidemics, human-to-human transmission of plague bacilli often occurred by respiratory droplets when septicemic infections progressed to the lungs. This highly contagious form of the disease therefore is called **pneumonic plague.** Lung symptoms are similar

to pneumonia, with headache, malaise, and extensive coughing. Hemorrhaging and fluid accumulation also are common. Many individuals suffer cardiovascular collapse with death common within 2 to 3 days of symptom onset. Mortality rates for pneumonic plague are near 100%. How the plague bacilli overcome the lung immune defenses is examined in **INVESTIGATING THE MICROBIAL WORLD 13**.

Treatment and Prevention

When detected early, plague can be treated with specific antibiotics, reducing mortality to less than 10%. A vaccine consisting of dead *Y. pestis* cells is available for high-risk groups.

Tularemia Has More Than One Disease Presentation

Tularemia is one of several microbial diseases first identified in the United States (others include St. Louis encephalitis, Rocky Mountain spotted fever, Lyme disease, and Legionnaires' disease). The CDC reported more than 240 cases in 2015.

The causative organism of tularemia is *Francisella tularensis*, a very small, aerobic, gram-negative rod that often, like *Y. pestis*, exhibits bipolar staining. It is extremely virulent because less than 50 cells can trigger disease onset. Tularemia is a zoonotic disease that can be transmitted by a broad variety of wild animals, especially rodents, and it is particularly prevalent in rabbits (in which case it is known as **rabbit fever**). Dogs can acquire the organisms during romps in the woods. Humans are incidentally infected by contact with an infected animal

Investigating the Microbial World 13

The Lungs as Microbial Playgrounds

When the human lungs become infected with a bacterial pathogen, the usual response is for the immune system to mount an immediate attack in an attempt to rid the infection. And, as healthy adults, should we encounter one of those pathogens, the immune system usually works fine to generate a restrictive environment in which the pathogens cannot survive—and are eliminated.

OBSERVATIONS: Infection of the lungs with *Yersinia pestis*, the agent of plague, produces a different result. The infection does not trigger an immediate immune response as detected by the lack of inflammation and disease symptoms. Rather, the plague bacilli can grow unmolested for more than 36 hours before an immune response is mounted. Somehow, the bacterial cells have produced a permissive environment in which they have ample time to grow and multiply to levels that the immune system usually cannot handle, and pneumonic plague will result.

QUESTION: *Why does* Y. pestis, *unlike other pathogens, not trigger early inflammation?*

HYPOTHESIS: The lack of an inflammatory response is because the plague bacilli suppress the early immune response, making the otherwise restrictive environment into a permissive one. If so, coinfecting the lungs with *Y. pestis* and other microbes will also produce a permissive environment in which those coinfected microbes, which are usually destroyed, can grow.

EXPERIMENTAL DESIGN: Rats were used as experimental organisms and were infected or coinfected intranasally with *Y. pestis, Y. pestis* mutants, or other bacterial species. Bacterial growth was then determined by counting the numbers of bacterial cells (as colony-forming units [CFUs]) in lung tissue.

EXPERIMENT 1: Separate groups of rats were infected or coinfected (inoculated) with 10^4 CFU of wild type (wt) *Y. pestis* and a nonpathogenic *Y. pestis* mutant (mut) that is normally destroyed in the lungs. The CFUs per lung were determined at 24, 48, and 60 hours postinoculation.

EXPERIMENT 2: Separate groups of rats were infected or coinfected with 10^4 CFU of wild type *Y. pseudotuberculosis* or *Klebsiella pneumoniae*, both of which generate lung infections similar to *Y. pestis*, or nonpathogenic mutants of these species that again normally are destroyed in the lungs. The CFUs per lung were determined at 48 hours post inoculation.

RESULTS:

EXPERIMENT 1: *see Figure A.*

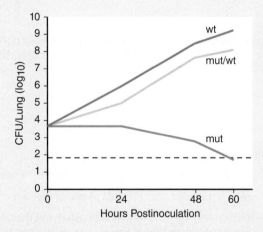

CFUs formed by wild type (wt) and a nonpathogenic mutant (mut) of *Y. pestis* and mut CFUs formed from a coinfection with wt (mut/wt).

Data from Price, P. A., Jin, J., and Goldman, W. E. 2012. *Proc Natl Acad Sci* USA 109(8):3083–3088.

(continues)

Investigating the Microbial World 13 (*Continued*)

The Lungs as Microbial Playgrounds

EXPERIMENT 2: *see Figure B.*

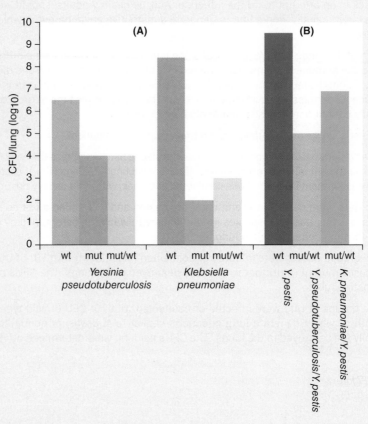

(A) CFUs formed by wt and mut strains of *Yersinia pseudotuberculosis* and *Klebsiella pneumoniae* and mut CFUs formed from a coinfection with the wt species. **(B)** CFUs formed by wt *Y. pestis* and by coinfection with mut *Y. pseudotuberculosis* and mut *K. pneumoniae* stains.

Data from Price, P. A., Jin, J., and Goldman, W. E. 2012. *Proc Natl Acad Sci* USA 109(8):3083–3088.

CONCLUSIONS:

QUESTION 1: *Was the hypothesis validated? Explain.*

QUESTION 2: *What information can you deduce from the results of experiment 1 (Figure A)?*

QUESTION 3: *What additional information is provided from the data obtained from experiment 2 (Figure B)?*

Y. pestis and *Y. pseudotuberculosis* are evolutionarily very closely related. Studying the genetic and molecular differences between these two species may identify the mechanism that *Y. pestis* uses to produce the permissive environment.

You can find answers online in **Appendix E**.

Modified from Price, P. A., Jin, J., and Goldman, W. E. 2012. *Proc. Natl. Acad. Sci* USA 109(8):3083–3088.

or by contact with infected ticks, biting flies, or mosquitoes. Other methods of transmission include consumption of infected rabbit meat, drinking contaminated water, or inhaling infectious aerosols.

Various forms of tularemia exist, depending on where the bacilli enter the body. A tick bite, for example, can lead to swollen lymph nodes (**glandular tularemia**) and a skin ulcer at the bite site (**FIGURE 13.9**). Individuals typically experience flu-like symptoms. Inhalation of *F. tularensis* cells can lead to respiratory disease (**inhalation tularemia**) and produce swollen lymph nodes, a dry cough, and pain under the breastbone.

Tularemia usually resolves on treatment with the antibiotic streptomycin, and few people die of the disease. Epidemics are unknown, and evidence suggests that tularemia is not communicable among humans despite the many modes of entry into the body. Avoiding infected animals, aerosols, and ticks are the best prevention methods.

Lyme Disease and Tickborne Relapsing Fever Are Transmitted by Spirochetes

Besides the spirochete that causes leptospirosis (discussed earlier in this chapter), other spirochetes, specifically in the genus *Borrelia*, are associated with human disease. The organisms often are found in damp soil, mud, and sewage-polluted water. The cells are slender, tightly coiled, gram-negative bacteria that can be quite long. The cells exhibit a twisting motility caused by endoflagella that run the length of the cell and are located in the periplasmic space between the bacterial cell membrane and outer membrane.

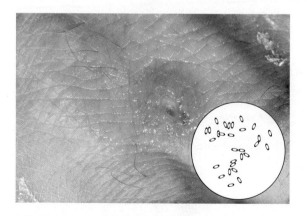

FIGURE 13.9 A Tularemia Lesion. The lesions of tularemia occur where the bacilli enter the body. Inset: Artist sketch of *F. tularensis* cells seen with the light microscope at 1000X.
»» What are the symptoms associated with lesion formation?

Courtesy of Dr. Brachman/CDC.

Lyme Disease

Borrelia burgdorferi is the causative agent of **Lyme disease**, currently the most commonly reported arthropod-borne illness in the United States. About 95% of cases occur in the northeastern, Mid-Atlantic, and north central states (**FIGURE 13.10**). More than 33,000 Lyme disease cases were reported in 2014.

The arthropod that transmits most cases of Lyme disease in the Northeast and Midwest is the deer tick *Ixodes scapularis*; in the West, the major vector is the western blacklegged tick *I. pacificus*. The 2-year life cycle of the ticks transmitting Lyme disease is shown in **FIGURE 13.11**. When an infected tick takes a blood meal, it also defecates into the wound, transmitting the spirochetes. Therefore, if the tick is observed on the skin, it should be removed with forceps or tweezers, and the area thoroughly cleansed with soap and water before applying an antiseptic.

Some 20% of infected individuals experience nothing more than flu-like symptoms. For others, Lyme disease has a variable incubation period of 3 to 31 days and, untreated, typically has three stages. The **early localized stage** involves fever, fatigue, and a slowly expanding red rash at the site of the tick bite. The rash is called **erythema** (red) **migrans** (expanding), or **EM**. Beginning as a small flat or raised lesion, the rash increases in diameter in a circular pattern over a period of weeks, and sometimes reaches a diameter of 10 to 15 inches (**FIGURE 13.12**). With an intense red border, it is termed a **bull's-eye rash**. It can vary in shape and is usually hot to the touch, but it need not be present in all cases of disease. Fever, aches and pains, and flu-like symptoms usually accompany the rash. The tick bite can be distinguished from a mosquito bite because the latter itches, whereas a tick bite does not. The reason is explained in MICROFOCUS 13.3.

Left untreated, an **early disseminated stage** begins weeks to months later with the spread of *B. burgdorferi* to the skin, heart, nervous system, and joints. On the skin, multiple smaller EMs develop, and invasion of the nervous system can lead to meningitis, facial palsy, and peripheral nerve disorders. Cardiac abnormalities are the most common, such as brief, irregular heartbeats. Joint and muscle pain also occur.

If still left untreated, a **late stage** occurs months to years later. About 10% of patients develop chronic arthritis with swelling in the large joints, such as the knee. Although Lyme disease is not known to have a

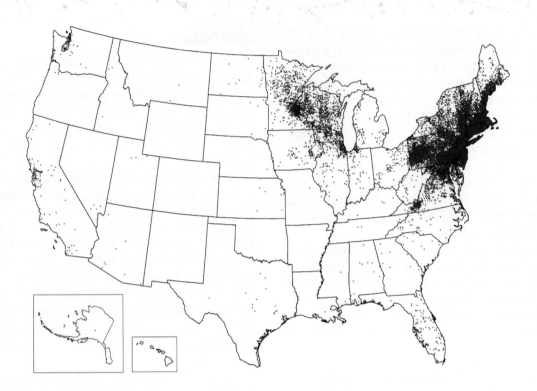

FIGURE 13.10 Reported Cases of Lyme Disease—by County of Residence. Although cases of Lyme disease have been reported in most every state, the vast majority of cases are reported from southern Maine to northern Virginia and from Wisconsin, northern Minnesota, and northern Illinois. »» *From the map, what types of habitats seem to be preferred for endemic Lyme disease?*

Courtesy of CDC.

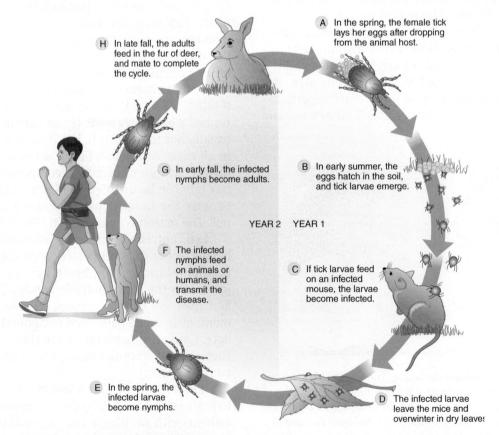

FIGURE 13.11 Life Cycle of the Blacklegged Ticks. The life cycle of *Ixodes scapularis* and *I. pacificus,* the ticks that transmit Lyme disease. »» *Why is Lyme disease called a zoonotic disease?*

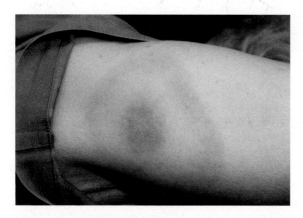

FIGURE 13.12 Erythema Migrans. Erythema migrans is the rash accompanying 80% of cases of Lyme disease. The rash consists of a large patch with an intense red border. It is usually hot to the touch, and it expands with time. *»» What is the common name for the rash?*

Courtesy of James Gathany/CDC.

high mortality rate, the overall damage to the body can be substantial.

Effective treatment depends on the stage of infection, but it can usually be treated with specific antibiotics. Avoiding ticks and wearing protective clothing are preventative measures. A vaccine for dogs (LymeVax®) has been licensed and is in routine use. No vaccine currently is available for humans.

Tickborne Relapsing Fever

Another spirochete, *B. hermsii*, is responsible for **tickborne relapsing fever** (**TBRF**), which is endemic to the mountains in the western United States, with 70% of the cases occurring in California, Colorado, and Washington. Cabins in these wilderness areas are favorable nesting sites for infected rodents and their ticks, especially rustic cabins where rodents have access, as **CLINICAL CASE 13** describes.

An individual often is bitten briefly (5 to 20 minutes) at night by an infected tick. The individual then goes through febrile periods of 3 to 5 days with relapses every 5 to 7 days as the spirochetes increase and decline in number in the blood (**FIGURE 13.13**). With this tickborne form of relapsing fever, up to 13 relapses can occur and, untreated, mortality can be 2% to 5%. A more severe form of TBRF was first identified in the northeastern United States in 2013. The bacterium, *B. miyamotoi*, is carried by deer ticks, and the disease often causes sepsis, which requires hospitalization.

TBRF can be treated with antibiotics to hasten recovery.

MICROFOCUS 13.3: Evolution

But It Doesn't Itch!

How many of us at some time have had a small red spot or welt on our skin that looks like an insect bite—but it didn't itch? When most of us are bitten by an insect, such as an ant, bee, or mosquito, the site of the bite begins itching almost immediately. Yet, there are cases in which bites do not itch, such as a tick bite. So, what's the difference?

Most insect bites trigger a typical inflammatory reaction. Take a mosquito as an example. When it bites and takes a blood meal, saliva injected into the skin contains substances so the blood does not coagulate. The bite causes tissue damage and activation of chemical mediators that trigger the typical characteristics of an inflammatory response: redness, warmth, swelling, and pain at the injured site. One of the major chemical mediators is histamine, released by the damaged tissue and immune cells. An itch actually is a form of pain and arises when histamine affects nearby nerves. Itching is the result.

A tick feeding head-down in human skin.

© IanRedding/Shutterstock.

In a tick bite, the scenario is slightly different. The bite triggers the same set of inflammatory reactions, and histamine is produced. However, in the tick saliva is another molecule called histamine-binding protein (HBP). HBP binds to histamine, preventing the mediator from affecting the nearby nerves—and no itching occurs.

So why is that an advantage for the tick? Probably because ticks take long blood meals that sometimes can last for days. By the bite being painless and not itching, there is less chance of the tick being discovered and removed.

Clinical Case 13

An Outbreak of Tickborne Relapsing Fever

In late July, a family reunion was planned at a remote, previously uninhabited cabin in the mountains of northern New Mexico. Three days before the reunion, three family members arrived to clean the cabin. After the reunion, about half of the 39 family members slept in the cabin overnight.

Four days after the reunion, one of the family members who cleaned the cabin arrived at a local hospital complaining of fever, chills, muscle aches, and a rash on the forearms. The symptoms had started 2 days previously.

A hospital lab technician took a blood sample from the patient and identified spirochetes on a blood smear (see figure). The immediate diagnosis was tickborne relapsing fever (TBRF). On August 2, the New Mexico Department of Health and the Indian Health Services were notified.

By August 7, 39 family members had sought medical care or they had been visited by a public health nurse. A study was begun to determine risk factors for infection, better understand the outbreak, and provide prevention measures.

A case of TBRF was defined. Based on the definition, 14 family members suffering the illness were identified.

Blood samples were taken from all 14 symptomatic patients. Spirochetes were identified in the blood smears of nine patients.

Samples from 13 patients were sent to the Centers for Disease Control and Prevention (CDC) in Atlanta. Eleven samples, including two samples negative for spirochetes by blood smear, were cultured and all grew *Borrelia hermsii*.

All 14 patients received antibiotic therapy and recovered.

Risk analysis revealed that all three family members who cleaned the cabin and eight of the other 36 were more likely to be TBRF patients.

Inspection of the cabin by Indian Health Service workers identified abundant rodent nesting materials and droppings within the walls of the cabin.

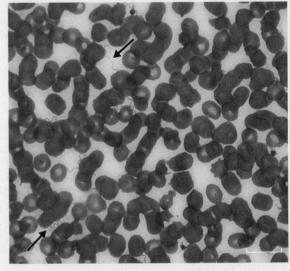

Tickborne relapsing fever spirochetes (arrows). (Bar = 8 μm.)

© Science Source.

Questions

a. Why was such a prompt diagnosis and notification of health services necessary?
b. How would you define a case of TBRF?
c. Why were spirochetes only found in 9 of the 14 patients?
d. Why did all three who cleaned the cabin become ill?
e. Of the other eight patients, what is most likely the reason for them contracting the disease?

You can find answers online in **Appendix E**.

For additional information, see www.cdc.gov/mmwr/preview/mmwrhtml/mm5234a1.htm.

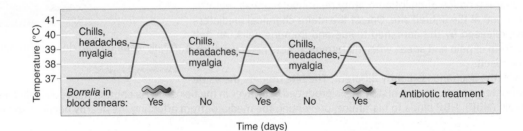

FIGURE 13.13 The Cycles of Relapsing Fever. In relapsing fever, chills, headache, and fever peak with high numbers of *Borrelia* spirochetes in the blood. »» *During which part of each cycle would specific antibodies against* Borrelia *be produced?*

Concept and Reasoning Checks 13.2

a. Explain how bubonic plague can develop into communicable pneumonic plague.
b. Describe the three clinical stages of untreated Lyme disease.
c. Explain the reason that infected individuals experience relapsing fevers.

Chapter **Challenge B**

Although the arthropod-borne pathogens discussed in this section are transmitted by fleas (plague) or ticks (tularemia, Lyme disease, and tickborne relapsing fever), they still include soil involvement.

QUESTION B: *Which of these arthropod-borne pathogens could be considered euedaphic pathogens and which are soil-transmitted pathogens based on the vector used to transmit the pathogen?*

You can find answers online in **Appendix F.**

■ KEY CONCEPT 13.3   Rickettsial and Ehrlichial Diseases Are Spread by Arthropods

In 1909, Howard Taylor Ricketts, a University of Chicago pathologist, described a previously unknown organism in the blood of patients with Rocky Mountain spotted fever and showed that ticks transmit the disease. A year later, he located a similar organism in the blood of animals infected with Mexican typhus and discovered that fleas were the important vectors in this disease. Unfortunately, in the course of his work, Ricketts fell victim to the disease and died. When later research indicated that Ricketts had described a unique group of microorganisms, the name "rickettsias" was coined to honor him.

Rickettsial Infections Often Involve a Characteristic Rash

The rickettsias are very small, gram-negative, obligate, intracellular parasites. Most infections are transmitted by ticks, lice, or fleas, and all illnesses can be treated effectively with the antibiotic doxycycline or chloramphenicol.

Rocky Mountain Spotted Fever (RMSF)

Contrary to its name, **Rocky Mountain spotted fever (RMSF)** is not commonly reported in western states any longer but is a problem in Oklahoma, Arkansas, Texas, and many southeastern and Atlantic Coast states. RMSF is caused by *Rickettsia rickettsii*

(**FIGURE 13.14A**) and is transmitted by hard ticks. Children are its primary victims because of their contact with ticks in wooded areas.

Following a bite from an infected tick, the hallmarks of RMSF are a high fever lasting for many days, severe headaches, and a skin rash reflecting damage to the small blood vessels. The rash begins as pink spots (macules) and progresses to pink-red papules (**FIGURE 13.14B**). Where the spots fuse, they form a flat area, forming a so-called **maculopapular** rash, which becomes dark red and then fades without evidence of scarring. As the disease progresses, the rash appears on the palms of the hands and soles of the feet and progressively spreads to the body trunk. Mortality rates of untreated cases are about 30%. Antibiotic treatment reduces this rate significantly, and prevention involves prompt tick removal.

Epidemic Typhus

One of the most notorious of all bacterial diseases is **epidemic typhus (typhus fever)**. It is caused by *Rickettsia prowazekii*, which is transmitted to humans by body lice where hygiene is poor. The rickettsias are excreted in feces from the lice, so scratching the bite will facilitate infection into the wound (rather than through the bite itself). Following an incubation period of 1 to 3 weeks, the characteristic fever and rash are particularly evident. The maculopapular

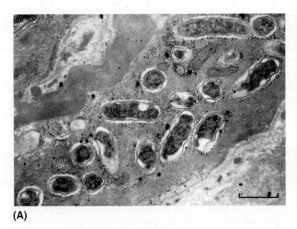 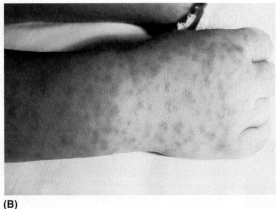

(A)   (B)

FIGURE 13.14 *Rickettsia rickettsii* **and Symptoms of Rocky Mountain Spotted Fever. (A)** *R. rickettsii* in the cytoplasm of a kidney cell. (Bar = 0.5 μm.) **(B)** Child's right hand and wrist displaying the characteristic spotted rash of RMSF. *»» What is happening in the host to produce a spotted rash?*

(A) Courtesy of Billie Ruth Bird/CDC. (B) Courtesy of the CDC.

rash, unlike the rash of RMSF, appears first on the body trunk and progresses to the extremities. Intense fever, sometimes reaching 40°C (104°F), remains for many days as the patient hallucinates and becomes delirious. Some patients suffer permanent damage to the blood vessels, heart, kidney, and lungs, and over 75% of sufferers die in epidemics. Again, antibiotic therapy reduces this percentage substantially, and the disease is rare in the United States.

Murine Typhus

The agent of **murine (endemic) typhus** (*murin* = "of the mouse") is *Rickettsia typhi* and the disease is prevalent in rodent populations where fleas abound

(such as rats and squirrels). Cats and their fleas are involved also, and lice can harbor the bacilli. Most recent cases have occurred in south Texas.

When an infected flea feeds on the human skin, it deposits the rickettsias into the wound. Symptoms include high fever, persistent headache, chills, nausea, and a maculopapular rash that spreads from trunk to extremities. Often the recovery is spontaneous, without the need of antibiotic therapy. However, lice can transport the rickettsias to other individuals and initiate an epidemic.

Having covered several arthropod-borne bacterial diseases, **MICROINQUIRY 13** describes an arthropod-borne disease called "war fever." See if you can identify the infectious disease from the historical details.

MICROINQUIRY 13

Dreams of Conquest Stopped by a Microbe

In the spring of 1812, Napoleon Bonaparte was at the height of his power. He was emperor to 45 million French and, by conquering much of Europe, had made France the wealthiest and most influential empire on the European continent. France's power was based on its Grande Armée, against which no opposing army could match. With such power, Napoleon set out to expand his European conquests by invading Russia. On June 22, 1812, Napoleon's troops entered Poland and began the march into Russia with 600,000 soldiers. Napoleon was confident that Tsar

Alexander's outnumbered Russian troops would collapse when faced with the Grande Armée. However, a third, but microscopic, enemy was lurking on the Russian frontier. This virulent and swift microscopic foe, this "war fever" bug, along with battle wounds and weather, would take the lives of all but a few thousand soldiers.

Once in Poland, progress became difficult for Napoleon's army as the roads were poor and getting needed supplies of food and water to the front lines was increasingly difficult. Because the Russians had destroyed much

Napoleon's retreat from Russia left many dead from "war fever." Drawing by Raffet and engraved by J Smith.

© Hulton Archive/Stringer/Getty Images.

of Poland in their retreat, Napoleon's troops were forced to search for food in the countryside, pillaging from the peasants who lived in filthy conditions and who were infested with bed bugs, fleas, and lice. Soon, many soldiers began to develop high fevers and pink rashes that developed on the body trunk and spread to the extremities. Some developed hallucinations and became delirious only to then feel like their heart had briefly stopped. The virulent microbe had begun its attack on the Grande Armée.

Within the first month of the campaign, Napoleon had lost more than 80,000 soldiers to the infection and, along with dysentery, the mortality and morbidity numbers soon reduced the Grande Armée to 130,000 soldiers. Napoleon's doctors could not cope with the magnitude of the disease and were powerless to explain what had struck down so many soldiers. Perhaps it was a miasma, as was a common belief for disease spread at the time.

Napoleon's forces entered Moscow in September with only 100,000 men. In Moscow, most Russian citizens had fled the city with their belongings and deliberately set fire to the city on their retreat. There was little habitable shelter in Moscow and no food—and Tsar Alexander refused to surrender. After another 10,000 soldiers died from "war

fever," Napoleon's remaining 90,000 soldiers began the 1,000-mile retreat as the storms of the Russian winter approached. The filthy, dirty clothing of the soldiers would be a warm home for the lice carrying the disease.

After infection by the "war fever" microbe, a soldier experienced a high fever that continued for some 2 weeks, during which time symptoms might also include severe headaches, bronchial disturbances, and mental confusion. After about 6 days, a red rash appeared on the torso and then on the hands, feet, and face. The soldiers also slept in large groups in confined spaces, and here the lice could quickly move to uninfected soldiers, spreading the disease. Soldiers would wake from sleeping with a horrendous tingling as lice covered and crawled over their bodies.

By the end of December, the remnants of the Grande Armée finally reached France. It is estimated that of the original 600,000 soldiers, more than 50% of the deaths might have been solely from "war fever." It should be mentioned that the Russian army also suffered from malnutrition, exposure, dysentery, and "war fever." Estimates suggest that more than 100,000 Russian soldiers died from wounds and disease. The actual number probably was much higher.

In the end, the clash between two imposing and remarkable forces (human and microbe) meeting on the road to Moscow resulted in a microbial victory. French military strength would never fully recover and Napoleon's illustrious career was crushed, partly the result of microbes. Then, the microbe disappeared—and by 1814 it was no longer a major epidemic killer in armed conflict.

From the description of the battle and the symptoms of infection, what was the disease that caused "war fever"?

Discussion Point

Why did "war fever" become an explosive epidemic in the Russian campaign, killing tens of thousands, if not hundreds of thousands of troops and citizens? With filthy clothing and uniforms, how did the lice transmit the "war fever" microbe?

You can find answers (and comments) online in **Appendix E**.

If you want to read the entire story, check out a copy of Stephan Talty's *The Illustrious Dead: The Terrifying Story of How Typhus Killed Napoleon's Army.* New York: Crown Publishers, 2009.

Other Tickborne Zoonoses Are Emerging Diseases in the United States

Ehrlichiosis and **anaplasmosis** are now recognized as two similar rickettsial tickborne diseases. **Human monocytic ehrlichiosis (HME)** is caused by *Ehrlichia chaffeensis* and **human granulocytic anaplasmosis (HGA)** is caused by *Anaplasma phagocytophilum*. Because *E. chaffeensis* reproduces in the body's monocytes (hence "monocytic") and *A.*

phagocytophilum reproduces in neutrophils (a type of granulocyte, hence "granulocytic"), both bring about a lowering of the white blood cell count (**leukopenia**).

The CDC reported almost 1,500 cases of HME and 2,800 cases of HGA in 2014. However, this might be an underestimate because identification and reporting of human cases are incomplete at the state level. HME cases predominate in New York, North Carolina, and the central Midwest, whereas

(A)   (B)

FIGURE 13.15 Reported Cases of Tickborne Ehrlichiosis and Anaplasmosis—2013. (A) Human monocytic ehrlichiosis (HME) is primarily found in the Midwest, South, and Northeast regions. **(B)** Human granulocytic anaplasmosis (HGA) is primarily found in the Upper Midwest and New England. Both diseases are reported as the number of cases by county for 2013. *»» Compare the map for anaplasmosis (B) with that for Lyme disease (Figure 13.10). What conclusions can you draw about these tickborne diseases?*

(A and B) Courtesy of CDC.

HGA cases predominate in the upper Midwest and Northeast (**FIGURE 13.15**).

HME is transmitted by the lone star tick (prevalent in the South), whereas HGA is transmitted by the blacklegged tick, which also transmits Lyme disease (prevalent in the Northeast). Patients with HME or HGA suffer from headache, malaise, and high fever, with some liver disease and, infrequently, a maculopapular rash. Indeed, both HME and HGA are quite similar to Lyme disease, except the symptoms come on faster in HGA. With HME, the symptoms clear more quickly, and a rash is infrequent.

Specific antibiotics can be used for both diseases.

TABLE 13.2 summarizes the systemic bacterial diseases.

In conclusion, we have learned about several soilborne and arthropod-borne bacterial diseases. As you might have noticed, many of the arthropod-borne diseases have a similar clinical presentation. Of the *Rickettsia*-caused diseases, RMSF is the most common and well known of the spotted fevers. However, other spotted fevers have been recognized in people for more than a hundred years. Among these are Mediterranean spotted fever, Helvetica spotted fever, and others that occur in Europe, Asia, and Africa. Often these spotted fevers are less pathogenic than RMSF but can be confused with RMSF clinically. Therefore, the CDC has combined cases of RMSF and other spotted fevers under a new category called "spotted fever rickettsiosis." Typhus is a group of similar diseases also caused by species in the genus *Rickettsia*. However, spotted fevers and typhus are different clinically, so epidemic typhus and murine typhus are not included as part of the spotted fever rickettsiosis group.

Concept and Reasoning Checks 13.3

a. What distinguishes the different rickettsial diseases from one another?

b. Draw a concept map for the rickettsial diseases indicating how the diseases differ from one another.

TABLE 13.2 A Summary of Arthropod-Borne Bacterial Diseases

Disease	Causative Agent	Signs and Symptoms	Transmission	Treatment	Prevention
Plague	*Yersinia pestis*	Bubonic: sudden onset of fever and chills, headache, fatigue, muscle aches, and buboes Septicemic: high fever, abdominal pain, diarrhea Pneumonic: headache, malaise, extensive coughing	Infected flea bite Inhaled infectious droplets from person or animal	Streptomycin and gentamicin given intravenously or intramuscularly	Avoiding contact with sick or dead animals Flea control
Tularemia	*Francisella tularensis*	Flu-like symptoms Own set of symptoms depending on body site	Several modes of transmission: ticks, sick or dead animals, airborne, contaminated food or water	Streptomycin	Avoiding and preventing tick bites Avoiding sick or dead animals
Lyme disease	*Borrelia burgdorferi*	Rash, flu-like symptoms, joint pain	Bite of infected deer tick	Amoxicillin	Avoiding and preventing tick bites
Tickborne/ Relapsing fever	*Borrelia hermsii*	Periods of fever and chills	Bite of infected tick	Doxycycline	Avoiding and preventing tick bites
Rocky Mountain spotted fever	*Rickettsia rickettsii*	High fever, severe headache, skin rash of red spots	Bite of infected hard tick	Doxycycline or tetracycline	Avoiding and preventing tick bites
Epidemic typhus	*Rickettsia prowazekii*	Fever and rash	Scratching bites from body lice	Doxycycline	Avoiding and preventing lice infestation
Murine typhus	*Rickettsia typhi*	Mild fever, persistent headache, rash	Bite of infected flea	Doxycycline	Avoiding and preventing flea bites
Ehrlichiosis and anaplasmosis	(HME) *Ehrlichia chaffeensis* (HGA)*Anaplasma phagocytophilum*	Headache, malaise, fever	Bite of infected lone star tick Bite of blacklegged tick	Tetracycline or doxycycline	Avoiding and preventing tick bites

Chapter **Challenge C**

The rickettsial and ehrlichial arthropod-borne pathogens discussed in this final section were transmitted by ticks (RMSF, ehrlichiosis, anaplasmosis), lice (epidemic typhus), and fleas (murine typhus).

QUESTION C: *Which of these arthropod-borne pathogens could be considered euedaphic pathogens and which are soil-transmitted pathogens, based on the vector used to transmit the pathogen?*

You can find answers online in **Appendix F**.

■ SUMMARY OF KEY CONCEPTS

Concept 13.1 Several Soilborne Bacterial Diseases Develop from Endospores

1. **Anthrax** is an acute infectious disease caused by the spore-former *Bacillus anthracis*. Inhalation produces symptoms of respiratory distress and causes a blood infection. Consumption and skin contacts lead to boil-like lesions. (Figure 13.3)

2. *Clostridium tetani* is the causative agent of **tetanus**. Symptoms of generalized muscle stiffness and **trismus** lead to convulsive contractions with an unnatural fixed smile. (Figure 13.4)

3. **Gas gangrene** is caused by the anaerobic endospore-forming *Clostridium perfringens*. Symptoms include intense pain, swelling, and a foul odor at the wound site. (Figure 13.5)

4. **Leptospirosis** is a **zoonosis** caused by *Leptospira interrogans*. Infected individuals have flu-like symptoms. (Figure 13.6)

Concept 13.2 Some Bacterial Diseases Can Be Transmitted by Arthropods

5. *Yersinia pestis* is the causative agent of **plague**. **Bubonic plague** is characterized by the formation of **buboes**. Spreading of the bacilli to the blood leads to **septicemic plague**. Localization in the lungs is characteristic of **pneumonic plague**, which can be spread person to person. (Figure 13.8)

6. **Tularemia** is caused by *Francisella tularensis*, which is highly infectious at low doses. Skin ulcers, eye lesions, and pulmonary symptoms can result. (Figure 13.9)

7. **Lyme disease** results from an infection by the spirochete *Borrelia burgdorferi*. It involves three stages: the **erythema migrans** (**EM**) rash; neurological and cardiac disorders of the central nervous system; and migrating arthritis. (Figure 13.12)

8. **Tickborne relapsing fever** results in recurring attacks of high fever caused *Borrelia hermsii*. Symptoms include substantial fever, shaking chills, headache, and drenching sweats. (Figure 13.13)

Concept 13.3 Rickettsial and Ehrlichial Diseases Are Spread by Arthropods

9. Rickettsial and ehrlichial infections usually involve a skin rash, fever, or both.

   ▶ **Rocky Mountain spotted fever** is caused by *Rickettsia rickettsii*, an intracellular parasite. Carried by ticks, the symptoms include a high fever for several days, severe headache, and a maculopapular skin rash. (Figure 13.14)

   ▶ **Epidemic typhus** is caused by *Rickettsia prowazekii* that is carried by body lice. A maculopapular rash progresses to the extremities, and an intense fever, hallucinations, and delirium are characteristic symptoms of the disease.

   ▶ **Murine typhus** is transmitted by fleas. The causative agent, *Rickettsia typhi*, causes a mild fever, headache, and maculopapular rash.

10. Ehrlichiosis has been recognized in two forms: **human monocytic ehrlichiosis** (*Ehrlichia chaffeensis*) and **human granulocytic anaplasmosis** (*Anaplasma phagocytophilum*).

■ CHAPTER SELF-TEST

For **Steps A–D**, you can find answers to questions and problems in **Appendix D**.

■ STEP A: REVIEW OF FACTS AND TERMS

Multiple Choice

Read each question carefully before selecting the *one* answer that best fits the question or statement.

1. Woolsorter's disease applies to the _____ form of _____.
   A. inhalation; tularemia
   B. toxic; myonecrosis
   C. intestinal; anthrax
   D. inhalation; anthrax

2. Which one of the following describes the mode of action of tetanospasmin?
   A. It inhibits muscle contraction.
   B. It damages and lyses red blood cells.
   C. It disrupts cell tissues.
   D. It inhibits muscle relaxation.

3. A crackling sound associated with myonecrosis is due to _____.
   A. respiratory distress due to plague
   B. nerve contractions due to tetanus
   C. lymph node swelling due to plague
   D. gas produced by *C. perfringens*

4. *Leptospira interrogans* has all the following characteristics *except* _____.
   A. endoflagella
   B. aerobic metabolism
   C. exotoxin production
   D. a hook at one end of the cell

5. A characteristic of *Y. pestis* is a/an _____ staining.
   A. gram-positive
   B. bipolar
   C. acid-fast
   D. gram-variable

6. Skin ulcers are a common lesion resulting from being bitten by _____.
   A. a tick infected with *B. burgdorferi*
   B. fleas infected with *Y. pestis*
   C. a tick infected with *F. tularensis*
   D. lice infected with *C. tetani*

7. Erythema migrans is typical of the _____ stage of Lyme disease.
   A. early localized
   B. early disseminated
   C. late
   D. recurrent

8. A brief tick bite and a small number of recurring periods of fever and chills are typical of _____.
   A. relapsing fever
   B. ehrlichiosis
   C. Rocky Mountain spotted fever
   D. epidemic typhus

9. Rocky Mountain spotted fever is most common in the _____.
   A. southeastern United States
   B. Rocky Mountains
   C. Pacific Northwest
   D. New England

10. A lowering of the white blood cell count is characteristic of _____.
    A. plague
    B. anthrax
    C. ehrlichiosis
    D. RMSF

11. In the accompanying figure, outline the life cycle (A–H) of the blacklegged tick responsible for the transmission of Lyme disease.

Matching

Match the statement on the left with the disease on the right by placing the correct letter in the available space. A letter can be used once, more than once, or not at all.

Statement

___ **12.** Accompanied by erythema migrans

___ **13.** Affects monocytes in the body; transmitted by ticks

___ **14.** Caused by a spore-forming rod that produces hyaluronidase and α-toxin

___ **15.** Maculopapular rash beginning on extremities and progressing to body trunk

___ **16.** Caused by a spirochete that infects kidney tissues in pets and humans

___ **17.** Occurs in small game animals, especially rabbits

___ **18.** Caused by a gram-negative rod with bipolar staining; transmitted by the rat flea

___ **19.** Up to 13 attacks of substantial fever, joint pains, and skin spots; *Borrelia* involved

___ **20.** Most commonly reported tickborne disease in the United States

Disease

A. Anthrax

B. Ehrlichiosis

C. Epidemic typhus

D. Gas gangrene

E. Leptospirosis

F. Lyme disease

G. Murine typhus

H. Plague

I. Rocky Mountain spotted fever

J. Tetanus

K. Tickborne relapsing fever

L. Tularemia

■ STEP B: CONCEPT REVIEW

21. Summarize the clinical significance of the three forms of **anthrax**. (**Key Concept 13.1**)

22. Explain why **gas gangrene** is referred to as **myonecrosis**, and identify the toxins and enzymes involved with the disease. (**Key Concept 13.1**)

23. Distinguish between the mild and systemic forms of **leptospirosis**. (**Key Concept 13.1**)

24. Compare **bubonic, septicemic**, and **pneumonic plague**. (**Key Concept 13.2**)

25. Summarize the clinical significance of glandular and inhalation **tularemia**. (**Key Concept 13.2**)

26. Discuss the characteristics of the two forms of **ehrlichiosis**. (**Key Concept 13.3**)

■ STEP C: APPLICATIONS AND PROBLEM SOLVING

27. In February 1980, a patient was admitted to a Texas hospital complaining of fever, headache, and chills. He also had greatly enlarged lymph nodes in the left armpit. A sample of blood was taken and Gram stained, whereupon gram-positive diplococci were observed. The patient was treated with cefoxitin, a drug for gram-positive organisms, but soon thereafter he died. On autopsy, *Yersinia pestis* was found in his blood and tissues. Why was this organism mistakenly thought to be diplococci, and what error was made in the laboratory? Why were the symptoms of plague missed?

28. Some estimates place epidemic typhus among the all-time killers of humans; one listing even has it in third place behind malaria and plague. In 1997, during a civil war, an outbreak of epidemic typhus occurred in the African country of Burundi. The outbreak was estimated to be the worst since World War II. What conditions might have led to this epidemic?

29. A young woman was hospitalized with excruciating headache, fever, chills, nausea, muscle pains in her back and legs, and a sore throat. Laboratory tests ruled out meningitis, pneumonia, mononucleosis, toxic shock syndrome, and other diseases. On the third day of her hospital stay, a faint pink rash appeared on her arms and ankles. By the next day, the rash had become darker red and spread from her hands and feet to her arms and legs. Can you guess the eventual diagnosis?

30. In Chapter 9 of the Bible, in the Book of Exodus, the sixth plague of Egypt is described in this way: "Then the Lord said to Moses and Aaron, 'Take a double handful of soot from a furnace, and in the presence of Pharaoh, let Moses scatter it toward the sky. It will then turn into a fine dust over the whole land of Egypt and cause festering boils on man and cattle throughout the land.'" Which disease in this chapter is probably being described?

■ STEP D: QUESTIONS FOR THOUGHT AND DISCUSSION

31. Although the tetanus toxin is second in potency to the toxin of botulism, many physicians consider tetanus to be a more serious threat than botulism. Would you agree? Why?

32. Murine typhus was observed in five members of a Texas household. On investigation, epidemiologists learned that family members had heard rodents in the attic, and, 2 weeks previously, they had used rat poison on the premises. Investigators concluded that both the rodents and the rat poison were related to the outbreak. Why?

33. Centuries ago, the habit of shaving one's head and wearing a wig probably originated in part as an attempt to reduce lice infestations in the hair. Why would this practice also reduce the possibilities of certain diseases? Which diseases?

34. At various times, local governments are inclined to curtail deer hunting. How might this lead to an increase in the incidence of Lyme disease?

35. In autumn, it is customary for homeowners in certain communities to pile leaves at the curbside for pickup. How might this practice increase the incidence of tularemia, Lyme disease, and Rocky Mountain spotted fever in the community?

CHAPTER 14

Sexually Transmitted and Contact Transmitted Bacterial Diseases

What do Abraham Lincoln, Adolf Hitler, Friedrich Nietzsche, Oscar Wilde, Ludwig van Beethoven, and Vincent van Gogh have in common? Very likely, they all suffered from syphilis. In 2003, Deborah Hayden wrote a book titled *Pox: Genius, Madness, and the Mysteries of Syphilis* (New York: Basic Books) in which she looked at 14 eminent figures from the 15th to 20th centuries whose behavior, careers, or personalities were more than likely shaped by a syphilis infection.

Syphilis originally was called the Great Pox to separate it from smallpox. Until the introduction of penicillin in 1943, syphilis was untreatable and caused a chronic and relapsing disease that could disseminate itself throughout the body, only to reappear later as

so-called tertiary syphilis. In this most dangerous and terminal form, the disease produces excruciating headaches, gastrointestinal pains, and eventually deafness, blindness, paralysis, and insanity.

Yet, sometimes ecstasy and fierce creativity were part of the "symptoms." As Deborah Hayden says, "one of the 'warning signs' of tertiary syphilis is the sensation of being serenaded by angels." In fact, writer Karen Blixen (Isak Dinesen) once said,

Computer-generated, false-color image of *Treponema pallidum,* the causative agent of syphilis.
© Science Picture Co/Science Source.

"Syphilis sold her soul to the devil for the ability to tell stories." Deborah Hayden believes such emotions provided much of the creative spark for the notable historical figures she describes in her book. Perhaps the most intriguing is the debated proposal that syphilis might have driven Hitler mad and that he was dying of syphilis when he committed suicide in his Berlin bunker in the final days of World War II.

If Deborah Hayden's arguments are true, it is amazing how a bacterial organism has affected the body and mind in different ways, shaping the thoughts of writers and philosophers, the creative genius of artists, composers, and scientists—and, yes, the madness of dictators.

Today, **sexually transmitted infections (STIs)** remain a health problem in the United States and around the world. Globally, more than one million people acquire an STI every day. In the United States, the Centers for Disease Control and Prevention (CDC) estimates that in 2015, more than 20 million Americans contracted an STI, with half of these cases reported among the 15- to 24-year age group. Among the STIs discussed in this chapter, three crown the list of the top 10 infectious diseases in the United States (**FIGURE 14.1**). Importantly, these represent only the notifiable cases, and, because many STIs go unreported, the true number of infections is substantially higher. Also, note in Figure 14.1 that AIDS (HIV diagnosis) also can be sexually transmitted. In fact, individuals infected with an STI are two to five times more likely to acquire HIV than are uninfected individuals if exposed to the AIDS virus through sexual contact.

The STIs are but one example of how changing social patterns can affect the incidence of other bacterial infections and diseases discussed in this chapter. We also will examine urinary tract infections, which are another common infection in people, and identify other contact diseases of the skin and eyes. We will begin with diseases associated with the reproductive system.

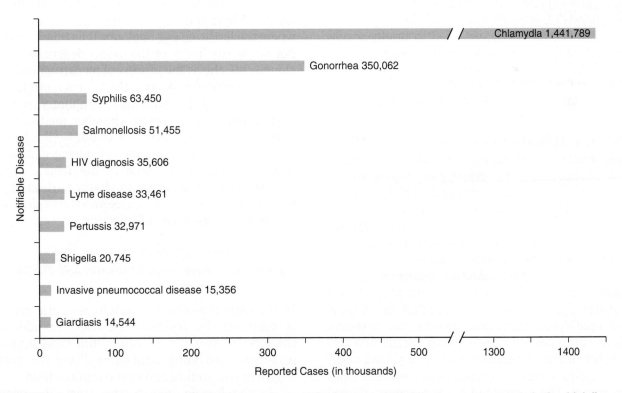

FIGURE 14.1 Reported Cases of Notifiable Diseases in the United States, 2014. The three most reported microbial diseases in the United States in 2014 were sexually transmitted. *»» How many of the top 10 diseases have bacteria as their causative agent?*

Data from Summary of Notifiable Infectious Diseases and Conditions—United States, 2014/CDC.

Chapter Challenge

This chapter on the infections and diseases caused by bacterial species brings together several different parts of human anatomy, including the female and male reproductive systems, the urinary system, the skin, and the eye. So, in this chapter, separate challenges will be offered for the different systems and organs discussed.

■ KEY CONCEPT 14.1 Portions of the Female and Male Reproductive Systems Contain a Resident Microbiome

Sexually transmitted infections (STIs) usually are transmitted through sexual contact, and they affect the male and female reproductive systems. STIs also are called **sexually transmitted diseases** (**STDs**). However, these infections often are considered "hidden epidemics" because many individuals might not realize they are infected and they may not exhibit any disease symptoms. Therefore, the term "sexually transmitted infection" is becoming more commonly used, and we will use the term here. However, look back again at Figure 14.1 to remind yourself of the significant numbers of reported infections causing chlamydia, gonorrhea, and syphilis.

To understand STIs, we need to briefly review the male and female reproductive systems.

The Male and Female Reproductive Systems Consist of Primary and Accessory Sex Organs

The **male reproductive system** produces, maintains, and transports sperm cells and is the source of male sex hormones. The **primary sex organs** are the testes, which produce sperm cells (**FIGURE 14.2A**). The testes also produce sex hormones called androgens, such as testosterone, that are important for the maturation, development, and changes in activity of the reproductive system organs.

The **accessory reproductive organs** include the epididymis, a coiled tube leading out of the testes that stores sperm until they are mature and motile. The epididymis terminates in the vas deferens, which combines with the seminal vesicle and terminates in the ejaculatory duct. The paired seminal vesicles secrete a viscous fluid that nourishes the sperm. The prostate gland is a walnut-shaped structure at the base of the urethra. It also contributes a slightly milky fluid that contains nutrients for sperm health and an antibacterial compound to combat urinary tract infections.

The external organs are the scrotum, containing the testes, and the penis, which contains the urethra and through which urine passes and semen is ejaculated.

The **female reproductive system** consists of the primary sex organs, which include the ovaries, and several accessory sex organs, specifically the fallopian tubes (also called uterine tubes), uterus, vagina, and vulva (**FIGURE 14.2B**). During ovulation, an egg, released from one of the ovaries, enters a fallopian tube. If sperm are present, fertilization can take place, after which the fertilized egg moves through the fallopian tube to the uterus. It is in the uterine lining where the fertilized egg implants. If fertilization does not occur, the lining of the uterus degenerates and sloughs off; this is the process of menstruation.

The terminal portion of the female reproductive tract is the vagina, which is a tube about 9 cm long. It represents the receptive chamber for the penis during sexual intercourse, the outlet for fluids from the uterus, and the passageway during childbirth. One very important tissue of the female reproductive tract is the cervix, the lower portion of the uterus and the part that opens into the vagina. The cervix is a common site of infection.

The Male and Female Reproductive Systems Have a Set of Natural Defenses Against Infection

In the male reproductive system, only the urethra is colonized by resident microbes. In addition, the reproductive system can mount local immune responses, providing antibodies along the entire length of the urethra and in the seminal fluid.

In the female reproductive tract, the vagina has an acidic pH because *Lactobacillus* and other acid-producing species produce lactic acid. The acidic environment discourages the growth of many potential pathogens. Because the cervix represents

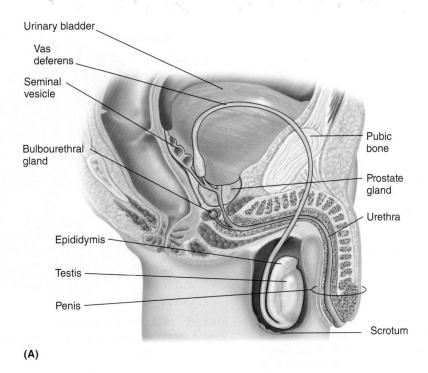

Urinary bladder
Vas deferens
Seminal vesicle
Bulbourethral gland
Epididymis
Testis
Penis
Pubic bone
Prostate gland
Urethra
Scrotum

(A)

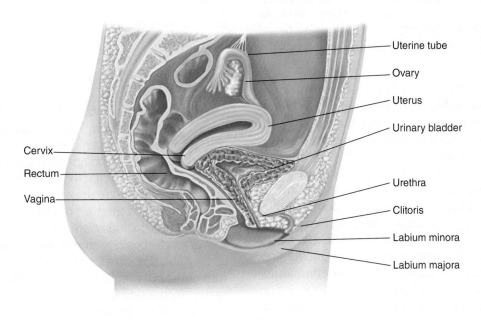

Uterine tube
Ovary
Uterus
Urinary bladder
Cervix
Rectum
Vagina
Urethra
Clitoris
Labium minora
Labium majora

(B)

FIGURE 14.2 Reproductive Systems of Males and Females. The **(A)** male reproductive system and **(B)** female reproductive system showing the primary and secondary sex organs. *»» Identify the specific primary and secondary sex organs in males and females.*

a potential portal of entry for microbes from the lower genital tract, the cervix contains several antimicrobial defense mechanisms, including a mucociliary escalator; mucus that contains a variety of antimicrobial chemicals, including lysozyme and lactoferrin; and antibacterial peptides that can kill or inhibit the growth of many bacterial species.

These chemicals and peptides fluctuate in concentration in the cervix and vagina during the menstruation cycle. As the concentrations of lactoferrin and IgA antibodies drop, the concentrations of human defensins and lysozyme rise (**FIGURE 14.3**). Thus, there is a constant chemical defense against potential microbial invasion.

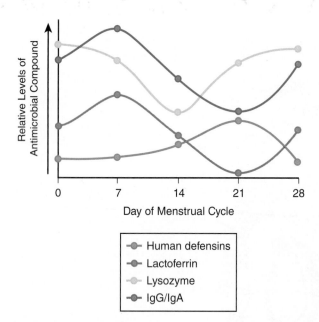

FIGURE 14.3 Fluctuations in Antimicrobial Chemicals and Proteins in the Vagina During the Menstrual Cycle. During the female menstrual cycle, as some antimicrobial compounds fall (IgG/IgA, lactoferrin), others rise (lysozyme, human defensins). *»» Why are there fluctuations in the levels of these antimicrobial compounds?*

Modified from Wilson, M. 2005. *Microbial Inhabitants of Humans: Their Ecology and Role in Health and Disease.* Cambridge University Press.

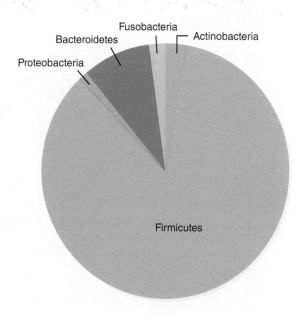

FIGURE 14.4 The Bacterial Composition of the Female Vagina. Members of the gram-positive phylum Firmicutes predominate and *Lactobacillus* is the primarily genus found. *»» Propose a reason why there are so few other phyla present in the vagina.*

Data from Cho, I. and Blaser, M. J. 2012. *Nat Rev Gen* 13(4):260–270.

The cervix and vagina harbor a resident **microbiome**, the vagina having been examined in most detail (**FIGURE 14.4**). Researchers believe the composition of the healthy vaginal microbiome varies by race, age, and perhaps even within an individual, so the composition shown in Figure 14.4 does not identify specific bacterial species. However, the vast majority of women harbor a bacterial community dominated by one of four *Lactobacillus* strains (phylum Firmicutes). During and after pregnancy, the microbiome remains relatively stable. However, women experiencing preterm births have an altered microbial community characterized by a drop in *Lactobacillus* species and an increase in a diverse population of anaerobic species. This microbial imbalance, called a **dysbiosis**, can last for up to one year. Shifts like these and other changes that coincide with hormonal and lifestyle changes illustrate the complexity of and factors influencing the vaginal microbiome.

Common Vaginal Infections Come from the Resident Microbiome

Not all bacterial infections of the reproductive system originate from outside pathogens. Some infections are the result of normal resident bacterial species.

Bacterial Vaginosis

The most common vaginal infection reported in women of childbearing age is **bacterial vaginosis (BV)**, which occurs when there is a dysbiosis of community microbes. Although most women are unaware of the condition, the lactobacilli that normally produce lactic acid and hydrogen peroxide in the vagina are replaced by a mixture of other resident bacterial genera. These include *Gardnerella vaginalis* and *Mobiluncus* (gram-positive rods of the phylum Actinobacteria) and *Prevotella* (very short, gram-negative rods of the phylum Bacteroidetes). The reason for the switch in dominance from lactobacilli to the other polymicrobial species is not clearly understood. Menstruation normally causes a rise in vaginal pH to about 7, and this rise perhaps is one factor why the population changes. In addition, the composition of the vaginal microbiome might help predict potential preterm delivery. Specifically, a high abundance of *Gardnerella* and a scarcity of *Lactobacillus* appear to be associated with preterm deliveries.

For 50% of women with BV, there are no outward symptoms of the condition. If a woman does have symptoms, they include a foul-smelling, grayish-white vaginal discharge that is thin and watery. Antibiotic therapy generally provides relief.

Concept and Reasoning Checks 14.1

a. Trace the path of (a) a sperm cell and (b) an egg as they move through their respective reproductive systems.

b. Explain how the resident microbiome protects the vagina from infection.

c. What factors appear to be responsible for the overgrowth of the vagina by other members of the resident microbiome?

Chapter **Challenge A**

In this section, it was stated that the female reproductive system is prone to more infections than is the male reproductive system.

QUESTION A: *Do you agree that the female reproductive system is more prone to infections? Provide evidence to support your answer.*

You can find answers online in **Appendix F**.

■ KEY CONCEPT 14.2   Many Sexually Transmitted Infections Are Caused by Bacteria

STIs continue to be a major public health challenge in the United States. In fact, the CDC reports that STIs as a whole are at an all-time high (**FIGURE 14.5**). The CDC estimates that almost 20 million new infections occur each year—and 50% of these infections occur in young people between the ages 15 to 24.

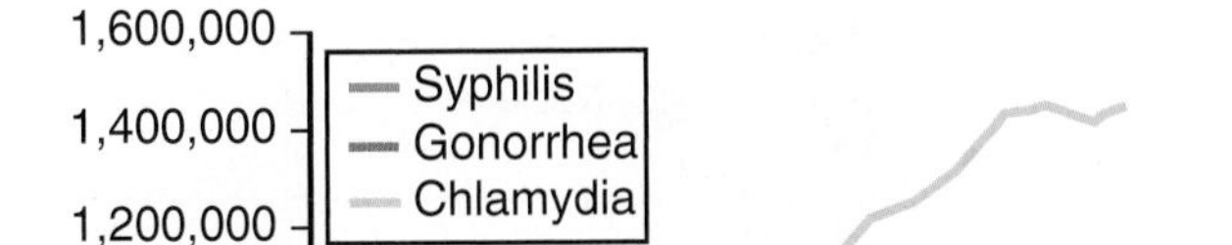

The bacterial species causing STIs have some distinctive characteristics. Because these organisms are dependent on sexual intercourse (which might be sporadic) for transmission, the pathogens must:

▶ Persist in an infective form until that person has sexual intercourse

▶ Hide themselves from recognition and elimination by the immune system

▶ Be ready to be transmitted when the infected individual has sexual intercourse

In this section, we consider the common STIs and the bacterial species that cause them. We begin with three that require mandatory reporting to the CDC—chlamydia, gonorrhea, and syphilis.

Chlamydial Urethritis Is the Most Frequently Reported STI

Chlamydial urethritis represents one of several diseases collectively known as **non-gonococcal urethritis**, or **NGU**. NGU is a general term for a condition in which people without gonorrhea have a demonstrable infection of the urethra. About 50% of NGUs are due to chlamydial urethritis, 25% to *Ureaplasma* urethritis, and another 25% are due to unidentified infectious agents.

FIGURE 14.5 Reported Cases of Chlamydia, Gonorrhea, and Syphilis—United States and U.S. Territories, 1995–2014. In 2014, the CDC reported the largest ever number of cases of chlamydia. **»» Why are STIs considered a health challenge?**

Data from the CDC.

Causative Agent and Epidemiology

The disease **chlamydia** or **chlamydial urethritis** is caused by *Chlamydia trachomatis*, an exceptionally small (0.35 µm), round to ovoid-shaped organism. Being an obligate, intracellular parasite, it has one of the smallest bacterial genomes, having about 600 genes (*Escherichia coli* has around 4,200 genes). Serotypes D–K of *C. trachomatis* are associated with NGU.

C. trachomatis has a biphasic and unique reproductive cycle (**FIGURE 14.6**). There is a nonreplicating, extracellular, infectious **elementary body (EB)** and a replicating, intracellular, noninfectious **reticulate body** (**FIGURE 14.7**). Humans appear to be the only host for the organism.

Chlamydial urethritis is the most common STI globally and the most frequently reported nationally notifiable infectious disease in the United States (see Figure 14.1). About 75% of these cases occur in women between 15 and 29 years of age. Besides an increase in incidence of chlamydial urethritis, the rise in reported cases is due to the increased number of screening programs available and the development of better diagnostic tests.

Clinical Presentation

C. trachomatis is transmitted by vaginal, anal, or oral sex. In fact, any sexually active individual can be infected through sexual contact with an infected individual. The disease has an incubation period of about 1 to 3 weeks. Chlamydia often is referred to as the "silent disease" because the organism does not cause extensive tissue injury directly. Thus, some 85% to 90% of infected individuals are asymptomatic, whereas others might not seek treatment and thus can unknowingly pass the disease on to others.

If symptoms do occur, females often note a slight cervical discharge (drainage of fluid from the

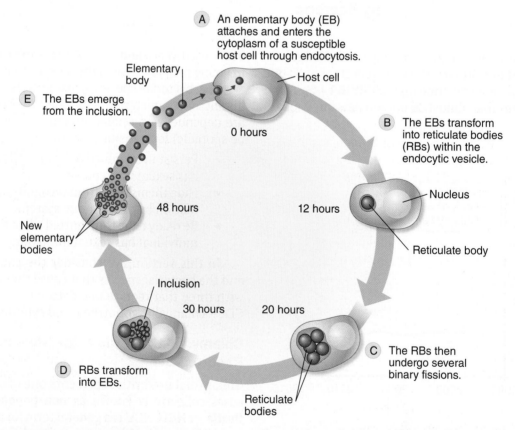

FIGURE 14.6 The Chlamydial Life Cycle. The reproductive cycle of the chlamydiae involves two types of cells. After infection, nonreplicating elementary bodies (EBs) reorganize into reticulate bodies (RBs), which divide to form additional RBs. Within 30 hours, the RBs begin to reorganize into EBs within an inclusion. The EBs are released directly from the inclusion body or by the spontaneous lysis of the inclusion and host cell. »» *Because* C. trachomatis *only reproduces inside cells, what cellular products might the bacterium require to survive and grow?*

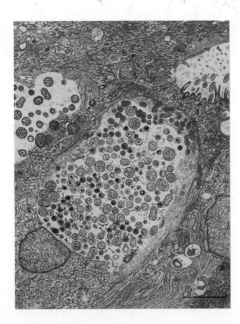

FIGURE 14.7 A Chlamydial Inclusion. False-color transmission electron micrograph of cells from a woman's fallopian tube infected with *Chlamydia trachomatis.* Spherical cells (green/brown) are elementary bodies and reticulate bodies seen inside an inclusion body (yellow). Cell cytoplasm is reddish-brown. (Bar = 0.5 μm.) *»» What complications can arise from an infection of the fallopian tubes?*

© Dr. R. Dourmashkin/Science Source.

vaginal wall and cervix), as well as inflammation of the cervix. Burning pain also is experienced during urination, reflecting infection in the urethra. In complicated cases, the disease can spread to, block, and inflame the fallopian tubes, causing **salpingitis**. About 40% of untreated infections progress to **pelvic inflammatory disease** (**PID**), which is an inflammation of the uterus, fallopian tubes, and/or ovaries. PID from chlamydia and gonorrhea is believed to affect about 50,000 women in the United States annually. Often, however, there are few symptoms of disease before the salpingitis manifests itself, thus adding to the danger of infertility and a life-threatening **ectopic pregnancy**, in which a fertilized egg begins to develop in a fallopian tube rather than the uterus, as CLINICAL CASE 14 describes. In fact, chlamydial urethritis is the number one cause of first trimester pregnancy-related deaths in the United States.

In males, chlamydia is characterized by painful urination and a discharge that is watery and copious. The discharge often is observed after urinating for the first time in the morning. Tingling sensations in the penis are generally evident. Inflammation of the epididymis can result in sterility, but this

complication is uncommon. Chlamydial pharyngitis or inflammation of the anus (**proctitis**) is possible through oral or anal intercourse.

Treatment and Prevention

Individuals who are asymptomatic remain infected. In symptomatic cases, the infection can be treated successfully (95% cure rate) with antibiotics. However, post-treatment rescreening is recommended to ensure reinfection has not occurred.

Gonorrhea Is Becoming Untreatable

One of the most common STIs in men and women is gonorrhea. Importantly, the ability to treat gonorrhea infections with antibiotics might soon become more challenging.

Causative Agent and Epidemiology

The agent of **gonorrhea** is *Neisseria gonorrhoeae*, a small, unencapsulated, nonmotile, gram-negative diplococcus named for Albert Neisser, who isolated it in 1879. The organism, commonly known as the **gonococcus**, has a characteristic double-bean shape (**FIGURE 14.8**). The great majority of cases of gonorrhea are transmitted during sexual intercourse. (Gonorrhea is sometimes called "the clap," from the French "clappoir" for "brothel.") The organism is found only in humans, where it can infect the cervix, uterus, fallopian tubes, and urethra, as well as the mouth, throat, and anus.

Gonorrhea is the second most frequently reported nationally notifiable infectious disease in the United States (see Figure 14.1) with the highest percentage of infections being in persons 15 to 24 years of age. The CDC estimates that 3 to 4 million cases go undetected or unreported each year. Globally, the World Health Organization (WHO) believes there are more than 78 million infections annually.

Clinical Presentation

Following attachment of *N. gonorrhoeae* by pili to the genital tract, the incubation period for gonorrhea ranges from 2 to 6 days. In females, the cervix becomes reddened, and a discharge might be expressed by pressure against the pubic area. Patients often report abdominal pain and a burning sensation on urination, and the normal menstrual cycle might be interrupted.

In some females, gonorrhea also spreads to the fallopian tubes. As these thin passageways become riddled with pouches and adhesions, salpingitis and

Clinical Case 14

A Case of Chlamydia with Tragic Consequences

An educated, professional woman met a man at a friend's house one evening. She was director of a New York law firm. The man was equally successful in his professional career.

The couple hit it off immediately. There were many evenings of quiet candlelit dinners, and soon, they became sexually intimate. Neither one used condoms or other means of protection.

Three days after having intercourse, the woman began experiencing fever, vomiting, and severe abdominal pains. She immediately made an appointment to see her doctor.

Assuming the illness was an intestinal upset, the physician prescribed appropriate medication. The woman was actually suffering from chlamydia, but the fact that she was sexually active did not come up during the examination.

Feeling better, the woman continued her normal routine. But 6 months later, with no apparent warning, she collapsed on a New York City sidewalk.

The woman was rushed to a local hospital, where doctors diagnosed a chlamydial infection by observing the dark inclusion bodies typical of chlamydia (see figure). They performed emergency surgery: her uterus and fallopian tubes were badly scarred. She recovered; however, the scarring left her unable to have children.

Questions

a. Would condom use block the transfer of *C. trachomatis* cells?

b. Do fever, vomiting, and abdominal pains indicate a chlamydial infection?

c. What "complications" were developing during the 6 months when the woman was in improved health?

d. What are inclusion bodies?

e. How would scarring lead to infertility?

You can find answers online in **Appendix E**.

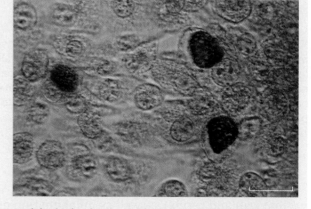

Dark inclusion bodies typical of a chlamydial infection. (Bar = 2 μm.)

Courtesy of Dr. E. Arum/Dr. N. Jacobs/CDC.

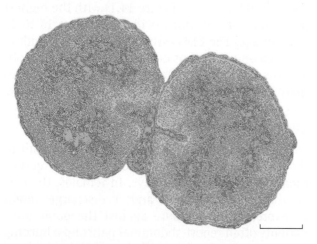

FIGURE 14.8 *Neisseria gonorrhoeae.* A false-color transmission electron micrograph of a *N. gonorrhoeae* diplococcus. (Bar = 0.3 μm.) *»» What does the blue area inside the cells represent?*

© Dr. David Phillips/Visuals Unlimited/Corbis, Inc.

PID might occur. Sterility can result from scar tissue remaining after the disease has been treated, or a woman might experience an ectopic pregnancy. It should be noted that symptoms are not universally observed in females, and an estimated 50% of affected women remain asymptomatic but can spread the disease unknowingly through sexual intercourse.

Gonorrhea is particularly dangerous to an infant born to an infected woman. The infant can contract gonococci during passage through the birth canal and develop neonatal conjunctivitis.

Symptoms of gonorrhea tend to be more acute in males than in females, and males thus tend to seek diagnosis and treatment more readily. When gonococci infect the mucus membranes of the urethra, symptoms include a tingling sensation in the penis, followed in a few days by pain when urinating. There is also a thin, watery discharge at first, followed later by more obvious yellow, thick fluid resembling semen (**FIGURE 14.9**). Frequent urination and an urge to urinate develop as the disease spreads further into the urethra. The lymph nodes of the groin can swell, and sharp pain might be felt in the testicles. Unchecked infection of the epididymis can lead to sterility.

Like chlamydia, gonorrhea does not restrict itself to the urogenital organs. **Gonococcal pharyngitis** can develop in the pharynx if bacterial cells are transmitted by oral–genital contact; patients complain of sore throat or difficulty in swallowing. Infection of the rectum, or **gonococcal proctitis**, also can ocur, especially in individuals performing anal intercourse.

Treatment and Prevention

Historically, gonorrhea has been treated effectively with antibiotics. However, today *N. gonorrhoeae* is becoming resistant to most all antibiotics used for treatment, and it might be only a matter of time before treatment failures occur. Such changes are transforming *N. gonorrhoeae* into a major superbug and a potential global public health threat.

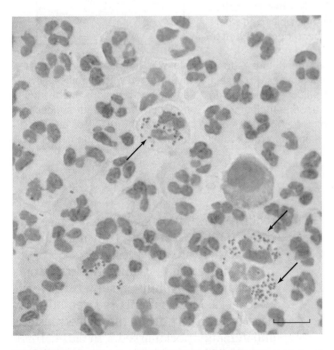

FIGURE 14.9 Gonococcal Urethral Smear. A gram-stained smear of discharge from the male urethra showing the gonococci (arrows) in the cytoplasm of white blood cells. (Bar = 10 μm.) »» *What is the Gram reaction for* N. gonorrhoeae?

Courtesy of Joe Miller/CDC.

Syphilis Is a Chronic, Infectious Disease

Over the centuries, Europeans have had to contend with four pox diseases: chickenpox, cowpox, smallpox, and the Great Pox, the latter now known as **syphilis**. The first European epidemic was recorded in the late 1400s, shortly after the conquest of Naples by the French army. Its origin remains uncertain, but some interesting clues have been found, as MICROFOCUS 14.1 uncovers.

Causative Agent and Epidemiology

Syphilis is caused by *Treponema pallidum*, (*trepo* = "turn"; *nema* = "thread"; *pallid* = "pale"; thus literally "pale turning thread"). This spirochete moves by means of endoflagella. Humans are the only host for *T. pallidum*, so the organism must spread by direct human-to-human contact, usually during sexual intercourse.

Syphilis is currently ranked among the top five most-reported microbial diseases in the United States (see Figure 14.1). In 2014, the rate of reported cases was the highest since 1994. Statistics indicate more than 12 million cases are reported each year worldwide, with more than 90% of cases being in men. Taken alone, these figures suggest the magnitude of the syphilis epidemic, but some public health microbiologists believe that for every case reported, as many as nine cases go unreported.

The cells of *T. pallidum* can evade immune defenses because the pathogen has very few surface proteins that the immune system can recognize. This invisibility allows the pathogen to persist in the body.

The pathogen penetrates the skin surface through the mucous membranes of the genitalia or via a wound, abrasion, or hair follicle. The variety of clinical symptoms accompanying the stages, and their similarity to other diseases, have led some physicians to call syphilis the "great imitator." Untreated, the disease can progress through a number of stages, which can overlap.

Clinical Presentation

The incubation period for syphilis varies greatly (10 to 90 days), but it averages about 3 weeks.

▶ **Primary Syphilis.** A lesion, called a **chancre**, which is a painless circular, purplish ulcer with a small, raised margin with hard edges is typical of **primary syphilis** (**FIGURE 14.10A**). The chancre develops at the site of entry of the spirochetes, often the genital organs. However, any area of the skin can be affected, including the pharynx, rectum, or lips. The

MICROFOCUS 14.1: History

The Origin of the Great Pox

Among the more intriguing questions in medical history is how and why syphilis suddenly emerged in Europe in the late 1400s. Writers of that period tell of a terrible new disease that swept over Europe and on to India, China, and Japan. But where did the actual disease originate?

For most of the past 500 years, Christopher Columbus has been alternately blamed or exonerated for bringing syphilis to Europe from the New World. In the 20th century, critics of the "Columbian theory" proposed that syphilis had always been present in the Old World, perhaps coming from Spain and Portugal through the slave trade with Africa.

Now, a genetic analysis of the disease-causing *Treponema pallidum* carried out by Kristin Harper and colleagues at Emory University in Atlanta suggests that today's syphilis is a close cousin to another South American disease called yaws, a tropical infection of the skin, bones, and joints.

To determine if Columbus and his men introduced syphilis to Europe after "catching yaws" in the Americas, Harper and her colleagues collected species of *T. pallidum* from across the globe (Africa, Europe, Asia, the Middle East, the Americas, and the Pacific Islands) to determine which microbes gave rise to which—and perhaps determine where syphilis originated. They also sequenced and compared genes from *Treponema pertenue*, which causes yaws, from Africa and South America. By comparing genetic sequences between the *Treponema* species, the researchers found the species that cause syphilis originated relatively recently, with their closest relatives causing yaws in South America.

However, yaws is spread through simple skin contact, not sexual intercourse. So, how did a simple skin disease in South America possibly become an STI in Europe? One suggestion is that syphilis became sexually transmitted only when it reached Europe, where the climate was not as hot and humid as it was in the tropics and where people wore more clothing, limiting skin contact. Therefore, the bacterium adapted to use sexual intercourse as the transmission mechanism.

To confirm the "New World" theory, more genetic analyses need to be done, including sequencing of the entire DNA of the related *Treponema* species. Professor George Armelagos, one of Harper's coworkers on the project said, "Understanding [*T. pallidum's*] evolution is important not just for biology, but for understanding social and political history. It could be argued that syphilis is one of the important early examples of globalization and disease, as globalization remains an important factor in emerging diseases today."

This woodcut (1484) depicts a man infected with syphilis as evidenced by the lesions on the face and legs. The woodcut suggests syphilis is the result of an unfavorable planetary alignment.

Mary Evans Picture Library/Alamy Stock Photo.

---

chancre teems with spirochetes and represents the most infectious stage. It persists for 3 to 6 weeks and then heals spontaneously. However, the infection has not been eliminated, as the spirochetes have spread through the blood and lymph to other body organs.

▶ **Secondary Syphilis.** Several weeks after the chancre of primary syphilis has healed, the patient develops a fever and a flu-like illness as well as swollen lymph nodes. With **secondary syphilis**, a skin rash develops, which can be mistaken for measles, rubella, or chickenpox. The rash appears as reddish-brown spots on the palms, face, and trunk (**FIGURE 14.10B**). Transmission can occur if there are moist lesions.

In untreated patients, the symptoms of secondary syphilis resolve after several weeks. Most patients recover, but they bear pitted scars from the healed lesions and remain "pockmarked." In 2014, there were 20,000 cases of primary and secondary syphilis reported to the CDC.

Many patients will have relapses of secondary syphilis during which time they remain infectious. Within 4 years, the relapses cease and the disease is no longer infectious (except in pregnant females). Patients either remain asymptomatic or slowly progress to the third stage.

▶ **Tertiary Syphilis.** About 40% of untreated patients develop **tertiary syphilis**. This stage occurs in many forms, but most commonly, it involves the skin, skeletal, or cardiovascular and nervous systems. The hallmark of tertiary syphilis is the **gumma**, a soft, painless, gummy noninfectious granular lesion (**FIGURE 14.10C**). In the cardiovascular system, gummas weaken the major blood vessels, causing them to bulge (aneurysm) and burst, whereas in the spinal cord and meninges, gummas lead to deterioration of the tissues and paralysis. In the brain, they alter the patient's personality and judgment

and cause insanity so intense that for many generations, people with tertiary syphilis were confined to mental institutions (but read the chapter introduction). Damage can be so serious as to cause death.

▶ **Congenital syphilis. Congenital syphilis** is a serious problem in pregnant women because the *Treponema* spirochetes penetrate the placental barrier after the third or fourth month of pregnancy. Infection in the fetus can lead to death (stillbirth); surviving infants can develop skin lesions and open sores. Affected children often suffer poor bone formation, meningitis, or **Hutchinson's triad**, a combination of deafness, impaired vision, and notched, peg-shaped teeth. In 2014, there were nearly 460 congenital syphilis cases reported to the CDC.

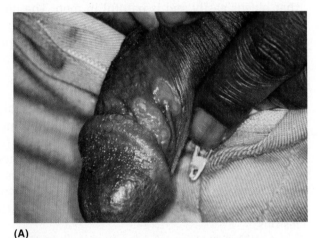

(A)

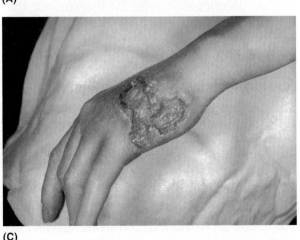

(C)

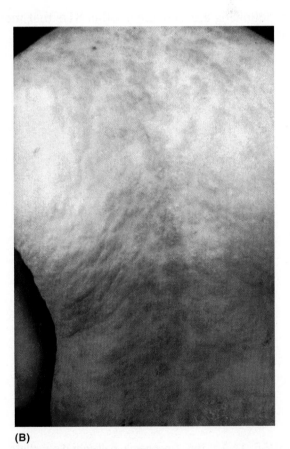

(B)

FIGURE 14.10 The Stages of Syphilis. (A) The chancre of primary syphilis as it occurs on the penis. The chancre has raised margins and is usually painless. **(B)** A skin rash is characteristic of secondary syphilis. **(C)** The gumma that forms in tertiary syphilis is a granular, diffuse lesion compared with the primary chancre. *»» From which of these stages could the spirochete be most contagious?*

Treatment and Prevention

There is no immunity that develops from a prior infection and no vaccine for syphilis. Therefore, the cornerstone of syphilis control is safe sex practices and the identification and treatment of the sexual contacts of patients. Penicillin is the drug of choice, and a single dose usually is sufficient to cure primary and secondary syphilis. Because *T. pallidum* cannot be cultivated on laboratory media, diagnosis of the primary infection depends on the observation of spirochetes from the chancre using fluorescence or dark-field microscopy (**FIGURE 14.11**). As the disease progresses, a number of tests to detect syphilis antibodies become useful, including the Venereal Disease Research Laboratory (VDRL) test for syphilis screening and the newer rapid plasma reagin (RPR) test for screening and monitoring of syphilis.

Other Sexually Transmitted Infections Also Exist

The following are a couple of other STIs that merit brief mention.

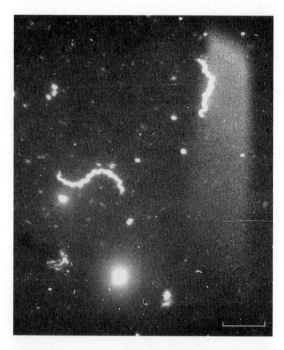

FIGURE 14.11 The *Treponema pallidum* Spirochete. A dark-field microscope view of the syphilis spirochete seen in a sample taken from the chancre of a patient. (Bar = 5 μm.) »» *Why are all photos of* T. pallidum *taken from patient samples and not from samples grown in culture?*

Courtesy of CDC.

Chancroid

An STI believed to be more prevalent worldwide (especially in tropical and semitropical climates) than gonorrhea or syphilis is **chancroid**. The disease is endemic in many developing nations with tropical climates and occurs where public health standards are low. The disease has been decreasing in the United States since 1987; there were 6 cases reported in 2014. However, the organism is difficult to culture and therefore the disease could be substantially underdiagnosed.

The causative agent of chancroid is *Haemophilus ducreyi*, a small gram-negative rod. The transmission of chancroid depends on contact with open ulcers or their fluid. After a 3- to 5-day incubation period, a tender, red papule forms at the entry site. The papule quickly becomes pus filled and then breaks down, leaving a shallow, saucer-shaped ulcer that bleeds easily and is painful in men. The ulcer has ragged edges and soft borders, a characteristic distinguishing it from the chancre formed in primary syphilis. For this reason, the disease is often called **soft chancre**.

The lesions in chancroid most often occur on the penis in males and the labia or clitoris in females (see Figure 14.2). Substantial swelling in the lymph nodes of the groin might be observed. However, the disease generally goes no further; in fact, in women, it often goes unnoticed. The clinical picture in chancroid makes the disease recognizable, but definitive diagnosis depends on isolating *H. ducreyi* from the lesions.

Antibiotics are useful for therapy, but the disease often disappears in 10 to 14 days without treatment.

Lymphogranuloma venereum

A systemic STI caused by serotypes L1–L3 of *C. trachomatis* is **lymphogranuloma venereum (LGV)**. LGV is more common in males than females and is accompanied by fever, malaise, and swelling and tenderness in lymph nodes of the groin. Females might experience proctitis if the chlamydial cells pass from the genital opening to the nearby rectum. LGV is prevalent in Southeast Asia and in Central and South America. Sexually active individuals returning from these areas can show symptoms of the disease, but treatment with antibiotics leads to rapid resolution.

TABLE 14.1 summarizes the STIs caused by bacterial agents.

TABLE 14.1 A Summary of Sexually Transmitted Illnesses (STIs) Caused by Bacteria

Disease	Causative Agent	Signs and Symptoms	Transmission	Treatment	Prevention
Chlamydia	*Chlamydia trachomatis* (serotypes D–K)	Pain on urination Watery discharge Salpingitis	Sexual intercourse	Doxycycline Erythromycin	Practicing safe sex Limiting sex partners Screening for STIs
Gonorrhea	*Neisseria gonorrhoeae*	Pain on urination Discharge Salpingitis	Sexual intercourse	Ceftriaxone	Practicing safe sex Avoiding oral sex
Syphilis	*Treponema pallidum*	Chancre Skin lesions Gumma	Sexual intercourse	Penicillin	Practicing safe sex
Chancroid	*Haemophilus ducreyi*	Soft chancre Erythema Swollen inguinal lymph nodes	Sexual intercourse	Azithromycin Erythromycin Ceftriaxone	Practicing safe sex
Lymphogranuloma venereum (LGV)	*Chlamydia trachomatis* (serotypes L1-L3)	Swollen lymph nodes Proctitis	Sexual intercourse	Doxycycline	Practicing safe sex

Concept and Reasoning Checks 14.2

a. Why is chlamydial urethritis referred to as the "silent disease"?
b. Contrast the symptoms of gonorrhea in females and males.
c. Describe the three stages of adult syphilis.
d. Propose an explanation as to why all the STI-causing walled bacteria are gram negative.

Chapter Challenge B

Most STIs have an acute phase that causes pain, produces a visible lesion, and generates an inflammatory reaction at the entry site. Such infections also can have a chronic phase during which the infection can spread to sites that are more distant and generate damage to host tissues.

QUESTION B: *Identify the acute phase and the chronic phase of the infection for chlamydia, gonorrhea, and syphilis.*

You can find answers online in **Appendix F**.

■ KEY CONCEPT 14.3 Urinary Tract Infections Are the Second Most Common Body Infection

Each year, more than 1 million **urinary tract infections (UTIs)** and 20 million STIs are diagnosed by hospital emergency room (ER) departments in the United States. A 2015 report found that such UTIs and STIs are misdiagnosed almost 50% of the time by those ERs. Consequently, about 50% of women diagnosed with a UTI do not have such an infection. In addition, STIs are misdiagnosed 37% of the time, most being identified as a UTI. Part of the reason for the diagnosis confusion is the overlapping symptoms, such as urination problems and frequency and urgency of urination, in both infections. So, let's examine the urinary system and the major groups of UTIs.

The Urinary System Removes Waste Products from the Blood and Helps Maintain Homeostasis

The organs of the urinary system consist of two kidneys, two ureters, a urinary bladder, and a urethra (FIGURE 14.12). The kidneys filter waste products from the bloodstream and convert the filtrate into urine. The ureters, urinary bladder, and urethra are collectively known as the **urinary tract** because they transport the urine from the kidneys through the ureters to the urinary bladder. **Urination** involves expulsion of urine from the bladder through the urethra to the exterior.

When considering infections and diseases of the urinary system, it is important to understand that the urethral opening in females is closer to the anus than in males. It also is close to the vaginal opening, making these heavily colonized sites important sources of potential microbial colonizers. These factors combine to generate significant differences not only in the types of microbes colonizing the urinary system of males and females but also their relative susceptibility to infection.

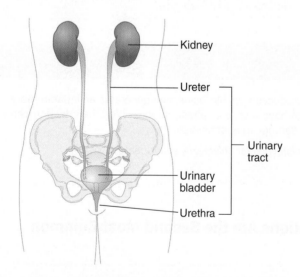

FIGURE 14.12 The Human Urinary Tract. After the kidneys remove excess liquid and waste from the blood as urine, the ureters carry the urine to the bladder in the lower abdomen. Stored urine in the bladder is emptied through the urethra. *»» Explain how an infection of the kidney would arise.*

Part of the Urinary Tract Harbors a Resident Microbiome

Until recently, it was believed that the kidneys and ureters of healthy individuals were sterile; no resident microbes were present. However, research now suggests that the bladder and urethra, at least in females, are not sterile, which means the urine also is not sterile. Rather, these areas are colonized by resident microbes. The main resident bacterial genera known to colonize both the female and male urethra are *Corynebacterium*, *Streptococcus*, *Staphylococcus*, *Bacteroides*, and a variety of other genera including *Peptostreptococcus*, *Prevotella*, and *Porphyromonas*.

Because the urethra of females is much shorter than that of males, ascending infections into the bladder and kidneys are more frequent in females. One of the main antimicrobial defense mechanisms in the urethra is the shedding of the outermost cells of the mucosa with their adherent microbial populations. In addition, the secretion of mucus forms a layer on the epithelial surface that entraps microbes as well as prevents their adhesion to the urethral epithelium. Microbial colonization of the urethra also is hindered by the flushing action of urine, and its normally slightly acidic pH (6.2) and high urea content can kill invading microbes or inhibit their growth. The urine also contains a number of antimicrobial proteins, including defensins and secretory antibodies.

Many of the defense mechanisms operating in the male urethra are similar to those in the female urethra, including shedding of epithelial cells, urinary flow, and the presence in urine of antimicrobial and anti-adhesive factors. In addition, the prostate secretes zinc, which in prostatic secretions can inhibit the growth of potential pathogens such as *E. coli.*

Elimination of bacterial cells from the urethra occurs mainly during urination. In females and males who do not have a urinary tract infection, the number of identifiable bacterial cells in the urine can be up to 10^5 cells per milliliter, with most individuals having between 10 and 1,000 cells per milliliter of urine. The diagnosis of a UTI depends on counting only the infecting bacterial cells in urine, not the resident population. This is accomplished by allowing the initial urinary flow to flush out the resident urethral population. Then a "mid-stream" urine sample is taken for analysis. However, traditional urine culture tests do not detect most bacterial species.

Although, like urine, semen is sterile when first produced, it will accumulate microbes during its passage through the urethra. In fact, the semen of approximately 80% of males contains between one and nine different bacterial species. Therefore, sexual intercourse is likely to provide an avenue by which urethral pathogens can be disseminated.

Most UTIs Involve the Lower Urinary Tract

It has been estimated that up to 50% of humans will suffer a UTI at some time during their lives. In the United States, the CDC reports that UTIs account for about 4 million ambulatory-care visits each year, representing about 1% of all outpatient visits. UTIs account for about 10 million doctor visits each year. **FIGURE 14.13** shows the major bacterial agents that cause UTIs.

Women are most at risk of developing a UTI. In fact, half of all women will develop a UTI during their lifetimes, and many will experience more than one. UTI infections in males remain rare through the fifth decade of life. Then, normal enlargement of the male prostate begins to interfere with emptying of the bladder, and infection rates rise 20%.

Most UTIs start in the lower urinary tract (**FIGURE 14.14**).

Urethritis

In an uncomplicated UTI, pathogens in the gut can contaminate the area around the urethra. In women, organisms like *E. coli*, which cause the majority of UTIs, can travel from the lower intestine and anus to the vagina. From there, due to the close proximity, the infecting organisms can travel to the urethra and cause an inflammation called **urethritis**. Infections in

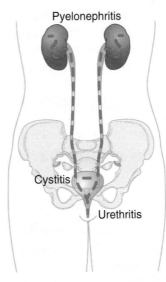

4. The uropathogen enters the kidneys where bacterial toxins damage the tissue.

3. The uropathogen ascends through the ureters to the kidneys.

2. In the bladder, the pathogen colonizes with biofilm formation. Bacterial toxins damage the epithelium.

1. The uropathogen (green rods) enters the urethra and migrates to the bladder.

FIGURE 14.14 Infection of the Urinary Tract. Colonization of the urethra can lead to infection of the bladder. Without treatment, infection can spread to the kidneys via the ureters. *»» How would an uropathogen like* E. coli *evade being washed away in the urine flow?*

men are far less frequent due to the greater distance between the urethra and anus. In addition, to establish an infection in the urinary tract, the uropathogen must withstand the flow force of the urine. For example, *E. coli* uses its pili to latch onto the urinary tract lining. Infections confined to the urethra are characterized by painful urination in women and men and discharge of mucoid or **purulent** (pus) material from the urethral orifice in men.

Treatment of urethritis depends on the cause of the infection. Antibiotics are given for bacterial infections. STIs that cause urethritis can be prevented by using a condom.

Cystitis

If the uropathogen ascends to the bladder, the infection is called **cystitis**. A host immune response (inflammation) ensues as white blood cells (neutrophils) attempt to clear the infection. If the uropathogen can evade the immune response, the bacterial cells will multiply and form a biofilm. Bacterial toxins then cause cell damage to the bladder tissue. Pregnancy can increase the chances of cystitis because the pregnancy can interfere with emptying of the bladder. The symptoms of cystitis are painful urination, increased urinary frequency and urgency, and presence of pus in the urine. The symptoms are clinically distinguished from urethritis by a more acute onset, more severe symptoms, the presence of

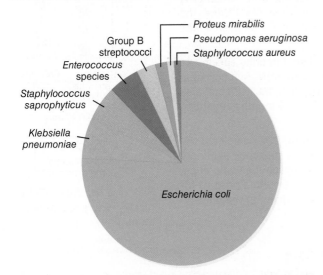

FIGURE 14.13 Common Causes of UTIs. Several bacterial species can cause UTIs. *E. coli* (phylum Proteobacteria) causes the vast majority of uncomplicated UTIs. It is normally found as part of the microbiome of the gastrointestinal tract. *»» How would* E. coli *get from the gastrointestinal tract to the urinary tract?*

bacteria in the urine, and, in approximately half of cases, the presence of red blood cells in the urine.

Symptoms of cystitis often disappear without treatment. People with frequent bladder infections caused by uropathogens like *E. coli* often take antibiotics regularly at low doses. INVESTIGATING THE MICROBIAL WORLD 14 looks into cranberry juice for preventing bladder infections.

Investigating the Microbial World 14

Does Cranberry Juice Cure Urinary Tract Infections?

Urinary tract infections (UTIs), such as cystitis, are an unpleasant illness. Besides the increased urge to urinate, there often is a burning sensation when one does urinate. Although the infection and symptoms can resolve without medical treatment, there is a 24% chance of a recurrence within 6 months.

OBSERVATIONS: Cranberry products, such as cranberry juice or cranberry capsules, have been considered or even touted by many as an effective home treatment for preventing recurring UTIs. Proponents say that cranberry products work by acidifying the urine, which would make the urinary tract less hospitable to pathogens like *Escherichia coli*, the most common cause of UTIs. Also, the sugar (fructose) and proanthocyanidins in cranberries might interfere with the ability of the pili on *E. coli* cells to adhere to the cells lining the urinary tract. Opponents say the evidence is less than compelling and too anecdotal. In addition, good quality, randomized, double-blind, and placebo-controlled studies on the effects of cranberries have not been undertaken.

QUESTION: *Does cranberry juice prevent recurrent UTIs?*

HYPOTHESIS: Regular drinking of cranberry juice cocktail (CJC) will reduce the likelihood of recurrent UTIs. If so, a randomized, double-blind comparison of the efficacy of CJC and placebo juice on women with an acute UTI should reduce the rate and duration of UTI symptoms.

EXPERIMENTAL DESIGN: Out of 419 college women enrolled, 319 had a positive urine culture for a UTI. The experimental juice consisted of a formulated low-calorie CJC (27% juice) with a standardized proanthocyanidins component. The placebo juice was formulated to imitate the flavor (sugar and acidity) and color of cranberry juice but without any cranberry or proanthocyanidin content.

EXPERIMENT: The 319 women were randomly split into two groups. One group (155 women) drank two 8 oz. glasses of CJC twice daily for 6 months. The other group (164 women) drank two 8 oz. glasses of placebo juice twice daily for 6 months. Neither the participants nor investigators knew which group was drinking which juice. Compliance was based on self-reporting.

The clinical assessment consisted of analyzing clean-catch, mid-stream urine specimens from the participants at the beginning of the study and at 3 and 6 months. Self-collected vaginal and rectal specimens were cultured for *E. coli* pathogens. Participants also completed questionnaires at the beginning of the study and at 3 and 6 months, regarding any UTI symptoms as well as other pertinent medical information.

RESULTS: See figure. Of the 319 participants that started the study, 230 completed the entire study (116 in the CJC group and 114 in the placebo group). The presence of urinary and vaginal symptoms over the course of the study was similar between the two groups. A positive UTI was based on a combination of symptoms and a urine culture positive for a known uropathogen. Gastrointestinal symptoms were twice as frequent in the placebo group as in the CJC group.

Risk of a recurring UTI by history and juice assignment.

Modified from Barbosa-Cesnik, C., et al. (2011). *Clin Infect Dis* 52 (1):23–30.

CONCLUSIONS:

QUESTION 1: *Was the hypothesis validated? Explain using the figure.*

QUESTION 2: *Explain why this was a (a) randomized, (b) double-blind, and (c) placebo-controlled study.*

QUESTION 3: *Can you think of any problems or caveats in the setup and performance of this study that could make the results questionable?*

You can find answers online in **Appendix E.**

Modified from Barbosa-Cesnik, C., et al. 2011. *Clin Infect Dis* 52(1):23–30.

Pyelonephritis

A uropathogen that passes up the ureters and into the kidneys can cause an infection and inflammation called **pyelonephritis**. Colonization again is damaging, this time to the kidney tissue. Symptoms of pyelonephritis often begin suddenly with chills, fever, pain in the lower part of the back, nausea, and vomiting. About one-third of people with pyelonephritis also have symptoms of cystitis. If not treated properly, a kidney infection can permanently damage the kidneys or spread to the bloodstream and cause a life-threatening infection. Thus, prompt medical attention and antibiotic therapy are essential.

Prostatitis

An inflammation and infection of the prostate, called **prostatitis**, typically develops in young men and in older men following placement of an indwelling catheter. In **acute bacterial prostatitis**, symptoms come on suddenly and can include fever and chills and the signs and symptoms of cystitis, including increased urinary urgency and frequency. In **chronic bacterial prostatitis**, signs and symptoms develop more slowly and usually are not as severe as are those of acute prostatitis. Antibiotics are usually the choice of treatment for both types of prostatitis.

Healthcare-acquired UTIs can occur in a medical care facility if patients have a urinary catheter placed through the urethra and into the bladder to collect urine. Although these infections can generate the same sequence of events in the urinary tract as just described, catheter-caused infections are often referred to as "complicated UTIs" because catheterization leads to a compromised bladder; that is, uropathogens use the catheter as a prime location for colonization, which induces an extremely strong immune response.

TABLE 14.2 summarizes the bacterial diseases of the urinary system.

TABLE 14.2 Urinary Tract Infections Caused by Bacteria

Disease	Causative Agent	Signs and Symptoms	Transmission	Treatment	Prevention
Urethritis	*Escherichia coli* *Chlamydia trachomatis* *Mycoplasma* Other bacterial species	Persistent urge to urinate Burning sensation when urinating Blood in the urine Cloudy, strong-smelling urine	Bacterial entry from the large intestine to the urethra Sexual intercourse	Antibiotics	Drinking plenty of liquids (water) Urinating frequently Keeping the genital area clean
Cystitis	*Escherichia coli*	Strong, persistent urge to urinate Burning sensation when urinating Blood in the urine Passing cloudy or strong-smelling urine	Bacterial entry from the large intestine through the urethra to the bladder Sexual intercourse	Antibiotics	Drinking plenty of liquids (water) Urinating frequently Keeping the genital area clean
Prostatitis	*Escherichia coli* *Staphylococcus* *Proteus* *Pseudomonas*	Fever and chills Flu-like symptoms Pain in the prostate gland, lower back, or groin Increased urinary urgency and frequency Pain when urinating Blood-tinged urine	Bacteria from the large intestine From bladder Sexual intercourse	Antibiotics	Practicing good hygiene Keeping the penis clean Drinking enough fluids to cause regular urination
Pyelonephritis	*Escherichia coli* *Enterococcus* *Klebsiella* *Pseudomonas*	Frequent urination Persistent urge to urinate Burning sensation or pain when urinating Abdominal pain or pressure Cloudy urine with a strong odor Pus or blood in urine	Bacteria from the large intestine enter urethra From bloodstream	Antibiotics	Practicing good hygiene Drinking enough fluids to cause regular urination

Concept and Reasoning Checks 14.3

a. How does the urinary system differ from the urinary tract and why is the urinary tract more likely to become infected?
b. Why is the urethra the only part of the urinary system normally colonized by resident microbiome?
c. How can you diagnose the different UTIs based on signs and symptoms?

Chapter Challenge C

An earlier challenge asked you to decide whether you believe the female reproductive system is prone to more infections than is the male reproductive system. How about the female urinary system?

QUESTION C: *Do you believe the female urinary system would be more prone to infections than the male urinary system? Again, back up your decision with evidence.*

You can find answers online in **Appendix F**.

■ KEY CONCEPT 14.4   Contact Diseases Can Be Caused by Resident Bacterial Species

The skin is the largest organ of the body, accounting for more than 10% of body weight and having a surface area of almost 2 square meters. Being the interface between environment and internal body tissues and organs, the skin plays a critical role as a barrier against microbial entry and infection. Should there be damage to, or wounding or puncturing of, the skin, invasive bacterial disease can occur.

The Skin Protects Underlying Tissues from Microbial Colonization

The skin, along with its accessory structures (nails, hair, sweat glands, and sebaceous glands) constitutes the **integumentary system** (FIGURE 14.15).

The upper, bloodless portion of skin is the **epidermis** (*epi* = "over"). It consists of layers of **keratinocytes** that produce a tough, fibrous protein called **keratin**. The keratinocytes are produced in the substratum basale and, over a two-week period, they are pushed toward the surface where they eventually die in the stratum corneum. They then are worn away and replaced by newer cells that are continually pushing up from below. This thickened, keratinized layer is dry and waterproof, and, along with epidermal antimicrobial peptides and immune defenses present in the epidermal sublayers, promotes pathogen clearance and prevents most bacterial cells, viruses, and other foreign substances from entering the underlying tissues.

Beneath the epidermis lies the **dermis**, a complex, thick layer of fibrous and elastic connective tissue. The dermis contains sweat glands, the sweat being composed of water, salt, and a few other antimicrobial substances. The salt and chemicals create an unwelcome environment for most pathogens. Our characteristic body odor comes from the microbial digestion of sweat, the smelly topic for MICROFOCUS 14.2.

Hair, which covers much of the skin surface, consists of dead keratinized cells emerging from hair follicles embedded in the dermis. Associated with these follicles are **sebaceous glands** that are primarily found on the face, scalp, and upper torso. These glands produce an oily substance, called **sebum**, which keeps the skin and hair soft and moist. Sebum has a low pH (4.5 to 5.5) and, along with the breakdown of the fatty acids in the oils to toxic end products, make the skin surface inhospitable to nonresident microbes.

The Skin Harbors a Resident Microbiome

The skin can be envisioned as an ecosystem composed of various habitats or specialized niches colonized by a resident microbiome, a unique community of normally harmless resident and diverse

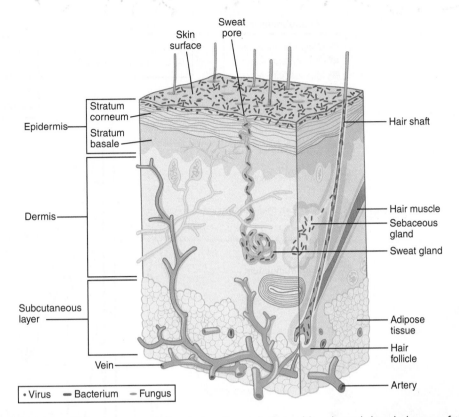

FIGURE 14.15 **Layers of the Skin.** A cross-sectional view through the epidermis and dermis layers of the skin. Microbes, including many bacteria, fungi, and viruses, cover the skin surface as well as colonizing the sebaceous glands, sweat glands, and the hair shafts and follicles. *»» Why are the sebaceous glands and the sweat glands so susceptible to infection?*

MICROFOCUS 14.2: Public Health

"Smelly Sweat"

Several years ago, a man was removed from an Air Canada flight due to his body odor. The man was escorted off the plane after multiple passengers complained about his smell. Sweating can make a person smell awful, as in the case of the airline passenger. You certainly have experienced this by being unlucky enough to walk past a "smelly" pedestrian or get a whiff of the "stinky" person next to you on a bus.

Body odor, called bromhidrosis (*brom* = "stench"; *hidro* = "sweat"; *sis* = "the act of") begins when the 3 to 4 million sweat glands on the face, palms, and soles of the feet—but especially in the apocrine glands of the armpits and genital regions—secrete sweat. When bacterial cells on the skin multiply in the sweat, they break down some of the sweat proteins. For example, the propionibacteria in the sebaceous glands break down amino acids into propionic acid, which has a pungent vinegar-like smell. Likewise, *Staphylococcus epidermidis* on the skin degrades the amino acid leucine into isovaleric acid, which has the typical "locker room" smell (see figure).

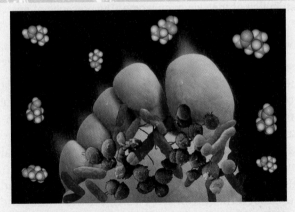

Foot odor is due to the bacterial microbiome (false-color scanning electron micrograph) and the acid products they produce (molecular models = isovaleric acid).
© Scimat/Science Source.

For most cases of sweat, just make sure you maintain good hygiene; take regular showers or baths and use a deodorant (designed to eliminate odor by turning the skin pH acidic, making for a more hostile environment for bacterial growth) or antiperspirant (designed to prevent odor and reduce sweat production by blocking the sweat pores). However, a word of caution: recent research suggests that antiperspirants and deodorants modify the armpit microbiome, and presently, we have no idea if that is beneficial or harmful to our skin microbiome and our health.

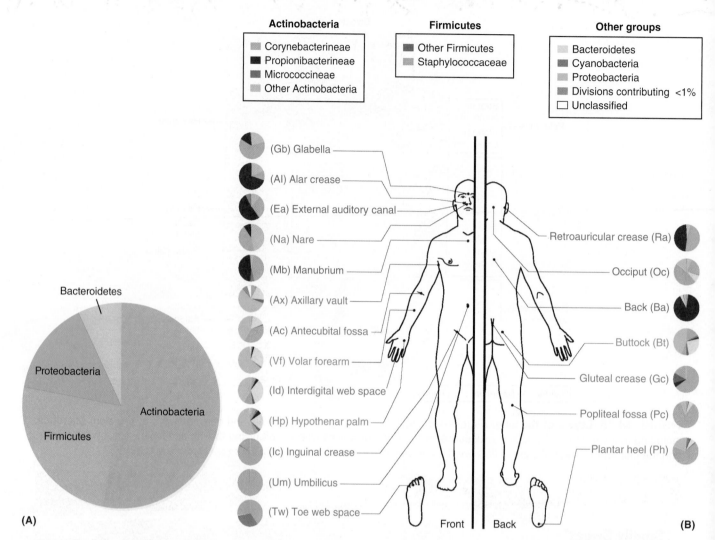

FIGURE 14.16 The Skin Microbiome. (A) Members of four phyla compose most of the skin microbiome. **(B)** A topographical distribution of the skin microbiome. The presence of specific bacterial taxa is governed by the microenvironment of each of the three skin sites. »» *What are the most common organisms on each of the three skin sites?*

(A) Data from Cho, I. and Blaser, M. J. 2012. *Nat Rev Gen* 13(4):260–270. (B) Courtesy of Darryl Leja, National Human Genome Research Institute (NHGRI).

microbial species, including viruses. The most recent studies suggest that the skin microbiome remains stable in healthy individuals despite constant exposure to the environment. These microbial communities are dominated by four bacterial phyla: the gram-positive Actinobacteria and Firmicutes and the gram-negative Bacteroidetes and Proteobacteria (**FIGURE 14.16A**). Furthermore, the diversity of these phyla and specific species varies by skin site (**FIGURE 14.16B**), as described here:

▶ **Sebaceous Sites.** Bacterial diversity appears lowest at sebaceous (oily) sites, such as the forehead and back. The predominant organisms are species of *Staphylococcus* (phylum Firmicutes) and *Propionibacterium*

(phylum Actinobacteria), the latter being capable of metabolizing the sebum into products that are toxic to many other bacterial species. *Propionibacterium acnes* chiefly inhabits hair follicles where the microaerophilic environment is optimal for growth. In higher numbers, as we will discuss in a moment, *P. acnes* can be associated with skin conditions like acne.

▶ **Moist Sites.** Analysis of the microbial communities in moist sites, such as the navel, groin, sole of the foot, back of the knee, and the inner elbow, indicate these sites also are dominated by *Staphylococcus* and *Corynebacterium* species. The gram-positive corynebacteria usually do not cause skin infections.

▶ **Dry Sites.** The areas with the highest diversity are the dry sites, which include the forearm, legs, and hand. These sites are dominated by *Staphylococcus epidermidis* and several gram-negative species in the Proteobacteria and Bacteroidetes phyla.

Lastly, the antagonistic activities of the resident microbiome toward exogenous or transient microbes also are the consequence of their community interactions. For example, *S. epidermidis*, which covers the skin surface, out competes and produces chemical compounds that inhibit its potentially pathogenic cousin, *S. aureus*. Although *P. acnes* plays a role in skin acne, the organism also produces short-chain fatty acids that inhibit invasion of the skin by exogenous pathogens.

Acne Is the Most Common Skin Condition in the Developed World

Whether you call it zits, blackheads, or pimples, **acne** can be an annoying and persistent malady that is especially stressful for some 85% of adolescents and 10% of adults (**FIGURE 14.17A**).

Causative Agent and Etiology

Many medical experts refer to acne as a nontransmissible "physiological condition" rather than a disease. The development of acne, referred to medically as "acne vulgaris," is not the result of poor hygiene or eating greasy foods. Rather, it involves several interacting factors.

The most common microbial factor is *P. acnes*, a small, slow-growing, aerotolerant, gram-positive rod (**FIGURE 14.17B**). It typically colonizes the sebaceous skin environment (e.g., face, upper chest, and back). However, it is not considered the cause of acne because *P. acnes* cells also are part of the stable skin microbiome of acne-free individuals (**FIGURE 14.18**). Thus, other factors also contribute to the development of acne. For example, during puberty, sex hormone changes (especially testosterone in males) cause the sebaceous glands to enlarge and produce excessive amounts of sebum. The *P. acnes* cells degrade the lipids in the sebum and adhere to the fatty acid products, aiding colonization and population growth. Other factors include excess sebum production, abnormal follicular development, and the individual's immune response to the buildup of *P. acnes* cells.

Clinical Presentation

The mildest form of acne occurs when hair follicles become swollen with sebum and keratinocytes. The *P. acnes* cells exacerbate the condition by causing the keratinocytes to become sticky, plugging the follicles. Such plugged sebaceous glands, called

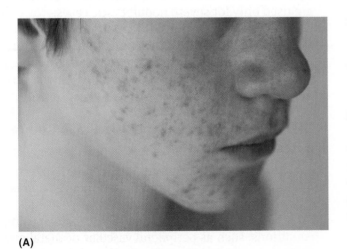
(A)

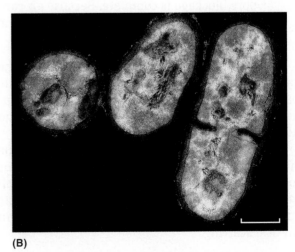
(B)

FIGURE 14.17 Acne Vulgaris. (A) Acne, a typical affliction of teenagers and young adults, primarily affects the face, upper torso, and back. **(B)** *Propionibacterium acnes* bacteria. This false-color transmission electron micrograph shows three of the rod-shaped bacterial cells. (Bar = 1 µm.) *»» Why are the face, upper torso, and back the sites most likely to develop acne?*

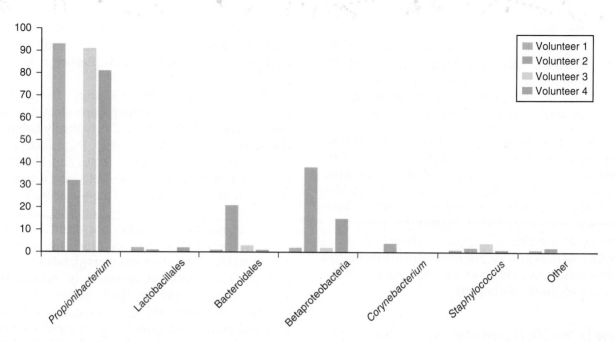

FIGURE 14.18 Variation in Skin Microbiome of the Back. The distribution of resident skin microbes (by bacterial taxa) differs between these four volunteers. However, *Propionibacterium* is the predominant organism. *»» What is unusual about volunteer 2?*

Data from Grice, E. A. and Segre, J. A. 2010. *Nature Rev Microbiol* 9:244–252.

comedones (sing., comedo), can exist in two noninflammatory forms:

▶ "Blackheads" represent follicles in which the sebum plug does not completely block the pore. A chemical reaction oxidizes melanin, giving the material in the follicle a black color.

▶ "Whiteheads" represent follicles in which pore blockage is complete. Because air cannot reach the melanin, it is not oxidized and the follicle remains white.

The overgrowth of *P. acnes* breaks down the trapped sebum into products irritating the skin, resulting in a local inflammation in the form of papules (small, fluidless, and often red skin bumps) and pustules (pus-containing skin bumps).

However, some 20% of individuals with acne will develop a severe condition in which numerous nodules or cysts develop deeper in the dermis. Such **nodulocystic acne** can be quite painful, and cysts can join to form large abscesses that can result in physical scarring.

Treatment and Prevention

Most mild cases can be treated effectively at home with good daily skin care using mild soap and, if necessary, over-the-counter treatments, such as benzoyl peroxide. More severe cases might require antibiotics, which can reduce the numbers of bacterial cells and the production of inflammatory substances. Topical retinoids, such as tretinoin (Retin-A®), are used to prevent blockage of follicles and to promote extrusion of the plugged material. Systemic retinoids, such as isotretinoin (Accutane®) are used only for severe nodulocystic cases.

Before leaving this skin condition, MICROFOCUS 14.3 identifies a possible microbial connection to **rosacea**, which is a condition that causes redness and often small, red, pus-filled bumps on the face that can resemble acne.

Skin Wound Infections Can Be the Result of a Resident Microbiome

Should a wound occur, not only could external microbes enter and cause an infection but also resident microbes.

Chronic wounds are the result of predisposing conditions, such as peripheral vascular or metabolic diseases in which there is a lack of blood flow to body tissues. In these cases, dead skin breaks and produces an open sore, resulting in a leg ulcer or pressure sores (bedsores) that become infected with resident bacteria. Skin microbiome studies have not identified

MICROFOCUS 14.3 Medical Microbiology

Rosacea

Rosacea is a skin condition that causes an inflammation of the skin around the nose, cheeks, and chin (see figure). The condition affects about 3% of individuals in a population, with fair-skinned females between 30 and 50 years of age being three times more susceptible than are men. It also occurs more often in individuals with skin changes due to aging or sunburn. As the figure shows, rosacea can be mistaken for acne or another skin condition based on the typical redness and the presence of small facial skin papules and pustules. Although antibiotics have been used to treat rosacea (although they are not a cure), a potential bacterial agent responsible for the condition had not been identified—until recently.

Skin mites (*Demodex folliculotum*) are normal inhabitants on the skin and live harmlessly around the hair follicles on the face. These arthropods have been associated with rosacea because rosacea patients have a higher number of the mites than non-rosacea individuals.

Recently, researchers in Ireland have discovered a bacterium that lives in the digestive tract of the mites. The bacterium, *Bacillus oleronius*, produces proteins that trigger an immune reaction. In fact, rosacea patients strongly react with these bacterial substances. *Staphylococcus epidermidis* has also been identified in pustules of some rosacea patients. Interestingly, both bacterial species are susceptible to the antibiotics normally used to treat rosacea patients.

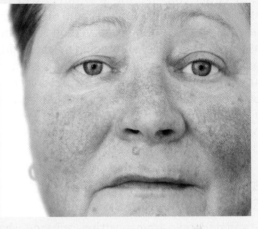

The red skin rash on the face of an individual suffering from rosacea.

© Lipowski Milan/Shutterstock.

Accordingly, one infection scenario suggests that when the mites die on the skin, they decompose and set free the bacteria. The bacterial cells then invade the surrounding skin tissue, producing the chemicals triggering the immune reactions leading to rosacea. In addition, the mites might transport *S. epidermidis* around the face, accounting for their presence in some rosacea patients. Although conclusive proof for a bacterial agent(s) remains to be shown, an alternative to antibiotic therapy might be to develop a chemical that would kill or control the mite population on the face.

a unique organism to a particular type of chronic wound. In fact, most are **polymicrobial**; that is, they are the result of more than one infectious agent.

An **acute wound** arises from cuts, lacerations, or surgical procedures on or to the skin. In such cases, resident microbes can enter the deeper layers of tissue and cause an infection. For example, each year in the United States intravenous catheters are used in about 50% of hospitalized patients. Because the catheter must be inserted through the skin, there is a potential that resident skin bacteria, such as

S. epidermidis, can be mechanically "injected" into the tissues when the catheter is inserted. *S. epidermidis* now acts as an opportunistic pathogen to colonize the catheter and form a **biofilm**. Implanted medical devices, such as hip and knee prostheses, pacemakers, and heart valves, also can be subject to the same biofilm scenario.

To try to limit such infections, rigorous cleaning and disinfection of the skin surface must be done to minimize the resident microbiome. In addition, the device or implant needs to be checked for sterility.

Concept and Reasoning Checks 14.4

a. How do the epidermis and dermis help protect the body from microbial colonization and infection?

b. Identify the three skin sites colonized by the resident microbiome and provide some examples of the microbes found at these sites.

c. How does *Propionibacterium acnes* contribute to acne?

d. How can resident bacterial species cause an infection from a chronic or acute wound?

■ KEY CONCEPT 14.5   Contact Diseases Also Can Be Caused by Bacterial Species That Originate Externally

When pathogens infect the skin from the outside, they either colonize the skin directly or reach the skin through the bloodstream. Such bloodborne involvement might be observed on the skin surface as a rash. Therefore, a physician must consider many possible diseases when evaluating a skin infection and might need to identify the presence or absence of specific signs and symptoms before coming to a diagnosis. FIGURE 14.19 identifies several of the bacterial diseases that we will encounter, based on their anatomical site on the skin.

Staphylococcal Contact Diseases Have Several Manifestations

Staphylococci, gram-positive spheres that form clustered cell arrangements, are normal inhabitants of the human skin, mouth, nose, and throat. Although they generally live in these areas without causing harm, they can initiate disease when they penetrate the skin barrier or the mucous membranes. *S. aureus* is the species often involved in these contact diseases.

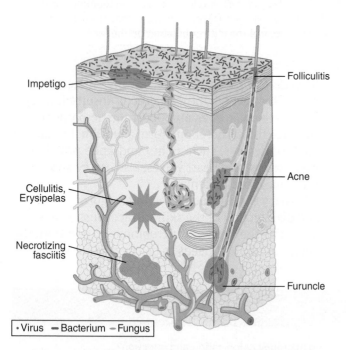

Impetigo

Folliculitis

Cellulitis, Erysipelas

Acne

Necrotizing fasciitis

Furuncle

• Virus — Bacterium — Fungus

FIGURE 14.19 Anatomical Sites Associated with Skin and Soft Tissue Infections. Although the symptoms of several of these diseases overlap, each is a distinct disease based on the pathogen causing the infection. *»» Which infections would you suspect are most likely to be severe? Explain.*

Localized Skin Infections

The hallmark of an *S. aureus* skin infection is **folliculitis,** which produces pus-filled pockets at the base of a hair follicle typically on the armpits, buttocks, face, or neck (**FIGURE 14.20A**). *Pseudomonas aeruginosa*, an aerobic gram-negative rod, also can cause folliculitis, especially with people who frequent hot tubs or whirlpool baths that are not kept clean and chlorinated.

A more serious infection of folliculitis by *S. aureus* involves the formation of an **abscess,** a confined pus-filled lesion. A **furuncle** (boil) is a warm, painful abscess that develops in the region of a hair follicle, whereas **carbuncles** are a group of connected, deeper abscesses (**FIGURE 14.20B**). Because bacterial cells can enter the blood, treatment might require antibiotic treatment and **debridement,** which involves the removal of dead, damaged, and infected tissue.

Toxin-Associated Contact Diseases

Several skin diseases are associated with toxins produced by *S. aureus*, such as those described here:

▶ **Impetigo.** A highly contagious but superficial staphylococcal skin disease that commonly affects children is **impetigo.** The production of toxins allows the bacterial cells to spread through the skin, which leads to an infection that develops as thin-walled blisters oozing a yellowish fluid before crusting over (**FIGURE 14.21A**). Usually, the blisters occur on the exposed parts of the body, especially on the face and limbs where skin trauma has occurred.

Impetigo is commonly treated topically, although antibiotic therapy might be needed for multiple lesions.

▶ **Scalded Skin Syndrome.** Another contact disease is **scalded skin syndrome,** which is usually seen in infants and young children under the age of five. The staphylococcal toxins produced cause the skin to become red, wrinkled, and tender to the touch, giving the skin a sandpaper appearance. The epidermis can then peel off just above the stratum basale, through a process called **desquamation** (**FIGURE 14.21B**). Antibiotic therapy should be instituted early even in a tentative diagnosis.

▶ **Toxic Shock Syndrome.** Another toxin-producing strain of *S. aureus* causes a rare illness called **toxic shock syndrome (TSS).** The staphylococci involved in TSS can be the result of a skin abscess, surgery, burn wounds, or barrier contraceptives (sponges and diaphragms).

TSS toxin-1 activates immune cells and triggers the release of immune proteins that cause inflammation. The earliest symptoms of disease are nonspecific with fever, chills, vomiting, and watery diarrhea. Patients then experience a sore throat, severe muscle aches, and a sunburn-like rash. This is followed by desquamation, especially on the soles of the feet and palms of the hands about 1 to 2 weeks after the onset of the rash (**FIGURE 14.21C**). Clinical symptoms can progress rapidly, leading to a sudden drop in blood platelets, renal failure, shock, and death. Antibiotics can be used to inhibit bacterial growth, but measures such as blood transfusions must be taken to control the shock. In 2015, 98 cases of TSS were reported to the CDC.

Skin Diseases Caused by Streptococcus pyogenes Can Be Mild to Severe

Streptococci are a large and diverse group of encapsulated, nonmotile, facultatively anaerobic, gram-positive cocci. Medically, the streptococci are classified in two ways. The first divides streptococci into hemolytic (*hemo* = "blood") groups, depending on how they affect sheep red blood cells when the streptococci are plated on blood agar. The

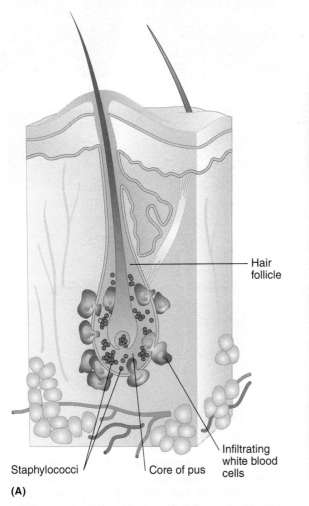

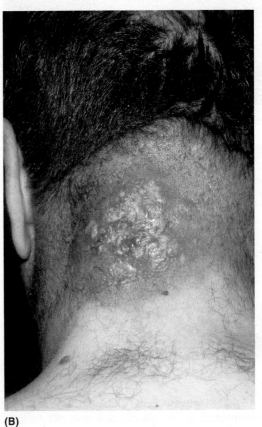

(A)   (B)

FIGURE 14.20 Staphylococci and Skin Abscesses. (A) Staphylococci infect the base of a hair follicle during the development of a skin lesion. White blood cells that engulf bacterial cells have begun to collect at the site as pus accumulates, and the skin has started to swell. **(B)** A severe carbuncle on the back of the head/neck. *»» How would a severe carbuncle be treated?*

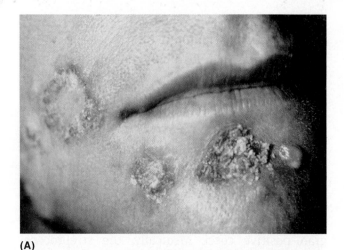

(A)

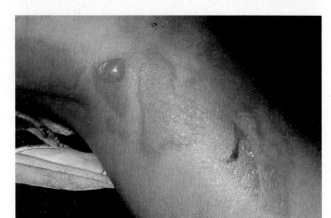

(B)

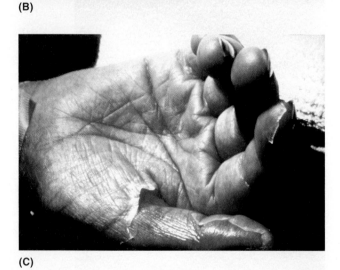

(C)

FIGURE 14.21 Staphylococcal Skin Diseases. (A) A patient with impetigo on the cheeks and chin. **(B)** An individual with scalded skin syndrome. **(C)** A patient with toxic shock syndrome. In all cases, note the peeling of the skin. *»» What chemical factors are responsible for the pathology observed?*

α-hemolytic streptococci turn blood agar an olive-green color as a secreted toxin partially destroys red blood cells in the medium; colonies of **β-hemolytic** streptococci produce clear, colorless zones around the colonies due to the complete destruction of red blood cells (**FIGURE 14.22A**); **nonhemolytic** streptococci cause no change in blood agar.

The second classification is based on variants of a carbohydrate located in the cell walls of some

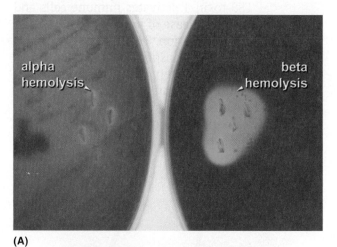

(A)

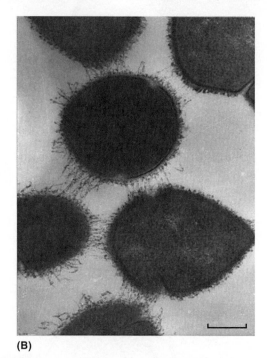

(B)

FIGURE 14.22 Streptococci. (A) An example of α- and β-hemolysis caused by hemolysins released from *Streptococcus pyogenes* cells. **(B)** A close-up view of *S. pyogenes* cells showing strands of M protein protruding through the capsule. (Bar = 0.5 μm.) *»» What function does M protein play in pathogenesis?*

β-hemolytic streptococci. Groups A and B are the most important to human disease and *S. pyogenes* is the most common species. This β-hemolytic organism is generally implied when physicians refer to **group A streptococci** (**GAS**).

The pathogenicity of *S. pyogenes* is enhanced by the presence of fibril-protruding **M proteins**, which are anchored in the cell wall (**FIGURE 14.22B**). With more than 100 chemical versions of these proteins, M protein gene (*emm*) typing has been used to identify *S. pyogenes* serotypes.

The CDC estimates that there are about 10,000 to 15,000 cases of invasive GAS disease every year, accounting for up to 1,600 deaths. There are about 600 million infections and 500,000 deaths globally. Transmission occurs by direct contact with the mucus from the nose or throat of individuals with *S. pyogenes* infections or through contact with infected wounds or sores on the skin. It also can be spread by asymptomatic carriers.

Most infections caused by *S. pyogenes* are relatively mild, such as **streptococcal pharyngitis**, popularly known as **strep throat**. Although *S. aureus* is the most common cause of impetigo, *S. pyogenes* also can cause a similar infection.

Some other streptococcal soft tissue infections can be more severe, as described below.

Cellulitis

S. pyogenes is a major cause of **cellulitis**, a severe, fast-spreading skin infection of the dermis and subcutaneous tissues resulting from minor trauma, penetrating wounds, eczema, or surgical incisions. Symptoms include a high fever, shaking and chills, and headache. The skin lesion enlarges rapidly, invading and spreading through the lymphatic vessels and appears as a fiery red, swollen, warm, and painful rash, usually on the lower limbs or face. A superficial form of cellulitis involving the dermis is **erysipelas** (*erysi* = "red"; *pela* = "skin"), which is often found in infants, children, and older adults (**FIGURE 14.23**). Cellulitis treatment usually involves oral or intravenous antibiotic therapy.

Although these strep infections tend to be mild, other GAS infections can cause potentially life-threatening diseases. Researchers believe that in these latter, potentially deadly cases, *S. pyogenes* cells respond to the body's immune attack by altering their gene expression. Among the changes is an increased rate of cell division, leading to rapid, possibly uncontrollable spread of the pathogen in the infected tissues.

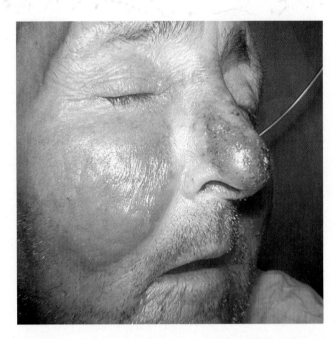

FIGURE 14.23 Facial Erysipelas. Infection of the skin and subcutaneous tissue with *Streptococcus pyogenes* produces a bright red rash of the affected areas, in this case the cheek and nose. »» *What is most likely the source of this patient's infection?*

Courtesy of Dr. Thomas F. Sellers, Emory University/CDC.

Streptococcal Toxic Shock Syndrome

GAS causes **streptococcal toxic shock syndrome** (**STSS**), which is caused by toxins structurally similar to those in staphylococcal TSS. Without treatment, more than half of STSS patients die, highlighting the need for prompt antibiotic therapy.

Necrotizing Fasciitis

Occasionally described in the news media as the "flesh-eating disease," **necrotizing fasciitis** is a rare but dangerous *S. pyogenes* infection that destroys muscles, fat, and soft skin tissue (**FIGURE 14.24**). The pathogens reach the subcutaneous tissue through a wound or trauma to the skin surface. There, the enzymes and toxins produced by the bacterial cells cause **necrosis** (cell death) of the subcutaneous tissue and lower skin layer. Without early diagnosis and treatment, damage to the surrounding tissue spreads rapidly and, along with blockage of small subcutaneous vessels, infection produces additional dermal cell death. Debridement and surgery often are needed to remove damaged tissue; in severe cases, amputation might be the only recourse to remove the infection.

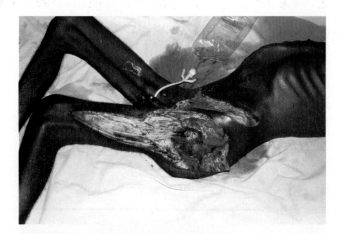

FIGURE 14.24 Necrotizing Fasciitis. The extensive loss of connective tissue can be seen in the leg of a 15-year-old AIDS patient. Extensive wound debridement was required. *»» What tissues are most affected by the GAS?*

© Dr. M.A. Ansary/Science Source.

Other Skin Punctures and Wound Trauma Also Can Lead to Skin Infections

The risk of infection due to a traumatic wound, such as a battlefield injury, compound fracture, or thermal burn, depends on several factors, including the extent of potential contamination, the contaminating dose of bacterial cells, and their virulence. The physical and physiological nature of the wound—that is, are there areas of necrosis or poor blood and oxygen supply—also are important factors affecting bacterial colonization and infection. **Gas gangrene** is often the result of a traumatic wound. It is a soil-borne disease, which is described in Chapter 13.

Burns are one of the most common and devastating forms of trauma. In a burn injury, the skin has been mechanically damaged and the underlying tissues are open to potential infection (**FIGURE 14.25**). Burn wounds can be classified as **wound cellulitis**, which involves the unburned skin at the margin of the burn, or as an **invasive wound infection**, which is characterized by microbial invasion of viable tissue beneath the burn.

Although a variety of bacterial species can cause infection (TABLE 14.3), one of the most likely infective agents in many burn centers is *P. aeruginosa*. This aerobic, gram-negative rod is widely distributed in soil, water, plants, and animals (including humans). *P. aeruginosa* can form mature biofilms within about 10 hours after colonization. Therefore, limiting burn wound infections and patient mortality requires rapid burn debridement and wound

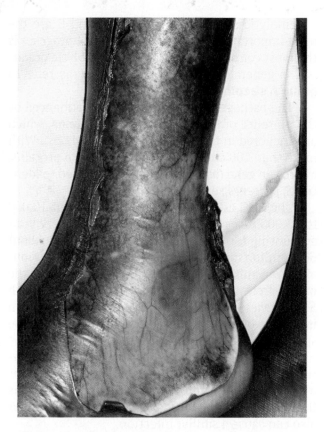

FIGURE 14.25 A Burn Trauma to the Leg. Burns to this patient's lower leg and ankle have reached underlying fat and muscle tissue, making the tissue extremely susceptible to infection. *»» What types of bacteria are likely to infect this patient?*

© Dr. M.A. Ansary/Science Source.

TABLE 14.3 Bacterial Genera Commonly Associated with Invasive Burn Wound Infections

Gram-Positive Bacterial Phyla and Genera	Gram-Negative Bacterial Phyla and Genera
Firmicutes • *Enterococcus* • *Staphylococcus*	**Proteobacteria** • *Acinetobacter* • *Enterobacter* • *Escherichia* • *Klebsiella* • *Proteus* • *Pseudomonas* • *Serratia* **Bacteroidetes** • *Bacteroides*

closure. In addition, patients with serious invasive wound infections require immediate topical and systemic antibiotic therapy to minimize morbidity and mortality.

Leprosy (Hansen Disease) Is a Chronic, Systemic Infection

You might believe that **leprosy** is a disease of the past. In fact, the disease still strikes many individuals today. For most of the past 2,600 years, leprosy has been considered a curse of the damned. It did not kill, but neither did it seem to end. Instead, it lingered for years, causing a chronic infectious disease and resulting in severe disabling deformities of the body.

The agent of leprosy, or **Hansen disease**, is *Mycobacterium leprae*, an acid-fast rod related to *M. tuberculosis*. Being an obligate intracellular parasite, it cannot be cultivated in artificial laboratory media. In 1960, researchers at the CDC succeeded in cultivating the bacillus in the footpads of mice, and in 1969, scientists discovered it would grow in the skin of nine-banded armadillos.

In the United States, there were 63 new cases reported in 2015 and there are some 3,600 patients being treated for the chronic condition. About 70% of the new cases occur in susceptible individuals who traveled to a leprosy-endemic area abroad. Of concern to health experts is a new strain of *M. leprae* that was first identified in 2011. This strain appears to be spread by zoonotic transfer, specifically from infected armadillos to native-born Americans living in the southern United States, especially Texas and Louisiana.

Leprosy is hard to transmit because about 95% of the world's population has a natural immunity to the disease. It is spread by contact with nasal secretions, which are taken up through the upper respiratory tract. The disease has an unusually long incubation period of 3 to 5 years, a factor making diagnosis very difficult. Because the organism's optimal growing temperature is 31°C, symptoms occur in the skin (e.g. hands, feet, face, and earlobes) and peripheral nervous system that represent cooler parts of the body.

The WHO categorizes leprosy by the type and number of skin areas affected. **Paucibacillary (tuberculoid) leprosy** is a more limited and mild disease characterized by a few (*pauci* = "few") flat or slightly raised skin lesions of various sizes that are typically pale or slightly red and numb to touch. **Multibacillary (lepromatous) leprosy** is a more severe, contagious disease, involving larger numbers of skin lesions. In both forms, the most important pathological feature is involvement and damage to the peripheral nerves. However, leprosy also can involve the kidneys, eyes, and nose (**FIGURE 14.26A**).

Until recently, one of the principal drugs for the treatment and cure of leprosy was a sulfur

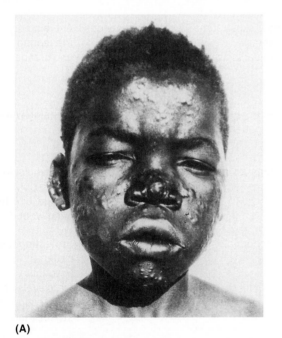
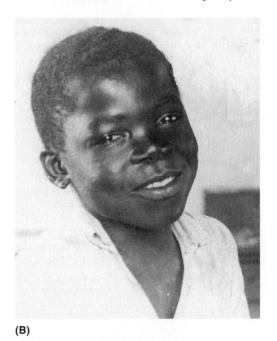

(A) (B)

FIGURE 14.26 Treating Leprosy. The young boy with multibacillary leprosy is pictured **(A)** before treatment with dapsone and **(B)** some months later, after treatment. »» *Why is the disfiguring caused by* Mycobacterium leprae *limited to the body extremities?*

(A and B) Courtesy of American Leprosy Missions, www.leprosy.org.

compound called Dapsone. In many cases, such as the one shown in **FIGURE 14.26B**, the results can be dramatic. Today, treatment involves multidrug therapy with dapsone, rifampin, and clofazimine.

TABLE 14.4 summarizes the bacterial contact diseases. **MICROINQUIRY 14** presents four cases for study involving sexually transmitted and contact diseases.

TABLE 14.4 A Summary of the Major Contact Diseases Caused by Bacterial Species

Inflammation or Disease	Causative Agent	Signs and Symptoms	Transmission	Treatment	Prevention
Acne	*Propionibacterium acnes*	Comedones (whiteheads and blackheads)	Part of resident skin microbiome	Benzoyl peroxide Antibiotics Isotretinoin	Gentle washing of affected skin Benzoyl peroxide
Furuncle and Carbuncle	*Staphylococcus aureus*	Painful single or cluster of boils	Autoinfection Contact with infected person	Drainage Debridement	Practicing good hygiene
Impetigo	*Staphylococcus aureus* *Streptococcus pyogenes*	Thin-walled blisters forming a crust	Direct or indirect contact	Skin cleansing Topical antibiotic	Practicing good hygiene
Scalded skin syndrome	*Staphylococcus aureus*	Red, wrinkled and tender skin Epidermis may peel	Direct or indirect contact	Cefazolin	Practicing good hygiene
Toxic shock syndrome (TSS)	*Staphylococcus aureus*	Fever, vomiting, watery diarrhea, sore throat, muscle aches, and sunburn-like rash	Vaginal tampons Skin wounds Surgery	Supportive care Antibiotics	Avoiding highly absorbent vaginal tampons
Erysipelas	*Streptococcus pyogenes*	High fever, shaking and chills, headache leading to fiery rash on lower limbs and face	Minor skin trauma Eczema Surgical incisions	Oral or intravenous antibiotics	Avoiding dry skin Preventing cuts and scrapes
Streptococcal TSS (STSS)	*Streptococcus pyogenes*	Fever, dizziness, confusion, and flat body rash	Direct contact with patients or carriers	Clean any wounds Debridement	Cleaning wounds
Necrotizing fasciitis	*Streptococcus pyogenes*	Fever with pain and swelling at the wound site	Trauma to skin surface	Broad spectrum antibiotics Debridement Surgery	Cleaning skin after a cut, scrape, or other deep wound
Burn infections	*Pseudomonas aeruginosa* Other bacterial species	Difficult to diagnose May be absent, minimal, or late developing	Nosocomial	Antibiotic therapy Debridement	Protecting burn patients Practicing high levels of disinfection and sterilization
Leprosy (Hansen disease)	*Mycobacterium leprae*	Disfiguring of skin and bones, loss of pain sensation, loss of facial features	Nasal secretions	Multidrug therapy with dapsone, rifampin, and clofazimine	Avoiding contact where endemic

MICROINQUIRY 14

Sexually Transmitted and Contact Disease Identification

The following are several descriptions of sexually transmitted, contact, and miscellaneous bacterial infections based on material presented in this chapter. Read the case history and then answer the questions posed. You can find answers online in **Appendix E**.

Case 1

The patient is a 17-year-old woman who comes to the clinic indicating that several days ago she started feeling nauseous but had not experienced any vomiting. She tells the physician that the day before coming to the clinic, she had a fever and chills; she also has been urinating more frequently, and the urine has a foul smell. She is diagnosed as having a urinary tract infection.

14.1a.   What bacterial species could be responsible for her illness?

14.1b.   Why are these types of diseases more prevalent in women than they are in men?

14.1c.   What types of urinary infections can occur?

14.1d.   How could this patient attempt to avoid another UTI?

14.1e.   What role do biofilms play in UTIs?

Case 2

A 19-year-old unwed mother arrives at the emergency room of the county hospital complaining of having cramps and abdominal pain for several days. She says she had never had a urinary tract infection and could not have gonorrhea because she was treated and cured of the STI 2 years ago. She has not experienced nausea or vomiting. When questioned, she tells the emergency room nurse that she has a single male sexual partner and condoms always are used. Based on further examination, the patient is diagnosed with pelvic inflammatory disease (PID). An endocervical swab is used for preparing a tissue culture. Staining results indicate the presence of cell inclusions.

14.2a.   What bacterial species can be associated with PID? What disease does she most likely have?

14.2b.   Why was a tissue culture inoculum ordered? Describe the reproductive cycle of this organism.

14.2c.   What other tests could be ordered for the patient's infection?

14.2d.   Why was the emergency room concerned about her sexual activity?

14.2e.   What misconception does the patient have about her past gonorrhea infection?

Case 3

A 17-year-old man comes to a free neighborhood clinic. He says that he noticed some white pus-like discharge and a tingling sensation in his penis. Since yesterday, he has had pain when urinating. He tells the physician that he has been sexually active with several female partners over the past eight months, but no one has had any sexually transmitted disease. Examination determines that there is no swelling of the lymph nodes in the groin or pain in the testicles. A Gram stain indicates the presence of gram-negative diplococci. The patient is given antibiotics, instructed to tell his female partners they both should be medically examined, and then he is released.

14.3a.   Based on the clinic findings, what disease does the patient have and what bacterial species is responsible for the infection?

14.3b.   Why is it important for his sexual partners to be medically examined, even if they experience no symptoms? What complications could arise if they are infected?

14.3c.   For which other organisms is this patient at increased risk? Why?

14.3d.   What significance can be drawn from the fact that the patient does not have any swelling of the lymph nodes in the groin or pain in the testicles?

14.3e.   What antibiotics most likely would be given to the patient?

Concept and Reasoning Checks 14.5

a.   Summarize the characteristics of toxigenic *S. aureus* infections.
b.   Assess the clinical significance of GAS in causing STSS and necrotizing fasciitis.
c.   Why is debridement necessary for most burn wounds?
d.   Distinguish between paucibacillary (tuberculoidal) and multibacillary (lepromatous) leprosy.

Chapter Challenge D

The human skin is the largest organ of the integumentary system. It weighs about 4.5 kilograms and has a surface area of almost two square meters. With such a large area, it is not surprising that there are numerous potential bacterial pathogens that can affect the human skin. Thus, one of the most important roles of the skin is to act as a barrier to these outside pathogens.

QUESTION D: *Describe the difference between resident skin diseases and those skin diseases caused by external pathogens, and explain why the exogenous ones are more dangerous to human health.*

You can find answers online in **Appendix F**.

■ KEY CONCEPT 14.6   Several Contact Diseases Affect the Eye

The eye has a multilayered defense against infection. The front of the eye and the lid of the eye are covered by a mucous membrane called the conjunctiva (**FIGURE 14.27**). The tear film over the conjunctiva protects the eye from infection by containing several antimicrobial agents, including lysozyme and antibodies. In addition, the blinking reflex clears away foreign material on the eye surface.

Microbiome research now reports that the surface of the healthy eye also has its own ocular microbiome. Although it is unclear if the microbial members are permanent or transient, preliminary studies suggest there are about a dozen bacterial genera on the conjunctiva and a different dozen on the corneal surface. When an eye disease does occur, the diversity of microbes in the eye microbiome is

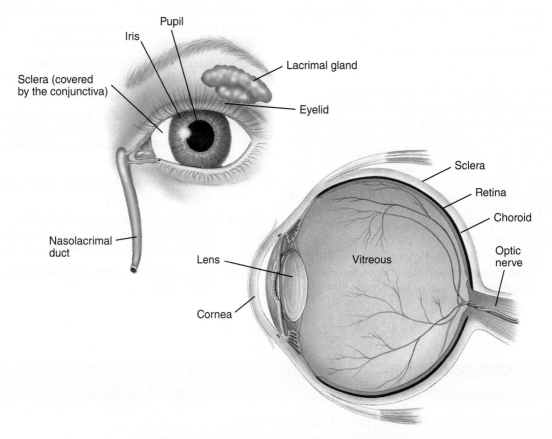

FIGURE 14.27 Eye Anatomy. External infections are most often associated with the eyelid, conjunctiva, or cornea. »» *Why would the conjunctiva and cornea be most susceptible to infection?*

reduced in half, with *Pseudomonas* strains becoming prominent members. Importantly, dysbiosis occurs before an eye infection can be diagnosed, suggesting changes to the eye microbiome might be an indicator of potential infection.

Eye infections typically involve inflammation of the eyelid, conjunctiva, or cornea.

Conjunctivitis Can Be Caused by a Variety of Pathogens

Several eye infections are caused by a few bacterial species. *S. aureus* can infect the eyelid and the cornea. When it infects the eyelid margin, a painful red inflammation called **blepharitis** develops. The inflammation sometimes leads to the formation of a **stye**. Treatment usually involves warm compresses and a topical antibiotic, such as one containing bacitracin.

Bacterial conjunctivitis, commonly called **pink eye** or **red eye**, is an inflammation that causes dilation of the conjunctival blood vessels (**FIGURE 14.28**). It is common in childhood, but the inflammation can occur in people of any age. The organisms that most commonly cause bacterial conjunctivitis are staphylococci, pneumococci, or streptococci that are picked up by contact with the eye. Bacterial conjunctivitis usually causes no long-term eye or vision damage.

Neonatal conjunctivitis is an inflammation of the conjunctiva of the newborn. The inflammation, also called **ophthalmia of the newborn**, results from contact with the bacterium during passage through the birth canal of a mother infected with *N. gonorrhoeae* or *C. trachomatis*. Infection with *N. gonorrhoeae* is the most serious and, if untreated, can lead to blindness. Prevention involves the use of antimicrobial drugs put into the eyes of all newborns after delivery.

Trachoma is the world's leading infectious cause of blindness. It occurs in hot, dry regions of the world, and it is prevalent in Mediterranean countries, parts of Africa and Asia, and in Native American populations in the southwestern United States. There are 500 million infections, mostly in children, worldwide, and 7 to 9 million individuals have been blinded by trachoma (**FIGURE 14.29**).

Trachoma is caused by serotypes A–C of *C. trachomatis*. These serotypes, unlike the others described in this chapter (see Table 14.1), are not sexually transmitted; rather, they are transmitted by personal contact with contaminated fingers, towels, and optical instruments. Face-to-face contact and flies also are important modes of transmission.

The pathogen multiplies in the conjunctiva, forming a series of tiny, pale nodules that are rough in appearance (*trach* = "rough"; *oma* = "tumor"). An initial infection typically heals without permanent damage. However, the initial infection sets up a hypersensitive state, such that repeated infections result in a chronic inflammation. Over 10 to 15 years, scarring of the conjunctiva occurs as the eyelashes turn inward and abrade the cornea, eventually leading to blindness.

Antibiotic therapy helps reduce the symptoms of trachoma, but for many patients relief is only temporary as reinfection can occur (**FIGURE 14.30**). In 1997, the WHO established the Alliance for Global Elimination of Trachoma by 2020 (GET 2020). Since then, 10 national programs, making up 50% of the global trachoma burden, have reduced acute infections in children by 50%. This has involved using a "**SAFE strategy**"; that is, **S**urgery of the eyelids; **A**ntibiotics for acute infections; **F**acial hygiene improvements; and **E**nvironmental access to safe water.

TABLE 14.5 summarizes the bacterial eye diseases.

In conclusion, we have now finished our survey of many bacterial infections and diseases that affect the human body. If you go to the CDC website and visit the diseases and conditions page (www.cdc.gov/diseasesconditions/), you will find an index of many more diseases than the ones described in this and other chapters. In addition, you can visit the weekly *Morbidity and Mortality Report* (*MMWR*) at (www.cdc.gov/mmwr/). This is the agency's "primary vehicle for scientific publication of timely, reliable, authoritative, accurate, objective, and useful public health information and recommendations." The information provided in the *MMWR* is

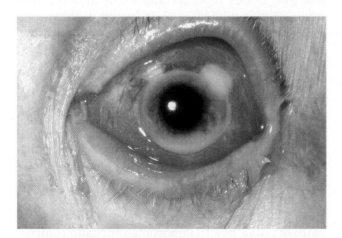

FIGURE 14.28 Conjunctivitis. The inflammation on the conjunctiva, also called pink eye, causes swelling on the surface. »» *What causes the redness in the sclera?*

Status of elimination of trachoma as a public health problem, 2016

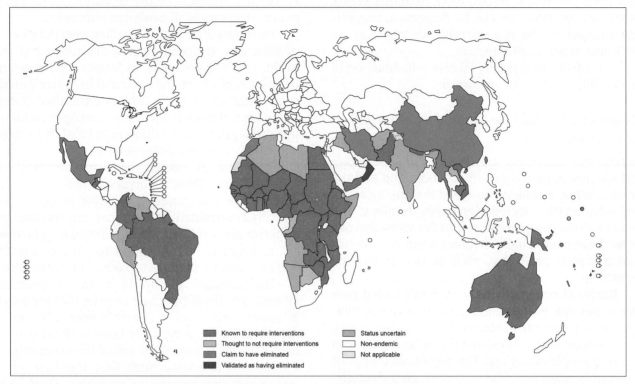

The boundaries and names shown and the designations used on this map do not imply the expression of any opinion whatsoever on the part of the World Health Organization concerning the legal status of any country, territory, city or area or of its authorities, or concerning the delimitation of its frontiers or boundaries. Dotted lines on maps represent approximate border lines for which there may not yet be full agreement. © WHO 2016. All rights reserved

Data Source: World Health Organization
Map Production: Control of Neglected Tropical Diseases (NTD)
World Health Organization

 World Health Organization

FIGURE 14.29 **Worldwide Trachoma.** Blinding or suspected blinding trachoma (red-colored areas) affects individuals worldwide but especially in Africa, Southeast Asia, Mexico, and parts of South America. »» *How can trachoma be treated?*

Map by World Health Organization.

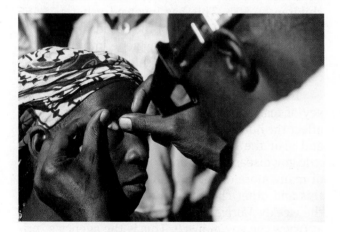

FIGURE 14.30 **Trachoma Management.** An ophthalmic surgeon in rural Malawi, Africa, attends to a trachoma patient. »» *How is trachoma spread?*

based on weekly reports compiled by the CDC from information supplied by state health departments. It is written at a level that you, as a "microbiology student," can comprehend. When you read about an infectious disease outbreak in the news media, more than likely there will be one or more articles soon appearing in the *MMWR*. Likewise, a similar publication put out by the WHO, called the *Weekly Epidemiological Record* (*WER*; www.who.int/wer/en/), has a more global perspective. It "serves as an essential instrument for the rapid and accurate dissemination of epidemiological information on cases and outbreaks of diseases under the International Health Regulations and on other communicable diseases of public health importance, including emerging or reemerging infections." If you want to keep up with the latest in disease outbreaks, epidemics, and sometimes pandemics, these are the places to go.

TABLE 14.5 A Summary of Infectious Bacterial Eye Diseases

Inflammation or Disease	Causative Agent	Signs and Symptoms	Transmission	Treatment	Prevention
Blepharitis (stye)	*Staphylococcus aureus*	Burning sensation in eye	Contaminated instruments Hands Shared towels Droplets	Warm compress Antibiotic medication	Practicing good hygiene
Bacterial conjunctivitis	*Staphylococcus aureus* *Streptococcus pyogenes* *Haemophilus influenzae* *N. gonorrhoeae*	Eye pain, swelling, redness, and a yellow or greenish discharge	Direct or indirect contact	Broad-spectrum antibiotic	Practicing good hygiene
Neonatal conjunctivitis	*N. gonorrhoeae* *Chlamydia trachomatis*	Eye swelling and pus discharge Watery discharge	Infected mother to child during childbirth	Topical and oral antibiotics	Using silver nitrate or antibiotics Screening mother
Trachoma	*Chlamydia trachomatis* (serotypes A–C)	Tiny, pale nodules on the conjunctiva Upper eyelid abrasion can cause blindness	Indirect contact Mechanical vector	Topical or oral antibiotics	Washing face Controlling flies Freshwater source

Concept and Reasoning Check 14.6

a. What are the challenges facing health officials trying to eliminate trachoma?

Chapter Challenge E

The human eye is exposed to the environment and all the microbes and pathogens that might meet the eye. The infectious agents that potentially can cause disease can be introduced to the eye either directly or indirectly.

QUESTION E: *For blepharitis, bacterial conjunctivitis (pink eye), and trachoma, identify if each infection is the result of direct or indirect contact. As such, what is the best way to protect your eyes from potential infections?*

You can find answers online in **Appendix F**.

■ SUMMARY OF KEY CONCEPTS

Concept 14.1 Portions of the Female and Male Reproductive Systems Contain a Resident Microbiome

1. The **primary sex organs** in the male are the testes, wheras the epididymis, vas deferens, seminal vesicles, prostate, and penis are **accessory reproductive organs**. In females, the primary sex organs are the ovaries, whereas the accessory organs consist of the fallopian tubes, uterus, vagina, and vulva. (Figure 14.2)
2. Antimicrobial defenses in the reproductive tracts include the urethral mucosa, the vagina, vulva, and cervix in females. *Lactobacillus* species in the vagina as well as antibodies and other antimicrobial products of the systems produce an environment not favorable for pathogen colonization. (Figure 14.3)
3. Nonsexually transmitted illnesses include **bacterial vaginosis**, involving:
   ▶ *Gardnerella vaginalis*
   ▶ *Prevotella*
   ▶ *Mobiluncus*
   ▶ Other resident bacterial species

Concept 14.2 Many Sexually Transmitted Infections Are Caused by Bacteria

▶ **Chlamydial urethritis**
   ① *Chlamydia trachomatis*
▶ **Gonorrhea**
   ② *Neisseria gonorrhoeae*
▶ **Syphilis**
   ③ *Treponema pallidum*
▶ **Chancroid**
   ④ *Haemophilus ducreyi*
▶ **Lymphogranuloma venereum**
   ⑤ *Chlamydia trachomatis*

Concept 14.3 Urinary Tract Infections Are the Second Most Common Body Infection

The organs of the **urinary system** that are susceptible to infection are the kidneys and the urinary tract (ureters, urinary bladder, and urethra). The kidneys, ureters, and bladder are normally sterile due to normal urine flow. However, the urethra is colonized by microbes either along its entire length or near the terminus. The bacterial infections include:

▶ **Urethritis**
   ⑥ *Escherichia coli*
   ⑦ *Chlamydia trachomatis*
   ⑧ *Neisseria gonorrhoeae*
▶ **Cystitis**
   ⑨ *Escherichia coli*
▶ **Prostatitis** (in males—not shown)
   – *Escherichia coli*
   – *Staphylococcus aureus*
   – *Proteus*
   – *Pseudomonas aeruginosa*

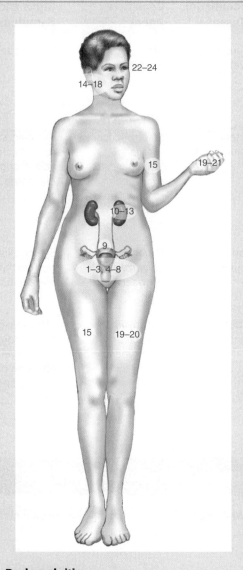

▶ **Pyelonephritis**
   ⑩ *Escherichia coli*
   ⑪ *Staphylococcus aureus*
   ⑫ *Klebsiella pneumoniae*
   ⑬ *Pseudomonas aeruginosa*

Concept 14.4 Contact Diseases Can Be Caused by Resident Bacterial Species

Many bacterial diseases are caused by contact with the skin, which normally protects the underlying tissues from bacterial colonization and infection. The bacterial infections include:

▶ **Acne**
   ⑭ *Propionibacterium acnes*
▶ Burn infections
   ⑮ *Pseudomonas aeruginosa*

Concept 14.5 Contact Diseases Also Can Be Caused by Bacterial Species That Originate Externally

▶ **Furuncles** (boils) and carbuncles
  16 *Staphylococcus aureus*
▶ **Impetigo**
  17 *Staphylococcus aureus*
  18 *Streptococcus pyogenes*
▶ **Scalded skin** and **toxic shock syndromes**
  19 *Staphylococcus aureus*
▶ **Erysipelas** and **necrotizing fasciitis**
  20 *Streptococcus pyogenes*

▶ **Leprosy** (Hansen disease)
  21 *Mycobacterium leprae*

Concept 14.6 Several Contact Diseases Affect the Eye

▶ **Conjunctivitis**
  22 *Neisseria gonorrhoeae*
  23 *Chlamydia trachomatis*
▶ **Trachoma**
  24 *Chlamydia trachomatis*

■ CHAPTER SELF-TEST

For **Steps A–D**, you can find answers online in **Appendix D**.

■ STEP A: REVIEW OF FACTS AND TERMS

Multiple Choice

Read each question carefully before selecting the *one* answer that best fits the question or statement.

1. The primary sex organs of the female reproductive system is/are the _____.
   A. uterus
   B. vagina
   C. fallopian tubes
   D. ovaries

2. What part or parts of the male and female reproductive systems are typically colonized by resident microbiome?
   A. Male: ureters; female: vagina and ovaries
   B. Male: testes and epididymis
   C. Male: urethra; female: vagina, vulva, and cervix
   D. Male: bladder and ureters; female: fallopian tubes and cervix

3. Which one of the following microbes is *not* associated with vaginosis?
   A. *Mobiluncus*
   B. *Gardnerella*
   C. *Prevotella*
   D. *Staphylococcus*

4. Which one of the following statements is *not* correct concerning the reproductive cycle of *Chlamydia*?
   A. Reticulate bodies are infectious.
   B. Reticulate bodies reorganize into elementary bodies.
   C. Elementary bodies infect host cells.
   D. Elementary bodies transform into reticulate bodies.

5. Salpingitis is associated with _____ and can lead to _____.
   A. syphilis; gumma formation
   B. gonorrhea; sterility
   C. chlamydia; ophthalmia
   D. chancroid; soft chancre

6. A chancre is typical of which stage of syphilis?
   A. Primary
   B. Secondary
   C. Tertiary
   D. Chronic, latent

7. Besides chlamydia urethritis, what other STI is associated with another serotype of *Chlamydia trachomatis*?
   A. Lymphogranuloma venereum (LGV)
   B. Genital warts
   C. Gonorrhea
   D. Chancroid

8. Which one of the following is *not* part of the urinary tract?
   A. Urethra
   B. Bladder
   C. Kidneys
   D. Ureters

9. What bacterial species is most often associated with cystitis?
   A. *Treponema pallidum*
   B. *Escherichia coli*
   C. *Chlamydia trachomatis*
   D. *Pseudomonas aeruginosa*

10. What type of immune defensive cell is found in the sublayers of the epidermis?
    A. Keratinocyte
    B. Dendritic (Langerhans) cell
    C. Neutrophil
    D. Basophil
11. The skin is _____.
    A. dominated by gram-negative bacterial cells
    B. free of bacterial cells
    C. without a microbiome
    D. dominated by gram-positive bacterial cells
12. Which one of the following statements is *not* true of acne?
    A. Acne is associated with *Propionibacterium acnes*.
    B. Plugged sebaceous glands are called erythemas.
    C. Whiteheads are completely blocked follicles.
    D. Acne is not a preventable disease.
13. An acute wound could be due to _____.
    A. surgical procedures
    B. cuts
    C. lacerations
    D. All the above (A–C) are correct.
14. In children, this skin disease is characterized by the production of thin-walled blisters oozing a yellowish fluid and forming yellowish-brown flakes.
    A. Toxic shock syndrome
    B. Scalded skin syndrome
    C. Erysipelas
    D. Impetigo

15. Which one of the following skin diseases is *not* caused by *Streptococcus pyogenes*?
    A. Necrotizing fasciitis
    B. Toxic shock syndrome
    C. Gas gangrene
    D. Erysipelas
16. The most common cause of an invasive wound infection, such as a burn, is _____.
    A. gram-positive bacterial species
    B. *Treponema pallidum*
    C. *Pseudomonas aeruginosa*
    D. *Escherichia coli*
17. Leprosy can be contracted by contact with _____.
    A. contaminated water
    B. insects
    C. nasal secretions
    D. contaminated food
18. The SAFE strategy has greatly reduced the global burden of what disease?
    A. Trachoma
    B. Neonatal conjunctivitis
    C. Leprosy
    D. Blepharitis

Fill-in

Answer each of the following by filling in the blank with the correct word or phrase.

19. Bacterial conjunctivitis is commonly called _____.
20. Lymphogranuloma venereum (LGV) is prevalent in _____ Asia.
21. *Gardnerella* is the cause of _____.
22. _____ are a group of connected, deep abscesses caused by *S. aureus*.
23. _____ is the most common agent of urinary tract infections.
24. Sufferers of UTIs have a _____ on urination.
25. Dapsone, rifampin, and clofazimine are drugs used to treat _____.

▮ STEP B: CONCEPT REVIEW

26. Describe the causes of **bacterial vaginosis**. (**Key Concept 14.1**)
27. Distinguish between the signs and symptoms of **chlamydial urethritis** in males and females. (**Key Concept 14.2**)
28. Describe (a) the possible complications resulting from **gonorrhea** in females, and (b) explain the danger of gonorrhea in pregnant females. (**Key Concept 14.2**)
29. Distinguish among the three possible stages of **syphilis**. (**Key Concept 14.2**)
30. Differentiate among the various forms of UTIs: **urethritis, cystitis, prostatitis**, and **pyelonephritis**. (**Key Concept 14.3**)
31. Assess the role of *Propionibacterium acnes* in triggering **acne** and describe the follicle-associated lesions. (**Key Concept 14.4**)

32. Assess the role of *Staphylococcus aureus* as an agent for contact diseases. (**Key Concept 14.5**)

33. Describe the skin infections caused by *Streptococcus pyogenes*. (**Key Concept 14.5**)

34. Summarize the types of diseases caused by **traumatic wounds**, including burns. (**Key Concept 14.5**)

35. Explain why **trachoma** is a major cause of blindness in many developing nations. (**Key Concept 14.6**)

■ STEP C: APPLICATIONS AND PROBLEM SOLVING

36. Suppose that a high incidence of leprosy existed in a particular part of the world. Why is it conceivable that there might be a correspondingly low level of tuberculosis?

37. An African patient reports to a local hospital with an upper lip swollen to about three times its normal size. Probing with a safety pin at facial points where major nerve endings terminate showed that the area to the left of the nose and above the lip was without feeling. When a biopsy of the tissue was examined, it revealed round reservoirs of immune system cells called granulomas within the nerves. On bacteriological analysis, acid-fast rods were observed in the tissue. What disease do all these data suggest?

38. Certain microscopes have the added feature of a small hollow tube that fits over the eyepieces or oculars. Viewers are encouraged to rest their eyes against the tube and thereby block out light from the room. Why is this feature hazardous to health?

39. A woman suffers two miscarriages, each after the fourth month of pregnancy. She then gives birth to a child, but impaired hearing and vision become apparent as it develops. Also, the baby's teeth are shaped like pegs and have notches. What medical problem existed in the mother?

■ STEP D: QUESTIONS FOR THOUGHT AND DISCUSSION

40. One of the major problems of the current worldwide AIDS epidemic is the possibility of transferring the human immunodeficiency virus (HIV) among those who have a sexually transmitted infection. Which diseases in this chapter would make a person particularly susceptible to penetration of HIV into the bloodstream?

41. At a specified hospital in New York City, hundreds of patients pay a regular visit to the "neurology ward." Some sign in with numbers; others invent fictitious names. All receive treatment for leprosy. Why do you think this disease still carries such a stigma?

42. Several years ago, the Rockefeller Foundation offered a $1 million prize to anyone who could successfully develop a simple and rapid test to detect chlamydia and/or gonorrhea. The test had to use urine as a test sample and be performed and interpreted by someone with a high school education. No one ever claimed the prize. Can you guess why?

PART V Viruses and Eukaryotic Microorganisms

The five chapters in Part V embrace several aspects of the six principal (overarching) concepts as described in the American Society of Microbiology (ASM) fundamental statements for a concept-based microbiology curriculum.

PRINCIPAL CONCEPTS

Evolution	• Mutations and horizontal gene transfer, with the immense variety of microenvironments, have selected for a huge diversity of microorganisms. Chapters 15–19
Cell Structure and Function	• The structure and function of microorganisms have been revealed by the use of microscopy (including bright field, phase contrast, fluorescent, and electron). Chapters 15–19 • While microscopic eukaryotes (for example, fungi, protozoa, and algae) carry out some of the same processes as bacteria, many of the cellular properties are fundamentally different. Chapters 18 and 19 • The replication cycles of viruses (lytic and lysogenic) differ among viruses and are determined by their unique structures and genomes. Chapters 15–17
Information Flow and Genetics	• Genetic variations can impact microbial functions (e.g., in biofilm formation, pathogenicity, and drug resistance). Chapters 15–19 • Although the central dogma is universal in all cells, the processes of replication, transcription, and translation differ in Bacteria, Archaea, and Eukaryotes. Chapters 18 and 19 • The regulation of gene expression is influenced by external and internal molecular cues and/or signals. Chapters 15–19 • The synthesis of viral genetic material and proteins is dependent on host cells. Chapters 15–17
Microbial Systems	• Microorganisms and their environment interact with and modify each other. Chapters 18 and 19 • Microorganisms, cellular and viral, can interact with both human and nonhuman hosts in beneficial, neutral, or detrimental ways. Chapters 15–19
Impact of Microorganisms	• Microorganisms provide essential models that give us fundamental knowledge about life processes. Chapters 15–19 • Because the true diversity of microbial life is largely unknown, its effects and potential benefits have not been fully explored. Chapters 15–19

CHAPTER 15

The Viruses and Virus-Like Agents

Under the Naica hills in northern Mexico lies the Cave of Crystals. Formed some 500,000 years ago, these 50-ton gypsum crystals (see the chapter opener image) are buried 300 meters below the desert surface, in caves having temperatures of 45°C to 50°C and saturating humidity. The caves, which have been cut off from the outside world for possibly millions of years, appeared devoid of life when first discovered by silver miners in 2000. Then, in 2009, scientists collected water samples from the cave pools for analysis. The scientists, led by Dr. Curtis Suttle, wanted to discover if any organisms could survive in the harsh conditions of the caves. When the water samples were taken back to the lab in British Columbia for analysis, the findings stunned the researchers.

In the water from these gigantic caverns, some the length of a football field and two stories high, were bacterial cells. Moreover, in these samples were tiny, geometric shapes—viruses! And they were numerous—perhaps up to 200 million per drop of water.

Almost halfway around the world in Pozzuoli, Italy, in acidic hot springs (85°C to 95°C; pH 1.5), microbiologists discovered that enrichment cultures of several archaeal organisms contained virus particles. Much further south in the Antarctic's Lake

Limnopolar, a freshwater lake that is frozen nine months of the year, scientists found bacterial cells, protists, and—yes—viruses. Again, vast numbers were present, representing perhaps 10,000 different types, making the lake one of the most diverse virus communities in the world. Even the world's oceans are filled with viruses. For example, Rachel Parsons and her collaborators have detected enormous virus populations in the western Atlantic (**FIGURE 15.1**).

Cave of Crystals, Naica Mine, Chihuahua, Mexico.
© Javier Trueba/MSF/Science Source.

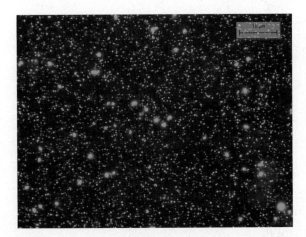

FIGURE 15.1 The Virosphere. Viruses can be found wherever life exists. This includes in huge, open expanses such as the world's oceans. The larger, fluorescent dots are bacterial cells and the smaller spots are viruses. *»» Does it seem strange to you that huge numbers of viruses are found in places like the Cave of Crystals and the open oceans? Explain.*

Courtesy of Rachel Parsons, Bermuda Institute of Ocean Science.

The number of bacterial viruses alone on Earth exceeds 10^{30}, and scientists and microbiologists are finding them everywhere. In fact, the planet's most abundant "coterrestrials" are the viruses,

which make up the so-called **virosphere**. Suttle's lab explains, "*wherever life is found; it is a major cause of* [morbidity and] *mortality, a driver of global geochemical cycles, and a reservoir of the greatest unexplored genetic diversity on Earth.*" Moreover, viruses are a major driving force for evolution because each virus infection has the potential to introduce new genetic information into just about any organism in the domains Bacteria, Archaea, and Eukarya or into their own progeny viruses. The virosphere is massive, it is incredibly diverse, and it has tremendous impact beyond infection and disease.

In this chapter, we study the properties of viruses, especially bacterial and clinically important animal (human) viruses, focusing on their unique structure and mechanism for replication. We will see how they are classified, how they are cultured and identified, and how some are associated with tumors and cancer. The chapter concludes by discussing another unusual virus-like agent that can cause disease in animals and humans.

So, are viruses alive? Let's consider this question as we study the viruses, their structure, and behavior in this chapter.

Chapter **Challenge**

Viruses are often described as being on the edge of life, in some limbo land between "life" and "nonlife." But on which side of the edge are they? Most biology textbooks use certain emergent properties of life to define something as living. These properties include the ability to:

- Grow and develop
- Reproduce
- Establish complex organization
- Regulate the internal environment (maintain homeostasis)
- Transform energy (light to chemical; chemical to cellular)
- Respond to the environment
- Evolve by adapting to a changing environment

■ KEY CONCEPT 15.1   Filterable Infectious Agents Cause Disease

The development of the germ theory in the 1880s recognized disease patterns associated with a specific bacterial species. However, some diseases, such as smallpox, measles, and rabies, resisted identification. Many of these diseases would turn out to be of viral origin.

Many Scientists Contributed to the Early Understanding of Viruses

In the late 1800s, tobacco growers in Europe noticed their tobacco plants were developing a mottled or mosaic pattern on the leaves, which came to be called

(A)

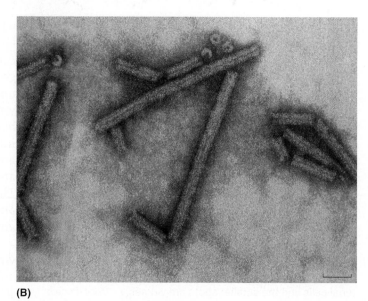

(B)

FIGURE 15.2 Tobacco Mosaic Disease and Its Virus. (A) An infected tobacco leaf exhibiting the mottled or mosaic appearance caused by the disease. **(B)** This false-color transmission electron micrograph of TMV shows the rod-shaped structure of the virus particles. (Bar = 80 nm). *»» Why was it impossible for Ivanowsky and his contemporaries to see viruses with the light microscope?*

(A) Courtesy of Clemson University - USDA Cooperative Extension Slide Series, Bugwood.org. (B) © Dennis Kunkel Microscopy/Science Source.

tobacco mosaic disease (TMD) (**FIGURE 15.2A**). Several scientists, including Dimitri Ivanowsky and Martinus Beijerinck, independently filtered the crushed leaves of a TMD-infected plant, hoping to trap what they believed was a bacterial infectious agent on the filter, as the bacterial cells would be too large to pass through. Rather, to their surprise, the clear liquid passing through the filter contained the infectious agent. When placed on healthy leaves, TMD would soon appear. Beijerinck believed TMD was caused by a filterable *contagium vivum fluidum* (contagious living fluid), which did not fit the prevailing germ theory. Throughout his studies, Beijerinck often referred to the mysterious, infectious agent as the "virus" (*virus* = "poison").

In 1898, foot-and-mouth disease (a contagious disease among cattle, sheep, and deer) also was suspected as being caused by a filterable virus, implying that there were viruses capable of infecting animals as well as plants. Three years later, the American Walter Reed and his group in Cuba provided evidence linking yellow fever with a virus, so viruses could be associated with human disease, as well.

In 1915, English bacteriologist Frederick Twort discovered viruses that infected bacterial cells. Two years later, such viruses were identified by French-Canadian scientist Felix d'Herrelle. He called them

bacteriophages (*phage* = "eat"), or simply **phages**, for their ability to destroy the bacterial cells they infected.

By the early 1930s, viruses were thought to be living microorganisms below the resolving power of light microscopes. However, in 1935 the tobacco mosaic virus (TMV) was crystallized, suggesting viruses might be nonliving agents of disease. Additional work with TMV revealed the virus was composed exclusively of nucleic acid and protein, and the viruses lacked the cytoplasmic structures typical of prokaryotic and eukaryotic cells. Soon, advances in staining of viruses, including TMV, made their study with the transmission electron microscope possible (**FIGURE 15.2B**). Such studies indicated they were not cellular.

Being acellular, viruses cannot be grown on a nutrient agar plate or in a broth the way many bacterial organisms will grow. Rather, virus cultivation requires that they be grown in cultures of cells or tissues, a technique that became well developed by the 1940s. With this technique, the extraordinary nature and behavior of the viruses could be investigated in detail; the field of **virology** was off and running, and it remains one of the most vibrant fields in microbiology, as the **MICROBIOLOGY PATHWAYS** highlights.

MICROBIOLOGY PATHWAYS

Virology

When Ed Alcamo, the original author of this textbook, was in college, he was part of a group of twelve biology majors. Each of them had a particular area of "expertise." One was going to be a surgeon, whereas another was interested in marine biology. A third was a budding dentist, and Ed was the local virologist. He was fascinated with viruses and wrote a term paper summarizing arguments for the living or nonliving nature of viruses. (At the time, neither side was persuasive, and even his professor gracefully declined to take a stand.)

Ed never quite made it to being a virologist, but if your fascination with these infectious particles is as keen as his was, you might like to consider a career in virology.

Virologists investigate dreaded diseases such as AIDS, polio, and rabies (see figure), whereas others (epidemiologists) investigate disease outbreaks. Virologists also concern themselves with many types of cancers; others study the chemical interactions of viruses with various tissue culture systems and animal models.

Working with dangerous viruses requires wearing special protective equipment.

Courtesy of James Gathany/CDC.

Virologists also are working to replace agricultural pesticides with viruses able to destroy mosquitoes and other pests. Some virologists are inserting viral genes into plants and are hoping the plants will produce viral proteins to lend resistance to disease. One particularly innovative group is trying to insert genes from hepatitis B viruses into bananas. They hope that one day we can vaccinate ourselves against hepatitis B by having a banana for lunch.

If you wish to consider the study of viruses, an undergraduate major in biology would be a good choice. Because the biology of viruses is related to the biology of cells, courses in biochemistry and cell biology will be required. You should have good observation, communication, and analysis skills and be able to think critically, reason, and problem solve. Science and medical computer skills would be helpful, as would basic skills using laboratory equipment, tools, and instruments.

Most virologists study for an MD or PhD degree following completion of college. MD students pursue virology research in the context of patients or disease and become investigators with an interest in infectious disease or epidemiology. Most PhD students pursue basic questions with academic institutions or work for industrial or governmental organizations.

A great way to find out if you have the "research bug" is to work in a college or university laboratory. Many colleges and universities with research programs employ undergraduate students (near minimum wage!) in the research laboratory. So, here is a good way to "get your feet wet." If you find it interesting, it also will enhance your chances to be accepted by a top-flight graduate school.

Not interested in lab bench research? Virologists also can find careers in full-time teaching. In addition, your knowledge can be used to pursue a career in communications, serving as a science writer or reporter. They also pursue careers in business administration or law, especially involving the pharmaceutical industry or patent law.

Concept and Reasoning Check 15.1

a.   Describe the major events leading to the recognition of viruses as agents of infectious disease.

Chapter **Challenge A**

We have learned that (1) viruses, like TMV, can be crystallized, and (2) viruses cannot be grown in culture by themselves; rather, they need to be cultivated in living cells.

QUESTION A: *Based on these observations and the characteristics of life, where would you place the viruses? Are they living or nonliving, and why?*

You can find a discussion online in **Appendix F**.

■ KEY CONCEPT 15.2   Viruses Have a Simple Structural Organization

Today more than 5,000 viruses have been identified and sequenced. However, this is only a small proportion of the estimated 1 million different viruses that virologists believe might exist—their total numbers making viruses, as we saw in the chapter introduction, the most abundant "biological entities" on Earth and the most extensive carriers of genetic information.

Viruses Are Tiny Infectious Agents

Viruses are small, obligate, intracellular particles (**FIGURE 15.3**); that is, most can be seen only with the electron microscope. However, there are always exceptions, as **MICROFOCUS 15.1** reveals. No matter their size, these acellular particles must infect and take over a host cell in order to

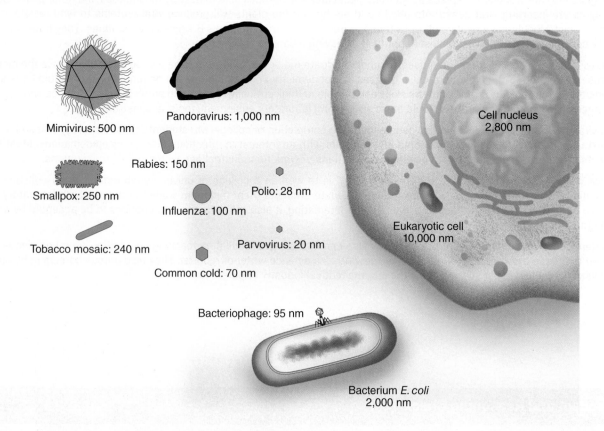

FIGURE 15.3 Size Relationships Among Cells and Viruses. The sizes (not drawn to scale) of various viruses relative to a eukaryotic cell, a cell nucleus, and the bacterium *Escherichia coli.* Most viruses range from the very small poliovirus to the much larger smallpox virus. Newly discovered mimiviruses and pandoraviruses are much larger »» *Propose a hypothesis to explain why most viruses are so small.*

MICROFOCUS 15.1: Evolution

Hypothesis: Giant Viruses Arose from Cellular Life

As in all sciences, dogma (beliefs) can change with new observations and discoveries. Take, for example, the viruses, which are traditionally viewed as very small particles with simple genome complexity. Times have changed.

In 2003, a virus the size of a very small bacterial cell (0.4 µm) was found infecting amoebas (see figure). Called mimivirus, this giant virus has a genome made of double-stranded DNA containing some 1,000 protein-coding genes (the influenza viruses and HIV each have nine genes). Soon after, other "giant viruses" related to the mimivirus were discovered. These include the mamavirus and the megavirus, the latter having a genome containing more than 1,200 protein-coding genes. Still, like all viruses, these viruses cannot convert energy or replicate on their own.

When one of these giant viruses infects an amoeba host, the virus constructs a gigantic "replication factory" to assemble hundreds of new virions. These factories contain the viral genes along with host enzymes, ribosomes, and mitochondria. To make matters more interesting, the giant viruses also have virophages, which are small viruses that attack and infect the giant viruses. One called "Sputnik" uses the mimivirus replication factory for its own replication. These viruses are beginning to sound like parasitic cells!

More recently, other giant viruses have been discovered. The pandoravirus also infects amoebas, and it has the most complex genome, containing more than 2,500 protein-coding genes. This genome is approaching the genome size of some cellular microorganisms that also lead a parasitic existence. The pandoravirus genes are unique because up to 94% of the genes are unknown to science. Most recently, another giant virus, called the pithovirus, was discovered. It represents the largest virus yet discovered (1.5 µm in length).

Could any of these giant viruses also cause disease in humans? One recent report suggests the mimivirus might be capable of causing pneumonia in humans. Antibodies against the virus have been detected in some individuals suffering from pneumonia, and the viruses have been detected in human blood. However, it is unclear how common such giant virus infections might be in causing acute human illness.

These discoveries of giant viruses appear to blur the line between viruses and single-celled organisms and between nonliving and living. Indeed, up to 90% of the genes the giant viruses contain are unique, and the genes do not have similar counterparts in any known prokaryotic or eukaryotic organisms. Consequently, some scientists hypothesize such giant viruses evolved billions of years ago from ancient, free-living cells in long-extinct domains of life. These ancestors gradually lost genes, which, like many parasitic prokaryotes, resulted in their parasitic lifestyle and their inability to reproduce (replicate) on their own. Some scientists propose a controversial idea—the giant viruses represent additional domains of life and should be included with the living organisms in the "Tree of Life." Obviously, much more study needs to be done to understand what role these viruses have played in the origin and evolution of life on Earth.

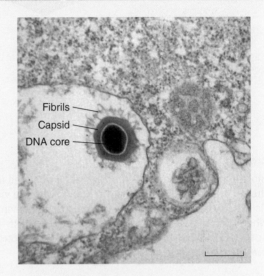

Transmission electron micrograph of the mimivirus. (Bar = 400 nm.)

Courtesy of Didier Raoult, Rickettsia Laboratory, La Timone, Marseille, France.

replicate (**FIGURE 15.4**). This dependence on a host cell is necessary because viruses lack the chemical machinery for generating energy and carrying out metabolism. In fact, viruses are missing many of the typical prokaryotic and eukaryotic features. They have no organelles, no cell wall, no cytoplasm, and no cell nucleus or nucleoid. Instead, they are comprised of two basic components: an indispensable nucleic acid core and a surrounding coat of protein.

The **viral genome** contains one or more molecules of DNA or RNA in either a double-stranded or a single-stranded form. Usually the nucleic acid is a linear or circular molecule, although in some viruses,

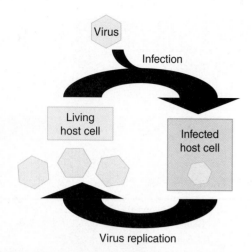

FIGURE 15.4 The Viral Replication Cycle. To produce more viruses, an inert virus must infect a host cell in which the viral genome controls the production of more virus particles. The virus particles are released into the environment, where each initiates a new infectious cycle with another host cell. »» *What do you predict happens to the infected host cell when the replicated viruses are released?*

like the influenzaviruses, the genome exists as separate, nonidentical nucleic acid segments. In most cases, the viral genome consists of relatively few genes, ranging from less than ten to a few hundred.

The protein coat of a virus particle, called the **capsid**, surrounds the viral genome while providing shape or symmetry to the virus (**FIGURE 15.5**).

The capsid is built from individual protein subunits called **capsomeres** that self-assemble into the virus's shape. Together, the capsid with its enclosed genome is referred to as a **nucleocapsid**.

The capsid also protects the viral genome against chemical and physical agents and other environmental fluctuations (e.g., temperature and pH changes). In a few viruses, such as the AIDS virus, a few inactive enzymes might be present along with the genome.

Viruses composed solely of a nucleocapsid are sometimes referred to as **nonenveloped viruses**. On these viruses (Figure 15.5), either special capsid proteins called **spikes** or other capsid proteins protrude from the surface. These external proteins help attach the viruses to and facilitate entry into specific host cells.

The nucleocapsids of many other viruses are surrounded by a flexible, loose-fitting membrane known as an **envelope**; such viruses are referred to as **enveloped viruses** (Figure 15.5). The envelope is composed of lipids and protein, similar to the host membranes. Enveloped viruses can lose their infectivity if the envelope is destroyed.

Many enveloped viruses also contain protein spikes projecting from the envelope. These proteins also function for attachment and host cell entry. Enveloped viruses can contain a layer or two of protein between the capsid and envelope called the **matrix** that holds the nucleocapsid to the envelope.

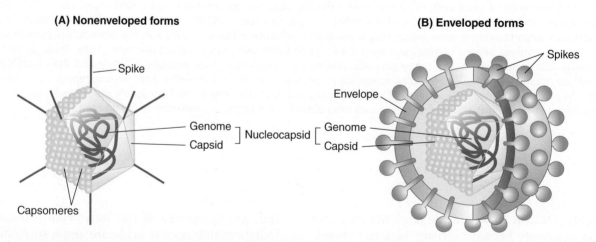

FIGURE 15.5 The Components of Viruses. (A) Nonenveloped viruses consist of a nucleic acid genome (either DNA or RNA) and a protein capsid. Capsomere units are shown on two faces of the capsid from which spikes may protrude. **(B)** Enveloped viruses have an envelope that surrounds the nucleocapsid. Again, spikes usually are present on the envelope. »» *What important role do spikes play in the infective behavior of viruses?*

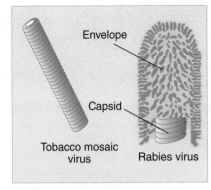

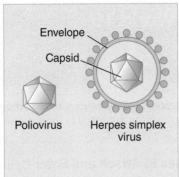

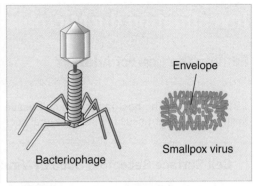

(A) Helical symmetry: The capsid forms a hollow, helical rod inside of which is the virus genome.

(B) Icosahedral symmetry: The capsid forms a 20-sided structure inside of which is the virus genome.

(C) Complex symmetry: The capsid forms a complex structure inside of which is the virus genome.

FIGURE 15.6 Viral Shapes. Viruses exhibit variations in shape. The capsid may have **(A)** helical symmetry, **(B)** icosahedral symmetry, or **(C)** complex symmetry. *»» Although the bacteriophages are classified as complex, what two symmetries do these viruses exhibit?*

A completely assembled and mature infectious virus particle outside its host cell is known as a **virion**.

Viruses Have Diverse Shapes

The self-assembly of the capsid generates the three-dimensional shape or morphology of a virus. For many viruses, including those infecting humans, there are three morphologies or symmetries:

▸ **Helical Symmetry.** Some virus capsids, such as those of the rabies and Ebola viruses, exist in the form of a rod or filament and are said to have **helical symmetry** (**FIGURE 15.6A**). The helix is a tightly wound coil resembling a corkscrew or spring.

▸ **Icosahedral Symmetry.** Other viruses, such as the herpesviruses (enveloped) and the polioviruses (nonenveloped), have capsids in the shape of a polyhedron with 20 triangular sides and hence, **icosahedral** (*icos* = "twenty," *edros* = "side") **symmetry** (**FIGURE 15.6B**).

▸ **Complex Symmetry.** Some viruses have capsids with a complex geometrical pattern, meaning they have several parts with different shapes (**FIGURE 15.6C**). Some bacteriophages, for example, have an icosahedral head with a collar and tail assembly in the shape of a helix. Poxviruses, by contrast, are brick shaped, with protein filaments occurring in a swirling pattern at the surface of the virus. In either case, these viruses demonstrate **complex symmetry**.

Viruses Have a Host Range and Tissue Specificity

As a group, viruses can infect almost any cellular organism in the domains Bacteria, Archaea, and Eukarya. A virus's **host range** refers to what organisms (hosts) the virus can infect, and it is based on a virus's capsid or envelope proteins.

Many viruses have a very narrow host range. The smallpox virus and polioviruses only infect humans. At the other extreme, a few viruses have a broad host range. For example, the rabies viruses infect humans and most warm-blooded animals (e.g., bats, skunks, and dogs).

Even within its host range, many plant and animal viruses only infect certain cell types or tissues within a multicellular host organism. This limitation is called the **cell/tissue specificity**. For example, the host range for the human immunodeficiency virus (HIV) is a human, but once in the host, HIV primarily infects specific groups of susceptible white blood cells called T lymphocytes and macrophages. This is because the envelope has protein spikes for specifically binding to receptor molecules on these cells. Therefore, if a potential host cell lacks the appropriate receptor or the virus lacks the complementary protein, the virus usually cannot infect that cell. How we know that susceptible host cells have receptors that bind specific viruses is explored in the **INVESTIGATING THE MICROBIAL WORLD 15**.

TABLE 15.1 lists some human viruses and their cell/tissue specificity.

Investigating the Microbial World 15

Finding the Correct Address

OBSERVATION: The host range and cell tissue tropism allow viruses to find the appropriate host and then the correct susceptible cell or tissue to infect. Several such receptors are listed in the accompanying table.

Cell Surface Receptors Used by Viruses to Attach and Enter Cells

Virus	Cell Surface Receptor
Influenza A	Sialic acid
HIV-1	CD4 and co-receptors (CXCR5, CCR4)
Hepatitis C	Low-density lipoprotein receptor
Rabies	Acetylcholine receptor, neural cell adhesion molecule, nerve growth factor, gangliosides, phospholipids
Rhinovirus	Intracellular adhesion molecule 1 (ICAM-1)
Hepatitis B	IgA receptor
Adenovirus	Integrins
Poliovirus	Immunoglobulin superfamily protein (CD155)

QUESTION: *Because one cannot see cell surface receptors with a microscope, how do we know these viruses actually bind to receptors on the surface of target host cells?*

HYPOTHESIS: Viruses cannot infect cells if the correct membrane receptor is absent on the host cell. If so, removing and blocking the receptor molecules, or altering the cell surface, on susceptible host cells will prevent virus infection.

EXPERIMENTAL DESIGN: Appropriate susceptible host cells are treated in one of three ways to remove, block, or alter their receptors:

EXPERIMENT 1: Susceptible respiratory host cells are treated with neuraminidase (removes sialic acid).

EXPERIMENT 2: Susceptible respiratory cells are treated with an antibody that binds to a cell surface receptor called ICAM-1.

EXPERIMENT 3: The gene (*PVR*) coding for the poliovirus receptor is genetically engineered into mouse cells that normally do not bind poliovirus.

RESULTS:

EXPERIMENT 1

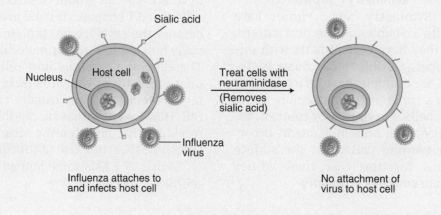

Influenza attaches to and infects host cell

No attachment of virus to host cell

EXPERIMENT 2

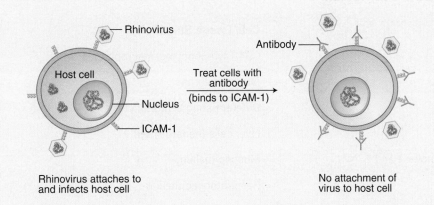

Rhinovirus attaches to
and infects host cell

No attachment of
virus to host cell

EXPERIMENT 3

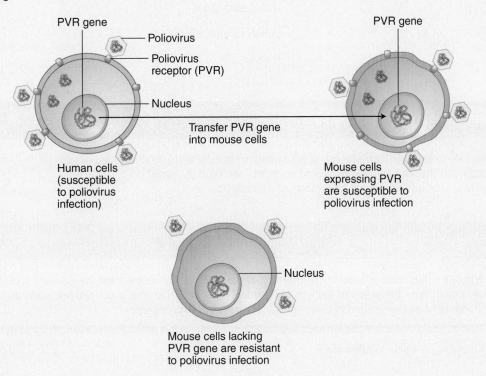

CONCLUSIONS:

QUESTION 1: *Is the hypothesis supported? Answer by analyzing the results from each of the three experiments.*

QUESTION 2: *Explain the relationship between a cell/tissue tropism and host cell receptors.*

QUESTION 3: *What are the controls in each of these three experiments?*

You can find answers online in **Appendix E**.

Modified from Shors, T. 2013. *Understanding Viruses*, Second Edition. Burlington, MA: Jones & Bartlett Learning.

TABLE 15.1 Viral Cell Tropism

Virus	Cell/Tissue Specificity
HIV	CD4 T lymphocytes, macrophages
Rabies	Muscle, neurons
Human papilloma	Differentiating keratinocytes
Hepatitis A, B, C	Liver cells (hepatocytes)
Human herpes simplex 1 and 2	Mucoepithelium
Influenza A	Respiratory epithelium
Norovirus	Intestinal epithelium
Rhinovirus	Nasal epithelium
Poliovirus	Intestinal epithelium
Epstein-Barr	B cells

Concept and Reasoning Checks 15.2

a. Identify the role of each structure found on (a) a nonenveloped and (b) an enveloped virus.
b. What shapes can viruses have and what structure determines that shape?
c. How does viral structure determine host range and tissue tropism?

Chapter Challenge B

Viruses often have a precise geometrical shape that is determined by the capsid. This crystalline-like architecture in the world of cells is quite unusual. In addition, the giant viruses, such as Mimivirus and Pandoraviruses described in MicroFocus 15.1 have been discovered.

QUESTION B: *How do viral architecture and the unusual properties of giant viruses affect your opinion of viruses being alive or not?*

You can find a discussion online in **Appendix F**.

■ KEY CONCEPT 15.3   Viruses Can Be Classified by Their Genome

An ongoing effort to classify viruses similar to that for cellular organisms has been complicated by the fact that viruses evolve rapidly and they can transfer their genes between viruses and with cellular organisms.

A Taxonomic Scheme for All Viruses Has Yet to Be Universally Adopted

Viral nomenclature has used a variety of virion features. The measles virus and poxviruses, for example, are named after the disease they cause; the Ebola and Marburg viruses after the location from which they were originally isolated; and the Epstein-Barr virus

after the researchers who studied it. Other viruses are named after morphological factors—the coronaviruses (*corona* = "crown") have a crown-like capsid, and the picornaviruses (*pico* = "small"; *rna* = "ribonucleic acid") are very small viruses with an RNA genome. Such characteristics, however, do not reflect the evolutionary history of the viruses.

A more encompassing classification system based on evolutionary relationships is being devised by the International Committee on Taxonomy of Viruses (ICTV). At this writing, higher-order taxa (phyla and classes) have not been completely developed. To date, six orders are recognized that comprise 87 families, each ending with -viridae (e.g., Herpesviridae). However, many other viruses have not yet been assigned to a family. Viruses have been categorized into hundreds of genera; each genus name ends with the suffix -virus (e.g., human herpes virus). In this text, we mostly use the family name (e.g., Herpesviridae) when referring to the whole family of viruses and the common name (e.g., herpes simplex virus) when discussing a specific virus in a family.

Animal viruses usually are classified into two groups based on their genome (DNA or RNA) and then split into separate families based on characteristics such as strand type (double stranded and single stranded) and presence or absence of an envelope. FIGURE 15.7 presents a selection of viral families affecting humans, together with some of their characteristics and diseases. Please refer to this figure as we describe the DNA viruses and the RNA viruses.

DNA Viruses

Many viruses contain either double-stranded (ds) or single-stranded (ss) DNA genomes with nucleocapsids that are nonenveloped or enveloped. Among the nonenveloped dsDNA viruses are the Adenoviridae, which includes viruses causing common colds, and the Papillomaviridae, which includes members causing common skin warts. The enveloped dsDNA viruses include the Herpesviridae whose members include viruses responsible for cold sores (fever blisters) and chickenpox; the Poxviridae that includes the smallpox virus; and the Hepadnaviridae that includes the hepatitis B virus.

The ssDNA viruses are nonenveloped. The Parvoviridae include one virus that causes a childhood rash called fifth disease. A separate canine parvovirus infects the lymph nodes and intestines of puppies.

RNA Viruses

A large number of viruses have RNA genomes consisting of either dsRNA or ssRNA (see Figure 15.7).

The viruses that have dsRNA include the nonenveloped Reoviridae, whose genomes are segmented and surrounded by a two-layered capsid. In this family, the rotaviruses can cause severe diarrhea and intestinal distress (gastroenteritis) in infants and young children.

The ssRNA viruses have nucleocapsids that are enveloped or nonenveloped. Their genomes are classified according to the sense or polarity of their RNA into positive sense and negative sense. Please refer to this figure as we describe the different groups of viruses by their type of genome.

▶ **Positive-Sense (+Sense) RNA Viruses.** The ssRNA in the +sense RNA viruses acts as mRNA and, on infection of a host cell, the RNA can be translated directly by the host cell ribosomes (thus called + sense). The non-enveloped Picornaviridae contain viruses responsible for many well-known human diseases, including common colds, polio, and hepatitis A. The enveloped families, such as the Flaviviridae, also include many viruses responsible for human diseases, such as yellow fever, West Nile disease, hepatitis C, and Zika virus infection. In fact, the +sense RNA viruses are the largest group of RNA viruses.

Although the Retroviridae are single-stranded, +sense RNA viruses, this family of viruses behaves in an unusual way. After infecting a host cell, the virus uses its own enzyme to produce dsDNA from its ssRNA genome through a process of reverse transcription (*retro* = "backwards"). Only after the new dsDNA has integrated into the host cell genome can the viral DNA genes be transcribed into mRNA and translated. The human immunodeficiency virus (HIV) that is responsible for AIDS is the most recognized virus in this family.

▶ **Negative-Sense (−Sense) RNA Viruses.** The ssRNA in −sense RNA viruses is complementary in base sequence to mRNA (thus called − sense). As a result, the RNA must be made into a "readable" form. This is accomplished when a RNA polymerase converts the −sense RNA into a +sense RNA prior to translation. The +sense RNA then is translated by host ribosomes. The virus families in this group are enveloped, and several members are the infectious agents for a considerable number of human diseases, including Ebola virus disease, influenza, measles, mumps, and rabies.

As a rule, the genomes of RNA viruses are smaller than those in DNA viruses, and they depend more heavily on host cell proteins and enzymes for replication. RNA virus genomes also are more error (mutation) prone when copying their RNA genome because the RNA polymerase lacks the efficient proofreading exhibited by DNA polymerase to correct replication errors (FIGURE 15.8). Thus, RNA viruses, such as the influenzaviruses, tend to "genetically drift," evolving more rapidly into new, potentially epidemic strains.

FIGURE 15.9 identifies the different types of viral genomes by their host domain and, for the Eukarya, also by animal, fungal, plant, and protist.

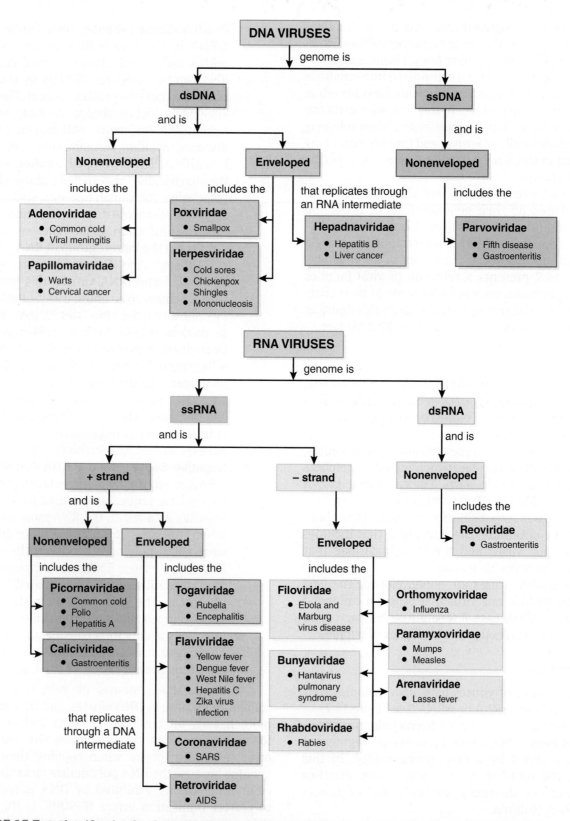

FIGURE 15.7 A Classification for the Medically Relevant Human Viruses. The virus families causing human diseases have been classified based on genome type, being enveloped or nonenveloped, and, for the ssRNA viruses, having + sense or − sense. In this concept map, some of the distinctive human diseases associated with each family are indicated. *»» According to this concept map, does it mean, for example, that the Herpesviridae are closely related to the Poxviridae? Explain.*

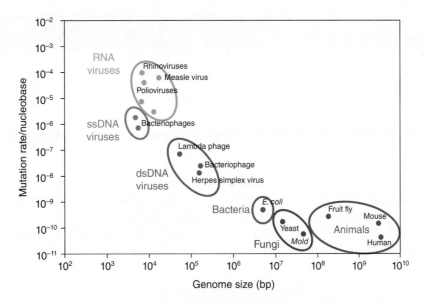

FIGURE 15.8 **Mutation Rate Versus Genome Size.** The mutation rate per nucleobase versus genome size is plotted for several viruses, a bacterial species, two fungal species, and three animals. *»» Why are RNA viruses so prone to mutations?*

Modified from Gago et al. 2009 *Science* 323:1308

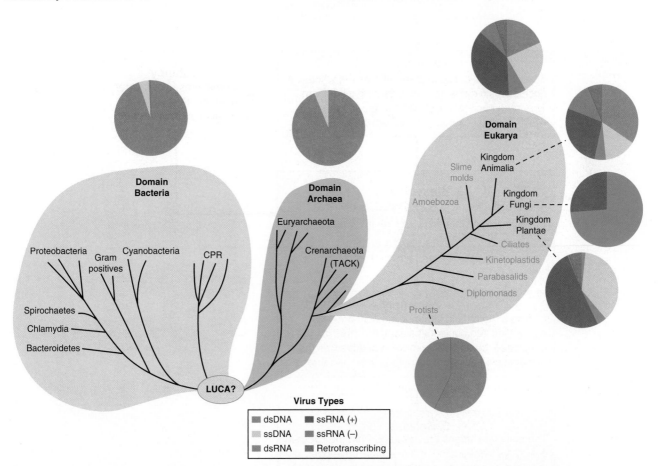

FIGURE 15.9 **Diversity of Viruses.** The pie charts superimposed on the "tree of life" show the types of viral genomes associated with the domains of life and within the animals, fungi, plants, and protists. CPR = Candidate Phyla Radiation; TACK = Thaumarchaeota, Aigarchaeota, Crenarchaeota, and Korarchaeota groups. *»» Which group of organisms has (A) mostly dsRNA genomes and (B) about 50% ssRNA (+) genomes? Which group lacks RNA viruses completely?*

Pie charts modified from *Small Things Considered* (January 2016). Viral Abundance and Diversity.

Concept and Reasoning Check 15.3

a. How do DNA viruses differ from RNA viruses?

■ KEY CONCEPT 15.4 Virus Replication Follows a Set of Common Steps

The process of viral replication is one of the most remarkable events in nature (**FIGURE 15.10**). In this five-step process, a virus (1) encounters and attaches to the appropriate host cell, (2) enters the host cell that is a thousand or more times its size, and (3) hijacks the metabolism of the cell to construct viral parts that are then (4) assembled into new virus particles. New virions then are (5) released from the cell, the host cell often being destroyed (lysed) in the process.

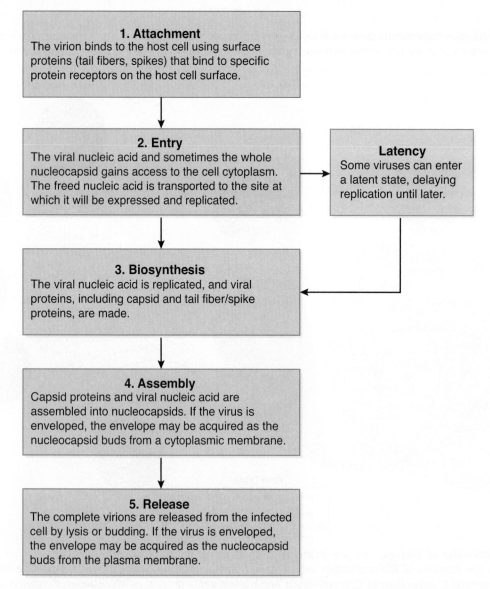

FIGURE 15.10 The Steps of Virus Replication. These five steps are common to all viruses. »» *Why can't viruses carry out these steps outside a host cell?*

Replication has been studied in a wide range of viruses and their host cells. Before we examine the animal (human) viruses, we will look at the ubiquitous bacteriophages because much of what we know about viruses comes from phage studies.

The Replication of Bacteriophages Can Follow One of Two Pathways

The bacteriophages are the most prevalent and numerous biological agents in the world. These viruses exhibit two strategies of infection: a lytic (virulent) pathway and a lysogenic (temperate) pathway.

The Lytic Pathway

One of the best-studied processes of replication is that carried out by bacteriophages of the T-even group (T for "type"). Bacteriophages T2, T4, and T6 are in this group. They are large, complex, DNA virions with a characteristic head and tail structure (**FIGURE 15.11**). They contain tail fibers, which, like the spikes on animal viruses, function to attach the phage to a bacterial cell. The T-even phages are

virulent viruses, meaning they lyse the host cell while carrying out a **lytic pathway** of replication.

We will use phage T4 replication in *Escherichia coli* as a model for the lytic pathway. An overview of the five-step process is presented in **FIGURE 15.12**.

1. **Attachment.** The first step in the replication pathway of a virulent phage occurs when phage and bacterial cells randomly collide. If sites on the phage's tail fibers match with a complementary receptor site on the cell envelope of the bacterial cell, attachment will occur. The actual attachment (adsorption) consists of a weak chemical union between phage and receptor site.

2. **Entry.** Following attachment, the phage DNA crosses the cell envelope (Figure 15.11). The tail of the phage releases lysozyme, an enzyme that dissolves a portion of the bacterial cell wall. The tail sheath then contracts, and, as the tail core drives through the cell envelope, the DNA is ejected through the core and into the bacterial cytoplasm. The ejection

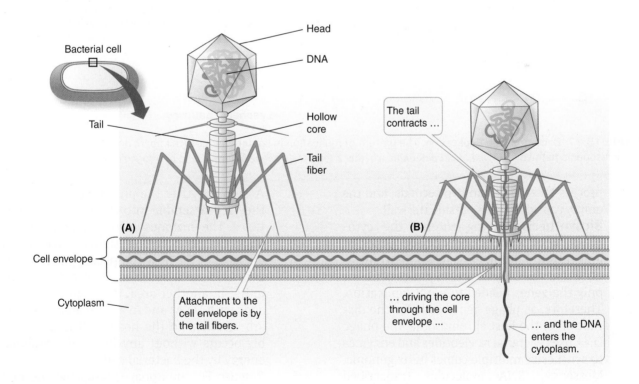

FIGURE 15.11 Bacteriophage Structure, Attachment, and Genome Entry. (A) The structure of a bacteriophage consists of the protein head, inside of which is the nucleic acid, and the protein tail. The hollow core of the tail allows transfer of the nucleic acid during infection while the tail fibers attach the phage to receptors on the cell envelope. **(B)** Following attachment, the tail core contracts and is driven through the cell wall. *»» Is the phage in this figure infecting a gram-positive or gram-negative cell? Explain.*

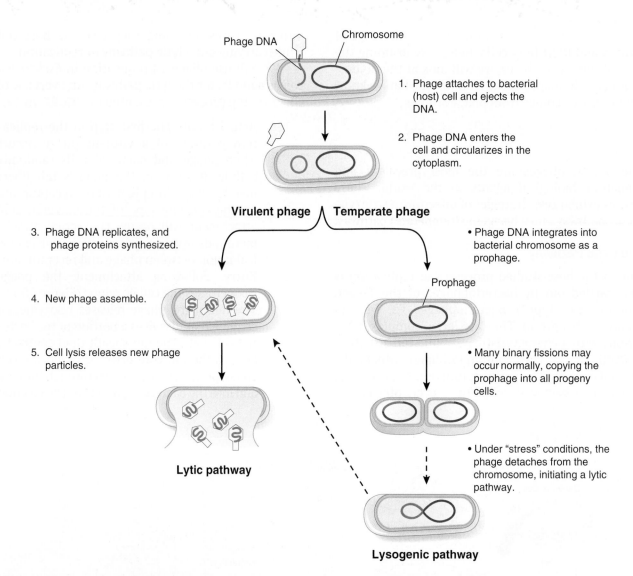

1. Phage attaches to bacterial (host) cell and ejects the DNA.

2. Phage DNA enters the cell and circularizes in the cytoplasm.

Virulent phage | **Temperate phage**

3. Phage DNA replicates, and phage proteins synthesized.

• Phage DNA integrates into bacterial chromosome as a prophage.

4. New phage assemble.

5. Cell lysis releases new phage particles.

• Many binary fissions may occur normally, copying the prophage into all progeny cells.

Lytic pathway

• Under "stress" conditions, the phage detaches from the chromosome, initiating a lytic pathway.

Lysogenic pathway

FIGURE 15.12 Bacteriophage Replication. The pattern of replication in bacteriophages (tail fibers not shown) can involve a lytic or lysogenic pathway. *»» What might determine whether a temperate phage follows a lytic or lysogenic pathway?*

process takes less than two seconds, and the empty capsid remains outside the wall.

3. **Biosynthesis.** Having entered the cytoplasm, production of new phage genomes and capsid parts begins. It is important to note that the DNA in the T4 phage contains only the genes needed for viral replication. Therefore, as phage genes code for the disruption of the host chromosome, the phage DNA uses bacterial nucleotides and enzymes to synthesize multiple copies of its genome. Messenger RNA molecules transcribed from phage genes appear in the cytoplasm, and the translation of phage enzymes and capsid proteins begins. The bacterial cells' ribosomes, amino acids, and enzymes are all enlisted for biosynthesis.

4. **Assembly.** After the phage parts are made, they self-assemble into complete virus particles. The enzymes encoded by viral genes guide the assembly in systematic fashion. In one area of the host cytoplasm, phage heads and tails are assembled from protein subunits; in another area, the heads are packaged with DNA; and in a third area, the tails are attached to the heads. All this assembly occurs without any input of metabolic energy by the bacterial cell.

5. **Release.** For some phages, lysozyme, encoded by the bacteriophage genes late in the replicative cycle, degrades the bacterial cell wall. Mature phage particles now burst out from the ruptured bacterial shell and are set free to infect more bacterial cells. The number of

phages released from an infected bacterial cell is called the **burst size**. Although this size depends on the volume of the bacterial cell, burst size typically ranges from 50 to a few hundred phages per infected cell.

The lytic pathway represents a **productive infection** because many virions are produced with little delay between attachment and release. Importantly, not every phage infection spells disaster for the bacterial cell. Bacterial cells can modify the receptors on their surface (through mutation) so the phages have no place to attach. Bacterial cells

also have a type of "immune defense," called the **CRISPR-Cas system**, which is capable of recognizing and destroying foreign (plasmid or viral) DNA that enters the cell. Of course, viral mutations can occur so phages can get past the bacterial "immune" defenses. In the end, it often becomes a molecular "arms race" between phage and bacterial cells as to whether the predator (phage) or prey (bacterial cell) comes out on top.

MICROFOCUS 15.2 describes the use of phages as a possible "therapy" to cure bacterial infections.

MICROFOCUS 15.2: History/Public Health/Biotechnology

Phage Therapy and Clinical Medicine

When bacteriophages were identified in the early 1900s, some scientists, including one of their discoverers, Felix d'Herrelle, began to promote phages as therapy for curing dreaded bacterial diseases such as cholera and bubonic plague. D'Herrelle and others assumed that if phages could destroy bacterial cells in culture (see figure), why not try using them to destroy bacterial cells causing diseases in the human body. Consequently, in 1919, they treated four children suffering from bacterial dysentery. All four began to recover within 24 hours. Phage therapy, however, never was deeply embraced in the United States, partly due to the rise and availability of antibiotics in the 1940s and 1950s. Today, as antibiotic resistance reaches a crisis level, phage therapy again is being looked at as a potential antibacterial therapy.

Let's examine a few examples of phage therapy in food production and clinical medicine.

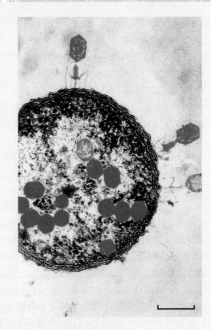

Phage infection of an *E. coli* cell. (Bar = 100 nm.)

© Lee D. Simon/Science Source.

- The U.S. Food and Drug Administration (FDA) has approved a "cocktail" of six purified phages (ListShield™) that can be safely sprayed on luncheon meats (cold cuts, hot dogs, sausages) to kill any of 170 strains of *Listeria monocytogenes*, which cause listeriosis, a serious, life-threatening foodborne disease to newborns, pregnant women, and immunocompromised individuals.
- A Polish microbiologist has injected a solution of phages into 550 patients with a blood infection. Each patient benefited from the treatment, and the researchers report that several patients appear completely cured.
- Austrian researchers have engineered bacteriophages designed to kill bacterial cells without lysing them. These phages were ten times more lethal than the natural phages.
- A Boston research team has engineered a phage to attack and destroy bacterial biofilms, which often are the cause of human infections and disease. In addition, a group of Italian researchers has identified a phage capable of curing 97% of mice infected with a deadly form of methicillin-resistant *Staphylococcus aureus* (MRSA).
- The FDA has approved the first clinical trial using phages with people having venous leg ulcers. No harm to the volunteers was reported.
- Several biotech companies are conducting clinical trials that use phage cocktails to treat *Pseudomonas aeruginosa*, MRSA, *Escherichia coli*, and other infections caused by bacterial pathogens.
- The European Phagoburn study is evaluating phages as treatment for infections resulting from burn wounds.

At a time when we are seeing increased bacterial resistance to antibiotics, phage therapy using "phage cocktails" or phage enzymes is on the rebound. Although many of these phage therapy studies are in early clinical trials, the use of phage therapy holds great potential as a complementary approach for the treatment of acute and chronic bacterial infections and in support of human health.

The Lysogenic Pathway

Other phages interact with bacterial cells in a slightly different way, called a **lysogenic pathway** (see Figure 15.12). For example, lambda (λ) phage also infects *E. coli* but might not immediately enter a lytic pathway. Instead, after attaching and entering the bacterial cell, the phage DNA integrates into the bacterial chromosome as a **prophage**. Bacteriophages participating in this cycle are known as **temperate** (self-controlled) phages because the phage does not immediately kill the host bacterial cell. Rather, the bacterial cell survives the infection and continues to grow and divide normally. However, as the bacterial cell goes through its cell cycle, the prophage is copied and vertically transferred to daughter cells as part of the replicated bacterial chromosome. Because the prophage remains "inactive," the infection is referred to as a **latent infection**.

Such binary fissions can continue for an undefined period with all daughter cells "infected" with the prophage. Usually at some point, the bacterial cells become "stressed" (e.g., lack of nutrients, presence of harmful chemicals). This triggers the prophage to detach itself from the bacterial chromosomes (see Figure 15.12) and initiate a lytic pathway, lysing the bacterial cells as new λ phages are released.

The CRISPR-Cas system mentioned above is more than a defense against foreign DNA entering the cell. The CRISPR-Cas system is actually a surveillance system; that is, the system can recognize whether the incoming DNA is potentially destructive or beneficial. For example, the bacterial pathogen *Corynebacterium diphtheriae* is the agent responsible for diphtheria. The toxin produced by these cells is coded by a gene that the organism picked up from a lysogenic infection. Because this gene is beneficial from the pathogen's perspective, it is not destroyed by CRISPR-Cas. We are only now beginning to understand this adaptive immune ability that bacteria possess.

Animal Virus Replication Often Results in a Productive Infection

Like bacteriophages, animal viruses often have brief but eventful "lives" as they infect cells and produce more virions. Such a productive infection retains the five steps of virus replication described for the bacteriophages. An overview of the dsDNA and ssRNA viruses is presented here, with emphasis on the differences in replication events from the phages.

1. **Attachment.** Animal viruses infect host cells by binding to receptors on the host cell's plasma membrane. This binding is facilitated by capsid proteins or spikes spread over the surface of the envelope (e.g., Herpesviridae) or capsid (e.g., Picornaviridae) (**FIGURE 15.13**).

2. **Entry.** Viral entry differs from that in phages, in that animal viruses often are taken into the cytoplasm as intact nucleocapsids. For the enveloped herpesviruses, the viral envelope fuses with the plasma membrane and releases the nucleocapsid into the cytoplasm. For other animal viruses, such as the non enveloped poliovirus, the virion is taken into the cell by endocytosis. At the attachment site, the cell engulfs the virus within a vacuole and brings the virus-containing vacuole into the cytoplasm (**FIGURE 15.13B**).

   Regardless of entry mechanism, once in the cytoplasm, the capsid disassembles from the genome in a process called **uncoating** and the genome is transported to the site (i.e., cell nucleus, cytoplasm) where transcription or replication will occur.

3. and 4. **Biosynthesis and Assembly.** One way we split the virus families is based on whether they have a DNA or RNA genome.

   ▶ **DNA Genome Expression.** The DNA of a DNA virus supplies (expresses) the genetic codes for enzymes that synthesize viral parts from the host's available building blocks. The Herpesviridae, for example, employ a division of labor: the DNA genome is replicated and transcribed in the host cell nucleus, whereas the capsid proteins are translated in the cytoplasm (**FIGURE 15.14**). The viral proteins are then transported to the nucleus where they join with the nucleic acid molecules for assembly.

   ▶ **RNA Genome Expression.** RNA viruses follow a slightly different pattern. Because the RNA in +sense RNA viruses acts as an mRNA, following uncoating, the RNA is directly translated by the host cell's ribosomes into viral proteins as genome replication occurs. However, the −sense RNA viruses use their RNA as a template to synthesize a complementary +sense of RNA. The synthesized +ssRNA then is used as an mRNA molecule for protein synthesis as well as the template to form the −ssRNA genome.

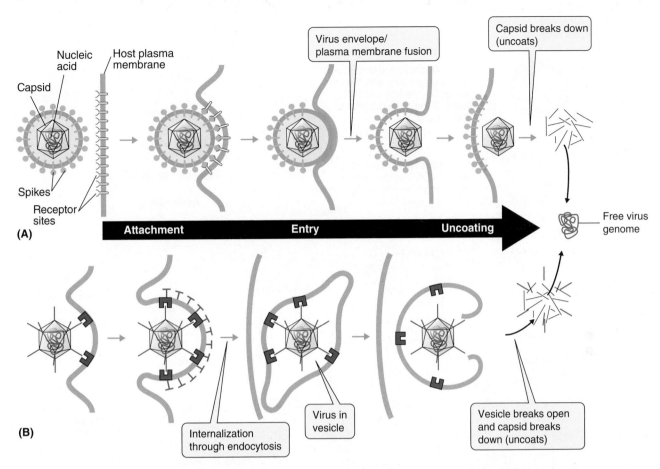

FIGURE 15.13 Animal Virus Entry Strategies. (A) An enveloped virus, such as herpesvirus, contacts the plasma membrane and the spikes interact with receptor sites on the membrane surface of the host cell. **(B)** For nonenveloped viruses, such as the poliovirus, entry into the cytoplasm occurs through endocytosis. »» *Propose a hypothesis to explain why two different entry mechanisms exist for virion entry into a host cell.*

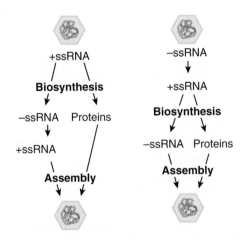

The final step of assembly for some enveloped viruses is the acquisition of an envelope. In this step, envelope proteins (spikes) are synthesized, and, depending on the virus, the spikes are incorporated into a nuclear or cytoplasmic membrane during virus assembly

or from the plasma membrane during virus release (see Figure 15.14).

5. **Release.** In the final stage, the nucleocapsids of some viruses push through the plasma membrane, forcing a portion of the membrane ahead of and around the virion, resulting in an envelope. Other enveloped viruses, like the Herpesviridae, fuse with the plasma membrane, releasing the virion. This process, called **budding**, need not always kill or damage the cell during release. The same cannot be said for nonenveloped viruses. They leave the cell when the cell membrane ruptures; this process generally results in cell lysis and death.

The number of virions released (burst size) from an infected host cell again depends on the volume of the host cell. In general, the burst size typically ranges from 500 to several thousand virions per infected host cell.

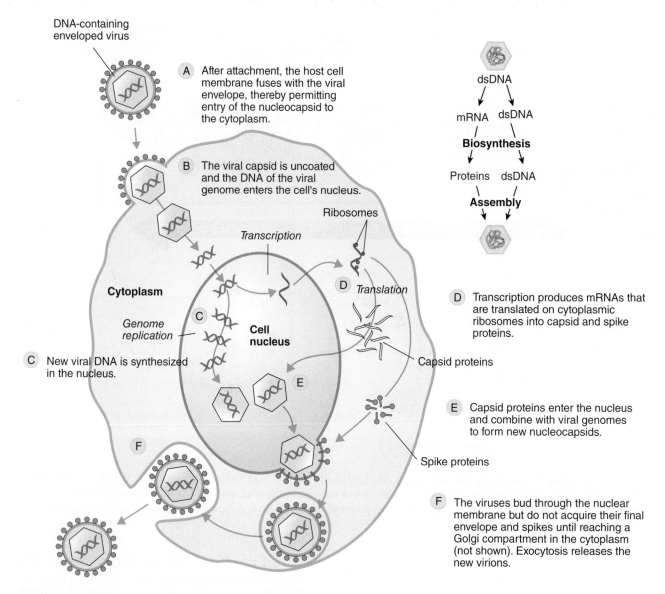

DNA-containing enveloped virus

A. After attachment, the host cell membrane fuses with the viral envelope, thereby permitting entry of the nucleocapsid to the cytoplasm.

B. The viral capsid is uncoated and the DNA of the viral genome enters the cell's nucleus.

Transcription

Ribosomes

Cytoplasm

D. *Translation*

Genome replication

Cell nucleus

C. New viral DNA is synthesized in the nucleus.

Capsid proteins

Spike proteins

dsDNA

mRNA   dsDNA

Biosynthesis

Proteins   dsDNA

Assembly

D. Transcription produces mRNAs that are translated on cytoplasmic ribosomes into capsid and spike proteins.

E. Capsid proteins enter the nucleus and combine with viral genomes to form new nucleocapsids.

F. The viruses bud through the nuclear membrane but do not acquire their final envelope and spikes until reaching a Golgi compartment in the cytoplasm (not shown). Exocytosis releases the new virions.

FIGURE 15.14 Replication of a Double-Stranded DNA Virus. The replication cycle for a DNA virus, such as a herpes simplex virus, involves genome replication and assembly in the cell nucleus and biosynthesis in the cytoplasm. *»» Why must the viral DNA enter the cell nucleus to replicate?*

Some Animal Viruses Produce a Latent Infection

Unlike most RNA viruses that go through a productive infection, many of the DNA viruses and the Retroviridae can establish a latent (or persistent) infection, characterized by repression of most viral gene expression; therefore, the virus lies "dormant." For example, some Herpesviridae, such as the herpes simplex virus (HSV), can generate a productive or latent infection. In an infected sensory neuron, HSV undergoes latency as the viral dsDNA enters the neuron's cell nucleus and circularizes. No virions are produced for months or years until some stress event reactivates the viral dsDNA and a new productive infection will be initiated.

Retroviridae, such as HIV, also carry out a latent infection. However, in this group of viruses, the virus carries an enzyme, called **reverse transcriptase**, which, as described earlier, is used to reverse transcribe its +ssRNA into dsDNA (**FIGURE 15.15**). The dsDNA then enters the host cell nucleus. Like the temperate phage DNA inserted into a bacterial chromosome, the dsDNA becomes integrated randomly into

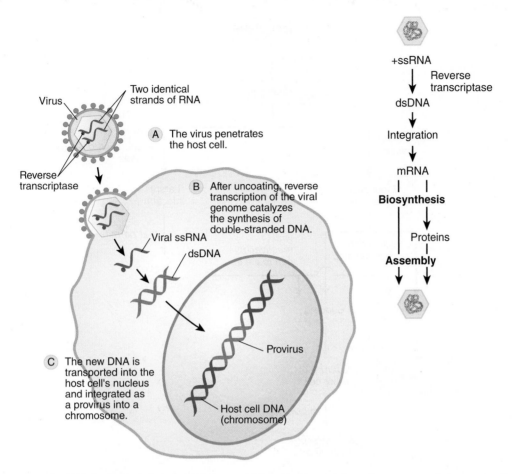

FIGURE 15.15 The Formation of a Provirus by HIV. Retroviridae, such as HIV, can incorporate their genome into a host cell chromosome. The single-stranded RNA genome is reverse transcribed into double-stranded DNA by a reverse transcriptase enzyme. The DNA then enters the cell nucleus and integrates into a chromosome. **»» *Propose a reason why genome integration is an advantage for the virus.***

the DNA of one chromosome. This integrated viral genome is referred to as a **provirus**, and it represents a unique and stable association between the viral DNA and the host genome. However, at any time, the provirus can be reactivated and a productive infection involving biosynthesis, assembly, and release of new virions will ensue. The advantage for the viral DNA to exist as a provirus is that every time the host cell divides, the provirus will be replicated along with the host genome and be present in all progeny cells.

Before leaving the topic of viral genomes, it is worth noting that the normal human genome contains almost 10% viral DNA sequences. That is, the human genome contains DNA sequences referred to as human **endogenous retroviruses** (ERVs). These ERVs are the result of ancient retroviral infections, and most of the DNA sequences are only fragments

of the original viral genome that integrated into the germline thousands of years ago. In addition, nonretroviral elements also have been discovered in vertebrate genomes, many of these coming from RNA viruses.

Viral Infections in Humans Can Be Treated but Not Cured

With the possible exception of hepatitis C, no human viral disease can be cured completely. Antibiotics do not work on viruses, although a number of antiviral drugs can ease the symptoms or shorten the duration of a viral infection. However, some viral diseases, such as smallpox, measles, and hepatitis B, can be prevented with the use of vaccines. **FIGURE 15.16** summarizes the outcomes for animal virus infections.

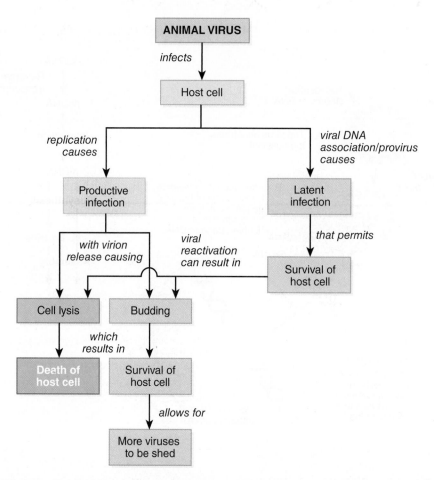

FIGURE 15.16 Potential Outcomes for Animal Virus Infection of a Host Cell. As depicted in this concept map, depending on the virus and host cell interaction, the host cell may be killed or survive. »» *Identify a specific virus that follows each of the three outcomes (productive infection; cell death or survival; latent infection).*

Concept and Reasoning Checks 15.4

a. Why would it be advantageous for a phage to carry out a lysogenic pathway rather than a lytic pathway?
b. Distinguish between the replication events in DNA and RNA viruses.
c. How do viruses cause disease?
d. Why don't RNA viruses, like the poliovirus or influenzavirus, undergo provirus formation?

Chapter Challenge C

Viruses do not replicate the way cellular organisms reproduce. There is no binary fission or mitosis whereby one cell divides into two daughter cells. Rather, a virus infection leads to the simultaneous replication of hundreds of new virions. The infected cell consists of multiple assembly lines, all cranking out new virus particles.

QUESTION C: *Considering the characteristics of life, the new information on virus replication, and the discovery of giant viruses, does any of this change, modify, or strengthen your assertion that viruses are alive or not alive?*

You can find a discussion online in **Appendix F**.

■ KEY CONCEPT 15.5   Viruses and Their Infections Can Be Detected in Various Ways

If an individual contracts a viral infection, there are various ways the viral agent can be detected for identification and eventual diagnosis.

Detection of Viruses Often Is Critical to Disease Identification

The diagnosis of viral infections, such as the flu and a cold, are usually straightforward and might not require further laboratory confirmation. In some cases, viral infections leave their mark on the infected individuals. Measles is accompanied by **Koplik spots**, which are a series of bright red patches with white pimple-like centers on the lateral mouth surfaces. Mumps and chickenpox are associated, respectively, with swollen salivary glands and teardrop-like skin lesions.

Virus detection, however, is not always so straightforward, and proper identification has a bearing on relating a particular virus to a particular disease. The clinical microbiology laboratory or reference laboratory diagnosis of a viral disease can be carried out by using light or electron microscopy to examine cells obtained from body tissues or fluids (**FIGURE 15.17**). Other common methods to detect viral infections use immunological procedures to search for viral antibodies in a patient's serum.

Because viruses need to get inside host cells to replicate, viruses cannot be grown by themselves in

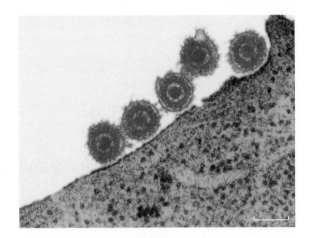

FIGURE 15.17 Cells and Viruses. A false-color transmission electron micrograph of tissue cells and associated herpes simplex viruses (orange). The envelope of the virus is seen as a ring at the viral surface, and the nucleocapsid of the virus is the darker center. (Bar = 200 nm.) *»» How might these viruses enter the host cell?*

© Dennis Kunkel Microscopy/Science Source.

culture. Rather, virologists use cultured human or animal cells to study the viral infection process and to produce large stocks of virus. One common method once used to cultivate and detect viruses was fertilized chicken eggs. A hole is drilled in the eggshell, and a viral suspension is introduced (**FIGURE 15.18A**). Today, only the influenzaviruses are routinely

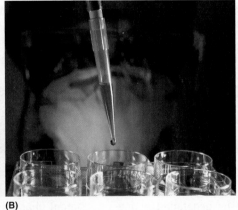
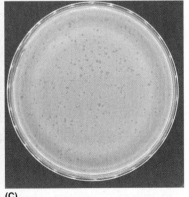

(A)   (B)   (C)

FIGURE 15.18 Infection of Cells in Embryonated Eggs and Cell Culture. (A) Inoculation of fertilized eggs is standard practice for growing the flu virus. **(B)** A masked researcher uses a pipette to infect a culture of human cells with a virus. The dishes contain a culture medium, which allows the cells to survive outside the body. The effect of the virus on the infected cells will then be studied. **(C)** Plaque formation in a cell culture produces clear areas (plaques) in the dish. *»» Why must a face guard or mask be worn when infecting embryonated eggs and cells in culture?*

(A) Courtesy of Greg Knobloch/CDC. (B) © James King-Holmes/Science Source. (C) Courtesy of Giles Scientific Inc., CA, www.biomic.com.

TABLE 15.2 Examples of Virus Cytopathic Effects

Virus Family	Cytopathic Effect
	Changes in cell structure
Picornaviridae	Shrinking of cell nucleus
Papovaviridae	Cytoplasmic vacuoles
Paramyxoviridae, Coronaviridae	Cell fusion (syncytia)
Herpesviridae	Chromosome breakage
Herpesviridae, Adenoviridae, Picornaviridae, Rhabdoviridae	Rounding and detachment of cells from culture
	Cell inclusions
Adenoviridae	Virions in nucleus
Rhabdoviridae	Virions in cytoplasm (Negri bodies)
Poxviridae	"Viral factories" in cytoplasm
Herpesviridae	Nuclear granules

cultivated by this method, usually to produce high virus concentrations for vaccine production.

The most common method for detecting animal viruses is to select an animal cell culture in which the virus of interest will replicate. The cell culture is usually grown as a single layer of cells called a **monolayer**. Viruses are then introduced into the culture (**FIGURE 15.18B**). When viruses replicate in a cell culture, they often cause a noticeable change to the cells (TABLE 15.2). This change, called the **cytopathic effect** (CPE), often is detected as microscopic alterations to cell structure or the formation of multinucleate giant cells called **syncytia**.

Viruses also can be detected by the formation of plaques in a cell culture. A **plaque** is a clear zone within the dense "lawn" of bacterial cells or monolayer of animal cells (**FIGURE 15.18C**). The viruses infect and replicate in the cells, thereby destroying them and forming the clear areas. Epidemiological investigations sometimes use **phage typing** to identify plaques characteristic of a specific strain of a bacterial disease, such as staphylococcal food poisoning. **MICROINQUIRY 15** uses the plaque assay to monitor intracellular virus production and extracellular virus release during a productive infection of animal cells in culture.

MICROINQUIRY 15

The One-Step Growth Cycle

In the research laboratory, we can follow the replication of viruses (productive infection) by generating a one-step growth curve. Realize that we are not really looking at growth but the replication and increase in number of virus particles. There are several periods associated with a virus growth (replication) cycle that you should remember because you will need to identify the periods in the growth curve. The "eclipse period" is the time when no virions can be detected inside cells (intracellular), and the "latent phase" is the time during which no extracellular virions can be detected. Also, the "burst size" is the number of virions released per infected cell.

To generate our growth curve, we start by inoculating our viruses onto a susceptible cell culture. There are 100,000 (10^5) cells in each of 10 cultures. We then add 10 times the number of viruses (10^6) in a small volume of liquid to each culture to ensure that all the cells will be infected rapidly. After 60 minutes of incubation, we wash each culture to remove any viruses that did not attach to the cells and add fresh cell growth medium. Then, at 0 hour and every 4 hours after infection, we remove one culture, pour off the growth medium, and lyse the cells. The number of viruses in the growth medium and within the lysed cells is determined. Because we cannot see viruses, we measure any viruses in the growth medium or in the lysed cells using a plaque assay (see text for details). We assume that one plaque (plaque-forming unit or PFU) resulted from one infected cell.

The curves drawn in the accompanying figure plot the results from our growth experiment.

Answer the following questions based on the one-step growth curve. You can find answers online in **Appendix E**.

15.1a. How long is the eclipse period for this viral infection? Explain what is happening during this period.

15.1b. How long is the latent period for this infection? Explain what is happening during this period.

15.1c. What is the burst size?

15.1d. Explain why the growth curve shows a decline in phages between 0 and 4 hours.

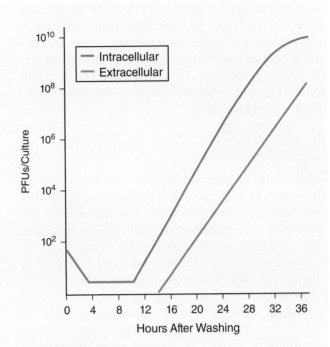

A One-Step Growth Cycle

Concept and Reasoning Check 15.5

a. Explain how some viruses can be identified simply by observing an infected cell culture.

■ KEY CONCEPT 15.6   Some Viruses Are Associated with Human Tumors and Cancers

Cancer is indiscriminate. It affects humans and animals, young and old, male and female, rich and poor. In the United States, some 560,000 people die of cancer annually, making the disease the third leading cause of death after heart disease and stroke. Worldwide, almost 8 million people die of cancer each year, two-thirds in developing nations. Cancer is a very complex topic, so we will only summarize the general concept and emphasize the important points related to viruses.

Cancer Is an Uncontrolled Growth and Spread of Cells

Cancer results, in part, from the uncontrolled reproduction (mitosis) of a single cell. The cell escapes the cell cycle's controlling factors and, as it continues to multiply, a mass of cells soon forms. Eventually, the cluster yields a clone of abnormal cells referred to as a **tumor**. Normally, the body will respond to a tumor by surrounding it with fibrous connective tissue called a capsule. Such a local tumor is designated **benign** because it usually does not spread to surrounding tissue.

Additional changes can occur to tumor cells that release them from their origins. The tumor cells might break out of the capsule and **metastasize**, a spreading of the cells to other tissues of the body. Such a tumor now is described as **malignant**, and the individual now has **cancer** (cancer = "crab"; a reference to the radiating spread of cells, which resembles a crab).

How can such a mass of cells bring illness and misery to the body? By their sheer numbers, cancer cells invade and erode local tissues, interrupt normal functions, and choke organs to death by robbing them of vital nutrients. Thus, the cancer patient will commonly experience weight loss even while maintaining a normal diet.

Oncogenic Viruses Influence Cancer Development

Some cancer researchers believe that about 65% of all human cancers are the result of genetic factors (FIGURE 15.19). Many of the remaining cancers are associated with **carcinogens**, which are chemical and physical agents that produce cellular changes leading to a tumor and cancer. The **Ames test**, discussed in Chapter 7, is designed to use bacterial cells to detect potential chemical carcinogens in humans.

Health experts suspect that about 12% of human cancers globally are associated with viruses, the majority caused by human papillomaviruses (HPVs) and the hepatitis B and C viruses. In all,

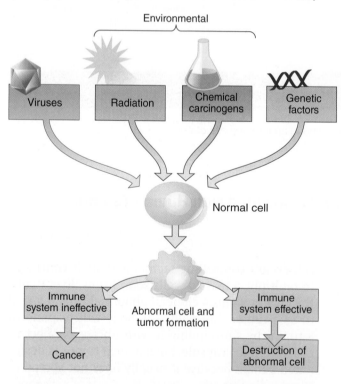

FIGURE 15.19 **The Onset of Cancer.** Viruses and other factors can induce a normal cell to become abnormal. When the immune system is effective, it destroys abnormal cells. However, when the abnormal cells evade the immune system, a tumor may develop and become malignant, spreading to other tissues in the body. »» *What factors are required for a tumor to become malignant?*

at least six viruses are associated directly or indirectly with human tumors and cancers (TABLE 15.3). When these **oncogenic (tumor-causing) viruses** are transferred to test animals or cell cultures, an observable change in the morphological, biochemical, or growth patterns of normal cells takes place. This cellular **transformation** involves a complex, multistep sequence of events, of which viruses play one part. Immune system activity, chronic inflammation, and mutagens usually are additional factors required for malignant transformation.

As Table 15.3 shows, oncogenic viruses are found only in some families composing the DNA viruses (DNA tumor viruses), which have dsDNA genomes, and the Retroviridae (RNA tumor viruses), which have dsDNA when integrated as a provirus in a host cell. The most common virus-associated malignancy among women is uterine cancer, caused by certain groups of human papillomaviruses (HPVs) in the Papoviridae. In men, liver cancer, caused by the hepatitis B virus (Hepadnaviridae), is the most common. One further point is that there might be other, yet unidentified viruses that also are associated with human cancers. On the positive side, individuals can be vaccinated to prevent infection by HPV and hepatitis B virus, and there are antiviral drugs available that reportedly can cure hepatitis C infections. Thus, the incidences of cancer caused by these viruses have been declining.

Oncogenic viruses are thought to cause tumors in three ways:

▸ **Oncogene Activity.** One way the cellular transformation process is mediated is by proteins encoded by **oncogenes**, which are viral genes that some members of the Retroviridae possess. On infection of a human cell, integration and expression of the viral oncogene convert a normal cell into a tumorous or cancerous cell (FIGURE 15.20A). These genes usually affect growth control, cell division, or chemical signaling involved in cell growth.

▸ **Gene Suppression.** Cellular transformation can occur through the loss of normal gene activity. Normal cells contain several **tumor suppressor genes (TSGs)**, the function of which is to inhibit transformation; that is, if a cell attempts to convert to an abnormal cell, the tumor suppressor proteins trigger a programmed cell death (cell suicide) called **apoptosis**. However, if a provirus integrates

TABLE 15.3 Human Oncoviruses and Their Effects on Cell Growth

Oncogenic Virus	Effect	Benign Disease	Associated Cancer
DNA Tumor Viruses			
Papillomaviridae • Human papillomavirus (HPV)	Encodes genes that inactivate cell growth regulatory proteins	Benign warts Genital warts	Cervical, penile, and oropharyngeal cancer
Papovaviridae • Merkel cell polyomavirus (MCPV)	Being investigated	Unknown	Merkel cell carcinoma
Herpesviridae • Epstein-Barr virus (EBV)	Stimulates cell growth and activates a host cell gene that prevents cell death	Infectious mononucleosis	Burkitt lymphoma Hodgkin lymphoma Nasopharyngeal carcinoma
• Herpesvirus 8 (HV8)	Forms lesions in connective tissue	Growths in lymph nodes	Kaposi sarcoma
• Cytomegalovirus (CMV)	Stimulates genes involved with tumor signaling	Mononucleosis syndrome	Salivary gland cancers
Hepadnaviridae • Hepatitis B virus (HBV)	Stimulates overproduction of a transcriptional regulator	Hepatitis B	Liver cancer
RNA Tumor Viruses			
Flaviviridae • Hepatitis C virus (HCV)	Chromosomal aberrations, including enhanced chromosomal breaks and sister chromatid exchanges	Hepatitis C	Liver cancer
Retroviridae • Human T-cell leukemia virus (HTLV-1)	Encodes a protein that activates growth-stimulating gene expression	Weakness of the legs	Adult T-cell leukemia/lymphoma

near a TSG, the close association with the TSG can cause the loss of tumor suppressor protein function, resulting in uncontrolled cell divisions (**FIGURE 15.20B**).

▶ **Gene Activation.** If a provirus integrates near a transcriptionally "silent" gene, the close association might activate the silent gene. If the gene normally helps control cell division or cell growth, its activation can permanently "turn on" the synthesis of growth-promoting proteins, again leading to uncontrolled cell division (**FIGURE 15.20C**).

This ability of viruses to get inside cells and deliver their viral information or influence host cell genetic information has been manipulated through genetic engineering to deliver genes to cure genetic illnesses. **MICROFOCUS 15.3** looks at the pros and cons of using viruses for this purpose.

Tumor Viruses Do Not Always Cause Cancer

Not everyone infected with an oncogenic virus develops cancer. Human cells have division controls and checkpoints that must be disrupted for malignant transformation to occur. In fact, a variety of factors, including environmental, diet, and lifestyle, as well as host genetics contribute to the transformation process. In addition, immune system status and surveillance are critical in determining if cancer might develop (see Figures 15.19 and 15.20). Thus, tumor virus infection is only one part of cancer development.

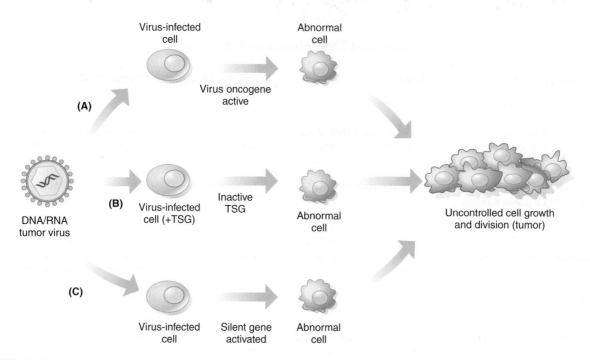

FIGURE 15.20 Tumor Viruses and Tumor Induction. (A) Some oncogenic viruses carry an oncogene whose protein product directly causes uncontrolled growth. **(B)** Other viruses integrate their DNA next to a tumor suppressor gene (TSG), resulting in inactivation of the TSG and leading to uncontrolled cell growth. **(C)** Yet other viruses integrate their DNA next to a silent growth control gene, resulting in activation of the genes and uncontrolled growth. *»» What advantage is it to the tumor virus to cause cells to divide endlessly?*

MICROFOCUS 15.3: Biotechnology

The Power of the Virus

Ever since the power of genetic engineering made it possible to transfer genes between organisms, scientists and physicians have wondered how they could use viruses to cure disease. One potential way is to use viruses against cancer. Many years ago, a mutant herpesvirus was produced that could only replicate in tumor cells. Because viruses make ideal cellular killers, perhaps these viruses would kill the infected cancer cells. Used on a terminal patient with a form of brain cancer, the virotherapy worked, as the patient was cured.

Additional "oncolytic" (cancer-killing) viruses are being developed. Some cause the cancer cells to commit cell suicide, whereas others deliver chemicals (e.g., radioactive materials, a toxic drug, or a gene that codes for an anticancer protein) that on infection are capable of destroying the cancer cells.

Perhaps the riskiest form of such virotherapy is trying to insert a gene correctly into a patient to reverse a genetic disorder. In 2002, four young French boys suffering with a nonfunctional immune system were given virotherapy designed to ferry into the immune cells the gene needed to "turn on" the immune system. The therapy worked, but the virus-inserted gene disrupted other cellular functions, leading to leukemia (cancer of the white blood cells) in two of the boys. Therefore, the idea of using virotherapy certainly works, but delivering the virus without complications and targeting the genes to the correct place can be a difficult assignment.

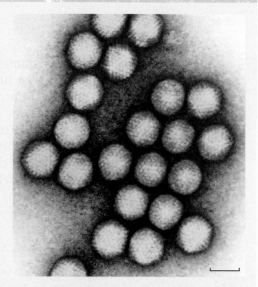

False-color transmission electron micrograph of adenoviruses. (Bar = 50 nm.)

Courtesy of Dr. G. William, Jr./CDC.

Another problem is the hurdle of overcoming a normal immune system response to the infecting virus. During virotherapy, the immune system mounts an attack on the injected viruses, which, therefore, might be destroyed before the viruses reach their target. In one gene therapy case, an 18-year-old patient died because of an immune reaction that developed in response to the adenovirus used to treat the patient's rare metabolic disorder. Importantly, 17 patients had been successfully treated using the same adenovirus prior to Gelsinger. One technique researchers have devised to overcome the immune system defenses is to coat the viruses with an "extracellular envelope" invisible to the immune system yet allowing the viruses to search out and find their target.

Today, virotherapy is in high gear with many early clinical trials in progress to examine the power of the virus to directly or indirectly kill cancer cells and to deliver genes that can correct genetic disorders.

Concept and Reasoning Check 15.6

a. Explain how the oncogene theory is related to virus infection by DNA and RNA tumor viruses.

■ KEY CONCEPT 15.7 Emerging Viruses Arise from Genetic Recombination and Mutation

Almost every year, a newly emerging influenzavirus descends upon the human population. Other viruses not even heard of a few decades ago, such as HIV, Ebolavirus, Hantavirus, West Nile virus, and Zika virus, are often in the news. Where are these viruses coming from?

Emerging Viruses Usually Arise Through Natural Phenomena

The human population is at greater risk than ever from **zoonotic diseases**, which are animal diseases capable of being transmitted from animals to humans (TABLE 15.4). In fact, 70% of human

TABLE 15.4 Examples of Emerging Zoonotic Viruses

Virus	Genome	Emergence Factor
HIV	+ssRNA	Contact with infected apes/monkeys in Africa
MERS	+ssRNA	Possibly spread by infected Egyptian tomb bats (via camels?)
SARS-associated	+ssRNA	Contact with infected horseshoe bats
West Nile	+ssRNA	Infected mosquitos transported unknowingly by global travel
Zika	+ssRNA	Contact with infected mosquitoes carried from Africa and South Pacific
Chikungunya	−ssRNA	Spread through new mosquito vectors and global travel
Ebola/Marburg	−ssRNA	Human contact with infected fruit bats
Influenza	−ssRNA	Mixed pig and duck agriculture, mobile population
Lassa	−ssRNA	Human contact with infected rodents
Nipah/Hendra	−ssRNA	Human contact with infected flying foxes (bats)
Sin Nombre (Hantavirus)	−ssRNA	Large, infected deer mice population and contact with humans

infections come from pathogens originally found in animals. Many of these **emerging infectious diseases** are the result of viruses appearing for the first time in a population or rapidly expanding their host range with a corresponding increase in detectable disease. Many are transmitted by insect vectors, such as ticks, fleas, or mosquitoes, as well as bats. One example is the West Nile virus that spread across the United States between 1999 and 2009. It now is endemic (prevalent) across the continental United States. As of this writing, another, the Zika virus, has appeared in parts of Florida, and many health experts expect this virus also will eventually become endemic in parts of the United States.

But no matter how these viruses are transmitted, what caused their emergence?

One way "new" viruses arise is through **genetic recombination**. Take, for example, influenza. The mixing of genes between different influenzaviruses generates new flu strains every season in both the Northern Hemisphere and in the Southern Hemisphere. The "swine flu" that broke out in Mexico in 2009 was the result of the reassortment of genome segments from a strain of avian flu virus, a human flu virus, and a swine flu virus.

Viruses also arise from a second force driving evolution—**mutation**. For example, when a single nucleotide is altered (point mutation) in a replicating RNA virus genome, the virus has no way to "proofread" and correct the mistake. Occasionally, one of these mutations might be advantageous. In the case of HIV and influenzaviruses, a beneficial mutation could generate a new virus strain resistant to an antiviral drug or to a vaccine. With a rapid replication rate and burst size, it does not take long for a beneficial mutation to establish itself within a population.

Even if a new virus has emerged, it must encounter an appropriate host in which to replicate and spread. It is believed that smallpox and measles both evolved from cattle viruses, whereas flu probably originated in ducks and pigs. HIV almost certainly evolved from a monkey (simian) immunodeficiency virus. Consequently, at some time such viruses had to make a species jump. What could facilitate such a jump?

Our proximity to wild and domestic animals and their pathogens makes such a jump possible. Today, population pressure is pushing the human population into more remote areas of the world where potentially virulent viruses might be endemic. Evidence shows the Machupo and Junin viruses (Arenaviridae viruses that cause hemorrhagic fevers in South America) jumped from rodents to humans because of increased agricultural practices that, for the first time, brought infected rodents into contact with humans.

An increase in the size of the animal host population carrying a viral disease also can "explode" as an emerging viral disease. The spring of 1993 in the American Southwest was a wet season, providing ample food for deer mice. The expanding deer mouse population brought them into closer contact with humans. Leaving behind mouse feces and dried urine containing the hantavirus made infection in humans likely. The deaths of 14 people with a mysterious respiratory illness in the Four Corners area that spring eventually were attributed to this newly recognized virus.

Therefore, emerging viruses are not new. They are simply evolving from existing viruses, and, through human changes to the environment, these viruses are given the "opportunity" to spread or to increase their host range. CLINICAL CASE 15 describes the transmission of viruses via the pet trade and the zoonosis that resulted.

Clinical Case 15

Multistate Outbreak of Monkeypox

On April 9, 2003, a Texas animal importer received a shipment of 800 small mammals from Accra, Ghana. This included rope squirrels, Gambian giant rats, and dormice. On April 21, an Illinois animal distributor received several Gambian rats and dormice from the Texas distributor. These animals were caged with prairie dogs the distributor had previously caught. All the animals were then distributed to six states (Illinois, Indiana, Kansas, Missouri, Ohio, and Wisconsin).

In June, a Wausau, Wisconsin, couple went to a local swap meet. At a stall selling exotic pets, they purchased two prairie dogs. Back home, the animals were introduced to their 3-year-old daughter. Two days later, while playing, the young girl was bitten on the finger by one of the prairie dogs.

Within several days, the young girl was taken to the local hospital with a fever of 40°C (104°F) that would not abate. She had developed skin pustules—small, round, raised areas of inflamed skin filled with pus (see figure). She was hospitalized, put on an IV, and started on antibiotics. A biopsy from the pustules failed to test positive for any known bacterial or viral pathogen.

Her mother soon began to get itchy, red pustules on her body. The illness progressed from bumps to cold sweats at night. A few days later, her husband also developed symptoms. Both recovered, and a biopsy from the mother's pustules was sent to the Centers for Disease Control and Prevention (CDC) in Atlanta for identification. Meanwhile, one of the purchased prairie dogs died.

After 7 days in the hospital, the daughter began improving and was sent home, where she recovered completely. The day after the daughter went home, the CDC report was received. A pox virus was identified, specifically monkeypox, a disease formerly found only in Central and West Africa.

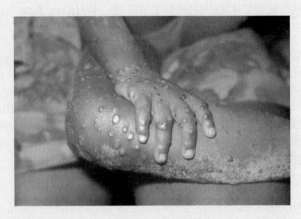

A photograph of an African child showing the pustules typical of a monkeypox infection.

Courtesy of CDC.

Alarmed by this potentially emerging virus, the CDC carried out an investigation to see how many other people exhibited the symptoms of monkeypox. The July 8 report identified 71 cases of monkeypox from Wisconsin (39), Indiana (16), Illinois (12), Missouri (2), Kansas (1), and Ohio (1). The number of cases increased from May 15 through the week ending June 8 and then declined with no additional cases reported after June 20. All patients recovered completely.

A traceback investigation was undertaken to determine the source of the confirmed human cases of monkeypox.

Questions:
  a. Why was the hospital unable to identify the pathogen causing the pustules in the daughter?
  b. How did the daughter become exposed to the monkeypox virus?
  c. How did the mother and father become exposed to the monkeypox virus?
  d. Based on the information provided, what did the traceback identify as the source of monkeypox cases?

You can find answers online in **Appendix E**.

For additional information, see www.cdc.gov/mmwr/preview/mmwrhtml/mm5227a5.htm.

Concept and Reasoning Check 15.7

a.   Distinguish the ways that emerging viral diseases arise.

Chapter Challenge D

The emergence of "new" viruses or the reemergence of others was described as the product of genetic recombination and mutation. As such, viruses can adapt to changing environmental conditions by inheriting characteristics that enhance their survival in a particular environment. Adaptation is one of the characteristics of life.

QUESTION D: *Because viruses evolve, is that enough to classify them as living? Can nonliving objects, like a car, or better yet, a computer virus, evolve? Explain.*

You can find answers online in **Appendix F**.

■ KEY CONCEPT 15.8   Prions Are Infectious, Virus-Like Agents

When viruses were discovered, scientists believed they were the ultimate infectious particles. It was difficult to conceive of anything smaller than viruses as agents of infectious disease. However, the perception was revised as scientists discovered new disease agents—the subviral particles referred to as **prions** (pronounced pree-ons).

Prions Are Infectious Proteins

In 1986, cattle in Great Britain began dying from a mysterious illness. The cattle experienced weight loss, became aggressive, lacked coordination, and were unsteady in their gait. These detrimental effects became known as "**mad cow disease**" and were responsible for the eventual death of these animals. A connection between mad cow disease and a similar human disease surfaced in Great Britain in 1995 when several young people died of a human brain disorder resembling mad cow disease. Symptoms included dementia, weakened muscles, and loss of coordination. Health officials suggested the human disease was caused by eating beef that had been processed from cattle with mad cow disease. It appeared the disease agent was transmitted from cattle to humans. This resulted in the identification of more than 180,000 infected cattle and the slaughter of 4.4 million animals in Great Britain.

Besides mad cow disease, similar neurological degenerative diseases have been discovered and studied in other animals and humans. These include scrapie in sheep and goats, chronic wasting disease in elk and deer, and Creutzfeldt-Jakob disease in humans. All are examples of a group of rare diseases called **transmissible spongiform encephalopathies** (**TSEs**) because, like mad cow disease (bovine spongiform encephalopathy; BSE), they can be transmitted to other animals of the same species and possibly to other animal species, including humans, and the disease causes the formation of "sponge-like" holes in brain tissue (**FIGURE 15.21A**).

Many scientists originally believed these agents were a new type of virus. However, in the early 1980s, Stanley Prusiner and colleagues isolated an unusual protein from scrapie-infected tissue, which they thought represented the infectious agent. Prusiner called the proteinaceous infectious particle a prion.

This led Prusiner to propose the **protein-only hypothesis**, which states that prions are pure protein and contain no nucleic acid. The protein-only hypothesis further proposes that there are two types of prion proteins (**FIGURE 15.21B, C**). Normal cellular prions (PrP^C) are found in normal brain and nerve tissue. These proteins have important roles to play in everything from neuron protection

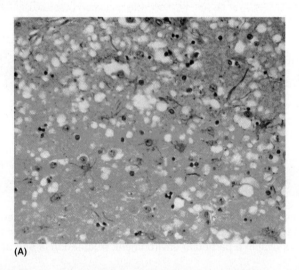

(A)

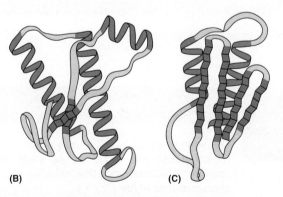

(B)   (C)

FIGURE 15.21 Prion "Infection" and Structure. (A) A photomicrograph showing the vacuolar ("spongy") degeneration of gray matter (clear areas) characteristic of human and animal prion diseases. **(B)** The tertiary structure of the normal prion protein (PrP^C). The helical regions of secondary structure are green. **(C)** A misfolded prion (PrP^{Sc}). This infectious form of the prion protein results from the helical regions in the normal protein unfolding and an extensive pleated sheet secondary structure (blue ribbons) forming within the protein. The misfolding allows the proteins to clump together and react in ways that contribute to disease. »» *How does a change in tertiary structure affect the functioning of a protein?*

(A) Courtesy of Dr. Al Jenny/CDC.

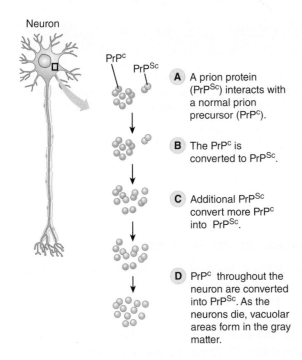

Neuron

PrP^C

PrP^Sc

A A prion protein (PrP^Sc) interacts with a normal prion precursor (PrP^C).

B The PrP^C is converted to PrP^Sc.

C Additional PrP^Sc convert more PrP^C into PrP^Sc.

D PrP^C throughout the neuron are converted into PrP^Sc. As the neurons die, vacuolar areas form in the gray matter.

FIGURE 15.22 Prion Formation and Propagation. Prion proteins (PrP^Sc) bind to normal prion precursors (PrP^C) in neurons, causing these proteins to the misfolded PrP^SC conformation. This process is repeated many times, leading to the accumulation of PrP^Sc protein fibers that cause progressive neuron cell death. *»» How could these areas of PrP^Sc protein fiber accumulation be observed in a brain autopsy from a person who has died from variant CJD?*

to learning during infant development. Only when they undergo mutation do they become abnormal, misfolded, disease-causing prions (PrP^Sc). Proof that is more definitive came in 2004 when Prusiner's group showed that purified prions could cause disease when injected into the brains of mice.

Today, many researchers believe prion diseases—that is, TSEs—are spread by the infectious PrP^Sc being taken up intact from the digestive tract and transported to the nervous tissue. The particles then bind to normal PrP^C, causing the latter to change shape (**FIGURE 15.22**). In a domino-like effect, the newly converted PrP^Sc proteins, in turn, would cause more PrP^C proteins to become abnormal. The PrP^Sc

proteins then form insoluble protein fibers that aggregate, forming sponge-like holes left where groups of nerve cells have died. Importantly, PrP^Sc does not trigger an immune response. Loss of life occurs from nerve cell damage and loss of brain function.

The human form of TSE appears similar to the classic and spontaneous form of Creutzfeldt-Jakob disease (CJD). The new prion-derived form of CJD, called **variant CJD (vCJD)**, is characterized clinically by neurological abnormalities such as dementia. Neuropathology shows a marked spongiform appearance throughout the brain, and death occurs within 3 to 12 months after symptoms appear. New noninvasive tests for vCJD in living patients are becoming available. These tests have almost 100% accuracy in detecting PrP^Sc.

The number of confirmed and probable deaths from vCJD in the United Kingdom peaked in 2000 with 28 reported victims. The latest figures report 177 vCJD deaths in the United Kingdom and another 27 cases in France since 1995. As such, BSE does not appear to be highly contagious considering the thousands of individuals who must have ingested contaminated beef. Experts believe BSE is almost extinct because the last reported case in cattle in the United Kingdom was in 2012.

In North America, the first mad cow was discovered in Canada in 2003, and in December of that year, the first such cow was reported in the United States. In all, there have been 20 BSE cattle in Canada and 4 in the United States. No human cases of vCJD have been reported as originating in the United States, although two cases have been documented in individuals exposed to BSE in the United Kingdom and a third case from exposure to an individual while living in Saudi Arabia. Protection measures include condemning those animals having signs of neurological illness and holding any cows suspected of having BSE until test results are known. In addition, "downer cattle"—those unable to walk on their own—cannot be used for human food or animal feed.

Concept and Reasoning Check 15.8

a. How are prions (a) similar to and (b) different from viruses?

Chapter Challenge E

Although the fossil record has been useful to study the origins of plants and animals, and even some members of the domain Bacteria, the record is not much help for the viruses. Even though no one can know with 100% certainty how viruses originated, scientists and virologists have put forward three hypotheses (read **In Conclusion**, coming up in just a moment).

QUESTION E: *From your experiences with viruses and virus-like agents in this chapter, would any of these hypotheses fit with the idea that viruses are living infectious agents?*

A discussion can be found online in **Appendix F**.

In conclusion, the fossil record has been useful to study the origins of plants and animals and even some members of the domain Bacteria. Even though there is no fossil record for the viruses, scientists and virologists have put forward three hypotheses as to how viruses might have originated. The best guess is that they appeared some 2 to 3 billion years ago, making them about as old as the oldest life forms on Earth. Because they depend on host cells for replication, one could argue they appeared soon after the first cells. That said, the following subsections present the three main hypotheses for the origin of viruses.

Regressive Evolution Hypothesis

Viruses are degenerate life forms; that is, they are derived from small cells that functioned as intracellular parasites. Over time, the cells lost many functions that could be supplied by their hosts, and the life forms retained only those genes essential for directing their parasitic way of life. The giant viruses, such as mimivirus, pandoravirus, and pithovirus could be "living" examples.

Cellular Origins Hypothesis

Viruses are derived from subcellular components (DNA and RNA) and functional assemblies of macromolecules that have escaped their cellular origins by being able to replicate separately in host cells. Modern-day bacterial plasmids that replicate independently from the bacterial chromosome and transposons, the jumping genes that can move about in a genome or be transferred to another genome, lend support to this hypothesis.

Coevolution Hypothesis

Viruses coevolved with cellular organisms from the self-replicating molecules believed to have existed on the primitive prebiotic Earth. This hypothesis would suggest viruses arose at the same time as cellular life and transitioned to a parasitic form by hijacking the cellular machinery of the newly evolving organisms.

Although each of these theories has its supporters, the topic generates strong disagreements among experts. In the end, it is not so much a matter of how viruses arose but rather how we become infected with them.

SUMMARY OF KEY CONCEPTS

Concept 15.1 Filterable Infectious Agents Cause Disease

1. The concept of a **virus** as a filterable agent of disease was first suggested by Ivanowsky, while Twort and d'Herrelle identified viruses as infective agents in bacterial organisms. In the 1940s, the electron microscope enabled scientists to see viruses, and innovative methods allowed researchers to cultivate them.

Concept 15.2 Viruses Have a Simple Structural Organization

2. Viruses are small, obligate, intracellular particles composed of nucleic acid as either DNA or RNA and in either a **single-stranded** or a **double-stranded** form. The genome is surrounded by a protein **capsid**, and many viruses have an **envelope** surrounding the **nucleocapsid**. **Spikes** protruding from the capsid or envelope are used for attachment to host cells. (Figure 15.5)
3. Viruses have **helical, icosahedral**, or **complex symmetry**. (Figure 15.6)
4. The **host range** of a virus refers to what organisms (hosts) the virus can infect. Many viruses infect only certain cell types or tissues within the host referred to as a **tissue tropism** (tissue attraction).

Concept 15.3 Viruses Can Be Classified by Their Genome

5. Two broad classes of viruses can be organized based on their genome and strand type: the **single-stranded** or **double-stranded DNA viruses** and the **single-stranded** or **double-stranded RNA viruses**. (Figure 15.7)

Concept 15.4 Virus Replication Follows a Set of Common Steps

6. **Bacteriophages** undergo a **lytic pathway**, involving attachment, penetration, biosynthesis, maturation, and release stages, or a **lysogenic pathway** where the phage genome integrates into the bacterial chromosome as a **prophage**. (Figures 15.10 and 15.12)
7. Animal viruses also progress through the same stages of replication. However, penetration can occur by membrane fusion or endocytosis, and the biochemistry of nucleic acid synthesis varies among DNA and RNA viruses. (Figure 15.13)

8. Many of the DNA viruses and the retroviruses can incorporate their viral DNA independently within the host's cell nucleus, or as a **provirus**. (Figure 15.15)

Concept 15.5 Viruses and Their Infections Can Be Detected in Various Ways

9. Various detection methods for viruses are based on the cultivation of animal viruses using cell cultures. The detection of viruses can be determined by characteristic cytological changes referred to as the **cytopathic effect** and by **plaque** formation.

Concept 15.6 Some Viruses Are Associated with Human Tumors and Cancers

10. **Tumor** formation is a complex condition in which cells multiply without control and possibly develop into **cancers**. (Figure 15.19)
11. There are at least seven **carcinogenic viruses** known to cause human tumors. Many of these tumors can develop into **cancer**. (Table 15.3)
12. Viruses can bring about tumors by converting **proto-oncogenes** into tumor **oncogenes**. If such genes are carried in a virus, they are called **viral oncogenes**. (Figure 15.20)

Concept 15.7 Emerging Viruses Arise from Genetic Recombination and Mutation

13. Viruses use **genetic recombination** and **mutations** as mechanisms to evolve. Human population expansion into new areas and increased agricultural practices in previously forested areas have exposed humans to existing animal viruses. (Table 15.4)

Concept 15.8 Prions Are Infectious, Virus-Like Agents

14. **Prions** are infectious particles made of protein. They are capable of causing a number of animal diseases, including **mad cow disease** in cattle and **variant Creutzfeldt-Jakob disease (vCJD)** in humans. Prions cause disease by folding improperly and, in the misfolded shape, cause other prions to misfold. These and other similar diseases in animals are examples of a group of rare diseases called **transmissible spongiform encephalopathies (TSEs)**. (Figures 15.21, 15.22)

CHAPTER SELF-TEST

You can find answers for **Steps A–D** online in **Appendix D**.

■ STEP A: REVIEW OF FACTS AND TERMS

Multiple Choice

Read each question carefully and then select the *one* answer that best fits the question or statement.

1. Which one of the following scientists was *not* involved with discovering viruses?
   A. Frederick Twort
   B. Dimitri Ivanowsky
   C. Robert Fleming
   D. Martinus Beijerinck
2. Viral genomes consist of ____.
   A. DNA only
   B. RNA only
   C. DNA or RNA
   D. DNA and RNA
3. A nucleocapsid can have ____ symmetry.
   A. radial
   B. icosahedral
   C. vertical
   D. bilateral
4. Viral cell/tissue specificity refers to ____.
   A. what tissues grow due to a viral infection
   B. what tissues are resistant to viral infection
   C. what organisms a virus infects
   D. what cells or tissues a virus infects
5. An RNA virus genome in the form of messenger RNA is referred to as a ____ RNA.
   A. +sense
   B. double-stranded
   C. −sense
   D. reverse strand
6. A virulent bacteriophage will ____.
   A. carry out a lytic cycle
   B. integrate its genome in the host cell
   C. remain dormant in a bacterial cell
   D. exist as a prophage
7. The release of the viral genome from the capsid is called ____.
   A. uncoating
   B. endocytosis
   C. penetration
   D. maturation

8. Provirus formation is possible in members of the ____.
   A. single-stranded DNA viruses
   B. Retroviridae
   C. double-stranded RNA viruses
   D. single-stranded (−sense) RNA viruses
9. Cytopathic effects would include all the following *except* ____.
   A. changes in cell structure
   B. plaque formation
   C. vacuolated cytoplasm
   D. syncytia formation
10. A benign tumor ____.
    A. will metastasize
    B. represents cancer
    C. is malignant
    D. is a clone of dividing cells
11. A viral oncogene ____.
    A. inactivates tumor suppressor genes
    B. produces proteins that cause tumors
    C. causes viruses to replicate rapidly
    D. All of the above (**A–C**) are correct.
12. Newly emerging viruses causing human disease can arise from ____.
    A. species jumping
    B. mutations
    C. genetic recombination
    D. All of the above (**A–C**) are correct.
13. Which one of the following statements about prions is *false*?
    A. Prions are infectious proteins.
    B. Prions have caused disease in Americans.
    C. Human disease is called variant CJD.
    D. Prions can be transmitted to humans from infected beef.

Identification

Use the following syllables to compose the term that answers each clue from virology. The number of letters in the term is indicated by the dashes, and the number of syllables in the term is shown by the number in parentheses. Each syllable is used only once.

A BAC CAP CO COS DRON EN GENE HE I I LENT MOR O ON ONS OPE PHAGE PRI SID TER TU U VEL VIR

14. Viral protein coat (2) __ __ __ __ __ __
15. Bacterial virus (5) __ __ __ __ __ __ __ __ __ __ __ __ __ __
16. A cancer gene (3) __ __ __ __ __ __ __ __
17. Phages that lyse bacterial cells (3) __ __ __ __ __ __ __ __
18. May surround the capsid (3) __ __ __ __ __ __ __ __
19. Infectious protein particles (2) __ __ __ __ __ __

■ STEP B: CONCEPT REVIEW

20. Identify the important historical developments that led to the discovery of viruses as infectious agents. (**Key Concept 15.1**)

21. Distinguish between the structures common to all viruses and those that separate **nonenveloped** from **enveloped** viruses. (**Key Concept 15.2**)

22. Explain the difference between **DNA viruses** and **RNA viruses**. (**Key Concept 15.3**)

23. Describe the five stages of phage replication. (**Key Concept 15.4**)

24. Describe how animal viruses cause (a) a **productive infection** and (b) a **latent infection**. (**Key Concept 15.4**)

25. Explain how the **cytopathic effect** aids in the detection of viruses. (**Key Concept 15.5**)

26. Identify three viruses that are involved in causing human tumors. (**Key Concept 15.6**)

27. Explain how viruses can cause tumors. (**Key Concept 15.6**)

28. Describe how "new" viruses emerge. (**Key Concept 15.7**)

29. Summarize the properties of and diseases resulting from **prions**. (**Key Concept 15.8**)

■ STEP C: APPLICATIONS AND PROBLEM SOLVING

30. As part of a lab exercise, you and your lab partner are to estimate the number of bacteriophages in a sample of sewer water using a plaque assay with *Escherichia coli*. You add the phages to a lawn of bacteria and let the plates incubate for 24 hours before counting the number of plaques. This then got your lab partner thinking. She says, "If all these bacteriophages are in the sewer, they must have come from human waste flushed into the sewer. If they indeed did come from humans, why haven't the phage wiped out all the microbiota in the human digestive tract?" How would you respond to her question?

31. A person appears to have died of rabies. As a coroner, you need to verify that the cause of death was rabies. What procedures could you use to confirm the presence of rabies virus in the brain tissue of the deceased?

32. As a young genetic engineer with a biotech company, you have been given a virotherapy project. The hepatitis B virus is a dsDNA virus that during its replicative cycle often produces different forms of virus particles. The ones that are filamentous or spherical completely lack a viral genome and are thus defective; they are just assembled capsids but with the same spikes as the infective virions. The biotech company wants you to come up with a plan to use these defective particles for biotechnology applications. What might you include on your list of applications?

■ STEP D: QUESTIONS FOR THOUGHT AND DISCUSSION

33. The Nobel-laureate and British biologist Sir Peter Medawar (1915–1987) once remarked that a virus is *"just a piece of bad news wrapped up in protein."* Explain what he meant by this statement.

34. Oncogenes have been described in the recent literature as "Jekyll and Hyde genes." What factors might have led to this label, and what does it imply? In your view, is the name justified?

35. Researchers studying bacterial species that live in the oceans have long been troubled by the question of why these microorganisms have not saturated the oceanic environments. What might be a reason?

36. When discussing the multiplication of viruses, virologists prefer to call the process replication, rather than reproduction. Why do you think this is so? Would you agree with virologists that replication is the better term?

37. How have revelations from studies on viruses and prions complicated some of the traditional views about the principles of biology?

38. Suppose that all viruses on Earth suddenly disappeared. How would life on the planet change? Would life have evolved on Earth if viruses had never existed?

CHAPTER 16

Viral Infections of the Respiratory Tract and Skin

Every year there are seasonal flu outbreaks that can become **epidemic**; that is, they spread quickly through a large population. However, occasionally there is a **pandemic**, a worldwide epidemic. The biggest influenza pandemic, called the "Spanish flu," occurred in 1918 to 1919. A fifth of the world's population was infected by a virus thought to have originated in birds (see the opening figure). Estimates suggest that between 20 and 100 million people died, young adults being the hardest hit. As we will see in this chapter, such pandemics can occur when a new infectious virus appears for which the human population has no immunity.

Unfortunately, influenza continues to be an ongoing problem in the 21st century. Health experts and microbiologists have been saying that we are overdue for another serious and perhaps very deadly flu pandemic. In fact, one might be "brewing." An avian flu virus has been circulating throughout Asia and into some parts of Europe and Africa. This influenza virus has killed more than 450 people of the 900 infected in Asia and Southeast Asia since 2003. At this time, it is still difficult for people to contract the disease because it is spread through contact with

infected poultry. Human-to-human transmission has not yet been documented. However, it might only take a few mutations to transform the avian flu virus into a form that can become highly contagious.

Another flu strain caught the world's attention in 2009. In April 2009, the Centers for Disease Control and Prevention (CDC) identified two children in California who had been infected with a new strain of influenza virus. It soon was discovered that this strain was causing a major flu outbreak in Mexico. By the end of April, cases of this new flu strain, originally called "swine flu" but eventually known as "novel H1N1," were being reported in cities around the world. The quick spread caused the World Health Organization (WHO) to raise the global flu pandemic alert to 5 (6 is the highest), and it appeared the first flu pandemic in more than 40 years was emerging.

As the cases continued to mount up in the spring and early summer of 2009, the actual symptoms for

False-color transmission electron microscope image of influenza viruses.
Courtesy of CDC/Dr. F. A. Murphy.

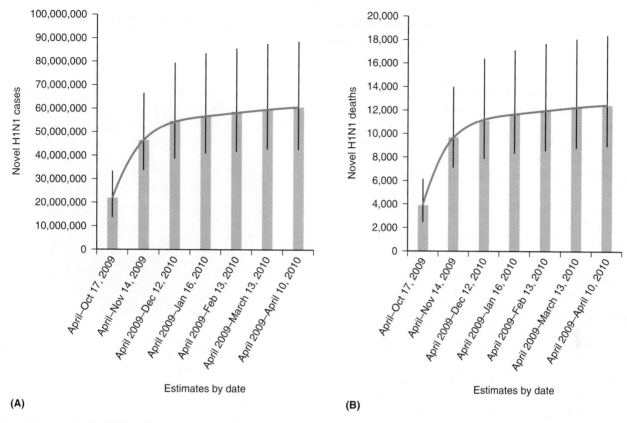

FIGURE 16.1 Cumulative Novel H1N1 Cases (A) and Deaths (B) in the United States by Date, April 2009−April 2010. The vertical bars indicate the range in estimates for each time period. *»» From these graphs, when did (a) the greatest increase in novel H1N1 cases and (b) greatest increase in novel H1N1 deaths occur?*

Reproduced from CDC Estimates of 2009 H1N1 Influenza Cases, Hospitalizations and Deaths in the United States.

most infected individuals remained relatively mild even though the WHO raised the pandemic alert to 6 in June. The fear was that there might be a second "flu wave" that would be much more devastating, which was the case for the Spanish flu. Luckily, that second wave never materialized. In fact, the death toll in the United States was estimated to be about 13,000 over the period, which was less than the 23,000 that usually die every year from the seasonal flu (**FIGURE 16.1**).

There were at least two major take-home lessons from the 2009 "mini-flu pandemic." One is that in this age of globalization, new viral threats can come from anywhere and spread with the speed of jet travel. The second lesson is that we need to be on

"virological alert" worldwide for new virus strains. Most flu outbreaks and epidemics originate in Southeast Asia—the 2009 pandemic surprised epidemiologists by coming from Mexico. So, in 2009 we dodged the bullet—next time we might not be so lucky.

In this chapter, we will focus on several viral diseases, including influenza, that affect the respiratory tract—the so-called "pneumotropic diseases." We also examine viral diseases affecting the skin—the "dermotropic diseases." Note: The division of the pneumotropic and dermotropic viral diseases is an artificial classification simply for grouping convenience. Therefore, the symptoms described might go beyond the respiratory tract or skin, respectively.

Chapter Challenge

Numerous viruses can cause infections in the human respiratory tract. Many of these viruses affect the upper respiratory tract only, whereas others first infect the upper respiratory tract before progressing to the lower respiratory tract. In addition, several viruses affect the skin. As we progress through this chapter, let's investigate some specific aspects of these viruses that are determined by their host range and tissue tropism.

■ KEY CONCEPT 16.1 Viruses Account for Many Upper Respiratory Tract Infections

The **upper respiratory tract** (**URT**) consists of the nose, sinuses, pharynx (throat), and larynx. The prominent location of the nose serves as a portal of entry, and the common cold viruses are a common infectious agent of the nasal cavities. Cold viruses account for more than 1 billion infections each year in the United States alone. In fact, it is the world's most common infectious disease.

Rhinovirus Infections Are a Frequent Cause of the Common Cold

Rhinoviruses (*rhino* = "nose") are small, nonenveloped, icosahedral viruses (**FIGURE 16.2**). These single-stranded (+ strand) viruses compose a broad group of more than 100 types within the family Picornaviridae (*pico* = "small"; hence, small RNA viruses). The genome contains only 10 genes.

Rhinoviruses thrive in the human nasopharynx, where the temperature is a few degrees cooler (33°C) than in the lower respiratory tract and the rest of the body. The viruses are transmitted directly through airborne respiratory droplets that are produced during coughing and sneezing, by contact with an infected person, or indirectly through contact with contaminated, nonliving objects (**fomites**). Rhinoviruses account for 30 to 50% of **common colds**, also called "head colds," although several other viruses also can cause colds (**FIGURE 16.3**). Adults typically suffer two or three colds and children six to eight colds per year. Rhinoviruses can be dangerous to people with pulmonary disorders (e.g., asthma, cystic fibrosis, chronic obstructive pulmonary disease),

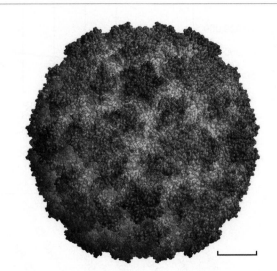

FIGURE 16.2 A Rhinovirus. A false-color, computer-generated image of the surface (capsid) of human rhinovirus 16. (Bar = 3 nm.) »» *What is unique about the rhinoviruses that make vaccine development impractical?*

Image by J. Y. Sgro, UW-Madison.
Reference: Hadfield, A. T., Lee, W. M., Zhao, R., Oliveira, M. A., Minor, I., Rueckert, R. R., & Rossmann, M. G. (1997). The refined structure of human rhinovirus 16 at 2.15 Å resolution: Implications for the viral life cycle. *Structure, 5,* 427–441.

for which a mild cold can trigger an uncontrolled lung inflammation.

Rhinoviruses infect the epithelial cells lining the URT (**FIGURE 16.4**). One to 3 days after infection, a set of medical signs and symptoms (common-cold syndrome) begins. These include sneezing, a sore throat, runny nose (mucus secretion) or stuffy nose (inflammation), mild aches and pains, and a mild-to-moderate hacking cough. The illness usually lasts 7 to 10 days and is self-limiting. One of the

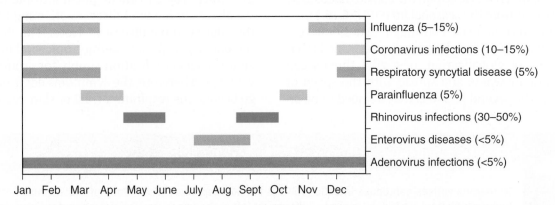

FIGURE 16.3 The Seasonal Variation of Viral Respiratory Diseases. This chart shows the months associated with various viral diseases of the respiratory tract (and their annual percentage). Enteroviruses cause diseases of the gastrointestinal tract as well as respiratory disorders. »» *Hypothesize why different cold viruses cause diseases at different times of the year.*

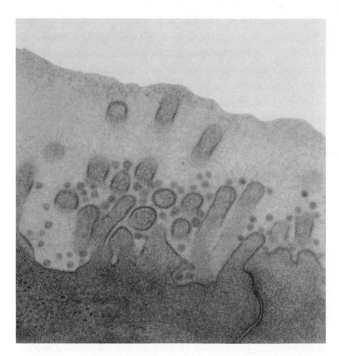

FIGURE 16.4 A Rhinovirus Infection. This false-color transmission electron micrograph shows rhinoviruses (green) near the surface of the epithelial cells (orange) of the nasopharynx. *»» What is the dark yellow material?*

© Dennis Kunkel Microscopy/Science Source.

old folk tales was that becoming chilled could cause colds. MICROFOCUS 16.1 investigates this claim.

Antibiotics will not prevent or cure a cold, and with more than 100 rhinovirus strains, development of a vaccine is difficult. Antihistamines and decongestants can be used to treat the symptoms of a cold; however, they do not shorten the length of the illness. The therapeutic value of consuming vitamin C is examined in INVESTIGATING THE MICROBIAL WORLD 16. In the end, the best way to try to prevent catching a cold is good hand hygiene.

Common cold viruses also can lead to **laryngitis**, an infection of the larynx. Symptoms are an unnatural change of voice, such as hoarseness, or even loss of voice that develops within hours to a day or so after infection. The throat might tickle or feel raw, and a person might have a constant urge to clear the throat. Fever, malaise, difficulty swallowing, and a sore throat can accompany severe infections. Treatment of viral laryngitis depends on the symptoms. Resting the voice (by not speaking), drinking extra fluids, and inhaling steam relieve symptoms and help healing.

TABLE 16.1 summarizes the viral diseases of the URT.

MICROFOCUS 16.1: Being Skeptical

Catching a Chill: Can It Cause a Cold?

How many times can you remember your mom or a family member saying to you, "Bundle up, or you will catch a cold!" Can you actually "catch" a cold from a body chill?

In 2005, a published scientific study reported that a drop in body temperature might allow a cold to develop. For this study, the researchers signed up 180 volunteers and split them into two groups. One group, dressed only in underwear, put their bare feet into bowls of ice-cold water for 20 minutes. The other group, also in underwear, put their bare feet in similar but empty bowls.

Over the next five days, 29% of individuals who had their feet chilled developed cold symptoms, while just 9% of the control group developed symptoms. Unfortunately, just being exposed to the cold may make more people believe they have the symptoms of a cold.

In 2013, a team at Yale University looked further into the cold temperature mystery. Using mice susceptible to a mouse-specific rhinovirus, they found that mice at warmer temperatures produced a

© fotografos/Shutterstock.

burst of anti-inflammatory immune signals (cytokines) to help fight off the viruses. Similarly, in human cells in culture, they found a rhinovirus infection caused the cells to commit cell suicide, which would limit the spread of the virus. At cooler temperatures, fewer cytokines were produced in infected mice, and the human cells in culture were not as likely to undergo cell suicide. However, this doesn't prove that cold temperatures would make a person more susceptible to catching a cold.

The verdict: although the science of these research projects is sound, and temperature might be one factor important to rhinovirus spread, the evidence that cold weather can make one more likely to catch a cold remains elusive.

Investigating the Microbial World 16

Vitamin C and the Common Cold

In this **Investigating the Microbial World**, rather than looking at one research study, a meta-analysis is presented. The purpose of a meta-analysis is to examine many previously published, peer-reviewed research studies with the aim of developing a single, concrete conclusion from all the study results.

OBSERVATION: Treating colds with vitamin C became very popular in the 1970s after Nobel laureate Linus Pauling suggested that vitamin C could prevent and lessen cold symptoms. Since then, numerous studies have looked at whether there are therapeutic benefits from taking vitamin C to prevent or reduce the length of a common cold syndrome.

QUESTION: *Does vitamin C reduce the severity, incidence, and/or duration of a common cold syndrome?*

OBJECTIVE: Use a meta-analysis to examine systematically studies assessing the incidence of colds with regular vitamin C intake among study subjects reporting one or more colds during the study period. Search criteria looked for trials using more than 0.2 g per day of vitamin C and having placebo controls. The term "incidence" refers to the percentage of participants experiencing one or more colds during the study period, and "duration" refers to the percentage of participants with a shorter duration of a common cold syndrome.

Data were extracted from electronic searches of: CENTRAL (a controlled trials registry), MEDLINE (biomedical literature resource), EMBASE (biomedical and pharmacological literature resource), CINAHL (Cumulative Index for Nursing and Allied Health Literature), LILACS (scientific and technical literature resource), and Web of Science (citation database). Searches were also done using the National Institutes of Health Clinical Trials registry and the World Health Organization CTRP (Clinical Trials Registry Platform).

SEARCH 1: Seven studies were found that looked at the therapeutic effect and severity of symptoms (3,249 cold episodes) while taking vitamin C regularly or a placebo during the study period.

SEARCH 2: Twenty-nine studies (11,306 participants) were found that examined the incidence (percentage of participants experiencing one or more colds during the study period) while taking vitamin C regularly or a placebo during the study period. These studies were separated into two subgroups: the general population and 598 participants exposed to short periods of severe physical exercise (marathon runners, skiers, and soldiers on subarctic winter exercises).

SEARCH 3: Thirty-one studies were found that examined the duration (percentage of participants experiencing a shorter duration of a common cold syndrome during the study period) while taking vitamin C or a placebo. These studies were separated into two subgroups: adults and children because (a) children often have more colds due to immune system immaturity and (b) children being smaller (less body weight) means a fixed dose of vitamin C corresponds to a higher dose per body weight.

RESULTS: See figures (A) and (B).

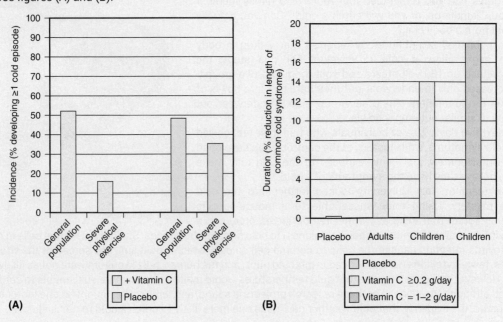

CONCLUSIONS: In Search 1, no consistent effect of vitamin C could be found among the 7 studies examining the therapeutic effect (reduced severity) among participants.

QUESTION 1. Looking at the figure (A): *Does vitamin C supplementation reduce the incidence of "catching a cold" in (i) the general population group and/or (ii) the severe physical exercise group? Explain.*

QUESTION 2. Looking at the figure (B):

(a) *Does vitamin C supplementation reduce the duration (length) of a common cold syndrome in adults and/or children? Explain.*
(b) *Does the dosage of vitamin C affect the duration?*

QUESTION 3: *What type of factors might bias the reporting of the incidence and duration of a common cold syndrome?*

QUESTION 4: *From the meta-analysis, would you take vitamin C to prevent and treat the common cold? Explain.*

You can find answers online in **Appendix E**.

Data from Hemilä, H. and Chalker, E. 2013. *Cochrane Database of Systematic Reviews* 2013, Issue 1. Art. No.: CD000980. DOI: 10.1002/14651858.CD000980.pub4.

TABLE 16.1   **A Summary of the Major Viral Diseases of the URT**

Disease	Causative Agent	Signs and Symptoms	Transmission	Treatment	Prevention
Common colds (rhinitis)	Rhinoviruses Adenoviruses Other viruses	Sneezing, sore throat, runny and stuffy nose, hacking cough	Respiratory droplets	Pain relievers Decongestants	Practicing good hygiene
Laryngitis	Cold viruses	Hoarseness Loss of voice	Respiratory droplets	Rest voice	Drinking plenty of water

---

Concept and Reasoning Check 16.1

a.   Why do we get colds repeatedly?

Chapter **Challenge A**

Most URT infections are caused by viruses. The result usually is a mild condition involving a cough and often fever, headache, and aches and pains.

QUESTION A: *What areas of the URT are most subject to infection and what is the best way to prevent such infections?*

You can find answers online in **Appendix F**.

■ KEY CONCEPT 16.2   Viral Infections of the Lower Respiratory Tract Can Be Severe

The **lower respiratory tract (LRT)** in humans consists of the lungs and the trachea (windpipe), which leads into the bronchial tubes of the lung. Each bronchus eventually branches into smaller bronchioles, which terminate in the air sacs called alveoli where gas exchange occurs. The anatomy of the human respiratory system is reviewed in Chapter 11.

Influenza Is a Highly Communicable Acute Respiratory Infection

Influenza (the "flu") is one of the most common infectious diseases, and it can be a serious respiratory illness.

Epidemiology

Influenza is a highly contagious and acute disease that is transmitted by airborne respiratory droplets. Although the illness starts in the URT, severe (complicated) cases also involve the LRT. Since the first recorded epidemic in 1510, scientists have identified 31 flu pandemics. As described in the chapter opener, the most notable pandemic of the twentieth century was the "Spanish" flu in 1918–1919 that took the lives of an estimated 20 million to 100 million people worldwide. Today, about 20% of the global population contracts the seasonal flu, and the WHO estimates that between 250,000 and 500,000 deaths occur from influenza complications each year. Ninety percent of these deaths are in older adults.

Causative Agent

There are three types of influenza viruses, fittingly named **influenza A**, **influenza B**, and **influenza C**. They all belong to the Orthomyxoviridae family and share similar structural features. Although clinical isolates often have an irregular morphology, the viruses are typically described as having a spherical shape in culture (**FIGURE 16.5**). Because influenza

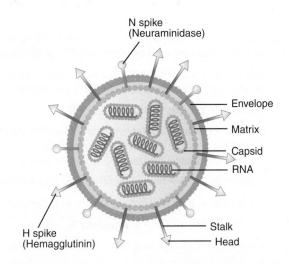

FIGURE 16.5 The Influenza A Virus. This diagram of the influenza A virus shows its eight segments of RNA, matrix, and envelope with hemagglutinin and neuraminidase spikes protruding. »» *What is the role of the hemagglutinin and neuraminidase spikes?*

A viruses are the most prevalent in the environment and are responsible for about 70% of human flu cases (**TABLE 16.2**), the remaining discussion will center on type A.

The genome of the influenza A virus consists of eight single-stranded (–sense) RNA segments (see Figure 16.5). Each helical segment is surrounded by a nucleocapsid and codes for one or two virus proteins. The entire nucleocapsid core is surrounded by a **matrix protein**, which in turn is surrounded by an envelope. Ten of the 12 genes of the influenza A

TABLE 16.2 Comparison of the Influenza Viruses

Characteristic	Influenza A	Influenza B	Influenza C
Reservoir	Humans, pigs, birds	Only humans	Only humans
Epidemiology • Epidemics • Pandemics	 Yes Yes	 Yes No	 No No
Occurrence	Seasonal	Seasonal	Nonseasonal
Antigenic changes	Drift/shift	Drift	Drift
Potential severity of illness	Severe	Moderate	Mild
Seasonal vaccine preparation	Two subtypes	One subtype	No vaccine
Genome	Eight segments	Eight segments	Seven segments

code for internal proteins needed for virus synthesis and assembly.

The remaining two genes code for the envelope spikes that project through the viral envelope. A single flu virus contains some 500 spikes.

- ▶ **H Spikes.** On an influenza A virus, about 80% of the spikes are **hemagglutinin (H)** proteins (**H spikes**). H spikes, all of which are identical on one flu virus, allow for attachment to and entry into the epithelial cells lining the nasopharynx and **tracheo-bronchial tree**, which consists of the airway structure involving the trachea and bronchi. Eighteen antigenically different types of H-spike proteins (classified as H1–H18) have been identified; most are naturally found in aquatic birds.

- ▶ **N Spikes.** The rest of the spikes on an influenza A virus are **neuraminidase (N)** proteins (**N spikes**). The N spikes assist in the release of the newly replicated virions from the host epithelial cells. There are 11 different N-spike proteins (N1–N11) and any one virus has but one type of N spike.

Thus, influenza A viruses are divided into subtypes based on composition of spike proteins. For example, of the 198 possible subtypes, the current seasonal flu subtypes in humans are H1N1 and H3N2.

Each year a slightly different seasonal flu strain evolves based, in part, on the changes to H and/or N spikes. Thus, there is a need for a new flu vaccine each year to protect individuals. In some years, the new subtype is quite mild (it might be a slightly changed version of H3N2), whereas in other years, a predominant subtype might be more dangerous because a completely new spike protein is present (e.g., H7N2).

An important question then is how do influenza A subtypes arise? **MICROINQUIRY 16** examines their evolution through mutation and genetic reassortment (gene swapping).

Clinical Presentation

Because some other viral agents and bacterial pathogens can produce an influenza-like respiratory illness, a diagnosis of influenza, if necessary, should be rapid. Today, there are several **rapid influenza diagnostic tests** (**RIDTs**) to detect influenza viral antigens in respiratory specimens. With results available in less than 15 minutes, an RIDT can be performed in the physician's office, allowing for prompt patient management.

The influenza A virus is extremely contagious. It can be spread person to person by way of expelled airborne respiratory droplets from coughing and sneezing. In addition, touching contaminated fomites and then touching one's eyes or nose can lead to infection.

After an incubation period of 1 to 4 days, an infected individual with an uncomplicated illness abruptly develops chills, fatigue, headache, and pain that are most pronounced in the chest, back, and legs. Over a 24-hour period, body temperature can rise to 40°C (104°F), and a severe cough develops. Individuals can experience sore throat, nasal congestion, sneezing, and tight chest, the latter a probable reflection of viral invasion of tissues of the trachea and bronchi of the LRT. Despite these severe symptoms, influenza is normally acute (short-lived) and usually resolves in 7 to 10 days. Often the public is confused about the difference in symptoms between the flu and a common cold. MICROFOCUS 16.2 highlights the similarities and differences between these illnesses.

Most of the annual deaths from seasonal influenza A are due to a secondary bacterial pneumonia. For example, the immune system's natural response to a flu infection can trigger *Streptococcus pneumoniae* cells to disperse from the biofilms normally found in the nasopharynx. These bacterial cells then turn on genes that make them more deadly, potentially causing pneumonia when they colonize the lung. *Staphylococcus aureus* is another bacterial species associated with secondary infections resulting from the flu. *S. aureus* can cause the patient's immune cells to damage their own lungs (pneumonia development) rather than attack the infecting bacteria.

In rare cases, influenza infections are associated with two other complications. **Guillain-Barré syndrome** (**GBS**) occurs (1 case per 100,000 infections) when the body's immune system mistargets the infection and damages its own peripheral nerve cells, causing muscle weakness and sometimes temporary paralysis. **Reye syndrome** usually makes its appearance in young people after they are given aspirin to treat fever or pain associated with influenza. It begins with nausea and vomiting, but progressive mental changes (such as confusion or delirium) can occur. Thankfully, very few children develop Reye syndrome (less than 0.03–1 case per 100,000 persons younger than 18 years).

MICROINQUIRY 16

Drifting and Shifting—How Influenza Viruses Evolve

Flu viruses circulate freely in the environment. They emerge from zoonotic reservoirs, and if flu viruses are passed back and forth between animal hosts, novel viral subtypes can arise. Aquatic birds are the natural reservoirs for most of the H-spike subtypes.

Influenza viruses evolve in two different ways. Both involve chemical and structural changes within the virus, especially to the hemagglutinin (H) and neuraminidase (N) spike proteins (antigens), thereby evolving new virus subtypes.

Antigenic drift involves unpredictable, small changes to the influenza A and B viruses (**Panel A**). These changes, involving minor point mutations resulting from RNA replication errors, will be expressed in the new virions produced. Although many such mutations can produce spike variants that make the virus unable to infect cells, on occasion a mutation might confer an advantage to the virus. For example, a spike protein might have a subtle change in shape (that is, the structural shape has "drifted") such that the proteins are not completely recognized by the host's immune system. This is what happens in most flu seasons. The virus spikes are different enough from the previous season that the host's antibodies fail to properly recognize the new subtype. This is why someone can still get the flu even though the person was vaccinated.

Antigenic shift is an abrupt, major change in structure to influenza A viruses only. Antigenic shift can give rise to new subtypes that can now jump to another species, including humans (**Panel B**); that is, the spike structure has "shifted" such that most everyone is immunologically defenseless and from which a pandemic can ensue (see figure).

Two mechanisms account for antigenic shift.

The "Spanish" flu was the introduction of a completely new flu strain (H1N1) that jumped from aquatic birds to humans. Once in the human host, the H1N1 subtype adapted quickly to replicate efficiently in humans (**Panels B, C**).

The second mechanism involves a process called "gene reassortment." If a host cell has been infected with two or more flu virus subtypes at the same time, a mixing (reassortment) of RNA segments from different viruses can occur (**Panel B**). Sometimes, this results in a hybrid virus with completely new surface spikes. For example, the 1957 influenza virus ("Asian" influenza; H2N2) was a reassortment, in which the human H1N1 subtype acquired new spike genetic segments (H2 and N2) from an avian species (**Panel D**). The 1968 influenza virus ("Hong Kong" influenza; H3N2) was the result of another reassortment; the human H2N2 subtype acquired a new hemagglutinin genetic segment (H3) from another avian species (**Panel E**).

In this second transmission mechanism, pigs usually are the reassortment "vessels" because they can be infected by both avian and human flu viruses. In fact, the 2009 swine flu is another example, involving reassortment of swine, avian, and human virus segments (**Panel F**). In late 2013, the first human cases of H7N9 and H5N1 (gene reassortment in poultry between domestic ducks and wild bird subtypes) were reported in China. Generating a virus with a functional reassortment is rare, but as you can see from the graph, they do occur with some regularity.

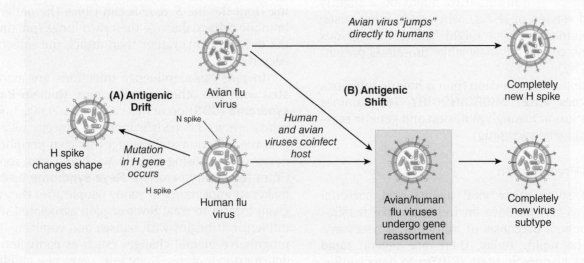

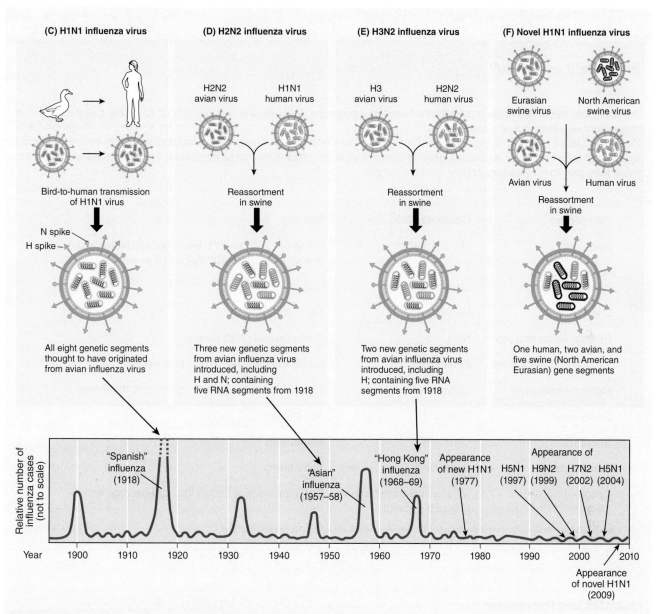

(C) H1N1 influenza virus

Bird-to-human transmission of H1N1 virus

N spike
H spike

All eight genetic segments thought to have originated from avian influenza virus

(D) H2N2 influenza virus

H2N2 avian virus   H1N1 human virus

Reassortment in swine

Three new genetic segments from avian influenza virus introduced, including H and N; containing five RNA segments from 1918

(E) H3N2 influenza virus

H3 avian virus   H2N2 human virus

Reassortment in swine

Two new genetic segments from avian influenza virus introduced, including H; containing five RNA segments from 1918

(F) Novel H1N1 influenza virus

Eurasian swine virus   North American swine virus

Avian virus   Human virus

Reassortment in swine

One human, two avian, and five swine (North American Eurasian) gene segments

Relative number of influenza cases (not to scale)

"Spanish" influenza (1918)

"Asian" influenza (1957–58)

"Hong Kong" influenza (1968–69)

Appearance of new H1N1 (1977)

H5N1 (1997)   H9N2 (1999)   H7N2 (2002)   H5N1 (2004)

Appearance of

Year   1900   1910   1920   1930   1940   1950   1960   1970   1980   1990   2000   2010

Appearance of novel H1N1 (2009)

The Major Influenza Pandemics and Novel H1N1 Appearance in Humans. The major influenza pandemics in humans. The four pandemics were the result of antigenic shifts.

Discussion Point

The current avian (or bird) flu subtype (H5N1 strain) is lethal to domestic fowl but is very inefficiently transmitted from fowl to human. Since 2003, of the approximately 850 reported human infections, 449 have died. Recently, two research groups, one at the University of Wisconsin, Madison and the other from the Erasmus Medical Center in the Netherlands, each identified the few mutations needed to make H5N1 readily transmissible to humans. Both reports were initially blocked from publication because critics feared that making the methodology and resulting mutation sequences available might give terrorists the "recipe" to make a "doomsday weapon" or the virus might escape from a research lab. Proponents for publication say knowing what mutations must occur would allow epidemiologists to "be on the lookout" for such naturally occurring mutations in H1N1 and to test the efficacy of existing vaccines and antiviral drugs. Both papers have been published. Was this a good idea?

MICROFOCUS 16.2: Public Health

Is It a Cold or the Flu?

Do you know the differences in symptoms between a common cold and the flu? As described in this chapter, both are respiratory illnesses but are caused by different viruses. Yet both often have some similar symptoms. In general, the flu has a sudden onset (3–6 hours), whereas a cold comes on more gradually. Flu symptoms are more severe than cold symptoms (see the table), and colds generally do not develop serious complications, such as pneumonia and bacterial infections, nor do they usually require hospitalization.

Symptoms	Common Cold	Flu
Fever	Uncommon	Common; 38°C to 39°C [occasionally higher, especially in young children (40°C)]; lasts 3 to 4 days
Headache	Uncommon	Common
Chills	Uncommon	Fairly common
General aches and pains	Mild	Common and often severe
Fatigue and weakness	Sometimes	Usual and can last up to 2 to 3 weeks
Extreme exhaustion	Uncommon	Usual and occurs at the beginning of the illness
Stuffy nose	Common	Sometimes
Sneezing	Common	Sometimes
Sore throat	Common	Sometimes
Cough	Mild to moderate hacking, productive cough	Dry, unproductive cough that can become severe

Data from National Institute of Allergy and Infectious Diseases. Retrieved from www.niaid.nih.gov/topics/Flu/Pages/coldOrFlu.aspx.

Treatment and Prevention

Treatment of uncomplicated cases of influenza involves bed rest, adequate fluid intake, and aspirin (or acetaminophen in children) for fever and muscle pain. Two antiviral drugs, zanamivir (Relenza®) and oseltamivir (Tamiflu®), are available by prescription. These drugs interfere with the neuraminidase spikes, blocking the release of new virions. If given to otherwise healthy adults or children early in disease onset, the drugs can reduce the duration of illness by one day, but significantly, the drugs might reduce the likelihood of complications. Importantly, these drugs should not be taken in place of an annual flu vaccination, which remains the best prevention strategy.

Each year's batch of flu vaccine is based on the previous year's predominant influenza A and B viruses. On average, the vaccine is about 75% effective. Today, several different forms of the vaccine are available, making flu vaccination available to almost anyone, as MICROFOCUS 16.3 illustrates.

Some Respiratory Viruses Cause Forms of Community-Acquired Pneumonia

Pneumonia refers to microbial disease of the bronchial tubes and lungs. **Community-acquired pneumonia (CAP)** is the most common type of pneumonia. It occurs outside of hospitals or other healthcare facilities, and viruses are the most frequent cause of CAP in children younger than 5 years. Although viral pneumonia usually is mild, on occasion it can become very serious.

Let's examine a few of the more common viruses and the pneumonia they cause. Several are members of the Paramyxoviridae family (**FIGURE 16.6**).

MICROFOCUS 16.3: Public Health

Which Flu Vaccine Is for Me?

Until recently, there were few options for getting vaccinated against the flu. You could get the standard shot in the arm, which is a vaccine containing two inactivated type A viruses and one type B virus. FluMist, an attenuated flu virus vaccine, was introduced as a nasal spray in 2003. Then, for the first time, in 2013 six different flu vaccines became available to meet just about anybody's need (see accompanying figure). These include:

- An **egg-free vaccine** (for people 18–49 years of age with egg allergies) is produced in caterpillar cells of insects rather than chicken eggs.
- The **nasal spray** (FluMist®) **vaccine** (for healthy individuals 2–49 years of age). Note: For the 2016–2017 season, the CDC did not recommended the nasal spray due to its reported poor or relatively low effectiveness between 2013 through 2016.
- The **microneedle vaccine** delivers the standard vaccine via a microneedle syringe to produce a less painful jab in the arm.
- The **high-dosage vaccine** (for people older than 65 years of age) provides a 24% stronger immune response (more antibodies produced).
- The **standard vaccine** (for anyone 6 months and older).

The standard and the nasal spray vaccines also come in a quadrivalent form (two type A and two type B subtypes). Hopefully, with the multiple types of vaccines, most Americans will get an annual vaccination. Most promising, it might not be long before there is a universal flu vaccine for protection against all strains of the flu viruses.

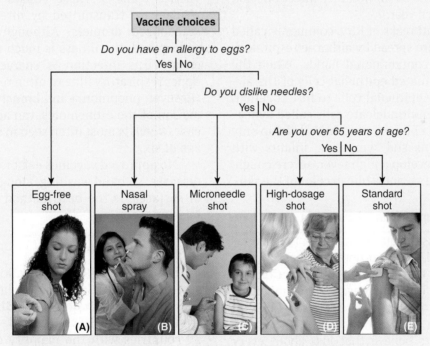

(A) © Brian Chase/Shutterstock. (B) Courtesy of James Gathany/CDC. (C) © Leah-Anne Thompson/Shutterstock. (D and E) © Alexander Raths/Shutterstock.

Respiratory Syncytial Disease

The respiratory syncytial virus (RSV), a member of the Paramyxoviridae, causes **respiratory syncytial (RS) disease**, which is the most common LRT illness affecting infants and children. Some virologists believe that up to 95% of all children have been infected by 2 years of age, and CDC epidemiologists estimate that RSV causes 51,000 to 82,000 hospitalizations and 4,500 deaths in infants and children each year in the United States. On a global

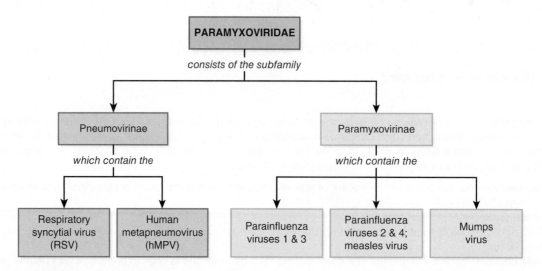

FIGURE 16.6 The Relationships Between the Human Paramyxoviruses. This concept map illustrates the relationships between the Paramyxoviridae that cause human respiratory and skin diseases. »» *Which virus is most closely related to the respiratory syncytial virus?*

scale, there are some 34 million RSV infections and 200,000 deaths each year.

Community outbreaks of RSV, commonly called **viral pneumonia**, are spread by airborne respiratory droplets or virus-contaminated hands. When the virus infects the ciliated epithelial cells of the airways, it causes the epithelial cells to fuse together, resulting in giant multinucleate cells called **syncytia**. Although most children and adults develop only mild cold-like signs and symptoms, infants with severe cases can develop a high fever, severe cough, rapid, short breathing, and a bluish color of the skin.

Human Metapneumovirus Pneumonia

An RSV-like illness is caused by the human metapneumovirus (hMPV), another member of the Paramyxoviridae. The symptoms of infection are mild, with cough, fever, and nasal congestion common. Some infections can progress to the LRT and lead to pneumonia.

Medical experts believe that just about every child in the world has been infected by age 10. The disease also can occur in adults, as evidenced by two outbreaks of hMPV in skilled nursing facilities in West Virginia and Idaho in 2011 and 2012. Among the 57 cases, six patients died.

Parainfluenza Infection

Forty percent of acute pneumonia cases in children are the result of infection by human parainfluenza viruses 1 and 3. These viruses in the Paramyxoviridae are transmitted by direct contact or via aerosolized droplets. Although the infection is widespread, the illness is much milder than influenza. Virus infection is characterized by minor upper respiratory illness, often referred to as a cold. However, pneumonia and **bronchiolitis**, an inflammation of the bronchioles, can accompany the disease, which is most often seen in children under the age of six.

No approved vaccines exist for these three viral diseases, so hand washing, along with disinfection of surfaces, is the best method to prevent transmission.

Respiratory Syndromes

In the spring of 2003, an infectious disease quickly spread through Southeast Asia and to Canada, as reported in **CLINICAL CASE 16**. During the epidemic, the WHO identified 8,100 cases from 29 countries with the majority of the 774 deaths being in China, Taiwan, Vietnam, Singapore, and Canada. The disease became known as **severe acute respiratory syndrome** (**SARS**).

SARS is caused by a coronavirus called SARS-CoV, which, being in the Coronaviridae family, is a single-stranded (+sense) RNA virus with helical symmetry and a spiked envelope (**FIGURE 16.7**). SARS is spread through close person-to-person contact by touching one's eyes, nose, or mouth after contact with SARS-CoV. Disease spread also

Clinical Case 16

The 2002–2003 Outbreak of SARS

February 11, 2003: The Chinese Ministry of Health notifies the World Health Organization (WHO) of a mystery respiratory illness that has been occurring since November 2002 in Guangdong province in southern China. However, Chinese officials refuse to allow WHO officials to investigate.

On February 21, a 64-year-old doctor from Guangdong comes to Hong Kong to attend a wedding. He stays at a Hong Kong hotel, unknowingly infecting 16 people who, on returning home, spread the disease to Hanoi, Vietnam; Singapore; and Toronto, Canada (see accompanying figure).

Two days later, a Canadian woman tourist checks out of the same Hong Kong hotel and returns to Toronto, where her family greets her. She dies 10 days later as five family members are hospitalized.

On February 28, Carlo Urbani, a WHO doctor in Hanoi, treats one of the people infected in Hong Kong and realizes this is a new disease, which he names "severe acute respiratory syndrome" (SARS). He dies of the disease 29 days later.

On March 15: The WHO declares SARS a worldwide health threat and issues an emergency travel advisory. To block the chain of transmission, isolation and quarantine are instituted. More than half of those infected are healthcare workers.

April 16: The identification is made by 13 laboratories around the world that a new, previously unknown, coronavirus (called SARS-CoV) is the causative agent of SARS.

June 30: The WHO announces there have been no new cases of SARS for 2 weeks. During the 114-day epidemic, more than 8,000 people from 29 countries were infected and 774 died.

Questions:
  a. Why were Chinese officials reluctant to allow a WHO investigation into the mystery illness?
  b. Justify the use of quarantine and isolation to break the chain of transmission for SARS-CoV.
  c. Why was there such a disproportionately high number of SARS-CoV infections in healthcare workers?
  d. How could such a large number of infections in healthcare workers have been prevented?
  e. Why does the SARS outbreak represent such a great example for an emerging infectious disease?

You can find answers online in **Appendix E**.

For additional information, see www.cdc.gov/mmwr/preview/mmwrhtml/mm5226a4.htm.

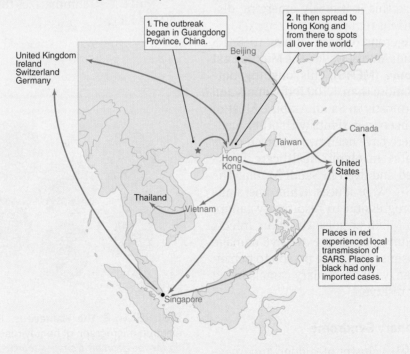

Spread of SARS.

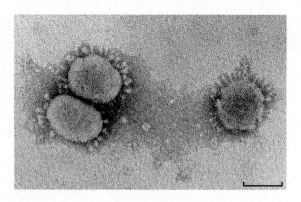

FIGURE 16.7 Coronaviruses. False-color transmission electron micrograph of three human coronaviruses. The spikes can be seen clearly extending from the viral envelope. (Bar = 100 nm.) »» *By looking at the micrograph, explain why the virus is referred to as a "corona" virus.*

© Gary Gaugler/Medical Images RM.

occurs when a SARS-infected individual contaminates fomites with infectious droplets from coughing or sneezing. Horseshoe bats are the primary reservoir of SARS-CoV.

Many people remain asymptomatic after being infected with SARS-CoV. However, in individuals who do develop symptoms, a dry cough and labored breathing are common. In those patients progressing to a severe respiratory illness, pneumonia develops. With insufficient oxygen reaching the blood, some 10% to 20% of patients require mechanical ventilation. Because this is a newly emerging disease, treatment options remain unclear.

In 2012, a dangerous disease similar to SARS was reported in Saudi Arabia. Called the **Middle East respiratory syndrome** (**MERS**), this ongoing outbreak has sickened more than 1,700 individuals and killed at least 500 (mostly in Saudi Arabia). Most of the fatalities have been in patients with other medical conditions. There have been two reported cases in the United States involving healthcare workers who had worked and lived in Saudi Arabia. The infection by MERS-CoV produces symptoms similar to SARS, but progression to respiratory failure is more rapid with MERS because the virus infects deeper regions of lungs, causing extensive damage to the alveoli. Transmission can be person to person and camel to human, with dromedary camels being the likely zoonotic reservoir.

Hantavirus Pulmonary Syndrome

In the spring of 1993, a cluster of sudden and unexplained deaths in previously healthy young adults occurred in the Four Corners region of the American Southwest. The CDC identified a hantavirus, now called Sin Nombre virus (SNV), as the infectious agent and termed the outbreak **hantavirus pulmonary syndrome** (**HPS**). Since 1993, there have been more than 2,000 cases throughout the Americas (600 in the western United States), and about 35% of reported cases have resulted in death. In 2012, an outbreak in tent cabins at Yosemite National Park in California might have exposed 1,700 campers and visitors to the disease. It did sicken at least six people and took the lives of three. In 2014, 34 cases of HPS were reported to the CDC.

The hantaviruses are enveloped, single-stranded (−sense) RNA viruses in the Bunyaviridae family (**FIGURE 16.8**). Infected deer mice, which are the zoonotic reservoir, shed the virus in saliva, urine, and feces. Humans usually are infected by breathing or directly contacting dried urine or feces containing the viruses or through direct contact with the infected rodents. Several weeks after exposure, the symptoms of infection occur and include fatigue, headache, fever, and muscle aches. About half of all HPS patients experience a severe illness with dizziness, difficulty breathing, and low blood pressure. A high rate of death is due to the development of a fatal pulmonary disease as the lungs fill with fluid. Prevention consists of eliminating rodent nests and minimizing contact with them. There is no vaccine available.

TABLE 16.3 summarizes the viral diseases affecting the LRT.

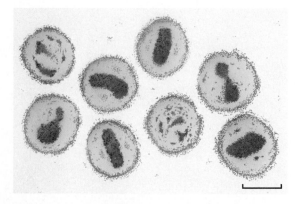

FIGURE 16.8 The Hantavirus. False-color transmission electron micrograph of hantaviruses. (Bar = 90 nm.) »» *What do the red geometric shapes represent in the hantaviruses?*

© Chris Bjornberg/Science Source.

TABLE 16.3 **A Summary of the Major Viral Diseases of the LRT**

Disease	Causative Agent	Signs and Symptoms	Transmission	Treatment	Prevention
Influenza	Influenza A, B, and C viruses	Chills, fatigue, headache Chest, back, and leg pain	Respiratory droplets	Bed rest and fluids Antiviral medications (oseltamivir or zanamivir)	Getting annual flu vaccination
Respiratory syncytial (RS) disease	Respiratory syncytial virus (RSV)	Influenza-like	Respiratory droplets Hand contact	Fever-reducing medications Ribavirin for severe cases	Practicing good hygiene
Parainfluenza	Human parainfluenza viruses 1 and 3	Cold-like	Respiratory droplets Direct contact	No specific therapy	Practicing good hygiene
RSV-like illness	Human metapneumovirus	Cold-like	Respiratory droplets Direct contact	Fever-reducing medications Ribavirin for severe cases	Practicing good hygiene
SARS MERS	SARS coronavirus (SARS-CoV) MERS coronavirus (MERS-CoV)	Fever, headache, body aches, dry cough, and breathing difficulty	Respiratory droplets and airborne particles Direct contact	No effective treatment	Practicing good hygiene
Hantavirus pulmonary syndrome	Hantavirus (Sin Nombre virus)	Fatigue, fever, muscle aches, headache, dizziness, breathing difficulty	Aerosolized droplets of rodent saliva, urine, feces	Supportive care	Eliminating rodent nests Minimizing contact

Concept and Reasoning Checks 16.2

a. From MicroInquiry 16, why is the influenza A virus the cause of most flu epidemics and pandemics?

b. Identify the common relationships between the paramyxoviruses.

c. How do the SARS coronavirus and the hantavirus differ in their spread between individuals?

Chapter Challenge B

During the 2009–2010 novel H1N1 flu pandemic, French scientists reported that the novel H1N1 influenza A virus did not become a major cause of disease among the French until after the fall rhinovirus season declined. Researchers found that the percentage of throat swabs from French individuals with respiratory illness that tested positive for novel H1N1 declined in September, while at the same time rhinovirus, which causes colds, rose. Then, in late October as the rhinovirus cases declined, the number of novel H1N1 cases again rose.

QUESTION B: *With viral cell/tissue specificity in mind, hypothesize why having a common cold might have protected these individuals from becoming infected with the influenza A virus.*

You can find answers online in **Appendix F**.

■ KEY CONCEPT 16.3   Herpesviruses Cause Several Human Skin Diseases

There are more than 100 viruses in the family Herpesviridae, and they are capable of infecting a broad range of animals. Presently, eight viruses within the family infect humans (**FIGURE 16.9**). Several viruses infect the skin. Some, such as the herpes simplex viruses, remain epidemic in contemporary times; others, such as varicella-zoster virus, are being brought under control through effective vaccination programs. However, there are relatively few antiviral drugs for these diseases, and prevention programs remain the major course of action.

All human Herpesviridae are large viruses with a nonsegmented, double-stranded DNA genome. The capsid exhibits icosahedral symmetry and is surrounded by an envelope with spikes. Another shared characteristic is that all the viruses can undergo both productive and latent infections. The word *herpes* is Greek for "creeping," referring to the spread of herpes infections through the body during productive and latent infection cycles.

Human Herpes Simplex Viruses Can Remain Latent in Sensory Nerve Ganglia

Two of the most prevalent viruses in the Herpesviridae are herpes simplex viruses 1 and 2 (HSV-1, HSV-2;

FIGURE 16.10A). Humans are the only reservoir for both viruses that formally are called human herpesvirus 1 (HHV-1) and human herpesvirus 2 (HHV-2). Each has a slight preference for the cells it infects.

Cold Sores

HSV-1 is primarily the agent responsible for **cold sores**. Worldwide, the WHO estimates that 67% of people younger than 50 (about 44% in the United States) are infected with HSV-1. However, only a minority of these individuals have visible symptoms. For cold sores, also called "fever blisters," symptoms include a tingling sensation, and then visible lesions appear as small, hard spots on the lip and edges of the mouth. Within a couple of days red, fluid-filled blisters appear (**FIGURE 16.10B**). The blisters eventually break, releasing the fluid containing infectious virions. A yellow crust forms and then peels off without leaving a scar on the skin. A person is most likely to transmit viruses from the time the blisters appear until they have completely dried and crusted over. Cold sores generally clear up without treatment in about two weeks.

After the acute (productive) infection, the viruses enter the trigeminal ganglion, where the viral genome becomes latent. Reactivation of the viral

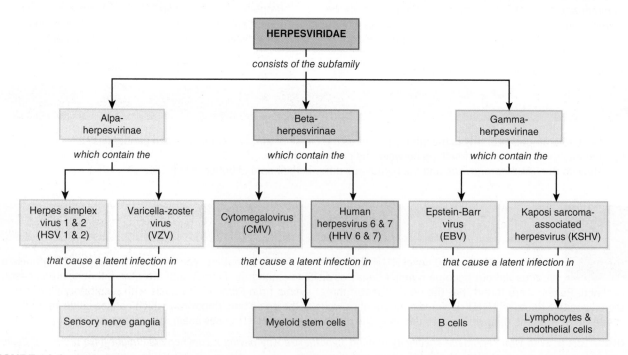

FIGURE 16.9 The Relationships Between the Human Herpesviruses. This concept map illustrates the relationships between six of the eight viral species in the family Herpesviridae that cause human disease. Only EBV and KSHV are known to be oncogenic. **»» What does it mean for a virus to be oncogenic?**

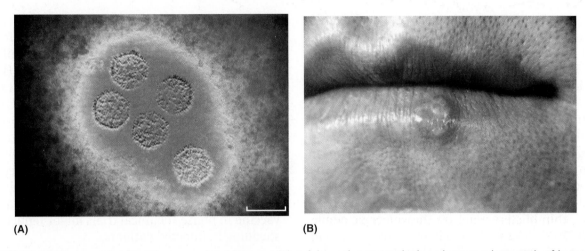

(A)   **(B)**

FIGURE 16.10 Herpes Simplex Viruses and Cold Sores. (A) A false-color transmission electron micrograph of herpes simplex viruses. (Bar = 150 nm.) **(B)** The cold sores (fever blisters) of herpes simplex erupting as tender, itchy papules and progressing to vesicles that burst, drain, and form scabs. *»» How would the release phase of the HSV replication cycle lead to blisters and vesicles on the lips?*

(A) © Medical Images RM/Claude Bucau. (B) Courtesy of Dr. Hermann/CDC.

genome often occurs after some form of trauma, often in response to a stressful trigger, such as fever, menstruation, or emotional disturbance. Even environmental factors like sunburn (exposure to ultraviolet light) can trigger reactivation (**FIGURE 16.11**). In rare complications, reactivation can lead to an infection of the eye (**ophthalmic herpes**) or brain (**herpes encephalitis**).

Not kissing others while the blisters are present can help prevent the transmission of cold sores. Washing one's hands often and avoiding shared items (utensils, towels, lipstick) used by a person with an active infection also will limit viral spread.

Genital Herpes

Although it is not a notifiable disease, the CDC estimates that more than 45 million Americans (primarily between the ages of 14 to 49) have **genital herpes**, with a slightly higher rate among women. The infection rate is continuing to increase with a 30% jump over the past three decades, accounting for more than 600,000 new cases being diagnosed each year. Alarmingly, about 80% of infected individuals do not know they are infected.

Most cases of genital herpes are a result of an HSV-2 infection, which is efficiently transmitted

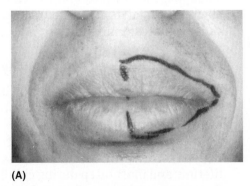
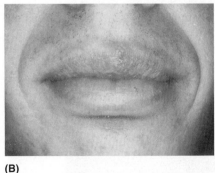

(A)   **(B)**

FIGURE 16.11 Ultraviolet Light and Cold Sores. These photos show the results of an experiment studying the effects of ultraviolet (UV) light on the formation of herpes simplex cold sores of the lips. This patient usually experienced sores on the left upper lip. **(A)** The patient was exposed to UV light from a retail cosmetic sunlamp on the left upper and lower lips in the area designated by the line. The remainder of the face was protected by a sunscreen. **(B)** Sores formed on the left upper lip. *»» What would you conclude from these experiments regarding the reactivation of herpesviruses?*

Reproduced from *J Clin Microbiol*, 1985, vol. 22, pp. 366–368, DOI and reproduced with permission from the American Society for Microbiology. Photo courtesy of Woody Spruance, MD, Professor of Internal Medicine, University of Utah.

through intimate genital contact with an infected individual. In fact, the WHO estimates that about 8% of the world population is infected with the virus. However, because of the high rate of HSV-1 infections, individuals, even if they don't show any signs or symptoms, can spread HSV-1 from the mouth to the genitals through oral sex and lead to genital herpes.

Signs of a productive infection generally appear within a few days of sexual contact, often as itching or throbbing in the genital area. This is followed by reddening and swelling of a small area where painful, thin blisters erupt on the vulvar or penile skin (**FIGURE 16.12**). The blisters crust over and the sores disappear, usually within about 3 weeks. The signs often are very mild and can go unnoticed. Fifty percent of infected individuals experience only one outbreak in their lifetime.

However, for many infected individuals, a latent period occurs during which time the virus remains in sensory nerve cells near the blisters. The latent genome can become reactivated again by stressful situations, and a recurrent productive infection results. This cycle of latency and recurrent infection can occur three to eight times a year.

Although genital herpes usually causes no serious complications, the presence of sores or breaks in the skin or lining of the vagina and rectum can lead to bleeding. When the sores come in contact with the mouth, vagina, or rectum during sex, there is an increased risk of transmitting or getting infected with HIV if the individual or the partner has HIV.

Prevention is similar to that for any sexually transmitted infection—abstain from sexual activity or limit sexual contact to a person who is not infected. Although there is no cure for genital herpes, antiviral drugs such as acyclovir can help heal the sores sooner, reduce the frequency of recurrent infections, and diminish the chances of passing the infection on to other individuals. An investigational vaccine in phase II clinical trials might reduce virus shedding in infected patients.

Neonatal Herpes

Although most pregnant women with genital herpes have healthy babies, a small number of infected mothers spread HSV to their newborns during labor and delivery. The result, **neonatal herpes**, can be a devastating and life-threatening disease causing blindness and persistent seizures in the newborn. If symptoms and a diagnosis of an active genital herpes infection in the mother are made prior to delivery, the healthcare provider might recommend a course of acyclovir before delivery, or birth by cesarean section, to reduce the risk for neonatal HSV infection.

The Varicella-Zoster Virus Also Generates a Latent Infection in Sensory Ganglia

In the centuries when pox diseases regularly swept across Europe and other parts of the world, people had to contend with the Great Pox (syphilis), the smallpox, the cowpox, and the chickenpox. Today, chickenpox and its associated disease, shingles, still cause significant morbidity.

Causative Agent and Epidemiology

Chickenpox and shingles are two clinical manifestations caused by the varicella-zoster virus (VZV), which is formally called human herpesvirus 3 (HHV-3). **Chickenpox (varicella)** is a highly contagious disease that usually attacks just once in a lifetime and most often during childhood. **Shingles (herpes zoster)** is usually an adult disease. Anyone who has had chickenpox will develop a latent VZV infection, which puts the person at risk of developing shingles later in life. However, should a person with an active case of shingles pass VZV to someone who has not had chickenpox, that person initially will develop chickenpox but not shingles.

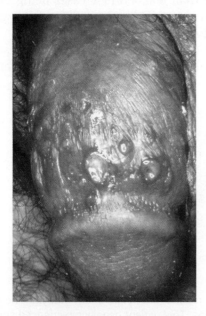

FIGURE 16.12 Genital Herpes Rash on the Penile Shaft. Signs of male genital herpes usually appear as blisters on or around the genitals or rectum. *»» How might this male patient have gotten genital herpes?*

Courtesy of Dr. N. J. Flumara and Dr. Gavin Hart/CDC.

Prior to the availability of a vaccine in 1995, about 4 million children contracted chickenpox each year in the United States. In 2014, the CDC reported about 10,000 cases. Shingles develops in about 30% of adults having a latent VZV infection. Still, the CDC reports that there are more than 1 million cases of shingles each year with about 50% of those cases occurring in men and women 60 years of age or older.

Clinical Presentations

VZV is transmitted by airborne respiratory droplets and skin contact, and it has an incubation period of about 2 weeks. The disease begins in the respiratory tract, with fever, headache, and malaise. Viruses then pass into the bloodstream and localize in the peripheral nerves and skin. As they multiply in the cutaneous tissues, VZV produces a red, itchy rash on the face, scalp, chest, and back, although it can spread across the entire body. The rash quickly turns into small, teardrop-shaped, fluid-filled vesicles (**FIGURE 16.13A**). The vesicles in chickenpox develop over 3 or 4 days in a succession of "crops." They itch intensely and eventually break open to yield highly infectious virus-laden fluid. In chickenpox, the vesicles form crusts that fall off. A person who has chickenpox can transmit the virus for up to 48 hours before the telltale rash appears and remains contagious until all spots crust over. Chickenpox usually lasts 2 weeks or less.

The development of chickenpox complications is rare. The most common complication in adults who contract chickenpox is a bacterial infection of the skin, pneumonia, or an inflammation of the brain (encephalitis). Reye syndrome can occur during the recovery period, so aspirin should not be used to reduce fever in either children or adults.

After having chickenpox, the VZV genome remains latent in the dorsal root ganglia (thoracic or lumbar nerves). Many years later, the virus genome might undergo replication in the sensory neurons. The viruses formed travel down the nerves to the skin of the body trunk and resurface as shingles (**FIGURE 16.13B**). Here, blisters with blotchy patches of red appear that encircle the trunk (*zoster* = "girdle"). Many sufferers also experience a series of headaches as well as facial paralysis and sharp "icepick" pains described as among the most debilitating known. The most common complication of shingles is **postherpetic neuralgia**, in which some 30% of suffers continue to have debilitating pain for months or years after the lesions disappear.

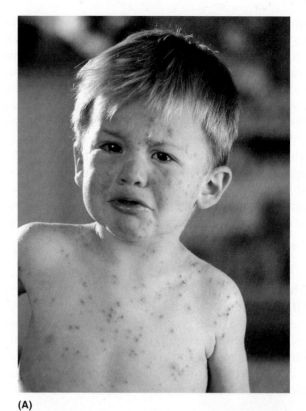

(A)

(B)

FIGURE 16.13 The Lesions of Chickenpox and Shingles. **(A)** A typical case of chickenpox. The lesions can be seen in various stages, with some in the early stage of development and others in the crust stage. **(B)** Dermal distribution of shingles lesions on the skin of the body trunk. The lesions contain less fluid than in chickenpox and occur in patches as red, raised blotches. *»» Summarize how varicella latency could lead to shingles.*

(A) © SW Productions/age fotostock. (B) © Science Photo Library/Science Source.

Treatment and Prevention

Acyclovir and valacyclovir have been used in patients to lessen the symptoms of chickenpox and shingles and to hasten recovery. A varicella-zoster immune globulin (VZIG), which contains antibodies to the chickenpox virus, also can be used. The chickenpox vaccine (Varivax®) is the safest, most effective way to prevent chickenpox and its possible complications. The varicella vaccine is 85% effective in disease prevention, and it has reduced hospitalizations for chickenpox by 93%. In 2006, a live, attenuated shingles vaccine, called Zostavax®, was approved. This is not a preventative vaccine but rather a therapeutic vaccine that can reduce the incidence of shingles by more than 50% and postherpetic neuralgia by more than 60% in people older than 60 years of age. Acyclovir therapy also lessens the symptoms.

Other Members of the Herpesviridae Cause Latent Infections in Blood Cells

The Epstein-Barr virus (human herpesvirus 4; HHV-4) and the cytomegalovirus (human herpesvirus 5; HHV-5) (see Figure 16.9) primarily infect and remain latent in various groups of white blood cells but do not present as a skin disease. Therefore, they will be discussed in the chapter dealing with infections of the blood.

Roseola

Human herpesvirus 6 (HHV-6) belongs to a different subfamily of the Herpesviridae (see Figure 16.9). HHV-6 primarily affects infants and young children 6 months to 2 years old (HHV-7 often acts together with HHV-6). The viruses cause **roseola**, an acute, self-limiting condition marked by high fever. This often is followed by a red body rash on the back and neck (**FIGURE 16.14**). Roseola can spread from person to person through contact with an infected person's respiratory secretions or saliva. The infection usually lasts about 1 week, and treatment requires bed rest, fluids, and fever-reducing (non-aspirin) medications.

HHV-6 can remain latent in myeloid stem cells, which give rise to monocytes and macrophages. This is reflected in the observation that 30% to 60% of bone-marrow transplant recipients suffer a reactivation of HHV-6 during the first few weeks after transplantation. The recurrence potentially can lead to pneumonia or encephalitis. No drugs have been approved for HHV-6 infections.

Kaposi Sarcoma

Human herpesvirus 8 (HHV-8) causes **Kaposi sarcoma (KS)**. The virus triggers the growth of new blood vessels, leading to a tumor of the blood vessels under the skin. The malignancy, which is marked by dark or purple skin lesions, has become one of the most common tumors in AIDS patients (**FIGURE 16.15**). Treatment of smaller skin lesions involves the use of liquid nitrogen, low-dose radiation, or chemotherapy applied directly to the lesion.

TABLE 16.4 summarizes the characteristics of the herpesviruses that affect the skin.

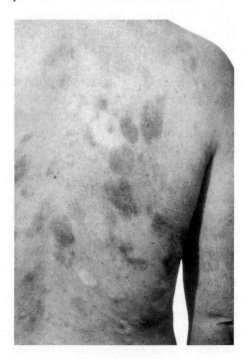

FIGURE 16.15 Kaposi Sarcoma. This man is suffering from AIDS. Kaposi sarcoma is an opportunistic disease, appearing as dark lesions on the skin. *»» What is meant by an opportunistic disease?*

Courtesy of the National Cancer Institute.

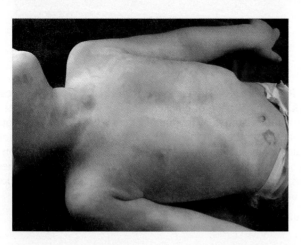

FIGURE 16.14 Roseola. Following a high fever, a red body rash may appear on the neck and trunk of infants and young children. *»» Why should aspirin not be given to infants or young children suffering a viral infection such as roseola?*

Courtesy of Carolyn Rogers Pershouse.

TABLE 16.4 **A Summary of Viral Diseases Caused by the Herpesviruses**

Disease	Causative Agent	Signs and Symptoms	Transmission	Treatment	Prevention
Cold sores (fever blisters)	Herpes simplex virus 1 (HSV-1)	Small, red hard spot on lip	Direct or indirect contact	Lidocaine Acyclovir	Avoiding skin contact Not sharing personal items
Genital herpes	Herpes simplex virus 2 (HSV-2)	Itching or throbbing in the genital area Reddening and swelling of a small area where painful, thin blisters erupt	Sexual intercourse	Antiviral drugs (acyclovir)	Abstaining from sexual activity Limiting sexual contact to only one person who is infection free
Neonatal herpes	Herpes simplex virus 2	Mental development can be delayed Blindness can occur Persistent seizures	From infected mother to newborn during childbirth	Acyclovir Supportive therapy	Cesarean section if the maternal infection is recognized Screening women considered to be at high risk
Chickenpox (varicella)	Varicella-zoster virus (VZV)	Fever, headache, malaise with red, itchy rash on face, scalp, chest, and back	Droplet contact	Supportive care Acyclovir	Chickenpox vaccine
Shingles (herpes zoster)	Varicella-zoster virus (VZV)	Blisters on body trunk with intense pain		Acyclovir	Shingles vaccine
Roseola	Human herpesvirus 6	Red rash on neck and trunk	Contact	Supportive care	Avoiding exposure to infected child
Kaposi sarcoma	Human herpesvirus 8	Dark lesion on skin	Contact (sexual and nonsexual)	Antiretroviral therapy	Using anti-HIV medications

Concept and Reasoning Checks 16. 3

a. Explain how a productive infection and a latent infection differ for herpes simplex virus 1 (HSV-1).
b. Assess the consequences if someone with shingles transmits the virus to a susceptible person who never had chickenpox.
c. Determine why a bone-marrow transplant recipient is at risk of a recurrent HHV-6 infection.

Chapter **Challenge C**

The most common virus infection of the lips is herpes labialis, more commonly called cold sores or fever blisters. Being capable of producing a latent infection, new and recurrent productive infections can occur, even though one does not need to have a cold or a fever to develop the herpes lesions.

QUESTION C: *Based on the mechanism for reactivation of latent HSV-1, provide an explanation as to how these oral lesions got the name of "cold sores" or "fever blisters."*

You can find answers online in **Appendix F**.

■ KEY CONCEPT 16.4 Several Other Viral Diseases Affect the Skin

A number of other viral infections also affect the skin. Some are common childhood diseases, such as measles, whereas others, like smallpox, were once a worldwide scourge.

A Few Viruses Cause Typical Childhood Illnesses

Measles (Rubeola)

Measles is caused by the measles virus, a single-stranded (–sense) RNA virus in the Paramyxoviridae family (see Figure 16.6). Prior to the development of a measles vaccine, some 4 million children in the United States came down with measles annually, resulting in about 500 deaths and another 1,000 children experiencing chronic, lifelong disabilities. In 2000, due to an aggressive vaccination campaign and a strong healthcare system, measles was declared eliminated from the United States, meaning there had been no continuous disease transmission for at least 12 months. In fact, measles is no longer endemic (constantly present) in the United States.

When measles does occur, the acute disease usually arises in young children for whom it is highly contagious, with an infection rate approaching 95% in susceptible individuals. In fact, on average, a person with measles infects 12 to 18 previously uninfected and unvaccinated (susceptible) individuals.

After an incubation period of about 10 days, symptoms appear. These include a hacking cough, runny nose, eye redness, and a high fever. After replicating in the lungs, the virus spreads to the regional lymph nodes, and a blood infection spreads the virus throughout the body. Diagnostic red patches called **Koplik spots** appear on the mucous membrane in the mouth 2 to 4 days after the onset of symptoms (**FIGURE 16.16A**). The characteristic red facial rash appears about 2 days after the first evidence of Koplik spots. Beginning as pink-red pimple-like spots (maculopapules), the rash breaks out at the hairline and then covers the face and spreads to the trunk and extremities (**FIGURE 16.16B**). Within 3 to 5 days, the rash turns brown and fades.

Measles usually is characterized by complete recovery. However, in immunocompromised patients the disease can be severe. In developing nations, a small percentage of cases develop an opportunistic bacterial ear infection or pneumonia. About 1 in 1,000 cases develops a dangerous complication called **subacute sclerosing panencephalitis (SSPE)**. This rare brain disease is characterized by a decrease in cognitive skills and loss of nervous function. This delayed complication can occur anytime from 1 month to 25 years after clinical measles (usually within 7 to 10 years).

Other than supportive care, there is no prescribed treatment for a measles virus infection. Prevention is accomplished with the measles vaccine,

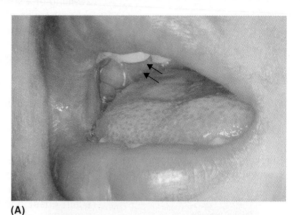

(A)

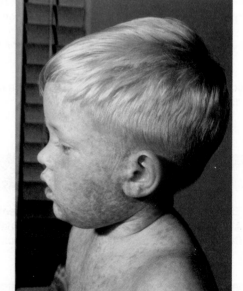

(B)

FIGURE 16.16 Koplik Spots and the Measles Rash. **(A)** Koplik spots on the inside cheek (arrows) of a child suffering from measles. The red spots with white centers are a frequent symptom. **(B)** A child with measles, showing the typical rash on face and torso. **»» Based on the rash, how would you distinguish measles from chickenpox?**

(A) Mediscan/Alamay Stock Photo. (B) Courtesy of CDC.

which usually is given as part of the measles, mumps, and rubella (MMR) vaccine.

If measles is no longer endemic in the United States, why are there still sporadic measles cases reported? Unfortunately, measles remains endemic in many other parts of the world, meaning importations of the virus by travelers can result in an outbreak (defined as a chain of transmission involving three or more cases). Although the vaccine is 97% effective, if small pockets of individuals are not vaccinated, those individuals are susceptible to infection by the measles virus. In 2008, 2011, and 2013–2015, there were substantial upticks in reported measles cases in the United States as compared with other years (**FIGURE 16.17**). These cases resulted from infected individuals bringing measles into the United States from foreign locales where measles has been more common. For example, during the 2015 outbreak, most measles cases were transmitted to unvaccinated children through individuals visiting or returning from the Philippines where a large outbreak was occurring.

Mumps

An outbreak of more than 40 mumps cases occurred among students at Harvard University in the first part of 2016. The disease is caused by the mumps virus, another member of the Paramyxoviridae (see Figure 16.6). Its name comes from the English "to mump," meaning to be sullen or to sulk. Although it is not a dermatotropic virus, the fact that it is a member of the Paramyxoviridae and is a component of the MMR vaccine makes it convenient for discussion here.

The characteristic sign of the disease is enlarged jaw tissues arising from swollen salivary glands, especially the **parotid glands. Infectious parotitis** is an alternate name for the disease. It is spread by respiratory droplets or contact with saliva or mucus from an infected individual. About 20% of people infected with the mumps virus have no signs or symptoms. When signs and symptoms do appear, swollen and painful salivary glands are typical, which causes a puffy cheek appearance. Obstruction of the ducts leading from the parotid glands retards the flow of saliva, which causes the characteristic swelling (**FIGURE 16.18**). The skin overlying the glands is usually taut and firm, and patients experience pain when the glands are touched as well as when chewing or swallowing.

In male patients, the mumps virus can pose a threat to the reproductive organs, causing swelling and damage to the testes, a condition called **orchitis** (*orchi* = "the testicles"). The sperm count can be reduced, but sterility is not common. An estimated 25% of mumps cases in post-adolescent males develop into orchitis.

Most children and adults recover from uncomplicated cases of mumps in about two weeks, so there is no specific treatment. Prevention of mumps is achieved in developed nations through vaccination with the MMR vaccine. Unfortunately, more than 40% of the world's nations have no mumps vaccine program.

Rubella

Rubella (*rube* = "reddish"; *ella* = "small") or **German measles** (*germanus* = "similar") is caused by the

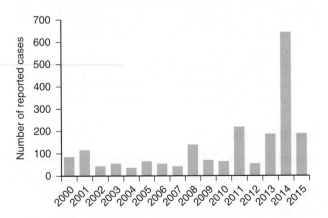

FIGURE 16.17 Reported Cases of Measles—United States, 2000–2015. Since the eradication of measles in 2000, there have been recent outbreaks linked to infected travelers coming from other countries. *»» What is meant by saying that measles has been eradicated?*

Data from CDC.

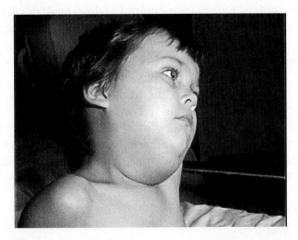

FIGURE 16.18 Mumps. Close-up of a young child with mumps (infectious parotitis). *»» What causes the swelling of the jaw region?*

Courtesy of NIP/Barbara Rice/CDC.

rubella virus, a single-stranded (+sense) RNA virus of the Togaviridae (*toga* = "cloak") family. In 1969, physicians reported 58,000 cases of rubella. Then, with the introduction of the MMR vaccine, reported cases dropped sharply. In 2014, there were only six reported cases, and rubella is no longer considered an endemic disease in the United States. In fact, the WHO has reported that rubella has been eliminated from the Americas.

Viral transmission generally occurs by airborne respiratory droplets, and infection is accompanied by fever with a pale-pink maculopapular rash beginning on the face and spreading to the body trunk and extremities. The rash develops rapidly, often within a day, and fades after another 2 days. Rubella symptoms are so mild that no treatment is required. However, rubella is dangerous to the developing fetus in a pregnant woman. Called **congenital rubella syndrome**, virus transmission from mother to fetus can lead to destruction of the fetal capillaries and blood insufficiency. The organs most often affected are the eyes, ears, and cardiovascular organs, and children can be born with cataracts, glaucoma, deafness, or heart defects. Today, there are some 120,000 children born (primarily in developing countries) each year with severe birth defects (deafness, blindness, and brain damage) due to rubella infection.

Fifth Disease

In the late 1800s, numbers were assigned to diseases accompanied by skin rashes. Disease I was measles, II was scarlet fever, III was rubella, and IV was Duke's disease (also known as roseola and now recognized as any rose-colored rash). The fifth such rash, called fifth disease, or **erythema infectiosum**, is caused by human parvovirus B19, which is a small, single-stranded DNA virus of the Parvoviridae (*parv* = "small") family.

Community outbreaks occur worldwide, primarily among children, and transmission appears to be by airborne respiratory droplets. Although most infections are asymptomatic, if symptoms occur, the outstanding characteristic is a red facial rash (**FIGURE 16.19**). The rash fades within 7 to 10 days.

Some Human Papillomavirus Infections Can Cause Cancer

Human papillomaviruses (HPVs) represent a collection of nearly 120 different types of small, icosahedral, naked, double-stranded DNA viruses of the Papovaviridae family. The viruses infect epithelial

FIGURE 16.19 Fifth Disease. The fiery red rash of a child with fifth disease (erythema infectiosum). »» *Why is fifth disease sometimes called "slapped cheek" disease?*

© Dr. P. Marazzi/Science Source.

cells, especially in the skin and anal/genital region. Some are of little consequence, whereas others are of clinical relevance.

Common Warts

Some HPV strains infect the skin and cause **common (dermal) warts**, which are small, usually benign skin growths. Specifically, HPV-1 to HPV-4 usually cause warts on the hands or fingers and deeper and more painful **plantar warts** on the soles of the feet. In most cases, these skin warts (also called papillomas) are white or pink and cause no pain. Common warts can be acquired through direct contact with HPV from another person or by direct contact with a towel or object used by someone who has the virus. Therefore, prevention requires maintaining proper cleanliness and not picking at the warts, which can lead to their spread.

Common warts can be difficult to treat and eradicate. A physician can try freezing (cryotherapy) or minor surgery or prescribe a chemical treatment.

Genital Warts

According to the CDC, there are some 1 million new cases of **genital warts** each year in the United States. About 90% of genital warts are caused by the HPV-6 and HPV-11 strains.

These HPV strains are most commonly transmitted through oral, vaginal, or anal sex with someone who has an HPV infection. Most people who acquire these types never develop warts or any other symptoms. When the warts do occur, they appear as small, flat, flesh-colored bumps or tiny, cauliflower-like

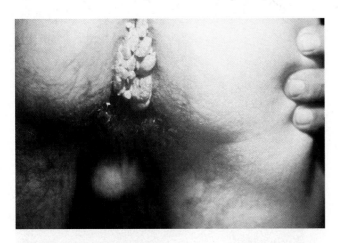

FIGURE 16.20 Genital Warts. Genital warts are caused by the human herpes simplex virus types 6 and 11. The warts typically appear as bumps on the genitalia or in this case on the anal region of this male patient. *»» What must the herpesvirus infection do to the infected skin and mucosal surfaces to produce the wart appearance?*

Courtesy of Dr. Wiesner/CDC.

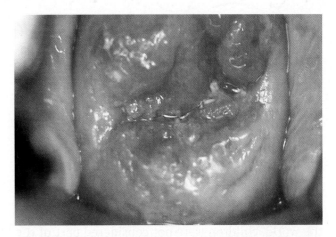

FIGURE 16.21 An Early Stage of Cervical Cancer. This photograph shows the precancerous cervix of a patient typified by erosion to her cervix. Such abnormal changes can lead to invasive cervical cancer. *»» Why is it most likely that this patient is in her 20s rather than her 50s?*

Courtesy of CDC.

bumps anywhere in the genital region or areas around the anus (**FIGURE 16.20**). Genital warts are sometimes called **condylomata** (*condylo* = "knob"), a reference to the bumpy appearance of the warts.

Although genital warts are not life threatening, there is no cure for the HPV infection, because it can remain dormant in a latent form. Visible genital warts can be removed by freezing or electrosurgical excision or by using specific chemicals available only to doctors. No one method is preferred. However, even after treatment, genital warts can come back. In fact, 25% of cases recur within 3 months.

HPV-Associated Cancers

Oncogenic (tumor-forming) HPVs are associated with most all cervical cancers and many vulvar, vaginal, penile, and oropharyngeal cancers. CDC analyses indicate that there are some 26,000 new HPV-associated cancer cases each year, with 18,000 occurring in women.

Although the HPV-6 and -11 strains are considered a "low risk" potential as cancer-causing agents, other HPV strains are strongly associated with precancerous changes and cervical cancers in young women (**FIGURE 16.21**). HPV-16 is responsible for about 50% of cervical cancers and together with strains 18, 31, and 45, account for 80% of the cancers. The cancer-causing HPV strains insert their genome into the host cell genome, deregulating cell division control and potentially resulting in tumor development. The development of HPV-associated

cancers typically unfolds over many years. These HPV viruses can result in abnormal Pap smears, so Pap tests are a critical screening procedure for all women, especially those who might be infected with HPV. In 2011, the cobas® HPV test was approved, which can identify HPV-16 and HPV-18, along with other high-risk strains, from a cervical cell sample from a female patient. This test helps physicians assess the need for additional diagnostic testing for cervical cancer.

A vaccine, called Gardasil®, is available for use in females and males 9 to 26 years of age. The vaccine is aimed at the most prominent types of HPV causing genital warts and cancer. In fact, the U.S. Food and Drug Administration (FDA) reports the vaccine to be essentially 100% effective against HPV strains 16 and 18, which cause approximately 70% of cervical cancers, and against HPV strains 6 and 11, which cause approximately 90% of genital warts. Since the introduction of the vaccine, health experts have seen a 64% drop in the vaccine-targeted HPV strains among teenage girls. Importantly, the vaccine is not a treatment because it does not eliminate the virus in individuals who already are infected.

Poxvirus Infections Have Had Great Medical Impacts on Human Populations

Though most skin diseases caused by viruses tend not to be life threatening, a few, such as smallpox, have exacted heavy tolls of human misery.

Smallpox

Smallpox is caused by a brick-shaped double-stranded DNA virus of the Poxviridae family (**FIGURE 16.22**). A unique characteristic of the smallpox (variola) virus is that the nucleocapsid is surrounded by a series of fiber-like rods with an envelope.

Smallpox is a contagious and sometimes fatal disease that, until recently, had ravaged people around the world since prebiblical times. Just in the 20th century, the death toll from smallpox exceeded 300 million. Few people escaped the pitted scars accompanying the disease, and, in some parts of the world, children were not considered part of the family until they had survived smallpox. Thanks to Edward Jenner, the first attempts at a vaccine were begun in the late 1700s. However, it would not be until 1979, after a coordinated global vaccination campaign in the 1960s and 1970s, that the WHO would report that smallpox and the virus had been eradicated as a naturally occurring virus.

Transmission of the smallpox virus is by airborne respiratory droplets or contact with the viruses from skin lesions. Following an incubation period of 7 to 17 days, the initial symptoms are high fever, headache, vomiting, and general body weakness. Pink-red spots, called **macules**, soon follow, first on the face and then on the body trunk. The spots become pink pimples, called **papules**, that develop into fluid-filled vesicles so large and obvious that the disease is also called variola (*varus* = "vessel"; **FIGURE 16.23**). The vesicles become deep **pustules**, which break open and emit pus. If the person survives, the pustules crust and scab over and leave pitted scars, or **pox**.

Smallpox vaccination has been hailed as one of the greatest medical and social advances because it was the first attempt to control an infectious disease on an international scale. It also was the first effort to protect the community rather than the individual. Today, there are two known stocks of smallpox virus, one at the CDC in Atlanta and the other at a similar facility in Russia. The destruction of these remaining smallpox stocks has been planned by the WHO. However, destruction has been postponed several times because of the controversy over the value of keeping smallpox stocks. MICROFOCUS 16.4 presents the pros and cons for smallpox stock retention/destruction.

Vaccinations against smallpox were discontinued in the United States in 1972 and elsewhere soon after. As a result, the majority of people around the world lack immunity to the disease. This makes smallpox one of the most dangerous weapons for

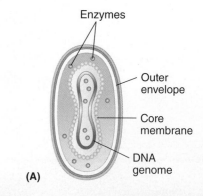

(A)

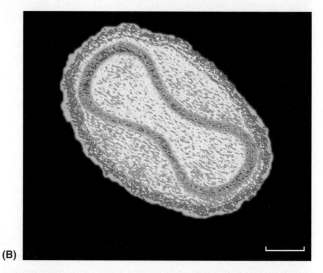

(B)

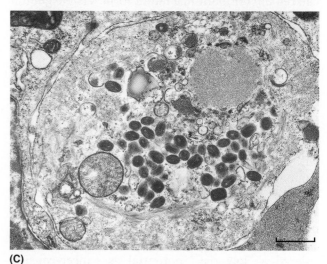

(C)

FIGURE 16.22 The Smallpox Virus. (A) A drawing of the smallpox virus, showing its structure. **(B)** A false-color transmission electron micrograph of the smallpox virus. (Bar = 30 nm.) **(C)** A false-color transmission electron micrograph of a cell infected with smallpox viruses (red). (Bar = 500 nm.) »» *Although variola virus has a DNA genome, it replicates totally in the cell cytoplasm. What enzyme must it carry to ensure proper DNA replication in the cytoplasm?*

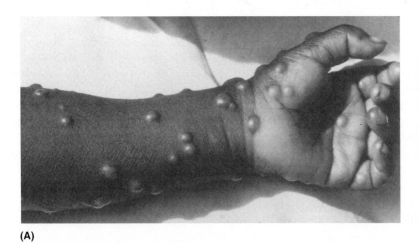
(A)

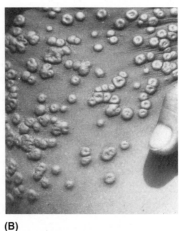
(B)

FIGURE 16.23 The Lesions of Smallpox. (A) The smallpox lesions are raised, fluid-filled vesicles similar to those in chicken-pox. Later, the lesions will become pustules **(B)** and then form pitted scars, the pocks. *»» What is the difference between a vesicle and a pustule?*

(A) Courtesy of World Health Organization Diagnosis of Smallpox Slide Series/CDC. (B) Courtesy of James Hicks/CDC.

MICROFOCUS 16.4: Public Health

"Should We or Shouldn't We?"

One of the liveliest global debates in microbiology is whether the last remaining stocks of smallpox viruses in Russia and the United States should be destroyed. Here are some of the arguments.

For Destruction

- People are no longer vaccinated, so if the virus should escape the laboratory, a deadly epidemic could ensue.
- The DNA of the virus has been sequenced, and many cloned fragments are available for performing research experiments; therefore, the whole virus is no longer necessary.
- The elimination of the remaining stocks of laboratory virus will eradicate the disease and complete the project.
- No epidemic resulting from the theft or accidental release of the virus can occur if the remaining stocks are destroyed.
- If the United States and Russia destroy their smallpox stocks, it will send a message that biological warfare will not be tolerated.

Against Destruction

- Future studies of the virus are impossible without the whole virus. Indeed, certain sequences of the viral genome defy deciphering by current laboratory means.
- Insights into how the virus causes disease and affects the human immune system cannot be studied without having the genome and whole virus. The virus research might identify better therapeutic options that can be applied to other infectious diseases.
- Mutated viruses could cause smallpox-like diseases, so continued research on smallpox is necessary in order to be prepared.
- No one actually knows where all the smallpox stocks are located. Smallpox virus stocks might be secretly retained in other labs around the world for bioterrorism purposes, so destroying the stocks could create a vulnerability in protecting the public. Smallpox viruses also can remain active in buried corpses.
- Destroying the virus impairs the scientists' right to perform research, and the motivation for destruction is political, not scientific.
- Today, it is possible to create the smallpox virus from scratch. So why bother to destroy it?
- Because the smallpox virus (vaccinia) might have evolved from camelpox, who is to say that such evolution could not happen again from camelpox?

Discussion Point

Now it's your turn. Can you add any insights to either list? Which argument do you prefer? Note: In 2011, the World Health Assembly of the World Health Organization (WHO) met to again consider the evidence for retention or destruction of the smallpox stocks. WHO decided to postpone a decision. In 2017, the debate continues.

bioterrorism, even though those who were vaccinated prior to 1972 still have some level of immunity. In addition, the United States has stated that it now has adequate stockpiles of smallpox vaccine to vaccinate every American, if necessary.

Molluscum contagiosum is another viral disease that forms mildly contagious, wart-like skin lesions. The virus of molluscum contagiosum is another member of the Poxviridae, and the virus is transmitted is by sexual contact.

The lesions are firm, waxy, and elevated with a depressed center. When pressed, they yield a milky, curd-like substance. Although usually flesh toned, the lesions can appear white or pink. Possible areas involved include the facial skin and eyelids in children and the external genitals in adults. The lesions can be removed by excising them (cutting them out) and pose no public health threat.

TABLE 16.5 presents a summary of these other dermotropic viral diseases.

TABLE 16.5 A Summary of Other Viral Diseases of the Skin

Disease	Causative Agent	Signs and Symptoms	Transmission	Treatment	Prevention
Measles (rubeola)	Measles virus	Cough, nasal discharge, eye redness, and high fever; Koplik spots	Droplet contact	Supportive care	MMR vaccine
Mumps	Mumps virus	Swollen and painful parotid glands Pain on chewing and swallowing	Person-to-person in infected saliva	Supportive care	MMR vaccine
Rubella (German measles)	Rubella virus	Fever with pale-pink maculopapular rash spreading to body trunk	Droplet contact	Supportive care	MMR vaccine
Fifth disease (erythema infectiosum)	Parvovirus 19	Maculopapular rash on cheeks and ears	Direct and indirect contact	Supportive hygiene	Practicing good care
Warts	Human papillomaviruses types 1–4 (HPV-1 to HPV-4)	White or pink skin growth on hands and feet	Direct and indirect contact	Freezing Minor surgery	Practicing proper care of affected areas
Genital warts	Human papillomavirus types 6 and 11 (HPV-6, HPV-11)	Small, flat, flesh-colored bumps or tiny, cauliflower-like bumps anywhere in the genital region or areas around the anus	Oral, vaginal, or anal sex with someone who has HPV	Removal by freezing, electro-surgical excision, or by using specific chemicals	Abstaining from sexual activity HPV vaccine
HPV-associated cancers	Human papillomavirus types 16 and 18 (HPV-16, HPV-18)	Vaginal bleeding and pelvic pain Precancerous growth on the cervix or penis	Oral, vaginal, or anal sex with someone who has HPV	Surgery Radiation therapy Chemotherapy	HPV vaccine Pap test
Smallpox (variola)	Variola virus	Fever, macules that became papules, then vesicles and pustules	Droplet contact Fomites	Immediate vaccination	Vaccine if available
Molluscum contagiosum	Molluscum contagiosum virus (MCV)	Flesh toned, wart-like lesions	Direct and sexual contact	Removal of papules	Avoiding touching papules Avoiding sexual contact

Concept and Reasoning Checks 16.4

a. How do measles and mumps differ in terms of signs and symptoms?
b. Why is rubella called three-day measles?
c. What is the unique feature of fifth disease?
d. Why would a woman be considered to be at risk of developing cervical cancer if she has genital warts?

Chapter Challenge D

Many of the viral infections discussed in this section (including measles, mumps, and rubella) primarily affect children. In fact, the first of the two doses of the MMR vaccine should be given to infants 12 to 15 months of age. Although adults can contract these diseases, it is rare.

QUESTION D: *Assuming no one has been immunized, what is it about adults that makes them more resistant to infection, whereas young children are most prone to contracting these diseases? Hint: Think about what vaccines (in general) are meant to do.*

You can find answers online in **Appendix F**.

In conclusion, you might wonder where all the new viral diseases come from. For AIDS, we know that it originated in chimpanzees and "jumped" to humans perhaps as early as the 1920s. This ability to jump species is not unique to HIV. Other viral diseases, including influenza and SARS that were covered in this chapter, also have emerged from an animal reservoir (e.g., influenza from wild birds and SARS from horseshoe bats). With this ability to jump species, a new viral disease could become pandemic quickly because of air travel or simply by jumping from wild animals to domestic ones and then on to humans. Therefore, today it is important to establish a global surveillance network to monitor viruses (and other microbial pathogens) in wild animals. Consequently, the Global Viral Forecasting Initiative (GVFI), as well as other virus identification projects around the world (primarily in tropical regions, which are reservoirs for many animal pathogens), has been set up to identify and track dangerous animal and human viruses, including their potential movement from animals to humans. And, importantly, this is not a one-way street. Human diseases can be transmitted back to animals. For example, measles has been transmitted to mountain gorillas, yellow fever to South American monkeys, and polio to chimpanzees—the latter two diseases being part of the viral community of pathogens we will discuss in Chapter 17.

■ SUMMARY OF KEY CONCEPTS

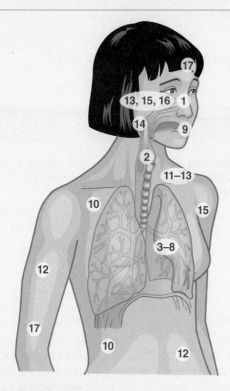

Concept 16.1 Viruses Account for Most Upper Respiratory Tract Infections

▶ **Common colds**
   1   Rhinoviruses and others
▶ **Laryngitis**
   2   Rhinoviruses

Concept 16.2 Viral Infections of the Lower Respiratory Tract Can Be Severe

▶ **Influenza**
   3   Influenza A and B viruses
▶ **Respiratory syncytial (RS) disease**
   4   Respiratory syncytial virus

▶ **Parainfluenza**
   5   Human parainfluenza viruses 1 and 3
▶ **RSV-like illness**
   6   Human metapneumovirus
▶ **SARS**
   7   SARS coronavirus
▶ **Hantavirus pulmonary syndrome (HPS)**
   8   Hantavirus

Concept 16.3 Herpesviruses Cause Several Human Skin Diseases

▶ **Cold sores**
   9   Herpes simplex 1
▶ **Genital herpes** (not shown)
      Herpes simplex 2
▶ **Chickenpox** and **shingles**
   10   Varicella zoster
▶ **Roseola** (children)
   11   Human herpesvirus 6
▶ **Kaposi sarcoma**
   12   Human herpesvirus 8

Concept 16.4 Several Other Viral Diseases Affect the Skin

▶ **Measles**
   13   Measles virus
▶ **Mumps**
   14   Paramyxoviruses
▶ **Rubella**
   15   Rubella virus
▶ **Fifth disease**
   16   Parvovirus B19
▶ **Warts** (extremities; not shown)
      Papillomaviruses
▶ **Smallpox**
   17   Variola

■ CHAPTER SELF-TEST

You can find answer for **Steps A–D** online in **Appendix D**.

STEP A: REVIEW OF FACTS AND TERMS

Multiple Choice

Read each question carefully, then select the *one* answer that best fits the question or statement.

1. There are more than _____ different rhinoviruses, which belong to the _____ family.
   A. 50; Orthomyxoviridae
   B. 100; Herpesviridae
   C. 30; Paramyxoviridae
   D. 100; Picornaviridae

2. Which one of the following statements is *not* true of the influenza viruses?
   A. They have a segmented genome.
   B. The genome is double-stranded DNA.
   C. The viruses have an envelope.
   D. There are three types of flu viruses.

3. Which one of the following is *not* a member of the Paramyxoviridae?
   A. RSV
   B. Human metapneumovirus
   C. SARS-CoV
   D. Parainfluenza virus

4. SARS is _____.
   A. a skin infection
   B. spread through close person-to-person contact
   C. a mild, respiratory infection
   D. most often seen in young children

5. Cold sores and genital herpes can be caused by _____.
   A. HSV-1
   B. HHV-6
   C. VZV
   D. HHV-8

6. A red, itchy rash that forms small, teardrop-shaped, fluid-filled vesicles is typical of _____.
   A. measles
   B. rubella
   C. chickenpox
   D. mumps

7. HHV-6 that causes roseola primarily affects _____.
   A. the elderly
   B. pregnant mothers
   C. infants
   D. teenagers

8. Kaposi sarcoma is a tumor of the _____.
   A. blood vessels
   B. liver
   C. lymph nodes
   D. kidneys

9. _____ are diagnostic for measles.
   A. Koplik spots
   B. Wart-like lesions
   C. Swollen lymph nodes
   D. Blisters on the body trunk

10. The characteristic sign of rubella is _____.
    A. orchitis
    B. pale-pink maculopapular rash
    C. fiery red rash on cheeks and ears
    D. high fever and sensitivity to light

11. Fifth disease _____.
    A. is hospital acquired
    B. causes white skin warts
    C. causes benign skin growths
    D. is transmitted by respiratory droplets

12. Some papillomaviruses are capable of causing _____.
    A. a "lacy" rash on the skin
    B. lung cancer
    C. pneumonia
    D. cervical cancer

13. Which one of the following statements applies to smallpox?
    A. The disease is associated with animal contact.
    B. The disease has been eradicated worldwide.
    C. It can be sexually transmitted.
    D. The virus can lie dormant in host cells.

Term Selection

For each choice, circle the word or term that best completes each of the following statements.

14. Rhinoviruses are a collection of (RNA, DNA) viruses having (helical, icosahedral) symmetry and the ability to infect the (air sacs, nose), causing (mild, serious) respiratory symptoms.

15. In children, the skin lesions of chickenpox occur (all at once, in crops) and resemble (teardrops, pitted scars), but in adults the lesions are known as (shingles, erythemas).

16. One of the early signs of (smallpox, measles) is a series of (Koplik spots, inclusion bodies) occurring in the (lungs, mouth) and signaling a (red, purple) rash is forthcoming.

17. Antigenic variation among (mumps, influenza) viruses seriously hampers the development of a highly effective (vaccine, treatment), and a life-threatening situation can occur if secondary infection due to (fungi, bacteria) complicates the primary infection.

18. Respiratory syncytial disease is caused by a (DNA, RNA) virus infecting the (lungs, intestines) of (adults, children) and inducing cells to (fuse together, cluster) and form giant cells called (syncytia, tumors).

19. SARS is caused by a (coronavirus, orthomyxovirus), a/an (naked, enveloped) virus spread by (sexual, person-to-person) contact.

20. Genital herpes is caused by a (helical, icosahedral) virus most often (HSV-1, HSV-2) and causes blisters with (thick, thin) walls that disappear in about 3 (days, weeks), only to reappear when (stress, physical injury) occurs.

◼ STEP B: CONCEPT REVIEW

21. Explain why a vaccine against **rhinoviruses** is not feasible. (**Key Concept 16.1**)

22. Identify the major **influenza viruses** and the structures involved in generating subtypes. (**Key Concept 16.2**)

23. Describe the infections caused by herpes simplex virus 1 and herpes simplex virus 2. (**Key Concept 16.3**)

24. Explain why the incidence of **chickenpox** has declined in the United States and how the varicella-zoster virus causes **shingles**. (**Key Concept 16.3**)

25. Distinguish between **common** and **genital warts**, including possible complications. (**Key Concept 16.4**)

◼ STEP C: APPLICATIONS AND PROBLEM SOLVING

26. The CDC reports an outbreak of measles at an international gymnastics competition. Some 700 athletes and numerous coaches and managers from 51 countries are involved. What steps would you take to avert a disastrous international epidemic?

27. A child experiences "red bumps" on her face, scalp, and back. Within 24 hours, the bumps have turned to tiny blisters that become cloudy, with some developing into sores. Finally, all become brown scabs. New "bumps" continue to appear for several days, and her fever reaches 39°C (102.2°F) by the fourth day. Then, the blisters cease and the fever drops. What disease has she had?

28. A man experiences an attack of shingles, and you warn him to stay away from children as much as possible. Why did you give him this advice?

◼ STEP D: QUESTIONS FOR THOUGHT AND DISCUSSION

29. Thomas Sydenham was an English physician who, in 1661, differentiated measles from scarlet fever, smallpox, and other fevers and set down the foundations for studying these diseases. How would you go about distinguishing the variety of look-alike skin diseases discussed in this chapter?

30. During a passenger flight to Kodiak, Alaska, a Boeing 737 developed engine trouble and was forced to land. While the airline rounded up another aircraft, the passengers sat for 4 hours in the unventilated cabin. One passenger, it seemed, was in the early stages of influenza and was coughing heavily. By the week's end, 38 of the 54 passengers on the plane had developed influenza. What lessons about infectious disease does this incident teach?

31. In the United Kingdom, the approach to rubella control is to concentrate vaccination programs on young girls just before they enter the childbearing years. In the United States, the approach is to immunize all children at the age of 15 months. Which approach do you believe is preferable? Why?

CHAPTER 17

Viral Infections of the Blood, Lymphatic, Gastrointestinal, and Nervous Systems

In early 2015, a pregnant 23-year-old woman in northeastern Brazil visited a local hospital for an ultrasound to see how her 18-week fetus was developing. The mother appeared healthy and had no symptoms of any illness, so there was no cause for concern. The ultrasound analysis proved otherwise. The examination indicated that the fetus was significantly underweight for its developmental age. A repeat ultrasound at 30 weeks revealed that the fetus now had a range of developmental defects. Therefore, two weeks later, labor was induced but it resulted in a stillbirth.

The fetus in this case exhibited **microcephaly**, a medical condition by which the fetus or infant has an unusually small head due to abnormal brain development. Laboratory tests for HIV, cytomegalovirus, rubella, and hepatitis C viruses that can affect fetal development came back negative. However,

the test for the Zika virus (ZIKV) was positive and alerted health officials that the virus had arrived in the Western hemisphere for the first time. A ZIKV outbreak was poised to trigger a multinational epidemic in the Americas.

ZIKV (see the opening image) was first identified in 1947 in monkeys inhabiting the Zika Forest of Uganda. In 1954, the first cases of human infection were reported in Nigeria. Since then, and up to 2007, small outbreaks involving ZIKV were reported in tropical Africa, Southeast Asia, and the Pacific Islands. There were no serious complications or deaths reported in these early cases. In fact, in almost 80% of cases, the infected individuals were asymptomatic, and, if symptoms did occur, they generally presented as flu-like and resolved within a week.

By October 2015, the Brazilian Ministry of Health confirmed an increase in the birth of infants with

False-color transmission electron microscope image of Zika viruses (blue).
Courtesy of CDC/ Cynthia Goldsmith.

microcephaly in northeast Brazil. By mid-November, the ministry reported a continued increase in microcephaly cases and noted a possible association of microcephaly with ZIKV infection during pregnancy. Then, on February 1, 2016, the World Health Organization (WHO) declared ZIKV a Public Health Emergency of International Concern, and local transmission now has been reported in more than 39 countries and territories of the Americas (**FIGURE 17.1A**). In April, the Centers for Disease Control and Prevention (CDC) reported a definite relationship between the ZIKV and brain abnormalities in newborns. In adults, the virus can cause Guillain-Barré syndrome, an immune system–triggered paralysis (usually temporary) of the peripheral nervous system.

ZIKV is believed to have "traveled" from the South Pacific to Brazil in late 2014. One question puzzling microbiologists and virologists is why ZIKV has suddenly emerged as a serious pathogen. Some virologists propose that internal genetic changes (mutations) might be responsible for increased infection and disease-causing abilities. Other virologists suggest ZIKV has always been capable of causing serious effects. However, having been relegated to relatively small populations, there would be few infections and

therefore few opportunities to cause serious illness. After its arrival in the Americas, with access to dense human populations, more infections can occur and many more cases that are serious are being reported. Right now, evidence is favoring the latter idea.

On August 1, 2016, the Florida Department of Health reported the presence of ZIKV in mosquitoes in a neighborhood of Miami, Florida. On November 28, 2016 the Texas Department of State Health Services reported the state's first case of local mosquito-borne Zika virus infection in Brownsville. As of March 2017, the CDC reported more than 5,100 ZIKV disease cases in the United States (**FIGURE 17.1B**). The CDC also reported that there were more than 1,100 completed pregnancies with laboratory evidence of possible ZIKV infection. Of those, 47 infants were born with birth defects, and 5 pregnancy losses were associated with birth defects.

In this chapter, we will examine several virus diseases that can be separated into three general categories. Some illnesses, such as hepatitis B, yellow fever, and dengue fever, are diseases of the blood, whereas others, including the noroviruses, affect the digestive system. Still others, such as ZIKV infection, polio, and West Nile disease, affect the nervous system.

Chapter Challenge

There is a large number of viruses that can infect various tissues and organs of the blood, gastrointestinal, and nervous systems. The diseases they cause can be very distinct. As you read this chapter, can you identify the disease and answer a few other questions based on the case histories provided in the chapter challenges?

■ KEY CONCEPT 17.1   Viral Infections Can Affect the Blood and the Lymphatic Systems

Several viral diseases are transmitted through the blood and the lymphatic systems. To reach these areas, the viruses generally are introduced into the body tissues by mechanical means or sexual contact.

Two Herpesviruses Cause Blood Diseases

Although most members of the human Herpesviridae primarily affect the skin (discussed in Chapter 16), two are associated with blood infections and disease.

Epstein-Barr Virus Disease

The Epstein-Barr virus (EBV) (also called human herpesvirus 4; HHV-4) is a relatively large,

enveloped, double-stranded DNA virus that has infected an estimated 95% of the adult population worldwide.

An EBV primary infection during childhood might trigger no symptoms, or the symptoms might be indistinguishable from other typical childhood illnesses. If the primary infection occurs during adolescence or young adulthood, the individual runs a 35% to 50% chance of developing **infectious mononucleosis** (or "mono"). This is commonly called the "kissing disease" because EBV is spread person to person orally via saliva or by saliva-contaminated objects (e.g., table utensils and drinking glasses). Infectious mononucleosis is a blood

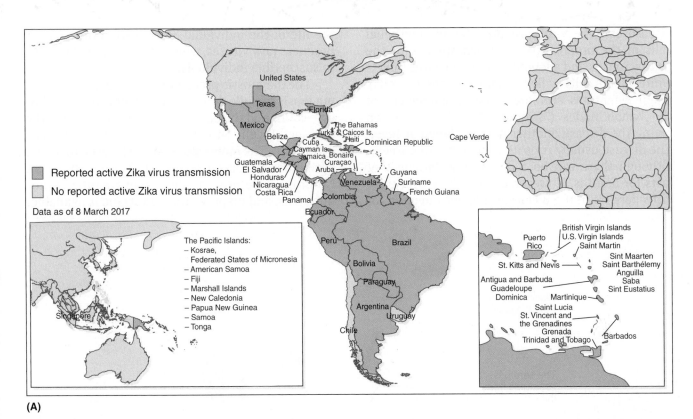

(A)

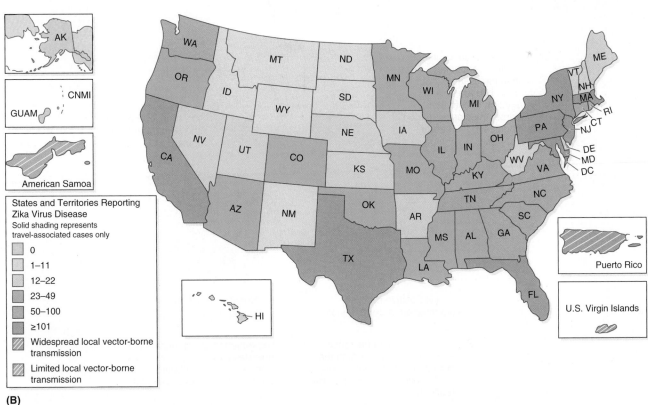

(B)

FIGURE 17.1 Areas with Active Zika Virus Transmission (as of March 8, 2017). (A) Countries and territories in the Western Hemisphere and Pacific Islands reporting Zika virus transmission. **(B)** Zika virus cases reported in the United States and its territories. *»» How might the Zika virus have crossed the Pacific Ocean from the South Pacific to South America?*

disease and the virus primarily targets epithelial cells of the oropharynx and then spreads to the nearby tonsils and adenoids, where the virus infects white blood cells called B lymphocytes (B cells). Major symptoms are sore throat, enlargement of the lymph nodes ("swollen glands"), and fever, giving the disease another name—**glandular fever**. The disease usually runs its course in 3 to 4 weeks, although there can be a debilitating fatigue that lasts for months. However, even with recovery, the virus persists in the B cells as a latent infection for the life of the individual. Reactivation of the virus produces new viruses, which can be spread by saliva to susceptible individuals.

The diagnostic procedures for infectious mononucleosis include detection of an elevated lymphocyte count and the observation of **Downey cells**, the damaged B cells with a vacuolated and granulated cytoplasm. The **Monospot test** can be used to detect antibodies produced against EBV (**FIGURE 17.2**). No drugs have proven effective in treating mononucleosis, and no preventative vaccine is available.

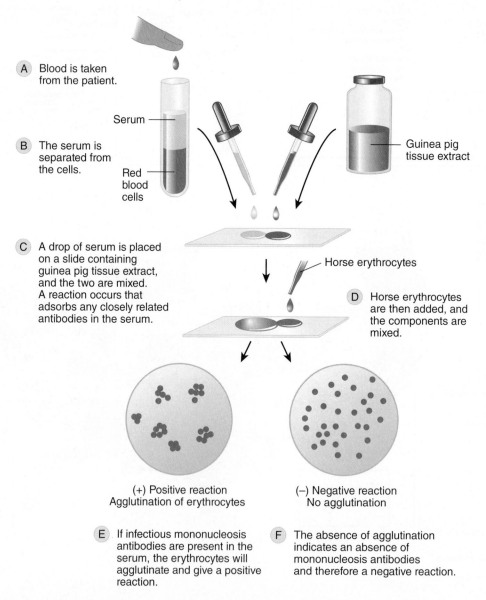

A Blood is taken from the patient.

Serum

B The serum is separated from the cells.

Red blood cells

Guinea pig tissue extract

C A drop of serum is placed on a slide containing guinea pig tissue extract, and the two are mixed. A reaction occurs that adsorbs any closely related antibodies in the serum.

Horse erythrocytes

D Horse erythrocytes are then added, and the components are mixed.

(+) Positive reaction
Agglutination of erythrocytes

(−) Negative reaction
No agglutination

E If infectious mononucleosis antibodies are present in the serum, the erythrocytes will agglutinate and give a positive reaction.

F The absence of agglutination indicates an absence of mononucleosis antibodies and therefore a negative reaction.

FIGURE 17.2 The Monospot Slide Test for Infectious Mononucleosis. About 1 week after the onset of infection by EBV, many patients develop heterophile antibodies. The antibodies peak at weeks two to five and can persist for several months to 1 year. The Monospot test is performed by first mixing samples of the patient's serum with a guinea pig tissue extract, which binds to and removes any closely related antibodies. Horse erythrocytes then are added, and the red blood cells are observed for agglutination. Suitable controls (not shown) must also be included. »» *What types of controls should be included?*

As mentioned, EBV causes a latent infection in the body. Reactivation in immunocompromised individuals can lead to the development of a tumor or cancer. In China and Southeast Asia, EBV causes a cancer of the nasopharynx called **nasopharyngeal carcinoma**. EBV also has been found in 10% of gastric (stomach) carcinoma tumors and 40% to 50% of **Hodgkin lymphoma**, a cancer of the B cells in the lymph nodes and spleen. In addition, evidence is mounting for an association between EBV and an increased risk of developing **multiple sclerosis** (MS), a chronic, muscle-weakening disease of the central nervous system. In all, excluding MS, there are more than 200,000 reported cases of EBV-associated tumors globally each year. Even though some 95% of the world's population has a lifelong EBV infection, why are EBV-associated tumors reported in only a very small fraction? Here is one explanation suggesting cofactors and other genetic changes are involved.

In areas of equatorial West Africa, children are especially susceptible to an EBV infection that causes **Burkitt lymphoma**, a tumor of the connective tissues and B cells of the jaw. Virologists and cancer specialists theorize the malaria parasite, common in central Africa, acts as a cofactor by acting as an irritant of the lymph gland tissue, thereby stimulating activation of the virus genome while inhibiting immune responses to EBV. In addition, in some 90% of Burkitt lymphoma cases the affected cells contain a chromosomal abnormality that activates an **oncogene**, which triggers uncontrolled mitosis and the tumorous condition.

Human Cytomegalovirus Disease

The human cytomegalovirus (HCMV) (human herpesvirus 5; HHV-5) is the largest member of the Herpesviridae. The virus takes its name from the enlarged cells (*cyto* = "cell"; *megalo* = "large") formed as a result of a CMV infection. Usually, these are cells of the salivary glands, epithelium, or liver. Transmission is through body fluid contact (e.g., blood, urine, saliva, feces, breast milk) from an infected person. After the primary infection, CMV undergoes lifelong latency.

Infection can be asymptomatic. In the case of **human cytomegalovirus disease**, a mononucleosis-like syndrome develops, involving fever and malaise. Most patients recover uneventfully.

However, in HCMV-infected women who are pregnant, the virus can pass into the fetal bloodstream and damage fetal tissues. This serious congenital disease can arise in 5% to 10% of infants. Of the 20,000 to 40,000 infants born each year with CMV infections, up to 8,000 develop permanent brain damage in the form of hearing loss, microcephaly, or cognitive deficits.

Antiviral drugs, including acyclovir and valacyclovir, provide effective treatment for transplant patients and as a preemptive therapy to HCMV reactivation.

Several Hepatitis Viruses Are Bloodborne

In the United States, between 3.5 and 5.3 million people are affected by untreated chronic viral hepatitis B and hepatitis C. Globally, deaths from hepatitis B and hepatitis C in 2014 surpassed those deaths due to HIV/AIDS to become the seventh leading cause of death worldwide. Therefore, the World Health Assembly of the WHO passed a resolution stating that urgent action must be taken to address this global health crisis.

Hepatitis is an acute viral disease affecting more than 700,000 Americans each year. The viruses causing hepatitis B and C are responsible for most liver infections, which are the reason for most liver transplants. Both forms of hepatitis go through the typical five stages characteristic of an virus infection (**FIGURE 17.3**). Hepatitis A affects the gastrointestinal tract, and we will discuss the disease later in this chapter.

Hepatitis B

Formerly called "serum hepatitis," **hepatitis B** is caused by the hepatitis B virus (HBV) that is in the family Hepadnaviridae (*hepa* = "liver"). The nucleocapsid is surrounded by an envelope containing a protein called the **hepatitis B surface antigen (HBsAg)**. HBV has an unusual genome consisting of partially double-stranded DNA and a short piece of single-stranded DNA. The genome is capable of coding for only four viral proteins.

Hepatitis B is a global health problem, accounting for 780,000 deaths every year. Two billion people, representing almost one-third of the world's population, have been exposed to HBV, and some 240 million have chronic HBV infections (HBV positive for more than six months). About 20% of these individuals are at risk of dying from HBV-related liver disease.

Transmission of HBV usually involves direct or indirect contact with infected body fluids such as blood. For example, transmission can occur by contact with blood-contaminated needles, such as hypodermic syringes or those used for tattooing, acupuncture, or ear piercing (**FIGURE 17.4**).

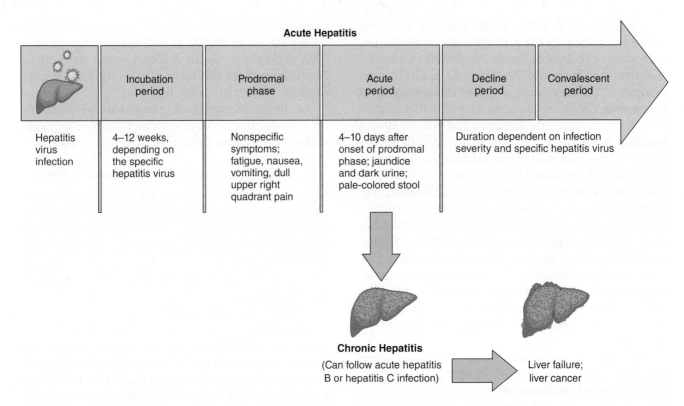

FIGURE 17.3 Hepatitis. Most infections by hepatitis viruses cause an acute disease. However, hepatitis B virus, and especially the hepatitis C virus, can cause a chronic illness sometimes involving liver failure and liver cancer. *»» In what organ would one most likely find these viruses replicating?*

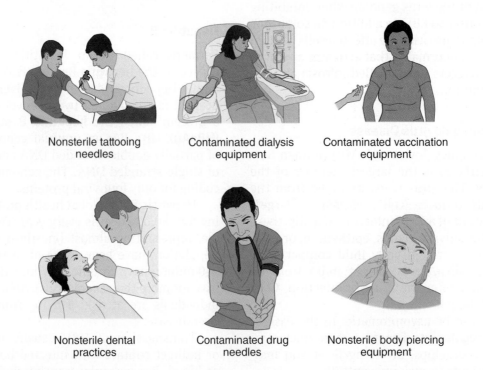

FIGURE 17.4 Transmission of Hepatitis B. There are several different ways that the hepatitis B virus can be transmitted. *»» What is the common denominator in all these methods of transmission?*

HBV-contaminated objects such as endoscopes, dental instruments, and renal dialysis tubing also are implicated as fomites involved with infection. HBV also can be transmitted from an infected partner through vaginal, anal, or oral sex.

Hepatitis B has a long incubation period of up to 12 weeks during which time the virus infects the liver but is not cytolytic. Thus, the primary (acute) infection might be asymptomatic. Symptoms are more likely in adults than in children and include fever, fatigue, loss of appetite, nausea and vomiting, and dark urine. Patients experience jaundice (yellowing of eyes or skin) weeks later. Abdominal pain and tenderness are felt in the upper-right quadrant of the abdomen (the liver is located on the upper-right side of the abdomen). Ninety percent of infected individuals clear the infection within six months, and these individuals develop immunity to reinfection.

About 10% of patients develop chronic infections (HBsAg present in blood samples for more than 6 months) with or without symptoms. In rare cases, **cirrhosis**, an extensive scarring of the liver, can occur due to immune system damage due to the infection. In addition, chronically infected carriers run a 100-times higher chance of developing liver cancer called **hepatocellular carcinoma (HCC)** than do noncarriers.

Hepatitis B can be prevented by immunization with a genetically engineered (recombinant) hepatitis B vaccine that is composed of HBsAg. Because of child and adolescent vaccinations, there has been a steady decline in reported cases of hepatitis B in the United States (slightly more than 2,800 acute and more than 12,000 chronic cases in 2014). For chronic hepatitis B, treatment involves injections of interferon alfa (Intron A®) along with multiple other antiviral agents.

HBV is also associated with another form of hepatitis, called **hepatitis D**. In this disease, HBV and another hepatitis virus, the hepatitis D virus (HDV), are involved. HDV consists of a protein fragment called the delta antigen and a segment of positive-sense, single-stranded RNA. However, HDV can cause liver damage only when HBV is present because HDV requires the HBsAg coat derived from HBV to infect cells. Therefore, one cannot become infected with hepatitis D unless the individual already is infected with HBV. Chronic liver disease with cirrhosis is 2 to 6 times more likely in a co-infection.

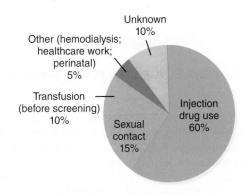

FIGURE 17.5 Hepatitis C Infections—United States, by Source. There has not been a reported case of transfusion-related hepatitis C in the United States since 1993. *»» In developed nations, what would you suspect are the two most likely sources of HCV infection? In developing nations?*

Data from CDC.

Hepatitis C

Another form of hepatitis, **hepatitis C**, is caused by the hepatitis C virus (HCV), an enveloped, positive-sense, single-stranded RNA virus of the Flaviviridae family. This chronic, bloodborne liver disease affects 150 million people worldwide, the highest proportion of cases being in Asia and Africa. In the United States, more than 3 million people are infected with HCV, most in baby boomers who received an HCV-contaminated blood transfusion before the mid-1970s. Thus, in 2014, there were almost 20,000 HCV-related deaths—an all-time high. However, new infections now are declining. The CDC reported slightly more than 2,200 new cases in 2014.

HCV is primarily transmitted via the blood, through injection drug use, or by blood transfusions (in countries without a blood-screening program; **FIGURE 17.5**). However, HCV also has been spread inadvertently through medical intervention, as MICROFOCUS 17.1 reports.

During an incubation period of up to 12 weeks, HCV is transmitted through the blood to the liver. Individuals with an **acute HCV** illness can remain asymptomatic or, in rare cases, develop fever, nausea, vomiting, and jaundice. No permanent immunity is generated. Up to 75% of patients develop a chronic HCV illness. Due to HCV's relentless destruction of liver cells, without treatment, about 25% of these

MICROFOCUS 17.1: History

What's Worse—the Disease or Medical Intervention?

Egypt has a population of 80 million and contains the highest prevalence of hepatitis C in the world. Physicians and researchers estimate a national prevalence rate of at least 12% (or 9.6 million people and more than 1.5 million in need of treatment). Chronic hepatitis C is the main cause of liver cirrhosis and liver cancer in Egypt and, indeed, one of the top five leading causes of death. The highest concentration of the HCV appears in farming people living in the Nile delta and rural areas. So, why does Egypt have such a high prevalence rate of HCV?

Schistosomiasis is a common parasitic disease in Egypt and can cause urinary or liver damage over many years. Farmers and rural populations are at greatest risk of acquiring the disease through swimming or wading in contaminated irrigation channels or standing water. Prior to 1984, the treatment for schistosomiasis was intravenous tartar emetic, which was used to induce vomiting. Between the 1950s and the 1980s, hundreds of thousands of Egyptians received this standard treatment, called parenteral antischistosomal therapy (PAT). Today, drugs for schistosomiasis are administered in pill form.

Unfortunately, inadequately sterilized needles (this was a time before disposable syringes and needles were available) used in the PAT campaign contributed to the transmission of HCV and set the stage for the world's largest transmission of bloodborne pathogens resulting from medical intervention. In addition, no one was aware of HCV prior to the 1980s or the dangers associated with blood exposure. Further evidence for the correlation between the PAT campaign and hepatitis C was the drop in the hepatitis C rate when PAT injections were replaced with oral medications.

Since late 2014, the Egyptian Ministry of Health has treated hepatitis C patients with recently approved antivirus drugs (see text) that can cure up to 90% of hepatitis C infections in 12 weeks. It is hoped that with these drugs the hepatitis C infections in Egypt will plummet quickly.

patients will develop cirrhosis and liver failure. Thus, each year almost 15,000 Americans (500,000 individuals globally) die from cirrhosis, HCC, or other complications. Indeed, cirrhosis from hepatitis C is the primary reason for liver transplants in the United States. In 2012, the CDC recommended that all baby boomers (people born between 1945 and 1965) be tested for hepatitis C because of the risk of having become infected with the virus from a blood transfusion or organ transplant prior to the start of blood supply screening for HCV in the early 1990s.

Although no vaccine is available to prevent HCV infection, a new hepatitis C blood test can screen for, detect, and confirm an HCV infection more rapidly than previous tests. In addition, new antiviral medicines taken in combination can cure (i.e., no detectable virus in the blood three months after treatment is completed) 90% of individuals with a chronic HCV infection within 12 weeks of treatment. Currently, some of these drugs are extremely expensive, and access to diagnosis and treatment is low in much of the developing world.

Hepatitis G

Another chronic liver illness is **hepatitis G**. The hepatitis G virus (HGV) is an enveloped, positive-sense, single-stranded RNA virus of the Flaviviridae. Like HBV and HCV, HGV is transmitted by blood and blood products. It appears to cause persistent liver infections in 15% to 30% of infected adults. The disease process is not completely understood.

TABLE 17.1 summarizes the viral diseases associated with the blood and the lymphatic systems.

Concept and Reasoning Checks 17.1

a. Identify the serious complications that can result from (i) EBV and (ii) CMV infection.
b. Summarize the similarities in symptoms between hepatitis B and C.

TABLE 17.1 Viral Diseases Associated with the Blood and Lymphatic Systems

Disease	Causative Agent	Signs and Symptoms	Transmission	Treatment	Prevention
Infectious mononucleosis	Epstein-Barr virus	Sore throat, fever, swollen lymph nodes in neck and armpits	Person-to-person via saliva or saliva-contaminated objects	Bed rest and adequate fluid intake	Avoiding kissing Not sharing food or personal objects
Cytomegalovirus disease	Cytomegalovirus	Asymptomatic Fever, malaise, swollen glands	Contact with the body fluids of an infected person	None	Practicing good hand hygiene Avoiding kissing Not sharing food or personal objects
Hepatitis B	Hepatitis B virus	Fatigue, loss of appetite, nausea, vomiting Jaundice and abdominal tenderness	Direct or indirect contact with body fluids	Interferon and other drugs	Receiving the hepatitis B vaccine
Hepatitis C	Hepatitis C virus	Early: no signs or symptoms May experience fatigue, loss of appetite, nausea, vomiting, low-grade fever, jaundice	Direct or indirect contact with body fluids Blood transfusions	Interferon and other drug therapies	Avoiding illegal drugs, tattooing

Chapter Challenge A

A 27-year-old male with a history of intravenous and over-the-counter drug abuse comes to the hospital emergency room complaining of nausea, vomiting, headache, and abdominal pain. He indicates that he has had these symptoms for 2 days. His vital signs are normal and he presents with slight jaundice. Physical examination indicates discomfort in the upper-right quadrant. Serologic tests for HBsAg and serum antibodies to HAV are negative.

QUESTION A:
a. *Based on these symptoms and clinical findings, what disease does the patient have, and what virus is the cause?*
b. *What is the probable source of the infection?*
c. *For what other diseases is the patient at risk?*

You can find answers online in **Appendix F**.

▇ KEY CONCEPT 17.2   Some Viral Diseases Cause Hemorrhagic Fevers

Viral hemorrhagic fever (VHF) refers to acute febrile illnesses caused by any of approximately 30 viruses. Because many VHFs are of public health concern, particularly in Africa and South America, the possibility of importation of these viruses to the United States is closely monitored. Several of the more prominent VHFs are described in the sections that follow.

Flaviviruses Can Cause a Terrifying and Severe Illness

The Flaviviridae are a family of enveloped, icosahedral, positive-sense, single-stranded RNA viruses. They are referred to as **arboviruses** because they are *ar*thropod-*bo*rne viruses, meaning the viruses are transmitted by arthropods, primarily through bites by infected ticks and mosquitoes. Two of the most dangerous bloodborne viruses in the family cause yellow fever and dengue fever. Zika virus infection will be discussed later in this chapter with viruses affecting the nervous system.

Yellow Fever

Between December 2015 and May 2016, the African country of Angola experienced the largest outbreak of **yellow fever** in 30 years. More than 2,200 cases and almost 300 deaths were reported. The fear was that exported cases to other parts of Africa and to Asia could trigger a yellow fever epidemic and a global health emergency. Fortunately, in May, the WHO reported that a yellow fever vaccination campaign in Angola, Uganda, and Congo had brought the disease under control.

The first human disease found to be associated with a virus was yellow fever. As a result of the slave trade from Africa, the disease spread rapidly in large regions of the Caribbean and tropical Americas and was common in the southern and eastern United States for many generations in the 18th and 19th centuries. In 1901, a group led by the American Walter Reed identified mosquitoes as the agents of transmission, as INVESTIGATING THE MICROBIAL WORLD 17 recounts. With widespread mosquito control, the incidence of the disease gradually declined in the United States. Still, yellow fever is endemic in 42 countries in Africa and South America and causes up to 170,000 cases and 60,000 deaths annually. In March 2017, infectious disease experts warned that a yellow fever outbreak in Brazil could bring the disease back to the United States.

Sylvatic (jungle) yellow fever occurs in monkeys and other jungle animals, where the virus is picked up by blood-feeding mosquitoes. Human epidemics occur when a person is bitten by an infected mosquito in the forest. If that person then travels to urban areas, the virus can be passed via a blood meal to another mosquito species, *Aedes aegypti*. This species then transmits the virus, now causing "urban yellow fever" among humans.

The *A. aegypti* mosquito injects the virus into the human bloodstream while taking a blood meal. After 3 to 6 days, acute phase symptoms of headache, fever, and muscle pain appear and last 3 to 5 days. Most patients recover at this stage. In 15% of patients, the illness reappears in a much more severe form, causing high fever, severe nausea, uncontrollable hiccups, and a violent, black vomit (blood hemorrhage). Liver

Investigating the Microbial World 17

How Is Yellow Fever Transmitted?

In 1900, neither the causative agent for nor the mode of transmission of yellow fever was known. In 1881, Carlos Finlay, a Cuban physician, had suggested from his investigations that mosquitoes transmitted the yellow fever agent. Unfortunately, he did not have good experimental technique, and his mosquito idea was not accepted. Rather, the medical profession believed yellow fever was transmitted by infected bedding used by yellow fever patients and thus could be transmitted through the air (like a miasma). In 1900, the American Surgeon General sent Major Walter Reed and a team of young doctors (the Reed Commission) to Cuba to investigate the disease that in the past decades had produced severe outbreaks in many port cities in the United States.

OBSERVATIONS: The Reed Commission was able to identify infected mosquitoes as the transmission agent for yellow fever through a series of experiments using American soldiers, Spanish immigrants, and even team members as volunteers. They also found that the yellow fever agent would pass through a filter, so it was probably a virus. But to really prove the case, Reed needed to show that the other modes of transmission (contaminated bedding and the air) were not valid.

QUESTION: *Is yellow fever transmitted by other methods than mosquitoes?*

HYPOTHESIS: Yellow fever is only transmitted by infected mosquitoes. If so, volunteers exposed only to yellow fever–contaminated bedding or "infected" air should not contract yellow fever.

EXPERIMENTAL DESIGN: Two small frame houses (1 and 2) were constructed. House 1 was used for testing yellow fever–contaminated bedding and house 2 for testing "infected" air and mosquitoes. House 2 was further divided by a wire screen into two rooms (A and B) through which air could pass but mosquitoes could not. Room A contained volunteers and infected mosquitoes, whereas room B had only volunteers.

EXPERIMENT 1: House 1 was filled with boxes containing sheets, pillowcases, blankets, and other materials that had been in contact with individuals who had yellow fever and contained "discharges" (black vomit, urine, fecal matter) from such cases. Three nonimmune volunteers (A–C) thoroughly handled and made contact with all the bedding materials daily for 20 days. The experiment was repeated twice more, each time with two more nonimmune volunteers (D and E, F and G) for 20 days each.

EXPERIMENT 2: House 2 contained only articles that had been steam disinfected. In room A, 15 infected mosquitoes were released and then one volunteer (H) entered the room for 15 minutes to be bitten by the mosquitoes. This was repeated with the same volunteer two more times during the day. During each visit by volunteer H, two nonimmune volunteers (J and K) were in room B where they stayed and slept for 18 nights.

RESULTS: If any volunteers exhibited typical signs of yellow fever, the board of experts would confirm or refute the presence of yellow fever in the volunteer.

EXPERIMENTS 1 AND 2: See table.

Results of Experiments 1 and 2 Concerning Non-Mosquito Modes of Transmission

Experiment (Volunteers)	Number of Volunteers	
	Diagnosed with Yellow Fever	Remaining Healthy
1 (A–C)	0	3
1 (D and E)	0	2
1 (F and G)	0	2
2 (H)	1	0
2 (J and K)	0	2

CONCLUSIONS:

QUESTION 1: *Was the hypothesis validated? Explain, using the table.*

QUESTION 2: *Is the size of the volunteer population large enough to be statistically significant? Remember, one is dealing with a disease that the volunteers know could be deadly. Would there be concerns trying to do similar experiments with human volunteers today?*

QUESTION 3: *Provide three conclusions you can make from the Reed Commission experiments on yellow fever transmission.*

You can find answers online in **Appendix E**.

Note: Although the mode of transmission of the yellow fever virus was identified, it wouldn't be until 1927 that the virus was finally isolated.

Modified from Reed, W. R., et al. 1901. *J Amer Med Assoc* 36(7):431–440.

damage produces jaundice (thus the name "yellow fever"), and major hemorrhaging occurs from the mouth, eyes, and nose of patients. Up to 50% of these patients go into a coma and die from internal bleeding. The remaining patients amazingly recover without signs of major organ damage.

Being a zoonosis, yellow fever cannot be eradicated and, except for supportive therapy, no treatment exists. The disease can be prevented by immunization with a live but weakened vaccine, as was done in the 2016 Angola outbreak.

Dengue Fever

Another VHF is **dengue fever**, which is thought to take its name from the Swahili phrase *ka dinga pepo*, meaning "cramp-like attack," a reference to the symptoms. Dengue fever is the most common arboviral disease of humans, and the WHO reports that dengue fever is the most important mosquito-borne viral disease in the world. It also is the most rapidly spreading, with a 30-fold increase in global incidence over the past 50 years (**FIGURE 17.6**). Globally, there are more than 100 million new infections and more than 25,000 deaths annually. In 2015–2016, Hawaii's big island experienced the largest outbreak of dengue since the 1940s. Most likely, the virus was imported by infected travelers.

In the Americas, outbreaks have occurred in Puerto Rico and Venezuela (2007) and Brazil (2008). Along the U.S.–Mexico border, dengue outbreaks have occurred in Brownsville, Texas (2005), in several Texas counties (2013), and Yuma County, Arizona (2014). In 2009, the CDC reported the first three locally acquired cases of dengue in Key West, Florida, and a further study found that nearly 5% of the Key West population tested positive for dengue exposure.

The four dengue virus (DENV) serotypes (DENV-1–4) are transmitted by mosquitoes. While taking a blood meal, infected mosquitoes inject the viruses into the blood. Most infections are asymptomatic or involve a mild fever with or without a rash. More severe cases of dengue fever present as a sudden high fever, with headache, joint and muscle pain, and a maculopapular rash. The joint and muscle pain gives patients the sensation that their bones are breaking; thus, the disease has been called **breakbone fever**. After about a week, the symptoms fade.

Some 500,000 people globally develop **dengue hemorrhagic fever** (**DHF**) that results from the individual being infected by a second DENV serotype. This more severe form of the disease presents as a rash from skin hemorrhages on the face and extremities and severe vomiting and shock (**dengue shock syndrome**) as blood pressure decreases dramatically. DHF patients usually recover if they are treated early and are provided with fluid replacements. Still, almost 3% of DHF patients die.

With so many dengue infections globally, disease prevention is of utmost concern. Mosquito control, using insecticides and draining standing water,

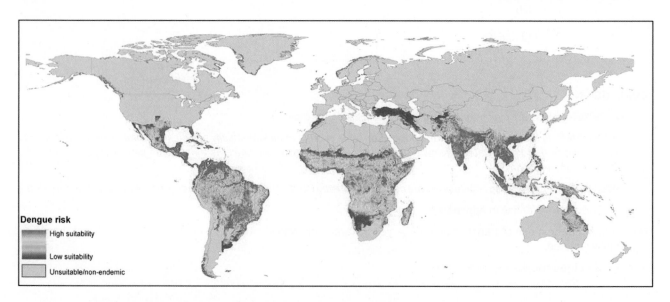

Dengue risk
High suitability
Low suitability
Unsuitable/non-endemic

FIGURE 17.6 Global Risk of Developing Dengue Fever. Much of the tropics and subtropics are at risk for dengue fever development and spread. *»» Why is dengue fever resident to the tropics and subtropics?*

Data from WHO. (2012). Global strategy for dengue prevention and control: 2012-2020. Geneva, Switzerland: Author. Data from WHO, CDC, Gideon online, ProMED, DengueMap, Eurosurveillance and published literature: Simmons, C.P. et al. 2012. *New England Journal of Medicine*, 366:1423–1432.

is the main environmental method. In addition, a vaccine against all four serotypes is needed. Several vaccines are in clinical trials and, in early 2016, the first attenuated vaccine, called Dengvaxia®, was approved by the Mexican, Philippine, and Brazilian health authorities.

A newly emerging and dangerous dengue-like hemorrhagic fever that is spreading and "knocking on" the U.S. borders is described in MICROFOCUS 17.2.

Members of the Filoviridae Produce Severe Hemorrhagic Lesions

Ebola and Marburg virus diseases are caused by viruses in the Filoviridae (*filo* = "thread") family. These viruses consist of long, thread-like, enveloped, negative-sense, single-stranded RNA viruses (FIGURE 17.7).

The first reported cases of **Ebola virus disease (EVD)** captured headlines in 1976 and 1979 when outbreaks occurred in Sudan and Zaire (now the Democratic Republic of Congo; DRC). (Ebola is the river in the DRC where the illness was first recognized.) When the outbreak was over, 88% (280) of infected people had died. Through 2012, there had been 21 confirmed outbreaks of EVD in Africa. Then, in 2014, the largest outbreak and epidemic of EVD occurred in West Africa, specifically in Guinea, Sierra Leone, and Liberia. Before the outbreak was brought under control, almost 29,000 cases and more than 11,000 deaths occurred. Flare-ups still were occurring in 2016.

MICROFOCUS 17.2: Public Health

A Newly Emerging Hemorrhagic Fever

In 2006, the Centers for Disease Control and Prevention (CDC) reported 37 cases of a unique hemorrhagic fever in U.S. travelers returning from destinations in the Indian Ocean and India—that is 34 more cases than had occurred in the previous 15 years. These travelers experienced fever, headache, fatigue, nausea, vomiting, muscle pain, and a skin rash—typical symptoms of dengue fever. However, unlike dengue, these patients also had incapacitating joint pain. The symptoms typically lasted a few days to a few weeks, although the joint pain sometimes lasted for many months. In the CDC cases, all recovered.

The disease experienced by these travelers was chikungunya (CHIK) fever (chikungunya means "to walk bent over," referring to the severe joint pain). CHIK fever is caused by the chikungunya virus (CHIKV; see the accompanying figure) that is endemic to tropical East Africa and regions rimming the Indian Ocean. CHIKV is an enveloped, positive-sense, single-stranded RNA virus in the family Togaviridae. It is transmitted by mosquitoes. The 2006 outbreak on Réunion Island in the Indian Ocean affected more than 300,000 of the 780,000 inhabitants and, for the first time, CHIK fever had claimed a substantial number of lives; 240 fatalities were attributed directly or indirectly to CHIKV. It then spread to India where more than 1.5 million cases were reported. Today, more than 37 countries have reported CHIK fever cases.

CHIKV spreads through the blood to the liver, muscles, brain, lymphatic tissues, and joints. There is no specific antiviral treatment for CHIK fever, and care is based on symptoms. Prevention consists of protecting individuals from mosquito bites and controlling the vector through insecticide spraying.

What makes this emerging disease especially worrisome is the spread of CHIKV to the Americas. Almost 80% of the CHIK fever cases have occurred in the Caribbean where 350,000 have been infected and 21 have died. In July 2014, the CDC reported that the virus had arrived in the United States. As of January 2016, cases of CHIK fever have been identified in 45 countries and territories in the Americas. More than 1.7 million suspected cases have been reported to the Pan American Health Organization. It is likely that CHIKV will spread to new areas in the Americas and locally transmitted cases are expected to grow.

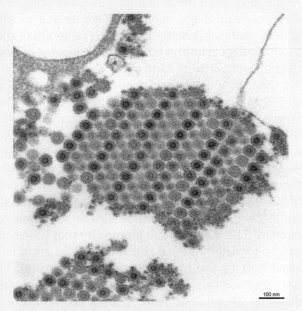

A false-color transmission electron micrograph of numerous Chikungunya viruses. (Bar = 100 nm.)

© Science Source. Colorization by Jessica Wilson.

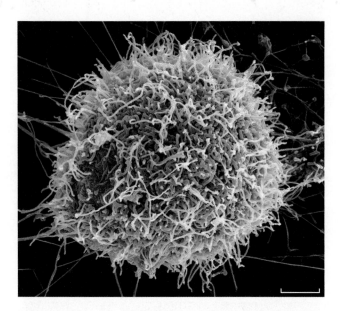

FIGURE 17.7 The Ebolavirus. False-color scanning electron micrograph of Ebola viruses (blue) being released from an infected cell. (Bar = 140 nm.) »» *Why are these viruses placed in the Filoviridae family?*

Courtesy of National Institute of Allergy and Infectious Diseases (NIAID).

EVD can be caused by infection with one of four infectious strains of ebolavirus. The 2014–2015 epidemic was caused by the *Zaire ebolavirus*. The ebolaviruses are zoonotic pathogens, and free-tailed fruit bats that are common in central and sub-Saharan Africa have been identified as a possible reservoir. Two research teams have reported that a single genetic mutation greatly increased the ability of the virus to infect cells and produce more viruses during the epidemic in West Africa.

Often, human disease spread begins with a single infection from the wildlife reservoir. The West African epidemic began when a young boy was bitten by an ebolavirus-infected bat. Uninfected individuals, including healthcare workers, also can be exposed to the virus by direct contact with body fluids/secretions (blood, vomit, urine, feces, sweat) of an infected/dead person or animal or through contact with objects, such as virus-contaminated needles. There is even the possibility that the virus can be spread by sexual transmission. In any case, after it is in the blood, the virus is captured by host immune cells. However, the virus escapes and destroys the cells as more viruses are replicated. These viruses then spread and infect more cells (see Figure 17.7).

Initial symptoms can appear anywhere from 2 to 21 days after exposure (average is 8 to 10 days) and include fever, severe headache, joint and muscle pain, sore throat, and weakness. This is followed in 2 to 3 days by diarrhea, vomiting, and stomach pain. A rash, red eyes, and hiccups might be seen in some patients and visible hemorrhaging occurs in less than half the cases. Because viral replication masks the viral RNA so that there is a slow immune response, the patient might bleed internally before a reasonable immunological defense can be mounted. Without treatment—and even in many cases with treatment—seizures are followed by loss of consciousness and death.

Of the more than 17,000 individuals who survived the West African epidemic, recovery has not been straightforward, with survivors experiencing various post-infection conditions. Four months after blood tests showed survivors were free of the virus, more than 50% still complained of neurological problems, joint pain, and muscle pain. Vision problems and sleeplessness also were common complaints. There also is evidence that in "recovered" individuals, the virus can linger in semen for months, and it can be transmitted to a sexual partner. The virus also can hide out in the eye and replicate months after a patient's recovery.

Named for Marburg, West Germany, where an outbreak occurred in 1967, **Marburg virus disease (MVD)** was first identified in tissues of green monkeys imported from Africa. The reservoir appears to be a different species of fruit bat from that for the ebolaviruses. After an incubation period of 5 to 10 days, MVD develops symptoms similar to EVD but usually has a lower fatality rate.

For both EVD and MVD, prevention measures include good infection control, safe burial practices for victims, and avoiding the hunting of bushmeat. The latter refers to animals, including monkeys and bats, which are hunted as food in some parts of Western Africa and can harbor the viruses. Processing the meat can spread the virus.

No approved vaccine or antiviral drugs are yet available for EVD or MVD. However, an experimental EVD vaccine in phase 3 clinical trials has been shown to be 100% effective in preventing the disease. For now, basic interventions for both diseases include providing intravenous fluids and electrolytes, maintaining oxygen status and blood pressure, and treating other infections, if they occur.

Members of the Arenaviridae Are Associated with Chronic Infections in Rodents

Lassa Fever

First reported in the town of Lassa, Nigeria, in 1969, **Lassa fever** is caused by a zoonotic, negative-sense, single-stranded RNA virus of the Arenaviridae family that has sandy-looking granules in the virion (*arena* = "sand"). There are more than 100,000 Lassa fever cases and about 5,000 deaths reported each year in West Africa. In 2015, a New Jersey man, who had traveled to West Africa, contracted Lassa fever. He was hospitalized when he returned to the United States but still died from the infection.

Transmission of the Lassa fever virus is shown in **FIGURE 17.8**. It involves rodents that carry and spread the virus to humans through contact with rodent urine and droppings. Infection leads to severe fever, exhaustion, and patchy blood-filled hemorrhagic lesions of the throat. The fever persists for weeks, and profuse internal hemorrhaging is common. The case fatality rate is 15% to 20%. An antiviral drug called ribavirin is effective in treating the disease. No vaccine is available.

Other Viral Hemorrhagic Fevers

There are additional VHFs found in various parts of the world. These include Crimean-Congo hemorrhagic fever, which occurs worldwide; Oropouche fever, which affects regions of Brazil; and Junin and Machupo, the hemorrhagic fevers of Argentina and Bolivia, respectively. The Sabia virus has caused hemorrhagic illnesses in Brazil, and the Guanarito virus is associated with Venezuelan hemorrhagic fever.

TABLE 17.2 summarizes the diseases caused by the hemorrhagic fever viruses.

FIGURE 17.8 Transmission of Lassa Fever Virus. Transmission of Lassa fever is through aerosol or direct contact with excretions from infected rodents, from contaminated food, or person to person. *»» Why might Lassa fever–infected rodents be a health hazard around homes or villages?*

TABLE 17.2 Viral Diseases Causing Hemorrhagic Fevers

Disease	Causative Agent	Signs and Symptoms	Transmission	Treatment	Prevention
Yellow fever	Yellow fever virus	Acute phase: headache, fever, muscle pain Toxic phase: severe nausea, black vomit, jaundice, hemorrhaging	Bite from an *Aedes aegypti* mosquito	No antiviral medications Supportive care	Vaccination Avoiding mosquito bites in endemic areas
Dengue fever	Dengue fever virus	Sudden high fever, headache, nausea, vomiting	Bite from an infected *Aedes aegypti* mosquito	No specific treatment available	Avoiding mosquito bites in endemic areas
Dengue hemorrhagic fever	A different serotype of dengue fever virus	Decrease in platelets, skin hemorrhage	Bite from an infected *Aedes* mosquito with another dengue virus	No specific treatment available	Avoiding mosquito bites in endemic areas
Ebola/Marburg viral disease	Ebola and Marburg viruses	Fever, headache, joint and muscle aches, sore throat, weakness Internal bleeding	Bite of infected fruit bat Blood transfer through cut, abrasion, or infected animal bite	No specific treatment available	Avoiding dead animals and bats in outbreak areas
Lassa fever	Lassa fever virus	Severe fever, exhaustion, hemorrhagic lesions on throat	Aerosol and direct contact with excreta from infected rodents	Ribavirin	Avoiding dead or infected rodents Maintaining good home sanitary conditions

Concept and Reasoning Checks 17.2

a. Compare the similarities and differences between yellow fever and dengue fever.
b. Why is EVD such a deadly infection?
c. How are humans most likely infected by the Lassa fever virus?

Chapter Challenge B

A 31-year-old woman comes to a neighborhood clinic with a fever, backache, headache, and bone and joint pain. She indicates these symptoms appeared about 2 days earlier. She also complains of eye pain. Questioning the woman reveals that she has just returned from a trip to Bangladesh where she was doing ecological research in the tropical forests. Skin examination shows remnants of several mosquito bites, which the woman confirms. She indicates this was her first trip to Southeast Asia. She also reports that she had been taking some antibiotics that she had been given.

QUESTION B:
a. *Based on the woman's symptoms, what viral disease does she most likely have? What clues lead you to this conclusion?*
b. *Explain why the antibiotics did not help the woman's condition.*
c. *Why might the woman want to seriously consider not returning to Southeast Asia to continue her ecological studies?*

You can find answers online in **Appendix F**.

■ KEY CONCEPT 17.3   Viral Infections of the Gastrointestinal Tract Are Major Global Health Challenges

The digestive tract is one of the major portals of entry for many infectious agents, including some viruses. As mentioned earlier in this chapter, hepatitis B and hepatitis C involve bloodborne transmission. However, hepatitis A is transmitted through contaminated food or water and causes only an acute or newly occurring infection; it does not become chronic.

Hepatitis Viruses A and E Are Transmitted by the Gastrointestinal Tract

Hepatitis A

As an acute inflammatory disease of the liver, **hepatitis A** is caused by the hepatitis A virus (HAV), which is a small, nonenveloped, positive-sense, single-stranded RNA virus belonging to the Picornaviridae family. Approximately 1.5 million cases of hepatitis A occur each year worldwide, and it remains one of the most frequently reported vaccine-preventable diseases. In 2014, there were more than 1,200 cases reported to the CDC. The largest-ever outbreak of hepatitis A in the United States occurred in 2003, as **CLINICAL CASE 17** recounts.

Transmission of hepatitis A (formerly referred to as "infectious hepatitis") often involves an infected food handler, with outbreaks also being traced to daycare centers where workers contacted contaminated feces (**fecal–oral route**). Interestingly, children often serve as the principal reservoir because their infections usually are asymptomatic. In addition, the disease can be transmitted by raw or undercooked shellfish such as clams and oysters because these animals filter feed and concentrate the viruses from virus-contaminated seawater.

The incubation period is usually about 4 weeks. Because the primary site of replication is the gastrointestinal tract, initial symptoms in older children and adults include anorexia, nausea, vomiting, and low-grade fever. The virus infection then spreads to the liver, where virus replication causes discomfort in the upper-right quadrant of the abdomen (the location of the liver). Considerable jaundice usually follows the onset of symptoms. Viruses from infected liver cells are shed through the intestines and into the feces. Convalescence can take 2 to 3 months, and relapses commonly occur in up to 20% of cases. Full recovery brings long-term immunity.

There is no treatment for hepatitis A except for prolonged rest and relieving symptoms. In those exposed to HAV, it is possible to prevent development of the disease by administering hepatitis A immune globulin within 2 weeks of infection. This preparation consists of anti-HAV antibodies obtained from blood donors. Today, in most developed countries the blood supply is routinely screened for hepatitis antibodies.

Maintaining high standards of personal hygiene (hand washing) and removing the source of contamination are essential to preventing the spread of hepatitis A. Two safe and highly effective vaccines, known commercially as Havrix® and Vaqta®, are available in the United States for people over 1 year of age. Also available for individuals over 18 years of age is Twinex®, a combination vaccine for hepatitis A and B.

Hepatitis E

An opportunistic, emerging disease caused by a nonenveloped, positive-sense, single-stranded RNA virus of the Caliciviridae family is **hepatitis E**. In developed nations, the hepatitis E virus (HEV) usually is spread through pork. Infected individuals seldom develop any symptoms, and they recover without medical intervention. This "silent epidemic" is infecting some 300,000 people in Germany alone every year.

In developing nations, HEV causes disease outbreaks through its spread in dirty water (fecal–oral route). This distinct form of hepatitis E takes the lives of almost 60,000 people every year in Africa and Asia. This form of hepatitis shares many clinical characteristics and symptoms with hepatitis A.

Several Unrelated Viruses Can Cause Viral Gastroenteritis

Viral gastroenteritis is an inflammatory condition caused by a variety of viruses that produce varying combinations of diarrhea, nausea, vomiting, low-grade fever, cramping, headache, and malaise. Illnesses are global, they occur in people of all ages, and, for many children, they can be life threatening. Viral gastroenteritis often is called the "stomach flu," although it has no association with the influenza viruses.

Clinical Case 17

Hepatitis A Outbreak

On November 5, 2003, an outbreak of hepatitis A was confirmed through lab analysis reported by the Pennsylvania Health Department. At that time, there were 34 cases confirmed at a mall restaurant (restaurant A) with 10 customers and 12 restaurant employees reporting symptoms of hepatitis A infection.

By November 7, 130 people had contracted hepatitis A (see the accompanying figure), and the health department provided injections of immunoglobulin as a precaution for anyone who had eaten at the restaurant between September 22 and November 2. Number of hepatitis A cases by date of eating at restaurant A and illness onset.

At this time, state officials and arriving CDC investigators suspected the virus was spread by an infected worker who failed to wash his or her hands before handling food.

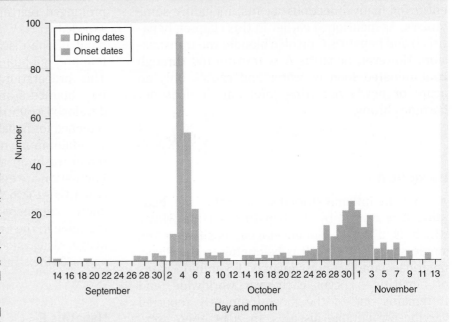

Number of hepatitis A cases by date of eating at restaurant A and illness onset.

On November 8, the first fatality from the hepatitis A outbreak was reported. The person died from liver failure. The outbreak had risen to 240 confirmed cases. Officials still believed the problem centered on an infected food worker.

Four days later, health officials announced the confirmed case count had risen to 340. Mention also was made of a recent multistate (Tennessee, Georgia, and North Carolina) outbreak of hepatitis A in late September and early October. These 250 cases resulted from eating contaminated green onions at a few local restaurants.

By November 13, 410 illnesses were reported. Transmission from an infected food worker was ruled out as all employees became ill after the outbreak began. Interviews of restaurant A patrons began in order to identify what and how much they ate at restaurant A.

By November 15, two more people had died from the hepatitis A outbreak and more than 500 illnesses had been reported.

Patron interviews and further menu item investigations pointed to Mexican salsa containing green and white onions as the prime source of illness. In fact, interview results showed that 98% of restaurant A patrons who became ill reported eating salsa containing raw green onions. Illness was not associated with eating salsa containing raw white onions.

In all, more than 550 people had been stricken during the hepatitis A outbreak. Genetic analysis of the virus implicated raw green onions imported from three firms in Mexico.

Questions:

  a. Why would the infections of restaurant employees be significant?
  b. Explain why immunoglobulin injections were recommended.
  c. How common are deaths from hepatitis A?
  d. Why was it important to discover what and how much food the affected restaurant customers ate?
  e. Provide some ways that the green onions might have initially become contaminated with the hepatitis A virus.

You can find answers online in **Appendix E**.

For additional information, see www.cdc.gov/mmwr/preview/mmwrhtml/mm5247a5.htm.

Rotavirus Gastroenteritis

Prior to 2009, the world's severest and deadliest form of gastroenteritis in children was **rotavirus gastroenteritis**. The diarrhea-related illness was associated with 125 million cases and more than 500,000 deaths worldwide among children younger than 2 years of age. In the United States, there are an estimated 75,000 hospitalizations and 20 to 40 childhood deaths each year. However, since 2008, two oral vaccines, RotaTeq® and Rotarix®, have been available in the United States. With the use of these vaccines, hospitalizations for severe diarrhea have been reduced by almost 95%.

The rotavirus (*rota* = "wheel") is a highly infective, nonenveloped, circular-shaped virus whose genome contains 11 segments of double-stranded RNA (**FIGURE 17.9**). It is a member of the Reoviridae family. Rotavirus infections tend to occur in the cooler months (October–April) in the United States, and the infections are often referred to as "winter diarrhea."

Transmission occurs by the fecal–oral route. After an incubation period of two days, the virus infects enterocytes in the small intestine and disrupts normal absorption. Thus, the patient can exhibit diarrhea, vomiting, and chills. The disease normally runs its course in 3 to 8 days. Recovery from the infection does not guarantee immunity;

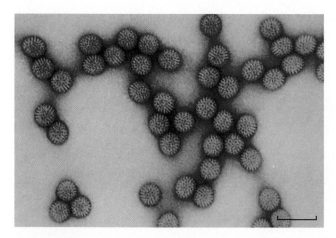

FIGURE 17.9 Rotaviruses. False-color transmission electron micrograph of rotaviruses. (Bar = 140 nm.) »» *From this micrograph, why are these viruses called rotaviruses?*

© Science Source.

indeed, many children have multiple rounds of reinfection. Severe water and electrolyte loss due to diarrhea can be life threatening.

Norovirus Gastroenteritis

Health experts believe that more than 20% of acute gastroenteritis cases ("food poisoning") globally and more than 50% of gastroenteritis cases in hospitals are caused by noroviruses, making **norovirus gastroenteritis** the most common cause of nonbacterial gastroenteritis.

Noroviruses (formerly called the Norwalk-like viruses) belong to the family Caliciviridae and are nonenveloped, icosahedral viruses with a positive-sense, single-stranded RNA genome. Currently, there are four known norovirus groups, although new strains seem to emerge every 2 to 3 years. All of them are highly contagious, and as few as 20 virus particles can lead to illness in healthy individuals. Humans are the only known reservoir. Outbreaks typically occur where people are in close contact, such as schools, hospitals, prisons, and dormitories. Norovirus outbreaks on cruise ships often make the news media reports.

Noroviruses are transmitted primarily through the fecal–oral route, either by consumption of contaminated food or water or by direct person-to-person spread. Contamination of surfaces also can be a source of infection because the viruses are extremely stable in the environment. The CDC estimates that almost 60% of all foodborne outbreaks are caused by norovirus infections. In the United States, this amounts to 21 million cases, 71,000 hospitalizations, and some 800 deaths each year.

The incubation period for norovirus gastroenteritis is 15 to 48 hours. After crossing the epithelium of the small intestine, typical gastrointestinal symptoms of fever, diarrhea, abdominal pain, and vomiting last about 24 to 48 hours. Although recovery is complete, dehydration is the most common complication.

As with rotavirus gastroenteritis, the only treatment for norovirus gastroenteritis is fluid and electrolyte replacement. Washing hands and having safe food and water are important prevention measures.

TABLE 17.3 summarizes the characteristics of the virally caused gastrointestinal illnesses.

TABLE 17.3 Viral Infections of the Gastrointestinal Tract

Disease	Causative Agent	Signs and Symptoms	Transmission	Treatment	Prevention
Hepatitis A	Hepatitis A virus	Nausea, vomiting, low-grade fever	Indirect through food and raw shellfish	No specific treatment	Receiving the hepatitis A vaccine Practicing good hygiene
Hepatitis E	Hepatitis E virus	Nausea, muscle pain, low-grade fever	Indirectly through water Zoonosis	No effective treatment	Avoiding untreated water
Viral gastroenteritis	Rotavirus	Diarrhea, nausea, vomiting, low-grade fever, stomach cramping, headache, malaise	Indirect through food or water Contaminated surfaces	Self-care	Practicing good hand hygiene Avoiding shared items
	Norovirus	Nausea, vomiting, diarrhea, stomach cramping	Indirect through food or water Person to person	Self-care Oral rehydration fluids	Practicing good hand hygiene Washing fruits and vegetables

Concept and Reasoning Checks 17.3

a. Compare and contrast hepatitis A and E.

b. Explain why oral rehydration salt solutions are the treatment for most forms of viral gastroenteritis.

Chapter Challenge C

A 59-year-old man visits the local hospital emergency room complaining of nausea, abdominal discomfort with vomiting, and fever. On physical examination, he has a temperature of 38°C and appears jaundiced. He indicates the symptoms appeared abruptly 2 weeks earlier after he had returned from a vacation tour to Egypt. The patient reveals that while he was there, he lived like the locals, ate in local restaurants, and often drank the local water. He did not have any vaccinations before leaving on his trip.

QUESTION C:

a. *Name two viral diseases discussed in this chapter that the patient might have.*

b. *What is this patient's most likely disease? Why would you make that diagnosis?*

c. *Identify a possible transmission route for this virus.*

d. *What treatment could have been prescribed if the patient had sought medical intervention earlier in the illness?*

You can find answers online in **Appendix F**.

■ KEY CONCEPT 17.4   Viral Diseases of the Nervous System Can Be Deadly

Several viral diseases affect the human nervous system, which can suffer substantial damage when viruses replicate in the tissue. Rabies, polio, and West Nile virus disease have been the most recognized diseases. Now Zika virus infection can be added to the list.

Zika Virus Infection Represents an Emerging Viral Pathogen

Until recently, ZIKV was thought to cause only mild to moderate infection, and no serious complications or deaths had been reported (see chapter opener). However, reports of a dramatic increase in the birth of microcephalic infants during the recent epidemic of ZIKV in Brazil led the WHO to declare Zika-associated microcephaly a global health emergency. However, as with any emerging infection, it will be some time before questions about transmission and the full spectrum of disease are fully answered.

ZIKV is a member of the Flaviviridae family and is an enveloped, icosahedral, positive-sense, single-stranded RNA virus. ZIKV is another example of an arbovirus (**FIGURE 17.10**), and the virus is spread by the same species of mosquitoes that spread chikungunya fever and dengue fever (described earlier in this chapter). In addition, ZIKV has been identified in amniotic fluid and the placenta, and it can spread from mother to fetus during the early perinatal period. Sexual transmission also has been documented. The virus has been detected in asymptomatic blood donors, so blood transfusions (and by extension, organ transplantation) provide another mode of transmission. There have been no reports of infants infected through breastfeeding. Infected humans and nonhuman primates appear to be the main reservoirs of the virus.

If a person is bitten by a ZIKV-infected mosquito, the incubation period can range from 3 to 12 days. Health experts report that 80% of individuals infected with ZIKV will not develop any symptoms. Of the 20% that do have symptoms, a rash, joint pain, and conjunctivitis (red eyes) can appear 2 to 7 days after the bite. Recovery from the infection is thought to confer lasting immunity.

ZIKV also can be spread from an infected pregnant mother to her fetus. In 1% to 10% of these fetal infections, the most severe outcomes were the result of infection during the first or second trimester. By far, the most alarming complication is microcephaly (small head size). In Brazil, where the infection was first identified in the Western hemisphere, more than 2,000 cases of microcephaly or other fetal abnormalities of the central nervous system have been reported. As yet, it is uncertain what the outcome will be in terms of neurological abnormalities as the infants continue to grow. In fact, scientists studying ZIKV suspect that an infection of the fetus also can lead to other neurological impairments and ocular anomalies and even have developmental consequences on the digestive, urogenital, and cardiac

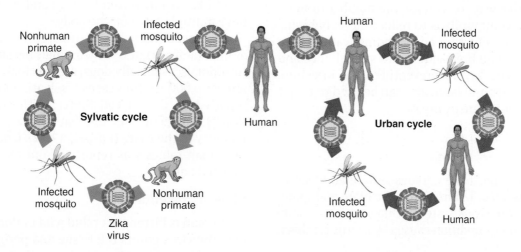

FIGURE 17.10 Zika Transmission Cycles. In Africa, the Zika virus normally cycles between nonhuman primates and mosquitoes (sylvan cycle). Should an infected mosquito bite a human, the virus can be carried into an urban setting, where the virus can now cycle between humans and mosquitoes (urban cycle). »» *Could a Zika urban cycle be the source to start a sylvan cycle? Explain.*

systems. Consequently, the fetal infection is becoming known as **congenital Zika syndrome**.

At this time, data on the pathophysiology of ZIKV are incomplete, and researchers are unsure how the virus breaches the placental barrier and leads to microcephaly. Some evidence suggests the virus targets cells that make new neurons. Likewise, it is not known how to prevent the neurological damage the virus might cause. It is thought that ZIKV first damages the placenta and then the fetus itself. One possibility is that the virus damages the blood capillaries that supply nourishment to the fetus. In addition, ZIKV might activate an immune response that leads to cell death. As the disease is further studied, more information will be gathered about the clinical aspects.

A small number of adults infected with ZIKV also might experience nervous system abnormalities. A neurological paralyzing condition, called **Guillain-Barré syndrome**, has been reported. This condition occurs if the immune system damages nerve cells, which can cause muscle weakness in the arms and legs. A temporary paralysis can affect breathing and might require the assistance of a breathing tube. The condition can last from a few weeks to several months. In the severest of cases, there is permanent damage, and in about 5% of these cases, death results.

Health authorities are still determining what prevention measures to take against ZIKV infection. Better and quicker diagnostic tests are being developed. Without a vaccine, the WHO suggests that people in regions where mosquitoes might be present (e.g., the tropics and subtropics) should use insect repellent and wear long-sleeved shirts and pants. In addition, the CDC emphasizes that the most effective way to decrease the number of mosquitoes in a community is to reduce larval habitats. That means eliminating any standing water where mosquitoes could breed. Mosquito surveillance and control are the key to preventing Zika infection. Pesticide sprays and larvicides can be used to target high–mosquito-density areas.

The Rabies Virus Is of Great Medical Importance Worldwide

Some viruses infect the nervous system and lead to **encephalitis**, an inflammation of the brain and spinal cord (central nervous system). One such potentially life-threatening condition caused by a virus is **rabies**.

Causative Agent and Epidemiology

The rabies virus is an enveloped, negative-sense, single-stranded RNA virus of the Rhabdoviridae

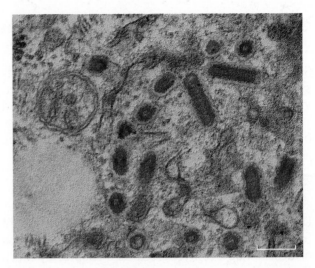

FIGURE 17.11 The Rabies Virus. An accumulation of rabies viruses (red) in the salivary gland tissue of a canine. In this false-color transmission electron micrograph, the viruses are red. (Bar = 100 nm.) »» *In this micrograph, identify the viruses that most closely have the shape characteristic of rhabdoviruses.*

© Eye of Science/Science Source.

family. It is rounded on one end, flattened on the other, and looks like a bullet (**FIGURE 17.11**). The rabies virus has only five genes in its genome.

Rabies (*rabies* = "madness") is notable for having the highest mortality rate of any human infectious disease after the symptoms have fully appeared; there are an estimated 55,000 deaths annually, mostly in rural areas of Africa and Asia. The most important global source of human rabies in developing countries is from rabid dogs. Children usually are at the greatest risk from rabies because they are more likely to be bitten by dogs and more likely to have multiple bites on the body.

In most developed countries, almost all cases of animal rabies occur in wild animals because domestic animals, especially dogs, are vaccinated. In 2014, almost 6,000 wildlife cases ("sylvatic rabies") were reported in the United States, with sheep/goats, skunks, domestic cats, and bats accounting for the majority of the cases (**FIGURE 17.12**). One fatal case of human rabies was reported from the bite of an infected bat.

Clinical Presentation

If a person is bitten by a rabid wild or domestic animal, the virus enters the tissue and peripheral nervous system through a skin wound contaminated with the saliva from the rabid animal. However, only 5% to 15% of individuals bitten by a rabid animal develop the disease. In these cases, the incubation

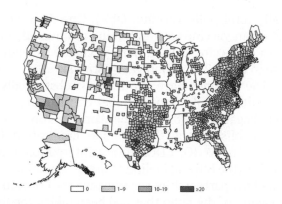

FIGURE 17.12 Reported Cases of Animal Rabies, by County—United States, 2014. This geographic map identifies the location of animal rabies reported in all states (except Hawaii). *»» What wild animal is most associated with rabies in your area?*

Reproduced from Adams, D.A., Thomas, K.R., Jajosky, R. et al. (2016). Summary of notifiable infectious diseases and conditions—United States, 2014. *MMWR Morb Mortal Wkly, 63*: 1–152. Data from the Division of High Consequence Pathogens and Pathology, National Center for Emerging and Zoonotic Infectious Diseases.

period for rabies varies according to the amount of virus entering the nerves and the wound's proximity to the central nervous system. For example, a bite on the face can begin producing symptoms in as few as six days, whereas a bite on the lower leg could take months before symptoms appear.

Rabies is a disorder in neurotransmission. Early signs of **rabies encephalitis** are abnormal sensations such as tingling, burning, or coldness at the site of the bite. Fever, nausea, and vomiting are additional symptoms. The acute (neurological) phase is characterized by clinical signs indicative of rabies. The patient has increased muscle tension and becomes hyperactive and aggressive. Soon there is paralysis, especially in the swallowing muscles of the pharynx, and saliva drips from the mouth. Brain inflammation, together with an inability to swallow, increases the violent reaction to the sight, sound, or thought of water. The disease therefore has been called **hydrophobia**—literally, the fear of water. The acute phase usually ends in 2 to 10 days. After clinical signs of rabies appear, the disease is nearly always fatal, and treatment is typically supportive. Death usually comes from respiratory paralysis. Interestingly, a few people have survived a full-blown, lethal infection through medical intervention, as MICROFOCUS 17.3 reports.

MICROFOCUS 17.3: Public Health

The "Milwaukee Protocol"

On September 12, 2004, 15-year-old Jeanna Giese of Fond du Lac, Wisconsin, was bitten on the finger as she tried to rescue a bat outside her church. Not realizing that the bat was rabid, Jeanna did not seek immediate treatment and became gravely ill a month later.

Normally, after the symptoms of rabies set in, recovery is not an option, and death occurs within a week. Jeanna was rushed to Children's Hospital in Milwaukee where she presented symptoms of fever, slurred speech, and profuse salivation. The next day, doctors administered an unproven treatment. She was placed in a chemically induced coma and given a combination of antiviral drugs (ketamine, midazolam, ribavirin, and amantadine), sedatives, and anesthetics in their effort to save her life.

Nine days later, Jeanna was brought out of the coma. She was paralyzed, unable to speak or walk, and without sensation, as she suffered from the effects of rabies on her nervous system. Physicians detected brain-wave activity but were unsure what was ahead after the drugs wore off. However, when they did wear off, Jeanna demonstrated some eye movements and reflexes—a real positive sign. Progress continued and on January 1, 2005, Jeanna was released from the hospital in a wheelchair. She had a long road to recovery, if indeed recovery would be complete. She would need to regain her faculties, including her ability to speak and walk.

What followed were two years of rehabilitation during which time she regained her ability to walk and talk, and nerves damaged by rabies recommunicated with muscles. Jeanna returned to finish high school and then went on to earn a bachelor's degree in 2011.

Jeanna's lead physician, Dr. Rodney E. Willoughby, Jr. at Children's Hospital and the Medical College of Wisconsin, has had his so-called "Milwaukee protocol" used more than 30 times with some success. Jeanna is now one of several who apparently has survived rabies through the Milwaukee protocol. The others were in Colombia, Brazil, Peru, Qatar, and California. Unfortunately, the survivors in Colombia and Peru died during rehabilitation.

Importantly, although the Milwaukee protocol might be a potential cure for human rabies, medical experts agree that the focus needs to remain on rabies prevention. To that end, in the summer of 2011, Jeanna and her family went to an island in the Philippines where rabies had recently been eradicated. There she visited schools and learned about the efforts used for eradication.

Treatment and Prevention

An estimated 10 million people worldwide receive **postexposure immunization** each year after being bitten by rabies-suspect animals. This relatively painless immunization entails four injections in the shoulder muscle on days 0, 3, 7, and 14 post exposure. These injections are preceded by thorough cleansing of the bite site and one dose of rabies immune globulin, which provides an immediate supply of antibodies to help neutralize the virus until the infected person can respond to the vaccine by producing his or her own antibodies.

The Polio Virus May Be the Next Infectious Disease Eradicated Globally

The name **polio** is a shortened form of **poliomyelitis** (*polio* = "gray"; *myelo* = for "spinal cord"), referring to the "gray matter," which is the nerve tissue of the spinal cord and brain that the virus infects.

The polioviruses, being in the Picornaviridae family, are among the smallest virions, measuring 27 nm in diameter. They are composed of a nonenveloped capsid containing a positive-sense, single-stranded RNA genome.

The polioviruses usually enter the body by the fecal–oral route through contaminated water and food. About 90% to 95% of cases are asymptomatic. The most common clinical presentation is an **abortive poliomyelitis** wherein the patient experiences fever, headache, nausea, and sore throat. There is no entry of the virus into the central nervous system, and complete recovery occurs. In perhaps 1% of cases, the poliovirus infection passes across the blood–brain barrier and localizes in the meninges. One result can be aseptic meningitis, which is a **nonparalytic poliomyelitis** characterized by fever, headache, and stiff neck. Complete recovery usually occurs.

The other outcome in the most severe infections is **paralytic poliomyelitis**, which occurs as the virus infects the brain. Paralysis of the arms, legs, and body trunk are affected. Swallowing is difficult as paralysis develops in the tongue, facial muscles, and neck. Paralysis of the diaphragm muscle causes labored breathing, which can end in death. Those who do recover often have permanent paralysis.

In the 1950s, a team led by Jonas Salk grew large quantities of the viruses and inactivated them with formaldehyde to produce the first polio vaccine (inactivated polio vaccine; IPV). Albert Sabin's group subsequently developed a vaccine containing attenuated (weakened) polioviruses. This vaccine was in widespread use by 1961 and could be taken orally (oral polio vaccine; OPV). One drawback of the Sabin vaccine was that being attenuated, a few cases of vaccine-caused polio occurred (1 reported case in every 2.4 million vaccinations).

The vaccines have contributed substantially to the reduction of polio. A global effort to eradicate polio began in 1988 when the World Health Assembly established the Global Polio Eradication Initiative. The WHO, CDC, Rotary International, and UNICEF (the Bill and Melinda Gates Foundation later joined the partnership) spearheaded the effort to rid the world of the polio that had been taking the lives of 350,000 children every year. Now, in 2017, eradication is in sight. Two of the three strains of poliovirus (types 2 and 3) have been eradicated. In 2016, only 37 wild (type 1) poliovirus cases were reported and these cases were all in Afghanistan, Pakistan, and Nigeria. The WHO believes that if all the remaining unvaccinated children in these countries can be immunized, polio eradication can finally be accomplished.

Enterovirus Infections

Another group of viruses in the Picornaviridae is the enteroviruses that can cause a form of gastroenteritis called **hand, foot, and mouth disease**. This is a moderately contagious disease typically affecting infants and young children in the spring to fall. Symptoms include fever, poor appetite, malaise, and a sore throat. Most patients recover in 7 to 10 days.

Recently, a more severe syndrome thought to be associated with other enteroviruses, specifically enterovirus 68-type D (EV-D68), has been identified that causes a polio-like syndrome. From August 2014 through December 2015, CDC and state public health laboratories confirmed more than 1,200 cases apparently caused by EV-D68. Almost all of the confirmed cases in the American Midwest and across North America were among children, many of whom had asthma. These individuals exhibited extremely weak muscles and loss of muscle tone (**flaccid paralysis**), and meningitis, encephalitis, myocarditis, or sepsis was reported in some patients. Of the almost 2,000 reported cases, EV-D68 was found in 14 patients who died in this outbreak. The precise cause of the condition and its infection are still being investigated.

Additional Viruses Also Can Cause Encephalitis

Some arboviruses can cause a rare primary encephalitis called **arboviral encephalitis**. Arboviral encephalitis is another example of a zoonosis, and again mosquitoes are the most common vector.

The primary transmission cycle involves a mosquito taking a blood meal from an infected bird (**FIGURE 17.13**). The infected mosquito can then infect other birds, or another animal such as a horse or human, when taking another blood meal. However, these latter infections represent an **incidental infection** because the illness occurs in what is called a **dead-end host**; that is, a host not producing significant number of viruses in the blood and thus not contributing to the primary transmission cycle. Because all these arboviral infections involve bites from infected mosquitoes, prevention includes using insect repellent (N, N-diethyl-meta-toluamide; DEET) and wearing protective clothing. **MICROINQUIRY 17** examines the proposal to rid the world of mosquitoes.

Arboviral encephalitis can be sporadic or epidemic, and the diseases have a global distribution (**FIGURE 17.14**). In the United States, the most common cause of arboviral disease is the West Nile virus.

West Nile Virus Disease

Along with Zika virus infection and dengue fever, one of the more prominent emerging arboviral diseases in the Western Hemisphere is **West Nile virus disease**.

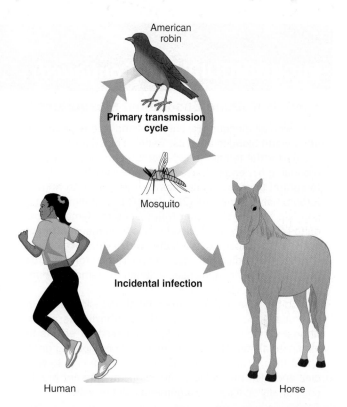

FIGURE 17.13 The Transmission of Arboviral Encephalitis. Shown here is the generalized pattern of viral encephalitis transmission among various animals, including humans. *»» Although it might be difficult to interrupt viral encephalitis transmission, why would mosquito elimination be the best bet for disease interruption?*

MICROINQUIRY 17

There Is Nothing Good About Mosquitoes

They have been with us for more than 5,000 years. They have adapted to humans and they love us, especially our blood. No, it isn't vampires of folklore (or of novels) but rather living insect bloodsuckers, the mosquitoes. These insects (and the infectious agents they can carry) cause more disease and death than any other animal on the planet. There are almost 3,000 different species of mosquitoes found around the world with 200 species found in the United States. Fortunately, only a small number transmit infectious diseases. For example, *Aedes aegypti* (see the accompanying figure) can carry and transmit Zika virus, yellow fever virus, dengue fever virus, and chikungunya fever virus. Any of 43 species can carry West Nile virus and *Anopheles gambiae* carries the protistan parasite causing malaria. The Bill and Melinda Gates Foundation estimates that mosquito-borne diseases kill 725,000 people (600,000 due to malaria) every year.

A female *Aedes aegypti* mosquito taking a blood meal from a human host. Note the distended abdomen that contains the ingested blood meal.

Courtesy of James Gathany/CDC.

(continues)

MICROINQUIRY 17 (Continued)

There Is Nothing Good About Mosquitoes

As with all mosquitoes, only the females bite people or other warm-blooded animals. Taking a blood meal is necessary for the female to get the nutrients necessary for producing her eggs. *A. aegypti* is especially troublesome to control because (1) it tends to breed close to human (urban) habitations, and (2) it doesn't bite at night, and it is a "sip feeder." This latter trait means the female mosquito takes many small blood meals from many different people. If the mosquito is infected with a virus, like Zika or dengue, sip feeding provides an opportunity for the virus to be injected into many people during the female's short, three-week life span.

With the Zika outbreak, the control and elimination of the mosquitoes that transmit these arboviruses are of utmost concern. Some people have called for efforts to eliminate all mosquitoes because there is "nothing good about them." However, with the number of mosquito species and their prevalence, total elimination is not practical. For any species, the arguments fall this way:

Recently, new ways to control mosquitoes and possibly eradicate a specific species, such as *A. aegypti*, have been proposed. It is possible using new molecular technologies to alter the genetic information and to modify genes in organisms. For *A. aegypti* (or any species), companies are generating sterile males in captivity. Releasing these males into the wild could result in mating with unsuspecting females, which would produce no eggs—and no new generation of mosquitoes. Even if this works, the number of sterile males released would have to be monumental. Consequently, it is probably not possible to even eliminate one species, although their numbers might be substantially reduced, which would substantially reduce the disease burden.

Discussion Point

Based on the given information, and any thoughts and information you might have, is it worthwhile in terms of time and money to try to eradicate a species of mosquito? If so, which species?

Eradicate Mosquitoes?	
Yes	**No**
• Will save thousands of human lives every year • Will become prevalent as human population and urbanization increase • If not eradicated, control mosquitoes (and the viruses they can carry and transmit)	• Some mosquito species pollinate flowers and are a vital food source for birds, bats, and other animals. • Attempts at eradication could result in selection of insecticide-resistant strains of mosquitoes.

It is caused by the West Nile virus (WNV), another member of the Flaviviridae (**FIGURE 17.15**). WNV is the major cause of domestically acquired arboviral disease, and it has a somewhat broad host range; it can infect humans, birds, mosquitoes, horses, and other mammals. After the first outbreak in New York City in 1999, each year WNV moved farther west across the United States. Today, WNV is endemic in the 48 contiguous states, and experts believe the virus now is endemic, as it has caused seasonal outbreaks every summer and fall (**FIGURE 17.16**).

Although up to 80% of infected individuals remain asymptomatic, the other 20% of infected people develop symptoms between 3 and 14 days after they are bitten by the infected mosquito. Most

of these infected individuals develop WNV disease, which is a nonneuroinvasive infection involving a fever, headache, body aches, and extreme fatigue. These people also might develop a skin rash on the chest, stomach, and back. Symptoms typically last a few days. In 2016, less than 50% of cases were a nonneuroinvasive infection.

In the neuroinvasive form, called **West Nile encephalitis** or **West Nile meningitis**, infection affects the central nervous system. Symptoms can include high fever, headache, stiff neck, stupor, disorientation, coma, convulsions, muscle weakness, vision loss, numbness, paralysis, and coma. These symptoms can last several weeks to longer than a year and the neurological effects can be permanent.

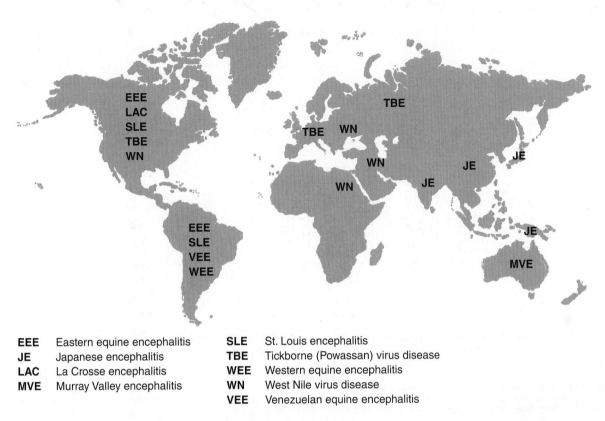

EEE	Eastern equine encephalitis	SLE	St. Louis encephalitis
JE	Japanese encephalitis	TBE	Tickborne (Powassan) virus disease
LAC	La Crosse encephalitis	WEE	Western equine encephalitis
MVE	Murray Valley encephalitis	WN	West Nile virus disease
		VEE	Venezuelan equine encephalitis

FIGURE 17.14 Global Distribution of Major Arboviral Encephalitides. The various diseases caused by arboviruses are shown.
»» Propose a reason why there is such a large variety of arboviral encephalitides in the Western Hemisphere.

Courtesy of CDC. Retrieved from http://www.cdc.gov/ncidod/dvbid/arbor/worldist.pdf.

About 9% of these patients died from the neuroinvasive disease in 2016.

There is no vaccine available nor is there any specific treatment for human WNV infections. In

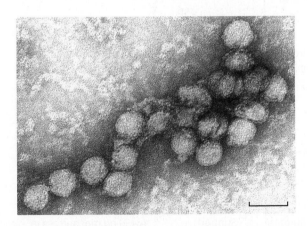

FIGURE 17.15 West Nile Virus. False-color transmission electron micrograph of West Nile viruses (WNVs). (Bar = 100 nm.) *»» How would you describe the shape of the West Nile virus?*

Courtesy of Cynthia Goldsmith/CDC.

cases of encephalitis and meningitis, people might need to be hospitalized so that they can receive supportive treatment including intravenous fluids, help with breathing, and nursing care. People who spend a lot of time outdoors are more likely to be bitten by an infected mosquito. These individuals should take special care to avoid mosquito bites. In addition, individuals older than 50 years of age are more likely to develop serious symptoms from a WNV infection if they do get sick, so they too should take special care to avoid mosquito bites.

TABLE 17.4 summarizes the neurotropic viral diseases.

In conclusion, we have examined several viral diseases that affect various body systems. In their studies, students often wonder where these viruses originated. Again, there is good evidence that many have "jumped" species from an animal to humans (a zoonosis) at some time in the past. For example, hepatitis B jumped from apes, yellow fever came from African primates, and dengue fever and Zika virus infection originated in Old World primates. Virologists are interested in understanding how

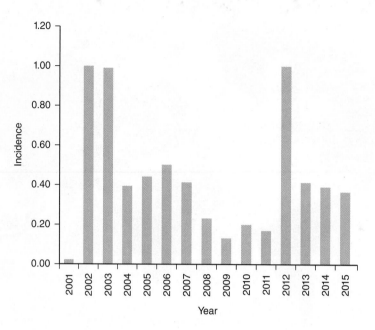

FIGURE 17.16 Incidence of West Nile Virus—United States, 2001–2015. This figure shows the incidence (reported cases per 100,000 individuals) of neuroinvasive disease since 2001. WNV incidence was stable between 2004 and 2007 and then decreased through 2009. »» *What factors might have brought about the sudden increase in incidence of WNV in 2012?*

Data from CDC. (2015). Summary of notifiable infectious disease and conditions--United States, 2013. *MMWR Morb Mortal Wkly, 62*(53):1–119; CDC. (2016). West Nile virus: Final annual maps & data for 1999–2015. Retrieved from http://www.cdc.gov/westnile/statsmaps/finalmapsdata/index.html; CDC. (2017). West Nile virus: Weste Nile virus disease cases and presumptive viremic blood donors by state--United States, 2016 (as of January 17, 2017). Retrieved from http://www.cdc.gov/westnile/statsmaps/preliminarymapsdata/histatedate.html..

TABLE 17.4 A Summary of the Viral Diseases of the Nervous System

Disease	Causative Agent	Signs and Symptoms	Transmission	Treatment	Prevention
Zika virus infection	Zika virus	Fever, maculopapular rash, joint pain, or conjunctivitis (red eyes); microcephaly (newborns) and Guillain-Barré (adults)	Bite from an infected *Aedes aegypti* mosquito	No effective treatment	Avoiding mosquito bites in endemic areas
Rabies	Rabies virus	Tingling, burning, coldness at bite site Fever, headache, increased muscle tension Paralysis and hydrophobia	Bite from rabid animal	Rabies immune globulin Rabies vaccine	Avoiding rabid animals Thoroughly washing the bitten area Pre-exposure vaccination when needed
Polio	Poliovirus	Asymptomatic Abortive Nonparalytic Paralytic	Fecal–oral route	Supportive treatments	Polio vaccine Practicing good personal hygiene
Arboviral encephalitis	Arboviruses (West Nile)	Sudden high fever, severe headache, stiff neck, disorientation, paralysis, mental disorders Convulsions and coma	Mosquitoes and ticks	Supportive treatment	Avoiding mosquitoes in endemic areas Insect repellant Covering exposed skin

these transformations from animal to human occur and have conceived a "five-step program" for the "evolution" of such diseases.

Stage 1: The virus is found in animals but not in humans.

Stage 2: The virus has evolved so it now can be transmitted to humans but not between humans (e.g., rabies).

Stage 3: The virus is transmitted to humans and briefly between humans (brief outbreak) before "burning out" (e.g., Ebola virus disease).

Stage 4: The virus is transmitted to humans and then between humans, which leads to extended outbreaks (e.g., dengue fever, Zika virus infection).

Stage 5: The virus has evolved into a pathogen exclusively in humans, and outbreaks are solely the result of human-to-human transmission (e.g., AIDS).

Concept and Reasoning Checks 17.4

a. Although the Zika virus is in the same family of viruses as the yellow fever virus, what viral structures might specifically target Zika to the nervous system?
b. Why is the incubation period so variable for the rabies virus?
c. Explain why the American polio vaccination schedule has dropped the OPV as part of the vaccination.
d. Assess the consequences of being infected with the WNV.

Chapter Challenge D

A 29-year-old man visited a neurologist with 2 days of increasing pain in the right arm and abnormal sensations. The neurologist diagnosed atypical neuropathy. The next day, the man returned, complaining of hand spasms and sweating on the right side of the face and trunk. The patient was discharged twice from an emergency department but symptoms worsened. After developing difficulties swallowing, having excessive saliva production, exhibiting agitation, and presenting generalized muscle twitching, the patient was admitted to a local hospital. Vital signs and blood tests were normal, but within hours, he became confused. The patient was placed on broad-spectrum antibiotics and mechanical ventilation. Renal failure developed and the patient died 3 days later.

QUESTION D:
a. *Based on the man's symptoms, what important signs did the doctor miss in trying to produce a diagnosis? What is the one important question that was not asked of the patient?*
b. *What disease did the man die from?*
c. *What medical treatment, if started immediately, might have saved the patient's life?*

You can find answers online in **Appendix F**.

■ SUMMARY OF KEY CONCEPTS

Concept 17.1 Viral Infections Can Affect the Blood and the Lymphatic Systems

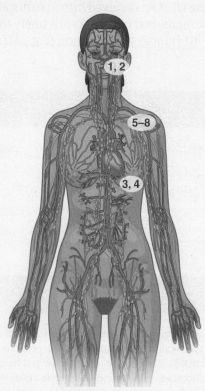

▶ **Infectious mononucleosis**
   1 Epstein-Barr virus
▶ **Cytomegalovirus disease**
   2 Cytomegalovirus
▶ **Hepatitis B**
   3 Hepatitis B virus
▶ **Hepatitis C**
   4 Hepatitis C virus

Concept 17.2 Some Viral Diseases Cause Hemorrhagic Fevers

▶ Hemorrhagic fevers
   5 Yellow fever virus
   6 Dengue fever virus
   7 Ebola virus/Marburg virus
   8 Lassa fever virus

Concept 17.3 Viral Infections of the Gastrointestinal Tract Are Major Global Health Challenges (not shown)

▶ **Hepatitis A**
   Hepatitis A virus
▶ **Hepatitis E**
   Hepatitis E virus
▶ **Viral gastroenteritis**
   Rotavirus, norovirus

Concept 17.4 Viral Diseases of the Nervous System Can Be Deadly

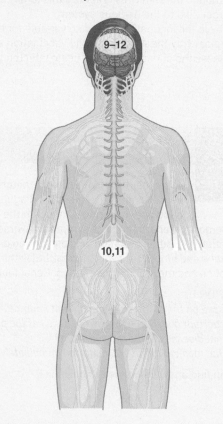

▶ **Zika virus infection**
   9 Zika virus
▶ **Rabies**
   10 Rabies virus
▶ **Polio**
   11 Poliovirus
▶ **Arboviral encephalitis**
   12 West Nile virus

■ CHAPTER SELF-TEST

For **Steps A–D**, you can find answers online in **Appendix D**.

STEP A: REVIEW OF FACTS AND TERMS

Multiple Choice

Read each question carefully before selecting the *one* answer that best fits the question or statement.

1. Mononucleosis is an infection of _____ cells by the _____.
   A. T; cytomegalovirus
   B. B; Epstein-Barr virus
   C. lung; cytomegalovirus
   D. red blood; Epstein-Barr virus

2. Which of the following is *not* a transmission mechanism for hepatitis B?
   A. Sexual contact
   B. Nonsterile body piercing equipment
   C. Fecal–oral route
   D. Blood-contaminated needles

3. Symptoms of headache, fever, and muscle pain lasting 3 to 5 days, followed by a 2 to 24 hour abating of symptoms are characteristics of _____.
   A. yellow fever
   B. hepatitis C
   C. dengue fever
   D. Ebola hemorrhagic fever

4. A long thread-like RNA virus is typical of the _____ viruses.
   A. hepatitis C
   B. Ebola
   C. polio
   D. West Nile

5. The reservoir for Lassa fever is _____.
   A. rats
   B. mosquitoes
   C. ticks
   D. sandflies

6. Which one of the following characteristics pertains to hepatitis A?
   A. Transmission is by the fecal–oral route.
   B. The incubation period is 2 to 4 weeks.
   C. It is an acute, inflammatory liver disease.
   D. All of the above (**A–C**) are correct.

7. _____ are the single most important cause of diarrhea in infants and young children admitted to American hospitals.
   A. Noroviruses
   B. Echoviruses
   C. Hepatitis A viruses
   D. Rotaviruses

8. Hydrophobia is a term applied to _____.
   A. rotavirus infections
   B. West Nile fever
   C. arboviral encephalitis
   D. rabies

9. These viruses multiply first in the tonsils and then the lymphoid tissues of the gastrointestinal tract.
   A. Rabies virus
   B. Rotavirus
   C. Polio virus
   D. Hepatitis A virus

10. A major neurological symptom of a Zika virus infection in adults is _____.
    A. Guillain-Barré syndrome
    B. flaccid paralysis
    C. microcephaly
    D. hydrophobia

11. Arboviral encephalitis is an example of a _____.
    A. disease causing gastroenteritis
    B. disease spread by the fecal–oral route
    C. zoonosis
    D. type of hepatitis

True-False

Each of the following statements is true (T) or false (F). If the statement is false, substitute a word or phrase for the underlined word or phrase to make the statement true.

12. _____ The infectious agent for both yellow fever and dengue fever is a DNA virus transmitted by mosquitoes.

13. _____ Eighty percent of people infected by West Nile virus experience flu-like symptoms.

14. _____ The term "hydrophobia" means "fear of water," and it is commonly associated with patients who have Zika virus infection.

15. _____ Norovirus and rotavirus are both considered to be agents of viral encephalitis.

16. _____ The Epstein-Barr virus is the cause of infectious mononucleosis.

17. _____ The Salk and Sabin vaccines have been used for immunizations against hepatitis.

18. _____ Hepatitis B is most commonly transmitted by contact with infected semen or infected blood.

19. _____ One of the most important causes of diarrhea in infants and young children admitted to hospitals is the Epstein-Barr virus.

20. _____ Filoviruses are long, thread-like viruses that cause hemorrhagic fevers and include the marburgvirus.

■ STEP B: CONCEPT REVIEW

21. Compare the similarities and differences between the nature of the hepatitis B and C viruses and the illnesses they cause. (**Key Concept 17.1**)
22. Summarize the symptoms of **Ebola virus disease** and **Marburg virus disease**. (**Key Concept 17.2**)
23. Describe how the hepatitis A virus is spread and prevented. (**Key Concept 17.3**)

24. Explain why rotavirus infections are so deadly in children and describe how noroviruses are transmitted. (**Key Concept 17.3**)
25. Describe the outcome to someone who has been bitten by a rabid animal, and recommend treatment if **rabies** symptoms have not yet appeared. (**Key Concept 17.4**)
26. Explain how the polioviruses cause disease, and identify the two types of **polio** vaccines. (**Key Concept 17.4**)

■ STEP C: APPLICATIONS AND PROBLEM SOLVING

27. Written on some blood donor cards is the notation "CMV." What do you think the letters mean, and why are they placed there?
28. Sicilian barbers are renowned for their skill and dexterity with razors (and sometimes their singing voices). French researchers studied a group of 37 Sicilian barbers and found that 14 had antibodies against hepatitis C, despite never having been sick with the disease. By comparison, when a random group of 50 blood donors was studied, none had

the antibodies. As an epidemiologist, what might account for the high incidence of exposure to hepatitis C among these barbers?
29. As a state health inspector, you suggest all restaurant workers should be immunized with the hepatitis A vaccine. Why would restaurant owners agree or disagree with your idea?
30. An epidemiologist notes that India has a high rate of dengue fever but a very low rate of yellow fever. What might be the cause of this anomaly?

■ STEP D: QUESTIONS FOR THOUGHT AND DISCUSSION

31. Health authorities panicked when an outbreak of Ebola hemorrhagic disease occurred among imported macaques in a quarantine facility in Reston, Virginia, in 1989. What sparks such a dramatic response when a disease like Ebola virus disease breaks out?
32. A diagnostic test has been developed to detect hepatitis C in blood intended for transfusion purposes. Obviously, if the test is positive, the blood is not used. However, there is a lively controversy as to whether the blood donor should be informed of the positive result. What is your opinion? Why?
33. In the southwestern United States, abundant rain and a mild winter often bring conditions that encourage

a burgeoning rodent population. Under these circumstances, what viral disease would health officials anticipate, and what precautions should they give residents?
34. Disney World and 20 swampy counties in Florida use "sentinel chickens" strategically placed on the grounds to detect any signs of viral encephalitis. Why do you suppose they use chickens? Why are Disney World and many Florida counties particularly susceptible to outbreaks of viral encephalitis? What recommendations might be offered to tourists if the disease broke out?

CHAPTER 18

Eukaryotic Microorganisms: The Fungi

In August 2005, Hurricane Katrina left unimaginable devastation everywhere along the south-central coast of the United States. In New Orleans, thousands of homes were flooded and left sitting in feet of water for weeks. Many health experts were concerned that outbreaks of infectious diseases like cholera, West Nile virus infection, and gastrointestinal illnesses might occur. Homes sitting in stagnant water could become breeding grounds for microorganisms, some of which could cause human disease.

Thankfully, most of these infectious disease scenarios did not occur. However, what did break out in many of the parishes in New Orleans was mold (see chapter opening photo). There were mold colonies on walls, ceilings, cabinets, clothes, and just about anything that provided a source of moisture and nutrients. Spore-forming colonies were everywhere (**FIGURE 18.1**).

If you see small spots of mold in your home, a bleach solution will do a great job to kill and eliminate the problem. Then again, homes flooded by Katrina were another problem. Many home items, such as beds, couches, or cabinets that were even

above the water level, contained mold due to the prolonged humidity. More than likely, most of these homes were demolished (or stripped to the framing), and the furniture and other home contents destroyed. All would have to be replaced. Most health officials told residents to follow the same slogan used for potentially spoiled food: "When in doubt, throw it out."

At Tulane Hospital, the first floor was covered with mold, which made cleanup and reopening of many wards a very difficult chore. In homes as well as offices and hospitals, molds were discovered growing in ventilation systems and the ventilation ducts. If the ventilation fans were turned on, literally hundreds of billions of fungal spores would be

The mold *Penicillium* growing as colonies on agar.
Courtesy of Dr. Jeffrey Pommerville.

(A)

(B)

FIGURE 18.1 **A Wall of Mold.** Every circular spot seen on these walls **(A)** and **(B)** represents a mold colony. Each colony started from a single spore; now each mature colony contains millions of spores. *»» How do you think molds reproduce?*

(A) © DarrenTownsend/iStock/Getty Images Plus/Getty. (B) Ken Hawkins/Alamy Stock Photo.

blown and spread to new areas to germinate and grow. In fact, six weeks after Katrina, the mold spore count in the air was still twice the number considered normal for New Orleans.

What about illnesses from breathing the mold spores? By November 2005, many New Orleans residents who had returned were suffering from upper respiratory problems—the residents called it the "Katrina cough." Residents with asthma, bronchitis, and allergies who had left New Orleans were asked not to return just yet because the high spore counts could worsen their medical condition.

This discussion illustrates that molds grow in many natural and constructed environments. They often break down organic matter, be it natural or fabricated, and some can act as pathogens and cause some dangerous and debilitating diseases. On the other hand, many mold species are of great importance to the natural recycling in the biosphere.

In this chapter, we will encounter some beneficial fungi, such as those used to produce antibiotics or put to work in commercial and industrial processes. We also will identify and discuss several human diseases caused by fungi.

Our study begins with a focus on the community of fungi in the human body and then on the structures, growth patterns, and life cycles of fungi—something quite unique from the other groups of microorganisms.

Chapter **Challenge**

This chapter on the fungi covers their basic characteristics and classification in the domain Eukarya and the major human diseases they cause. Because these are diverse topics, let's try to solve a set of mini-challenges as we progress through this chapter.

■ KEY CONCEPT 18.1   The Kingdom Fungi Includes the Molds and Yeasts

The **fungi** (sing., **fungus**) are a diverse group of microorganisms in the domain Eukarya. Some 75,000 species have been described, although as many as 5 million might exist. Like the prokaryotic

world, the fungi are found in every ecosystem on Earth. The majority of fungi are active in degrading and recycling biomass, that is, dead animal and plant matter. Some fungi have been put to use as sources of antibiotics, such as penicillin, whereas other fungi make useful products for medicine and the food industry. The latter are discussed in Chapter 25 (online). There are about 200 fungal species associated with human disease, although only about 12 species are of serious medical consequence.

The Human Fungal Microbiome Is an Emerging Field of Study

The human microbiome refers to the tremendous number of microbes that normally live in and on the human body. This community is composed primarily of resident bacterial species, which play important roles in health and disease. However, the human microbiome also includes resident viruses, protists, and fungi. With regard to the fungi, they have been less well studied, but they are equally important to human health. Thus, the fungal microbiome, or so-called **human mycobiome**, consists of fungal organisms that are resident on and in the human body, although the fungal communities are much smaller populations than those comprising the bacterial communities (**FIGURE 18.2**).

To date, most of the studies investigating the human mycobiome have been descriptive. For example, some 100 different fungal species (about 280 bacterial species by comparison) can inhabit the oral cavity, each person having between 9 and 23 of these fungal species. In the bacterial microbiome, when one bacterial community disappears, or is reduced in number, another bacterial community spreads. The result of this imbalance, called **dysbiosis**, can result in the development of a disease condition. This community change also appears to be true for the mycobiome. Researchers have found that the oral mycobiome of HIV-infected people is different from the mycobiome in uninfected individuals. Likewise, studies on the bacterial and fungal communities in the gastrointestinal tract have shown that a lack of community diversity influences obesity. In addition, like the bacterial microbiome, the mycobiome helps modulate immune function, such that dysbiosis can make a person more susceptible to inflammatory bowel diseases, such as ulcerative colitis and Crohn's disease.

The study of the human mycobiome is still in its infancy and, for now, the impact of the resident

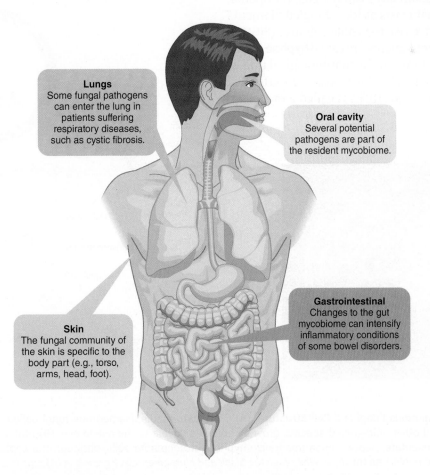

Lungs
Some fungal pathogens can enter the lung in patients suffering respiratory diseases, such as cystic fibrosis.

Oral cavity
Several potential pathogens are part of the resident mycobiome.

Skin
The fungal community of the skin is specific to the body part (e.g., torso, arms, head, foot).

Gastrointestinal
Changes to the gut mycobiome can intensify inflammatory conditions of some bowel disorders.

FIGURE 18.2 The Human Fungal Mycobiome. Diverse species of fungi inhabit the skin and other anatomical sites in the human body. *»» Why don't people commonly develop illnesses or diseases to these resident fungi?*

fungal organisms on human health remains to be clearly understood. So, let's examine some features of those fungi that we understand much better.

The Fungal Body Plan Is Distinctive

Being eukaryotic organisms, fungi contain a variety of cellular organelles, including cell nuclei with linear chromosomes, mitochondria, an endomembrane system, ribosomes, and a cytoskeleton. The cell wall is composed of large amounts of the carbohydrate chitin. **Chitin** is a polymer of glucose molecules modified with amino acid and acetyl groups (called acetylglucosamine units). The cell wall provides rigidity and strength, which, like the cell wall of all microorganisms, allows the cells to resist osmotic lysis (bursting) due to high internal water pressure.

The fungal body plan (form) can be broken into one of two types—the multicellular filamentous fungi (molds) and the unicellular yeasts.

Filamentous Molds

One major group of fungi, the **molds**, which you have come to appreciate already from the chapter opener, has a body plan that exists as long, tangled filaments that branch and give rise to visible colonies (**FIGURE 18.3A**). The tubular filaments are called **hyphae** (sing., **hypha**), and they represent the morphological unit of a filamentous fungus. The thick mass of intertwined hyphae you see in Figure 18.3A is called a **mycelium**

(pl., mycelia). Whereas many of the hyphae composing the mycelium remain in the soil or within the nutrient medium, other hyphae can extend upward, and it is on these aerial hyphae that reproductive structures and their spores usually form.

The hyphae of filamentous fungi have a distinctive cellular arrangement. In many species, cross walls, called **septa** (sing., **septum**), divide the hyphal cytoplasm into separate "cells" (**FIGURE 18.3B**). Notice that each septum contains a pore that allows adjacent cell cytoplasms to mix. In other molds, the filaments are without septa. Such nonseptate hyphae are considered **coenocytic**, meaning the "cell" contains a large, random number of cell nuclei in a common cytoplasm. No other organisms have this unique filamentous arrangement of cells and cell nuclei.

Unicellular Yeasts

The other group of fungi is the **yeasts**, whose body plan is a single cell (**FIGURE 18.4**). When they grow to high numbers on agar, they produce colonies that visually resemble bacterial colonies. Yeasts can be found in environments as diverse as water, soil,

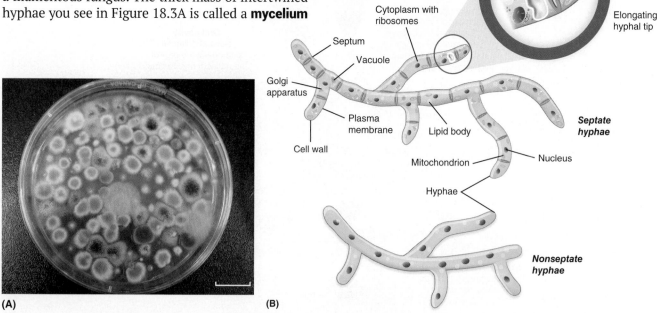

(A) **(B)**

FIGURE 18.3 Filamentous Fungi and Cell Structure. (A) On growth media, filamentous fungi called molds, such as *Aspergillus*, *Penicillium*, and other unidentified species, grow as colonies visible to the naked eye. **(B)** Molds have hyphae that are either septate or nonseptate. Septa compartmentalize hyphae into separate cells, although the septa have a pore through which cytoplasm and nuclei can move. »» *What does the black and blue-green fuzzy growth in (A) represent?*

(A) Courtesy of Ginger Chew, Sc.D./CDC. (B) © Designua/Shutterstock.

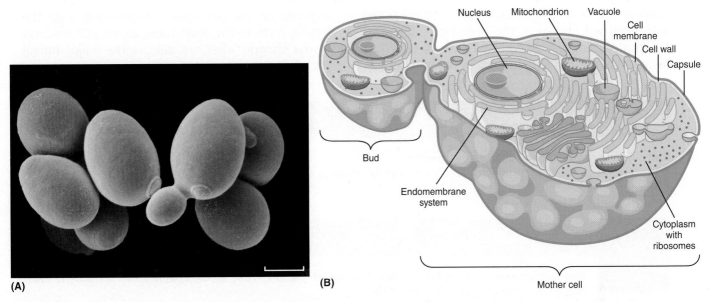

(A)   **(B)**

FIGURE 18.4 Yeasts and Yeast Cell Structure. (A) False-color scanning electron micrograph of yeast cells. Yeasts represent fungal cells that often grow as a unicellular organism. (Bar = 1 µm.) **(B)** Yeast cell structure contains the typical eukaryotic organelles.
»» *What cell process must occur for both the mother cells and bud to have a cell nucleus?*

(A) © SCIMAT/Science Source.

and the human skin. There are some 1,500 known species of yeasts, but perhaps the most familiar to us is the genus *Saccharomyces*. *Saccharomyces* (*saccharo* = "sugar") has all the typical eukaryotic cell structures found in the filamentous fungi (**FIGURE 18.4B**). The most commonly used species are *S. cerevisiae* and *S. ellipsoideus*, the former used for bread baking (baker's yeast) and alcohol production, the latter for alcohol production through fermentation. The extent of the wine, beer, and spirit industries globally is evidence to the significance of the fermentation yeasts.

S. cerevisiae probably is the most understood eukaryotic organism at the molecular and cellular levels. The genome contains about 6,000 genes, about 20% of which are common to the human genome. Today, *Saccharomyces* and its relatives have been used in many areas of research. For example, the chemical and signaling process by which an animal or plant cell prepares itself for asexual cell division (mitosis) was either first discovered using yeast cells or major contributions to the understanding came from research with yeast cells. Being similar to human cells in several ways, the potential value of many experimental drugs for human disease treatment can be screened (tested) by using yeast cells. Yeast cells are also biochemical factories used in biotechnology for the production of medical therapeutics. Modified yeast cells, for instance, have been used to produce insulin to treat diabetes and to create a hepatitis B vaccine.

Some fungi are **dimorphic**; that is, they can switch back and forth between the filamentous and yeast growth forms. Usually at ambient temperature (23°C [73°F]) and with limited nutrients, the fungus grows as a saprophytic, filamentous mold, which lives off dead organic matter (**FIGURE 18.5**).

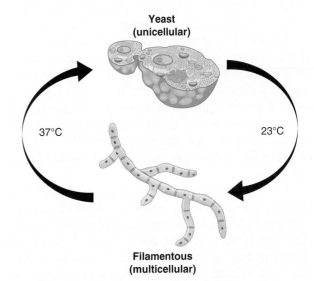

**Yeast
(unicellular)**

37°C   23°C

**Filamentous
(multicellular)**

FIGURE 18.5 Dimorphic Growth Flexibility. Many fungi have the ability to alternate between a filamentous-type growth at ambient temperature and a yeast-type growth at human body temperature. »» *What advantage is it for a fungal environmental pathogen to be able to grow at human body temperature?*

Filamentous image © Designua/Shutterstock.

However, at body temperature (37°C [98.6°F]) and with enriched nutrients, the organism transforms to a unicellular, pathogenic yeast. In fact, in just about all cases, fungal virulence in the human host depends on the pathogen transitioning from the hyphal form to the yeast form, as we will see later in the chapter when we discuss the major fungal pathogens of humans.

Concept and Reasoning Checks 18.1

a. Assess the role of hyphae to fungal growth.
b. Summarize the importance of yeasts to commercial interests and research.

Chapter Challenge A

This section examined the basic characteristics of the fungi. One of the unique characteristics is the filamentous (mycelial) arrangement of cells.

QUESTION A: *How can many molds exist as a coenocytic (multinucleate) structure? Most organisms, humans included, would die if cells contained more than one cell nucleus.*

You can find answers online in **Appendix F**.

■ KEY CONCEPT 18.2   Fungal Growth and Reproduction Are Dependent on Spores

Generally, fungal growth, be it filamentous or unicellular, involves two phases: a growth (vegetative) phase and a reproductive (spore-forming) phase.

Most molds grow by elongation and branching at the hyphal tips (see Figure 18.3B), which in the natural environment allows the mycelium to function as a very efficient surface for absorbing nutrients. Often, one cubic centimeter of soil can contain up to a kilometer of intertwined hyphae. For the yeasts, nutrients are absorbed across the cell surface, similar to the way bacterial cells obtain their nutrients.

For efficient nutrient uptake, the molds and yeasts secrete enzymes into the surrounding environment to break down (hydrolyze) complex organic compounds into simpler ones. Thus, all fungi are chemoheterotrophs, and most of these are saprobes that feed on dead plant and animal matter. Together with many bacterial species, these fungi make up the **decomposers**, recycling vast quantities of organic matter, as depicted in MICROFOCUS 18.1. By living off the remains of dead organisms, they break down complex biological compounds into simpler raw materials (e.g., carbon, minerals) that can then be recycled and used by other organisms.

Fungi that feed on living plants or animals represent potential pathogens that in the host might cause disease. As we will see later in the chapter, many are **opportunistic**; that is, they only cause serious disease in people with a weakened immune system, that is, individuals who are **immunocompromised**.

Fungal growth is influenced by many factors. Normally, high concentrations of sugars stimulate fungal growth, and laboratory media for fungi usually contain extra glucose; examples include Sabouraud dextrose (glucose) agar and potato dextrose agar.

Besides the availability of chemical nutrients, oxygen, temperature, and pH also influence growth.

▶ **Oxygen.** The majority of fungi are aerobic organisms, with the notable exception of the facultative yeasts like *Saccharomyces*, which can grow in the presence of oxygen or under fermentation conditions.

▶ **Temperature.** Most fungi act as decomposers at ambient temperatures. Psychrophilic fungi grow (though slowly) at lower temperatures, such as the 5°C (41°F) found in a normal refrigerator. However, as mentioned a moment ago, dimorphic fungi have the

MICROFOCUS 18.1: Evolution

When Fungi Ruled the Earth

About 250 million years ago, a catastrophe of epic proportions occurred on Earth. Scientists believe that more than 90% of animal species in the seas vanished. The great Permian extinction, as it is called, also wreaked havoc on land animals and cleared the way for dinosaurs to inherit the planet. However, many land plants managed to survive, and before the dinosaurs came, these plants spread and enveloped the world. At least, that is what paleobiologists traditionally believed.

Now, however, they are revising their theory and finding a significant place for the fungi. Many scientists now believe that land plants were decimated by the Permian extinction, and, as a result, for a brief geological span, dead wood covered the planet. During this period, they suggest the fungi emerged, and wood-rotting species (decomposers) experienced a powerful spike in their populations (see the accompanying figure). Support for this theory is offered by numerous findings of fossil fungi from the post-Permian period. The fossils are bountiful, and they come from all corners of the globe. Significantly, they contain fungal hyphae—the active feeding forms rather than the dormant spores.

A mushroom species growing on a rotting log.

© Izatul Lail bin Mohd Yasar/Shutterstock.

In the post-Permian period, fungal decomposers proliferated wildly and entered a period of feeding frenzy during which they were the dominant eukaryotic form of life on Earth. It's something worth considering next time you kick over a mushroom growing on a rotting log.

metabolic flexibility to grow well at 37°C. Such morphological transitions are a method that the human fungal pathogens have evolved to adapt and grow in the human or animal body.

▶ **pH.** Many fungi thrive under mildly acidic conditions at a pH between 5 and 6. Thus, mold contamination frequently occurs in acidic foods such as sour cream, yogurt, citrus fruits, and many vegetables, with molds purposely being added to numerous types of cheese (**FIGURE 18.6**).

Some Fungi Require or Benefit from a Partner for Growth

Many fungi in the environment live and grow in a mutually beneficial relationship with other species, especially plants, through a symbiotic association called **mutualism**. Fungi called **mycorrhizae** (*rhiza* = "root") live in association with plants, where the hyphae of these fungi invade or encase the plant roots (**FIGURE 18.7A**). Mycorrhizae have been found in plants from salt marshes to deserts and pine forests. In fact, mycorrhizae are believed to live in and around the roots of 95% of plant species.

FIGURE 18.6 Roquefort Cheese. Roquefort cheese is made from cow's milk and is purposely produced with the mold *Penicillium roqueforti.* »» *Why would a mold be added to the ripening process of cheeses such as this?*

© Jones & Bartlett Learning. Photographed by Kimberly Potvin.

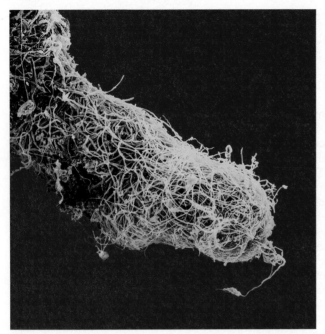

(A)

(B)

FIGURE 18.7 Mycorrhizae and Their Effect on Plant Growth. (A) Mycorrhizae surround a root section from a *Eucalyptus* tree in this false-color scanning electron micrograph. These fungi are involved symbiotically with their plant host, such as aiding in mineral metabolism. **(B)** An experiment analyzing the mycorrhizal effects on plant growth. *»» Which plant or plants (CK, GM, GE) represent(s) the control and experimental set ups?*

(A) © Dr. Gerald Van Dyke/Visuals Unlimited/Getty. (B) © Science VU/R. Roncadori/Visuals Unlimited.

In the mutualistic relationship, the plant provides the fungus with carbohydrates that the fungus needs for growth. In return, the mycorrhizae act as a second, extensive root system, supplying the plant with up to 80% of the essential minerals (nitrogen and phosphorus) and water the plant needs for growth (**FIGURE 18.7B**). In addition, the filamentous fungal network also acts as a communication system, linking separate plants into a common chemical communication system. Without mycorrhizae, few trees, food crops, shrubs, and herbs would thrive.

Reproduction in Fungi Involves Spore Formation

From the description of the situation in New Orleans after Hurricane Katrina, you can appreciate fungi as producers of massive numbers of **spores**, which represent microscopic cells for disseminating the fungal organisms. **Sporulation** is the process of spore formation. After they are formed, the spores are released, and they are often blown far distances on air currents or carried away by insects to new environments.

The reproductive structures producing spores are usually aerial. They can be asexual and invisible to the naked eye or sexual structures, called **fruiting bodies**, such as the macroscopic mushrooms.

Asexual Reproduction

Asexual reproductive structures develop at the ends of specialized aerial hyphae. Here, mitotic divisions produce thousands of genetically identical spores.

Many asexual spores develop within sacs or vessels called **sporangia** (sing., sporangium; *angio* = "vessel") (**FIGURE 18.8A**). Appropriately, the spores are called **sporangiospores**. Other fungi produce spores on supportive structures called **conidiophores** (**FIGURE 18.8B**). These unprotected, dustlike spores are known as **conidia** (sing., conidium; *conidio* = "dust"). They are passively released into the air and, being extremely light, the spores are blown about by wind currents. In yet other fungi, asexual spores form simply by fragmentation of the hyphae, yielding **arthrospores** (*arthro* = "joint") (**FIGURE 18.8C**). The fungi that cause athlete's foot multiply in this manner.

After they are free of the reproductive structure, spores landing in an appropriate environment have the capability of germinating to reproduce a new hypha (**FIGURE 18.9A**). Continued growth and branching quickly produce a mycelium.

Many yeasts reproduce asexually by **budding**. In this process, the cell becomes swollen at one edge, and a new cell, called a **blastospore** (*blasto* = "bud"), develops (buds) from the parent cell (**FIGURE 18.9B**). Although the spore usually breaks free to live as an independent cell, sometimes a chain of cells, called a **pseudohypha**, will form.

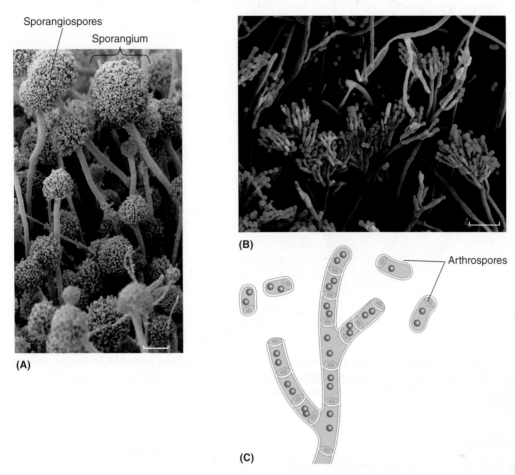

FIGURE 18.8 Fungal Reproductive Structures. False-color scanning electron micrographs of sporangia and conidia. **(A)** Sporangia of the common bread mold *Rhizopus*. Each round sporangium contains thousands of sporangiospores. (Bar = 20 µm.) **(B)** The conidiophores and conidia in the mold-like phase of *Penicillium roqueforti*. Many conidiophores are present within the mycelium. Conidiophores (orange) containing conidia (blue) are formed at the end of specialized hyphae (green). (Bar = 20 µm.) **(C)** Some fungi produce arthrospores by simple fragmentation of the hyphae *»» Why are fungal spores often elevated on the tips of hyphae?*

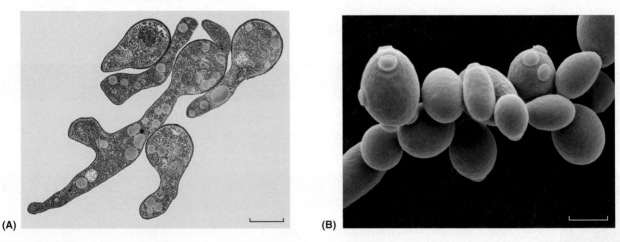

FIGURE 18.9 Germinating Fungal Spores. (A) A false-color transmission electron micrograph of germinating fungal spores. (Bar = 3 µm.) **(B)** False-color scanning electron micrograph of the unicellular yeast *Saccharomyces cerevisiae*. (Bar = 2 µm.) *»» Propose what the circular areas represent on the parent cell at the left and right center of the micrograph in (B).*

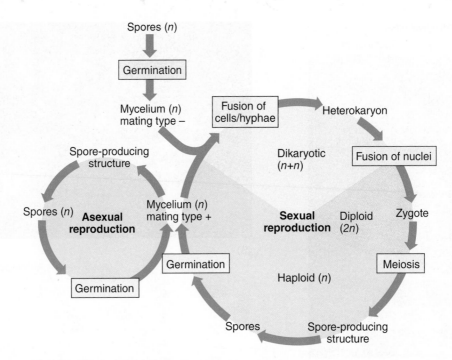

FIGURE 18.10 A Typical Life Cycle of a Mold. Many fungi have both an asexual and sexual reproduction cycle characterized by spore formation. The unique phase in sexual reproduction is the presence of a heterokaryon where nuclei from two different mating types remain separate (dikaryotic) in a common cytoplasm. *»» How could an organism like a mold survive without a sexual cycle?*

Sexual Reproduction

Many fungi also produce spores through sexual reproduction. In this process, hyphae of opposite **mating types** come together and fuse (**FIGURE 18.10**). Because the nuclei are genetically different in each mating type, the fusion cell represents a **heterokaryon** (*hetero* = "different"; *karyo* = "nucleus"), that is, a cell with genetically dissimilar nuclei existing in a **dikaryotic** (two nuclei) cytoplasm. Eventually the nuclei fuse, and a diploid cell is formed. The chromosome number soon is halved by meiosis, returning the cell or organism to a haploid condition.

Sexual reproduction usually occurs in the fruiting body, which represents the aerial part of a fungus in which spores are formed and from which they are released. The spores are forcibly ejected from the fruiting body and are carried away by air currents. In fact, estimates have suggested that mushrooms release billions of spores into the atmosphere every day, amounting to more than 50 million tons of spores each year.

Concept and Reasoning Checks 18.2

a. Estimate the value of mycorrhizae to plant survival.
b. Assess the role of asexual and sexual reproduction in fungi.

Chapter Challenge B

Both fungal molds and plants are dependent on winds for dissemination of spores (fungi) and some seeds (green plants).

QUESTION B: *How are fungal spores and plant seeds similar and different?*

You can find answers online in **Appendix F**.

KEY CONCEPT 18.3   Fungi Have Evolved into a Variety of Forms

Historically, fungal classification was based on structural, physiological, and biochemical differences between organisms. However, with the advent of DNA analysis and genome sequencing, new evolutionary relationships among fungal groups are being developed (**FIGURE 18.11**).

Fungi Can Be Classified into Several Major Groups

Fungi can be classified into one of several phyla, including the Chytridiomycota, Glomeromycota, Zygomycota, Ascomycota, and Basidiomycota. If a fungal organism lacks a recognized sexual reproductive cycle and it appears to only reproduce asexually, it is placed into an informal group sometimes called the mitosporic fungi. We begin, though, with two other groups whose relationships to the fungi are uncertain.

The Cryptomycota and Microsporidia

In 2011, scientists in England, Spain, and the United States discovered a diverse, widespread, and previously undescribed group of eukaryotic organisms. Tentatively called the **Cryptomycota** (*crypto* = "hidden"), these small, flagellated cells found in soil, freshwater, and marine habitats are unique in lacking a chitin cell wall. Being motile, the researchers believe the Cryptomycota must be a very ancient group of microorganisms, but how they are related to the other fungal groups requires more study.

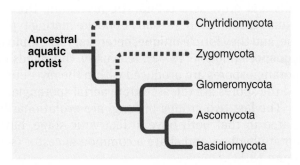

FIGURE 18.11 A Proposed Evolutionary Relationship for the Kingdom Fungi. This generalized scheme for the relationships between fungal groups is based on molecular findings. The dashed lines represent unclear lineages for the chytrids and zygomycetes. *»» Because the chytrids are the only fungal group with flagella, indicate on the evolutionary tree where you would place the phrase "Loss of flagella."*

Another group of eukaryotic microbes, the **Microsporidia**, is thought to be a sister group of the fungi. These organisms are unicellular parasites that live within the cells or tissues of protists and animals. They are among the smallest known eukaryotes, some cells being only 1 μm in diameter. The cells have chitin cell walls but lack true mitochondria. As with the Cryptomycota, more studies are under way to better clarify the evolutionary relationships within the group and with other fungal groups.

The Chytridiomycota and Glomeromycota

The oldest known fungi are ancestors to the modern day **Chytridiomycota**, commonly called **chytrids**. Members of the phylum can be found in moist soil and water. They produce flagellated asexual spores, called **zoospores**. Most chytrids act as decomposers, although a few chytrids have had a significant disease impact on some vertebrates, as MICROFOCUS 18.2 explains.

The **Glomeromycota** form what some consider the most extensive symbiosis on Earth. These fungi represent the mycorrhizae described earlier that exist within the roots of most land plants. In fact, some mycologists believe that plant evolution onto land more than 400 million years ago depended on symbiosis with the ancestral Glomeromycota, which delivered plants with needed nutrients by decomposing organic matter and building up richer soils.

The Zygomycota

The Zygomycota is of uncertain evolutionary origin, and its members might represent members that taxonomically belong with other fungal groups. However, for now, the Zygomycota is a useful category to describe these organisms.

The **Zygomycota** ("zygomycetes") consists of primarily saprophytic fungi that inhabit terrestrial environments. Familiar representatives include fast-growing bread molds and other powdery molds typically growing on spoiled fruits with high sugar content or on acidic vegetables (**FIGURE 18.12**). On these and similar materials, the mycelium grows inside their food, dissolving the substrate with extracellular enzymes and taking up nutrients by absorption through the hyphae. Many members are decomposers, whereas some can cause infections and disease.

MICROFOCUS 18.2: Environmental Microbiology

The Day the Amphibians Died

Amphibians (frogs, toads, salamanders) are the oldest class of land-dwelling vertebrates, having survived the effects that brought the extinction of the dinosaurs. Now amphibians are facing their own potential extinction from a different, infectious effect.

In the early 1990s, researchers in Australia and Panama reported massive declines in the number of amphibians in ecologically pristine areas. As the decade progressed, massive die-offs occurred in dozens of frog species, and a few species even became extinct. Once filled with frog song, the forests were quiet. "They're just gone," said one researcher. By 2012, scientists estimated that more than 300 frog species had gone nearly extinct worldwide, and many others were facing the same outcome. Why were the frogs dying?

One of the reasons for the decline was due to an infectious disease. Many amphibian species on four continents were being infected with a chytrid called *Batrachochytrium dendrobatidis*. This fungus accumulates in the frog's skin and secretes a chemical that cripples the frog's immune system. In addition, the fungus infection causes the

A species of harlequin frog, one of many species being wiped out by a chytrid infection.

© Anneka/Shutterstock.

frog to overstimulate production of keratin. This overproduction alters the frog's skin permeability, which leads to a fatal osmotic imbalance and heart failure. Roughly one-third of the world's 6,000 amphibian species are now considered under threat of extinction due to the disease chytridiomycosis (see the accompanying figure).

It now appears that crayfish are reservoirs for the fungus and are spreading the disease on to the amphibians. But to make matters worse, in 2014 researchers identified two viruses infecting toads and newts and another species of *Batrachochytrium* affecting European newts and salamanders.

Where the chytrid came from remains a mystery. The global amphibian trade might be one origin. Are these amphibian deaths a sign of a yet unseen shift in the ecosystem, much like a canary in a coal mine? Some believe this type of "pathogen pollution" might be as serious as chemical pollution. It represents an alarming example of how emerging pathogens (fungal and viral) could potentially wipe out an entire species and, eventually, an entire ecosystem.

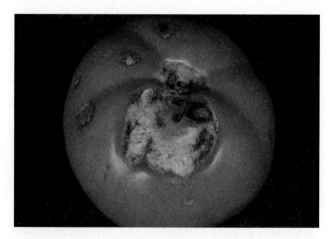

FIGURE 18.12 A Mold Growing on a Tomato. Zygomycete molds typically grow and reproduce on overripe fruits or vegetables, such as this tomato. *»» Identify the black structures and the gray fuzzy growth on the tomato.*

© Jones & Bartlett Learning. Photographed by Kimberly Potvin.

The Zygomycetes make up about 1% of the described species of fungi. They have chitinous cell walls and grow as a coenocytic mycelium. During sexual reproduction, sexually opposite mating types fuse, and they form a unique, heterokaryotic, diploid **zygospore** (FIGURE 18.13). Asexually, thousands of sporangiospores are produced within the mycelium, and the spores are released from aerial sporangia.

The last two groups of fungi are evolutionarily related in that both have a dikaryotic stage, have septate hyphae, and share a common ancestor (see Figure 18.11).

The Ascomycota

Members of phylum **Ascomycota** (*asco* = "sac"; "ascomycetes") are very diverse and account for about 75% of all known fungi. The phylum contains many saprophytic species that serve as decomposers.

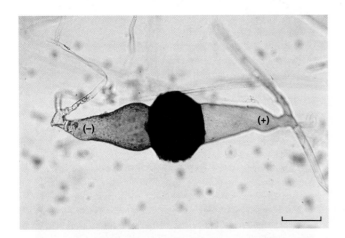

FIGURE 18.13 Sexual Reproduction in the Zygomycota. Sexual reproduction between hyphae of different mating types (+ and −) produces a darkly pigmented zygospore. (Bar = 30 μm.) *»» When the zygospore germinates, what will be produced from the structure?*

© Carolina Biological Supply Com/Medical Images RM.

Perhaps the best-known ascomycetes are *S. cerevisiae* (the yeast used in baking and wine/beer production), *Penicillium chrysogenum* (the mold that produces the antibiotic penicillin), and *Morchella esculentum* (the edible morel) (**FIGURE 18.14**).

The phylum also has several members associated with illness and disease. *Aspergillus flavus* produces **aflatoxin**, a toxic metabolite found in *Aspergillus*-contaminated nuts and stored grain. Aflatoxin is the most potent natural **carcinogen**, that is, a substance capable of causing cancer. *Claviceps purpurea* grows as hyphae on kernels of rye, wheat, and barley where the fungus produces another powerful toxin.

FIGURE 18.14 A Representative Ascomycete. The edible morel (*Morchella esculentum*) represents the fruiting body of an ascomycete prized for its delicate taste. *»» Have you ever eaten a morel? Are they as tasty as their price suggests?*

© Liane Matrisch/Dreamstime.com.

The toxin actually is a group of peptide derivatives called alkaloids that are deposited in the grain as a substance called **ergot**. Products, such as bread made from contaminated rye grain, can cause ergot rye disease, or **ergotism**. MICROFOCUS 18.3 relates a possible effect of ergotism in American history.

MICROFOCUS 18.3: History

"The Work of the Devil"

As an undergraduate, Linda Caporael was missing a critical history course for graduation. Little did she know that through this class she was about to provide a possible answer for one of the biggest mysteries of early American history: the cause of the Salem Witch Trials. These trials in 1692 led to the execution of 20 people who had been accused of being witches in Salem, Massachusetts (see the accompanying figure).

Linda Caporael, now a behavioral psychologist at New York's Rensselaer Polytechnic Institute, in preparation of a paper for her history course had read a book in which the author could not explain the hallucinations among the people in Salem during the witchcraft trials. Caporael made a connection between the "Salem witches" and a French story of ergot poisoning in 1951. In Pont-Saint-Esprit, a small village in the south of France, more than 50 villagers went insane for a month after eating ergotized rye flour. Some people had fits of insomnia, others had hallucinogenic visions, and still others were reported to have fits of hysterical laughing or crying. In the end, three people died.

Caporael noticed a link between these bizarre symptoms, those of Salem witches, and the hallucinogenic effects of drugs like LSD, which is a derivative of ergot. Could ergot possibly have been the perpetrator in Salem, too?

(continues)

MICROFOCUS 18.3: History (*Continued*)

"The Work of the Devil"

During the Dark Ages, Europe's poor lived almost entirely on rye bread. Between 1250 and 1750, ergotism, then called "St. Anthony's fire," caused miscarriages, chronic illnesses in people who survived, and mental illness. Importantly, hallucinations were considered "the work of the devil."

Toxicologists know that eating ergotized food can cause violent muscle spasms, delusions, hallucinations, crawling sensations on the skin, and a host of other symptoms—all of which Linda Caporael found in the records of the Salem witchcraft trials. Ergot thrives in warm, damp, rainy springs and summers, which were the exact conditions Caporael says existed in Salem in 1691. Add to this that parts of Salem village consisted of swampy meadows that would be the perfect environment for fungal growth and that rye was the staple grain of Salem—and it is not a stretch to suggest that the rye crop consumed in the winter of 1691–1692 could have been contaminated by large quantities of ergot.

A witch trial in Massachusetts, 1692.

North Wind Picture Archives/Alamy Stock Photo.

Caporael concedes that ergot poisoning can't explain all of the events at Salem. Some of the behaviors exhibited by the villagers probably represent instances of mass hysteria. Still, as people reexamine events of history, it seems that just maybe ergot poisoning did play some role—and, hey, not bad for an undergraduate history paper!

Candida albicans, the agent responsible for "yeast infections," will be described later in the chapter, along with other ascomycetes that cause serious diseases in humans.

Some ascomycete pathogens cause nonhuman diseases. Most notably is *Cryphonectria parasitica*, the agent responsible for chestnut blight, a devastating disease that has wiped out more than 4 billion American chestnut trees since the early 1900s. Another ascomycete has a more recent association with infectious disease called white nose syndrome in bats, as **INVESTIGATING THE MICROBIAL WORLD 18** explores.

The mycelial ascomycetes have the ability to form conidia through asexual reproduction or **ascospores** through sexual reproduction. Ascospores are formed within a reproductive fruiting body called an ascocarp (see Figure 18.14). Each **ascus** (pl., asci) within the ascocarp produces eight haploid ascospores (**FIGURE 18.15**). Once formed, the spores are forcibly discharged from the asci.

Investigating the Microbial World 18

Bats, White Noses, and Koch's Postulates

White-nose syndrome (WNS) is a disease associated with the deaths of up to 6 million North American bats. The disease causes a distinctive white lesion visible on the nose (and muzzle), wings, tail, and ears of hibernating bats. According to the U.S. Fish and Wildlife Service, since it was first detected in 2006 in New York, WNS has spread rapidly and has now been found among bats in more than 115 caves and mines from Canada to Indiana and Missouri and south to Alabama. In 2016, the disease was found in bat populations on the Pacific coast of the United States. As far as is known, there is no evidence of a health threat to humans.

OBSERVATIONS: An ascomycete fungus, called *Pseudogymnoascus* (formerly *Geomyces*) *destructans*, has been isolated from infected bats. However, its role in causing WNS has been uncertain because the characteristic white lesions are the only consistent pathological finding associated with the disease. In addition, the same fungus has been found to colonize the skin of European bats without causing any reported deaths.

QUESTION: *Is P. destructans the cause of WNS?*

HYPOTHESIS: *P. destructans* is the cause of WNS. If so, using Koch's postulates should confirm that susceptible bats contract WNS from *P. destructans*.

EXPERIMENTAL DESIGN: Twenty-five little brown bats that were naturally infected with WNS were collected in New York. Additional healthy little brown bats collected from an area in Wisconsin outside the known range for WNS were housed in the lab under hibernation conditions. Pure cultures of the fungus were grown and the conidia collected from the plates. Test samples in a phosphate-buffered solution (PBS) contained 5×10^5 conidia.

EXPERIMENT 1: A group of 29 healthy bats had a sample of conidia applied to one wing and another sample applied to the fur between the eye and the ear. The control group consisted of 34 healthy bats that had PBS only applied to the same regions. The experiment was run for up to 102 days.

EXPERIMENT 2: The group of 25 infected bats (positive control) was placed in physical contact with 18 healthy bats for up to 102 days.

EXPERIMENT 3: A group of 36 healthy bats was placed in mesh cages next to the 25 infected bats for up to 102 days.

RESULTS: Any bats that died and all surviving bats after 102 days were studied histologically by examination of the muzzle and skin from each wing. The *P. destructans* fungus from infected bats was reisolated in culture from wing skin and compared genomically to the applied or natural strain. All strains were identical.

EXPERIMENTS 1–3: See table.

Development of WNS in Experimental Bat Groups

Experiment	Group	Number with WNS lesions	Number Without WNS lesions	Total
1	Conidia treated	29	0	29
	PBS treated	0	34	34
2	Contact exposure	16	2	18
	Positive control	25	0	25
3	Airborne exposure	0	36	36

Data from Lorch, J. M. et al. (2011). *Nature* 480(7377):376–378.

CONCLUSIONS:

QUESTION 1: *Was the hypothesis validated? Explain, using the table.*

QUESTION 2: *What is the important epidemiological and disease management implication that can be drawn from experiments 2 and 3?*

QUESTION 3: *Using experiment 1, reconstruct the four steps of Koch's postulates for WNS.*

You can find answers online in **Appendix E**.

Data from Lorch, J. M., et al. 2011. *Nature* 480(7377):376–378.

Ascomycetes are the most frequent fungal partner in lichens. Until recently, a **lichen** was thought to be a mutualistic association between a fungus and a photosynthetic partner (**photobiont**), such as a cyanobacterium or green alga (**FIGURE 18.16A**). In 2016, it was reported that many common lichens appear to have a third partner, a basidiomycete yeast.

Most of the lichen's visible body is the fungus, with the hyphae receiving carbohydrate nutrients by penetrating the core of the lichen, which contains the photobiont and yeast partner. The photobiont receives water and protection from the fungus, whereas the function of the yeast partner remains to be determined. Together, the organisms form a symbiotic association that readily grows in environments where the separate organisms could not survive by themselves (e.g., rock surfaces; **FIGURE 18.16B**).

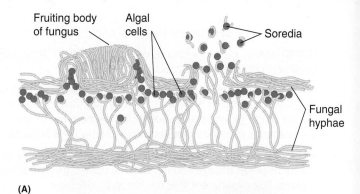

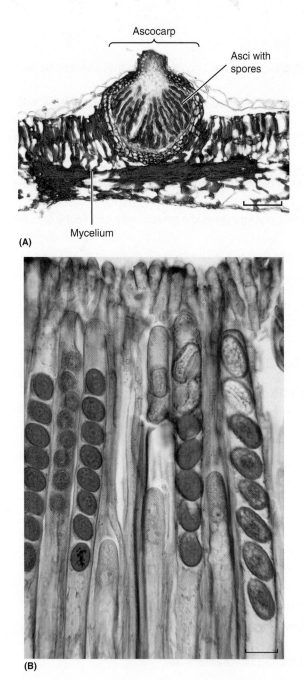

FIGURE 18.15 Ascomycetes and Ascospores. (A) Cross section of a stained fruiting body on an apple leaf. (Bar = 20 µm.) **(B)** A higher magnification of several asci, each containing eight ascospores. (Bar = 5 µm.) »» *Ascospores are representative of what stage of the ascomycete life cycle?*

(A) © Biodisc/Visuals Unlimited, Inc. (B) © Dr. John D. Cunningham/Visuals Unlimited/Corbis.

FIGURE 18.16 Lichens. (A) A cross section of a lichen, showing the upper and lower surfaces where tightly coiled fungal hyphae enclose photosynthetic algal cells. On the upper surface, a fruiting body has formed. Airborne clumps of algae and fungus called soredia are dispersed from the ascocarp to propagate the lichen. Loosely woven fungi at the center of the lichen permit the passage of nutrients, fluids, and gases. **(B)** A typical lichen growing on the surface of a rock. The brown fruiting bodies can be seen. »» *What attributes of the fungus and alga permit the lichen to withstand extreme environmental conditions?*

(B) Courtesy of Dr. Jeffrey Pommerville.

The Basidiomycota

Members of the phylum **Basidiomycota** ("basidiomycetes") contain about 30,000 identified species. The basidiomycetes include unicellular "yeast-like" species and multicellular filamentous fungi. The most recognized members are the mushrooms and puffballs (**FIGURE 18.17A, B**). Some basidiomycetes are important saprobes, decomposing wood and other plant products. Other members form mycorrhizae, whereas still others are important plant parasites, such as the so-called "rusts" and "smuts" that infect corn, cereal plants, and other

(A) **(B)**

FIGURE 18.17 The Basidiomycota. The basidiomycetes are characterized by sexual reproductive structures that usually are macroscopic. These include **(A)** the mushrooms, such as this species of *Amanita muscaria*, and **(B)** the puffballs. »» *Being a fruiting body, where on the mushroom do the spores form?*

(A and B) Courtesy of Dr. Jeffrey Pommerville.

commercial crops. A few Basidiomycota can cause serious diseases in animals and humans.

In the soil, the basidiospores germinate and grow as haploid mycelia. When hyphae of different mating types come in contact, they fuse into a heterokaryon containing genetically different haploid nuclei (**FIGURE 18.18**). Under appropriate environmental conditions, some of the hyphae become tightly compacted, and they force their way to the surface and grow into an aerial fruiting body (basidiocarp)

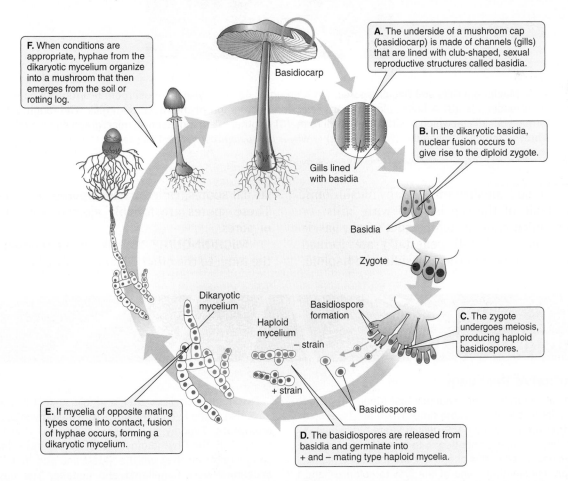

F. When conditions are appropriate, hyphae from the dikaryotic mycelium organize into a mushroom that then emerges from the soil or rotting log.

Basidiocarp

A. The underside of a mushroom cap (basidiocarp) is made of channels (gills) that are lined with club-shaped, sexual reproductive structures called basidia.

Gills lined with basidia

B. In the dikaryotic basidia, nuclear fusion occurs to give rise to the diploid zygote.

Basidia

Zygote

Dikaryotic mycelium

Haploid mycelium

Basidiospore formation

– strain

+ strain

C. The zygote undergoes meiosis, producing haploid basidiospores.

Basidiospores

E. If mycelia of opposite mating types come into contact, fusion of hyphae occurs, forming a dikaryotic mycelium.

D. The basidiospores are released from basidia and germinate into + and – mating type haploid mycelia.

FIGURE 18.18 The Sexual Reproduction Cycle of a Typical Basidiomycete. The basidiomycetes include the mushrooms from which basidiospores are produced. »» *Identify the portion of the cycle that exists in a dikaryotic state.*

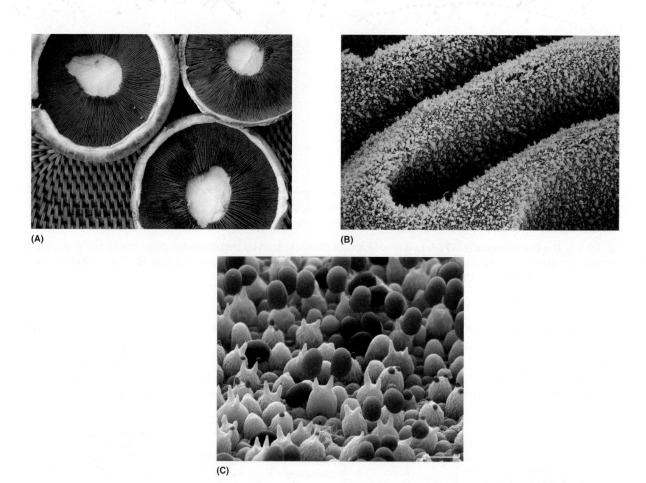

FIGURE 18.19 **Mushroom Gills and Basidiospores. (A)** A group of *Agaricus bisporis* mushroom (Portobello) caps, showing the gills on their underside. **(B)** A false-color scanning electron micrograph of the mushroom gills. The rough appearance represents forming basidia on which basidiospores will form. **(C)** Another scanning electron micrograph showing the basidiospores (dark brown) attached to basidia. (Bar = 20 μm.) *»» Why is the specific epithet for this species of* Agaricus *called* bisporis?

(A) Courtesy of Dr. Jeffrey Pommerville. (B) © Dr. Jeremy Burgess/Science Source. (C) © Eye of Science/Science Source.

typically called a **mushroom**. In many mushrooms, the underside of the cap is lined with "gills" or pores on which club-shaped **basidia** (sing., basidium; *basidium* = "small pedestal") are formed (**FIGURE 18.19**). Within these basidia, the haploid, sexual spores, called **basidiospores**, are produced. These spores are forcibly ejected from the gills or pores.

MICROINQUIRY 18 looks at the relationship of the fungi to the other eukaryotic kingdoms.

MICROINQUIRY 18

Evolution of the Fungi

The fossil record provides evidence that fungi and plants "hit the land" about the same time. There are fossilized fungi that are more than 450 million years old, which is about the time that plants began to colonize the land. In fact, as described in this chapter, some fossilized plants, perhaps representing some of the first to colonize land, appear to have mycorrhizal associations.

Taxonomists believe that a single ancestral species gave rise to the five fungal phyla (Chytridiomycota, Glomeromycota, Zygomycota, Ascomycota, and Basidiomycota). Only the Chytridiomycota (chytrids) have flagella, suggesting that they are the oldest line and that fungal ancestors were flagellated and aquatic. The chytrids might have evolved from a protistan ancestor, and, as fungi colonized the land, they evolved different reproductive styles, which taxonomists have separated into the four nonflagellated phyla described in this chapter.

So, what evolutionary relationships do the fungi have with the eukaryotic plant and animal kingdoms? The accompanying table has some data that you need to analyze and from which you need to make a conclusion as to what relationships are best supported by the evidence. You do not need to understand the function or role for every characteristic listed.

You can find answers online in **Appendix E**.

18.1. Based on the data presented, are fungi more closely related to animals or plants? Justify your answer.

18.2. Draw a taxonomic scheme showing the relationships between plant, animal, and fungal kingdoms.

Comparisons of Organismal, Cellular, and Biochemical Characteristics

Characteristic	Fungi	Plants	Animals
Protein sequence similarities	✓		✓
Elongation factor 3 protein (translation) similarities	✓		✓
Types of polyunsaturated fatty acids			
Alpha linolenic		✓	
Gamma linolenic	✓		✓
Cytochrome system similarities	✓		✓
Mitochondrial UGA codes for tryptophan	✓		✓
Plate-like mitochondrial cristae	✓		✓
Presence of chitin	✓		some
Type of glycoprotein bonding			
O-linked	✓		✓
N-linked	✓		✓
Absorptive nutrition	✓		some
Glycogen reserves	✓		✓
Pathway for lysine biosynthesis			
Aminoadipic acid pathway	✓		
Diaminopimelic pathway		✓	
Type of sterol intermediate			
Lanosterol	✓		✓
Cycloartenol		✓	
Mitochondrial ribosome 5S RNA present		✓	
Presence of lysosomes	✓		✓

TABLE 18.1 Comparisons of the Fungal Phyla Important to Human Disease

Phylum	Common Name	Cross Walls (Septa)	Sexual Structure	Sexual Spore	Asexual Spore	Representative Genera
Zygomycota	Zygomycetes	No	Zygospore	Zygospore	Sporangiospore	*Rhizopus*
Ascomycota	Ascomycetes	Yes	Ascus	Ascospore	Conidia	*Saccharomyces Aspergillus Penicillium*
Basidiomycota	Basidiomycetes	Yes	Basidium	Basidiospore	Conidia Hyphal fragments	*Agaricus Amanita*

The Mitosporic Fungi

Certain fungi lack a known sexual cycle of reproduction; consequently, they are labeled with the term **mitosporic fungi** because the asexual spores are the product of mitosis. Many mitosporic fungi are reclassified when a sexual cycle is observed or comparative genomics identifies a close relationship to a known fungal phylum.

Among the mitosporic fungi, a few are pathogenic for humans. These fungi usually reproduce by asexual spores, budding, or fragmentation.

TABLE 18.1 compares the three phyla of fungi important to human health.

Concept and Reasoning Checks 18.3

a. What are the unique features of the Chytridiomycota and Glomeromycota?
b. Describe the unique properties of the Zygomycota.
c. Summarize the properties of the Ascomycota.
d. Identify the properties of the Basidiomycota.
e. Distinguish the characteristics of the mitosporic fungi.

Chapter Challenge C

The fungi, together with the plants and animals, make up the domain Eukarya and, with the domains Bacteria and Archaea, form the so-called "tree of life."

QUESTION C: *In the historical versions of the "tree of life," the fungi were thought to be part of the plant kingdom. Why might scientists at the time belive that? What information eventually lead to the fungi being more closely allied with the animal kingdom? Consider MicroInquiry 18 when answering the question.*

You can find answers online in **Appendix F**.

■ KEY CONCEPT 18.4   Some Fungi Can Invade the Skin

Fungal diseases, called **mycoses** (sing., **mycosis**), often affect different regions of the human body. Some of these pathogens affect the skin or body surfaces. For example, several diseases, including ringworm and athlete's foot, involve the skin areas. Other mycoses can occur in the oral cavity, intestinal tract, or vaginal tract.

Cutaneous Mycoses Affect the Skin, Hair, and Nails

Dermatophytosis (*dermato* = "skin"; *phyto* = "plant," referring to the days when fungi were grouped with plants) is a general name for a fungal disease of the hair, skin, or nails. The diseases are commonly known as **tinea infections** (*tinea* = "worm") because in ancient times, worms were thought to be the cause. These inflammation-causing infections of the living layers of skin and nails comprise several forms of **ringworm**, including athlete's foot (tinea pedis); head ringworm (tinea capitis); body ringworm (tinea corporis); groin ringworm or "jock itch" (tinea cruris); and nail ringworm (tinea unguium).

The infectious agents of dermatophytosis are a group of fungi called **dermatophytes**. Three of the most common genera, *Epidermophyton, Trichophyton,* and *Microsporum* (**FIGURE 18.20A**), can survive for weeks on wooden floors of shower rooms or on mats. Thus, transmission can be by contact and by sharing contaminated towels, combs, hats, and numerous other **fomites** (inanimate objects). Because tinea infections affect cats and dogs, these animals also can transmit the fungi to humans (**FIGURE 18.20B**).

These cutaneous mycoses are commonly accompanied by blister-like lesions. Often, a thin fluid exudes when the blisters are scratched or irritated, and as the blisters dry, they leave a scaly ring. Infection is limited due to the immune system's defensive inflammatory response.

Treatment of dermatophytosis often is directed at changing the conditions of the skin environment. Topical antifungal agents, such as undecylenic acid (Desenex®) and mixtures of acetic acid and benzoic acid (Whitfield's ointment), are active against the dermatophytes. In addition, tolnaftate (Tinactin®) and miconazole (Micatin®) are useful as topical agents for infections of the living skin.

A Mucocutaneous Mycosis Is Represented by Candidiasis

Yeasts in the Ascomycota that are a genus separate from *Saccharomyces* can cause opportunistic infections. The most common pathogen is *Candida albicans*, which is frequently present on the mucous membranes of the skin, mouth, vagina, and intestinal tract of healthy humans. Although *C. albicans* is dimorphic, at human body temperature the organism grows as filaments called **pseudohyphae** rather than as the yeast-like form. When immune system defenses are compromised, or when changes occur in the resident mycobiome on the body,

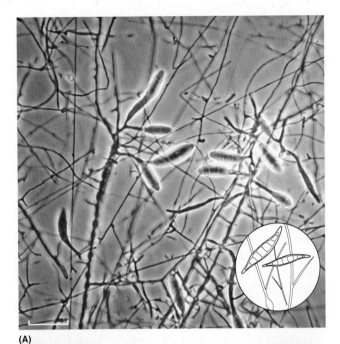

(A)

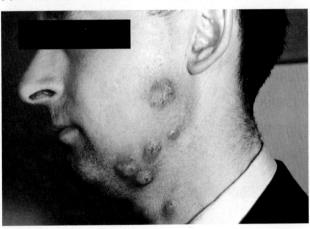

(B)

FIGURE 18.20 Ringworm. Ringworm of the skin or body (tinea corporis). **(A)** Light microscope photograph of *Microsporum*, one of the fungi causing ringworm on the scalp and body. (Bar = 15 µm.) **(B)** A case of body ringworm on the face and neck that was "caught" from a cat. »» *Why is the shape of each ringworm skin lesion in (B) circular?*

(A) Courtesy of Dr. Libero Ajello/CDC. (B) Courtesy of the CDC.

C. albicans flourishes on the skin and mucous membranes. Depending on the pathogen's body location, the result can be a mucocutaneous infection called **candidiasis**.

Vaginal Candidiasis

C. albicans is part of the resident vaginal mycobiome and, in opportunistic conditions, can cause **vaginal candidiasis** (vulvovaginitis). In the United States, every year there are some 20 million cases of vaginal candidiasis, often referred to as a "yeast infection." In immunocompromised individuals, a toxin normally produced by the hyphae of *C. albicans* cannot be neutralized by the immune system. As a result, the toxic protein causes membrane damage to the surrounding epithelial cells at the mucosal surface, and this injury results in cell leakage. Moreover, studies have shown that excessive antibiotic use destroys the bacterial lactobacilli normally present in the vaginal environment. Without lactobacilli as competitors, *C. albicans* flourishes.

Symptoms include an itching sensation, burning internal pain, and a white "cheesy" discharge. Reddening (erythema) and swelling of the vaginal tissues also occur. Diagnosis is performed by observing *C. albicans* in a sample of vaginal discharge or on a vaginal smear (**FIGURE 18.21A**) or by cultivating the organisms on laboratory media. Treatment is usually successful with nystatin (Mycostatin®) applied as a topical ointment or suppository. Miconazole, clotrimazole, and ketoconazole are useful alternatives.

Thrush

The oral cavity is another location where *C. albicans* can cause an opportunistic infection. **Oral candidiasis**, commonly called **thrush**, involves hyphal invasion of the mucosal tissue. This is manifested as small, white patches that appear on the mucous membranes of the oral cavity and then grow together to form soft, crumbly, milk-like curds (**FIGURE 18.21B**). When scraped off, a red, inflamed base

is revealed. Oral suspensions of gentian violet and nystatin (oral technique called "swish and spit") are effective for therapy. The disease is common in newborns, who acquire it during passage through the birth canal of infected mothers. Children also can contract thrush from contact with nursery utensils, toys, or the handles of shopping carts. The illness also can be an early sign of AIDS in an adult patient.

Other Mucocutaneous Forms

Candidiasis in the intestinal tract is closely tied to the use of antibiotics that destroy many of the bacterial cells normally found in the intestine. As with vaginal disease, the dysbiosis allows *C. albicans* to flourish. In the hospital or healthcare facility, the use of *Candida*-contaminated catheters and other medical devices can lead to bloodstream infections. Mortality rates can be as high as 50% in immunocompromised individuals.

Before leaving the discussion of *Candida* infections, mention should be made of the recent identification of *C. auris*, an emerging, multidrug-resistant species that can cause invasive infections with high mortality rates. Since 2009, the fungus has been identified in at least 12 countries on four continents. As of March 2017, the Centers for Disease Control and Prevention (CDC) had identified 35 cases of *C. auris* infection in the United States, mostly in New York State. All patients had a history of serious underlying medical conditions and the organism appears resistant to all three major classes of antifungal drugs. Preliminary analysis suggests

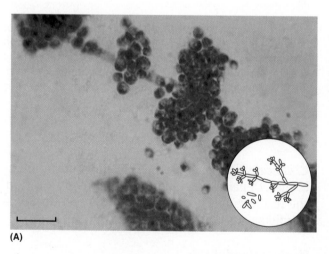

(A)

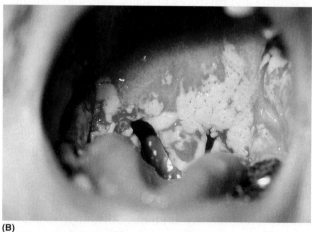

(B)

FIGURE 18.21 *Candida albicans.* **(A)** A light micrograph of stained *C. albicans* cells from a vaginal swab. (Bar = 60 μm.) **(B)** A severe case of oral candidiasis, showing a thick, creamy coating over the back of the oral cavity. *»» From photo (A), how do you know this specimen of* C. albicans *was from a human infection and not from the natural environment?*

(A) Courtesy of Dr. Godon Roberstad/CDC. (B) Courtesy of CDC.

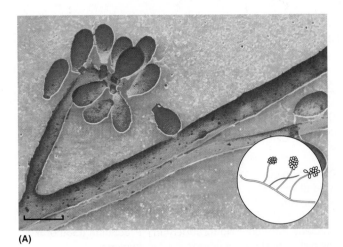

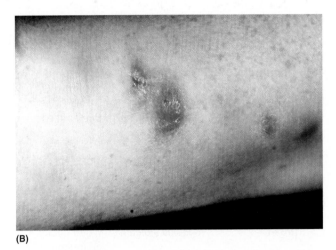

(A)   **(B)**

FIGURE 18.22 *Sporothrix schenckii.* **(A)** A false-color scanning electron micrograph of hyphae (orange) and conidia (purple) formed on conidiophores. (Bar = 8 μm.) **(B)** A patient showing the lesions of sporotrichosis on an infected arm. Characteristic "knots" can be felt under the skin. *»» Explain how a rose thorn harboring* S. schenckii *conidia can cause a skin disease.*

the fungus was transmitted in healthcare facilities, emphasizing the importance of maintaining strict infection control measures.

Sporotrichosis Is a Subcutaneous Mycosis

People who work with animals or with soil, thorny plants, and vegetation (e.g., sphagnum moss or bales of hay) can develop a subcutaneous mycosis in which the infection has established itself in the cutaneous and subcutaneous tissues. The infection, called **sporotrichosis**, is caused by the dimorphic fungus *Sporothrix schenckii* (**FIGURE 18.22A**). In one American outbreak, 84 cases of sporotrichosis occurred in people who handled conifer seedlings packed with sphagnum moss containing *S. schenckii*

conidia. Small cuts or scrapes on their hands were the portal of entry for the pathogen.

Symptoms of infection include pus-filled red or purple lesions at the site of entry, and "knots" can be felt under the skin (**FIGURE 18.22B**). Pneumonia-like lung infections can occur if airborne conidia are inhaled, and dissemination though the lymphatic system can spread the infection to bones, the joints, and the central nervous system. Usually, these types of disseminated infections occur only in immunocompromised patients. *S. schenckii* infections are controlled with oral itraconazole for 3 to 6 months, but severe systemic infections might require amphotericin B therapy.

TABLE 18.2 summarizes the fungal diseases of the skin.

Concept and Reasoning Checks 18.4

a. Describe the characteristics of dermatophytosis.
b. Describe the different forms of candidiasis based on body location.
c. Describe the characteristics of sporotrichosis.

 Chapter Challenge D

Dermatophytes comprise a group of fungi that can cause skin, hair, and nail infections. Infections caused by these fungi are also known as "ringworm" or "tinea" infections.

QUESTION D: *Suppose there is a ringworm outbreak in your child's school/daycare center. What should you do?*

You can find answers online in **Appendix F**.

TABLE 18.2 A Summary of the Fungal Skin Diseases

Inflammation or Disease	Causative Agent	Signs and Symptoms	Transmission	Treatment	Prevention
Dermatophytosis (Tineas)	*Epidermophyton Microsporum Trichophyton*	Blister-like lesions	Direct or indirect contact Soil	Antifungal medications	Keeping skin dry Not sharing personal items
Vaginal candidiasis	*Candida albicans*	Vaginal itching, irritation, white, thick discharge	Disruption of normal vaginal microbiota	Antifungal medications	Avoiding baths and hot tubs Avoiding douches
Thrush	*Candida albicans*	White flecks on mucous membranes	Passage through the birth canal	No treatment in children	Practicing good oral hygiene Limiting sugar intake
Sporotrichosis	*Sporothrix schenckii*	Pus-filled, purplish lesions	From plant material harboring the fungus	Itraconazole Amphotericin B	Wearing long sleeves and gloves while working with suspect materials or vegetation

■ KEY CONCEPT 18.5   Several Fungal Pathogens Cause Lower Respiratory Tract Diseases

Although humans are exposed to fungal spores constantly, only a small number of species cause deep mycoses of the lungs. Unfortunately, many of these can become systemic. Pathogens, such as *Histoplasma*, *Blastomyces*, and *Coccidioides*, are considered environmental pathogens. Others, like *Cryptococcus* and *Pneumocystis*, can exist at low levels in the body for long periods without causing disease. Should an individual become immunocompromised, the pathogens act opportunistically to cause dangerous, often life-threatening diseases.

Most of the fungal respiratory pathogens discussed below are dimorphic, and the pathogenic form is a yeast-like growth. The pathogens cause disease through direct invasion of tissues, where the organisms secrete enzymes that cause cell and tissue damage. Injury also arises from the damaging effects of the immune system's inflammatory response to the infection.

Cryptococcosis Usually Occurs in Immunocompromised Individuals

Cryptococcosis is caused by *Cryptococcus neoformans*, a basidiomycete that produces yeast-like cells

coated in a thick, gelatinous capsule (**FIGURE 18.23A**). Cryptococcosis is among the most dangerous fungal diseases in humans. Worldwide, there are an estimated 1 million new cases and 625,000 deaths each year from cryptococcal meningitis. *C. neoformans* is found in the soil of urban environments around the world, and the fungus grows actively in the droppings of pigeons but not within the pigeon tissues.

The spores of the fungus become airborne when dried bird droppings are stirred up by gusts of wind, and the spores subsequently enter the respiratory passageways of humans when inhaled (**FIGURE 18.23B**). Infection in healthy individuals usually produces a mild or asymptomatic pneumonia. However, in immunocompromised patients, the cryptococci cause an opportunistic mycosis. The encapsulated spores can pass into the bloodstream and localize in the meninges and brain, causing the patient to experience piercing headaches, stiffness in the neck, and paralysis. Left untreated, the ensuing meningoencephalitis can be fatal. However, intravenous treatment with the antifungal drug amphotericin B is usually successful. A new diagnostic test can detect within 10 minutes both early

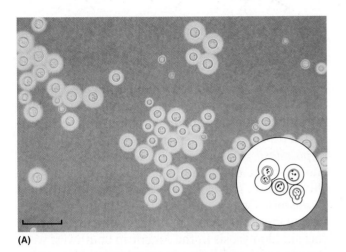

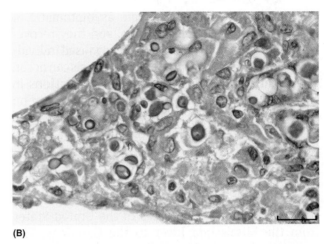

(A)

(B)

FIGURE 18.23 *Cryptococcus neoformans.* **(A)** A light microscope photomicrograph of *C. neoformans* cells. The white halo surrounding the cells is the capsule. (Bar = 20 μm.) **(B)** A stained photomicrograph of *C. neoformans* cells (red) from lung tissue of an AIDS patient. The capsule surrounding the cells provides resistance to phagocytosis and enhances the pathogenic tendency of the fungus. (Bar = 20 μm.) »» *Why would* C. neoformans *be a serious health threat to AIDS patients?*

and advanced cryptococcal infections with better than 95% accuracy.

Another species of *Cryptococcus, C. gattii,* emerged in British Columbia in 1999 and soon spread to the Pacific Northwest (Washington, Idaho, Oregon, and California) and western Canada. This subtropical fungus, which might have originated in Brazil, has adapted to a more temperate climate. Unlike *C. neoformans, C. gattii* is not associated with bird droppings; rather, it is found in the soil around eucalyptus and Douglas fir trees. After a person inhales the fungal spores, the incubation period can be as long as 13 months. Respiratory symptoms in immunocompromised patients include chest pain, shortness of breath, a persistent cough, and fever. Should the infection become systemic and spread from the lungs to the brain, symptoms include fever, headache, neck pain, vomiting and nausea, and confusion or behavioral changes.

Although antifungal drugs (amphotericin B) are available and must be taken for at least 6 months, up to 25% of individuals who develop systemic infections of *C. gatti* die, so public health officials must exert increased vigilance for this potentially serious, life-threatening infection. Presently there is no vaccine or preventive medications for either form of cryptococcosis.

Histoplasmosis Can Produce a Systemic Disease

Histoplasmosis is a lung disease that occurs worldwide. It is endemic in the Ohio and Mississippi River valleys where it is often called "summer flu." The causative agent is *Histoplasma capsulatum.* In the Midwest, estimates are that as many as 25,000 people develop histoplasmosis each year.

The lungs are the primary portal of entry, and infection usually occurs from the inhalation of spores present in dry, dusty soil contaminated with bird or bat droppings (**FIGURE 18.24**). Being a dimorphic fungus, it grows as a yeast form when infecting humans. Human-to-human transmission does not occur.

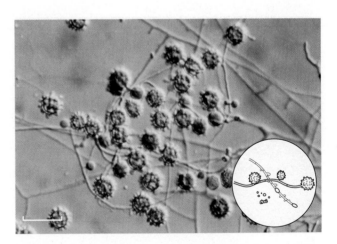

FIGURE 18.24 *Histoplasma* **Mycelium and Spores.** The spores are produced from the hyphal tips. (Bar = 20 μm.) »» *Which dimorphic form of the fungus is associated with growth in the soil?*

Most people infected remain asymptomatic or experience a mild flu-like illness, and they recover without treatment. In immunocompromised individuals, a disseminated form of histoplasmosis can occur. These patients develop tuberculosis-like lesions in the lungs and other internal organs. Depending on the severity of the disease, a patient might need to be on antifungal drugs for up to one year.

Blastomycosis Is an Uncommon but Serious Respiratory Infection

Blastomycosis occurs principally in Canada, the Great Lakes region, and areas of the United States from the Mississippi River to the Carolinas. The largest outbreak in United States history occurred in central Wisconsin in 2010 when 55 people were infected and two patients died.

The infectious agent, *Blastomyces dermatitidis*, is a dimorphic fungus typically found in moist soil, particularly in those soils near barns and sheds and in decomposing organic matter, such as wood and leaves (**FIGURE 18.25**). The fungus produces airborne conidia that are inhaled. Within the lungs, the spores germinate and grow as yeast-like cells.

About 50% of people infected with the fungus experience no symptoms. If symptoms do develop, they often are flu like. In immunocompromised patients, chronic pneumonia is the most common clinical manifestation. However, the pathogen can spread from the lungs and involve many internal organs (bones, liver, spleen, or central nervous system) and can be fatal. For severe cases, amphotericin B treatment can last for up to a year.

Coccidioidomycosis Can Become a Potentially Lethal Infection

Coccidioidomycosis, known more commonly as "valley fever" or "desert fever," is caused by *Coccidioides immitis* and *C. posadasii*. These fungal species naturally inhabit the arid soil in the southwestern United States (*C. immitis* in the central valley of California, *C. posadasii* in southern Arizona), parts of Mexico, and in parts of Central and South America. New clusters of *C. immitis* are spreading from California to eastern Oregon and eastern Washington State. During most of the 1980s, about 450 annual cases of coccidioidomycosis in the American Southwest were reported to the CDC. Due to recent population movements in this region, better reporting of cases, and increased lab testing, the number of reported cases in 2014 was more than 8,200, the majority occurring in California and Arizona. **CLINICAL CASE 18** describes an outbreak in northeastern Utah.

Coccidioides fungi are dimorphic. In the soil, the ascomycete fungus grows as a mycelium, and the hyphae produce arthrospores (**FIGURE 18.26**). In the soil, the arthrospores germinate and form another mycelium, which perpetuates the saprobic cycle.

When the soil is disrupted by farming, construction, dust storms, or even earthquakes, arthrospores can become airborne. If inhaled into the human lungs, the arthrospores become lodged in the terminal bronchioles where they develop into large multicellular spherules (**FIGURE 18.27**). These structures eventually rupture and release many endospores that can develop into more spherules, which perpetuates

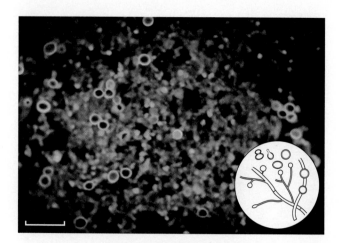

FIGURE 18.25 *Blastomyces* **Spores.** In this stained sample from infected lung tissue, *B. dermatitidis* spores are scattered throughout the tissue. (Bar = 10 μm.) »» *Which dimorphic form of the fungus is associated with human infections?*

Courtesy of CDC/Dr. Leo Kaufman.

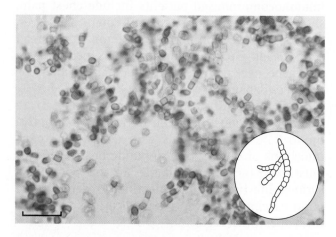

FIGURE 18.26 *Coccidioides* **Arthrospores.** Numerous barrel-shaped spores can be seen in this light microscope photo. (Bar = 20 μm.) »» *Where would one normally find these spores in the environment?*

Courtesy of CDC/Lucille Georg.

Clinical Case 18

An Outbreak of Coccidioidomycosis

On June 18, 2001, two archaeologists with the National Park Service (NPS) directed a team of six student volunteers and two leaders at an archaeological site in Dinosaur National Monument (DNM), a 515–square mile area in northeastern Utah and northwestern Colorado.

The volunteers and leaders laid stone steps, built a retaining wall, and sifted dirt for artifacts. Peak dust exposure occurred on June 19, the day most sifting occurred. Workers did not wear protective face masks.

Between June 29 and July 3, all of the eight team members and the two NPS archaeologists sought medical care at a local hospital

Modified from Dr. Errol Reiss/CDC.

emergency department for a flu-like illness with fever, cough, headache, rash, and muscle aches. Chest X-rays showed that all 10 individuals had fluid filling the pulmonary airspaces; eight individuals were hospitalized with pneumonia.

NPS closed the work site to all visitors and staff, and the TriCounty Health Department alerted the public. On July 2, the TriCounty Health Department, the Utah Department of Health, and the Centers for Disease Control and Prevention (CDC) initiated an investigation to identify the risk factors, cause, and extent of the outbreak.

During July 2–4, 18 persons (8 team members and 10 archaeologists) with potential exposure to dust at the work site in June were interviewed to determine symptoms and previous activities. Hospital records were reviewed to ascertain clinical information. A case definition was defined for a person working at DNM.

Results of blood cultures from the hospitalized persons were negative for bacterial pathogens. Initial serological tests were negative for antibodies to *Francisella tularensis, Yersinia pestis, Mycoplasma* species, *Histoplasma capsulatum*, and *Blastomyces dermatitidis.* Further serological tests to detect IgM antibodies (typically found in an early immunological response) to *Coccidioides immitis* showed positive precipitin lines indicative of IgM antibodies in 9 of the 10 acute serum specimens from patients, confirming the diagnosis of acute coccidioidomycosis (see the accompanying figure). All hospitalized patients were treated with fluconazole.

A serosurvey of 40 other park employees was conducted to identify other infected persons and to guide prevention and control measures. None of the 40 had detectable antibodies to *C. immitis.*

The DNM site, which reopened on September 28, is located approximately 200 miles north of areas where *C. immitis* is normally found.

Questions:

a. What would be present in the dust from the sifted dirt that would lead to the described illness?

b. A case definition is the method public health professionals use to define who is included as a "case" in an outbreak investigation. Provide a case definition for the illness described in this clinical case.

c. Why was a serosurvey seen as an important part of the outbreak analysis? What would health officials be looking for in a serosurvey, that is, in serum collected from the 40 other individuals?

d. What conclusions can you draw from the fact that this coccidioidomycosis outbreak was much farther north than where the disease is usually found?

e. How could these cases of coccidioidomycosis have been prevented?

You can find answers online in **Appendix E**.

For additional information, see www.cdc.gov/mmwr/preview/mmwrhtml/mm5045a1.htm.

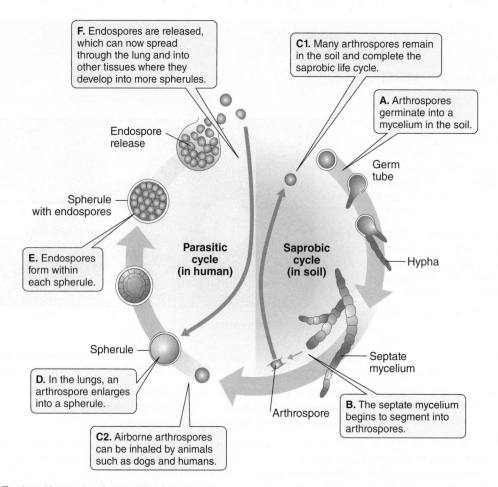

F. Endospores are released, which can now spread through the lung and into other tissues where they develop into more spherules.

C1. Many arthrospores remain in the soil and complete the saprobic life cycle.

A. Arthrospores germinate into a mycelium in the soil.

Germ tube

Endospore release

Spherule with endospores

Parasitic cycle (in human)

Saprobic cycle (in soil)

Hypha

E. Endospores form within each spherule.

Spherule

Septate mycelium

D. In the lungs, an arthrospore enlarges into a spherule.

Arthrospore

B. The septate mycelium begins to segment into arthrospores.

C2. Airborne arthrospores can be inhaled by animals such as dogs and humans.

FIGURE 18.27 The Life Cycle of *Coccidioides*. Outside the body, the fungus goes through a saprobic cycle. However, the arthrospores in the respiratory tract go through a parasitic cycle, producing endospores capable of forming more spherules and infecting other tissues such as the skin, bone, and central nervous system. »» *Explain why* Coccidioides *is considered a dimorphic fungus.*

the pathogenic cycle. Human-to-human transmission does not occur.

Most people who breathe in the spores remain asymptomatic. About 40% of infections develop a flu-like illness, with a dry, hacking cough, chest pains, and high fever. These individuals will recover on their own within several weeks to a few months. About 5% to 10% of cases involve immunocompromised patients. These individuals are at risk of developing a self-limiting pulmonary illness (community-acquired pneumonia). In about 1% of cases, the fungus becomes disseminated, and the systemic infection spreads from the lungs to the skin, bone, and the central nervous system, including the meninges of the spinal cord. Although most patients recover and have lifelong immunity, the CDC estimates that there are about 200 deaths each year from coccidioidomycosis.

Besides the clinical symptoms, confirmation of the diagnosis is done by a blood test for antibodies.

Therapy with amphotericin B for 6 months is prescribed for severe cases. Another drug, nikkomycin, and a vaccine are in the early stages of development.

Pneumocystis Pneumonia Can Cause a Lethal Infection

The most common form of nonbacterial pneumonia in Americans with AIDS is ***Pneumocystis* pneumonia (PCP)**. The causative organism, *Pneumocystis jirovecii* (previously called *Pneumocystis carinii*), is an obligate parasite (**FIGURE 18.28**). It is transmitted person to person by airborne respiratory droplets, although transmission from the environment also can occur. A wide cross section of individuals normally harbors the organism without symptoms, mainly because of the control imposed by the immune system.

P. jirovecii has a complex life cycle that takes place entirely in the alveoli of the lung. There, an active trophic (feeding) occurs in immunocompromised

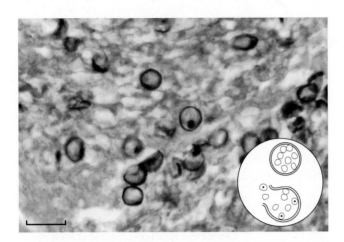

FIGURE 18.28 *Pneumocystis jirovecii* Infection. This biopsy is from lung tissues from an AIDS patient. (Bar = 10 μm.) *»» How is the organism transmitted person-to-person?*

Courtesy of CDC/ Dr. Edwin P. Ewing, Jr.

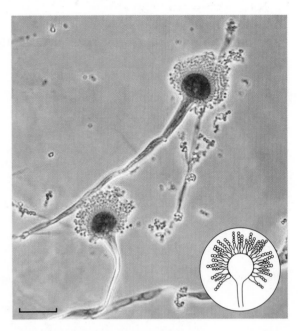

FIGURE 18.29 *Aspergillus* Structure. Atop the filamentous conidiophore are chains of spherical conidiospores that represent the asexual reproductive structures of this organism. (Bar = 10 μm.) *»» When the spores germinate and grow in the lungs, what type of structure results?*

Courtesy of CDC.

individuals with the *Pneumocystis* cells filling the alveoli and occupying the air spaces. A nonproductive cough develops, with fever and difficult breathing. Progressive deterioration leads to a dense mass of cells in the alveoli and, eventually, death through respiratory failure.

The current treatment for severe PCP is a 3-week therapy with trimethoprim-sulfamethoxazole (co-trimoxazole).

Other Fungi Also Cause Mycoses

A few other ascomycete fungal diseases deserve brief mention because they are important in certain parts of the United States, or they affect individuals in certain professions. Generally, the diseases are mild, although complications can lead to serious tissue damage in severely immunocompromised patients.

Invasive **aspergillosis** is a unique disease because the fungus enters the body as conidia and then transforms into a mycelium in the lungs. Disease usually occurs in immunocompromised individuals or those in whom an overwhelming number of conidia have entered the tissue. The most common agent of aspergillosis is *Aspergillus fumigatus,* which is found in decaying leaves, rotting vegetables, and stored grain (**FIGURE 18.29**). An opportunistic infection of the lung can yield a round-shaped ball of mycelium called a **pulmonary aspergilloma** ("fungus ball") that causes a bloody cough, chest pain, and wheezing with shortness of breath. The fungus does not disseminate to other areas of the body.

The most deadly form of aspergillosis is **invasive aspergillosis**. This form of the disease occurs when the fungal infection spreads beyond the lungs and disseminates to the other organs, such as the skin, heart, kidneys, or brain. Signs and symptoms depend on which organs are affected but, in general, include headache, fever with chills, bloody cough, shortness of breath, and chest or joint pain. Patients with severe cases of invasive aspergillosis might require surgery.

Treatment usually involves antifungal drugs such as voriconazole. It can be difficult to avoid *Aspergillus* spores in the environment. Staying away from obvious sources of mold, such as compost piles and damp places, can help prevent infection in susceptible individuals.

Although most fungal disease outbreaks come from environmental exposure, a multistate outbreak of **fungal meningitis** in 2012 occurred among patients being treated with injections of steroidal medications for back and neck pain. It was discovered that some of these steroidal medications were contaminated with *Exserohilum rostratum,* a common mold found in soil and on plants. In one case, *A. fumigatus* was identified in a contaminated preparation. In all, more than 500 individuals were infected, 35 of whom died from the infection.

TABLE 18.3 A Summary of the Major Fungal LRT Diseases

Disease	Causative Agent	Signs and Symptoms	Transmission	Treatment	Prevention
Cryptococcosis	*Cryptococcus neoformans* *Cryptococcus gattii*	Asymptomatic Opportunistic infection leads to severe headache, stiff neck, paralysis	Airborne yeast cells	Amphotericin B	Maintaining strong immune system
Histoplasmosis	*Histoplasma capsulatum*	Mild influenza-like illness Can disseminate to other organs	Airborne spores	Amphotericin B or ketoconazole for systemic disease	Wearing face mask in contaminated areas
Blastomycosis	*Blastomyces dermatitidis*	Persistent cough Chest pains Chronic pneumonia	Airborne spores	Amphotericin B	Wearing face mask in contaminated areas
Coccidioidomycosis	*Coccidioides immitis* *Coccidioides posadasii*	Dry, hacking cough Chest pains High fever	Airborne arthrospores	Amphotericin B	Limiting exposure where infection is highest
Pneumocystis pneumonia	*Pneumocystis jirovecii*	Nonproductive cough Fever Breathing difficulty	Airborne droplets	Trimethoprim-sulfamethoxazole	Maintaining strong immune system
Aspergillosis	*Aspergillus fumigatus*	Bloody cough Chest pain Wheezing Shortness of breath	Airborne spores	Voriconazole	Staying away from sources of mold

TABLE 18.3 summarizes the fungal diseases of the lower respiratory tract.

In conclusion, we have learned about several of the fungal infections and diseases that affect humans. In the global picture, fungi infect billions of people every year, yet their significance to worldwide infectious disease remains mostly unrecognized. Although there is no accurate record keeping for the diseases caused by the fungal pathogens of humans, experts say that, taken as a whole, these pathogens kill as many people every year as tuberculosis or malaria. Part of the increasing incidence of fungal diseases is due to immunosuppressive diseases, like AIDS, that weaken the immune system's response to opportunistic infections such as cryptococcosis and other respiratory diseases we discussed in this chapter. Importantly, this increased incidence of fungal diseases is not limited to humans. In the past 20 years, an extraordinary number of fungal diseases have caused severe die-offs and extinctions. As described in this chapter, frog populations, especially in Central America, are being decimated by a fungal disease, and the bat population in North America could be greatly diminished by a deadly fungus causing white-nose syndrome. In addition, collapse of honeybee colonies in 2010 was linked to a co-infection of fungus and virus that might have been responsible for 40% to 60% of disappearing beehives since 2005. So, other animals, as well as humans, are at risk for life-threatening fungal infections.

Concept and Reasoning Checks 18.5

a. Explain why cryptococcosis is such a dangerous fungal disease.
b. How does histoplasmosis differ in "healthy" individuals versus in immunocompromised individuals?
c. Why is blastomycosis a dangerous disease in immunocompromised individuals?
d. What is the unique feature of a coccidioidomycosis infection?
e. What are the unique features of *P. jirovecii*?
f. Summarize the unique features of aspergillosis.

Chapter Challenge E

Coccidioidomycosis (valley or desert fever) is a fungal disease that local and state health departments must report to the CDC. As mentioned in this section, in 2014 there were more than 8,200 reported cases of coccidioidomycosis, the majority of which were located in Arizona and California.

QUESTION E: *Based on the accompanying graph, provide (a) some reasons why there was a steady increase in reported cases through 2011 and (b) a decrease in 2012 through 2014.*

You can find answers online in **Appendix F**.

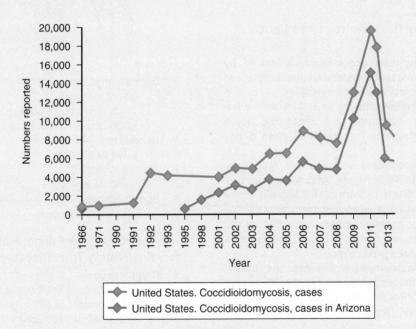

Coccidioidomycosis in the United States.

Courtesy of GIDEON (www.gideononline.com), Coccidioidomycosis in the United States. Data from the CDC.

■ SUMMARY OF KEY CONCEPTS

Concept 18.1 The Kingdom Fungi Includes the Molds and Yeasts

1. Fungi are eukaryotic microorganisms with **heterotrophic** metabolism. Most fungi consist of masses of **hyphae** that form a **mycelium. Cross walls** separate the cells of hyphae in many fungal species. Most fungi are **coenocytic**. (Figure 18.3)

Concept 18.2 Fungal Growth and Reproduction Are Dependent on Spores

2. Fungi secrete enzymes into the surrounding environment and absorb the breakdown products. Tremendous absorption can occur when there is a large mycelium surface. Most fungi are aerobic, grow best around 25°C (77°F), and prefer slightly acidic conditions.
3. Reproductive structures generally occur at the tips of hyphae. Masses of asexually or sexually produced **spores** are formed within or at the tips of **fruiting bodies**. (Figure 18.9)

Concept 18.3 Fungi Have Evolved into a Variety of Forms

4. The phylum **Chytridiomycota** is characterized by motile cells, whereas the **Glomeromycota** represent fungi living symbiotically with land plants.
5. In the phylum **Zygomycota**, the sexual phase is characterized by the formation of a **zygospore**, which releases haploid spores that germinate into a new mycelium.
6. The phylum **Ascomycota** includes the unicellular yeasts and filamentous molds. **Ascospores** are produced that germinate to form a new haploid mycelium, whereas asexual reproduction is through the dissemination of **conidia. Lichens** are a mutualistic association between an ascomycete and either a cyanobacterium or a green alga.
7. The phylum **Basidiomycota** includes the **mushrooms**. Within these fruiting bodies, **basidiospores** are produced. On germination, they produce a new haploid mycelium. "Rusts" and "smuts" that cause many plant diseases are additional members of the phylum.
8. The **mitosporic fungi** lack a sexual phase. Many human fungal diseases involve fungi in this informal group.
9. *Saccharomyces* is a notable unicellular ascomycete **yeast** involved in baking and brewing and scientific research.

Concept 18.4 Some Fungi Can Invade the Skin

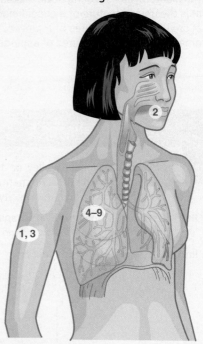

▶ **Dermatophytosis**
  1 *Microsporum*
    Trichophyton (nail plate; not shown)
    Epidermophyton (webs of toes; not shown)
▶ **Candidiasis (vulvovaginitis)** (not shown)
    Candida albicans
▶ **Thrush** (children)
  2 *Candida albicans*
▶ **Sporotrichosis**
  3 *Sporothrix schenckii*

Concept 18.5 Many Fungal Pathogens Cause Lower Respiratory Tract Diseases

▶ **Cryptococcosis**
  4 *Cryptococcus neoformans, C. gatti*
▶ **Histoplasmosis**
  5 *Histoplasma capsulatum*
▶ **Blastomycosis**
  6 *Blastomyces dermatitidis*
▶ **Coccidioidomycosis**
  7 *Coccidioides immitis, C. posadasii*
▶ ***Pneumocystis* pneumonia**
  8 *Pneumocystis jirovecii*
▶ **Aspergillosis**
  9 *Aspergillus fumigatus*

■ CHAPTER SELF-TEST

You can find answers for **Steps A–D** online in **Appendix D**.

■ STEP A: REVIEW OF FACTS AND TERMS

Multiple Choice

Read each question carefully, and then select the *one* answer that best fits the question or statement.

1. Which one of the following statements about fungi is *not* true?
   A. Some fungi are dimorphic.
   B. Fungi have cell walls made of chitin.
   C. Fungi are photosynthetic organisms.
   D. Fungi consist of the yeasts and molds.

2. Which one of the following best describes the growth conditions for a typical fungus?
   A. pH 3; 23°C; no oxygen gas present
   B. pH 8; 30°C; no oxygen gas present
   C. pH 6; 30°C; oxygen gas present
   D. pH 3; 23°C; oxygen gas present

3. All the following are examples of asexual spore formation except _____.
   A. arthrospores
   B. conidia
   C. sporangiospores
   D. basidiospores

4. An organism without a known sexual stage would be classified in the _____.
   A. mitosporic fungi
   B. Zygomycota
   C. Basidiomycota
   D. Ascomycota

5. Yeasts of the species *Saccharomyces* _____.
   A. are used in bread making
   B. reproduce by budding
   C. are members of the ascomycetes
   D. All of the above (**A–C**) are correct.

6. Aflatoxin is produced by _____ and is _____.
   A. *Sporothrix schenckii*; a hallucinogen
   B. *Aspergillus flavus*; carcinogenic
   C. *Claviceps purpurea*; a hallucinogen
   D. *Aspergillus niger*; a mycotoxin

7. This fungal disease can cause a blister-like lesion on the scalp.
   A. Candidiasis
   B. Dermatophytosis
   C. Cryptococcosis
   D. Histoplasmosis

8. _____ causes more than 20 million cases each year, and symptoms include an itching sensation and burning internal pain.
   A. Thrush
   B. "Jock itch"
   C. Vulvovaginitis
   D. Sporotrichosis

9. From the following, sporotrichosis would most likely be transmitted from _____.
   A. peat moss
   B. bat caves
   C. mushrooms
   D. dusty soil

10. Which one of the following fungi would most likely be found in pigeon droppings?
    A. *Pneumocystis*
    B. *Cryptococcus*
    C. *Coccidioides*
    D. *Sporothrix*

11. Moving to the Ohio or Mississippi River valleys might make one susceptible to _____.
    A. PCP
    B. valley fever
    C. dermatophytosis
    D. histoplasmosis

12. This dimorphic fungus produces conidia that are inhaled from dusty soil or bird droppings.
    A. *Aspergillus fumigatus*
    B. *Pneumocystis jirovecii*
    C. *Candida albicans*
    D. *Blastomyces dermatitidis*

13. The formation of arthrospores and spherules is characteristic of _____.
    A. coccidioidomycosis
    B. histoplasmosis
    C. candidiasis
    D. aspergillosis

14. This fungal disease is the most common cause of nonbacterial pneumonia in immunocompromised individuals.
    A. Blastomycosis
    B. *Pneumocystis* pneumonia
    C. Aspergillosis
    D. Coccidioidomycosis

15. The deadliest form of aspergillosis is _____.
    A. a pulmonary form
    B. a toxigenic form
    C. a blood form
    D. an invasive form

Matching

Match the statement on the left to the organism on the right by placing the correct letter in the available space. You may use a letter once, more than once, or not at all.

Statement

_____ 16. Produces a widely used antibiotic
_____ 17. Used for bread baking
_____ 18. Causes valley fever in the southwestern United States
_____ 19. Agent of rose thorn disease
_____ 20. Associated with the droppings of pigeons
_____ 21. Cause of vaginal yeast infections
_____ 22. One of the causes of dermatophytosis
_____ 23. Often found in chicken coops and bat caves
_____ 24. Reproduction includes a spherule

Organism

a. *Aspergillus fumigatus*
b. *Blastomyces dermatitidis*
c. *Candida albicans*
d. *Coccidioides immitis*
e. *Cryptococcus neoformans*
f. *Epidermophyton* species
g. *Histoplasma capsulatum*
h. *Penicillium notatum*
i. *Pneumocystis jirovecii*
j. *Saccharomyces cerevisiae*
k. *Saccharomyces ellipsoideus*
l. *Sporothrix schenckii*

STEP B: CONCEPT REVIEW

25. Differentiate between **molds, yeasts**, and **dimorphic fungi. (Key Concept 18.1)**
26. Differentiate the forms of asexual and sexual spores produced by fungi, and describe the generalized sexual life cycle of a mold. **(Key Concept 18.2)**
27. Identify the key characteristics of the **Zygomycota, Ascomycota**, and **Basidiomycota. (Key Concept 18.3)**

28. Summarize the symptoms and treatment of **dermatophytosis. (Key Concept 18.4)**
29. Summarize the symptoms and complications of **cryptococcosis. (Key Concept 18.5)**
30. Describe the symptoms of and the complications arising from **coccidioidomycosis. (Key Concept 18.5)**
31. Explain the potential seriousness of a Pneumocystis **pneumonia** infection. **(Key Concept 18.5)**

STEP C: APPLICATIONS AND PROBLEM SOLVING

32. You decide to make bread. You let the dough rise overnight in a warm corner of the room. The next morning you notice a distinct beer-like aroma in the air. What did you smell, and where did the aroma come from?
33. In a Kentucky community, a crew of five workers demolished an abandoned building. Three weeks later, all five required treatment for acute respiratory illness, and three were hospitalized. Cells obtained from the patients by lung biopsy revealed oval bodies, and epidemiologists found an accumulation of

bat droppings at the demolition site. As the head epidemiologist, can you discern what disease the workers contracted?
34. A woman comes to you complaining of a continuing problem of ringworm, especially of the lower legs in the area around the shins. Questioning her reveals she has five very affectionate cats at home. What disease does she have, and what would be your suggestion to her?

35. Using the diagram below, explain why these fungal diseases are considered systemic mycoses.

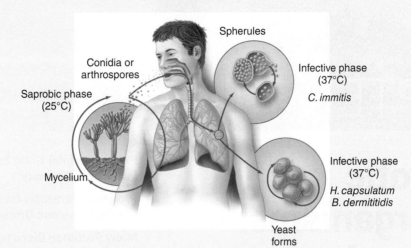

■ STEP D: QUESTIONS FOR THOUGHT AND DISCUSSION

36. In a suburban community, a group of residents obtained a court order preventing another resident from feeding the flocks of pigeons that regularly visited the area. Microbiologically, was this action justified? Why?

37. Mr. A and Mr. B live in an area of town where the soil is acidic. Oak trees are common, and azaleas and rhododendrons thrive in the soil. In the spring, Mr. A spreads lime on his lawn, but Mr. B prefers to save the money. Both use fertilizer, and both have magnificent lawns. Come June, however, Mr. B notices that mushrooms are popping up in his lawn and that brown spots are beginning to appear. By July, his lawn has virtually disappeared. What is happening in Mr. B's lawn, and what can Mr. B learn from Mr. A?

38. Residents of a New York community, unhappy about the smells from a nearby composting facility and concerned about the health hazard posed by such a facility, had the air at a local school tested for the presence of fungal spores. Investigators from the testing laboratory found abnormally high levels of *Aspergillus* spores on many inside building surfaces. Is there any connection between the high spore count and the composting facility? Is there any health hazard involved?

39. Many years ago, a serious earthquake struck the Northridge section of Los Angeles County in California. From that date through March 15, 170 cases of coccidioidomycosis were identified in adjacent Ventura County. This number was almost four times the previous year's number of cases. What is the connection between the two events?

CHAPTER 19

Eukaryotic Microorganisms: The Parasites

Every year, more than 600,000 people die from malaria, an infection caused by a protistan parasite that is carried from person to person by mosquitoes. The disease is one of the most severe public health problems worldwide and is a leading cause of death and disease in many developing nations. Yet in 1957, a global program to eradicate the parasite commenced only to end in failure 21 years later. What happened?

You might be surprised to learn that malaria once was an infectious killer in the United States. In the late 1880s, malaria was common in the American plains and southeast with epidemics reaching as far north as Montreal, Canada. Malaria was a major source of casualties in the American Civil War, and malaria remained endemic in the southern states until the 1930s (see photo at top).

To eliminate malaria, American officials established the National Malaria Eradication Program on July 1, 1947. This was a cooperative undertaking between the newly established Communicable Disease Center (the original CDC) and state and local health agencies of the 13 malaria-affected southeastern states. In 1947, 15,000 malaria cases were reported. However, by 1950, after more than 4,650,000 homes had been sprayed with the pesticide DDT (dichlorodiphenyltrichloroethane) to

kill the mosquitoes, only 2,000 malaria cases were reported. By 1952, the United States was malaria free, and the program ended.

Encouraged by the success of the American eradication effort, in 1957 the World Health Organization (WHO) began a similar effort to eradicate malaria worldwide. These efforts involved house spraying with insecticides, antimalarial drug treatment, and surveillance. Successes were made in some countries, but the emergence of drug resistance, widespread mosquito resistance to DDT, wars, and massive population movements made the eradication efforts unsustainable. The eradication campaign was eventually abandoned in 1978.

Now, nearly 40 years later, a substantial portion (3.3 billion) of the world's population is still at risk

---

An aircraft spraying an insecticide during malarial control operations in the southern United States in the 1930s.
Courtesy of CDC.

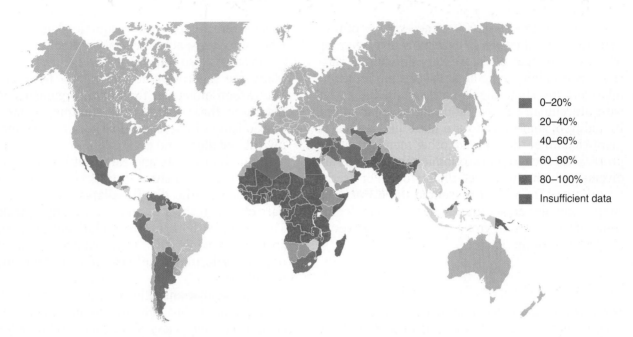

FIGURE 19.1 Percent of Population at Risk of Malaria—2016. Malaria occurs in 97 countries and puts an estimated 3.3 billion people at risk. »» *How can almost 50% of the world's population be at risk?*

Data from World Malaria Report. Medicines for Malaria Venture/WHO Global Malaria Programme.

of malaria (**FIGURE 19.1**). The good news is that another global campaign, initiated by the WHO, several United Nations agencies, and the World Bank, is in progress. As we will see, there is great hope for that program.

Malaria is just one of a number of human diseases caused by **parasites**, organisms that must live in or on a different species to get their nourishment and to reproduce. There are two different groups of eukaryotic parasites of concern to microbiology because of their ability to cause infectious disease. One group contains single-celled **protists**. Some of the diseases they cause are familiar to us, such as malaria. Others affect the intestine, blood, or other organs of the body. The second group are multicellular parasites, referred to as **helminths** (*helminth* = "worm"). These include the flatworms and roundworms, which together probably affect more people worldwide than any other group of organisms. In the strict sense, flatworms and roundworms are animals, but they often are studied in microbiology because of the helminth's ability to cause disease. Together with the parasitic protists, they are the subject of study of the biological discipline known as **parasitology**.

Chapter Challenge

The protists are by far the most diverse of any eukaryotic group of organisms, which illustrates the broad evolutionary relationships among these microorganisms. Therefore, the term "protist" does not describe a close-knit taxonomic group but rather it is an "umbrella" term for any eukaryotic organism that is not plant, animal, or fungus. In this chapter's challenges, let's see if we can solve a few problems about the protists and then finish with a challenge that includes the parasitic helminths.

■ KEY CONCEPT 19.1   Protists Exhibit Great Structural and Functional Diversity

The protists were first seen by Antony van Leeuwenhoek more than 300 years ago when he wrote in a letter to the Royal Society, "*No more pleasant sight has met my eye than this.*" Indeed, many scientists and microbiologists have continued to study this structurally and functionally diverse collection of eukaryotes.

There are about 200,000 named species of protists. However, they are an extremely diverse group of eukaryotes that often are very difficult to classify. For example, some algal species are multicellular, and the slime molds have unique life cycles with unicellular, colonial, and multicellular stages. In addition, some species are more like animal or plant cells than they are like other members of the protists. As such, their taxonomic relationships are diverse and not always well understood. For us, the evolutionary framework presented in **FIGURE 19.2** simply provides an accepted scheme for cataloging some of the protistan groups that we will survey in the first part of this chapter.

Most Protists Are Unicellular and Nutritionally Diverse

Many protists are unicellular, free living, and thrive where there is water. They can be located in damp soil, mud, drainage ditches, and puddles. Some species remain attached to aquatic plants or rocks, whereas other species swim about freely.

Protists also are very diverse nutritionally. Many are heterotrophic and obtain their energy and carbon molecules from other organisms, often obtaining these materials by forming a parasitic relationship with a susceptible host. Others, including the **red algae** and **green algae** (Archaeplastida), contain chloroplasts and carry out photosynthesis similar to green plants (**FIGURE 19.3A**). Some protists, such as the **dinoflagellates** (Alveolata), are part of the freshwater and marine **phytoplankton**. This includes both heterotrophic and autotrophic species, making the phytoplankton important organisms forming the base of food webs in the world's oceans.

The **radiolarians** (Rhizaria) also are part of the phytoplankton but are unique in having highly sculptured and glassy silica plates with radiating cytoplasmic arms to capture prey (**FIGURE 19.3B**).

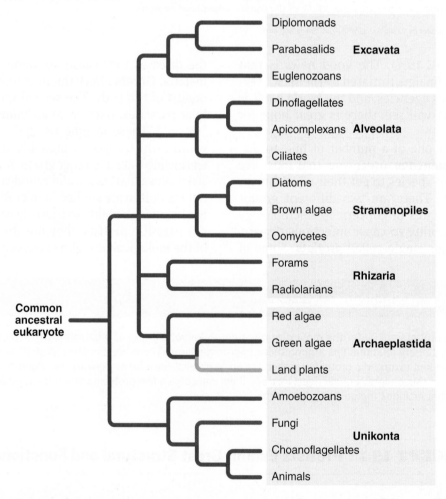

FIGURE 19.2 A Tentative Phylogeny for the Domain Eukarya. The phylogeny includes the six eukaryotic "supergroups" (in boldface). Everything besides the land plants and animals can be considered microbial, and everything besides the land plants, animals, and fungi would be considered protists. »» *What kingdom of organisms is most closely related to the animals?*

Among the most abundant eukaryotic microbes are the **foraminiferans** (**forams**; Rhizaria). They are marine protists that have chalky skeletons, often in the shape of snail shells with openings between sections (*foram* = "little hole"). The shells of dead forams form oceanic sediments of limestone that are hundreds of meters thick. When brought to the surface by geological upthrusting, massive white cliffs are exposed (**FIGURE 19.3C**).

The **diatoms** are the most diverse group (Stramenopiles) of single-celled protists. By being encased in a two-part, hard-shelled silica wall (*diatomos* = "cut in half"), the cells can withstand great pressures and are not easily crushed or destroyed by predators (**FIGURE 19.3D**). Diatoms carry out photosynthesis and compose an important part of the phytoplankton found in marine and freshwater environments. The massive accumulations of fossilized diatom walls are mined and ground up

into **diatomaceous earth**, which has many useful applications ranging from a filtering agent in swimming pools to a mild abrasive in household products, including toothpastes and metal polishes. Diatomaceous earth also can be used as a pesticide, because it grinds holes in the exoskeleton of crawling insects, causing the animals to desiccate.

Yet another group of protists is the **oomycetes** (Stramenopiles). These protists resemble fungi because they produce filamentous growth characteristic of the molds, and they are completely heterotrophic, absorbing extracellularly digested food materials. Some oomycetes are plant pathogens. *Phytophthora ramorum*, for example, infects the California coastal live oak and causes sudden oak death. *P. infestans* is the agent of late blight in potatoes. Being responsible for the Irish famine of the 19th century, its infection had a great impact on humans and society, as MICROFOCUS 19.1 recalls.

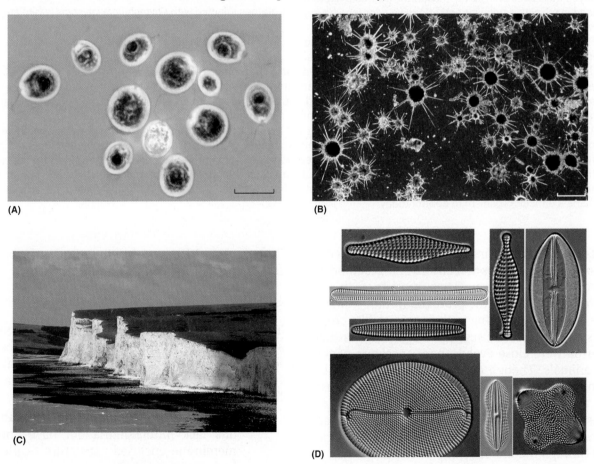

(A)

(B)

(C)

(D)

FIGURE 19.3 Algae and Phytoplankton. (A) A light micrograph of the green alga *Chlamydomonas.* (Bar = 10 μm.) **(B)** Light micrograph of an assortment of radiolarians. Radiolarians build hard skeletons made of silica around themselves as they float in seas with other plankton. (Bar = 120 μm.) **(C)** The white chalk cliffs at Beachy Head in Sussex, England, consist of the ancient shells of foraminiferans. **(D)** The diversity in shape of the diatoms. »» *What characteristics are similar in the algae and phytoplankton?*

MICROFOCUS 19.1: History

The Great Famine

Ireland of the 1840s was an impoverished country of 8 million people. Most were tenant farmers paying rent to land-lords who were responsible, in turn, to the English property owners. The sole crop of Irish farmers was potatoes, grown season after season on small tracts of land. What little corn was available was usually fed to the cows and pigs.

Early in the 1840s, heavy rains and dampness portended calamity. Then, on August 23, 1845, *The Gardener's Chronicle and Agricultural Gazette* reported, "A fatal malady has broken out amongst the potato crop. On all sides we hear of the destruction. In Belgium, the fields are said to have been completely desolated."

The potatoes had suffered before. However, nothing was quite like this new disease. Beginning as black spots, it decayed the leaves and stems and left the potatoes a rotten, mushy mass with a peculiar and offensive odor (see figure). Even the harvested potatoes rotted.

The winter of 1845 to 1846 was a disaster for Ireland. Farmers left the rotten potatoes in the fields, and the disease

Potatoes infected with *Phytophthora infestans* (white areas).

Tony Cunningham/Alamy.

spread. Facing starvation, the farmers ate the animal feed and then the animals themselves. They also consumed the seed potatoes, leaving nothing for spring planting. As starvation spread, the English government attempted to help by importing corn and establishing relief centers. In England, the potato disease had few repercussions because English agriculture included various grains. In Ireland, however, famine spread quickly.

After two years, the potato rot seemed to slacken, but in 1847 ("Black '47") it returned with a vengeance. Despite relief efforts by the English, over 2 million Irish people died from starvation. At least 1 million Irish people left the land and moved to cities or foreign countries. During the 1860s, great waves of Irish immigrants came to the United States—and in the next century, their Irish American descendants would influence American culture and politics (today, 11% of the American population is of Irish descent). And it all resulted from a filamentous protist, *Phytophthora infestans*, which remains a difficult organism to control today.

The protists also include many motile, predatory or parasitic species that absorb or ingest food by infecting living organisms. These protists traditionally have been called the **protozoa** (*proto* = "first"; *zoo* = "animal"), referring to their animal-like properties that incorrectly suggested to biologists that protozoa were close evolutionary ancestors of the first animals. Though often studied by zoologists, protozoa also interest microbiologists because they are unicellular, most have a microscopic size, and several are responsible for infectious disease. Therefore, we will emphasize these microbial parasites through the first half of this chapter.

The Protists Encompass a Variety of Parasitic Lifestyles

Based on comparative studies involving genetic analyses and genomics, a tentative taxonomy places the human protist parasites in one of three informal

"supergroups." Here, we briefly consider the biological features of the three groups.

Supergroup Excavata

The supergroup Excavata (see Figure 19.2) contains species that are single celled and possess flagella for motility. Some members in the group might represent organisms whose ancestors were the earliest forms of eukaryotes.

▶ **Parabasalids.** Members of the **parabasalids** lack mitochondria but have another membrane-enclosed structure called a **hydrogenosome**. This organelle can make ATP through a fermentation-like process that produces acetate and hydrogen gas as end products. As such, these protists live in low oxygen or anaerobic environments. Several species, including *Trichonympha*, are found in the guts of termites where the symbionts

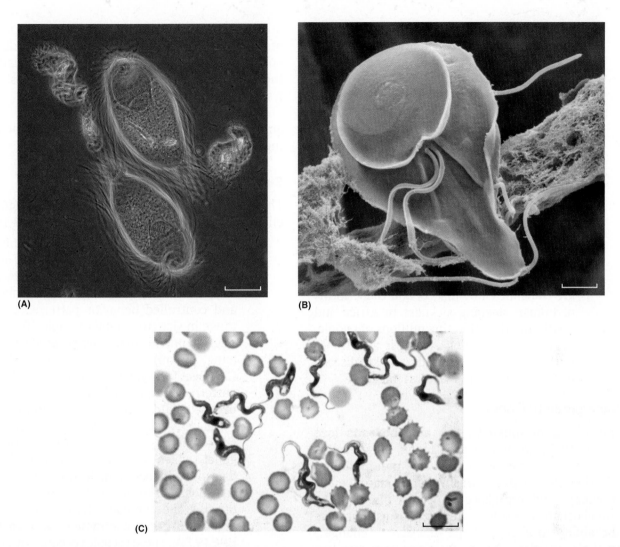

FIGURE 19.4 Parabasalid, Diplomonad, and Kinetoplastid Parasites. (A) A light micrograph of *Trichonympha*, a parabasalid parasite found in the gut of termites. Each thin, wispy line represents a flagellum used for motility. (Bar = 25 μm.) **(B)** A false-color scanning electron micrograph of *Giardia intestinalis*. The pear-shaped cell body of this diplomonad and flagella are typical features. (Bar = 5 μm.) **(C)** Light micrograph of *Trypanosoma* in a blood smear. The wavy cell appearance is due to the wavy membrane and flagellum. (Bar = 10 μm.) **»» *If the parabasalids and diplomonads lack (or have modified) mitochondria, what metabolic pathway remains for synthesizing ATP?***

participate in a mutualistic relationship (**FIGURE 19.4A**). The cells of *Trichonympha* contain hundreds of flagella with the characteristic arrangement of microtubules found in all eukaryotic flagella. Another species, *Trichomonas vaginalis*, is a human parasite and is transmitted through sexual intercourse.

▶ **Diplomonads.** The **diplomonads** have two haploid nuclei and three pairs of flagella at the anterior end and one pair of flagella at the posterior end, giving the cell bilateral symmetry. These protists only reproduce asexually by binary fission, they have mod-

ified mitochondria, and they are thought to represent one of the earliest branches of eukaryotic evolution.

The most notable species is *Giardia intestinalis* (**FIGURE 19.4B**). This potential pathogen can contaminate water and, thus, affects the gastrointestinal tract. The diplomonad can survive outside the anaerobic environment of the intestine by forming a **cyst**, which is a dormant, highly resistant stage. Many lakes and rivers in the United States are contaminated with the cysts, requiring hikers and campers to boil or filter the water before drinking.

▶ **Euglenozoans.** A third group of protists in the Excavata is the **Euglenozoans**. Among these are the **kinetoplastids**, another ancient lineage of heterotrophic species. A unique characteristic of these protists is a single, posterior flagellum that is attached to the cell's wavy, undulating membrane (**FIGURE 19.4C**). The kinetoplastids have a single mitochondrion containing the **kinetoplast**, which possesses thousands of copies of mitochondrial DNA.

Some 60% of the kinetoplastid species are trypanosomes (*trypano* = "hole"; *soma* = "body"), referring to the hole the organism bores through the skin to enter and infect the host. Two *Trypanosoma* species are transmitted by insects and cause forms of human sleeping sickness in Africa and South America, affecting millions of people. Species of *Leishmania*, which is also transmitted by insects, can produce a skin disease or a potentially fatal visceral infection.

Supergroup Unikonta

One group of unikonts, the **amoebozoans** (see Figure 19.2), are mostly free-living, single-celled organisms that can be as large as 1 mm in diameter. They usually live in freshwater or marine environments and reproduce by binary fission. The amoebozoans are soft-bodied organisms that have the ability to change shape (*amoeba* = "change") as their cytoplasm flows into temporary formless projections called **pseudopods** (*pseudo* = "false"; *pod* = "a foot"); thus, the motion is called **amoeboid motion**. Pseudopods also capture bacterial cells and

other protistan cells through the ingestive process of **phagocytosis** (**FIGURE 19.5A**). The pseudopods enclose the particles to form an organelle called a **food vacuole**, the contents of which are digested by lysosomal enzymes.

The species *Entamoeba histolytica* can cause amebic dysentery from drinking water or consuming food contaminated with amoebal cysts.

Supergroup Alveolata

This group is very diverse and includes the dinoflagellates described earlier. Two more groups are the

▶ **Ciliates.** Within the **Alveolata** are the ciliated protists, or **ciliates**, that can be found in almost any pond water sample. They have a variety of shapes and can exhibit elaborate and controlled behavior patterns. Ciliates range in size from a microscopic 10 μm to a huge 3 mm. All ciliates are covered with **cilia** (sing., cilium) in longitudinal or spiral rows. Cilia beat in a synchronized and coordinated pattern, the organized "rowing" action moving the ciliate along in one direction.

Ciliates, such as *Paramecium*, are heterotrophic by ingestion through a primitive gullet, which sweeps in food particles for digestion. In addition, freshwater protozoa continually take in water by the process of osmosis and eliminate the excess water via organelles called **contractile vacuoles** (**FIGURE 19.5B**). These vacuoles expand with water drawn from the cytoplasm and then appear to "contract" as they release water through a temporary opening in the cell membrane.

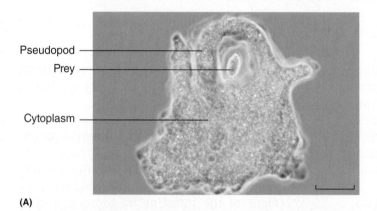

Pseudopod ⎯⎯⎯⎯
Prey ⎯⎯⎯⎯

Cytoplasm ⎯⎯⎯⎯

(A)

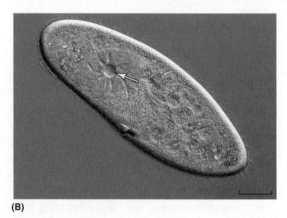

(B)

FIGURE 19.5 An Amoeba and a Ciliate. (A) A light micrograph of *Amoeba proteus.* A pseudopod is extending around the prey, beginning the process of phagocytosis. (Bar = 100 μm.) **(B)** This light micrograph of the ciliate *Paramecium* shows the contractile vacuole (arrow). (Bar = 50 μm.) »» *What is the function of the contractile vacuole?*

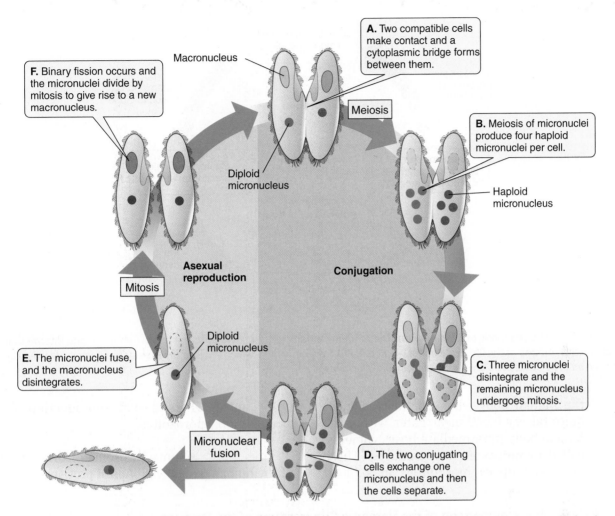

FIGURE 19.6 Conjugation and Reproduction in *Paramecium*. Ciliates, such as *Paramecium*, reproduce sexually by the process of conjugation. In this process, an exchange of micronuclei gives rise to new macronuclei. »» *Provide a hypothesis to explain why ciliates, unlike most other organisms, must have two types of nuclei.*

Asexual reproduction in the ciliates occurs by binary fission. However, the complexity of ciliates is illustrated by the nature of sexual reproduction. Ciliates have two types of nuclei: a single, large **macronucleus** that has multiple copies of genes for cell metabolism and a **micronucleus** that undergoes a form of sexual reproduction called **conjugation**. **FIGURE 19.6** outlines the conjugation process and nuclear exchange.

▶ **Apicomplexans.** The last group of alveolates we will mention is the **Apicomplexans**, so named because the *api*cal tip of the cell contains a *complex* of organelles used for penetrating host cells. Adult apicomplexan cells have no cilia or flagella and are heterotrophic parasites in animals.

The apicomplexans have a complex life cycle including alternating sexual and asex-

ual reproductive phases. These phases often occur in different hosts. Two parasitic species, *Toxoplasma* and *Plasmodium*, are of special significance because the first is associated with AIDS and the second is the causative agent of malaria, one of the most prolific killers of humans.

Although many parasitic protists require only one host for the completion of their life cycle, most apicomplexans require two or more different hosts to complete their life cycle. The host organism in which the parasite completes its sexual cycle is called the **definitive host**, whereas the host in which the parasite completes its asexual cycle is the **intermediate host** (**FIGURE 19.7**). For example, the parasite *Plasmodium* produces infective **sporozoites** in mosquitoes (definitive host). These parasitic cells enter

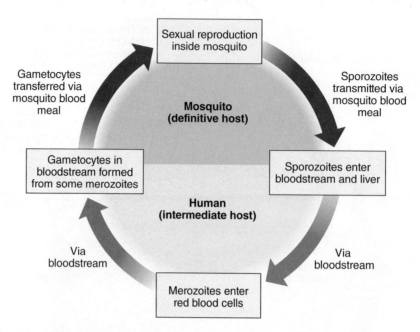

FIGURE 19.7 The Life Cycle of *Plasmodium*. The malaria parasite requires two different hosts (definitive and intermediate) in which specific stages of the life cycle occur. »» *In trying to prevent malaria, which stage of the* Plasmodium *life cycle would appear to be the most amenable to blocking transmission of the parasite?*

the human body when an infected mosquito takes a blood meal. Later, within the human body (intermediate host), gametes, called **gametocytes**, are produced. These are taken up by mosquitoes in a blood meal where sexual reproduction forms more sporozoites.

TABLE 19.1 summarizes some of the attributes of the human parasitic groups of protists.

TABLE 19.1 Comparison of the Parasitic Supergroups of the Protists

Supergroup	Motility	Other Characteristics	Representative Genera
Excavata Parabasalids Diplomonads Kinetoplastids	Flagella Flagella Flagella	Thousands of flagella; no mitochondria Two nuclei; modified mitochondria Kinetoplast DNA	*Trichomonas* *Giardia* *Trypanosoma*, *Leishmania*
Unikonta Amoebozoans	Amoeboid movement	Pseudopodia	*Entamoeba*
Alveolata Ciliates Apicomplexans	Cilia Flagella (gametes only)	Macro- and micronuclei Apical complex; multiple hosts	*Paramecium* *Plasmodium*, *Toxoplasma*

Concept and Reasoning Checks 19.1

a. Summarize the characteristics of the plant-like and fungal-like protists.
b. What features can be used to separate the parabasalids, diplomonads, and kinetoplastids?
c. What unique cellular structures and behaviors are associated with the Unikonta?
d. How does conjugation in ciliates differ from that in bacterial cells?
e. What are some unique features of the apicomplexans?

Chapter Challenge A

As you have seen, the protists are a diverse group of eukaryotic microorganisms. Until recently, they had been placed in the kingdom Protista, which was a catchall for any eukaryotic organism that was not a plant, animal, or fungus. Some taxonomists today want to split the protists into a loose grouping of 30 or more dissimilar phyla based on structure and function.

QUESTION A: *Provide some examples of how structure and function could be used to assign these microorganisms to separate and unique groups. What structures and functions do they have in common?*

You can find answers online in **Appendix F**.

■ KEY CONCEPT 19.2   Protistan Parasites Attack the Skin and the Digestive and Urinary Tracts

Protistan diseases can occur at a variety of anatomical sites in the human body. For example, some diseases, such as amebiasis and giardiasis, take place in the digestive system, whereas others, such as trichomoniasis, develop in the urogenital tract. We begin with a cutaneous protistan disease.

Leishmania Can Cause a Cutaneous Infection

Leishmaniasis is a vector-borne disease that occurs in 88 countries with an at-risk population of 350 million people. About 12 million people are affected and leishmaniasis carries the ninth highest infectious disease burden globally. The responsible protists are in the kinetoplastid group of the Excavata and include several species of *Leishmania*, including *L. tropica* and *L. donovani* (**FIGURE 19.8A**). Transmission

is zoonotic, by the bite of an infected female sandfly that is only about one-third the size of a mosquito. Usually, the fly becomes infected during a blood meal by ingesting infected cells of an animal, such as a rodent, dog, another human, or another mammal.

One species, *L. tropica*, produces a disfiguring **cutaneous leishmaniasis** that globally causes over 1 million new cases each year. When an infected sandfly takes a blood meal, it injects the flagellated parasites. Within a few weeks after being bitten, the person develops a sore on the skin at the bite site as the parasites multiply. The sore then expands and ulcerates to resemble a volcano with a raised edge and central crater (**FIGURE 19.8B**). A scab eventually forms and leaves a scar when healed. Many American soldiers were infected during the Iraq conflicts, as described in MICROFOCUS 19.2.

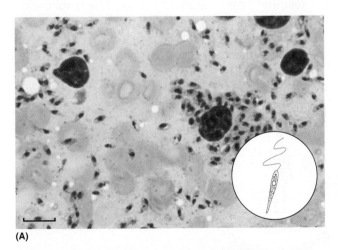

(A)

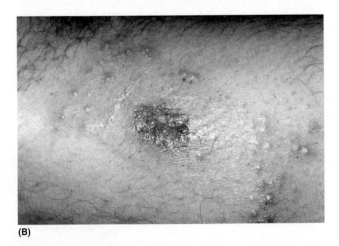

(B)

FIGURE 19.8 *Leishmania* and Cutaneous Leishmaniasis. (A) A stain preparation of *Leishmania* cells, as seen with the light microscope. (Bar = 20 μm.) **(B)** Skin lesion due to cutaneous leishmaniasis. *»» Explain how the parasite can bring about the physical skin lesion.*

MICROFOCUS 19.2: Public Health

The "Baghdad Boil"

The United States Department of Defense (DOD) identified and treated hundreds of cases of cutaneous leishmaniasis (CL) among military personnel serving in Afghanistan and Iraq. The disease was "affectionately" called the "Baghdad boil" (see figure). Between 2003 and 2008, more than 2,000 laboratory-confirmed cases were reported.

Leishmania major, which is endemic in Southwest/Central Asia, was the parasitic species identified in the 176 cases analyzed. Patients were treated with sodium stibogluconate.

The DOD implemented prevention measures to decrease the risk of CL. These procedures included improving hygiene conditions, instituting a CL awareness program among military personnel, using permethrin-treated clothing and bed nets to kill or repel sandflies, and applying insect repellent containing 30% DEET to exposed skin.

These measures, according to the Department of Defense's Medical Surveillance Monthly Report, reduced the number of reported cases from 52 per month to less than three per month in fall 2009.

Cutaneous leishmaniasis.
© AP Photo/Leslie E. Kossoff.

Another species, *L. braziliensis*, causes a disfiguring **mucocutaneous leishmaniasis** primarily in the Americas, whereas *L. donovani* causes a systemic **visceral** (body organ) **leishmaniasis** called **kala azar**. With the visceral disease, symptoms do not appear until several months after being bitten by a sandfly. Infection of white blood cells leads to irregular bouts of fever, swollen spleen and liver, progressive anemia, and emaciation. If not treated, about 90% of cases are fatal. There are about 1.5 million new cases globally each year, with 90% occurring in India, Bangladesh, Brazil, Nepal, and Sudan.

The antimony compound, sodium stibogluconate, has been used to treat established cases. In 2014, the U.S. Food and Drug Administration (FDA) approved a new oral medication called miltefosine (Impavido®) to treat all three forms of the disease.

Control of the sandfly remains the most important method for preventing outbreaks of leishmaniasis.

Several Protistan Parasites Cause Diseases of the Digestive System

Many of the protistan parasites have at least two stages in their life cycle (**FIGURE 19.9**). One is the **cyst** stage. During this period, the organism remains in a dormant, resistant state. The cyst has a thick wall to protect it from the environment and, as such, the cyst is the mechanism for transmission between hosts. Excystment leads to the **trophozoite** (*tropho* = "feeding"; *zoi* = "animal") stage. The trophozoite emerges from the cyst and during this active, infective period in the host, the parasite reproduces and can cause disease. Encystment of trophozoites to cysts returns the parasite to a dormant stage.

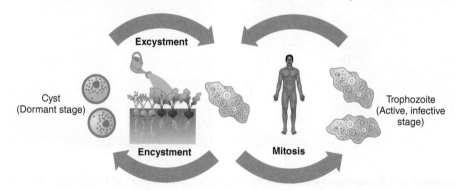

FIGURE 19.9 The Cyst and Trophozoite Stages. Many parasitic protists have a dormant cyst stage that survives outside the host in water and food. Excystment leads to the development of infective trophozoites that actively feed and multiply in the host. Encystment once again produces dormant cysts. »» *What special characteristics must a cyst have to survive in the environment?*

There are several parasitic diseases associated with the gastrointestinal (GI) tract. Giardiasis, cryptosporidiosis, and cyclosporiasis cause an inflammatory gastroenteritis, whereas amebiasis causes an invasive infection of the intestines. All display symptoms of **chronic diarrhea**, which is defined as daily loose bowel movements that last for several weeks.

Giardiasis

In 2013, the CDC received reports of almost 13,000 cases of **giardiasis**, making it the most common intestinal parasitic infection reported in the United States. Because not all cases are reported to the CDC, the disease is estimated to cause more than 1 million infections annually. The disease is sometimes mistaken for viral gastroenteritis and is considered a type of traveler's diarrhea.

The causative agent of giardiasis is the flagellated diplomonad *Giardia intestinalis*. This organism is distinguished by four pairs of anterior flagella and two nuclei that stain darkly to give the appearance of eyes on a face. The cell can be divided equally along its longitudinal axis in what is referred to as bilateral symmetry (**FIGURE 19.10**).

Giardiasis is commonly transmitted by food or water containing *Giardia* cysts or by the fecal–oral route. Acute giardiasis develops after an incubation period of about 7 days. The cysts pass through the

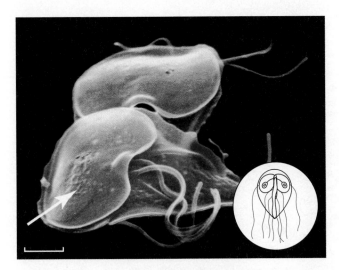

FIGURE 19.10 *Giardia intestinalis.* False-color scanning electron micrograph of *G. intestinalis* in the human small intestine. The pear-shaped cell body and sucker device (arrow) are evident, as are the flagella of this diplomonad. (Bar = 1 μm.) »» *What is the function of the sucker device?*

© Jerome Paulin/Visuals Unlimited, Inc.

stomach and, in the duodenum of the small intestine, the flagellated trophozoites emerge from the cyst. These cells multiply in the small intestine, where some adhere to the intestinal lining using a sucking disk located on the lower cell surface (see Figure 19.10).

The patient feels nauseous, experiences gastric cramps and flatulence, and emits a foul-smelling watery diarrhea sometimes lasting for up to 3 weeks. Trophozoites encyst, and the cysts are excreted in the feces. As such, person-to-person transmission is possible.

Treatment of giardiasis can be administered with drugs such as metronidazole or tinidazole. However, these drugs have side effects, and the physician might want to let the disease run its course without treatment. When individuals recover, they often become carriers and excrete cysts for years.

Cryptosporidiosis

Globally, outbreaks of **cryptosporidiosis** occur every year. The most remarkable outbreak in the United States occurred in 1993 in Milwaukee, Wisconsin, where more than 400,000 people were affected (and 54 died), making it the largest waterborne infection ever recorded in the United States. The CDC estimates there are almost 750,000 cases annually, although less than 2% are reported. The highest number of reported cases of cryptosporidiosis is in young children and older adults.

Human cryptosporidiosis is usually caused by the apicomplexans *Cryptosporidium parvum* and *C. hominis*. Transmission occurs mainly through ingestion of thick-walled, fertilized gametes called **oocysts** that are found in fecally contaminated food or water (such as drinking water and recreational water). The pathogen is extremely tolerant to chlorine, so communal swimming pools also are a potential source of infection.

Cryptosporidiosis has an incubation period of about 1 week. Following ingestion, sporozoites emerge from the oocysts, and the sporozoites infect epithelial cells of the digestive or respiratory tract. Patients with competent immune systems appear to suffer limited diarrhea lasting 2 to 3 weeks, during which time newly formed infectious oocysts are excreted in the feces. However, in immunocompromised individuals, such as AIDS patients, infection causes a cholera-like, profuse, non-bloody diarrhea that can be severe and irreversible. These patients become dehydrated and emaciated and can die without oral rehydration therapy.

Cyclosporiasis

Another single-celled apicomplexan, *Cyclospora cayetanensis*, causes the disease **cyclosporiasis**. In a recent outbreak, hundreds of individuals in several states were sickened by imported *Cyclospora*-contaminated cilantro. In fact, fresh produce and water usually serve as vehicles for transmission of oocysts, which are ingested in what is usually feces-contaminated food or water.

Cyclosporiasis has an incubation period of 2 to 14 days. In the gastrointestinal tract, mature oocysts transform into infectious sporozoites that infect the epithelial cells of the small intestines. Symptoms of the disease include watery diarrhea and abdominal cramping. Recovery can take a few days to a month. Sporocysts mature into oocysts that are shed in the feces.

Cyclospora and *Cryptosporidium* can be differentiated based on the size of the oocysts (*Cyclospora* oocysts are larger) and the ability of *Cyclospora* oocysts to autofluoresce with UV light. A differential diagnosis is important because *Cyclospora* is susceptible to the drug combination of trimethoprim-sulfamethoxazole, whereas *Cryptosporidium* is not.

Amebiasis

The causative agent of **amebiasis** (amebic colitis) is *Entamoeba histolytica* (Amoebozoans). Worldwide, 40,000 to 100,000 people die annually from amebiasis, and it is only surpassed by malaria in terms of deaths from parasitic disease. Amebiasis primarily affects children and adults who are undernourished and living in unsanitary conditions.

In nature, the parasite exists in the cyst form, which enters the body by food or water contaminated with human or animal feces, or by direct contact with cyst-containing feces (fecal–oral route). The cysts pass through the stomach and emerge as trophozoites in the ileum of the small intestine before they migrate to the colon (**FIGURE 19.11**). Infections can be asymptomatic and produce mild

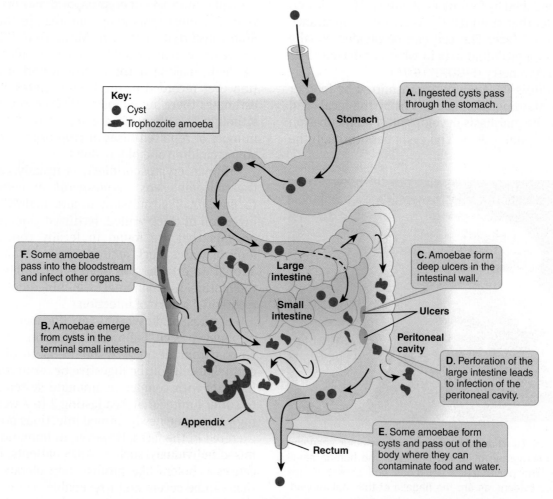

Key:
● Cyst
🦠 Trophozoite amoeba

Stomach

A. Ingested cysts pass through the stomach.

F. Some amoebae pass into the bloodstream and infect other organs.

C. Amoebae form deep ulcers in the intestinal wall.

Large intestine

Small intestine

Ulcers

Peritoneal cavity

B. Amoebae emerge from cysts in the terminal small intestine.

D. Perforation of the large intestine leads to infection of the peritoneal cavity.

Appendix

Rectum

E. Some amoebae form cysts and pass out of the body where they can contaminate food and water.

FIGURE 19.11 The Course of Amebiasis Due to *Entamoeba histolytica*. The parasite enters the body as a cyst, which develops into an amoeboid form that causes deep ulcers. »» *What is the advantage for the parasite to enter the body as a cyst?*

symptoms including watery diarrhea. The trophozoites produce cysts that are passed in the feces. Metronidazole is used to treat the trophozoite stage.

About 10% of people who are infected develop an invasive, intestinal disease. By secreting enzymes, *E. histolytica* cells tear off chunks of host cells and digest the tissue, which produces lesions and deep ulcers in the colon. Patients experience stomach pain, bloody stools, and fever. In rare cases, the parasites invade the blood and cause an invasive infection in the liver, lung, or brain, where fatal abscesses can develop.

Another Protistan Parasite Causes a Sexually Transmitted Infection

Trichomoniasis is among the most common parasitic diseases in men and women in industrialized countries, including the United States, where an estimated 7.4 million new cases occur annually. The disease is transmitted primarily by sexual contact and is considered a sexually transmitted infection (STI). It is more common than all bacterial STIs combined.

The causative agent of trichomoniasis, *Trichomonas vaginalis*, is a pear-shaped, flagellated protist of the parabasalids (**FIGURE 19.12**). The organism has only a trophozoite stage, and its only host is humans, where it thrives and replicates by binary fission in the slightly acidic environment of the female vagina and the male urethra.

The incubation period is 5 to 28 days. In females, trichomoniasis is accompanied by intense itching (pruritus) and discomfort during urination and sexual intercourse. Usually, a yellow-green, frothy discharge with a strong odor also is present. The symptoms frequently are worse during menstruation, and erosion of the cervix can occur. In males, the disease can be asymptomatic or cause

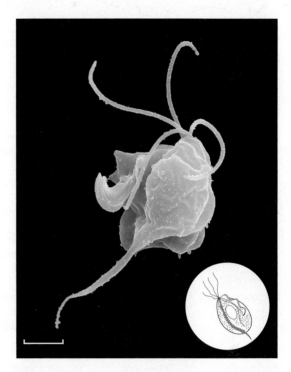

FIGURE 19.12 *Trichomonas vaginalis.* False-color scanning electron micrograph of a *T. vaginalis* trophozoite. (Bar = 5 μm.) *»» What are the whip-like structures evident in the micrograph?*

© Dr. Dennis Kunkel/Visuals Unlimited, Inc.

inflammation of the urethra (urethritis), which is accompanied by slight pain on urination and a thin, mucoid discharge.

The drug of choice for treatment is orally administered metronidazole or tinidazole, and both patient and sexual partners should be treated concurrently to prevent transmission or reinfection. Drug resistance has become of increasing concern.

TABLE 19.2 summarizes the protistan diseases of the skin and the digestive and urinary tracts.

TABLE 19.2 **A Summary of Protistan Diseases of the Skin and the Digestive and Urinary Tracts**

Disease	Causative Agent	Signs and Symptoms	Transmission	Treatment	Prevention
Cutaneous leishmaniasis	*Leishmania major*	Skin sore that ulcerates	Sandfly bites	Stibogluconate	Protecting skin from sandfly bites
Visceral leishmaniasis (kala azar)	*Leishmania donovani*	Fever, swollen spleen and liver, anemia, emaciation	Sandfly bites	Stibogluconate	Protecting skin from sandfly bites

(continues)

TABLE 19.2 A Summary of Protistan Diseases of the Skin and the Digestive and Urinary Tracts (*Continued*)

Disease	Causative Agent	Signs and Symptoms	Transmission	Treatment	Prevention
Giardiasis	*Giardia intestinalis*	Nausea, gastric cramps, flatulence, foul-smelling watery diarrhea	Fecal–oral route	Metronidazole	Practicing good hand hygiene Avoiding untreated water
Cryptosporidiosis	*Cryptosporidium parvum* *Cryptosporidium hominis*	Watery diarrhea, dehydration, vomiting, nausea, stomach cramps, fever, malaise	Contact with contaminated water	Nitazoxanide, paromomycin Fluid replacement	Practicing good hand hygiene Avoiding untreated water
Cyclosporiasis	*Cyclospora cayetanensis*	Watery diarrhea, nausea, abdominal cramping, bloating, vomiting	Contaminated fresh produce or water	Trimethoprim-sulfamethoxazole Fluid replacement	Avoiding untreated water and contaminated fresh produce
Amebiasis	*Entamoeba histolytica*	Bloody stools, stomach pain and cramping	Indirect through food or water Direct contact with feces	Metronidazole Paromomycin	Avoiding contaminated food, unpasteurized milk
Trichomoniasis	*Trichomonas vaginalis*	Intense itching Discomfort on urination and sexual intercourse Yellow-green frothy discharge with strong odor	Oral, vaginal, or anal sex with someone who is infected	Metronidazole or tinidazole	Abstaining from sexual activity Limiting sexual partners Patients and partner treatment

Concept and Reasoning Checks 19.2

a. Summarize the two types of leishmaniasis and their health consequences.
b. Identify why most cases of giardiasis are reported in the summer to early fall.
c. Compare and contrast cryptosporidiosis and cyclosporiasis.
d. Explain why amebiasis is considered an example of "invasive" gastroenteritis.
e. Why is trichomoniasis such a common protozoan disease?

Chapter Challenge B

Here is a map showing the incidence of giardiasis by state. A *Giardia* infection can be transmitted in several ways. Infection can come from swallowing cysts picked up from surfaces (such as bathroom handles, changing tables, diaper pails, or toys) that contain stool from an infected person. It can come from drinking *Giardia*-contaminated water (for example, untreated or improperly treated water from lakes, streams, or wells). Infection also can develop after swallowing water while swimming or playing in water where *Giardia* might live, especially in lakes, rivers, ponds, and streams.

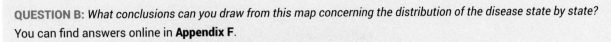

QUESTION B: *What conclusions can you draw from this map concerning the distribution of the disease state by state?* You can find answers online in **Appendix F**.

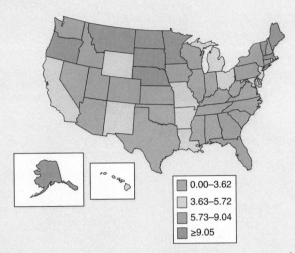

Incidence of Giardiasis Cases, by State—2013. Incidence is the number of reported cases per 100,000 individuals.

Data from Summary of Notifiable Infectious Diseases and Conditions — United States, 2013. *Morbidity and Mortality Weekly Report*, 62(53):1–119. 2013.

■ KEY CONCEPT 19.3 Many Protistan Diseases of the Blood and Nervous System Can Be Life Threatening

We finish our discussion of the protist parasites by examining those that cause infections in the blood or in the nervous system. This includes two of the most prevalent diseases—malaria and sleeping sickness.

Malaria Remains a Major Human Disease

According to the WHO, in 2015 some 3.3 billion people, representing almost half the world's population, were at risk of contracting malaria. In that year, 97 countries and territories reported cases of malaria transmission.

Causative Agent and Epidemiology

Malaria is caused by any of six species of the apicomplexan genus *Plasmodium*; *P. falciparum* is the most serious life-threatening species and the one we will focus on in this discussion.

Malaria has been infecting humans for more than 5,000 years. In 2015, about 200 million people were suffering from malaria, which exacts its greatest toll in Africa. The WHO estimates that more than 600,000 people die from malaria every year, with younger children most susceptible (450,000 deaths)

because their immune system has not yet had time to develop a degree of protective immunity. No infectious disease of contemporary times can claim such a dubious distinction. Even the United States is involved in the ongoing malaria pandemic—over 1,600 imported cases were reported in 2014.

All plasmodial species are transmitted by the female *Anopheles* mosquito, the definitive host. Humans represent the intermediate host. The limits of transmission are primarily determined by atmosphere temperatures at which mosquitoes can sustain development.

Clinical Presentation

The life cycle of *P. falciparum* and its clinical presentation are described in **FIGURE 19.13**. Note that the cycle alternates between mosquitoes and human hosts and that the parasite has three important stages: the sporozoite, the merozoite, and the gametocyte/gamete.

▶ **Sporozoite Stage.** The presence of sporozoites in the salivary glands of female mosquitoes leads to an increase in biting

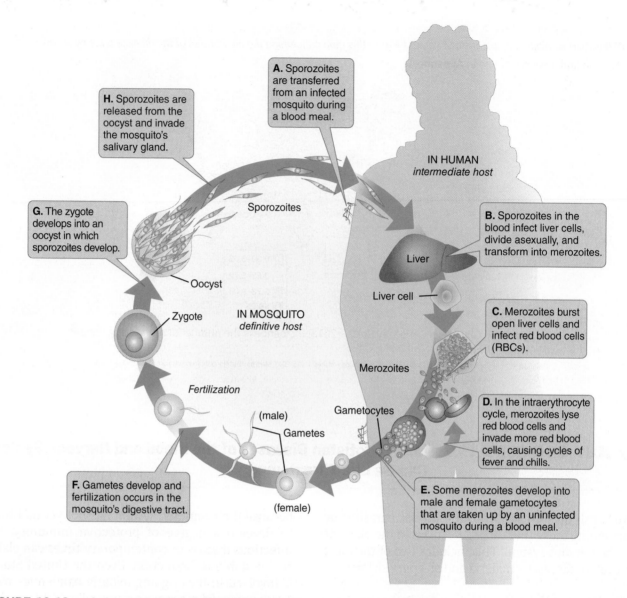

H. Sporozoites are released from the oocyst and invade the mosquito's salivary gland.

A. Sporozoites are transferred from an infected mosquito during a blood meal.

G. The zygote develops into an oocyst in which sporozoites develop.

IN HUMAN
intermediate host

Sporozoites

Liver

B. Sporozoites in the blood infect liver cells, divide asexually, and transform into merozoites.

Oocyst

Liver cell

Zygote

IN MOSQUITO
definitive host

C. Merozoites burst open liver cells and infect red blood cells (RBCs).

Merozoites

Fertilization

Gametocytes

D. In the intraerythrocyte cycle, merozoites lyse red blood cells and invade more red blood cells, causing cycles of fever and chills.

(male)

Gametes

F. Gametes develop and fertilization occurs in the mosquito's digestive tract.

(female)

E. Some merozoites develop into male and female gametocytes that are taken up by an uninfected mosquito during a blood meal.

FIGURE 19.13 Malaria and the *Plasmodium* Life Cycle. The propagation of the *Plasmodium* parasite requires two hosts, mosquitoes and humans. **»» *Why are two hosts required to complete the* Plasmodium *life cycle?***

frequency (i.e., taking a blood meal). As a result, several hundred sporozoites enter the person's bloodstream and quickly migrate to the liver. After a few days, the transformation of sporozoites to merozoites is completed, and up to 40,000 merozoites emerge from each infected liver cell.

▶ **Merozoite Stage.** After they are released, the free merozoites invade red blood cells (RBCs). Within RBCs, over a 48h-hour period each merozoite undergoes asexual reproduction to produce 16 to 32 merozoites. In response to a parasite biochemical signal, thousands of infected RBCs rupture simultaneously releasing the merozoites and

their toxins. The released merozoites now infect more RBCs, and the 48-hour cycle occurs repeatedly.

This chronic cycle of RBC infection and cell lysis called the **intraerythrocyte cycle** is responsible for the symptoms of malaria. First, there is intense cold, with shivers and chattering teeth. Body temperature then rises rapidly to 40°C, and the sufferer develops intense fever, headache, and delirium. After 2 to 3 hours, massive perspiration ends the fever stage, and the patient often falls asleep, exhausted. With repeated cycles of RBC infection and cell lysis, the symptoms repeat themselves.

In a fraction of untreated individuals, the infection is not controlled and severe complications can lead to death. Death from malaria is due to a number of factors related to the loss of RBCs. Substantial anemia develops, and the hemoglobin from ruptured blood cells darkens the urine; malaria is sometimes called **blackwater fever**. Cell fragment and parasite biomass adherence accumulate on small blood vessels of the brain, kidneys, heart, liver, and other vital organs, which impairs blood flow and causes clots to form. Heart attacks, respiratory distress, cerebral hemorrhages, and kidney failure are common.

▶ **Gametocyte/Gamete Stage.** During the merozoite stage, a proportion of the merozoites are developmentally reprogrammed to undergo a change into male and female gametocytes. In the peripheral circulation, the gametocytes will be ingested when an uninfected mosquito takes a blood meal. In the mosquito's midgut, the gametocytes emerge as male and female gametes. Fertilization occurs and eventually motile sporozoites develop that pass into the salivary glands.

Treatment and Prevention

Since its discovery around 1640, quinine has been the mainstay for treating malaria. During World War II, American researchers developed the drug chloroquine, which remained an important mode of therapy until recent years, when drug resistance began emerging in *Plasmodium* species. Another drug, **artemisinin**, is effective in curing malaria especially when combined with other drugs to limit the development of drug resistance. However, the fear is that the current rise of artemisinin resistance in *Plasmodium* in Southeast Asia might spread to India and Africa, which would threaten their efficacy to eliminate the parasite and the disease.

The prevention of malaria includes the use of insecticide-treated mosquito nets, indoor spraying with insecticides, better diagnostic testing for infections, and treatment with effective antimalarial medicines. Since 1989, mefloquine has been recommended as a preventative drug for individuals entering malaria regions of the world. However, serious medical side effects, including cognitive disorders, have been associated with some people taking the drug. Malavone is a newer antimalarial drug now recommended for malaria caused by

P. falciparum. Experimental vaccines directed against the sporozoite or merozoite stage are being tested.

The WHO Global Technical Strategy for Malaria 2016 has been developed with the aim of guiding and supporting regional and country programs that are trying to control and eliminate malaria. The goal of the malaria strategy is to lower the global malaria burden by 90% by 2030.

Babesia Causes a Malaria-Like Illness

Babesiosis, found in the Northeast United States, is a malaria-like disease caused by the apicomplexan parasite *Babesia microti*. These protists live in ticks, and they are transmitted when the arthropods feed in human skin. Tickborne transmission occurs primarily during the spring and summer. However, the geographic distribution is spreading such that it might soon rival Lyme disease as the most common tickborne illness in the United States.

After they are in the human blood, sporozoites of *B. microti* penetrate human red blood cells. As the cells disintegrate, a mild anemia develops. Piercing headaches accompany the disease and, in rare cases, meningitis occurs. A suppressed immune system appears to favor establishment of the disease. However, babesiosis is rarely fatal, and drug therapy is not recommended. Humans are dead-end hosts, and tick control is considered the best method of disease prevention.

The *Trypanosoma* Parasites Can Cause Life-Threatening Systemic Diseases

Trypanosomiasis is the general term for two diseases, African trypanosomiasis and Chagas disease, caused by parasitic species of the kinetoplastid *Trypanosoma* (**FIGURE 19.14**).

African Trypanosomiasis

African trypanosomiasis (human **African sleeping sickness**) is endemic in 36 African countries and exerts a level of mortality greater than that of HIV disease/AIDS. The WHO estimates there are more than 500,000 new cases every year with some 50,000 deaths.

The *Trypanosoma* parasite's life cycle alternates between humans and its vector, the tsetse fly. If an infected fly takes a blood meal from a person, the portal of entry in the human becomes painful and swollen and a chancre similar to that in syphilis develops. The trypanosomes are carried to the lymphatic system before migrating to the bloodstream. In the blood, the parasites are carried to other sites in the body including the central nervous system, where meningoencephalitis can occur.

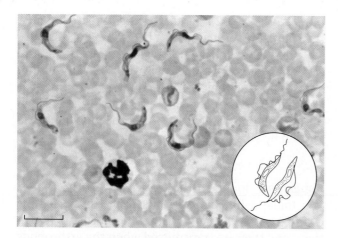

FIGURE 19.14 *Trypanosoma.* A light micrograph of stained *Trypanosoma* cells among red blood cells. (Bar = 15 μm.) *»» What is unique about the* Trypanosoma *cells?*

Courtesy of Dr. Mae Melvin/CDC.

Two forms of African sleeping sickness exist. A chronic form, common in central and western Africa, is caused by *Trypanosoma brucei* variety *gambiense.* It is accompanied by chronic bouts of fever, as well as severe headaches, changes in sleep patterns and behavior, and a general body wasting. As the trypanosomes invade the brain, the patient slips into a coma (hence the name, "sleeping sickness"). The second form, common in eastern and southern Africa, is due to *Trypanosoma brucei* variety *rhodesiense.* The disease is more acute with high fever and rapid coma preceding death. The trypanosomes are transmitted back to the tsetse fly when the fly takes a blood meal from an infected individual.

Prevention involves clearing brush lands and treating areas where the tsetse flies breed. Patients are treated with the drug pentamidine, melarsoprol, or eflornithine, depending on the form and stage of the disease.

Chagas Disease

Chagas disease (American trypanosomiasis) is caused by *Trypanosoma cruzi* and is transmitted by the reduviid (triatomine) bug. The insect feeds at night and bites individuals where the skin is thin, such as on the forearms, face, or lips. For this reason, the vector is called the "kissing bug." It is endemic in rural areas of Latin America where there are an estimated 8 million cases. There are approximately 200,000 new cases every year and more than 15,000 deaths. In 2007, the American Red Cross and Blood Systems started screening the U.S. blood supply for Chagas disease and *T. cruzi.* About one in 30,000 donors test positive, so it is estimated that there are

at least 330,000 infected individuals in the United States, primarily among immigrants coming from Chagas-endemic countries.

The parasites are carried in the gut of the reduviid bug. Transmission of *T. cruzi* occurs when the infected bug takes a blood meal and defecates on the host. If the host then scratches the feces into the bite site, the trypanosomes enter the blood and invade many cell types. During this acute phase, the individual experiences high parasite numbers even though most infections are asymptomatic. Following the acute phase, the parasite number declines. In 20% to 30% of infected individuals, a chronic irreversible disease occurs that in 10 to 30 years can develop clinical symptoms, which vary by geographic region. Individuals can experience widespread tissue damage including intestinal tract abnormalities and extensive cardiac nerve destruction (cardiomyopathy) that is so thorough the victim experiences sudden heart failure. Benznidazole and nifurtimox have proved useful for acute disease, and a vaccine is being developed.

Toxoplasma Causes a Relatively Common Blood Infection

Perhaps the most common human parasite is *Toxoplasma gondii,* which is responsible for **toxoplasmosis** (**FIGURE 19.15**). The disease affects up to 50% of

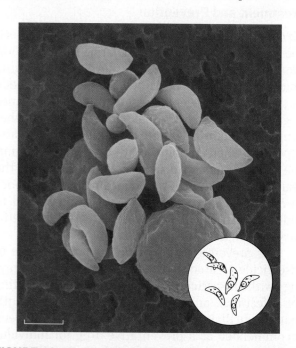

FIGURE 19.15 *Toxoplasma gondii.* A false-color scanning electron micrograph of numerous crescent-shaped *T. gondii* trophozoites (orange). (Bar = 5 μm.) *»» What infective stage is represented by these cells?*

© Dennis Kunkel Microscopy/Science Source.

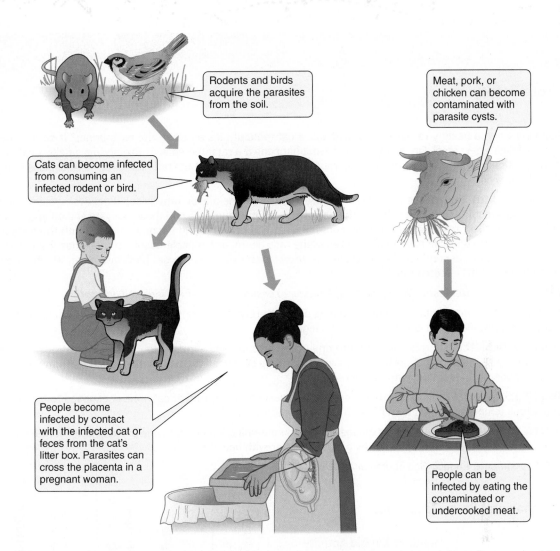

Rodents and birds acquire the parasites from the soil.

Meat, pork, or chicken can become contaminated with parasite cysts.

Cats can become infected from consuming an infected rodent or bird.

People become infected by contact with the infected cat or feces from the cat's litter box. Parasites can cross the placenta in a pregnant woman.

People can be infected by eating the contaminated or undercooked meat.

FIGURE 19.16 The Routes of Infection Causing Toxoplasmosis. Humans become infected through contact with the feces of an infected cat or consumption of undercooked, contaminated beef. »» *What form of the* Toxoplasma *parasite is passed to humans?*

the world's population, including 60 million Americans. In immunocompromised patients, the parasite can attack brain tissue, causing encephalitis. Left untreated, toxoplasmosis often results in cerebral lesions, seizures, and death.

T. gondii, an apicomplexan parasite, exists in three forms: the trophozoite, the cyst, and the oocyst (**FIGURE 19.16**). In nature, the parasite exists in cyst and oocyst forms. Grazing animals (beef cattle, pigs, and sheep) acquire the cysts from the soil and pass them to humans who eat the undercooked meat.

Domestic cats acquire the cysts from the soil or from infected birds or rodents. Oocysts then form in the cat. Should the pet owner forget to wash his or her hands after contacting cat feces, changing the cat litter, or simply touching the cat, oocysts can

be brought to the hands. Contact with the mouth (fecal–oral route) then transfers the oocysts

Toxoplasmosis develops after trophozoites are released from the cysts or oocysts in the host's gastrointestinal tract. The trophozoites rapidly invade the intestinal lining and spread throughout the body via the blood. For most healthy individuals, the parasite causes no serious illness and is nonneuroinvasive. Pregnant women, however, are at risk of developing a dangerous toxoplasmosis infection because the parasite can cross the placenta and infect the fetal tissues (see Figure 19.16). Such congenital infections can lead to disabling complications, such as hearing loss, mental disability, and blindness. INVESTIGATING THE MICROBIAL WORLD 19 takes a fascinating look at how *Toxoplasma* manipulates one of its natural intermediate hosts, the rat.

Investigating the Microbial World 19

Love Your Enemy

What happens when a healthy rat smells its nemesis, a cat (actually, it's smelling the cat's urine)? It scrams before it becomes the feline's next meal. On the other hand, what happens if a rat infected with *Toxoplasma* smells a cat? It sticks around and continues to sniff the odors—a dangerous decision on the part of the rat! Moreover, *Toxoplasma* is controlling the rat's risky behavior.

OBSERVATIONS: Often the parasite *Toxoplasma* uses a cat's (domestic or otherwise) digestive system for sexual reproduction. As a result, oocysts are shed in the cat's feces in large numbers and these oocysts end up in soil or water. Intermediate hosts, like rats, ingest the oocyst-contaminated soil or water and become infected with the oocysts. The oocysts transform into a motile form that localizes in the rat's neural and muscle tissue. In those tissues, they develop into cysts. Now, in order to continue the life cycle, the *Toxoplasma* cysts have to get back into a cat, which means cats have to catch and eat the infected rats.

QUESTION: *How does* Toxoplasma *"manipulate" rat behavior to ensure the parasite is ingested by a cat?*

HYPOTHESIS: *Toxoplasma* affects the limbic system (a part of the brain that controls emotions) of the rat brain. If so, then cat urine should show increased activity in this brain region.

EXPERIMENTAL DESIGN: Male rats were put into one of four groups and exposed to cat urine or estrous female odor. Immediately after a 20-minute exposure to the urine/odor, rats were sacrificed and brain activity analyzed immunochemically for c-Fos activation (an indirect indicator of neural activity).

EXPERIMENT 1: Uninfected male rats were exposed to a towel containing 1 ml of cat urine or exposed to estrous female rat odor (estrous female separated from males by a plastic barrier, with holes).

EXPERIMENT 2: Infected male rats were exposed to a towel containing 1 ml of cat urine or exposed to estrous female rat odor (estrous female separated from males by a plastic barrier, with holes).

RESULTS: Experiments 1 and 2: See figures and table.

Rat Behavioral Brain Responses

Experimental Treatment	Relative Limbic Activity Involved with	
	Defensive Behavior	**Reproductive Behavior**
Uninfected exposure to • Urine • Estrous odor	Increased Diminished	No increase Increased
Infected exposure to • Urine • Estrous odor	Unchanged Diminished	Increased Increased

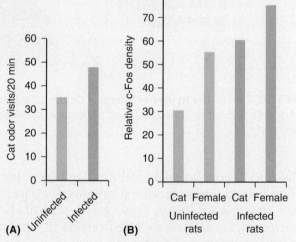

(A) Time spent exploring cat urine. **(B)** Neural activity (c-Fos activity) in rats exposed to cat urine (bars 1 and 3) or to female estrous odor (bars 2 and 4).

Modifed from House, P.K., Vyas, A., and Sapolsky, R. (2011). *PLoS ONE* 6(8): e23277.

CONCLUSIONS:

QUESTION 1: *Was the hypothesis validated? Explain using the figures and table.*

QUESTION 2: *How does brain activity in uninfected male rats exposed to estrous female odor compare to brain activity in infected male rats exposed to cat urine?*

QUESTION 3: *What can you conclude about the effect of* Toxoplasma *on the innate fear response?*

Note: About 33% of the human population has been exposed to *Toxoplasma*, and several studies find that the infection increases the risk for schizophrenia and obsessive-compulsive disorder.

You can find answers online in **Appendix E.**

Data from House, P. K., et al. 2011. *PLoS ONE* 6(8):e23277.

Naegleria Can Infect the Central Nervous System

Primary amebic meningoencephalitis (PAM) is a rare disease with less than 200 cases reported worldwide since 1965 (34 cases in the United States since 2004). However, it is the most deadly disease of the central nervous system after rabies. More than 97% of patients (primarily children and young adults) die within 4 to 5 days of infection, as **CLINICAL CASE 19** describes.

PAM is caused by several species of thermophilic amoebal parasites in the genus *Naegleria* (Excavata), especially *N. fowleri* that lives in fresh water (**FIGURE 19.17**). It often is referred to as the "brain-eating" amoeba.

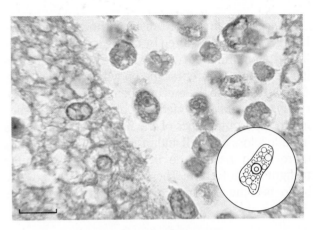

FIGURE 19.17 *Naegleria fowleri.* A light micrograph of brain tissue showing the infection with *N. fowleri* amoebae (round cells). (Bar = 25 µm.) »» *How might these aquatic amoebae infect brain tissue?*

Courtesy of Dr. Govinda S. Visvesvara/CDC.

Clinical Case 19

Primary Amebic Meningoencephalitis Associated with a Hot Spring

In late June, a previously healthy 11-year-old boy was evaluated in a local emergency department for a 2-day history of headache, stiff neck, nausea, and vomiting. He had a low-grade fever and was lethargic.

Two days later, the boy was hospitalized with presumptive viral meningitis. There were no signs of meningitis, and a lumbar puncture identified no infectious organisms in the cerebral spinal fluid (CSF). All other routine tests were negative.

The boy's condition continued to deteriorate, and he developed slurred speech, altered mental status, and seizures. Two days later, the boy was intubated and provided mechanical ventilation. In a second lumbar puncture sample, motile amoebae were identified.

The next day, the Florida Department of Health (FDH) was notified about a suspected case of primary amebic meningoencephalitis (PAM). The FDH immediately contacted the Centers for Disease Control and Prevention (CDC) to obtain miltefosine, an antiparasitic drug for treating PAM.

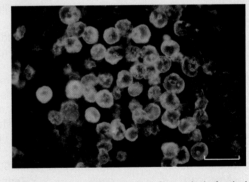

Fluorescent antibody stain of *Naegleria fowleri* amoebae. (Bar = 40 µm.)

Courtesy of the CDC.

Unfortunately, before the drug arrived the patient's condition worsened, with progressive neurological deterioration, and the boy died.

A postmortem CSF sample molecularly identified *Naegleria fowleri* (see figure).

An epidemiological investigation by the FDH discovered that the boy and his family had visited Costa Rica. The boy's symptoms developed on the day they returned. In Costa Rica, the boy had participated in water activities that included swimming and water sliding at a resort's hot springs four days before symptoms appeared.

Questions:
  a. Would the absence of microbes in the initial lumbar puncture mean the young boy did not have meningitis? Explain.
  b. Why was a follow-up lumbar puncture ordered?
  c. How might the patient have become infected with the parasite?
  d. Why were no other children infected?
  e. What precautions should be taken to prevent another incident?

You can find answers online in **Appendix E**.

For additional information, see http://www.cdc.gov/mmwr/preview/mmwrhtml/mm6443a5.htm.

Naegleria is an opportunistic pathogen of humans. If forcefully inhaled, often by diving into or water skiing in contaminated warm surface water, the free-living trophozoites enter the body through the nose. After attaching to the mucous membranes in the nasal cavity, the parasites infect the olfactory tracts leading to the frontal lobes of the brain. Early symptoms include a loss of smell and taste. As *N. fowleri* digests more brain tissue, the immune system attempts to fight the infection. The surrounding areas of the brain become inflamed, which leads to symptoms resembling those in other forms of encephalitis and acute meningitis, that is, rapid onset, piercing headaches, fever, delirium, stiff neck, and occasional seizures. Death usually results from extreme brain pressure in the skull caused by the inflammation and swelling. This severs the brain–spinal cord connection, and the patient dies from respiratory failure. Treatment with miltefosine has had limited success even if infections are diagnosed very early.

TABLE 19.3 summarizes the protistan diseases of the blood and nervous system.

TABLE 19.3 A Summary of Protistan Diseases of the Blood and Nervous System

Disease	Causative Agent	Signs and Symptoms	Transmission	Treatment	Prevention
Malaria	4 *Plasmodium* species	Moderate to severe shaking chills, high fever, profuse sweating with chills, malaise	Bite of infected *Anopheles* mosquito	Chloroquine, quinine, mefloquine, malarone	Mefloquine, malarone
Babesiosis	*Babesia microti*	Mild anemia, piercing headache	Bite of infected tick	Azithromycin and atovaquone	Avoiding ticks in endemic areas
African trypanosomiasis (African sleeping sickness)	*Trypanosoma brucei*	Chronic bouts of fever, severe headache, change in sleep patterns and behavior	Bite of infected tsetse fly	Pentamidine, melarsoprol, eflornithine	Clearing areas where tsetse flies breed
Chagas disease (American trypanosomiasis)	*Trypanosoma cruzi*	Acute phase: redness and swelling at the site of infection, asymptomatic Chronic phase: wide-spread tissue damage	Bite of an infected reduviid bug	Acute phase: benznidazole and nifurtimox Chronic phase: depends on signs and symptoms	Avoiding endemic residences Insecticides
Toxoplasmosis	*Toxoplasma gondii*	Usually asymptomatic	Contact with contaminated cat feces, ingestion of contaminated food, water, or fomites	Pyrimethamine and sulfadiazine for acute disease	Practicing good hygiene Avoiding raw or undercooked meat and unpasteurized goat's milk
Primary amebic meningoencephalitis (PAM)	*Naegleria fowleri*	Piercing headache, fever, delirium, neck rigidity	Inhalation of contaminated water	Early with amphotericin B, miconazole, and rifampin	Avoiding swimming in contaminated water

Chapter **Challenge C**

A number of protistan diseases are transmitted by "flying insects." This includes malaria (mosquitoes), African sleeping sickness (tsetse flies), and leishmaniasis (sandflies). In the next section, you will see that lymphatic filariasis is also transmitted by mosquitoes.

QUESTION C: *Why have these diseases evolved to use a flying insect as the means of transmission? Hint: Think about how the parasite is carried through the human host.*

You can find answers online in **Appendix F**.

■ KEY CONCEPT 19.4   Parasitic Helminths Cause Substantial Morbidity Worldwide

The **helminths** are among the world's most common animal parasites. For example, 2 billion people—approximately 33% of the human population—are infected with soil-transmitted helminths! In concluding this chapter on the parasites, although we are concerned with medical helminthology and the diseases caused by the parasitic worms, it is worth knowing that most intestinal parasite infections do not trigger disease in healthy people as long as the worm load remains low.

Parasitic worms are of interest to microbiologists because of the parasites' ability to cause an enormous level of morbidity worldwide. However, different from many of the bacterial and fungal pathogens, most parasitic helminths are dependent on the host or hosts for nourishment, so it is to the helminth's benefit that the host stays alive. Therefore, the helminths tend to cause infections of debilitation and chronic morbidity resulting from physical factors related to the **helminthic load** (number of worms present) or location in the body.

There Are Two Groups of Parasitic Helminths

The helminths of medical significance are the flatworms and the roundworms.

Flatworms

Animals in the phylum Platyhelminthes (*platy* = "flat"; *helmin* = "worm") are the **flatworms**. These free-living organisms live in marine and freshwater environments and in damp soil. As multicellular animals, they have tissues functioning as organs in organ systems. However, they have no specialized respiratory or circulatory structures, and they lack a digestive tract. The gut, called the gastrovascular cavity, simply consists of a sac with a single opening, thus placing the worm in close contact with its surroundings. Complex reproductive systems are found in many species within the phylum, and a large number of species are **hermaphroditic**; that is, the flatworms have both male and female reproductive organs. Two groups of flatworms are of concern regarding human disease.

▶ **Trematodes.** The **trematodes (flukes)** have flattened, broad bodies with bilateral symmetry (**FIGURE 19.18A**). They have a complex life cycle that includes encysted egg stages and temporary larval forms. Sucker devices are commonly present; one is used to take in nutrients and eliminate waste products, whereas the other is used to attach to its host. In many cases, two hosts exist: an intermediate host, which harbors the larval (asexual) form, and a definitive host, which harbors the mature adult (sexual) form. In this chapter, we are concerned with parasites whose definitive host is a human.

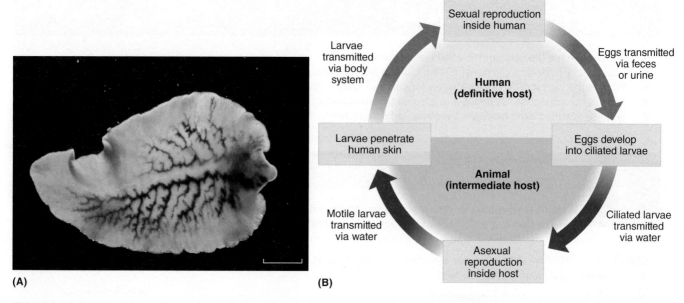

FIGURE 19.18 The Fluke Life Cycle. (A) Light micrograph of a liver fluke, *Fasciola hepatica*, a parasite infecting sheep, cattle, and humans. (Bar = 50 µm.) **(B)** The blood fluke *Schistosoma* has alternating sexual and asexual generations that require different hosts—human and snail. »» *What added pressures are placed on the blood flukes infecting the human bloodstream that would not be present as free-living larvae in the water?*

(A) © Sinclair Stammers/Science Source.

The life cycle of a fluke often contains several phases (**FIGURE 19.18B**). In the human host, the parasite produces fertilized eggs generally released in the feces. When the eggs reach water, they hatch and develop into tiny ciliated larvae. The larvae penetrate the intermediate host and go through a series of asexual reproductive stages. The larvae produced are released into the environment where they can be picked up by another human host.

The trematode lifestyle requires the parasite to evade the host's immune system.

It accomplishes this by having its surface resemble the surface of the host cells, so the immune system "sees" the worm as a "normal" cell, not an invader. This mimicry is quite effective as some flukes can remain in a human host for 40 years or more.

MICROINQUIRY 19 examines some amazing behaviors of how parasitic flukes can manipulate their host for their own benefit—and survival.

▸ **Cestodes.** The other group of flatworms is the **cestodes** (**tapeworms**). These parasitic worms are widespread parasites infecting

MICROINQUIRY 19

Parasites as Manipulators

Over the last several chapters, we have been discussing viral, bacterial, and fungal pathogens and protist and multicellular parasites. For the latter parasites, you might conclude they make their living "sponging off" their hosts with the ultimate aim to reproduce. In fact, for a long time scientists thought these organisms simply evolved so they could take advantage of what their hosts had to offer. In reality, the parasites might be quite a bit more sophisticated than previously thought. Two

cases involving multicellular parasites are submitted for your inquiry.

Case 1

The lancet fluke, *Dicrocoelium dendriticum*, infects cows. Because the infected cow is the definitive host, the fluke needs to get out of the cow to form cysts in another animal. If the cysts can then be located on grass blades, another cow might eat those cyst-containing grass blades, and another round of infection by *D. dendriticum* can occur.

19.1a. As a parasite, how do you ensure cysts will be on grass blades that cows eat?

Infected cows excrete dung containing fluke eggs. Snails (intermediate host #1) forage on the dung and, in the process, ingest the fluke eggs. The eggs hatch and bore their way from the snail gut to the digestive gland where fluke larvae are produced. In an attempt to fight off the infection, the snail smothers the parasites in balls of slime coughed up into the grass.

Ants (intermediate host #2) come along and swallow these nutrient-rich slime balls. Now, here is the interesting part, because the lancet fluke has yet to get cysts onto grass blades so they will be eaten by more cows. In infected ants, the larvae travel to the ant's head and specifically the nerves that control the ant's mandibles. Although most of the larvae then return to the abdomen where they form cysts, a few flukes remain in the head and control the ant's behavior. As evening comes and the temperature cools, the head-infected ants with a belly full of cysts leave their colony, climb to the tip of grass blades, and hold on tight with their mandibles (see figure). The lancet fluke has taken control of the ants and placed them in a position most favorable to be eaten by a grazing cow.

19.1b. What happens if no cow comes grazing that evening?

If a grazing cow does not come by, the next morning the flukes release their influence on the ants, which return to their "normal duties." But, the next evening, the flukes

An ant clutching a grass blade with its feet and mandibles.

© Eye of Science/Science Source.

again take control, and the ants march back up to the tips of grass blades. If a cow does eat the grass with the ants, the fluke cysts in the ant abdomen quickly hatch, and another cycle of parasite reproduction begins in the cow. Now, that is quite a behavioral driving force to control your destiny—in this case, to ensure fluke cysts are positioned so they will be eaten by the cow.

Case 2

(This case is based on research carried out in a coastal salt marsh by Kevin Lafferty's group at the University of California at Santa Barbara.) Another fluke, *Euhaplorchis californiensis*, uses shorebirds as its definitive host. Infected birds drop feces loaded with fluke eggs into the marsh. Horn snails (intermediate host) eat the droppings containing the eggs. The eggs hatch and castrate the snails. Larvae are produced in the water, and they then latch onto the gills of their second intermediate host, the California killifish. In the killifish, the larvae move from the blood vessels to a nerve that carries the larvae to the brain where the larvae form a thin layer on top of the brain. There they stay, waiting for the fish to be eaten by a shorebird. After they are in the bird's gut, the adults develop and produce another round of fertilized eggs. Here, we have two questions.

19.2a. What is the purpose for the larvae castrating the snails?

An experiment was set up to answer this question. Throughout the salt marsh, Lafferty set up cages with uninfected snails and other cages with infected ones. The results were as expected—the uninfected snails produced many more offspring than did the infected snails. Lafferty believes a marsh full of uninfected snails would reproduce so many offspring as to deplete the algae and increase the population of crabs that feed on the snails. By castrating the snails, the parasite actually is controlling the snail population and keeping the salt marsh ecosystem balanced—again for its benefit.

19.2b. Why do the larvae in the killifish take up residence on top of the fish's brain?

Through a series of experiments, Lafferty's group discovered that infected fish underwent a swimming behavior not observed in the uninfected fish. The infected killifish darted near the water surface, a very risky behavior that makes it more likely the fish would be caught by shorebirds. In fact, experiments showed infected fish were 30 times more likely to be caught by shorebirds than uninfected fish. Now, from the parasite's perspective, they want to be caught so another round of reproduction can begin in the shorebird. So again, the fluke has manipulated the situation to its benefit.

Parasites might cause disease, but they also maximize their chances for ensuring infection. Pretty amazing!

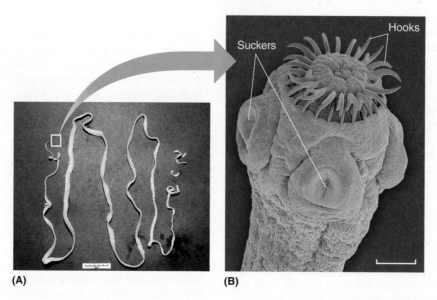

FIGURE 19.19 The Tapeworms. (A) Photograph of a beef tapeworm, *Taenia saginata*, which has grown to several meters in length. **(B)** A false-color scanning electron micrograph of the head, scolex, of *T. saginata* showing the suckers and hooks. (Bar = 1 mm.) *»» What is the purpose of the suckers and hooks on the scolex head of the tapeworm shown in (B)?*

(A) Mediscan/Alamy Stock Photo. (B) © Dennis Kunkel Microscopy/Science Source.

practically all mammals, as well as many other vertebrates. With rare exceptions, tapeworms require at least two hosts. Humans often acquire the larvae by eating undercooked meat containing tapeworm cysts, which then develop into mature adult worms.

Tapeworms generally live in the intestines of a host vertebrate. The parasites have a head region, called the **scolex**, that is used for attachment and a ribbon-like body consisting of reproductive segments called **proglottids** (**FIGURE 19.19**). The proglottids most distant from the scolex are filled with fertilized eggs resulting from sexual reproduction by the parasite.

Although they lack a digestive tract, tapeworms in the intestines are constantly bathed by nutrient-rich fluid, from which they absorb food already digested by the host. Morbidity often arises from the large tapeworm load in the intestine, which can rob the host of nutrients and lead to nutrient deficiencies. In an infected individual, the eggs produced are spread through the host's feces, as the proglottids break free from the tapeworm.

Roundworms

Among the most prevalent animals are the **roundworms** in the phylum Nematoda (*nema* = "thread").

These helminths have a thread-like cylindrical body and live in oceans, freshwater, and soil from the polar regions to the tropics. Good topsoil, for example, can contain billions of nematodes per acre. They parasitize every conceivable type of plant and animal, causing both economic crop damage and serious disease in animals. Conversely, they also appear to have a beneficial effect for humans, as MICROFOCUS 19.3 conveys.

Roundworms range in size from less than one millimeter to several meters in length. The body is covered by a tough coat called a **cuticle**. Roundworms lack a circulatory system, have a group of longitudinal muscles, and have separate sexes. Following fertilization of the female by the male, the eggs hatch to larvae that resemble miniature adults. Growth then occurs by cellular enlargement and mitosis. Damage in hosts is generally caused by large worm burdens in the blood vessels, lymphatic vessels, or intestines. In addition, the infestation can result in nutritional deficiency or damage to the muscles.

Trematodes Can Cause Human Illness

Schistosomiasis is caused by several species of blood flukes that are found in different parts of the world, including *Schistosoma mansoni* (Africa and South America), *S. japonicum* (Asia), and *S. haematobium* (Africa and India). In some regions, the term **bilharziasis** is still used for the disease; it

MICROFOCUS 19.3: Being Skeptical

Eat Worms and Cure Your Ills—and Allergies!

Doctors and researchers have shown that eating worms (actually drinking worm eggs) can fight disease. What! Are you nuts? Drink worm eggs!

Joel Weinstock, a gastroenterologist at the University of Iowa, discovered that as allergies and other diseases have increased in Western countries, infections by roundworm parasites have declined. However, in many other countries, allergies are rare, and worm infections are quite common. For that reason, Weinstock wondered if there was a correlation between allergy increase and parasite decline.

To test his hypothesis, Weinstock "brewed" a liquid concoction consisting of thousands of pig whipworm eggs (ova). The whipworms are called *Trichuris suis*, so his product with ova is called TSO. This roundworm (see figure) was chosen because, after the ova hatch, they will not survive for long in the human digestive system, and they will be passed out in the feces.

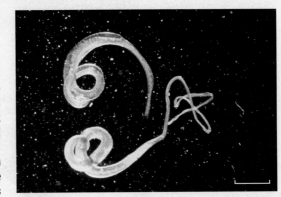

Adult whipworms (male top; female bottom). (Bar = 1 cm.)

© CNRI/Science Source.

One woman in Iowa was suffering from incurable ulcerative colitis, which is caused by the immune system overreacting. Immune cells begin attacking the person's own gut lining, making it bleed. The symptoms are severe cramps and acute, intense diarrhea. In a trial run, Weinstock gave the woman a small glass of his TSO. Every 3 weeks she downed another glass. Guess what? Her ulcerative colitis went into remission, and she no longer suffers any disease symptoms.

Further trials have involved 100 people suffering ulcerative colitis and a further 100 suffering Crohn's disease, which is another type of inflammatory bowel disease related to immune function. In this study, 50% of the volunteers suffering ulcerative colitis and 70% of those suffering Crohn's disease went into remission, as identified by no symptoms of abdominal pain, ulcerative bleeding, and diarrhea after TSO therapy.

Weinstock believes some parasites are so intimately adapted with the human gut that if they are eradicated, bad things might happen, such as the bowel disorders mentioned. Since then, several other studies by separate scientists have documented similar reversals of disease and further showed that the worm therapy restored the normal mucosal microbiota (resident bacteria) of the colon.

Is there something to this? Alan Brown, an academic researcher in the United Kingdom (UK) picked up a hookworm infection while on a field trip outside the UK. Being that he was a well-nourished Westerner, the 300 hookworms in his gut caused no major problem. However—since being infected, his hay fever allergy has disappeared!

The verdict: more studies are showing the therapeutic value of worm treatment. If I had an incurable inflammatory bowel condition, would I drink work eggs? Hmm…

---

comes from the older name for the genus, *Bilharzia*. The WHO estimates that 250 million people in 74 countries suffer from schistosomiasis, which kills approximately 200,000 people every year.

The adult flukes of *Schistosoma* measure about 10 mm in length. The eggs hatch in freshwater to produce larvae, which then infect snails (intermediate host) (**FIGURE 19.20**). In the snails, the larvae convert into a second larval form, which then escape from the snails and attach themselves to the bare skin of humans (definitive host) wading in larvae-contaminated water. The larvae infect the blood and mature into adult flukes, which cause fever and chills. However, the major effects of disease are due

to eggs produced through sexual reproduction of the parasite. The eggs are carried by the bloodstream to the liver or bladder, where they cause substantial liver damage, bloody urine, and pain on urination. In the intestines, ulceration, diarrhea, and abdominal pain occur. Male and female species of *Schistosoma* mate in the human liver and produce eggs that are released in the feces or urine. Once again, in water, the eggs will hatch into larvae. The anti-helminthic drug praziquantel is used for treatment.

Other species of *Schistosoma* live in the blood of waterfowl and animals that live near water. The eggs of these parasites are released in feces from the host, and the larvae first infect a species of aquatic

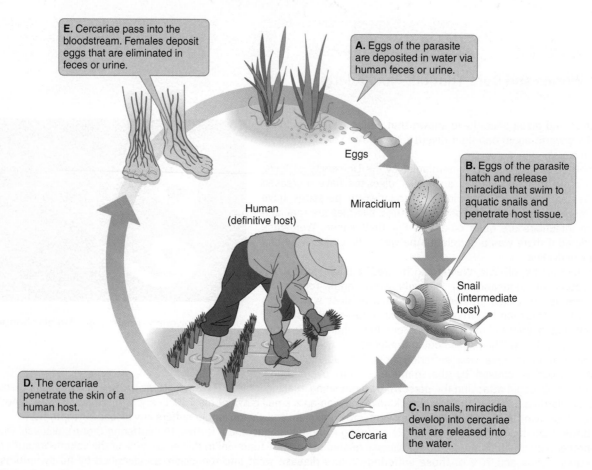

FIGURE 19.20 The Life Cycle of the Blood Fluke *Schistosoma mansoni*. The life cycle of *S. mansoni* involves an intermediate host, the snail, and a definitive host, a human. »» *What defines a definitive and intermediate host?*

snails that live in shallow water near the shoreline. The larvae released from the snail then can infect a bird or animal or a human swimming or wading in the schistosome-contaminated water. However, the parasite penetrates no farther than the skin because humans are not a suitable long-term host; in fact, the larvae are attacked and destroyed by the body's immune system. However, the larvae release allergenic substances that cause an itchy body rash, commonly known as **swimmer's itch**. The rash might last about a week, but it will gradually go away. Approximately 400,000 individuals in the United States, primarily in the northern lakes region, suffer from this mild form of the disease.

Tapeworms Survive in the Human Intestines

Tapeworm infection involves the infestation of the digestive tract, and several tapeworm diseases affect humans.

> ▶ **Beef and Pork Tapeworm Diseases.** Globally, each year approximately 50,000 people die from beef and pork tapeworm diseases.

Such diseases in the United States are rare because most cattle and pigs are free of the parasite.

Humans are the definitive hosts for both the beef tapeworm *Taenia saginata* and the pork tapeworm *Taenia solium*. Humans acquire the tapeworm cysts by eating raw or poorly cooked beef or pork infected with the parasite. In the infected individual, over a two-month period, adult tapeworms grow and can reach 8 meters or more in length. The tapeworms attach to the small intestine via the scolex, and obstruction of this organ can result. In most cases, however, there are few symptoms other than mild diarrhea, and a mutual tolerance between parasite and host can develop. Each tapeworm can have up to 2,000 proglottids, and infected individuals will expel numerous egg-containing proglottids daily. The proglottids accumulate in the soil where they are consumed by cattle or pigs. Embryos from the eggs travel to the animal's muscle, where they form cysts. In the case of *T. solium*, cysts can accumulate in the eyes, brain, muscle, or lung. This complication is called **cysticercosis**.

▶ **Human Echinococcosis.** Sheep, dogs, and other canines such as wolves, foxes, and coyotes are the reservoir and definitive hosts for the comparatively tiny tapeworm *Echinococcus granulosus* (**FIGURE 19.21A**). Eggs reach the soil in feces and spread to numerous intermediary hosts, one of which is humans. Contact with a dog also might account for transmission. In humans, the parasites travel by the blood to the liver, where they form thick-walled **hydatid cysts**

capable of surviving for decades in the individuals (**FIGURE 19.21B**). Common symptoms include abdominal and chest pain and coughing up blood. Treatment might require an antiparasite drug or surgical removal of the cysts. Prevention includes deworming dogs and practicing good hygiene.

Humans Are Hosts to at Least 50 Roundworm Species

Here, we discuss a few significant parasitic infections caused by roundworms.

Pinworm Disease

The most prevalent helminthic infection in the United States is **pinworm disease**, where an estimated 30% of children and 16% of adults are infected. Pinworm disease is caused by *Enterobius vermicularis*. Ingested eggs hatch into larvae in the small intestine before the male and female worms establish an infection in the large intestine. In an infected individual, female worms migrate to the anal region at night and release a considerable number of eggs. The anal area itches intensely, and scratching contaminates the hands and bed linens with eggs. Reinfection can take place if the hands are brought to the mouth or if eggs are deposited in foods by the hands. When the eggs are swallowed, they hatch and start another cycle of infection.

Diagnosis of pinworm disease can be made accurately by applying the sticky side of cellophane tape to the area about the anus and examining the tape microscopically for pinworm eggs (**FIGURE 19.22**).

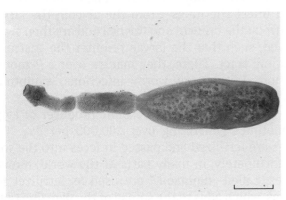

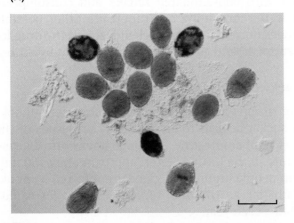

(A)

(B)

FIGURE 19.21 The Dog Tapeworm. (A) A light micrograph of the dog tapeworm, *Echinococcus*. A large section of the worm (right) contains numerous eggs. (Bar = 2 mm.) **(B)** In humans, the tapeworm migrates to the liver or lungs, where it forms slow-growing hydatid cysts (red). (Bar = 200 μm.) *»» Identify the scolex in the light micrograph of* **Echinococcus.** *What was the identifying feature?*

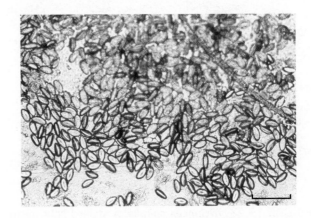

FIGURE 19.22 Diagnosing Pinworm Disease. The transparent tape technique is used in the diagnosis of pinworm disease. (Bar = 2 mm.) *»» What time of day might be best to use the tape technique?*

Mebendazole is effective for controlling the disease, but even without medication, the worms will die in a few weeks, and the infection will disappear as long as reinfection is prevented.

Trichinellosis

Most of us are familiar with the term **trichinellosis** (also called trichinosis) because packages of pork often contain warnings to cook the meat thoroughly to avoid this disease. Today, the disease is rare in the United States due to laws enacted to prevent pork infection.

Trichinellosis often is caused by the small roundworm *Trichinella spiralis*, which globally is among the most widespread of zoonotic pathogens. When contaminated raw or poorly cooked pork is consumed, the cysts pass into the human intestines where larvae are released. In the mucosa of the small intestine, the larvae develop into adult worms, causing diarrhea and abdominal pain. Through sexual reproduction, the females release larvae that migrate to striated muscle where they encyst (**FIGURE 19.23**). Fever and muscle pain accompany this muscular phase of the infection.

Complications of trichinellosis occur when the female-produced larvae migrate to the muscles primarily in the tongue, eyes, and ribs where the cysts eventually die and calcify. The patient commonly experiences pain in the breathing muscles of the ribs and loss of eye movement.

Drugs have little effect on cysts, although albendazole can be used to kill larvae.

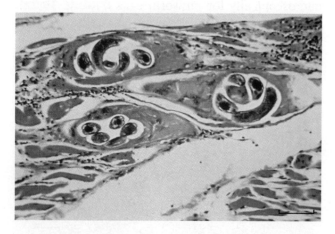

FIGURE 19.23 *Trichinella spiralis*. A stained light micrograph of *T. spiralis* larvae (dark red) coiled inside cysts in muscle tissue (red). (Bar = 2 mm.) »» *What is the purpose of forming a cyst in muscle, which represents a good nutrient source for a parasite?*

Courtesy of the CDC.

Soil-Transmitted Diseases

These diseases are caused by the most significant parasites in humans. They are associated with poverty, lack of adequate sanitation and hygiene, and overpopulation. One parasite, caused by whipworms, was mentioned in MICROFOCUS 19.3.

Ascariasis

Ascaris lumbricoides, a parasite that is the second most prevalent multicellular parasite in the United States, is responsible for **ascariasis**. Globally, the WHO estimates that 1 billion people worldwide are infected, leading to about 60,000 deaths every year, especially in tropical and subtropical regions.

When ingested in contaminated food, eggs hatch in the small intestine. Larvae then penetrate the intestinal wall, enter the venous bloodstream, and infect the lungs. From the alveoli, the larvae crawl up the bronchi and trachea and are then swallowed, such that the larvae re-enter the gastrointestinal tract. There, they mature over a 2-month period. With large worm infections, symptoms include abdominal pain and vomiting.

A female *Ascaris* is a prolific producer of eggs, sometimes generating over 200,000 per day. The eggs are fertilized and passed in feces into the soil. Unfortunately, in many parts of the world, human feces, called "nightsoil," are used as fertilizer for crops, which adds to the spread of the parasite. Contact with contaminated fingers and consumption of water containing soil runoff are other possible modes of transmission. Mebendazole is the drug for treatment, and good hygiene can prevent infection.

Hookworm Disease

One group of roundworms possesses a set of hooks or sucker devices to attach firmly to tissues of the host's upper intestine. Although **hookworm disease** is rare in the United States, approximately 750 million people around the globe are infected by hookworm larvae (**FIGURE 19.24**), resulting in 65,000 deaths each year.

Two hookworms, both about 10 mm in length, can be involved in human disease. The first is the Old World hookworm, *Ancylostoma duodenale*, which is found in Europe, Asia, and the United States; the second is the New World hookworm, *Necator americanus*, which is prevalent in the Caribbean islands.

The life cycle of a hookworm involves only a single host (**FIGURE 19.25**). Female hookworms can release 5,000 to 20,000 eggs, which are excreted into

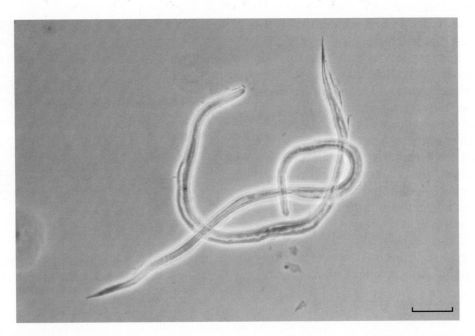

FIGURE 19.24 Hookworm Larvae. Light micrograph of larvae of the hookworm *A. duodenale*. (Bar = 100 μm.) »» *Why are these parasites called hookworms?*

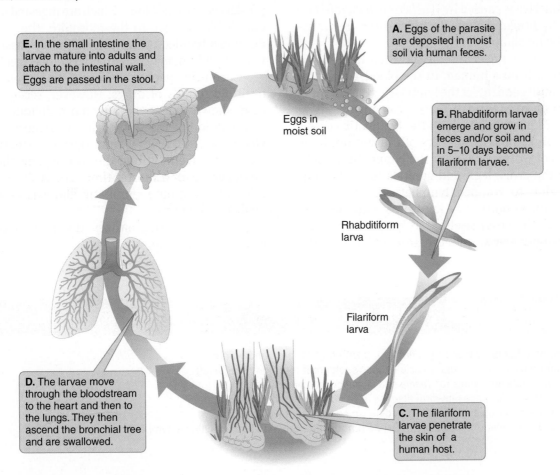

E. In the small intestine the larvae mature into adults and attach to the intestinal wall. Eggs are passed in the stool.

A. Eggs of the parasite are deposited in moist soil via human feces.

B. Rhabditiform larvae emerge and grow in feces and/or soil and in 5–10 days become filariform larvae.

Eggs in moist soil

Rhabditiform larva

Filariform larva

D. The larvae move through the bloodstream to the heart and then to the lungs. They then ascend the bronchial tree and are swallowed.

C. The filariform larvae penetrate the skin of a human host.

FIGURE 19.25 The Life Cycle of the Hookworms *Ancylostoma duodenale* **and** *Necator americanus*. The filariform larvae are the infective form. »» *What other helminth can be coughed up and swallowed?*

the soil where they remain viable for months. If the thread-like larvae that hatch from the eggs contact a human's bare feet, the larvae penetrate the skin layers causing a skin rash called "ground itch." They then enter the bloodstream and soon localize in the lungs. From there, they are carried up to the bronchi and trachea and are swallowed into the intestines. The parasites live in the human intestine, where they suck blood from ruptured capillaries. Hookworm disease therefore is accompanied by blood loss and is generally manifested by anemia. Cysts also can become lodged in the intestinal wall, and ulcer-like symptoms might develop.

Mebendazole can be used to reduce the worm burden. Infection can be prevented by wearing shoes and gloves and properly disposing of human waste.

Roundworms Also Infect the Lymphatic System

Lymphatic filariasis is a parasitic disease affecting over 130 million people in 80 countries throughout the tropics and subtropics. The disease in Africa and South America is caused by *Wuchereria bancrofti* and in Asia by *Brugia malayi*. Both parasites are transmitted by mosquitoes.

Mosquitoes inject the larvae when they take a blood meal from a human. In the blood, the larvae grow to adults and infect the lymphatic system where they can survive for up to 7 years, causing extensive inflammation and damage to the lymphatic vessels and lymph glands. After years of infestation, large numbers of worms can block the lymphatic vessels, causing the arms, legs, and genitals to swell enormously due to trapped lymphatic fluid (**FIGURE 19.26**). This condition is known as **elephantiasis** because of the gross swelling of lymphatic tissues, called **lymphedema**, and the resemblance of the

FIGURE 19.26 Elephantiasis. The severely swollen legs of a patient with lymphatic filariasis (elephantiasis) caused by *Wuchereria bancrofti*. »» *Why is the disease called elephantiasis?*

Courtesy of CDC.

skin to elephant hide. The infection often results in permanent disability.

The adult worms mate and release millions of pre-larvae into the blood, which are ingested in a mosquito's blood meal. In the mosquito, the pre-larvae develop into infective larvae, ready to be passed along to another human during the next blood meal.

Currently, people suffering the disease take multiple rounds of drugs. However, recent studies have shown that if the affected individual takes a combination of the three drugs (ivermectin, diethylcarbamazine, and albendazole), the treatment is more effective in killing the worms, and the treatment can be for a shorter time. The WHO is making efforts to eliminate lymphatic filariasis as a global health problem by 2020.

TABLE 19.4 summarizes the human diseases caused by the helminths.

Concept and Reasoning Checks 19.4

a. Compare the body plan of the flatworms with that of the roundworms.
b. Summarize how schistosomiasis causes such great morbidity worldwide.
c. Draw simple life cycles for *Taenia* and *Echinococcus* parasites.
d. Why do you suppose pinworm disease is so common in the United States?
e. Why are ascariasis and hookworm disease so prevalent worldwide?
f. What is the relationship between the nematode infection and limb swelling in lymphatic filariasis?

TABLE 19.4   A Summary of Human Helminthic Diseases

Disease	Causative Agent	Signs and Symptoms	Transmission	Treatment	Prevention
Schistosomiasis	*Schistosoma* species	Ulceration, diarrhea, abdominal pain Bloody urine and pain on urination	Larval spread through bloodstream	Praziquantel	Avoiding wading in snail-contaminated water
Beef tapeworm disease Pork tapeworm disease	*Taenia saginata* *Taenia solium*	Mild diarrhea	Eating poorly cooked beef or pork	Praziquantel	Eliminating livestock contact with tapeworm eggs Thoroughly cook or freeze all meat and pork
Human echinococcosis	*Echinococcus granulosus*	Abdominal chest pain, coughing up blood	Transmitted from domestic dogs and livestock	Praziquantel	Avoiding infected animals
Pinworm disease	*Enterobius vermicularis*	Anal or vaginal itching, irritability and restless-ness, intermittent abdominal pain and nausea	Ingestion of pinworm eggs from contaminated food, drink, or hands	Albendazole	Practicing good hygiene and household cleaning
Ascariasis	*Ascaris lumbricoides*	Vague abdominal pain, nausea, vomiting, diarrhea or bloody stools	Ingestion of contaminated food	Mebendazole Albendazole	Practicing good hygiene
Trichinellosis	*Trichinella spiralis*	Intestinal pain, vomiting nausea, constipation	Consumption of raw or undercooked pork	Mebendazole	Avoiding under-cooked pork or wild animal meat
Hookworm disease	*Ancylostoma duodenale* *Necator americanus*	Raised rash at skin entry site Abdominal pain, loss of appetite, diarrhea, weight loss	Larvae in soil	Mebendazole	Improving sanitation Avoiding contact with contaminated soil
Lymphatic filariasis	*Wuchereria bancrofti*	Swelling of arms, legs, scrotum	Lymphatic vessels from bite of infected mosquito	Diethyl-carbamazine Albendazole	Avoiding mosquitoes in endemic areas

Chapter Challenge D

Helminths are anything but microorganisms. Some tapeworms are meters long, and many are visible with the naked eye, making them macroscopic.

QUESTION D: *Why are the human helminth diseases discussed with other true "microscopic diseases" (e.g., viral, bacterial, and protistan diseases)?*

You can find answers online in **Appendix F**.

In conclusion, we have now finished the chapters describing the major pathogens (bacterial, viral, fungal, protistan, and helminthic) that affect humans. One of medicine's goals has been to eradicate as many of the highly dangerous and deadly diseases as possible. So far, that goal has only been reached for smallpox—although in 2011 a viral disease called rinderpest (cattle plague) became the first animal disease to be eradicated. Other human diseases are being targeted by global health workers. Polio is near its end. All these diseases have been targeted with vaccines and regional or global vaccination programs. Others have to be eradicated in a different fashion. Take for example Guinea worm (dracunculiasis), which has been widespread across Africa, the Middle East, and parts of Asia. People are infected from drinking stagnant water contaminated with the roundworm. After about a year, the infection causes paralysis and a gruesome affliction that produces a skin blister that feels like one has been stabbed with a red-hot needle—all this as the worms try to emerge from the skin. Therefore, global eradication efforts have focused on providing clean water. The results have been astounding. In the mid-1980s, when the program began, there were 3.5 million Guinea worm cases globally. In 2016, only 25 cases were reported, so it should not be long before Guinea worm is eradicated. It is hoped that other parasitic diseases such as lymphatic filariasis will not be far behind.

■ SUMMARY OF KEY CONCEPTS

Concept 19.1 Protists Exhibit Great Structural and Functional Diversity

1. Protists are a diverse group. The majority are unicellular and free-living organisms inhabiting moist areas or water. Some protists, such as the **green algae**, are photosynthetic, and other protists, like the **dinoflagellates, radiolarians**, and **foraminiferans**, are part of the marine **phytoplankton**. Other protists are heterotrophic and have a fungus-like structure.
2. The human parasitic protists have many lifestyles, but the parasites can be cataloged into one of three super groups that contain human parasites.

Concept 19.2 Protistan Parasites Attack the Skin and the Digestive and Urinary Tracts

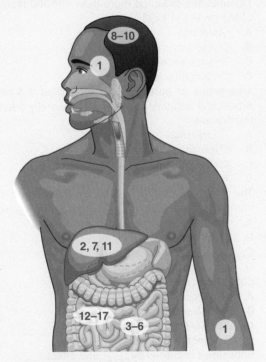

▶ **Cutaneous leishmaniasis**
　1　*Leishmania major*
▶ **Visceral leishmaniasis**
　2　*Leishmania donovani*
▶ **Amebiasis**
　3　*Entamoeba histolytica*
▶ **Giardiasis**
　4　*Giardia intestinalis*

▶ **Cryptosporidiosis**
　5　*Cryptosporidium parvum, C. hominis*
▶ **Cyclosporiasis**
　6　*Cyclospora cayetanensis*
▶ **Trichomoniasis** (not shown)
　　Trichomonas vaginalis

Concept 19.3 Many Protistan Diseases of the Blood and Nervous System Can Be Life Threatening

▶ **Malaria**
　7　*Plasmodium* species
▶ **Babesiosis** (not shown)
　8　*Babesia microti*
▶ **African Trypanosomiasis (human African sleeping sickness)**
　　Trypanosoma brucei
▶ **Chagas disease (American trypanosomiasis)** (not shown)
　　Trypanosomiasis cruzi
▶ **Toxoplasmosis**
　9　*Toxoplasma gondii*
▶ **Primary amebic meningoencephalitis**
　10　*Naegleria fowleri*

Concept 19.4 Parasitic Helminths Cause Substantial Morbidity Worldwide

Among the **helminths**, there are two groups responsible for parasitic diseases in humans—**flatworms** and **roundworms**.

▶ **Schistosomiasis**
　11　*Schistosoma mansoni*
▶ **Beef/pork tapeworm disease**
　12　*Taenia saginata, T. solium*
▶ **Echinococcosis**
　13　*Echinococcus granulosus*
▶ **Pinworm disease**
　14　*Enterobius vermicularis*
▶ **Trichinellosis**
　15　*Trichinella spiralis*
▶ **Ascariasis**
　16　*Ascaris lumbricoides*
▶ **Hookworm disease**
　17　*Ancylostoma duodenale, Necator americanus*
▶ **Filariasis** (not shown)
　　Wuchereria bancrofti, Brugia malayi

◼ CHAPTER SELF-TEST

You can find answers for **Steps A–D** online in **Appendix D**.

◼ STEP A: REVIEW OF FACTS AND TERMS

Multiple Choice

Read each question carefully before selecting the *one* answer that best fits the question or statement.

1. The _____ are members of the _____.
   A. green algae; dinoflagellates
   B. dinoflagellates; phytoplankton
   C. radiolarians; fungus-like protists
   D. ciliates; photosynthetic protists

2. This group of protists has a single mitochondrion with a mass of DNA.
   A. Kinetoplastids
   B. Ciliates
   C. Apicomplexans
   D. Diplomonads

3. The _____ are a group of protists that have food vacuoles and pseudopods.
   A. amoebas
   B. parabasalids
   C. kinetoplastids
   D. ciliates

4. Which one of the following is NOT found in the ciliates?
   A. A contractile vacuole
   B. Macronuclei and micronuclei
   C. A complex of organelles in the tip
   D. Mitochondria

5. An intermediate host is _____.
   A. where parasite asexual cycle occurs
   B. always a nonhuman host
   C. where parasite sexual cycle occurs
   D. the human host between two other animal hosts

6. The vector transmitting leishmaniasis is the _____.
   A. mosquito
   B. sandfly
   C. tsetse fly
   D. sand flea

7. _____ enters the human body as a cyst and develops into a trophozoite in the small intestine; a severe form of dysentery can occur.
   A. *Cryptosporidium parvum*
   B. *Entamoeba histolytica*
   C. *Giardia intestinalis*
   D. *Cyclospora cayetanensis*

8. Sucker-like devices allow this protist to adhere to the intestinal lining.
   A. *Cyclospora cayetanensis*
   B. *Entamoeba histolytica*
   C. *Giardia intestinalis*
   D. *Cryptosporidium parvum*

9. This disease sickened more than 400,000 residents of Milwaukee in 1993.
   A. Giardiasis
   B. Cryptosporidiosis
   C. Trypanosomiasis
   D. Cyclosporiasis

10. This genus of parabasalid affects over 7.4 million Americans annually and causes a sexually transmitted disease.
    A. *Toxoplasma*
    B. *Cryptosporidium*
    C. *Cyclospora*
    D. *Trichomonas*

11. The _____ form of the malarial parasite enters the human blood while the _____ enters the mosquito.
    A. sporozoites; merozoites
    B. merozoites; gametocytes
    C. merozoites; sporozoites
    D. sporozoites; gametocytes

12. Chagas disease is caused by _____.
    A. *Trypanosoma cruzi*
    B. *Toxoplasma gondii*
    C. *Babesia microti*
    D. *Trypanosoma brucei*

13. Babesiosis is carried by _____ and infects _____.
    A. mosquitoes; RBCs
    B. fleas; the kidneys
    C. ticks; RBCs
    D. mosquitoes; the liver

14. This opportunistic protist causes primary amebic meningoencephalitis (PAM).
    A. *Plasmodium falciparum*
    B. *Toxoplasma gondii*
    C. *Naegleria fowleri*
    D. *Paragonimus westermani*

Matching

Read the statement concerning the helminths, and then select the answer or answers that best apply to the statement. Place the letter(s) next to the statement.

_____ 15. Transmitted by mosquitoes
    A. Filariasis
    B. Trichinosis
    C. Hookworm disease

_____ 16. Beef tapeworm species
    A. *Echinococcus granulosus*
    B. *Schistosoma mansoni*
    C. *Taenia saginata*

_____ 17. Cause of pinworm disease
    A. *Trichinella*
    B. *Enterobius*
    C. *Ascaris*

_____ 18. Type of tapeworm
    A. *Taenia*
    B. *Echinococcus*
    C. *Necator*

_____ 19. Acquired by consuming contaminated pork
    A. *Taenia solium*
    B. *Echinococcus*
    C. *Necator americanus*

_____ 20. Causes inflammation and damage to the lymphatic vessels
    A. *Echinococcus*
    B. *Ascaris*
    C. *Wuchereria*

Label Identification

21. Label the two parasite life cycles for (1) malaria and (2) schistosomiasis, using the term lists provided.

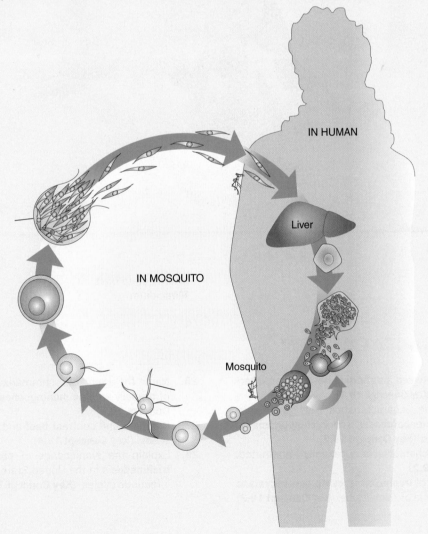

IN HUMAN

Liver

IN MOSQUITO

Mosquito

(1) *Plasmodium* Life Cycle

Term List

Definitive host
Gametes
Gametocyte
Intermediate host

Merozoites
Oocyst
Sporozoites
Zygote

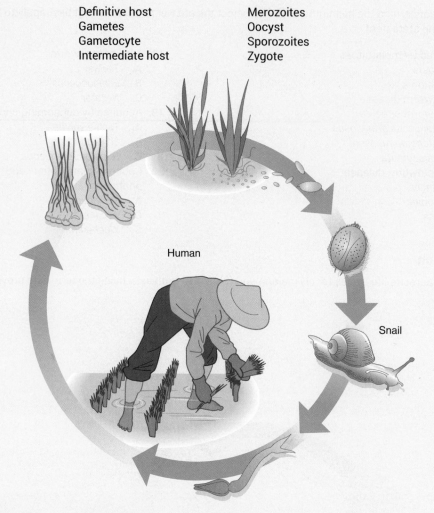

Human

Snail

(2) *Schistosoma* Life Cycle

Term List

Cercaria
Definitive host
Eggs

Intermediate host
Miracidium

■ STEP B: CONCEPT REVIEW

22. Differentiate between a **definitive host** and an **intermediate host**. (**Key Concept 19.1**)

23. Summarize the characteristics of **amebiasis, giardiasis, cryptosporidiosis**, and **cyclosporiasis** as human diseases. (**Key Concept 19.2**)

24. Explain how **trichomoniasis** is sexually transmitted. (**Key Concept 19.2**)

25. Identify the role of **trophozoites, cysts**, and **oocysts** to the infection cycle of *Toxoplasma*. (**Key Concept 19.3**)

26. Name the agent, and summarize the characteristics of **primary amebic meningoencephalitis (PAM)**. (**Key Concept 19.3**)

27. Compare and contrast **beef** and **pork tapeworm diseases**. (**Key Concept 19.4**)

28. Explain the significance of **pinworm disease** and **trichinellosis** in the United States, and describe their infectious cycles. (**Key Concept 19.4**)

■ STEP C: APPLICATIONS AND PROBLEM SOLVING

29. You and a friend who is 3 months pregnant stop at a hamburger stand for lunch. Based on your knowledge of toxoplasmosis, what helpful advice can you give your friend? On returning to her home, you notice she has two cats. What additional information might you share with her?

30. Cardiologists at a local hospital hypothesized that a few patients with a certain protozoal disease easily could be lost among the far larger population of heart disease sufferers patronizing county clinics. They proved their theory by finding 25 patients with this protozoal disease among patients previously diagnosed as having coronary heart disease. Which protozoal disease was involved?

31. Federal law stipulates that food scraps fed to pigs must be cooked to kill any parasites present. It also is known that feedlots for swine are generally more sanitary than they have been in the past. Because of these and other measures, the incidence of trichinellosis in the United States has declined, and the acceptance of "pink pork" has increased. Do you think this is a dangerous situation? Why?

■ STEP D: QUESTIONS FOR THOUGHT AND DISCUSSION

32. Many years ago, a newspaper article asserted that parasitology is a "subject of low priority in medical schools because the diseases are exotic infections only occurring in remote parts of the world." Would you agree with that statement? Explain.

33. It has been said that until recent times, many victims of a particular protozoal disease were buried alive because their life processes had slowed to the point where they could not be detected with the primitive technology available. Which disease was probably present?

34. Some restaurants offer a menu item called steak tartare, which is a dish served with raw ground beef. What hazard might this meal present to the restaurant patron?

35. Many of the parasitic diseases described in this chapter, including schistosomiasis, ascariasis, filariasis, trypanosomiasis, leishmaniasis, and some bacterial diseases mentioned in previous chapters, such as trachoma and leprosy, are often considered "neglected diseases." Why do you think these diseases have been somewhat "neglected" by the medical community, especially in the developed world?

PART VI Interactions and Impact of Microorganisms with Humans

The five chapters in Part VI embrace several aspects of the six principal (overarching) concepts as described in the American Society for Microbiology (ASM) fundamental statements for a concept-based microbiology curriculum.

PRINCIPAL CONCEPTS

Evolution	• Human impact on the environment influences the evolution of microorganisms (e.g., emerging diseases and the selection of antibiotic resistance). Chapter 20
Cell Structure and Function	• Bacteria and Archaea have specialized structures (e.g., flagella, endospores, and pili) that often confer critical capabilities. Chapter 20
Metabolic Pathways	• The survival and growth of any microorganism in a given environment depends on its metabolic characteristics. Chapter 20
Information Flow and Genetics	• Genetic variations can impact microbial functions (e.g., in biofilm formation, pathogenicity and drug resistance). Chapters 20–24
Microbial Systems	• Microorganisms are ubiquitous and live in diverse and dynamic ecosystems. Chapter 20 • Most bacteria in nature live in biofilm communities. Chapter 20 • Microorganisms and their environment interact with and modify each other. Chapters 20–22 • Microorganisms, cellular and viral, can interact with both human and nonhuman hosts in beneficial, neutral, or detrimental ways. Chapter 20
Impact of Microorganisms	• Microorganisms provide essential models that give us fundamental knowledge about life processes. Chapter 20

CHAPTER 20

The Host-Microbe Relationship and Epidemiology

By late July 1999, crows were literally dropping out of the sky in New York City, and dead crows were found in surrounding areas, as well. By early September, officials at the Bronx Zoo discovered that birds at the zoo, including a cormorant, two red Chilean flamingos and an Asian pheasant, had died of the same brain inflammation (encephalitis) as found in the crows.

On August 23, an infectious disease physician at a hospital in northern Queens reported to the New York City Department of Health that two patients had been admitted with encephalitis. In fact, on further investigation, the health department identified a cluster of six patients with encephalitis. Blood samples from the initial cases were given to the Centers for Disease Control and Prevention (CDC) for testing for common North American **arboviruses**—viruses transmitted by arthropods, such as insects. The results came back positive for St. Louis encephalitis (SLE) virus, which is carried by mosquitoes. These findings prompted the New York City Health Department to begin aerial and ground application of insecticides.

News of SLE and spraying caught the attention of the Bronx Zoo officials. If the birds were dying from the same encephalitis disease as in humans, it could not be caused by the SLE virus because SLE does not infect birds.

Consequently, a reinvestigation by the CDC of virus samples taken from humans, birds, and mosquitoes was carried out. Results identified West Nile virus (WNV), a virus never before detected in the Western hemisphere, as the source of the infection (see the chapter opening image).

By early fall, mosquito activity waned and the number of human cases declined. In all, 61 people were infected and seven died. Although New York City and the surroundings could breathe a sigh of relief, the outbreak was only the beginning of a seven-year spread of WNV across the continental United States (**FIGURE 20.1**).

The outbreak of West Nile encephalitis is one example highlighting the **epidemiology** of infection and disease, that is, the scientific (and medical)

False-color transmission electron micrograph of West Nile viruses.
Courtesy of Cynthia Goldsmith/CDC.

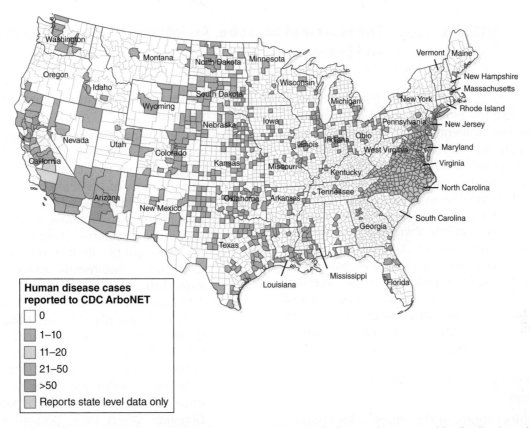

FIGURE 20.1 West Nile Virus Activity, United States—2015. The map indicates the geographic distribution of human disease cases by county in 2015 (a county is shaded no matter in what area of the county the disease case(s) occurred). *»» On the map, what was the activity in your county in 2015?*

CDC/USGS.

study of the causes, transmission, and prevention of disease within populations. Today, the CDC, the World Health Organization (WHO), and other agencies throughout the world have built on the historical work of John Snow, Ignaz Semmelweis, Joseph Lister, Louis Pasteur, Robert Koch, and many others. Yet, many of the epidemiologists employed by these organizations seek the same goals as John Snow: to identify and investigate disease outbreaks, conduct research to enhance prevention, and then devise prevention strategies.

In this chapter, we discuss the mechanisms underlying the spread and development of infectious disease. Our purpose is to bring together many concepts of disease and synthesize an overview of the host–microbe relationship. We summarize much of the important terminology used in medical microbiology and outline some of the factors used by microorganisms to establish themselves in tissues. An understanding of the topics concerning the host–microbe relationship will be essential preparation for the detailed discussion of host resistance (immune) mechanisms.

Chapter Challenge

Patients, especially older adults, receiving medical care can contract serious, sometimes life-threatening infections called healthcare-associated infections (HAIs). One very serious HAI is caused by *Clostridium difficile* (commonly called *C. diff*), which produces lower gastrointestinal disease. In mild cases, this gram-positive, anaerobic, spore-forming rod causes watery diarrhea three or more times a day for two or more days along with mild abdominal cramping and tenderness. In severe cases, *C. diff* causes pseudomembranous colitis (an inflamed colon). Because the bacterial cells are found in the feces, poor hygiene can result in items or surfaces being contaminated with *C. diff*. Contacting such items, and then touching the mouth or mucous membranes, can lead to a *C. diff*-associated infection (CDAI). Healthcare workers can spread the bacterial cells to patients or contaminate surfaces through hand contact. In fact, severe cases of *C. diff* infections account for up to 30,000 American deaths each year. So, what is it about *C. diff* that makes it a "poster microbe" for infection and disease, the subject of this chapter? Let's find out!

■ KEY CONCEPT 20.1   The Host and Microbe: An Intimate Relationship in Health and Disease

By the early 1970s, some health experts claimed we could "close the books on infectious diseases" because the development and use of antibiotics and vaccines would make the threat of infectious disease of little consequence. However, antibiotic resistance and new emerging diseases, including Legionnaires' disease, AIDS, Lyme disease, hantavirus pulmonary syndrome (HPS), severe acute respiratory syndrome (SARS), and, most recently, Zika virus disease have thwarted such optimism. In 2016, of the approximately 55 million humans who died worldwide, more than 25% (14 million) died from infectious diseases, making them the second leading cause of death (behind cardiovascular disease; **FIGURE 20.2**). In fact, infectious diseases are the leading cause of death in children younger than 5 years of age.

Colonization Can Lead to Infection and Disease

When microbes (note: in this chapter, for simplicity of discussion, "microbe" includes the viruses)

interact with a host, such as a human, several outcomes can occur:

▶ **Colonization.** If a healthy person is exposed to a bacterial or fungal pathogen, the microbe might attempt to establish a foothold on or in some part of the body. However, **colonization** will be challenging because of the individual's strong immune defenses and resident microbiome. Most of the entering microbes will be eliminated from or assume a transient relationship with the host.

▶ **Infection.** Should the pathogen establish a foothold, it now can begin to invade the body and multiply. With an **infection**, there is now competition between the host and the invading microbe. Again, additional immune system defenses might eliminate the pathogen but in the process cause inflammation at the site of the infection.

▶ **Disease.** Often in a person with weak immune resistance (immunocompromised),

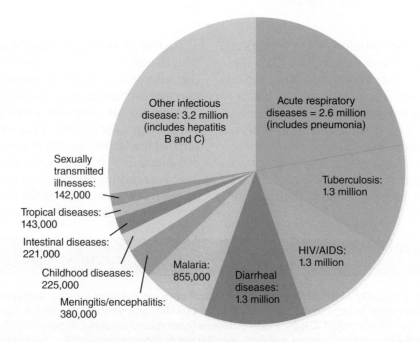

FIGURE 20.2 Infectious Disease Deaths Worldwide—2013. This pie chart depicts the leading causes of infectious diseases and the number of worldwide deaths in 2013. Childhood diseases: diphtheria, measles, pertussis, chickenpox, and tetanus. Tropical diseases: dengue fever, yellow fever, leishmaniasis, African trypanosomiasis, Chagas disease, schistosomiasis, filariasis, and onchocerciasis. »» *Only one of the "pie slices" has grown explosively in mortality numbers since 1993. Identify the disease and explain why that is so.*

Data from Global Burden of Disease Study 2013 (*The Lancet* 385:117–171. 2015).

infection leads to **disease**, which represents any change from the general state of good health. For an infectious disease, tissue or organ damage or organ dysfunction can occur. Immune system defenses and/or antibiotic therapy might eliminate the pathogen or there might be a chronic condition, where the pathogen persists even though it cannot grow and spread. Of course, the other outcome could be death of the individual due to widespread organ or body system failure due directly or indirectly to the pathogen.

It is important to note though that infection and disease are not synonymous; a person can be infected without suffering a disease.

The Human Body Maintains a Symbiosis with Its Microbiome

For colonization, whether host or pathogen gets the upper hand is due in part to trillions of microbes normally found on and in the human body. This community of beneficial microbes (primarily bacterial species), called the **human microbiome**, normally resides in the body without directly causing infection or disease (TABLE 20.1). These resident microbes establish a permanent relationship with the surfaces of those cells and tissues having a direct or indirect connection to the environment. Thus, the nose, mouth, eyes, upper respiratory tract, openings to the urogenital tract, and the intestines are colonized by hundreds, if not thousands, of microbial species (**FIGURE 20.3**). In the large intestine, for example, bacterial species such as *Escherichia coli*, *Bacteroides fragilis*, *Lactobacillus casei*, and the fungal species *Candida albicans* are a few of the hundreds of species composing the gut community. The different microbiomes found at different body sites are discussed in the relevant chapters dealing with bacterial, viral, and eukaryotic diseases.

Most other tissues of the body remain **sterile**; the blood, cerebrospinal fluid, joint fluid, and internal organs, such as the kidneys, liver, muscles, bone, and brain, do not contain any microbes unless an infection has occurred.

One of the important roles that our microbiome plays is to shield us from pathogen invasion. By forming a cellular barrier over the epithelium of the upper respiratory and digestive tracts, the microbiome prevents pathogens from establishing a foothold for colonization. In addition, the microbiome successfully competes for the available nutrients and secretes antimicrobial compounds that make the environment less hospitable for pathogens. Thus, the relationship between the body and its microbiome is an example of a **symbiosis**, or living together. If the symbiosis is beneficial to both the host and the microbe, the relationship is called **mutualism**. For example, species of *Lactobacillus* live in the female vagina and derive nutrients from the environment while producing acid to prevent the overgrowth of other potentially infectious organisms.

Commensalism represents a type of symbiosis in which the microbe benefits and the host is unaffected. *E. coli* is generally presumed to be a commensal in the human intestine because even though it produces small amounts of vitamins B and K for our diet, we can obtain all these vitamins from other sources.

TABLE 20.1   The Human Microbiome—The Numbers

30 trillion	The approximate number of microbes (primarily bacterial cells) on and in a human body, which is about 3 times the number of human cells building the body
3 pounds	The average weight of the human microbiome in a 150-pound person (= 1%–3% of total body weight)
3 million	The estimated number of unique protein-coding genes in the human microbiome, which is more than 100 times the number of human genes coding for proteins (20,000)
35,000	An estimated upper limit on the number of bacterial species in the human intestine
500	The estimated number of bacterial species in the human mouth
0	The number of microbial cells on and in the fetus prior to birth

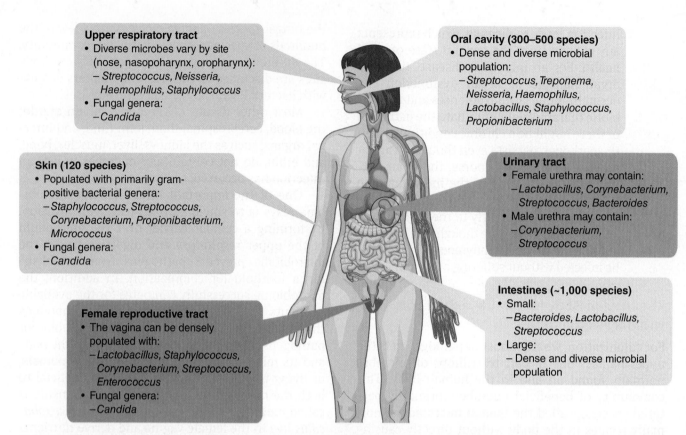

FIGURE 20.3 A Sampling of Organisms Composing the Human Microbiome. In reality, there are thousands of microbial species on and within some human body systems. »» *Explain why some body systems (e.g., circulatory system, nervous system) normally remain sterile.*

Although our microbiome is capable of maintaining itself, physical, chemical, or biological injury to the resident microbes can sometimes overextend its repair abilities. Taking antibiotics for an infection is one example of a process that can destroy and injure the gut microbiome. In some cases, this allows opportunistic pathogens, such as antibiotic-resistant *Clostridium difficile* (*C. diff*), to overrun the remaining bowel microbiome and become a **super-infection**, that is, a new infection (*C. diff*) occurring in a patient having a preexisting infection.

Many health experts suggest that ingesting live microbes can help reestablish the normal microbiome. Called **probiotics**, these products are composed of avirulent microbes similar to those normally found in the gut. One of the most familiar food products that might help as a probiotic is yogurt. Those yogurts containing *Lactobacillus bulgaricus*, *Streptococcus thermophilus*, and *Bifidobacterium lactis* are said to contain "live, active cultures." Could bacterial species like these in adequate amounts compete with pathogens and help restore the normal microbiome? INVESTIGATING THE MICROBIAL WORLD 20 takes a look.

Researchers are just now beginning to comprehend some of the roles the human microbiome plays in human health. In fact, more than 50% of the bacterial species and the majority of the fungal species and viruses compose a so-called "microbial dark matter," that is, species, organisms, and viruses whose function and behavior are completely unknown at present. In 2016, a new National Microbiome Initiative (NMI) was introduced and funded by several national funding agencies with support from more than 100 external institutions. The aim of the NMI is to advance the understanding of microbiome behavior across all communities of microbes, including studies that could have applications for better protection and restoration of a healthy human microbiome.

The Human Microbiome Begins at Birth

As we just saw, humans (and other animals, too) have coevolved with a diverse set of resident

Investigating the Microbial World 20

Can a Yogurt a Day Keep the Doctor Away?

Anecdotal evidence represents nonscientific observations, usually based solely on people's personal experiences. Such stories do not provide proof, yet they often are used in place of clinical or scientific evidence to support a claim. Take, for example, the probiotic properties of yogurt.

OBSERVATION: For more than a century, people have touted the health benefits of soured milk and then cultured dairy products for human digestion. In reality, the validity of such claims to human health remains unknown. However, what such claims often do is stimulate research to verify or refute the claim.

QUESTION: *Can the bacterial species in probiotic yogurt affect the diverse population of normal microbes in the human gut?*

HYPOTHESIS: Daily consumption of yogurt significantly alters the composition of the gut's microbiome. If so, test subjects will show a changed bacterial composition after consuming a daily cup of yogurt.

EXPERIMENTAL DESIGN:

EXPERIMENT 1: The fecal bacterial composition in seven healthy, adult female identical twins was analyzed before, during, and after consuming a specific brand of yogurt (with five known bacterial strains) referred to as a fermented milk product (FMP) over a 4-month period. All participants consumed two servings of FMP per day for 7 weeks.

EXPERIMENT 2: The fecal bacterial composition in gnotobiotic mice was analyzed before, during, and after being inoculated with the same five bacterial strains found in the FMP. Note: gnotobiotic mice are born with germ-free guts and then "humanized" by rearing them such that their guts only contain 15 members of a typical human gut's microbiome (representing the three major bacterial phyla found in the human gut).

RESULTS: In the twins experiment, consumption of the FMP did not change the bacterial species of the gut microbial community. Using the "humanized" mice, again there was no change in the composition of the 15 bacterial species, and the five FMP bacterial species did not take up residence in the gut of the mice. However, in both the twins experiment and the "humanized" rat experiment, study analysis showed that consumption of the FMP or exposure to the five bacterial species in the FMP did result in a significant change in the way the gut microbiome metabolically processed carbohydrates.

CONCLUSIONS:

QUESTION 1: *Do these two experiments support the hypothesis? Explain.*

QUESTION 2: *Why were the mice in experiment 2 "humanized"?*

QUESTION 3: *Propose what might have happened if these experiments were done using individuals that had a gut microbiome that was out of balance, say with a significant* Clostridium difficile *population.*

You can find answers online in **Appendix E**.

Data from McNulty, N. P., et al. 2011. *Science Translational Medicine* 3:1–14.

---

microorganisms. Where did all these microbes come from that inhabit the adult human body?

Although there is growing evidence that microbes from a mother's mouth might colonize the placenta during fetal development, the newborn and infant will acquire the vast majority of microbes that eventually become the adult human microbiome during three phases (**FIGURE 20.4**).

▶ **Method of Birth.** When the newborn passes through the mother's birth canal (vaginal delivery), the newborn picks up the

community of microbes (bacteria and fungi) present in the mother's vaginal microbiome. However, if it is a **cesarean birth** (C-section), the newborn has no contact with the mother's vaginal microbes. Therefore, the microbes colonizing the skin, nose, mouth, and rectum of the newborn are usually from the skin of other individuals that had contact with the newborn. These microbes tend to be quite different from those picked up through a vaginal birth.

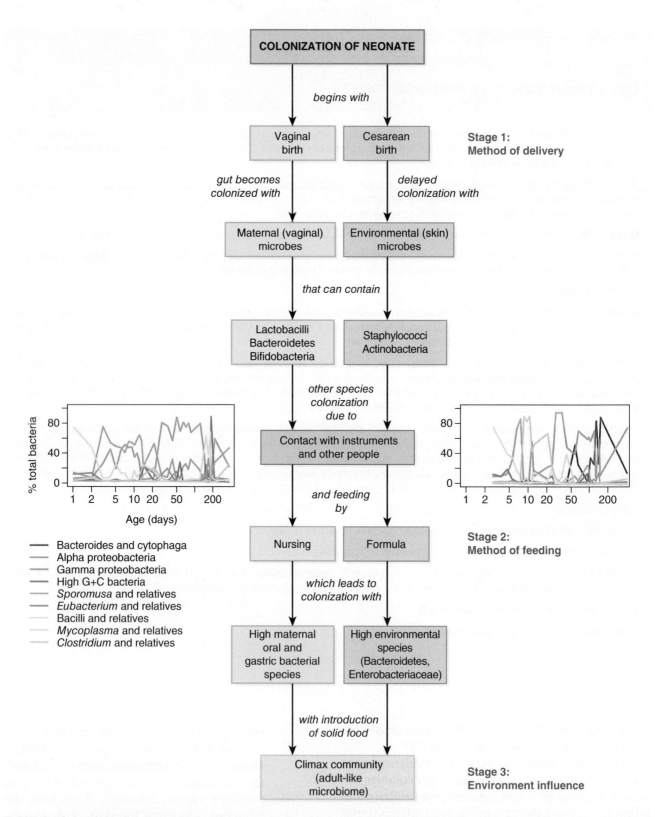

FIGURE 20.4 Bacterial Colonization of the Neonate. This concept map diagrams possible gut colonization of two newborns based on the three major stages of microbiome acquisition. The graphs (insets) plot fluctuations in the bacterial microbiome (left: vaginal birth; right: cesarean birth) over the first 300 days. *»» How would each form of birth and feeding introduce new microbes into the newborn?*

Graphs modified from Palmer, C., et al. 2007. *PLoS Biology* 5(7), el77 doi:10.1371/journal.pbio.0050177.

▶ **Method of Feeding.** Additional organisms enter upon first feeding where the carbohydrates coming from breast milk or formula can influence what microbes dominate in the newborn's gut. Breast milk is quite diverse in the microbial community it contains. In fact, each woman has a slightly different milk microbiome. In addition, milk contains antibodies and nutrients (fats, proteins, vitamins, and carbohydrates) that a baby needs. The carbohydrates are particularly interesting because these short chain sugars are only digestible by the microbes in the newborn's gut. Because formula-dominant feeding often lacks these needed nutrients, breastfeeding can be critical for normal infant development.

▶ **Environmental Contacts.** During the weeks following birth, additional contact with the mother, family, friends, medical staff, other individuals, and home pets will expose the infant to additional groups of microbes. Importantly, the administration of antibiotics early in life might lead to a less diverse microbial community. If a child undergoes repeated courses of antibiotics, the treatment might interfere with the normal development of the intestinal microbiome.

Although more work needs to be done to understand the role of the infant's microbiome, life's early events might predispose a child to noncommunicable disease such as asthma, allergies, and obesity later in childhood. In addition, the infant's microbiome plays a key role in the development of the immune system. Changes from the "normal" community of microbes, called **dysbiosis**, can bring on possible immunological-based diseases later in life, as well.

By three years of age, the infant's encounters with parents, family, other people, foods, pets, and the environment have established the unique composition of the individual's microbiome. This community of microbes remains stable throughout life, undergoing small changes in response to the internal and external environment of the individual.

Then again, on occasion, true pathogens gain the advantage and disease might occur.

Pathogens Differ in Their Ability to Cause Disease

There also is a symbiotic relationship, called **parasitism**, in which the pathogen benefits and causes damage to the host; disease can result. Microbiologists once believed microbes were either pathogenic or nonpathogenic; they caused disease or they did not. We now know that is not true from the preceding discussion.

Pathogenicity refers to the ability of a microorganism to gain entry to the host's tissues and bring about a physiological or anatomical change, resulting in altered health and leading to disease. Certain pathogens, sometimes referred to as **primary pathogens**, are well known for their ability to cause disease in healthy individuals. Examples would include the pathogens responsible for the common cold, plague, and typhoid fever. Others, called **opportunistic pathogens**, often are commensals taking advantage of a shift in the body's delicate balance to one that favors the microbe. For example, if the resident microbiome is damaged or the host's immune system is weakened, some commensals seize the "opportunity" to invade tissues and cause disease. AIDS is an example wherein crippling of the immune system makes the patient highly susceptible to a variety of potential opportunistic organisms.

Whether a disease is mild or severe depends on the pathogen's ability to do harm to a susceptible host. Thus, the degree of pathogenicity, called **virulence**, depends on the host–microbe interaction. For example, a microbe invariably causing disease, such as the typhoid bacillus, is said to be "highly virulent." By comparison, a microbe sometimes causing disease, such as *C. albicans*, is labeled "moderately virulent." The level of virulence then depends partly on the structural and/or chemical factors, called **virulence factors**, that the microbe possesses. In general, the more factors present, the more dangerous (virulent) the microbe. We will learn about some virulence factors later in this chapter.

The level of virulence also depends on the host because host immunity through vaccination can negate virulence. Thus, someone who has been vaccinated against hepatitis B is protected even if the person is exposed to the virus. In addition, certain organisms, described as **avirulent**, are not regarded as disease agents because they usually lack the virulence factors capable of infecting or causing disease in that host.

Concept and Reasoning Checks 20.1

a. Identify the importance of the human microbiome in human health and disease.
b. How might the route of birth affect a newborn's health?
c. Distinguish between pathogenicity and virulence.

Chapter Challenge A

People most at risk of developing a CDAI are those who take antibiotics as part of their medical care. In fact, *C. diff* is resistant to many of the antibiotics normally used to treat other HAIs. Therefore, standard antibiotic therapy could provide the opportunity for a CDAI and contribute to the outbreaks that have been widely reported in hospitals and other healthcare settings.

QUESTION A: *Suppose that a healthy person carrying* C. diff *transmits the pathogen to a patient on antibiotic therapy. The healthy person does not become ill, but the patient soon contracts a CDAI. Based on the material presented in Key Concept 20.1, how can that be if the patient is on antibiotics? In addition, what have the antibiotics done to the patient's gut that gave the pathogen an even easier passage into the intestinal cells for both colonizing the tissue and establishing an infection?*

You can find answers online in **Appendix F**.

■ KEY CONCEPT 20.2   The Host and Microbe: Establishment of Infection and Disease

Disease is the result of a dynamic series of events originating from the competition between host and pathogen.

A Diagnosis of an Infectious Disease Depends on Signs and Symptoms

When a health professional makes a **diagnosis** of an infectious disease, she or he takes a complete patient history, oversees a comprehensive physical exam of the patient, and, as a result, makes an evaluation of the patient's signs and symptoms. **Signs** of a disease represent some type of evidence of disease as detected by an observer (e.g., physician or other healthcare professional). Abnormal blood pressure, bacterial cells in the blood, or lab test results would be examples.

A correct diagnosis also can depend on **symptoms**, which represent changes in body function sensed by the patient. For example, aches, pains, nausea, and headaches could be symptoms of a **clinical disease**, which represents a situation in which an infected patient is **symptomatic**; that is, the patient is experiencing symptoms. On the other hand, a **subclinical disease** represents a circumstance in which an infected patient might be **asymptomatic** and is not experiencing any symptoms.

Diseases also can be characterized by a specific collection of signs and symptoms called a **syndrome**. AIDS (acquired immunodeficiency syndrome) is an example in which the individual exhibits over time a set of signs and symptoms that point to a specific disease, typically a set of opportunistic infections or what are referred to as AIDS-defining conditions.

Several Events Must Occur for Disease to Develop in the Host

From the standpoint of the pathogen, several factors come into play for an infection to be "successful" in causing a disease (**FIGURE 20.5**).

Portal of Entry

Pathogens typically have a characteristic route by which they enter a susceptible host. This **portal of entry** may be one of the following:

- ▶ **Skin.** Being the largest organ of the body, the skin represents a large surface over which pathogens might gain entry into the host. Although the skin normally is impenetrable to most pathogens, a break in, or wound to, the skin can serve as a portal of entry. Natural openings in the skin (e.g., sweat glands and hair follicles) also can serve as portals of entry.

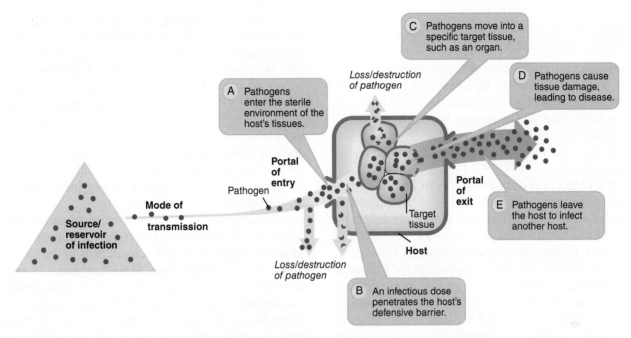

FIGURE 20.5 The Sequence of Events Leading to an Infectious Disease. The infectious dose and adhesion to cells or tissues are required to initiate infection and disease. *»» Identify which events would not occur if an infection but not a disease occurred.*

▶ **Respiratory Tract.** Inhalation of pathogens that are carried on airborne droplets, such as the bacterial pathogens that cause pneumonia and tuberculosis, or the viruses that cause colds and the flu, can enter the upper respiratory tract and progress to the lungs.

▶ **Gastrointestinal Tract.** Ingestion of contaminated foods or water, such as bacterial agents in fecally contaminated ground beef or bacterial/viral pathogens in contaminated water, can give the pathogen access to the mucous membranes of the intestines. This gastrointestinal portal is often referred to as the **fecal–oral route** of infection.

▶ **Reproductive/Urinary Tracts.** Sexual intercourse with a partner having a sexually transmitted infection might give the pathogen access to the reproductive tissues. Likewise, pathogens that cause urinary tract infections can enter through the mucous membranes of the urethra.

▶ **Other Parenteral Route.** Any pathogen entering the body in a manner other than through the gastrointestinal tract is called a **parenteral route**. Accordingly, a break or piercing of the skin due to animal/insect bites, wounds, or injections represents parenteral routes by which a pathogen gains

access to the blood and body tissues. Medical equipment, such as catheters and intubation tubes, also provide parenteral routes to the urinary or respiratory tracts.

Many pathogens are adapted to a specific portal of entry, so if the pathogen does not find the appropriate portal, no growth or infection will occur. On the other hand, some pathogens, such as *Staphylococcus*, can invade through any of several portals, including the skin, gastrointestinal tract, and urogenital tract.

Reaching the appropriate portal of entry is not sufficient to cause an infection. The development of an infection usually depends on the presence of a sufficient number of pathogenic organisms. This **infectious dose** refers to the minimal number of pathogens needed before the signs and symptoms of an infection are apparent. For example, the consumption of a few hundred *Campylobacter* cells will lead to food poisoning, whereas 100,000 *Salmonella* bacilli must be ingested to initiate food poisoning. One explanation for the difference is the higher resistance of the *Campylobacter* cells to the acidic conditions in the stomach, in contrast to the low resistance of *Salmonella* bacilli. Insufficient numbers will not cause an infection, and the pathogen will be destroyed or eliminated by the immune system or other body defenses.

Infection Locations Depend on the Pathogen's Tissue Preferences

Should the pathogen gain entry in sufficient numbers, an infection might be imminent unless additional host defenses eliminate the pathogen. In other cases, the pathogen and host might reach a stalemate by which neither has the advantage. The latent form of tuberculosis is an example of such a chronic state.

If an infection does occur, it can be one of two types. A **primary infection** occurs in an otherwise healthy individual, whereas a **secondary infection** develops in an individual weakened by the primary infection. In the influenza pandemic of 1918, hundreds of millions of individuals contracted influenza as a primary infection. The disease and immune defenses caused so much injury that the damaged respiratory tissue provided an easy portal of entry for opportunistic, pneumonia-causing bacterial species. In fact, some secondary infections can be more serious than the primary infection due to tissue damage or immune system dysfunction caused by the primary infection.

The location of the infection depends on the nature of the invading pathogen and its preferences for growth (see Figure 20.5). Many pathogens are restricted to a single area of the body. Examples of such **local infections** include boils and abscesses on the skin caused by *Staphylococcus*. **Systemic diseases** are those whereby the infection has spread via the blood to deeper organs and body systems. Thus, a staphylococcal skin boil beginning as a localized skin lesion can become more serious if the staphylococci spread via the blood to the bones, meninges, or heart tissue. In this case, the boil is not only a local infection but also the **focus of infection**, that is, the origin of a systemic infection.

Pathogens and toxins often are spread through the body via the blood. The transient appearance of small numbers of living bacterial cells in the blood is referred to as **bacteremia**. More dangerous is **septicemia** or blood poisoning, by which there is a proliferation and spread of large numbers of bacterial cells in the blood. **Sepsis**, an inflammatory condition arising from septicemia, can be a life-threatening condition as outlined in MICROFOCUS 20.1.

Other infectious microbes also are disseminated. **Fungemia** refers to the spread of fungi, **viremia** to the spread of viruses, and **parasitemia** to the spread of protists and multicellular worms through the blood.

MICROFOCUS 20.1: Clinical Microbiology

Sepsis and Septic Shock

Because the blood is sterile, any bacterial cells detected in the blood are cause for alarm. The presence of living, transient bacteria in the bloodstream is called **bacteremia**. In most bacteremic situations, however, only a small number of bacterial cells gain entry and no symptoms develop because the transients are rapidly removed by immune blood cells.

When more cells enter the bloodstream than can be effectively removed, **septicemia** develops in which the infectious agent spreads through the bloodstream (see the accompanying figure for types of pathogens often found). Septicemia can result from a local infection in the body (such as pneumonia) or from surgery on infected tissue.

For reasons that are not completely understood, **sepsis** arises if the body's immune response to the infection becomes unregulated, resulting in physiological, biochemical, and pathological abnormalities. The response has the potential to damage and overwhelm other healthy tissues and organs.

Septic shock brings about overwhelming circulatory, cellular, and metabolic abnormalities. It is one of the most dangerous killers in hospitals, yet one of the least understood medical conditions. According to medical experts, in the United States there are approximately 750,000 cases of septic shock every year, and between 15% and 30% of those patients die. Globally, septic shock is a leading cause of mortality, taking the lives of 18 million people every year.

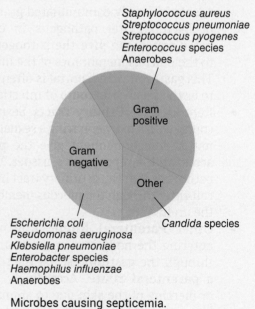

Staphylococcus aureus
Streptococcus pneumoniae
Streptococcus pyogenes
Enterococcus species
Anaerobes

Gram positive

Gram negative

Other

Escherichia coli
Pseudomonas aeruginosa
Klebsiella pneumoniae
Enterobacter species
Haemophilus influenzae
Anaerobes

Candida species

Microbes causing septicemia.

Septic shock then refers to those cases of sepsis that are likely to have poor outcomes due to the collateral damage inflicted by the immune system response and that are at greater risk of mortality. In an effort to identify the condition quickly, the medical community has established the quick Sequential (Sepsis-related) Organ Failure Assessment (qSOFA). Using qSOFA, a doctor would identify a potential septic shock patient by noting at least two of the following three clinical signs:

- An increased respiratory rate greater than 22 breaths/min (normal = 12–16 bpm)
- A drop in blood pressure (systolic) to less than 100 mm Hg (normal systolic = 90–120 mm Hg)
- Altered mental activity (ranges from slight confusion to total disorientation and coma)

If a patient falls in the qSOFA criteria, multiple organ failure (kidneys, brain, and lungs) and death can result. Therefore, such patients need to be identified early and given antibiotics that can eliminate the infection and large amounts of intravenous fluids to counteract the drop in blood pressure. Hopefully, these measures will calm any overactive immune response that is beginning. Unfortunately, many doctors miss the early signs of septic shock. It has been said, "For every hour delay in giving antibiotics, there is about a 7% increase in risk of death."

How successful can the guidelines be? In Phoenix, Arizona, one hospital system has targeted sepsis and septic shock by being watchful for potential qSOFA signs. In these cases, the system's death rate has decreased 27%.

Pathogen Persistence in the Host Often Depends on Structural Virulence Factors

After a pathogen has successfully entered the host in sufficient numbers, it must establish and maintain itself in the host (see Figure 20.5). Here, we encounter two structural virulence factors that help establish an infection and, at the same time, avoid the host's immune defenses.

Capsules and Pili

Many pathogens can establish an infection and increase the chance of causing disease if they can adhere to tissues or cell surfaces. A variety of structural virulence factors can facilitate adhesion. These include **capsules** on some bacterial, fungal, and protistan pathogens. The capsule, discussed in Chapter 4, allows the pathogen to stick to the host cell surface.

The capsule also is a defensive structure. During infection, the host mobilizes defensive immune cells, such as **macrophages**, to engulf and destroy the invaders (**FIGURE 20.6**). This process, called **phagocytosis**, is less effective on encapsulated pathogens because the macrophages cannot easily stick to the pathogens. Other bacterial pathogens, such as *Shigella*, *Listeria*, and *Mycobacterium*, have evolved escape strategies such that even though they are engulfed by immune cells, they are not destroyed by these cells. In fact, the pathogens might actually multiply in the immune cells and then use the immune cells as agents to spread the invaders through the body.

Other cell structures also can provide for adhesion. Many gram-negative bacterial species have **pili** that stick the cells to specific receptor sites found only on the appropriate target cells of the host

(**FIGURE 20.7**). Some gram-positive bacterial species, such as *C. diff*, have surface proteins that accomplish the same function. Many viruses have spikes on the capsid or envelope, allowing for attachment. By attaching to host cells or tissues, pathogens can avoid elimination, making infection and disease development more likely.

For some pathogens, only adhesion is necessary for infection and disease development. For example, the cells of *Vibrio cholerae* that cause cholera adhere to the mucous membranes of the colon. While adhering to the surface, the cells secrete toxins (described

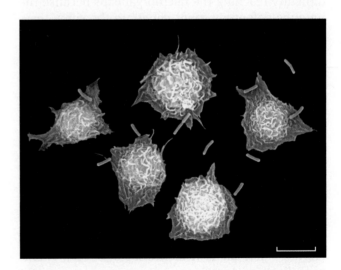

FIGURE 20.6 Macrophages Undergoing Phagocytosis. In this false-color scanning electron micrograph, macrophages, a type of immune cell, are capturing bacterial cells (blue) for phagocytosis. (Bar = 10 μm.) »» *Why might phagocytosis be an important activity of the human immune system when infection occurs?*

© Dr. K.G. Murti/Visual Unlimited/Getty.

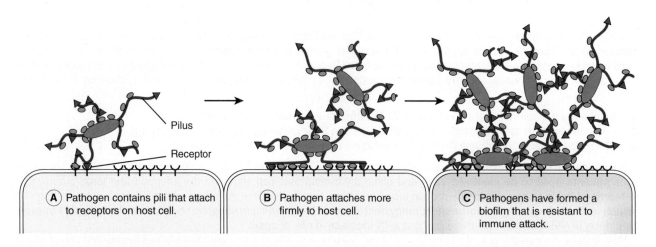

A | Pathogen contains pili that attach to receptors on host cell.

B | Pathogen attaches more firmly to host cell.

C | Pathogens have formed a biofilm that is resistant to immune attack.

FIGURE 20.7 Pathogen Adhesion to a Host Cell. Many gram-negative bacterial species contain pili that can adhere to a specific host cell or tissue. *»» What is the advantage of the pathogen forming a biofilm on the host cell surface?*

Modified from Telford, J. L., et al. 2006. *Nature Reviews Microbiology* 4(7):509–519.

later in this chapter) responsible for the massive water loss symptomatic of the disease. There is no invasion of host tissues by the bacterial cells.

The ability of many microbes to form an adhesive biofilm also makes the establishment of an infection more likely. CDC officials have estimated that 65% of human infections involve biofilms. A **biofilm** is a sticky layer of extracellular polysaccharides and proteins enclosing a community of microbial cells. If an invading pathogen forms a biofilm by adhering to host tissue, immune cells and antibodies have difficulty reaching the microorganisms because the immune defenses cannot penetrate the slimy mass of armor-like material composing the biofilm. Moreover, microorganisms often survive without dividing in a biofilm and are unaffected by antibiotics, which preferentially attack dividing cells.

Successful Infection Often Depends on a Variety of Enzyme Virulence Factors

When a pathogen enters a host cell, the pathogen is confronted with a profoundly different environment in which to survive. The pathogen must adapt and spread while withstanding the immune attack put forward by the host. To deal with these challenges, some bacterial species secrete extracellular enzymes that act as virulence factors. A few examples illustrate how bacterial enzymes act on host cells and interfere with cell functions or the barriers meant to retard invasion.

Coagulase and Streptokinase

Clot formation is one defensive measure used by the host to isolate an infected area and prevent

further spread of the infectious agent. However, some pathogens also can produce a plasma clot. For example, *Staphylococcus aureus* secretes the enzyme **coagulase** that coats the surface of the bacterial cells and allows the cells to remain clustered together in a clot and thus hidden (walled off) from the host's immune defenses (**FIGURE 20.8A**). An example is a boil, which represents a localized staphylococcal infection. In the diagnostic identification of staphylococci, the **coagulase test** is used to distinguish between pathogenic *S. aureus* (coagulase-positive) and other species like *S. epidermidis* (coagulase-negative).

Many streptococci and staphylococci have the ability to produce another enzyme called **streptokinase** (or staphylokinase) that dissolves the plasma clot produced by the pathogen or by the host. In either case, dissolving the clot allows further tissue invasion by the pathogen. Today, streptokinase can be produced through biotechnology and the enzyme used medically as an effective clot-dissolving medication for some cases of heart attacks and pulmonary blockage.

Hyaluronidase

Sometimes called the "spreading factor" because it enhances penetration (spreading) of a bacterial pathogen through tissues, **hyaluronidase** digests hyaluronic acid, a polysaccharide that binds cells together in a tissue (**FIGURE 20.8B**). The enzyme is an important virulence factor for pneumococci and certain species of streptococci and staphylococci. The result is a widespread infection that often involves destruction of connective tissue.

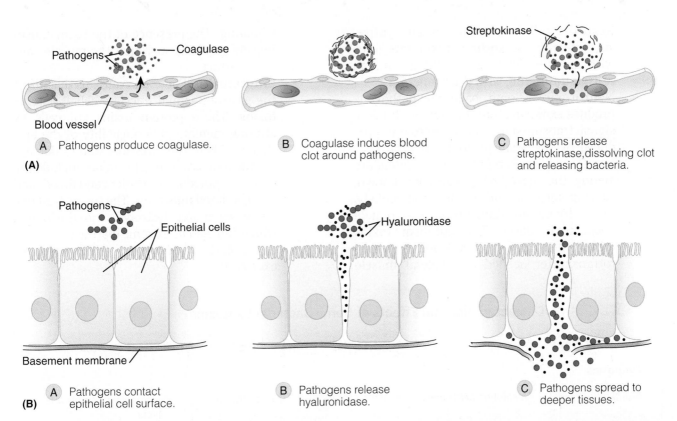

FIGURE 20.8 Enzyme Virulence Factors. (A) Some bacterial cells produce the enzyme coagulase, which triggers clotting of blood plasma. Bacterial cells within the clot can break free by producing streptokinase. **(B)** Some invasive bacterial species produce the enzyme hyaluronidase, which degrades the cementing polymer holding cells of the intestinal lining together. *»» From these examples, how do virulence factors protect bacterial cells and increase their virulence?*

Toxins Are Virulence Factors That Cause Damage or Dysfunction in Host Cells and Tissues

Microbial poisons, called **toxins**, represent another set of virulence factors that can profoundly affect the course and seriousness of an infectious disease. The ability of pathogens to produce toxins is referred to as **toxigenicity**, whereas the needed number of toxins to cause disease is called the **toxic dose**. The presence of toxins in the blood is referred to as **toxemia**, and a person is considered "intoxicated" if that individual is affected by a bacterial toxin. Two groups of bacterial toxins are recognized: exotoxins and endotoxins.

Exotoxins

Many gram-positive and gram-negative bacterial species produce and secrete unique **exotoxins**. These poisons are heat-sensitive protein molecules that are manufactured during bacterial metabolism. Depending on the bacterial species, exotoxins are coded by genes found on the bacterial chromosome,

on plasmids, or on prophages. The exotoxins are released into the host environment or, with some gram-negative bacteria, injected directly into the host cell. Toxins often damage host membranes or interfere with host cell function. Exotoxins are grouped into one of three types based on their primary target.

► **Cytotoxins.** Some staphylococci, streptococci, and pneumococci produce cytotoxins that kill cells. One cytotoxin, called **leukocidin**, forms pores in and lyses those immune white blood cells (leukocytes) whose job it is to phagocytize and destroy the pathogens. **Hemolysin**, produced by staphylococci and streptococci, lyse both white and red blood cells. In red blood cells (erythrocytes), lysis provides the pathogens with a source of iron (in hemoglobin) that the bacterial cells need for metabolism. In the clinical (diagnostic) laboratory, hemolysin producers can be detected by **hemolysis**, a destruction of sheep blood cells by bacteria growing on a blood agar medium.

Diphtheria and anthrax are diseases caused by other cytotoxins. The diphtheria toxin kills cells by blocking protein synthesis, whereas the anthrax toxin interferes with immune defenses by killing immune cells or interfering with their function.

▶ **Neurotoxins.** Some bacterial pathogens produce exotoxins that interfere with nerve signal transmission. The neurotoxins produced by the gram-positive bacterial species *Clostridium botulinum* and *C. tetani* are among the most lethal exotoxins known, causing botulism and tetanus, respectively. These toxins are dangerous in extremely small toxic doses. The botulism toxins inhibit the release of acetylcholine, a neurotransmitter needed for skeletal muscle signaling. The presence of the toxin causes muscle paralysis such that the muscle cannot contract.

▶ **Enterotoxins.** Bacterial exotoxins that affect the gastrointestinal tract are called **enterotoxins**. These protein toxins kill cells by altering membrane permeability of the epithelial cells of the intestinal wall or damaging the mucosal lining. Examples include the toxins responsible for cholera and *E. coli*- and *C. diff*-related infections. The toxins lead to a loss of water and electrolytes into the lumen, which can lead to extreme diarrhea.

TABLE 20.2 identifies other types of exotoxins.

TABLE 20.2 Characteristics and Effects of Some Bacterial Exotoxins

Exotoxin (Type)	Organism	Host Target	Disease	Effect on Host
Cytotoxins				
Alpha toxin	*Clostridium perfringens*	Plasma membrane	Gas gangrene	Hemolysis; membrane lysis
Anthrax toxin	*Bacillus anthracis*	Cellular chemical communication	Anthrax	Hemorrhaging and pulmonary swelling
Diphtheria toxin	*Corynebacterium diphtheriae*	Ribosomes and protein synthesis	Diphtheria	Inhibits protein synthesis; cell death
Erythrogenic toxin	*Streptococcus pyogenes*	Immune response	Scarlet fever	Capillary destruction; rash
Neurotoxins				
Botulism toxin	*Clostridium botulinum*	Cellular exocytosis	Botulism	Respiratory paralysis
Tetanus toxin	*Clostridium tetani*	Cellular exocytosis	Tetanus	Rigid paralysis; respiratory failure
Enterotoxins				
Cholera toxin	*Vibrio cholerae*	Cellular chemical communication	Cholera	Severe diarrhea
Clostridium difficile toxins A and B	*Clostridium difficile*	Cellular chemical communication	Colitis Pseudomembranous colitis	Mucosal lining destruction; diarrhea
Enterotoxin A	*Staphylococcus aureus*	Immune response	Food poisoning	Diarrhea and nausea
Pertussis toxin	*Bordetella pertussis*	Cellular chemical communication	Whooping cough (pertussis)	Interferes with host cell communication
Pyrogenic toxin	*Staphylococcus pyogenes*	Immune system	Toxic shock syndrome	Fever, shock
Shiga toxin	*Escherichia coli/Shigella dysenteriae*	Ribosomes and protein synthesis	Diarrhea	Diarrhea

TABLE 20.3 **A Comparison of Exotoxins and Endotoxins**

Characteristic	Exotoxins	Endotoxins
Source	Living gram-positive and gram-negative bacterial cells	Lysed gram-negative bacterial cells
Location	Released from living cells	Part of cell wall
Chemical composition	Protein	Lipopolysaccharide
Heat sensitivity	Unstable (above 60°C [140°F])	Stable (at 60°C)
Immune reaction	Strong	Weak
Conversion to toxoid	Possible	No
Fever	No	Yes
Toxicity	High	Low
Toxic dose	Small	Large
Representative diseases	Botulism Diphtheria Cholera	Salmonellosis Typhoid fever Meningococcal meningitis

The body responds to exotoxins by producing antibodies called **antitoxins**. When toxin and antitoxin molecules combine with each other, the toxin is neutralized. This process represents an important defensive measure in the body. Therapy for people who have botulism, tetanus, or diphtheria often includes injections of antitoxins (immune globulin) to neutralize the toxins.

Endotoxins

The outer membrane of the cell wall of many gram-negative bacteria contains **lipopolysaccharide (LPS)**. The lipid portion of the LPS represents an **endotoxin**, which is usually released upon lysis (or during binary fission) of the bacterial cells. Endotoxins do not stimulate a strong immune response in the body.

Endotoxins are very different from exotoxins. All endotoxins have comparable toxic effects and similar signs and symptoms. Usually an individual experiences a fever, chills, substantial body weakness, and muscle aches. At high toxic doses, damage to the circulatory system can lead to **endotoxin shock**. In this case, the permeability of the blood vessels changes, and blood leaks into the intercellular spaces. The tissues swell, the blood pressure drops, and the patient can lapse into a coma.

TABLE 20.3 summarizes the characteristics of the bacterial toxins.

Pathogens Must Be Able to Leave the Host to Spread Disease

To be transmitted to another susceptible host, pathogens must exit the host through some suitable **portal of exit** (**FIGURE 20.9**). Often these portals are the same as the portals of entry. Respiratory

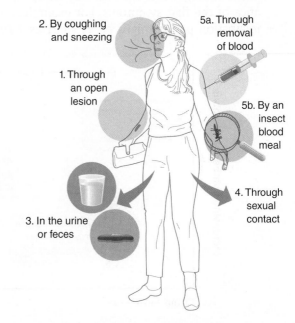

FIGURE 20.9 Six Different Portals of Exit from the Body.
Pathogens may differ in the way they exit the host. *»» Why do pathogens need specific portals of exit?*

pathogens are typically expelled through coughing and sneezing, which easily spread nasal secretions, saliva, and sputum as droplets or aerosols. Intestinal tract pathogens exit the body in feces, usually as a watery diarrhea.

Many **incidental infections** involve an infected host from which the pathogen cannot be transmitted to another susceptible host. Therefore, the infected individual is called a **dead-end host**. WNV disease, valley fever, and many helminthic diseases represent some examples in which humans are dead-end hosts for these diseases.

Finally, some diseases leave aftereffects, called **sequelae** (sing. sequela), resulting from the disease. For example, there might be paralysis after polio, deafness from meningitis, arthritis from Lyme disease, and skin scars (pox) following smallpox.

Diseases Progress Through a Series of Stages

In most instances, there is a recognizable pattern in the progress of an infectious disease following the entry of the pathogen into the host. Let's summarize the entire process of colonization, infection, and disease by examining the stages typically seen for an infectious disease. This disease progression occurs in five stages (**FIGURE 20.10**):

▶ **Incubation Period.** The **incubation period** reflects the time elapsing between the entry of the pathogen into the host and the appearance of the first symptoms. For example, an incubation period might be as short as 2 to 4 days for the flu, 1 to 2 weeks for measles, or 3 to 6 years for leprosy. Such factors as the number of organisms, their generation time, virulence, and the level of host resistance determine the incubation period's length. The location of entry also can be a determining factor. For instance, the incubation period for rabies can be as short as several days or as long as a year, depending on how close the infecting viruses are to the central nervous system.

▶ **Prodromal Phase.** The **prodromal phase** is a time of mild signs or symptoms. For many diseases, this period is characterized by indistinct and general symptoms such as headache and muscle aches, which indicate the competition between host and microbe has begun. Making a diagnosis at this time can be difficult due to the lack of specific signs and symptoms.

▶ **Acute Period.** The stage of the disease when signs and symptoms are of most intense and specific to the disease is the **acute period** (or climax). For example, patients with the flu suffer high fever and chills, dry skin and a pale complexion, a headache, cough, body and joint aches, and loss of appetite. The length of the acute period can be quite variable, depending on the body's immune response to the pathogen and the virulence of the pathogen. Although the patient feels miserable, there is evidence that some signs and symptoms can be beneficial, as demonstrated in MICROFOCUS 20.2.

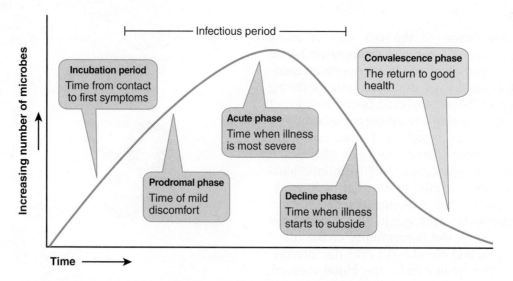

FIGURE 20.10 The Course of an Infectious Disease. Most infectious diseases go through a five-stage process. The length of each stage would depend on the specific pathogen and the host response. »» *How would the graph differ in the case of a chronic disease?*

MICROFOCUS 20.2: Public Health

Illness Might Be Good for You

For most of the 20th century, medicine's approach to infectious disease was relatively straightforward: note the signs and symptoms, and eliminate them. However, that approach might change in the future, as Darwinian medicine gains a stronger foothold. Proponents of Darwinian medicine ask why the body has evolved its symptoms and question whether relieving the symptoms might leave the body at greater risk.

As an example, consider coughing. In the rush to stop a cough, we might be neutralizing the body's mechanism for clearing pathogens from the respiratory tract. In addition, it might not be in our best interest to stifle a fever (at least a low-grade fever), because fever enhances the immune response to disease. Many physicians view iron insufficiency in the blood as a symptom of disease, yet many bacterial species (e.g., tuberculosis bacilli) require this element, and as long as iron is sequestered out of the blood in the liver, the bacterial cells cannot grow well. Even diarrhea can be useful—it helps propel pathogens from the intestine and assists the elimination of the toxins responsible for the illness.

Darwinian biologists point out that disease symptoms have evolved over the vast expanse of time and probably have other benefits waiting to be understood. However, these biologists are not suggesting a major change in how doctors treat their patients, but they are pushing for more studies on whether symptoms are part of the body's natural defenses. So, don't throw out the Nyquil®, Tylenol®, or Imodium® quite yet.

© Comstock Images/Jupiterimages.

▶ **Period of Decline.** As the signs and symptoms begin to subside, the host enters a **period of decline** that indicates the end of the illness. With the flu, sweating can be common as the body releases excessive amounts of heat, and the normal skin color soon returns as the blood vessels dilate. It is often during this period that a person can acquire a secondary infection, which, in the case of the flu, might be an opportunistic bacterial pneumonia. Some individuals can remain carriers of a disease for some time. A person who has had typhoid fever might appear asymptomatic, but the individual can still readily transmit the agent (*Salmonella typhi*), causing new disease cases.

▶ **Period of Convalescence.** The disease sequence concludes after the body passes through a **period of convalescence**. During this time, the body's systems return to normal.

This final period might not bring disease resolution. With an **acute disease** like the flu, severe symptoms develop rapidly, come to a climax, and then fade rather quickly during convalescence. A **chronic disease**, by contrast, often lingers for long periods; although it is controlled, it is not cured. The symptoms are slower to develop, an acute period is rarely reached, and convalescence might continue for months or years. Hepatitis B, hepatitis C, and latent tuberculosis are examples of chronic diseases. Finally, the disease might overwhelm the body and the immune system, resulting in the patient's death. Active tuberculosis, Ebola virus disease, and malaria are examples.

Concept and Reasoning Checks 20.2

a. Assign the following signs and symptoms of the flu to the appropriate disease stage (fever, headache, chills, cough, sore throat, fatigue, and muscle aches).
b. What are portals of entry, and how do these portals play a role in infection and disease?
c. Evaluate the role of enzymes and toxins as important virulence factors in the establishment of disease.
d. Discuss whether the portals of entry and exit are always the same. Explain.

Chapter Challenge B

So, our patient has a CDAI. Although the organism is highly resistant to many frontline antibiotics, it also is a highly infectious pathogen (high virulence). Therefore, it must have an array of virulence factors that makes it a dangerous pathogen.

QUESTION B: *From your reading in Key Concept 20.2, identify what enzymatic and toxin virulence factors* C. diff *possesses. How does each factor contribute to the organism's overall virulence?*

You can find answers online in **Appendix F**.

■ KEY CONCEPT 20.3 Epidemiology Is the Foundation of Public Health

Infectious diseases were the most serious health threat in the world until the middle of the twentieth century, when cardiovascular diseases, cancers, diabetes, and chronic lung disorders began to dominate human illnesses, at least among the developed nations. Even so, infectious diseases today remain serious health threats. Many of these infectious diseases are **communicable diseases**, that is, diseases such as tuberculosis and AIDS that are transmissible among susceptible individuals in a population. If a communicable disease spreads rapidly between susceptible individuals, the disease is said to be **contagious**. Chickenpox and measles fall into this category.

Noncommunicable diseases are singular events in which the agent is acquired directly from the environment and the agent cannot be transmitted to other susceptible individuals. With tetanus, the *C. tetani* spores in the soil could be transmitted on

a sharp object that punctures the skin. If the individual then develops tetanus, the person cannot transmit the agent to another person. As mentioned earlier in the chapter, that individual would be a dead-end host.

Epidemiology is at the forefront in understanding how infectious diseases are distributed in a population and in identifying the factors that influence or establish that distribution. In this final section, we examine the factors putting population groups at risk of contracting infectious disease. We also look at special environments, such as healthcare settings, and the public agencies saddled with the job of emerging disease identification, control, and prevention, much of which is the foundation of public health.

MICROBIOLOGY PATHWAYS describes the training of an epidemiology intelligence officer, sometimes referred to as a "disease detective."

MICROBIOLOGY PATHWAYS

Epidemiology

Flying to an impoverished African country on your second day of work to battle Ebola, one of the deadliest viruses, isn't most people's idea of a dream assignment. But it was for Marta Guerra. In fact, the trip to Uganda was the assignment she had been coveting. "I wasn't that worried," Guerra says. "This particular strain has only a 65% death rate instead of the Congo strain, which is 85%."

Guerra is a disease epidemiologist, popularly known as a "disease detective," with the Epidemic Intelligence Service (EIS) of the Centers for Disease Control and Prevention (CDC) in Atlanta.

Growing up in the multicultural environments of Havana, Cuba, and Washington, D.C., Marta Guerra developed a keen interest in other cultures and teaching people about health risks. She was fascinated

Disease detectives processing patient samples during an outbreak.

Courtesy of Ethleen Lloyd/CDC.

by stories her professors told about working overseas. After seeing the movie *Outbreak,* "I thought, that's what I want to do—help contain a deadly epidemic," recalls Guerra. So, she obtained a master's in public health and a PhD in tropical medicine.

Like all EIS officers, Guerra's job is to isolate the cause of an outbreak, prevent its spread, and get out public health messages to people who could have been exposed. When Guerra flew to Uganda, the Ebola outbreak had already been identified, so her task was to go from village to village, trying to locate family members and friends, and educate them about symptoms and treatment. "In every corner of Africa, people know the word Ebola, and they are terrified of it," Guerra explains. "Sometimes they hide sick family members; sometimes they're frightened of survivors."

The job of a disease detective can be difficult—and dangerous (see figure). Although Guerra seldom considered she might acquire a disease she was investigating, she was concerned about political violence. In Uganda, Guerra's team needed military escorts on their travels through villages. In Ethiopia, while on a polio eradication mission the previous summer she recalls, "There was rebel activity in all the areas we traveled through—plus land mines. It was pretty scary."

Perhaps you might be interested in a similar, and sometimes adventurous, career. "You have to be highly motivated, with the ability to think fast on your feet and make quick decisions. You have to be able to walk into a chaotic situation and deal with whatever is thrown at you," Guerra says. "Sometimes I barely drop my bags at home before I'm called out again. Being adaptable is really essential."

To get started, you need a bachelor's degree in a biological science. In addition to required courses in chemistry and biology, undergraduates should study microbiology, mathematics, and computer science. A master of science in epidemiology or public health also is required; many have a PhD or a medical or veterinary degree. Most American disease epidemiologists then apply to EIS's two-year, post-graduate program of service and on-the-job training, where they work with mentors like Marta. She says, "I like the fact that I am contributing to science in the sense that what I do will affect people far into the future."

Data from *Disease Detective* by Carol Sonenklar in www.MedHunters.com.

Epidemiologists Often Need to Identify the Reservoir of an Infectious Disease

Pathogens seldom survive for long periods in the open environment. Rather, the pathogen must find a suitable place in which it can survive and multiply and from which it can be transmitted to a suitable host (**FIGURE 20.11**). This suitable location is called

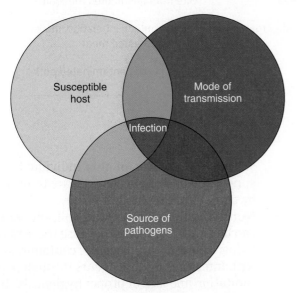

FIGURE 20.11 Infectious Disease Elements. Assuming there is a susceptible host, to cause an infection the pathogen must have a way to be transmitted to that susceptible host. **»» *What is meant by a susceptible host?***

a **reservoir** of infection. There are three types of potential reservoirs:

▸ **Nonhuman Animal Reservoirs.** Wild and domesticated animals are one type of reservoir. A rabid bat, for example, would be a reservoir for the rabies virus. The bat also would be a **vector** because the bat is an organism that transmits the pathogen (i.e., rabies virus) from reservoir to host. If a human were bitten by the rabid bat, the person would not be a reservoir because that individual cannot pass on the virus; in this case, the human again is a dead-end host.

An infectious disease transmitted from animals to humans is called a **zoonosis** (pl. zoonoses). The CDC reports that about 75% of recently emerging infectious diseases that affect humans are zoonotic, and some 60% of all human pathogens are zoonotic. TABLE 20.4 lists some common zoonotic diseases and the animals that act as reservoirs.

▸ **Human Reservoirs.** Not all infectious diseases arise in nonhuman reservoirs. Infected humans also can be reservoirs, and in some cases, they might be the only reservoir. The smallpox virus only infects humans, so humans are its sole reservoir. As such, in the 1960s and 1970s, the WHO and other agencies stopped the spread of the smallpox

TABLE 20.4 Some Common Zoonotic Infections and Their Reservoirs

Zoonotic Disease	Pathogen	Reservoirs	Mode of Transmission
Viral			
Hantavirus pulmonary syndrome	Hantavirus	Deer mice	Inhalation of contaminated dry feces or urine
Influenza	Influenza A	Birds, swine, bats	
Rabies	Rabies virus	Many mammals	Bite of infected mammal
West Nile virus disease	West Nile virus	Birds, horses	Bite of an infected mosquito
Ebola virus disease	Ebola viruses	Fruit bats	Bite from an infected bat
Zika virus infection	Monkeys (historically)	Mosquitoes	Bite from infected mosquito
Bacterial			
Anthrax	*Bacillus anthracis*	Domestic animals	Direct contact with/inhalation from infected animal/animal product
Bubonic plague	*Yersinia pestis*	Rats	Bite from an infected flea
Lyme disease	*Borrelia burgdorferi*	Deer	Bite from an infected tick
Salmonellosis	*Salmonella enterica* serotypes	Poultry, reptiles	Fecal–oral route
Other Microbes and Parasites			
Ringworm (fungal)	Dermatophytes	Domestic mammals	Direct contact
Malaria (protist)	*Plasmodium* species	Monkeys	Bite from an infected mosquito
Toxoplasmosis (protist)	*Toxoplasma gondii*	Cats, other mammals	Direct contact, inhalation, ingestion of contaminated meat
Pork tapeworm (helminth)	*Taenia solium*	Pigs	Ingestion of contaminated pork

virus by locating all human reservoirs and instituting a global vaccination campaign. Smallpox was eradicated because the virus no longer had a susceptible reservoir in which to replicate.

Sometimes, a person who has recovered from a disease and has no signs or symptoms (asymptomatic) continues to shed the disease agents. For instance, an individual who has recovered from typhoid fever or amebiasis becomes a **carrier** for many weeks after the symptoms of disease have subsided. The feces of this individual can spread the bacterial cells or cysts to others by contaminating food or water.

▶ **Nonliving Reservoirs.** Food, soil, and water can be another reservoir because these nonliving sources can become contaminated with infectious disease agents through poor sanitation and lack of proper hygiene. Water is perhaps the most dangerous, acting as a reservoir for the agents of many diseases including polio, hepatitis A, cholera, and traveler's diarrhea.

Infectious Agents Can Be Transmitted in Several Ways

No matter what the reservoir of infection, for a person to be infected there must be contact between the infectious agent and the host. Public health officials typically put the transmission of infectious agents into one of two groups: contact transmission (direct and indirect) and airborne transmission.

Contact Transmission

Contact transmission can be direct or indirect. With **direct contact transmission**, there must be close or personal (direct) contact with someone who is infected or who has a communicable disease (**FIGURE 20.12A**). Activities such as handshaking or kissing an infected person would be examples of contact through **horizontal transmission**. For zoonotic diseases, including rabies and Lyme disease, a bite or scratch by an infected animal can transmit the viral or bacterial agent to an uninfected person. The exchange of body fluids, such as through sexual intercourse, is another example of direct contact transmission typical for sexually transmitted infections like gonorrhea and syphilis.

One other form of direct contact transmission, called **vertical transmission**, includes the spread of pathogens, such as HIV, from a pregnant mother to her unborn child (see Figure 20.12A). In addition, vertical transmission of gonorrhea from mother to newborn can occur during labor or delivery.

Indirect contact transmission can be the result of contact with a nonliving object, a vehicle, or a vector (**FIGURE 20.12B**). **Fomites** are inanimate (lifeless) objects on or in which disease organisms linger for some period. For instance, doorknobs or drinking glasses can be contaminated with flu or cold viruses; syringe needles can be contaminated with hepatitis B virus or HIV. These and similar agents can be transmitted to another person when that person touches or uses the fomite.

Indirect spread of disease through contaminated food and water is referred to as **vehicle transmission**. Foods can be contaminated during processing or handling (fecal–oral route), or they might be dangerous when prepared from diseased animals. Poultry products, for example, are often a reservoir for salmonellosis because *Salmonella* species frequently infect chickens and can lead to cross contamination during processing of the food product.

Vector transmission represents another indirect, zoonotic method of disease transmission.

Typical disease vectors include arthropods such as mosquitoes, ticks, and fleas, which act as reservoirs and carry disease agents from one infected host to another susceptible host. **Mechanical vectors** are arthropods that passively transport microbes on their legs and other body parts. For example, houseflies can carry diseases picked up on their feet and then deposit them on another object, such as food, when the fly lands. In other cases, arthropods represent **biological vectors**, whereby the pathogen must multiply in the insect before the pathogen is capable of infecting another host. The bacterial organism responsible for plague infects and reproduces in ticks and the malarial protist infects and reproduces in mosquitoes. When the infected arthropod takes a blood meal, the pathogens are introduced into the host's blood.

Airborne Transmission

The simple act of talking, sneezing, or coughing can generate airborne particles called **aerosols**; from an infected individual, these particles can harbor infectious agents (**FIGURE 20.12C**). This form of disease transmission can involve the violent expulsion of airborne **respiratory droplets** through sneezing or coughing (**FIGURE 20.13**). This form of transmission through the air requires the "recipient" be near an infected individual because the droplets are large (5–100 µm) and fall out of the air within about 1 meter of their source. If an uninfected person is within that distance, respiratory droplets from the ill person can carry the pathogens to mucosal portals of entry, such as the eyes, mouth, and upper respiratory tract. These larger particles also can settle on objects, which would allow for indirect contact transmission via fomites.

Pathogens also can be transmitted through the air on smaller particles (<5 µm) called **droplet nuclei** (see Figure 20.12B). These particles can remain suspended in the air for indefinite periods, during which time they can be transmitted some distance by air currents from the infected individual. Being much smaller particles, droplet nuclei could be deposited in the lower respiratory tract.

In some cases, there can be long-distance, global airborne transmission, as described in MICROFOCUS 20.3.

Epidemiologists Also Track How Infectious Diseases Spread Within Populations

When epidemiologists investigate an infectious disease, they need to determine if it is localized or

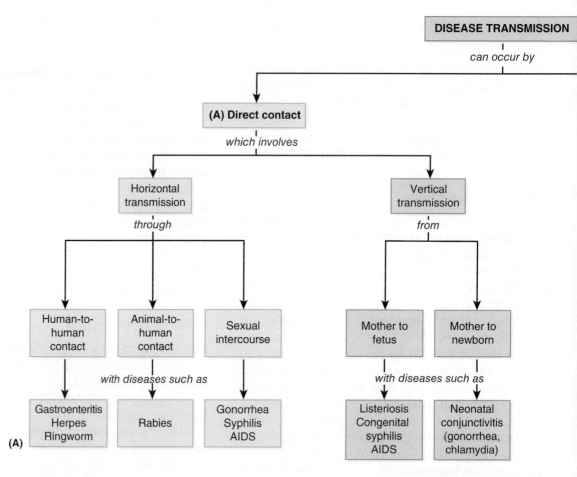

FIGURE 20.12 Transmission of Pathogens. (A) Pathogens can be transmitted by direct contact involving horizontal or vertical transmission. **(B)** Pathogens also can be transmitted indirectly by fomites, contaminated food and water, and vectors. **(C)** Pathogens can be carried airborne as respiratory droplets and droplet nuclei. *»» Explain whether indirect contact transmission represents horizontal or vertical transmission.*

FIGURE 20.13 Airborne Transmission. High-speed video imaging of a sneeze. The larger respiratory droplets (lower half of aerosol) can extend about one meter from the person sneezing. Smaller droplet nuclei (upper half of aerosol) remain suspended for longer periods in the moist air of the sneeze. *»» Based on this image, what should you do when you feel a sneeze coming on?*

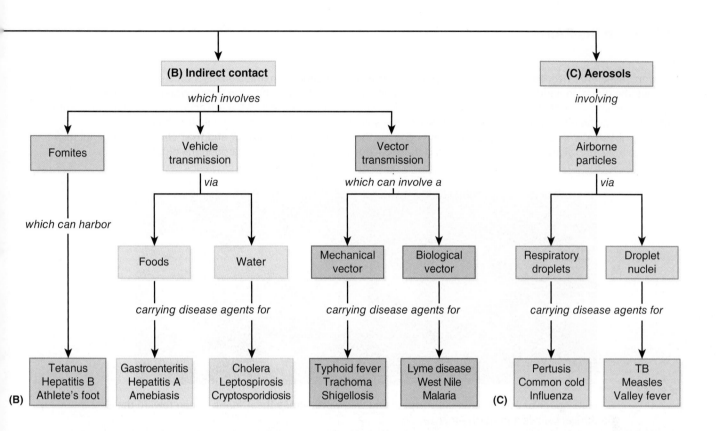

MICROFOCUS 20.3: Environmental Microbiology

Riders on the Storm

Dust storms can be a relatively common meteorological phenomenon in arid and semi-arid regions of the world. Sometimes called sandstorms, haboobs, or siroccos, they arise when a gust front passes or when the wind force is strong enough to remove loose sand and dust from the dry soil surface. The result can be an awesome wall of sand that can obscure visibility within seconds. On a more global scale, the major dust storms arise from the Gobi Desert in Asia, the Sahara Desert in North Africa, and arid lands around the Arabian Peninsula. Annually, such storms carry more than 3 billion metric tons of dust aloft into the atmosphere.

The long-range movement of dust and suspended particles can certainly have an impact on air quality. In February and March, dust from the Gobi Desert moves eastward across the Pacific Ocean to North America. Between May and October, strong winds blow off the Sahara Desert and the west coast of North Africa, carrying soil and dust westward across the Atlantic Ocean (see the accompanying figure). Although much of the dust settles in the ocean, large amounts remain suspended and make the journey to the Caribbean islands and Florida. It is not unusual to wake up in the morning and see a fiery orange sunrise—what the locals call a "tequila sunrise." The orange is the African dust. In fact, traces of such dust have been detected as far west as New Mexico!

Today, these Asian and African storms are becoming of more concern as worldwide deforestation, overgrazing, and climate change combine to generate massive dust clouds that can carry particles aloft. Besides causing respiratory distress, such as asthma, these storms can have another impact on human health—pathogens can be "riders on the storm."

Scientists have shown that in dust storms, microbes less than 20 μm can travel 15,000 feet into the atmosphere and thousands of kilometers in distance. Recent studies have shown that hundreds of bacterial and fungal species can be cultured from samples of dust clouds moving across the mid-Atlantic Ocean, and 20% to 30% of these microbes are animal or plant pathogens.

(continues)

MICROFOCUS 20.3: Environmental Microbiology (*Continued*)

Riders on the Storm

More research is needed to define the role of dust storms as an indirect transmission mechanism spreading pathogens across the globe. In addition, the dust is rich in iron minerals. When the iron hits the ocean waters, the trace metal can become so abundant that it encourages massive explosions in microbial growth, what are called microbial blooms. Some of these could be blooms of bacterial pathogens, such as *Vibrio cholerae* that is responsible for cholera.

More observations and studies are important because these "riders on the storm" and their chemical nutrients might become more prevalent and more frequent as climate change expands arid regions around the globe.

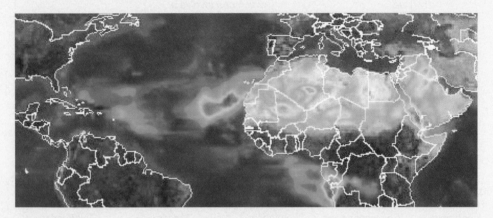

Satellite image of African dust storm spreading westward across the Atlantic Ocean toward the Caribbean islands and North America.

Courtesy of Jay Herman/NASA.

spreading through a community or region. **Endemic** refers to a disease that is persistent at a low level in a certain geographic area. Plague in the American Southwest is an example.

By comparison, an **epidemic** refers to a disease that occurs in excess (i.e., above endemic levels) of what is normally expected within a defined community, population, or geographic area. Influenza often causes widespread epidemics each flu season. This should be contrasted with an **outbreak**, which is a more contained epidemic. A cluster of measles cases in one American city would be classified as an outbreak. Not only do epidemiologic investigations look at current outbreaks but they also consider future outbreaks, including those artificially caused, such as bioterrorism. The topic is discussed in MICROFOCUS 20.4.

A **pandemic**, according to the internationally accepted definition, is an epidemic occurring worldwide or spreading over a very wide area and crossing international boundaries to affect a large number of

people. The most obvious example here would be AIDS and the 2009 H1N1 flu.

The late and former U.S. Surgeon General C. Everett Koop once stated, "*Health care matters to all of us some of the time, public health matters to all of us all of the time.*" Therefore, maintaining vigilance against infectious disease is extremely important. For this reason, national and international public health organizations, such as the CDC and the WHO, learn a great deal about diseases by analyzing disease data reported to them. The CDC, for example, has a list of infectious diseases that physicians must report to state health departments, who then compile the data and send it on to the CDC (see the **Table in Appendix B**). These are published in the *Morbidity and Mortality Weekly Report.*

MICROINQUIRY 20 explores some basics when using epidemiological data as a tool for understanding the occurrence and transmission of infectious diseases.

MICROFOCUS 20.4: History/Public Health

Bioterrorism: The Weaponization and Purposeful Transmission of Human Pathogens

The anthrax attacks that occurred on the East Coast of the United States in October 2001 confirmed what many health and governmental experts had been saying for many years—it is not *if* bioterrorism would occur, but *when* and *where*. Bioterrorism represents the intentional or threatened use of primarily microorganisms or their toxins to cause fear in or actually inflict death or disease upon a large population for political, religious, or ideological reasons.

Is Bioterrorism Something New?

Bioterrorism has a long history, beginning with infectious agents that were used for biowarfare. In the United States, during the aftermath of the French and Indian Wars (1754–1763), British forces, under the guise of goodwill, gave smallpox-laden blankets to rebellious tribes sympathetic to the French. The disease decimated the Native Americans, who never had been exposed to it before and therefore had no immunity. Between 1937 and 1945, the Japanese established Unit 731 to carry out experiments designed to test the lethality of several microbiological weapons as biowarfare agents on Chinese soldiers and civilians. In all, some 10,000 "subjects" died of bubonic plague, cholera, anthrax, and other diseases. After years of their own research on biological weapons, the United States, the Soviet Union, and more than 100 other nations in 1973 signed the Biological and Toxin Weapons Convention, which prohibited nations from developing, deploying, or stockpiling biological weapons. Unfortunately, the treaty provided no way to monitor compliance. As a result, in the 1980s the Soviet Union developed and stockpiled many microbiological agents, including the smallpox virus, and anthrax and plague bacteria. After the 1991 Gulf War, the United Nations Special Commission (UNSCOM) analysts reported that Iraq had produced 8,000 liters of concentrated anthrax solution and more than 20,000 liters of botulinum toxin solution. In addition, anthrax and botulinum toxin had been loaded into SCUD missiles.

In the United States, several biocrimes have been committed. **Biocrimes** are the intentional introduction of biological agents into food or water, or by injection, to harm or kill groups of individuals. One of the most publicized biocrimes occurred in Oregon in 1984 when the Rajneeshee religious cult, in an effort to influence local elections, intentionally contaminated salad bars of several restaurants with the bacterium *Salmonella*. The unsuccessful plan sickened more than 750 citizens and hospitalized 40. Whether biocrime or bioterrorism, the 2001 events concerning the anthrax spores that were mailed to news offices and to two U.S. congressional representatives, only increase our concern over the use of microorganisms or their toxins as bioterror agents.

What Microorganisms Are Considered Bioterror Agents?

A considerable number of human pathogens and toxins have potential as microbiological weapons. These "Tier 1 agents" include bacterial organisms, bacterial toxins, and viruses. The seriousness of the agent depends on the severity of the disease it causes (virulence) and the ease with which it can be disseminated. Tier 1 agents can be spread by aerosol contact, such as anthrax and smallpox, or added to food or water supplies, such as the botulinum toxin (see the accompanying table).

Why Use Microorganisms?

Some Tier 1 Agents and Perceived Risk of Use

Type of Agent	Disease (Microbe Species or Virus Name)	Perceived Risk
Bacterial	Anthrax (*Bacillus anthracis*) Plague (*Yersinia pestis*) Tularemia (*Francisella tularensis*)	High Moderate Moderate
Viral	Smallpox (variola virus) Hemorrhagic fevers (Ebola, Marburg virus)	Moderate Low
Toxin	Botulinum toxin (*Clostridium botulinum*)	Moderate

At least 12 nations are believed to have the capability of producing bioweapons from microorganisms. Such microbiological weapons offer clear advantages to these nations and terrorist organizations in general. Perhaps most important, biological weapons represent "The Poor Nation's Equalizer." Microbiological weapons are cheap to produce compared to chemical and nuclear weapons and provide those nations with a deterrent every bit as dangerous and deadly as the nuclear weapons possessed by other nations. For those who pursue biological weapons, there is the advantage of achieving high impact and getting the most "bang for the buck." In addition, microorganisms can be deadly in minute amounts to a defenseless (nonimmune) population. They are odorless, colorless, and tasteless, and unlike conventional and nuclear weapons, microbiological weapons do not damage infrastructure, yet they can contaminate such areas for extended periods. Without rapid medical treatment, most of the select agents can produce high numbers of casualties

(continues)

MICROFOCUS 20.4: History/Public Health (*Continued*)

Bioterrorism: The Weaponization and Purposeful Transmission of Human Pathogens

that would overwhelm medical facilities. Lastly, the threatened use of microbiological agents creates panic and anxiety, which often are at the heart of terrorism.

How Would Microbiological Weapons Be Used?

All known microbiological agents (except smallpox) represent organisms naturally found in the environment. For example, the bacterium causing anthrax is found in soils around the world (see left accompanying figure). Assuming that one has the agent, the microorganisms can be grown (cultured) easily in large amounts. However, most of the select agents must be "weaponized"; that is, they must be modified into a form that is deliverable, stable, and has increased infectivity and/or lethality. Nearly all of the microbiological agents in category A are infective as an inhaled aerosol. Weaponization, therefore, requires the agents be small so that inhalation would bring the organism deep into the respiratory system and prepared so that the particles do not stick together or form clumps. Several of the anthrax letters of October 2001 involved such weaponized spores.

Dissemination of biological agents by conventional means would be a difficult task. Aerosol transmission, the most likely form for dissemination, exposes microbiological weapons to environmental conditions to which they are usually very sensitive. Excessive heat, ultraviolet light, and oxidation would limit the potency and persistence of the agent in the environment. Although anthrax spores are relatively resistant to typical environmental conditions, the bacterial cells causing tularemia become ineffective after just a few minutes in sunlight. The possibility also exists that some nations have developed or are developing more lethal bioweapons through genetic engineering and biotechnology. The former Soviet Union might have done so. Commonly used techniques in biotechnology could create new, never-before-seen bioweapons, making the resulting "designer diseases" true doomsday weapons.

Conclusions

Ken Alibek, a scientist and defector from the Soviet bioweapons program, has suggested the best biodefense is to concentrate on developing appropriate medical defenses that will minimize the impact of bioterrorism agents. If these agents are ineffective, they will cease to be a threat; therefore, the threat of using human pathogens or toxins for bioterrorism, like that for emerging diseases such as 2009 H1N1 flu and West Nile fever, is being addressed by careful monitoring of sudden and unusual disease outbreaks. Extensive research studies are being carried out to determine the effectiveness of various antibiotic treatments (see right accompanying figure) and how best to develop effective vaccines or administer antitoxins. To that end, vaccination perhaps offers the best defense. The United States has stated that it has stockpiled sufficient smallpox vaccine to vaccinate the entire population if a smallpox outbreak occurred. Other vaccines for other agents are in development.

This primer is not intended to scare or frighten; rather, it is intended to provide an understanding of why microbiological agents have been developed as weapons for bioterrorism. We cannot control the events that occur in the world, but by understanding bioterrorism, we can control how we should react to those events—should they occur in the future.

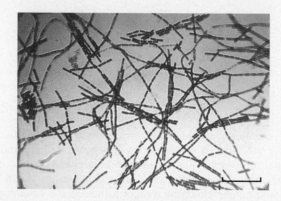

Light micrograph of gram-stained *Bacillus anthracis*, the causative agent of anthrax. (Bar = 20 μm.)

Courtesy of CDC.

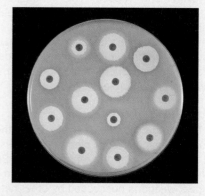

Antibiotic drugs in paper discs are used to test the sensitivity of anthrax bacteria (*Bacillus anthracis*) cultured on an agar growth medium. The clear zone surrounding each disc indicates the bacterial cells are sensitive to the antibiotic.

Courtesy of CDC/Gilda L. Jones.

MICROINQUIRY 20

Epidemiological Investigations

Infectious disease epidemiology is a scientific study from which health problems are identified. In this inquiry, we are going to look at just a few of the applications and investigative strategies for analyzing the patterns of illness. You can find answers online in **Appendix E**.

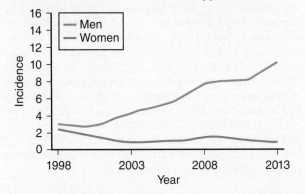

FIGURE A Incidence of Primary and Secondary Syphilis— United States, 1998–2013. Incidence numbers are per 100,000 population.

Reproduced from CDC. (2015). Summary of notifiable infectious diseases and conditions—United States, 2013. *MMWR Morb Mortal Wkly 62*(53): 1-119.

One of the important measures is to assess disease occurrence. The **incidence** of a disease is the number of new cases within a set population in a given period. **FIGURE A** is a line graph showing the incidence of primary and secondary syphilis per year in men and women in the United States since 1998. The **prevalence** of a disease refers to the total number of infected individuals (new and old) within a population at a given time.

20.1a.   What was the incidence of syphilis in 1998 and 2013?

20.1b.   How has the incidence of syphilis changed between 2000 and 2013?

Descriptive epidemiology describes studies analyzing the characteristics of an infectious disease. Often a comprehensive description can be provided by showing the disease trend over time, its geographic extent (place), and the populations (people) affected by the disease.

Characterizing by Time

Traditionally, drawing a graph of the number of cases by the date of onset shows the time course of an outbreak or epidemic. An epidemic curve, or "epi curve," is a histogram providing a visual display of the magnitude and time trend of a disease.

Look at the epi curve in **FIGURE B** for the 2015 MERS-CoV outbreak in the Republic of Korea and China. One important aspect of a bar graph is to consider its overall shape. An epi curve with a single, rapidly rising peak indicates a **common source** (or **"point source"**) **epidemic** (or **outbreak**) in which people are exposed to the same source over a short time. If the duration of exposure is prolonged

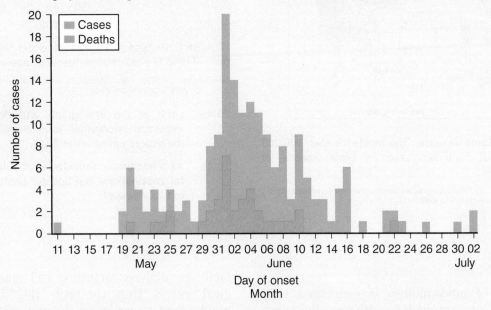

FIGURE B Confirmed Cases of MERS-CoV—Republic of Korea and China, 2015. Number of cases of, and deaths from, Middle East respiratory syndrome by day from May 11 to July 2, 2015.

Reproduced from WHO. (2017). Middle East respiratory syndrome coronavirus (MERS-CoV) maps and epicurves. Retrieved from http://www.who.int/csr/disease/coronavirus_infections/maps-epicurves-20-july-26-july/en/

(continues)

MICROINQUIRY 20 (*Continued*)

Epidemiological Investigations

such that the numbers of cases rises gradually, the epidemic (or outbreak) is called a **propagated epidemic**, and the epi curve will have a plateau instead of a peak. Person-to-person transmission is likely, and its spread might have a series of plateaus, one incubation period apart. The first case in such an outbreak or epidemic is referred to as the **index case**.

20.2a.   Identify the type of epi curve drawn in Figure B. Provide a reason for your decision.

20.2b.   Explain the significance of knowing the total number of cases and the number of individuals that died.

Characterizing by Place

Analysis of a disease or outbreak by place provides information as to where the disease was acquired, on the geographic extent of a problem, and possible clusters or patterns that provide clues to the identity and origins of the infectious agent. **FIGURE C** shows a geographic distribution for Lyme disease in 2013. This is a county map, for which each reported case of a disease in a state is shown as a highlighted county.

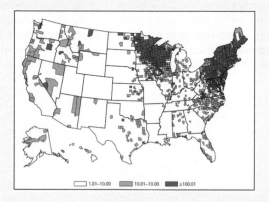

1.01–10.00   10.01–10.00   ≥100.01

FIGURE C Lyme Disease. The incidence (per 100,000 population) of confirmed cases of Lyme disease by county in 2013.

Reproduced from CDC. (2015). Summary of notifiable infectious diseases and conditions—United States, 2013. *MMWR Morb Mortal Wkly 62*(53): 1-119.

20.3.   From this map, what conclusions can you draw with regard to the reported cases of Lyme disease?

Characterizing by Person

Populations at risk for a disease can be determined by characterizing a disease or outbreak by person. Persons also refer to populations identified by personal characteristics (e.g., age, race, gender) or by exposures (e.g., occupation, leisure activities, drug intake). Such a profile of those who became ill is important because the people or population might be related to disease susceptibility and to opportunities for exposure. Age and gender often are the characteristics most strongly related to exposure and to the risk of disease. For example, **FIGURE D** is a line graph showing the number of pertussis (whooping cough) cases reported in the United States in 2013.

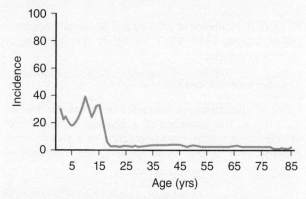

FIGURE D Pertussis. The incidence (per 100,000 population) of reported cases of pertussis by age in 2013.

Reproduced from CDC. (2015). Summary of notifiable infectious diseases and conditions—United States, 2013. *MMWR Morb Mortal Wkly 62*(53): 1-119.

20.4a.   Look at the line graph and describe what important information is conveyed in terms of the majority of cases in 2013.

20.4b.   As a healthcare provider, what role do you see for vaccinations and booster shots with regard to this disease?

HAIs Have Moved Beyond Hospitals

One aspect of epidemiology is concerned with tracking and preventing HAIs. Historically, these infections were contracted in a hospital and called **nosocomial infections**. Today, many infectious diseases are transmitted and spread in long-term care facilities, dialysis facilities, and ambulatory surgical centers. Thus, the term "HAI" is more commonly used to cover all the potential healthcare locations for infectious disease transmission. The number of HAIs is alarming. They account for an estimated 1.7 million infections (1 in 20 patients receiving medical care) and, according to the CDC, some 99,000 deaths each year in the United States. Most hospitals have an infection control officer who is the on-site epidemiologist charged with tracking any HAI outbreaks and identifying breaches in proper safety practices. The officer also is charged with training healthcare staff, including nurses, in proper aseptic procedures.

Like all infections, HAIs involve three elements: a compromised host (the hospital patient), a source of pathogens, and a chain of transmission (**FIGURE 20.14**).

The Compromised Host

Many healthcare patients are immunocompromised; that is, their immune system is weakened or impaired, rendering these patients very susceptible to opportunistic infections. Immunocompromised individuals would include people with AIDS or undergoing chemotherapy, patients taking broad-spectrum antibiotics, burn and transplant patients, newborns, surgical patients, and older adult patients. Should a pathogen enter the body of such a patient, the individual's immune system might be unable to mount an attack to eliminate the pathogen. The most common sites of infection are identified in **FIGURE 20.15**.

The Pathogens

Healthcare personnel attempt to maintain a sanitary and clean working environment. Still, the facility can be a reservoir for human pathogens coming from other patients being treated for an infectious disease, from patient visitors, or by transmission from healthcare staff. Many HAIs are caused by opportunistic agents that, given the "opportunity," can infect an immunocompromised patient. More than 50% of surgical infections come from bacterial cells that patients normally carry as opportunists on their skin. However, public health officials state that proper methods of patient screening, scrubbing with skin disinfectants, and pretreating with an antibiotic ointment can reduce infections by 60%.

Patients also can become infected from contaminated indwelling medical devices such as central lines and catheters. *S. epidermidis*, a common bacterial species found on the skin surface, can form biofilms because the microbe can be introduced with the insertion of an indwelling catheter. Again, proper skin degerming and extensive cleaning of medical devices before and after use can greatly reduce these sources of HAI pathogens. In fact, in 2015, the awareness of proper sanitation has brought a reduction in infection rates for central-line bloodstream infections, surgical site

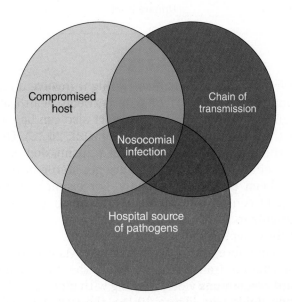

FIGURE 20.14 Healthcare-Associated Infection Elements. For a healthcare-associated infection (HAI) to occur, there needs to be a susceptible (immunocompromised) host, pathogenic organisms within the hospital setting, and a chain of transmission for the pathogen. *»» How does this HAI figure compare to that for an infectious disease (see Figure 20.11)?*

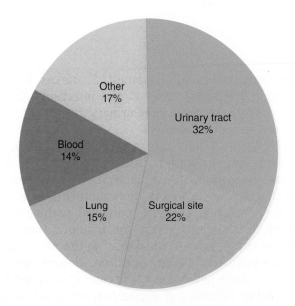

FIGURE 20.15 Sites of Healthcare-Associated Infections. This pie chart shows the sites for the estimated 1.7 million HAIs in American hospitals. *»» What might be the primary source for the infections among the top four sites?*

Data from CDC. Division of Healthcare Quality Promotion. (2010).

TABLE 20.5 Top 10 Infectious Agents Involved in Nosocomial Infections

Microorganism	% of Total Infections	% Antibiotic Resistant[1]	Nosocomial Infections
Gram-positive bacteria			
Coagulase-negative staphylococci	15.3%	Not reported (historically greater than 80%)	Blood and surgical site
Staphylococcus aureus	14.4%	56%	Urinary, blood, lung, and surgical site
Enterococcus species	12.1%	3%–90%	Urinary tract, blood, and surgical site
Gram-negative bacteria			
Escherichia coli	9.6%	1%–30%	Urinary and surgical site
Pseudomonas aeruginosa	7.9%	6%–33%	Urinary, blood, and surgical site
Klebsiella pneumoniae	5.8%	3%–27%	Lung, blood, and surgical site
Enterobacter species	4.8%	Not reported	Urinary and surgical site
Acinetobacter baumannii	2.7%	26%–37%	Urinary, blood, lung, and surgical site
Klebsiella oxytoca	1.1%	3%–17%	Urinary, blood, lung, and surgical site
Fungal			
Candida species	10.7%	Not reported	Urinary tract

infections, methicillin-resistant *S. aureus* infections, and hospital-onset *C. diff* infections. Only cases of catheter-associated urinary tract infections showed an increased rate.

TABLE 20.5 lists the most common microorganisms responsible for HAIs. More than 30% are caused by gram-negative bacteria. While examining this table, note that there is another dimension to the virulence of these potential pathogens—antibiotic resistance. With the patient's immune system compromised, the use of antibiotics often is necessary to fight an infection. Unfortunately, many of these pathogens are becoming resistant to increasing numbers of antibiotics, meaning more toxic and expensive drugs must be used.

The Chain of Transmission

The key to HAIs and their prevention stems from the way the agents are transmitted to the patient.

These **chains of transmission** might involve direct contact between patients or between healthcare staff and patient. Indirect contact also can be part of the chain of transmission (**FIGURE 20.16**). One of the most common chains of transmission is by indwelling instruments that are not sterile or have not been cleaned thoroughly. For example, some 41,000 central-line bloodstream infections strike patients in the United States each year. About 25% of these septicemic patients die.

Therefore, the key to reducing HAIs is to break the chain of transmission. Besides the use of **standard precautions** when working with blood or other body fluids (see Figure 20.16), the CDC has published preferred methods for cleaning, disinfecting, and sterilizing patient-care medical devices and general methods for cleaning and disinfecting the healthcare environment. The proper use of chemical disinfectants, and sterilization when possible,

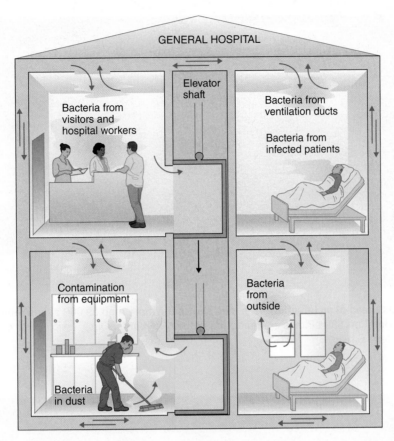

FIGURE 20.16 Microbe Transmission in a Hospital and Standard Precautions. Potential pathogens can be spread through several means within the hospital or healthcare environment. Protecting patients and other healthcare workers also means using standard precautions if working with a patient who may be infected. The precautions when handling blood or other body fluids that may harbor pathogens (HIV, hepatitis B or C) include:

- Washing hands
- Wearing personal protective equipment (gloves, mask, and eye protection)
- Handling and disposing of sharps (hypodermic needles) properly
- Disposing of all hazardous and contaminated materials in approved and labeled biohazard containers
- Cleaning up all spills with disinfectant or diluted bleach solution to kill any pathogens present

»» Why is hand washing always at the top of the list for preventing disease transmission?

are essential. These chemical agents and physical methods must be used properly to reduce the risk for infections associated with both invasive and noninvasive medical and surgical devices. Importantly, it all starts with good hand hygiene on the part of healthcare staff and visitors.

Infectious Diseases Continue to Challenge Public Health Organizations

In an era of supersonic jet travel and international commerce, it is not possible to adequately protect the health of any nation without focusing on diseases and epidemics elsewhere in the world (**FIGURE 20.17**).

Emerging infectious diseases are caused by newly identified pathogens or by pathogens appearing in new areas of the world for the first time. **CLINICAL CASE 20** describes a recent outbreak in Key West, Florida. **Re-emerging (resurgent) infectious diseases** are those that previously were controlled or declining but now are showing increased occurrence. Globally, there have been more than 40 new emerging diseases and at least 20 resurgent diseases identified in the last 40 years (**FIGURE 20.18**).

There are more than 1,400 known human pathogens, and more than half represent zoonoses. However, fewer than 100 are specialized within humans. Some 335 (24%) represent emerging or resurgent

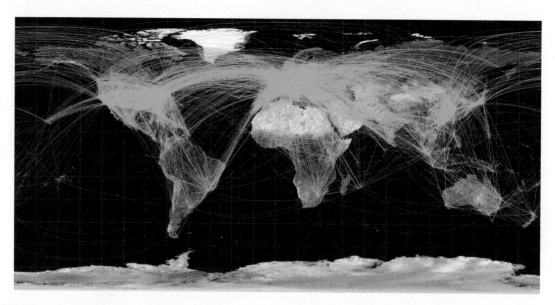

FIGURE 20.17 Potential Spread of an Infection. This map shows the more than 59,000 airline routes (green lines) between some 3,200 airports (red dots) on 530 airlines spanning the globe. »» *From this map, which parts of the world might be most susceptible to potential disease spread?*

Courtesy of openflights.org/Contentshare.

Clinical Case 20

Locally Acquired Dengue Fever

In mid-August, a previously healthy, 34-year-old woman in Rochester, New York, went to her primary care physician complaining of fever, headache, malaise, and chills. She told the physician that the symptoms had appeared 24 hours earlier. A urine sample was taken for analysis.

Two days later, the patient returned to her physician. Her fever had abated, but she had a more severe headache, severe pain behind the eyes that worsened on eye movement, and a feeling of light-headedness. Her urinalysis report indicated bacterial cells and red blood cells were present in the urine. She was referred to a local hospital emergency department.

The emergency room evaluation showed all vital signs were normal. A complete blood cell workup revealed a low white blood cell and platelet count and a normal hematocrit (percent of red blood cells in the blood). A CT scan and cerebral spinal fluid (CSF) from a lumbar puncture were normal. Because her light-headedness disappeared, she was discharged from the emergency department.

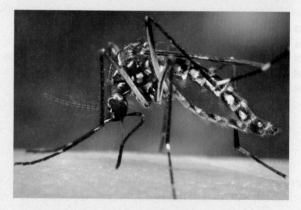

A mosquito taking a human blood meal.

Courtesy of Prof. Frank Hadley Collins, Director, Center for Global Health and Infectious Diseases, University of Notre Dame/CDC.

Four days later, the patient returned to her primary care physician expressing the feeling that she just didn't feel good. Although all vital signs were normal, petechiae (tiny purplish-red spots due to blood hemorrhages) were noted on her lower extremities.

A consultation with an infectious disease specialist suggested the patient could have dengue fever. Questioning the patient, it was determined that she had not traveled to any dengue-endemic area in the world. She did state that prior to the onset of symptoms she had just returned from a trip to Key West, Florida, and, while there, had been bitten several times by mosquitoes. A serum sample from the patient was tested for antibodies to dengue fever virus. The results were positive. Confirmatory testing of serum and CFS samples was done by the Centers for Disease Control and Prevention (CDC) in Atlanta. Both samples were positive for antibodies against dengue fever virus.

Two days later, the patient reported to her physician that she was feeling much better. She had completely recovered when interviewed by the Monroe County (Florida) Health Department.

Further investigation identified another 24 cases of dengue fever, all locally acquired in the Key West area.

Questions:
a. Why was a urine sample taken for analysis?
b. Why didn't the original CSF sample taken on August 13 indicate a dengue fever infection?
c. What sign indicated to the infectious disease specialist that the patient might have dengue fever?
d. Considering the number of dengue fever cases in Key West, what measures should be taken to lessen and control the outbreak?

You can find answers online in **Appendix E**.

For additional information, see www.cdc.gov/mmwr/preview/mmwrhtml/mm5919a1.htm.

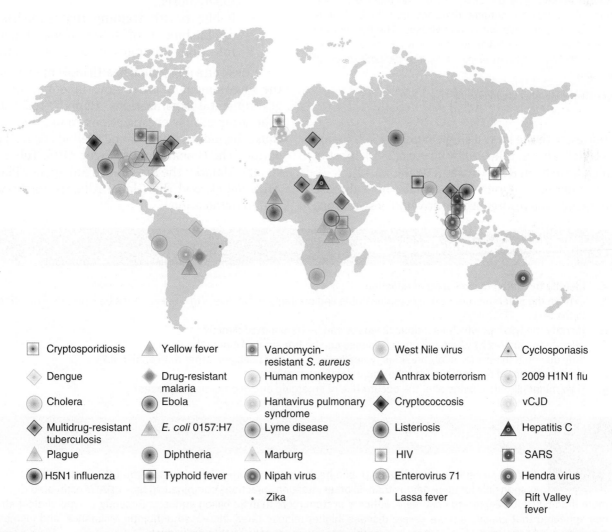

Cryptosporidiosis | Yellow fever | Vancomycin-resistant *S. aureus* | West Nile virus | Cyclosporiasis

Dengue | Drug-resistant malaria | Human monkeypox | Anthrax bioterrorism | 2009 H1N1 flu

Cholera | Ebola | Hantavirus pulmonary syndrome | Cryptococcosis | vCJD

Multidrug-resistant tuberculosis | *E. coli* 0157:H7 | Lyme disease | Listeriosis | Hepatitis C

Plague | Diphtheria | Marburg | HIV | SARS

H5N1 influenza | Typhoid fever | Nipah virus | Enterovirus 71 | Hendra virus

Zika | Lassa fever | Rift Valley fever

FIGURE 20.18 The Global Occurrence of Emerging and Re-Emerging Diseases. These two maps identify the global scope of emerging and re-emerging diseases over the past four decades. *»» Can you suggest a reason why many of the diseases have appeared in North America? (Note: see Figure 20.17).*

Data from American College of Microbiology. *Clinical Microbiology in the 21st Century: Keeping the Pace.* ASM Press, 2008, Washington, D.C.

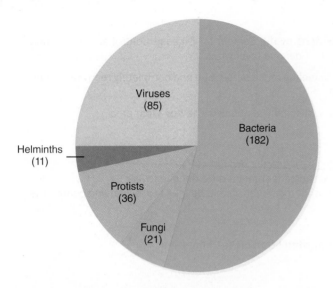

FIGURE 20.19 The Diversity of Emerging/Re-Emerging Disease Agents. The pie chart shows the relative proportions of emerging/resurgent diseases caused by bacteria, fungi, protists, helminths, and viruses. The numbers in parentheses are the known number. *»» Name several prominent emerging/resurgent diseases caused by each group of microbes.*

Modified from Woolhouse, M. E. J. 2008. *Nature* 451:898–899.

diseases, with the largest single number (54%) being bacterial species, followed by viruses (25%), most of which have been RNA viruses (**FIGURE 20.19**).

Numerous reasons help explain how disease emergence and resurgence are driven. These reasons can be placed in one of three categories, which are listed in **TABLE 20.6**.

By understanding these drivers, health organizations, such as the CDC and WHO, can develop plans to limit or stop emerging disease threats. The CDC has had a history of serving American public health. Regarding emerging and resurgent diseases, the CDC's priority areas include the following:

▶ International outbreak assistance to host countries to maintain control of new pathogens when an outbreak is over.
▶ A global approach to disease surveillance by establishing a global "network of networks" for early warning of emerging health threats.
▶ Global disease control through initiatives to reduce HIV disease/AIDS, malaria, and tuberculosis.
▶ Public health training that supports the establishment of International Emerging Infections Programs in developing nations.

However, a new global health movement requires the involvement of more governmental and nongovernmental organizations. Happily, today there is an array of new organizations committed to this cause, including: the Bill and Melinda Gates Foundation; The Global Fund to Fight AIDS, Tuberculosis, and Malaria; the President's Emergency Plan for AIDS Relief; and the Global Alliance for Vaccines and Immunization.

Concept and Reasoning Checks 20.3

a. Identify the different reservoirs of infection.
b. Explain the difference between a communicable and contagious disease. Provide examples beyond those mentioned in the text.
c. Identify the ways by which infectious diseases can be transmitted directly.
d. Identify the modes by which infectious disease can be transmitted indirectly.
e. Why do you think the term "outbreak" is typically used in news releases rather than epidemic?
f. How do standard precautions limit the chains of transmission?
g. How might climate change alter the occurrence of emerging and re-emerging diseases?

Chapter Challenge C

CDAI represents a considerable public health hazard. In the United States, the CDC reports that *C. diff* is responsible for more deaths than all other intestinal infections combined. In fact, CDAI is estimated to affect more than 500,000 people and cost the healthcare system more than $3 billion annually. Recently, a hypervirulent strain has appeared that causes a more severe disease, higher relapse rates, increased antibiotic resistance, and increased mortality due to the production of greater quantities of toxins—even in healthy individuals.

QUESTION C: *Would you consider CDAI an emerging disease? A re-emerging disease? How about the hypervirulent strain? As a healthcare provider, what standard precautions should you take when treating CDAI patients?*

You can find answers online in **Appendix F**.

TABLE 20.6 Drivers of Disease Emergence and Re-Emergence

Disease Driver		Examples (B = bacterium; V = virus; P = protist; H = helminth)
Environmental change	**Changes in land use or agriculture practices.** Urbanization, deforestation, and water projects can bring new or re-emergent diseases.	Dengue fever (V) and schistosomiasis (H)
	Pathogen evolution. Pathogens have developed resistance to antibiotics and antimicrobial drugs.	Tuberculosis (B) and yellow fever (V)
	Climate change. Global changes in weather patterns bring new diseases to new latitudes and elevations.	Hantavirus pulmonary syndrome (V)
Economic/ social/ political change	**International agricultural trade.** The global transport of produce can introduce new or resurgent foodborne diseases.	Hepatitis A (V) and cyclosporiasis (P)
	Animal trade. Wildlife trade (legal and illegal) provides new mechanisms for disease transmission.	Monkeypox (V)
	Poor population health. In many developing nations, large numbers of people suffer from malnutrition or poor public health infrastructure, making disease eruption much more likely.	Cholera (B) and diphtheria (B)
	Contamination of food sources and water supplies. Substandard application of, or lapses in, sanitation practices can bring a resurgence of disease.	Listeriosis (B)
	Failure of public health systems. Failure of immunization programs can bring a resurgence of disease.	Diphtheria (B)
Demographic change	**Human migration.** The migration of many peoples or whole societies from agrarian to dense, urban settings has brought new diseases that can be easily transmitted to a susceptible population.	Malaria (P)
	International travel. The number of people traveling internationally can spread diseases quickly to other parts of the globe.	SARS (V) and West Nile fever (V)

In conclusion, with the appearance of emerging and resurgent infectious diseases and the declining power of antibiotics, our ability to live a healthy, germ-free life seems harder to maintain. One of the luminaries in microbiology, the French-American microbiologist René Dubos (1901–1982), devoted much of his professional life to the study of microbial diseases and the mechanisms of resistance to infection. Dubos also considered the environmental and social factors that affect human health. In 1959, he wrote *Mirage of Health*, in which he argues, *"Complete freedom from disease ... is almost incompatible with the process of living."* He argues that the possibility of a germ-free existence is a dream because science and medicine have ignored the dynamic process of adaptation that every microorganism (and every living organism) must face (e.g., antibiotic resistance) in a constantly changing environment (medicine's use of antibiotics). Dubos argues that infectious disease is not simply a battle between the "good guys" (humans) fighting the "bad guys" (pathogens). Now, almost 60 years after the publication of Dubos' book, we are losing the battle against many pathogens. We might not have lost the war, but we need to be aware of other factors, besides the causative organism, to effect good health. One factor is our immune system, which, along with sound sanitary practices, perhaps gives us our best chance to shield ourselves from infectious diseases.

■ SUMMARY OF KEY CONCEPTS

Concept 20.1 The Host and Microbe: An Intimate Relationship in Health and Disease

1. **Infection** refers to competition between host and microbe for supremacy, whereas **disease** results when the microbe wins the competition. However, the human body contains large populations of resident microbes (**human microbiome**) that usually outcompete invading pathogens. (Figure 20.3)
2. The human microbiome is established at birth and is influenced by the method of birth, feeding, and environmental factors. (Figure 20.4)
3. **Parasitism** is the symbiotic relationship occurring when the microbe does harm to the host. **Pathogenicity**, the ability of the pathogen to cause disease, and resistance go hand in hand. **Virulence** refers to the degree of pathogenicity a microbe displays.

Concept 20.2 The Host and Microbe: Establishment of Infection and Disease

4. A medical diagnosis is based on a patient's **signs** and **symptoms**.
5. If the immune system is compromised, microbes normally acting as commensals might cause **opportunistic infections**. A **primary infection** is an illness caused by a pathogen in an otherwise healthy host, whereas a **secondary infection** involves the development of other diseases as a result of the primary infection. **Local diseases** are restricted to a specific part of the body, whereas **systemic diseases** spread to several parts of the body and deeper tissues. To cause disease, most pathogens must enter the body through an appropriate **portal of entry**. The **infectious dose** represents the number of pathogens taken into the body causing a disease.
6. Bacterial pili or virus spikes allow for attachment to specific cells. (Figure 20.7)
7. The possibility of disease is enhanced if a microbe can penetrate host tissues. The pathogen's ability to penetrate tissues and cause damage is called **invasiveness**. Virulence and invasiveness are strongly dependent on the spectrum of **virulence factors** a pathogen possesses. Adhesion or attachment may lead to **phagocytosis** of the pathogens.

8. Many bacterial species produce enzymes to overcome the body's defenses. These include **coagulase**, **streptokinase**, and **hyaluronidase**. In addition, some species produce **exotoxins** that are proteins released by gram-positive and gram-negative cells. Their effects depend on the host cells or tissues affected. **Endotoxins** are the lipopolysaccharides released from dead, gram-negative cell walls. Their effects on the host are universal. (Figure 20.8 and Table 20.3)
9. For a disease to spread efficiently to other hosts, the pathogen also must leave the body through an appropriate **portal of exit**. (Figure 20.9)

Concept 20.3 Epidemiology Is the Foundation of Public Health

10. **Reservoirs** include humans, who represent carriers of a disease, arthropods, and any food and water in which some parasites survive.
11. A disease can be **communicable**, such as measles, or **noncommunicable**, such as tetanus.
12. Diseases can be transmitted by **direct** or **indirect** contact. Indirect methods include consumption of contaminated food or water, contaminated inanimate objects (**fomites**), and arthropods. Arthropods can be **mechanical** or **biological vectors** for the transmission of disease. (Figure 20.12)
13. The occurrence of diseases falls into three categories: **endemic**, **epidemic**, and **pandemic**. An "outbreak" is essentially the same as an epidemic, although usually an outbreak is more confined in terms of disease spread.
14. **HAIs** are acquired as a result of the patient being treated for some other injury or medical problem. **Standard precautions** limit the **chain of transmission**. (Figures 20.14, 20.15)
15. Public health organizations, such as the CDC, are charged with the duty of limiting or stopping disease threats, including **emerging** and **re-emerging diseases**. (Figure 20.18)

■ CHAPTER SELF-TEST

You can find answers for **Steps A–D** online in **Appendix D**.

STEP A: REVIEW OF FACTS AND TERMS

Multiple Choice

Read each question carefully and then select the *one* answer that best fits the question or statement.

1. Which one of the following is *not* part of the resident human microbiome?
   A. *Lactobacillus*
   B. *Candida*
   C. *Clostridium*
   D. *Escherichia*

2. A newborn ____.
   A. contains resident microbiome before birth
   B. remains sterile for many weeks after birth
   C. becomes colonized soon after conception
   D. is colonized with many common microbes within minutes after birth

3. Factors affecting virulence may include ____.
   A. the presence of pili
   B. their ability to penetrate the host
   C. the infectious dose
   D. All the above (**A–C**) are correct.

4. A healthy person can be diagnosed as having a ____ infection with ____, the multiplication of bacterial cells in the blood.
   A. primary; bacteremia
   B. primary; viremia
   C. primary; septicemia
   D. secondary; parasitemia

5. Changes in body function sensed by the patient are called ____.
   A. symptoms
   B. syndromes
   C. prodromes
   D. signs

6. In the body, bacterial invasiveness can be limited by ____.
   A. fever
   B. phagocytosis
   C. enzyme production
   D. toxin production

7. Which one of the following is *not* true of exotoxins?
   A. They are proteins.
   B. They are part of cell wall structure.
   C. They are released from live bacterial cells.
   D. They trigger antibody production.

8. A portal of exit would be ____.
   A. the feces
   B. an insect bite
   C. blood removal
   D. All of the above (**A–C**) are correct.

9. All of the following are examples of communicable diseases *except* ____.
   A. chickenpox
   B. measles
   C. the common cold
   D. tetanus

10. If a person has recovered from a disease but continues to shed disease agents, that person is a ____.
    A. vector
    B. fomite
    C. vehicle
    D. carrier

11. Which one of the following is an example of transmission by indirect contact?
    A. Coughing
    B. Aerosols
    C. Sexual intercourse
    D. Mother to fetus

12. Fifty cases of hepatitis A during one week in a community would most likely be described as a/an ____.
    A. outbreak
    B. pandemic
    C. endemic disease
    D. epidemic

13. The most common HAI involves ____.
    A. the blood
    B. the lungs
    C. the urinary tract
    D. a surgical site

14. Emerging infectious diseases can be the result of ____.
    A. pathogen evolution
    B. international trade
    C. agricultural practices
    D. All of the above (**A–C**) are correct.

True-False

Each of the following statements is true (T) or false (F). If the statement is false, substitute a word or phrase for the underlined word or phrase to make the statement true.

____ 15. An <u>epidemic</u> disease occurs at a low level in a certain geographic area.

____ 16. <u>Commensalism</u> is a form of symbiosis where the microbe benefits and causes no damage to the host.

____ 17. A <u>biological</u> vector is an arthropod that carries pathogenic microorganisms on its feet and body parts.

_____ **18.** Organisms causing disease when the immune system is depressed are known as <u>opportunistic</u> organisms.

_____**19.** Few symptoms are exhibited by a person who has a <u>subclinical disease</u>.

_____ **20.** A <u>chronic</u> disease develops rapidly, is usually accompanied by severe symptoms, and comes to a climax.

■ STEP B: CONCEPT REVIEW

21. Differentiate between **infection** and **disease**. (**Key Concept 20.1**)

22. Explain how the human microbiome is first established. (**Key Concept 20.1**)

23. Distinguish between (a) **primary** and **secondary infections**, (b) **local** and **systemic diseases**, and (c) **bacteremia** and **septicemia**. (**Key Concept 20.1**)

24. Explain the differences between **signs, symptoms,** and **syndromes**. (**Key Concept 20.2**)

25. Name three enzymes and describe their roles as **virulence factors**. Summarize the differences between **exotoxins** and **endotoxins** as virulence factors associated with disease. (**Key Concept 20.2**)

26. Distinguish between the **direct** and **indirect contact** methods of disease transmission. (**Key Concept 20.3**)

27. Explain how **nosocomial infections** can be controlled or eliminated through using the **standard precautions** to break the **chain of transmission**. (**Key Concept 20.3**)

■ STEP C: APPLICATIONS AND PROBLEM SOLVING

28. The transparent covering over salad bars is commonly called a "sneeze guard" because it helps prevent nasal droplets from reaching the salad items. As a community health inspector, what other suggestions might you make to prevent disease transmission at the salad bar?

29. While slicing a piece of garden hose, your friend cut himself with a sharp knife. The wound was deep, but it closed quickly. Shortly thereafter, he reported to the emergency room of the community hospital, where he received a tetanus shot. What did the tetanus shot contain, and why was it necessary?

30. After reading this chapter, you decide to make a list of the ten worst "hot zones" in your home. The title of your top ten list will be "Germs, Germs Everywhere." What places will make your list, and why?

31. As a state epidemiologist responsible for identifying any disease occurrences, would an epidemic disease or an endemic disease pose a greater threat to public health in the community? Explain.

■ STEP D: QUESTIONS FOR THOUGHT AND DISCUSSION

32. In 1840, Great Britain introduced penny postage and issued the first adhesive stamps. However, politicians did not like the idea because it deprived them of the free postage they were used to. Soon, a rumor campaign was started, saying that these gummed labels could spread disease among the population. Can you see any wisdom in their contention? Would their concern "apply" today?

33. In 1892, a critic of the germ theory of disease named Max von Pettenkofer sought to discredit Robert Koch's work by drinking a culture of cholera bacilli diluted in water. Von Pettenkofer suffered nothing more than mild diarrhea. What factors might have contributed to the failure of the bacilli to cause cholera in von Pettenkofer's body?

34. When Ebola fever broke out in Africa in 1995, disease epidemiologists noted how quickly the responsible virus killed its victims and suggested the epidemic would end shortly. Sure enough, within 3 weeks it was over. What was the basis for their prediction? What other conditions had to apply for them to be accurate in their guesswork?

35. A woman takes an antibiotic to relieve a urinary tract infection caused by *E. coli*. The infection resolves, but in 2 weeks, she develops a *C. albicans* ("yeast") infection of the vaginal tract. What conditions might have caused this yeast infection?

CHAPTER 21

Resistance and the Immune System: Innate Immunity

Edward Jenner's epoch-making moment in history came in 1798 when he demonstrated the protective action of vaccination against smallpox. As great as Jenner's discovery was, it did not advance the development or understanding of **immunity**, which is how the human body can generate resistance to a particular disease, whether by recovering from the disease or being protected through vaccination, such as that first devised by Jenner.

Although Pasteur introduced vaccines for chicken cholera, anthrax, and rabies in the 1880s, it was the Russian zoologist and immunologist Elie Metchnikoff who devised the first experiments to study immunity (see the chapter opener). Through landmark experiments with invertebrates, Metchnikoff suggested that certain types of cells could attack, engulf, and destroy foreign material, including infectious microbes. His work culminated in the theory of phagocytosis as a way the immune system can capture and destroy disease-causing microbes that had penetrated the body.

Also in Pasteur's lab, Émile Roux and Alexandre Yersin identified a bacterial toxin as the disease-causing agent of diphtheria. Then, Koch's coworker Emil von Behring modified the diphtheria toxin to produce a substance with "antitoxic activity" capable of protecting animals for a short period against diphtheria. The substance gave the body the power to resist infection.

With these early discoveries, the science of immunology was born. Today, Metchnikoff's work forms part of the basis for "innate immunity," which consists of several nonspecific defenses present in most animals. The discoveries of Roux, Yersin, and von Behring are the basis for part of what is referred to today as "adaptive immunity." This form of protective immunity or resistance is found only in vertebrates and is a response directed only against a specific invading

Ilya Ilyich (Elie) Metchnikoff.
© Science Source.

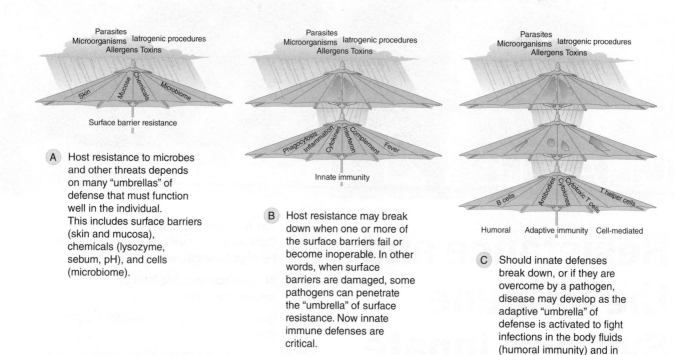

A Host resistance to microbes and other threats depends on many "umbrellas" of defense that must function well in the individual. This includes surface barriers (skin and mucosa), chemicals (lysozyme, sebum, pH), and cells (microbiome).

B Host resistance may break down when one or more of the surface barriers fail or become inoperable. In other words, when surface barriers are damaged, some pathogens can penetrate the "umbrella" of surface resistance. Now innate immune defenses are critical.

C Should innate defenses break down, or if they are overcome by a pathogen, disease may develop as the adaptive "umbrella" of defense is activated to fight infections in the body fluids (humoral immunity) and in infected cells (cell-mediated).

FIGURE 21.1 The Relationship Between Host Resistance and Disease. Host resistance can be likened to a three-layered "microbiological umbrella" that forms a barrier or defense against infectious disease. *»» Why does the microbiological umbrella require three levels of defense?*

pathogen. Together, innate and adaptive immunity are equally important arms that form a "microbiological umbrella" protecting us against the torrent of potential microbial pathogens and other substances to which we are continually exposed (**FIGURE 21.1**).

This chapter examines innate immunity, the body's immediate response to pathogens, toxins, and other foreign substances. It will serve as the foundation for a discussion on adaptive immunity, which is described in Chapter 22.

Chapter Challenge

As described in Chapter 16, influenza commonly called the flu, is a contagious respiratory illness caused by influenza viruses that infect the upper respiratory tract (nose and throat) and eventually the lower respiratory tract (the lungs). The type A flu virus has two types of surface protein spikes, called hemagglutinin (H) and neuraminidase (N). The H spikes are needed for the virus to attach to and penetrate host cells of the respiratory tract, whereas the N spikes are needed for the release of the virus from the infected cells.

So, what happens immunologically when we are exposed to the type A flu virus? Let's examine the innate immune response as we proceed through this chapter. Warning: we won't have recovered from the flu by the end of this chapter! We will have such a severe case that we will need to continue the investigation into the chapter on adaptive immunity before we get better. So, let's catch the flu!

■ KEY CONCEPT 21.1   The Immune System Is a Network of Cells and Molecules to Defend Against Foreign Substances

The work of Metchnikoff, Roux, Yersin, and von Behring, as well as that of their contemporaries and followers, opened up the entire field of **immunology**,

the scientific study of how the immune system functions in the body to prevent or destroy foreign material, including pathogens.

Blood Cells Form an Important Part of Immune Defense

In the circulatory system, blood consists of three major components: fluid, clotting agents, and formed elements—the blood cells and platelets. The fluid portion, called **serum**, is an aqueous solution of minerals, salts, proteins, and other organic substances. When clotting agents, such as fibrinogen and prothrombin, are present, the fluid is referred to as **plasma**.

All blood and immune cells originate in the bone marrow from what are called **multipotential hematopoietic stem cells** (**FIGURE 21.2**). When these undifferentiated cells divide, some maintain a reserve population of multipotent hematopoietic stem cells, whereas the others differentiate into myeloid and lymphoid cell types. The **myeloid stem cells** differentiate into most of the white blood cells (basophils, mast cells, eosinophils, neutrophils, monocytes), whereas the **lymphoid stem cells** differentiate into B and T lymphocytes. A separate cell line gives rise to the red blood cells and platelets (megakaryocytes).

Leukocytes (White Blood Cells)

As their name suggests, **leukocytes** (*leuko* = "white") have no pigment in their cytoplasm and therefore

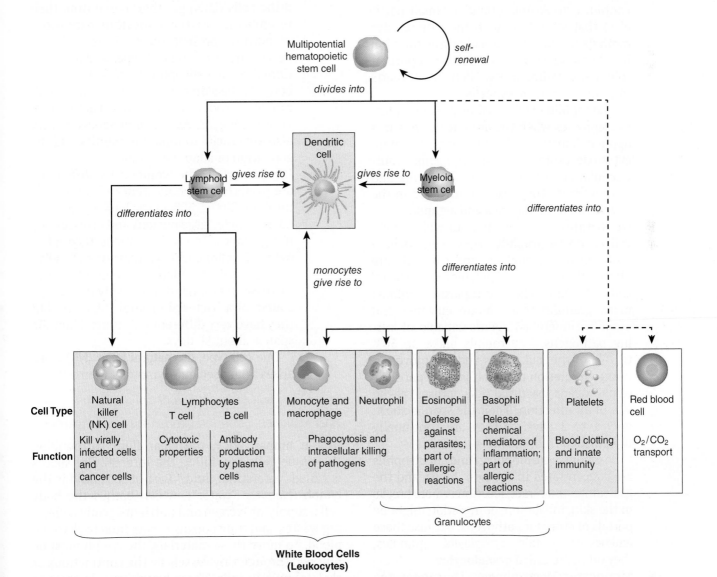

FIGURE 21.2 Major Types of Leukocytes and Their Origin. All blood cells, including the white blood cells, arise from multipotential hematopoietic stem cells in the bone marrow. Each derived cell line becomes more specialized down to the final cell type. The dendritic cells are involved with phagocytosis and antigen presentation. The dashed line represents a separate line of stem cells. »» *Explain what could occur if an individual lacked neutrophils, basophils, eosinophils, and monocytes/macrophages.*

appear gray when observed with the light microscope. They number about 4,000 to 12,000 per microliter of blood and have different life spans, depending on the cell type. Most can be identified in a blood smear or tissue sample by the size and shape of the cell nucleus or by the staining of visible cytoplasmic granules, if present, when observed with the light microscope.

There are several different cell types working together as a team to produce an effective immune defense in the body:

▶ **Basophils and Eosinophils.** Two types of leukocytes can be identified by staining their cytoplasmic contents. **Basophils** contain cytoplasmic granules (small particles) that stain blue with the cationic dye **methylene blue**. These leukocytes function in allergic reactions and play an important role in innate immunity. Their counterparts in tissues are the **mast cells**.

Eosinophils exhibit red or pink cytoplasmic granules when the anionic dye **eosin** is applied. Substances in the granules contain cytotoxic proteins to defend against multicellular parasites such as flatworms and roundworms. They also are involved in the development of allergies and asthma.

▶ **Neutrophils.** Another type of white blood cell is the **neutrophils**. These cells have a multilobed cell nucleus, and the cells are often referred to as "polymorphonuclear" cells (PMNs). Their cytoplasm contains many granules that contain enzymes and other antimicrobial agents capable of killing pathogens. Neutrophils make up 50% to 70% of the leukocytes and can move out of the circulation to engulf foreign particles or pathogens (bacterial cells, viruses) at the site of an infection. Their life span is short, only 1 to 2 days, so they are continually replenished from the bone marrow.

Basophils, eosinophils, and neutrophils are concentrated in the skin, lungs, and the gastrointestinal tract, locations where breaks in the skin, inhalation, or ingestion might be portals of entry for pathogens. Because these leukocytes contain cytoplasmic granules, they often are called **granulocytes**.

▶ **Monocytes/Macrophages.** The **monocytes** circulate in the blood, they have a single, bean-shaped cell nucleus, and they lack visible cytoplasmic granules. In tissues, monocytes mature into **macrophages**. Composing 5% to 7% of the leukocytes, monocytes spend 1 to 2 weeks in the circulation before they enter body tissues. There, they can live for several months while "patrolling" for any invading pathogens.

Macrophages are one of the key defensive cells in both innate and adaptive immunity. As such, they are tissue residents in the spleen, lymph nodes, and thymus and are common immune cells in the lungs, skin, and nervous system. Because both neutrophils and macrophages carry out phagocytosis, they are called **phagocytes**.

▶ **Dendritic Cells.** Phagocytes called **dendritic cells** (DCs) get their name from their resemblance to the long, thin extensions (dendrites) present on neurons. DCs are derived from several cell types and are positioned at specific portals of entry in the body. At the skin and mucosal surfaces, DCs rapidly respond to invading bacterial or viral pathogens. As active phagocytes, they also are crucial to innate immunity and the activation of adaptive immunity.

▶ **Lymphocytes.** The lymphocytes differentiate from the lymphoid stem cells and make up about 20% to 30% of the leukocytes. They have a single, large nucleus and no cytoplasmic granules. The two cell types, **B lymphocytes (B cells)** and **T lymphocytes (T cells)**, are key cells of adaptive immunity. They increase in number dramatically during the course of a bacterial or viral infection, but they have very different functions. Their life span is about 90 days.

One cell type, the **natural killer (NK) cells**, arises from a separate lymphoid stem cell line.

The Lymphatic System Is Composed of Cells and Tissues Essential to Immune Function

In the human body, the clear fluid surrounding the tissue cells and filling the intercellular spaces is called "interstitial fluid." Part of that fluid is the **lymph** (*lymph* = "water"), which bathes the body cells, supplying oxygen and nutrients, while collecting wastes, and transporting excess fluid back to the blood. To move these materials, the lymph must be pumped through tiny vessels by the contractions of skeletal muscle cells. These lymph vessels unite to form larger lymphatic vessels. The afferent vessels carry lymphatic fluid to lymph nodes, which filter the fluid before it is returned via efferent vessels to the thoracic duct and the bloodstream.

In the lymphatic tissues, lymphocytes mature, differentiate, and divide. The **primary lymphoid tissues** consist of the **thymus**, which lies behind the breastbone, and the **bone marrow**. Both are sites where T and B cells form or mature (**FIGURE 21.3A**).

The **secondary lymphoid tissues** are sites where mature immune cells interact with pathogens and carry out the adaptive immune response. The **spleen**, a flattened organ at the upper left of the abdomen, contains immune cells to monitor and fight infectious microbes entering the body. Likewise, the **lymph nodes**, prevalent in the neck, armpits, and groin, are bean-shaped structures containing macrophages and dendritic cells that engulf pathogens in the lymph and lymphocytes, which respond specifically to foreign substances brought in from the circulation (**FIGURE 21.3B**). Because resistance mechanisms are closely associated with the lymph nodes, it is not surprising they become enlarged (often called "swollen glands") during infections.

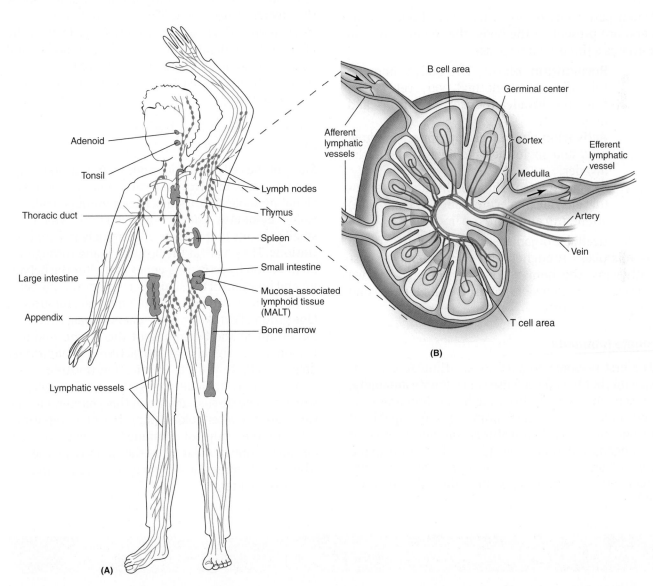

FIGURE 21.3 The Human Lymphatic System. (A) The human lymphatic system consists of lymphocytes, lymphatic organs, lymph vessels, and lymph nodes located along the vessels. The lymphatic organs are illustrated, and the preponderance of lymph nodes in the neck, armpits, and groin is apparent. **(B)** A lymph node is a filtering site for circulation of lymph. Any microbes present are removed by macrophages and dendritic cells before the "cleaned" lymph exists at an efferent lymphatic vessel. *»» Why are there so many lymph nodes distributed throughout the body?*

Other organ systems in the body contain additional secondary lymphoid tissues. The appendix and the Peyer patches in the ileum of the small intestine form part of the **mucosa-associated lymphoid tissue** (**MALT**), whereas secondary lymphoid tissues associated with the respiratory tract are the **tonsils**. Specialized lymphocytes and dendritic cells of the MALT help defend against pathogen infection.

Innate and Adaptive Immunity Compose a Fully Functional Human Immune System

When pathogens or abnormal cells (tumors, cancers) are present in the body, the immune system initiates a three-step response:

▶ **Recognition phase.** The body's immune cells distinguish normal body cells (self) from the invader (nonself organisms and molecules);

▶ **Activation phase.** The appropriate immune cells and molecules mobilize to fight the invader;

▶ **Effector phase.** The mobilized cells (effector cells) and molecules (effector molecules) attempt to eliminate the invader.

Multicellular organisms are constantly exposed to microbes throughout life. To defend against pathogens, the human body is protected by two overlapping systems of immunity called innate and adaptive immunity.

Innate Immunity

The first response to a pathogen stimulates a nonspecific, but flexible defense called **innate immunity**. Present in the body from birth, the immune cells and molecules of innate immunity are capable of recognizing microbial features common to diverse pathogens and foreign substances. Therefore, in the case of an infection, innate immunity represents the host's early-warning system that, although not specific to a particular pathogen, can respond within minutes of, or a few hours after, infection.

Although the effector cells and molecules of innate immunity are capable of eliminating or holding an infection in check, the innate system also sends additional chemical signals to tissues involved with activating adaptive immunity.

Adaptive Immunity

The response to a specific pathogen in the body is called **adaptive immunity**, so-called because the system must first adapt to (acquire) information about the invading pathogen. Therefore, the response is slow compared to the innate response; in fact, the effector phase of adaptive immunity (antibodies and lymphocytes) takes several days to more than a week to generate an effective response and to mount a protective defense.

Cytokines Are Chemical Messengers Used in Innate and Adaptive Immune Responses

Many of the cell interactions during innate and adaptive immunity involve chemical communication. The signaling molecules used between immune cells are called **cytokines** (*kine* = "movement"), which are small proteins that operate at extremely low concentrations. They work at the beginning and throughout an infection by activating genes that alter the behavior and function of the signal-receiving immune cells.

Cytokines can be separated into several groups. One group of cytokines is the **chemokines**, which are important at the beginning of the innate immune response to signal the attraction of additional defense cells to the site of infection (**FIGURE 21.4**). Other groups include pyrogens and interferons, which will be discussed later in this chapter. Finally, there are the **interleukins** (**ILs**) that play important roles in the adaptive immune response. We will become more familiar with the actions of the ILs when we examine the adaptive immune defenses in the next chapter.

Concept and Reasoning Checks 21.1

a. In a blood smear, how could you tell if a patient was deficient in neutrophils?
b. Assess the role of the lymphatic system to defense against pathogens.
c. Summarize the need for chemical communication between immune defensive cells.

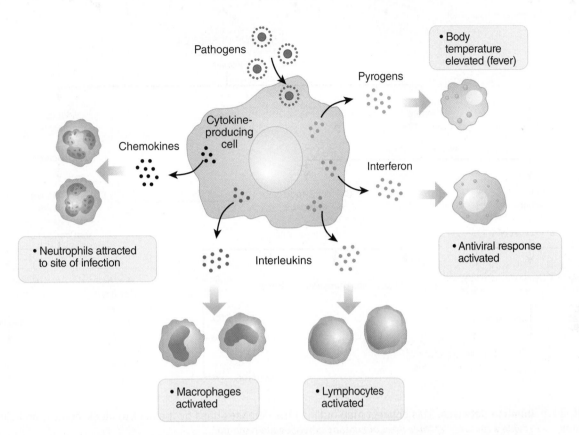

FIGURE 21.4 An Example of Cytokine Secretion by Immune Cells. In this summary diagram, an activated immune cell can respond to pathogen presence by secreting several cytokines that, depending on the secreting cell, can have one of several effects. Pyrogens, interferon, and chemokines are discussed in this chapter. »» *Why are so many different types of cytokines needed to produce an immune response?*

Chapter **Challenge A**

We have now been infected with a sufficient number (infectious dose) of flu viruses such that contracting the flu is likely. However, at this point in the incubation period, we are not aware that we are infected. While we are still healthy and not showing any symptoms, consider the following question.

QUESTION A: *Most viral infections, including the flu, are controlled by the innate and adaptive arms of the immune system. Make a checklist of the types of white blood cells discussed in this section of the chapter that would most likely be involved in identifying flu viruses and would be activated to fight the infection. What role will the lymphatic system play in optimizing the response?*

You can find answers online in **Appendix F**.

■ KEY CONCEPT 21.2   Surface Barriers Are Part of Innate Immunity

The innate defenses to infection first involve a broad group of nonspecific surface barriers through which a pathogen would have to pass in order to cause an infection (**FIGURE 21.5**).

Host Defensive Barriers Limit Pathogen Entry

Pathogens can enter the body through several portals of entry. Two anatomical (physical) portals are

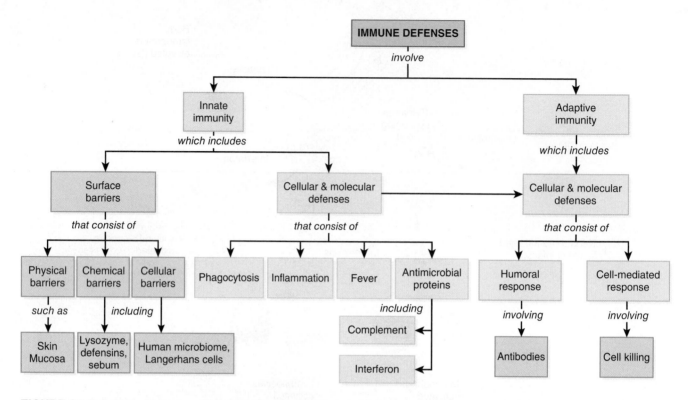

FIGURE 21.5 Immune Defenses. This concept map outlines the defenses used by the host to block, control, and eliminate pathogens. »» *Provide a reason why three lines of immune defenses are required.*

the intact skin and the mucous membranes, which extend over the body surface or into the body cavities. These barriers provide physical, chemical, and cellular defenses against pathogen colonization and invasion of deeper body tissues.

Physical Barriers

One of the first hurdles for many pathogens is penetrating the skin and mucous membranes.

▶ **Skin.** The skin, composed of the epidermis and dermis, is a multilayered tissue and, as such, forms a tough, semi-watertight, and impenetrable barrier to infection by most microorganisms (**FIGURE 21.6**). In addition, as the thin outer epidermis is sloughed off, any attached microorganisms are removed with the shed epidermal cells. The skin also is a poor source of nutrients for invading pathogens, and its low water content presents a veritable desert to pathogens. Unless pathogens can penetrate the skin barrier, infection and disease are rare, although some superficial skin ailments caused by fungal pathogens (**dermatophytes**), like athlete's foot and ringworm, can establish localized infections.

▶ **Mucous Membranes.** The moist epithelial tissue and underlying connective tissue of the gastrointestinal, urogenital, and respiratory tracts compose the **mucous membranes (mucosa)**. Because these exposed

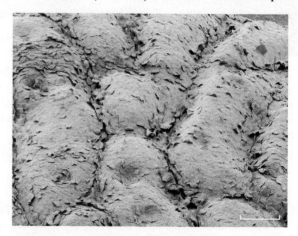

FIGURE 21.6 The Outer Surface of the Human Skin. This false-colored scanning electron micrograph shows the top layer (epidermis) of the skin. It is composed of dead cells that are continuously sloughed off and replaced with new cells from below. (Bar = 10 μm.) »» *From a potential colonization standpoint, what is the advantage of dead skin cells being sloughed off?*

© Steve Gschmeissner/Science Source.

sites represent potential portals of entry, the mucosa lining these cavities needs to provide protection from pathogen invasion. Goblet cells of the mucous membranes secrete **mucus**, a sticky secretion of glycoproteins that traps heavy particles, microorganisms, and viruses. It also separates physically the pathogens from the actual epithelium. If the pathogens are taken into the lower respiratory tract, the epithelial cells are covered with **cilia**. By beating as rhythmic waves, these hair-like structures propel the microbe-laden mucus up to the throat (called the **mucociliary elevator**), where the material is swallowed or expelled through coughing or sneezing.

Some upper respiratory infections, such as the common cold, can cause excessive mucus production, as described in MICRO-FOCUS 21.1.

Chemical Barriers

Besides being an anatomical barrier, the physical barriers secrete and employ antimicrobial chemicals to defend against pathogen invasion. These chemical effector molecules inhibit microbial growth or viral replication. Resistance in the vaginal tract is enhanced by the low pH where resident *Lactobacillus* species break down glycogen to various acids. Many researchers believe the disappearance of lactobacilli during antibiotic treatment allows opportunistic diseases such as candidiasis (vaginal yeast infection) to develop because of the lack of an acidic environment.

A natural barrier to the gastrointestinal tract is provided by stomach acid, which, with a pH of approximately 2.0, prevents colonization and the passage of most pathogens into the intestines. A notable survivor in the stomach is *Helicobacter pylori*, a major cause of peptic ulcers.

Bile from the gallbladder enters the digestive system at the duodenum and, besides emulsifying fats, is toxic to many pathogens. Bile also helps maintain the resident microbiome of the small intestine. In addition, duodenal enzymes can hydrolyze the proteins, carbohydrates, lipids, and other large molecules of microorganisms.

The chemical barriers also consist of numerous small, cationic, antimicrobial peptides called

MICROFOCUS 21.1: Miscellaneous

Going with the Flow

Every time we get a cold, the flu, or a seasonal allergy, we often end up with the sniffles or a truly raging runny nose. When this happens, it is simply your body's response to what it believes is an infection—or a hypersensitivity to cold temperatures or spicy food. As a defensive barrier against infection, innate immunity uses mucus flow as the best way to wash respiratory pathogens out of the airways.

You always are producing—and swallowing—mucus. However, most people are not aware of their mucus production until they get an infection like the common cold. When infected, the body revs up mucus secretion in response to the virus. So how much mucus is produced?

In a healthy individual, glands in the nose and sinuses are continually producing clear and thin mucus—often more than 200 milliliters (a little less than a cup) each day! An individual is not aware of this production because the mucus flows down the throat and is swallowed. Now, if

© nazira_g/Shutterstock.

that individual comes down with a cold or the flu, the nasal passages often become congested, forcing the mucus to flow out through the nostrils of the nose. This requires clearing by blowing the nose or (to put it nicely) expectorating from the throat. With a serious cold or flu, the mucus may become thicker and gooier and have a yellow or green color. The revved up mucus flow in such cases can amount to about 200 milliliters every hour; if you blow your nose 20 times an hour, each blow could amount to anywhere from 2 to 10 milliliters of mucus. If you have watery eyes as well, then that tear liquid can enter the nasal passages and combine with the mucus, producing an even larger "flow per blow."

So, although a runny nose is usually just an annoyance, make sure you drink plenty of water to make up for the lost mucus from the runny nose.

defensins that are produced by neutrophils, macrophages, and epithelial cells of the skin as well as the respiratory, gastrointestinal, and urogenital tracts. The defensins are active against many gram-negative and gram-positive bacterial cells, fungi, and some viruses where the peptides damage membranes and bring about cell lysis. The **sebum** on the skin surface contains several defensins. In the body, some defensins are produced constantly, whereas others are induced by microbial products. INVESTIGATING THE MICROBIAL WORLD 21 examines how fruit flies are used to determine the specificity of these defensins.

In a somewhat similar manner, **lysozyme** is a chemical inhibitor found in human tears, mucus, and saliva. The enzyme disrupts the cell walls of gram-positive bacterial cells, resulting in cell lysis

and bacterial cell death. Other soluble proteins, such as the complement components and interferon, are an important part of innate immunity and will be described later in this chapter.

Cellular Barriers

Innate defensive barriers also include the normal microbiome of the body surfaces. These resident, nonpathogenic microbes form a cellular barrier preventing colonization by pathogens. The human microbiome also can outcompete pathogens for nutrients and attachment sites on the skin and mucous membranes.

Langerhans cells, which are a specialized type of dendritic cell, are found under the skin and mucosa surfaces. These defensive cells recognize pathogens that have made it past the skin surface.

Investigating the Microbial World 21

Who Goes There?

Pathogens that manage to breach the body's surface barriers comprising the skin and mucous membranes are now in a position to initiate colonization and infection. To prevent this from happening, innate defenses must recognize the intruder as being an invader and activate a response to eliminate the pathogen. In other words, the host defenses must know "who goes there"—friend or foe.

OBSERVATION: The identification of a pathogen by innate immunity is often accomplished by recognizing unique molecular markers on the surface of the pathogen that are not found on the surface of host cells. This recognition results in the production and secretion of several types of antimicrobial peptides.

QUESTION: *Will the same antimicrobial peptide defend against different pathogens?*

HYPOTHESIS: The antimicrobial peptides function against a diverse set of pathogen molecular markers. If so, a single antimicrobial peptide should be effective against two different types of pathogens.

EXPERIMENTAL DESIGN: Fruit flies are excellent experimental organisms because mutant fruit flies can be produced with just about any specific phenotype. In this case, mutant flies were generated that can recognize pathogens but cannot respond by producing any antimicrobial peptides (AMP$^-$).

Four groups of fruit flies were used. Group 1 consisted of wild type flies that can respond by producing antimicrobial peptides (AMP$^+$) and group 2 was the mutant strain (AMP$^-$) mentioned above. The other two groups were AMP$^-$ flies that had been genetically engineered to produce significant amounts of the antimicrobial peptide defensin (Group 3; Def$^+$) or drosomycin (Group 4; Dro$^+$).

EXPERIMENT 1: One set of 20 flies from all four groups were infected with the fungus *Neurospora crassa* and monitored over 5 days.

EXPERIMENT 2: Another set of 20 flies from all four groups were infected with the bacterial species *Micrococcus luteus* and monitored over 5 days.

RESULTS:

EXPERIMENT 1: See figure A.

EXPERIMENT 2: See figure B.

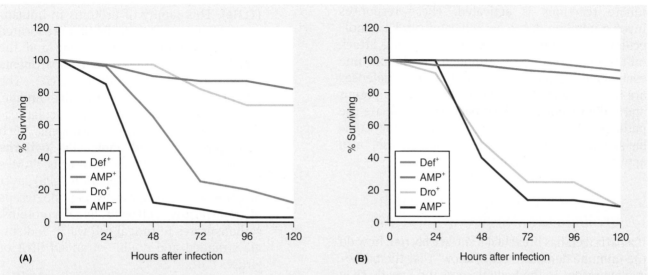

(A) Percent fly survival after infection with Neurospora crassa. **(B)** Percent fly survival after infection with Micrococcus luteus.

(A and B) Modified from Tzou, P., et al. 2002. *Proc Natl Acad Sci* USA 99:2152–2157.

CONCLUSIONS:

QUESTION 1: *In experiments 1 and 2, what do the results with AMP⁺ (group 1) and AMP⁻(group 2) flies tell you about the flies' ability to mount an antimicrobial peptide defense?*

QUESTION 2: *In experiments 1 and 2, what do the results with Def⁺ (group 3) and Dro⁺ (group 4) flies tell you about the flies' ability to use the same antimicrobial peptide against two different pathogens?*

QUESTION 3: *Was the hypothesis supported? Explain.*

You can find answers online in **Appendix E**.

Data from Tzou, P., et al. 2002. *Proc Natl Acad Sci USA* 99:2152–2157.

Concept and Reasoning Check 21.2

a. Explain how the skin, mucous membranes, and cellular barriers provide a defense against infection.

Chapter **Challenge B**

So, we have been exposed to an infectious dose (100 to 1,000 particles) of flu virus transmitted via air droplets. As such, the droplet-containing flu viruses can infect the mucosa of the respiratory tract.

QUESTION B: *What surface barriers of innate immunity attempt to keep the flu viruses out of the respiratory tract? What mechanisms are involved in the attempted elimination?*

You can find answers online in **Appendix F**.

■ **KEY CONCEPT 21.3** **Coordinated Cellular Defenses Respond to Pathogen Invasion**

As just described, the surface barriers of innate immunity guard against pathogens trying to invade and colonize the skin and mucous membranes.

Should one of these barriers be breached by a pathogen, say by a cut finger, a skin puncture, a burn, or a bite from an infected insect, a group of coordinated innate reactions is activated. These responses involve immune defensive cells and cytokine molecules to generate an effective and strong attack on the infectious agent. Although we will examine each defense separately, realize that these defenses are interconnected and flexible to provide a potent and rapid response to beat back and eliminate the pathogen. The well-known sports and military combat saying, "The best defense is a good offense," also applies to innate immunity.

Innate Immunity Depends on Recognition of Unique Pathogen-Associated Molecules

If a pathogen has breached a surface barrier, how do the immune defensive cells "know" that the pathogen is present in the body tissues, and how do they then react? Earlier, we mentioned that an effective immune reaction involves recognition, activation, and effector responses. For innate immunity, the responses go like this:

▶ **Recognition.** Recognition of an invading pathogen rests with the macrophages, neutrophils, dendritic cells, and nonimmune epithelial cells. These cells contain membrane proteins called **toll-like receptors**

(TLRs). This family of proteins in humans consists of at least 10 TLRs that are located on the cell's plasma membrane and the membrane of cytoplasmic compartments called "endosomes" (**FIGURE 21.7**). The job of the TLRs is to recognize unique, highly conserved (invariant) molecular features in or on invading pathogens. Such **pathogen-associated molecular patterns (PAMPs)** are absent from host cells. PAMPs include the lipopolysaccharide (LPS) layer of gram-negative cell walls, peptidoglycans of gram-positive cell walls, flagella proteins, carbohydrates in fungal cell walls, and single-stranded and double-stranded RNA of viruses (TABLE 21.1).

▶ **Activation.** After a TLR binds to a PAMP, the TLR triggers (activates) several processes (phagocytosis and inflammation) by the immune cells. In addition, specific antimicrobial chemicals (complement and interferon) are synthesized and released (see Figure 21.7).

▶ **Effector Outcomes.** Because of activation, the phagocytosis and inflammation processes attempt to limit or destroy

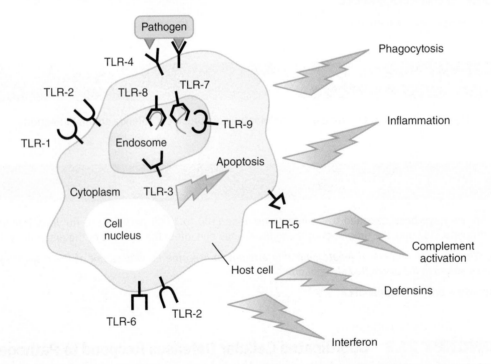

FIGURE 21.7 Pattern Recognition Receptors and the Innate Response. Proteins called toll-like receptors (TLRs) are located on the plasma membrane and endosome membrane. Binding to a pathogen-associated molecular pattern (PAMP) (e.g., bacterial wall PAMP on plasma membrane with TLR-4 or viral single-stranded RNA [purple lines] in endosome with TLR-7 and -8) initiates (lightening bolts) several processes associated with the innate immune response. »» *Why do some TLRs have to exist in the endosome?*

TABLE 21.1 The Recognized Human Toll-Like Receptors

Toll-Like Receptor	Location	PAMP (Molecular Sequences)
1	Cell surface	Bacterial cells (lipoprotein)
2	Cell surface	Gram-positive bacterial cell walls (peptidoglycan) Gram-positive bacterial cell walls (lipoteichoic acid) Fungal cell walls (zymosan)
3	Endosome membrane	Viruses (double-stranded RNA)
4	Cell surface	Gram-negative bacterial cell walls (LPS)
5	Cell surface	Bacterial flagella (flagellin)
6	Cell surface	*Mycoplasma* (lipoprotein)
7	Endosome membrane	Bacteria and viruses (single-stranded RNA)
8	Endosome membrane	Bacteria and viruses (single-stranded RNA)
9	Endosome membrane	Bacteria and viruses (DNA)
11 and 12	Endosome membrane	*Toxoplasma* cytoskeletal protein (profilin)
13	Endosome membrane	Bacterial ribosomes (RNA sequence)

the pathogens. Complement enhances the phagocytosis and inflammatory responses, while interferon (if the infection is viral) aims to destroy the viruses. Each of these effector functions is described next.

Phagocytosis Attempts to Clear Microbes from Infected Tissues

From the chapter opener, one of Metchnikoff's key observations involved the larvae of starfish. When jabbed with a splinter of wood, motile cells in the starfish gathered around the splinter. From observations like these, Metchnikoff suggested that cells could actively seek out and engulf (ingest) foreign particles in the body of the starfish larvae. Metchnikoff's theory of phagocytosis laid the foundation for one of the most important functions of innate and adaptive immunity. Let's examine the process, using a bacterial pathogen as an example.

Phagocytosis (*phago* = "eat") is the capturing, engulfing, and digesting of foreign particles by phagocytes (neutrophils, macrophages, and dendritic cells). Handling the pathogen is a four-step process (**FIGURE 21.8B**):

▶ **Attachment.** Phagocytic cells, like macrophages, are always on the lookout for foreign material in the body. Macrophages extend thin cytoplasmic protrusions, called **filopodia**, that wave like a fishing lure (**FIGURE 21.8A**). If a filopodium "catches" its victim (e.g., bacterial pathogen), the membrane of the filopodium attaches onto a PAMP, and the filopodium rapidly retracts and brings the "catch" to the immune cell surface.

On binding the pathogen, the phagocytes release chemokines that attract to the infection site neutrophils and then monocytes, which mature into macrophages, to help eliminate the pathogen.

▶ **Ingestion.** After attachment, the plasma membrane of the macrophage surrounds the microbe and an invagination, or infolding, of the plasma membrane produces a free, internalized vacuole called a **phagosome**.

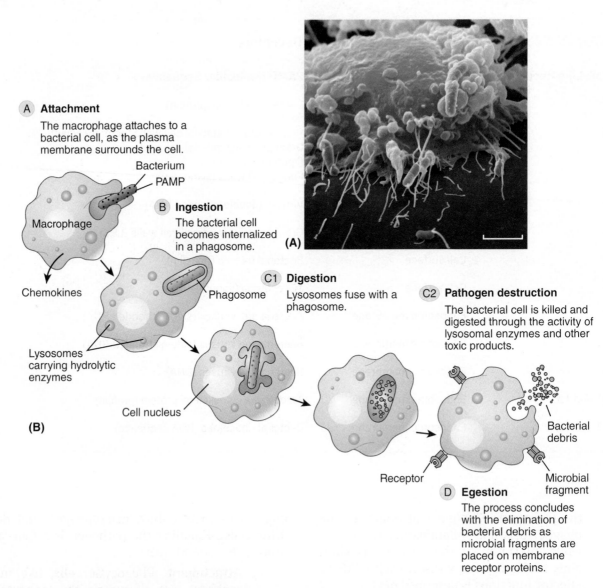

A Attachment
The macrophage attaches to a bacterial cell, as the plasma membrane surrounds the cell.

Bacterium
PAMP

Macrophage

(A)

Chemokines

B Ingestion
The bacterial cell becomes internalized in a phagosome.

Phagosome

C1 Digestion
Lysosomes fuse with a phagosome.

C2 Pathogen destruction
The bacterial cell is killed and digested through the activity of lysosomal enzymes and other toxic products.

Lysosomes carrying hydrolytic enzymes

Cell nucleus

(B)

Receptor

Bacterial debris

Microbial fragment

D Egestion
The process concludes with the elimination of bacterial debris as microbial fragments are placed on membrane receptor proteins.

FIGURE 21.8 The Mechanism of Phagocytosis. (A) A false-color scanning electron micrograph of a macrophage (gold) attaching to *E. coli* cells (pink) by means of filopodia. (Bar = 5 μm.) **(B)** The stages of phagocytosis. *»» What might the bacterial debris or waste products be in (D)?*

(A) © Eye of Science/Science Source.

▶ **Digestion and Pathogen Destruction.** The phagosome now becomes acidified and the lowered pH aids in killing or inactivating the pathogen. Lysosomes fuse with the phagosome and the lysosome's digestive enzymes, including lysozyme and acid hydrolases, digest the pathogen. In addition, phagocytosis produces a so-called **respiratory burst** that yields toxic reactive oxygen (RO) metabolites capable of damaging the pathogen membrane. These RO metabolites include hydrogen peroxide (H_2O_2), nitric oxide (NO), superoxide anions (O_2^-), and

hypochlorous acid (OCl^-) that is the active compound in bleach.

During digestion, macrophages and dendritic cells place some of the digested fragments from the pathogen on proteins located on the phagocyte's cell surface. This "presenting of antigen" will be critical to the activation of adaptive immunity.

The host–microbe relationship is always evolving. This includes the pathogen's attempts at evading, manipulating, or defeating phagocytosis, which is the topic of MICROFOCUS 21.2.

MICROFOCUS 21.2: Evolution

Avoiding the "Black Hole"

In astronomy, a black hole is a region of space that sucks in nearby matter and from which gravity prevents anything, including light, from escaping. In the immune world, the phagosome is the equivalent of a black hole. Once sucked into the phagosome, the pathogen cannot escape—well, usually. In the back and forth battles between microbe and host, some pathogens have evolved ways to avoid being killed by phagocytosis. Here are a few examples.

- Some species, such as *Streptococcus pneumoniae*, have a thick capsule, which interferes with the attachment stage of phagocytosis. More interesting, a few foodborne pathogens, such as *Shigella* and *Listeria*, make themselves invisible to the phagocyte by masking themselves with host proteins, so even if they are taken up into the phagocyte, they are not targeted for destruction.
- Once in the phagosome, the respiratory pathogen *Legionella* secretes proteins that prevent the phagosome from maturing. Therefore, the digestion stage of phagocytosis does not occur, and no destruction of the bacterial cells results.
- Lysosomes contain enzymes, proteins, and other substances capable of efficiently digesting, or being toxic to, almost any pathogen (see accompanying figure). *Yersinia*

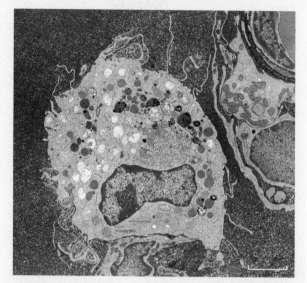

False-color transmission electron micrograph of lysosomes (purple) in a macrophage. (Bar = 4 μm.)

© MedImage/Science Source.

pestis, the bacterium causing plague, has a capsule that is very resistant to the lysosomal enzymes. As a result, the pathogen actually multiplies in the vacuole. When respiratory pathogens, such as *Mycobacterium*, *Legionella*, and *Staphylococcus*, are internalized by phagocytes, the bacterial cells prevent fusion of lysosomes with the phagosome. Without the enzymes, no digestion can occur and, like *Y. pestis*, the bacterial cells actually reproduce within the phagosome. *Toxoplasma gondii*, the protist parasite responsible for toxoplasmosis, evades the lysosomal "black hole" by enclosing itself in its own membrane vesicle that does not fuse with lysosomes.

The process of pathogenesis continues to evolve, pitting the genetic flexibility of microbes against the innate immune mechanisms of the host. As microbiologists, immunologists, and other scientists better understand these processes, the better we can devise strategies to help immune defenses make the black hole truly a death trap for all pathogens.

▸ **Egestion.** After the pathogen has been destroyed, the macrophage separates out any useful molecules for internal use and eliminates (egests) the remaining cellular debris by exocytosis.

Extracellular Traps Also Help Eliminate Pathogens

Besides phagocytosis, many neutrophils can combat infections by forming extracellular fiber-like networks called **neutrophil extracellular traps (NETs)**. When such a neutrophil binds a PAMP, the neutrophil undergoes a loss of nuclear continuity, disintegration of the nuclear envelope, and mixing of the nuclear proteins and DNA with cytoplasmic antimicrobial proteins and enzymes (**FIGURE 21.9**).

This mixing forms the NETs that are released as the neutrophil plasma membrane breaks and the cell dies. The released NETs have antimicrobial activity and, by binding to and killing the pathogens, help prevent further spreading of the pathogen in the body.

Platelets also have the ability to bind to pathogens. By recognizing PAMPs by TLRs, platelets bind to and activate neutrophils, which then quite rapidly (in minutes) form NETs. The accumulation of platelets and neutrophils at a wound form a clot, which, along with NETs, again can serve as a physical barrier preventing further spread of the pathogen. Therefore, besides chemokine communication and phagocytosis, neutrophils have a third, independent weapon to attack pathogens.

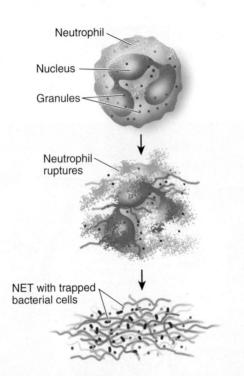

FIGURE 21.9 Formation of NETs. A neutrophil becomes activated, leading to the release of DNA and granule proteins that, on rupture, form into NETs. *»» How is innate immunity improved by generating NETs?*

Modified from Lee, Warren, L. and Grinstein, S. 2004. *Science* 303:1477–1478.

Inflammation Is the Body's Response to Injury

Acute inflammation is an immediate, short-term, but nonspecific response by the body to trauma. It is a normal process that occurs after a mechanical injury, such as a wound or blow to the skin, or from exposure to a chemical agent, such as acid or bee venom. Microbiologically, inflammation also can be a normal response to an infection. The process usually is beneficial because it eliminates invading pathogens and assists with tissue repair.

In the case of an infection, the pathogen's presence and tissue injury set into motion an innate process to limit and impede the spread of the pathogen and to localize the infection (**FIGURE 21.10**). In this example of a bacterial infection, the three-step inflammation process involves white blood cells and the cytokines and other chemical factors they release.

▸ **Trauma.** At the infection site, bacterial cells invade the injury. Resident macrophages start to engulf bacterial cells and mast cells are activated.

▸ **Inflammation.** At the infection site, the inflamed area exhibits four cardinal signs: redness (*rubor*; from increased blood flow);

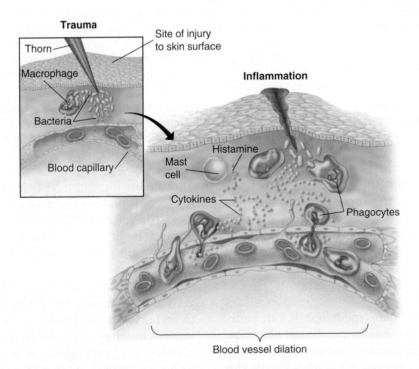

FIGURE 21.10 The Process of Inflammation in Response to Infection. A series of nonspecific steps composes the inflammatory response to trauma caused by an infection, such as being pierced by a plant thorn harboring bacterial cells. Histamine causes blood vessel dilation, which increases blood flow and defense cell influx. Neutrophils and macrophages release chemokines to attract additional defense cells. The phagocytic cells engulf the bacterial cells, helping the tissue repair process. *»» Why does the inflammatory response generate the same set of events no matter what the nature of the foreign substance?*

heat (*calor*; from the warmth of the blood); swelling (*tumor*; from the accumulation of fluid); and pain (*dolor*; from swelling, which puts pressure on nerve endings).

Specifically, resident tissue macrophages secrete several cytokines triggering local (swelling) dilation of blood vessels (**vasodilation**) and increasing blood flow (capillary permeability). Basophils and mast cells release **histamine** that also increases vascular permeability and capillary dilation. These events permit the flow of plasma into and excess fluid accumulation (**edema**) at the site of infection. Chemokines secreted by the immune cells attract more phagocytes (neutrophils and monocytes) to the infected tissue to help in the elimination of the bacterial cells. The arriving phagocytes adhere to the blood vessels near the infection site and then migrate between capillary cells (**diapedesis**) into the infected tissue.

Often during the inflammation process, pus forms. Pus consists of dead immune cells, tissue debris, pathogens, and fluid. When the area becomes enclosed in a wall of fibrin through the clotting mechanism, an **abscess**, or boil, might form.

▶ **Tissue Repair.** At the infection site, additional neutrophils augment phagocytosis as new monocytes differentiate into more macrophages. To limit tissue damage, the body tries to regulate the number of phagocytes and the level of cytokine production. Therefore, the inflammatory response must be managed so that it is short-lived. After the infection has been cleared, the body uses an elaborate mechanism to shut down inflammation and promote tissue repair.

Should the pathogens not be cleared, **chronic inflammation** can occur. The persisting cytokines can then produce excessive tissue damage, leading to a number of chronic degenerative diseases, including rheumatoid arthritis, coronary heart disease, atherosclerosis (hardening of the arteries), and cancer. With a pathogen like *Mycobacterium tuberculosis*, macrophages and lymphocytes form a granuloma (tubercle) in an attempt to "wall off" the infected and inflamed tissue.

A consequence of an excessive (overwhelming) inflammatory response, called a "cytokine storm," is **septic shock**, which is characterized by a collapse of the circulatory and respiratory systems. In some gram-negative bacterial infections, large amounts of cell wall lipopolysaccharides (LPS) are released, which are referred to as endotoxins. The systemic spread of LPS through the body triggers macrophages to secrete the "overdoses" of cytokines. Shock can lead to death.

Moderate Fever Benefits Host Defenses

Fever is an abnormally high body temperature that remains elevated above the normal oral temperature of 37°C (98.6°F). Fever supports the immune system response by slowing the reproduction or replication of the pathogen. On the other hand, fever is metabolically costly; keeping the body temperature elevated is an energy drain and makes the heart work harder.

A fever is caused by fever-producing substances, called **pyrogens**. These substances can be specific cytokines, such as **interleukin-1** that is produced by activated macrophages (see Figure 21.4), or they can be external substances like bacterial toxins or fragments from pathogens. These pyrogens move through the systemic circulation and affect the **hypothalamus**, resulting in elevated body temperature. As this takes place, cell metabolism increases and blood vessels constrict, thus denying blood to the skin and keeping its heat within the body. Patients thus experience cold skin and chills along with the fever.

Most health experts suggest a low to moderate fever in adults can be beneficial to immune defense because the elevated temperature inhibits the rapid growth of many microbes, inactivates bacterial toxins, encourages rapid tissue repair, and enhances phagocytosis. Fever gives the immune system the opportunity to get the "upper hand" on the infection. However, if the temperature rises above 42°C (107.5°F), convulsions and death can result from protein denaturation and inhibition of nerve impulses. Infants with a fever above 38°C (100°F), or older children with a fever of 39°C (102°F), might need medical attention. MICRO-FOCUS 21.3 examines the old proverb that it is good to "feed a cold and starve a fever."

Natural Killer Cells Recognize and Kill Abnormal Cells

Besides the phagocytes, natural killer (NK) cells are another type of defensive leukocyte (see Figure 21.2). NK cells are formed in the bone marrow before they migrate to the tonsils, lymph nodes, and spleen. On stimulation, they secrete several cytokines, triggering adaptive immune responses by macrophages and other immune cells. While those events are mobilized, the NK cells move into the blood and peripheral tissues where they act as potent effector cells that can destroy tumor (cancer) cells. They also

MICROFOCUS 21.3: Being Skeptical

Feed a Cold, Starve a Fever?

One of the typical symptoms of the flu is a high fever (39 to 40°C [102.2 to 104°F]), whereas a cold seldom produces a fever. So, is there any truth to the old adage, "You should feed a cold and starve a fever?" The definite answer is—yes!

There are three reasons you should reduce food intake with a fever. First, food absorbed by the intestines could be misidentified by the body as an allergen, worsening the illness and fever. In addition, with the body already under stress, heavy eating can contribute on rare occasions to body seizures, collapse, and delirium.

Third, eating stimulates digestion of the foods. During times of increased physiological stress, digestion can overstimulate the parasympathetic nervous system, when the sympathetic nervous system (involved in constricting blood vessels and conserving heat = fever) already is active. In other words, eating tends to drop body temperature, possibly eliminating fever as a protective mechanism.

So, with the flu or other illness-induced fever, stick to bed rest and just drink plenty of liquids.

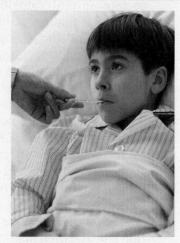

© Blend Images/Jupiterimages.

destroy pathogen-infected cells through the secretion of cytotoxic proteins that cause the death of the infected cells. The killing is precise and, like a military "surgical strike," NK cells target only abnormal or pathogen-infected cells; the surrounding normal cells usually are not damaged.

Concept and Reasoning Checks 21.3

a. Justify the need for the spectrum of toll-like receptors found on and in human immune cells.
b. Discuss the importance of phagocytosis to innate immunity.
c. Explain how inflammation serves as an innate immune defense.
d. Assess the advantage of a low-grade fever during an infection.
e. How do natural killer cells assist the other innate defense mechanisms during infection by a pathogen?

Chapter Challenge C

Unfortunately, the surface barriers did not hold, and the flu viruses have breached the mucous membranes of the nose and upper respiratory tract. In fact, we are now beginning to feel the signs and symptoms: sudden onset of fever, body aches, headache, and fatigue.

QUESTION C: *What cellular defenses of innate immunity are now activated in trying to limit the spread of the infection? How is this related to headaches and muscle aches?*

You can find answers online in **Appendix F**.

▄ KEY CONCEPT 21.4   Effector Molecules Damage Invading Pathogens

Besides the cellular processes of phagocytosis and inflammation, innate defenses also produce and release a variety of antimicrobial proteins to attack pathogens or infected cells. We have already mentioned the defensin peptides. In this last section, we examine the complement proteins and interferons, which, unlike the defensins, are unique molecules specific to vertebrate immune systems.

Complement Marks Pathogens for Destruction

Complement (also called the complement system) is a group of nearly 30 normally inactive proteins that are produced in the liver and that circulate in the bloodstream and tissues. If a microbe penetrates the body, complement proteins become sequentially activated through a cascade of steps that enhance the inflammatory response and phagocytosis. Additional complement proteins in the cascade also help in the destruction of the invading bacterial cells and some viruses through cell or virus lysis. The three pathways for complement activation are illustrated in **FIGURE 21.11**.

All these pathways—the classical, alternative, and lectin—culminate in the activation of an enzyme that splits complement protein C3 into

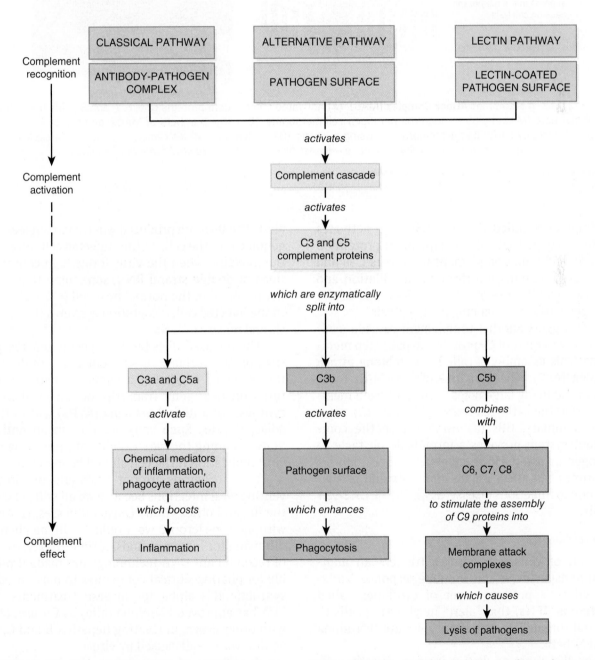

FIGURE 21.11   A Concept Map of the Complement Pathways. The early events of all three pathways of complement activation culminate in the splitting of C3 and C5. »» *What is the advantage of having three pathways leading to complement activation?*

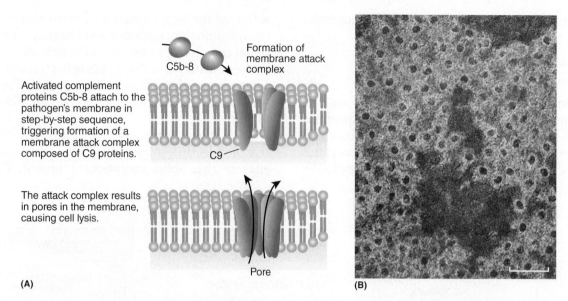

FIGURE 21.12 A Membrane Attack Complex (MAC). (A) Activated complement proteins C5b, 6, 7, and 8 guide the formation of a tubular pore (MAC) by C9 through the pathogen cell membrane. **(B)** Large numbers of MACs, as seen by transmission electron microscopy, perforate the membrane, causing cell lysis. (Bar = 50 nm.) *»» Would bacterial species that are gram positive or gram negative, and viruses that are enveloped or nonenveloped, be most sensitive to MAC formation? Explain.*

two fragments called C3a and C3b. C3b activates another enzyme that splits complement protein C5 into C5a and C5b. Complement fragments C3a and C5a boost inflammation through vasodilation and increased capillary permeability. C3b binds to the pathogen surface, enhancing phagocytosis.

C5b triggers another complement cascade leading to the assembly of C9 proteins on pathogen membranes. This assembly is called a **membrane attack complex (MAC)**. When hundreds of these MACs form, there are so many large holes in the pathogen membranes that the pathogen lyses (**FIGURE 21.12**).

In summary, this teamwork within the complement system provides a formidable obstacle to pathogen invasion. However, genetic abnormalities sometimes can affect innate immunity and predispose one to infectious disease, as **CLINICAL CASE 21** details.

Interferon Puts Cells in an Antiviral State

Infection of cells with viruses triggers an additional nonspecific, innate immune response. Virus-infected cells produce a set of cytokines called **interferons (IFNs)** that "alert" neighboring cells to the viral threat. Two important IFNs are IFN-alpha and IFN-beta.

The IFN proteins do not interact directly with viruses; rather, they interact with the cells that they

alert. The IFNs are produced when a virus releases its genome into the cell. A virus-infected cell recognizes the infection when the virus forms high concentrations of double-strand RNA, something that would not be found in the normal host cell (**FIGURE 21.13**). In the infected cell, recognition activates the synthesis and release of IFNs.

The released IFNs bind to specific IFN receptor sites on the surfaces of adjacent cells. Binding has the effect of triggering a signaling pathway that turns on new gene transcription involved in the synthesis of **antiviral proteins (AVPs)** within those adjacent cells. Such cells are now in an **antiviral state**, meaning they are capable of interfering with viral replication if the cell should be infected.

The interferons are not 100% effective in preventing viral infections because we all still get colds, the flu, and other virally caused illnesses. However, without interferons, we would be ill much more often and for longer periods of time. Today, the use of recombinant DNA technology has made it possible for pharmaceutical companies to mass-produce synthetic IFN-alpha for disease treatments. The FDA has approved FN-alpha (alfa) 2a for use, along with other drugs, in treating hepatitis B and C, and genital warts—all caused by viruses.

FIGURE 21.14 summarizes the innate barrier mechanisms.

Clinical Case 21

A Clinical History of Bacterial Infections

A 4-year-old boy was taken by his parents to his pediatrician. The boy was suffering another bacterial infection, a recurrent problem over the last few years, including two hospitalizations for bacterial pneumonia. The pediatrician took several specimens for testing, placed the young boy on antibiotics, and asked the parents to bring back their son in 3 days for the test results.

On his return, the pediatrician informed the parents that their son had a *Haemophilus influenzae* type b infection but that the antibiotics once again should take care of the infection. However, the recurrence of infections, all due to gram-negative bacterial pathogens, was unusual so the pediatrician asked an infectious disease (ID) specialist to consult.

A clinical history report for the boy was given to the ID specialist, and she was told the boy's bacterial infections had always been cleared with long-term antibiotic therapy and that he was up to date on his vaccinations.

The ID specialist asked if the boy had suffered any increased susceptibility to viral or fungal infections. The answer was no; they had all been gram-negative bacterial infections. She then asked the parents if any other members in or relatives of the family had ever had a similar history of recurrent infections. The parents remarked that

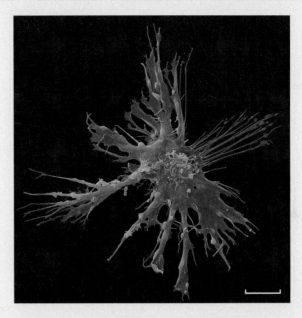

A false-colored scanning electron micrograph of a dendritic cell showing the long cell surface projections. (Bar = 20 μm.)

© David Scharf/Science Source.

two relatives, one female and one male, had had a similar condition. The ID specialist was unsure about a definitive cause for the boy's illnesses but suggested a battery of blood and immunological tests be carried out.

A blood cell count, serum antibody test, and cytokine production analysis were carried out. All results were normal, suggesting all his blood cells, including neutrophils, monocytes, and lymphocytes, were in the normal range and appeared to be functioning properly. The ID specialist was intrigued by the lack of viral or fungal infections, so she ordered tests to examine the ability of the boy's immune system to recognize infectious agents.

Functional tests for dendritic cells (see photo) and macrophages were run. The results confirmed a lack of a toll-like receptor (TLR) on the cell surface of these blood cells, thus "blinding" the white blood cells in recognizing gram-negative cells.

Questions:
a. Why hadn't the pediatrician brought in an ID specialist earlier in the boy's medical history?
b. Why would the ID specialist ask about previous viral and fungal infections?
c. What was the significance in asking about other family relatives having a similar recurrent condition?
d. In the analysis of dendritic cells and macrophages, which TLR (using Table 21.1) is most likely missing?

You can find answers online in **Appendix E**.

For additional information, see www.nature.com/nrmicro/journal/v3/n1/full/nrmicro1068.html.

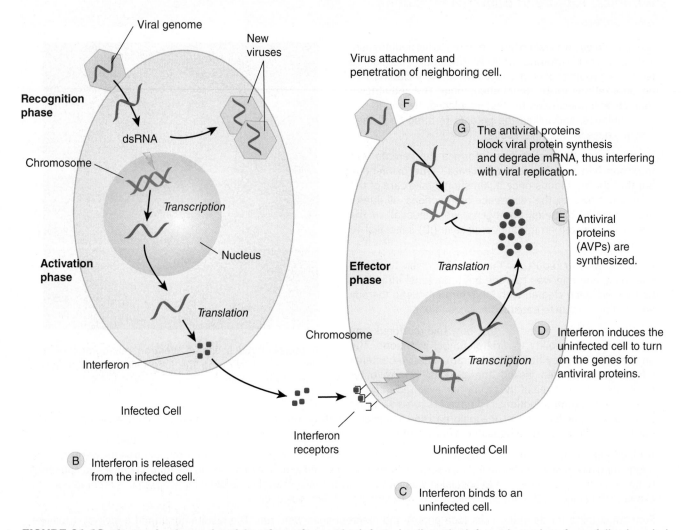

FIGURE 21.13 The Production and Activity of Interferon. The infected cell on the left produces interferon following viral infection. The neighboring cell on the right reacts to interferon binding at the cell surface by synthesizing antiviral proteins. The uninfected cell now is in an activated, antiviral state capable of inhibiting viral replication if the cell is infected by a virus. *»» Why would interferon itself not be considered an antiviral compound?*

Concept and Reasoning Checks 21.4

a. Explain the outcome to someone who has an infectious disease but has a genetic defect making the person unable to produce C5 complement protein.

b. Distinguish between the role of interferon and antiviral proteins.

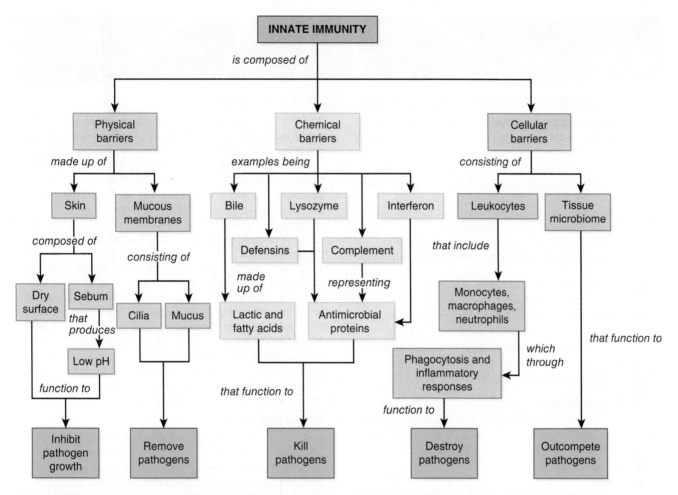

FIGURE 21.14 Innate Defense Barriers to Pathogens. This concept map summarizes the ways innate immunity is poised to inhibit, eliminate, destroy, and compete against an invading pathogen. *»» Why does there need to be such a diversity of defense mechanisms?*

Chapter **Challenge D**

The cellular defenses are having a difficult time limiting virus spread in the upper respiratory tract, and the virus infection is progressing to the lungs. In fact, we now have the classic symptoms: fever/chills, cough, sore throat, runny or stuffy nose, muscle or body aches, headache, and fatigue.

QUESTION D: *What other innate molecules are at work trying to assist the cellular defenses in stopping the spread of the flu viruses? How are these molecules attempting to work?*

Unfortunately, the innate defenses will not be sufficient to eliminate the flu viruses from our bodies. The last line of defense, adaptive immunity, will need to strike the decisive blow. That part of the challenge is addressed in Chapter 22.

You can find answers online in **Appendix F**.

In conclusion, physical barriers, immune cells, antimicrobial substances, and cytokines are needed to protect the human body, generate an effective innate immune response, and induce the adaptive immune response. **MICROINQUIRY 21** presents an overview showing the connections between innate and adaptive immunity.

MICROINQUIRY 21

Visualizing the Bridge Between Innate and Adaptive Immunity

If the physical barriers of skin or mucous membranes are breached, innate immunity involves several cells and chemical mediators:

- Macrophages, neutrophils, and dendritic cells are involved with phagocytosis and cytokine production and secretion;
- Natural killer cells can search out and destroy virus-infected cells using cytotoxic proteins;
- Defensive proteins (defensins, complement, and interferon) can lyse pathogens or interfere with virus replication; and
- Signaling receptors (the toll-like receptors) and immune system proteins (the cytokines) are instrumental in stimulating and coordinating a sustained immune defense.

Still, many pathogens in just sheer numbers, or by having appropriate virulence factors, can win the battle against the innate immune response. Therefore, it is critical for innate immunity to muster the apparatus to activate the last and ultimate defense against pathogens—the adaptive immune response. But how does the innate response do this? How does it bridge these two overlapping components of the immune system into a seamless and directed immune response?

The major link between innate and adaptive immunity actually is quite simple—and therefore rather easy to subvert if you are the human immunodeficiency virus (HIV).

The link involves the toll-like receptors and phagocytosis of the pathogen during the innate immune response (see figure, part **A**). During phagocytosis, the process not only destroys the pathogen, it also "presents" to the adaptive immune system cells (T lymphocytes) fragments of the pathogen on the phagocyte's cell surface (see figure, part **B**; also see Figurer 21.8). This fragment presentation is primarily carried out by dendritic cells, macrophages, and monocytes, so these cells are often called "antigen presenting cells" (APCs). The environment in which the T cell recognizes a pathogen fragment on an APC, the amount of pathogen fragments presented, and for how long the fragments are presented will affect the nature and strength of the T-cell response. The activated dendritic cells are the most effective triggers of T-cell and B-cell activity because the dendritic cell population has a high expression of pathogen fragment presentation and generates stimulatory molecules necessary for maximal T-cell stimulation—and adaptive immunity in general (see the accompanying figure, part **C**).

Discussion Point

Innate immunity is integral to adaptive immunity. The HIV virus can "disconnect" the bridge between the two components of the immune system by infecting and destroying T cells. How does this affect the overall function of the immune system, and what does it tell us about the ability of the innate immune system to control and eliminate viral infections?

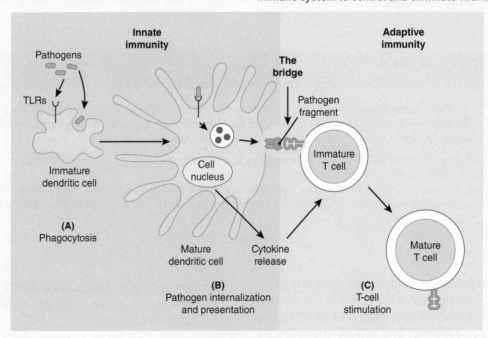

The Bridge Between Innate and Adaptive Immunity. Pathogen recognition receptors and phagocytosis are important in establishing the bridge to activation of adaptive immunity.

SUMMARY OF KEY CONCEPTS

Concept 21.1 The Immune System Is a Network of Cells and Molecules to Defend Against Foreign Substances

1. In terms of infection, the most important cells are the **leukocytes**, which consist of the **neutrophils, eosinophils, basophils, macrophages, dendritic cells**, and **lymphocytes** (**NK cells, B lymphocytes**, and **T lymphocytes**). (Figure 21.2)
2. The **lymphatic system** functions in disease to filter and trap pathogens in **lymph nodes** or other secondary lymphoid tissues populated with B cells, T cells, macrophages, and dendritic cells. (Figure 21.3)
3. A fully functional immune defense system consists of the **innate immune response** that is genetically encoded from birth and the **adaptive immune responses** that one develops during one's lifetime through exposure to pathogens. Communication within and between responses is dependent on **cytokines**. (Figure 21.4)

Concept 21.2 Surface Barriers Are Part of Innate Immunity

4. The **skin** and **mucous membranes** are physical barriers, whereas chemical defenses include the acidic environment in the urinary tract and stomach. The normal **microbiota** limits pathogen spread and infection, while **defensins** damage cell membranes and **lysozyme** destroys bacterial cells.

Concept 21.3 Coordinated Cellular Defenses Respond to Pathogen Invasion

5. **Toll-like receptors** are responsible for the ability of innate immunity to recognize **pathogen-associated molecular patterns** (**PAMPs**). (Figure 21.7)
6. **Phagocytosis** by **phagocytes** (neutrophils, macrophages, dendritic cells) internalizes and kills pathogens. (Figure 21.8)
7. **Inflammation** can be a defensive mechanism to tissue injury by pathogens. Chemical mediators, including cytokines, trigger **vasodilation** and increase capillary permeability at the site of infection. The inflamed area exhibits redness, warmth, swelling, and pain. (Figure 21.10)
8. Moderate **fever** inhibits the growth of pathogens while increasing metabolism for tissue repair and enhancing phagocytosis.
9. **Natural killer** (**NK**) **cells** are lymphocytes acting nonspecifically on virus-infected cells and abnormal host cells.

Concept 21.4 Effector Molecules Damage Invading Pathogens

10. **Complement** proteins enhance innate defense mechanisms and lyse cells with **membrane attack complexes**. (Figures 21.11, 21.12)
11. **Interferon** represents a group of antiviral proteins naturally produced by cells when infected by a virus. Release of interferon stimulates uninfected neighboring cells to enter an **antiviral state** by producing antiviral proteins. (Figure 21.13)

CHAPTER SELF-TEST

You can find answers for **Steps A–D** online in **Appendix D**.

STEP A: REVIEW OF FACTS AND TERMS

Multiple Choice

Read each question carefully and then select the *one* answer that best fits the question or statement.

1. Which pair of cells represents granulocytes?
   A. Basophils and lymphocytes
   B. Neutrophils and eosinophils
   C. Eosinophils and monocytes
   D. Lymphocytes and monocytes

2. The secondary lymphoid tissues include the ____ and ____.
   A. thymus; bone marrow
   B. bone marrow; tonsils
   C. spleen; thymus
   D. spleen; lymph nodes

3. Which one of the following statements is *not* true of innate immunity?
    A. It is an early-warning system against pathogens.
    B. It is a form of immunity found only in vertebrates.
    C. It is a nonspecific response.
    D. It responds within minutes to many infections.

4. The stomach is a chemical barrier to infection because the stomach _____.
    A. contains bile
    B. harbors *Helicobacter pylori*, a member of the host microbiota
    C. possesses defensive cells
    D. has an acid pH

5. _____ bind to _____ on microbial invaders.
    A. Toll-like receptors; PAMPs
    B. Mast cells; histamine
    C. Toll-like receptors; complement
    D. Macrophages; defensins

6. Which one of the following is the correct sequence for the events of phagocytosis?
    A. Cell attachment, acidification, phagosome formation, phagolysosome formation
    B. Cell attachment, phagosome formation, acidification, phagolysosome formation
    C. Phagosome formation, cell attachment, acidification, phagolysosome formation
    D. Cell attachment, phagosome formation, phagolysosome formation, acidification

7. Which characteristic sign of inflammation is *not* correctly associated with its cause?
    A. Edema—nerve damage
    B. Heat—blood warmth
    C. Swelling—fluid accumulation
    D. Redness—blood accumulation

8. Pyrogens are _____.
    A. proteins affecting the hypothalamus
    B. bacterial fragments
    C. fever-producing substances
    D. All the above (**A–C**) are correct.

9. Natural killer (NK) cells kill by secreting _____.
    A. lysozymes
    B. cytotoxic proteins
    C. defensins
    D. interferons

10. Which one of the following is *not* a function of complement?
    A. Stimulation of inflammation
    B. Stimulation of antibody formation
    C. Formation of membrane attack complexes
    D. Heightened level of phagocytosis

11. Which one of the following statements about interferon is *false*?
    A. Interferon is produced in response to a viral infection.
    B. Interferon is a naturally produced protein.
    C. Interferon puts uninfected cells in an antiviral state.
    D. Interferon is a protein that binds to RNA virus genomes.

Matching

Match the statement on the left to the term on the right by placing the correct letter in the available space. You can use a term once, more than once, or not at all.

Statement

_____ 12. Recognizes cancer tumors and virus-infected cells.
_____ 13. Produced in response to viral infection.
_____ 14. Small protein released by various defensive cells in response to an activating substance.
_____ 15. A monocyte matures into one of these.
_____ 16. Also called a PMN.
_____ 17. Some form holes in a target cell membrane.
_____ 18. A phagocyte that circulates in the blood and survives for several months.
_____ 19. Refers to a substance causing fever.
_____ 20. A short-lived phagocyte.

Term

A. Complement
B. Cytokine
C. Dendritic cell
D. Granulocyte
E. Interferon
F. Lymphocyte
G. Macrophage
H. Natural killer (NK) cell
I. Neutrophil
J. Pyrogen

STEP B: CONCEPT REVIEW

21. Summarize the functions for the six groups of **leukocytes**. (**Key Concept 21.1**)
22. Compare and contrast the **primary** and **secondary lymphoid tissues** in terms of function. (**Key Concept 21.1**)
23. Distinguish between **innate** and **adaptive immunity**. (**Key Concept 21.1**)
24. Describe how physical, chemical, and cellular barriers protect against pathogen invasion. (**Key Concept 21.2**)

25. Explain the importance of **phagocytosis** as a nonspecific defense mechanism. (**Key Concept 21.3**)
26. Explain why **inflammation** is a key factor in the identification of and defense against an invading microbe. (**Key Concept 21.3**)
27. Diagram the pathway showing how **interferon** stimulates an **antiviral state**. (**Key Concept 21.4**)

STEP C: APPLICATIONS AND PROBLEM SOLVING

28. A roommate cuts his finger and develops an inflammation at the cut site. Having taken microbiology, he asks you to explain exactly what is causing the throbbing pain and the warmth at the cut site. How would you reply to your roommate's question?

29. On a windy day, some dust blows in your eye and your eye waters. Why does your eye water when something gets into your eye, and how does this relate to immune defenses?
30. A friend is ill with an infection and asks you why she has "swollen glands" behind the jaw. As a microbiology student, what would you tell her?

STEP D: QUESTIONS FOR THOUGHT AND DISCUSSION

31. The opening of this chapter suggests that for many diseases, a penetration of the mechanical barriers surrounding the human body must take place. Can you think of any diseases where penetration is not a prerequisite to illness?
32. It has been said that no other system in the human body depends and relies on signals as greatly as the immune system. From the discussion of innate immunity, what evidence can you offer to support or reject this concept?

33. Phagocytes have been described as "bloodhounds searching for a scent" as they browse through the tissues of the body. The scent they usually seek is a chemotactic factor, a peptide released by bacterial cells. Does it strike you as unusual that bacterial cells would release a substance to attract the "bloodhound" that will eventually lead to the bacterial cell's demise? Explain.

CHAPTER 22

Resistance and the Immune System: Adaptive Immunity

In 1981, an immunodeficiency disease among gay men was reported. After there was an alarming rise in the number of new cases of what became known as **acquired immunodeficiency syndrome (AIDS)**, the race was on to discover its cause and how the disease was spread.

Great breakthroughs in any field of science are infrequent. Although most experiments advance our understanding of a phenomenon, more often than not experiments result in a "yes" or "no" answer to the hypothesis being tested. Then again, sometimes an experiment or observation not only gives a yes or no answer but also makes a giant leap forward in scientific understanding. The observations and work carried out by Pasteur and Koch on germ theory in the late 1800s are just one historical breakthrough that comes to mind. Another modern breakthrough was the discovery of the virus that causes AIDS and how the virus infects the body.

The eventual appearance of AIDS in distinctly different populations including homosexual and heterosexual individuals, children, intravenous drug users, hemophiliacs, and blood transfusion recipients suggested it must be an infectious agent. Moreover, as the number of AIDS cases continued to rise,

so did the hypotheses about its possible causation.

Luc Montagnier, Françoise Barré-Sinoussi, and their colleagues at the Pasteur Institute in Paris were the first to report the discovery of the virus now called the **human immunodeficiency virus (HIV)** associated with AIDS. Coupled with this breakthrough was the work of Robert Gallo and his research team at the National Cancer Institute in Maryland. Gallo's group showed that the virus identified by Montagnier was the cause of AIDS.

So, what was the virus doing to cause the disease? As the clinical descriptions of AIDS solidified, it became apparent that the virus infected T lymphocytes, a group of white blood cells key to the control of immune function (see the chapter opening image). HIV infection caused a dramatic decrease in the T-cell number (**FIGURE 22.1**).

Finding the cause of AIDS led to the development of a blood test to identify HIV, which prevented millions of people from being infected through the transfusion of tainted blood. Knowing how the virus interacts and replicates in T cells stimulated the

A T cell (blue-green) being infected by HIV (yellow).
© Science Source.

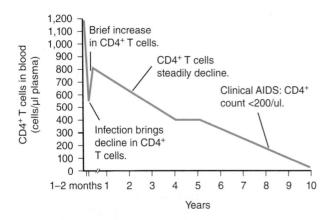

FIGURE 22.1 T-Cell Decline with HIV. HIV infection brings about an abrupt drop and rebound in T-cell number. As HIV disease progresses, the T-cell number slowly declines. Once the cell count falls below 200/ul, the person is said to have AIDS. *»» Using your knowledge of virus replication, how might HIV lead to a decline in the T-cell population?*

identification and design of many antiretroviral drugs to combat the disease, allowing infected individuals to live longer and to have more productive lives.

Most scientists today would agree that the discoveries made by Montagnier and Gallo rank as one of the major scientific breakthroughs of the 20th century. In 1987, both Montagnier and Gallo were given equal credit for the discovery. However, in 2008, the Nobel Prize in Physiology or Medicine went to Barré-Sinoussi and Montagnier.

If one set out to design a virus to cripple the immune system and the very cells designed to defend against the virus, one could not do better than the naturally evolved HIV. By destroying T cells, the entire immune system is crippled and unable to respond to infectious disease agents.

How is this possible? How can a simple virus with only nine genes inactivate a complex and powerful immune defense? This chapter provides the answer as we examine how the actions of innate immunity supply the adaptive immune response with the needed boost to fight pathogens.

Chapter **Challenge**

Recall from Chapter 21 that to better illustrate the immune system's response, we contracted the flu. For our discussion here, let's assume that we have the flu, and the cellular defenses associated with innate immunity (refer to the text in previous chapter) were unable to limit virus spread in the upper respiratory tract. The infection now has progressed to the lungs. We are exhibiting the classic signs and symptoms: fever/chills, cough, sore throat, runny or stuffy nose, muscle or body aches, headache, and fatigue. Because innate immunity did not stem the spread of the flu viruses, the last line of defense, adaptive immunity, needs to provide the force to eradicate the disease. How does adaptive immunity allow us to recover and return to normal good health? Let's find out as we recover from the flu!

■ KEY CONCEPT 22.1   The Adaptive Immune Response Targets a Specific Invading Pathogen

With regard to infectious disease, innate immunity is designed to nonspecifically eliminate any of a diverse group of pathogens—or at least limit their spread until the other half of the immune response, **adaptive immunity**, can mobilize the body's cells and molecules to eliminate the identified pathogen. Before we examine the components and events of adaptive immunity, let's consider its key features.

The Ability to Recognize and Eliminate Pathogens Requires a Multifaceted Approach

Adaptive immunity must defend against a tremendous variety of potential pathogens. This

defense is based on the following four important properties:

▶ **Specificity**. Adaptive immunity has the ability to recognize a diverse group of foreign (nonself) substances, including pathogens. Such microbes and their molecular parts that are capable of mobilizing the immune system and provoking an immune response are called **antigens** (also called **immunogens**).

Adaptive immunity does not recognize an entire pathogen. Rather, antigens are present on the surface of the microbe or virus to which immune cells or antibodies can bind

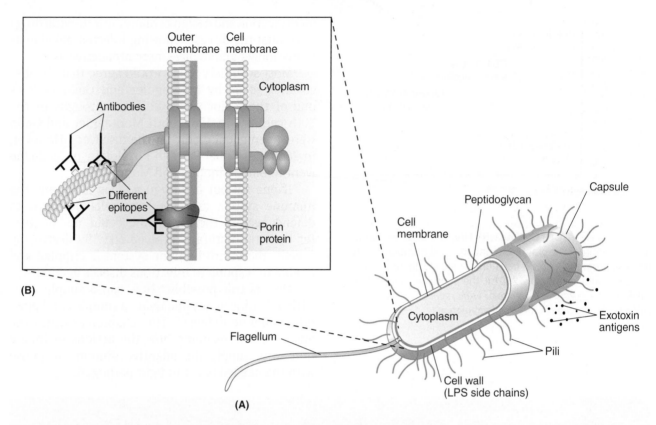

FIGURE 22.2 The Various Epitopes Possible on a Bacterial Cell. (A) Each identified cell structure in this idealized bacterial cell (antigen) is capable of stimulating an immune response. **(B)** Examples of epitopes include sites on the flagellum and cell membrane to which specific antibodies can bind. »» *What organic molecule is common to most of the bacterial structures identified in (A)?*

(**FIGURE 22.2A**). More specifically, there are discrete regions of the antigen called **epitopes** (or **antigenic determinants**) that combine with immune cells or antibodies. For example, a structure such as a bacterial flagellum (the antigen) on *E. coli* can have several epitopes, each with a characteristic and distinct three-dimensional shape (**FIGURE 22.2B**). It is the job of adaptive immunity to recognize these specific microbial "fingerprints" and generate the appropriate adaptive immune response. Importantly, an adaptive response to one pathogen is specific to that pathogen only; that is, a response to an *E. coli* infection would not provide immunological protection against a flu virus infection.

Proteins are the most potent at stimulating an immune response; that is, they are strongly **immunogenic** because the amino acids have the greatest array of building blocks. Carbohydrates also are immuno-

genic, but lipids and nucleic acids are less so because they lack chemical diversity and rapidly break down in the body. It is estimated that the human immune system can respond to about 10^{14} diverse epitopes. If the body loses its ability to respond to antigens and epitopes, an **immune deficiency** will occur. AIDS is one example.

▶ **Tolerance of "Self."** Under normal circumstances, one's own cells and molecules ("self") with their own distinct three-dimensional shapes do not stimulate an adaptive immune response. This **self-tolerance** can be established in several ways. Among them is **immune suppression**, whereby **regulatory T cells (T_{reg})** prevent autoreactive T cells from attacking self.

A second mechanism of self-tolerance is through **clonal deletion**. Should any autoreactive cells bind to a self-antigen, those immune cells are genetically programmed to undergo self-inflicted cell death.

So, the immune system develops a tolerance of "self" and remains extremely sensitive to "nonself" epitopes present on pathogens, tumor cells, and, in the case of an organ transplant, to cells of other individuals of the same species. However, if some autoreactive T cells escape suppression or elimination, self-tolerance can break down, causing an **autoimmune disorder**. Lupus erythematosus and rheumatoid arthritis are two familiar examples.

Many nucleotides, hormones, peptides, and other molecules are recognized by immune cells but are not immunogenic. However, when these molecules, called **haptens** (*hapt* = "fastened") are linked (fastened) to a carrier protein that is immunogenic, the larger combination can be recognized as "nonself" and trigger an immune response to both the hapten and the carrier protein. Examples of haptens include penicillin molecules, chemicals in poison ivy plants, and molecules in certain cosmetics and dyes. These often are the cause of allergies.

▶ **Minimal "Self" Damage.** An adaptive immune response must be strong enough to eliminate the pathogen yet controlled so as not to cause extensive collateral damage to the neighboring healthy tissues and organs of the body. Therefore, most encounters between pathogen and host occur at local sites, such as the actual infection site and in the secondary lymphatic tissues, such as the lymph nodes. In fact, many of the symptoms of an infectious disease are due to collateral damage caused by the innate and adaptive immune responses and not directly caused by the invading pathogen. If the immune system does overreact, a severe and potentially fatal allergic reaction called **anaphylactic shock** can occur.

▶ **Immunological Memory.** Most of us during our lives have become immune to certain diseases such as chickenpox and measles, from which we have recovered or to which we have been immunized. This precise long-term ability to "remember" past pathogen exposures is called **immunological memory** and is a hallmark of adaptive immunity. Thanks to immunological memory, a second or ensuing exposure to the same antigenic variation of a pathogen produces such a rapid and vigorous immune response that

the person usually does not know they were exposed. Loss of immunological memory is one reason why some people suffer frequent and repeated infections.

Adaptive Immunity Generates Two Complementary Responses to Most Pathogens

The cornerstones of adaptive immunity are the **lymphocytes**. Although all lymphocytes look similar when viewed in the microscope, two types of lymphocytes can be distinguished based on their complementary adaptive immune responses, cellular function, and unique surface proteins.

Overview of Humoral Immunity

Part of adaptive immunity comprises the **humoral immune response** (*humor* = "a fluid"). This response involves the **B lymphocytes** (**B cells**), which arise from lymphoid stem cells in the bone marrow (**FIGURE 22.3**). After maturing in the bone marrow,

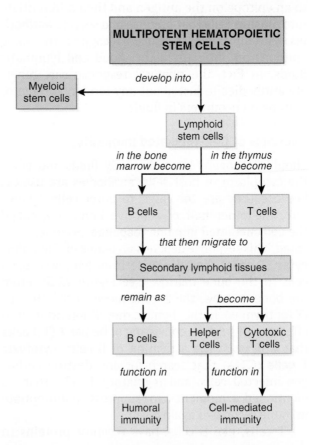

FIGURE 22.3 The Fate of Lymphoid Stem Cells. Lymphocytes arise in the bone marrow and mature in the bone marrow (B cells) or thymus (T cells). *»» Why are two different lineages of lymphoid progenitors needed?*

the B cells move through the circulation to colonize the secondary lymphoid tissues, such as the lymph nodes.

B cells recognize epitopes by means of surface receptors anchored in the plasma membrane. Within one individual's B-cell population, there is an enormous variety of receptors. However, on any one B cell, there is only one type of receptor, although there are some 100,000 identical receptors on that one cell.

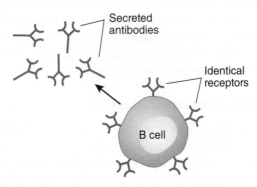

In response to an antigen, B-cell receptors bind to an epitope on the antigen and then differentiate into **plasma cells** that produce and secrete **antibodies**. The antibodies produced recognize the same epitopes on antigens in the blood and lymphatic fluids. In fact, the humoral response can generate antibodies to just about any foreign antigen or epitope it encounters in fluids.

Overview of Cell-Mediated Immunity

Should pathogens leave the body fluids and enter the cytoplasm of host cells, antibodies are useless because they are too large to enter cells. Therefore, the other half of adaptive immunity, called the **cell-mediated immune response**, becomes activated. This response depends on teams of **T lymphocytes** (**T cells**) that also arise from lymphoid stem cells in the bone marrow (see Figure 22.3). From the bone marrow, the T cells move to the thymus (T for thymus). In the thymus, they differentiate into effector cell subsets, consisting of **helper T (T_H) cells** that will assist the activation of B cells, **cytotoxic T cells** (CTLs) that search for and destroy pathogen-infected cells, and regulatory T cells that, as mentioned a moment ago, suppress inappropriate immune responses against self.

T cells, like B cells, have receptor proteins in the plasma membrane. Again, there can be some 100,000 identical receptors on a single T cell, and the entire T-cell population within an individual contains an enormous variety of receptors.

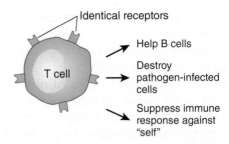

The presence of diverse and highly specific receptor proteins on the lymphocyte cell surface ensures that even before an antigen enters the body, naïve B and T cells, which are lymphocytes that have not yet encountered a pathogen (antigen), are present and waiting "in the wings" to recognize epitopes. For example, although an individual might never contract malaria, that individual has the immune system capability of recognizing and binding to the epitopes of malaria parasites should the parasites infect the body.

Clonal Selection Is a Response to an Almost Infinite Diversity of Antigens

Adaptive immunity represents a response to a specific antigen, because out of the millions of naïve B and T cells in the body, only those cells having receptors that recognize a specific epitope on an antigen (or pathogen) are activated. The stimulated cells then carry out their effector role, whereas the vast majority of the naïve B and T cells remain naïve; that is, unreactive. This immune process is called **clonal selection** and is diagrammed for B cells in **FIGURE 22.4**. A similar set of events occurs to activate T cells, which will be described in Section 22.3. As in most immune processes, clonal selection involves recognition, activation, and effector phases.

▶ **Recognition.** When antigens enter a lymphoid organ, such as a lymph node, they will eventually encounter a naïve B cell with the matching receptor. Interaction of the antigen's epitope with the appropriate B cell leads to binding of the antigen (epitope) to the receptor proteins on the B-cell surface.

▶ **Activation.** Binding of epitope, often along with additional stimulation from helper T cells, activates the B cell to divide and form a **clone**, that is, a population of genetically identical B cells.

▶ **Effector.** Most of the cells in the clone differentiate into effector cells called **plasma cells**, which synthesize and release antibodies that have an identical binding specificity

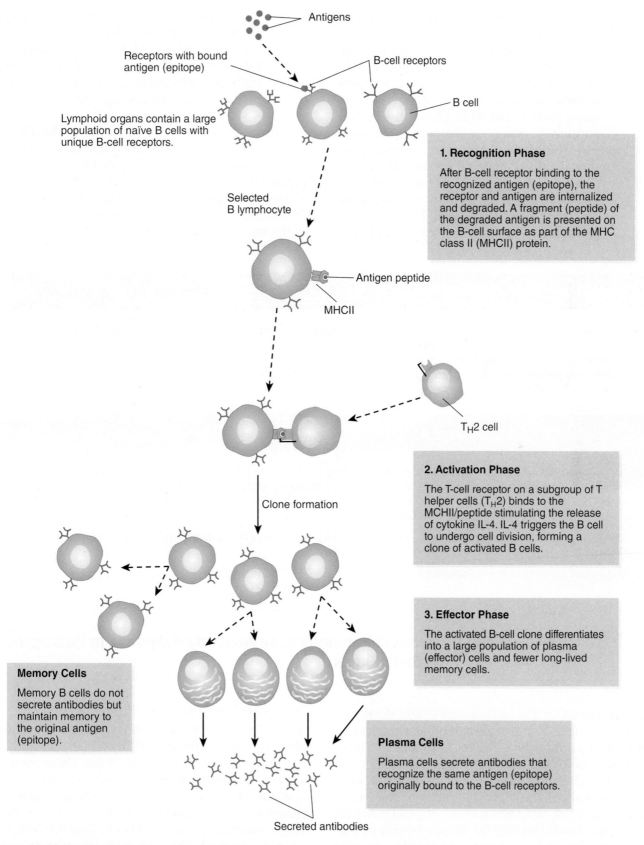

FIGURE 22.4 Clonal Selection of B Cells. In clonal selection, only those B cells that have a surface receptor that matches the shape of an epitope (not shown on antigen drawn) will be stimulated to divide, producing a clone of identical B cells. »» *Why are the plasma cells called effector cells?*

to that of the B-cell surface receptor that first bound to the epitope.

Although the majority of B cells in the clone become plasma cells, about 10% of the B cells differentiate into **memory B cells**, which are capable of surviving for many years in the body. Should the person encounter the same pathogen (antigen) with the identical epitopes a second time, the adaptive immune response will kick in much faster because of the presence of B memory cells. These memory cells are ready to respond and quickly differentiate into plasma cells that release antibodies against the infecting agent. This explains why most people are immune to many diseases after they have recovered from those diseases. Likewise, immunization against a specific pathogen stimulates an adaptive immune response and memory cell formation.

On first exposure to an antigen, the clonal selection process, from binding epitope to secreting detectable antibody, takes about 14 days. This period is called **seroconversion** time. However, once memory cells are produced, the time to seroconvert on a second or ensuing exposure of memory B cells to the same antigen can be as short as 2–3 days.

Concept and Reasoning Checks 22.1

a. Justify the statement, "The four characteristics of adaptive immunity are simply variations of specificity."
b. How do the two arms of adaptive immunity (humoral and cell-mediated) differ from each other?
c. Describe the fates for the lymphoid stem cells.
d. What are the effector cells in the clonal selection of B cells? Why are they called effector cells?

Chapter Challenge A

Adaptive immunity consists of two balancing responses to infection. Humoral immunity attacks antigens (pathogens) in body fluids, whereas adaptive immunity attacks cells that have become infected with the pathogen. Both responses aim to destroy the infectious agent. At the height of the flu, we have millions of flu viruses being replicated in and released from infected respiratory tissues.

QUESTION A: *What is the response that (a) humoral immunity and (b) adaptive immunity will mount to try to eliminate the infection of flu viruses?*

You can find answers online in **Appendix F**.

■ KEY CONCEPT 22.2   Humoral Immunity Is an Antibody Response to Pathogens in Body Fluids

In the late 1800s, the mechanisms of specific resistance to infectious disease were unknown because no one was sure how the body responded when infected. However, Paul Ehrlich, the father of antimicrobial chemotherapy, and others knew that certain proteins of the blood interact with chemical compounds of microorganisms. **INVESTIGATING THE MICROBIAL WORLD 22** illustrates one of the most significant experiments in infectious disease medicine that identified antibodies as the important factor in humoral immunity.

Antibodies Share a Common Structure

Antibodies are a class of proteins called **immunoglobulins**. The basic antibody unit (monomer) consists of four polypeptide chains (**FIGURE 22.5**). There are two identical **heavy (H) chains** (each heavy chain consists of about 400 amino acids) and two identical **light (L) chains** (each light chain has about 200 amino acids). These chains are joined together by disulfide bridges (-S–S-) to form a Y-shaped flexible structure.

Each light and heavy chain has both a constant and variable domain. The **constant domain** contains nearly identical amino acids in both light and heavy chains. Depending on the antibody class, the tail or stem of the heavy chains, called the **Fc segment** (for "fragment crystallizable"), performs various functions (e.g., recognition by phagocytes, activation of the complement system, or attachment to certain cells in allergic reactions).

The amino acids comprising the **variable domain** are different across the hundreds of thousands of antibodies produced in response to

Investigating the Microbial World 22

It's in the Blood

From the experiments of Jenner and Pasteur, in which immunity to a disease (smallpox and rabies) could be generated through vaccination, scientists in the late 1800s thought that blood in some way could render toxic substances harmless; that is, the blood provided immunity to diseases like tetanus and diphtheria. In 1890, Emil von Behring and Shibasaburo Kitasato (both in Koch's research group) published a set of experiments that today is considered one of the most important contributions that microbiology has made to medicine. These experiments provided the foundation for the concept of antibodies as the neutralizing agent in blood.

OBSERVATION: Experiments indicated that the blood of immune animals could render the tetanus and diphtheria toxins harmless, making the animals immune to the diseases.

QUESTION: *What is it about the blood that confers the immunity? Specifically, is it the blood cells or something in the serum (blood fluid without white or red blood cells)?*

HYPOTHESIS: The blood cells of immunized animals neutralize the tetanus toxin. If so, the blood serum from immunized animals will not protect against a later exposure to the same toxin because the blood cells are absent.

EXPERIMENTAL DESIGN: Immune and nonimmune rabbits and nonimmune mice were used as the test animals, and the tetanus toxin was used as the test substance. Immune rabbits were resistant to the tetanus toxin.

EXPERIMENT 1: Nonimmune mice were injected with a potentially lethal dose of the tetanus toxin. The mice were then observed for the effect of the toxin.

EXPERIMENT 2: Immune rabbits were injected with a potentially lethal dose of tetanus toxin. Blood from surviving rabbits was injected into nonimmune mice, which were then injected with the same dose of tetanus toxin as used in experiment 1. The mice were then observed for the effect of the toxin.

EXPERIMENT 3: Serum from the surviving mice in experiment 2 was injected into a group of nonimmune mice, which were then injected with either the same dose of tetanus toxin as used in experiment 1 or an equally lethal dose of diphtheria toxin. The mice were then observed for the effect of the two toxins.

RESULTS: See the accompanying figure.

CONCLUSIONS:

QUESTION 1: *What affect did the tetanus toxin have on the nonimmune mice (experiment 1)?*

QUESTION 2: *Explain why the rabbits and mice in experiment 2 survived the lethal dose of tetanus toxin.*

QUESTION 3: *Explain why the mice in experiment 3 survived or did not survive the lethal dose of tetanus and diphtheria toxins.*

QUESTION 4: *Was the hypothesis supported? Explain.*

You can find answers online in **Appendix E**.

Note: Emil von Behring won the first-ever Nobel Prize in Physiology or Medicine in 1901 for his sole work with the diphtheria toxin. Although the tetanus work with Kitasato was the key investigation, diphtheria was a more serious threat to children of the time—thus the award to von Behring.

Data from Brock, T. D. 1999. in *Milestones in Microbiology: 1546 to 1940.* ASM Press: Washington, D.C.

(continues)

Investigating the Microbial World 22 (*Continued*)

It's in the Blood

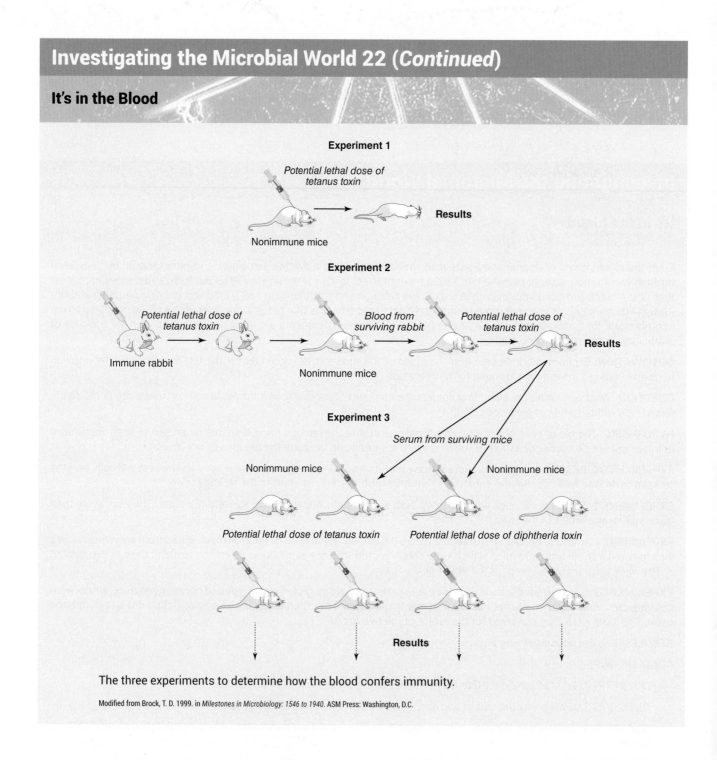

The three experiments to determine how the blood confers immunity.

Modified from Brock, T. D. 1999. in *Milestones in Microbiology: 1546 to 1940*. ASM Press: Washington, D.C.

antigens (epitopes). The variable domains in a light and heavy chain form a highly specific, three-dimensional pocket called the **antigen-binding site**. The antigen (with its epitope) binds to this region on the antibody.

The antigen-binding site is uniquely shaped to "fit" a specific epitope. Moreover, the two "arms" of the antibody are identical, so a single monomeric antibody can combine with two identical epitopes on the same or separate pathogens. As we will see, the combination of antibody and antigen becomes a target for elimination by phagocytes.

There Are Five Immunoglobulin Classes

The five classes of immunoglobulins each have unique structural characteristics. Using the abbreviation Ig (immunoglobulin), the five classes are designated IgG, IgM, IgA, IgE, and IgD. Each class

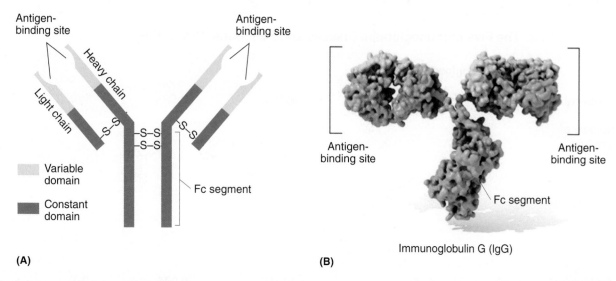

FIGURE 22.5 Structure of an Antibody. (A) An immunoglobulin molecule consists of two shorter light polypeptide chains and two longer heavy polypeptide chains. The variable domains in each light and heavy chain form a pocket called the antigen-binding site. **(B)** A three-dimensional molecular model of the antibody immunoglobulin G (IgG). *»» Using the illustration in (A), identify the light and heavy chains in (B).*

(B) Genetics Home Reference. [Internet] Bethesda (MD): National Library of Medicine (US); 2003- [updated 2004 Aug 4; cited 2006 Aug 08] Immunoglobulin G (IgG) Available from: http://ghr.nlm.nih.gov/handbook/illustrations/igg.

also has a specific function, as some antibodies are secreted into the bloodstream, others are attached onto a cell as a receptor, and yet others are deposited in body secretions. TABLE 22.1 provides an outline of each of the five classes; the list that follows offers more detailed descriptions:

▶ **IgG.** The antibody commonly referred to as "gamma globulin" represents the **IgG** class. This antibody is the major circulating antibody in serum. IgG primarily is produced in response to protein antigens, and the antibody coats the antigen (pathogen) to make the antigen more susceptible to phagocytosis. Booster injections of a vaccine raise the level of this antibody considerably in the serum. IgG is the only immunoglobulin that can cross the placenta to confer passive immunity to the fetus.

▶ **IgM.** The **IgM** (M stands for macroglobulin; *macro* = "large") class is the largest antibody molecule and, in serum, consists of a pentamer (five monomers) whose tail segments point inward. The antibody also can be found as a monomer (receptor) on the surface of B cells. IgM primarily is produced in response to polysaccharides and is the first, although short-lived, immunoglobulin to appear in the circulation after antigenic stimulation.

Thus, the presence of IgM in the serum of a patient indicates a very recent infection. Because of its large size, most IgM remains in circulation and cannot cross the placenta.

▶ **IgA.** The dimeric form (two monomers) of IgA, called **secretory IgA**, is the predominant immunoglobulin secreted. This secretion occurs through specialized epithelial cells at mucosal surfaces such as the respiratory and gastrointestinal tracts. Thus, IgA is an important part of **mucosal immunity** where it provides a first-line defense against potential pathogens, as explained in **MICROINQUIRY 22**. IgA also is located in tears and saliva and in colostrum, the first breast milk secreted by a nursing mother. When consumed by an infant, the antibodies in colostrum provide added resistance to potential gastrointestinal pathogens.

CLINICAL CASE 22 describes an immunodeficiency involving IgA.

▶ **IgE and IgD.** The **IgE** class is another monomeric immunoglobulin. It plays a major role in allergic reactions by sensitizing mast cells and basophils to certain antigens referred to as allergens. The result is the secretion of histamine and other chemical mediators responsible for the watery eyes and runny nose of hay fever and other allergies.

TABLE 22.1   The Five Immunoglobulin Classes and Functional Properties

Property	Immunoglobulin Class				
	IgG	IgM	IgA	IgE	IgD
Number of monomers	1 	5 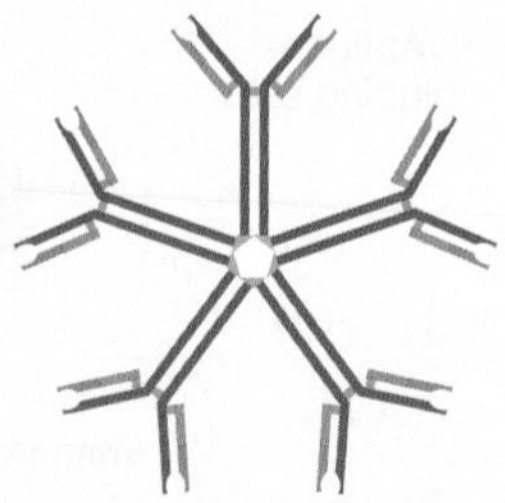	2 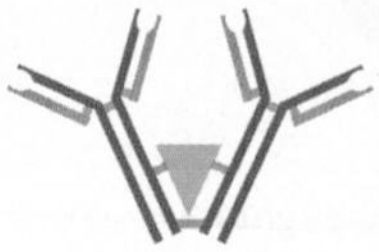	1 	1 
Antibody in serum (%)	80	5–8	10–15	0.004	0.2
Half-life[1] in serum (days)	7–24	5–10	4–7	1–5	2–8
Activates complement	Yes	Yes	No	No	No
Crosses placenta	Yes	No	No	No	No
Neutralizes bacterial toxins	Yes	Yes	Yes	No	No
Function	Principal antibody of secondary antibody response	First antibody formed in a primary antibody response	Monomer in serum; dimer secreted onto mucosal surfaces	Role in allergic reactions; effective against parasitic worm infections	Receptor on B-cell surface to initiate humoral response

[1]The time for 50% of the antibodies to be eliminated.

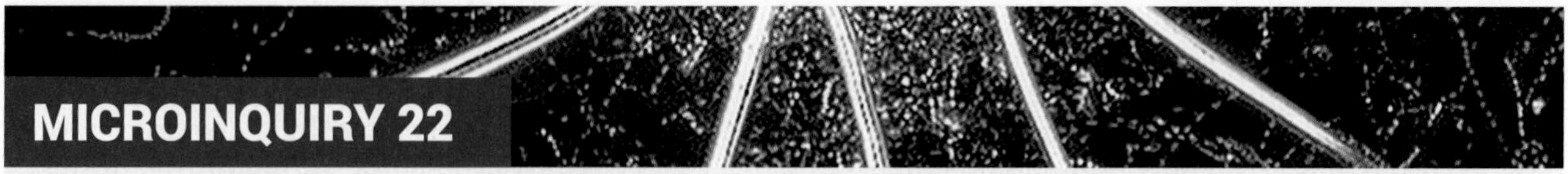

MICROINQUIRY 22

Mucosal Immunity

Most of this chapter discusses the systemic immune system, that is, the body's immune cells and chemical signals located in the blood and lymphatic systems. Until recently, however, not much was known about immunity at the body surfaces, where specialized cells and antibodies protect the body at its vulnerable portals of entry, such as the mucous membranes (mucosa) lining the respiratory, urinary, and gastrointestinal tracts. This defense is called mucosal immunity. Let's briefly look at the structure and function of mucosal immunity, using the gastrointestinal (GI) tract as our example.

The Mucosal Surface. Mucosal immunity represents an integrated network of tissues, cells, and effector molecules to protect the host from infection of the mucous membrane surfaces (see the accompanying figure and numbers as we proceed). If the epithelium is damaged by gastrointestinal pathogens such as *Vibrio cholerae*, enterotoxigenic *Escherichia coli* (ETEC), *Shigella* spp., or noroviruses, the pathogen can gain access to the mucosal tissues ①. Therefore, the mucosal surface must protect itself against such possibilities.

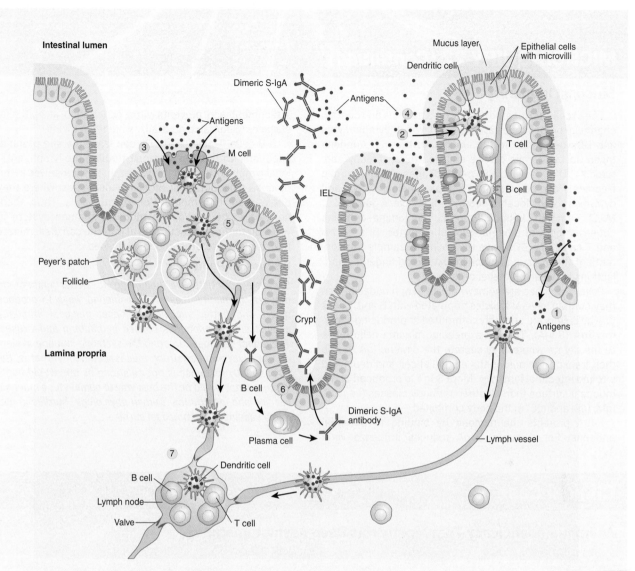

The Mucosa of the Digestive Tract. Mucosal immunity helps protect the body from colonization and infection by pathogenic and commensal microorganisms.

Modified from Wells, Jerry M. and Mercenier, A., *Nat Rev Microbiol* 6 (2008):349–362.

In the GI tract, mucosal immunity involves the **mucosa-associated lymphoid tissue (MALT)**, the largest immune organ of the body. The intestinal epithelium contains specialized **M cells** that are important to uptake, transport, processing, and possibly presentation of microbial antigens within the intestinal lumen. The epithelium also contains **intraepithelial lymphocytes (IELs)**. The lamina propria includes secondary lymphoid structures, such as the Peyer's patches, and a large concentration of macrophages, dendritic cells (DCs), plasma cells, and B and T lymphocytes. In a healthy human adult, MALT houses almost 80% of all immune cells.

Although not discussed here, the mucosa must exhibit "oral tolerance." In an environment in which immune cells are continuously exposed to large numbers of foreign but nonpathogenic antigens from foods as well as from the resident gut microbiome, inflammatory responses would damage and destroy the intestinal lining. The phenomenon of oral tolerance is a unique feature of the mucosal immune system.

MALT Function: Epithelial Layer. If a pathogen makes it through the acidic environment of the stomach, most often the pathogen ends up in the intestinal lumen or trapped in the mucus layer ②. The MALT, therefore, forms a defensive line against the pathogenic microbes, which tend to invade intestinal spaces devoid of microbiota, such as epithelial crypts and epithelial cells. The epithelial cells make up most of the cells of the intestinal lining, so they have the most contact with pathogens in the lumen. The epithelial cells can ingest and present antigen fragments to IELs scattered through the epithelium and secrete cytokines to stimulate development of the IELs. The IELs contribute to the removal of injured or infected epithelial cells.

(continues)

MICROINQUIRY 22 (*Continued*)

Mucosal Immunity

The M cells are located over Peyer's patches and contain channels or passageways through which lymphocytes and DCs get closer to the luminal surface ③. Antigens taken up by M cells can be shuttled to, or directly captured by, DCs, B cells, and macrophages. These antigen fragments can be presented to typical helper T cells and cytotoxic T cells located nearby in the Peyer's patches.

MALT Function: Lamina Propria. The lamina contains large numbers of plasma cells, phagocytes, DCs, and B and T cells. The DCs have the ability to sample the contents of the intestinal lumen by extending finger-like projections between epithelial cells ④.

Peyer's patches are somewhat like lymph nodes in that they contain several follicles populated with B and T cells ⑤. The B cells are primarily committed to producing secretory IgA (S-IgA). The IgA-expressing plasma cells move to the crypts where they secrete the dimeric IgA that is then transported across the epithelial cell and deposited into the intestinal lumen ⑥. More S-IgA is produced at the mucosal surface than all other antibody classes (i.e., IgG, IgM, IgD, and IgE) in the body combined.

S-IgA protects the mucosa by binding to potential antigens. For example, S-IgA reduces influenza virus attachment and prevents virus penetration at epithelial surfaces. S-IgA also neutralizes bacterial toxins.

⑦ B cells, T cells, and DCs can leave the site of initial encounter with antigen, transit through the lymph, enter local lymph nodes, and then reestablish themselves in the mucosa of origin, or at other mucosal sites, where they differentiate into memory or effector cells. Thus, there appears to be a common mucosal immune system in which immune cells activated at one site can disseminate that immunity to other mucosal tissues.

Discussion Point

Because scientists are beginning to better understand mucosal immunity, they are considering ways to produce newer and better vaccines to boost mucosal immunity against enteric infections caused by bacteria and viruses. Traditional vaccines targeting the systemic immune system do not always confer strong mucosal immunity; that is, the IgA antibody response is not as strong as one would hope. Discuss what would be the best way to administer a mucosal vaccine and identify the "human physiology" hurdles in getting a vaccine to the mucosal surface.

Clinical Case 22

An Immunodeficiency That Mostly Has Been Asymptomatic

An 8-year-old girl complaining of extreme and severe gastrointestinal distress is taken by her parents to her doctor. In fact, over the years the girl has experienced many but less severe cases of diarrhea and sporadic upper respiratory tract infections.

Having had all childhood vaccinations with no reported side effects, the doctor suspects there must be some underlying cause for the girl's illnesses. The mother and father have been without illness, and their daughter is an only child, so there are no other immediate family members for comparison. However, the doctor learns that two close relatives living in another part of the United States also have had recurring severe sinusitis along with gastrointestinal and upper respiratory tract infections.

Suspecting there might be an immunological disorder, the doctor orders tests for IgG antibodies to childhood vaccines and a complete blood count. IgG titers to childhood vaccines are normal and all blood cell counts (phagocytes, T cells, B cells, plasma cells) are within the normal range.

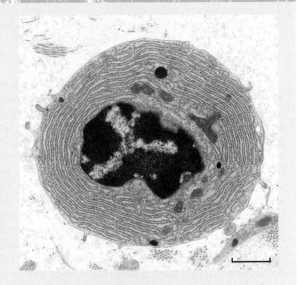

An antibody-secreting plasma cell. (Bar = 2 µm.)

© Steve Gschmeissner/Science Source.

Still believing there must be an immunological explanation, the doctor orders a complete serum immunoglobulin analysis. The results show normal IgG and IgM levels, but no IgA antibodies could be detected.

The decision is to simply treat each of the girl's future infections as it occurs. No therapy should be attempted to replace the missing IgA antibodies.

Questions:
  a. Why was it important to know if the girl had all her childhood vaccinations?
  b. Why test for antibodies to childhood vaccines?
  c. What important information is provided by the identification of other relatives with similar disease symptoms?
  d. How does the IgA deficiency correlate with the sites (respiratory, sinus, gastrointestinal) where infections are occurring?
  e. Why would any attempt to replace the missing IgA antibodies be unsafe?

You can find answers online in **Appendix E**.

For additional information, see www.pediatrics.about.com/od/primaryimmunodeficiency/a/iga_deficiency.htm.

The **IgD** class, which also exists as a cell surface receptor on B cells, has no known effector role in serum.

Antibody Responses to Pathogens Are of Two Types

We can summarize the humoral immune response by briefly examining the two two types of antibody responses to a pathogen. A **primary antibody response** represents the first time the immune system (naïve B cells) encounters a pathogen or antigen (**FIGURE 22.6**). B cells are activated and effector cells, the plasma cells, begin producing and secreting antibodies. As mentioned earlier, the time to seroconvert is about 14 days, and the measurable quantity of antibody in the bloodstream is called the **titer**. The IgM antibodies are the first to appear, but they soon are replaced by a longer-lasting IgG response.

During the primary antibody response, memory cells also are produced, which provide the immunological memory needed for subsequent encounters with

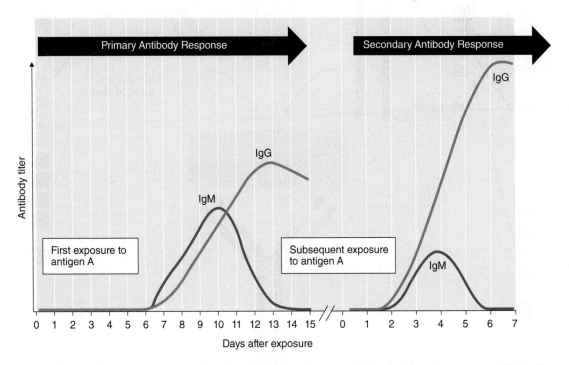

FIGURE 22.6 The Primary and Secondary Antibody Responses. After the initial antigenic stimulation, IgM is the first antibody to appear in the circulation. Later, IgM is supplemented by IgG. On a subsequent exposure to the same antigen (antigen A), the production of IgG is more rapid, and the concentration in the serum reaches a higher level than in the primary response. *»» Hypothesize why the IgM titer in the secondary antibody response is weaker than in the primary antibody response.*

the same antigen. Therefore, a second or subsequent infection by the same pathogen or antigen produces a more powerful and longer lasting **secondary antibody response**. Due to the presence of memory cells, a rapid differentiation into plasma cells generates IgG, the principal antibody. The secondary antibody response has a seroconversion time of only 2–3 days.

The secondary antibody response also is the basis for vaccination. By receiving a vaccine (immunization), the immune system responds by generating a primary immune response and the production of memory cells. Later, if the vaccinated individual encounters the "live" pathogen, memory cells are ready to trigger a secondary antibody response, and the pathogen is rapidly eliminated before any symptoms arise.

Antibody Diversity Is a Result of Gene Rearrangements

For decades, immunologists were puzzled as to how antibodies could recognize millions of different epitopes from the limited number of genes associated with the immune system—the human genome has only about 20,000 genes for all cellular functions.

The antibody diversity problem has been solved by showing that developing B cells contain about 300 genetic segments, which are "mixed and matched" and combined in each B cell as a unique arrangement of segments. This process is known as **somatic recombination**, and **FIGURE 22.7** illustrates how the process occurs.

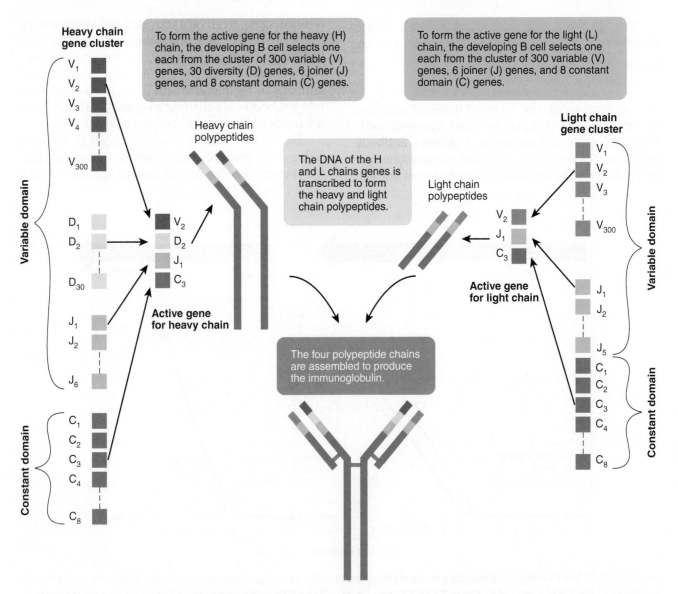

FIGURE 22.7 The Source of Antibody Diversity. Almost any substance can stimulate the production of a specific antibody. Immunoglobulin gene rearrangements provide the mechanism for this diversity. Note: Not all potential heavy and light chain segments are shown in the diagram. *»» How can gene arrangements generate such antibody diversity?*

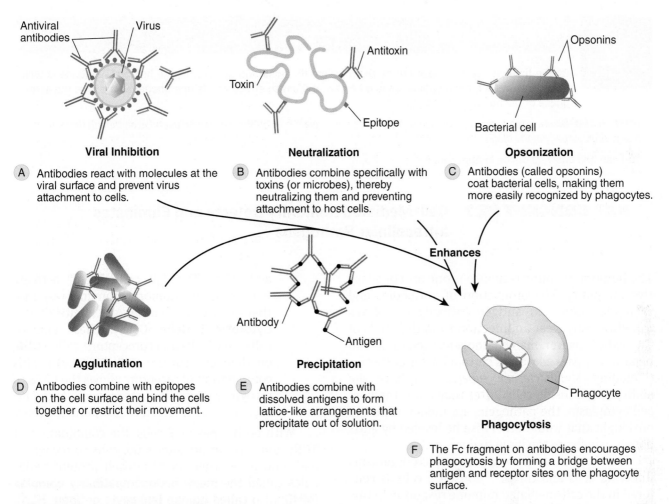

Viral Inhibition

A Antibodies react with molecules at the viral surface and prevent virus attachment to cells.

Neutralization

B Antibodies combine specifically with toxins (or microbes), thereby neutralizing them and preventing attachment to host cells.

Opsonization

C Antibodies (called opsonins) coat bacterial cells, making them more easily recognized by phagocytes.

Enhances

Agglutination

D Antibodies combine with epitopes on the cell surface and bind the cells together or restrict their movement.

Precipitation

E Antibodies combine with dissolved antigens to form lattice-like arrangements that precipitate out of solution.

Phagocytosis

F The Fc fragment on antibodies encourages phagocytosis by forming a bridge between antigen and receptor sites on the phagocyte surface.

FIGURE 22.8 Mechanisms of Antigen Clearance. In all cases, the antibody-antigen complex facilitates phagocytosis by phagocytes. *»» Reorganize these mechanisms for antigen clearance into two basic strategies (i.e., clumping of antigen versus blocking of antigen).*

Antibody Interactions Mediate the Disposal of Antigens (Pathogens)

Finally, how does the formation of **antigen-antibody complexes** result in elimination of the pathogen? **FIGURE 22.8** summarizes the "red flagging" of pathogens for disposal.

A final example of antigen-antibody interaction involves the three pathways of the complement system, described in the chapter on innate immunity. One set of reactions leads to lysis of bacterial cells through a complement cascade at the pathogen's cell surface, resulting in the formation of membrane attack complexes. These complexes form pores in the membrane, increasing cell membrane permeability and inducing the cell to undergo lysis through the unregulated flow of salts and water.

Concept and Reasoning Checks 22.2

a. How is an antibody's antigen binding site similar to the active site on an enzyme?
b. Why are there five different classes of antibodies in the body?
c. Justify the need for the two antibody responses (primary and secondary) to pathogens.
d. Why don't the constant genes for an antibody show the same diversity as the variable genes?
e. To this point, summarize the roles that phagocytes have had in the adaptive immune response.

Chapter Challenge B

By the time we are experiencing the most severe flu signs and symptoms, the humoral response is well under way. In fact, the response should be near 100% with effector cells operational by the time the acute phase of the flu infection has occurred.

QUESTION B: *Diagram the humoral response to flu viruses and explain how antibodies can be useful even though there is little, if any, virus actually in the blood.*

You can find answers online in **Appendix F**.

KEY CONCEPT 22.3   Cell-Mediated Immunity Detects and Eliminates Intracellular Pathogens

The humoral immune response is one arm of adaptive immunity. The production of antibodies and their association in antigen-antibody complexes can effectively clear an infection from the fluids of the body. However, many pathogens such as flu and hepatitis viruses as well as certain bacterial species (including *Mycobacterium tuberculosis*) have the ability to get inside cells. After they enter the host cell cytoplasm, the pathogens are hidden from the onslaught that would otherwise be leveled by antibodies, which are too large to enter cells.

The body's defense against microorganisms infecting cells and tumor ("nonself") cells is centered in the cell-mediated immune response. In this final section, we describe the roles for the T cells in recognizing and eliminating virus-infected cells and tumor cells.

Cell-Mediated Immunity Relies on T-Lymphocyte Receptors and Recognition

As we have seen, B-cell receptors directly recognize epitopes on an antigen. However, the interaction between antigen and T cells depends on sets of surface receptors capable of recognizing a multitude of antigenic (peptide) fragments. T-cell receptors are capable of such recognition because, like B cells and antibodies, they too are the result of somatic recombination.

As mentioned earlier, there are two major subpopulations of T cells involved in immune defenses:

▸ **Helper T (T$_H$) Cells.** One group of T cells, the so-called helper T cells, have T-cell receptors (TCRs) and coreceptors called **CD4** (**FIGURE 22.9A**). These cells mature into two effector classes. The **T$_H$1 cells** control many cell-mediated processes, such as cytotoxic T-cell, macrophage, and neutrophil activation. The **T$_H$2 cells** "help" activate B cells in the humoral immune response that we discussed earlier (see Figure 22.4).

▸ **Cytotoxic T Cells (CTLs).** These effector cells have TCRs and coreceptors, called **CD8**, on their cell surface (**FIGURE 22.9B**). This combination of receptors allows the CTLs to recognize and eliminate nonself antigens, such as virus-infected cells and tumor cells.

With both types of T cells, the combination of TCRs and coreceptors allows the cells to recognize and bind to another set of nonself protein molecules called the **major histocompatibility complex** (**MHC;** also called **human leukocyte antigen**; **HLA**). MHC proteins are embedded in the membranes of most cells of the body. Due to the number of MHC genes that exist, the variety of MHC proteins existing in the human population is enormous, and the chance of two people having the same MHC proteins is incredibly small. (The notable exception is in identical twins.) The MHC proteins define the uniqueness of the individual and play a role in the immune response.

Recent evidence also suggests MHC proteins might play more "amorous" roles, as MICRO-FOCUS 22.1 points out.

There are two classes of MHC proteins (see also TABLE 22.2):

▸ **MHC Class I (MHC-I).** The MHC-1 molecules are found on all nucleated cells of the body. The MHC-I molecules fold into a shape that can bind a small, internalized antigen peptide. If these proteins contain an antigen fragment from an infecting virus, the TCR/CD8 receptor complex on CTLs will bind to the "presented" MHC-I/antigen fragment (see Figure 22.9B).

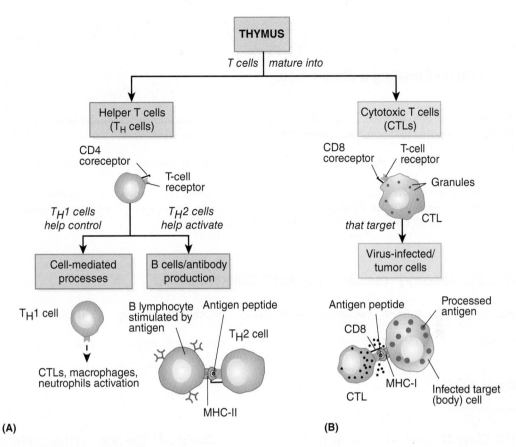

FIGURE 22.9 T-Cell Receptors. In the thymus, **(A)** helper T cells gain T-cell receptors and CD4 coreceptors, one class helping eliminate pathogens from infected antigen-presenting cells and the other class helping activate B cells and antibody production. **(B)** Cytotoxic T cells gain T-cell receptors and CD8-coreceptors that recognize antigen peptides on virus-infected cells and tumor cells. *»» What is common to the receptors of all T cells?*

MICROFOCUS 22.1: Being Skeptical

Of Mice and Men—Smelling a Mate's MHC

Would you pick a mate by sight—or smell? At least in the nonhuman world, it might be by smell.

The MHC is a group of immune system molecules essentially unique to each individual. Recent research suggests mammals, including humans, might unconsciously (or consciously?) select a mate who has a different MHC than themselves. What's the evidence?

In mice, there is a region of the nose that detects chemicals in the air that are important to mouse reproduction. Frank Zufall and colleagues at the University of Maryland at College Park have discovered that mice also use the nasal region to recognize MHC molecules found in mouse urine. The researchers believe mice might use such "markers" to identify an individual through smell—and perhaps a suitable mate, as well. In fact, mice prefer to smell mice that have a different MHC from themselves and prefer mating with mice of another MHC.

Now to the human "experience." In a human study, people were given sweaty t-shirts to smell. In recorded responses, the participants indicated they preferred smells that the researchers then linked back to different MHCs from those of the person smelling the t-shirt.

Zufall's theory is that in both mice and humans, individuals preferring a different MHC are preventing inbreeding. Therefore, mating between two individuals with different MHCs will produce offspring having a unique set of MHCs that might make that individual more resistant to factors like infectious disease.

However, most humans don't pick their mates by smelling sweaty t-shirts. So, Craig Roberts at the University of Liverpool wondered if there was a relationship between human faces and MHC. After determining the MHC for a large group, he

(continues)

MICROFOCUS 22.1: Being Skeptical (*Continued*)

Of Mice and Men—Smelling a Mate's MHC

selected 92 women and showed them photographs of six men, three with somewhat similar MHCs and three with very different MHCs as compared to the women. The women were asked if they would prefer a long- or short-term relationship with each.

Roberts was surprised to find that the women preferred the faces of men who had similar MHCs for a long-term relationship.

The verdict? There certainly is much, much more research to be done. However, Roberts proposes two mechanisms might apply in the selection of a mate. First, select a mate facially with a similar MHC. Then, smell will make sure the mate is not too similar, so more offspring that are "successful" will be produced.

TABLE 22.2 Antigen-Presenting Cells and T-Cell Recognition

Cell Type	Antigen Presented	MHC Class	T-Cell Recognizing	T-Cell Surface Coreceptor
Nucleated cells	Protein fragments	I	Cytotoxic T cell (CTL)	CD8
Macrophages, dendritic cells, B cells	Protein fragments	II	Helper T cell (Th cell)	CD4

▶ **MHC Class II (MHC-II).** The MHC-II proteins are found primarily on the surface of B cells, macrophages, and dendritic cells. Again, the MHC-II molecules have a shape in which small antigen peptides can bind. If these proteins contain an antigen fragment from an infecting pathogen, the TCR/CD4 receptor complex on the T_H2 cells will bind to them (see Figure 22.9A).

Confused? The next section explains how these responses occur and what they accomplish. As a break, read MICROFOCUS 22.2, which talks about the possibility of "thinking healthy."

Naïve T Cells Mature into Effector T Cells

The actual adaptive immune response originates with the innate immune response and the entry of antigens into the body. At their site of entry or

MICROFOCUS 22.2: Being Skeptical

Can Thinking "Well" Keep You Healthy?

In 2010, psychological scientists at the University of Kentucky and the University of Louisville wanted to see how optimism affects the immune system. Studying law students, they were interested in knowing if student expectations about their future (optimistic or not) were reflected in their immune responses. The 6-month study showed an optimistic disposition made no difference in their immune responses. However, as each student had highs and lows in law school, his or her immune response showed a similar response. In optimistic times, their immune system was quicker to respond to an immunological challenge, and in pessimistic times, their immune system was slower to respond to a similar challenge. The scientists suggest that an optimistic outlook may promote a stronger immune system.

The idea that mental states can influence the body's susceptibility to, and recovery from, disease has a long history. The Greek physician Galen thought cancer struck more frequently in melancholy women than in cheerful women. During the 20th century, the concept of mental state and disease was researched more thoroughly, and a firm foundation was established linking the nervous system and the immune system. Because of these studies, a new field called "psychoneuroimmunology" has emerged.

The outcome of these discoveries is the emergence of a strong correlation between a patient's mental attitude and the progress of disease. Rigorously controlled studies conducted in recent years have suggested that the aggressive determination to conquer a disease can increase the life span of those afflicted. Therapies consist of relaxation techniques as well as using mental imagery suggesting that disease organisms are being crushed by the body's immune defenses. Behavioral therapies of this nature can amplify the body's response to disease and accelerate the mobilization of its defenses.

In a 2009 study, scientists at Ohio State University concluded that women who practiced yoga regularly had lower levels of interleukin-6 (IL-6) in their blood than those women who did not practice yoga. Besides the benefits of IL-6 signaling as a proinflammatory cytokine produced by macrophages to stimulate an immune response to trauma and the acute phase response, when unregulated (under stress conditions), IL-6 signaling has been shown to contribute to the onset of rheumatoid arthritis, inflammatory bowel disease, osteoporosis, multiple sclerosis, and some types of cancer. In the yoga study, after a stressful experience, the women doing yoga showed smaller increases in IL-6 than their nonpractitioners.

Few reputable practitioners of behavioral therapies believe such therapies should replace drug therapy. However, the psychological devastation associated with many diseases, such as AIDS, cannot be denied, and it is this intense stress that the "thinking well" movement attempts to address. Very often, for instance, a person learning of a positive HIV test goes into severe depression, and because depression can adversely affect the immune system, a double dose of immune suppression ensues. Perhaps by relieving the psychological trauma, the remaining body defenses can adequately handle the virus.

As with any emerging treatment method, there are numerous opponents of behavioral therapies. Some opponents argue that naïve patients might abandon conventional therapy; another argument suggests therapists might cause enormous guilt to develop in patients whose will to live cannot overcome failing health. Proponents counter with the growing body of evidence showing that patients with strong commitments and a willingness to face challenges—signs of psychological hardiness—have relatively greater numbers of T cells than passive, nonexpressive patients. To date, no study has proven that mood or personality has a life-prolonging effect on immunity. Still, doctors and patients are generally inspired by the possibility of using one's mind to help stave off the effects of infectious disease.

within the lymphatic system, antigens are phagocytized by dendritic cells and macrophages. As the cells migrate to the lymph nodes or other secondary lymphoid tissues, the antigens are broken down in the cytoplasm, and antigen fragments (peptides) are displayed on the surface as MHC-II/peptide complexes (**FIGURE 22.10A**). Such cells are referred to as **antigen-presenting cells** (**APCs**) and the cell-mediate response also involves recognition, activation, and effector phases.

▶ **Recognition.** When the APCs enter the lymphoid tissue, they mingle among the myriad groups of naïve T cells (T_H0), which, like naïve B cells, have not encountered an appropriate antigen. The APCs in the tissue search for the naïve T cells having the surface receptors that recognize the MHC-II/peptide. This process requires considerable time and energy because only one cluster of T lymphocytes might have the appropriate matching T cell/CD4 receptors.

▶ **Activation.** Binding to the APC stimulates T-cell activation (**FIGURE 22.10B**). The T cell bound to APC secretes cytokines, including **interleukin-2** (**IL-2**), which stimulate cell division of that T cell and others activated by the APCs. The result is the production of a clone (clonal selection) of antigen-specific activated T helper cells. Some of these mature into memory T helper cells and await a future encounter with the same MHC-II/peptide.

▶ **Effector.** Activated T cells mature into one of several types of effector T cells, depending on the influence of specific cytokines (**FIGURE 22.10C**). Some mature into helper T2 (T_H2) cells that "help" in the activation of humoral immunity, which we discussed earlier. Other activated T cells mature into helper T1 (T_H1) cells. These effector cells control other adaptive and innate immune responses; that is, they trigger macrophage and neutrophil cell division as well as boost CTL division. Therefore, T_H cells are the master controllers, ensuring that an appropriate immune response is generated to the invading pathogen.

Activated Cytotoxic T Cells Destroy Virus-Infected Cells

Because body cells also can become infected with viruses, the immune system has evolved a way to attack virus-infected cells. During the infection,

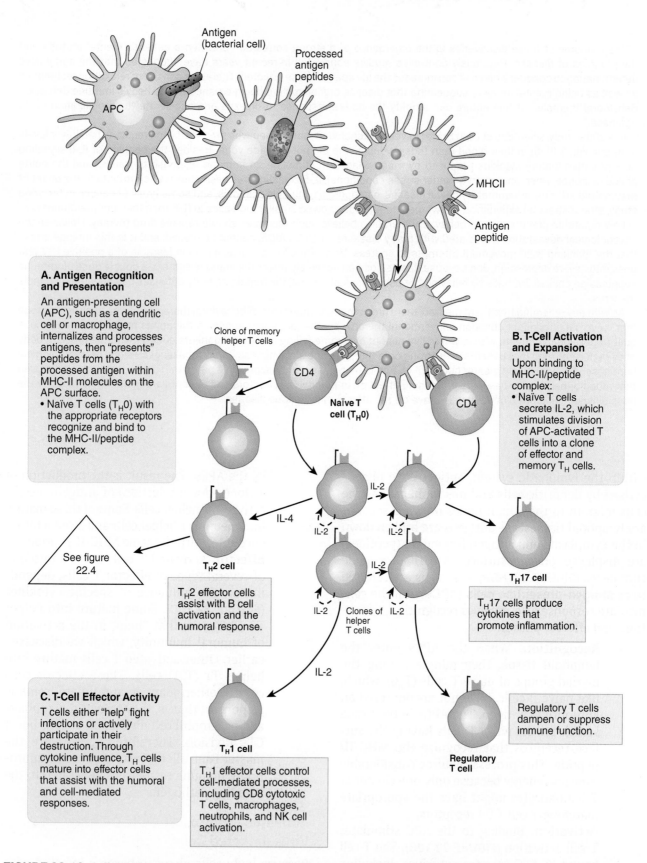

A. Antigen Recognition and Presentation

An antigen-presenting cell (APC), such as a dendritic cell or macrophage, internalizes and processes antigens, then "presents" peptides from the processed antigen within MHC-II molecules on the APC surface.
• Naïve T cells (T_H0) with the appropriate receptors recognize and bind to the MHC-II/peptide complex.

B. T-Cell Activation and Expansion

Upon binding to MHC-II/peptide complex:
• Naïve T cells secrete IL-2, which stimulates division of APC-activated T cells into a clone of effector and memory T_H cells.

C. T-Cell Effector Activity

T cells either "help" fight infections or actively participate in their destruction. Through cytokine influence, T_H cells mature into effector cells that assist with the humoral and cell-mediated responses.

T_H2 effector cells assist with B cell activation and the humoral response.

T_H17 cells produce cytokines that promote inflammation.

T_H1 effector cells control cell-mediated processes, including CD8 cytotoxic T cells, macrophages, neutrophils, and NK cell activation.

Regulatory T cells dampen or suppress immune function.

See figure 22.4

FIGURE 22.10 Cell-Mediated Immunity: Helper T Cell Activation and Effector Activity. Triggered by antigen peptide presentation, naïve T cells become activated and develop into specialized subtypes of Th cells important to adaptive (humoral and cell-mediated) immunity and innate immunity. *»» How do T_H1 cells and T_H2 cells differ in their control of immune responses?*

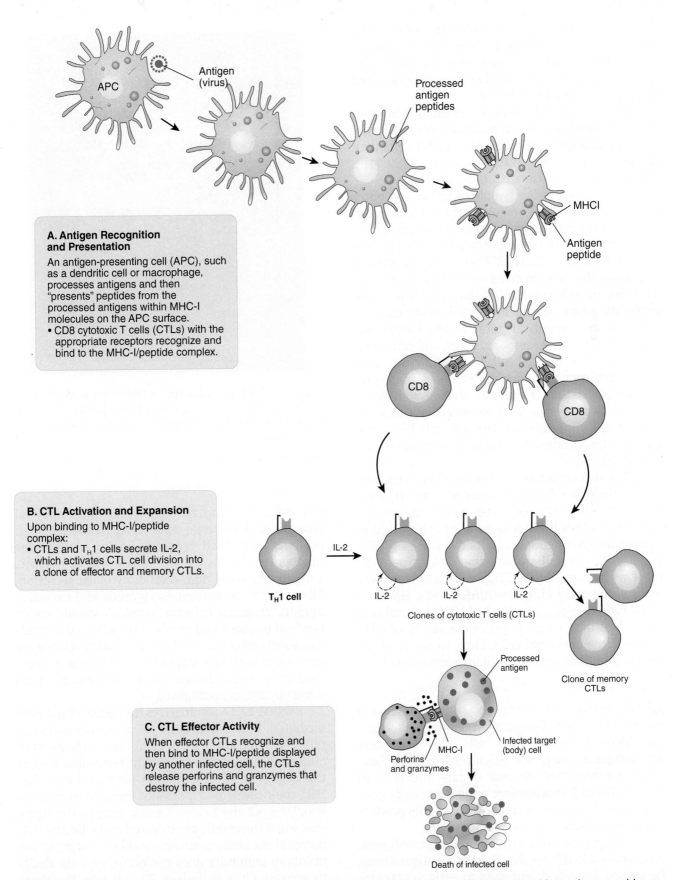

A. Antigen Recognition and Presentation

An antigen-presenting cell (APC), such as a dendritic cell or macrophage, processes antigens and then "presents" peptides from the processed antigens within MHC-I molecules on the APC surface.
• CD8 cytotoxic T cells (CTLs) with the appropriate receptors recognize and bind to the MHC-I/peptide complex.

B. CTL Activation and Expansion

Upon binding to MHC-I/peptide complex:
• CTLs and T$_H$1 cells secrete IL-2, which activates CTL cell division into a clone of effector and memory CTLs.

C. CTL Effector Activity

When effector CTLs recognize and then bind to MHC-I/peptide displayed by another infected cell, the CTLs release perforins and granzymes that destroy the infected cell.

Antigen (virus)

Processed antigen peptides

MHCl

Antigen peptide

CD8

CD8

T$_H$1 cell

IL-2

IL-2   IL-2   IL-2

Clones of cytotoxic T cells (CTLs)

Clone of memory CTLs

Processed antigen

Infected target (body) cell

MHC-I

Perforins and granzymes

Death of infected cell

APC

FIGURE 22.11 Cell-Mediated Immunity: Cytotoxic T Cell Activation and Effector Activity. Triggered by antigen peptide presentation, naïve cytotoxic T cells (CTLs) become activated and primed to destroy virus-infected or tumor cells. **»» *Why are CTLs also called killer T cells?***

the host cells manage to degrade some of the viral antigens into small peptide fragments. These are attached onto MHC-I proteins and transported to the cell surface where they are displayed like "red flags" to denote an infected cell.

Activation of CD8 T cells occurs through interaction with dendritic cells or macrophages displaying a MHC-I/peptide (**FIGURE 22.11A**). CD8 T cells then divide into a clone of activated CTLs (**FIGURE 22.11B**). CTL memory cells also are produced.

The CTLs leave the lymphoid tissue and enter the lymph and blood vessels. They circulate until they come upon their target cells—the infected cells displaying the telltale MHC-I/peptide on their surface (**FIGURE 22.11C**). The CTLs bind to the MHC-I/peptide on the virus-infected cell surface and release a number of active molecules. Toxic proteins, called **perforins**, insert into the membrane of the infected cell, forming cylindrical pores in the membrane. This "lethal hit" releases ions, fluids, and cell structures from the infected cell. In addition, the CTLs release **granzymes**, which are enzymes that enter the target cell and trigger cell suicide. Cell death not only deprives the viral pathogen of a place to survive and replicate but, through cell lysis, also exposes the pathogen to antibodies in the extracellular fluid.

CTLs also are active against tumor cells because these cells often display distinctive (nonself) molecules on their surfaces. The molecules are not present in other body cells, so they are viewed as foreign antigen peptides. Harbored within MHC-I proteins at the cell surface, the antigen peptides react with receptors on CTLs, and the tumor cells are subsequently killed (**FIGURE 22.12**). However, some tumors reduce the level of MHC-I proteins at the cell surface, which impedes the ability of CTLs to "find" the abnormal cells. Thus, tumor cells can escape immunological surveillance and survive.

Some Antigens Are T-Cell Independent

All the examples of antigens we have discussed so far are called **T-dependent antigens** because they depend on T helper cells. However, there are antigens, such as polysaccharides in bacterial capsules that do not require the help of T cells. These **T-independent antigens** usually generate a weaker immune response, primarily producing IgM antibodies and no memory cells.

Another group of antigens is the **superantigens**. As described in this section, "regular" antigens must be broken down and processed to antigen peptides before they are presented on the APC's cell surface

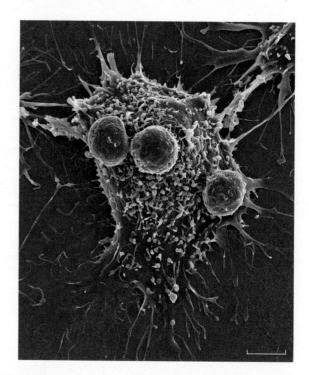

FIGURE 22.12 A "Lethal Hit." A false-color scanning electron micrograph of three cytotoxic T cells (dark pink) attacking a tumor cell. (Bar = 10 μm.) »» *How do cytotoxic T cells "find" a tumor cell?*

© Steve Gschmeissner/Science Source.

(**FIGURE 22.13**). Superantigens, such as certain viral proteins and bacterial exotoxins, cross-link MHC proteins and T-cell receptors. As a result, large numbers of T cells become active and proliferate, with an unusually high level of cytokine secretion. The result is an extremely vigorous and excessive immune response called a "cytokine storm," which can lead to shock and death of the affected individual. Superantigens and massive cytokine release are associated with the staphylococcal toxins causing food poisoning and toxic shock syndrome and with some viruses, as spotlighted in MICROFOCUS 22.3.

FIGURE 22.14 summarizes the humoral and cell-mediated immune responses of adaptive immunity. Note that the T helper cells are at the heart of almost all responses. You might remember that at the beginning of this chapter we mentioned that HIV could destroy a person's immune system by knocking out the helper T cells. Note in the figure that when these cells are infected and killed by HIV, humoral immunity is adversely affected, stimulation of innate immunity does not occur, and the ability to activate CTLs is limited. That is why the chapter opener referred to HIV as the perfectly designed

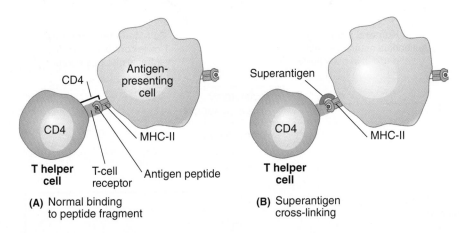

(A) Normal binding to peptide fragment

(B) Superantigen cross-linking

FIGURE 22.13 Superantigen Binding. (A) Normal binding between APC and T cell requires an antigen peptide that was processed within the APC. **(B)** The presence of a superantigen cross-links the T-cell receptor and the MHC-II receptor on the APC, triggering T-cell activation. **»» *What type of molecule (carbohydrate, lipid, protein, or nucleic acid) is a superantigen? Explain.***

virus to cripple the immune response. As the T-cell population drops, cell-mediated immunity becomes less able to respond to pathogens. Eventually, there are so few T cells that the entire immune response collapses. Chapter 24 discusses much more about HIV and AIDS.

In conclusion, Elie Metchnikoff and Paul Ehrlich were jointly awarded the Nobel Prize in Physiology or Medicine in 1908 *"in recognition of their* [groundbreaking] *work on immunity."* In the ensuing 110 years, great progress has been made in

understanding how the body fights infectious disease. It has been more than 200 years since the first vaccine was developed by Edward Jenner to provide immunity to smallpox and some 130 years since Pasteur applied his immunization method to protect animals against anthrax and people against rabies. In both cases, Jenner and Pasteur had no idea how their vaccines worked or how the immune system protected vaccinated individuals against these diseases. Today, vaccines are one of the backbones of infectious disease prevention.

Concept and Reasoning Checks 22.3

a. How can the different T cells be separated based on receptors and function?
b. Justify the need for two populations of "helper T cells."
c. Summarize the events leading to cell death of infected cells.
d. What are T-independent antigens, and how are they similar to superantigens?

Chapter **Challenge C**

Besides the activation of the humoral response to the flu virus, the cell-mediated arm of adaptive immunity also is well under way.

QUESTION C: *What two types of effector T cells will be critical in the attempt to rid the body of the flu viruses? Explain how each cell type works.*

Moreover, thanks to our immune system we have made it through the acute phase of infection and disease triggered by the flu virus. We are well on our way through the period of convalescence and we'll be back in class on Monday!

You can find answers online in **Appendix F**.

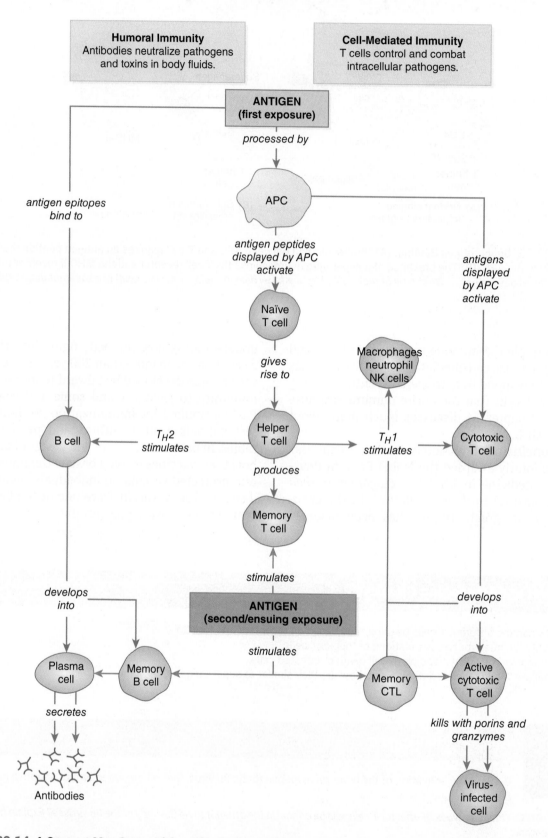

FIGURE 22.14 A Concept Map Summarizing Adaptive Immunity. Humoral immunity produces antibodies that respond and bind to antigens, while cell-mediated immunity stimulates helper T cells that activate B cells and cytotoxic T cells that bind to and kill infected cells or other abnormal cells. Memory B and T cells are important in a second or ensuing exposure to the same antigen (red lines). »» *How would an HIV infection affect the immune response illustrated in this figure?*

MICROFOCUS 22.3: Infectious Disease

The "Over-Perfect" Storm

You get a cold or the flu, or a bacterial infection, and you suffer for several days. We think of pathogens as our worst enemy during an outbreak of influenza or pneumonia, and we rely on our immune system to get us over the illness. However, sometimes the immune system is worse than the pathogen when it overreacts to an infection. It can result in a potentially fatal immune response—an "over-perfect storm" in terms of disease elimination.

For reasons scientists do not completely understand, too many immune cells triggered by innate immunity can be sent to the infection site, and the cells produce more cytokines, which brings even more immune cells to the site of infection. Along with T-cell activation of cytokines, this uncontrolled immune response propagates what is referred to as a "cytokine storm" in which excessive numbers of immune cells are caught in an endless positive feedback loop that brings more and more immune cells to fight the infection. This "over-perfect" storm causes inflammation in the tissue surrounding the infection, and this is what makes the immune reaction potentially so deadly.

For example, a cytokine storm in response to a lung infection can potentially cause permanent lung damage, resulting in multiple organ system failure due, in part, to an inadequacy of oxygen reaching the body's tissues. It explains why the 1918 Spanish flu pandemic, which killed an estimated 50 to 100 million people worldwide, took the lives of a dispro-

How the H5N1 influenzavirus triggers a cytokine storm.

Modified from Osterholm, M. T., *N Engl J Med* 352 (2005):1839–1842.

portionate number of young adults—their healthy immune systems triggered a cytokine storm. A large number of the 774 individuals killed during the 2003 SARS epidemic also were the result of a cytokine storm.

Two more examples involve flu illnesses: the 2009–2010 pandemic of H1N1 (swine) flu, which did not trigger a cytokine storm, and the H5N1 avian flu that does trigger a flood of cytokines (see the accompanying figure). The latter has been responsible for 450 reported deaths among the more than 851 people (53%) that have been infected with H5N1 as of February 2017. For some reason, H5N1 viruses are "more potent inducers" of cytokines in infected epithelial cells, macrophages, and T cells than are the H1N1 viruses responsible for seasonal pandemic flu. In fact, autopsies of H5N1 flu victims in Southeast Asia have revealed that these victims had lungs filled with debris from the excessive inflammation triggered by the virus and cytokines. In lab studies, the H5N1 virus produced 10 times higher levels of cytokines from human cells in culture than did seasonal flu viruses.

Research is continuing to understand the nature of this "over-perfect" storm and how best to treat affected individuals.

■ SUMMARY OF KEY CONCEPTS

Concept 22.1 The Adaptive Immune Response Targets the Specific Invading Pathogen

1. **Adaptive immunity** requires specificity for **epitopes**, tolerance of and minimal damage to "self," and generates memory of past infections. (Figure 22.2)
2. Adaptive immunity consists of **humoral immunity** maintained by **B cells** and **antibodies** and **cell-mediated immunity** controlled by **T cells**.
3. Immune system cells arise from stem cells in the bone marrow. **Lymphoid progenitor cells** give rise to T cells that mature in the thymus and B cells that mature in the bone marrow. These cells colonize the lymphoid organs. (Figure 22.3)
4. Appropriate B- and T-cell populations are activated through **clonal selection** involving recognition of specific epitopes on antigens or antigen fragments. (Figure 22.4)

Concept 22.2 Humoral Immunity Is an Antibody Response to Pathogens in Body Fluids

5. Monomeric antibodies consist of two identical **light chains** and two identical **heavy chains**. The **variable regions** of the light and heavy chains form two identical **antigen-binding sites** where the epitope (antigen) binds. (Figure 22.5)
6. Of the five classes of antibodies, **IgM** and **IgG** are primary disease fighters, whereas secretory **IgA** is found on body (mucosal) surfaces. (Table 22.1)
7. A **primary antibody response** produces IgM and IgG antibodies and memory cells, the latter providing immunity in a **secondary antibody response**. (Figure 22.6)

8. Antibodies can be produced that recognize almost any antigen. To do this, **somatic recombination** between a large number of gene clusters (variable, diversity, joiner, constant) for the heavy and light chains occurs. (Figure 22.7)
9. The actual clearance of antigen is facilitated by antibody-antigen interactions (**inhibition, neutralization, opsonization, agglutination, precipitation**) that result in phagocytosis by phagocytes. In addition, the formation of **membrane attack complexes** by complement directly lyses and kills pathogens. (Figure 22.8)

Concept 22.3 Cell-Mediated Immunity Detects and Eliminates Intracellular Pathogens

10. Two subpopulations of T cells carry T-cell receptors. The **cytotoxic T cells** have a **CD8** coreceptor, whereas the **helper T cells** have a **CD4** coreceptor. The receptors are responsible for recognition of antigen fragments presented in **major histocompatibility complex (MHC)** proteins. (Figure 22.9)
11. Cellular reactions and cytokines mediated by T_H2 cells stimulate B cells and humoral immunity (Figure 22.10), whereas T_H1 cells triggers the activation of CTLs, macrophages, and neutrophils. (Figure 22.10)
12. **Cytotoxic T cells** are activated by APC/CD4 T-cell combination. Active cells recognize abnormal cells (virus-infected cells or tumors) presented as MHC-I proteins with bound antigen peptide. Binding of cytotoxic T cells triggers the release of **perforins** and **granzymes**, which lyse and kill the abnormal cells. (Figure 22.11)
13. Immune responses can be generated in a **T-independent** manner and by **superantigens**, for which help from T cells is not required. (Figure 22.13)

■ CHAPTER SELF-TEST

You can find answers for **Steps A–D** online in **Appendix D**.

■ STEP A: REVIEW OF FACTS AND TERMS

Multiple Choice

Read each question carefully and then select the *one* answer that best fits the question or statement.

1. All the following are immunogenic *except* ____.
   A. bacterial flagella
   B. haptens
   C. bacterial pili
   D. viral spikes

2. ____ cells are associated with ____ immunity while ____ cells are part of ____ immunity.
   A. B; cell-mediated; T; innate
   B. T; humoral; B; cellular
   C. T; cell-mediated; B; humoral
   D. T; humoral; B; nonspecific

3. Which one of the following cell types is *not* derived from lymphoid progenitors?
   A. Macrophages
   B. B cells
   C. Helper T cells
   D. Cytotoxic T cells
4. Clonal selection includes ____.
   A. antigen-receptor binding on B cells
   B. antibody secretion recognizing same epitope as on B cell receptors
   C. differentiation of B cells into plasma cells and memory cells
   D. All the above (**A–C**) are correct.
5. An antigen binding site on the IgG antibody is a combination of ____.
   A. one variable region from a light chain and one variable region from a heavy chain
   B. two variable regions from two light chains
   C. two variable regions from two heavy chains
   D. one variable region from a constant region and one from a variable region
6. This dimeric antibody class often occurs in secretions of the respiratory and gastrointestinal tracts.
   A. IgE
   B. IgM
   C. IgA
   D. IgG
7. The presence of IgM antibodies in the blood indicates ____.
   A. an early stage of an infection
   B. a chronic infection
   C. an allergic reaction is occurring
   D. humoral immunity has yet to start

8. Antibody diversity results from ____.
   A. mutation
   B. antigenic shift
   C. somatic recombination
   D. complement binding
9. A/an ____ mechanism facilitates the clearance of toxins from the body.
   A. opsonization
   B. precipitation
   C. agglutination
   D. neutralization
10. MHC class I proteins would be found on ____, whereas MHC class II proteins would be found on ____.
    A. nucleated cells; plasma cells
    B. nucleated cells; macrophages
    C. dendritic cells; neutrophils
    D. only white blood cells; red blood cells
11. T_H1 cells activate ____.
    A. B cells
    B. CTLs, macrophages, and NK cells
    C. antibodies
    D. humoral immunity
12. Perforins and granzymes are found in ____.
    A. helper T cells
    B. antigen-presenting cells
    C. cytotoxic T cells
    D. B cells
13. T-independent antigens include ____.
    A. bacterial capsules
    B. superantigens
    C. bacterial flagella
    D. All the above (**A–C**) are correct.

True-False

Each of the following statements is true (T) or false (F). If the statement is false, substitute a word or phrase for the underlined word or phrase to make the statement true.

____ 14. <u>Cell-mediated</u> immunity involves T lymphocyte activity.

____ 15. <u>IgM</u> consists of two monomers.

____ 16. A secondary antibody response primarily involves <u>IgG</u>.

____ 17. Antibodies are transported in the <u>blood</u>.

____ 18. There are <u>two</u> polypeptide chains in a monomeric antibody molecule.

____ 19. The <u>humoral</u> immune response depends on antibody activity.

____ 20. <u>Lysozyme</u> is secreted by cytotoxic T cells.

▉ STEP B: CONCEPT REVIEW

21. Identify and describe the four characteristics of an **adaptive immune response**. (**Key Concept 22.1**)
22. List the steps in **clonal selection** of B and T cells. (**Key Concept 22.1**)
23. Draw the structure of a monomeric **antibody** and label the parts. (**Key Concept 22.2**)
24. Summarize the characteristics for each of the five **immunoglobulin** classes. (**Key Concept 22.2**)

25. Distinguish between the antibody-specific mechanisms used to clear antigens (pathogens) from the body. (**Key Concept 22.2**)
26. Summarize how naïve T cells are activated, and identify how these effects contribute to cell-mediated immunity. (**Key Concept 22.3**)
27. Discuss how cytotoxic T cells are activated and how they eliminate virus-infected cells. (**Key Concept 22.3**)

■ STEP C: APPLICATIONS AND PROBLEM SOLVING

28. A microbiology professor suggests that an epitope arriving in the lymphoid tissue is like a parent searching for the face of a lost child in a crowd of a million children. Do you agree with this analogy? Why or why not?

29. In the book and classic movie, *Fantastic Voyage*, a group of scientists is miniaturized in a submarine (the Proteus) and sent into the human body to dissolve a blood clot. The odyssey begins when the miniature submarine carrying the scientists is injected into the bloodstream. Today, microscopic robots called nano-robots are being designed that would be injected into the body to fight diseases, including infectious ones. What do you think about this future microscopic robot technology?

30. As a consultant for the company Acme Nanobots, what immunological hurdles need to be considered before such microscopic robots could be fully developed?

■ STEP D: QUESTIONS FOR THOUGHT AND DISCUSSION

31. The ancestors of modern humans lived in a sparsely settled world where communicable diseases were probably very rare. Suppose that by using some magical scientific invention, one of those individuals was thrust into the contemporary world. How do you suppose he or she would fare in relation to infectious disease? What is the immunological basis for your answer?

32. Your brother's high school biology text contains the following statement: "T cells do not produce circulating antibodies. Rather, they carry cellular antibodies on their surface." What is fundamentally incorrect about this statement?

33. Some time ago, an immunologist reported that cockroaches injected with small doses of honeybee venom develop resistance to future injections of venom that would ordinarily be lethal. Does this finding imply that cockroaches have an immune system? What might be the next steps for the research to take? What does this research tell you about the cockroach's ability to survive for 3 or 4 years, far longer than most other insects?

CHAPTER 23

Immunization and Serology

Prior to the age of modern vaccines, the only way one could become immune to a disease was to catch the disease and hope for recovery. Unfortunately, the symptoms often were severe and perhaps disfiguring. There also was the risk of complications; if the disease did not kill, a complication might. In addition, ill individuals could be contagious and spread the disease to other unsuspecting and susceptible people, leading to epidemics or even pandemics.

As early as the 11th century, Chinese doctors ground up smallpox scabs and blew the powder into the noses of healthy people to protect them against the ravages of smallpox. Jenner improved on this technique by using a preparation from cowpox that was scratched into the recipient's arm.

With the development of modern vaccines, these and many other infectious diseases have been eradicated (smallpox), almost eliminated (polio; see the chapter opening image), or brought under control (diphtheria) in much of the world. Equally important, vaccination is much safer and cheaper than trying to cure a disease after it has already struck.

Today, vaccination is the most cost-effective medical intervention. In fact, the reduction in mortality and morbidity associated with vaccine-preventable diseases in the United States represents one of the greatest public health achievements (**FIGURE 23.1A, B**). Outside the United States, equally stunning results have been reported. In Africa, meningitis A has been a

major killer of children for decades. Through a collaborative immunization program to vaccinate everyone between 1 and 29 years of age, the MenAfriVac vaccine has brought the number of meningitis A cases to zero (**FIGURE 23.1C**).

Unfortunately, many people live in regions of the world where vaccines are not readily available. In 2010, the Bill and Melinda Gates Foundation pledged $10 billion over the next 10 years to help research, develop, and deliver vaccines, such as MenAfriVac, to the poorest developing nations. Unfortunately, some individuals and groups have elected not to use the opportunity of vaccination to protect themselves and their family. The result has been outbreaks of disease.

Thanks to vaccination against pertussis (whooping cough), cases of the disease in the United States reached an all-time low in 1981 (**FIGURE 23.1D**). However, since then, and most strikingly in the last few years, whooping cough cases have been

A public health field worker preparing supplies for polio vaccinations in central Africa.
Courtesy of CDC/Benjamin Dahl, PhD, M.P.H.

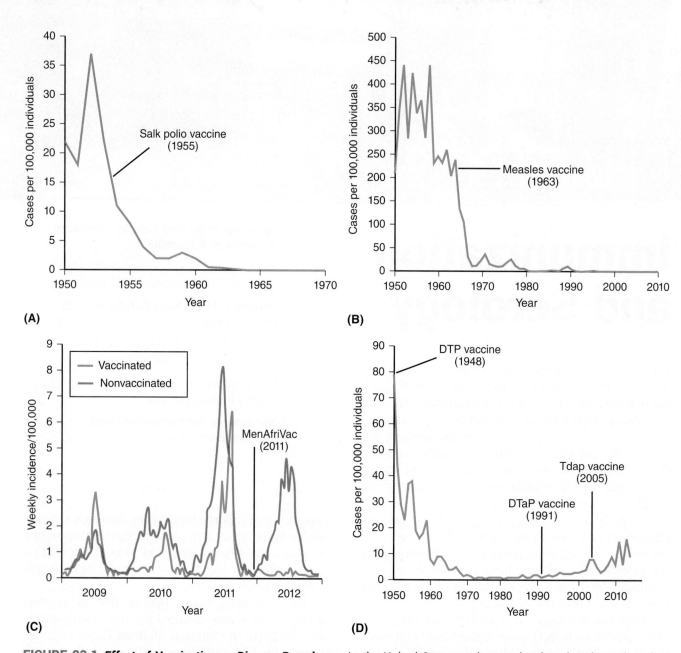

FIGURE 23.1 Effect of Vaccination on Disease Prevalence. In the United States and most developed nations, there has been a swift decline in infections following the introduction of vaccines. **(A)** Reduction of polio cases and **(B)** measles cases in the United States with the introduction of polio vaccine (1955) and measles vaccine (1963). **(C)** Reduction in cases of meningococcal meningitis in Chad (Africa) with the introduction of the meningococcus A vaccine (MenAfriVac). **(D)** Reduction of cases of pertussis (whooping cough) with the introduction of vaccines to protect against the disease. *»» Provide a rebuttal to counter the argument that measles was on the decline before the introduction of the measles vaccine in 1963.*

(C) Data from Dagula, D.M. (2013). The impact of MenAfriVac on serogroup A invasive meningococcal disease and carriage in Chad. Presented at the Meningitis Research Foundation 2013 conference, Meningitis & Septicaemia in Adults.

increasing. Reports from other parts of the world also are describing similar upticks in pertussis cases.

In this chapter, we study the role and safety of vaccines in the immune response. We also examine the serological tests used to identify antibodies in the serum or other fluids of a patient. Physicians can use such tests for diagnostic purposes when an infection is suspected.

Chapter Challenge

Children have always been more susceptible to infectious diseases than have their parents. The major reason for this is that kids have had limited exposure to the pathogens causing the diseases, and they have not yet built the immunologic memory required for protection. However, many of the historical childhood infections, such as polio, diphtheria, pertussis, rubella (German measles), mumps, chickenpox, and measles, are no longer major threats, at least in developed nations, because childhood vaccines are available to generate the needed immunity. But suppose that you can't remember if you ever had one of these diseases (let's take measles), you have no immunization record of having been vaccinated, and your parents do not remember if you ever had measles during your childhood. Knowing that 90% of susceptible individuals will get the disease if exposed to the virus, what should you do now that you are a 25-year-old adult? Let's determine a course of action as we proceed through this chapter on vaccines and the blood tests available to detect infectious diseases.

■ KEY CONCEPT 23.1   Immunity to Disease Can Be Generated Naturally or Artificially

In biology, **immunity** refers to a condition under which an individual is protected from particular infectious diseases. However, it does not mean one is immune to all diseases, but only those that the individual has been exposed to and recovered from or has been vaccinated against. Immunity depends on **adaptive immunity** and on the presence of antibodies, T lymphocytes, and other factors originating from the immune system's response to a specific foreign (nonself) antigen. Although the types of adaptive immunity discussed here focus on humoral immunity, remember that cell-mediated immunity also is an important and essential arm of resistance to infectious disease.

Adaptive Immunity Can Result by Actively Producing Antibodies to an Antigen

Active immunity occurs when antigens enter the body and the individual's immune system actively responds by producing antibodies to epitopes on those antigens. This type of immunity usually lasts for many years, if not a lifetime. This exposure to antigens can be unintentional, as when one becomes infected with a pathogen, or intentional, when one is purposely exposed to a weakened or inactive antigen or antigens via a vaccine.

▶ **Naturally Acquired Active Immunity.** One way to become immune to a disease is

by experiencing the disease and recovering from it. Thus, such **naturally acquired active immunity** follows a bout of illness and occurs in the "natural" scheme of events (**FIGURE 23.2A**). Also, **subclinical diseases** might result in acquiring active immunity. For example, many people develop inapparent or unnoticeable infections to mumps or fungal diseases such as valley fever. In either the clinical infection or subclinical infection, the individual produces antibodies as part of the primary antibody response.

Following the primary antibody response, the **memory cells** produced circulate in the blood and reside in the bone marrow for years. If the individual again is exposed to the active pathogen, the memory cells replicate and almost immediately produce antibodies (primarily IgG) to reestablish immune protection.

▶ **Artificially Acquired Active Immunity.** The other way to generate active immunity is by **vaccination** (**FIGURE 23.2B**). This form of active immunity is called **artificially acquired active immunity** because the antigens are contained in a vaccine. Vaccination is less risky than naturally acquired active immunity because the recipient of the

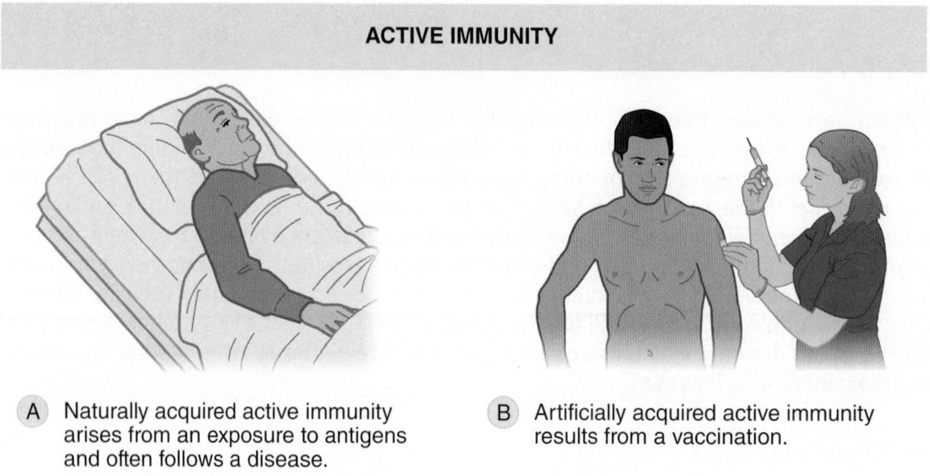

ACTIVE IMMUNITY

A   Naturally acquired active immunity arises from an exposure to antigens and often follows a disease.

B   Artificially acquired active immunity results from a vaccination.

FIGURE 23.2 Two Ways to Acquire Immunity. Immunity can be generated by natural or artificial means and acquired in an active or passive form. *»» Why do these forms of immunity represent an adaptive response rather than an innate response to an antigen?*

vaccine does not (usually) contract the disease or potential disease complications. The vaccination process still produces antibodies and memory cells following exposure to the vaccine antigens. If the vaccinated person is exposed to the same infective pathogen later, the memory cells act swiftly and produce a secondary antibody response, stopping the infection before it can make the individual ill. Vaccines then are simply "educating" the body's natural infection fighting system.

Thus, **vaccines** are substances that prepare the immune system to recognize and respond to a pathogen (antigen), resulting in protection (immunity) to that pathogen. They are composed of treated microorganisms or viruses, chemically altered toxins, or molecular parts of microorganisms, and the vaccines usually are administered by injection or oral consumption.

Most vaccines represent **preventative vaccines**, that is, vaccines that protect an individual from contracting a particular infectious disease. A few vaccines represent **therapeutic vaccines**, which act to lessen the severity of the illness that might appear or already exists. For example, shingles can be a very painful and debilitating disease. Getting the shingles vaccine will not protect one from getting shingles (the viral DNA already is present in the body), but the vaccine will lessen the severity and length of symptoms should shingles occur.

Although we will use the term "vaccination" in much of this chapter, the terms "vaccination" and "immunization" are not quite the same thing. **Vaccination** is the process of administering a vaccine that produces immunity in the body. **Immunization** is the process by which an individual becomes protected from an infectious disease.

Let's now examine the different types of vaccines to find out how they are made and how they work.

Whole-Agent Vaccines Contain Weakened or Inactivated Antigens

TABLE 23.1 summarizes the viral and bacterial vaccines currently in use in the United States. Most of these vaccines were developed by using the whole bacterium, virus, or bacterial toxin as the antigen.

Attenuated Vaccines

Some viral and bacterial pathogens can be weakened in the pharmaceutical lab such that they should not cause disease. These **attenuated** pathogens multiply only at low rates in the body, increasing the dose of antigen to a level that will stimulate an immune response. Such vaccines are the closest to the natural pathogens and, because they most closely mimic a natural infection, they generate the strongest immune response. Often, the person vaccinated will have lifelong immunity.

TABLE 23.1 The Major Bacterial and Viral Vaccines Currently in Use

Bacterial Disease	Route of Administration	Recommended Vaccine Usage/Comments
Contain Killed Whole Cells		
Cholera	Subcutaneous (SQ) injection	For travelers to endemic areas; short-term protection
Typhoid fever	SQ and intramuscular (IM)	For travelers to endemic areas
Contain Attenuated Cells		
Tuberculosis (BCG)	Intradermal (ID) injection	For high-risk occupations only; protection variable
Subunit Vaccines (Capsular Polysaccharides)		
*Meningitis (meningococcal)	SQ	Part of childhood immunization schedule
Meningitis (*H. influenzae*)	IM	Part of childhood immunization schedule; may be administered with DTaP
*Pneumococcal pneumonia	IM or SQ	Part of childhood immunization schedule; moderate protection
Pertussis	IM	Part of childhood immunization schedule
Subunit Vaccine (Protective Antigen)		
Anthrax	SQ	For lab workers and military personnel
Toxoids (Formaldehyde-Inactivated Exotoxins)		
Diphtheria	IM	Part of childhood immunization schedule
Tetanus	IM	Part of childhood immunization schedule
Viral Disease	**Route of Administration**	**Recommended Vaccine Usage/Comments**
Contain Inactivated Whole Particles		
Polio (Salk)	IM	Part of childhood immunization schedule; highly effective
Rabies	IM	For individuals sustaining animal bites or otherwise exposed
Influenza	IM	Part of childhood immunization schedule
*Hepatitis A	IM	Part of childhood immunization schedule; protection for travelers to endemic areas
Contain Attenuated Particles		
Measles (rubeola)	SQ	Part of childhood immunization schedule; highly effective
Mumps (parotitis)	SQ	Part of childhood immunization schedule; highly effective

(continues)

TABLE 23.1 **The Major Bacterial and Viral Vaccines Currently in Use (Continued)**

Bacterial Disease	Route of Administration	Recommended Vaccine Usage/Comments
Polio (Sabin)	Oral	Possible vaccine-induced polio
Smallpox (vaccinia)	Pierce outer layers of skin	For lab workers, military personnel, healthcare workers
Rubella	SQ	Part of childhood immunization schedule; highly effective
Chickenpox (varicella)	SQ	Part of childhood immunization schedule; immunity can diminish over time
*Shingles (zoster)	SQ	Therapeutic vaccine for individuals 60 years and older
Yellow fever	SQ	For travelers, military personnel in endemic areas
Influenza	IM	Nasal spray form
*Rotavirus infection	Oral	Part of childhood immunization schedule
Recombinant Vaccines		
Hepatitis B	IM	Part of childhood immunization schedule; medical, dental, laboratory personnel
*Genital warts/cervical cancer	IM	Preventative vaccine for individuals 9–26 years old

SQ = subcutaneous injection; IM = intramuscular injection; ID = intradermal injection
*Vaccines licensed since 2000.

The downside of attenuated vaccines results from the slow but continued multiplication of the bacterium or virus composing the vaccine. Because the vaccines contain dividing bacterial cells or replicating viruses, there is a remote chance one of them could increase in population size to cause a mild form of the disease. Usually a healthy person with a fully functioning immune system (**immunocompetent**) clears the infection without serious consequence. However, infants, older adults, and individuals with a compromised immune system, such as patients with AIDS or pregnant women, should not be given attenuated vaccines, if possible. Today, many vaccines, such as the Sabin oral polio vaccine and the measles, mumps, and chickenpox vaccines, contain attenuated viruses.

To avoid multiple injections of immunizing agents, it sometimes is advantageous to combine vaccines into a **single-dose vaccine**. The **measles-mumps-rubella** (**MMR**) vaccine is one example. Attenuated vaccines can be destroyed by heat, so they must be refrigerated to retain their effectiveness. Consequently, on a global scale, attenuated vaccines might not be the vaccine strategy of choice, because many developing nations lack widespread refrigeration facilities.

Inactivated Vaccines

Another strategy for preparing vaccines is to "kill" the bacterial or viral pathogen. These **inactivated vaccines** are relatively easy to produce because the pathogen is killed by simply using specific chemicals and/or heat. Although the agents in these vaccines cannot reproduce or replicate, the inactivation process alters the antigen's structure and shape. This results in a weaker humoral immune response and

MICROFOCUS 23.1: Public Health/Tools

Preparing for Battle

Each flu season up to 20% of Americans get the flu. Of this number, some 200,000 are hospitalized, and anywhere from 6,000 to 15,000 die from complications of a flu infection. Many of these hospitalizations and deaths could be prevented with a yearly flu shot, especially for those people at increased risk (people older than 50, immunocompromised individuals, and healthcare workers in close contact with flu patients). Because influenza viruses change often, the influenza vaccine is updated each year to ensure that it is as effective as possible. How is each year's vaccine designed and manufactured?

Each flu season, information on circulating influenza virus strains and epidemiological trends are gathered by the World Health Organization (WHO) Global Influenza Surveillance Network. The network consists of 128 influenza centers in 99 countries and 4 WHO Collaborating Centres for Reference and Research on Influenza located in Atlanta, United States; London, United Kingdom; Melbourne, Australia; and Tokyo, Japan. The national influenza centers sample patients with influenza-like illness and submit representative isolates to WHO Collaborating Centres for immediate strain identification.

Twice a year (February: northern hemisphere; September: southern hemisphere), WHO meets with the directors of the Collaborating Centres and representatives of key national laboratories to review the results of their strain identifications and to recommend the composition of the influenza vaccine for the next flu season.

In the United States, the Food and Drug Administration (FDA) or the Centers for Disease Control and Prevention (CDC) provides the identified viral strains to vaccine manufacturers in February. Each virus strain is grown separately in chicken eggs (see the accompanying figure). After the virus has replicated many times, the fluid containing the viruses is removed and the viruses purified before being attenuated or inactivated. Then, the appropriate strains are mixed together with a carrier fluid. Production usually is completed in August and ready for shipment by September.

Chicken eggs being inspected prior to being "inoculated" with a flu virus.

Courtesy of James Gathany/CDC.

little or no cell-mediated immunity. For this reason, most inactivated vaccines require multiple doses to generate a protective response, and periodic supplemental doses, or booster shots, might be needed throughout life to maintain immunity.

The Salk polio vaccine and the hepatitis A vaccine typify such preparations of inactivated whole viruses. In the case of influenza, the virus changes genetically from year to year, so a different vaccine must be provided annually. MICROFOCUS 23.1 describes how the components of the flu vaccine are selected each year.

Compared to attenuated vaccines, inactivated vaccines are safer because they cannot cause the disease in a vaccinated individual. The vaccines can be stored in a freeze-dried form at room temperature, making them a vaccine of choice in developing nations.

Toxoid Vaccines

For some bacterial diseases, such as diphtheria and tetanus, a bacterial exotoxin is the main cause of illness. For these diseases, a third immunization strategy is to inactivate these toxins and use them as a vaccine. Such toxins can be inactivated with formalin (an aqueous solution of formaldehyde) and the resulting inactivated toxin is called a **toxoid**. Immunity induced by a toxoid vaccine allows the body to generate antibodies and memory cells to recognize the natural toxin, should the individual be exposed to the active toxin. Because toxoid vaccines are inactivated products, booster shots are necessary (see the section later in the chapter).

Single-dose vaccines include **diphtheria-pertussis-tetanus** (**DPT**) vaccine and the newer diphtheria-tetanus-acellular pertussis (DTaP) vaccine. For other vaccines, however, a combination single-dose vaccine might not be useful because the antibody response might be lower for the combination than for each vaccine taken separately.

Newer Vaccines Contain Only Subunits or Fragments of Antigens

So-called second-generation vaccines have been developed that contain only a subunit or fragment of the bacterial cell or virus. With only a small piece of the antigen, a vaccinated individual cannot contract the disease from the vaccine, but these vaccines might fail to generate a cytotoxic T-cell response as part of adaptive immunity.

Subunit Vaccines

Unlike the whole-agent vaccines, the strategy for a **subunit vaccine** is to have the vaccine contain only those parts or subunits of the antigen that stimulate a strong immune response. For example, the subunit vaccine for pneumococcal pneumonia contains 23 different polysaccharides from the capsules of 23 strains of *Streptococcus pneumoniae*.

One way of producing a subunit vaccine is to use recombinant DNA technology, in which the resulting vaccine is called a **recombinant subunit vaccine**. The hepatitis B vaccine (Recombivax HB® and Engerix-B®) are examples. Several hepatitis B virus genes are isolated and inserted into yeast cells, which then synthesize the hepatitis antigens. These antigens are collected and purified to make the vaccine. Adverse reactions to such subunit vaccines are very rare, and the subunits cannot produce disease in the person vaccinated.

Conjugate Vaccines

Haemophilus influenzae b (Hib), which is responsible for a form of childhood meningitis, produces a capsule. Materials, like the capsular polysaccharides, represent **haptens**, which refers to substances that are not strongly immunogenic; that is, they do not provoke a strong immune response. Therefore, the strategy is to conjugate (attach) the haptens to a protein molecule, such as the tetanus or diphtheria toxoid that will stimulate a strong immune response. The result is the Hib **conjugate vaccine**, which has been instrumental in reducing the incidence of *Haemophilus* meningitis in young American children (younger than 5 years of age) from 18,000 cases annually in 1986 to 22 cases in 2015.

FIGURE 23.3 presents the 2016 immunization schedule for children 0 to 6 years of age as

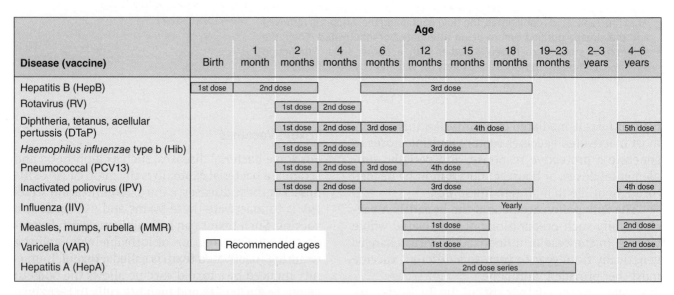

FIGURE 23.3 Childhood (0–6 years) Immunization Schedule–2017. Each year the Advisory Committee on Immunization Practices of the CDC reviews the childhood immunization schedule. This schedule indicates the recommended ages for routine administration of licensed vaccines. RV = rotavirus vaccine; PCV13 = pneumococcal conjugate vaccine; IPV = inactivated polio vaccine; IIV = inactive influenza vaccine; VAR = varicella vaccine. Additional recommendations concerning the schedule are available from the CDC at www.cdc.gov/vaccines/schedules/downloads/child/0–6yrs-schedule-pr.pdf. **»» *If you have children, or younger brothers or sisters, are they current on these vaccinations?***

Modified from CDC. (2017). Immunization schedules for infants and children. Retrieved from https://www.cdc.gov/vaccines/schedules/easy-to-read/child.html#print.

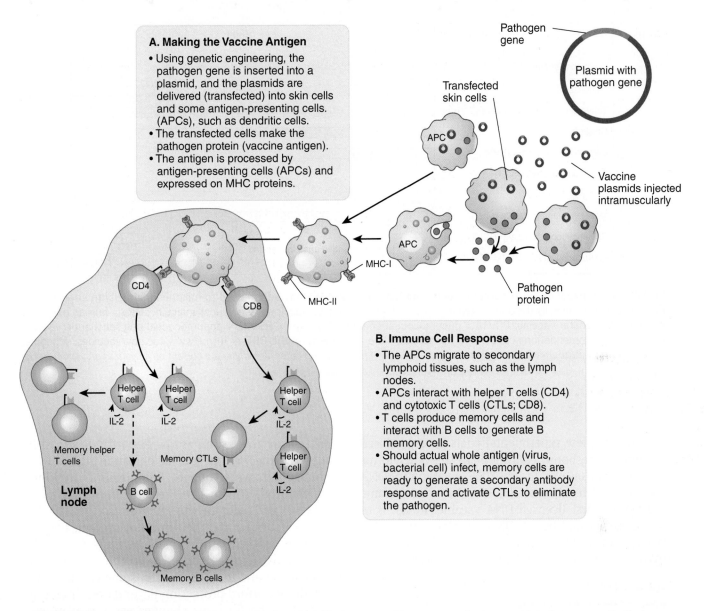

A. Making the Vaccine Antigen
- Using genetic engineering, the pathogen gene is inserted into a plasmid, and the plasmids are delivered (transfected) into skin cells and some antigen-presenting cells. (APCs), such as dendritic cells.
- The transfected cells make the pathogen protein (vaccine antigen).
- The antigen is processed by antigen-presenting cells (APCs) and expressed on MHC proteins.

B. Immune Cell Response
- The APCs migrate to secondary lymphoid tissues, such as the lymph nodes.
- APCs interact with helper T cells (CD4) and cytotoxic T cells (CTLs; CD8).
- T cells produce memory cells and interact with B cells to generate B memory cells.
- Should actual whole antigen (virus, bacterial cell) infect, memory cells are ready to generate a secondary antibody response and activate CTLs to eliminate the pathogen.

FIGURE 23.4 Making a DNA Vaccine. A DNA plasmid with the gene of interest can be inserted into immune cells that can trigger an immune response. The memory cells produced within the vaccinated individual are capable of quickly responding to an actual pathogen infection. *»» Describe the memory cells' response if such a vaccinated individual were infected with the actual pathogen.*

recommended by the Centers for Disease Control and Prevention (CDC).

DNA Vaccines

Part of the renaissance in vaccine technology is a strategy to use DNA as a vaccine. These investigational **DNA vaccines** consist of plasmids engineered to contain one or more protein-encoding genes from a viral or bacterial pathogen (**FIGURE 23.4**). By being injected intramuscularly in a saline solution, cells in the body will take up the injected foreign DNA, express (transcribe and translate) the proteins encoded by the DNA, and display these antigens on the cell surface, much like an infected cell presents antigen peptides to T cells. Such a display should stimulate a strong antibody (humoral) and cell-mediated immune response. If the vaccinated individual encounters the actual pathogen later, the individual should be immune to contracting the disease.

Disease (Vaccine)	Age Group				
	19–21 years	22–26 years	27–59 years	60–64 years	65+ years
Influenza (IIV)	1 dose every year				
Tetanus, diphtheria, pertussis (Td/Tdap)	Get a Tdap once, then a Td booster every 10 years				
Chickenpox (VV)	2 doses				
Cervical/penile cancer (HPV)	3 doses				
Penile cancer (HPV)	3 doses				
Shingles (HZV)				1 dose	
Measles, mumps, rubella (MMR)	1 or 2 doses				
Pneumococcal 13-valent (PCV13)	Recommended ages for vaccine				1 dose
Pneumococcal 23-valent (PPSV23)					1 dose

FIGURE 23.5 Recommended Adult Immunization Schedule—2017. Many adults need to maintain immunity to several infectious diseases, especially the ones listed here. IIV = inactive influenza vaccine; VV = varicella vaccine; HPV = human papilloma vaccine; ZV = zoster vaccine; PCV13 = pneumococcal conjugate vaccine; PPSV23 = pneumococcal polysaccharide vaccine. Additional recommendations concerning the schedule are available from the CDC at http://www.cdc.gov/vaccines/schedules /downloads/adult/adult-schedule-easy-read.pdf. *»» Look at the table and determine if you are current on all your adult vaccinations.*

Modified from CDC. (2017). Immunization schedules. Retrieved from https://www.cdc.gov/vaccines/schedules/easy-to-read/adult.html.

Thus far, few experimental trials have produced an immune response equivalent to the other forms of vaccines. However, in July 2005, a veterinary DNA vaccine to protect horses from West Nile virus became the world's first licensed DNA vaccine. In terms of humans, plasmid-based DNA vaccines are being studied for a variety of infections, including AIDS, influenza, hepatitis C, and for several cancers.

Some Vaccines Are Specifically Recommended for Adults

Many adults are unaware (or forget) that they need vaccines or booster shots throughout their lives. As a result, adult vaccination rates in the United States are unacceptably low. In fact, more than 40,000 adults in the United States die each year from diseases that are vaccine preventable. **FIGURE 23.5** presents the CDC's immunization schedule for adults (19 years and older).

There are several reasons why adult vaccinations are needed. First, some adults might not have experienced a childhood disease such as chickenpox or measles, or perhaps they were never vaccinated as children. Adult chickenpox can lead to serious pneumonia complications, whereas adult measles can produce serious complications, including arthritis and encephalitis (brain inflammation). Second,

newer vaccines, such as the hepatitis B and shingles vaccines, might not have been available when some adults were younger. Shingles is a rare illness in children, so the CDC recommends that adults 60 years and older get the shingles vaccine. Third, as adults get older and their immune systems begin to decline in strength, they become more susceptible to serious diseases caused by common infections such as flu and tetanus. Fourth, some vaccines simply do not afford protection beyond one "infectious season." The flu virus changes every season, so a vaccine taken one season might not protect for the next flu season.

One other reason for adult vaccinations stems from the fact that many vaccines do not provide lifelong immunity, and the immune system's memory begins to fade after several years. Therefore, **booster shots** provide additional doses of a vaccine given periodically to "refresh" the immune system's antibodies and memory cells. Consequently, some form of either the tetanus-diphtheria (Td) vaccine or the tetanus-diphtheria-acellular pertussis (Tdap) vaccine is recommended for adults every 10 years.

In some cases, killed pathogens do not stimulate a strong immune response, and substances need to be added to increase their effectiveness, as **MICRO-FOCUS 23.2** describes.

MICROFOCUS 23.2: Tools

Enhancing Vaccine Effectiveness

Vaccines are like any other medical product. They contain active and inactive ingredients that usually are suspended in water and salt (a saline solution). **Active Ingredients.** The main component in a vaccine is the antigen that is prepared in such a way as to stimulate the immune response. Sometimes, a vaccine is made more effective by adding another component, called an **adjuvant** (*adjuvare* = "help"), which increases the immune response and antibody production. By provoking a more sustained immune response, adjuvants also can allow the vaccine manufacturer to reduce the actual amount of antigen needed in each dose. The adjuvants licensed for general use in human vaccines in the United States are aluminum salts, specifically aluminum hydroxide or aluminum phosphate formulations, simply called alum.

A pregnant woman receiving an influenza vaccine in the Lao People's Democratic Republic (Laos).

Courtesy Athit Perawongmetha, United Parcel Service (UPS)/CDC.

Inactive Ingredients. Vaccines can contain stabilizers that act like preservatives. These are added to ensure that the vaccine does not break down and remains effective until ready to use. Some vaccines might contain antibiotics that are added during the manufacturing process to prevent potential bacterial contamination. Low levels of formaldehyde that are safe for humans might also be used. Other stabilizers, such as sugars or amino acids, might be added to protect the vaccine from damage that can be caused by light, heat, or humidity.

One preservative used since the 1930s in some vaccines (not MMR) is **thimerosal**, a mercury-containing compound. Although thimerosal was safe at the low levels used in vaccines, the American Academy of Pediatrics, the U.S. Public Health Service, and vaccine manufacturers stated in 1999 that, as a precautionary measure, thimerosal would be eliminated in all vaccines except tetanus-diphtheria (Td) vaccine and some flu vaccines. The seasonal flu vaccine often is packaged as multidose vials, and these vials contain thimerosal to protect the vaccine from possible contamination after it is opened. The single-dose vaccine units and the nasal spray vaccine do not contain thimerosal. The preservative is not found in any of the pediatric vaccines given to children younger than 6 years of age.

However, 120 developing nations depend on thimerosal-containing vaccines, which prevent an estimated 1.4 million child deaths each year. Therefore, health experts recommend leaving thimerosal in multidose vials of children's vaccines. In fact, the tiny amount of ethyl mercury in thimerosal is harmless, according to numerous recent research studies.

Adaptive Immunity Also Can Result by Passively Receiving Antibodies to an Antigen

So far, we have been discussing how an individual develops adaptive immunity through vaccination, specifically by artificially acquired active immunity. We finish this section by examining the ways in which an individual can acquire antibodies in a passive and temporary manner.

Passive immunity develops when antibodies enter the body from an outside source (in contrast to active immunity, in which individuals produce their own antibodies). Again, the source of antibodies can be unintentional, such as a fetus receiving antibodies from the mother, or intentional, such as the transfer of antibodies to an individual through injection (see **FIGURE 23.6A** and **B**). However, passive immunity is temporary, as the protection declines after several weeks or a few months.

Naturally Acquired Passive Immunity

The most common form of passive immunity is **naturally acquired passive immunity**, which is also called "congenital immunity." This form of passive immunity develops when, during the last 1 to 2 months of pregnancy, IgG antibodies pass into the fetal circulation from the mother's bloodstream via the placenta and umbilical cord. The process occurs in the "natural" scheme of events. The maternally passed IgG antibodies remain with the child for up to one year after birth and play an important role during these

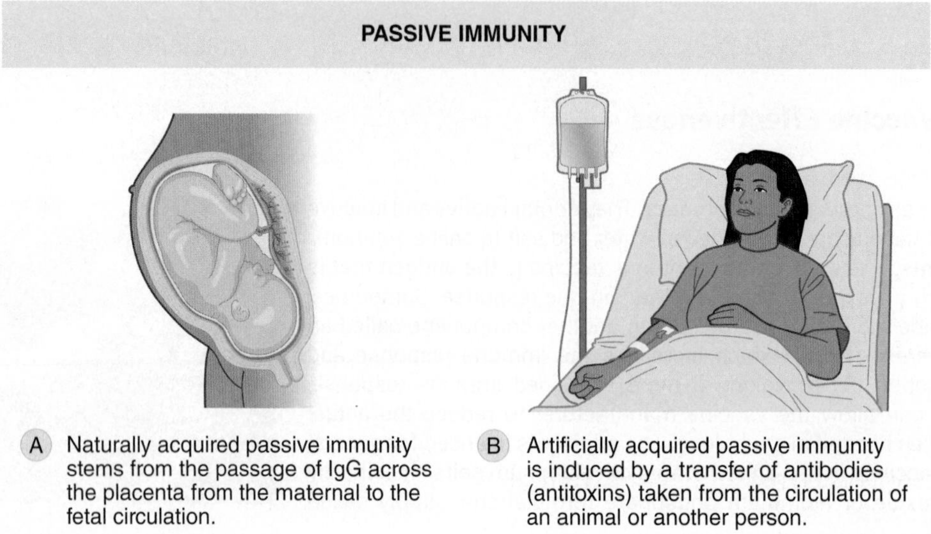

PASSIVE IMMUNITY

A. Naturally acquired passive immunity stems from the passage of IgG across the placenta from the maternal to the fetal circulation.

B. Artificially acquired passive immunity is induced by a transfer of antibodies (antitoxins) taken from the circulation of an animal or another person.

FIGURE 23.6 Two Ways to Acquire Passive Immunity. Immunity can be generated by natural or artificial means and acquired in an active or passive form. »» *Why do these forms of immunity represent an adaptive response rather than an innate response to an antigen?*

months of life in providing additional resistance to certain diseases such as measles and rubella.

Maternal antibodies also pass to the newborn through the first milk, or **colostrum**, of a nursing mother as well as during future breast feedings. In this instance, IgA is the predominant antibody, although IgG and IgM also might be found in the milk. The antibodies accumulate in the respiratory and gastrointestinal tracts of the child and support increased disease resistance.

Artificially Acquired Passive Immunity

The second form of passive immunity is called **artificially acquired passive immunity** because immunity arises from the intentional transfer of antibody-rich serum (blood products) from one human (or other animal) into the patient's circulation. The exposure to antibodies is thus "artificial." In the decades before the development of antibiotics, serum injection was an important therapeutic tool for the treatment of disease. The practice still is used for viral diseases such as Lassa fever and arthropod-borne encephalitis. In fact, in the Ebola epidemic in West Africa in 2013–2016, attempts were made to collect antibodies from survivors of Ebola virus disease and inject the antibodies into patients suffering the disease. These attempts were of questionable value.

Artificially acquired passive immunity makes use of the following three major types of **antisera** (antibodies) to treat human diseases:

▶ **Gamma Globulin.** One form, called **gamma (immune) globulin**, takes its name from the fraction of blood serum in which most antibodies are found. Gamma globulin consists of a pool of sera that are combined from many different human donors and thus contains a mixture of IgG antibodies, including those for the disease to be treated. Gamma globulin is primarily used for postexposure treatment of hepatitis A and measles.

▶ **Hyperimmune Globulin.** Another type of serum product is called **hyperimmune globulin**. It contains high concentrations of specific antibodies that are prepared from donor plasma having the antibodies of interest. Like gamma globulin, these hyperimmune preparations are used for postexposure treatment of infectious diseases such as hepatitis B, rabies, chickenpox, and tetanus.

▶ **Antitoxins.** Another type of antiserum is an **antitoxin** blood product. This serum is produced in animals, especially horses, and it contains antibodies against only one antigen. Antitoxins are typically used to treat botulism and diphtheria. One caution with this blood product is that the horse antibodies are "seen" as foreign (nonself) antigens by the recipient's immune system. The ensuing interaction causes an allergic reaction and forms **immune complexes** (antigen-antibody complexes) with the activation of complement. The recipient develops a condition called **serum sickness**, which is characterized by a hive-like rash at the injection site, labored breathing, and swollen joints.

Herd Immunity Results from Effective Vaccination Programs

In 2000, measles was declared eliminated from the United States. However, in 2014 there were 23 outbreaks affecting almost 670 individuals. This was much higher than the usual number of outbreaks (four outbreaks) and cases (60 cases) in the past decade (**FIGURE 23.7**). According to the CDC, most of these cases were in children and teens whose

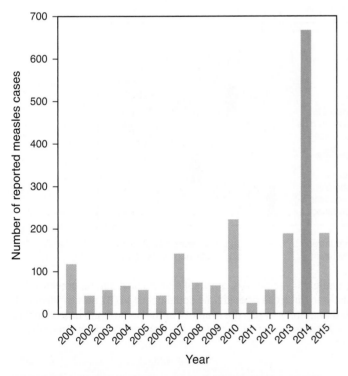

FIGURE 23.7 Measles Cases, by Year—United States, 2001–2015. In 2014, 667 confirmed measles cases were reported to CDC's National Center for Immunization and Respiratory Diseases. This is the highest number of reported cases since measles elimination was documented in 2000. *»» Were the majority of the people who contracted measles vaccinated or unvaccinated? Explain.*

Modified from CDC. (2016). Measles data and statistics. Retrieved from https://www.cdc.gov/measles/downloads/measlesdataandstatsslideset.pdf.

parents had exempted them from being vaccinated as part of the mandatory school vaccinations with the MMR vaccine. Many of the cases in 2014 were the result of viral infections linked to importation, that is, to American citizens who were infected in a foreign country or to foreign visitors in the United States who spread the infection to unvaccinated or unprotected Americans. In fact, many of the cases in the United States in 2014 were associated with travel to the Philippines, which was experiencing a large measles outbreak.

This example shows what can happen if small numbers of individuals or pockets of unvaccinated individuals encounter a pathogen. A disease outbreak is likely. Although immunizations are never meant to reach 100% of the population, if the vast majority of the population is immunized, it is unlikely that an infectious individual will be exposed to a susceptible person and spread the disease. This indirect protection, called **herd (community) immunity**, implies that if enough people in a population are immunized against certain infectious diseases, it is very difficult for those diseases to spread (**FIGURE 23.8**).

According to health experts and disease epidemiologists, when greater than 85% of the population (around 95% for some diseases like measles) is vaccinated, the spread of the disease is effectively stopped. This percentage is called the **herd immunity threshold**. Although the rest of the "herd" or population remains susceptible, there are so many vaccinated people in the "herd" that it is unlikely an infected person could easily spread the disease.

Herd immunity can be affected by several factors. One is the environment. People living in crowded cities are more likely to catch a disease if they are not vaccinated than nonvaccinated people living in rural areas because of the constant close contact with other people in the city.

Another factor is the strength of an individual's immune system. People whose immune systems are compromised, either because they currently have a disease or because of medical treatment (anticancer or antirejection drug therapy), often cannot be immunized because they are at a greater risk of catching the disease to which others are immunized.

MICROFOCUS 23.3 looks at the consequences of declining vaccinations and herd immunity.

Vaccine Side Effects Are Rare and Minor

Many people today are too young to remember the prevaccine era, the large number of infectious cases, and the terrible damage some infectious diseases caused—birth defects from German measles;

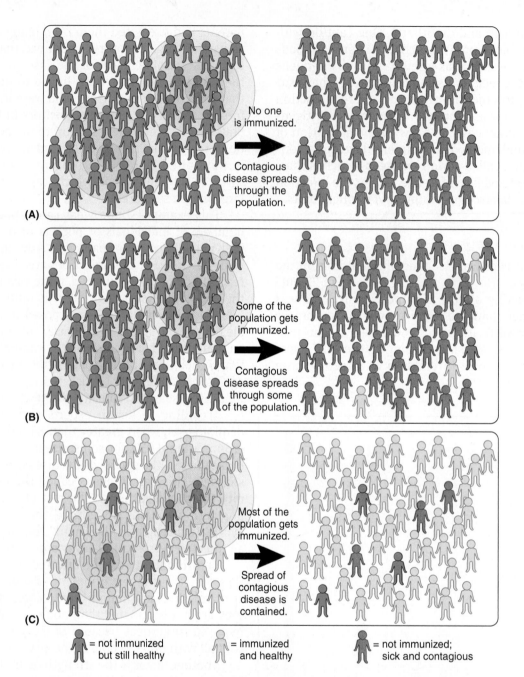

FIGURE 23.8 Herd Immunity. If a large percentage of a population is vaccinated, outbreaks of disease can be prevented. In **(A)**, no individuals in a community are vaccinated, making it likely an outbreak or epidemic could occur. In **(B)**, some individuals in the community are vaccinated but not enough to confer herd immunity. In **(C)**, the majority of the community has been vaccinated, which protects most individuals in the community. »» *What groups of individuals in the community, who are not eligible for certain vaccines, would get protection from the herd immunity?*

Courtesy National Institute of Allergy and Infectious Diseases.

MICROFOCUS 23.3: Public Health

The Risk of Not Vaccinating: Community and Healthcare Settings

Data from epidemiological studies clearly show that community vaccinations essentially eliminate outbreaks of infectious disease by maintaining high herd immunity. In fact, the Centers for Disease Control and Prevention (CDC) reported in 2014 that approximately 90% of American children receive all the recommended vaccines against childhood diseases before entering kindergarten. Yet, about half the states in the United States allow parents to apply for vaccination waivers prior to enrolling their children in school. There are some valid medical reasons why a few children should not be vaccinated, including children with a history of allergic reactions or who are immunocompromised. Of great concern is the increasing number of nonmedical exemptions granted, based on religious, personal, or philosophical grounds. In 2011–2012, 80% of vaccination exemptions fell in the nonmedical group. In addition, only a small percentage of adults in the community receive booster shots for diseases like tetanus and diphtheria. All these examples represent "holes" in herd immunity, and, as a result, there has been an upsurge in childhood diseases such as pertussis and measles.

A young girl receiving a vaccination.

Courtesy of James Gathany, CDC.

Likewise, vaccination of healthcare workers (HCWs) at risk for exposure to serious, and sometimes deadly, diseases like influenza and hepatitis is just as important to maintain patient safety and quality-of-care. For example, HCWs (i.e., physicians, nurses, emergency medical personnel, pharmacists, medical and nursing students, dental professionals and students, laboratory technicians, hospital volunteers, and administrative staff) are frequently exposed to influenza, and even if they are asymptomatic, they can still transmit the virus to patients, especially the vulnerable older adults and immunocompromised individuals. Influenza vaccination of HCWs not only diminishes the disease burden by maintaining high herd immunity but also reduces the rate of influenza disease and overall mortality in the patients cared for by HCWs. Therefore, since 1981, the CDC has recommended priority vaccinations for all HCWs, unless they have had a previous adverse reaction to the inactivated influenza vaccine.

Yet, despite local and national efforts to encourage influenza vaccination, the overall vaccination rate among HCWs in the United States remains too low at slightly more than 70%. Unvaccinated HCWs surveyed said they did not get the influenza vaccine because they feared the side effects and needles, they were skeptical of vaccine efficacy, they believed in their own innate ability to resist infection, or they reported an inability to access the vaccine. As with community vaccinations, unless there are extenuating circumstances, HCWs, as part of their "duty of care," should be vaccinated against influenza to reduce the chance of illness to themselves and transmission to others.

central nervous system paralysis from polio; and the uncontrollable, violent coughing and breathing struggles from whooping cough (TABLE 23.2). Therefore, it is imperative that children develop a strong immune system to combat these diseases and other potential infections. To accomplish this, vaccines are the answer.

If you look back at Figure 23.3, notice the current immunization schedule in the United States suggests that by age two, children should have had up to 30 shots to provide natural immunity to infectious diseases. Records show that such vaccinations have been very successful at eliminating or greatly reducing the incidence of childhood diseases (e.g.,

pertussis, chickenpox, pneumonia, and measles) in the United States. Importantly, the vast majority of these children experienced few, if any, side effects to these vaccines. Yet, some parents request that their physician spread out the childhood vaccines over a longer vaccine schedule. In most cases, this is ill advised because underimmunization often puts the unvaccinated (or incompletely vaccinated) children at risk of contracting a disease, and it runs the potential of generating an outbreak.

Still, many parents wonder if this number of vaccines can have adverse effects. The overwhelming evidence from numerous studies shows that some vaccines (in children and adults) may cause

TABLE 23.2   Decline in Disease Cases, United States

Disease	Mean U.S. Cases per Year		% Decrease
	Pre-Vaccine	2015	
Diphtheria	21,053	0	100
Measles	530,217	189	>99
Mumps	162,344	1,057	99
Pertussis (whooping cough)	200,752	32,971 (for 2014)	84
Polio	16,316	0	100
Rubella	47,745	9 (for 2013)	>99
Smallpox	29,005	0	100

minor swelling and redness at the injection site. These effects pass in a day or two. However, the parents or adults need to realize that these minor side effects are an indication that the vaccine worked; the symptoms are due to the immune system recognizing and responding to the microbial agent in the vaccine.

The chance of experiencing a serious side effect, such as an allergic reaction from a vaccine, is extremely rare (about 1 in a million). Usually, these cases can be treated with standard medications. Even rarer is the possibility of experiencing a serious "adverse event" such as a seizure. To "catch" any adverse events, the U.S. Food and Drug Administration (FDA) and the CDC have established the **Vaccine Adverse Events Reporting System** (**VAERS**) to which anyone, including doctors, patients, and parents, can report adverse vaccine reactions. FDA and CDC personnel regularly monitor the system for any suspicious trends.

The FDA requires vaccine manufacturers to follow extensive safety procedures when producing vaccines used in the United States to ensure they are safe and effective. After lab and animal tests have been completed, a promising vaccine must undergo clinical trials before being licensed by the FDA for use.

Perhaps the vaccine of most concern regarding side effects and childhood development is the MMR vaccine. In 1998, a medical paper (now discredited) was published suggesting that there might be a link between the MMR vaccine and autism in children. The idea was put forward that perhaps the sheer number of vaccinations could overwhelm a child's immune system and cause neurological damage. To make a long story short, as of 2017, more than a dozen studies carried out by other scientists and researchers around the world have unanimously found no evidence to support the autism claim. The MMR vaccine does not lead to increased rates of autism in children.

Through vaccination and booster shots, vaccines have been essential to maintaining public health and controlling infectious diseases. Although a few of the millions of people vaccinated each year suffer a serious consequence from vaccination, the risks of contracting a disease (especially in infants and children) from not being vaccinated are thousands of times greater than the risks associated with any vaccine. In fact, the chances of developing a seizure from the MMR vaccine are less likely than developing a seizure from an actual measles infection. In addition, the licensed vaccines are always being reexamined for ways to improve their safety and effectiveness.

From a historical perspective, INVESTIGATING THE MICROBIAL WORLD 23 recounts the field studies carried out to test the effectiveness of the Salk polio vaccine.

Investigating the Microbial World 23

The Salk Polio Vaccine Field Trials

Experiments are not always done in the lab. Many science investigations take place outside the lab and compose what are called field studies. These research projects still use science inquiry methods to collect data about some phenomenon or behavior. They also may involve clinical trials that involve observational studies concerning the effects of some intervention on the health of a population. In the 1950s, the field trials for the Salk polio vaccine were one of the most famous and largest statistical studies ever conducted.

OBSERVATION: The graph on the right provides the approximate number of polio cases in the United States between 1930 and 1955, the year the Salk polio vaccine was released. The need for a vaccine was urgent and two labs, those of Jonas Salk and Albert Sabin, were involved in the search. Looking at the Salk investigation, a vaccine was eventually prepared from a heat- and formaldehyde-killed sample of the polio virus. The key in the making of the vaccine was to ensure the viruses were "killed" (cannot replicate in host cells) but the treatment was not so severe that it destroyed the immunogenic properties of the antigens. After the vaccine was thought to be safe according to lab experiments, it needed to be field tested for its effectiveness.

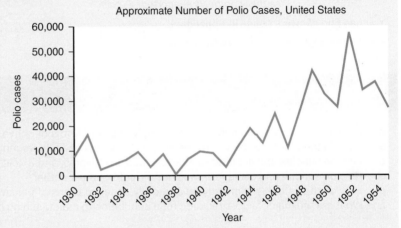

Approximate number of polio cases, United States.

Modified from Francis, T., et al. 1955. *Amer J Pub Health* 45 (suppl.):1–50.

QUESTION: *Was the vaccine effective? Did it prevent polio?*

HYPOTHESIS: The vaccine is effective. If so, administering the vaccine to a treatment group should show a statistically significant reduction in polio cases compared to a nontreatment (control) group.

EXPERIMENTAL DESIGN: Designing epidemiological field studies often are fraught with problems. The treatment and control groups had to be as identical as possible. The individuals in each group needed to come from the same region, and the studies needed to be done at the same time of year. Large numbers of individuals were needed because the incidence of paralytic polio in the United States was about 50 cases per 100,000 individuals. In addition, diagnosis of mild polio cases (nonparalytic) was tricky because the symptoms of fever and weakness also could be typical of other diseases.

The field trials, carried out in 1954, involved two distinct designs:

FIELD TRIAL 1: Almost 950,000 first- through third-graders in 127 selected schools in 33 states were enlisted (with parent's consent). The second-graders at selected schools across the country were vaccinated (treatment group), whereas the first- and third-graders in the same schools were not vaccinated (control group).

FIELD TRIAL 2: More than 400,000 children in 84 selected schools in 11 states were enlisted (with parent's consent). One group was given the vaccine while a second group was given a placebo injection (a saline solution).

RESULTS:

Field Trial	Group	Number of Students	Paralytic Polio Cases	Incidence per 100,000
1	Treatment	221,998	38	17.1
	Control	725,173	330	45.5
2	Treatment	200,745	33	16.4
	Placebo	201,229	115	57.1

Data from Francis, T. et al. 1955. *Amer J Pub Health* 45 (suppl.):1–50.

(continues)

Investigating the Microbial World 23 (*Continued*)

The Salk Polio Vaccine Field Trials

CONCLUSIONS:

QUESTION 1: *Were the number of paralytic polio cases in the control groups about equal to the actual incidence in the United States? Explain.*

QUESTION 2: *Did treatment with the Salk polio vaccine reduce the incidence of paralytic polio? Explain.*

QUESTION 3: *Was the hypothesis supported? Explain.*

You can find answers online in **Appendix E**.

Modified from Francis, T. et al. 1955. *Amer J Pub Health* 45 (suppl.):1–50.

Concept and Reasoning Checks 23.1

a. How do the naturally acquired and artificially acquired forms of active immunity differ?
b. What is the difference between a preventative vaccine and a therapeutic vaccine?
c. List the benefits and drawbacks of attenuated, inactivated, and toxoid vaccines.
d. How do subunit and conjugate vaccines differ from one another?
e. How do the naturally acquired and artificially acquired forms of passive immunity differ?
f. What are the major factors required for a population to maintain high herd immunity?

Chapter Challenge A

You cannot remember if you ever had measles or the vaccination as a child. Therefore, you do some background research.

QUESTION A: *What is the schedule for childhood measles vaccination, and what is the combination formulation used for measles vaccination? What are the antigens normally found in the vaccine?*

You can find answers online in **Appendix F**.

■ KEY CONCEPT 23.2   Serological Reactions Can Be Used to Diagnose Disease

Antigen-antibody interactions analyzed in the immunology section of the clinical microbiology laboratory (CML) are known as **serological reactions** because they commonly involve serum from a patient. The principle behind these interactions is simple and straightforward: if the patient has an abnormal level of a specific antibody in the serum, that antibody probably was produced in response to an infecting disease agent. Consequently, **serology**, the study of blood serum and its constituents, has great diagnostic significance as well as more broad-ranging applications. For example, serological reactions are used to confirm identifications made by other procedures and to detect pathogens in body tissues. In addition, the results of the reactions help the physician follow the course of disease and determine the immune status of the patient. The serological reactions can even aid in determining the subspecies (serotypes) of taxonomic groupings of microorganisms.

Serological Reactions Have Certain Characteristics

Diagnostic microbiology focuses on applying microbiology to medical diagnosis and makes extensive use of the serological reactions between antigen and antibody. Thus, immunodiagnostic procedures can be used to identify infectious diseases by detecting antigens, or the antigens can be used to detect antibodies present in a blood (serum) sample.

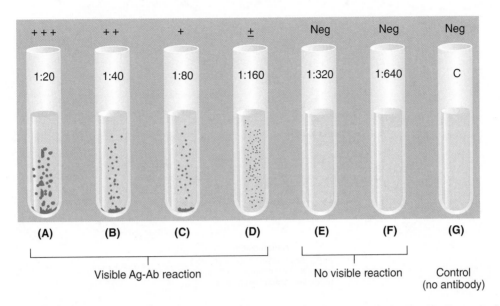

FIGURE 23.9 **The Determination of Titer.** A sample of antibody (Ab)-containing serum was diluted in saline solution to yield the dilutions shown. An equal amount of antigen (Ag) was then added to each tube, and the tubes were incubated. An antigen-antibody interaction may be seen in tubes **(A)** through **(D)** but not in tubes **(E)** or **(F)** or the control tube **(G)**. The titer of antibody is the highest dilution of serum antibody in which a reaction is visible. *»» What is the titer of antibody in this example?*

Successful serological reactions require the antigen or antibody solution to be greatly diluted in order to reach a concentration at which a reaction will be most favorable. The process, called **titration**, can be used to the physician's advantage because the dilution series is a valuable way of determining the titer of antibodies. The **titer** is the most dilute concentration of serum antibody yielding a detectable reaction with its specific antigen (**FIGURE 23.9**). This number is expressed as the denominator of the dilution (i.e., a 1:50 dilution, titer = 50). Thus, one might say the titer of influenza antibodies rose from 20 to 320 as influenza progresses, then continues upward and stabilizes at 1,280 as the disease reaches its peak. A rise in the titer is evidence that an individual has a disease, an important factor in diagnosis. Serology has become a highly sophisticated—and often automated—branch of immunology, as we will see as the following tests are described.

Neutralization Involves the Binding of Antibodies to Antigens

Neutralization is a serological reaction used to identify bacterial toxins and antitoxins as well as viruses and viral antibodies. Normally, little or no visible evidence of a neutralization reaction is present, so the test mixture of toxin (or virus) and antibodies must be injected into a laboratory animal to determine whether neutralization has occurred.

Suppose that several people in a community have become ill and have the signs and symptoms of botulism: weak muscles, double vision, nausea, and vomiting. The identification of botulism toxin can be made by using a neutralization test, which is described in **FIGURE 23.10**. The key to the test is the use of botulism antitoxins (antibotulism toxin antibodies) to neutralize the toxin, if present.

A similar test for diphtheria is the **Schick test**, in which a person's immunity to diphtheria can be determined by injecting diphtheria toxin intradermally. No skin reaction will occur if the person has neutralizing antibodies in the blood. However, a local red, swollen area will develop if no antidiphtheria antibodies are present, indicating the person is susceptible to diphtheria.

Precipitation Requires the Formation of a Lattice Between Soluble Antigen and Antibody

Precipitation reactions are serological reactions involving thousands of soluble antigen and antibody molecules cross-linked to form antigen-antibody (Ag-Ab) complexes (lattices). The lattices are so large and insoluble that they precipitate from a solution or in a gel.

Precipitation tests are performed in either fluid media or gels. In fluids, the antibody and antigen solutions are layered over each other in a thin tube.

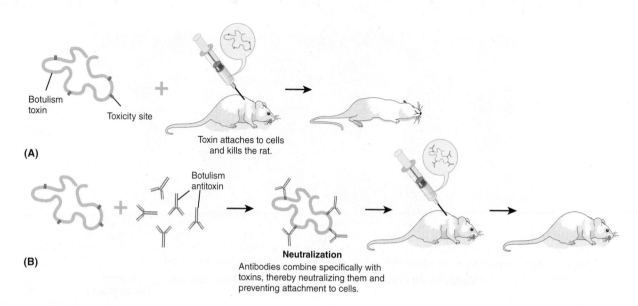

FIGURE 23.10 Neutralization Test. The neutralization test can be used to detect bacterial toxins, such as the botulism toxin. **(A)** If a food is believed to be contaminated with the botulism toxin, injecting the food into a rat will allow the toxin to attach to nerve cells, and the released toxin will kill the rat. **(B)** To make sure it is the botulism toxin and not another foodborne toxin, the food material is mixed with botulism antitoxin (antibotulism antibodies). If the food does contain botulism toxin, the antibodies will block binding of the toxin (neutralization), and on injection of the mix, the rat will survive. »» *What would be the conclusion if the mixture in (B) did kill the rat?*

The molecules then diffuse through the fluid until they reach a **zone of equivalence**—the ideal concentration of antigen and antibody that forms visible, insoluble lattices.

In **immunodiffusion**, the diffusion of antigens and antibodies takes place through a semisolid gel, such as agar. As the molecules diffuse through the gel, they eventually reach the zone of equivalence, where they interact and form a visible precipitate (lattice). A variation of this technique is called the double diffusion technique, so named because both reactants diffuse.

In immunodiffusion, soluble antigen and antibody solutions are placed in wells cut into the agar gel in Petri dishes. The wells are filled with appropriate antigen and antibodies and incubated. When examined, precipitation lines form at the zone of equivalence (**FIGURE 23.11**).

In the antigen-detecting procedure known as **immunoelectrophoresis**, gel electrophoresis and diffusion are combined. A mixture of antigens is placed in a reservoir on an **agarose**-coated slide (agarose is a polysaccharide isolated from agar), and an electrical field is applied to the ends of the slide. The different antigens move through the agarose at different rates of speed, depending on their electrical charges (**FIGURE 23.12**). A trough is then cut into the agarose along the same axis, and a known antibody solution is added. During incubation, antigens and antibodies diffuse toward each other and precipitation lines form, as in the immunodiffusion technique.

Agglutination Involves the Clumping of Antigens

The amount of antigen or antibody needed to form a visible reaction can be reduced if either is attached to the surface of an object. The result is a clumping together, or **agglutination**, of the linked product. Agglutination procedures are performed on slides or in tubes. For bacterial agglutination, emulsions of unknown bacterial species are added to drops of known antibodies on a slide, and the mixture is observed for clumping.

Direct agglutination involves IgM antibodies. Most often, antigens are adsorbed onto the surface of latex spheres or polystyrene particles (**FIGURE 23.13A**). IgM antibodies can be detected rapidly by observing agglutination of the carrier particles. Bacterial infections caused by a number of pathogens can be detected rapidly by mixing the bacterial sample with latex spheres attached to streptococcus antibody (**FIGURE 23.13B, C**).

Hemagglutination refers to the clumping of red blood cells. This process is particularly important in the determination of blood types prior to blood transfusion (**FIGURE 23.13D**). In addition, certain viruses, such as the measles and mumps viruses, agglutinate red blood cells. Antibodies for these viruses can be detected by a procedure in which the serum is first

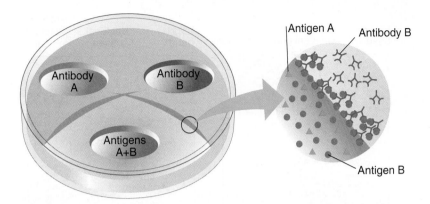

FIGURE 23.11 Double Diffusion Technique. Wells are cut into a plate of agar. Different known antibodies then are placed into the two upper wells, and a mixture of unknown antigens is placed into the lower well. During incubation, the reactants diffuse outward from the wells, and cloudy lines of precipitate form. The lines cross each other because each antigen has reacted only with its complementary antibody. »» *Why does the precipitin line form where it does?*

combined with laboratory-cultivated viruses and then added to the red blood cells. If serum antibodies neutralize the viruses, agglutination fails to occur. This test is called the **hemagglutination inhibition (HAI) test** and is explained in MICROFOCUS 23.4. A version of the hemagglutination test, called the **Coombs test**, is used to detect Rh antibodies involved in hemolytic disease of the newborn. A slide agglutination test, called the **Venereal Disease Research Laboratory (VDRL) test**, is used for the rapid screening of patients to detect syphilis.

Complement Fixation Can Detect Antibodies to a Variety of Pathogens

The **complement fixation test** is performed in two parts. The first part—the test system—heats the patient's serum to destroy any complement present in the serum. Next, carefully measured amounts of antigen and guinea pig complement are added to the serum (**FIGURE 23.14A**). This test system is incubated at 37°C (98.6°F) for 90 minutes. If antibodies specific for the antigen are present in the patient's

A Gel electrophoresis:
On a gel-coated slide, an antigen sample is placed in a central well. An electrical current is run through the gel to separate antigens by their electrical charge (electrophoresis). Unlike the depiction in the diagram, the separate antigens cannot be detected visually at this point.

B Addition of antibodies:
A trough is made on the slide, and a known antibody solution is added.

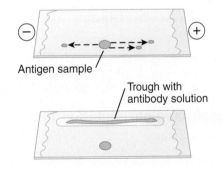

C Diffusion of antigens and antibodies:
As antigens and antibodies diffuse toward one another through the gel, precipitin lines are seen where optimal concentrations of antigen and antibodies meet.

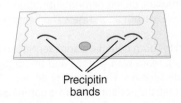

FIGURE 23.12 Immunoelectrophoresis. By applying an electrical field, immunoelectrophoresis separates antigens by their electrical charge before antibodies are added. »» *In (A), what is the electrical charge (+ or −) on the three antigens shown?*

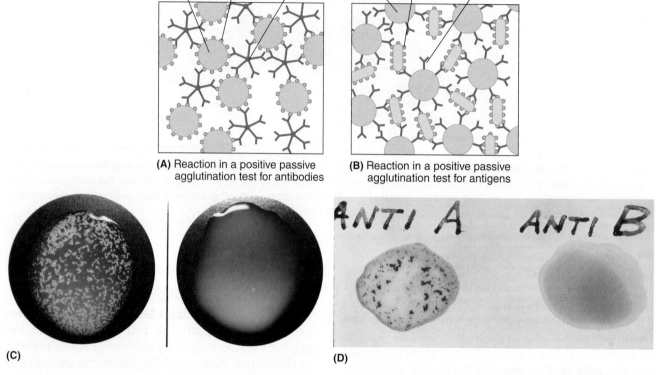

(A) Reaction in a positive passive agglutination test for antibodies

(B) Reaction in a positive passive agglutination test for antigens

(C)

(D)

FIGURE 23.13 Direct Agglutination Tests. (A) When particles are bound with antigens, agglutination indicates the presence of antibodies, such as the IgM shown here. **(B)** When particles are bound with antibodies, agglutination indicates the presence of antigens. **(C)** Latex agglutination test for *Staphylococcus*. If a specific staph antigen is present, it will bind to antibody-coated latex beads and agglutinate them for a positive test (left). A negative test is shown for comparison on the right. **(D)** Blood type can be determined using the agglutination test. Red blood cells are diluted with saline and mixed with anti-A agglutinin (at left) and anti-B agglutinin (at right). For some minutes, they are allowed to react. The clumping (agglutination) is due to antibody-antigen reactions. These tests use antigens or antibodies adsorbed onto the surface of latex spheres. »» *What is the blood type in the example shown in (D)?*

MICROFOCUS 23.4: Tools

The Hemagglutination-Inhibition Test

An indirect method for detecting viruses is to search for viral antibodies in a patient's serum. This can be done by combining serum (the blood's fluid portion) with known viruses.

Certain viruses—such as those of influenza, measles, and mumps—have the ability to agglutinate (clump) red blood cells. This phenomenon, called hemagglutination (HA), will produce a thin layer of cells over the bottom of a plastic dish (Figure A). If the cells do not agglutinate (no hemagglutination), they fall to the bottom of the well accumulating as a small "button."

HA can be used for both detection and identification purposes. It is possible to detect antibodies against certain viruses because antibodies react with viruses and tie up the reaction sites that otherwise would bind to red blood cells, thereby inhibiting hemagglutination. This property allows a laboratory test called the hemagglutination-inhibition (HAI) test to be performed. In the test, the patient's serum is combined with known viruses, to which red blood cells are then added (Figure B). Hemagglutination inhibition indicates antibodies are present in the serum. Such a finding implies the patient has been exposed to the virus.

Figure C presents a scenario for detecting exposure to the measles virus.

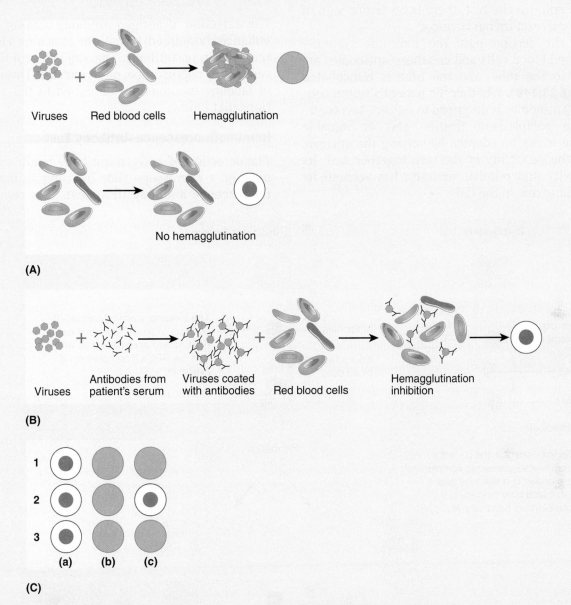

Tests for Detecting Viruses (A) In hemagglutination, viruses of certain diseases, such as measles and mumps, can agglutinate (clump) red blood cells, forming a film of cells over the bottom of a plastic well. No agglutination produced a small pellet or "button" at the bottom of the cell. **(B)** In the hemagglutination-inhibition (HAI) test, known viruses are combined with serum from a patient. If the serum contains antibodies for that virus, the antibodies coat the virus, and when the viruses are then combined with red blood cells, no agglutination will take place. **(C)** In December, three young children (1, 2, 3) who attended the same daycare center were admitted to the hospital with nonbacterial pneumonia. All three were tested for the presence of influenza A antibodies using the HAI test. Which child or children likely had the flu? (a) = positive control; (b) = negative control; (c) = patient's serum.

serum, an antibody-antigen interaction takes place, and the complement is used up, or "fixed." However, at this point in the test, there is no visible sign of whether a reaction has occurred.

In the second part—the indicator system— sheep red blood cells and antisheep antibodies are added to the tube, and the tube is reincubated (**FIGURE 23.14B**). Whether the patient's serum contained antibodies is described in **FIGURE 23.14C, D**.

The complement fixation test is valuable because it can be adapted by varying the antigen. Thus, the versatility of the test, together with its sensitivity and relative accuracy, has secured its continuing role in the CML.

Labeling Methods Are Used to Detect Antigen-Antibody Binding

The detection of antigen-antibody binding can be enhanced (amplified) visually by attaching a label to the antigen or antibody. This tag can be a fluorescent dye, a radioisotope, or an enzyme. These types of labeling methods are described in the subsections that follow.

Immunofluorescence Antibody Test

The detection of antigen-antibody binding can be done on a microscope slide by using an **immunofluorescence antibody (IFA) test**. Two commonly

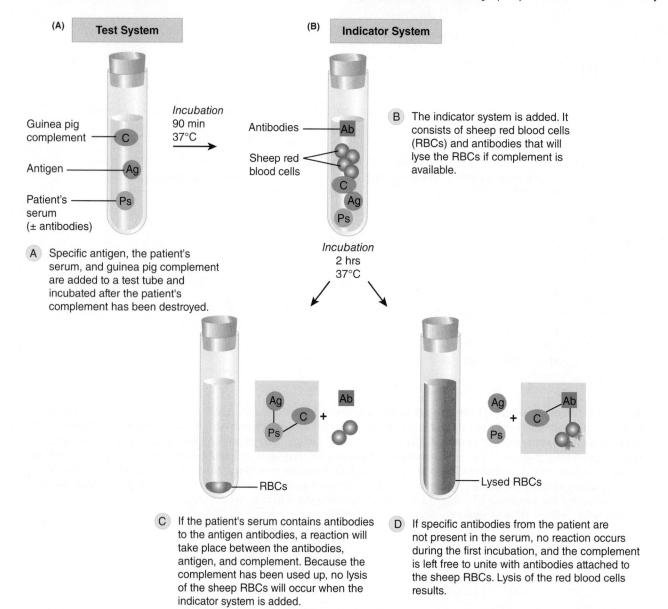

FIGURE 23.14 The Complement Fixation Test. Complement fixation tests are carried out in two stages involving **(A)** a test system and **(B)** an indicator system. *»» What would happen if the patient's complement proteins were not inactivated?*

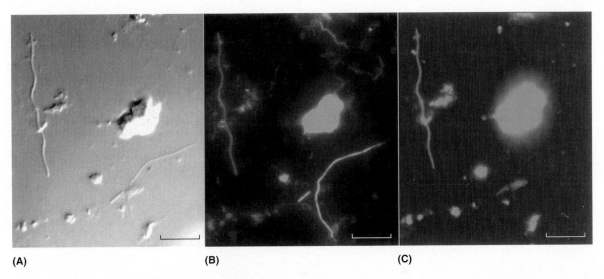

FIGURE 23.15 Direct Immunofluorescence Antibody (IFA) Test. Three views of spirochetes in the hindgut of a termite identified by a direct IFA test. **(A)** An unstained area displayed by phase contrast microscopy. (Bar = 10 μm.) **(B)** The same viewing area seen by fluorescence microscopy using rhodamine B as a stain. (Bar = 10 μm.) **(C)** The same area viewed after staining with fluorescein. (Bar = 10 μm.) »» *Why is this method referred to as direct fluorescent antibody technique?*

Reproduced from *Appli. Environ. Microbiol*, 1996, vol. 62(2), p. 347–352, DOI and reproduced with permission from the American Society for Microbiology. Photo courtesy of Doctor Bruce J. Paster, Senior Member at Department of Molecular Genetics, The Forsyth Institute.

used dyes are fluorescein, which emits an apple-green glow, and rhodamine, which gives off orange-red light when exposed to ultraviolet (UV) light.

The IFA test can be direct or indirect. In the direct method, the fluorescent dye is linked directly to a known antibody. After combining with particles having complementary antigens, the labeled antibodies fluoresce (glow) when viewed with a fluorescence microscope using UV light.

For example, a microbiologist wants to know if spirochetes are present in termites. **FIGURE 23.15** explains the process and microscopic identification.

With the indirect IFA test, the presence of pathogens can be identified. Suppose that a clinical technician needs to determine if spirochetes (typical of a syphilis infection) are present in a patient's serum. A fluorescent dye can be linked to an antibody recognizing human antibodies in a patient's serum. **FIGURE 23.16** shows its setup and result interpretation.

The IFA test is adaptable to a broad variety of antigens and antibodies and is widely used in serology.

Radioimmunoassay

An extremely sensitive serological procedure used in the CML is the **radioimmunoassay (RIA)**. The RIA measures the concentration of very small molecules, such as haptens. One of its major advantages is that it can detect nanograms of a substance.

The **radioallergosorbent test (RAST)** is an extension of the radioimmunoassay. This indirect RIA can be used, for example, to detect IgE antibodies in the serum of a person possibly allergic to compounds like penicillin. **FIGURE 23.17** outlines the procedure and results.

Enzyme-Linked Immunosorbent Assay

Another test to detect antigens or antibodies is the **enzyme-linked immunosorbent assay (ELISA)**, also called the **enzyme immunoassay (EIA)**. It has the same sensitivity as RIA and RAST but does not require expensive equipment or involve working with radioactivity. ELISA is used to detect antibodies to a variety of pathogens. **FIGURE 23.18** is an example of using the ELISA to determine if someone is HIV positive; that is, does the individual have antibodies to HIV?

The availability of inexpensive ELISA kits has brought the procedure into the doctor's office and CML. Importantly, a positive ELISA for HIV or for any pathogen being tested only indicates that antibodies to the pathogen are present. A positive ELISA does not necessarily mean the person has the disease. The presence of antibodies only says the person has been exposed to the pathogen. **MICROINQUIRY 23** looks at the ELISA test, and **CLINICAL CASE 23** uses serological tests to detect a case of tickborne encephalitis.

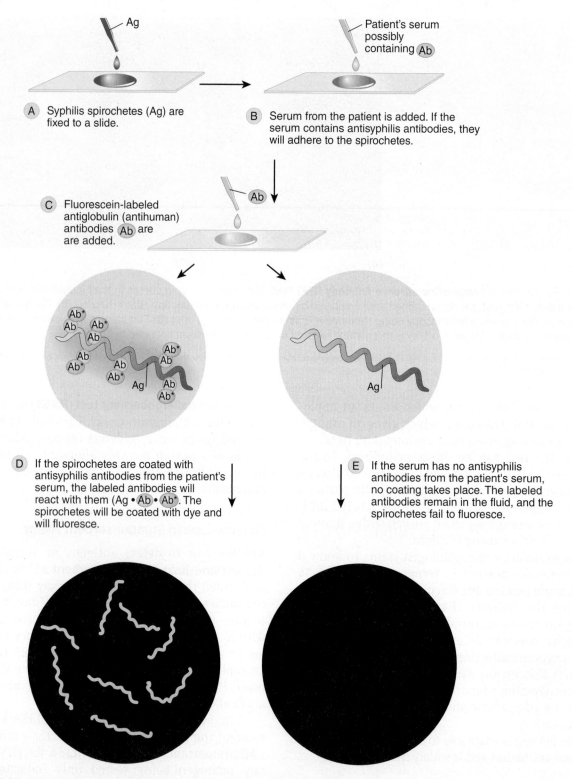

FIGURE 23.16 Indirect Immunofluorescence Antibody Test (IFA) for Diagnosing Syphilis. With the indirect IFA test, a larger visible signal (fluorescence) is observed. *»» Why is this method referred to as the indirect fluorescent antibody technique?*

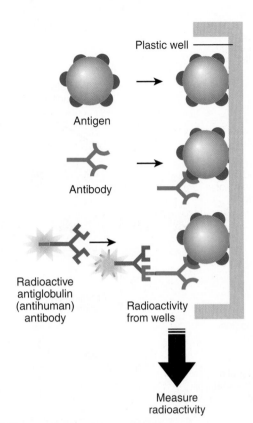

(A) The antigen to the suspected antibody is bound to a plastic device, such as a plastic plate with wells.

(B) The test solution, such as serum, is added to the device. The antibody may or may not be present in the solution.

(C) An antibody that reacts with a human antibody (an antiglobulin antibody) is added. The antiglobulin antibody carries a radioactive label. If radioactivity is detected in the wells, the technician concludes that the solution contained the suspected antibody.

Plastic well

Antigen

Antibody

Radioactive antiglobulin (antihuman) antibody

Radioactivity from wells

Measure radioactivity

FIGURE 23.17 The Indirect Radioimmunoassay (RIA). This technique provides an effective way of detecting very tiny amounts of antibody in a preparation. A known antigen is used. The objective is to determine whether the complementary antibody is present. »» *Would this test represent direct or indirect radioactive antibody labeling? Explain.*

Concept and Reasoning Checks 23.2

a. What does titration accomplish?
b. Summarize the uses for neutralization, precipitation, and agglutination tests.
c. Describe the uses for the complement fixation test.
d. Assess the need to label antigen or antibody in the RAST and ELISA procedures.

Chapter Challenge B

The fact that you have no evidence of measles immunity becomes more of concern because you are planning a trip to Europe. You read that in recent years, the number of measles cases in Americans has increased, and many of these individuals were infected with the measles virus while traveling abroad. As you plan to travel, you read a report that says recently there have been more than 6,500 cases of measles in Europe. You become more alarmed when you read that adults can have a rare but serious measles brain inflammation called acute measles encephalitis. The report says that people of all ages should be up to date on MMR vaccination before traveling abroad. Therefore, you wonder: Are there any blood (serological) lab tests that can be used to determine if a person has ever had measles or been vaccinated?

QUESTION B: *Identify three serological tests that could be used to determine if you have antibodies to the measles virus. How does each test work?*

You can find answers online in **Appendix F**.

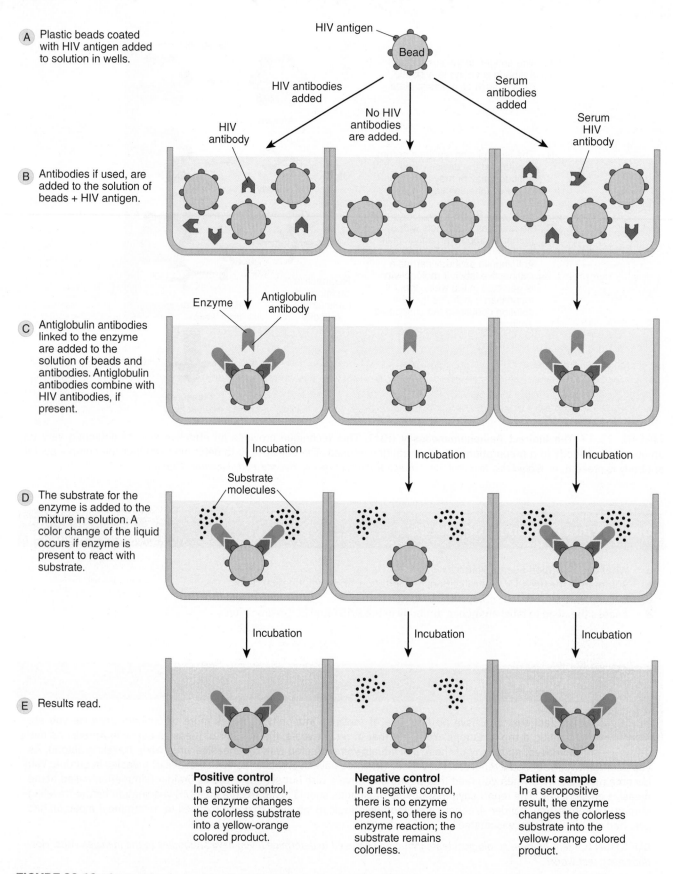

FIGURE 23.18 The ELISA. The enzyme-linked immunosorbent assay (ELISA) as it is used to detect antibodies (in this case HIV) in a patient's serum. *»» What are positive and negative controls, and why are they needed in the test?*

MICROINQUIRY 23

Applications of Immunology: Disease Diagnosis

Ancient Egyptian medical papyri from 1500 BC refer to many different disease symptoms and treatments. Some of these symptoms can still be used to identify diseases today. Thus, one of the most traditional "tools" of diagnosis over the centuries has been a patient's signs and symptoms. However, there are problems when relying solely on these signs and symptoms. Initially, many diseases display common symptoms. For example, the initial symptoms of a hantavirus or anthrax infection are very similar to those of the flu. Some diseases do not display symptoms for perhaps weeks, months—or years in the case of AIDS. Yet, it is important to identify these diseases rapidly so that appropriate treatment, if possible, can be started.

Serological (blood) tests have been used in the United States since about 1910 to both diagnose and control infectious disease. Today's understanding of immunology has brought newer tests that rely on identifying antibody-mediated (humoral) immune responses, that is, antibody reactions with antigens. Serology laboratories work with serum or blood from patients suspected of having an infectious disease. The lab tests look for the presence of antibodies to known microbial antigens. Serological tests that are seropositive indicate antibodies to the microbe were detected, whereas a seronegative result means no antibodies were detected in a patient's serum.

Let's use a hypothetical person, named Pat, who "believes" that 7 days ago he might have been exposed to the hepatitis C virus through unprotected sex. Afraid to go to a neighborhood clinic to be tested, Pat goes to a local drugstore and purchases an over-the-counter, FDA-approved, hepatitis C home testing kit (see the accompanying figure). Using a small spring-loaded device that comes with the kit, Pat pricks his little finger and puts a couple of drops of blood onto a paper strip included with the kit. He fills out the paperwork and mails the paper strip in a pre-paid mailer (supplied in the kit) to a specific blood-testing facility where the ELISA test is performed. In 10 business days, Pat can call a toll-free number anonymously, identify himself by a unique 14-digit code that came with the testing kit, and ask for his test results. If necessary, he can receive professional post-test counseling and medical referrals.

Later in the day, Pat runs into a good friend of his who is a nurse. Pat explains his "predicament" and that he is anxious about having to wait 10 days for the test results. He asks his friend some questions. If you were the nurse, how would you respond to Pat's questions? You can find answers online in **Appendix E**.

23.1a. What is ELISA, and how is it performed?

23.1b. You mention positive and negative controls in your explanation of the ELISA process. Pat asks, "What are positive and negative controls, and why are they necessary?"

After your explanations, Pat still has more questions for you to answer.

23.1c. Pat says, "So suppose I am seropositive. Does it mean I have hepatitis C?"

23.1d. Pat says, "Okay, so if I am seronegative, then I must be free of the virus. Right?"

23.1e. If, indeed, Pat was seronegative, what advice would you give him?

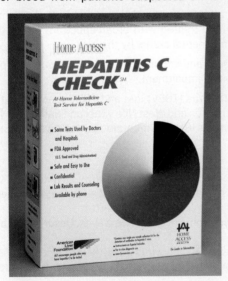

An FDA-approved home testing kit for hepatitis C.

Courtesy of Home Access Health Corporation.

Clinical Case 23

A Case of Tickborne Encephalitis

A 57-year-old man in previous good health came to a Utah hospital emergency room complaining of fever and right arm tremors. Further examination indicated a fever of 39.3°C (102.7°F), an excessively rapid heartbeat, and right-side rigidity (stiffness of the stomach muscles). He appeared disoriented and showed a lack of mental alertness. He was immediately admitted into the hospital.

On questioning the patient, he indicated he had traveled to eastern Russia in June. Further questioning discovered that the patient had been bit multiple times by ticks. Two weeks before coming to the emergency room, the patient had developed fever, muscle pain, and nausea, but the symptoms improved with a broad-spectrum antibiotic prescribed by his doctor.

Then, 4 days before coming to the ER, the patient stated that he had developed a headache, stiff neck, and confusion, which became worse over the next few days. The worsening symptoms are what triggered the patient's coming to the ER.

Ixodes ricinus taking a blood meal on the human skin.

Graham Ella/Alamy Stock Photo.

A serum sample was taken. Examination of a cerebrospinal fluid sample showed abnormal numbers of lymphocytes (112/mm^3; normal = 0–5/mm^3) and elevated protein (84 mg/dl; normal = 15–45 mg/dl). These findings were indicative of an infectious condition, so further microbiological testing was ordered. Bacterial cultures were negative, as were tests for herpes simplex virus and enterovirus infections.

An MRI was ordered, which revealed left-sided cerebral edema with a restriction in blood supply in the thalamus and striatum. A diagnosis of encephalitis of unknown cause was made and treated with broad-spectrum antibiotics and corticosteroids. The patient exhibited no fever and demonstrated normal mental status by the next day.

The patient was released after 14 days with residual right-sided weakness and impaired cognition. Although the patient's motor skills returned within 6 months, he continued to have mild cognitive impairment.

Serum collected on admission tested positive for tickborne encephalitis virus (TBEV) IgM and TBEV-specific neutralizing antibodies. Serological tests for other arboviruses [St. Louis encephalitis virus and West Nile virus] and pathogens [*Borrelia burgdorferi* (Lyme disease), *Leptospira* (leptospirosis), and *Rickettsia* (typhus, Rocky Mountain spotted fever; RMSF)] were negative.

Questions:
  a. What might the right arm tremors, right-sided rigidity, disorientation, and lack of mental alertness indicate?
  b. For the patient, a diagnosis of encephalitis of unknown cause was made. What was missed such that the encephalitis was "of unknown cause"?
  c. Tickborne encephalitis is called a biphasic disease, meaning the symptoms occur in two waves. Describe the biphasic nature of tickborne encephalitis in this patient.
  d. Serum collected on admission tested positive for tickborne encephalitis virus (TBEV) IgM and TBEV-specific neutralizing antibodies. What does the presence of IgM antibodies indicate, and how would the presence of neutralizing antibodies identify TBEV?
  e. What types of serological tests might have been done to rule out the other arboviruses and pathogens?

You can find answers online in **Appendix E**.

For additional information, see www.cdc.gov/mmwr/preview/mmwrhtml/mm5911a3.htm.

■ KEY CONCEPT 23.3   Monoclonal Antibodies Are Used for Immunotherapy

The variety of laboratory tests used to diagnose infectious diseases continues to expand as new biochemical and molecular techniques are adapted to special clinical situations. Monoclonal antibodies represent one of the promising developments.

Monoclonal Antibodies Are Becoming a "Magic Bullet" in Biomedicine

Usually a pathogen (antigen) or vaccine contains several different epitopes. Therefore, a different B-cell population is activated for each epitope, and different antibodies specific to each of the epitopes are produced. Serum from such a patient would contain **polyclonal antibodies** because many different antibodies are derived from different clones

of activated B cells. However, in the laboratory it is possible to produce populations of identical antibodies that bind to only one epitope.

In 1975, César Milstein and Georges J. F. Köhler developed a method for the laboratory production of **monoclonal antibodies** (**MAbs**), that is, antibodies recognizing only one epitope. The key was using **myelomas**, cancerous tumors that arise from the uncontrolled division of plasma cells. Although these myeloma cells have lost the ability to produce antibodies, they can be grown and will divide indefinitely ("immortalized") in culture.

There are two basic steps to generating MAbs. In the first step, a mouse is injected with the antigen of interest (**FIGURE 23.19**), triggering humoral

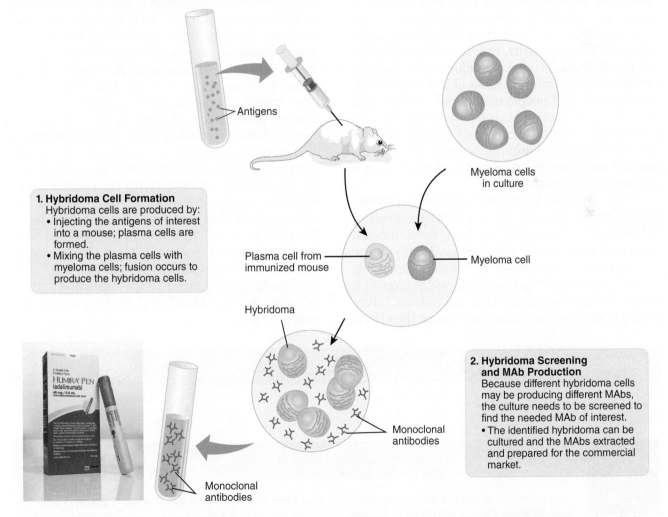

1. Hybridoma Cell Formation
Hybridoma cells are produced by:
• Injecting the antigens of interest into a mouse; plasma cells are formed.
• Mixing the plasma cells with myeloma cells; fusion occurs to produce the hybridoma cells.

Antigens

Myeloma cells in culture

Plasma cell from immunized mouse

Myeloma cell

Hybridoma

2. Hybridoma Screening and MAb Production
Because different hybridoma cells may be producing different MAbs, the culture needs to be screened to find the needed MAb of interest.
• The identified hybridoma can be cultured and the MAbs extracted and prepared for the commercial market.

Monoclonal antibodies

Monoclonal antibodies

FIGURE 23.19 The Production of Monoclonal Antibodies. The production of a monoclonal antibody (MAb) requires fusing a myeloma cell with a plasma cell. After screening for the MAb needed, those clones can be cultivated to produce more of the same MAb. (inset) Humira® was the first fully human MAb approved for use in the United States. »» *What are the unique features of the plasma cell and the myeloma cell?*

immunity. Antibody-secreting plasma cells are then removed from the mouse's spleen and mixed with myeloma cells from a cell culture. By forcing the fusion of plasma cells with myeloma cells, hybrid cells are formed. These so-called **hybridomas** can multiply indefinitely (a myeloma-cell characteristic), producing antibodies that recognize but one epitope (a B-cell characteristic).

Because the antigen originally injected might have several epitopes, different hybridomas might be producing different MAbs. Therefore, the second step is to **screen** (e.g., identify and select) each hybridoma clone for the synthesis of the desired antibody. The hybridoma producing the desired MAb can then be propagated (cloned) in cell culture.

Currently, more than 40 MAbs have been approved for use in the United States. The most prominent MAb targeted at infectious disease is palivizumab (Synagis®), which is designed to prevent respiratory syncytial virus infections in high-risk groups (e.g., infants, children and adults with weakened immune systems, adults with asthma, congestive heart failure, chronic obstructive pulmonary disease, and people with immunodeficiencies such as AIDS). Another MAb, raxibacumab, has been approved for the treatment, and possibly prevention, of inhalational anthrax. Monoclonal antibodies also are in the pipeline for treating hepatitis B and hepatitis C infections.

Most approved MAbs are being used for **immunotherapy**, specifically for controlling overactive inflammatory responses. In fact, today several MAb products, such as adalimumab (Humira®), are directed against inflammatory responses responsible for rheumatoid arthritis, Crohn's disease, or ulcerative colitis. MAbs, such as ustekinumab (Stelara®), represent immunosuppressants that reduce the effects of plaque psoriasis or psoriatic arthritis. Other MAbs function as anticancer treatments. MAbs have been approved for treating malignant lymphoid tumors, leukemia, both colorectal and pancreatic cancers, and breast cancer.

Techniques have been devised to add a "payload" to an MAb. For example, scientists can link a toxin onto the Fc segment of an MAb. When the antibodies are injected back into the patient, they act like "stealth missiles." They react specifically with the tumor cells, and the toxin kills the cells without destroying surrounding tissue.

The use of MAbs represents one of the most elegant expressions of modern biotechnology—and their potential uses in the clinic and medicine remain to be fully appreciated.

In conclusion, we have now learned how vaccines work to protect an individual from infection by many human pathogens. It would seem that we should be less susceptible to disease as more vaccines are developed and made available. In fact, it was through a worldwide vaccine campaign in the 1960s and 1970s that smallpox was eradicated globally. However, here is an interesting twist. Can removal of one pathogen through global vaccination produce the opportunity for other pathogens to flourish? In 2010, researchers at Virginia's George Mason University suggested that the smallpox vaccine gave some protection against HIV and the development of AIDS. With the eradication of smallpox, no one is being vaccinated against the pathogen, and this inadvertently has helped boost the AIDS pandemic. In tests with white blood cells, the investigators report that the smallpox vaccine somehow cut HIV replication fivefold. A link? Both the smallpox virus and HIV use the same surface receptor on white blood cells for attachment and host cell entry.

Concept and Reasoning Check 23.3

a. Explain how MAbs are produced.

Chapter **Challenge C**

All the serological tests to detect measles antibodies were negative. Therefore, it appears you did not ever have measles. You still plan to go to Europe.

QUESTION C: *To be on the safe side, before traveling to Europe, what should you do to ensure you will be protected should you be exposed to someone infected with the measles virus?*

You can find answers online in **Appendix F**.

■ SUMMARY OF KEY CONCEPTS

Concept 23.1 Immunity to Disease Can Be Generated Naturally or Artificially

1. In **naturally acquired active immunity**, antibodies are produced, and lymphocytes activated, in response to an infectious agent. Immunity arises from both contracting the disease and recovering, or from a subclinical infection. **Artificially acquired active immunity** is when antibodies and lymphocytes are produced as a result of a **vaccination**, and immunity is established for some length of time. (Figure 23.2)

2. There are several strategies for producing vaccines, including **attenuated, inactivated, toxoid, subunit**, and **conjugate vaccines**.

3. Adult vaccines can return immunological memory and immune cells to protective levels. (Figure 23.5)

4. **Naturally acquired passive immunity** comes from the passage of antibodies from the mother to fetus or mother to newborn through **colostrum**, which confers immunity for a short period of time. **Artificially acquired passive immunity** is when one receives an **antiserum** (antibodies) produced in another human or animal. This form confers immunity for a short period. Serum sickness can develop if the recipient produces antibodies against the antiserum.

5. **Herd immunity** results from a vaccination program that lowers the number of susceptible members within a population capable of contracting a disease. If there are few susceptible individuals, the probability of disease spread is minimal. (Figure 23.8)

6. Established vaccines are quite safe, although some people do experience side effects that can include mild fever, soreness at the injection site, or malaise after the vaccination. Few people suffer serious consequences.

Concept 23.2 Serological Reactions Can Be Used to Diagnose Disease

7. Serological reactions consist of antigens and antibodies (serum). Successful reactions require the correct dilution (**titer**) of antigens and antibodies. (Figure 23.9)

8. **Neutralization** is a serological reaction in which antigens and antibodies neutralize each other. Often there is no visible reaction, so injection into an animal is required to see if the reaction has occurred. (Figure 23.10)

9. In a **precipitation** reaction, antigens and antibodies react to form a matrix that is visible to the naked eye. Different microbial antigens can be detected using this technique. (Figure 23.11)

10. **Agglutination** involves the clumping of antigens. The cross-linking of antigens and antibodies causes the complex to clump. Some infectious agents can be detected readily by this method. (Figure 23.13)

11. **Complement fixation** can detect antibodies to a variety of pathogens. This serological method involves antigen-antibody complexes that are detected by the fixation (binding) of complement. (Figure 23.14)

12. The **fluorescent antibody technique** involves the addition of fluorescently tagged antibodies to a known pathogen to a slide containing unknown antigen (pathogen). If the antibody binds to the antigen, the pathogen will glow (fluoresce) when observed with a fluorescence microscope, and the pathogen is identified. The **radioimmunoassay (RIA)** uses radioactivity to quantitate radioactive antigens through competition with nonradioactive antigens for reactive sites on antibody molecules. The **enzyme-linked immunosorbent assay (ELISA)** can be used to detect if a patient's serum contains antibodies to a specific pathogen or to detect or measure antigens in serum. A positive ELISA is seen as a colored reaction product. (Figures 23.15–22.18)

Concept 23.3 Monoclonal Antibodies Are Used for Immunotherapy

13. **Monoclonal antibodies** result from the fusion of a specific B cell (plasma cell) with a **myeloma cell** to produce a **hybridoma** that secretes an antibody recognizing a single epitope. (Figure 23.19)

■ CHAPTER SELF-TEST

You can find answers for **Steps A–D** online in **Appendix D**.

■ STEP A: REVIEW OF FACTS AND TERMS

Multiple Choice

Read each question carefully, then select the *one* answer that best fits the question or statement.

1. Exposure to the flu virus, contracting the flu, and recovering from the disease would be an example of _____.
   A. artificially acquired passive immunity
   B. naturally acquired active immunity
   C. artificially acquired active immunity
   D. naturally acquired passive immunity

2. An attenuated vaccine contains _____.
   A. inactive toxins
   B. living, but slow-growing (replicating) antigens
   C. killed bacteria
   D. noninfective antigen subunits

3. Which one of the following vaccinations would most likely require booster shots periodically throughout life?
   A. Tetanus
   B. Measles
   C. Hepatitis B
   D. Mumps

4. Immune complex formation and serum sickness are dangers of _____.
   A. artificially acquired passive immunity
   B. naturally acquired active immunity
   C. artificially acquired active immunity
   D. naturally acquired passive immunity

5. Herd immunity is affected by _____.
   A. the percentage of a population that is vaccinated
   B. the strength of an individual's immune system
   C. the number of susceptible individuals
   D. All the above (**A–C**) are correct.

6. Approximately _____ of 100,000 vaccinated individuals are likely to suffer a serious reaction to the vaccination.
   A. 1
   B. 50
   C. 100
   D. 500

7. Titer refers to _____.
   A. the most concentrated antigen-antibody concentration showing a reaction
   B. the first diluted antigen-antibody concentration showing a reaction
   C. the precipitation line formed between an antigen-antibody reaction
   D. the most dilute antigen-antibody concentration showing a reaction

8. _____ is a serological reaction that produces little or no visible evidence of a reaction.
   A. Precipitation
   B. ELISA
   C. Neutralization
   D. Agglutination

9. The serological reaction where antigens and antibodies form an extensive lattice of large particles is called _____.
   A. fixation
   B. precipitation
   C. neutralization
   D. agglutination

10. When antigens are attached to the surface of latex beads and then reacted with an appropriate antibody, a/an _____ reaction occurs.
    A. inhibition
    B. agglutination
    C. neutralization
    D. precipitation

11. What serological test requires sheep red blood cells and a preparation of antibodies that recognizes the sheep red blood cells?
    A. ELISA
    B. Radioimmunoassay
    C. Immunodiffusion
    D. Complement fixation test

12. In an ELISA, the primary antibody represents _____.
    A. the patient's serum
    B. the antibody recognizing the secondary antibody
    C. the enzyme-linked (labeled) antibody
    D. the antibodies having been washed away

13. A hybridoma cell _____.
    A. secretes monoclonal antibodies
    B. presents antigens on its surface
    C. secretes polyclonal antibodies
    D. is an antigen-presenting cell

Completion

On completing Section 23.1 (Immunity to Disease Can Be Generated Naturally or Artificially), test your comprehension of the section's contents by filling in the following blanks with two terms that best answer the description.

14. Two general forms of immunity: _____ and _____

15. Two types of natural immunity: _____ and _____

16. Two types of passive immunity: _____ and _____

17. Two names for antibody-containing serum: _____ and _____

18. Two ways newborns have acquired maternal antibodies: _____ and _____

19. Two types of viruses in viral vaccines: _____ and _____

20. Two bacterial diseases for which toxoids are used: _____ and _____

STEP B: CONCEPT REVIEW

21. Explain how **artificially acquired active immunity** produces immunity. (**Key Concept 23.1**)

22. Explain why **booster shots** are needed for some vaccines. (**Key Concept 23.1**)

23. Compare and contrast (a) **neutralization**, (b) **precipitation**, and (c) **agglutination** as examples of serological reactions. (**Key Concept 23.2**)

24. Differentiate between **fluorescent antibody, RAST**, and **ELISA** as labeling methods to detect antigen-antibody binding. (**Key Concept 23.2**)

25. Summarize the characteristics of **monoclonal antibodies**. (**Key Concept 23.3**)

STEP C: APPLICATIONS AND PROBLEM SOLVING

26. A friend is heading on a trip to a foreign country where hepatitis A outbreaks are quite common. For passive immunity, he gets a shot of an antiserum containing IgG antibodies. Why do you suppose IgM is not used, especially given that the immunoglobulins are the important components of the primary antibody response?

27. A complement fixation test is performed with serum from a patient with an active case of syphilis. In the process, however, the technician neglects to add the syphilis antigen to the tube. Would lysis of the sheep red blood cells occur at the test's conclusion? Why?

28. Given a choice, which of the four general types of adaptive immunity would it be safest to obtain? Why?

29. An immunocompromised child is taken by his mother to the local clinic to receive a belated measles-mumps-rubella (MMR) vaccination. As the nurse practitioner, you decline to give the child the vaccine. Explain to the mother why you refused to vaccinate the child.

STEP D: QUESTIONS FOR THOUGHT AND DISCUSSION

30. It is estimated that when at least 90% of the individuals in a given population have been immunized against a disease, the chances of an epidemic occurring are very slight. The population is said to exhibit "herd immunity" because members of the population (or herd) unknowingly transfer the immunizing agent to other members and eventually immunize the entire population. What are some ways by which the immunizing agent can be transferred?

31. When children are born in Great Britain, they are assigned a doctor. Two weeks later, a social services worker visits the home, enrolls the child on a national computer registry for immunization, and explains immunization to the parents. When a child is due for an immunization, a notice is automatically sent to the home, and if the child is not brought to the doctor, the nurse goes to the home to learn why. Do you believe a method similar to this can work (or should be used) in the United States to achieve uniform national immunization?

32. Since 1981, the incidence of pertussis has been increasing in the United States, with the greatest increase found in adolescents and adults. Why are adolescents and adults targets of the bacterial pathogen, considering these individuals were usually considered immune to the disease?

33. Children between the ages of 5 and 15 are said to pass through the "golden age of resistance" because their resistance to disease is much higher than that of infants and adults. What factors might contribute to this resistance?

CHAPTER 24

Immune Disorders and AIDS

We are more than 35 years into the pandemic of acquired immunodeficiency syndrome (AIDS), and health officials are still reporting slow progress toward finding a cure for the disease. The Joint United Nations Programme on HIV/AIDS (UNAIDS) reported that in 2016 there are some 37 million people globally living with AIDS (**FIGURE 24.1**), and only about half know they are infected. More than 35 million people out of the 78 million infected have died of AIDS since the start of the epidemic. UNAIDS also reported that worldwide about 5,700 people still become infected with the human immunodeficiency virus (HIV) each day (see the chapter opening figure), which amounts to almost 240 new infections every hour.

In 2016, news that is more positive was reported concerning the pandemic. Globally, since 2010 there has been a 6% reduction in new HIV infections, with new childhood infections dropping by as much as 50%. More than 17 million infected individuals are on antiretroviral therapy, which is up from 7.5 million in 2010. Further, in 2016 almost 50% of children living with HIV were on antiretroviral therapy, which is up 130% from 2010. In all, more than 60 million lives have been saved with drug therapy. Still, for every one person put on antiretrovirals,

two new people become infected and, according to UNAIDS, 20% of patients on drug therapy stop taking the drugs within 1 year.

Although there is no safe and effective vaccine yet developed for AIDS, it is possible that a combination of vaccine and antiviral drugs will just do the trick—and eradicate HIV.

In general, human host defenses (innate and adaptive immune responses) continually try to protect the body from, among other things, infectious diseases. These responses work together to provide a powerful and intricate system that usually works without fail. However, sometimes due to genetic errors or an infection, the immune system defenses

Human immunodeficiency virus (HIV) particles (yellow) attached to a T lymphocyte (red).
Courtesy National Institute of Allergy and Infectious Diseases (NIAID).

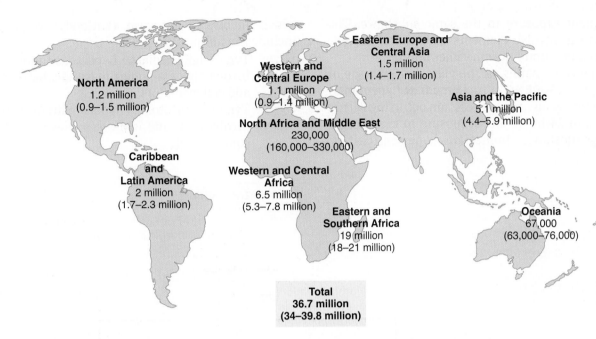

FIGURE 24.1 Adults and Children Living with HIV—2016. The numbers of HIV infections shown by World Health Organization (WHO) region is an estimate. *»» Provide some reasons why these numbers represent only estimates.*

Data from WHO.

do not function properly, leading to some type of immune disorder.

Even though AIDS is the major topic in this chapter, it is by no means the only immune disorder. Hypersensitivities and allergies represent immune disorders that many believe can be at least partly the result of

children not fighting the types of infectious diseases experienced by past generations. Also included in the broad category of immune disorders are the autoimmune diseases that often have a microbial component associated with them. We will survey each type of immune disorder in the following sections.

Chapter Challenge

Carla is a 20-year-old long-distance runner for her college track team. Recently she has begun to develop what appears to be an allergic reaction to many unrelated events. As a 17-year-old high school student, she began to develop hives over her skin surface every time she ran. Hives, also called urticaria, is a sudden outbreak of swollen, pale red bumps (wheals and flares) on the skin that most often is an allergic reaction to a drug or food item. But an allergy to exercise? For Carla, the itchy bumps would disappear in a couple of hours after running but would reappear every time she went running. Now in college, the same reaction still occurs. What has made her more concerned is that while taking an anatomy and physiology exam, hives began appearing all over her body, just like when she ran. However, this time they lasted for 5 months and itched intensely. What is going on with Carla's body? Let's find out as we progress through this chapter on immune disorders.

■ KEY CONCEPT 24.1 Type I Hypersensitivity Represents a Familiar Allergic Response

Hypersensitivity is a multistep phenomenon triggered by the immune system from exposure to substances like pollen, peanuts, or jewelry. The

reactions consist of a dormant (latent) stage, during which time an individual becomes sensitized, and an active (hypersensitive) stage following a subsequent exposure to the same substance. The process can involve elements of humoral or cell-mediated immunity or sometimes both.

Hypersensitivities can be classified into one of four types (**FIGURE 24.2**). An **immediate hypersensitivity** refers to an immune reaction occurring within minutes to 24 hours after exposure to the allergy-causing substance. Immediate hypersensitivities, which involve a humoral (antibody) response, include the following:

▶ Type I IgE-mediated hypersensitivity: this involves IgE antibodies, mast cells, basophils, and cell mediators.

▶ Type II cytotoxic hypersensitivity: this involves IgG and IgM antibodies as well as complement.

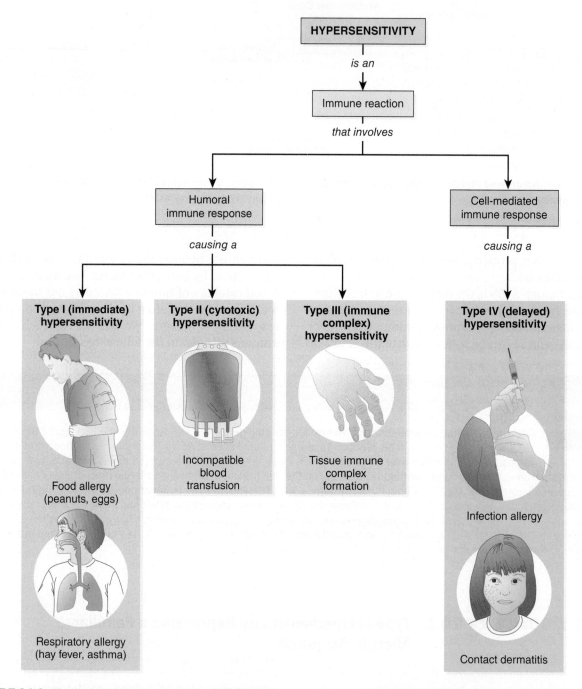

FIGURE 24.2 The Various Forms of Hypersensitivities. Immune system hypersensitivities may be related to B lymphocytes or T lymphocytes. Stimulation of the immune system is the starting point for all these disorders. *»» Why are all these examples of immune stimulation referred to as hypersensitivities?*

▶ Type III immune complex hypersensitivity: this involves IgG and IgM antibodies, complement, and the formation of antigen-antibody immune complexes in tissues.

The fourth type is a **delayed hypersensitivity**, in which a cell-mediated immune response develops over 2 to 3 days. This hypersensitivity involves cytokines and T lymphocytes.

Type I Hypersensitivity Is Induced by Allergens

A **type I hypersensitivity** (commonly called an **allergy**) is an exaggerated (hypersensitive) or inappropriate immune response triggered by IgE antibodies. One explanation for type I hypersensitivity reactions is that noninfectious, but often noxious, chemical substances we are exposed to in the environment are identified by the immune system as something that must be eliminated. These environmental antigens, called **allergens**, include insect venoms, mosquito and tick salivary excretions, and irritants like dust. An allergic response to bee venom might have evolved as a protection against the often severe, tissue damaging and life-threatening effects caused by enzymes in the venom. Unfortunately, an overreaction, if triggered systemically, can also be life threatening.

But what about allergens like plant pollens? Certainly, they are not noxious or dangerous substances yet in an allergic individual, a pollen allergen will provoke an "expulsion response" that can involve coughing, sneezing, and watery eyes. Unfortunately, a complete understanding of the hypersensitivity response remains elusive, so let's just examine the reactions typically occurring with an allergy. The process consists of two steps, described in the following subsections.

Sensitization

A type I hypersensitivity has all the characteristics of a humoral immune response. In a susceptible individual, the hypersensitivity begins with the recognition of an allergen, which can be any of a wide variety of materials such as plant pollen, certain foods, bee venom, serum proteins, or a drug, such as penicillin. Allergists refer to this first exposure to the allergen as the **sensitizing dose**, and the process is called **sensitization**.

In this first exposure, the immune system responds to the allergen as if it were a dangerous antigen, such as a pathogen. The allergen is taken up by antigen-presenting cells (APCs) and fragments presented to **helper T2 (T$_H$2) cells** (FIGURE 24.3). As in humoral immunity, T$_H$2 cells stimulate B cells to mature into plasma cells, which in this case are stimulated to produce IgE antibodies. The antibodies enter the circulation and attach by the Fc tail segment to the surface of mast cells and basophils. Both types of white blood cells can be found in the respiratory and gastrointestinal tracts and nearby blood vessels. Importantly, they are filled with "granules" containing histamine and other physiologically active substances.

As IgE antibodies attach to mast cells and basophils, the individual becomes "sensitized." Multiple stimuli by allergen molecules might be required to sensitize a person fully. This is why penicillin often must be taken several times before a penicillin allergy reveals itself.

Sensitization usually requires a minimum of 1 week, during which time millions of molecules of IgE attach to thousands of mast cells and basophils. Because attachment occurs at the Fc fragment of the antibody, the allergen (antigen) binding sites point outward from the cell (see Figure 24.3).

Degranulation and Allergy Development

After a person is sensitized, a subsequent exposure to the same allergen allows the allergen molecules to bind to the allergen binding sites on the IgE antibodies. This binding triggers **degranulation**, a release of granule contents at the cell surface. Degranulation can occur within minutes of or a few hours after allergen exposure.

As granules fuse with the plasma membrane of the basophils and mast cells, an **immediate allergic response** releases a number of preformed chemical mediators having substantial pharmacological activity (FIGURE 24.4). The best-known preformed mediator of allergic reactions is **histamine**. After it enters the bloodstream, histamine and other mediators like serotonin attach to the receptors present on most body cells, triggering vascular (capillary) permeability, tissue swelling, redness, mucus secretion, and smooth muscle contraction that can lead to breathing difficulties.

Still other mediators must be synthesized after the antigen-IgE reaction, triggering a **late-phase allergic response** (see Figure 24.4). These mediators, along with cytokines, are extremely potent at causing smooth muscle constriction and bronchial tube constriction, which further intensifies the inflammatory response and the final allergic outcome.

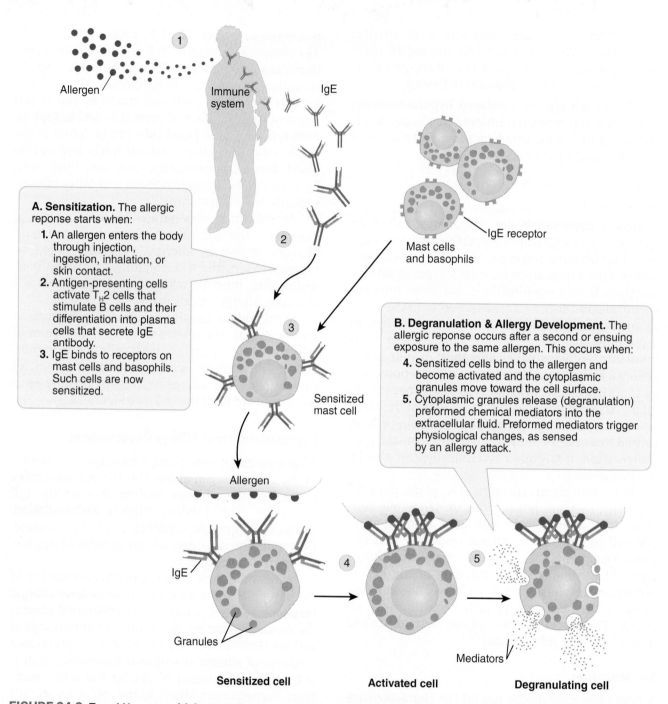

A. Sensitization. The allergic reponse starts when:
1. An allergen enters the body through injection, ingestion, inhalation, or skin contact.
2. Antigen-presenting cells activate T_H2 cells that stimulate B cells and their differentiation into plasma cells that secrete IgE antibody.
3. IgE binds to receptors on mast cells and basophils. Such cells are now sensitized.

B. Degranulation & Allergy Development. The allergic reponse occurs after a second or ensuing exposure to the same allergen. This occurs when:
4. Sensitized cells bind to the allergen and become activated and the cytoplasmic granules move toward the cell surface.
5. Cytoplasmic granules release (degranulation) preformed chemical mediators into the extracellular fluid. Preformed mediators trigger physiological changes, as sensed by an allergy attack.

FIGURE 24.3 Type I Hypersensitivity Reaction. Basophils, mast cells, and IgE antibodies are involved in typical allergies like hay fever. *»» Does a person experience the symptoms of an allergy on the first or ensuing exposure to an allergen? Explain.*

Type I Hypersensitivities Can Be Localized or Systemic

There are two forms of type I hypersensitivities. The often less severe form is **atopy** (*atopo* = "out of place"), which refers to any localized and chronic allergy such as hay fever or even asthma. A more severe allergic response is called **anaphylaxis** (*ana* = "without"; *phylaxi* = "protection"), which involves a systemic and potentially life-threatening reaction. Let's take a closer look at each.

Atopy

The vast majority of hypersensitivities involve a localized sensitization of mast cells and a controlled production of IgE antibody. According to public health estimates, more than 60 million Americans

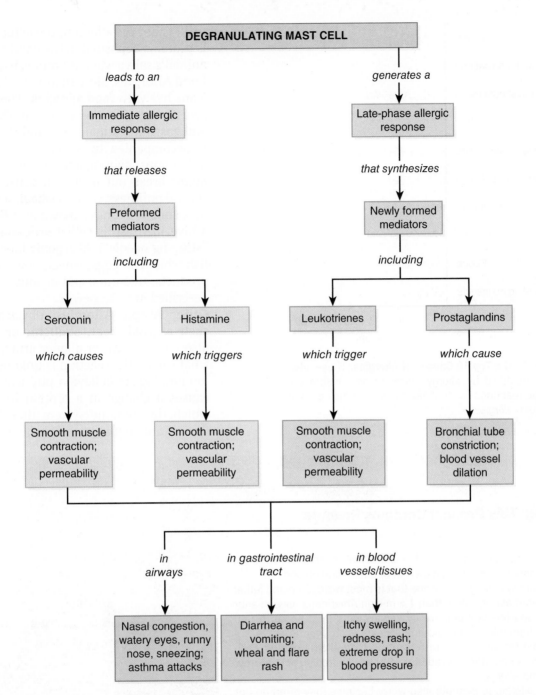

FIGURE 24.4 Concept Map for Type I Hypersensitivity and Mediator Substance Production. Several chemical mediators are stimulated upon IgE cross-linking on mast cells and basophils that have various effects in the body. *»» Because IgE is an antibody normally targeting parasitic helminths, what might these mediator substances have done if it was a parasite infection rather than an allergen?*

test positive for some form of an atopic disorder (**FIGURE 24.5**).

▶ **Allergic Rhinitis.** Hay fever is an example of what is called **allergic rhinitis**. It affects between 10% and 30% of all adults and up to 40% of children in the United States. These seasonal allergies, mostly in the nose and eyes, occur when allergens, such

as dust or pollen, are breathed in by a sensitized individual.

There also are year-round allergies. These perennial hypersensitivities usually result from chronic exposure to such substances as house dust, mold spores, dust mites, detergent enzymes, and the particles of animal skin and hair ("dander"). Actually, dander

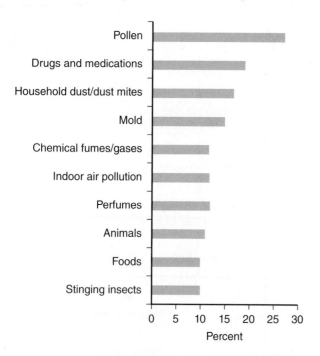

FIGURE 24.5 The Top 10 Causes of Allergies. These allergens were identified by allergy sufferers as causing the most allergic reactions. *»» Of these allergens, which ones, if any, cause your allergies?*

Data from Marist Institute for Public Opinion.

itself is not the allergen; the actual allergen is proteins deposited in the dander from the animal's saliva when it grooms itself.

▶ **Food Allergies.** More than 15 million Americans live with **food allergies**. These hypersensitivities usually occur in the mouth and gastrointestinal tract, and the reaction is accompanied by swollen lips, abdominal cramps, nausea, and diarrhea. The skin might break out in a rash consisting of a central puffiness, called a **wheal**, surrounded by a zone of redness known as a **flare**. Such a hive-like rash is called **urticaria** (*urtica* = "stinging needle"). **Allergenic** foods include fish, shellfish, eggs, wheat, cow's milk, soy products, tree nuts, and peanuts, the latter described in MICROFOCUS 24.1.

▶ **Other Atopic Allergies.** Physical factors such as cold, heat, sunlight, or sweating also can lead to an allergy attack. Exactly what causes the reaction is unknown. Some immunologists believe a physical stimulus causes a change in a protein in the skin, which the immune system then "sees" as foreign and mounts an immune response.

MICROFOCUS 24.1: Miscellaneous

Warning: This Product Contains Peanuts

Americans love peanuts—salted, unsalted, oil roasted, dry roasted, Spanish, honey crusted, in shells, out of shells, and on and on.

Yet of all food allergies, the one that is most worrisome and lethal is the peanut allergy. More than 1.5 million Americans have peanut allergies, and hospital emergency departments treat about 30,000 cases of food-related allergic anaphylaxis each year. Peanuts are responsible for about 50% to 60% of all food allergy deaths each year in the United States. In addition, 80% of children do not outgrow their peanut allergy.

People allergic to peanuts also worry because they often do not need to come into direct contact with peanuts to develop an allergic reaction. Peanut particle allergens can travel by air or be "trace contaminants" in other foods processed on the same manufacturing line used for peanuts. Although medical experts believe it is unlikely that sensitized individuals would go into anaphylactic shock merely from inhaling or ingesting these minute quantities of peanut particles, the sensitized individuals often do develop allergic symptoms—like a runny nose or itchy, watery eyes—in the presence of these particles.

Courtesy of USDA-ARS/Scott Bauer.

The good news for peanut allergy sufferers is that recent immunotherapy trials are helping kids beat their allergies to peanuts. These oral treatments involved ingestion of very small doses of peanut products (powder, peanut protein, or probiotic form) for a year or more. Most of the children tolerated the therapy without developing an allergic reaction. For up to 90% of the children in a therapy program, after the therapy stopped, they could now tolerate peanuts in their regular diet; they were said to have sustained unresponsiveness. However, the therapy did not work for all children, as some still maintained their hypersensitivity to peanuts.

Obviously, much more research must be done. However, the research is looking promising and some researchers hope that soon there will be some form of immunotherapy generally available for individuals suffering from peanut allergies.

There also can be immediate hypersensitivity reactions to chemicals, specifically to proteins found in natural latex. Although latex gloves are effective in preventing the transmission of many infectious diseases to healthcare workers, more than 10% of healthcare workers exhibit a latex allergy from wearing latex gloves or contacting medical products that contain latex. Symptoms include skin redness, itching, or a rash, whereas more severe reactions can cause sneezing, a runny nose, itchy eyes, and wheezing. Rarely, life-threatening shock can occur. Today, most products containing latex have suitable alternatives.

Systemic Anaphylaxis

One of the potentially dangerous reactions of type I hypersensitivity involves allergens, such as bee venom, penicillin, nuts, and seafood. If these allergens are deposited directly into the bloodstream or via the digestive tract, they can trigger degranulation of basophils throughout the body. In affected individuals, the main activity of the released mediators is to contract smooth muscles in the small veins and increase vascular permeability, forcing fluid into the tissues (see Figure 24.4). The resulting drop in blood pressure can lead to unconsciousness and **anaphylactic shock**.

In addition, the skin might become swollen around the eyes, wrists, and ankles, a condition called **edema**. The edema is accompanied by a hive-like rash, along with burning and itching in the skin, as the sensory nerves are excited. Contractions also occur in the gastrointestinal tract and bronchial muscles, leading to sharp cramps and shortness of breath, respectively. The individual inhales rapidly without exhaling and traps carbon dioxide in the lungs, an ironic situation in which the lungs are fully inflated but lack oxygen. Being an immediate hypersensitivity, if prompt action is not taken, death from asphyxiation can occur in 10 to 15 minutes. Many doctors suggest that people subject to systemic anaphylaxis should carry a self-injecting syringe containing epinephrine (adrenalin), which stabilizes mast cells and basophils, dilates the airways, and constricts the capillary pores, keeping fluids in the circulation.

Allergic Reactions Can Initiate Many Cases of Asthma

Asthma is a chronic condition characterized by wheezing, coughing, and stressed breathing. Some

FIGURE 24.6 Using an Asthma Inhaler. For some asthma sufferers, simple exercise might bring on an asthma attack. Using an inhaler that delivers medication directly to the lungs allows sufferers to live active lives without fear of an attack. *»» How could physical exercise trigger an asthma attack?*

© Michael Donne/Science Source.

300 million people live with asthma worldwide (20 million adults, 7 million children in the United States), and the disorder is responsible for more than 3,400 deaths annually.

Asthma has been blamed on airborne allergens such as pollen, mold spores, or products from insects like dust mites. In fact, about 50% of asthma cases have an environmental origin. Attacks also can be induced by physical exercise or cold temperatures—situations that irritate and inflame the airways (**FIGURE 24.6**). Importantly, microbes can influence the onset of asthma, as INVESTIGATING THE MICROBIAL WORLD 24.1 explores.

Asthma primarily is an inflammatory disorder that occurs in two stages. First, there is an **immediate allergic response** to allergen exposure in which T_H2 cells control the IgE response. However, unlike a common allergy, the mediators are not released in

Investigating the Microbial World 24.1

Microbes and the Allergic Response

Allergies and asthma have long been suffered by many individuals in various populations. The causes of the disorders have also been studied for decades, and yet we still have an incomplete picture of how the process is triggered even as the numbers of allergy and asthma cases continue to increase. Recently though, researchers have been getting some very interesting clues to the causes.

OBSERVATION: Children on farms appear to suffer many fewer allergic attacks and less asthmatic lung disease than children who live in more urban areas and large cities (see MicroInquiry 24). In addition, it has been proposed that the overuse of antibiotics could be a predisposing factor leading to allergies and asthma. Therefore, some researchers are investigating whether the bacterial populations (microbiome) in our airways and gut influence the onset of an allergic airway disease (AAD) such as asthma. It is known that germ-free animals have defects in their immune responses that produce an allergic airway response that mimics the clinical features seen in humans.

QUESTION: *Does the removal of natural bacterial populations (microbiome) from the gut lead to the development of AAD?*

HYPOTHESIS: Established bacterial populations in the gut lessen the chances of developing AAD. If so, in test animals the removal of the gut microbial population and exposure to a set of airway allergens will cause AAD response.

EXPERIMENTAL DESIGN: Groups of pathogen-free mice were split into three groups. An AAD response was characterized by increased numbers of immune cells (eosinophils, mast cells, neutrophils) and immune molecules (IgE antibody and cytokines).

EXPERIMENTAL GROUP (GROUP A): One group of mice was treated for 5 days with a broad-spectrum antibiotic to decrease the total bacterial microbiome and then given a single oral dose of the yeast *Candida albicans* to establish a low level of yeast in the microbiome, which typically occurs as a result of human antibiotic therapy. On days 2 and 9, the mice were exposed intranasally to mold spores from *Aspergillus fumigatus*, a common indoor allergen affecting humans. Mouse tissues were examined on day 12.

CONTROL GROUPS (GROUPS B AND C): One group of mice (group B) received the antibiotic and yeast exposures but no exposure to mold spores. The other group (group C) was not exposed to antibiotic or yeast but was exposed to mold spores on days 2 and 9.

RESULTS: See the accompanying figure. Note: the intranasal exposure to fungal spores did not cause any type of infection or fungal disease in any of the three groups of mice.

CONCLUSIONS: Based on the results, clinical features mimicking an AAD were noted.

> **QUESTION 1:** *Did the experimental study support the hypothesis being tested? Explain.*
>
> **QUESTION 2:** *Which group(s) of mice exhibited the characteristics of AAD? Explain your selection.*
>
> **QUESTION 3:** *Only animals with active helper T cells show the characteristics of AAD. Why were active T cells only found in Group A?*

You can find answers online in **Appendix E**.

Fecal analysis:
- Group A: Bacterial microbiota reduced 99.99% in the gut

C. albicans analysis:
- Group A: More than 99.95% of yeast in the gut; no yeast or bacterial cells detected in or cultured from the lungs

Pulmonary analysis:
- Group A: Increase in eosinophils, mast cells, neutrophils, IgE, and cytokines
- Increase in helper T-cell response

Group A: Antibiotic (in water) + C. albicans (oral route) + A. fumigatus spores (intranasal) + A. fumigatus spores (intranasal) — days −4, 0, 1, 2, 6, 9, 12

Group B: Antibiotic (in water) + C. albicans (oral route) — No exposure to spores — days −4, 0, 1, 2, 6, 9, 12

Group C: Sterile water — No exposure to C. albicans spores — A. fumigatus spores (intranasal) — A. fumigatus spores (intranasal) — days −4, 0, 1, 2, 6, 9, 12

Fecal analysis:
- Group B: Bacterial microbiota reduced by 99.99% in the gut
- Group C: No change in gut microbiota

C. albicans analysis:
- Group B: More than 99.95% of yeast were in the gut; no yeast or bacterial cells detected in or cultured from the lungs
- Group C: No yeast cells detected

Pulmonary analysis:
- Group B: No AAD response detected
- Group C: Only cytokines and neutrophils increase
- Group B and C: No increase in helper T-cell response

Data from Noverr, M. C., et al. 2004. *Infect Immun* 72(9):4996–5003.

Data from Noverr, M. C. et al. (2004). *Infect. Immun.* 72 (9): 4996–5003.

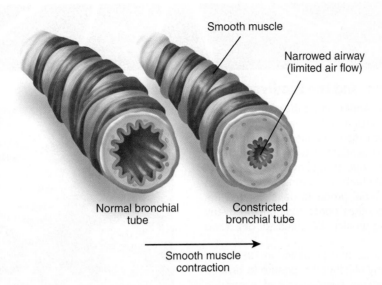

Smooth muscle

Narrowed airway
(limited air flow)

Normal bronchial
tube

Constricted
bronchial tube

Smooth muscle
contraction

FIGURE 24.7 Airway Narrowing During an Asthma Attack. During an asthma attack, chemical mediators (cytokines) from mast cells cause the smooth muscle layers to spasm, constrict, and form thick mucus that can block the airway. *»» What types of cells are responsible for releasing the asthma-causing chemical mediators?*

the nose or eyes but rather from mast cells in the lower respiratory tract. The resulting bronchial constriction, vasodilation, and mucus buildup can make breathing difficult (**FIGURE 24.7**).

In the second stage, the synthesis and release of specific cytokines result in the recruitment of eosinophils and neutrophils into the lower respiratory tract. Being part of a **late-phase allergic response**, the presence of the eosinophils and neutrophils can cause tissue injury and potentially cause blockage of the airways (bronchioles). Bronchodilators can be used to open the airways by widening the bronchioles. More recently, the use of anti-inflammatory agents such as inhaled steroids or nonsteroidal cromolyn sodium have been prescribed. Cromolyn sodium blocks degranulation by mast cells.

Allergies Are Not Universal

Not everyone suffers from allergies. Research suggests that IgA antibodies in nonallergic individuals block the production of IgE antibodies. In fact, in atopic individuals the IgA level is greatly reduced to that in nonallergic individuals. Immunologists believe the B lymphocytes and plasma cells that produce IgA in nonallergic individuals somehow shield lymphocytes producing IgE antibodies.

Another theory of atopic disorders involves the control of IgE synthesis by **suppressor T cells**. In nonallergic individuals, a rising IgE concentration causes suppressor T cells to shut off IgE synthesis. However,

in atopic individuals the mechanism malfunctions, possibly because the T cells are defective, and IgE therefore is produced in massive quantities. Allergic people are known to possess almost 100 times the IgE level of people who do not have allergies.

Most interestingly, hygiene appears to affect the development of allergies and asthma, especially in young children, but not in the way you might think, because microbes might play a critical role in preventing allergies. **MICROINQUIRY 24** looks into the microbial connection.

Therapies Sometimes Can Control Allergies

The best way to avoid hypersensitivities is to identify and avoid contact with the allergen. Let's look at a couple of the tests to identify allergens and the available treatments.

Skin Tests

An allergy skin test can be performed to determine to what allergens a person is sensitive. The skin test can be applied to an individual's forearm or back. In the example shown in **FIGURE 24.8**, the nurse first prepares the individual's back by wiping it with alcohol and then makes a series of number marks on the back with a pen that correspond to a series of numbered allergens that are to be tested. The test allergens include grasses; molds; common tree, plant, and weed pollens; cat and animal dander;

MICROINQUIRY 24

Allergies, Microbes, and the Biodiversity Hypothesis

Today, about 20 million Americans suffer from asthma, including more than 7 million children and adolescents. The reasons for this increase—and allergies in general—are uncertain. Many scientists have suggested the increase is in large part due to our overly clean urban lifestyle. We use disinfectants for almost everything in the home, and antibacterial products have flooded the commercial markets. In other words, maintaining overly good hygiene is making us sick. We need to "get down and dirty"!

The "hygiene hypothesis," first proposed in 1989, suggests that a lack of early childhood exposure to diverse microbes and other infectious agents can lead to immune system weakness and an increased risk of developing asthma and allergies. In past centuries, infants and their immune systems had to battle all sorts of infectious diseases—from typhoid fever and polio to diphtheria and tuberculosis—as well as ones that were more mundane. Such interactions and recovery "pumped up" the immune system and prepared it to act in a controlled manner. Today, most children in developed nations are exposed to far fewer pathogens, and their immune systems remain "wimpy," often unable to respond properly to non-pathogenic substances like pollen and cat dander. Their immune systems have not had the proper "basic training."

In 2003, researchers at the National Jewish Medical and Research Center in Denver reported that mice infected with the bacterium *Mycoplasma pneumoniae* had less severe immunological responses when challenged with an allergen. However, if mice were exposed to allergens first, they developed more severe allergic responses. Also, the allergy-producing mediators were at lower levels in mice first exposed to *M. pneumoniae*. So, early "basic training" of the immune system seems to temper allergic responses.

Many other studies also bolster the now so-called "biodiversity hypothesis." One study used data from the Third National Health and Nutrition Examination Survey conducted by the Centers for Disease Control and Prevention (CDC). The survey included 34,000 American residents ranging from 1 year in age to older than 90. The CDC analysis concluded that humans who were seropositive for hepatitis A virus, *Toxoplasma gondii*, and herpes simplex

Dirt may be good for you.

virus type 1—that is, markers for previous microbial exposures—were at a decreased risk of developing hay fever, asthma, and other atopic diseases.

Another study used data collected from 812 European children ages 6 to 13 who either lived on farms or did not live on farms. Using another marker—an endotoxin found in dust samples from bedding—the investigators reported that children who did not live on farms were more than twice as likely to have asthma or allergies than were children growing up on farms. Presumably, on farms, children are exposed to more and a wider range of "immune-strengthening" microbes.

Moreover, not just children can be affected. A 2013 report suggested an adult's immune system could also be positively affected by exposure to a farming environment to dampen down hypersensitivities. Don't live on a farm? The presence of a dog or other pet in the home before birth, attending day care during the first year of life, and simply allowing children to play together and get dirty might just bring a downward spiral to allergies.

Discussion Point

Do you suffer from common allergies? Discuss whether your allergies and your childhood fit the description in this MicroInquiry.

---

foods; and other common substances that people encounter. Each allergen is applied by pricking the skin with a separate sterile lancet containing a drop of the allergen.

It takes about 15 to 20 minutes for possible reactions to develop (immediate allergic response).

Positive reactions appear as small circles of inflammation (wheal and flare reactions), which itch and look like mosquito bites. The nurse then measures the diameter of any wheals with a millimeter ruler. To be sensitive, a wheal must be at least 3 mm larger than the reaction to the negative control. Based on

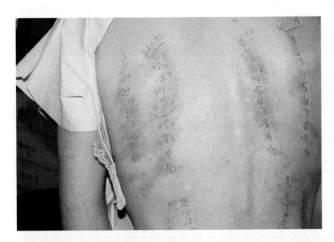

FIGURE 24.8 An Allergy Skin Test. The back of the individual is marked and then pricked with a large variety of test allergens. In about 20 minutes, the nurse notes if a skin wheal has developed in response to each allergen. The diameter of the wheal is then measured. »» *Would you consider this patient to have an overall skin reaction that is mild, moderate, or severe?*

Courtesy of Dr. Jeffrey Pommerville.

wheal size, allergen sensitivity can be assessed as mild, moderate, or severe. After all the test allergens are tabulated, the specific allergens to which the person is sensitive can be deduced and possible treatments discussed with the allergist.

Blood Tests

Allergy blood tests often are used for people who cannot have skin tests or for those who have had severe allergic reactions (anaphylaxis) in the past. The most common type of blood test used is the **enzyme-linked immunosorbent assay** (**ELISA**). It measures the level of IgE in a person's blood in response to certain allergens. IgE levels four times the normal level usually indicate sensitivity to the allergen or set of allergens.

Clearly, avoiding the allergen or allergens is the most effective way of preventing an allergic reaction. However, often avoidance is not possible, so various medications are available that "cover up" allergic symptoms or prevent them temporarily. There are numerous over-the-counter (OTC) and prescription drugs used to relieve the symptoms of allergies or prevent the release of the chemical mediators that cause the allergies.

Antihistamines include OTC and prescription drugs, such as Allegra® and Claritin®. The drugs block the effects of histamine, and they are used to relieve or prevent the symptoms of hay fever and other types of atopic reactions. Steroids—specifically **corticosteroids**—help reduce inflammation associated with allergies. These medicines, such as Flonase® and Advair® (for asthma), are sprayed or inhaled into the nose or mouth. Mast cell stabilizers, such as cromolyn, help control allergic symptoms by inhibiting the release of chemical mediators from mast cells. These products, therefore, are only effective where they are applied (nasal passages or eyes). Other drugs block the effects of those chemical substances that cause smooth muscle contractions.

When medications fail to control allergy symptoms, an allergist may recommend **allergen immunotherapy** (commonly called **allergy shots**) to reduce sensitivity to allergens. It is especially helpful for individuals sensitive to seasonal allergens, indoor allergens, and insect stings. As described in MICROFOCUS 24.1, therapy involves a series of shots containing the allergen extract given regularly for several years. The belief is that the shots will allow the individual's body to stop producing as much IgE antibody, and, therefore, the individual will not have as severe an allergic response when exposed to the natural allergen in the environment. This process is called **desensitization**. After a course of allergy shots, 80% to 90% of patients have less severe allergic reactions and may even have their allergies completely resolved.

Concept and Reasoning Checks 24.1

a. Summarize the events occurring during a first and second exposure to an allergen.
b. Distinguish between the different forms of atopic disorders.
c. Explain why systemic anaphylaxis is life threatening.
d. How can asthma be a life-threatening allergic reaction?
e. Evaluate the medications and treatments available for allergies.

Chapter Challenge A

Carla's outbreak of hives continues. She decides to visit an allergist. The allergist asks her if she is on any kind of medication or taking antibiotics. He also asks Carla if she can remember eating any specific type of food before the onset of hives. Carla says no and this condition has been recurring for 3 years, anyway.

QUESTION A: *What type of localized anaphylaxis does the allergist suspect Carla might have? Does this hypersensitivity seem to match her symptoms? What might you suspect could be the cause of Carla's skin reaction?*

You can find answers online in **Appendix F**.

■ KEY CONCEPT 24.2   Other Types of Hypersensitivities Represent Immediate or Delayed Reactions

Immunological responses involving IgG antibodies or T cells also can lead to adverse hypersensitivity reactions.

Type II Hypersensitivity Involves Cytotoxic Reactions

A **type II hypersensitivity** is a cytotoxic, cell-damaging response that occurs when IgG or IgM antibodies react with antigens on the surfaces of cells (**FIGURE 24.9**). Complement often is activated, but IgE does not participate nor is there any degranulation of mast cells. Rather, antibody-bound cells are subject to the formation of complement-stimulated **membrane attack complexes** (**MACs**).

A well-known example of cytotoxic hypersensitivity is the transfusion reaction arising from the mixing of incompatible blood types (TABLE 24.1). A person receiving blood of the incorrect type can have a series of reactions capable of destroying red blood cells (RBCs; erythrocytes) and producing an inflammatory response that can lead to kidney failure or even death. For example, if a person with type A blood donates to a recipient with type O blood, the A antigens on the donor's erythrocytes will react with anti-A antibodies in the recipient's plasma, and the cytotoxic effect will be expressed as agglutination of donor erythrocytes and activation of complement in the recipient's circulatory system. Therefore, blood banks cross-match the donor's RBCs with the recipient's serum to ensure compatibility.

MICROFOCUS 24.2 examines a microbial reason as to why there are different human blood types.

When some antibiotic drugs, such as penicillin, are taken by patients, the drug can adsorb nonspecifically to membrane proteins on erythrocytes (see Figure 24.9). The complex in some patients leads to a rare **drug-induced hemolytic anemia** if the complex triggers antibody formation. Binding to the attached antibodies induces complement-mediated lysis and thus progressive anemia.

Another expression of cytotoxic hypersensitivity is **hemolytic disease of the newborn**, or **Rh disease**. This problem arises from the fact that erythrocytes of approximately 85% of Caucasian Americans contain a surface antigen, first described in rhesus monkeys and therefore known as the Rh antigen. Such individuals expressing the antigen are said to be Rh positive. The 15% who lack the antigen are Rh negative.

The ability to produce the Rh antigen is a genetically inherited trait. When an Rh-negative woman marries an Rh-positive man, there is a 3 to 1 chance (or 75% probability) that the trait will be passed to the child, resulting in an Rh-positive child. During the birth process, a woman's circulatory system is exposed to her child's blood, and if the child is Rh positive, the Rh antigens enter the woman's blood and stimulate her immune system to produce Rh antibodies (**FIGURE 24.10**). If a succeeding pregnancy results in another Rh-positive child, IgG antibodies (from memory cell activation) will cross the placenta (along with other resident IgG antibodies) and enter the fetal circulation. These antibodies will react with Rh antigens on the fetal erythrocytes and cause complement-mediated lysis of the cells. The fetal circulatory system rapidly releases immature erythroblasts to replace the lysed blood cells, but these cells are also destroyed. The result can be stillbirth or, in a less extreme form, a baby with jaundice.

Modern treatment for hemolytic disease of the newborn consists of the mother receiving an

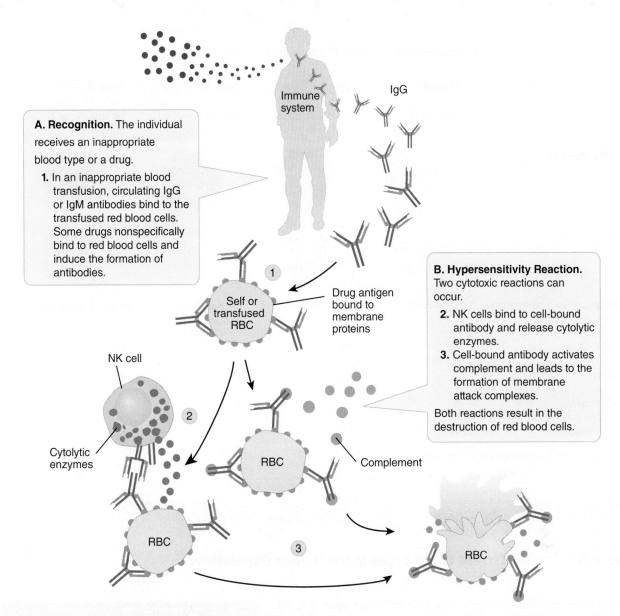

A. Recognition. The individual receives an inappropriate blood type or a drug.

1. In an inappropriate blood transfusion, circulating IgG or IgM antibodies bind to the transfused red blood cells. Some drugs nonspecifically bind to red blood cells and induce the formation of antibodies.

B. Hypersensitivity Reaction. Two cytotoxic reactions can occur.

2. NK cells bind to cell-bound antibody and release cytolytic enzymes.
3. Cell-bound antibody activates complement and leads to the formation of membrane attack complexes.

Both reactions result in the destruction of red blood cells.

Immune system

IgG

Self or transfused RBC

Drug antigen bound to membrane proteins

NK cell

Cytolytic enzymes

RBC

Complement

RBC

RBC

FIGURE 24.9 Type II Hypersensitivity Reaction. Type II hypersensitivities involve the lysis of red blood cells due to incorrect blood cell typing or to antibody formation to an antibiotic. *»» In an antibiotic-sensitive patient, what will happen when the drug is removed?*

injection of Rh antibodies (RhoGAM®). The injection is given within 72 hours of delivery of an Rh-positive child (no injection is necessary if the child is Rh negative). Antibodies in the preparation interact with Rh antigens on any fetal red blood cells in the mother and remove them from the circulation, thereby preventing the cells from stimulating the woman's immune system.

Type III Hypersensitivity Involves an Immune Complex Reaction

A **type III hypersensitivity** occurs when IgG antibodies combine with antigens, forming soluble immune complexes that accumulate in blood vessels or on organ surfaces (**FIGURE 24.11**). These immune complexes activate complement proteins capable of increasing vascular permeability and exerting a chemotactic effect on phagocytic neutrophils. At the target site, neutrophils release lysosomal enzymes, which cause tissue damage. Local inflammation is common, and fibrin clots can complicate the problem.

Artificially acquired passive immunity involves giving a patient a serum of antibodies that were collected from numerous other animals. Receiving this serum can lead to one of two syndromes.

TABLE 24.1 Some Characteristics of the Major Blood Types

	Type A	Type B	Type AB	Type O
	Antigen A	Antigen B	Antigens A and B	Neither A nor B antigens
Red blood cells				
Serum				
Approximate percentage, U.S. population*	B antibody	A antibody	Neither A nor B antibody	A and B antibodies
African-American	26	19	4	51
Asian	28	25	7	40
Caucasian	40	11	4	45
Hispanic	31	10	2	57

*Percentages from the American Red Cross.

MICROFOCUS 24.2: Evolution

Why Are There Different Blood Types in the Human Population?

The human population contains four different blood types: A, B, AB, and O. Is this just a matter of divergent evolution producing different populations with different blood types? Or is there another reason for the difference? Remember, as shown in Table 24.1, type A people have anti-B antibodies, and type B people have anti-A antibodies; Type AB have neither, whereas type O have both antibodies in the circulation.

Robert Seymour and his group at University College London believe that blood types are an evolutionary response to balancing defenses against viruses and bacteria. For example, when measles viruses break out of infected cells, they carry on their envelope the chemical group (antigen) identifying the blood type of that individual. Therefore, measles viruses from people with blood type A or B would be neutralized by type O blood because O blood has antibodies to antigens A and B. In reverse, measles viruses emerging from someone with blood type O carry neither the A nor B antigen,

A photograph of blood types being stored.

© John Foxx/Stockbyte/Getty.

so the viruses would not be neutralized by the blood of people with blood types A, B, or AB. Therefore, people with blood type O are better prepared to defend against viruses coming from people with other blood types—and they are better at transmitting viruses to those blood types.

To make matters more complex, Alexandra Rowe and her colleagues at the University of Edinburgh have reported that, at least in Africa, children suffering from mild forms of malaria are three times more likely to be blood type O than those with severe malaria. To cause an infection, the malaria parasite needs to attach to specific sugars on the surface of red blood cells, and type O blood, compared to A and B, has far fewer of these sugars.

If this is all true, then why isn't type O blood universal? Apparently, because of other pathogens. Seymour's group suggests that because there would be more type O individuals, probability says they would more likely be attacked by other pathogens, especially bacterial. This idea is supported by Rowe's group who says that type O individuals are more susceptible to other diseases, such as stomach ulcers and cholera. In fact, in Latin America where the majority of people are type O, people suffering from severe cholera are eight times more likely to be type O than those suffering a milder form of the disease.

If all this is validated with much more work, here is yet another way that microbes and viruses have affected the human species. Once again, they do rule the world!

Serum sickness develops when the immune system produces IgG against residual proteins in a serum preparation. The IgG then reacts with the proteins, and immune complexes gather in the kidney over a period of days. The problem is compounded if IgE antibodies, also from the immune system, attach to mast cells and basophils, thereby inducing a type I hypersensitivity. The sum total of these events is kidney damage, along with hives and swelling in the face, neck, and joints.

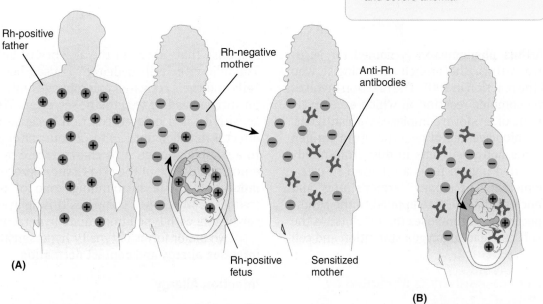

(A) First Pregnancy. An Rh-positive man and an Rh-negative woman have an Rh-positive baby.
- At delivery, Rh antigens enter the mother's circulation through breaks in the placenta.
- The Rh antigens in the mother trigger the production of anti-Rh antibodies.

(B) Second or Subsequent Pregnancy. If any future pregnancy produces an Rh-positive fetus:
- The mother's anti-Rh antibodies will attach to fetal erythrocytes.
- Rh/blood cell-antibody complex will trigger a cytotoxic hypersensitivity, resulting in erythrocyte destruction and severe anemia.

Rh-positive father

Rh-negative mother

Anti-Rh antibodies

Rh-positive fetus

Sensitized mother

(A)

(B)

FIGURE 24.10 Hemolytic Disease of the Newborn. If a mother and a second child are Rh incompatible, maternal antibodies cause the lysis and removal of the fetal red blood cells. *»» Why is this type II hypersensitivity also called erythroblastosis fetalis?*

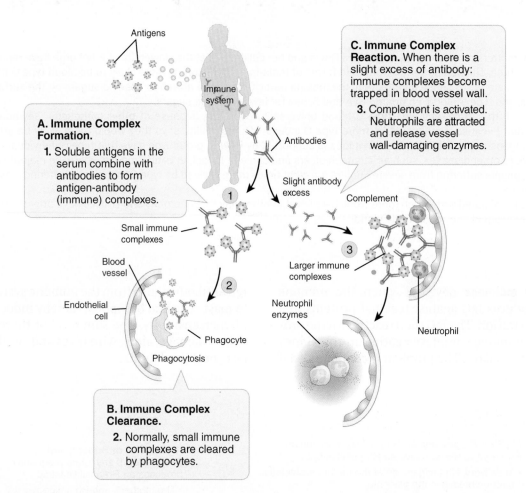

FIGURE 24.11 Type III Hypersensitivity Reaction. Type III hypersensitivities occur when excess antibodies combine with antigens to form aggregates (immune complexes) that accumulate in blood vessels or on tissue surfaces. *»» What types of physical problems can develop from immune complex formation?*

The **Arthus phenomenon** is named for Nicolas Maurice Arthus, the French physiologist who described the reaction in 1903. The reaction involves an immune complex reaction in which excessively large amounts of IgG form complexes with antigens, either in the blood vessels or near the site of antigen entry into the body. Antigens in dust from moldy hay and in dried pigeon feces are known to cause this phenomenon. The names "farmer's lung" and "pigeon fancier's disease" are applied to the conditions, respectively. Thromboses (blood clots) in the blood vessels can lead to oxygen starvation and cell death.

Type IV Hypersensitivity Is Mediated by Antigen-Specific T Cells

A **type IV hypersensitivity** or "delayed-type hypersensitivity" is an exaggerated cell-mediated immune response that occurs in the absence of antibodies. The adjective "cell-mediated" is used because T_H1 cells and their activated macrophages interact with an antigen leading to cytokine secretion. The hypersensitivity is a delayed reaction because maximal effect is not seen until 24 to 72 hours after exposure to a soluble antigen. It is characterized by a thickening and drying of the skin tissue, a process called **induration** and a surrounding zone of **erythema** (redness). TABLE 24.2 compares this delayed hypersensitivity with type I immediate hypersensitivity.

Two major forms of type IV hypersensitivity are **infection allergy** and **contact dermatitis**.

Infection Allergy

When the immune system responds to certain microbial agents, sensitized T_H1 cells migrate to the antigen site and release cytokines. The cytokines

TABLE 24.2 Immediate and Delayed Hypersensitivities Compared

	Type I Immediate Hypersensitivity	Type IV Delayed Hypersensitivity

Clinical state	Hay fever
	Asthma
	Urticaria
	Allergic skin conditions
	Anaphylactic shock
		Drug allergies
		Infectious allergies
		    Tuberculosis
		    Rheumatic fever
		    Contact dermatitis

Onset	Immediate	Delayed
Duration	Short: hours	Prolonged: days or longer
Allergens	Pollen Molds House dust Danders Drugs Antibiotics Soluble proteins and carbohydrates Foods	Drugs Antibiotics Microorganisms: bacteria, viruses, fungi, animal parasites Poison ivy/oak and plant oils Plastics and other chemicals Fabrics, furs Cosmetics
Passive transfer of sensitivity	With serum	With cells or cell fractions of lymphoid series

attract phagocytes and encourage phagocytosis. Sensitized lymphocytes then remain in the tissue and provide immunity to successive episodes of infection. The classic delayed hypersensitivity reaction is demonstrated by injecting an extract of the microbial agent into the skin of a sensitized individual. As the immune response takes place, the area develops induration and erythema.

An important application of infection allergy is the **tuberculin test** for diagnosing tuberculosis (**FIGURE 24.12**). A purified protein derivative (PPD) of *Mycobacterium tuberculosis* is applied to the skin by intradermal injection (the Mantoux test) or multiple punctures (tines). Individuals sensitized by a previous exposure to *Mycobacterium* species develop a vesicle, erythema, and induration. However, a positive result does not constitute a final diagnosis. Rather, sensitivity might have developed from a subclinical exposure to *M. tuberculosis*, a vaccination, from clinical disease years before, or from a former screening test in which the test antigens elicited a T-cell response.

Contact Dermatitis

After exposure to a variety of antigens, a delayed-type hypersensitivity reaction occurs. In poison ivy, the allergen has been identified as urushiol, a low-molecular-weight hapten on the surface of the leaf (**FIGURE 24.13**). In the body, urushiol attaches to tissue proteins to form allergenic compounds. Within 48 hours, a rash appears that consists of very itchy pinhead-sized blisters.

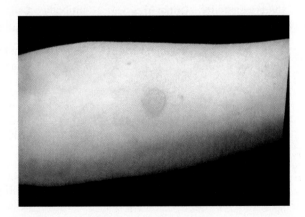

FIGURE 24.12 A Positive Tuberculin Test. The raised induration and zone of inflammation indicate that antigens have reacted with T cells, probably sensitized by a previous exposure to tubercle bacilli. *»» Does this test mean the person has tuberculosis? Explain.*

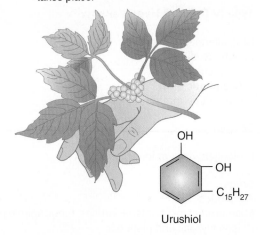

A When a sensitized person touches poison ivy, a substance called urushiol stimulates T lymphocytes in the skin; within 48 hours, a type IV reaction takes place.

Urushiol

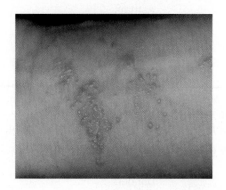

B The reaction is characterized by pinhead-sized blisters that usually occur in a straight row.

FIGURE 24.13 Contact Dermatitis. When sensitized skin makes contact with substances like urushiol in poison ivy, the chemical combines with tissue proteins to form allergenic compounds. The result is a rash on the skin surface. *»» Why is this form of hypersensitivity called delayed hypersensitivity?*

(B) © Bill Beatty/Visuals Unlimited/Getty.

The course of contact dermatitis is typical of the type IV reaction. Repeated exposures cause a drying of the skin, with erythema and scaling. Examples of other type IV responses occur on the scalp when allergenic shampoo is used, on the hands when contact is made with detergent enzymes, and on the wrists when an allergy to costume jewelry develops. Contact dermatitis also can occur on the face where contact is made with cosmetics, on areas of the skin where chemicals in permanent-press fabrics have accumulated, and on the feet when there is sensitivity to dyes in leather shoes. The list of possibilities is endless.

TABLE 24.3 summarizes the four types of hypersensitivities.

TABLE 24.3 Overview of the Hypersensitivity Reactions

Hypersensitivity Type	Origin of Hypersensitivity	Antibody Involved	Cells Involved	Mediators Involved	Evidence of Hypersensitivity	Examples
Type I IgE-mediated	B lymphocytes	IgE	Mast cells Basophils	Histamine Serotonin Leukotrienes Prostaglandins	30 minutes or less	Hay fever Systemic anaphylaxis Asthma
Type II Cytotoxic	B lymphocytes	IgG IgM	RBC WBC	Complement	5–8 hours	Transfusion reactions Hemolytic disease of newborns
Type III Immune complex	B lymphocytes	IgG	Host tissue cells	Complement	2–8 hours	Serum sickness Arthus phenomenon SLE Rheumatic fever Rheumatoid arthritis
Type IV Cellular	T lymphocytes	None	Host tissue cells	Cytokines	1–3 days	Contact dermatitis Infection allergy

Concept and Reasoning Checks 24.2

a. Summarize the factors responsible for type II hypersensitivities.
b. Summarize the factors responsible for type III hypersensitivities.
c. Summarize the factors responsible for type IV hypersensitivities.

Chapter **Challenge B**

Carla's outbreak of hives has become a chronic condition that baffles the allergist, who has even called in an infectious disease specialist to consult. Allergy skin tests were unable to pinpoint a specific cause. In the meantime, the itching has become so bad that at times that Carla cannot sleep. Over-the-counter anti-itch (antihistamine) creams and combinations of prescription antihistamines provide little relief.

QUESTION B: *Do any of the other forms of hypersensitivity, especially type IV, appear to correlate with Carla's symptoms? Explain.*

You can find answers online in **Appendix F**.

■ KEY CONCEPT 24.3   Autoimmune Disorders and Transplantation Are Immune Responses to "Self"

One of the properties of the immune system is tolerance of "self"; that is, one's own cells and molecules with their antigenic determinants do not stimulate an immune response. Should self-tolerance break down, an **autoimmune disorder** can occur.

Autoimmune Disorders Are a Failure to Distinguish Self from Nonself

Each individual's immune system must have a mechanism to differentiate "self" from "nonself," so the system will only react to foreign antigens, such as pathogens. Such tolerance is thought to develop in a number of ways (see also **FIGURE 24.14**):

▶ **Clonal Deletion Theory.** This theory says that self-reactive lymphoid cells are destroyed through a programmed cell death during the development of the immune system in an individual.

▶ **Clonal Anergy Theory.** This theory proposes that self-reactive T cells become inactivated in the normal individual, and these cells cannot differentiate into effector cells when presented with antigen.

▶ **Regulatory T Cell Theory.** According to this theory, specific regulatory T cells function to suppress exaggerated immune responses.

In fact, all of these theories might be correct and, as such, several mechanisms could actively contribute to the development of immunological tolerance.

Up to 8% of the American population (24.5 million individuals) suffers from an autoimmune disorder, and far more women than men are affected. So, what causes the loss of tolerance? Part of the answer lies in human heredity. Various gene mutations have been identified that affect cell division and programmed cell death—and therefore clonal deletion, anergy, and regulatory T-cell activity. There also are several other ways autoimmune disorders can be triggered:

▶ **Access to Privileged Sites.** Certain parts of the body (e.g., brain, eyes) limit access of immune cells and molecules. If such sites become "accessible" to the immune system through injury, an immune response will be mounted.

▶ **Antigenic Mimicry.** A foreign substance or microbe entering the body might closely resemble (mimic) a similar body substance. The immune system then responds to the "self" substance as it attacks the foreign substance or pathogen. Rheumatoid arthritis and type I diabetes can be examples.

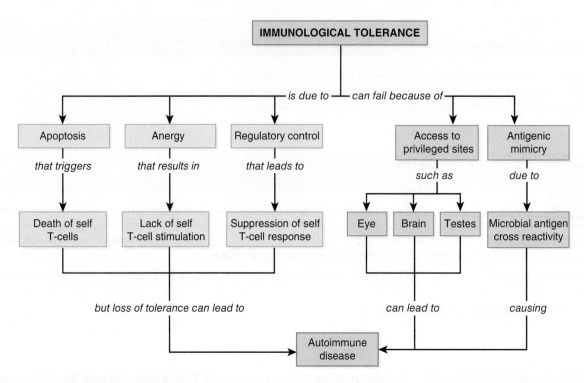

FIGURE 24.14 **Mechanisms of Tolerance and Failure.** Immunological tolerance involves killing self T cells, blocking stimulation, or suppressing a self T-cell response. Autoimmune diseases result from a loss of tolerance, T-cell access to privileged sites, or antigenic mimicry through cross reaction between and exogenous and self-antigen. »» *Why might T cells have gained access to privileged sites?*

Because of the loss of tolerance, the immune system attempts to mount an immune response against its own cells and tissues. Although many such disorders cause fever, specific symptoms depend on the disorder and the part of the body affected. The resulting inflammation and tissue damage can cause pain, deformed joints, jaundice, chronic itching and urticaria, breathing difficulties, and even death. To date, some 100 clinically distinct autoimmune disorders have been identified (TABLE 24.4).

Among the more notable organ-specific disorders are myasthenia gravis, type I diabetes, systemic lupus erythematosus (SLE) (FIGURE 24.15), and rheumatoid arthritis.

Treatment of most autoimmune disorders requires suppressing the immune system, which means interrupting the system's ability to fight infectious disease. Immunosuppressants include corticosteroids, such as prednisone. Some disorders resolve as spontaneously as they appear, whereas others become chronic, life long disorders. Thus, the prognosis depends on the particular disorder.

Transplantation of Tissues or Organs Is an Important Medical Therapy

Transplantation of tissues and organs is a common medical procedure today. However, the techniques are plagued by the immune system's natural

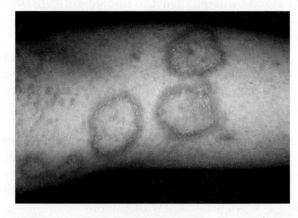

FIGURE 24.15 **Skin Lesions of Systemic Lupus Erythematosus.** This autoimmune disorder can affect the skin and other body organs. »» *What causes the inflammation seen with lupus?*

Scott Camazine/Alamy Stock Photo.

TABLE 24.4 **A Summary of Some Autoimmune Disorders**

Disorder	Tissue/Target	Effect	Female:Male Ratio
Autoantibody Mechanism			
Thrombocytopenia	Blood platelets/extracellular matrix protein	Red blood cell lysis and anemia	3:1
Goodpasture syndrome	Kidney/extracellular matrix collagen	Glomerulonephritis	1:6
Myasthenia gravis	Skeletal muscle/acetylcholine receptor	Muscle weakness	3:1
Graves disease	Thyroid/thyroid stimulating hormone	Hyperthyroidism	5–10:1
Systemic lupus erythematosus	Skin, kidneys, joints/nuclear antigens	Skin rash, glomerulonephritis, arthritis	9:1
Self T-Cell Mechanism			
Multiple sclerosis	Nerve myelin/myelin proteins	Partial or complete paralysis	Children—3:1 Adults over 50—1:1
Self T-Cell/Autoantibody Mechanism			
Hashimoto disease	Thyroid/thyroglobulin	Destruction of thyroid gland	10–20:1
Type 1 (insulin-dependent) diabetes	Pancreatic islet B cells/cell products	Failure to produce insulin	1:1
Rheumatoid arthritis	Synovium of joints/unknown	Joint inflammation Loss of movement	3–5:1

response to destroy foreign antigens (the transplant). Most of the damage results from cytotoxic T-cell activity, so transplantation can be considered similar to a type IV hypersensitivity.

Four types of transplantations, or grafts, are recognized, depending upon the genetic relationship between donor and recipient. A graft taken from one part of the body and transplanted to another part of the same body is called an **autograft**. This graft is never rejected because it is the person's own tissue. A tissue taken from an identical twin and grafted to the other twin is an **isograft**. This, too, is not rejected because the genetic constitutions of identical twins are very similar.

Rejection mechanisms might be more vigorous as the genetic compositions of donor and recipient become more varied. For instance, grafts between brother and sister, or between fraternal twins, might lead to only mild rejection because of their genetic similarity. Grafts between cousins might be rejected more rapidly, and as the relationship becomes more distant, the vigor of rejection increases proportionally. Most transplants today are **allografts**, or grafts between two unrelated individuals. **Xenografts**, or grafts between members of different species, such as a pig and a human, are rarely successful without a high degree of immunosuppression.

Mechanism of Allograft Rejection

With an allograft, the transplanted donor tissue or organ is rejected by the recipient if the immune system of the recipient interprets the transplant as "nonself." This recognition of nonself is stimulated by the recipient's immune system recognizing foreign **major histocompatibility complex** (MHC) proteins on the surface of the transplant. These proteins, typically referred to as **human leukocyte antigens (HLAs)** in the transplantation field, confer uniqueness to each individual. In the case of an allograft, the HLAs of donor and recipient can be distinct.

The mechanism of rejection is defined by the speed of the rejection mechanism. **Hyperacute rejection** can occur within minutes after transplantation as preformed antibodies attack the foreign HLA transplant. The antibodies activate complement, and they immediately block and destroy blood vessels in the donor transplant. In **acute rejection**, over a period of 10 to 30 days, T_H1 cells respond to the foreign HLA transplant by activating cytotoxic T cells (CTLs) that search out and destroy transplant cell tissue. T_H2 cells activate B cells and antibody production, which, as in hyperacute rejection, activate complement and lead to transplant destruction.

The closer the match between the donor's and the recipient's HLA proteins, the greater the chance of a successful transplant. To determine which HLA proteins are present and how closely those proteins match the HLA protein of the donor, tissue typing is carried out. Even with tissue matching, chronic rejection can occur over a period of months to years as T cells and antibodies respond to minor HLA differences.

A rejection mechanism of a completely different sort is sometimes observed in bone marrow transplants. Suppose that an individual receives a bone marrow transplant from another individual. Often, the transplanted marrow contains immune system cells that form immune products against the host after the host's immune system has been suppressed during transplant therapy. Essentially, the graft is rejecting the host; it is a "reverse rejection." This phenomenon, called a **graft-versus-host reaction (GVHR)**, sometimes can lead to fatal consequences in the host body.

Immunosuppressive Agents Prevent Allograft Rejection

To inhibit rejection, it is necessary to suppress activity of the immune system. Therefore, all transplant patients require treatment with **immunosuppressive agents**, which are drugs that in some way suppress or inhibit T cells and antibody production. Because these drugs are injected and spread systemically through the body, they all have drawbacks that are related either to their toxicity and side effects in the body or to a failure to produce sufficient immune system suppression. In addition, immunosuppression can leave the body open to tumor (cancer) formation and opportunistic infections, so often transplant patients need to be on antibiotics to minimize the latter risk.

Concept and Reasoning Checks 24.3

a. Identify the common attributes to all autoimmune disorders.
b. Assess the need for tissue typing before a transplant is performed.
c. Explain why immunosuppression leaves a transplant patient susceptible to opportunistic infections.

Chapter Challenge C

On further conversations with Carla, the allergist discovers that Carla's mother has recently experienced similar episodes of hives when she swims in her pool that has quite cold water. Carla tells her allergist that she learned from her mother that even her grandmother and one of her aunts experienced hives to sun exposure earlier in their lives, but such reactions have since disappeared. The allergist says "Oh, now I know what your disorder is! The good news is that you have chronic idiopathic urticaria. The bad news," he said, "is that we do not know its precise cause, which is what idiopathic means." The allergist prescribes a drug that seems to relieve the itch in others with Carla's condition and says he hopes the condition will spontaneously resolve as it did in her grandmother and aunt.

QUESTION C: *What type of disorder do Carla and her mother have that eventually resolved in her grandmother and aunt? Provide a hypothesis as to how the disorder developed.*

You can find answers online in **Appendix F**.

■ KEY CONCEPT 24.4   Immunodeficiency Disorders Can Be Inherited or Acquired

If there is an immune system malfunction or developmental abnormality in immune system function, an **immunodeficiency disorder** can result. Such disorders hamper the immune system's ability to provide a strong response to viral and microbial infections. The spectrum of immunodeficiency disorders ranges from relatively minor deficiency states to major abnormalities that are life threatening.

Immunodeficiencies Can Involve Any Aspect of the Immune System

Immunodeficiency disorders are identified by the part of the innate or adaptive immune system affected. Thus, the disorders might involve B cells and antibodies, T cells, both B and T cells, phagocytes, or complement proteins. The affected immune component might be absent, present in reduced numbers or amounts, or functioning abnormally.

Immunodeficiencies can be congenital (**primary immunodeficiency**), meaning the immunodeficiency is the result of a genetic abnormality. Such disorders present from birth are rare, although more than 70 different congenital disorders have been documented. People with primary immunodeficiencies have immune systems that do not work properly. Such individuals are subject to multiple infections, especially recurring respiratory infections. **FIGURE 24.16** provides several examples.

An immunodeficiency also can be acquired later in life (**secondary immunodeficiency**). Cases often result from immunosuppressive treatments such as chemotherapy and radiation therapy that reduce or eliminate populations of white blood cells.

Secondary immunodeficiencies also can be the result of an infection. **Acquired immunodeficiency syndrome (AIDS)**—caused by the human immunodeficiency virus (HIV)—is the most recognizable, and the disease and virus are the subject of the last section in this chapter.

The Human Immunodeficiency Virus (HIV) Is Responsible for HIV Infection and AIDS

CLINICAL CASE 24 details a syndrome first described by American physicians in 1981. The illness involved the development of certain opportunistic infections, including fungal pneumonia and cytomegalovirus infections, and immune system deficiency. Their report presented the signs and symptoms for a disease that would become known as **acquired immunodeficiency syndrome (AIDS)**. In 1983, the infectious agent, a virus, was discovered. It would be named the **human immunodeficiency virus (HIV)**.

The origin of HIV had been debated and investigated for decades. Today, it origins are quite certain, as detailed in MICROFOCUS 24.3.

Structure of HIV

HIV is a member of the Retroviridae. The virus contains two copies of a single-stranded (positive sense) RNA (**FIGURE 24.17**). Packed with these RNA viruses are several enzymes, one of which is called **reverse transcriptase**, which is needed to copy the single-stranded RNA (ssRNA) into double-stranded DNA (dsDNA). This reversal of the usual mode of gene expression (transcription) gives the virus its name, retrovirus (*retro* = "backward") and the enzyme its name, reverse transcriptase.

Like other enveloped viruses, the HIV envelope contains protein spikes for attachment and entry into the host cell. One spike protein, gp120, is used to attach to an appropriate host cell (see the following subsection), whereas the other, gp41, promotes fusion of the viral envelope with the host plasma membrane after attachment.

Replication of HIV

The viral replication cycle begins when gp120 proteins on the HIV spikes contact CD4 receptors present on the plasma membrane of a host cell (**FIGURE 24.18**). If the infection is transmitted through sexual intercourse, the initial cells encountered are the dendritic cells (DCs) and CD4 (helper) T cells associated with the mucosa of the genital tract. At least 50% of the body's CD4 T cells are located in the mucosal lining.

Following entry of the capsid into the host cell cytoplasm, which is facilitated by the gp41 spike proteins, uncoating occurs and the reverse transcriptase synthesizes a molecule of dsDNA from the viral ssRNA. Within 72 hours of infection, the DNA molecule integrates into one of the host chromosomes as a **provirus**. The exact site of integration determines how quickly the disease arises. If

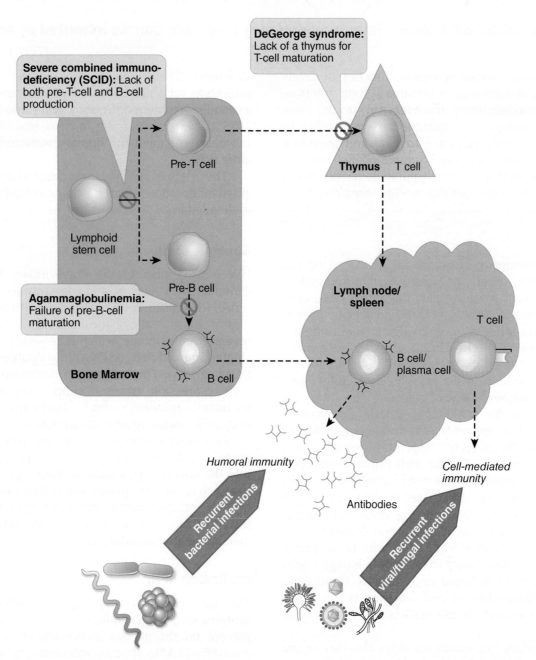

FIGURE 24.16 Examples of Some Immunodeficiencies. Lymphoid stem cells normally give rise to pre-T and pre-B cells in the bone marrow. The pre-T cells move to the thymus where they mature. Pre-B cells mature in the bone marrow. Both T cells and B cells then migrate to the lymph nodes and spleen. Immunodeficiencies, such as the one indicated, block normal T-cell and B-cell formation or maturation. *»» Why are recurrent bacterial infections associated with errors in humoral immunity and recurrent viral and fungal infections with cell-mediated errors?*

the infected cell is actively dividing, the provirus initiates DNA transcription and translation, resulting in the biosynthesis of new virus particles. The new, immature particles then "bud" from the host cell and rearrange themselves into the mature form that is able to infect other cells. If the infected cell is not dividing, the provirus remains dormant. In either situation, the infected individual now has a "primary HIV infection."

Typically, within 2 to 3 weeks after infection, half of the immune cell population, including CD4 T cells in the mucosal lining, have been depleted. If transmission occurs via the blood, CD4 T cells, macrophages, and DCs are among the first cells infected,

Clinical Case 24

Pneumocystis Pneumonia

Note: On June 5, 1981, the CDC published a report in *Morbidity and Mortality Weekly Report* about homosexual men in Los Angeles who had contracted *Pneumocystis carinii* pneumonia. This was the first published report of what, a year later, would become known as acquired immunodeficiency syndrome (AIDS). These descriptions are taken almost verbatim from the report (see the reference at end of the case).

Between October 1980 and May 1981, five young men, all sexually active homosexuals, arrived at three different hospitals in Los Angeles, California, and were treated for biopsy-confirmed *Pneumocystis carinii* (now called *P. jirovecii*) pneumonia. All five patients also had laboratory-confirmed cases of a previous or current infection with cytomegalovirus (CMV) and with the fungus *Candida albicans* (mucosal candidiasis infection).

- The patients did not know one another, had no known common contacts or knowledge of sexual partners who had had similar illnesses, and had no comparable histories of sexually transmitted infections.
- Two of the five patients reported having frequent homosexual contacts with various partners.
- All five patients reported using inhalant drugs, and one reported injection drug abuse.

Patient 1: A previously healthy 33-year-old man developed *P. carinii* pneumonia and oral mucosal candidiasis in March 1981 after a 2-month history of fever associated with elevated liver enzymes, low leukocyte count, and CMV in the urine. The serum complement-fixation CMV titer in October 1980 was 256; in May 1981, it was 32. The patient's condition deteriorated despite courses of treatment with the drugs trimethoprim-sulfamethoxazole (TMP/SMX), pentamidine, and acyclovir. He died May 3, 1981, and postmortem examination showed residual *P. carinii* and CMV pneumonia.

Patient 2: A previously healthy 30-year-old man developed *P. carinii* pneumonia in April 1981 after a 5-month history of fever each day and of elevated liver-function tests, CMV in the urine, and documented antibodies to CMV in immunofluorescence tests. He also had a low white blood cell count and mucosal candidiasis. His pneumonia responded to a course of intravenous TMP/SMX, but, as of the latest reports, he continues to have a fever each day.

Patient 3: A 30-year-old man was well until January 1981 when he developed esophageal and oral candidiasis that responded to amphotericin B treatment. He was hospitalized in February 1981 for *P. carinii* pneumonia that responded to oral TMP/SMX. His esophageal candidiasis recurred after the pneumonia was diagnosed, and he was again given amphotericin B. The CMV complement-fixation titer in March 1981 was 8. Material from an esophageal biopsy was positive for CMV.

Patient 4: A 29-year-old man developed *P. carinii* pneumonia in February 1981. He had had Hodgkin's disease 3 years earlier but had been successfully treated with radiation therapy alone. He did not improve after being given intravenous TMP/SMX and corticosteroids and died in March. Postmortem examination showed no evidence of Hodgkin's disease, but *P. carinii* and CMV were found in lung tissue.

Patient 5: A previously healthy 36-year-old man with a clinically diagnosed CMV infection in September 1980 was seen in April 1981 because of a 4-month history of fever, breathing difficulties, and a cough. On admission, he was found to have *P. carinii* pneumonia, oral candidiasis, and CMV retinitis. A complement-fixation CMV titer in April 1981 was 128. The patient has been treated with two short courses of TMP/SMX that have been limited because of a sulfa-induced neutropenia. He is being treated for candidiasis with topical nystatin.

Questions:
  a. If the patients did not know one another, how did they all end up with essentially the same set of infectious diseases?
  b. What does the huge reduction in the serum complement-fixation CMV titer indicate in patient 1?
  c. Do you believe patient 2 had been cured of his illness? Explain.
  d. Knowing today that all the infectious diseases described were due to HIV weakening their immune system, what is the most likely way that the patients became infected with HIV?

You can find answers online in **Appendix E**.

For additional information, see www.cdc.gov/mmwr/PDF/wk/mm4534.pdf.

MICROFOCUS 24.3: Evolution

HIV's Family Tree

For years, scientists and researchers have debated the origin of the human immunodeficiency virus (HIV). Did it come from contaminated polio vaccine? Was it a government secret project that went awry? Where did the virus arise?

HIV is a member of the genus *Lentivirus*, all of which produce slow (*lent* = "slow"), incessant infections of the immune system. These viruses have been found in several animals, including cats, sheep, horses, cattle, and most important, monkeys, from which the simian immunodeficiency virus (SIV) has been isolated. It is now generally accepted that HIV is a descendant of SIV because certain strains of SIV bear very close gene sequences to HIV-1 and HIV-2, the two types of HIV. HIV-2 corresponds to the SIV strain found in the sooty mangabey (or green monkey), which is indigenous to western Africa. It is believed that in the 1960s SIV jumped nine times from sooty mangabeys to humans.

Until recently, the origin for the more virulent HIV-1 was more difficult to place. The closest counterpart to SIV was found in chimpanzees. However, chimps are infected with many different SIVs. Through close study of the viruses, one SIV from forest chimps in Cameroon was found to be a close match to HIV-1. Based on additional work, the most likely scenario is that sometime in the early 1900s HIV-1 jumped on 13 separate occasions to humans in Cameroon where it continued to evolve into a form that today can infect only humans. Infected humans living in the Cameroon forest brought the virus to Leopoldville (now Kinshasa), which was a rapidly growing city with an active sex trade. From there, in the 1960s the virus spread around the world.

Gene sequencing analysis suggests that HIV arrived in the United States around 1969. Considering that it takes about 10 years for an infection to produce symptoms, it is not surprising that the first cases did not show up until 1981. In fact, some researchers believe that by then there were already some 100,000 infections.

A young chimpanzee.

© Dana Ward/Shutterstock.

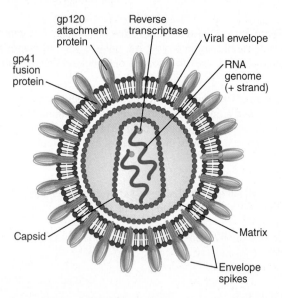

FIGURE 24.17 A Diagram of the Human Immunodeficiency Virus (HIV). The virus consists of two molecules of RNA and reverse transcriptase. A protein capsid surrounds the genome, and an envelope with spikes of protein lies outside the capsid and matrix. »» *What is the most unique feature about the structure of HIV?*

primarily in the lymphoid organs. How the CD4 T cells are killed is the subject in **INVESTIGATING THE MICROBIAL WORLD 24.2.**

Clinical Disease Progression

Although the rate of disease progression varies among individuals, realize the following stages are a continuum for someone who is not on antiviral drug therapy (**FIGURE 24.19**):

▶ **I: Primary (Acute) HIV Stage.** Many people are asymptomatic when they first become infected with HIV. However, about 70% experience an **acute HIV infection**, which produces a nonspecific, flu-like illness within a month or two of HIV exposure. The symptoms of fever, chills, rashes, and night sweats usually last no longer than a few days. About 20% of these individuals have symptoms serious enough to require a doctor's exam. However, being that the symptoms are nonspecific, a correct diagnosis often is missed.

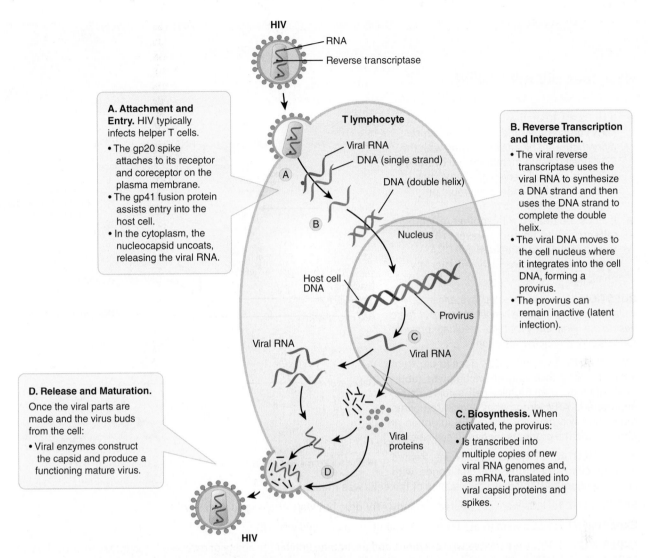

A. Attachment and Entry. HIV typically infects helper T cells.

- The gp20 spike attaches to its receptor and coreceptor on the plasma membrane.
- The gp41 fusion protein assists entry into the host cell.
- In the cytoplasm, the nucleocapsid uncoats, releasing the viral RNA.

B. Reverse Transcription and Integration.

- The viral reverse transcriptase uses the viral RNA to synthesize a DNA strand and then uses the DNA strand to complete the double helix.
- The viral DNA moves to the cell nucleus where it integrates into the cell DNA, forming a provirus.
- The provirus can remain inactive (latent infection).

C. Biosynthesis. When activated, the provirus:

- Is transcribed into multiple copies of new viral RNA genomes and, as mRNA, translated into viral capsid proteins and spikes.

D. Release and Maturation.

Once the viral parts are made and the virus buds from the cell:

- Viral enzymes construct the capsid and produce a functioning mature virus.

FIGURE 24.18 The Replication Cycle of the Human Immunodeficiency Virus (HIV). Replication is dependent on the presence and activity of the reverse transcriptase enzyme. *»» What is the role of the reverse transcriptase enzyme?*

Investigating the Microbial World 24.2

How Does HIV Kill T Cells?

After investigators determined that AIDS was caused by a virus (HIV) that infected CD4 T cells, the scientists wondered how the virus killed and brought about a progressive depletion in T-cell number as the years of infection passed. Certainly, many of the infected cells were killed directly as a result of a virus productive infection within the infected cells. But is that the only way?

OBSERVATIONS: More than 98% of CD4 T cells are found in lymphoid organs, such as the tonsils and spleen, and it is within these organs that HIV replicates in an infected individual. However, using human lymphoid cultures (HLCs) prepared from tonsillar tissue, which closely mimic the cellular environment normally experienced by HIV in humans, the following was observed:

■ Less than 5% of the CD4 T cells show a productive infection (produce HIV virions), yet 99% of all the CD4 T cells in those cultures die. Therefore, direct killing cannot explain all the T-cell deaths. For some reason, the 94% of surrounding cells, called "bystander cells," are also dying even though they produce few, if any, virus particles.

(continues)

Investigating the Microbial World 24.2 (*Continued*)

How Does HIV Kill T Cells?

- If HLCs are treated with the antiretroviral (ARV) drug enfuvirtide (see the accompanying figure) to block virus fusion with the CD4 T-cell plasma membrane, no bystander cell deaths are observed after HIV challenge, so the bystander cells must be "infected" for death to occur.
- If HLCs are treated with another ARV drug, raltegravir (see figure) to prevent HIV DNA from inserting into the host cell DNA, bystander deaths are observed after HIV challenge.

QUESTION: *How are infected bystander CD4 T cells being killed in these lymphoid organs if there has to be viral entry but no resulting provirus formation or productive infection?*

HYPOTHESIS: Bystander CD4 T cells die from indirect killing involving an incomplete (abortive) infection that triggers cell suicide. If so, appropriate ARV drugs that block retrovirus replication will pinpoint where HIV infection is aborted and the products triggering cell suicide.

EXPERIMENTAL DESIGN: HCLs were treated with specific ARV drugs and challenged with HIV (see figure). After 9 days, the viability of the cells was measured.

Data from Doitsh, G. et al. 2010. *Cell* 143(5): 789–801.

EXPERIMENT 1: HLCs were not treated with any ARV drug but were challenged with HIV.

EXPERIMENT 2: HLCs were treated with AZT and then challenged with HIV.

EXPERIMENT 3: HLCs are treated with efavirenz and nevirapine (inhibit the ability of reverse transcriptase to copy the viral ssRNA into dsDNA) and then challenged with HIV.

RESULTS:

Experiment	ARV Drug: Mode of Action	Result
1	None used	HLCs depleted of CD4 T cells after 9 days
2	AZT: blocks the elongation stage of DNA replication but allows up to 250 base pairs in the early initiation stage of DNA replication to occur	HLCs depleted of CD4 T cells after 9 days
3	Efavirenz and nevirapine: inhibit the ability of reverse transcriptase to copy the viral ssRNA into dsDNA	HLCs not depleted of CD4 T cells after 9 days

Modified from Doitsh, G. et al. 2010. *Cell* 143 (5): 789–801.

CONCLUSIONS: The bystander cells, which are the major cause of CD4 T-cell depletion in lymphoid organs, become infected, but due to an abortive infection, the accumulated products trigger cell death.

QUESTION 1: *By adding different ARV drugs that target precise steps of the HIV replicative cycle, where does the abortive step of viral replication occur?*

QUESTION 2: *What are the accumulated products that trigger CD4 T-cell death?*

You can find answers online in **Appendix E**.

Data from Doitsh, G. et al. 2010. *Cell* 143(5):789–801.

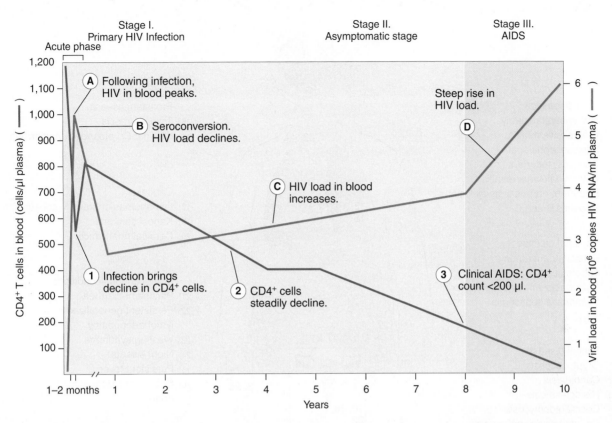

FIGURE 24.19 **HIV Infection and AIDS.** Upon being infected with HIV, an immune response brings about an abrupt drop in the blood HIV population, which rises as the immune system fails (**A–D**). Without antiretroviral therapy, the T-cell count slowly drops (1–3). Once the T-cell count is below 200/µl, the person is said to have AIDS. »» *Explain why it takes so long for the T-cell number to drop.*

During an acute phase, the immune system is at war with HIV as the virus infects lymph nodes ①, replicates, and releases virions (A) into the bloodstream. The immune system begins responding to the infection by producing antibodies and T cytotoxic cells directed against HIV. Such a process in the body is referred to as **seroconversion**, meaning HIV antibodies can now be detected in the blood. This usually occurs within 1 to 3 months after infection (B).

▶ **II: Clinical Asymptomatic Stage.** Without drug treatment, the infected individual often remains asymptomatic and free of major diseases for many years. However, early signs of immune system decline might involve periods of swollen lymph nodes. Even with extremely low numbers of HIV particles in the peripheral blood, the individual can infect others. During this stage, which can last more than 7 years ("clinical latency"), the level of HIV in the blood slowly rises (C) as the number of CD4 T cells steadily declines ②.

As the immune system continues to decline, the affected individual will begin to experience mild **HIV disease** symptoms, such as skin rashes, night sweats, fatigue, and significant weight loss. As the disease progresses, more serious conditions might occur, including recurrent herpes blisters on the mouth, diarrhea, and fever.

▶ **III: AIDS.** Over time, the immune system loses the fight against HIV. At full bore, HIV can produce more than 10 billion new virions per day (D). Thus, lymph nodes and tissues become damaged from the battles, and the immune system can no longer replace T cells at the rate they are destroyed. In addition, HIV might have mutated into a form that is more pathogenic and more aggressive in destroying T cells.

Infected individuals are said to have AIDS when their CD4 T cell count is below 200 cells per microliter of blood (healthy adults usually have CD4 T cell counts of 1,000 or higher) ③ and/or have one or more of what

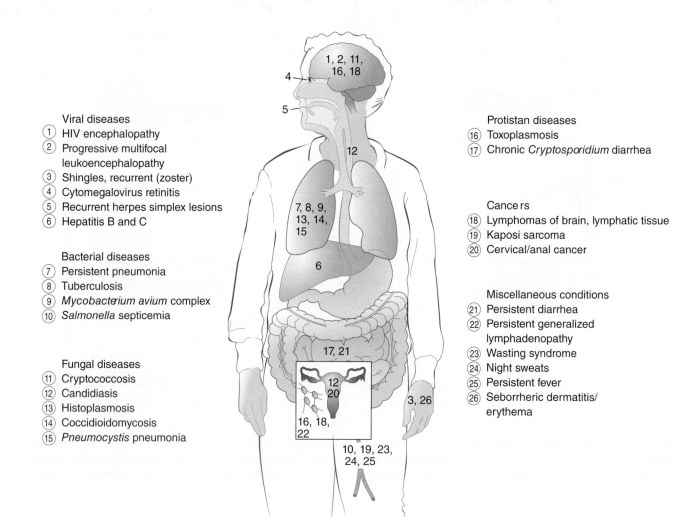

Viral diseases
(1) HIV encephalopathy
(2) Progressive multifocal leukoencephalopathy
(3) Shingles, recurrent (zoster)
(4) Cytomegalovirus retinitis
(5) Recurrent herpes simplex lesions
(6) Hepatitis B and C

Bacterial diseases
(7) Persistent pneumonia
(8) Tuberculosis
(9) *Mycobacterium avium* complex
(10) *Salmonella* septicemia

Fungal diseases
(11) Cryptococcosis
(12) Candidiasis
(13) Histoplasmosis
(14) Coccidioidomycosis
(15) *Pneumocystis* pneumonia

Protistan diseases
(16) Toxoplasmosis
(17) Chronic *Cryptosporidium* diarrhea

Cancers
(18) Lymphomas of brain, lymphatic tissue
(19) Kaposi sarcoma
(20) Cervical/anal cancer

Miscellaneous conditions
(21) Persistent diarrhea
(22) Persistent generalized lymphadenopathy
(23) Wasting syndrome
(24) Night sweats
(25) Persistent fever
(26) Seborrheric dermatitis/erythema

FIGURE 24.20 AIDS Defining Conditions. The variety of opportunistic illnesses that affect the body as a result of infection with HIV. Note the various systems that are affected and the numerous organisms involved. *»» Why is there such a diverse group of potential pathogens that can cause opportunistic infections?*

are referred to as **AIDS defining conditions** (**FIGURE 24.20**). The AIDS defining conditions represent a set of opportunistic infections that can occur in many parts of the body. Some of the most common are *Pneumocystis* pneumonia, cytomegalovirus infections, toxoplasmosis, and candidiasis. People with AIDS also are likely to develop various cancers, especially those caused by viruses, such as Kaposi sarcoma and cervical cancer, or cancers of the lymphatic system known as lymphomas. These infections can be severe and eventually fatal because the immune system is so damaged by HIV that the body cannot fight off multiple pathogen infections.

Scientists have been studying those few individuals who have been infected with HIV for 10 or more years and yet HIV is virtually undetectable in their bodies. The researchers are trying to discover what factors allow these "**long-term nonprogressors**" to keep the viral load down. One intriguing finding concerns the way the virus enters cells. The HIV gp120 spike needs to bind to two receptors on host helper T cells: the CD4 receptor and a coreceptor called CCR5 (**FIGURE 24.21**). HIV-infected individuals having helper T cells with both receptors tend to develop AIDS. However, about 1% of the Caucasian population lacks the CCR5 receptor so the virus cannot attach and enter the cells. These individuals would have natural resistance to HIV infection.

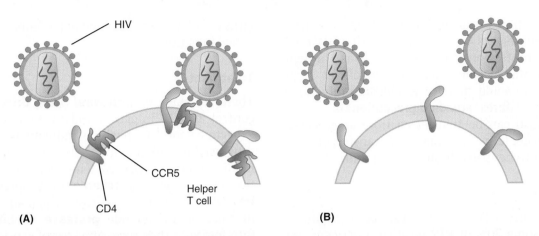

FIGURE 24.21 HIV Attachment. (A) To infect a host cell, like a helper T cell, the HIV spikes must recognize and bind to two receptors, CD4 and CCR5, on the host cell plasma membrane. **(B)** A few individuals naturally lack the CCR5 coreceptor and, as such, the virus cannot attach to the cells. *»» What would happen if the helper T cells lacked the CD4 receptor?*

Transmission

HIV can be transmitted by "risky behaviors," such as sharing of blood-contaminated needles with an HIV-infected person; having unprotected sexual contact, including vaginal, anal, or oral with an infected individual; or having unprotected intercourse with a person of unknown HIV status (**FIGURE 24.22**).

Donated red blood cells and blood factor concentrates can contain the virus, but extensive blood tests now are performed in the United States and many other countries to preclude HIV being transmitted

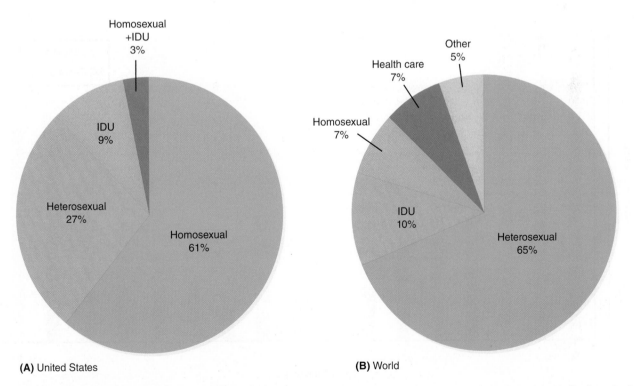

(A) United States

(B) World

FIGURE 24.22 Number of New HIV Infections, by Mode of Transmission. (A) Estimated number of new HIV infections in the United States has been roughly stable since the 1990s. **(B)** Globally, most cases are transmitted heterosexually. *»» In (B), what might be a transmission mechanism in healthcare settings?*

Data from (A) CDC. Retrieved from www.cdc.gov/nchhstp/newsroom/docs/HIV-infections_2006-2009.pdf. (B) Avert. Retrieved from www.avert.org/worlstatinfo.htm.

through a blood transfusion or blood products. Healthcare workers are at risk of acquiring HIV during their professional activities, such as through an accidental needle stick. Individuals at high risk always should practice established infection-control procedures (**standard precautions**).

HIV also can be transferred from an infected mother to her fetus (transplacental transfer) or baby at birth or to her baby through breast milk.

Diagnosis

Because early HIV infection can be asymptomatic and some 20% of HIV-positive Americans are unaware of their infection (greater than 50% in the 13- to 24-year-old group), increased opportunities for HIV testing are vital. Six FDA-approved, rapid HIV antibody screening tests are available. Most of these are ELISA-type tests that can be performed inexpensively and confidentially.

Other tests include the **APTIMA assay**, which amplifies and detects HIV RNA, as a diagnosis of primary HIV-1 infection, and the **viral load test**, which detects HIV RNA and can monitor HIV-1

virus circulating in the blood of patients with established infections.

Treatment

There are many **antiretroviral (ARV) drugs** used to control an HIV infection and to increase the life expectancy of HIV-positive individuals. The first drug used for treatment was azidothymidine, commonly known as AZT. AZT interferes with reverse transcriptase activity (**reverse transcriptase inhibitor**). Other antiretroviral drugs, which are discussed in Chapter 15, include **protease inhibitors** that interfere with the processing step of capsid production; **entry inhibitors**, which work by blocking viral entry into the CD4 cells; and **integrase inhibitors** that block provirus formation.

During replication, HIV can undergo a high rate of error, producing some mutations that can make the virus resistant to ARV drugs if the drug is used singly, that is, as **monotherapy** (FIGURE 24.23). Cases have been reported in which an HIV-infected individual contained 28 different strains of HIV. Therefore, a more effective treatment requires a

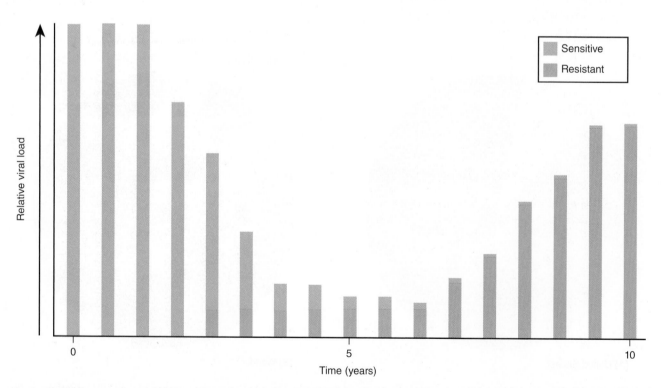

FIGURE 24.23 Viral Load and Drug Resistance. Because HIV replicates at a high rate and the reverse transcriptase enzyme is error prone, mutations will occur. After several years of drug monotherapy, the antiretroviral drug may have eliminated most of the original strain. However, some mutations will have conferred drug resistance, and these viruses can now replicate free of any inhibition to that drug. »» *What is the most logical way to combat drug resistance to monotherapy treatment?*

combination or "cocktail" of drugs. When three or more drugs are used together, the combination is referred to as **antiretroviral therapy** (**ART**). Although ART is not a cure, it has been significant in reducing the risk of HIV transmission as well as saving the lives of untold numbers of infected individuals. In fact, many patients are being told that they can expect a near normal life expectancy if they stay on their medications.

Prevention

Until recently, the only way to prevent HIV infection was to avoid behaviors putting a person at risk, such as sharing needles or having unprotected sex. However, there is now a medication to "prevent" HIV infection. The drug, called Truvada®, consists of two reverse transcriptase inhibitors that can be taken once daily by adults who are HIV negative but are at risk of becoming infected (e.g., certain groups of healthcare workers). Importantly, the drug combination, along with other safe-sex practices, reduces substantially (up to 75%) the risk of acquiring an HIV infection; still, it is not a 100% effective prevention strategy.

There is no vaccine for AIDS. Two types of vaccines are being considered. **Preventive vaccines** (such as Truvada®) for HIV-negative individuals could be given to prevent infection. **Therapeutic vaccines**, for HIV-positive individuals, would help the individual's immune system control HIV either by blocking virus entry, provirus formation, or replication, which should stop the progression of the disease and its transmission to others.

There are two major reasons why a successful vaccine has been lacking. First, HIV continually mutates and recombines. Such behavior by the virus means that a vaccine has to protect individuals against a moving target. In addition, because HIV infects CD4 T cells, the vaccine needs to activate the very cells infected by the virus.

Although at times it appears that vaccine research is an uphill battle, scientists are optimistic that a safe, effective, affordable, and stable HIV vaccine can be produced. Importantly, progress in basic and clinical research is moving forward, and scientists are inching closer to identifying products suitable for a successful HIV vaccine.

In conclusion, the introduction and spread of antiretroviral therapy has served as a highly effective "therapeutic treatment" approach. By treating infected individuals with antiretroviral drugs, today more than 65% of patients are living beyond five years after being infected with HIV. By lowering the viral load, treatment also is making prevention a possibility. Unfortunately, 35% of infected individuals are still dying within five years, so infections and lack of antiviral drugs remain a problem. The World Health Organization (WHO) now recommends that all HIV-infected people immediately be provided therapy as soon as they are diagnosed. The world now has the tools to stop the HIV pandemic and eliminate the threat.

Concept and Reasoning Checks 24.4

a. Compare and contrast the symptoms of the primary immunodeficiency disorders.
b. Distinguish between the three stages of HIV disease and AIDS.

■ SUMMARY OF KEY CONCEPTS

Concept 24.1 Type I Hypersensitivity Represents a Familiar Allergic Response

1. **Type I hypersensitivity** is caused by **IgE antibodies** produced in response to certain antigens called **allergens**. The antibodies can attach to the surface of **mast cells** and **basophils** and trigger the release of the mediators on subsequent exposure to the same allergen. (Figure 24.3)
2. Type I hypersensitivities can involve **systemic anaphylaxis**, a whole-body reaction in which a series of mediators induces vigorous and life-threatening contractions of the smooth muscles of the body.
3. **Atopic diseases** involve **localized anaphylaxis** (common allergies), such as hay fever or a food allergy. (Figure 24.5)
4. **Asthma** is an inflammatory disease involving an early response to an allergen and a late response by eosinophils and neutrophils that cause tissue injury and airway blockage.
5. Several ideas have been put forth concerning the nature of allergies. These include the shielding of IgE-secreting B cells (plasma cells) by IgA-secreting plasma cells and feedback mechanisms where IgE limits its own production in nonatopic individuals.
6. **Desensitization therapy** attempts to limit the possibility of an anaphylactic reaction through the injection of tiny amounts of allergen over time. Therapy also can involve the presence of IgG antibodies as **blocking antibodies**.

Concept 24.2 Other Types of Hypersensitivities Represent Immediate or Delayed Reactions

7. In **type II cytotoxic hypersensitivity**, the immune system produces IgG and IgM antibodies, both of which react with the body's cells and often destroy the latter. (Figure 24.9)
8. No cells are involved in **type III hypersensitivity**. Rather, the body's IgG and IgM antibodies interact with dissolved antigen molecules to form visible **immune complexes**. The accumulation of immune complexes in various organs leads to local tissue destruction in such illnesses as **serum sickness, Arthus phenomenon**, and **systemic lupus erythematosus**. (Figure 24.11)

9. **Type IV cellular hypersensitivity** involves no antibodies but is an exaggeration of cell-mediated immunity based in T lymphocytes. **Contact dermatitis** and **infection allergies** are manifestations of this hypersensitivity.

Concept 24.3 Autoimmune Disorders and Transplantation Are Immune Responses to "Self"

10. **Autoimmune disorders** can occur through defects in **clonal deletion, clonal anergy**, or **regulatory T-cell activity**. Human heredity as well as access to **privileged sites** and **antigenic mimicry** also can trigger an autoimmune response. (Figure 24.14)
11. Four types of grafts or transplants can be performed: **autografts, isografts, allografts**, and **xenografts**. Allografts are the most common. Rejection of grafts or transplants involves cytotoxic T cells and antibodies. The graft also can be rejected by immune cells in the graft that reject the recipient (**graft-versus-host reaction**).
12. Prevention of rejection is strengthened by using immunosuppressive agents, including antimetabolites, immune cell inhibitors, anti-inflammatory drugs, and monoclonal antibodies.

Concept 24.4 Immunodeficiency Disorders Can Be Inherited or Acquired

13. **Immunodeficiency disorders** can be congenital (**primary immunodeficiencies**) or acquired later in life (**secondary immunodeficiencies**).
14. **AIDS (acquired immunodeficiency syndrome)** is the final stage of the HIV (human immunodeficiency virus) infection. HIV infects and destroys CD4 T cells, eventually leading to an inability of the immune system to fend off opportunistic diseases. Transmission is through blood, blood products, contaminated needles, or unprotected sexual intercourse. Many **antiretroviral drugs** are available to slow the progression of disease. (Figures 24.18, 24.19)

■ CHAPTER SELF-TEST

You can find answers for **Steps A–D** online in **Appendix D**.

STEP A: REVIEW OF FACTS AND TERMS

Multiple Choice

Read each question carefully, then select the *one* answer that best fits the question or statement.

1. All the following are types of immediate hypersensitivities *except* _____.
   A. asthma
   B. contact dermatitis
   C. food allergies
   D. hay fever

2. Systemic anaphylaxis is characterized by _____.
   A. contraction of smooth muscles
   B. a red rash
   C. blood poisoning
   D. hives

3. Which of the following is *not* a type I hypersensitivity?
   A. Food allergies
   B. Contact dermatitis
   C. Allergic rhinitis
   D. Exercise-induced allergies

4. The early response in asthma is due to _____ activity.
   A. cytotoxic T cell
   B. basophil
   C. T_H2 cell and NK cell
   D. dendritic cells

5. What type of immune cell might control IgE-mediated hypersensitivities?
   A. Suppressor T cells
   B. Plasma cells
   C. Cytotoxic T cells
   D. Neutrophils

6. Desensitization therapy can involve _____.
   A. the use of blocking antibodies
   B. injections of small amounts of allergen
   C. allergen injections of several months
   D. All the above (**A–C**) are correct.

7. A cytotoxic hypersensitivity would occur if blood type _____ is transfused into a person with blood type _____.
   A. A; AB
   B. O; AB
   C. A; O
   D. O; B

8. Serum sickness is a common symptom of _____.
   A. contact dermatitis
   B. hemolytic disease of the newborn
   C. immune complex hypersensitivity
   D. food allergies

9. Which one of the following allergens is *not* associated with contact dermatitis?
   A. Foods
   B. Cosmetics
   C. Poison ivy
   D. Jewelry

10. Immunological tolerance to "self" is established by _____.
    A. destruction of self-reactive lymphoid cells
    B. clonal anergy
    C. clonal deletion
    D. All the above (**A–C**) are correct.

11. A _____ is a graft between genetically different members of the same species.
    A. xenograft
    B. autograft
    C. allograft
    D. isograft

12. Immunosuppressive agents used in preventing transplant rejection primarily affect _____.
    A. macrophages
    B. neutrophils
    C. dendritic cells
    D. T cells

13. What immunodeficiency disorder is associated with a lack of T and B cells and complete immune dysfunction?
    A. DiGeorge syndrome
    B. Severe combined immunodeficiency disease
    C. Agammaglobulinemia
    D. Type 1 diabetes

14. A patient with a CD4 T-cell count of 1,000/µl would be in the _____ of HIV disease/AIDS.
    A. HIV disease stage
    B. primary (acute) HIV stage
    C. AIDS stage
    D. clinical asymptomatic stage

Rearrangement

This chapter has summarized some of the disorders associated with the immune system. To gauge your understanding, rearrange the scrambled letters to form the correct word for each of the spaces in the statements.

15. The compound _____ is one of the major chemical mediators released during allergy reactions.

    I T M E H I A S N

16. In a _____ hypersensitivity, antibodies unite with cells and trigger a reaction that results in cell destruction.

    X Y O T C C I T O

17. Hay fever is an example of an _____ disease, one in which a local allergy takes place.

    O C T I A P

18. Immune complex hypersensitivities develop when antibody molecules interact with _____ molecules and form aggregates in the tissues.

    E N I G N A T

19. Mast cells and _____ are the two principal cells that function in anaphylactic responses.

    S S B I O H P A L

20. A key element in transplant acceptance or rejection is a set of molecules abbreviated as _____ proteins.

    L A H

21. The HIV DNA integrated into a chromosome is called a _____.

    S R V R U P O I

■ STEP B: CONCEPT REVIEW

22. Summarize the events occurring in and the role of humoral immunity in **type I hypersensitivities**. (**Key Concept 24.1**)
23. Summarize the characteristics of **type II hypersensitivity** and its relationship to blood transfusions and **hemolytic disease of the newborn**. (**Key Concept 24.2**)
24. Summarize the characteristics of **type IV hypersensitivity** and its relationship to **infection allergies** and **contact dermatitis**. (**Key Concept 24.2**)
25. Identify the ways in which an **autoimmune disease** can arise. (**Key Concept 24.3**)

26. Assess the usefulness of **immunosuppressive drugs** by transplant patients. (**Key Concept 24.3**)
27. Contrast **primary** and **secondary immunodeficiencies**, and list several primary disorders and their accompanying immunological deficiencies. (**Key Concept 24.3**)
28. Diagram how the **human immunodeficiency virus** (**HIV**) infects a cell, and identify the prevention and treatment methods used for **HIV disease** and **AIDS**. (**Key Concept 24.4**)

■ STEP C: APPLICATIONS AND PROBLEM SOLVING

29. During war and under emergency conditions, a soldier whose blood type is O donates blood to save the life of a fellow soldier with type B blood. The soldier lives and after the war becomes a police officer. One day he is called to donate blood to a brother officer who has been wounded and finds that it is his old friend from the war. He gladly rolls up his sleeve and prepares for the transfusion. Should it be allowed to proceed? Why?
30. Coming from the anatomy lab, you notice that your hands are red and raw and have begun peeling in several spots. This was your third period of dissection.

What is happening to your hands, and what could be causing the condition? How will you solve the problem?

31. "He had a history of nasal congestion, swelling of his eyes, and difficulty breathing through his nose. He gave a history of blowing his nose frequently, and the congestion was so severe during the spring he had difficulty running." The person in this description is former President Bill Clinton, and the writer is an allergist from Little Rock, Arkansas. What condition (technically known as allergic rhinitis) is probably being described?

■ STEP D: QUESTIONS FOR THOUGHT AND DISCUSSION

32. As part of an experiment, one animal is fed a raw egg while a second animal is injected intravenously with a raw egg. Which animal is in greater danger? Why?

33. A woman is having the fifth injection in a weekly series of hay fever shots. Shortly after leaving the allergist's office, she develops a flush on her face, itching sensations of the skin, and shortness of breath. She becomes dizzy and then faints. What is taking place in her body, and why has it not happened after the first four injections?

34. You may have noted that brothers and sisters are allowed to be organ donors for one another but that a person cannot always donate to his or her spouse. Many people feel badly about being unable to help a loved one in time of need. How might you explain to someone in such a situation the basis for becoming an organ donor and why it may be impossible to serve as one?

35. The immune system is commonly regarded as one that provides protection against disease. This chapter, however, seems to indicate that the immune system is responsible for numerous afflictions. Even the title is "Immune Disorders." Does this mean that the immune system should be given a new name? On the other hand, is it possible that all these afflictions are actually the result of the body's attempts to protect itself? Finally, why can the phrase "immune disorder" be considered an oxymoron?

36. In many diseases, the immune system overcomes the infectious agent, and the person recovers. In other diseases, the infectious agent overcomes the immune system, and death follows. Compare this broad overview of disease and resistance to what is taking place with AIDS, and explain why AIDS is probably unlike any other disease encountered in medicine.

APPENDIX A

■ Metric Measurement

	Fundamental Unit	Quantity	Symbol	Numerical Unit	Scientific Notation
Length	Meter (m)				
		kilometer	km	1,000 m	10^3
		centimeter	cm	0.01 m	10^{-2}
		millimeter	mm	0.001 m	10^{-3}
		micrometer	μm	0.000001 m	10^{-6}
		nanometer	nm	0.000000001 m	10^{-9}
Volume (liquids)	Liter (l)				
		milliliter	ml	0.001 l	10^{-3}
		microliter	μl	0.000001 l	10^{-6}
Mass	Gram (g)				
		kilogram	kg	1,000 g	10^3
		milligram	mg	0.001 g	10^{-3}
		microgram	μg	0.000001 g	10^{-6}

■ Temperature Conversion Chart

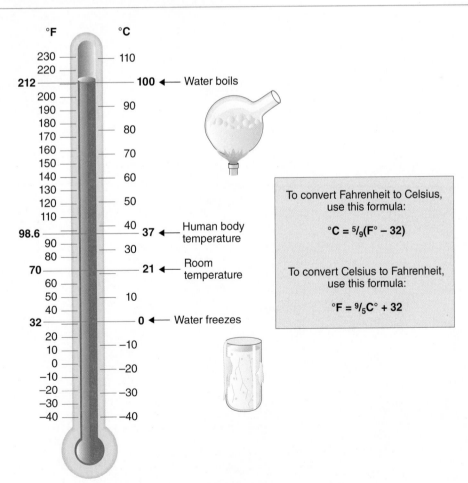

To convert Fahrenheit to Celsius, use this formula:

$$°C = \tfrac{5}{9}(F° - 32)$$

To convert Celsius to Fahrenheit, use this formula:

$$°F = \tfrac{9}{5}C° + 32$$

Water boils — 100 °C / 212 °F

Human body temperature — 37 °C / 98.6 °F

Room temperature — 21 °C / 70 °F

Water freezes — 0 °C / 32 °F

APPENDIX B

CDC Summary of Notifiable Diseases in the United States

■ **Infectious Diseases and Conditions Designated by Council of State and Territorial Epidemiologists (CSTE) and CDC as Nationally Notifiable During 2014**

(http://www.cdc.gov/mmwr/volumes/63/wr/mm6354a1.htm?s_cid=mm6354a1_w)

A notifiable infectious disease or condition is one for which regular, frequent, and timely information regarding individual cases is considered necessary for the prevention and control of the disease or condition.

- Anthrax
- Arboviral diseases, neuroinvasive and nonneuroinvasive
  - California serogroup viruses
  - Eastern equine encephalitis virus
  - Powassan virus
  - St. Louis encephalitis virus
  - West Nile virus
  - Western equine encephalitis virus
- Babesiosis
- Botulism
  - Foodborne
  - Infant
  - Other (includes wound and unspecified)
- Brucellosis
- Chancroid
- *Chlamydia trachomatis* infection
- Cholera (*Vibrio cholerae* O1 or O139)
- Coccidioidomycosis
- Cryptosporidiosis
- Cyclosporiasis
- Dengue virus infections
  - Dengue fever
  - Dengue hemorrhagic fever
  - Dengue shock syndrome

- Diphtheria
- Ehrlichiosis/Anaplasmosis
  - *Anaplasma phagocytophilum* infection
  - *Ehrlichia chaffeensis* infection
  - *Ehrlichia ewingii* infection
  - Undetermined human ehrlichiosis/anaplasmosis
- Giardiasis
- Gonorrhea
- *Haemophilus influenzae*, invasive disease
- Hansen's disease (Leprosy)
- Hantavirus pulmonary syndrome
- Hemolytic uremic syndrome, postdiarrheal
- Hepatitis, viral
  - Hepatitis A, acute
  - Hepatitis B, acute
  - Hepatitis B, chronic
  - Hepatitis B, perinatal infection
  - Hepatitis C, acute
  - Hepatitis C, past or present
- Human Immunodeficiency Virus (HIV) diagnoses
- Influenza-associated pediatric mortality
- Invasive pneumococcal disease (*Streptococcus pneumoniae*, invasive disease)

- Legionellosis (Legionnaire's Disease or Pontiac fever)
- Leptospirosis
- Listeriosis
- Lyme disease
- Malaria
- Measles
- Meningococcal disease (*Neisseria meningitidis*)
- Mumps
- Novel influenza A virus infections
- Pertussis
- Plague
- Poliomyelitis, paralytic
- Poliovirus infection, nonparalytic
- Psittacosis
- Q fever
  - Acute
  - Chronic
- Rabies
  - Animal
  - Human
- Rubella
- Rubella, congenital syndrome
- Salmonellosis
- Severe acute respiratory syndrome-associated coronavirus disease (SARS-CoV)

- Shiga toxin-producing *Escherichia coli* (STEC)
- Shigellosis
- Smallpox
- Spotted fever rickettsiosis
- Streptococcal toxic shock syndrome
- Syphilis
- Syphilis, congenital
- Tetanus
- Toxic shock syndrome (other than streptococcal)
- Trichinellosis
- Tuberculosis
- Tularemia

- Typhoid fever (caused by *Salmonella enterica* serotype *Typhi*)
- Vancomycin-intermediate *Staphylococcus aureus* (VISA) infection
- Vancomycin-resistant *Staphylococcus aureus* (VRSA) infection
- Varicella (morbidity)
- Varicella (mortality)
- Vibriosis (any species of the family *Vibrionaceae*, other than toxigenic *Vibrio cholerae* O1 or O139)

- Viral Hemorrhagic Fever
  - Crimean-Congo Hemorrhagic fever virus
  - Ebola virus
  - Lassa virus
  - Lujo virus
  - Marburg virus
  - New World Arenaviruses (Guanarito, Machupo, Junin, and Sabia viruses)
- Yellow fever

APPENDIX C

Pronouncing Organism Names

Some of the scientific names for microorganisms, which have Latin or Greek roots, can be hard to pronounce. As an aid in pronouncing these names, the primary microorganisms used in this textbook are listed alphabetically below, followed by the pronunciation. The following pronunciation key will aid you in saying these names. The accented syllable (ʻ) is placed directly after the syllable being stressed.

Pronunciation Key					
a add	ch check	g go	o odd	ou out	u put
a¯ ace	e end	i it	o¯ open	sh rush	ü rule
ã care	e¯ even	ı¯ ice	ô order	th thin	u¯ use
ä father	eˑ term	ng ring	oi oil	u up	

Acanthamoeba castellani a-kan-thä-me¯ʻbä kas-tel-än'e¯

Acetobacter aceti a-se¯ʻto¯-bak-teˑr a-set'e¯

Acinetobacter baumannii a-si-ne'to¯-bak-teˑr bou-mä'ne¯-e¯

Actinobacillus muris ak-tin-o¯-bä'cil-lus mu¯ʻris

Agaricus bisporis ä-gãr'i-kus bı¯-spo¯ʻr'us

Agrobacterium tumefaciens ag'ro¯-bak-ti're¯-urn tü'me-fa¯sh-enz

Ajellomyces dermatitidis ä-jel-lo¯-mı¯ʻse¯s deˑr-mä-tit'i-dis

Alcaligenes viscolactis al'kä-li-gen-e¯s vis-co-lak'tis

Amanita muscaria am-an-ı¯ʻtä mus-kãr'e-ä

A. phalloides fal-loi'dez

Amoeba proteus ä-me¯ʻbä pro¯ʻte–-us

Anaplasma phagocytophilum an'ä-plaz-mä fäg'o-sı¯-to¯-fil-um

Ancylostoma duodenale an-sil-o¯ʻsto¯-mä du¯-o¯-de'näl-e¯

Aquifex ä'kwe¯-feks

Armillaria är-mil-lãr'e¯-ä

Arthrobacter är-thro¯-bak'teˑr

Arthroderma är-thro¯-deˑr'mä

Ascaris lumbricoides as'kar-is lum-bri-koi'de¯z

Aspergillus favus a-speˑr-jil'lus fla¯ʻvus

A. fumigatus fü-mi-gä'tus

A. niger nı¯ʻjeˑr

A. oryzae ô'ri-zı¯

A. parasiticus pãr-ä-si-ti-kus

Azotobacter ä-zo'to-bak-teˑr

Babesia bigemina ba-be¯ʻse¯-ä big-em-e¯ʻna

B. microti mı¯-kro¯ʻte–

Bacillus amyloliquefaciens bä-sil'lus am-i-lo¯-li-kwä-fäs'e¯-enz

B. anthracis an-thra¯ʻsis

B. cereus se're¯-us

B. sphaericus sfe'ri-kus

B. subtilis su'til-us

B. thuringiensis thur-in-je¯-en'sis

Bacteroides fragilis bak-te'-roi'de¯z fra'gil-is

B. thetaiotaomicron tha¯-tä-ı¯-o¯-täw-mi'kron

Bartonella henselae bär-to¯-nel'lä hen'sel-ı¯

Beggiatoa bej'je¯-ä-to¯-ä

Blastomyces dermatitidis blas-to¯-mı¯'se¯z de'r-mä-tit'i-dis

Bordetella bronchiseptica bor-de-tel'lä bron-ke¯-sep'ti-kä

B. parapertussis pär'ä-pe'r-tus-sis

B. pertussis pe'r-tus'sis

Borrelia burgdorferi bôr-rel'e¯-ä burg-dôr'fe'r-e¯

B. hermsii he'rm-se¯'-e¯

B. recurrentis re¯-cür-ren'tis

B. turicatae te'r-i-kät'-ı¯

Botrytis cinerea bo-trı¯'tis cin-e'r-e¯'ä

Brevibacterium bre-vi-bak-ti're¯-um

Brucella abortus brü'sel-lä ä-bôr'tus

B. canis can'is

B. melitensis me-li-ten'sis

B. suis sü'is

Brugia malayi brü'-ge¯-ä mä-la¯'e¯

Burkholderia cepacia berk'ho¯ld-e'r-e¯-ä se-pa¯'se¯-ä

Campylobacter coli kam'pi-lo¯-bak-te'r ko¯'lı¯ (or ko¯'le¯)

C. jejuni je¯-ju¯'ne¯

Candida albicans kan'did-ä al'bi-kanz

Caulobacter crescentus ko¯-lo¯-bak'te'r kre-sen'tus

Cellulomonas sel-u-lo-mo¯'näs

Cephalosporium acremonium sef-ä-lo¯-spô're¯-um ac-re-mo¯'ne¯-um

Chlamydia trachomatis kla-mi'de¯-a trä-ko¯'mä-tis

Chlamydomonas klam-i-do¯-mo¯'näs

Chlamydophila pneumoniae kla-mi'dof-i-la nü-mo¯'ne¯-ı¯

C. psittaci sit'a-se¯

Chromobacter violaceum kro¯-mo¯-bak'-te'r vı¯-o¯-la¯'se¯-um

Claviceps purpurea kla'vi-seps pür-pü-re¯'ä

Clostridium acetobutylicum klôs-tri'de¯-um a¯-se¯-to¯-bu¯-til'i-kum

C. botulinum bot-u¯-lı-'num

C. difficile dif'fi-sil-e¯

C. perfringens pe'r-frin'jens

C. tetani te'tän-e¯

Coccidioides immitis kok-sid-e¯-oi'de¯z im'mi-tis

C. posadasii po¯-sä-da'se¯-e¯

Corynebacterium diphtheriae kôr'e¯-ne¯-bak-ti-re¯-um dif-thi're¯-ı-

Coxiella burnetii käks'e¯-el-lä be'r-ne'te¯-e¯

Cryptococcus neoformans krip'to¯-kok-kus ne¯-o¯-fôr'manz

Cryptosporidium coccidi krip'to¯-spô-ri-de¯-um kok'sid-e-

C. hominis ho¯'mi-nis

C. parvum pär'vum

Cyclospora cayetanensis sı¯'klo¯-spô-rä kı¯'e¯-tan-en-sis

Deinococcus radiodurans dı¯'no¯-kok-kus ra¯-de¯-o¯-dür'anz

Desulfovibrio de¯'sul-fo¯-vib-re¯-o¯

Desulfuromonas de¯'sul-für-o¯-mo¯-näs

Echinococcus granulosus e¯-kı¯n-o¯-kok'kus gra-nu¯-lo¯'sis

Ehrlichia chaffeensis e'r'lik-e¯-ä chäf-fen'sis

E. phagocytophila fa¯-go¯-cı¯-to'fı¯-lä

Emmonsiella capsulata em'mon-se¯-el-lä cap-sül-ä'tä

Entamoeba histolytica en-tä-me¯'bä his-to¯-li'ti-kä

Enterobacter aerogenes en-te-ro¯-bak'te'r ã-rä'jen-e¯z

E. cloacae klo¯-a¯'ki

Enterobius vermicularis en-te-ro¯'be¯-us ver-mi-ku¯-lar'is

Enterococcus faecalis en-te˙-ro¯-kok'kus fe¯-ka¯'lis

E. faecium fe¯'se¯-um

Epidermophyton ep-e¯-der-mo¯-fı–'ton

Erysipelothrix rhusiopathiae a¯r-e¯-sip'e-lo¯-thriks rü'sı–-o¯-pa-the¯

Escherichia coli esh-e˙r-e¯'ke¯-ä ko¯'lı¯ (or ko¯'le¯)

Euglena u¯-gle¯'nä

Filobasidiella neoformans fı¯-lo-ba-si-de¯-el'lä ne¯-o-fôr'mäns

Francisella tularensis fran'sis-el-lä tü'lä-ren-sis

Fusobacterium fu¯-so¯-bak-ti're¯-um

Gambierdiscus toxicus gam'be¯-e˙r-dis-kus toks'i-kus

Gardnerella intestinalis gärd-ne˙-rel'lä in-tes-ti-nal'is

G. vaginalis va-jin-al'is

Geobacillus stearothermophilus je¯-o¯-bä-sil'lus ste-är-o¯-the˙r-mä'fil-us

Giardia lamblia je¯-är'de¯-ä lam'le¯-ä

Gluconobacter glü'kon-o˙-bak-te˙r

Gonyaulax catanella gon-e¯-o-'laks kat-ä-nel'lä

Gymnodinium jim-no¯-din'e¯-um

Haemophilus ducrcyi he¯-mä'fil-us dü-krä'e–

H. influenzae in-flü-en'zı¯

Halobacterium salinarum ha-lo¯-bak-ti're˙-um sal-i-när'um

Hartmannella vermiformis hart-mä-nel'lä vêr-mi-fôr'mis

Helicobacter pylori he¯'lik-o¯-bak-te˙r pı–'lo¯-re–

Histoplasma capsulatum his-to¯-plaz'mä kap-su-lä'tum

Klebsiella pneumoniae kleb-se¯-el'lä nü-mo-'ne-ı–

Lactobacillus acidophilus lak-to¯-bä-sil'lus a-sid-o'fil-us

L. bulgaricus bul-gā'ri-kus

L. caseii ka¯'se¯-e¯

L. plantarum plan-tär'um

L. sanfranciscensis san-fran-si-sen'-sis

Lactococcus lactis lak-to¯-kok'kus lak'tis

Lagenidium giganteum la-je-ni'de¯-um jı–-gan'te¯-üm

Legionella pneumophila le¯-jä-nel'lä nü-mo¯'fi-lä

Leishmania donovani lish'mä-ne¯-ä don'o–-vän-e¯

L. tropica trop'i-kä

Leptospira interrogans lep-to¯-spı–'rä in-te˙r'ro¯-ganz

Leuconostoc citrovorum lü-ku¯-nos'tok sit-ro¯-vôr'um

L. mesenteroides mes-en-ter-oi'de–z

Listeria monocytogenes lis-te're¯-ä mo-no¯-sı–-tô'je-ne–z

Methanobacterium meth-a-no¯-bak-te˙r'e¯-um

Methanococcus jannaschii meth-a-no¯-kok'kus jan-nä'she¯-e¯

Micrococcus luteus mı¯-kro¯-kok'kus lu¯'te¯-us

Micromonospora mı¯-kro¯-mo¯-nos'po¯r-ä

Microsporum racemosum mı¯-kro¯-spô'rum ras-e¯-mo¯s'um

Morchella esculentum môr-che'lä es-kyu¯-len'tum

Mucor mu¯'kôr

Mycobacterium avium mı¯-ko¯-bak-ti're¯-um a¯'ve¯-um

M. bovis bo¯'vis

M. cheloni ke¯-lo¯'e¯

M. haemophilum he¯-mo'fil-um

M. kansasii kan-sä-se¯'ı¯

M. leprae lep'rı¯

M. marinum mār'in-um

M. tuberculosis tü-be˙r-ku¯-lo-'sis

Mycoplasma genitalium mı¯-ko¯-plaz'mä jen'i-tä-le¯-um

M. pneumoniae nu-mo¯-'ne¯-ı¯

Myxococcus xanthus micks-o¯-kok'kus zan'thus

Naegleria fowleri nı¯-gle're¯-ä fou'le˙r-e¯

Nannizzia nan'ne˙-ze¯-ä

Necator americanus ne-ka¯'tôr ä-me-ri-ka'nus

Neisseria gonorrhoeae nı̄-se're̅-ä go-nôr-re̅'ı̄

N. meningitidis me-nin ji'ti-dis

Neurospora nu̅-ros'po–r-ä

Nitrobacter nı̄-tro̅-bak'te˙r

Nitrosomonas nı̄-tro̅-so̅-mo̅'näs

Nocardia asteroids no̅-kär'de̅-a as-te˙r-oi'de–z

Nostoc nos'tok

Paramecium pãr-ä-me̅'se̅-um

Pasteurella multocida pas-tye˙r-el'lä mul-to̅'si-dä

Pelagibacter ubique pel-aj'e̅-bak-te˙r u̅-be̅k

Penicillium camemberti pen-i-sil'le̅-um kam-am-be˙r'te–

P. chrysogenum krı̄-so'gen-um

P. griseofulvin gris-e̅-o̅-fül'vin

P. notatum no̅-tä'tum

P. roqueforti ro̅-ko̅-fôr'te̅

Peptostreptococcus pep-to̅-strep'to̅-kok'kus

Pfiesteria piscicida fes-ter'e̅-ä pis-si-se̅'dä

Photorhabdus luminescens fo̅-to̅-rab'dus lü-mi-nes'senz

Phytophthora infestans fı̄-tof'thô-rä in-fes'tans

P. ramorum rä-môr'-um

Plasmodium falciparum plaz-mo̅'de̅-um fal-sip'är-um

P. malariae mä-la̅'re̅-ı̄

P. ovale o̅'va'le̅

P. vivax vı̄'vaks

Pneumocystis jiroveci nü-mo̅-sis'tis je˙r-o̅-vek'e̅

Porphyromonas gingivalis pôr'fı̄-ro̅-mo̅'näs jin-ji-val'is

Prochlorococcus pro̅-klôr-o̅-kok'kus

Propionibacterium acnes pro̅-pe̅-on'e̅-bak-ti-re̅-um ak'ne–z

Proteus mirabilis pro̅'te̅-us mi-ra'bi-lis

Pseudomonas aeruginosa su̅-do̅-mo̅'näs ã-rü ji-no̅'sä

P. cepacia se-pa̅'se̅-ä

P. marginalis mär-gin-al'is

Rhizobium rı̄-zo̅'be̅-um

Rhizopus stolonifer rı̄'zo-pus sto̅-lon-i-fe˙r

Rhodospirillum rubrum ro̅-do̅-spı̄-ril'um ru̅b'rum

Rickettsia akari ri-ket'se̅-ä ä-kãr'ı̄

R. prowazekii prou-wa-ze'ke̅-e̅

R. rickettsii ri-ket'se̅-e̅

R. tsutsugamushi tsü-tsü-gäm-ü'she̅

R. typhi tı̄'fe̅

Saccharomyces carlsbergensis sak-ä-ro̅-mı̄'se̅s kä-rls-be˙r-gen'sis

S. cerevisiae se-ri-vis'e̅-ı̄

S. ellipsoideus e̅-lip-soi'de̅-us

Saccharopolyspora erythraea sak-kãr-o̅-pol'e̅-spo-rä e̅-rith'rä-e̅

Salmonella enterica säl-mo̅n-el'lä en-te˙r-i'kä

S. enterica serotype Enteritidis en-te˙r-i-tı̄'dis

S. enterica serotype Typhi tı̄'fe̅

S. enterica serotype Typhimurium tı̄-fi-mur'e̅-um

Schistosoma haematobium shis-to̅-so̅'mä he̅-mä-to̅'be̅-um

S. japonicum ja-po̅'ne̅-kum

S. mansoni man'-so̅-ne̅

Serratia marcescens ser-rä'te̅-ä mär-ses'sens

Shigella dysenteriae shi-gel'lä dis-en-te're̅-ı̄–

S. sonnei son'ne̅-e̅

Spirillum minus spı̄'ril-lum mı̄'nus

Sporothrix schenkii spô-ro̅'thriks shen'ke̅-e̅

Staphylococcus aureus staf-i-lo̅-kok'kus ô-re̅-us

S. epidermidis e-pi-der'mi-dis

S. saprophyticus sa-pro̅-fi'ti-kus

Streptobacillus moniliformis strep-to̅-bä-sil'lus mon-i-li-fôr'mis

Streptococcus agalactiae strep-to̅-kok'kus a-gal-ac'te̅-ı̄

S. cremoris kre-mo̅'r'is

S. lactis lak'tis

S. mutans mu̅'tans

S. pneumoniae nü-mo⁻'ne⁻-ı⁻

S. pyogenes pı⁻-äj'en-e⁻z

S. sobrinus so⁻'bri-nus

S. thermophilus the˙r-mo'fil-us

Streptomyces cattley strep-to⁻-mı⁻'se⁻s kat-tel'-e⁻

S. coelicolor ko⁻'le⁻-ku-le˙r

S. erythraeus er-i-thra⁻-us

S. griseus gri'se⁻-us

S. lincolnensis lin-ko⁻l-nen'sis

S. mediterranei me-di-te˙r-rä'ne⁻-e⁻

S. nodosus no⁻-do⁻'sus

S. noursei ner'se⁻-e⁻

S. orientalis o⁻r-e⁻-en-tal'is

S. venezuelae ve-ne-zü-e'lı⁻

Sulfolobus acidocaldarius sul'fo⁻-lo⁻-bus as-i-do⁻-käl- där'e⁻-us

Taenia saginata te'ne⁻-ä sa-ji-nä'tä

T. solium so⁻'le⁻-um

Tetrahymena pyriformis tet-rä-hı⁻'me-nä pir-i-fôr'mis

Thermotoga maritima the˙r'mo⁻-to⁻-gä mar-i-te⁻'mä

Thiobacillus thı⁻-o⁻-bä-sil'lus

Thiomargarita namibiensis thı⁻'o⁻-mär-gä-re⁻-tä na'mi- be⁻-n-sis

Thiothrix thı⁻'o⁻-thriks

Toxoplasma gondii toks-o⁻-plaz'mä gon'de⁻-e⁻

Treponema pallidum tre-po⁻-ne⁻'mä pal'li-dum

T. pertenue pe˙r-ten'u⁻-e⁻

Trichinella spiralis trik-in-el'lä spı⁻-ra'lis

Trichoderma major trik'o⁻-de˙r-ma ma⁻'jôr

Trichomonas vaginalis trik-o⁻-mo–n'äs va-jin-al'is

Trichonympha trik-o⁻-nimf'ä

Trichophyton trik-o⁻-fı⁻'ton

Trypanosoma brucei gambiense tri-pa'no⁻-so⁻-mä brüs'e⁻ gam-be⁻-ens'

T brucei rhodesiense ro⁻-de⁻-se⁻-ens'

T. cruzi krüz'e–

Ureaplasma urealyticum u⁻-re⁻-ä-plaz'mä u⁻-re⁻-ä-lit'i-kum

Vibrio cholerae vib're⁻-o⁻ kol'e˙r-ı⁻

V. parahaemolyticus pa-rä-he⁻-mo⁻-li'ti-kus

V. vulnificus vul-ni'fi-kus

Wuchereria bancrofti vu⁻-ke˙r-är'e⁻-ä ban-krof'te⁻

Yersinia enterocolitica ye˙r-sin'e⁻-ä en'te˙r-o⁻-ko⁻l-it-ik-ä

Y. pestis pes'tis

Y. pseudotuberculosis su⁻-do⁻-tu⁻-be˙r-kyu-lo⁻'sis

Zoogloea ramigera zo⁻'o⁻-gle⁻-ä ram-i-ge˙r'ä

GLOSSARY

This glossary contains concise definitions for microbiological terms and concepts only. **Please refer to the index for specific infectious agents, infectious diseases, anatomical terms, specific antimicrobial drugs, and immune disorders.**

A

abortive poliomyelitis   The most common clinical presentation of polioviruses. the patient experiences fever, headache, nausea, and sore throat.

abscess   A circumscribed pus-filled lesion characteristic of staphylococcal skin disease; also called a boil.

abyssal zone   The environment at the bottom of oceanic trenches.

accessory digestive organs   An organ that helps with digestion but is not part of the digestive tract.

accessory reproductive organs Include the epididymis, a coiled tube leading out of the testes that stores sperm until they are mature and motile.

acetyl CoA   One of the starting compounds for the citric acid cycle.

acid   A substance that releases hydrogen ions (H$^+$) in solution; *see also* **base**.

acid-fast technique   A staining process in which mycobacteria resist decolorization with acid alcohol.

acidic dye   A negatively charged colored substance in solution that is used to stain an area around cells.

acidophile   A microorganism that grows best at acidic pHs below 4.

Acinetobacter **pneumonia**   occurs in outbreaks and, like *P. aeruginosa*, is usually associated with contaminated respiratory-support equipment or fluids.

Actinobacteria   A phylum in the domain Bacteria that exhibits fungus-like properties when cultivated in the laboratory.

actinomycete   A soil bacterium that exhibits fungus-like properties when cultivated in the laboratory.

activated sludge   Aerated sewage containing microorganisms added to untreated sewage to purify it by accelerating its bacterial decomposition.

activation energy   The energy required for a chemical reaction to occur.

active immunity   The immune system responds to antigen by producing antibodies and specific lymphocytes.

active site   The region of an enzyme where the substrate binds.

active transport   An energy-requiring movement of substances from an area of lower concentration across a biological membrane to a region of higher concentration by means of a membrane-spanning carrier protein.

acute disease   A disease that develops rapidly, exhibits substantial symptoms, and lasts only a short time.

acute HCV   An acute illness with a discrete onset of any sign or symptom consistent with acute viral hepatitis.

acute HIV infection   A flu-like syndrome that occurs immediately after a person contracts HIV.

acute inflammation   An immediate, but nonspecific innate immune defense to trauma.

acute period   The phase of a disease during which specific symptoms occur and the disease is at its height.

acute phase protein   A defensive blood protein secreted by liver cells that elevates the inflammatory and complement responses to infection.

acute rejection   A transplant rejection response that takes 10 to 30 days.

acute wound   One that arises from cuts, lacerations, bites, or surgical procedures.

adaptive immunity   A response to a specific immune stimulus that involves immune defensive cells and frequently leads to the establishment of host immunity.

adenosine diphosphate (ADP)   A molecule in cells that is the product of ATP hydrolysis.

adenosine triphosphate (ATP)   A molecule in cells that provides most of the energy for metabolism.

adhesin   A protein in bacterial pili that assists in attachment to the surface molecules of cells.

adjuvant   An agent added to a vaccine to increase the vaccine's effectiveness.

ADP   *See* **adenosine diphosphate**.

aerobe   An organism that uses oxygen gas (O$_2$) for metabolism.

aerobic respiration   The process for transforming chemical energy to ATP in which the final electron acceptor in the electron transport chain is oxygen gas (O$_2$).

aerosol   A small particle transmitted through the air.

aerotolerant   A bacterium not inhibited by oxygen gas (O$_2$).

aflatoxin   A toxin produced by *Aspergillus flavus* that is cancer causing in vertebrates.

African trypanosomiasis (African sleeping sickness)   An insect-borne parasitic disease of humans and other animals.

agar   A polysaccharide derived from marine seaweed that is used as a solidifying agent in many microbiological culture media.

agarose   A substance that is the main constituent of agar and is used especially in gels for electrophoresis. It is a polysaccharide mainly containing galactose residues.

agglutination   A type of antigen-antibody reaction that results in visible clumps of organisms or other material.

agranulocyte   A white blood cell lacking visible granules; includes the lymphocytes and monocytes; *see also* **granulocyte**.

alcoholic fermentation   A catabolic process that forms ethyl alcohol during the reoxidation of NADH to NAD$^+$ for reuse in glycolysis to generate ATP.

aldehyde   Highly active agents containing a–CHO (carbonyl) functional group.

alga (pl. **algae**)   An organism in the kingdom Protista that performs photosynthesis.

algal bloom   An excessive growth of algae on or near the surface of water, often the result of an oversupply of nutrients from organic pollution.

alginate   A sticky substance used as a thickener in foods and beverages.

alkaliphile   An organism that grows best at pH 9 or higher.

allergen   An antigenic substance that stimulates an allergic reaction in the body.

allergenic   A type of antigen that produces an abnormally vigorous immune response in which the immune system fights off a perceived threat that would otherwise be harmless to the body.

allergen immunotherapy   A procedure involving a series of allergen shots over several years to reduce the sensitivity to allergens.

allergic rhinitis   Also known as hay fever, is a type of inflammation in the nose which occurs when the immune system overreacts to allergens in the air.

allergy   A damaging immune response by the body to a substance to which it has become hypersensitive.

allergy shot   *See* **allergen immunotherapy**.

allograft   A tissue graft between two members of the same species, such as between two humans (not identical twins).

alpha helix   A spiral portion of a polypeptide consisting of amino acids stabilized by hydrogen bonds.

alpha (α) hemolytic   Referring to those bacterial species that when plated on blood agar cause a partial destruction of red blood cells as seen by an olive green color in the agar around colonies.

Alveolata   A group of protists, considered a major clade and superphylum within Eukarya, and are also called Alveolata.

amatoxin   A group of toxic compounds found in several genera of poisonous mushrooms.

amebiasis   Infection with amoebas, especially as causing dysentery.

Ames test   A diagnostic procedure used to detect potential cancer-causing agents in humans by the ability of the agent to cause mutations in bacterial cells.

amino acid   An organic acid containing one or more amino groups; the monomers that build proteins in all living cells.

aminoglycoside   An antibiotic that contains amino groups bonded to carbohydrate groups that inhibit protein synthesis; examples are gentamicin, streptomycin, and neomycin.

aminoquinolone   Any of various quinoline derivatives notable as antimalarial drugs.

amoeboid motion   A crawling type of movement caused by the flow of cytoplasm into plasma membrane projections; typical of the amoebas.

amoebozoan   A member of the super group Unikonta characterized by amoeboid movement and pseudopodia.

amylase   An enzyme, found chiefly in saliva and pancreatic fluid, that converts starch and glycogen into simple sugars.

anabolism   An energy-requiring process involving the synthesis of larger organic compounds from smaller ones; *see also* **catabolism**.

anaerobe   An organism that does not require or cannot use oxygen gas (O_2) for metabolism.

anaerobic respiration   The production of ATP where the final electron acceptor is an inorganic molecule other than oxygen gas (O_2); examples include nitrate, sulfate, and carbonate.

anaphylactic shock   A sudden drop in blood pressure, itching, swelling, and difficulty in breathing typical of some allergic reactions.

anaphylaxis   Severe allergic reaction.

anaplasmosis   One of two rickettsial tickborne diseases.

animalcule   A tiny, microscopic organism observed by Leeuwenhoek.

anion   An ion with a negative charge; *see also* **cation**.

anoxic   Without oxygen gas (O_2).

anoxygenic photosynthesis   A form of photosynthesis in which molecular oxygen (O_2) is not produced.

antibiogram   A collection of data usually in the form of a table summarizing the percent of individual bacterial pathogens susceptible to different antimicrobial agents.

antibiotic   A substance naturally produced by a few bacterial or fungal species that inhibits or kills other microorganisms.

antibody   A highly specific protein produced by the body in response to a foreign substance, such as a bacterium or virus, and capable of binding to the substance.

anticodon   A three-base sequence on the tRNA molecule that binds to the codon on the mRNA molecule during translation.

antigen   A chemical substance that stimulates the production of antibodies by the body's immune system.

antigen-antibody complex   The interaction of an antigen with antibodies.

antigen-binding site   The region on an antibody that binds to an antigen.

antigen-presenting cell (APC)   A macrophage or dendritic cell that exposes antigen peptide fragments on its surface to T cells.

antigenic determinant   A section of an antigen molecule that stimulates antibody formation and to which the antibody binds; also called epitope.

antigenic drift   A minor variation over time in the antigenic composition of influenza viruses.

antigenic shift   A major change over time in the antigenic composition of influenza viruses.

antigenic variation   Referring to changes in the spike proteins on the flu viruses.

antihistamine   A drug that blocks cell receptors for histamine, preventing allergic effects such as sneezing and itching.

antimicrobial   An agent that kills microorganisms or inhibits their growth. Antimicrobial medicines can be grouped according to the microorganisms they act primarily against.

antimicrobial agent (drug)   A chemical that inhibits or kills the growth of microorganisms.

antimicrobial resistance (AMR)   The ability of a microorganism (like bacteria, viruses, and some parasites) to stop an antimicrobial (such as antibiotics, antivirals and antimalarials) from working against it.

antimicrobial spectrum   The range of antimicrobial drug action.

antiretroviral (ARV) drug   Medications that treat HIV.

antiretroviral therapy (ART)   The use of HIV medicines to treat HIV infection.

antisepsis   The use of chemical methods for eliminating or reducing the growth or replication of infectious agents.

antiseptic   A chemical used to reduce or kill pathogenic microorganisms on a living object, such as the surface of the human body.

antiserum (pl. antisera)   A blood-derived fluid containing antibodies and used to provide temporary immunity.

antitoxin   An antibody produced by the body that circulates in the bloodstream to provide protection against toxins by neutralizing them.

antiviral protein (AVP)   A protein made in response to interferon and that blocks viral replication.

antiviral state   A cell capable of inhibiting viral protein synthesis due to interferon activation.

apicomplexan   A protist containing a number of organelles at one end of the cell that are used for host penetration; no motion is observed in adult forms.

aplastic anemia   An inability to produce red blood cells.

apoptosis   A programmed cell death.

APTIMA assay   A test that amplifies and detects HIV RNA, as a diagnosis of primary HIV-1 infection.

aqueous solution One or more substances dissolved in water.

arboviral encephalitis A rare primary encephalitis. It is an example of a zoonosis, and mosquitoes are the most common vector.

arbovirus A virus transmitted by arthropods (e.g., insects).

Archaea The domain of living organisms that excludes the Bacteria and Eukarya.

artemisinin An effective drug used in curing malaria; especially when combined with other drugs to limit the development of drug resistance.

arthrospore An asexual fungal spore formed by fragmentation of a septate hypha.

Arthus phenomenon An immune complex hypersensitivity when large amounts of IgG antibody for complexes with antigens in blood vessels or near the site of antigen entry.

artificially acquired active immunity The production of antibodies by the body in response to antigens in a vaccination; or the passive transfer of antibodies formed in one individual or animal to another susceptible person.

artificially acquired passive immunity The temporary immunity resulting from the transfer of antibodies from one individual or animal to another.

ascariasis A type of roundworm infection. Accounts for a major burden of disease worldwide.

Ascomycota A phylum of fungi whose members have septate hyphae and form ascospores within saclike asci, among other notable characteristics.

ascospore A sexually produced fungal spore formed by members of the ascomycetes.

ascus (pl. **asci**) A saclike structure containing ascospores; formed by the ascomycetes.

asepsis The process or method of bringing about a condition in which no unwanted microbes are present.

aseptic technique The practice of transferring microorganisms to a sterile culture medium without introducing other contaminating organisms.

asexual reproduction The form of reproduction that maintains genetic constancy while increasing cell numbers.

aspergillosis A condition in which certain fungi infect the tissues.

asthma A respiratory condition marked by spasms in the bronchi of the lungs, causing difficulty in breathing. It usually results from an allergic reaction or other forms of hypersensitivity.

atom The smallest portion into which an element can be divided and still enter into a chemical reaction.

atomic nucleus The positively charged core of an atom, consisting of protons and neutrons that make up most of the mass.

atomic number The number of protons in the nucleus of an atom.

atopy The genetic tendency to develop the classic allergic diseases: atopic dermatitis, allergic rhinitis (hay fever), and asthma.

ATP See **adenosine triphosphate**.

ATP/ADP cycle The cellular processes of synthesis and hydrolysis of ATP.

ATP synthase The enzyme involved in forming ATP by using the energy in a proton gradient.

attenuated Referring to the reduced ability of bacterial cells or viruses in a vaccine to do damage to the exposed individual.

attractant A substance that attracts cells through motility.

autoantibody An antibody made against a self antigen.

autoclave An instrument used to sterilize microbiological materials by means of high temperature using steam under pressure.

autograft Tissue taken from one part of the body and grafted to another part of the same body.

autoimmune disorder A reaction in which antibodies react with an individual's own chemical substances and cells.

autotroph An organism that uses carbon dioxide (CO_2) as a carbon source; see also **chemoautotroph** *and* **photoautotroph**.

autotrophy The process by which an organism makes its own food.

auxotroph A mutant strain of an organism lacking the ability to synthesize a nutritional need; *see also* **prototroph**.

avermectin Effective drugs used against a wide variety of roundworms. The drugs affect the roundworm's nervous system, causing muscle paralysis.

avirulent Referring to an organism that is not likely to cause disease.

B

babesiosis Aa malaria-like disease caused by the apicomplexan parasite *Babesia microti*. These

Bacillé Calmette-Guérin (BCG) A strain of attenuated *Mycobacterium bovis* used for immunization against tuberculosis and, on occasion, leprosy.

bacillary dysentery A type of dysentery, and is a severe form of shigellosis. Characterized by bloody diarrhea.

bacillus (pl. **bacilli**) (1) Any rod-shaped bacterial or archaeal cell. (2) When referring to the genus *Bacillus*, it refers to an aerobic or facultatively anaerobic, rod-shaped, endospore-producing, gram-positive bacterial genus.

bacitracin A cyclic polypeptide that is bactericidal and interferes with the transport of the NAG-NAM disaccharides through the cell membrane. Bacitracin is available in ointments for topical treatment of superficial skin infections.

bacteremia The presence of live bacterial cells in the blood.

Bacteria The domain that includes all organisms not classified as Archaea or Eukarya.

bacterial growth curve The events occurring over time within a population of growing and dividing bacterial cells.

bactericidal Referring to any agent that kills bacterial cells.

bacteriochlorophyll A pigment located in the membrane systems of purple sulfur bacteria that upon excitement by light, loses electrons and initiates photosynthetic reactions.

bacteriocin One of a group of bacterial proteins toxic to other bacterial cells.

bacteriology The scientific study of prokaryotes in the domain Bacteria and Archaea.

bacteriophage (phage) A virus that infects and replicates within bacterial cells.

bacteriostatic Referring to any substance that prevents the growth of bacteria.

bacterium (pl. **bacteria**) A single-celled microorganism lacking a cell nucleus and membrane-enclosed compartments, and often having peptidoglycan in the cell wall.

barophile A microorganism that lives under conditions of high atmospheric pressure.

base   A chemical compound that accepts hydrogen ions (H⁺) in solution; *see also* **acid**.

base analog   A nitrogenous base with a similar structure to a natural base but differing slightly in composition.

basic dye   A positively charged colored substance in solution that is used to stain cells.

Basidiomycota   The phylum of fungi whose members have septate hyphae and form basidiospores on supportive basidia, among other notable characteristics.

basidiospore   A sexually produced fungal spore formed by members of the basidiomycetes.

basidium (pl. **basidia**)   A club-like structure containing basidiospores; formed by the basidiomycetes.

basophil   A type of white blood cell with granules that functions in allergic reactions.

batch method   A method for milk pasteurization that involves heating the liquid at 63°C for 30 minutes.

B cell   *See* **B lymphocyte**.

benign   Referring to a tumor that usually is not life threatening or likely to spread to another part of the body.

benthic zone   The environment at the bottom of a deep river, lake, or sea.

benzoic acid   A chemical preservative used to protect beverages, catsup, and margarine.

beta (β) **hemolytic**   Referring to those bacterial species that when plated on blood agar completely destroy the red blood cells as seen by a clearing in the agar around the colonies.

beta-lactamase   The enzyme that converts the beta-lactam antibiotics (penicillins, cephalosporins, and carbapenems) into inactive forms.

beta-oxidation   The breakdown of fatty acids during cellular metabolism through the successive removal from one end of two carbon units.

bile   A mixture of cholesterol, phospholipids, acids, and immunoglobulins that are stored in the gallbladder.

bilharziasis   Another name for Schistosomiasis. An infection caused by parasitic flukes of the genus Schistosoma.

binary fission   An asexual process in bacterial and archaeal cells by which a cell divides to form two new cells while maintaining genetic constancy.

bioaugmentation   A form of bioremediation where bacteria are added to an area is referred.

bioburden   The estimated number of viable microbes in or on a medical device or surface before sterilization.

biocatalysis   The metabolic reactions of microorganisms carried out at an industrial scale.

biochemical oxygen demand (BOD)   A number referring to the amount of oxygen used by the microorganisms in a sample of water during a 5-day period of incubation.

biocrime   The intentional introduction of biological agents into food or water, or by injection, to harm or kill individuals.

biofilm   A complex community of microorganisms that form a protective and adhesive matrix that attaches to a surface, such as a catheter or industrial pipeline.

biofuel   Renewable biological materials that store potential energy in forms capable of providing useful energy.

biogenesis   The idea that life arises only from life.

biogeochemical cycle   The ways chemical elements, such as carbon, nitrogen, sulfur, and phosphorus, are recycled in Nature and converted from one chemical form to another.

biological pollution   The presence of microorganisms from human waste in water.

biological safety cabinet   A cabinet or hood used to prevent contamination of biological materials.

biological vector   An infected arthropod, such as a mosquito or tick, that transmits disease-causing organisms between hosts; *see also* **mechanical vector**.

biome   A specific environment that's home to living things suited for that place and climate.

biomass   The total weight of all organisms in a defined area or environment.

bioreactor   A large fermentation tank for growing microorganisms used in industrial production; also called a fermentor.

bioremediation   The use of microorganisms to degrade toxic wastes and other synthetic products of industrial pollution.

biosphere   The regions of the surface, atmosphere, and hydrosphere of the earth (or analogous parts of other planets) occupied by living organisms.

biostimulation   A process encouraging bacterial growth.

biotechnology   The commercial application of genetic engineering using living organisms.

bioterrorism   The intentional or threatened use of biological agents to cause fear in or actually inflict death or disease upon a large population.

biotype   A group of individuals having the same genotype

bipolar staining   A characteristic of *Yersinia* and *Francisella* species in which stain gathers at the poles of the cells, yielding the appearance of safety pins.

bisphenol   A combination of two phenol molecules used in disinfection.

blanching   A process of putting food in boiling water for a few seconds to destroy enzymes.

blastomycosis   A disease caused by infection with parasitic fungi affecting the skin or the internal organs.

blastospore   A fungal spore formed by budding.

B lymphocyte (B cell)   A white blood cell that matures into memory cells and plasma cells that secrete antibody.

booster shot   A repeat dose of a vaccine given some years after the initial course to maintain a high level of immunity.

B period   The growth phase in a bacterial cell cycle when the cells are increasing in size and mass.

bright-field microscopy   An optical configuration of the light microscope that magnifies an object by passing visible light directly through the lenses and object.

broad spectrum   Referring to an antimicrobial drug useful for treating many groups of microorganisms, including gram-positive and gram-negative bacteria; *see also* **narrow spectrum**.

bronchiolitis   Inflammation of the bronchioles.

bronchopneumonia   Scattered patches of infection in the respiratory passageways.

broth   A liquid growth medium.

bubo   A swelling of the lymph nodes due to inflammation.

bubonic plague   One of the most devastating infectious diseases in human history.

budding   (1) An asexual process of reproduction in fungi, in which a new cell forms as a swelling at the border of the parent cell and then breaks free to live independently. (2) The controlled release of virus particles from an infected animal cell.

buffer   (1) A compound that minimizes pH changes in a solution by neutralizing added acids and bases. (2) Refers to a solution containing such a substance.

bull's-eye rash   A circular lesion on the skin with a red border and central clearing; characteristic of Lyme disease.

Burkitt lymphoma   A tumor of the connective tissues of the jaw that is especially prevalent in children.

burst size   The number of virus particles released from an infected bacterial cell.

C

campylobacteriosis   One of the most commonly reported bacterial causes of invasive gastroenteritis in the United States.

cancer   A disease characterized by the radiating spread of malignant cells that reproduce at an uncontrolled rate.

candidiasis   An infection caused by a species of the yeast Candida, usually *Candida albicans*.

capnophilic   Referring to a prokaryotic cell requiring low oxygen gas (O_2) and a high concentration of carbon dioxide gas (CO_2) for metabolism.

capsid   The protein coat that encloses the genome of a virus.

capsomere   Any of the protein subunits of a capsid.

capsule   A layer of polysaccharides and small proteins covalently bound some prokaryotic cells; *see also* **slime layer** *and* **glycocalyx**.

carbapenem   A broad spectrum of beta-lactam drugs that are one of the most important groups of clinically useful antibiotics.

carbapenem-resistant Enterobacteriaceae (CRE)   Gram-negative bacteria that are resistant to the carbapenem class of antibiotics, considered the drugs of last resort for such infections.

carbohydrate   An organic compound consisting of carbon, hydrogen, and oxygen that is an important source of carbon and energy for all organisms; examples include simple sugars, starch, and cellulose.

carbon cycle   A series of interlinked processes involving carbon compound exchange between living organisms and the nonliving environment.

carbon-fixing reaction   A chemical reaction in the second part of photosynthesis in which carbohydrates are formed.

carbuncle   An enlarged abscess formed from the union of several smaller abscesses or boils.

carcinogen   Any physical or chemical substance that causes tumor formation.

carrier   An individual who has recovered from a disease but retains the infectious agents in the body and continues to shed them.

casein   The major protein in milk.

casing soil   A non-nutritious soil used to provide moisture for mushroom formation.

catabolism   An energy-liberating process in which larger organic compounds are broken down into smaller ones; *see also* **anabolism**.

cation   A positively charged ion; *see also* **anion**.

cationic dye   *See* **basic dye**.

CD4   A coreceptor on the cell surface of a T helper cell.

CD8   A coreceptor on the cell surface of a cytotoxic T cell.

cell cycle   The periods through which a cell passes between one division and the next.

cell envelope   The cell wall and cell membrane of a bacterial or archaeal cell.

cell line   A group of identical cells in culture and derived from a single cell.

cell-mediated immune response   The body's ability to resist infection through the activity of T-lymphocyte recognition of antigen peptides presented on macrophages and dendritic cells and on infected cells.

cell membrane   A thin bilayer of phospholipids and proteins that surrounds the prokaryotic cell cytoplasm. *See also* **plasma membrane**.

cell metabolism   The sum of all chemical reactions in cells, including dehydration and hydrolysis reactions.

cell morphology   The shape and arrangement of cells with respect to one another.

cell theory   The tenet that all organisms are made of cells and arise from preexisting cells.

cell (tissue) tropism   Refers to the specific cells or tissues that a virus infects.

cellular chemistry   The chemical study of cells and cellular molecules.

cellular oncogene (*c-onc*)   A mutated form of a normal gene controlling cell growth.

cellular respiration   The process of converting chemical energy into cellular energy in the form of ATP.

cellulose   The tough material composing plant cell walls, and most of the weight of a switchgrass plant is cellulose.

cell wall   A carbohydrate-containing structure surrounding fungal, algal, and most bacterial and archaeal cells.

central dogma   The doctrine that DNA codes for RNA through transcription and RNA is converted to protein through translation.

centrosome   The microtubule-organizing center of a eukaryotic cell.

cephalosporin   Any of a group of semisynthetic broad-spectrum antibiotics resembling penicillin.

cercaria (pl. cercariae)   A tadpole-like larva form in the life cycle of a trematode.

cesspool   Concrete cylindrical rings with pores in the walls that is used to collect human waste.

cestode   A flatworm, commonly known as a tapeworm, that lives as a parasite in the gut of vertebrates.

CFU   *See* **colony-forming unit**.

chain of transmission   How infectious diseases can be spread from human to human (or animal to human).

Chagas disease   A disease caused by *Trypanosoma cruzi*. It is transmitted by the reduviid bug and is found in Mexico and 17 countries in Central and South America.

chancroid   A sexually transmitted disease characterized by painful genital ulceration and inflammatory inguinal adenopathy.

chancre   A painless, circular, purplish hard ulcer with a raised margin that occurs during primary syphilis.

chaperone   A protein that ensures a polypeptide folds into the proper shape.

chemical bond   A force between two or more atoms that tends to bind those atoms together.

chemical element   Any substance that cannot be broken down into a simpler one by a chemical reaction.

chemically defined medium A substance in which the form and quantity of each nutrient is known; *see also* **complex medium**.

chemical pollution The presence of inorganic and organic waste in water.

chemical reaction A process that changes the molecular composition of a substance by redistributing atoms or groups of atoms without altering the number of atoms.

chemiosmosis The use of a proton gradient across a membrane to generate cellular energy in the form of ATP.

chemoautotroph An organism that derives energy from inorganic chemicals and uses the energy to synthesize nutrients from carbon dioxide gas (CO_2).

chemoheterotroph An organism that derives energy from organic chemicals and uses the energy to synthesize nutrients from carbon compounds other than carbon dioxide gas (CO_2).

chemokine A protein that prompts specific white blood cells to migrate to an infection site and carry out their immune system functions.

chemotaxis A movement of a cell or organism toward a chemical or nutrient.

chemotherapeutic agent A chemical compound used to treat diseases and infections in the body.

chemotherapeutic index A number that represents the highest level of an antimicrobial drug tolerated by the host divided by the lowest level of the drug that eliminates the infectious agent.

chemotherapy The process of using chemical agents to treat diseases and infections, or other disorders, such as cancer.

chitin A polymer of acetylglucosamine units that provides rigidity in the cell walls of fungi.

Chlamydiae A phylum of extremely small bacterial cells that can be cultured only in living host cells.

chloramphenicol An antibiotic used against serious infections such as typhoid fever.

chlorhexidine A compound used as a surgical hand scrub and superficial skin wound cleanser.

chlorination The process of treating water with chlorine to kill harmful organisms.

chlorophyll A green or purple pigment in algae and some bacterial cells that functions in capturing light for photosynthesis.

chloroplast A double membrane-enclosed compartment in algae that contains chlorophyll and other pigments for photosynthesis.

CHNOPS The acronym CHNOPS, which stands for carbon, hydrogen, nitrogen, oxygen, phosphorus, sulfur, represents the six most important chemical elements whose covalent combinations make up most biological molecules on Earth.

cholera One of many illnesses affecting the digestive system.

cholera toxin An enterotoxin that triggers an unrelenting loss of fluid.

Chromalveolata One of the super groups of the protists that includes the dinoflagellates, diatoms, ciliates, and the apicomplexans.

chromosome A structure in the nucleoid or cell nucleus that carries hereditary information in the form of genes.

chronic disease A disease that develops slowly, tends to linger for a long time, and requires a long convalescence.

chronic inflammation A long-lasting, but nonspecific innate immune response to trauma.

chronic rejection A form of transplant rejection that takes months to years to develop.

chronic wound An open sore, such as a leg ulcer or bed sore.

chytrid A fungus in the phylum Chytridiomycota.

Chytridiomycota A phylum of predominantly aquatic fungi.

ciliate A protistan organism that moves with the aid of cilia.

cilium (pl. **cilia**) A hair-like projection on some eukaryotic cells that along with many others assist in the motion of some protozoa and beat rhythmically to aid the movement of a fluid past the respiratory epithelial cells in humans.

cirrhosis Extensive injury of cells of the liver.

citric acid cycle A metabolic pathway in which acetyl groups are completely oxidized to carbon dioxide gas and some ATP molecules are formed. Also called Krebs cycle.

class A category of related organisms consisting of one or more orders.

classification The arrangement of organisms into hierarchal groups based on relatedness.

climax *See* **acute period**.

climate change A change in global or regional climate patterns, in particular a change apparent from the mid to late 20th century onwards and attributed largely to the increased levels of atmospheric carbon dioxide produced by the use of fossil fuels.

clinical disease A disease in which the symptoms are apparent.

clonal deletion A process by which the immune system eliminates autoreactive cells.

clonal selection The theory that certain lymphocytes are activated from the mixed population of B or T lymphocytes when stimulated by antigen or antigen peptide fragments.

clone A population of genetically identical cells (or plasmids).

cloning vector A plasmid used to introduce genes into a bacterial cell.

coagulase An enzyme produced by some staphylococci that catalyzes the formation of a fibrin clot.

coagulase test A diagnostic procedure to distinguish between *Staphylococcus aureus* and *S. epidermidis*.

coccidioidomycosis Known more commonly as "valley fever" or "desert fever," is caused by *Coccidioides immitis* and *C. posadasii*.

coccus (pl. **cocci**) A spherical-shaped bacterial or archaeal cell.

codon A three-base sequence on the mRNA molecule that specifies a particular amino acid insertion in a polypeptide.

coenocytic Referring to a fungus containing no septa (cross-walls) and multinucleate hyphae.

coenzyme A small, organic molecule that forms the nonprotein part of an enzyme molecule.

coenzyme A (CoA) A small, organic molecule of cellular respiration that functions in release of carbon dioxide gas (CO_2) and the transfer of electrons and protons to another coenzyme.

cofactor A inorganic substance that acts with and is essential to the activity of an enzyme; examples include metal ions and some vitamins.

cohesion The action or property of like molecules sticking together, being mutually attractive.

coliform bacterium   A gram-negative, nonsporeforming, rodshaped cell that ferments lactose to acid and gas and usually is found in the human and animal intestine; high numbers in water is an indicator of contamination.

colony   A visible mass of micro-organisms of one type growing on or in a solid growth medium.

colony-forming unit (CFU)   A measure of the viable cells by counting the number of colonies on a plate; each colony presumably started from one viable cell.

colostrum   The yellowish fluid rich in antibodies secreted from the mammary glands of animals or humans prior to the production of true milk.

comedo (pl. **comedones**)   A plugged sebaceous gland.

commensalism   A close and permanent association between two species of organisms in which one species benefits and the other remains unharmed and unaffected.

commercial sterilization   A canning process to eliminate the most resistant bacterial spores.

communicable disease   A disease that is readily transmissible between hosts.

community-acquired pneumonia (CAP)   One of the most common infectious diseases and is an important cause of mortality and morbidity.

comparative genomics   The comparison of DNA sequences between organisms.

competence   Referring to the ability of a cell to take up naked DNA from the environment.

competitive inhibition   The prevention of a chemical reaction by a chemical that competes with the normal substrate for an enzyme's active site; *see also* **noncompetitive inhibition**.

complement   A group of blood proteins that functions in a cascading series of reactions with antibodies to recognize and help eliminate certain antigens or infectious agents.

complement fixation test   A serological procedure to detect anti-bodies to any of a variety of pathogens by identifying antibody-antigen-complement complexes.

complex   Referring to one form of symmetry found in some viral capsids.

complex medium   A chemically undefined medium in which the nature

and quantity of each component has not been identified; *see also* **chemically defined medium**.

compound   A substance made by the combination of two or more different chemical elements.

compound microscope   *See* **light microscope**.

condensation reaction   *See* **dehydration synthesis reaction**.

congenital rubella syndrome   Virus transmission from mother to fetus that can lead to destruction of the fetal capillaries and blood insufficiency.

congenital syphilis   A severe, disabling, and often life-threatening infection seen in infants. A pregnant mother who has syphilis can spread the disease through the placenta to the unborn infant.

congenital Zika syndrome   A pattern of birth defects found among fetuses and babies infected with **Zika** virus during pregnancy.

conidiophore   The supportive structure on which conidia form.

conidium (pl. **conidia**)   An asexually produced fungal spore formed on a supportive structure without an enclosing sac.

conjugate vaccine   An antigen preparation consisting of the antigen bound to a carrier protein.

conjugation   (1) In Bacteria and Archaea, a unidirectional transfer of genetic material from a live donor cell into a live recipient cell during a period of cell contact. (2) In the ciliates, a sexual process involving the reciprocal transfer of micronuclei between cells in contact.

conjugation pilus (pl. **pili**)   A protein filament essential for conjugation between donor and recipient bacterial cells.

constant domain   The invariable amino acids in the light and heavy chains of an antibody.

contact dermatitis   A delayed hypersensitivity reaction occurring after contact with an allergen.

contagious   Referring to a disease whose agent passes with particular ease among susceptible individuals.

contractile vacuole   A membrane-enclosed structure within a cell's cytoplasm that regulates the water content by absorbing water and then contracting to expel it.

contrast   Refers to the ability to see an object against the background in the microscope.

control   That part of an experiment not exposed to or treated with the factor being tested.

Coombs test   An antibody test used to detect Rh antibodies involved in hemolytic disease of the newborn.

corticosteroid   A synthetic drug used to control allergic disorders by blocking the release of chemical mediators.

covalent bond   A chemical linkage formed by the sharing of electrons between atoms or molecules.

C period   The DNA replication phase of a bacterial cell cycle.

Crenarchaeota   A group within the domain Archaea that tend to grow in hot or cold environments.

CRISPR-Cas system   A type of "immune defense" which is capable of recognizing and destroying foreign (plasmid or viral) DNA that enters the cell.

critical control point (CCP)   In the food processing industry, a place where contamination of the food product could occur.

Cryptomycota   A group of fungi with small, flagellated cells without a chitin cell wall.

cryptosporidiosis   An illness caused by tiny, one-celled cryptosporidium parasites.

cutaneous anthrax   The most common and least dangerous form of anthrax. Cutaneous anthrax is typically caused when B. anthracis spores enter through cuts on the skin.

cutaneous (skin) leishmaniasis   This is the most common form of leishmaniasis, both in general and in U.S. travelers. Unless otherwise specified, cutaneous leishmaniasis refers to localized cutaneous leishmaniasis, rather than to much less common forms, such as diffuse cutaneous leishmaniasis and disseminated cutaneous leishmaniasis.

Cyanobacteria   A phylum of oxygen-producing, photosynthetic bacterial cells.

cyst   A dormant and very resistant form of a protozoan and multicellular parasite.

cysticercosis   A parasitic tissue infection caused by larval cysts of the tapeworm *Taenia solium*.

cytochrome   A compound containing protein and iron that plays a role as an electron carrier in cellular respiration and photosynthesis; *see also* **electron transport chain**.

cytokine Small proteins released by immune defensive cells that affects other cells and the immune response to an infectious agent.

cytokinesis The splitting of a cell in half.

cytopathic effect (CPE) Visible effect that can be seen in a virus-infected host cell.

cytoplasm The complex of chemicals and structures within a cell; in plant and animal cells excluding the nucleus.

cytoskeleton (1) The structural proteins in a prokaryotic cell that help control cell shape and cell division. (2) In a eukaryotic cell, the internal network of protein filaments and microtubules that control the cell's shape and movement.

cytosol The fluid, ions, and compounds of a cell's cytoplasm excluding organelles and other structures.

cytotoxic hypersensitivity A cell-damaging or cell-destroying hypersensitivity that develops when IgG reacts with antigens on the surfaces of cells.

cytotoxic T cell (CTL) The type of T lymphocyte that searches out and destroys infected cells.

cytotoxin A chemical that is poisonous to cells.

D

dark-field microscopy An optical system on the light microscope that scatters light such that the specimen appears white on a black background.

dead-end host A host that cannot transmit a pathogen.

deamination A biochemical process in which amino groups are enzymatically removed from amino acids or other organic compound.

decimal reduction time (D valve) The time required to kill 90% of the viable organisms at a specified temperature.

decline phase The final portion of a bacterial growth curve in which environmental factors overwhelm the population and induce death; also called death phase.

decomposer A bacterial or fungal organism that breaks down and recycles dead or decaying matter.

defensin An antimicrobial peptide present in white blood cells that plays a role in the prevention or elimination of infection.

definitive host An organism that harbors the adult, sexually mature form of a parasite.

degerm To mechanically remove organisms from a surface.

degranulation The release of cell mediators from mast cells and basophils.

dehydration synthesis reaction A process of bonding two molecules together by removing a water molecule and joining the open bonds.

delayed hypersensitivity An immune response to an allergen that takes 2 to 3 days to develop.

denaturation A process caused by heat or pH in which proteins lose their function due to changes in their 3-D structure.

dendritic cell A white blood cell having long finger-like extensions and found within all tissues; it engulfs and digests foreign material, such as bacterial cells and viruses, and presents antigen peptides on its surface.

dental plaque A biofilm consisting of salivary proteins, food debris, and bacteria attached to the tooth surface.

deoxyribonucleic acid (DNA) The genetic material of all cells and many viruses.

dermatophyte A pathogenic fungus that affects the skin, hair, or nails.

dermatophytosis An infection of the hair, skin, or nails caused by any one of the dermatophytes.

dermis A complex, thick layer of fibrous and elastic connective tissue beneath the epidermis.

desensitization A process in which minute doses of antigens are used to remove antibodies from the body tissues to prevent a later allergic reaction.

desiccation The process of extreme drying.

desquamation The process buy which the epidermis peels off just above the stratum basale,

detergent A synthetic cleansing substance that dissolves dirt and oil.

dextran A water-soluble glucose polymer.

diagnostic microbiology A field that involves applying microbiology to medical diagnosis.

diapedesis A process by which phagocytes move out of the blood vessels by migrating between capillary cells.

diatom One of a group of microscopic marine algae that performs photosynthesis.

debridement The removal of dead, damaged, or infected tissue.

diatomaceous earth A soft, crumbly, porous sedimentary deposit formed from the fossil remains of diatoms.

dichotomous key A method of deducing the correct species assignment of a living organism by offering two alternatives at each juncture, with the choice of one of those alternatives determining the next step.

differential medium A growth medium in which different species of microorganisms can be distinguished visually.

differential staining procedure A technique using two dyes to differentiate cells or cellular objects based on their staining; *see also* **simple stain technique**.

dikaryotic Referring to fungal cytoplasm in which two genetically different haploid nuclei closely pair.

dimorphic Referring to pathogenic fungi that take a yeast form in the human body and a filamentous form when cultivated in the laboratory.

dinoflagellate A microscopic photosynthetic marine alga that forms one of the foundations of the food chain in the ocean.

dipicolinic acid An organic substance that helps stabilize the proteins and DNA in a bacterial spore, thereby increasing spore resistance.

diplobacillus (pl. diplobacilli) A pair of rod-shaped bacterial or archaeal cells.

diplococcus (pl. diplococci) A pair of spherical-shaped bacterial or archaeal cells.

diploid Having two sets of genes.

diplomonad A protist that contains four pair of flagella, two haploid nuclei, and live in low oxygen or anaerobic environments; most members are symbiotic in animals.

direct contact The form of disease transmission involving close association between hosts; *see also* **indirect contact**.

direct microscopic count Estimation of the number of cells by observation with the light microscope.

direct observation treatment (DOT) A treatment strategy whereby a healthcare worker administers and ensures that a patient has taken prescribed TB drugs.

disaccharide A sugar formed from two single sugar molecules; examples include sucrose and lactose.

disease Any change from the general state of good health.

disinfectant A chemical used to kill or inhibit pathogenic microorganisms on a lifeless object such as a tabletop.

disinfection The process of killing or inhibiting the growth of pathogens.

disk diffusion method A procedure for determining bacterial susceptibility to an antibiotic by determining if bacterial growth occurs around an antibiotic disk; also called the Kirby-Bauer test.

disseminated intravascular coagulation (DIC) A systemic clotting of blood in small blood vessels.

disulfide bridge A covalent bond between sulfur-containing R groups in amino acids.

DNA *See* **deoxyribonucleic acid**.

DNA helicase An enzyme that unwinds and separates the two polynucleotide strands during DNA replication.

DNA ligase An enzyme that binds together DNA fragments.

DNA microarray A series of DNA segments used to study changes in gene expression.

DNA polymerase An enzyme that catalyzes DNA replication by combining complementary nucleotides to an existing strand.

DNA probe A short segment of single stranded DNA used to locate a complementary strand among many other DNA strands.

DNA replication The process of copying the genetic material in a cell.

DNA vaccine A preparation that consists of a DNA plasmid containing the gene for a pathogen protein.

domain The most inclusive taxonomic level of classification; consists of the Archaea, Bacteria, and Eukarya.

double diffusion assay Another name for the diffusion of antibodies and antigens.

double helix The structure of DNA, in which the two complementary strands are connected by hydrogen bonds between complementary nitrogenous bases and wound in opposing spirals.

Downey cell A swollen lymphocyte with foamy cytoplasm and many vacuoles that develops as a result of infection with infectious mononucleosis viruses.

D period The binary fission phase of a bacterial cell cycle.

droplet nucleus Particles 1–10 mcm in diameter, implicated in spread of airborne infection.

droplet transmission The movement of an airborne particle of mucus and sputum from the respiratory tract.

drug synergism Referring to two antimicrobial drugs that work better together than seperately.

dysbiosis A change to the normal (healthy) composition of the gut microbiota.

dysentery A type of gastroenteritis that results in diarrhea with blood.

E

Ebola virus disease (EVD) A rare and deadly viral illness that is reportable to the National Notifiable Disease Surveillance System (NNDSS) in all U.S. states and territories.

eclipse period The period of a viral infection when no viruses can be found inside the infected cell.

ecotype A subgroup of a species that has special characteristics to survive in its ecological surroundings.

ectopic pregnancy Development of a fertilized egg outside the womb, often in a fallopian tube.

edema A swelling of the tissues brought about by an accumulation of fluid.

effector cell An activated immune cell targeting a pathogen.

ehrlichiosis One of two rickettsial tickborne diseases.

electrolyte An ion in cells, blood, or other organic material.

electron A negatively charged particle with a small mass that moves around the nucleus of an atom.

electron micrograph Images recorded on electron-sensitive film.

electron shell An energy level surrounding the atomic nucleus that contains one or more electrons.

electron transport chain A series of proteins that transfer electrons in cellular respiration to generate ATP.

elementary body (EB) An infectious form of *Chlamydia* in the early stage of reproduction.

ELISA *See* **enzyme-linked immunosorbent assay**.

elongation (1) The addition of complementary nucleotides to a parental DNA strand. (2) The addition of addition of amino acids onto the forming polypeptide during translation.

emerging infectious disease A new disease or changing disease that is seen within a population for the first time; *see also* **reemerging infectious disease**.

encephalitis an inflammation of the brain.

endemic Referring to a disease that is constantly present in a specific area or region.

endergonic reaction A chemical process that requires energy; *see also* **exergonic reaction**.

endocytosis The process by which many eukaryotic cells take up substances, cells, or viruses from the environment.

endoflagellum (pl. endoflagella) A microscopic fiber located along cell walls in certain species of spirochetes; contractions of the filaments yield undulating motion in the cell.

endogenous infection A disorder that starts with a microbe or virus that already was in or on the body as part of the microbiota; *see also* **exogenous infection**.

endogenous retrovirus (ERV) DNA sequences within a viral genome.

endogenous viral element (EVE) Nucleic acid sequences within the viral genome.

endomembrane system A cytoplasmic set of membranes that function in the transport, modification, and sorting of proteins and lipids in eukaryotic cells.

endophyte A fungus that lives within plants and does not cause any known disease.

endoplasmic reticulum (ER) A network of membranous plates and tubes in the eukaryotic cell cytoplasm responsible for the synthesis and transport of materials from the cell.

endospore An extremely resistant dormant cell produced by some gram-positive bacterial species.

endosymbiont theory A theory that states states that at some time in the very distant past free-living respiratory bacteria and free-living photosynthetic bacteria were engulfed and retained by an evolving eukaryotic cell.

endosymbiosis The idea that mitochondria and chloroplasts originated from free-living bacterial cells and cyanobacteria that took up residence in a primitive eukaryotic cell.

endotoxin   A metabolic poison, produced chiefly by gram-negative bacteria that are part of the bacterial cell wall and consequently are released on cell disintegration; composed of lipid-polysaccharide-peptide complexes.

endotoxin shock   A drop in blood pressure due to an endotoxin.

energy-fixing reaction   A reaction in the first part of photosynthesis where light energy is converted into chemical energy in the form of ATP.

enriched medium   A growth medium in which special nutrients must be added to get an species to grow.

enterotoxin   A toxin that is active in the gastrointestinal tract of the host.

enterotype   Refers to a specific composition of bacterial species in the gut.

entry inhibitor   Antiretroviral drug that works by blocking viral entry into the CD4 cells.

envelope   The flexible membrane of protein and lipid that surrounds many types of viruses.

enveloped virus   A virus that has an outer wrapping or envelope.

environmental microbiology   The scientific study of microorganisms in the environment.

enzootic   Refers to a disease endemic to a population of animals.

enzyme   A reusable protein molecule that brings about a chemical change while itself remaining unchanged.

enzyme immunoassay (EIA)   *See* **enzyme-linked immunosorbent assay**.

enzyme-linked immunosorbent assay (ELISA)   A serological test in which an enzyme system is used to detect an individual's exposure to a pathogen.

enzyme-substrate complex   The association of an enzyme with its substrate at the active site.

eosin   A red anionic (acidic) dye.

eosinophil   A type of white blood cell with granules that stains with the dye eosin and plays a role in allergic reactions and the body's response to parasitic infections.

epidemic   Referring to a disease that spreads more quickly and more extensively within a population than normally expected.

epidemic typhus   One of the most notorious of all bacterial diseases caused by *Rickettsia prowazekii*, which is transmitted to humans by body lice where hygiene is poor.

epidemiology   The scientific study of the source, cause, and transmission of disease within a population.

epidermis   The upper, bloodless portion of skin that is visible.

epitope   *See* **antigenic determinant**.

ergotism   A condition inducing convulsions and hallucinations from fungal toxins in rye.

erysipelas   A superficial form of cellulitis involving the dermis.

erysipeloid   An infection in humans, usually on the hands or feet, caused by *Erysipelothrix rhusiopathiae*.

erythema   A zone of redness in the skin due to a widening of blood vessels near the skin surface.

erythema infectiosum   A rash caused by human parvovirus B19, which is a small, single-stranded DNA virus of the Parvoviridae (parv = "small") family.

erythema migrans (EM)   An expanding circular red rash that occurs on the skin of patients with Lyme disease.

erythrogenic   Referring to a streptococcal poison that leads to the rash in scarlet fever.

eschar   A black, necrotic (dying) scar that can be caused by antrhrax infection.

E test   An antibiotic sensitivity test using a paper strip containing a marked gradient of antibiotic.

ethylene oxide   A chemical gas that is used to sterilize many objects and instruments.

etiology   Disease causation.

Euglenozoa   A group of protists in the Excavata.

Eukarya   The taxonomic domain encompassing all organisms having a cell nucleus and membrane-bound compartments.

eukaryote   An organism whose cells contain a cell nucleus with multiple chromosomes, a nuclear envelope, and membrane-bound compartments; *see also* **prokaryote**.

eukaryotic   Referring to a cell or organism containing a cell nucleus with multiple chromosomes, a nuclear envelope, and membrane-bound compartments.

Euryarchaeota   A group within the domain Archaea that contains the methanogens and extreme halophiles.

exanthema   A maculopapular rash occurring on the skin surface.

Excavata   One of the super groups of the protists, many lacking true mitochondria.

exergonic reaction   A chemical process releasing energy; *see also* **endergonic reaction**.

exogenous infection   A disorder that starts with a microbe or virus that entered the body from the environment; *see also* **endogenous infection**.

exon   The coding sequence in a split gene; *see also* **intron**.

exotoxin   A bacterial metabolic poison composed of protein that is released to the environment; in the human body, it can affect various organs and systems.

experiment   A test or trial to verify or refute a hypothesis.

extreme acidophile   An archaeal organism living at an extremely acidic pH.

extreme halophile   A microbial species living at high salt (NaCl) concentration.

extreme thermophile   microorganisms that grow optimally at temperatures greater than 80°C.

extremophile   A microorganism that lives in extreme environments, such as high/low temperature, high acidity, or high salt.

extrinsic factor   An environmental characteristic that influences the growth of food microbes. *See also* **intrinsic factor**.

evolutionary medicine   A new field of medicine providing new strategies to slow the evolution of antibiotic resistance, to treat infectious diseases, and to better understand the role of microbes in human health.

F

facilitated diffusion   The movement of substances from an area of higher concentration across a biological membrane to a region of lower concentration by means of a membrane-spanning channel or carrier protein.

facultative anaerobe   Referring to an organism that grows in the presence or absence of oxygen gas (O_2).

family   A category of related organisms consisting of one or more genera.

Fc segment The stem portion of an antibody molecule that combines with phagocytes, mast cells, or complement.

fecal-oral route Pathogens in fecal material passed from one infected individual to the oral cavity of another individual.

fecal transplant The process of transferring fecal microbiota from a healthy individual into a recipient suffering a stubborn intestinal disease.

feedback inhibition The slowing down or prevention of a metabolic pathway when excess end product binds noncompetitively to an enzyme in the pathway.

fermentation A metabolic pathway in which carbohydrates serve as electron donors, the final electron acceptor is not oxygen gas (O_2), and NADH is reoxidized to NAD^+ for reuse in glycolysis for generation of ATP.

fermentor *See* **bioreactor**.

fever An abnormally high body temperature that is usually caused by a bacterial or viral infection.

F factor A plasmid containing genes for plasmid replication and conjugation pilus formation.

field The circular area seen when looking in the light microscope.

filopodium (pl. **filopodia**) A thin protrusion from a cell, such as a phagocyte.

filtration A mechanical method to remove microorganisms by passing a liquid or air through a filter.

Firmicutes A phylum in the domain Bacteria that contains many of the gram-positive species.

flaccid paralysis A loss of voluntary movement in which the limbs have little tone and become flabby.

flagellum (pl. **flagella**) A long, hair-like appendage composed of protein and responsible for motion in microorganisms; found in some bacterial, archaeal protistan, algal, and fungal cells.

flare A spreading zone of redness around a wheal that occurs during an allergic reaction; *also see* **wheal**.

flash pasteurization method A treatment in which milk is heated at 71.6°C for 15 seconds and then cooled rapidly to eliminate harmful bacteria; also called HTST ("high temperature, short time") method.

flatworm A multicellular parasite with a flattened body; examples include the tapeworms.

flavin adenine dinucleotide (FAD) A coenzyme that acts as an electron carrier for ATP production.

floc A jelly-like mass that forms in a liquid and made up of coagulated particles.

flocculation (1) A serological reaction in which particulate antigens react with antibodies to form visible aggregates of material. (2) The formation of jelly-like masses of coagulated material in the water-purification process.

fluid mosaic model The representation for the cell (plasma) membrane where proteins "float" within or on a bilayer of phospholipid.

fluke *See* **trematode**.

fluorescence microscopy An optical system on the light microscope that uses ultraviolet light to excite dye-containing objects to fluoresce.

fluorescent antibody technique A diagnostic tool that uses fluorescent antibodies with the fluorescence microscope to identify an unknown organism.

fluoroquinolone One of the most prescribed antibiotics in the United States used to treat urinary tract infections, gonorrhea and chlamydia, and intestinal tract infections. Examples include levofl oxacin and ciprofl oxacin (Cipro).

flush A burst of mushroom growth.

folic acid The organic compound in bacteria whose synthesis is blocked by sulfonamide drugs.

folliculitis Inflammation of the hair follicles.

fomite An inanimate object, such as clothing or a utensil, that carries disease organisms.

food vacuole A membrane-enclosed compartment in some eukaryotic that results from the intake of large molecules, particles, or cells, for digestion.

foraminiferan (foram) A shell-containing amoeboid protozoan having a chalky skeleton with window-like openings between sections of the shell.

formalin A solution of formaldehyde used as embalming fluid, in the inactivation of viruses, and as a disinfectant.

F plasmid A DNA plasmid in the cytoplasm of an F^+ bacterial cell that may be transferred to a recipient bacterial cell during conjugation.

fruiting body The general name for a reproductive structure of a fungus from which spores are produced.

functional genomics The identification of gene function from a gene sequence.

functional group A group of atoms on hydrocarbons that participates in a chemical reaction.

fungemia The dissemination of fungi through the circulatory system.

Fungi One of the four kingdoms in the domain Eukarya; composed of the molds and yeasts.

fungicidal Referring to any agent that kills fungi.

fungistatic Referring to any substance that inhibits the growth of fungi.

furuncle An infection of a hair follicle.

G

gametocyte The stage in the life cycle of the malaria parasite during which it reproduces sexually in the blood of a mosquito.

gamma (immune) globulin A general term for antibody-rich serum.

gamma ray An ionizing radiation that can be used to sterilize objects.

gangrene A physiological process in which the enzymes from wounded tissue digest the surrounding layer of cells, inducing a spreading death to the tissue cells.

gastroenteritis Infection of the stomach and intestinal tract often due to a virus.

gastrointestinal anthrax An acute inflammation of the intestinal tract caused by the consumption of endospore-contaminated and undercooked meat.

gene A segment of a DNA molecule that provides the biochemical information for a polypeptide or for a functioning RNA molecule.

generalized transduction A process by which a bacteriophage carries a bacterial chromosome fragment from one cell to another; *see also* **specialized transduction**.

generation time The time interval for a cell population to double in number.

genetically modified organism (GMO) An organism produced by genetic engineering.

genetic code   The specific order of nucleotide sequences in DNA or RNA that encode specific amino acids for protein synthesis.

genetic engineering   The use of bacterial and microbial genetics to isolate, manipulate, recombine, and express genes.

genetic recombination   The process of bring together different segments of DNA.

genome   The complete set of genes in a virus or cell.

genome annotation   The identification of gene locations and function in a gene sequence.

genomic island   A series of up to 25 genes absent in other strains of the same prokaryotic species.

genomics   The study of an organism's gene structure and gene function in viruses and organisms.

genus (pl. **genera**)   A rank in the classification system of organisms composed of one or more species; a collection of genera constitute a family.

germicide   Any agent that kills microorganisms.

germ theory of disease   The principle formulated by Pasteur and proved by Koch that microorganisms are responsible for infectious diseases.

giardiasis   A common intestinal parasitic infection.

gingivitis   One of the most common forms of early-stage periodontal disease. Bacterial cells in plaque buildup multiply and build up between the teeth and gums.

Glomeromycota   A group of mycorrhizal fungi that exist within the roots of most land plants.

glucose   A six-carbon sugar used as a major energy source for metabolism.

glutamic acid (**glutamate**)   A valuable food supplement for humans and animals, and its sodium salt, monosodium glutamate, is used in food preparations.

glutaraldehyde   A liquid chemical used for sterilization.

glycocalyx   A viscous polysaccharide material covering many prokaryotic cells to assist in attachment to a surface and impart resistance to desiccation; *see also* **capsule** *and* **slime layer**.

glycogen   A substance deposited in bodily tissues as a store of carbohydrates.

glycolysis   A metabolic pathway in which glucose is broken down into two molecules of pyruvate with a net gain of two ATP molecules.

glycopeptide   A class of antibiotics that inhibit bacterial cell wall synthesis.

Golgi apparatus   A stack of flattened, membrane-enclosed compartments in eukaryotic cells involved in the modification and sorting of lipids and proteins.

graft-versus-host (GVH) reaction A phenomenon in which a tissue graft or transplant produces immune substances against the recipient.

gram-negative   Referring to a bacterial cell that stains red after Gram staining.

gram-positive   Referring to a bacterial cell that stains purple after Gram staining.

Gram stain technique   A staining procedure used to identify bacterial cells as gram-positive or gram-negative.

granulocyte   A white blood cell with visible granules in the cytoplasm; includes neutrophils, eosinophils, and basophils; *see also* **agranulocyte**.

granuloma   A small lesion caused by an infection.

granzyme   A cytotoxic enzyme that causes infected cells to undergo a programmed cell death.

groundwater   Water originating from deep wells and subterranean springs; *see also* **surface water**.

group A streptococci (GAS)   Contains those streptococcal species that are β-hemolytic organisms.

growth factor   A substance, such as a vitamin or hormone, that is required for the stimulation of growth in living cells.

Guillain-Barré syndrome (GBS) A complication of influenza and chickenpox, characterized by nerve damage and polio-like paralysis.

gumma   A soft, granular lesion that forms in the cardiovascular and/or nervous systems during tertiary syphilis.

H

HACCP   *See* **hazard analysis critical control point**.

halogen   A chemical element whose atoms have seven electrons in their outer shell; examples include iodine and chlorine.

halophile   An organism that lives in environments with high concentrations of salt.

halotolerant   A microbe that grows best without salt but can tolerate low concentrations.

hand, foot, and mouth disease   A moderately contagious disease typically affecting infants and young children in the spring to fall. Symptoms include fever, poor appetite, malaise, and a sore throat. Most patients recover in 7 to 10 days.

haploid   Having a single set of genetic information.

hapten   A small molecule that combines with tissue proteins or polysaccharides to form an antigen.

Hazard Analysis Critical Control Point (HACCP)   A set of federally enforced regulations to ensure the dietary safety of seafood, meat, and poultry.

healthcare-acquired pneumonia (HAP) An inflammation of one or both lungs that develops primarily in older adults and immunosuppressed patients at least 48 hours after administration to a hospital or other healthcare facility.

healthcare-associated infection (HAI)   An infection resulting from treatment for another condition while in a hospital or healthcare facility.

heat fixation   The use of warm temperatures to prepare microorganisms for staining and viewing with the light microscope.

heavy (H) chain   The larger polypeptide in an antibody.

heavy metal   A chemical element often toxic to microorganisms; examples include mercury, copper, and silver.

helical   One form of symmetry found in some viral capsids.

helminth   A term referring to a multicellular parasite; includes roundworms and flatworms.

helminthic load   The number of worms in the human body.

helper T (T$_H$) cell   A type of lymphocyte involved in regulation of cell-mediated immunity and in helping B cells to produce antibody.

helper T1 (T$_H$1) cell   A T lymphocyte that enhances the activity of B lymphocytes.

helper T2 (T$_H$2) cell   A T lymphocyte that stimulates destruction of macrophages infected with bacterial cells.

hemagglutination   The formation of clumps of red blood cells.

hemagglutination-inhibition (HAI) test   A test using a patient's serum to detect the presence of antibodies against a specific infectious agent.

hemagglutinin (H)   (1) An enzyme composing one type of surface spike on influenza viruses that enables the viruses to bind to the host cell. (2) An agent such as a virus or an antibody that causes red blood cells to clump together.

hemolysin   A bacterial toxin that destroys red blood cells.

hemolysis   The destruction of red blood cells.

hepatitis B core antigen (HBcAG)   An antigen located in the inner lipoprotein coat enclosing the DNA of a hepatitis B virus.

hepatitis B surface antigen (HBsAg)   An antigen located in the outer surface coat of a hepatitis B virus.

herd (community) immunity   The proportion of a population that is immune to a disease; also called community immunity.

herd immunity threshold   The percent of the population that must be vaccinated to effectively stop disease spread.

hermaphroditic   Refers to an organism having both male and female reproductive organs.

heterokaryon   A fungal cell that has two or more genetically different nuclei.

heterotroph   An organism that requires preformed organic matter for its energy and carbon needs; *see also* **photoheterotroph** *and* **chemoheterotroph.**

heterotrophy   The process by which an organism uses preformed organic compounds for metabolism.

high-efficiency particulate air (HEPA) filter   A fibrous filter that traps particles, microbes, and spores in the air.

high frequency of recombination (Hfr)   Referring to a bacterial cell containing an F factor incorporated into the bacterial chromosome.

highly active antiretroviral therapy (HAART)   The combination of several (typically three or four) antiretroviral drugs for the treatment of infections caused by retroviruses, especially the human immunodeficiency virus.

highly perishable   Referring to foods that spoil easily.

histamine   A mediator in type I hypersensitivity reactions that is released from the granules in mast cells and basophils and causes the contraction of smooth muscles.

histone   A type of basic protein that forms the unit around which DNA is coiled in the nucleosomes of eukaryotic chromosomes.

histoplasmosis   A lung disease that occurs worldwide. It is endemic in the Ohio and Mississippi River valleys where it is often called "summer flu." The causative agent is *Histoplasma capsulatum.*

Hodgkin lymphoma   A cancer of the lymph nodes.

holding (batch) method   A pasteurization process that exposes a liquid to 63°C for 30 min.

homeostasis   The tendency of an organism to maintain a steady state or equilibrium with respect to specific functions and processes.

hops   The dried petals of the vine *Humulus lupulus* that impart flavor, color, and stability to beer.

horizontal gene transfer (HGT)   The movement of genes from one organism to another within the same generation; also called lateral gene transfer.

horizontal transmission   The spread of disease from one person to another.

host   An organism on or in which a microorganism lives and grows, or a virus replicates.

host range   The variety of species that a disease-causing microorganism can infect.

human cytomegalovirus disease   A mononucleosis-like syndrome involving fever and malaise.

human genome   The complete set a genetic information in a human cell.

human leukocyte antigen (HLA)   Cell surface antigens involved with tissue transplantation.

human microbiome   Refers to the tremendous number of microbes that normally live in and on the human body.

human monocytic ehrlichiosis (HME)   One of two similar rickettsial tickborne diseases. It is caused by *Ehrlichia chaffeensis.*

humic substance   Major component of the natural organic matter (NOM) in soil and water.

humoral immune response   The immune reaction of producing antibodies directed against antigens in the body fluids.

hyaluronidase   An enzyme that digests hyaluronic acid and thereby permits the penetration of pathogens through connective tissue.

hybridoma   The cell resulting from the fusion of a B cell and a cancer cell.

hydatid cyst   A thick-walled body formed in the human liver by *Echinococcus granulosus* and containing larvae.

hydrocarbon   An organic molecule containing only hydrogen and carbon atoms that are connected by a sharing of electrons.

hydrogen bond   A weak attraction between a positively charged hydrogen atom (covalently bonded to oxygen or nitrogen) and a covalently bonded, negatively charged oxygen or nitrogen atom in the same molecule or separate molecules.

hydrolysis reaction   A process in which a molecule is split into two parts through the interaction of H^+ and $(OH)^-$ of a water molecule.

hydrophilic   Referring to a substance that dissolves in or mixes easily with water; *see also* **hydrophobic.**

hydrophobia   An emotional condition ("fear of water") arising from the inability to swallow as a consequence of rabies.

hydrophobic   Referring to a substance that does not dissolve in or mix easily with water; *see also* **hydrophilic.**

hyperacute rejection   An immediate rejection response to a transplant.

hyperimmune globulin   A type of serum product that contains high concentrations of specific antibodies that are prepared from donor plasma having the antibodies of interest.

hypersensitivity   An immunological response to exposure to an allergen or antigen.

hyperthermophile   A prokaryote that has an optimal growth temperature above 80°C.

hypertonic   A solution with more dissolved material (solutes) than the surrounding solution.

hypha (pl. hyphae)   A microscopic filament of cells representing the vegetative portion of a fungus.

hypothalamus   The part of the brain controlling involuntary functions.

hypothesis   An educated guess or answer to a properly framed question.

I

icosahedral   Referring to a symmetrical figure composed of 20 triangular faces and 12 points; one of the forms of symmetry found in some viral capsids.

IgA   The class of antibodies found in respiratory and gastrointestinal secretions that help neutralize pathogens.

IgD   The class of antibodies found on the surface of B cells that act as receptors for binding antigen.

IgE   The class of antibodies responsible for type I hypersensitivities.

IgG   The class of antibodies abundant in serum that are major diseases fighters.

IgM   The first class of antibodies to appear in serum in helping fight pathogens.

immediate allergic response   An allergic reaction that occurs within seconds or minutes.

immediate hypersensitivity   Immune reactions to an allergen that occur within minutes to a few hours.

immune complex   A combination of antibody and antigen capable of complement activation and characteristic of type III hypersensitivity reactions.

immune complex hypersensitivity   An immune response involving antibodies combining with antigens, the complex accumulating in blood vessels or on tissue surfaces.

immune deficiency   The lack of an adequate immune system response.

immune suppression   A process whereby the immune system does not respond to self antigens.

immunity   The body's ability to resist infectious disease through innate and acquired mechanisms.

immunization   The process of making an individual resistant to a particular disease by administering a vaccine; *see also* **vaccination.**

immunocompetent   The ability of the body to develop an immune response in the presence of a disease-causing agent.

immunocompromised   Referring to an inadequate immune response as a result of disease, exposure to radiation, or treatment with immunosuppressive drugs.

immunodeficiency disorder   The inability, either inborn or acquired, of the body to produce an adequate immune response to fight disease.

immunodiffusion   The movement of antigen and antibody toward one another through a gel or agar to produce a visible precipitate.

immunoelectrophoresis   A laboratory diagnostic procedure in which antigen molecules move through an electric field and then diffuse to meet antibody molecules to form a precipitation line; *see also* **electrophoresis.**

immunofluorescence antibody (IFA) test   The use of a fluorescently labeled antibody to detect an antigen.

immunogen   *See* "**Antigen**".

immunogenic   Capable of generating an immune response.

immunoglobulin (Ig)   The class of immunological proteins that react with an antigen; an alternate term for antibody.

immunological memory   The long-term ability of the immune system to remember past pathogen exposures.

immunology   The scientific study of the structure and function of the immune system.

immunosuppressive agent   A drug given to block transplant rejection by the immune system.

immunotherapy   The use of monoclonal antibodies to control an overactive inflammatory response.

inactivated vaccine   A vaccine consisting of virus particles, bacteria, or other pathogens that have been grown in culture and then killed using specific chemicals and/or heat.

incidence   The number of reported disease cases in a given time frame.

inclusion   (1) A granule-like storage structure found in the prokaryotic cell cytoplasm. (2) A virus in the cytoplasm or nucleus of an infected cell.

incubation period   The time that elapses between the entry of a pathogen into the host and the appearance of signs and symptoms.

index of refraction   A measure of the light bending ability of a medium through which light passes.

indicator organism   A microorganism whose presence signals fecal contamination of water.

indirect contact   The mode of disease transmission involving nonliving objects; *see also* **direct contact.**

induced mutation   A change in the sequence of nucleotide bases in a DNA molecule arising from a mutagenic agent used under controlled laboratory conditions.

induration   A thickening and drying of the skin tissue that occurs in type IV hypersensitivity reactions.

infection   The relationship between host and pathogen and the competition for supremacy that takes place between them.

infection allergy   A type IV hypersensitivity reaction in which the immune system responds to the presence of certain microbial agents.

infectious disease   A disorder arising from a pathogen invading a susceptible host and inducing medically significant symptoms.

infectious dose   The number of microorganisms needed to bring about infection.

inflammation   A nonspecific defensive response to injury; usually characterized by redness, warmth, swelling, and pain.

influenza   A highly contagious viral infection of the respiratory passages. There are three types of influenza viruses: influenza A, influenza B, and influenza C .= They all belong to the Orthomyxoviridae family and share similar structural features. Although clinical isolates often have an irregular morphology, the viruses are typically described as having a spherical shape in culture.

initiation   (1) The unwinding and separating of DNA strands during replication. (2) The beginning of translation.

injection anthrax   A recently reported version of cutaneous anthrax where drug users injecting heroin develop massive edema (swelling) due to heroin contaminated with anthrax spores.

innate immunity   An inborn set of the preexisting defenses against infectious agents; includes the skin, mucous membranes, and secretions.

insertional inactivation   Referring to the "turning off" of a gene by a provirus.

insertion sequence (IS)   A segment of DNA that forms a copy of itself, after which the copy moves into areas of gene activity to interrupt the genetic coding sequence.

integrase inhibitor   Drugs that prevent HIV genetic code from integrating with the human DNA.

integumentary system The organ system that protects underlying tissues from various kinds of disease. It is made up of the skin, along with its accessory structures (nails, hair, sweat glands, and sebaceous glands).

interferon An antiviral protein produced by body cells on exposure to viruses and which trigger the synthesis of antiviral proteins.

interleukin (IL) A chemical cytokine produced by white blood cells that causes other white blood cells to divide.

intermediate host The host in which the larval or asexual stage of a parasite is found.

intermittent sterilization A method to kill all living microbes and spores over three successive days using free-flowing steam. Also called tyndallization.

intoxication The presence of microbial toxins in the body.

intraepithelial lymphatocyte (IEL) An intestinal epithelial cell that removes injured or infected cells.

intraerythrocyte cycle The chronic cycle of RBC infection and cell lysis.

intraterrestrial Microbes living in sediment and rock.

intrinsic factor A characteristic of a food product that influences microbial growth. *See also* **extrinsic factor**.

intron A non-coding sequence in a split gene; *see also* **exon**.

in use test A procedure used to determine the value of a disinfectant or antiseptic.

invasiveness The ability of a pathogen to spread from one point to adjacent areas in the host and cause structural damage to those tissues.

invasive wound infection Microbial invasion of living tissue beneath a burn.

invertase An enzyme from yeast.

iodophor A complex of iodine and detergents that is used as an antiseptic and disinfectant.

ion An electrically charged atom.

ionic bond The electrical attraction between oppositely charged ions.

ionizing radiation A type of radiation such as gamma rays and X rays that causes the separation of atoms or a molecule into ions.

irritable bowel syndrome A bowel disorder that involves recurrent pain with constipation or diarrhea or alternating attacks of both.

isograft Tissue taken from one identical twin and grafted to the other twin.

isomer A molecule with the same molecular formula but different structural formula.

isotope An atom of the same element in which the number of neutrons differs; *see also* **radioisotope**.

K

keratin The tough, fibrous protein produced by keratinocytes.

keratinocyte The most common cell type of the epidermis.

kinetoplast The mass of DNA found in the mitochondria of the kinetoplastids.

kinetoplastid A protozoan with one flagellum; most are parasitic in aerobic or anaerobic environments.

Kirby-Bauer test *See* **disk diffusion method**.

Koch's postulates A set of procedures by which a specific pathogen can be related to a specific infectious disease.

Koplik spots Red patches with white central lesions that form on the gums and walls of the pharynx during the early stages of measles.

Krebs cycle *See* **citric acid cycle**.

L

lactic acid fermentation A catabolic process that produces lactic acid during the reoxidation of NADH to NAD^+ for reuse in glycolysis to generate ATP.

lactone A chemical compound used as a flavoring ingredient in foods and beverages.

lagering The secondary aging of beer.

lagging strand During DNA replication, the new strand that is synthesized discontinuously; *see also* **leading strand**.

lag phase A portion of a bacterial growth curve encompassing the first few hours of the population's history when no growth occurs.

latency A condition in which a virus integrates into a host chromosome without immediately causing a disease.

latent infection A viral infection where the viral DNA remains "dormant" within a host chromosome.

latent phase The period during a viral infection when virus particles cannot be found outside the infected cells.

late-phase allergic response After an acute IgE mediated reaction, a second response occurs many hours after the initial response, and appears to be based on the activity of eosinophils.

leading strand During DNA replication, the new strand that is synthesized continuously; *see also* **lagging strand**.

leukocidin A bacterial enzyme that destroys phagocytes, thereby preventing phagocytosis.

leukocyte Any of a number of types of white blood cells.

leukopenia A condition characterized by an abnormal drop in the normal number of white blood cells.

leukotriene A substance that acts as a mediator during type I hypersensitivity reactions; formed from arachidonic acid after the antigen-antibody reaction has taken place.

lichen An association between a fungal mycelium and a cyanobacterium or alga.

light (L) chain A smaller polypeptide in an antibody.

light microscope An instrument that uses visible light and a system of glass lenses to produce a magnified image of an object; also called a compound microscope.

lipid A nonpolar organic compound composed of carbon, hydrogen, and oxygen; examples include triglycerides, phospholipids, and sterols.

lipid A A component in the outer membrane of the gram-negative cell wall.

lipopolysaccharide (LPS) A molecule composed of lipid and polysaccharide that is found in the outer membrane of the gram-negative cell wall of bacterial cells.

littoral zone The environment along the shoreline of an ocean.

local infection A disease restricted to a single area of the body.

localized anaphylaxis A type I hypersensitivity that remains in one part of the body; *see also* **systemic anaphylaxis**.

lockjaw *See* **trismus**.

logarithmic (log) phase The portion of a bacterial growth curve during which active growth leads to a rapid rise in cell numbers.

looped domain structure The term used to describe organization and packing of the prokaryotic chromosome.

lymph The tissue fluid that contains white blood cells and drains tissue spaces through the lymphatic system.

lymphatic filariasis   A parasitic disease affecting over 130 million people in 80 countries throughout the tropics and subtropics.

lymphedema   An abnormal swelling due to the loss of normal lymph vessel drainage of the affected part; typical of elephantiasis.

lymph node   A bean-shaped organ located along lymph vessels that is involved in the immune response and contains phagocytes and lymphocytes.

lymphocyte   A type of white blood cell that functions in the immune system.

lymphogranuloma venereum (LGV)   A systemic STD caused by serotype of *C. trachomatis*.

lymphoid stem cell   A bone marrow cell that gives rise to lymphocytes; *see also* **myeloid stem cell**.

lyophilization   A process in which food or other material is deep frozen, after which its liquid is drawn off by a vacuum; also called freeze-drying.

lysis   The rupture of a cell and the loss of cell contents.

lysogenic pathway   The events of a bacterial virus infection that result in the integration of its DNA into the bacterial chromosome.

lysosome   A membrane-enclosed compartment in many eukaryotic cells that contains enzymes to degrade or digest substances.

lysozyme   An enzyme found in tears and saliva that digests the peptidoglycan of gram-positive bacterial cell walls.

lytic cycle pathway   A process by which a bacterial virus replicates within a host cell and ultimately destroys the host cell.

M

macrolide   An antibiotic that blocks protein synthesis.

macromolecule   A large, chemically-bonded substance built from smaller building blocks.

macronucleus   The larger of two nuclei in most ciliates that is involved in controlling metabolism; *see also* **micronucleus**.

macrophage   A large cell derived from monocytes that is found within various tissues and actively engulfs foreign material, including infecting bacterial cells and viruses.

macule   A pink-red skin spot associated with infectious disease.

maculopapular   Refers to a lesion with a broad base that slopes from a raised center.

magnetosome   A cytoplasmic inclusion body in some bacterial cells that assists orientation to the environment by aligning with the magnetic field.

magnification   The increase in the apparent size of an object observed with a microscope.

major histocompatibility complex (MHC)   A set of genes that controls the expression of MHC proteins; involved in transplant rejection; *see also* **MHC class I** *and* **MHC class II**.

malignant   Referring to a tumor that invades the tissue around it and may spread to other parts of the body.

malting   The process whereby barley grains are treated to digest starch to maltose.

Mantoux test   The infection of a tuberculosis purified protein derivative into the forearm.

mashing   A fermentable mixture of hot water and barley grain from which alcohol is distilled.

mass   The amount of matter in a sample.

mass number   The total number of protons and neutrons in an atom.

mast cell   A type of cell in connective tissue which releases histamine during allergic attacks.

mating type   Separate mycelia of the same fungus or separate hyphae of the same mycelium.

matrix protein   A protein shell found in some viruses between the genome and capsid.

matter   Any substance that has mass and occupies space.

M cell   An intestinal epithelial cell that processes foreign antigens.

mechanical vector   A living organism, usually an insect, that transmits disease agents on its surface; *see also* **biological vector**.

membrane attack complex (MAC)   A set of complement proteins that forms holes in a bacterial cell membrane and leads to cell destruction.

membrane filter   A pad of cellulose acetate or polycarbonate.

memory cell   A cell derived from B lymphocytes or T lymphocytes that reacts rapidly upon re-exposure to antigen.

meninges   The membranes surrounding and protecting the brain and spinal chord.

meningitis   A general term for inflammation of the covering layers of the brain and spinal cord due to any of several bacteria, fungi, viruses, or protozoa.

merozoite   A stage in the life cycle of the malaria parasite that invades red blood cells in the human host.

mesophile   An organism that grows in temperature ranges of 20°C to 40°C.

messenger RNA (mRNA)   An RNA transcript containing the information for synthesizing a specific polypeptide.

metabolic engineering   An inter-disciplinary field that designs and optimizes ways to alter a set of enzymes (metabolic pathway) to produce a desired product.

metabolic pathway   A sequence of linked enzyme-catalyzed reactions in a cell.

metabolism   The sum of all biochemical processes taking place in a living cell; *see also* **anabolism** *and* **catabolism**.

metabolite   A substance formed in or necessary for metabolism.

metachromatic granule   A polyphosphate-storing granule commonly found in *Corynebacterium diphtheriae* that stains deeply with methylene blue; also called volutin.

metagenome   The collective genomes from a population of organisms.

metagenomics   The study of genes isolated directly from environmental samples.

metastasize   Referring to a tumor that spreads from the site of origin to other tissues in the body.

methanogen   An archaeal organism that lives on simple compounds in anaerobic environments and produces methane during its metabolism.

methanotroph   Prokaryotes that metabolize methane as their only source of carbon and energy.

MHC class I   A group of glycoproteins found on all nucleated cells to which cytotoxic T cells can bind.

MHC class II   A group of glycoproteins found on B cells, Macrophages, and dendritic cells that are recognized by helper T cells.

miasma   An ill-defined idea of the 1700s and 1800s that suggested diseases were caused by an altered chemical quality of the atmosphere.

microaerophile   An organism that grows best in an oxygen-reduced environment.

microaspiration   The passage of fluids, secretions, and microbes from the oropharynx.

microbe   *See* **microorganism**.

microbial antagonism   The method of using established cultures of microorganisms to prevent the intrusion of foreign strains.

microbial ecology   Involves the study of microbial dynamics in the environment and their interactions with each other and with other organisms.

microbial evolution   Refers to the genetically driven changes that occur in microorganisms and that are retained over time.

microbial forensics   The discipline involved with the recognition, identification, and control of a pathogen.

microbial genomics   The discipline of sequencing, analyzing, and comparing microbial genomes.

microbicidal   Referring to any agent that kills microbes.

microbiology   The scientific study of microscopic organisms and viruses, and their interactions with other organisms and the environment.

microbiome   The community of microorganisms in an environment or body location.

microbiome-induced therapy   Today, microbiome-induced therapy is being used to reintroduce microbes into the colon of CDI patients to stimulate the repopulation of lost resident microbes. The therapy, also called fecal transplantation, involves using fecal bacteria from a close relative or healthy donor.

microbiostatic   Referring to any agent that inhibits microbial growth.

microbiota   The population of microorganisms that colonize various parts of the human body and do not cause disease in a healthy individual.

microcephaly   A medical condition by which the fetus or infant has an unusually small head due to abnormal brain development.

microcompartment   A region in some bacterial cells surrounded by a protein shell.

microenvironment   A cell's or organism's physical and chemical surroundings.

micrometer (µm)   A unit of measurement equivalent to one millionth of a meter; commonly used in measuring the size of microorganisms.

micronucleus   The smaller of the two nuclei in most ciliates that contains genetic material and is involved in sexual reproduction; *see also* **macronucleus**.

microorganism (microbe)   A microscopic form of life including bacterial, archaeal, fungal, and protistan cells.

Microsporidia   A sister group of fungi that are parasitic in protists and animals.

mineralization   The conversion of organic compounds to inorganic compounds and ammonia.

minimum inhibitory concentration (MIC)   The lowest concentration of an antimicrobial agent that will inhibit its growth.

miracidium (pl. miracidia)   A ciliated larva representing an intermediary stage in the life cycle of a trematode.

mismatch repair   A mechanism to correct mismatched bases in the DNA; *see also* **excision repair**.

missense mutation   A base substitution that codes for an incorrect amino acid.

mitochondrion (pl. mitochondria)   A double membrane-enclosed compartment in eukaryotic cells that carries out aerobic respiration.

mitosporic fungus   A fungus without a known sexual stage of reproduction.

mold   A type of fungus that consists of chains of cells and appears as a fuzzy mass of thin filaments in culture.

molecular formula   The representation of the kinds and numbers of each atom in a molecule.

molecular taxonomy   The systematized arrangement of related organisms based on molecular characteristics, such as ribosomal DNA nucleotide sequences.

molecular weight   The sum of the atomic masses of all atoms in a molecule.

molecule   Two or more atoms held together by a sharing of electron pairs.

molluscum contagiosum   A viral disease that forms mildly contagious, wart-like skin lesions. The virus of molluscum contagiosum is another

member of the Poxviridae, and the virus is transmitted is by sexual contact.

monoclonal antibody (MAb)   A type of antibody produced by a clone of hybridoma cells, consisting of antigen-stimulated B cells fused to myeloma cells.

monocyte   A circulating white blood cell with a large bean-shaped nucleus that is the precursor to a macrophage.

monolayer   (1) A single layer of cultured cells; (2) A single layer of phospholipids.

monomer   A simple organic molecule that can join in long chains with other molecules to form a more complex molecule; *see also* **polymer**.

monosaccharide   A simple sugar that cannot be broken down into simpler sugars; examples include glucose and fructose.

Monospot test   A method used to detect the presence of heterophile antibodies, which is indicative of infectious mononucleosis.

monotherapy   The use of a single drug to treat a disease.

morphology   Refers to the form (shape) and structure of cells and organisms.

most probable number (MPN)   A laboratory test in which a statistical evaluation is used to estimate the number of bacterial cells in a sample of fluid; often employed in determinations of coliform bacteria in water.

M protein   A protein that enhances the pathogenicity of streptococci by allowing organisms to resist phagocytosis and adhere firmly to tissue.

mucin   The major glycoprotein in mucus.

mucociliary clearance   A defensive mechanism to clear microbes or foreign material from the airways via the ciliated epithelium.

mucociliary elevator   A major barrier against infection. Microorganisms hoping to infect the respiratory tract are caught in the sticky mucus and moved by the mucociliary escalator.

mucosa-associated lymphoid tissue (MALT)   A class of lymphoid tissue comprising nodular aggregates found in association with the wet mucosal surfaces of the body such as those of the respiratory, digestive, and urinary systems.

mucosal immunity   The immune defenses on the mucosal surfaces, such as the gastrointestinal tract.

mucous membrane (mucosa)   A moist lining in the body passages of all mammals that contains mucus-secreting cells and is open directly or indirectly to the external environment.

mucus   A sticky secretion of glycoproteins.

multidrug-resistant (MDR)   Referring to microbes that are resistant to the effects of multiple antibiotics.

multidrug-resistant tuberculosis (MDR-TB)   A form of tuberculosis infection caused by bacteria that are resistant to treatment with at least two of the most powerful first-line anti-tuberculosis medications, isoniazid and rifampin.

multiple sclerosis   A chronic, typically progressive disease involving damage to the sheaths of nerve cells in the brain and spinal cords, whose symptoms may include numbness, impairment of speech and of muscular coordination, blurred vision, and severe fatigue.

mushroom   A spore-bearing fruiting body typical of many members of the basdiomycetes.

must   The juice resulting from crushing grapes.

mutagen   A chemical or physical agent that causes a mutation.

mutant   An organism carrying a mutation.

mutation   A permanent alteration of a DNA sequence.

mutualism   A close and permanent association between two populations of organisms in which both benefit from the association.

mycelium (pl. **mycelia**)   A mass of fungal filaments from which most fungi are built.

mycetism   Mushroom poisoning.

mycobiome   The combined genetic material of the microorganisms in a particular environment.

mycolic acid   A waxy lipid composing the cell wall of mycobacterial species.

mycology   The scientific study of fungi.

mycorrhiza (pl. **mycorrhizae**)   A close association between a fungus and the roots of many plants.

mycosis (pl. **mycoses**)   A disease caused by a fungus.

mycotoxin   A poison produce by a fungus that adversely affects other organisms.

myeloid stem cell   A bone marrow cell that gives rise to red blood cells and all white blood cells except lymphocytes; *see also* **lymphoid stem cell**.

myeloma   A malignant tumor that develops in the blood-cell–producing cells of the bone marrow; the cells are used in the procedure to produce monoclonal antibodies.

myonecrosis   *See* **trismus**.

myxobacteria   A group of soil-dwelling bacterial species that exhibit multicellular behaviors.

N

nanometer (nm)   A unit of measurement equivalent to one billionth of a meter; the unit used in measuring viruses and the wavelength of energy forms.

narrow spectrum   Referring to an antimicrobial drug that is useful for a restricted group of microorganisms; *see also* **broad spectrum**.

natural killer (NK) cell   A type of defensive body cell that attacks and destroys cancer cells and infected cells without the involvement of the immune system.

naturally acquired active immunity   A host response resulting in antibody production as a result of experiencing the disease agent, or a passive response resulting from the passage of antibodies to the fetus via the placenta or the milk of a nursing mother.

naturally acquired passive immunity   The form of immunity that results from the passage of IgG antibodies from mother to fetus.

natural selection   The process whereby organisms better adapted to their environment tend to survive and produce more offspring

necrosis   Cell or tissue death.

necrotizing fasciitis   An acute disease in which inflammation of the fasciae of muscles or other organs results in rapid destruction of overlying tissues.

negative control   A form of gene regulation where a repressor protein binds to an operator and blocks transcription.

negative selection   A method for identifying mutations by selecting cells or colonies that do not grow when replica plated.

negative-sense RNA virus   Aso known as an antisense-strand RNA virus, a virus whose genetic information consists of a single strand of RNA that is the negative or antisense strand which node not encode mRNA.

negative stain technique   A staining process that results in colorless bacterial cells on a stained background when viewed with the light microscope.

negative strand   Referring to those RNA viruses whose genome cannot be directly transcribed into protein.

Negri body   A cytoplasmic inclusion that occurs in brain cells infected with rabies viruses.

neuraminidase (N)   An enzyme composing one type of surface spike of influenza viruses that facilitates viral release from the host cell.

neuraminidase inhibitor   A compound that inhibits neuraminidase, preventing the release of flu viruses from an infected cell.

neurotoxin   A poison that interferes with nerve transmission.

neutralization   A type of antigen-antibody reaction in which the activity of a toxin is inactivated.

neutron   An uncharged particle in the atomic nucleus.

neutrophil   The most common type of white blood cell; functions chiefly to engulf and destroy foreign material, including bacterial cells and viruses that have entered the body.

neutrophil extracellular traps (NETS)   The nuclear DNA and antimicrobial proteins of dead neutrophils that form a fibrous mesh trapping pathogens.

neutrophile   An organism that grows best at pH 7.

next-generation DNA sequencing (NGS)   also known as high-throughput sequencing, is a catch-all term used to describe a number of different modern sequencing technologies including Illumina (Solexa) sequencing, Roche 454 sequencing, Ion torrent, and SOLiD sequencing.

nicotinamide adenine dinucleotide (NAD⁺)   A coenzyme that acts as an electron carrier for ATP production.

nicotinamide adenine dinucleotide phosphate (NADP⁺)   A cofactor used in anabolic reactions, such as lipid and nucleic acid synthesis, which require NADPH as a reducing agent.

nitrogen cycle   The processes that convert nitrogen gas (N_2) to nitrogen-containing substances in soil and living organisms, then reconverted to the gas.

nitrogen fixation   The chemical process by which microorganisms convert nitrogen gas (N₂) into ammonia.

noncommunicable disease   A disease whose causative agent is acquired from the environment and is not transmitted to another individual.

noncompetitive inhibition   The prevention of a chemical reaction by a chemical that binds elsewhere than to active site of an enzyme; *see also* **competitive inhibition**.

noneveloped ("naked") virus   Utilizing capsid proteins to mediate binding to host cells.

non-gonococcal urethritis (NGU)   An inflammation of the urethra that is not caused by gonorrheal infection.

nonhalophile   A microorganism that cannot grow in the presence of added sodium chloride.

nonhemolytic   Referring to those bacterial species that when plated on blood agar cause no destruction of the red blood cells.

nonparalytic poliomyelitis   A type of polio that does not lead to paralysis (abortive polio). This usually causes the same mild, flu-like signs and symptoms typical of other viral illnesses.

nonperishable   Referring to foods that are least likely to spoil.

nonpolar covalent bond   A type of bond that occurs when two atoms share a pair of electrons with each other.

nonpolar molecule   A covalently bonded substance in which there is no electrical charge; *see also* **polar molecule**.

nonsense mutation   A base substitution that codes for a stop codon.

nosocomial infection   A disorder acquired during an individual's stay at a hospital or chronic care facility.

N spike   Neuraminidase proteins that assist in the release of the newly replicated virions from the host cells.

nuclear envelope   The double-layered membrane enclosing the nucleus of a eukaryotic cell which has pores that allow the passage of materials into and out of the nucleus.

nucleic acid   A high-molecular-weight molecule consisting of nucleotide chains that convey genetic information and is found in all living cells and viruses; *see* **deoxyribonucleic acid** *and* **ribonucleic acid**.

nucleobase   Any of five nitrogen-containing compounds found in nucleic acids, including adenine, guanine, cytosine, thymine, and uracil.

nucleocapsid   The combination of genome and capsid of a virus.

nucleoid   The chromosomal region of a bacterial and archaeal cell.

nucleoid-associated protein (NAP)   A protein that causes the DNA double helix to twist (supercoil).

nucleoside (base) analog   A chemical compound having a structure similar to that of another but differing from it in respect to a certain component; it may have similar or opposite action metabolically.

nucleosome   A structural unit of a eukaryotic chromosome, consisting of a length of DNA coiled around a core of histones.

nucleotide   A component of a nucleic acid consisting of a carbohydrate molecule, a phosphate group, and a nitrogenous base.

nucleotide excision repair   A process that removes a thymine dimer, along with adjacent nucleotides, and replaces them with the correct sequence.

O

obligate aerobe   An organism that requires oxygen gas (O₂) for metabolism.

obligate anaerobe   An organism that cannot use oxygen gas (O₂) for metabolism.

observation   The use of the senses or instruments to gather information on which science inquiry is based.

Okazaki fragment   A segment of DNA resulting from discontinuous DNA replication.

oligotroph   An organism that can live in an environment that offers very low levels of nutrients.

oncogene   A segment of DNA that can induce uncontrolled growth of a cell if permitted to function.

oncogenic   Referring to any agent such as viruses that can cause tumors.

oocyst   A cyst containing a zygote formed by a parasitic protozoan such as the malaria parasite.

oomycete   Any of various algaelike fungi constituting the phylum Oomycota of the kingdom Fungi, characterized by the formation of oospores.

operator   A sequences of bases in the DNA to which a repressor protein can bind.

operon   The unit of bacterial DNA consisting of a promoter, operator, and a set of structural genes.

opisthotonus   An arching of the back that is characteristic of tetanus.

opportunist   A microorganism that invades the tissues when body defenses are suppressed.

opportunistic infection   A disorder caused by a microorganism that does not cause disease but that can become pathogenic or life threatening if the host has a low level of immunity.

opportunistic pathogen   An infectious microorganism that is normally a commensal or does not harm its host but can cause disease when the host;s resistance is low.

opsonin   An antibody or complement component that encourages phagocytosis.

opsonization   Enhanced phagocytosis due to the activity of antibodies or complement.

optimal growth temperature   The temperature at which a plant or other organism will grow at the most stable healthy rate.

oral rehydration therapy   A procedure that uses a mixture of blood salts and glucose in water to restore their normal levels in the body.

order   A category of related organisms consisting of one or more families.

organelle   A specialized compartment in cells that has a particular function.

organic acid   A short-chain chemical used as a preservative.

organic compound   A substance characterized by chains or rings of carbon atoms that are linked to atoms of hydrogen and sometimes oxygen, nitrogen, and other elements.

origin of transfer   The fixed point on an F plasmid (factor) where one strand is nicked and transferred to a recipient cell.

osmosis   The net movement of water molecules from where they are in a high concentration through a semipermeable membrane to a region where they are in a lower concentration.

osmotic lysis   The bursting of a cell due to the inflow of water.

osmotic pressure   The force that must be applied to a solution to inhibit the inward movement of water across a membrane.

outbreak   A small, localized epidemic.

outer membrane   A bilayer membrane forming part of the cell wall of gram-negative bacteria.

oxazolidinone   Any of a class of synthetic antibiotics that inhibit protein synthesis, used against Gram-positive bacteria.

oxidation   A chemical change in which electrons are lost by an atom; *see also* **reduction**.

oxidation lagoon   A large pond in which sewage is allowed to remain undisturbed so that digestion of organic matter can occur.

oxidative phosphorylation   A series of sequential steps in which energy is released from electrons as they pass from coenzymes to cytochromes, and ultimately, to oxygen gas (O_2); the energy is used to combine phosphate ions with ADP molecules to form ATP molecules.

oxygenic photosynthesis   A form of photosynthesis in which molecular oxygen (O_2) is produced.

P

pandemic   A worldwide epidemic.

papule   A pink pimple on the skin.

para-aminobenzoic acid (PABA)   A crystalline acid that is widely distributed in plant and animal tissue.

parabasalid   A protist that contains numerous of flagella and lives in low oxygen or anaerobic environments; most members are symbiotic in animals.

paralytic poliomyelitis   Acute onset of a flaccid paralysis of one or more limbs with decreased or absent tendon reflexes in the affected limbs, wihout other apparent cause, and without sensory or cognitive loss.

parasite   A type of heterotrophic organism that feeds on live organic matter such as another organism.

parasitemia   The spread of parasitic protists and multicellular worms through the circulatory system.

parasitism   A close association between two organisms in which one (the parasite) feeds on the other (the host) and may cause injury to the host.

parasitology   The scientific study of parasites.

parenteral route   An organism gets access to the tissues underneath the mucous membranes or the skin.

parotid gland   Either of a pair of large salivary glands situated just in front of each ear.

paroxysm   A sudden intensification of symptoms, such as a severe bout of coughing.

particle transmission   The movement of particles suspended in the air.

passive agglutination   An immuno-logical procedure in which antigen molecules are adsorbed to the surface of latex spheres or other carriers that agglutinate when combined with antibodies.

passive immunity   The temporary immunity that comes from receiving antibodies from another source.

pasteurization   A heating process that destroys human pathogens in a fluid such as milk and lowers the overall number of bacterial cells in the fluid.

pasteurizing dose   The amount of irradiation used to eliminate pathogens.

pathogen   A microorganism or virus that causes illness or disease.

pathogen-associated molecular pattern (PAMP)   A unique microbial molecular sequence recognized by innate immune system receptors.

pathogenicity   The ability of a disease-causing agent to gain entry to a host and bring about a physiological or anatomical change interpreted as disease.

pathogenicity island   A set of adjacent genes that encode virulence factors.

pattern recognition receptor   A protein sensor that recognizes pathogen-associated molecular patterns.

pectinase   Any enzyme that breaks down pectin, a polysaccharide substrate found in the cell wall of plants, into simple sugars and galacturonic acid.

penicillin   Any of a group of antibiotics derived from *Penicillium* species or produced synthetically; effective against gram-positive bacteria and several gram-negative bacteria by interfering with cell wall synthesis.

peptide bond   A covalent linkage between the amino group on one amino acid and the carboxyl group on another amino acid.

peptidoglycan   A complex molecule of the bacterial cell wall composed of

alternating units of *N*-acetylglucosamine and *N*-acetylmuramic acid cross linked by short peptides.

perforin   A protein secreted by cytotoxic T lymphocytes and natural killer cells that forms holes in the plasma membrane of a targeted infected cell.

period of convalescence   The phase of a disease during which the body's systems return to normal.

period of decline   The phase of a disease during which symptoms subside.

periodontal disease (PD)   An inflammatory disease that affects the soft and hard structures that support the teeth.

periodontitis   A serious disease of the soft tissue and bone supporting the teeth; bone resorption occurs and the periodontal ligament may be lost.

periplasm   A metabolic region between the cell membrane and outer membrane of a gram-negative bacterial cell.

peroxide   A compound containing an oxygen–oxygen single bond.

persister cell   A bacterial cell that has temporarily ceased cell division but maintains a very low metabolic rate.

pH   An abbreviation for the power of the hydrogen ion concentration [H^+] of a solution.

phage   *See* **bacteriophage**.

phage typing   A procedure of using specific bacterial viruses to identify a particular strain of a bacterial species.

phagocyte   A white blood cell capable of engulfing and destroying foreign materials, including bacterial cells and viruses.

phagocytosis   A process by which foreign material or cells are taken into a white blood cell and destroyed.

phagolyososme   A membrane-enclosed compartment resulting from the fusion of a phagosome with lysosomes and in which the foreign material is digested.

phagosome   A membrane-enclosed compartment containing foreign material or infectious agents that the cell has engulfed.

phallotoxin   A group of chemical compounds present in the mushroom *Amanita phalloides*.

phase-contrast microscopy   An optical system on the light microscope that uses a special condenser and objective lenses to examine cell structure.

phenol   A chemical compound that has one or more hydroxyl groups attached to a benzene ring and derivatives are used as an antiseptic or disinfectant; also called carbolic acid.

phenol coefficient (PC)   A number that indicates the effectiveness of an antiseptic or disinfectant compared to phenol.

phenotype   The visible (physical) appearance of an organism resulting from the interaction between its genetic makeup and the environment.

phospholipid   A water-insoluble compound containing glycerol, two fatty acids, and a phosphate head group; forms part of the membrane in all cells.

phospholipid bilayer   A two-layered arrangement of phosphate and lipid molecules that form a cell membrane, the hydrophobic lipid ends facing inward and the hydrophilic phosphate ends facing outward.

phosphorus cycle   the cycling of phosphorus between the biotic and abiotic components of the environment.

phosphorylation   The addition of a phosphate group to a molecule.

photoautotroph   An organism that uses light energy to synthesize nutrients from carbon dioxide gas (CO_2).

photobiont   The photosynthetic partner in a lichen.

photoheterotroph   An organism that uses light energy to synthesize nutrients from organic carbon compounds.

photophobia   Sensitivity to bright light.

photophosphorylatyion   The generation of ATP through the trapping of light.

photoreactivation   A process that repairs DNA damaged by ultraviolet light using an enzyme that requires visible light.

photosynthesis   A biochemical process in which light energy is converted to chemical energy, which is then used for carbohydrate synthesis.

photosystem   A group of pigments that act as a light trapping system for photosynthesis.

pH scale   A range of values that extends from 0 to 14 and indicates the degree of acidity or alkalinity of a solution.

phycology   The scientific study of algae.

phylogenetic tree   A diagram illustrating the inferred relationships among organisms.

phylogeny   The evolutionary history of a species or group of species.

phylum (pl. phyla)   A category of organisms consisting of one or more classes.

physical mutagen   A mutation agent which is in the form of physical substances, such as short wave (ultraviolet and radiation ray such as alpha, beta, and gamma).

physical pollution   The presence of particulate matter in water.

phytoplankton   Microscopic free-floating communities of cyanobacteria and unicellular algae.

pilus (pl. pili)   A short hair-like structure used by bacterial cells for attachment.

planktonic bacteria   Referring to bacterial cells that live as individual cells.

plaque   A clear area on a lawn of bacterial cells where viruses have destroyed the bacterial cells.

plasma   The fluid portion of blood remaining after the cells have been removed; *see also* **serum**.

plasma cell   An antibody-producing cell derived from B lymphocytes.

plasma membrane   The phospholipid bilayer with proteins that surrounds the eukaryotic cell cytoplasm; *see also* **cell membrane**.

plasmid   A small, closed-loop molecule of DNA apart from the chromosome that replicates independently and carries nonessential genetic information.

platelet   A small disc-shaped cell without a cell nucleus that functions in blood clotting and innate immunity.

pleated sheet   The zig-zag secondary structure of a polypeptide in a flat plane.

pluripotent stem cell   An undifferentiated cell from which specialized cells arise.

pneumonia   An inflammation of the bronchial tubes and one or both lungs.

point mutation   The replacement of one base in a DNA strand with another base.

polar covalent bond   A type of bond between two or more atoms in which atoms do not share their pair of electrons equally.

polar molecule   A substance with electrically-charged poles; *see also* **nonpolar molecule**.

point source epidemic   Exposure to the same source over a brief time, such as through a single meal or event. The number of cases rises rapidly to a peak and falls gradually. Most cases occur within one incubation period.

polychlorinated biphenyl (PCB)   Any of a class of toxic aromatic compounds, often formed as waste in industrial processes, whose molecules contain two benzene rings in which hydrogen atoms have been replaced by chlorine atoms.

polyclonal antibodies   Antibodies produced from different clones of B cell.

polymer   A substance formed by combining smaller molecules into larger ones; *see also* **monomer**.

polymerase chain reaction (PCR)   A technique used to replicate a fragment of DNA many times.

polymicrobial disease   A clinical or pathological condition caused by more than one infectious agent.

polymixin   Any of a group of polypeptide antibiotics that are active chiefly against Gram-negative bacteria.

polynucleotide   A chain of linked nucleotides.

polypeptide   A chain of linked amino acids.

polysaccharide   A complex carbohydrate made up of simple sugars linked into a branched or chain structure; examples include starch and cellulose.

polysome   A cluster of ribosomes linked by a strand of mRNA and all translating the mRNA.

porin   A protein in the outer membrane of gram-negative bacterial cells that acts as a channel for the passage of small molecules.

portal of entry   The site at which a pathogen enters the host.

portal of exit   The site at which a pathogen leaves the host.

positive selection   A method for selecting mutant cells by their growth as colonies on agar.

positive-sense RNA virus   A virus whose genetic information consists of a single strand of RNA that is the posititve (or sense) strand which encodes mRNA and protein.

positive strand   Referring to the RNA viruses whose genome consists of a mRNA molecule.

post-exposure immunization   The receiving of a vaccine after contracting the pathogen.

post-polio syndrome   A condition that affects polio survivors years after recovery from an initial acute attack by the polio virus.

potability   Referring to water that is safe to drink because it contains no harmful material or microbes.

pour-plate method   A process by which a mixed culture can be separated into pure colonies and the colonies isolated; *see also* **streak-plate method**.

praziquantel   A synthetic anthelmintic drug used in the treatment of schistosomiasis and other infestations of humans and animals with parasitic trematodes or cestodes.

precipitation   A type of antigen-antibody reaction in which thousands of molecules of antigen and antibody cross-link to form visible aggregates.

prevalence   The percentage of the population affected by a disease.

prevacuum autoclave   An instrument that uses saturated steam at high temperatures and pressure for short time periods to sterilize materials.

preventative vaccine   A vaccine that could be given to HIV-negative individuals to prevent infection to HIV.

primary antibody response   The first contact between an antigen and the immune system, characterized by the synthesis of IgM and then IgG antibodies; *see also* **secondary antibody response**.

primary cell culture   Animal cells separated from tissue and grown in cell culture.

primary immunodeficiency   An autoimmune disorder present from birth. *See also* **secondary immunodeficiency**.

primary infection   A disease that develops in an otherwise healthy individual.

primary lymphoid tissue   Includes the thymus, which lies behind the breast bone, and the bone marrow. Both are sites where the T and B lymphocytes form and mature.

primary metabolite   A small molecule essential to the survival and growth of an organism; *see also* **secondary metabolite**.

primary pathogen   A virus, bacteria, fungi, or any other biological entity that causes a disease when it gains entry into a victim's body.

primary structure   The sequence of amino acids in a polypeptide.

primary treatment   A plain sedimentation process to remove suspended organic solids from the sewage. Chemicals are sometimes used to remove finely divided and colloidal solids.

prion   An infectious, self-replicating protein involved in human and animal diseases of the brain.

probiotic   Living microbes that help reestablish or maintain the human microbiota of the gut.

prodromal phase   The phase of a disease during which general symptoms occur in the body.

producer   Organism that produces organic compounds from carbon dioxide gas.

product   A substance or substances resulting from a chemical reaction.

productive infection   The active assembly and maturation of viruses in an animal cell.

progenote   A hypothetical first life form.

proglottid   One of a series of segments that make up the body of a tapeworm.

prokaryote   A microorganism in the domain *Bacteria* or *Archaea* composed of single cells having a single chromosome but no cell nucleus or other membrane-bound compartments.

prokaryotic   Referring to cells or organisms usually having a single chromosome but no cell nucleus and few subcompartments.

promoter   The region of a template DNA strand or operon to which RNA polymerase binds.

propagated epidemic   Outbreaks in which the disease propagates in one or more initial cases and then spreads to others, a relatively slow method of spread.

prophage   The viral DNA of a bacterial virus that is inserted into the bacterial DNA and is passed on from one generation to the next during binary fission.

propionic acid   A chemical preservative used in cheese, breads, and other bakery products.

prostaglandin   A hormone-like substance that acts as a mediator in type I hypersensitivity reactions.

protease   An enzyme that breaks down proteins and peptides.

protease inhibitor   A compound that breaks down the enzyme protease, inhibiting the replication of some viruses, such as HIV.

protein   A chain or chains of linked amino acids used as a structural material or enzyme in living cells.

protein-only hypothesis   The idea that prions are composed solely of protein and contain no nucleic acid.

protein synthesis   The process of forming a polypeptide or protein through a series of chemical reactions involving amino acids.

Proteobacteria   A phylum of gram-negative, chemoheterotrophic species in the domain Bacteria that are defined primarily in terms of their ribosomal RNA (rRNA) sequences; examples include *Escherichia coli*, *Salmonella*, and the rickettsiae.

protist   An informal term used to describe a eukaryote that is not animal, plant, or fungus.

proton   A positively charged particle in the atomic nucleus.

proto-oncogene   A region of DNA in the chromosome of human cells; they are altered by carcinogens into oncogenes that transform cells.

prototroph   An organism that contains all its nutritional needs; *see also* **auxotroph**.

protozoan (pl. **protozoa**)   A term formally used to describe a single-celled eukaryotic organism that lacks a cell wall and usually exhibits chemoheterotrophic metabolism.

protozoology   The scientific study of protozoa.

provirus   The viral DNA that has integrated into a eukaryotic host chromosome and is then passed on from one generation to the next through cell division.

pseudohypha (pl. **pseudohyphae**)   A chain of easily disrupted fungal cells that is intermediate between a chain of budding cells and a true hypha, marked by constrictions rather than septa at the junctions.

pseudomembrane   An accumulation of mucus, leukocytes, bacteria, and dead tissue in the respiratory passages of diphtheria patients.

pseudopeptidoglycan   A complex molecule of some archaeal cell walls composed of alternating units of *N*-acetylglucosamine and *N*-acetyltalosamine uronic acid.

pseudopod   A projection of the plasma membrane that allows movement in members of the amoebozoans.

psychrophile   An organism that lives at cold temperature ranges of 0°C to 20°C.

psychrotolerant   Referring to microorganisms that grow at 0°C but have a temperature optima of 20° to 40°C.

psychrotroph   *See* **psychrotolerant**.

pure culture   An accumulation or colony of microorganisms of one species.

pus   A mixture of dead tissue cells, leukocytes, and bacteria that accumulates at the site of infection.

pyrogen   A fever-producing substance.

pyruvate   The end product of the glycolysis metabolic pathway.

Q

quat   *See* **quaternary ammonium compound**.

quaternary ammonium compound (quat)   A positively charged detergent with four organic groups attached to a central nitrogen atom; used as a disinfectant.

quaternary structure   The association of two or more polypeptides in a protein.

question   A statement written from an observation and used to formulate a hypothesis.

quinolone   Any of a class of antibiotics used in treating a variety of mainly Gram-negative infections, and thought to be responsible for antibiotic resistance in some microbes.

quorum sensing   The ability of microbial cells to chemically communicate and coordinate behavior via signaling molecules.

R

radioallergosorbent test (RAST)   A type of radioimmunoassay in which antigens for the unknown antibody are attached to matrix particles.

radioimmunoassay (RIA)   An immunological procedure that uses radioactive-tagged antigens to determine the identity and amount of antibodies in a sample.

radioisotope   A unstable form of a chemical element that is radioactive; *see also* **isotope**.

radiolarian   A single-celled marine organism with a round silica-containing shell that has radiating arms to catch prey.

reactant   A substance participating in a chemical reaction.

recombinant DNA molecule   A DNA molecule containing DNA from two different sources.

recombinant DNA technology   A series of procedures that are used to join together (recombine) DNA segments, constructed from segments of two or more different DNA molecules.

recombinant F⁻ cell   A recipient cell that received a few chromosomal genes and partial F factor genes from a donor cell during conjugation.

recombinant subunit vaccine   The synthesis of antigens in a microorganism using recombined genes for the purpose of producing a vaccine.

red algae   A large group of algae that includes many seaweeds that are mainly red in color. Some kinds yield useful products or are used as food.

reduction   The gain of electrons by a molecule; *see also* **oxydation**.

redundancy   Referring to multiple codons coding for the same amino acid.

reemerging (resurgent) infectious disease   A disease showing a resurgence in incidence or a spread in its geographical area; *see also* **emerging infectious disease**.

regulatory gene   A DNA segment that codes for a repressor protein.

regulatory T cell   A population of lymphocytes that prevent other T lymphocytes from attacking self.

replication factory   The location in a cell where DNA synthesis occurs.

replication fork   The point where complementary strands of DNA separate and new complementary strands are synthesized.

replication origin   The fixed point on a DNA molecule where copying of the molecule starts.

replisome   Any of the sites on the matrix of a cell nucleus that contains a series of enzyme complexes in which DNA replication is considered to occur.

repressor protein   A protein that when bound to the operator blocks transcription.

reservoir   The location or organism where disease-causing agents exist and maintain their ability for infection.

resolution   The numerical value of a lens system indicating the size of the smallest object that can be seen clearly when using that system.

respiratory burst   A phagocytic response producing toxic metabolites.

respiratory droplet   Small liquid droplets expelled by sneezing or coughing.

restriction endonuclease   A type of enzyme that splits open a DNA molecule at a specific restricted point; important in genetic engineering techniques.

resurgent infectious disease   *see* **reemerging infectious disease**.

reticulate body   The replicating, intracellular, noninfectious stage of *Chlamydia trachomatis*.

reverse transcriptase   An enzyme that synthesizes a DNA molecule from the code supplied by an RNA molecule.

reverse transcriptase inhibitor   A compound that inhibits the action of reverse transcriptase, preventing the viral genome from being replicated.

revertant   Referring to a mutant organism or cell that has reacquired its original phenotype or metabolic ability.

R group   The side chain on an amino acid (–R).

rhabditiform   Referring to the elongated, rod-like shape of the larvae of the hookworm.

rhinosinusitis   Inflammation of the paranasal sinuses and nasal cavity.

rhinovirus   Any of a group of picornaviruses, including those that cause some forms of the common cold.

RhoGAM   Rh-positive antibodies.

riboflavin (vitamin B²)   A yellow vitamin of the B complex that is essential for metabolic energy production.

ribonucleic acid (RNA)   The nucleic acid involved in protein synthesis and gene control; also the genetic information in some viruses.

ribosomal RNA (rRNA)   An RNA transcript that forms part of the ribosome's structure.

ribosome   A cellular structure made of RNA and protein that participates in protein synthesis.

ribozyme   An RNA molecule that can catalyze a chemical reaction in a cell.

rice-water stool   A colorless, watery diarrhea containing particles of intestinal tissue in cholera patients.

rickettsiae   Short gram-negative rods that are obligate intracellular human pathogens.

RNA   *See* **ribonucleic acid**.

RNA polymerase   The enzyme that synthesizes an RNA polynucleotide from a DNA template.

rolling circle mechanism   A type of DNA replication in which a strand of DNA "rolls off" the loop and serves as a template for the synthesis of a complementary strand of DNA.

rose spots   A bright red skin rash associated with diseases such as typhoid fever and relapsing fever.

rotavirus gastroenteritis   A virus that infects the bowels and is most commonly found in children and in a genus of double-stranded RNA virus in the family Reoviridae.

roundworm   A multicellular parasite with a round body; examples include the nematodes.

R plasmid   A small, circular DNA molecule that occurs frequently in bacterial cells and carries genes for drug resistance.

S

SAFE strategy   A global effort to reduce trachoma, using Surgery of the eyelids; Antibiotics for acute infections; Facial hygiene improvements; and Evironmental access to safe water.

salmonellosis   Foodborne infection caused by salmonella serotypes and treated with antibiotics and fluid replacement.

salpingitis   Blocked and inflamed fallopian tubes caused by untreated chlamydia.

salt   An ionic compound resulting from the electrostatic attraction between positively and negatively charged atoms.

Sanitary Movement   New hygiene methods introduced in the mid-1800s that helped create the infrastructure for the public health systems we have today.

sanitization   To remove microbes or reduce their populations to a safe level as determined by public health standards.

sanitize   Referring to the reduction of a microbial population to a safe level.

saprobe   A type of heterotrophic organism that feeds on dead organic matter, such as rotting wood or compost.

sarcina (pl. **sarcinae**)   (1) A packet of eight spherical-shaped prokaryotic cells. (2) A genus of gram-positive, anaerobic spheres.

saturated   Referring to a water-insoluble compound that cannot incorporate any additional hydrogen atoms; *see also* **unsaturated**.

scalded skin syndrome   Exotoxins produced at the infection site by some strains of *S. aureus* travel through the bloodstream to the skin, resulting in red, wrinkled skin that is tender to the touch.

scanning electron microscope (SEM)   The type of electron microscope that allows electrons to scan across an object, generating a three-dimensional image of the object.

scarlet fever   A disease that can cause a pink-red skin rash, fever, sore throat, and inflamed tongue.

Schick test   A skin test used to determine the effectiveness of diphtheria immunization.

schistosomiasis   A parasite disease resulting in ulceration, diarrhea, abdominal pain, bloody urine, and pain on urination.

schmutzdecke   In water purification, a slimy layer of microorganisms that develops in a slow sand filter.

science   The organized body of knowledge that is derived from observations and can be verified or tested by further investigation.

scientific inquiry   The way a science problem is investigated by formulating a question, developing a hypothesis, collecting data about it through observation, and experiment, and interpreting the results; also called scientific method.

sclerotium (pl. **sclerotia**)   A hard purple body that forms in grains contaminated with *Claviceps purpurea*.

scolex   The head region of a tapeworm.

sebaceous glands   These glands produce an oily substance called sebum that keeps skin and hair soft and moist.

sebum   An oily substance produced by the sebaceous glands that keep the skin and hair soft and moist.

secondary antibody response   A second or ensuing response triggered by memory cells to an antigen and characterized by substantial production of IgG antibodies; *see also* **primary antibody response**.

secondary immunodeficiency   An autoimmune disorder acquired later in life. *See also* **primary immunodeficiency**.

secondary infection   A disorder caused by an opportunistic microbe as a result of a primary infection weakening the host.

secondary lymphoid tissue   Sites where mature immune cells interact with pathogens and carry out the adaptive immune response.

secondary metabolite   A small molecule not essential to the survival and growth of an organism; *see also* **primary metabolite**.

secondary syphilis   Symptoms include fever, flu-like symptoms, and a skin rash.

secondary structure   The region of a polypepetide folded into an alpha helix or pleated sheet.

secondary treatment   Sewage treatment used to degrade the biological content of the fluid (effluent) coming from the primary treatment.

secretory IgA (S-IgA)   The form of IgA antibodies secreted into the intestinal lumen.

sedimentation   The removal of soil particulates from water.

selective medium   A growth medium that contains ingredients to inhibit certain microorganisms while encouraging the growth of others.

selective toxicity   A property of many antimicrobial drugs that harm the infectious agent but not the host.

self-tolerance   The ability of the immune system to not respond to one's own molecular determinants.

semiconservative replication   The DNA copying process where each parent (old) strand serves as a template for a new complementary strand.

semiperishable   Referring to foods that spoil less quickly.

semisynthetic drug   A chemical substance synthesized from natural and lab components used to treat disease.

sense codon   A nucleotide sequence that specifies an amino acid.

sensitization   The process of the first exposure to an allergen.

sensitizing dose   The first exposure to an allergy-causing antigen.

sepsis   The growth and spreading of bacteria or their toxins in the blood and tissues.

septate   Referring to the cross-walls formed in the filaments of many fungi.

septic shock   A collapse of the circulatory and respiratory systems caused by an overwhelming immune response.

septicemia   A growth and spreading of bacterial cells in the bloodstream.

septicemic plague   The bacilli from the lymph nodes spreads to the bloodstream.

septic tank   An enclosed concrete box that collects waste from the home.

septum (pl. **septa**)   A cross-wall in the hypha of a fungus.

sequela (pl. **sequelae**)   Aftereffects resulting from a disease.

seroconversion   The time when antibodies to a disease agent can be detected in the blood.

serological reaction   An antigen-antibody reaction studied under laboratory conditions and involving serum.

serology   A branch of immunology that studies serological reactions.

serotype   A closely related group of microorganisms or structures distinguished by a specific set of immune-stimulating structures.

serovar   A closely related group of microorganisms or structures distinguished by a specific set of immune-stimulating structures.

serum (pl. **sera**)   The fluid portion of the blood consisting of water, minerals, salts, proteins, and other organic substances, including antibodies; contains no clotting agents; *see also* **plasma**.

serum sickness   A type of hypersensitivity reaction in which the body responds to proteins contained in foreign serum.

sexually transmitted disease (STD)   A disease such as gonorrhea or chlamydia that is normally passed from one person to another through sexual activity.

sexually transmitted infection (STI)   *See* **sexually transmitted disease**.

Shiga toxin   A bacterial poison that inhibits protein synthesis in target cells.

Shiga toxin-producing *E. coli* (STEC)   Foodborne illness through ingestion of undercooked meat or contaminated fruits and vegetables

shigellosis   An illness of variable severity characterized by diarrhea, fever, nausea, cramps, and tenesmus. Asymptomatic infections may occur.

Shingles (herpes zoster)   A viral disease characterized by a painful skin rash with blisters in a localized area.

sign   An indication of the presence of a disease, especially one observed by a doctor but not apparent to the patient; *see also* **symptom**.

silent mutation   A change in a base sequence that produces no change to the protein made.

simple stain technique   The use of a single cationic dye to contrast cells; *see also* **differential stain technique**.

single-dose vaccine   The combination of several vaccines into one measured quantity.

S-layer   The cell wall of most archaeal species consisting of protein or glycoprotein assembled in a crystalline lattice.

slide agglutination test   *See* **VDRL** test.

slime layer   A thin, loosely bound layer of polysaccharide covering some prokaryotic cells; *see also* **capsule** *and* **glycocalyx**.

sludge   The solids in sewage that separate out during sewage treatment.

sludge tank   The area in which secondary water treatment occurs.

smallpox   A virus that is caused by a brick-shaped double-stranded DNA virus.

sodium nitrite   Used to keep meat looking pink, to add flavor and texture, and to serve as antioxidants.

soft chancre   An STI resulting in n ulcer with ragged edges and soft borders

solute   Salt crystal, often dissolved in water.

somatic recombination   The reshuffling of antibody genetic segments in a B lymphocyte as it matures.

soredium (pl. **soredia**)   The disseminated group of fungal and photosynthetic cells formed by a lichen.

sour curd   The acidification of milk, causing a change in the structure of milk proteins.

spawn   A mushroom mycelium used to start a new culture of the fungus.

specialized transduction   The transfer of a few bacterial genes by a bacterial virus that carries the genes to another bacterial cell; *see also* **generalized transduction**.

species   The fundamental rank in the classification system of organisms.

specific epithet   The second of the two scientific names for a species.

spike   A protein projecting from the viral envelope or capsid that aids in attachment and penetration of a host cell.

spiral   A shape of many bacterial and archaeal cells.

spirillum (pl. **spirilla**)   A bacterial cell shape characterized by a twisted or curved rod.

Spirochaetes   A phylum in the domain Bacteria whose members possess a helical cell shape.

spirochete   A twisted bacterial rod with a flexible cell wall containing endoflagella for motility.

spontaneous generation   The doctrine that nonliving, decaying matter could spontaneously give rise to living organisms.

spontaneous mutation   A mutation that arises from natural phenomena in the environment.

sporangiospore   Asexual spore produced by many fungi.

sporangium (pl. **sporangia**)   The structures in fungi in which asexual spores are formed.

spore   (1) A reproductive structure formed by a fungus. (2) A highly resistant dormant structure formed from vegetative cells in several genera of bacteria, including *Bacillus* and *Clostridium*; *see also* **endospore**.

sporozoite   A stage in the life cycle of the malaria parasite that enters the human body.

sporulation   The process of spore formation.

sputum   Thick, expectorated matter from the lower respiratory tract.

stabilizing protein   A protein that keeps the DNA template strands separated during DNA replication.

standard plate count procedure   A direct method to estimate the number of cells in a sample dilution spread on an agar plate.

standard precautions   Using those measures to avoid contact with a patient's bodily fluids; examples include wearing gloves, goggles, and proper disposal of used hypodermic needles.

staphylococcus (pl. **staphylococci**)   An arrangement of bacterial cells characterized by spheres in a grapelike cluster.

start codon   The starting nucleotide sequence (AUG) in translation.

stationary phase   The portion of a bacterial growth curve in which the reproductive and death rates of cells are equal.

sterile   Free from living microorganisms, spores, and viruses.

sterilization   The removal of all life forms, including bacterial spores.

sterol   A type of lipid containing several carbon rings with side chains; examples include cholesterol.

stop codon   The nucleotide sequence that terminates translation.

streak-plate method   A process by which a mixed culture can be streaked onto an agar plate and pure colonies isolated; see also **pour-plate method**.

streptobacillus (pl. **streptobacilli**) A chain of bacterial rods.

streptococcus (pl. **streptococci**)   A chain of bacterial spheres.

streptokinase   An enzyme that dissolves blood clots; produced by virulent streptococci.

stridor   A high-pitched wheezing sound during breathing.

structural formula   A chemical diagram representing the arrangement of atoms and bonds within a molecule.

structural gene   A segment of a DNA molecule that provides the biochemical information for a polypeptide.

subacute sclerosing panencephalitis (SSPE)   A rare brain disease that can occur anytime from 1 month to 25 years after clinical measles.

subclinical disease   A disease in which there are few or inapparent symptoms.

substrate   The substance or substances upon which an enzyme acts.

substrate-level phosphorylation   The formation of ATP resulting from the transfer of phosphate from a substrate to ADP.

subunit vaccine   A vaccine that contains parts of microorganisms, such as capsular polysaccharides or purified fimbriae.

sulfonamide   An agent that interferes with bacterial metabolism.

sulfur cycle   The processes by which sulfur moves through and is recycled in the environment.

sulfur dioxide   A chemical preservative used in dried fruits.

superantigen   An antigen that stimulates an immune response without any prior processing.

supercoiled domain   A loop of wound DNA consisting of 10,000 bases.

supercoiling   The process by which a chromosome is twisted and packed.

superinfection   The overgrowth of susceptible strains by antibiotic resistant ones.

suppressor T cell   A group of lymphocytes that regulate IgE antibody production.

surface water   The water in lakes, streams, and shallow wells.

surfactant   A synthetic chemical, such as a detergent, that emulsifies and solubilizes particles attached to surfaces by reducing the surface tension.

sylvatic plague   It occurs in rural wildlife and accounts for most American and global cases today.

sylvatic (jungle) yellow fever   A virus that occurs in monkeys and other jungle animals and it picked up by blood-feeding mosquitos.

symbiosis   An interrelationship between two populations of organisms where there is a close and permanent association.

symptom   An indication of some disease or other disorder that is experienced by the patient; see also **sign**.

syncytium (pl. **syncytia**)   A giant tissue cell formed by the fusion of cells infected with respiratory syncytial viruses.

syndrome   A collection of signs or symptoms that together are characteristic of a disease.

synthetic biology   A field of study that attempts to "build" new living organisms by combining parts of other species.

synthetic drug (agent)   A substance made in the lab to prevent illness or treat disease.

systematics   The study of the diversity of life and its evolutionary relationships.

systemic anaphylaxis   The release of cell mediators throughout the body. See also **localized anaphylaxis**.

systemic disease   A disorder that disseminates to the deeper organs and systems of the body.

systemic inflammatory response (SIRS)   See **sepsis**.

T

tapeworm   See **cestode**.

taxonomy   The science dealing with the systematized arrangements of related living things in categories.

T cell   See **T lymphocyte**.

T-dependent antigen   An antigen that requires the assistance of $T_H 2$ lymphocytes to stimulate antibody-mediated immunity.

T-independent antigen   A foreign substance not requiring T helper cells for an immune response.

teichoic acid   A negatively charged sugar-alcohol polymer in the cell wall of gram-positive bacterial cells.

temperate   Referring to a bacterial virus that enters a bacterial cell and then the viral DNA integrates into the bacterial cell's chromosome.

termination   (1) The completion of DNA or RNA synthesis. (2) The release of a polypeptide from a ribosome during translation.

termination factor   A protein that triggers the release of a polypeptide from a ribosome.

terminator   A set of nucleobases that stops RNA synthesis.

terminus   The point where RNA synthesis stops.

tertiary structure   The folding of a polypeptide back on itself.

tertiary treatment   Advanced treatment of wastewater.

tetanospasmin   An exotoxin produced by *Clostridium tetani* that acts at synapses, thereby stimulating muscle contractions.

tetracyclines   Broad-spectrum, bacteriostatic antibiotics that inhibit translation by blocking attachment to the tRNA to the 30S subunit.

tetrad   An arrangement of four spherical bacterial cells in a cube shape.

theory   A scientific explanation supported by many experiments done by separate individuals.

therapeutic dose   The concentration of an antimicrobial drug that effectively destroys an infectious agent.

therapeutic vaccine   A vaccine that could be given to HIV-positive individuals to control HIV disease.

therapeutic window   The concentration range or dosage of the drug in the serum that is tolerated by the host but which will eliminate the infection or disease agent.

thermal death point (TDP)   The temperature required to kill a bacterial population in a given length of time.

thermal death time (TDT)   The length of time required to kill a bacterial population at a given temperature.

thermocline   The transition layer between the surface water and deep water layers.

thermoduric   Referring to an organism that tolerates the heat of the pasteurization process.

thermophile   An organism that lives at high temperature ranges of 40°C to 90°C.

thimerosal   A stabilizer put in some vaccines as a preservative.

thioglycollate broth   A microbiological medium containing a chemical that binds oxygen from the atmosphere and creates an environment suitable for anaerobic growth.

three domain system   The classification scheme placing all living organisms into one of three groups based, in part, on ribosomal RNA sequences.

thrush   Hyphal invasion of the mucosal tissue.

thylakoid membrane   The phospholipid bilayer where photosynthesis occurs in cyanobacteria.

thymus   A lymphoid tissue, which lies behind the breastbone.

tinea infection   A fungal disease of the hair, skin, or nails.

T-independent antigen   A foreign substance not requiring T helper cells for an immune response.

titer   A measurement of the amount of antibody in a sample of serum that is determined by the most dilute concentration of antibody that will yield a positive reaction with a specific antigen.

titration   A method of calculating the concentration of a dissolved substance, such as an antibody, by adding quantities of a reagent of known concentration to a known volume of test solution until a reaction occurs.

T lymphocyte (T cell)   A type of white blood cell that matures in the thymus gland and is associated with cell-mediated immunity.

toll-like receptor (TLR)   A signaling molecule on immune cells that recognizes a unique molecular pattern on an infectious agent.

toxemia   The presence of toxins in the blood.

toxic dose   (1) The amount of toxin need to cause a disease. (2) The amount of an antimicrobial drug that causes harm to the host.

toxigenicity   The ability of an organism to produce a toxin.

toxin   A poisonous chemical substance produced by an organism.

toxoid   A preparation of a microbial toxin that has been rendered harmless by chemical treatment but that is capable of stimulating antibodies; used as vaccines.

toxoplasmosis   A relatively common blood infection that affects 50% of the world's population.

trachoma   The world's leading infectious cause of blindness and occurs in hot, dry regions of the world.

transcription   The biochemical process in which RNA is synthesized according to a code supplied by the template strand of a gene in the DNA molecule.

transduction   The transfer of a few bacterial genes from a donor cell to a recipient cell via a bacterial virus.

transfer RNA (tRNA)   A molecule of RNA that unites with amino acids and transports them to the ribosome in protein synthesis.

transformation   (1) The transfer and integration of DNA fragments from a dead and lysed donor cells to a recipient cell's chromosome. (2) The conversion of a normal cell into a malignant cell due to the action of a carcinogen or virus.

transient microbiota   The microbial agents that are associated with an animal for short periods of time without causing disease; *see also* **indigenous microbiota**.

translation   The biochemical process in which the code on the mRNA molecule is converted into a sequence of amino acids in a polypeptide.

transmissible spongiform encephalopathy (TSE)   A group of progressive conditions that affect the brain and nervous system.

transmission electron microscope (TEM)   The type of electron microscope that allows electrons to pass through a thin section of the object, resulting in a detailed view of the object's structure.

transposable element (TE)   A fragment of DNA called an insertion sequence or transposon that can cause mutations.

transposase   An enzyme that moves insertion sequences to a new DNA location.

transposon   A segment of DNA that moves from one site on a DNA molecule to another site, carrying information for protein synthesis.

traumatic wound   A deep cut, compound fracture, or thermal burn.

trematode   A flatworm, commonly known as a fluke, that lives as a parasite in the liver, gut, lungs, or blood vessels of vertebrates.

triclosan   A broad-spectrum antimicrobial agent that destroys bacterial cells by disrupting cell membranes.

trismus   A sustained spasm of the jaw muscles, characteristic of the early stages of tetanus; also called lockjaw.

trivalent vaccine   A vaccine consisting of three components, each of which stimulates immunity.

trophozoite   The feeding form of a microorganism, such as a protozoan.

trypanosomiasis   The general term for two diseases, African trypanosomiasis and Chagas disease, caused by parasitic species or the kinetoplastid *Trypanosoma*.

tube (broth) dilution method   A procedure for determining bacterial susceptibility to an antibiotic by determining the minimal amount of the drug needed to inhibit growth of the pathogen; *see* **minimal inhibitory concentration**.

tubercle   A hard nodule that develops in tissue infected with *Mycobacterium tuberculosis*.

tuberculin test   A procedure performed by applying purified protein derivative from *Mycobacterium tuberculosis* to the skin and noting if a thickening of the skin with a raised vesicle appears within a few days; used to establish if someone has been exposed to the bacterium.

tuberculosis (TB)   A communicable infection that has been with us for thousands of years and continues to evolve.

tumor   An abnormal uncontrolled growth of cells that has no physiological function.

tumor suppressor gene (TSGs)   A normal gene that inhibits tumor formation.

turbid (turbidity)   The cloudiness of a broth culture due to bacterial growth.

tyndallization   *See* **intermittent sterilization**.

type I hypersensitivity   An exaggerated or inappropriate immune response triggered by IgE antibodies.

type II hypersensitivity   A cytotoxic, cell-damaging response that occurs when igG or igM antibodies react with antigens on the surfaces of cells.

type III hypersensitivity   igG antibodies combine with antigens, forming soluble immune complexes that accumulate in blood vessels or on organ surfaces.

type IV hypersensitivity   An exeggerated cell-mediated immune response that occurs in the absence of antibodies.

U

ultra-high-temperature (URT) method   A treatment in which milk is heated at 140°C for 3 seconds to destroy pathogens.

ultrastructure   The detailed structure of an cell, virus, or other object when viewed with the electron microscope.

ultraviolet (UV) light   A type of electromagnetic radiation of short wavelengths that damages DNA.

uncoating   Referring to the loss of the viral capsid inside an infected eukaryotic cell.

Unikonta   A group of protists that includes the amoebozoans.

unique gene   These necleotide sequences are unique to one strain of a species

universal solvent   solutes dissolved in water.

unsaturated   Referring to a water-insoluble compound that can incorporate additional hydrogen atoms; *see also* **saturated**.

upper respiratory tract (URT)   Composed of the nose, sinus cavity, pharynx (throat), and larynx.

urban plague   Disease among rodents that live in close association with human in urban areas.

urticaria   A hive-like rash of the skin.

use-dilution test   A method to evaluate the effectiveness of a chemical agent on microbes dried on steel cylinders.

V

vaccination   Inoculation with weakened or dead microbes, or viruses, in order to generate immunity; *see also* **immunization**.

vaccine   A preparation containing weakened or dead microorganisms or viruses, treated toxins, or parts of microorganisms or viruses to stimulate immune resistance.

Vaccine Adverse Events Reporting System (VAERS)   A reporting system designed to identify any serious adverse reactions to a vaccination.

valence shell   An unfilled outer electron shell of an atom.

variable   That part of an experiment exposed to or treated with the factor being tested.

variable domain   The different amino acids in different antibody light and heavy chains.

variant CJD (vCJD)   Characterized clinically by neurological abnormalities such as dementia.

variolation   An obsolete method to protect a susceptible person from smallpox by infecting the person with dried material from a smallpox vesicle.

vasodilation   A widening of the blood vessels, especially the arteries, leading to increased blood flow.

VBNC   *See* **viable but noncultured**.

VDRL test   A screening procedure used in the detection of syphilis antibodies.

vector   (1) An arthropod that transmits the agents of disease from an infected host to a susceptible host. (2) A plasmid used in genetic engineering to carry a DNA segment into a bacterium or other cell.

vehicle transmission   The spread of disease through contaminated food and water.

venereal disease   *See* **sexually transmitted disease**.

Venereal Disease Research Laboratory (VDRL) test   Used for the rapid screening of patients to detect syphilis.

vertical gene transfer   The passing of genes from one cell generation to the next; *see also* **horizontal gene transfer**.

vertical transmission   The spread of disease from mother to fetus or newborn.

vesicle   Small, membrane-bound spheres.

viable but noncultured (VBNC)   Referring to microbes that are alive but not dividing.

viable count   The living cells identified from a standard plate count.

vibrio   (1) A prokaryotic cell shape occurring as a curved rod.

vibriosis   Illnesses caused by *Vibrio* species other than *V.cholerae*.

viral gastroenteritis   An inflammatory condition caused by a variety of viruses that produce varying combinations of diarrhea, nausea, vomiting, fever, cramping, and malaise.

viral genome   Contains one or more molecules of DNA or RNA in either a double-stranded or single-stranded form.

viral hemorrhagic fever (VHF)   Acute febrile illness caused by any of approximately 30 viruses.

viral inhibition   The prevention of a virus infection by antibodies binding to molecules on the viral surface.

viral load test   A method used to detect the RNA genome of HIV.

viral pneumonia   Community outbreaks of RSV.

viremia   The presence and spread of viruses through the blood.

virion   A completely assembled virus outside its host cell.

viroid   An infectious RNA segment associated with certain plant diseases.

virology   The scientific study of viruses.

virosphere   Refers to all places where viruses are found or interact with their hosts.

virulence   The degree to which a pathogen is capable of causing a disease.

virulence factor   A structure or molecule possessed by a pathogen that increases its ability to invade or cause disease to a host.

virulent   Referring to a virus or microorganism that can be extremely damaging when in the host.

virus   An infectious agent consisting of DNA or RNA and surrounded by a protein coat; in some cases, a membranous envelope surrounds the coat.

volutin   *See* **metachromatic granule**.

W

wheal   An enlarged, hive-like zone of puffiness on the skin, often due to an allergic reaction; *see also* **flare**.

white blood cell   *See* **leukocyte**.

wild type   The form of an organism or gene isolated from nature.

Woolsorters' disease   A form of pneumonia caused by workers who process hides or wool, or shear sheep, inhaling the spores.

wort   A sugary liquid produced from crushed malted grain and water to which is added yeast and hops for the brewing of beer.

wound botulism   Caused by toxins produced in the anaerobic tissue of a wound infected with *C. botulinum*.

X

xenograft   A tissue graft between members of different species, such as between a pig and a human.

X-ray   An ionizing radiation that can be used to sterilize objects.

Y

yeast   (1) A type of unicellular, nonfilamentous fungus that resembles bacterial colonies when grown in culture. (2) A term sometimes used to denote the unicellular form of pathogenic fungi.

Z

zone of equivalence   The region in a precipitation reaction where ideal concentrations of antigen and antibody occur.

zone of inhibition   The area around a chemically soaked paper disk where growth is inhibited.

zoonosis (pl. **zoonoses**)   An animal disease that may be transmitted to humans.

zoonotic disease   A disease spread from another animal to humans.

zoospore   A flagellated asexual cell in the Chytridiomycota.

Zygomycota   A phylum of fungi whose members have coenocytic hyphae and form zygospores, among other notable characteristics.

zygospore   A sexually produced spore formed by members of the Zygomycota.

INDEX

Note: Page numbers followed by *f* or *t* indicate material in figures or tables respectively. Entries with **bold** page references are from online chapters.